The Clinical Anatomy of the Vascular System

Stephen J. Bordes, Jr. • Joe Iwanaga
Marios Loukas • R. Shane Tubbs
Editors

The Clinical Anatomy of the Vascular System

Springer

Editors
Stephen J. Bordes, Jr.
Department of Surgery
Louisiana State University Health
Sciences Center
New Orleans, LA, USA

Marios Loukas
Department of Anatomical Sciences
St. George's University
Grenada, Grenada

Joe Iwanaga
Department of Neurosurgery
Tulane University School of Medicine
New Orleans, LA, USA

R. Shane Tubbs
Department of Neurosurgery
Tulane University School of Medicine
New Orleans, LA, USA

ISBN 978-3-031-78328-9 ISBN 978-3-031-78326-5 (eBook)
https://doi.org/10.1007/978-3-031-78326-5

This Springer imprint is published by the registered company Springer Nature Switzerland AG
The registered company address is: Gewerbestrasse 11, 6330 Cham, Switzerland

Preface

Understanding the intricate anatomy of the blood vessels (Fig. 1) is fundamental to the practice of clinical medicine and surgery. This book, "Clinical Anatomy of the Blood Vessels," offers a detailed exploration of the vascular system, aiming to bridge the gap between anatomical knowledge and clinical application. It is a comprehensive resource for clinicians, surgeons, and students in the medical/dental and health-related sciences who seek to deepen their understanding of vascular architecture and its clinical significance.

The vascular system is the lifeline of the human body and develops from an intricate array of branching vessels within the embryo (Fig. 2). Any disruption in this network can lead to severe, often life-threatening conditions. For clinicians, precise knowledge of blood vessels is crucial for accurate

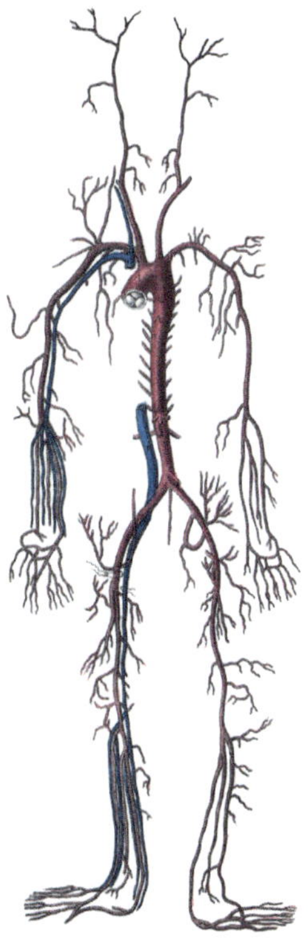

Fig. 1 Schematic drawing of the main blood vessels of the human body

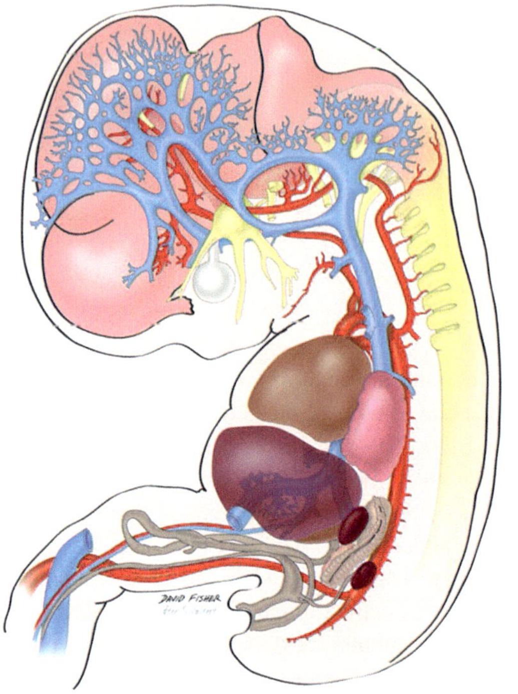

Fig. 2 Schematic drawing of the developing vascular system in the human embryo

diagnosis and effective intervention. It enhances the ability to interpret imaging studies, guides surgical planning, and improves patient outcomes. By understanding the common and rare anatomical variations, clinicians can anticipate potential complications and tailor their approaches to individual patient needs.

We hope this book will serve as a valuable resource, inspiring confidence and proficiency in the anatomical and clinical understanding of the body's blood vessels.

New Orleans, LA, USA — Stephen J. Bordes, Jr.
New Orleans, LA, USA — Joe Iwanaga
St. George's, Grenada — Marios Loukas
New Orleans, LA, USA — R. Shane Tubbs

Contents

Part V Upper Limb

Part I

Head and Neck

1 Intracranial Internal Carotid Artery and Vertebrobasilar System

Bilal Ibrahim, Diego Morales Roccuzzo, Edinson Najera, Hamid Borghei-Razavi, Michal Obrzut, and Badih Adada

1.1 Introduction

The encephalic circulation can be thought of as a double inlet vascular system with two parallel arterial pairs, the internal carotid arteries (ICA) and the vertebral arteries (VA), which together form the vertebrobasilar system, the former anterior and the latter posterior; and a single outlet for the foremost venous drain, the pair of internal jugular veins (IJVs). This complex layout is engineered so that each group of arteries, the ICA and the VA, respectively, is responsible for supplying the supratentorial space and the posterior fossa. In the latter, a "Rule of 3" of the posterior fossa has been described to correlate three portions of the brainstem, three fissures, three cerebellar peduncles, three cerebellar cortical surfaces, and three cranial nerve groups, all supplied by three respective arteries, forming three complexes (upper, middle, and lower) (Basma and Sorenson 2018). In other words, this so-called double inlet vascular system comprises simply the anterior and the posterior circulations, both of which are crucial components of the circle of Willis, each circuit with its territory, topography, singularities, and anatomical nuances. The circle of Willis lies in the basal aspect of the telencephalon and is responsible for supratentorial arterial blood supply.

B. Ibrahim · D. M. Roccuzzo · E. Najera (✉)
H. Borghei-Razavi · M. Obrzut · B. Adada
Department of Neurosurgery and Neurointerventional Radiology, Cleveland Clinic Florida,
Weston, FL, USA

1.2 Anterior Circulation

Internal Carotid Artery

The ICA has been subject of numerous studies and has long been recognized as among the most crucial vessels for neuroanatomy, neurosurgery, and neuroradiology. Several classifications divide the ICA into various numbers of segments (Fig. 1.1) (Shapiro et al. 2014). Keller, Rhoton, and Bouthellier are among the large pool of scholars who have analyzed its intrinsic anatomy, microsurgical anatomy, histology, and relationships with surrounding structures, thus contributing to the best-known classifications, of which Rhoton's is the most widely used (Gibo et al. 1981a). This classification divides the ICA into four portions, C1 to C4, that follow the blood flow from the heart from proximal to distal, which can be considered the natural, intuitive direction. These main portions are (Fig. 1.2):

1. Cervical ICA (C1).
2. Petrous ICA (C2)
3. Cavernous ICA (C3)
4. Supraclinoidal ICA (C4). This was further subdivided by Gibo et al. into the ophthalmic,

S. J. Bordes, Jr. et al. (eds.), *The Clinical Anatomy of the Vascular System*,
https://doi.org/10.1007/978-3-031-78326-5_1

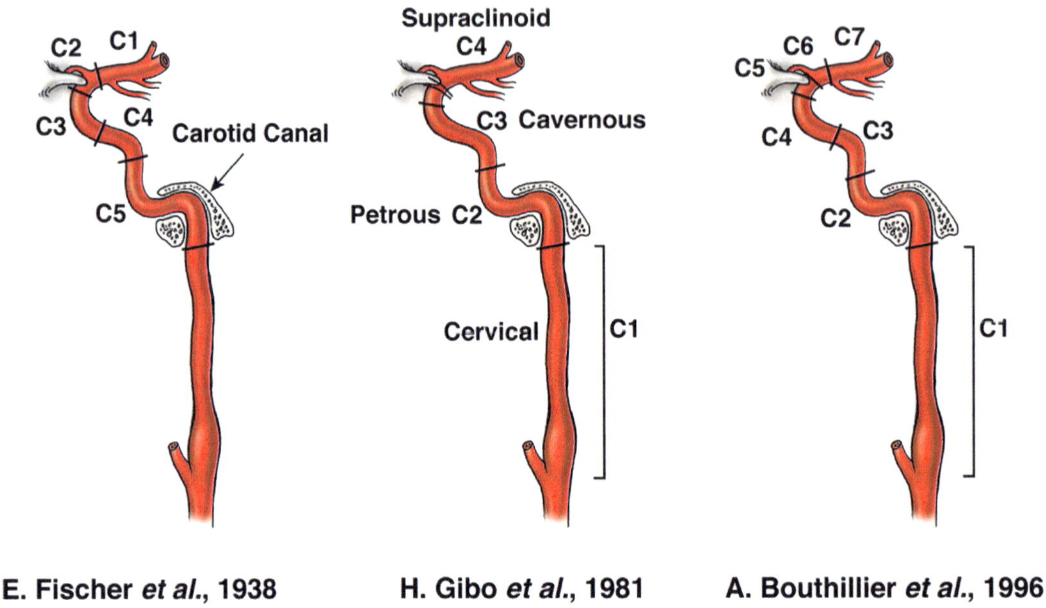

Fig. 1.1 Three systems of subclassifying the segments of the internal carotid artery

communicating, and choroidal segments (Gibo et al. 1981a). There are several collateral branches, often inconstant in various ways, but there are two constant terminal branches, the anterior and middle cerebral arteries.

Cervical ICA: C1

The cervical segment of the ICA is the longest segment and usually commences at the bifurcation of the common carotid artery (CCA) at the level of the fourth cervical vertebra. It lies deep to the medial border of the sternocleidomastoid muscle and relates laterally to the IJV and posteriorly to the cranial nerve (CN) X, wrapped along with other structures in the carotid sheath. It ascends posterior and lateral to the external carotid artery. Along it, postganglionic sympathetic fibers (PGSNs) ascend to form the internal carotid plexus. This contains the carotid sinus, a crucial baroreceptor that senses changes in systemic blood pressure and is located in the adventitia of the carotid bulb of the ICA. It is intimately related to the carotid body, a chemoreceptor found at the bifurcation of the CCA, but is a distinct organ. C1 typically has no branches. This segment is discussed in more detail in the Neck section of the book.

Clinical Significance: The most common causes of stroke in the young population are arterial dissections, which are caused mainly by stretching and tearing of the arterial walls. Carotid artery dissections are commonly seen after head trauma or can occur spontaneously. The most common location of carotid artery dissection is the cervical segment, 2–3 cm distal to the bifurcation. Symptoms include pain (headache and/or neck pain), CN deficits, especially the hypoglossal nerve, partial Horner's syndrome, or thromboembolism, and the results of stroke (Blum and Yaghi 2015).

Petrous ICA: C2

This segment begins when the ICA enters the carotid canal of the petrous part of the temporal bone and ends in the internal orifice of the carotid canal. It passes below the petrolingual ligament before entering the cavernous sinus, so the petrolingual ligament demarcates the ending of the petrous segment. Three subdivisions can be identified: (a) a shorter one (vertical) that ascends and then bends in a rostral direction; (b) a posterior

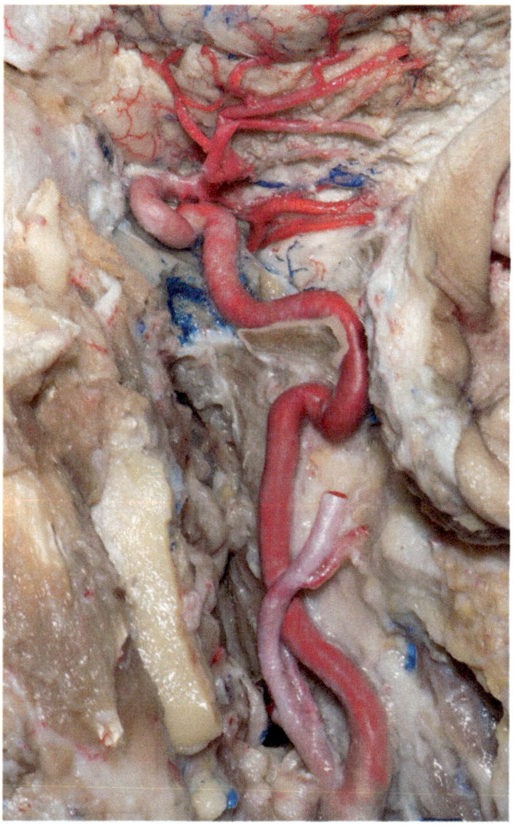

Fig. 1.2 Cadaveric dissection of the left internal carotid artery from its origin in the neck to its cortical branches. Compare with Fig. 1.1

loop anterior to the cochlea; and (c) a longer one (horizontal) (Fig. 1.3). The origins of two branches from this portion can be identified:

1. The caroticotympanic artery (which is a controversial branch)
2. The pterygoid branch (present in approximately 30%) (Quisling and Rhoton Jr 1979)

Cavernous ICA: C3

This segment starts from the petrolingual ligament, a fibrous band extending from the petrous apex to the lingual apophysis of the sphenoid bone and enters the cavernous sinus. It thus becomes the intracavernous ICA. It ends at the distal dural ring, a dural fold that extends medially from the upper surface of the anterior clinoid process to the tuberculum sella. This segment can

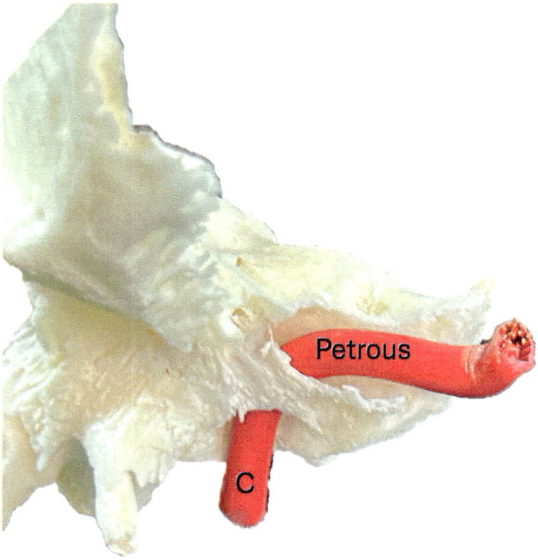

Fig. 1.3 Petrous part of the internal carotid artery traveling through the carotid canal with the overlying venous plexus. Note the cervical portion (C) of the artery

be subdivided into five portions: posterior ascending, posterior loop, horizontal, anterior loop, and anterior ascending (Fig. 1.4a). It is directed anteriorly and then superomedially to bend posteriorly and thus forms the posterior loop of the ICA. The ICA continues horizontally and afterward bends rostrally (part of the anterior loop of the ICA), heading toward the anterior clinoid process. It then touches the proximal dural ring (which encircles the ICA but not completely) and pierces the dura at the distal dural ring (which completely encircles it), where it becomes intradural (Fig. 1.5).

Typically, there are many branches, usually grouped as trunks, the main ones being:

(a) Meningohypophysial trunk (Fig. 1.6), which is the largest and most proximal, emerging at the level of the posterior loop. Branches of this trunk include the following:

   - Posterolateral branch, tentorial artery (or artery of Bernasconi and Cassinari); provides the blood supply for petroclival meningiomas
   - Posteromedial branch, dorsal meningeal artery (or dorsal clival)

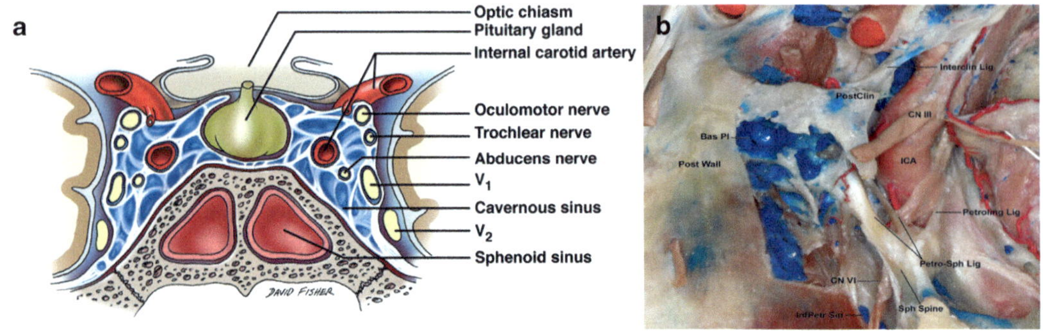

Fig. 1.4 (**a**) Coronal section through the cavernous sinus, noting the relationship of the internal carotid artery and the various cranial nerves (After Netter 2023). (**b**) Cadaveric dissection illustrating the cavernous part of the internal carotid artery and regional structures

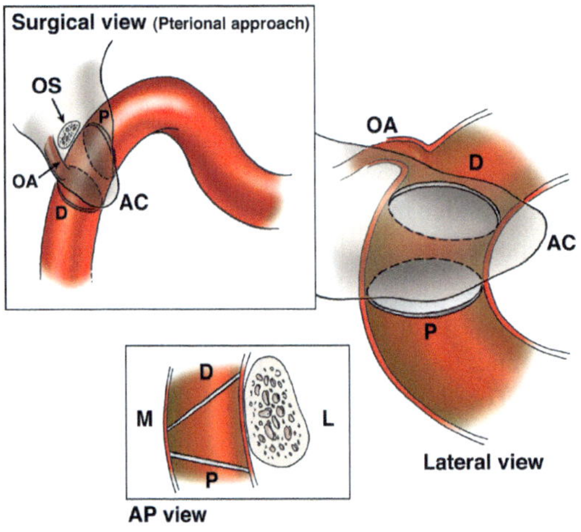

Fig. 1.5 Relationship of the internal carotid artery and the optic strut (OS), ophthalmic artery (OA), anterior clinoid process (AC), and proximal (P) and distal (D) dural rings

- Medial branch, inferior hypophysial artery (supplies posterior lobe of the hypophysis)

It is worth noting that the branches of this trunk can emerge directly from the ICA.

(b) Anterior meningeal artery.
(c) Inferolateral trunk, another constant collateral (present in >90%) that supplies the cranial nerves that travel through the inner layer of the lateral wall of the cavernous sinus.
(d) Finally, a medial branch is found in less than 10% of cases: the capsular artery of McConnell, which supplies the capsule of the pituitary gland.

Clinical Significance: The cavernous segment of the ICA can develop an abnormal connection with the cavernous sinus, which results in a carotid-cavernous fistula (CCF). The most common cause is a skull base fracture that tears the intracavernous ICA. Aneurysmal rupture of the cavernous ICA is an important spontaneous cause. The classical presentation is a triad of chemosis, pulsatile exophthalmos, and ocular bruit. Treatment is usually by occluding the fistula with detachable endovascular coils. Untreated cases can result in permanent visual loss or ocular motility deficits (Henderson and Miller 2018).

Supraclinoidal ICA: C4

Once the ICA runs through the distal dural ring, it is known as the supraclinoidal ICA. This segment is the intradural segment. It is directed upward, posterior and laterally, to end in a bifurcation just below the anterior perforated substance into two terminal branches, one anterior and medial, the anterior cerebral artery, and the other lateral, the middle cerebral artery. According to Rhoton's classification, C4 is segmented into three subdivisions: ophthalmic, communicating, and choroidal.

- *Ophthalmic Segment:* The largest portion of C4 commences immediately after the distal dural ring has been passed and ends just proxi-

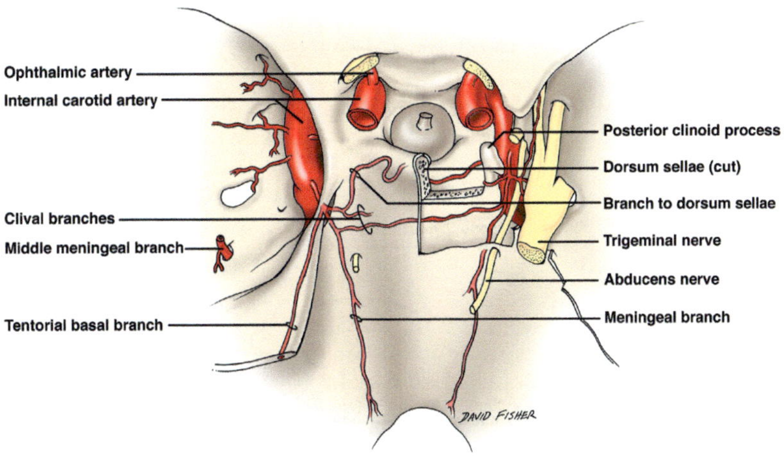

Fig. 1.6 Skull base illustrating the branches of the meningohypophysial trunk

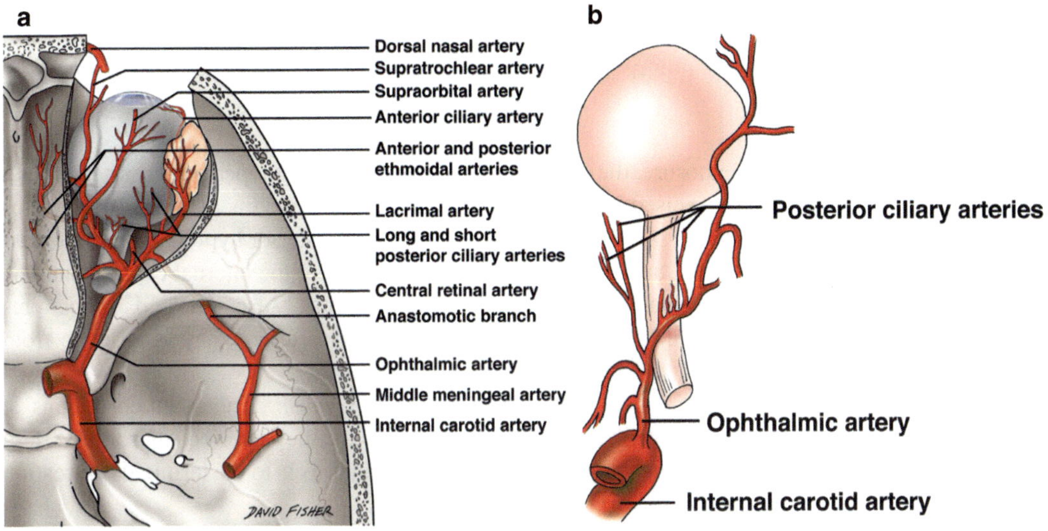

Fig. 1.7 (**a**) Major branches of the ophthalmic artery. (**b**) Ophthalmic segment of the internal carotid artery

mal to the PComA. Many important perforating arteries arise from this portion to supply the optic nerve. Two main collaterals are typically found:

(a) Ophthalmic artery: This emerges from the ICA distal to the cavernous sinus in more than 96% of cases, but in the remaining few it arises from the cavernous segment, the first intradural branch of the ICA. It runs through the optic canal into the orbit along with CN II, inferomedial to it. A characteristic "kink" can be seen on a lateral angiogram (Fig. 1.7).

(b) Superior hypophysial artery: this supplies the pituitary stalk and the anterior lobe of the gland (Fig. 1.8).

Clinical Significance: Ophthalmic artery aneurysms arise from the supraclinoidal ICA segment at the point where the ophthalmic artery branches. As they grow, they present with visual field defects, most commonly ipsilateral superior nasal quadrantanopsia, caused by optic nerve compression over its lateral aspect. Surgical clipping these aneurysms is challenging due to limited exposure and the critical structures packed

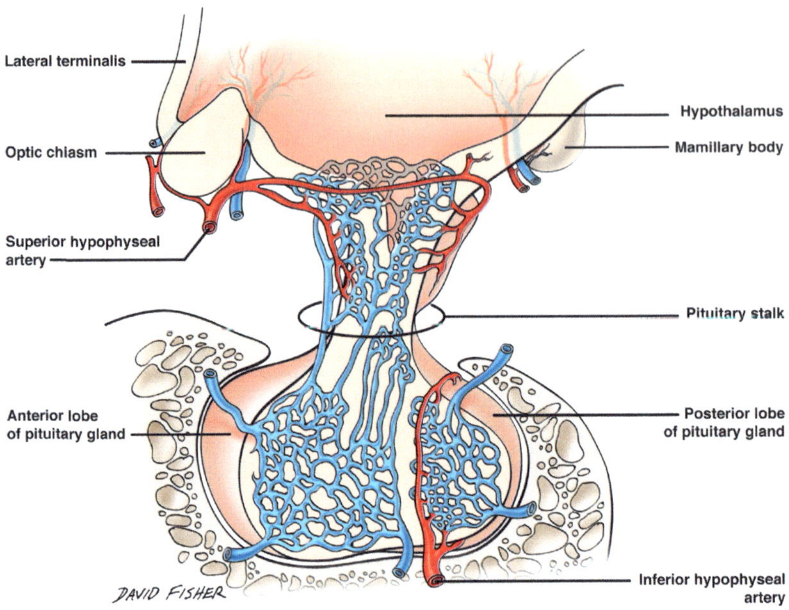

Fig. 1.8 Drawing of the hypophysial arteries. (After Netter 2023)

into this region. For adequate exposure, the anterior clinoid process needs to be excised, and the distal dural ring sometimes needs to be transected. This allows the ophthalmic artery to be identified early and prevents accidental inclusion of the artery in the clip (Cawley et al. 1998).

- *Communicating Segment:* This begins just proximal to the PComA origin and traverses CN II and CN III to terminate in the origin of the anterior choroidal artery. Its branches are as follows:
  - (a) Posterior communicating artery (PComA) (Fig. 1.9): This arises from the posterior medial wall of the ICA, and from there, it travels dorsal and slightly medial to encounter the posterior cerebral artery. The PComA can also be found as a variant in nearly 18% of cases. It is known as a fetal PcomA, which has a greater diameter than the ipsilateral posterior cerebral artery, thus providing blood flow to its respective territory instead (Golshani et al. 2010). Several perforators emerge from this collateral, which are destined for the hypothalamus, ventral thalamus, posterior perforated substance, and subthalamic nucleus.
  - (b) Anterior thalamoperforator arteries that supply the optic tract (Fig. 1.10), the chiasm, the posterior hypothalamus, the posterior limb of the internal capsule, and the subthalamus. Also referred to as premamillary arteries, they often arise from the PComA.
- *Choroidal Segment:* This segment begins at the origin of the anterior choroidal artery (AChA) and travels to the terminal bifurcation of the ICA.
- Anterior Choroidal Artery (Fig. 1.11): This arises from the posterior lateral wall of ICA 2–4 mm distal to the PComA. It mainly but not exclusively supplies the crucial structures of the diencephalon, mesencephalon, central core, and mesial temporal lobe, such as the portion of the optic tract, medial globus pallidus, genu of the internal capsule, inferior half of the posterior limb of the IC, uncus, retrolenticular fibers (optic radiation), lateral geniculate body, dentate gyrus, fornix, globus pallidus, red nucleus, and amygdala.

Clinical Significance: AChA syndrome is caused by AChA occlusion, which consists of hemiparesis (injury to the posterior limb of the internal capsule), hemiparesthesia (injury to the

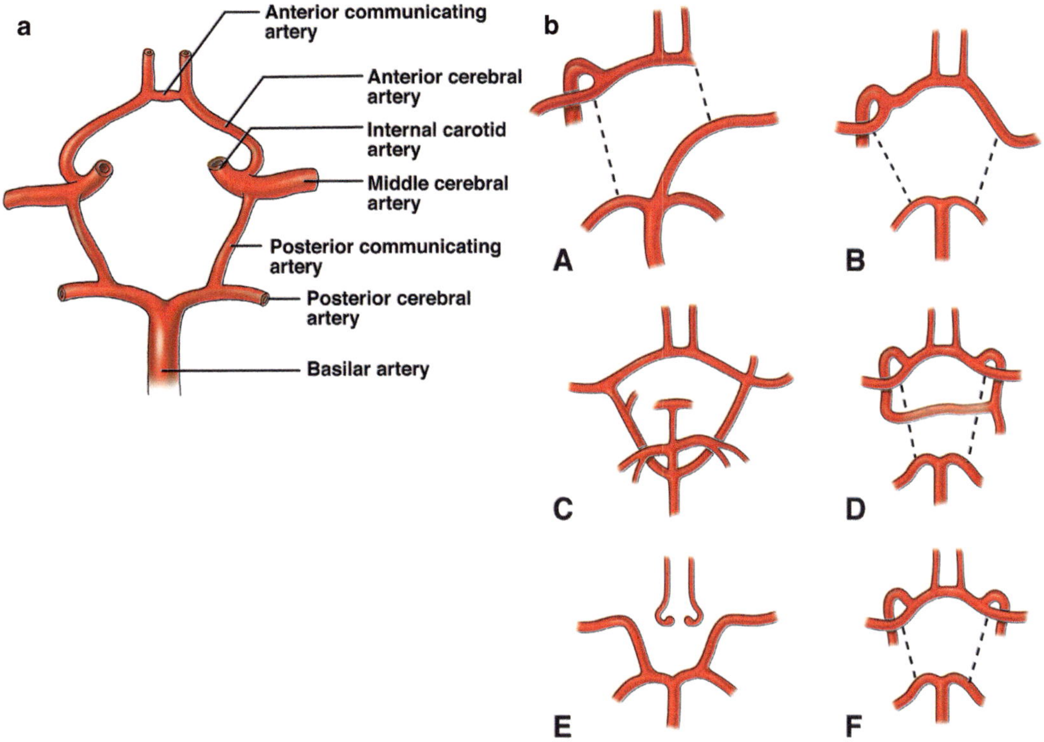

Fig. 1.9 (**a**) Schematic drawing of the circle of Willis. (**b**) Common and uncommon variants of the composition of the circle of Willis and its constituent vessels

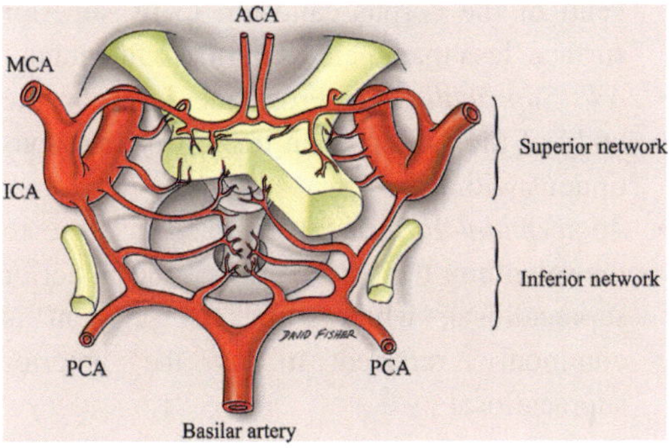

Fig. 1.10 Contributions to the optic tract from the posterior communicating artery

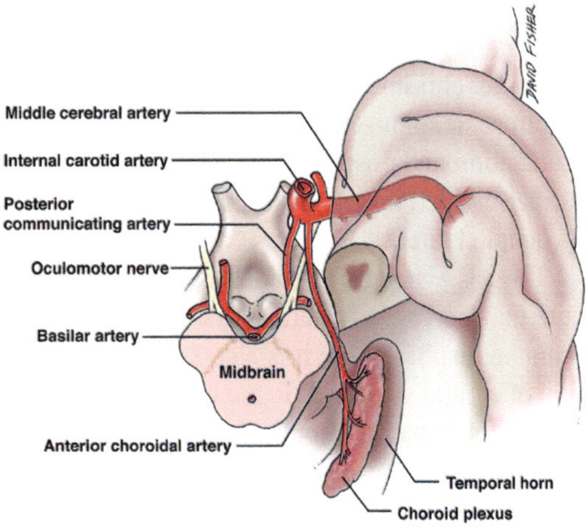

Fig. 1.11 Drawing noting the posterior communicating artery and anterior choroidal artery

ventral posterolateral nucleus of the thalamus), and homonymous hemianopia (injury to the lateral geniculate body). Owing to its location near the PComA, the AChA is at risk for occlusion when PComA aneurysms are clipped (Helgason and Caplan 2012).

One of the historical treatments of Parkinson's disease was AChA ligation. The blood supply to the pallidum by this artery made it a target for treatment. However, the destruction extended beyond the desired location and resulted in significant morbidity (Cooper 1953).

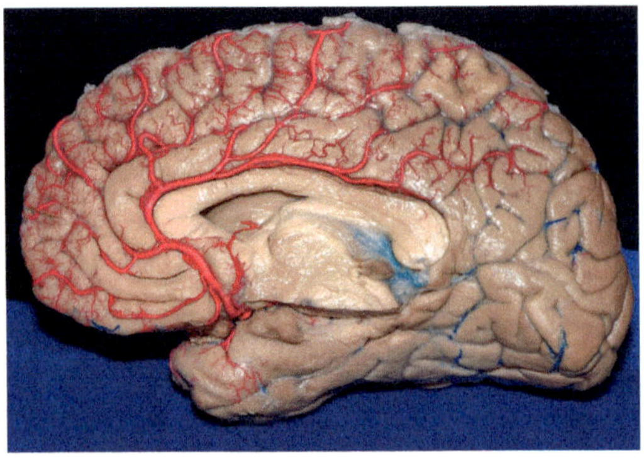

Fig. 1.12 Sagittal section of a cadaver brain with the right anterior cerebral artery injected with latex

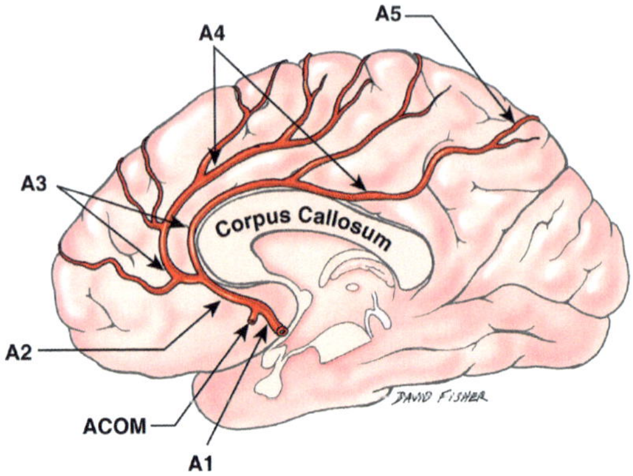

Fig. 1.13 Right anterior cerebral artery noting its segments

Anterior Cerebral Artery (ACA)

The ACA (Fig. 1.12) is one of the two paired terminal branches of the ICA, emerging from its anteromedial wall and traversing anteriorly over the optic chiasm, below the anterior perforated substance. It then bends over in a characteristic loop, adopting a curve of anterior convexity, and travels along the corpus callosum toward a more posterior situation. The ACA is one of the vessels that form the anterior half of the Circle of Willis, and each is connected to its contralateral homologue by the anterior communicating artery (AComA), which links them immediately above the optic chiasm and pituitary stalk. One can, therefore, recognize two large portions: precommunicating and postcommunicating. The latter is further divided into several segments, which will be detailed later.

The ACA mainly provides blood flow to the anterior and medial (parafalcine) aspects of the brain hemispheres, but it is also key to the irrigation of certain structures in the central core (parts of the globus pallidus, putamen, and internal capsule anterior limb, among others), the diencephalon, the corpus callosum, the olfactory tract, the optic chiasm, and a fairly important portion of the basal aspect of the frontal lobe. It has little expression on the convexity, but it correlates directly with Penfield's Homunculus sensory-motor functional areas facing the interhemispheric fissure, such as the lower extremities and the genitals, a clinically relevant fact when the numbers of cerebrovascular nosological entities and their ever-growing spread among the adult population are considered. It can be segmented as follows (Fig. 1.13):

- *A1 (Precommunicating):* From origin to anterior communicating artery
- *A2 (Postcommunicating):* From AComA to the site at which the callosomarginal artery emerges
- *A3 (Precallosal):* From the branching of the callosomarginal artery, curving around the genu of the corpus callosum to its superior surface, less than 5 cm posterior to the genu
- *A4 (Supracallosal):* From the level of the body of the CC up to an imaginary line continuous with the coronal suture
- *A5 terminal branch (Postcallosal):* Close to the splenium. It is also known as the posterior supracallosal, whereas the A4 segment is commonly referred to as the anterior supracallosal

The constant branches usually arising from the ACA are as follows:

(a) Anterior communicating artery (Fig. 1.14)
(b) The recurrent artery of Heubner (also known as the "medial striate artery"); the most conspicuous perforating vessel, this runs through the anterior third of the anterior perforated substance and heads to the basal ganglia and the internal capsule
(c) Orbitofrontal medial artery

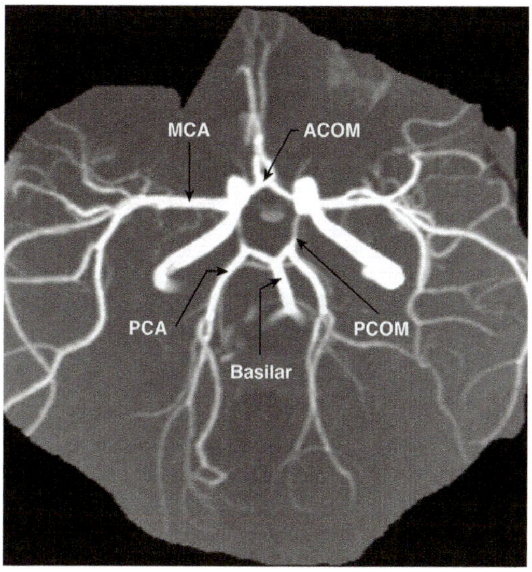

Fig. 1.14 Angiogram noting the circle of Willis and its major parts

(d) Frontopolar artery
(e) Callosal marginal artery:
   (a) Internal frontal branches
   (b) Paracentral artery
(f) Pericallosal artery
   (a) Internal superior parietal artery
   (b) Internal inferior parietal artery
   (c) Perforating arteries: numerous perforating arteries arise from the pre- and post-communicating segments and run through the anterior perforated substance. These are directed toward the optic tract, the floor of the third ventricle, and the basal ganglia.

Clinical Significance: The AcomA is the most common location of intracranial aneurysms. Rupture of these causes subarachnoid hemorrhage, IVH, and/or ICH, with devastating outcomes (Keedy 2006).

Middle Cerebral Artery (MCA)

The MCA (Fig. 1.15) is the remaining terminal branch of the ICA. In most cases, it originates from the anterolateral wall of the ICA in the carotid cistern and heads laterally to the Sylvian cistern, which is horizontally and parallel to the axial plane. From there, it splits into two, three, or four trunks and turns backward and upward to traverse the Sylvian fissure and through the insular lobe surface. At its terminus, it gives several opercular branches that head in cephalad and caudate directions to provide the convexity of the brain with blood flow.

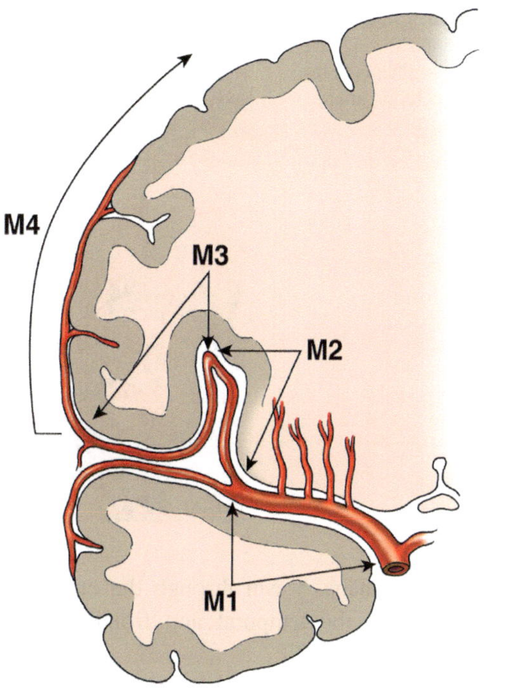

Fig. 1.15 Coronal section showing the segments of the middle cerebral artery

The MCA is usually divided into *four* portions:

- *M1 (Sphenoidal)*: From its origin to the point where it bends toward the insular surface at a 90° angle at the level of the limen insulae. It bifurcates into inferior and superior trunks in approximately 80% of cases. Trifurcation is found in 15% and a tetrafurcation or division into multiple lesser trunks in fewer than 5%.
- *M2 (Insular):* MCA trunks from the bifurcation to the emergence from the Sylvian fissure. This segment traverses across the inferior circular sulcus of the insula.
- *M3 (Opercular):* After several branches emerge and run dorsally across the sulci of the insula, another 90° turn is described as the

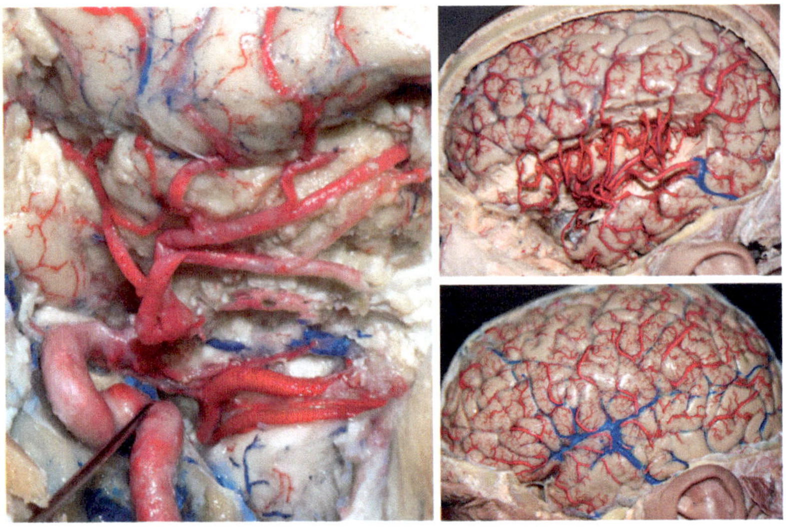

Fig. 1.16 Cadaveric dissection showing the cortical segments of the middle cerebral artery

MCA branches travel through the opercular surface of the temporal and fronto-parietal lobes.

- *M4 (Cortical):* The most distal segment, this distributes itself all along the convexity of the brain, supplying blood to the superolateral and temporal surfaces.

According to Rhoton, 12 cortical areas are irrigated by the MCA, and each of them is supplied by a stem artery that also supplies one or two more cortical areas; only the temporo-occipital, angular, and central areas have exclusive arterial stems (Gibo et al. 1981b). They can be listed as follows:

1. Orbitofrontal area
2. Prefrontal area
3. Precentral area
4. Central area
5. Anterior parietal area
6. Posterior area
7. Angular area
8. Temporo-occipital area
9. Posterior temporal area
10. Middle temporal area
11. Anterior temporal area
12. Temporo-polar area

The MCA provides an average of ten perforating branches per hemisphere (Fig. 1.16) divided into medial lenticulostriate, intermediate lenticulostriate, and lateral lenticulostriate. These are responsible for supplying part of the internal capsule and a large portion of the basal ganglia.

Clinical Significance: Most of the lateral cerebral hemisphere is supplied by the MCA. Occlusion of this artery, especially the M1/M2 segments, can result in malignant MCA infarction, a lethal condition with a mortality rate of up to 80% in untreated patients. It is characterized by a large ischemic infarction that results in significant surrounding brain edema. This exerts progressive compression and brain herniation. Management is mainly surgical by performing decompressive hemicraniectomy to relieve the effects of the increased intracranial pressure (Treadwell and Thanvi 2010).

1.3 Posterior Circulation

Vertebral Artery

The vertebral artery (VA) (Figs. 1.17 and 1.18) is a paired artery and is the origin of the posterior circulation, also known as the vertebrobasilar system. Each of the VAs merges into a single, larger trunk, known as the basilar artery (BA), and provides blood to the posterior cranial fossa with all its anatomical complexity. The VA is the first and usually the largest branch of the subcla-

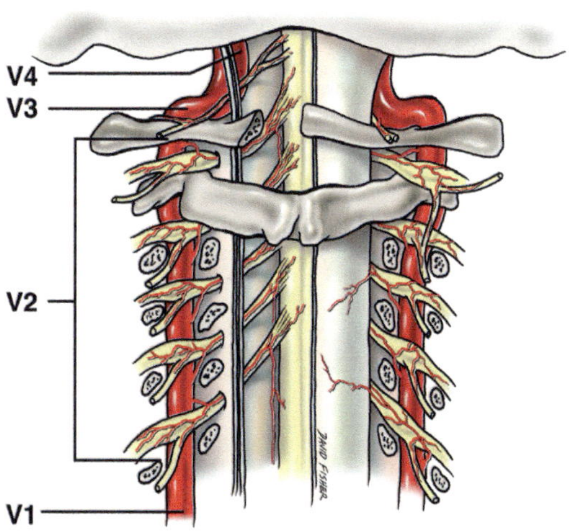

Fig. 1.17 Segments of the vertebral artery

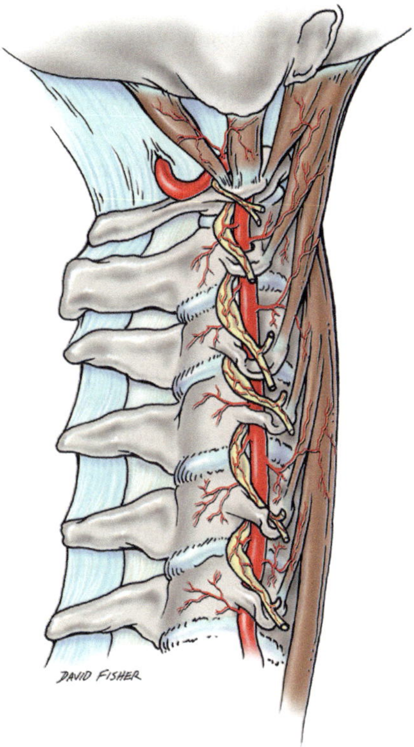

Fig. 1.18 Lateral view of the extracranial vertebral artery

vian artery although it may also arise from the aortic arch (Fig. 1.19). It has an average diameter of 3 mm and the left VA is dominant in more than 60%; the right VA is hypoplastic in 10% (Campero et al. 2011).

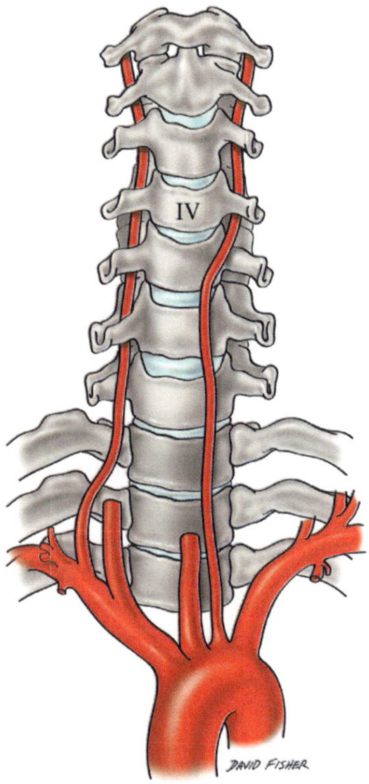

Fig. 1.19 Origin and extracranial course of the vertebral arteries. Note the left vertebral artery arising from the aortic arch and the right artery arising from the subclavian artery

Four segments are distinguished (Fig. 1.17).

- *V1—Prevertebral:* Starts from where the VA emerges from the subclavian artery and ends at the foramen transversarium of the sixth vertebral body.
- *V2—Intraforaminal:* This ascends cephalad through the transverse foramen of the sixth cervical vertebra surrounded by sympathetic fibers derived from the stellate ganglion, along a venous plexus that intimately surrounds it. It is located anterior to the cervical nerve roots. Finally, it bends laterally to enter the foramen within the transverse process of the axis (C2) in a superolateral trajectory.
- *V3—Suboccipital:* V3 leaves the foramen of the axis and curves posteriorly and medially in a groove on the upper aspect of the posterior arch of the atlas, and then enters the foramen magnum medial to the atlanto-occipital joint.

Table 1.1 Symptoms and signs of lateral medullary syndrome

Ipsilateral to infarct	Contralateral to infarct
Vertigo and nystagmus (lesion of inferior vestibular nucleus)	Pain and temperature sensation impaired in arm and leg (lesion of spinothalamic tract)
Dysarthria, dysphagia, and ipsilateral gag impairment (lesion of nuclei and fibers of CN IX and X)	General symptoms
Horner's syndrome (lesion of sympathetic fibers)	Headache, nausea, and vomiting
Ipsilateral ataxia (lesion of inferior cerebellar hemisphere or inferior cerebellar peduncle)	Hiccups (lesion of nucleus ambiguus)
Hemifacial numbness and paresthesia (lesion of descending trigeminal tract)	
Taste sensation impairment (lesion of nucleus tractus solitarius)	

- *V4—Intradural:* The final division is the intradural segment. Once it pierces the dura, it immediately enters the subarachnoid space. Finally, it joins the contralateral homolog at the vertebral confluence on the lower pontine border to form the basilar artery (BA).

Among the collateral branches, the PICA is the most relevant intracranially. It has the most complex and tortuous course, but the following can be listed as constant branches of the VA:

(a) Anterior meningeal artery
(b) Posterior meningeal artery
(c) Posterior spinal artery
(d) Anterior spinal artery
(e) Posterior inferior cerebellar artery (PICA)

Clinical Implication: Lateral medullary syndrome, aka Wallenberg syndrome, is caused by VA occlusion or, less commonly, by PICA occlusion. It is associated with various signs and symptoms related to injury to the lateral part of the medulla (Table 1.1). No pyramidal sign is usually found. Close neuromonitoring is needed for possible significant cerebellar edema and herniation, in which event suboccipital decompressive craniectomy is warranted (Shetty et al. 2012).

Posterior Inferior Cerebellar Artery (PICA)

The PICA (Figs. 1.20 and 1.21) is the largest, most important, and most complex branch. It usually arises 10 mm distal to the point where the VA becomes intradural, less than 20 mm proximal to the vertebrobasilar junction.

There are five segments:

(a) Anterior Medullary: From its origin from the VA to the inferior olivary eminence. This provides one or two short medullary circumflex branches destined to irrigate the ventral medulla oblongata.
(b) Lateral Medullary: From the most prominent point of the medullar olive to the origin of CN IX, X, and XI, giving up to five perforating branches that supply the brainstem.
(c) Tonsillomedullary: This starts where the PICA passes posterior to the glossopharyngeal, vagus, and accessory nerves and ends at the midpoint of the tonsillar medial surface. It has a caudal loop on lateral angiography.
(d) Telovelotonsillar: The most complex segment, this commences where the PICA ascends through the medial surface of the tonsil and then reaches the inferior medullary velum. It ends where the vessel leaves the fissure between the tonsil and the cerebellar hemisphere, where it projects a new loop, the cranial loop, which coincides with the fastigium. This segment supplies the fourth ventricle, the choroid plexus, and the tela choroidea.
(e) Cortical Segment: This starts where the PICA exits the fissure between the vermis and the tonsil and provides medial branches to the vermis and lateral branches to the suboccipital surface of the cerebellar hemisphere.

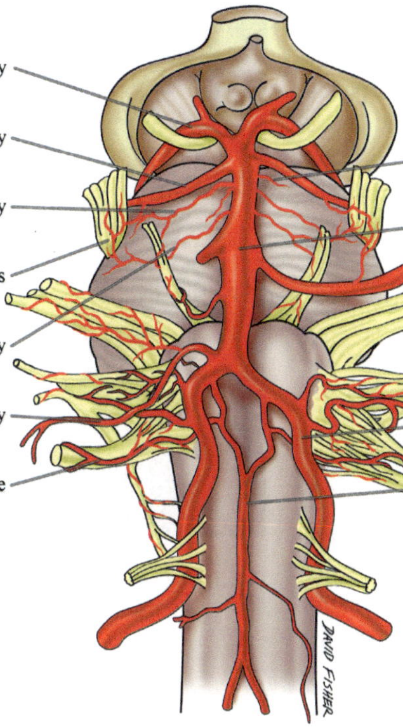

Fig. 1.20 Anterior view noting the posterior inferior cerebellar artery

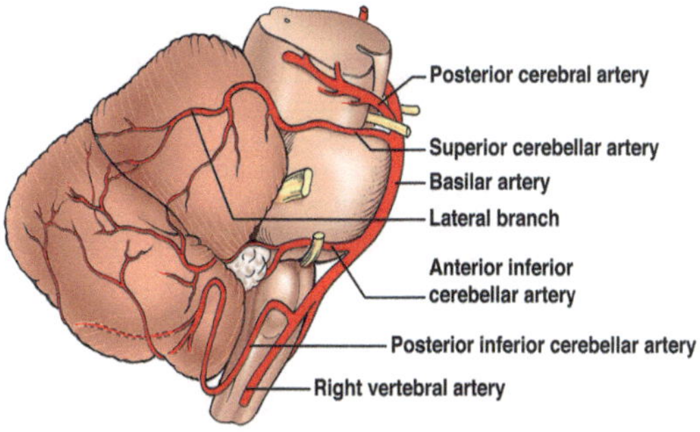

Fig. 1.21 Lateral view of the posterior inferior cerebellar artery

It typically provides three constant branches:

(a) Choroidal artery
(b) Tonsillohemispheric artery
(c) Inferior vermian artery

Clinical Implications: Occlusion of the PICA affects the posterior and inferior surfaces of the cerebellum, the lateral aspect of the medulla, and the inferior vermis. Nausea and vomiting, truncal ataxia, and lower cranial nerve deficits are the most common presentation. Lateral medullary syndrome, as mentioned previously, is commonly caused by VA occlusion, but it is also seen in 30% of PICA strokes (Miao et al. 2020).

This PICA may arise from the basilar or may be absent unilaterally. When arising from the basilar it originates from its upper third in 2% of cases, from the middle third in 46%, and from the lower third in 52% of cases. The vessel is duplicated in about 10% of individuals.

Extradural origin of the PICA is found in 5–20% of cases. Origins of the PICA from the ICA, the posterior meningeal artery, a persistent hypoglossal artery, and a persistent proatlantal artery have been reported. The PICA is duplicated in some 2.5–6% of cases. The PICA is hypoplastic in 5–16% of hemispheres. The PICA is sometimes absent. An AICA-PICA trunk is a normal variant. In 0.2% of the cases, the vertebral artery terminates in PICA. The posterior inferior cerebellar artery may provide the posterior spinal artery.

The posterior inferior cerebellar artery may arise from the internal carotid artery and the embryonic explanation postulated is the persistence of a primitive communicating vessel (presegmental artery) between the anterior and posterior circulation. The posterior inferior cerebellar artery may also arise from the posterior meningeal artery.

1.4 Basilar Artery

The basilar artery (BA) (Figs. 1.22 and 1.23) is a common trunk formed by the union of the left and right VAs. It begins at the medullary-pontine sulcus and along its course, it provides several pontine branches that irrigate numerous structures of the pons. It terminates at a bifurcation right in front of the pontomesencephalic sulcus, the point at which it gives off its two symmetrical terminal branches, the left and right posterior cerebral arteries.

The branches of the BA are as follows:

(a) Anterior inferior cerebellar artery (AICA)
(b) Pontine perforating branches: responsible for supplying numerous and crucial pontine nuclei
(c) Superior cerebellar artery (SCA)
(d) Posterior cerebral artery (PCA)

The basilar artery may be internally septated (Fig. 1.24). Very rarely the two vertebral arteries fail to unite to form the basilar artery, hence the basilar may appear to be doubled. The two longitudinal trunks may be united by anastomoses between them. Fenestration of the basilar artery is found in 1.33% of dissections and 0.12% of angiograms. The basilar artery may give rise to the middle meningeal artery.

The labyrinthine artery, which enters the internal acoustic meatus, can arise from multiple sources including the anterior inferior cerebellar artery, the basilar artery, the vertebral artery, or the posterior inferior cerebellar artery. Additionally, the vessel may be duplicated with each branch arising from a different source.

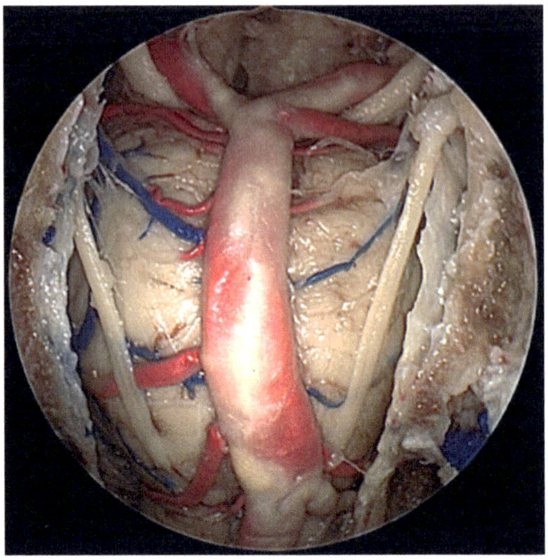

Fig. 1.22 Anterior view of the basilar artery in a cadaveric specimen. Note the laterally positioned abducens nerves

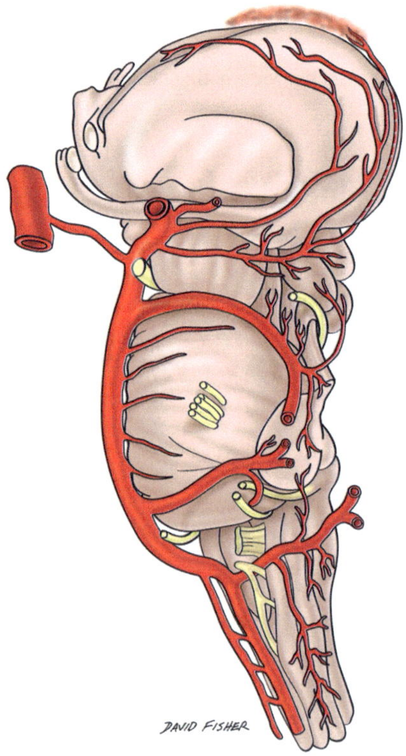

Fig. 1.23 Lateral view of the basilar artery and its main branches. From superior to inferior, note the posterior cerebral artery, superior cerebellar artery, anterior inferior cerebellar artery, and posterior inferior cerebellar artery. (After Haines 2008)

Anterior Inferior Cerebellar Artery (AICA)

The AICA (Figs. 1.21 and 1.25) is the first important collateral of the BA. It courses across the central portion of the cerebellopontine angle and is part of one of the three complexes described by Rhoton. It is intimately associated with the pons, the lateral recess of the fourth ventricle, the middle cerebellar peduncle, CN VI, VII, and VIII, the cerebellopontine fissure, and the cerebellar petrosal surface. It originates from the lower part of the BA and then runs posterolaterally anterior to CN VI, VII, and VIII. Very often it gives off a loop that runs across the internal acoustic meatus. It commonly divides into rostral and caudal trunks near the CN VII-VIII complex.

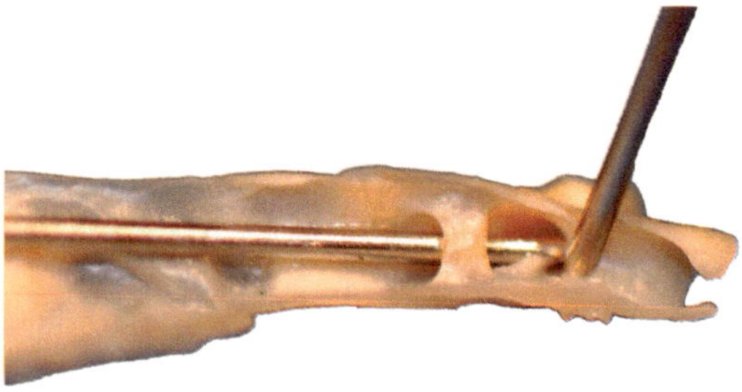

Fig. 1.24 Septation (over horizontally oriented needle) in an opened cadaveric basilar artery near the vertebrobasilar junction

The AICA supplies the anteroinferior surface of the cerebellum, the middle cerebellar peduncle, the flocculus, and the inferolateral surface of the pons. The origin of the vessel is single in 72% of cases, a duplicate in 26%, and triplicate in 2%. Origin of the AICA from the ICA has been reported.

Clinical Implication: Occlusion of the AICA results in lateral pontine syndrome, aka AICA syndrome or Marie-Foix syndrome. Signs and

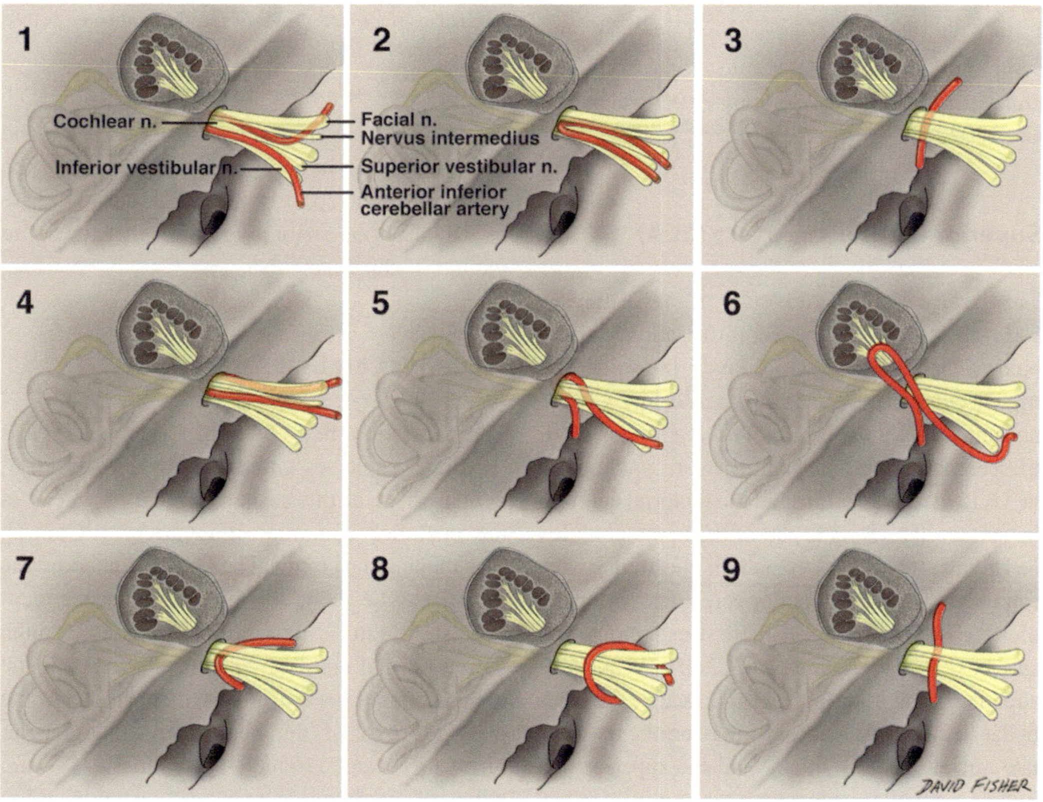

Fig. 1.25 The various orientations of the anterior inferior cerebellar artery and the nerves of the internal acoustic meatus

Table 1.2 Signs and symptoms of lateral pontine syndrome

Ipsilateral to lesion	Contralateral to lesion
Limb ataxia (lesions of middle and inferior cerebellar peduncle)	Hemiparesis (lesions of corticospinal tract)
Facial hemianesthesia (lesions of spinal trigeminal tract and nucleus)	Pain and temperature sensation impairment in arm and leg (lesions of spinothalamic tract)
Unilateral hearing loss and nystagmus (lesions of vestibular and cochlear nuclei)	
Horner's syndrome (lesions of sympathetic fibers)	
Unilateral facial weakness and loss of taste sensation on the anterior two-thirds of the tongue (lesions of the facial nucleus and nerve)	

symptoms of this are related to injury to the lateral aspect of the pons (Table 1.2) (Carvalho and Cruz 2020).

Superior Cerebellar Artery (SCA)

This artery (Figs. 1.21 and 1.26) originates 2–3 mm below the basilar bifurcation, travels below CN III, goes around the pontine-mesencephalic union, and passes below CN IV and above CN V in all cases. Afterward, it reaches the cerebellar-mesencephalic fissure and finally heads toward the vermis and the tentorial cerebellar surface. The SCA is part of the upper complex and relates to the upper half of the roof of the fourth ventricle, the superior cerebellar peduncle, the cerebellar-mesencephalic fissure, the tentorial surface of the cerebellum, and CN III, IV, and V. It supplies blood to the superior surface of the cerebellar hemispheres, the superior vermis, the dentate nucleus, the midbrain, and the superior and middle cerebellar peduncles.

The superior cerebellar artery (SCA) is duplicated in 14% of the hemispheres; in these cases,

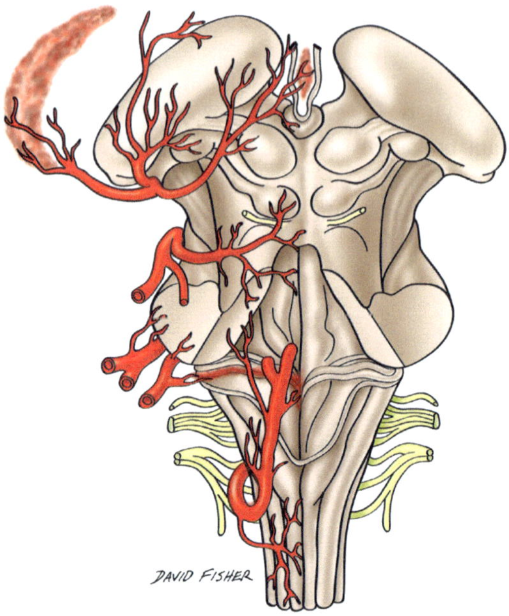

Fig. 1.26 Posterior view of the basilar artery and its main branches. From superior to inferior, note the posterior cerebral artery, superior cerebellar artery, anterior inferior cerebellar artery, and posterior inferior cerebellar artery. (After Haines 2008)

the duplicate vessels correspond to the rostral and caudal trunks. Although rare, the absence of the SCA has been reported. The SCA can give rise proximally to a perforating artery that travels to the interpeduncular fossa.

Posterior Cerebral Artery (PCA)

The posterior half of the Circle of Willis (Fig. 1.27) reaches completion when the PCA, which arises at the basilar bifurcation, receives the posterior communicating artery (PComA), thus closing the circuit. The PCA is joined by the PComA at the lateral margin of the interpeduncular cistern, circles the brainstem passes through the crural and ambient cisterns, and reaches the quadrigeminal cistern, finally distributing to the posterior part of the hemisphere. The basilar bifurcation can be found as far caudal as 1.3 mm below the pontomesencephalic connection and as far rostral as the mammillary bodies, sometimes even adjacent to the floor of the third ventricle.

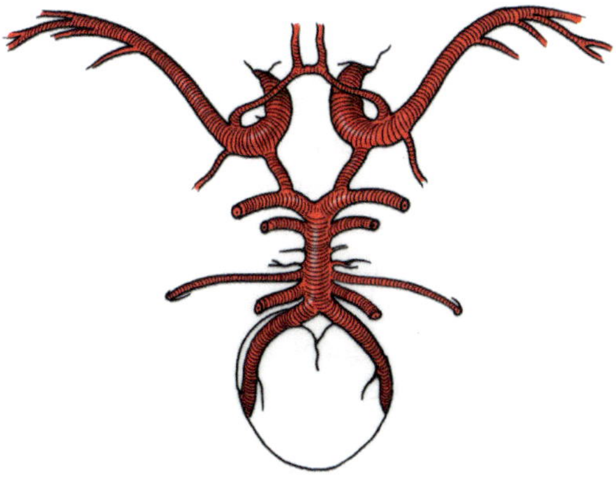

Fig. 1.27 Schematic drawing of the basilar artery being formed by the vertebral arteries at the foramen magnum and ascending to bifurcate into the left and right posterior cerebral arteries

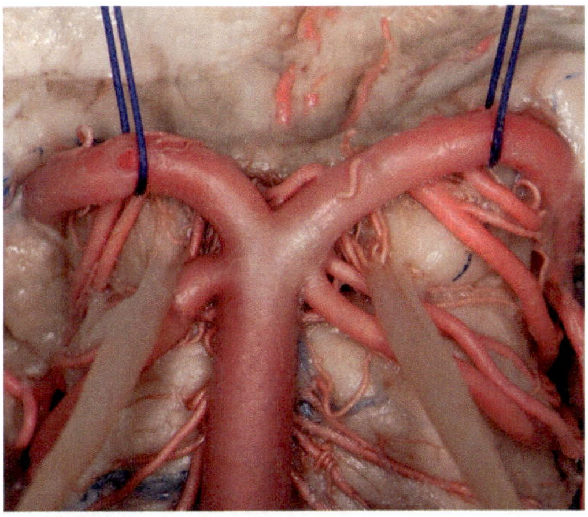

Fig. 1.28 Cadaveric specimen illustrating the basilar artery bifurcating into the left and right posterior cerebral arteries (sutures)

The PCA usually bifurcates opposite the interpeduncular fossa, but some variations have been described.

The PCA can be divided into four parts, P1 through P4:

- *P1 (Precommunicating):* Also referred to as the precommunicating segment (Fig. 1.28), this extends from the basilar termination at its bifurcation to the point at which the PCA is joined by the PComA. An anatomical variant in one-third of hemispheres is known as the fetal configuration, in which P1 is narrower than the PComA and the PCA emerges from the ICA. *Branches:* mesencephalic perforating arteries to the tectum, cerebral peduncles, and these nuclei: oculomotor, Edinger-Westphal, and trochlear; interpeduncular thalamoperforator artery, and medial posteriorchoroidal artery.
- *P2 Segment (Post-communicating):* This portion begins when the PCA is joined by the PComA. It lies within the crural and ambient cisterns and terminates laterally to the midbrain. *Branches:* lateral posterior choroidal artery, thalamogeniculate thalamoperforators to geniculate bodies and pulvinar, interior temporal, posterior temporal, parieto-occipital, and calcarine arteries.
- *P3 Segment (Quadrigeminal):* P3 (Fig. 1.26) courses dorsally from the posterolateral aspect of the midbrain and ambient cistern to reach the quadrigeminal cistern and then terminates at the anterior limit of the calcarine fissure. *Branches:* Quadrigeminal and geniculate branches, posterior pericallosal artery.
- *P4 Segment (Cortical):* It mainly comprises the branches distributed to the cortical surface and it commences at the anterior end of the calcarine sulcus.

Clinical Implications: As mentioned previously, thalmoperforating arteries are multiple and branch directly from P1 and P2 of the PCA to supply the thalamus and midbrain. The artery of Percheron is an anatomical variant of the thalamic blood supply seen in 30% of the population; it branches as a single trunk from the PCA and then branches to supply the thalamus and midbrain. The clinical significance of this variant is that occlusion of this artery results in a classic triad of altered mental status, memory impairment, and vertical gaze palsy. The affected structures are the rostral anteromedial portion of the midbrain and the bilateral mediodorsal and anterior nuclei of the thalamus (Lamot et al. 2015).

Occasionally, the proximal part of one of the posterior cerebral arteries is greatly reduced in size. The blood reaches the terminal region of the vessel from the internal carotid via the posterior communicating artery.

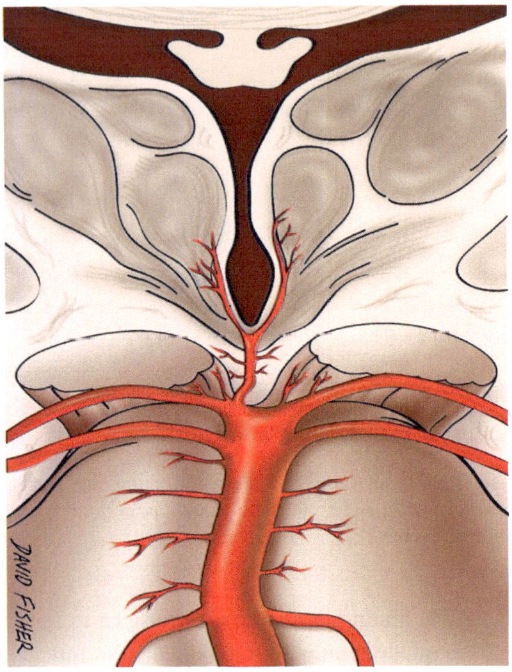

Fig. 1.29 The artery of Percheron arising from the right P1 segment of the posterior cerebral artery. (After Lazzaro et al. 2010)

The posterior cerebral may course below, rather than over, the oculomotor nerve. It may be absent and replaced by an accessory contralateral vessel. The posterior cerebral may provide the anterior cerebral artery. The posterior cerebral artery may arise on one side, from the internal carotid, by an enlargement of the posterior communicating branch.

Side-to-side asymmetry of the P1 segments is common being present on about half of angiograms. When a fetal P-comm artery is present, the ipsilateral P1 is typically hypoplastic.

In some persistent carotid-vertebrobasilar anastomoses, the PCA may be supplied by branches from the carotid system.

Duplication, fenestration, and a bilateral shared origin of the PCA and SCA may be seen involving the P1 segment of the posterior cerebral artery. Absence of the P1 is rare.

There may be a prominent perforating branch (artery of Percheron) (Fig. 1.29) that supplies portions of both the ipsilateral and contralateral thalamus and potentially the midbrain. Large perforating arteries can arise from a hypoplastic P1 segment. The P2 segment can give rise to a quadrigeminal artery, which may give off a branch to the superior cerebellum.

In rare cases, the parieto-occipital, posterior temporal, or calcarine arteries may arise directly from the ICA. Similarly, anomalous anterior choroidal artery supply to temporal, parietal, and occipital cortical regions normally supplied by branches of the PCA can occur. In the ambient cistern, the P2 segment can give rise to an anterior temporal branch that supplies the uncus. This latter branch may give rise to a posterior lateral choroidal artery to the temporal horn.

The calcarine artery is duplicated in 10% of the cases and arises from the parieto-occipital artery in 10% of cases. It can arise from the posterior cerebral artery and can travel deep in the calcarine sulcus or superficial over the medial edge of the calcarine cortex. The calcarine and parieto-occipital arteries can give rise to posterior temporal branches that extend out laterally and inferiorly.

The splenial artery (posterior pericallosal artery) arises from the parieto-occipital artery in 62% of cases but may arise from the calcarine (12%), medial posterior choroidal (8%), posterior temporal (6%), P2 or P3 segments (4% each), or the lateral posterior choroidal artery (4%). The parieto-occipital artery arises from the P3 segment in about half of the population. This vessel can arise from the calcarine, posterior lateral choroidal, or splenial arteries. The lateral posterior choroidal artery arises from the P3 segment in 11% of cases. The posterior medial choroidal artery can arise from the distal P1 segment and can be duplicated. The posterior communicating artery may be absent, or the branch representing it may fail to join the posterior cerebral artery.

References

Basma J, Sorenson J (2018) Surgical anatomy of the posterior fossa. In: Couldwell W (ed) Skull base surgery of the posterior fossa. Springer, Cham. https://doi.org/10.1007/978-3-319-67038-6_1

Blum CA, Yaghi S (2015) Cervical artery dissection: a review of the epidemiology, pathophysiology, treatment, and outcome. Arch Neurosci 2(4):e26670. https://doi.org/10.5812/archneurosci.26670

Campero A, Rubino PA, Rhoton AL (2011) Anatomy of the vertebral artery. In: Pathology and surgery around the vertebral artery. Springer, Paris. https://doi.org/10.1007/978-2-287-89787-0_4

Carvalho V, Cruz VT (2020) Clinical presentation of vertebrobasilar stroke. Porto Biomed J 5(6):e096. https://doi.org/10.1097/j.pbj.0000000000000096

Cawley CM, Zipfel GJ, Day AL (1998) Surgical treatment of paraclinoid and ophthalmic aneurysms. Neurosurg Clin N Am 9(4):765–783

Cooper IS (1953) Ligation of the anterior choroidal artery for involuntary movements; parkinsonism. Psychiatry Q 27(2):317–319. https://doi.org/10.1007/BF01562492

Duane EH (2008) Neuroanatomy: An Atlas of Structures, Sections, and Systems. 7th Ed. Lippincott: Williams & Wilkins

Gibo H, Lenkey C, Rhoton AL Jr (1981a) Microsurgical anatomy of the supraclinoid portion of the internal carotid artery. J Neurosurg 55(4):560–574. https://doi.org/10.3171/jns.1981.55.4.0560

Gibo H, Carver CC, Rhoton AL Jr, Lenkey C, Mitchell RJ (1981b) Microsurgical anatomy of the middle cerebral artery. J Neurosurg 54(2):151–169. https://doi.org/10.3171/jns.1981.54.2.0151

Golshani K, Ferrell A, Zomorodi A, Smith TP, Britz GW (2010) A review of the management of posterior communicating artery aneurysms in the modern era. Surg Neurol Int 1:88. https://doi.org/10.4103/2152-7806.74147

Helgason C, Caplan L (2012) Anterior choroidal artery territory strokes. In: Caplan L, Van Gijn J (eds) Stroke Syndromes, 3rd edn. Cambridge University Press, Cambridge, pp 375–386. https://doi.org/10.1017/CBO9781139093286.032

Henderson AD, Miller NR (2018) Carotid-cavernous fistula: current concepts in aetiology, investigation, and management. Eye (Lond) 32(2):164–172. https://doi.org/10.1038/eye.2017.240

Keedy A (2006) An overview of intracranial aneurysms. McGill J Med 9(2):141–146

Lamot U, Ribaric I, Popovic KS (2015) Artery of Percheron infarction: review of literature with a case report. Radiol Oncol 49(2):141–146. https://doi.org/10.2478/raon-2014-0037

Miao HL, Zhang DY, Wang T, Jiao XT, Jiao LQ (2020) Clinical importance of the posterior inferior cerebellar artery: a review of the literature. Int J Med Sci 17(18):3005–3019. https://doi.org/10.7150/ijms.49137

Quisling RG, Rhoton AL Jr (1979) Intrapetrous carotid artery branches: radioanatomic analysis. Radiology 131(1):133–136. https://doi.org/10.1148/131.1.133

Shapiro M, Becske T, Riina HA, Raz E, Zumofen D, Jafar JJ, Huang PP, Nelson PK (2014) Toward an endovascular internal carotid artery classification system. AJNR Am J Neuroradiol 35(2):230–236. https://doi.org/10.3174/ajnr.A3666

Shetty SR, Anusha R, Thomas PS, Babu SG (2012) Wallenberg's syndrome. J Neurosci Rural Pract 3(1):100–102. https://doi.org/10.4103/0976-3147.91980

Treadwell SD, Thanvi B (2010) Malignant middle cerebral artery (MCA) infarction: pathophysiology, diagnosis and management. Postgrad Med J 86(1014):235–242. https://doi.org/10.1136/pgmj.2009.094292

Common Carotid Artery

2

Dhaval Patel, Corey Leskanic, and Stephen J. Bordes, Jr.

2.1 Embryology

All the great arteries of the head and neck are derived from the six pharyngeal arches (PAA), which are composed of endothelial cells derived from the mesoderm and develop jointly with the six pharyngeal arches. The third PAA is responsible for the formation of the carotid artery system (Frisdal and Trainor 2014).

The ventral pharyngeal artery forms when the ventral portion of the second pharyngeal arch is detached from the dorsal aorta. At the 3 mm embryo stage, the internal carotid artery arises from fusion of the third branchial arch arteries with the distal segment of the dorsal aorta. The common carotid artery (CCA) forms when the ventral pharyngeal artery fuses proximally with the internal carotid arteries (Menshawi et al. 2015).

D. Patel
Department of Anatomical Sciences, University of Queensland School of Medicine, Herston, QLD, Australia

C. Leskanic
Department of Radiology, HCA Florida Trinity, Trinity, USA

S. J. Bordes, Jr. (✉)
Department of Surgery, Louisiana State University Health Sciences Center, New Orleans, LA, USA

2.2 Anatomy

The left and right CCAs differ in origin (Fig. 2.1). The right CCA originates from the brachiocephalic trunk traversing behind the right sternoclavicular joint. The left CCA originates posterolateral to the brachiocephalic trunk, directly from the aorta (Dungan and Heiserman 1996). The left CCA is therefore typically longer than its exclusively cervical counterpart on the right; it is 13.4–14.4 cm long while the right measures 9.4 cm (Dungan and Heiserman 1996; Huber 1982). The diameter of the CCA lumen ranges from 4.3 mm to 7.7 mm, and the intima-media thickness ranges from 0.4 mm to 0.8 mm. It is worth noting that men tend to have larger lumen diameters than women; however, intima-media thickness does not differ between the sexes (Limbu et al. 2006).

As the CCA ascends past the first rib on the sternum it is encompassed by the carotid sheath of the deep cervical fascia, along with the other great vessels in the neck. The arteries are part of the geometric formation known as the carotid triangle, which is formed by the posterior portion of the digastric muscle, the superior portion of the omohyoid muscle, and the sternocleidomastoid muscle on the anterior side. It includes the CCA and the external and internal CAs. It also contains the superior thyroid, lingual, facial, occipital, and ascending pharyngeal arteries and veins (Kikuta et al. 2019).

S. J. Bordes, Jr. et al. (eds.), *The Clinical Anatomy of the Vascular System*,
https://doi.org/10.1007/978-3-031-78326-5_2

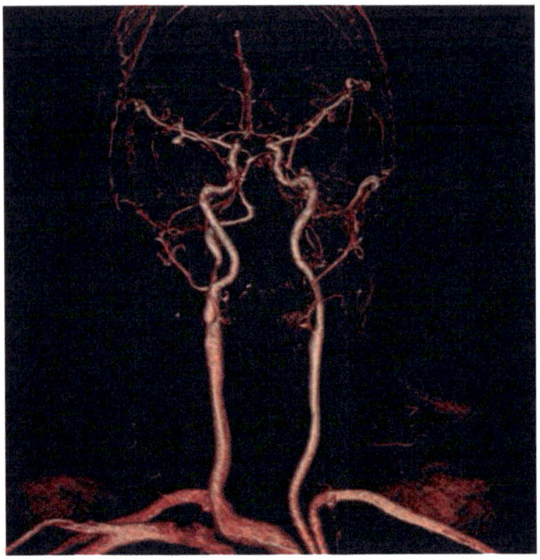

Fig. 2.1 CTA noting the left and right common carotid arteries

In the majority of individuals, at the upper edge of the thyroid cartilage, between the third and fifth vertebrae, the CCA bifurcates into the internal and external carotid arteries. In 90% of cases, the ICA is posterolateral to the ECA. In the other 10%, it has a medial position (Woodward et al. 2017).

2.3 Histology

The CCA comprises three layers: the tunica intima (inner layer), tunica media (middle layer), and tunica adventitia/externa (outer layer). The layers consist of endothelial cells in the tunica intima, longitudinal smooth muscle cells surrounded by connective tissue in the tunica media, and the circumjacent tissue creating the tunica adventitia. There can be adventitial thickening by inflammatory processes and increased vascularization via vasa vasorum in atherosclerosis of the CCA (Falk et al. 2009). Beyond the adventitia, perivascular adipose tissue can be deposited (Falk et al. 2009).

2.4 Imaging

Under ultrasound imaging, the carotid intima media thickness (CIMT) is measured to assess the burden of atherosclerosis. During the early phase of the systemic inflammatory process of atherosclerosis, medial cells have a hypertrophic response. This response can increase the CIMT in the CCA more than plaque formation in the internal carotid artery or carotid bulb. An increased CIMT is often related to stroke and hypertension.

Additionally, the carotid extra-media thickness (EMT) is visualized by ultrasound and is defined as the distance between the carotid media-adventitia interface and the jugular lumen (Falk et al. 2009). The carotid EMT is obtained approximately 1–1.5 cm proximal to the carotid bulb, where the carotid is the narrowest (Falk et al. 2009). The measurement encompasses part of the arterial media, arterial adventitia, perivascular adipose tissue, and the venous wall (Falk et al. 2009). The carotid EMT is greater in obesity, where there tends to be more perivascular adipose tissue (Falk et al. 2009).

While ultrasound is often the first best imaging test for evaluating carotid stenosis, an angiogram is the gold standard technique for visualizing the entire anatomical course of the vessels (Zach et al. 2012). For this, arterial access is gained and a catheter is advanced just proximal to the vessels. Contrast dye is then injected under CT visualization. However, computed tomographic angiograpy (CTA) is quickly growing in popularity among hospitals, especially in situations involving trauma. CT angiography is significantly cheaper, less invasive, and faster than angiography. The patient is typically placed on a CT scanner and scanned without and then with IV iodinated contrast. Once the contrast has been delivered, there is a brief delay before the CT machine scans the patient.

2.5 Variations

The bifurcation level of the CCA can vary greatly, a higher bifurcation being more common. A study showed that the right CCA bifurcated at the C3 level in 50% of cases, at C4 in 40%, and at C2 in 10% (Anu et al. 2007). The same study showed that the left CCA bifurcated at the C3 level in 55% of cases, at C4 in 35%, at C2 in 9%, and at C5 in 1% (Anu et al. 2007).

The origin of the CCAs can also vary greatly. Natis et al. created a classification system from I to VIII for variations of the aortic arch (Natsis et al. 2009). The smaller numbers denoted greater frequency. Type I, the normal branching pattern, was found in 83% of cases (Natsis et al. 2009). In type II (15%), the aortic branch divided into two branches, one of which gave rise to the right subclavian artery and the right and left CCAs (Natsis et al. 2009). Type III (0.79%) was not directly related to the CCA. In type IV (0.16%) there was a common origin for the left and right CCAs, so the aortic arch had three branches: the right and left subclavian arteries, and a bicarotid trunk from which the left and right CCAs originated (Natsis et al. 2009). In type V (0.16%), the left and right CCAs had a common origin with the right subclavian artery. In type VI (0.16%) the aortic arch had two branches, one bifurcating into the left and right CCAs and the other bifurcating into the right and left subclavian arteries (Natsis et al. 2009). In type VII (0.16%), the brachiocephalic trunk was absent, and both the right and left CCAs originated from the aortic arch (Natsis et al. 2009). Type VIII (0.16%) was not directly related to the CCA (Menshawi et al. 2015).

The point at which the left and right CCAs originate can vary depending on the type of arch variation. The absence of the brachiocephalic trunk, or a lower bifurcation point, means the right CCA originates lower than usual; a higher bifurcation point causes its origin to be higher than usual. When the right CCA originates from the aorta arch, it can run behind the trachea or the esophagus and form loops in the neck (Tubbs et al. 2016).

In some cases, the CCA is absent and the external and internal CAs originate directly from the aorta (Tubbs et al. 2016). In other cases, the termination of the CCA varies in that the external or the internal CA is missing. Cases of the superior thyroid artery emerging from the CCA instead of the external CA have also been reported (Tubbs et al. 2016; Lo et al. 2006).

2.6 Pathology

The CAs facilitate the delivery of blood to the brain. In carotid artery disease, blockages can decrease the blood supply to the brain. Such blockages are due to the buildup of cholesterol leading to narrowing of the artery, or stenosis. Risk factors include smoking, high blood pressure, high cholesterol, diabetes, and advanced age (Sobieszczyk and Beckman 2006). The narrowing of the artery can disrupt the blood supply to the brain leading to stroke.

Carotid sinus hypersensitivity (CSH) is a cause of syncope in the elderly. It is characterized by bradycardia, vasodilation, and hypotension. When patients show the characteristic symptoms due to carotid sinus massage, this defines CSH. It results from the stimulation of the baroreceptors located in the region where the CCA bifurcates into the internal and external CAs, known as the carotid sinus. There are three subtypes of CSH: cardioinhibitory, vasodepressor, and mixed. The pathophysiology of CSH is not well established but a few mechanisms have been investigated to explain how changes in baroreceptor function could cause it. These explanations include atherosclerotic noncompliance, sternocleidomastoid proprioceptive denervation, and general autonomic dysfunction (Amin and Pavri 2015). In patients with CSH, the central baroreceptor response can be impaired due to the accumulation of hyperphosphorylated tau in tyrosine hydroxylase-containing neurons (Miller et al. 2008).

Carotid artery aneurysms are rare, accounting for only 0.4–4% of all peripheral artery aneurysms (Fankhauser et al. 2015). They can be classified as true, in which all the layers of the artery are involved, or false, also known as pseudoaneurysm. The etiology of aneurysms affecting the extracranial CCA (ECCA) is very broad; true aneurysms due to atherosclerosis accounted for 37–41% of the total in one series (Fankhauser et al. 2015). Another series showed 35–51% of aneurysms being linked to trauma (Fankhauser et al. 2015). Lastly, pseudoaneurysms were linked to prior endarterectomy sites in 26–57% in another series (Fankhauser et al. 2015). It is important to note that the ECCA also includes the internal and external carotid arteries, and most aneurysms are linked to them, not to the CCA. Less common etiologies include true mycotic aneurysm, trauma, fibromuscular dysplasia, spontaneous dissection, connective tissue disorders, prior radiation, and congenital defects (Kraemer and Zhou 2019).

CCA dissection is a rare disease that can lead to stroke. In a dissection, the layers of the artery walls dissociate, causing a tear. The dissociation can be spontaneous or due to trauma. It is important to recognize CCA dissection as it was associated with up to 41% of aortic arch dissections in one series (Zach et al. 2012).

2.7 Clinical and Surgical Significance

Atherosclerosis in the CCA can be assessed by ultrasonography to determine the common carotid artery intima-media thickness (CCA-IMT). An increase in the CCA-IMT correlates with an increased risk of potential adverse cardiovascular events and myocardial infarction. Risks of stroke and transient ischemic attacks are greater in individuals with higher CCA-IMT values. In addition, higher CCA-IMT scores are associated with more severe neurological impairment from ischemic stroke.

While internal carotid artery dissection etiologies are most often spontaneous, or due to traumatic or iatrogenic events, dissections of a common carotid artery for those reasons are very rare (Zach et al. 2012). A common carotid artery dissection (CCAD) is more often associated with aortic arch dissection (AAD). CCAD has been associated with as many as 41% of AAD cases (Zach et al. 2012). While most of the symptoms can be associated with an AAD, patients in whom CCAD is the primary pathology can present with headache, neck pain, hemiparesis, hemiplegia, aphasia, monocular visual deficits, and/or altered level of consciousness. CCAD can progress to common carotid artery occlusion (Zach et al. 2012). There can be unilateral absence of a carotid pulse on physical exam (Zach et al. 2012). Early diagnosis is crucial in CAD occlusion events in order to improve neurological outcomes (Zach et al. 2012). While cerebral angiography is the gold standard for diagnosis of CCAD, CTA is more widely available and is less invasive (Zach et al. 2012). However, urgent carotid duplex ultrasonography can be used for emergency cases (Zach et al. 2012). Treatment of CCAD varies with the etiology, but available treatments include antiplatelet therapy, anticoagulation therapy, endovascular angioplasty and stenting, CEA, and bypass surgery (Zach et al. 2012).

Extracranial carotid artery aneurysms, ECAAs, are rare but can rupture leading to life-threatening hemorrhage. Etiologies of ruptured ECAAs include infection, connective tissue disorders, atherosclerosis, and trauma. The most common location for ECAAs is in zone II of the neck (41%), compared to zone III (23%) and zone I (8%). While internal carotid artery involvement is most common, ECAAs can also involve the CCA. Open reconstructive surgery is frequently reported as the method of treatment, especially for ECAAs located in zone II. An endovascular approach can also be performed in cases with a common etiology of posttraumatic injury, while ligation has been reported to be effective in cases of infection. In most unruptured ECAAs, open reconstructive surgery is the preferred method of treatment, but conservative non-operative treatments can be used for patients with minor or no symptoms.

References

Amin V, Pavri BB (2015) Carotid sinus syndrome. Cardiol Rev 23(3):130–134. https://doi.org/10.1097/CRD.0000000000000041

Anu VR, Pai MM, Rajalakshmi R, Latha VP, Rajanigandha V, D'Costa S (2007) Clinically-relevant variations of the carotid arterial system. Singapore Med J 48(6):566–569

Dungan DH, Heiserman JE (1996) The carotid artery: embryology, normal anatomy, and physiology. Neuroimaging Clin N Am 6(4):789–799. PMID: 8824131

Fankhauser GT, Stone WM, Fowl RJ et al (2015) Surgical and medical management of extracranial carotid artery aneurysms. J Vasc Surg 61(2):389–393. https://doi.org/10.1016/j.jvs.2014.07.092

Frisdal A, Trainor PA (2014) Development and evolution of the pharyngeal apparatus. Wiley Interdiscip Rev Dev Biol 3(6):403–418. https://doi.org/10.1002/wdev.147

Huber P (1982) Cerebral angiography, 2nd edn. Thieme Medical Publishers, New York

Kikuta S, Iwanaga J, Kusukawa J, Tubbs RS (2019) Triangles of the neck: a review with clinical/surgical applications. Anat Cell Biol 52(2):120–127. https://doi.org/10.5115/acb.2019.52.2.120

Kraemer CJK, Zhou W (2019) Carotid aneurysm review. Int J Angiol 28(1):17–19. https://doi.org/10.1055/s-0039-1677675

Limbu YR, Gurung G, Malla R, Rajbhandari R, Regmi SR (2006) Assessment of carotid artery dimensions by ultrasound in non-smoker healthy adults of both sexes. Nepal Med Coll J 8(3):200–203. PMID: 17203830

Lo A, Oehley M, Bartlett A, Adams D, Blyth P, Al-Ali S (2006) Anatomical variations of the common carotid artery bifurcation. ANZ J Surg 76(11):970–972. https://doi.org/10.1111/j.1445-2197.2006.03913.x

Menshawi K, Mohr JP, Gutierrez J (2015) A functional perspective on the embryology and anatomy of the cerebral blood supply. J Stroke 17(2):144–158. https://doi.org/10.5853/jos.2015.17.2.144

Miller VM, Kenny RA, Slade JY, Oakley AE, Kalaria RN (2008) Medullary autonomic pathology in carotid sinus hypersensitivity. Neuropathol Appl Neurobiol 34(4):403–411. https://doi.org/10.1111/j.1365-2990.2007.00903.x

Natsis KI, Tsitouridis IA, Didagelos MV, Fillipidis AA, Vlasis KG, Tsikaras PD (2009) Anatomical variations in the branches of the human aortic arch in 633 angiographies: clinical significance and literature review. Surg Radiol Anat 31(5):319–323. https://doi.org/10.1007/s00276-008-0442-2

Sobieszczyk P, Beckman J (2006) Carotid artery disease. Circulation 114(7):e244–e247. https://doi.org/10.1161/CIRCULATIONAHA.105.542860

Tubbs RS, Shoja MM, Loukas M (2016) Bergman's comprehensive encyclopedia of human anatomic variation. In: Common carotid and cervical part of the internal carotid arteries, pp 475–476. https://doi.org/10.1002/9781118430309

Woodward PJ, Griffith JF, Antonio GE, Ahuja AT, Wong KT, Kamaya A, Wong-You-Cheong J (2017) Imaging anatomy: ultrasound e-book. Elsevier Health Sciences

Zach V, Zhovtis S, Kirchoff-Torres KF, Weinberger JM (2012) Common carotid artery dissection: a case report and review of the literature. J Stroke Cerebrovasc Dis 21(1):52–60. https://doi.org/10.1016/j.jstrokecerebrovasdis.2010.05.001

External Carotid Artery

3

Patrick J. McCarty

3.1 Embryology

The pharyngeal arch arteries (PAAs) form during the third to fifth weeks of development when mesoderm-derived angiogenic mesenchymal cells migrate from the pharyngeal arches toward the aortic sac to form the endothelial lining of the developing vessels (Eichmann et al. 2005). The six paired branches emerge in a cranial to caudal sequence and terminate in the dorsal aortae. Neural crest cells within the pharyngeal arches give rise to the vascular smooth muscle cells and pericytes that contribute to the outer layer of the developing vasculature. The proximal portion of the common carotid artery (CCA) arises from the ventral aortic root connecting the third and fourth PAA while the homologous structure of the dorsal aorta regresses. As this ventral root elongates, the third PAA remains attached and ultimately forms the proximal portion of the internal carotid artery (ICA). The distal portion of the ICA is formed by the cranial segment of the dorsal aortae. The external carotid artery (ECA) then develops de novo as a branch of the CCA via angiogenesis (Frisdal and Trainor 2014).

The ECA branches develop between the first and second PAA through a complex series of events involving the formation and remodeling of primitive carotid-vertebrobasilar anastomoses. Briefly, the ascending pharyngeal artery is partially formed by the primitive hypoglossal artery, the occipital artery is partially formed from the primitive proatlantal artery, and the maxillary artery is formed from the maxillomandibular division of the transient stapedial artery, itself a remnant of the first and second arch arterial plexus. Partial regression of the second aortic arch forms a transient ventral pharyngeal artery from which both the lingual and facial arteries arise (Kiyosue 2020; Standring 2016).

3.2 Anatomy and Topography

The ECA typically begins at the superior border of the thyroid cartilage at the level of the fourth cervical vertebra (Figs. 3.1 and 3.2) and originates anteromedially from the bifurcation of the common carotid artery. It originates within the carotid triangle, which is bound posteriorly by the anterior margin of the sternocleidomastoid muscle, anteroinferiorly by the superior belly of the omohyoid muscle, and superiorly by the stylohyoid and the posterior belly of the digastric muscle. Within the carotid triangle, the ECA is covered by the skin, superficial fascia, cervical branch of the facial nerve, platysma muscle, and the investing layer of the deep cervical fascia. Portions of the thyrohyoid, hyoglossus, and

P. J. McCarty (✉)
Tulane University School of Medicine,
New Orleans, LA, USA
e-mail: pmccarty2@tulane.edu

S. J. Bordes, Jr. et al. (eds.), *The Clinical Anatomy of the Vascular System*,
https://doi.org/10.1007/978-3-031-78326-5_3

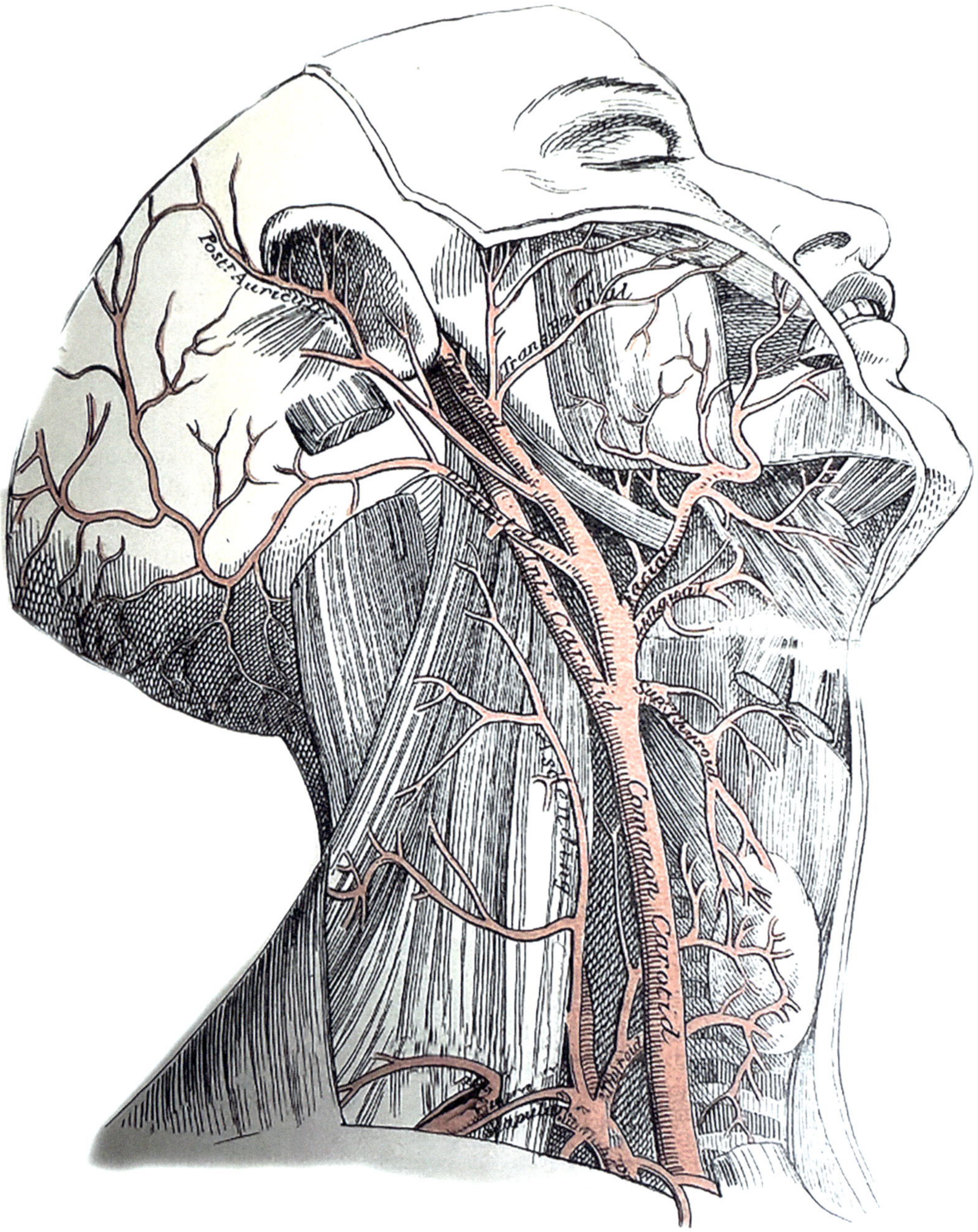

Fig. 3.1 Schematic drawing of the right carotid arteries noting the external carotid artery and its superior thyroid and lingual branches traveling anteriorly and posterior auricular and occipital branches traveling posteriorly. (After Allen 1882)

inferior and middle pharyngeal constrictor muscles make up the floor of the carotid triangle.

At a level just distal to the common carotid bifurcation (CCB), the ECA is crossed anterolaterally by the hypoglossal nerve running with the digastric muscle and the lingual and facial veins. At this level, the ECA is positioned lateral to the pharyngeal wall, superior laryngeal nerve, and the ascending laryngeal artery. As it ascends and exits the carotid triangle it is crossed by the pos-

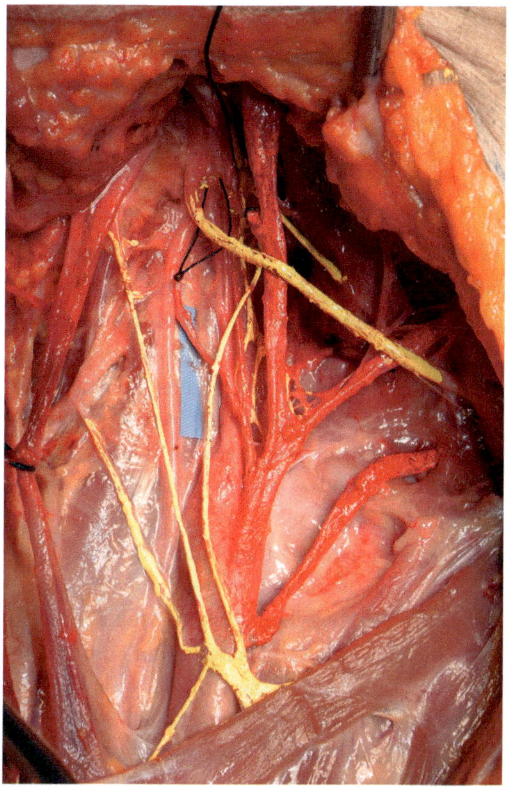

Fig. 3.2 Right-sided cadaveric dissection demonstrating branches of the external carotid artery traveling deep to the hypoglossal and ansa cervicalis nerves

terior belly of the digastric and stylohyoid muscles before entering the posteromedial aspect of the parotid viscera, wherein it lies medial to the retromandibular vein and facial nerve. Within the parotid viscera, approximately 2 cm proximal to the condylar head and deep to the posterior ramus of the mandible, the terminal branches of the ECA bifurcate with the superficial temporal artery coursing superiorly and crossing the root of the zygoma, and the maxillary artery coursing anteriorly within the infratemporal fossa between the neck of the mandibular condyle and the sphenomandibular ligament toward the pterygopalatine fossa (Brennan et al. 2020). At this level, the ECA is separated from the ICA by the styloid process and is crossed by and superficial to the styloglossus and stylopharyngeus muscles, the glossopharyngeal nerve, and the pharyngeal branch of the vagus nerve.

As the ECA ascends from the common carotid bifurcation it courses slightly anterior before heading toward the posterior neck of the mandible and the parotid viscera. It is positioned anteromedially to the ICA at its origin, but along its course becomes anterior and then finally lateral to the ICA near the level of separation by the styloglossus and stylopharyngeus muscles, the glossopharyngeal nerve, and the pharyngeal branch of the vagus nerve. Its size diminishes along its course as it gives off eight named branches, including its two terminal branches. These branches can be classified by the position in which they originate from the ECA as anterior, posterior, medial, and terminal.

Anterior branches:

- Superior thyroid artery
- Lingual artery
- Facial artery

Posterior branches:

- Occipital artery
- Posterior auricular artery

Medial branch:

- Ascending pharyngeal artery

Terminal branches:

- Maxillary artery
- Superficial temporal artery

As a major distributing vessel, the ECA supplies blood to the face, scalp, tongue, upper and lower teeth and gingivae, palatine tonsil, paranasal sinuses and nasopharyngeal tube, external and middle ears, pharynx, larynx, superior pole of the thyroid gland, and a portion of the meninges (Sadler 2018; Kiyosue 2020; Standring 2016).

3.3 Imaging

The ECA is often imaged as a component of the extracranial carotid and vertebral arterial systems. Doppler ultrasonography (US) is a readily accessible modality for the initial evaluation of carotid artery disease owing to its ability to capture velocity measurements, which is particularly useful for assessing carotid stenosis. However, US alone is limited in some situations by its inability to define anatomical details and its small field of view relative to other available imaging methods. Additionally, it can be impossible to obtain a complete study containing the CCA and both the ICA and ECA when the carotid bifurcation is high. Thus, US is typically most valuable as a general screening modality for suspected carotid vessel disease and as a complement to other more robust imaging studies.

Conventional contrast angiography has been the standard for extracranial carotid visualization both for evaluation and during intervention. Digital subtraction angiography (DSA) uses a series of contrast and noncontrast X-ray images that can represent the vasculature accurately in real-time and requires a lower dose of contrast media than some other angiographic imaging techniques. DSA is often used for interventions such as endovascular aneurysm repair, arterial balloon angioplasty, arterial stenting, endovascular embolization, and thrombectomy. However, there are important post-procedural considerations regarding recovery and monitoring of potential adverse events that typically are not required in other angiographic investigations.

The development of multidetector spiral CT scanners allowed for quicker acquisition times and improved visualization of blood vessel morphology and size (Kaufman and Lee 2014). Aided by advances in imaging reconstruction and post-processing techniques such as multiplanar reformation (MPR), maximum intensity projection (MIP), and volume rendering (VR), CT angiography (CTA) has been recognized as a cost-effective and time-sensitive imaging technique for evaluating and diagnosing numerous vascular pathologies of the carotid artery system (Calhoun et al. 1999; Lell et al. 2006). CTA is particularly useful for locating calcified atheromatous lesions precisely and determining the extent of carotid stenosis.

Magnetic resonance angiography (MRA) has also been used extensively for evaluating the cervical vessels and the extracranial carotid system. Noncontrast-enhanced MRA includes additional sequencing techniques such as time of flight (TOF) and phase contrast, which rely on flow phenomena to quantify blood flow. Some vascular territories such as the cerebral vasculature provide more advantageous flow conditions (i.e., laminar flow) than others. The application of TOF-MRA to the extracranial carotid systems is potentially limited by motion artifacts from pulsatile flow and loss of signal either from stenosis-induced turbulent flow (via spin-dephasing) or during instances of severe stenosis where distal flow is significantly reduced. Contrast-enhanced MRA can be used to reduce or eliminate motion artifacts in instances where pulsatile flow is of concern.

Although each of these imaging modalities is clinically useful on its own, a multimodal imaging strategy, when clinically feasible, typically provides a more accurate and comprehensive representation of the vascular anatomy and associated pathology. Johnston et al. (2002) showed that the rate of misclassification of surgically amenable carotid stenosis was lower with multimodal Doppler US/contrast-enhanced MRA (17%) than with Doppler US (36%) or contrast-enhanced MRA (24%) alone. Although intra-arterial angiography remains the standard, a meta-analysis of noninvasive imaging modalities showed that contrast-enhanced MRA was more sensitive and specific than Doppler US, noncontrast-enhanced MR angiography, and CT angiography for patients with 90–99% stenosis (Wardlaw et al. 2006). Each imaging modality has its own set of benefits, risks, and limitations that require careful consideration in the diagnostic and preprocedural settings (Calhoun et al. 1999; Johnston et al. 2002; Cappabianca et al. 2012; Kiyosue 2020; Kaufman and Lee 2014).

3.4 Histology

The ECA is a muscular artery that receives blood directly from the common carotid artery and acts as a major distributor of blood to the face, head, and neck. As in all blood vessels, its wall comprises three histologically distinct layers: the tunica intima, tunica media, and tunica adventitia. The tunica intima is the innermost layer and consists of a layer of simple squamous endothelium resting on a basal lamina with a subendothelial layer of connective tissue. A prominent feature of muscular arteries such as the ECA is a distinct internal elastic lamina that separates the intima from the media. This lamina is typically folded or wavy in appearance owing to the contraction of the smooth muscle within the tunica media. The ECA has a thick tunica media containing multiple concentric layers of smooth muscle cells that communicate with each other via gap junctions. Dynamic contraction and relaxation of this smooth muscle layer confer the ability to maintain blood pressure across a range of physiological perturbations and functional demands. The tunica media of the ECA also contains elastin fibers arranged as laminae that run parallel to the smooth muscle cells, although this feature is much less prominent than in large elastic arteries such as the aorta. The ECA also contains an external elastic lamina that is less compact than the internal lamina and separates the tunica media from the tunica adventitia. Although not easily distinguishable with traditional hematoxylin and eosin stains, the elastic fibers of the internal and external elastic laminae and those within the tunica media can be better visualized with Verhoeff stain. A modified version of this stain known as the Masson trichrome-Verhoeff stain permits elastin fibers, muscle fibers, and collagen fibers to be discriminated, which makes it particularly useful for studying vascular pathologies (Buk 1984). The tunica adventitia is the outermost layer of the ECA and consists primarily of fibroelastic connective tissue, with a higher relative density of elastin fibers than in the inner adventitia. It is much thinner than the tunica media in muscular arteries and typically blends with the surrounding connective tissue. It also contains lymphatic capillaries, nerves, and microvasculature known as the vasa vasorum, which extend into the outermost part of the tunica media to provide the blood vessel wall cells with oxygen and nutrients (Mescher 2018; Standring 2016).

3.5 Variations

Several radiological and cadaveric studies have indicated a relatively common incidence of variations in ECA anatomy and topography. The vast majority of these anatomical variations concern anomalous origins of branches, although the ECA itself is also subject to variation, particularly regarding its level of origin (Gluncic et al. 2001; Ozgur et al. (2008). The relatively common incidence of anatomical variation is probably attributable to the complex embryological remodeling processes discussed in the previous section. There have been many attempts to classify the most commonly observed branching variants of the ECA (Yamamoto et al. 2019; Jitpun et al. 2019; Cobiella et al. 2021). However, these studies are not uniform in classification parameters. Bergman's Comprehensive Encyclopedia of Human Anatomical Variation (Tubbs et al. 2016) provides an in-depth review of the reported variations in ECA anatomy.

The ECA can be absent unilaterally or, less commonly, bilaterally. Its origin is most commonly located at the level of the superior border of the thyroid cartilage anteriorly and the fourth cervical vertebra posteriorly, although it is not uncommon for it to be at a higher level. Variations in its branching pattern can be associated with its level of origin (Yamamoto et al. 2019). Most branching variation involves the proximal anterior branches, namely the superior thyroid, lingual, and facial arteries (Elsllabi et al. 2021). In commonly reported anomalous origins of the lingual and facial arteries, these branches arise from a common linguofacial trunk, the incidence ranging from 8% to 29.2% across various cadaveric and angiographic studies (Devadas et al. 2018; Navakalyani et al. 2016; Jitpun et al. 2019), possibly higher on the right side than the left (Heltzel and Jaynes 2011). A thyrolingual or thyrolinguofacial trunk is rare, each ranging from 0 to 2.5% (Devadas et al. 2018; Cobiella et al. 2021). Intra-individual asymmetries in branching variations, level of origin, and vessel diameter have also been reported. The carotid artery system also undergoes age-related remodeling from childhood to adulthood.

Variations in the Level of Origin

When they deviate from standard anatomy, the CCB and the origin of the ECA are more likely to be seen at a higher-than-normal than a lower-than-normal level. In approximately 30%, the CCB occurs at the C3–4 intervertebral level vs. approximately 12% at C4–5 (Standring 2016; Tubbs et al. 2016). Michalinos et al (2016). reviewed the literature regarding the level of the CCB in relation to various anterior and posterior anatomical landmarks. They reported that the incidence of a CCB located at or above the superior margin of the thyroid cartilage ranged from 72.5 to 95.7% (Lucev et al. 2000, 72.5%; Lo et al. 2006, 94%; Al-Rafiah et al. 2011, 94.9%; McNamara et al. 2015 95.7%).

Branching Variation

In an anatomical study of 40 cadavers by Devadas et al. (2018) the incidence of a linguofacial trunk was 20%; this was the most commonly observed branching variant. Similarly, Zumre et al. found a 20% incidence of a linguofacial trunk in their study of 40 fetal cadavers. Navakalyani et al. (2016) observed a linguofacial trunk in 8% of fetal cadavers. Jitpun et al. analyzed CTAs from 100 patients and found the incidence to be 29.2%.

In a large clinical study of 532 ECAs from 302 patients using digital subtraction angiography, Yamamoto et al. (2019) classified ECA branching variants into three types on the basis of the number of branches arising in proximity from a common stem of the proximal ECA: Type A, B, and C variants were defined as two, three, or ≥four branches arising at a common point from the proximal ECA, respectively. Type A was defined as the classical branching pattern in which all ECA branches arise individually. The authors reported the frequency of types A, B, and C to be 64.6%, 25.2%, and 10.2%, respectively. They also reported that 96.3% of patients with the type C branching variant had a high CCB at or above the C3/4 vertebral level, indicating a potential relationship between the height of the CCB and the incidence of a more proximally clustered ECA branching variant. Furthermore, 24% of the type C group had a linguofacial trunk.

An anomalous origin of the occipital artery is rare. Uchino et al. (2011) reported an incidence of 0.21% in a retrospective review of MRA studies of 2866 patients. In that review, variant origins of the occipital artery most commonly involved the ICA ($n = 4$); origins from the CCB ($n = 1$) and vertebral artery ($n = 1$) were less common. The ascending pharyngeal artery can arise directly from the CCA (65–80%), and less commonly from the occipital (20%), internal carotid (8%), or facial (2%) arteries (Standring 2016).

Intraindividual Variation

Heltzel and Jaynes (2011) observed contralateral differences in the origin of the superior thyroid arteries (STA) in their anatomical study of 73 cadavers. Specifically, the STA more frequently arose from the ECA than the CCA on the right side (28 and 20, respectively), whereas on the left side, the STA more frequently arose from the CCA than the ECA (34 and 19, respectively). They also observed that the left side had a higher prevalence of individually branching lingual and facial arteries than the right (81% vs 72%, respectively), suggesting an increased prevalence of right-sided linguofacial trunks. Yamamoto et al. observed bilaterality in types B and C branching variants in 10.4% and 2.2%, respectively, of the 230 patients in their study where bilateral angiograms were available. Unilateral types B and C branching variants with a normal (type A) contralateral vessel were observed in 25.2% and 10.4% of patients, respectively. Devadas et al. also observed bilateral linguofacial trunks in 12.5% of specimens.

Jitpun et al. observed no significant difference in the level of the CCB between the two sides in their study of 200 carotid systems from 100 patients. Schulz and Rothwell (2001) analyzed the angiograms of bilateral carotid artery systems in 755 patients obtained from the European Carotid Surgery Trial and identified intra-individual variation in the vessel diameter and area ratios of the carotid bifurcation, particularly in the ratio of outflow to inflow area, which differed by ≥25% between sides in 42% of patients.

Age-Related Variations

A study of 95 angiograms conducted by Seong et al. (2005) demonstrated that the angle split between the ICA and ECA relative to the CCA undergoes age-related morphological remodeling from childhood to adulthood, particularly during puberty. They reported that the ECA initially appears as a continuation of the CCA, but later during adolescence, as the carotid sinus and ICA develop in response to increased flow demand from the brain, the system is remodeled so that both the ECA and ICA appear as relatively equal continuations of the CCA.

The relatively common incidence of anatomical variations in branching patterns and the origin of the ECA have implications for all interventionists conducting vascular manipulations of the ECA and its related structures in the head and neck. As will be discussed, pre-emptive knowledge of these variations in anatomy is important in presurgical procedural planning (Tubbs et al. 2016; Kiyosue 2020; Stranding 2016).

3.6 Pathology

The most common pathology affecting the ECA is atherosclerosis. It is well known that atherosclerosis typically occurs at bifurcation points where the arterial geometry influences local hemodynamics. Atheromatous plaque buildup in the CCB can extend and stenose or obstruct the ECA. Atherosclerotic pathogenesis occurs with hyperlipidemia and locally disturbed hemodynamics, ultimately leading to endothelial dysfunction through a well-defined mechanism of leukocyte infiltration, lipid particle entrapment and oxidation, myointimal cell proliferation, collagen deposition, and ultimately necrosis. Once formed, plaques are susceptible to further complications including calcification, rupture, erosion, intraplaque hemorrhage, and mural thrombosis, each with its own related but distinct clinical sequelae. Plaques involving the ECA either in isolation or by invasion from the ICA or CCA can disrupt its ability to function in providing collateral blood flow.

Variation in the angle of the CCB can contribute to atherogenesis by disrupting normal hemodynamics. Lee and Kwak (2021) examined the association between morphological factors and plaque vulnerability associated with an intraplaque hemorrhage, lipid-rich necrotic core, and/or thin fibrous cap, and found that patients with vulnerable plaques had a smaller angle of the CCB than those with stable plaques (39.47 degrees vs. 47.60 degrees). Nevertheless, there was no significant difference between the groups in the ECA diameter or area and plaque stability, implicating the morphology of the CCB, rather than the ECA, in atherogenesis.

Spontaneous dissection of the ECA alone is rare, but it can occur indirectly via an extension of a common carotid vessel dissection. Traumatic dissection of the ECA can result from blunt trauma due to motor vehicle accidents, assaults, and sports injuries (Wang et al. 2021). Individuals with certain structural abnormalities and connective tissue disorders such as fibromuscular dysplasia and Ehlers-Danlos syndrome could be predisposed to dissections of conducting arteries including the ECA. Pseudoaneurysm of the ECA is also rare, with pathologies ranging from traumatic injury, irradiation, and tonsillitis/tonsillectomy (Nadig et al. 2009).

3.7 Clinical and Surgical Significance

Carotid endarterectomy (CEA) and carotid angioplasty and stenting (CAS) are the most frequent carotid revascularization procedures for reducing the risk of stroke, and considerable work has been done to generate guidelines for procedure selection. (Bagley and Priest, 2020). CEA has demonstrated clinical benefit as the first-line therapy for revascularization in patients with symptomatic stenosis, although CAS is an acceptable alternative for patients with various risk factors including a variant carotid system anatomy. The specific variables involved in periprocedural risk stratification, and the details of long-term efficacy studies, are beyond the scope of this chapter. However, it should be noted that the selection of CEA vs. CAS in asymptomatic stenosis is less clear and remains an active area of study.

Inclusion/exclusion of the ECA during CEA is contingent upon its involvement and thus the overall plaque morphology. However, the ECA is often affected and therefore routinely involved in the general CEA procedure. Patient-specific surgical approaches to, and procedural considerations for, CEA can vary depending on the anatomy of the carotid artery system. Eversion CEA can require the extension of the arteriotomy and division of the superior thyroid branch of the ECA. The level of the CCB can also influence the incision type (longitudinal vs. transverse vs. combined) and the degree of head positioning.

The ECA can also be the sole target vessel for endarterectomy, as in selective ECA endarterectomy in patients with isolated ECA occlusion or ipsilateral ICA occlusion. Selective ECA endarterectomy for advanced ICA stenosis highlights the importance of the ECA in providing collateral cerebral blood flow. Extracranial collateral flow occurs via ipsilateral ECA-derived middle meningeal and facial artery branches and the ICA-derived ophthalmic artery. While protective in cases of extracranial ICA occlusion, collateral flow from the ECA can supply arterial blood to skull base tumors and can be involved in arteriovenous malformations, namely carotid-cavernous fistulas (Hans 2018; Standring 2016).

This critical role in providing collateral cerebral blood flow has prompted investigations into the natural history of the ECA following endarterectomy and angioplasty with stenting (de Borst et al. 2007; Woo et al. 2007). Brown et al. (2015) measured the rate of ECA occlusion after CAS (3.8%) vs CEA (0.4%). They attributed the difference to stent-induced flow disturbances in the CAS group and routine treatment of stenotic ECA during CEA. Willfort-Ehringer et al. (2003) prospectively studied both the ipsilateral and contralateral ECA after ICA stenting in 112 patients. They found that after 24 months, the ipsilateral ECA had a significantly higher flow ratio than the contralateral ECA, indicating a progression of arterial occlusive disease, similar to the other reports (Reichmann et al. 2013).

The intraoperative and post-procedural risks for these carotid revascularization techniques include stroke and myocardial infarction. Cranial nerve injury is of particular concern in CEA and most commonly involves the vagus and hypoglossal nerves. Transient cranial nerve injury is more common than permanent injury. While numerous risk factors have been identified, a meta-analysis revealed that urgent procedures, and neurological or bleeding events requiring a return to the operating room, were the only factors that significantly increased the risk of cranial nerve injury after CEA (Kakisis et al. 2017).

Transcarotid artery revascularization (TCAR) is a relatively new approach to carotid revascularization that employs a hybrid open-endovascular technique to CAS. While yet to be evaluated in large randomized controlled trials, TCAR has proved inferior to neither CEA nor CAS. Using the Society of Vascular Surgery's Vascular Quality Initiative, which prospectively collects hundreds of patient- and procedure-specific variables from 701 centers in the United States and Canada, Dakour-Aridi et al. (2020) demonstrated that TCAR reduced cranial nerve injury more than CEA, and the difference was statistically significant. Others have shown that TCAR provides similar results to CAS (Malas et al. 2019; Schermerhorn et al. 2019). The TCAR technique minimizes certain transfemoral CAS procedural maneuvers that ostensibly increase the risk of microembolic events and employs a unique flow reversal mechanism that achieves proximal neuroprotection throughout the procedure (Schermerhorn et al. 2019; Sujijantarat et al. 2021).

Detailed knowledge of the ECA branches is imperative for certain procedures involving embolization or chemoembolization. The ascending pharyngeal (APA) branch of the ECA has numerous intracranial and extracranial anastomoses. These robust connections make it very useful in vascular studies and embolization of various dural vascular lesions. Stemming from the neuromeningeal division of the APA, the hypoglossal and jugal branches anastomose with branches of the meningohypophyseal trunk and inferolateral trunk, respectively, of the petrous segment of the ICA. The pharyngeal division of the APA also anastomoses with the inferolateral trunk of the petrous ICA via its foramen lacerum branch. In addition to ICA anastomoses, the descending branch of the hypoglossal artery exits the foramen magnum to anastomose with the posterior circulation via the odontoid arcade of the vertebral artery. A clinically important variant of this connection is the persistent hypoglossal artery. In this variant, the hypoglossal artery does not diminish normally during fetal life and instead persists as a robust intracranial connection to the vertebral artery (Meckel et al. 2011). The neuromeningeal trunk of the APA can originate from the occipital artery separately from the pharyngeal trunk and this is probably because of the difference in embryological origin of these branches. While most branches of the ECA participate in collateral anastomoses, the lingual artery is an important exception (Herrera-Núñez et al. 2020). It exhibits poor collateral flow to distal branches of the tongue and embolization can lead to tongue necrosis.

Procedures that involve using the ECA or its branches for bypass revascularization rely on its patency and overall health to achieve adequate perfusion of the target tissue. Encaphaloduroarteriosynangiosis (EDAS) is the most commonly used extracranial–intracranial bypass procedure for patients with a moyamoya-related middle cerebral artery (MCA) infarct (Yeh et al. 2017) and typically uses the superficial temporal artery for direct anastomosis to the MCA to revascularize the infarct territory. Similarly, the occipital artery is typically used as the donor vessel in posterior inferior cerebellar artery (PICA) territory infarcts. However, indirect bypass involves placing the donor artery directly on the brain without anastomosis to a cerebral artery and has been the preferred revascularization bypass technique for pediatric patients with moyamoya disease (Couldwell et al. 2020). In head and neck reconstruction, microsurgical free tissue transfers typically use the facial or superior thyroid artery as the donor vessel. However, when these branches are untenable owing to previous neck dissection or irradiation, a distally ligated ECA main trunk can be transposed for end-to-end anastomosis (Garg et al. 2017).

Carotid body paragangliomas, while rare, can receive blood flow from entrapped branches of the ECA and disrupt its anatomy and hemodynamics. These tumors are derived from neural crest cell chemoreceptors located in the carotid body, and although they tend to be highly vascular they are relatively slow growing. Larger carotid body paragangliomas characteristically displace the ECA anteriorly and the ICA posteriorly and can encapsulate either of those vessels, often necessitating more distal vessel exposure via elevation of the parotid gland and/or diversion of the digastric or stylohyoid muscles. These procedural considerations can increase the risk of injury to the ECA or its branches because they are close to the aforementioned structures. Furthermore, some cases require reconstruction of the ICA with transposition of the ECA. The most common complication for carotid body tumor resection is similar to the risk associated with CEA and includes injury to the vagus and hypoglossal nerves (Hans 2018).

The ECA can be injured during open reduction of fractures involving the mandibular ramus and/or condyle, and during procedures in which the parotid capsule/parenchyma is either dissected or removed, such as parotidectomy or temporomandibular joint (TMJ) surgery. Regarding the latter, a retromandibular approach to the TMJ to treat bony or fibrous ankylosis permits access to the ECA, which can be ligated if there is uncontrolled bleeding (Kasper et al. 2015).

References

Al-Rafiah A, El-Haggagy AA, Aal IH et al (2011) Anatomical study of the carotid bifurcation and origin variations of the ascending pharyngeal and superior thyroid arteries. Folia Morphol (Warsz) 70:47–55

Bagley JH, Priest R (2020) Carotid revascularization: current practice and future directions. Semin Intervent Radiol 37(2):132–139. https://doi.org/10.1055/s-0040-1709154

Brennan PA, Standring S, Wiseman SM (2020) Gray's surgical anatomy. Elsevier

Brown K et al (2015) Carotid artery stenting has increased risk of external carotid artery occlusion compared with carotid endarterectomy. J Vasc Surg 61(1):119–125

Buk SJ (1984) Simultaneous demonstration of connective tissue elastica and fibrin by a combined Verhoeff's elastic-Martius-scarlet-blue trichrome stain. Stain Technol 59(1):1–5. https://doi.org/10.3109/10520298409113822. PMID: 6206618.

Calhoun PS, Kuszyk BS, Heath DG, Carley JC, Fishman EK (1999) Threedimensional volume rendering of spiral CT data: theory and method. Radiographics 19(3):745–764. https://doi.org/10.1148/radiographics.19.3.g99ma14745. PMID: 10336201.

Cappabianca S, Scuotto A, Iaselli F et al (2012) Computed tomography and magnetic resonance angiography in the evaluation of aberrant origin of the external carotid artery branches. Surg Radiol Anat 34:393–399. https://doi.org/10.1007/s00276-011-0926-3

Cobiella R, Quinones S, Konschake M et al (2021) The carotid axis revisited, Sci Rep. 11(1):13847. https://doi.org/10.1038/s41598-021-93397-0

Couldwell MW, Cheshier S, Taussky P, Mortimer V, Couldwell WT (2020) Right frontotemporal craniotomy for ECA-to-MCA direct and indirect bypass and occipital artery indirect bypass to the posterior circulation: case report. J Neurosurg Pediatr 27(2):180–184. Published 2020 Nov 20. https://doi.org/10.3171/2020.7.PEDS20181

Dakour-Aridi H, Ramakrishnan G, Zarrintan S, Malas MB (2020) Outcomes of transcarotid revascularization with dynamic flow reversal versus carotid endarterectomy in the TCAR Surveillance Project. Semin Vasc Surg 33(1–2):24–30. https://doi.org/10.1053/j.semvascsurg.2020.10.001

de Borst GJ, Vos JA, Reichmann B, Hellings WE, de Vries JP, Suttorp MJ et al (2007) The fate of the external carotid artery after carotid artery stenting. A follow-up study with duplex ultrasonography. Eur J Vasc Endovasc Surg 33:657–663. [PubMed: 17337347]

Devadas D, Pillay M, Sukumaran TT (2018) A cadaveric study on variations in branching pattern of external carotid artery. Anat Cell Biol 51(4):225–231. https://doi.org/10.5115/acb.2018.51.4.225

Eichmann A, Yuan L, Moyon D, Lenoble F, Pardanaud L, Breant C (2005) Vascular development: from precursor cells to branched arterial and venous networks. Int J Dev Biol 49(2–3):259–267. https://doi.org/10.1387/ijdb.041941ae

Elsllabi M (2021) 865 Variation in the branching pattern of the anterior branches of the external carotid artery. Br J Surg 108(Supplement_2)

Frisdal A, Trainor PA (2014) Development and evolution of the pharyngeal apparatus. Wiley Interdiscip Rev Dev Biol 3(6):403–418. https://doi.org/10.1002/wdev.147

Garg RK, Poore SO, Wieland AM, Sanchez R, Baskaya MK, Hartig GK (2017) Recipient vessel selection in the difficult neck: outcomes of external carotid artery transposition and end-to-end microvascular anastomosis. Microsurgery 37(2):96–100. https://doi.org/10.1002/micr.22448

Gluncic V, Petanjek Z, Marusic A, Gluncic I (2001) High bifurcation of common carotid artery, anomalous origin of ascending pharyngeal artery and anomalous branching pattern of external carotid artery. Surg Radiol Anat 23:123–125

Hans SS (2018) Extracranial carotid and vertebral artery disease contemporary management. Springer, Cham

Heltzel S, Jaynes D (2011) Variation in the branching pattern of the four major caudal vessels emerging from the external carotid artery. FASEB J 25(S1):869.6–869.6

Herrera-Núñez M, Menchaca-Gutiérrez JL, Pinales-Razo R et al (2020) Origin variations of the superior thyroid, lingual, and facial arteries: a computed tomography angiography study. Surg Radiol Anat 42:1085–1093. https://doi.org/10.1007/s00276-020-02507-6

Jitpun E et al (2019) Do Asians have higher carotid bifurcation? A computed tomographic angiogram study of the common carotid artery bifurcation and external carotid artery branching patterns. Asian J Neurosurg 14(4) Wolters Kluwer India Pvt. Ltd:1082–1088

Johnston DCC et al (2002) Contrast-enhanced magnetic resonance angiography of carotid arteries: utility in routine clinical practice. Stroke (1970) 33(12):2834–2838., Lippincott Williams & Wilkins. https://doi.org/10.1161/01.STR.0000043632.51378.24

Kakisis JD, Antonopoulos CN, Mantas G, Moulakakis KG, Sfyroeras G, Geroulakos G (2017) Cranial nerve injury after carotid endarterectomy: incidence, risk factors, and time trends. Eur J Vasc Endovasc Surg 53(3):320–335. https://doi.org/10.1016/j.ejvs.2016.12.026

Kasper DL, Fauci AS, Hauser SL, Longo DL, Jameson JL, Loscalzo J (2015) Harrison's principles of internal medicine, 19th edn. McGraw Hill

Kaufman JA, Lee MJ (2014) Vascular and interventional radiology, 2nd edn. Elsevier Saunders

Kiyosue H (2020) External carotid artery: imaging anatomy atlas for endovascular treatment. Springer

Lee UY, Kwak HS (2021) Evaluation of plaque vulnerability via combination of hemodynamic analysis and simultaneous non-contrast angiography and intraplaque hemorrhage (SNAP) sequence for carotid intraplaque hemorrhage. J Pers Med 11(9):856. Published 2021 Aug 28. https://doi.org/10.3390/jpm11090856

Lell MM, Anders K, Uder M, Klotz E, Ditt H, Vega-Higuera F, Boskamp T, Bautz WA, Tomandl BF (2006) New techniques in CT angiography. Radiographics 26(Suppl 1):S45–S62. https://doi.org/10.1148/rg.26si065508

Lo A, Oehley M, Bartlett A, Adams D, Blyth P, Al-Ali S (2006) Anatomical variations of the common carotid artery bifurcation. ANZ J Surg 76(11):970–972. https://doi.org/10.1111/j.1445-2197.2006.03913.x

Malas MB, Dakour-Aridi H, Wang GJ et al (2019) Transcarotid artery revascularization versus transfemoral carotid artery stenting in the Society for Vascular Surgery Vascular Quality Initiative. J Vasc Surg

McNamara JR, Fulton GJ, Manning BJ. Three-dimensional computed tomographic reconstruction of the carotid artery: identifying high bifurcation. European Journal of Vascular and Endovascular Surgery. 2015;49(2):147–153. https://doi.org/10.1016/j.ejvs.2014.10.006.

Meckel S, Spittau B, McAuliffe W (2011) The persistent trigeminal artery: development, imaging anatomy, variants, and associated vascular pathologies. Neuroradiology 55(1):5–16

Michalinos A, Chatzimarkos M, Arkadopoulos N, Safioleas M, Troupis T (2016) Anatomical considerations on surgical anatomy of the carotid bifurcation. Anat Res Int 2016:6907472. https://doi.org/10.1155/2016/6907472

Nadig S, Barnwell S, Wax MK, Rosenthal EL (2009) Hoboken: Wiley Subscription Services, Inc., A Wiley Company. Head Neck 31(1):136–139

Navakalyani T, Janaki V, Sumalatha (2016) Variant branching patterns of external carotid artery – pharyngo—occipital trunk & occipito—auricular trunk. J Dental Med Sci 15(8):44–47

Ozgur Z, Govsa F, Ozgur T (2008) Anatomic evaluation of the carotid artery bifurcation in cadavers: implications for open and endovascular therapy. Surg Radiol Anat 30(6):475–480. https://doi.org/10.1007/s00276-008-0351-4. Epub 2008 May 6

Reichmann BL, Hellings WE, van der Worp HB, Algra A, Brown MM, Mali WP et al (2013) Flow velocities in the external carotid artery following carotid revascularization. Eur J Vasc Endovasc Surg 46:411–417. [PubMed: 23954165]

Schermerhorn ML, Liang P, Eldrup-Jorgenson et al (2019) Association of Transcarotid artery revascularization vs transfermoralcartid artery stenting with stroke or death among patients with carotid artery stenosis. JAMA 322:2313–2322

Schulz UG, Rothwell PM (2001) Major variation in carotid bifurcation anatomy: a possible risk factor for plaque development? Stroke 32(11):2522–2529. https://doi.org/10.1161/hs1101.097391

Seong J, Lieber BB, Wakhloo AK (2005) Morphological age-dependent development of the human carotid bifurcation. J Biomech 38(3):453–465., ISSN 0021-9290. https://doi.org/10.1016/j.jbiomech.2004.04.022

Standring S (2016). Gray's anatomy: the anatomical basis of clinical practice, 41st edn. Elsevier

Sujijantarat N, Antonios J, Koo A et al (2021) Transcarotid artery revascularization (TCAR): a technical video. J Neuro Interv Surg. https://doi.org/10.1136/neurintsurg-2021-018024

Tubbs RS, Shoja MM, Loukas M (2016) External carotid artery. In: Bergman's comprehensive encyclopedia of human anatomic variation. essay, Wiley Blackwell, pp 477–486

Uchino A, Saito N, Mizukoshi W, Okada Y (2011) Anomalous origin of the occipital artery diagnosed by magnetic resonance angiography. Neuroradiology 53:853–857

Wang GH, Shen HP, Chu ZM, Shen JG, Zhou HH (2021) Traumatic pseudoaneurysms of external carotid artery branch: case series and treatment considerations. Chin J Traumatol 24(6):368–373. https://doi.org/10.1016/j.cjtee.2021.04.003. Epub 2021 Apr 21. PMID: 33941432; PMCID: PMC8606599

Wardlaw JM, Chappell FM, Best JJ, Wartolowska K, Berry E, NHS Research and Development Health Technology Assessment Carotid Stenosis Imaging Group (2006) Non-invasive imaging compared with intra-arterial angiography in the diagnosis of symptomatic carotid stenosis: a meta-analysis. Lancet 367(9521):1503–1512. https://doi.org/10.1016/S0140-6736(06)68650-9

Willfort-Ehringer A, Ahmadi R, Gruber D et al (2003) Effect of carotid artery stenting on the external carotid artery. J Vasc Surg 38(5):1039–1044. https://doi.org/10.1016/s0741-5214(03)00550-0

Woo EY, Karmacharva J, Velazquesz OC, Carpenter JP, Skelly CL, Fairman RM (2007) Differential effects of carotid artery stenting versus carotid endarterectomy on external carotid artery patency. J Endovasc Ther 14:208–213. [PubMed: 17488178] 20. Zhou W, Lin PH, Bush RL, Peden EK, Guerrero MA, Kougi

Yamamoto D, Koizumi H, Ishima D, Kuroda H, Shibahara I, Niki J, Miyasaka K, Watanabe T, Kondo R, Kumabe T (2019) Angiographic characterization of the external carotid artery: special attention to variations in branching patterns. Tohoku J Exp Med 249(3):185–192

Yeh S-J, Tang S-C, Tsai L-K, Lee C-W, Chen Y-F, Liu H-M et al (2017) Color Dopplerultrasonography as an alternative tool forpostoperative evaluation of collaterals after indirectrevascularization surgery in Moyamoya disease. PLoS One 12(12):e0188948

Ascending Pharyngeal Artery

4

Charles J. Peper, Stephen J. Bordes, Jr.,
and R. Shane Tubbs

4.1 Introduction

The ascending pharyngeal artery (Fig. 4.1) has a complex course as it supplies many different areas of the head and neck including the posterior fossa of the skull base, the viscerocranium, the pharynx, the middle ear, and also the cervical spine. Furthermore, it forms several clinically significant anastomoses distally with other vascular territories.

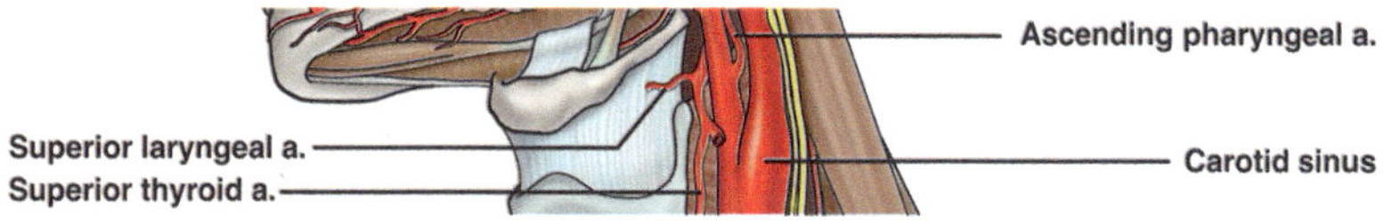

Fig. 4.1 Schematic drawing of the proximal ascending pharyngeal artery

C. J. Peper
Tulane University School of Medicine,
New Orleans, LA, USA
e-mail: cpeper@tulane.edu

S. J. Bordes, Jr.
Department of Surgery, Louisiana State University
Health Sciences Center, New Orleans, LA, USA

R. S. Tubbs (✉)
Department of Neurosurgery, Tulane University
School of Medicine,
New Orleans, LA, USA
e-mail: rtubbs@tulane.edu

S. J. Bordes, Jr. et al. (eds.), *The Clinical Anatomy of the Vascular System*,
https://doi.org/10.1007/978-3-031-78326-5_4

4.2 Embryology

The vascular territory of the carotid system arises from the third pharyngeal arch. Between weeks four and eight of development, the major branches of the external carotid artery form via angiogenesis (Weinzweig et al. 2010).

Arteries that pierce the skull base maintain their relative positions prior to the ossification of the occipital and sphenoid bones. Unlike most bones in the viscero- and neuro-cranium, which undergo intramembranous ossification, bones of the skull base ossify via endochondral ossification from multiple "primary" centers. The primary centers of the skull base develop in a caudal to rostral pattern beginning in the occipital bone at 12 weeks of development and finishing in the alisphenoid at 17 weeks. The vessels entering the cranium determine the position of the developing foramina, and reciprocally, the ossification of the skull base fixes those vessels in their final positions (Nemzek et al. 2000).

4.3 Anatomy

The ascending pharyngeal artery typically arises from the posterior or medial wall of the external carotid, slightly superior to the origin of the occipital artery. It ascends superiorly following a posterolateral course relative to the pharyngeal constrictor muscles, giving rise to posterior muscular branches before dividing into two main trunks, the pharyngeal and the neuromeningeal (Lasjaunias et al. 2006a; Hacein-Bey et al. 2002; Lang and Heilek 1984; Cavalcanti et al. 2009).

The neuromeningeal (posterior) trunk immediately gives rise to the inferior tympanic artery before continuing superiorly and posteriorly toward the skull base, where it divides into two consistent branches that enter the cranial vault and are named for the foramina through which they travel: the jugular and hypoglossal branches. The hypoglossal branch gives rise to a clival branch, which runs up the clivus toward the dorsum sellae and anastomoses thrice with the contralateral clival branch. From the most caudal anastomosis, bilateral arteries supplying the odontoid arch system travel caudally through the foramen magnum en route to anastomosing around the dens with a network of other arteries. We call this artery exiting through the foramen magnum the "Descending Odontoid Artery." (Lasjaunias et al. 2006a, b; Lasjaunias and Moret 1976).

The clival branch from the hypoglossal branch continues up the clivus to form an end-to-side anastomosis with the medial clival artery across the posterior surface of the dorsum sellae, and therefore contributes to the blood supply to the posterior lobe of the hypophysis (pars nervosa). The jugular branch produces an anterior branch traveling in the inferior petrosal sinus lateral to the clival branch, and forms an end-to-end anastomosis with the lateral clival artery from the internal carotid artery. The jugular branch also produces a posterior branch to the dura of the sigmoid sinus on its way to anastomosing with the transmastoid branches of the occipital artery (Lasjaunias et al. 2006a, b).

The musculospinal artery comes off the posterior wall of the neuromeningeal trunk and anastomoses with the ascending cervical artery at the level of C3. A prevertebral artery from the neuromeningeal trunk ascends the ventral surface of the C2 and C1 vertebrae before making a characteristic "inverted U" arch above the odontoid process. The posterior branch of the odontoid arch, also termed the ascending meningeal artery, arises from the vertebral artery, thereby establishing a vertebro-external carotid anastomosis. Note that the odontoid artery descending from the hypogloassal branch also contributes to this arch system, and that the connection to the ascending cervical artery provides a second route of vertebro-external carotid anastomosis (Lasjaunias et al. 2006a; Hacein-Bey et al. 2002; Geibprasert et al. 2009).

Important anastomotic connections of the inferior tympanic artery include the petrosquamal branch of the descending middle meningeal artery (MMA) and the corticotympanic branch of the internal carotid artery (Hacein-Bey et al. 2002).

The pharyngeal (anterior) trunk mainly supplies the mucosa of the naso- and oro-pharynx.

The inferior pharyngeal artery branches off the proximal middle pharyngeal artery or from the proximal pharyngeal trunk. The middle pharyngeal artery anastomoses with the internal maxillary artery and facial artery through the descending and ascending palantine arteries, respectively. The superior pharyngeal artery connects to the maxillary artery through the pterygovaginal artery. The superior pharyngeal artery also gives a posterior ramus connecting with the mandibular artery from the petrous internal carotid. An arterial branch supplying the eustachian tube is described as arising from the superior pharyngeal artery, middle pharyngeal artery, or pharyngeal trunk. At the midline, each pharyngeal branch connects to the contralateral branch (Lasjaunias et al. 2006a; Hacein-Bey et al. 2002; Geibprasert et al. 2009).

The superior pharyngeal artery gives off one more branch of note. The carotid ramus leaves the superior pharyngeal artery to travel through the foramen lacerum and then travels with the internal carotid artery to the point of the cavernous sinus. At this juncture, it anastomoses with the internal carotid artery through the inferolateral trunk and with the recurrent artery of the foramen lacerum (Lasjaunias et al. 2006a; Lasjaunias and Moret 1976).

Cranial nerve supply: The ascending pharyngeal artery is responsible for supplying the vasa nervorum of the lower cranial nerves. "Danger lurks in tiny vessels," and knowledge of the vascular supply to these nerves continue to be useful for the neurointerventionalist (Hendrix et al. 2013).

The CN IX vasa nervorum are supplied by the jugular branch and, more distally, by the inferior tympanic artery. Both typically arise from the neuromeningial trunk, but an important variation is addressed later.

The CN X vasa nervorum are supplied by the jugular branch.

The CN XI vasa nervorum are supplied by several arteries as the nerve both enters and exits the skull, having nerve roots from inside and outside the cranium. Near the jugular foramen, the jugular branch is responsible for the arterial supply to the nerve. However, its spinal roots are supplied by the musculospinal artery and also the neuromeningeal trunk.

The CN XII vasa nervorum arise from the hypoglossal branch.

The CN VI vasa nervorum are supplied proximally by the jugular branch (Lasjaunias et al. 2006a; Hacein-Bey et al. 2002).

The inferior tympanic artery has a questionable role in supplying the cranial nerves; it has been reported to anastomose with the stylomastoid artery, which supplies vasa vasorum to CN V and to the auricular branch of CN X.

The internal carotid vasa vasorum are supplied by the inferior tympanic artery (Hacein-Bey et al. 2002).

A major vascular collateral is notable; it travels through the odontoid arch system with the ascending cervical, deep cervical, occipital, and vertebral arteries (Lasjaunias et al. 2006a; Geibprasert et al. 2009).

4.4 Topography

Of the 13 angiosome territories of the head and neck, three are described as having no cutaneous involvement and therefore only supply deeper tissues in the region. The ascending pharyngeal artery is one of these three, along with the lingual and vertebral arteries. However, its territory in the posterior nasopharynx can be viewed easily (Hacein-Bey et al. 2002).

4.5 Variation

The ascending pharyngeal artery is known to have a variable origin from the external carotid. It can arise from orientations of the ECA other than the posterior wall and can also arise at various heights. Dual roots or twigs forming the APA have been described. Occasionally, it has been known to arise from the internal carotid artery, or as a third branch from the common carotid, forming a carotid trifurcation (Lang and Heilek 1984; Cavalcanti et al. 2009; Al-Rafiah et al. 2011).

Almost every branch of the neuromeningial trunk has been described in case studies as arising from the occipital artery, including the entire neuromeningial trunk itself. Indeed, the whole ascending pharyngeal artery can arise from the occipital artery, albeit rarely (Aggarwal et al. 2006). There are descriptions of atypical anastomoses with the vertebral artery not involving the hypoglossal branch, but connecting directly. These abnormal connections are associated with hypoplasia of one or both vertebral arteries (Shimamura et al. 1986; Nierling et al. 1966; Sato et al. 2011).

The inferior tympanic branch can arise from either the anterior (pharyngeal) or posterior (neuromengial) trunk, or even as a third trifurcating branch along with the pharyngeal and neuromeningeal branches. The eustachian branch has been described as arising from either the superior pharyngeal branch or the pharyngeal trunk (Hacein-Bey et al. 2002; Geibprasert et al. 2009).

Rarely, the posterior inferior cerebellar artery (PICA) arises from the ascending pharyngeal artery. Within this set of patients, the PICA can also supply the territory of the anterior inferior cerebellar artery (Lasjaunias et al. 2006a).

4.6 Clinical Significance

ECA Occlusion

At the craniocervical junction, there is an important anastomotic network termed the pharyngo-occipital collateral network. Strategic connections between the deep cervical, ascending cervical, vertebral, occipital, and ascending pharyngeal arteries present in the upper cervical region continue to be useful for the practicing neurointerventionalist. When occlusions or ligations hinder flow into a main artery of the head and neck, alternative collateral pathways within the pharyngo-occipital collateral network are "activated." When a proximal ECA occlusion is present, the cervico-pharyngeal pathway is typically activated. Although the C-3 vertebropharyngeal pathway can also be activated, the connection from the acending cervical to the ascending pharyngeal through the musculospinal artery is a more effective channel for endovascular interventions and has also been reported to be active during proximal subclavian occlusions or ligations (Lasjaunias et al. 2006a; Geibprasert et al. 2009).

Dural Arteriovenous Fistulas

After traversing the jugular foramen and entering the posterior fossa of the cranial vault, the jugular branch produces a rostral branch supplying the dura of the inferior petrosal sinus on its way to anastomosing with the lateral clival artery. It also produces a caudal branch to supply the dura of the sigmoid sinus before anastomosing with the transmastoid branches from the occipital artery. Because of this role, the ascending pharyngeal artery is the third most commonly implicated artery in creating dural arteriovenous fistulas within the cranial vault (Hacein-Bey et al. 2002; Lasjaunias et al. 2006b).

Epistaxis

Recurrent epistaxis following ligation of the internal maxillary artery has been known to result from atypical vessels from the ascending pharyngeal artery. It can also result from the typical anastomotic channel from the superior pharyngeal to the pterygovaginal artery (Kurata et al. 1993; Hacein-Bey et al. 1997).

Cancer

Meningiomas of the posterior fossa receive their vascular supply mainly from the ascending pharyngeal artery, particularly in the area of the jugular fossa (Hacein-Bey et al. 2002).

4.7 Complications

Cranial Nerve Ischemia

Cranial nerve ischemic arterial syndromes have been reported in CN V, IX, X, XI, and XII, related to disturbance of the ascending pharyngeal system. It is conceivable that CN VI could also exhibit an ischemic palsy after disturbance to the jugular branch of the neuromeningial trunk (Hendrix et al. 2013; Lapresle and Lasjaunias 1986).

References

Aggarwal NR et al (2006) Variant origin of superior thyroid artery, occipital artery and ascending pharyngeal artery from a common trunk from the cervical segment of internal carotid artery. Surg Radiol Anat 28(6):650–653. https://doi.org/10.1007/s00276-006-0145-5

Al-Rafiah A et al (2011) Anatomical study of the carotid bifurcation and origin variations of the ascending pharyngeal and superior thyroid arteries. Folia Morphol (Warsz) 70(1):47–55

Cavalcanti DD et al (2009) The ascending pharyngeal artery and its relevance for neurosurgical and endovascular procedures. Oper Neurosurg 65(6):114–120. https://doi.org/10.1227/01.neu.0000339172.78949.5b

Geibprasert S et al (2009) Dangerous extracranial–intracranial anastomoses and supply to the cranial nerves: vessels the neurointerventionalist needs to know. Am J Neuroradiol 30(8):1459–1468. https://doi.org/10.3174/ajnr.a1500

Hacein-Bey L et al (1997) Anastomoses in recurrent epistaxis. J Vasc Interv Radiol 8(4):535–538. https://doi.org/10.1016/s1051-0443(97)70605-5

Hacein-Bey L et al (2002) The ascending pharyngeal artery: branches, anastomoses, and clinical significance. Am J Neuroradiol 23(7):1246–1256

Hendrix P et al (2013) Arterial supply of the lower cranial nerves: a comprehensive review. Clin Anat 27(1):108–117. https://doi.org/10.1002/ca.22318

Kurata A et al (1993) Superselective embolization for severe traumatic epistaxis caused by fracture of the skull base. Am J Neuroradiol 14(2):343–345

Lang J, Heilek E (1984) Anatomo-clinical findings on the ascending pharyngeal artery. Anat Anz 156(3):177–207

Lapresle J, Lasjaunias P (1986) Cranial nerve ischaemic arterial syndromes. Brain 109(1):207–215. https://doi.org/10.1093/brain/109.1.207

Lasjaunias P, Moret J (1976) The ascending pharyngeal artery: normal and pathological radioanatomy. Neuroradiology 11(2):77–82. https://doi.org/10.1007/bf00345017

Lasjaunias P et al (2006a) Surgical neuroangiography, vol 1. Springer-Verlag

Lasjaunias P et al (2006b) Surgical neuroangiography, vol 3. Springer-Verlag

Nemzek WR et al (2000) MR, CT, and plain film imaging of the developing skull base in fetal specimens. Am J Neuroradiol 21(9):1699–1706

Nierling DA et al (1966) Ascending pharyngeal-vertebral anastomosis. Am J Roentgenol 98(3):599–601. https://doi.org/10.2214/ajr.98.3.599

Sato Y et al (2011) Ascending pharyngeal-vertebral anastomosis demonstrated by computed tomography angiography of the ascending pharyngeal artery: a case report. Acta Radiol 52(9):951–953. https://doi.org/10.1258/ar.2011.100308

Shimamura Y et al (1986) Ascending pharyngeal vertebral anastomosis with bilateral absence of vertebral arteries. Neurol Med Chir 26(7):552–555. https://doi.org/10.2176/nmc.26.552

Weinzweig J et al (2010) Head and neck embryology and anatomy. In: Plastic surgery secrets plus. Mosby/Elsevier, Philadelphia, pp 347–354

Occipital Artery

5

Nilo Alvarez Toledo

5.1 Embryology

The main arteries of the head and neck are derived from the pharyngeal arch arteries or aortic arches, which are six (I through VI) bilateral vascular structures, derived from mesoderm, that are part of the pharyngeal apparatus (Kathuria et al. 2011; Klostranec and Krings 2022). The second pharyngeal arch artery forms between 3 and 4 mm (24–28 days) (Menshawi et al. 2015) stage and is composed of a ventral and a dorsal portion (Kathuria et al. 2011). Close to the internal carotid artery origin, the ventral portion disconnects from the dorsal aorta and becomes the ventral pharyngeal artery (Kathuria et al. 2011). The ventral pharyngeal artery fuses with the internal carotid artery proximally to form the common carotid artery (Menshawi et al. 2015). The distal portion of the ventral pharyngeal artery develops into the external carotid artery (Kathuria et al. 2011; Klostranec and Krings 2022; Menshawi et al. 2015). Lasjaunias et al. (1978) hypothesized the origin of the occipital artery and its variants from "the remnants of the first two segmental arteries" (Lasjaunias et al. 1978; Vasović et al. 2009). The processes of vasculogenesis followed by angiogenesis are essential in this process (Klostranec and Krings 2022). Vasculogenesis is defined as "the differentiation of precursor cells (angioblasts) into endothelial cells and the novo formation of a primitive vascular network" (Vailhé et al. 2001). Angiogenesis consists of the growth of new vessels from others that already exist through two mechanisms, splitting (intussusception) and sprouting (Beets et al. 2013).

5.2 Anatomy

Origin

The occipital artery (Fig. 5.1) emerges from either the posterior or the lateral wall of the external carotid artery (Rhoton Jr. 2000; Alvernia et al. 2006), usually proximal to the emergence of the facial artery. Its origin is located 3–45 mm above the carotid bifurcation (Alvernia et al. 2006; Bademci and Yaşargil 2006; Small et al. 2014). Different authors have described a range of origins and different reference points. A meta-analysis found the occipital artery emerging among the first five branches of the external carotid artery in 100% of cases, most commonly as the fourth branch (Ostrowski et al. 2023). Other authors have described the occipital artery origin at the same level as the facial artery (Kawashima et al. 2005; Touré et al. 2010). Alvernia et al. (2006) found the origin within 5 mm above or below the emergence of the facial

N. Alvarez Toledo (✉)
Department of Neurosurgery, Icahn School of Medicine at Mount Sinai, New York, United States
e-mail: nilo.alvareztoledo@mountsinai.org

S. J. Bordes, Jr. et al. (eds.), *The Clinical Anatomy of the Vascular System*,
https://doi.org/10.1007/978-3-031-78326-5_5

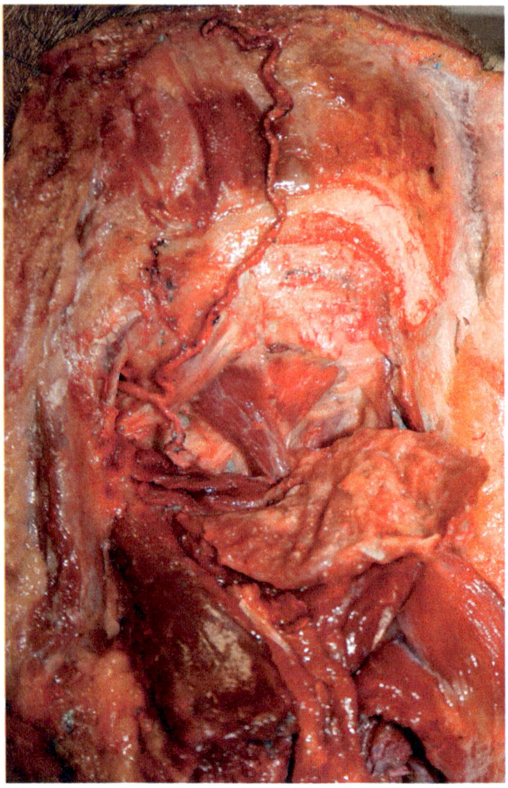

Fig. 5.1 Posterior dissection of the left occipital artery coursing just lateral to the suboccipital triangle and then ascending onto the scalp

artery, usually proximal (Alvernia et al. 2006). As related to the points of reference, some authors have described the emergence of the occipital artery at the level of the gonion (Rhoton Jr. 2000; Martins et al. 2005; Matsuo et al. 2020). Dumitru et al. (2023) found the point of emergence between the levels of the hyoid bone and the gonion, including both, in 74% of cases (Dumitru et al. 2023).

Classifications

Several classifications for the occipital artery anatomy have been proposed through the years (see table below). Three of those classifications are essential to fully understand the anatomy of the occipital artery from different perspectives. The classification proposed by Lasjaunias, in 1978, is based on neuroangiography and basically describes the course of the artery in three portions: ascending or cervical, horizontal or cervico-occipital, and ascending or occipital (Lasjaunias et al. 1978). The classification of Alvernia et al. (2006) is based on a microanatomical study and uses key relationships dividing the vessel into three segments: digastric, horizontal or cervico-occipital, and occipital (Alvernia et al. 2006). Lastly, the classification proposed by Fukuda et al. (2014) is more practical for the procedure of harvesting the occipital artery in bypass surgery. It is based on the three-dimensional relationship between the different muscle layers and the artery. This classification divides the occipital artery, distal to the posterior belly of the digastric muscle, into three segments: intramuscular, transitional, and subcutaneous (Fukuda et al. 2014).

Authors	Classification of the occipital artery
Lasjaunias et al. (1978)	Ascending or cervical Horizontal or cervico-occipital Ascending or occipital
Alvernia et al. (2006)	Digastric segment Inferior third Middle third Superior third Horizontal or suboccipital segment Subgaleal or occipital segment
Fukuda et al. (2014)	Subcutaneous Transitional Subcutaneous

Course, Segments, Branches, and Anatomical Relationships

The occipital artery first ascends posterosuperiorly from its origin, medial to the posterior belly of the digastric muscle, until the occipital groove, a small depression located posteromedial to the styloid process and medial to the mastoid process, in the temporal bone (Alvernia et al. 2006). At this point, the artery changes its course to posteromedial for a few centimeters, running first between the splenius capitis and the semispinalis capitis muscles, inside an extension of the styloid diaphragm (Fukuda et al. 2014). At the superior border of the splenius capitis muscle, the artery bends 90 degrees, piercing the tendon of the sternocleidomastoid muscle and the galea aponeurotica (Fukuda et al. 2014). Finally, above the superior nuchal line, the artery runs superficially, in the superficial fascia of the scalp to end in one or two terminal branches that create an extensive vascular network with other branches from the posterior auricular artery, the superficial temporal artery, and the contralateral occipital artery (Rhoton Jr. 2000; Alvernia et al. 2006; Martins et al. 2005).

Alvernia et al. (2006) divided the occipital artery into three segments: digastric, horizontal or suboccipital, and occipital (Alvernia et al. 2006). The digastric segment extends from the origin of the occipital artery to the occipital groove in the temporal bone. The horizontal or subgaleal segment extends from the exit of the occipital groove until the superior nuchal line. Finally, the occipital segment extends above the superior nuchal line (Alvernia et al. 2006).

Digastric Segment

The digastric is the deepest segment of the occipital artery. It ascends posterosuperiorly from the origin of the occipital artery to the occipital groove. It is subdivided into three thirds: inferior, middle, and superior (Alvernia et al. 2006). The inferior third runs between the origin of the occipital artery and the point where the spinal accessory nerve crosses the occipital artery posteriorly. The middle third extends from the crossing point of the spinal accessory nerve until the cranial base. The superior third runs in the occipital groove, sometimes a true bony canal (Alvernia et al. 2006). In some cases, the occipital groove can be absent and the artery runs superficial to the longissimus capitis muscle (Rhoton Jr. 2000). The mean diameter of the occipital artery at its origin is between 2.20 and 2.93 mm (Alvernia et al. 2006; Di et al. 2019). This segment gives off muscular, meningeal, and neuronal collateral branches (see table below) (Alvernia et al. 2006).

Occipital artery	Branches	Blood supply
Digastric segment	Muscular branches	3 to the sternocleidomastoid muscle
		1–3 to the digastric muscle
		1 to the rectus lateralis muscle
		Common trunk or single branches to the inferior oblique, superior oblique, and longissimus capitis muscles
	Meningeal branches	Stylomastoid trunk to the posterior cranial fossa and the facial nerve
	Neuronal branches	1–2 branches to the facial, spinal accessory, and hypoglossal nerves

The muscular branches of this segment provide blood supply to the sternocleidomastoid, digastric, rectus lateralis, inferior oblique, superior oblique, and longissimus capitis muscles. The stylomastoid trunk is the meningeal branch of this segment. It provides blood supply to the dura of the posterior cranial fossa and the facial nerve through small branches that reach the jugular foramen and the stylomastoid foramen (Alvernia et al. 2006). The neuronal branches provide blood supply to the facial, spinal accessory, and hypoglossal nerves (Alvernia et al. 2006).

This segment has important muscular, neural, and vascular relationships.

Muscular relationships:

- Digastric muscle, posterior belly
- Rectus capitis lateralis
- Superior oblique muscle

The digastric segment of the occipital artery runs medially to the posterior belly of the digastric muscle (Alvernia et al. 2006). It first arises between the posterior belly of the digastric muscle and the rectus capitis lateralis and then goes up between the posterior belly of the digastric muscle and the superior oblique muscle before reaching the occipital groove (Rhoton Jr. 2000; Di et al. 2019). The mastoid notch in the temporal bone, the site where "the posterior belly of the digastric muscle arises," is located lateral to the occipital groove, where the upper third of the digastric segment is located (Rhoton Jr. 2000).

Neural relationships:

- Hypoglossal nerve
- Ansa cervicalis
- Spinal accessory nerve
- Facial nerve

The extracranial segment of the hypoglossal nerve crosses superficially the occipital artery at a point located 7–9 mm above the origin of the artery (Bademci and Yaşargil 2006; Martins et al. 2005). The hypoglossal triangle, formed superiorly by the inferior border of the stylohyoid muscle, laterally by the descending hypoglossi, and medially by the transverse hypoglossi, contains the occipital artery (Bademci and Yaşargil 2006). The ansa cervicalis was found at the same level or above (0–15 mm) the origin of the occipital artery in all the specimens by Alvernia et al. (2006) (Alvernia et al. 2006). The spinal accessory nerve crosses the occipital artery between the inferior and the middle thirds of the digastric segment, posteriorly. The superior third runs medially to the exit of the facial nerve, at the stylomastoid foramen (Alvernia et al. 2006).

Vascular relationships:

- External carotid artery
- Internal carotid artery
- Internal jugular vein

The inferior and middle thirds of the digastric segment are surrounded by important vascular structures. The external carotid artery runs parallel and lateral, the internal carotid artery runs posteriorly, and the jugular vein runs medially (Rhoton Jr. 2000; Alvernia et al. 2006; Di et al. 2019).

Horizontal or Suboccipital Segment

The horizontal or suboccipital segment of the occipital artery runs from the occipital groove until the superior nuchal line (Alvernia et al. 2006). The segment runs posteromedially from its origin, passing, first, posterior to the superior oblique muscle, and then posterior to the splenius capitis muscle, from lateral to medial. At the exit from the occipital groove, the longissimus capitis muscle covers the artery posteriorly (Alvernia et al. 2006). In some cases, the artery runs superficially to the longissimus capitis muscle and the occipital groove is absent (Rhoton Jr. 2000; Martins et al. 2005). The splenius capitis muscle covers the artery posteriorly for about 6 centimeters (Fukuda et al. 2014). Fukuda et al. (2014) called this portion of the occipital artery, distal to the digastric muscle, "intramuscular" segment. This intramuscular segment runs between the semispinalis capitis and the splenius capitis muscle "in a single fatty connective tissue layer," a posterior extension of "the styloid diaphragm," from the infratemporal fossa (Fukuda et al. 2014; Janis et al. 2010). At the superior border of the splenius capitis muscle ends the intramuscular segment. Then, the artery leaves the "styloid diaphragm" (Fukuda et al. 2014; Janis et al. 2010) and bends at an angle of "45°–135°" (Fukuda et al. 2014). It pierces the sternocleidomastoid muscle tendon (Fukuda et al. 2014), or the tendinous arch formed by the insertion of both the sternocleidomastoid and trapezius muscles (Touré et al. 2010), and the galea aponeurotica from proximal to distal until the level of the superior nuchal line. Fukuda et al. (2014) called this small portion of 1 cm of the occipital artery "transitional" segment because it traverses the tendon and the galea aponeurotica (Fukuda et al. 2014). Some cases present a more medial course of the horizontal segment, and the transitional portion of the vessel runs under the belly of the trapezius muscle before piercing its tendon of insertion at the superior nuchal line (Fukuda et al. 2014). The greater occipital nerve joins the occipital artery below the superior nuchal line (Janis et al. 2010). Both the artery and nerve pierce together the tendon or the tendinous arch near the superior nuchal line. Usually, the occipital artery runs lateral to the greater occipital nerve (Won et al. 2018). Above the superior nuchal line, the horizontal segment becomes the occipital segment (Alvernia et al. 2006) (discussed below).

A table with results from different authors regarding the diameter, the length, and the distance from the inion to the horizontal segment of the occipital artery is shown below. The mean length of this segment has been reported between 65 and 82 mm (Kawashima et al. 2005; Fukuda et al. 2014; Di et al. 2019; Ateş et al. 2008; Benet et al. 2018). The intramuscular segment, "from the digastric groove to the superior edge of the splenius capitis muscle" (Fukuda et al. 2014), has a length of about 6 centimeters. The length of the transitional segment, "from the lateral edge of the splenius capitis muscle until the superior nuchal line" (Fukuda et al. 2014), is about 1 centimeter (Fukuda et al. 2014). Its mean diameter, at the exit of the occipital groove, is between 1.9 and 2.2 mm (Alvernia et al. 2006; Kawashima et al. 2005; Di et al. 2019). The mean diameter at the superior nuchal line ranges between 1.4 and 2.0 mm (Alvernia et al. 2006; Kawashima et al. 2005; Di et al. 2019; Ateş et al. 2008). The distance from the artery, at the superior nuchal line, and the inion varies in different studies from 33 to 45 mm (Alvernia et al. 2006; Ateş et al. 2008; Benet et al. 2018).

This horizontal segment is excellent to use in posterior circulation bypass surgery (Di et al. 2019). Particularly, the dissection of the transitional segment of the horizontal occipital artery is an essential step during the procedure of harvesting the occipital artery (Fukuda et al. 2014).

Authors	Mean diameter at the exit of occipital groove	Mean diameter at the superior nuchal line	Mean length of the segment	Distance occipital artery at the superior nuchal line to inion
Alvernia et al. (2006)	1.9 mm	1.4 mm	–	35.0 mm
Kawashima et al. (2005)	2.05 mm	2.01 mm	81.9 mm	–
Ateş et al. (2008)	–	1.9 mm	79.3 mm	45.0 mm
Di et al. (2019)	2.20 mm	1.77 mm	65.26 mm	–
Fukuda et al. (2014)	–	–	70.0 mm	–
Benet et al. (2018)	–	–	81.0 mm	33.0 mm
Range	**1.9–2.20 mm**	**1.4–2.0 mm**	**65.26–81.9 mm**	**33–45 mm**

This segment gives off muscular, neuronal, transosseous or transforaminal meningeal, and osteoarticular branches (see table below) (Alvernia et al. 2006).

Occipital artery Horizontal or suboccipital segment	Muscular branches	Superficial descending branch to the muscles of the occipital region
		Deep descending branch to the muscles of the occipital region
		1–3 horizontal branches to the sternocleidomastoid and the splenius capitis muscles
	Neuronal branches	Posterior deep occipital branch (branch from the deep descending branch) to the occipital nerve of Arnold
	Transosseous or transforaminal meningeal branches	1–3 branches to the dura of the posterior cranial fossa, the mastoid bone, and the occipital bone
	Osteoarticular branches	1–2 branches to the occipital condyle and the craniocervical junction

The muscular branches include a superficial descending branch, a deep descending branch, and a few small branches to the sternocleidomastoid and the splenius capitis muscles. The superficial and deep descending branches are constant and important in occipitovertebral anastomosis (Alvernia et al. 2006). These superficial and deep branches originate from a common trunk or separately (Alvernia et al. 2006), at a point where the occipital artery crosses the superior oblique muscle (Martins et al. 2005). The stylomastoid artery frequently arises from this segment of the occipital artery when it does not originate from the posterior auricular artery or the ascending pharyngeal artery (Martins et al. 2005). The stylomastoid artery ascends anteriorly to the mastoid and enters into the stylomastoid foramen to supply the facial nerve and the structures of the middle ear (Martins et al. 2005). The transforaminal or transosseous meningeal branches include branches to the dura of the posterior cranial fossa and the mastoid and occipital bones (Alvernia et al. 2006). The transmastoid branch or artery of the mastoid foramen is an inconstant transforaminal branch. If present, it arises at the level of the cranial insertion of the semispinalis capitis muscle, at a midpoint between the inferior and superior nuchal lines (Martins et al. 2005). It enters the skull through the mastoid foramen and communicates with the upper part of the sigmoid sinus (Martins et al. 2005). In addition, this segment gives off neuronal and osteoarticular branches that provide blood supply to the greater occipital nerve and the occipital condyle, respectively (Alvernia et al. 2006).

This segment has important muscular and neural relationships.

Muscular relationships:

- Longissimus capitis muscle
- Superior oblique muscle
- Semispinalis capitis muscle
- Splenius capitis muscle
- Sternocleidomastoid muscle
- Trapezius muscle

At the exit of the occipital artery from the occipital groove, the longissimus capitis muscle covers the artery posteriorly (Alvernia et al. 2006). In some cases, the artery runs superficial to the longissimus capitis muscle and the occipital groove is absent (Rhoton Jr. 2000; Martins et al. 2005). From its exit, the segment runs posteromedially, posterior to the superior oblique muscle first (Alvernia et al. 2006) and then between the semispinalis capitis (anterior) and the splenius capitis (posterior) muscles, "in a single fatty connective tissue layer," the superficial styloid diaphragm (Fukuda et al. 2014; Bejjani et al. 1998). At the level of the superior border of the splenius capitis muscle, the artery bends at an angle of approximately "45°–135°" (Fukuda et al. 2014), piercing the tendon of the sternocleidomastoid muscle and the galea aponeurotica from proximal to distal (Fukuda et al. 2014). In some cases, the artery runs more medial. In these particular cases, it runs under the belly of the trapezius muscle after leaving the styloid diaphragm and before piercing the trapezius tendon and the galea aponeurotica, at the superior nuchal line (Fukuda et al. 2014).

Neural relationships:

- Greater occipital nerve

The greater occipital nerve joins the occipital artery at the medial border of the occipital insertion of the sternocleidomastoid muscle tendon (Di et al. 2019). Both the artery laterally and the nerve medially pierce the sternocleidomastoid tendon and continue a vertical course above the superior nuchal line (Won et al. 2018). This nerve and its branches accompany the terminal branches of the occipital segment to innervate the occipital muscle (Alvernia et al. 2006). The direct anatomical relationship between the artery and the nerve has been described as a "single intersection point" or "a helical intertwining" (Janis et al. 2010; Shimizu et al. 2007).

Occipital Segment

The occipital segment extends above the superior nuchal line in the superficial fascia of the scalp (Rhoton Jr. 2000). Its course is vertical (Touré et al. 2010) and extends as a trunk above the superior nuchal line up to 50 mm, approximately 20–40 mm lateral to the midline (Alvernia et al. 2006). The occipital segment ends in one or two terminal branches (Fukuda et al. 2014). The occipital segment is accompanied by the greater occipital nerve and its branches (Alvernia et al. 2006; Martins et al. 2005). Fukuda et al. (2014) named this segment "subcutaneous," due to its anatomical position (Fukuda et al. 2014). At the superior nuchal line, the distance between the occipital artery and the inion ranges between 30 and 52 mm (Alvernia et al. 2006; Ateş et al. 2008; Benet et al. 2018). During its trajectory, it gives off collateral muscular and neuronal branches to supply the occipital muscle (Alvernia et al. 2006). It is important to describe the anastomosis with the branches from the neighbor arteries that create a rich vascular network. The medial terminal branch anastomoses with the branches from the contralateral occipital artery (Touré et al. 2010). The lateral terminal branch anastomoses with branches from the posterior auricular artery and the parietal branch of the superficial temporal artery (Touré et al. 2010). A meningeal branch from this occipital segment that traverses the parietal foramen to supply the supratentorial dura has been described (Martins et al. 2005).

Subgaleal or occipital segment	Muscular branches	15–30 branches to the occipital muscle
	Neuronal branches	Multiple branches accompanying the greater occipital nerve
	1–2 terminal branches (medial and lateral)	Anastomose with the contralateral occipital artery, and the ipsilateral posterior auricular artery and parietal branch of the superficial temporal artery, creating dense vascular network

This segment has important muscular, vascular, and neural relationships.

Muscular relationships:

- Occipital muscle

Above the superior nuchal line, the occipital segment provides multiple branches to the occipital muscle (Alvernia et al. 2006).

Vascular relationships:

- Posterior auricular artery
- Superficial temporal artery
- Contralateral occipital artery

The terminal branches of the occipital artery anastomose with the branches from the posterior auricular artery, the parietal branch of the superficial temporal artery, and with the contralateral occipital artery (Touré et al. 2010).

Neural relationships:

- Greater occipital nerve

Branches of the greater occipital nerve follow the branches of the occipital segment of the occipital artery to innervate the occipital muscle (Alvernia et al. 2006).

Anastomosis Occipital Artery–Vertebral Artery

The knowledge of the occipitovertebral anastomosis is very important for the endovascular procedures. These small anastomotic channels that connect the occipital artery with the vertebral artery are often not visible with angiography due to their small caliber. However, under certain hemodynamic situations, they may enlarge and become functional (Ostrowski et al. 2023). Obviously, the microanatomical techniques are more effective in studying such anastomosis. Schechter, in 1964, was the first to describe occipitovertebral anastomosis (Schechter 1964). Alvernia et al. described the anastomosis in 11 out of 12 specimens. The same author classified the anastomoses into three types according to the anatomical pattern (see table below), being the most common the anastomosis between the muscular branches of the superficial descending artery, from the horizontal segment, and the posterior radicular branch of the dorsal root of C2 (type II) (Alvernia et al. 2006).

Occipitovertebral anastomosis classification (adapted from Alvernia et al. 2006)

Occipitocervical anastomosis	Occipital artery branch	Vertebral artery branch	Frequency
Type I	One or two branches of the deep descending branch (horizontal segment)	Posterior radicular branch, related the dorsal root of C1	18%
Type II	Muscular branch from the superficial descending branch (horizontal segment)	Posterior radicular artery, related to the dorsal root of C2	72%
Type III	Distal muscular branch from the superficial descending branch (horizontal segment)	Posterior radicular artery related to the dorsal root of C3	9%

Other Anastomoses of the Occipital Artery

The occipital artery can form anastomosis also with the subclavian system, particularly with the deep cervical artery, a branch of the costocervical trunk (Ostrowski et al. 2023).

5.3 Anatomical Variations

Occipital Artery Originating from the External Carotid Artery as a Common Trunk

The origin of the occipital artery from a common trunk with other branches of the external carotid artery has been reported in the literature (Touré et al. 2010). An occipital artery arising from a common trunk with the ascending pharyngeal artery or with the posterior auricular artery has been reported (Touré et al. 2010).

Occipital Artery Originating from the Internal Carotid Artery

The variant of the occipital artery emerging unilaterally from the internal carotid artery is well known (Small et al. 2014). It has been reported in the literature since the nineteenth century with the work of Hyrtl and the early twentieth century with the report of Adachi (Newton and Young 1968). Small et al. (2014) in a large series of CTA images from 2602 patients, reported this type of origin in 0.49% of the sample (Small et al. 2014). Using MRA, other authors have reported the variant in 0.08% of cases (Uchino et al. 2011). The difference in percentage is the lower sensitivity of the MRA technique compared with the CTA (Small et al. 2014). Aggarwal et al. (2006) reported the occipital artery originating from a common trunk, with the superior thyroid and ascending pharyngeal arteries, from the internal carotid artery (Aggarwal et al. 2006).

Bilateral occipital arteries originating from both the right and left internal carotid arteries are extremely rare and only three cases have been reported (Uchino et al. 2011; Matsuda et al. 1977; Çetin et al. 2015). In 2015, Çetin et al. reported in their case left vertebral artery stenosis compensated with collateral blood flow from the ipsilateral occipital artery (Çetin et al. 2015).

Other Variants

Other variants of the occipital artery have been reported less commonly in the literature. These variants include bilateral origin from the thyrocervical trunk (Uchino and Ishihara 2022), occipital artery as a continuation of the ascending cervical artery (Tubbs et al. 2004), origin from the vertebral artery bilaterally (Demirbas et al. 2021), and occipital artery as the origin for the vertebral artery on both sides (Öner et al. 2017).

5.4 Pathology

Occipital Artery and Intracranial Dural Arteriovenous Fistula

Intracranial dural arteriovenous fistulas, also known as dural arteriovenous malformations (Baharvahdat et al. 2019), are defined as "pathological connections located within the dura between the pial arteries and the veno vasora, comprising the walls of the dural sinuses, bridging veins, or transosseous emissary veins' '(Alkhaibary et al. 2022). The most common location is at the transverse-sigmoid sinus (Alkhaibary et al. 2022). The middle meningeal artery and the occipital artery are usually the main feeding vessels (Gross et al. 2017). The fact that the occipital artery gives off meningeal branches for the posterior fossa and that those branches provide blood supply to the transverse-sigmoid sinus area explains why the artery is usually involved (Gross et al. 2017). Transarterial approach and embolization through the occipital artery are treatment options when there is no middle meningeal artery supplying the fistula (Stiefel et al. 2009).

Occipital Artery, Aneurysms and Pseudoaneurysms

Aneurysms of the occipital artery are extremely rare (Chaudhry et al. 2017). Most of the reported cases are pseudoaneurysms, secondary to blunt, penetrating, surgical head trauma, infection, or autoimmune diseases (Kim et al. 2010). True aneurysms, "defined as an abnormal, localized and permanent dilation, of all three layers of an arterial wall," are rarer with just a few reported cases (Chaudhry et al. 2017; Kim et al. 2010; Gum et al. 1987; Słoniewski et al. 2008; Kanematsu et al. 2011; Rao et al. 2012; Nguyen et al. 2016; Kawasaki et al. 2017; Illuminati et al. 2018; Bissacco et al. 2018; Lafford et al. 2020; Aljuboori et al. 2021). Three of those cases have been reported with intracranial location of the aneurysm (Słoniewski et al. 2008; Nguyen et al. 2016; Aljuboori et al. 2021).

Occipital Artery, Giant Cell Arteritis

Involvement of the occipital artery in giant cell arteritis has been reported (Wilkinson and Russell 1972). Despite the fact that superficial temporal artery biopsy is considered the gold standard for diagnosis, occipital artery biopsy has been successfully used as an alternative site for biopsy in patients with predominant occipital manifestations (Weems Jr. 1992; Pfadenhauer and Weber 2003).

Occipital Artery, Neurovascular Compression of the Occipital Nerve

Due to the anatomical relationship between the occipital artery and the greater occipital nerve, vascular pulsations of the aforementioned artery have been hypothesized as a cause of migraine headaches in the occipital region (Janis et al. 2010; Caviggioli et al. 2012).

5.5 Clinical/Surgical Topics

Occipital Artery—Posterior Circulation Bypass

Extracranial–intracranial bypass surgery for the posterior circulation is an option in some clinical scenarios including ischemic diseases, tumors of the skull base that involve the vertebral artery, and complex giant aneurysms (Ateş et al. 2008). The occipital artery, particularly the suboccipital segment, is an excellent donor artery due to its superficial location and the size of the vessel (Di et al. 2019; Ateş et al. 2008). At the superior nuchal line, the diameter is about 1.4–2.0 mm (Alvernia et al. 2006; Kawashima et al. 2005; Di et al. 2019; Ateş et al. 2008), which matches well with the caudal loop of the posteroinferior cerebellar artery (PICA) (Ateş et al. 2008). Therefore, the occipital artery–PICA bypass is an excellent option in the bypass surgery of the posterior circulation (Ateş et al. 2008). The 65–81 mm of the horizontal segment, plus the 50 mm of the occipital segment, of the occipital artery are enough to harvest a good donor vessel (Alvernia et al. 2006; Kawashima et al. 2005; Ateş et al. 2008).

The occipital artery–posterior cerebral artery bypass has been reported for revascularization in the territory of the posterior cerebral artery, particularly when the superficial temporal artery is unavailable (Touho et al. 1995; Korinth et al. 2000; Kazumata et al. 2013).

Occipital Artery—Middle Cerebral Artery Bypass

The bypass between the occipital artery and the middle cerebral artery is an option when the superficial temporal artery is hypoplastic or absent (Spetzler and Chater 1974; Kimura and Morita 2017; Hirano et al. 2018). This type of bypass is one of the direct techniques used in the revascularization of patients with Moyamoya disease (Reis et al. 2006). The use of occipital artery–saphenous vein–middle cerebral artery bypass variant has also been reported (Little et al. 1983).

Intraoperative Selective Angiography of the Occipital Artery

The use of the occipital artery for intraoperative angiography has been reported (Horiuchi et al. 2008). This direct cannulation method could be an alternative to transradial or transfemoral catheterization, particularly, when the patient is in a prone position during occipital or suboccipital craniotomies (Guo et al. 2019).

Occipitovertebral Anastomosis: Consequences and Applications

Complications during embolization related to the opening of occipitovertebral anastomosis have been reported (Spetzler et al. 1980). Patients who require external carotid ligation may develop collateral circulation via occipitovertebral anastomosis as reported by different authors (Takeuchi et al. 1994; Milnerowicz et al. 2019). The occipitovertebral anastomosis can be used in some cases for the delivery of thrombolytic drugs in case of vertebral artery occlusion (Alvernia et al. 2006; Macchi and Catini 1993).

Occipital Artery-Based Flaps

Occipital artery-based flaps have been used in defect reconstruction in the back of the neck, in the scalp of the parietal region, in the lateral face, and in ear reconstruction (Reinisch et al. 2019). It has been used in cases of hypopharynx reconstruction after tumor ablation (de Magalhães et al. 1998). A cadaveric study showing the favorable anatomic characteristics for possible use in skull base reconstruction using the prevertebral corridor was published by Durmaz et al. (2011).

References

Aggarwal NR, Krishnamoorthy T, Devasia B, Menon G, Chandrasekhar K (2006) Variant origin of superior thyroid artery, occipital artery and ascending pharyngeal artery from a common trunk from the cervical segment of internal carotid artery. Surg

Radiol Anat 28(6):650–653. https://doi.org/10.1007/s00276-006-0145-5

Aljuboori Z, McGrath M, Jadhav R, Ghodke B, Sekhar L (2021) A rare case of ruptured intracranial aneurysm arising from the retro-mastoid branch of the occipital artery. Surg Neurol Int 12:458. Published 2021 Sep 13. https://doi.org/10.25259/SNI_740_2021

Alkhaibary A, Alnefaie N, Alharbi A et al (2022) Intracranial dural arteriovenous fistula: a comprehensive review of the history, management, and future prospective [published online ahead of print, 2022 Nov 14]. Acta Neurol Belg. https://doi.org/10.1007/s13760-022-02133-6

Alvernia JE, Fraser K, Lanzino G (2006) The occipital artery: a microanatomical study. Neurosurgery 58(1Suppl):ONS114–ONS122. https://doi.org/10.1227/01.NEU.0000193519.00443.34

Ateş O, Ahmed AS, Niemann D, Başkaya MK (2008) The occipital artery for posterior circulation bypass: microsurgical anatomy. Neurosurg Focus 24(2):E9. https://doi.org/10.3171/FOC/2008/24/2/E9

Bademci G, Yaşargil MG (2006) Microsurgical anatomy of the hypoglossal nerve. J Clin Neurosci 13(8):841–847. https://doi.org/10.1016/j.jocn.2005.12.028

Baharvahdat H, Ooi YC, Kim WJ, Mowla A, Coon AL, Colby GP (2019) Updates in the management of cranial dural arteriovenous fistula. Stroke Vasc Neurol 5(1):50–58. Published 2019 Nov 21. https://doi.org/10.1136/svn-2019-000269

Beets K, Huylebroeck D, Moya IM et al (2013) Robustness in angiogenesis: notch and BMP shaping waves. Trends Genet 29:140–149

Bejjani GK, Sullivan B, Salas-Lopez E et al (1998) Surgical anatomy of the infratemporal fossa: the styloid diaphragm revisited. Neurosurgery 43(4):842–853. https://doi.org/10.1097/00006123-199810000-00072

Benet A, Tabani H, Ding X et al (2018) The transperiosteal "inside-out" occipital artery harvesting technique. J Neurosurg 130(1):207–212. https://doi.org/10.3171/2017.6.JNS17518

Bissacco D, Domanin M, Romagnoli S, Martelli E, Civelli V, Gabrielli L (2018) Spontaneous rupture of multiple occipital artery aneurysms in a patient with neurofibromatosis type 1. Vasc Endovasc Surg 52(1):86–88. https://doi.org/10.1177/1538574417742285

Caviggioli F, Giannasi S, Vinci V et al (2012) Neurovascular compression of the greater occipital nerve: implications for migraine headaches. Plast Reconstr Surg 129(2):353e–354e. https://doi.org/10.1097/PRS.0b013e31823aedc2

Çetin N, Akkan K, Uçar M, Onal B, Ilgit E (2015) Bilateral occipital arteries of internal carotid origin: report of a case and review of the literature. J Belg Soc Radiol 99(1):69–71. Published 2015 Sep 15. https://doi.org/10.5334/jbr-btr.828

Chaudhry NS, Gaynor BG, Hussain S, Dernbach PD, Aziz-Sultan MA (2017) Etiology and treatment modalities of occipital artery aneurysms. World Neurosurg 102:697.e1–697.e4. https://doi.org/10.1016/j.wneu.2017.02.121

de Magalhães RP, Brandão LG, Magalhães MG, Ferraz AR (1998) Galeal pedicle flap of the occipital region for pharynx reconstruction: anatomic and clinical considerations. Plast Reconstr Surg. 102(6):2124–2128. https://doi.org/10.1097/00006534-199811000-00049

Demirbas AT, Demirtas I, Sonmez Topcu F, Karasu S, Ayyıldiz B (2021) A rare case report: bilateral occipital artery arising from the vertebral artery. Surg Radiol Anat 43(11):1901–1904. https://doi.org/10.1007/s00276-021-02842-2

Di G, Fang X, Hu Q, Zhou W, Jiang X (2019) A microanatomical study of the far lateral approach. World Neurosurg 127:e932–e942. https://doi.org/10.1016/j.wneu.2019.04.004

Dumitru CC, Hostiuc S, Vrapciu AD, Rusu MC (2023) Vertical levels of the occipital artery origin. Medicina (Kaunas) 59(2):317. Published 2023 Feb 8. https://doi.org/10.3390/medicina59020317

Durmaz A, Fernandez-Miranda J, Snyderman CH, Rivera-Serrano C, Tosun F (2011) Prevertebral corridor: posterior pathway for reconstruction of the ventral skull base. J Craniofac Surg 22(3):848–853. https://doi.org/10.1097/SCS.0b013e31820f7d86

Fukuda H, Evins AI, Burrell JC, Stieg PE, Bernardo A (2014) A safe and effective technique for harvesting the occipital artery for posterior fossa bypass surgery: a cadaveric study. World Neurosurg 82(3–4):e459–e465. https://doi.org/10.1016/j.wneu.2013.09.015

Gross BA, Albuquerque FC, Moon K, McDougall CG (2017) Evolution of treatment and a detailed analysis of occlusion, recurrence, and clinical outcomes in an endovascular library of 260 dural arteriovenous fistulas. J Neurosurg. 126(6):1884–1893. https://doi.org/10.3171/2016.5.JNS16331

Gum GK, Numaguchi Y, Nadell JM, Robinson AE (1987) Aneurysm of the occipital artery: development after surgical ligation of the internal carotid artery. AJNR Am J Neuroradiol 8(6):1155–1156

Guo Y, Chen H, Chen X, Yu J (2019) Clinical importance of the occipital artery in vascular lesions: a review of the literature. Neuroradiol J 32(5):366–375. https://doi.org/10.1177/1971400919857245

Hirano T, Mikami T, Suzuki H et al (2018) Occipital artery to middle cerebral artery bypass in cases of unavailable superficial temporal artery. World Neurosurg 112:101–108. https://doi.org/10.1016/j.wneu.2018.01.103

Horiuchi T, Nitta J, Uehara T, Tanaka Y, Hongo K (2008) Intraoperative angiography through the occipital artery and muscular branch of the vertebral artery: technical note. Surg Neurol 70(6):645–648. https://doi.org/10.1016/j.surneu.2007.06.089

Illuminati G, Cannistrà M, Pizzardi G, Pasqua R, Frezzotti F, Calio' FG (2018) True aneurysm of the proximal occipital artery: case report. Int J Surg Case Rep 43:32–35. https://doi.org/10.1016/j.ijscr.2018.01.011

Janis JE, Hatef DA, Reece EM, McCluskey PD, Schaub TA, Guyuron B (2010) Neurovascular compression of the greater occipital nerve: implications for migraine

headaches. Plast Reconstr Surg. 126(6):1996–2001. https://doi.org/10.1097/PRS.0b013e3181ef8c6b
Kanematsu M, Kato H, Kondo H et al (2011) Neurofibromatosis type 1: transcatheter arterial embolization for ruptured occipital arterial aneurysms. Cardiovasc Intervent Radiol 34(Suppl 2):S131–S135. https://doi.org/10.1007/s00270-010-9850-x
Kathuria S, Gregg L, Chen J, Gandhi D (2011) Normal cerebral arterial development and variations. Semin Ultrasound CT MR 32(3):242–251. https://doi.org/10.1053/j.sult.2011.02.002
Kawasaki T, Yoshida K, Kikuchi T, Ishii A, Takagi Y, Miyamoto S (2017) Ruptured aneurysms of the occipital artery associated with congenital occipital bone defect. World Neurosurg 97:759.e13–759.e15. https://doi.org/10.1016/j.wneu.2016.09.116
Kawashima M, Rhoton AL Jr, Tanriover N, Ulm AJ, Yasuda A, Fujii K (2005) Microsurgical anatomy of cerebral revascularization. Part II: posterior circulation. J Neurosurg 102(1):132–147. https://doi.org/10.3171/jns.2005.102.1.0132
Kazumata K, Yokoyama Y, Sugiyama T, Asaoka K (2013) Occipital-posterior cerebral artery bypass via the occipital interhemispheric approach. Surg Neurol Int 4:90. Published 2013 Jul 12. https://doi.org/10.4103/2152-7806.114975
Kim HS, Son BC, Lee SW, Kim IS (2010) A rare case of spontaneous true aneurysm of the occipital artery. J Korean Neurosurg Soc 47(4):310–312. https://doi.org/10.3340/jkns.2010.47.4.310
Kimura T, Morita A (2017) Occipital artery to middle cerebral artery bypass: operative nuances. World Neurosurg 108:201–205. https://doi.org/10.1016/j.wneu.2017.08.126
Klostranec JM, Krings T (2022) Cerebral neurovascular embryology, anatomic variations, and congenital brain arteriovenous lesions. J Neurointerv Surg 14(9):910–919. https://doi.org/10.1136/neurintsurg-2021-018607
Korinth MC, Thron A, Bertalanffy H, Gilsbach JM (2000) Coil embolization of an incidental posterior cerebral artery aneurysm after initial OA-PCA bypass surgery. Zentralbl Neurochir 61(3):158–161. https://doi.org/10.1055/s-2000-11000
Lafford G, Sharp O, Rosich-Medina A (2020) True occipital artery aneurysm: a surgical case report and literature review. JPRAS Open 25:40–45. Published 2020 Jun 4. https://doi.org/10.1016/j.jpra.2020.05.005
Lasjaunias P, Théron J, Moret J (1978) The occipital artery. Anatomy–normal arteriographic aspects–embryological significance. Neuroradiology 15(1):31–37. https://doi.org/10.1007/BF00327443
Little JR, Furlan AJ, Bryerton B (1983) Short vein grafts for cerebral revascularization. J Neurosurg 59(3):384–388. https://doi.org/10.3171/jns.1983.59.3.0384
Macchi C, Catini C (1993) The anatomy and clinical importance of the collateral circles between the vertebral arteries and the cervical, costo-cervical, and occipital branches in 52 living subjects. Ital J Anat Embryol 98(3):153–163
Martins C, Yasuda A, Campero A, Ulm AJ, Tanriover N, Rhoton A Jr (2005) Microsurgical anatomy of the dural arteries. Neurosurgery 56(2 Suppl):211–251. https://doi.org/10.1227/01.neu.0000144823.94402.3d
Matsuda I, Handa J, Handa H, Mizuno H (1977) Bilateral anomalous occipital artery of internal carotid origin: case report. Nihon Geka Hokan 46(1):57–61
Matsuo S, Komune N, Akiyama O, Amano T, Nakamizo A (2020) Surgical anatomy of the donor arteries for extracranial-intracranial bypass surgery: an anatomic and radiologic study. World Neurosurg 136:e447–e459. https://doi.org/10.1016/j.wneu.2020.01.033
Menshawi K, Mohr JP, Gutierrez J (2015) A functional perspective on the embryology and anatomy of the cerebral blood supply. J Stroke 17(2):144–158. https://doi.org/10.5853/jos.2015.17.2.144
Milnerowicz M, Garcarek J, Bladowska J et al (2019) Bilateral thalamic stroke after tonsillectomy in patient with collateral extracranial anastomosis—case report. Pol J Radiol 84:126–130. https://doi.org/10.5114/pjr.2019.83004
Newton TH, Young DA (1968) Anomalous origin of the occipital artery from the internal carotid artery. Radiology 90:550–552
Nguyen HS, Doan N, Shabani S, Gelsomino M, Zaidat O (2016) Pure tentorial subdural hematoma from rupture of aneurysm along the transmastoid branches of the occipital artery. Surg Neurol Int 7(Suppl 18):S499–S501. Published 2016 Aug 1. https://doi.org/10.4103/2152-7806.187490
Öner Z, Öner S, Kahraman AS (2017) The right vertebral artery originating from the right occipital artery and the absence of the transverse foramen: a rare anatomical variation. Surg Radiol Anat 39(12):1397–1400. https://doi.org/10.1007/s00276-017-1882-3
Ostrowski P, Bonczar M, Plutecki D et al (2023) The occipital artery: a meta-analysis of its anatomy with clinical correlations. Anat Sci Int 98(1):12–21. https://doi.org/10.1007/s12565-022-00693-4
Pfadenhauer K, Weber H (2003) Giant cell arteritis of the occipital arteries–a prospective color coded duplex sonography study in 78 patients. J Neurol 250(7):844–849. https://doi.org/10.1007/s00415-003-1104-2
Rao VY, Hwang SW, Adesina AM, Jea A (2012) Thrombosed traumatic aneurysm of the occipital artery: a case report and review of the literature. J Med Case Rep. 6:203. Published 2012 Jul 17. https://doi.org/10.1186/1752-1947-6-203
Reinisch JF, van Hövell Tot Westerflier CVA, Tahiri Y et al (2019) The occipital artery–based fascial flap for ear reconstruction. Plast Reconstr Surg 143(3):592e–601e. https://doi.org/10.1097/PRS.0000000000005263
Reis CV, Safavi-Abbasi S, Zabramski JM, Gusmão SN, Spetzler RF, Preul MC (2006) The history of neurosurgical procedures for moyamoya disease. Neurosurg Focus 20(6):E7. Published 2006 Jun 15. https://doi.org/10.3171/foc.2006.20.6.7
Rhoton AL Jr (2000) The far-lateral approach and its transcondylar, supracondylar, and paracondylar exten-

sions. Neurosurgery 47(3 Suppl):S195–S209. https://doi.org/10.1097/00006123-200009001-00020
Schechter MM (1964) The occipital-vertebral anastomosis. J Neurosurg 21:758–762. https://doi.org/10.3171/jns.1964.21.9.0758
Shimizu S, Oka H, Osawa S et al (2007) Can proximity of the occipital artery to the greater occipital nerve act as a cause of idiopathic greater occipital neuralgia? An anatomical and histological evaluation of the artery-nerve relationship. Plast Reconstr Surg 119(7):2029–2034. https://doi.org/10.1097/01.prs.0000260588.33902.23
Słoniewski P, Dzierzanowski J, Och W (2008) Aneurysm of the meningeal branch of the occipital artery connecting with the distal portion of the posteroinferior cerebellar artery by the dural fistula. Folia Morphol (Warsz) 67(4):292–295
Small JE, Harrington J, Watkins E (2014) Prevalence of arterial branches arising from the extracranial internal carotid artery on CT angiography. Surg Radiol Anat 36(8):789–793. https://doi.org/10.1007/s00276-013-1246-6
Spetzler R, Chater N (1974) Occipital artery–middle cerebral artery anastomosis for cerebral artery occlusive disease. Surg Neurol 2(4):235–238
Spetzler RF, Modic M, Bonstelle C (1980) Spontaneous opening of large occipital-vertebral artery anastomosis during embolization. J Neurosurg 53(6):849–850. https://doi.org/10.3171/jns.1980.53.6.0849
Stiefel MF, Albuquerque FC, Park MS, Dashti SR, McDougall CG (2009) Endovascular treatment of intracranial dural arteriovenous fistulae using Onyx: a case series. Neurosurgery 65(6 Suppl):132–140. https://doi.org/10.1227/01.NEU.0000345949.41138.01
Takeuchi Y, Suzuki H, Numata T et al (1994) Hemodynamic changes in the head and neck after ligation of the unilateral carotid arteries: a study using color doppler imaging. Ann Otol Rhinol Laryngol 103(1):41–45. https://doi.org/10.1177/000348949410300107
Touho H, Karasawa J, Ohnishi H, Kobitsu K (1995) Anastomosis of occipital artery to posterior cerebral artery with interposition of superficial temporal artery using occipital interhemispheric transtentorial approach: case report. Surg Neurol 44(3):245–250. https://doi.org/10.1016/0090-3019(95)00052-6
Touré G, Méningaud JP, Vacher C (2010) Arterial vascularization of occipital scalp: mapping of vascular cutaneous territories and surgical applications. Surg Radiol Anat 32(8):739–743. https://doi.org/10.1007/s00276-010-0673-x
Tubbs RS, Salter G, Oakes WJ (2004) Continuation of the ascending cervical artery as the occipital artery in man. Anat Sci Int 79(1):43–45. https://doi.org/10.1111/j.1447-073x.2004.00062.x
Uchino A, Ishihara S (2022) Bilateral occipital arteries arising from the thyrocervical trunks (ascending cervical artery-occipital artery anastomosis) diagnosed by magnetic resonance angiography. Radiol Case Rep 17(7):2299–2303. Published 2022 May 5. https://doi.org/10.1016/j.radcr.2022.04.010
Uchino A, Saito N, Mizukoshi W, Okada Y (2011) Anomalous origin of the occipital artery diagnosed by magnetic resonance angiography. Neuroradiology 53(11):853–857. https://doi.org/10.1007/s00234-010-0825-z
Vailhé B, Vittet D, Feige JJ (2001) In vitro models of vasculogenesis and angiogenesis. Lab Invest 81(4):439–452. https://doi.org/10.1038/labinvest.3780252
Vasović L, Mojsilović M, Andelković Z et al (2009) Proatlantal intersegmental artery: a review of normal and pathological features. Childs Nerv Syst 25(4):411–421. https://doi.org/10.1007/s00381-008-0765-7
Weems JJ Jr (1992) Diagnosis of giant cell arteritis by occipital artery biopsy. Am J Med 93(2):231–232. https://doi.org/10.1016/0002-9343(92)90057-i
Wilkinson IM, Russell RW (1972) Arteries of the head and neck in giant cell arteritis. A pathological study to show the pattern of arterial involvement. Arch Neurol. 27(5):378–391. https://doi.org/10.1001/archneur.1972.00490170010003
Won HJ, Ji HJ, Song JK, Kim YD, Won HS (2018) Topographical study of the trapezius muscle, greater occipital nerve, and occipital artery for facilitating blockade of the greater occipital nerve. PLoS One 13(8):e0202448. Published 2018 Aug 15. https://doi.org/10.1371/journal.pone.0202448

6 Posterior Auricular Artery

Aaron Yu and Kyle Biggins

6.1 Embryology

Early in development (5–6 mm stage), a transient artery known as the ventral pharyngeal artery (VPA) originates from the ventral aortic roots cranial to the third aortic arch (Bertulli and Robert 2021; Hu et al. 2018). When the first and second aortic arches regress, the VPA extends directly from the distal ventral aorta, its distal segment merging with the maxillomandibular branch of the stapedial artery from the regressing second aortic arch to form the primitive external carotid artery (ECA) (Bertulli and Robert 2021; Hu et al. 2018; Menshawi et al. 2015; Tanoue et al. 2013; Rao et al. 2014). The more proximal portion of the VPA fuses with the internal carotid artery (ICA) to form the common carotid artery (CCA) (Menshawi et al. 2015).

Subsequently, when the definitive CCA emerges, several branches originate from the ECA, beginning around the 16–18 mm stage (Bertulli and Robert 2021). This intermediate stage of development can witness the angiogenesis of one such branch, the posterior auricular artery (PAA), at the level of the mandibular angle either as a direct, independent branch from the ECA or, rarely, as a common trunk with the occipital artery (Jinkins 2000).

A. Yu (✉) · K. Biggins
Tulane School of Medicine, New Orleans, LA, USA
e-mail: ayu2@tulane.edu; kbiggins@tulane.edu

6.2 Anatomy/Topography

The PAA (Fig. 6.1) is a small branch of the ECA primarily responsible for supplying an area of skin posterior to the auricle and the auricle itself, though it also contributes to supplying other structures through its anastomotic branches. It usually arises just superior to the occipital artery as the third posterior branch of the ECA near its distal termination and travels toward the stylomastoid foramen, often crossing the facial nerve trunk along its path (Kiyosue 2020; Kolhe and Leonard 1987; Standring et al. 2008; Liu et al. 2018).

Coursing posterosuperior from its origin along the superior border of the intimately related digastric and stylohyoid muscles, the PAA traverses beneath the parotid gland and styloid process (Kiyosue 2020; Kolhe and Leonard 1987). It gives off several small branches along its proximal course:

- *Muscular Branches*: supply the digastric, stylohyoid, and sternocleidomastoid muscles (Kiyosue 2020; Kolhe and Leonard 1987; Elvan et al. 2019).
- *Parotid Branches*: supply the inferior aspect of the parotid gland. These branches can anastomose with ECA branches supplying the parotid gland and the superficial temporal artery (STA) (Lasjaunias and Berenstein 1987).

S. J. Bordes, Jr. et al. (eds.), *The Clinical Anatomy of the Vascular System*,
https://doi.org/10.1007/978-3-031-78326-5_6

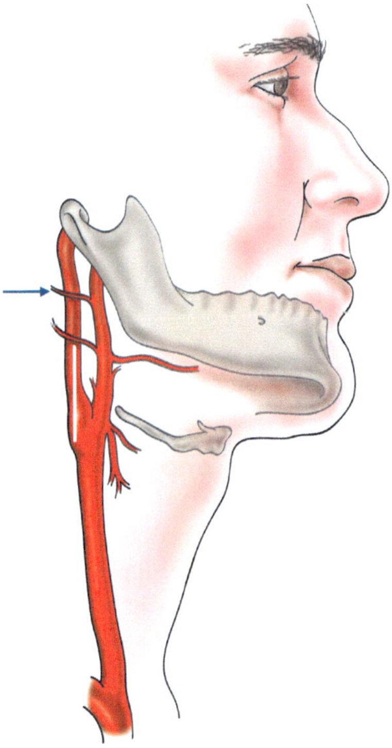

Fig. 6.1 Schematic view of the right posterior auricular artery (arrow)

At the stylomastoid foramen, the PAA ascends in a groove between the auricular cartilage and the mastoid process to reach the outer opening of the facial canal near the facial nerve (CN VII) (Kolhe and Leonard 1987; Liu et al. 2018; Elvan et al. 2019; Lasjaunias and Berenstein 1987). Closely accompanying the PAA is the posterior auricular nerve (PAN), a small ramus of the facial nerve that originates below the stylomastoid foramen and innervates the postauricular muscles (Liu et al. 2018). Inferior to the stylomastoid foramen, the facial nerve trunk crosses the PAA. It has therefore been suggested that the PAA is a novel anatomical landmark for identifying the facial nerve (Liu et al. 2018).

The two most consistent branches of the PAA are the auricular and occipital branches:

- *Occipital Branch*: arises superior to the groove between the auricle and mastoid process and forms an anastomosis with the occipital artery, providing a cutaneous supply to the scalp behind the ear and a muscular supply to the occipitofrontal muscle (Kiyosue 2020; Kolhe and Leonard 1987).
- *Auricular Branch*: typically the main terminal branch of the PAA as it ascends deep to the auricularis posterior muscle (Kolhe and Leonard 1987; Lasjaunias and Berenstein 1987). The auricular branch further divides into smaller perforating branches to supply the medial and lateral cranial surfaces of the auricle and can continue to ascend toward the parietotemporal area (Kolhe and Leonard 1987; Pinar et al. 2003). A potential anastomosis can form with the auricular branch of the PAA and the anterior auricular artery of the STA, or even its parietal branch (Kiyosue 2020; Lasjaunias and Berenstein 1987).

Arising variably from the PAA, the stylomastoid artery is a significant branch, accounting for the main contributions of the PAA outside the postauricular territory. The stylomastoid artery has been described as supplying the tympanic cavity, mastoid antrum, semicircular canals, chorda tympani, and the facial nerve through a notable anastomosis with the petrosal branch of the middle meningeal artery (petrosal artery) (Kolhe and Leonard 1987; Elvan et al. 2019). This anastomotic arch between the stylomastoid and petrosal arteries over the petrous part of the temporal bone is regarded as the facial nerve arterial arcade, the dominant supply to the facial nerve (Geibprasert et al. 2009). More specifically, this arcade is the vasa nervosum to the geniculate ganglion and the intraosseous segments of the facial nerve (Bhatia et al. 2021). The anastomotic arch accompanies the intrapetrous facial nerve and its mastoid and tympanic segments as it courses along the facial canal to exit through the stylomastoid foramen (Lasjaunias and Berenstein 1987; Kouo et al. 2018).

Among several further anastomotic branches with the ascending pharyngeal artery, meningohypophyseal trunk, posterior meningeal artery, and other MMA branches, the stylomastoid artery can supply the meninges of the posterior fossa (Geibprasert et al. 2009). The occipital artery can form an additional anastomosis with the stylomastoid branch of the PAA through its own corresponding branch that reaches the stylomastoid foramen (Lasjaunias and Berenstein 1987).

6.3 Variations

The PAA varies considerably in its angiographic dimensions, level of termination, origin, branching pattern, and relationship with the facial nerve. According to the classification system of Tokugawa et al. (2015), based on cerebral angiographical data (*n = 234*), it displays four types of variation in length and level of termination:

1. *Type A* (15.1%): PAA terminates between the origin and center of the external acoustic meatus.
2. *Type B* (34.9%): PAA terminates between the origin and apical helix of the auricle.
3. *Type C* (48.8%): PAA terminates between the apical helix and cranial vertex.
4. *Type D* (1.2%): PAA terminates at the cranial vertex (Tokugawa et al. 2015).

The diameter of the PAA varies in size (0.7–1.2 mm) (Gómez Díaz and Cruz Sánchez 2016). If the STA has no parietal branch, an extremely rare variant of the PAA can appear. Tokugawa et al. (2014) reported an unusually well-developed PAA, thought to take over the territory of the absent STA, the parietal scalp (Tokugawa et al. 2014). According to Wang et al. (2008), the PAA is usually (93%) the dominant arterial supply to the posterior ear, lobule, and retro-auricular skin; however, a dominant occipital artery supplying the posterior ear was found instead in 7% (Wang et al. 2008). A correspondingly small posterior auricular artery could be expected given a dominant occipital or STA.

Although the PAA was previously said to originate from the ECA just superior to the occipital artery, it arises from the occipital artery in 10–15% after a common occipitoauricular trunk (Elvan et al. 2019; Zümre et al. 2005). A PAA arising from the CCA has also been observed (Zümre et al. 2005).

The PAA can variably give rise to a stylomastoid branch toward the stylomastoid foramen into the third segment of the facial canal, ending in the tympanic cavity (Lasjaunias and Berenstein 1987; Liu et al. 2018). This so-called stylomastoid artery can originate from either the PAA, the occipital artery, or, more infrequently, directly from the ECA (Elvan et al. 2019; Lasjaunias and Berenstein 1987). It can also arise from a common trunk of the PAA and occipital artery (Geibprasert et al. 2009). Rarely, when the auricular branch of the PAA is either hypoplastic or absent, the stylomastoid artery can present as its main terminating branch instead (Kolhe and Leonard 1987).

Branches of the PAA participating in the helical rim arcade can also have variable branching patterns. Zilinsky et al. (2017) observed that three (31%) to seven (6%) branches were presented by the PAA as it traversed along the posterior aspect of the auricular cartilage on its path toward the helical rim (Zilinsky et al. 2017). It gave off two (38%) to four (31%) perforating branches that pierced the auricular cartilage, reaching the auricle's anterior aspect.

The variable relationship between the PAA, along with its stylomastoid branch (if present), and the facial nerve also merits attention. The PAA reliably crosses the main trunk of the facial nerve at a level inferior to the stylomastoid foramen (Elvan et al. 2019; Liu et al. 2018). Recently, Elvan et al. (2019) showed that it can cross the facial nerve from its lateral (79%) or medial (17%) spect or puncture the nerve itself proximally (2%) (Elvan et al. 2019). The anatomical positional relationship of this crossing usually showed bilateral symmetry, though asymmetry was also observed (Elvan et al. 2019). Thin extensions from the facial nerve trunk were seen to cross over the artery to rejoin the nerve in its course (Elvan et al. 2019). The stylomastoid artery has also been recorded as crossing the facial nerve either medially or laterally; the incidence of this has been contended (Elvan et al. 2019).

Trost et al. (2008) classified the variable collateral branches of the PAA dedicated to the facial nerve in the prestylian space (i.e., within the parotid gland) into three types:

1. *Type 1*: PAA presents one branch in the prestylian space.
2. *Type 2*: PAA presents several branches in the prestylian space.
3. *Type 3*: PAA presents several branches in both the prestylian space and the superficial retroauricular area (Trost et al. 2008).

6.4 Imaging

Certain radiological modalities accompanying cerebrovascular interventions or during a cadaveric study can help to identify ECA branches and, by extension, the PAA. Computerized tomography angiography (CTA) and digital subtraction angiography (DSA) reliably visualize the PAA in such investigations. Magnetic resonance angiography (MRA) is a more recent tool for assessing arteries, potentially an alternative to CTA as a radiation-free modality in preoperative microvascular mapping. MRA is also useful for determining graft patency. This all says that CTA lacks some of the limitations of MRA (Kramer et al. 2014). Kramer et al. (2014) reported that CTA was superior to MRA, especially for visualizing all ECA branches. Because its spatial resolution is inferior to that of CTA, MRA frequently failed to detect the smallest ECA branches, notably the ascending pharyngeal artery and the PAA (Kramer et al. 2014). CTA is arguably more reliable than DSA for planning microvascular reconstruction (Kramer et al. 2014).

Regardless, angiography on lateral projection has been standard for describing the dimensions, course, origin, and anastomotic branches of the PAA in both anatomical and clinical settings. The aforementioned Tokugawa et al. (2015) classification of the variable length and level of termination (Types A, B, C, D) of the PAA was based solely on the vessel's angiographical appearance. Often, with respect to its angiographical length, the PAA seems a slender vessel among the other small branches emerging from the ECA on conventional cerebral angiography. It is therefore usually not considered as a graft for arterial bypass and can be overlooked on imaging (Kramer et al. 2014). While type D (which involves an enlarged PAA terminating at the cranial vertex) is rare (1.2%), proper identification of this variation through lateral DSA visualization can have considerable clinical implications; for example, as an indication to use an overlarge PAA as a donor artery for revascularization surgery (Tokugawa et al. 2015).

Clinically, two studies emphasized the importance of proper radiological identification of a prominent PAA. Torazawa et al. (2017) presented a case involving an extracranial–intracranial (EC–IC) bypass in a patient with adult-onset ischemic moyamoya disease. As reported here, a prominent PAA has been observed on angiography to compensate for a correspondingly underdeveloped STA (Tokugawa et al. 2014; Torazawa et al. 2017). Given its anatomical similarity to the STA, especially if it has a large enough diameter, the PAA can be a surrogate graft. The combination STA–MCA and PAA–MCA bypass was successfully used in the direct double surgical revascularization performed in this case. In addition to preoperative ECA angiography, CTA was used on postoperative day 1 along with MRA to confirm the long-term patency of the PAA graft. The PAA–MCA bypass proved to be a feasible option for initial revascularization or as a rescue for recurrent ischemia provided its vessel diameter was $\geq$1.0 mm (Torazawa et al. 2017). Similarly, Germans and Regli (2014) presented the operative and radiological features of a viable PAA for EC–IC bypass in three cases (Germans and Regli 2014). Preoperative lateral DSA was again used to discern the size of the PAA and STA along with postoperative CTA and MRA to assess the patency of the bypasses. The prevalence of a PAA suitable for use as a donor vessel for EC–IC bypass was reported as 5.7% (Germans and Regli 2014).

6.5 Pathology

Pseudoaneurysm (PA) of the ECA and its branches is extremely rare, arising from trauma or iatrogenic complications. Other possible causes include inflammation, secondary infection, and arteriosclerosis; there are even spontaneous cases of idiopathic etiology (Wang et al. 2015; Sokol et al. 2017). PA of ECA branches is relatively more uncommon than PA of the ECA itself, presumably because the soft tissue buffer protects ECA branches in most locations (Wang et al. 2015). PAs in branches of the ECA result from total transection of the arterial vessel wall, leading to persistent leakage that develops later into an expanding mass (Wang et al. 2015).

To date, only a handful of cases in the literature present PA of the PAA. Among the earliest references to such a PA, Hashemi et al. (1978) describe a case of hemorrhage from the external acoustic meatus (otorrhagia) of the left ear, confirmed on angiography to arise from the PAA alone (Hashemi et al. 1978). Interestingly, aside from persistent headache in the left mastoid region associated with pulsating tinnitus, there was no history of trauma or other relevant causes of the PA in this case. The symptoms disappeared when the lesion was ligated at both ends. More recently, in 2015, Wang et al. studied 17 patients with head and neck PAs; only one of these PAs originated from the PAA viewed on DSA. Coil embolization of the PAs proved an effective alternative to standard ligation in this study (Hashemi et al. 1978).

Sokol et al. (2017) reported the only case of life-threatening bleeding from the external auditory meatus resulting from posttraumatic PA of the PAA (Sokol et al. 2017). Blunt head trauma led to the exposure of the PAA during the injury, causing the PA. It is suggested that the injured PA leaked through the skull base fracture via the ruptured tympanic membrane, causing massive bleeding into the external auditory canal (Sokol et al. 2017). Coil embolization of the distal segment of the PAA proved a successful treatment, immediately stopping the bleeding. There was no resultant facial weakness, though there was ipsilateral hearing loss. Other potential treatment options include resection and ligation of the aneurysm. However, open approaches to such lesions could entail increased risk owing to the distorted edematous tissue and bacterial colonization; thus, coil embolization was indicated as the most effective procedure here (Sokol et al. 2017).

Another unusual pathological presentation of the PAA cited in the literature is an auricular arteriovenous malformation (AVM). AVMs commonly arise from the intracranial area, rarely in the auricular area as abnormal connections between arteries and veins bypassing the capillary system (Rajesh and Saxena 2006). They are thought to be present normally at birth; however, some can remain quiescent, manifesting later in adulthood, their size increasing secondary to infection, trauma, previous excision attempts, or hormonal influences (pregnancy or puberty) (Rajesh and Saxena 2006). Such late-onset auricular AVMs have been reported in a few cases, with the principal AVM-feeding vessel originating from the PAA (Rajesh and Saxena 2006; Kim et al. 2017; Meher et al. 2008). Feeders from the auricular branches of the STA along with the PAA can contribute to auricular AVMs (Meher et al. 2008). Preoperative embolization of the feeder vessel followed by total surgical excision of the lesion has proved to be the treatment of choice for symptomatic auricular AVMs originating from the PAA (Rajesh and Saxena 2006; Kim et al. 2017; Meher et al. 2008).

6.6 Clinical and Surgical Significances

Clinically, it is important to note that lacerations within the region of the PAA, for instance in the ear or the region of the scalp posterior to the ear, cause significant bleeding. In the event of a laceration, knowledge of the PAA allows the clinician to control the hemorrhaging (Nguyen and Duong 2022). A strong understanding of the PAA and its anastomoses in the ear and posterior scalp region is critical for success in numerous surgical procedures.

Understanding the PAA is particularly important in reconstructive plastic surgery. In ear reconstruction, the surgeon can dissect the pedicle of tissue supplied by the PAA posterior to the ear. By rotating this tissue, he/she can construct a new ear (Nguyen and Duong 2022). Furthermore, diagnosis of giant cell temporal arteritis requires a biopsy of the superficial temporal artery. Postoperatively, owing to its anastomosis with the PAA, areas previously supplied by the superficial temporal artery now receive their primary vasculature from the PAA (Nguyen and Duong 2022; Kobayashi et al. 1997). The PAA is also a potential candidate for graft procedures, provided its diameter is approximately 1 mm. During the treatment of recurring stroke patients or those with adult-onset ischemic moyamoya disease, direct extracranial–intracranial bypass grafts using the PAA provided successful treatment of recurrent ischemia (Torazawa et al. 2017). Likewise, in procedures in which the surgeon seeks to identify the facial nerve, the PAA serves as an important anatomical landmark. The facial nerve travels inferior to the PAA within the stylomastoid foramen approximately 5 mm deep to the external meatal cartilage (Nguyen and Duong 2022; Liu et al. 2018). An in-depth understanding of the anatomy of the PAA and its anastomoses is beneficial in numerous surgical procedures within the ear and posterior scalp.

6.7 Complications

Owing to the anastomotic nature of the ear and the area of the scalp posterior to the ear, complications in this region are rare (Carpenter et al. 2017). However, if the PAA is compromised, the areas of the ear and posterior scalp will lack perfusion, which can result in ischemia and tissue necrosis (Nguyen and Duong 2022; Carpenter et al. 2017). During routine procedures, branches of the ECA typically provide a sufficient collateral blood supply. As surgical times increase, so does the likelihood that arterial injury will result in tissue necrosis and permanent harm to the patient.

When the PAA is injured during long surgical procedures within the posterior skull base, the pinna is at risk of ischemia and necrosis. Tissue necrosis can also impair quality of life through the negative aesthetic appearance within the ear and posterior regions of the scalp and neck (Carpenter et al. 2017). Furthermore, because of its proximity to the facial nerve, procedures that injure the PAA put patients at increased risk of acute idiopathic facial paralysis (Trost et al. 2008; Gordin et al. 2015).

Finally, in victims of stab wound injuries, injury to the PAA is relatively common. In these instances, the patient experiences serious profuse bleeding and needs immediate attention to prevent death or other serious outcomes. While these injuries can be extremely serious, cauterizing the PAA typically resolves 99% of the hemorrhaging and results in favorable patient outcomes (Carpenter et al. 2017; Casal et al. 2017).

References

Bertulli L, Robert T (2021) Embryological development of the human cranio-facial arterial system: a pictorial review. Surg Radiol Anat 43(6):961–973. https://doi.org/10.1007/s00276-021-02684-y

Bhatia KD, Lee H, Kortman H et al (2021) Endovascular management of intracranial dural arteriovenous fistulas: transarterial approach [published online ahead of print, 2021 Oct 7]. AJNR Am J Neuroradiol 43:324. https://doi.org/10.3174/ajnr.A7296

Carpenter PS, Burgette RC, Leonetti JP, Marzo SJ (2017) Auricular complications in parotid, temporal bone, infratemporal fossa, and lateral skull base surgery. Ear Nose Throat J 96(2):E27–E31. https://doi.org/10.1177/014556131709600216. PMID: 28231373

Casal D, Pelliccia G, Pais D, Carrola-Gomes D, Angélica-Almeida M, Videira-Castro J, Goyri-O'Neill J (2017) Stab injury to the preauricular region with laceration of the external carotid artery without involvement of the facial nerve: a case report. J Med Case Rep 11(1):205. https://doi.org/10.1186/s13256-017-1361-9

Elvan Ö, Bobuş A, Erdoğan S, Aktekin M, Olgunus ZK (2019) Fetal anatomy of the facial nerve trunk and its relationship with posterior auricular artery. Surg Radiol Anat 41(2):153–159. https://doi.org/10.1007/s00276-018-2126-x

Geibprasert S, Pongpech S, Armstrong D, Krings T (2009) Dangerous extracranial-intracranial anastomoses and supply to the cranial nerves: vessels the neurointerventionalist needs to know. AJNR Am J Neuroradiol 30(8):1459–1468. https://doi.org/10.3174/ajnr.A1500

Germans MR, Regli L (2014) Posterior auricular artery as an alternative donor vessel for extracranial-intracranial bypass surgery. Acta Neurochir 156:2095–2210. https://doi.org/10.1007/s00701-014-2206-6

Gómez Díaz OJ, Cruz Sánchez MD (2016) Anatomical and clinical study of the posterior auricular artery angiosome: in search of a rescue tool for ear reconstruction. Plast Reconstr Surg Glob Open 4(12):e1165. Published 2016 Dec 27. https://doi.org/10.1097/GOX.0000000000001165

Gordin E, Lee TS, Ducic Y, Arnaoutakis D (2015) Facial nerve trauma: evaluation and considerations in management. Craniomaxillofac Trauma Reconstr 8(1):1–13. https://doi.org/10.1055/s-0034-1372522

Hashemi HA, Mazhari A, Radbod K, Kamran B (1978) 'Otorrhagia'. A symptom of posterior auricular artery aneurysm. J Laryngol Otol 92(12):1127–1132. https://doi.org/10.1017/s0022215100086588

Hu CC, Lai YJ, Lai WJ (2018) Nonbifurcating carotid artery: a case report with a review of embryogenesis. Am J Case Rep 19:891–895. https://doi.org/10.12659/AJCR.910019

Jinkins JR (2000) Atlas of neuroradiologic embryology, anatomy, and variants. Lippincott Williams and Wilkins, Philadelphia

Kim SH, Han SH, Song Y, Park CS, Song JJ (2017) Arteriovenous malformation of the external ear: a clinical assessment with a scoping review of the literature. Braz J Otorhinolaryngol 83(6):683–690. https://doi.org/10.1016/j.bjorl.2016.09.004

Kiyosue H (2020) Superficial arteries from the distal ECA. In: Kiyosue H (ed) External carotid artery. Springer, Singapore

Kobayashi S, Nagase T, Ohmori K (1997) Colour Doppler flow imaging of postauricular arteries and veins. Br J Plast Surg 50(3):172–175. https://doi.org/10.1016/s0007-1226(97)91365-5. PMID: 9176003

Kolhe PS, Leonard AG (1987) The posterior auricular flap: anatomical studies. Br J Plast Surg 40(6):562–569. https://doi.org/10.1016/0007-1226(87)90149-4

Kouo T, Morales RE, Raghavan P (2018) Chapter 9 – Imaging of the facial nerve. In: Chong V (ed) Skull base imaging. Elsevier, pp 197–213

Kramer M, Schwab SA, Nkenke E et al (2014) Whole body magnetic resonance angiography and computed tomography angiography in the vascular mapping of head and neck: an intraindividual comparison. Head Face Med 10:16. Published 2014 May 12. https://doi.org/10.1186/1746-160X-10-16

Lasjaunias P, Berenstein A (1987) Surgical neuroangiography: 1 functional anatomy of craniofacial arteries, 1st edn. Springer, Berlin Heidelberg

Liu M, Wang SJ, Benet A, Meybodi AT, Tabani H, Ei-Sayed IH (2018) Posterior auricular artery as a novel anatomic landmark for identification of the facial nerve: a cadaveric study. Head Neck 40(7):1461–1465. https://doi.org/10.1002/hed.25127. Epub 2018 Mar 22. PMID: 29566447

Meher R, Varshney S, Pant HC (2008) Arteriovenous malformation related to the pinna. Hong Kong Med J 14(2):157–159

Menshawi K, Mohr JP, Gutierrez J (2015) A functional perspective on the embryology and anatomy of the cerebral blood supply. J Stroke 17(2):144–158. https://doi.org/10.5853/jos.2015.17.2.144

Nguyen JD, Duong H (2022) Anatomy, head and neck, posterior auricular artery. [Updated 2021 Jun 18]. In: StatPearls [Internet]. Treasure Island (FL): StatPearls Publishing. Available from: https://www.ncbi.nlm.nih.gov/books/NBK546687/

Pinar YA, Ikiz ZA, Bilge O (2003) Arterial anatomy of the auricle: its importance for reconstructive surgery. Surg Radiol Anat 25(3–4):175–179. https://doi.org/10.1007/s00276-003-0128-8

Rajesh BSS, Saxena RK (2006) Arteriovenous malformation and Color Doppler: posterior auricular artery. Indian J Otolaryngol Head Neck Surg 58(3):318–320. https://doi.org/10.1007/BF03050860

Rao NS, Manivannan K, Gaṅgādhara, Rao HR (2014) Trifurcation of external carotid artery and variant branches of first part of maxillary artery. Int J Anat Res 2(3):561–565

Sokol B, Stanislavski K, Jankowski R, Liebert W, Juszkat R (2017) Unusual presentation of traumatic pseudoaneurysm of the posterior auricular artery manifesting by massive bleeding from the external auditory canal: a case report. J Clin Case Rep 7:1–3. https://doi.org/10.4172/2165-7920.1000937

Standring S, Borley NR, Collins P, Crossman AR, Gatzoulis MA, Healy JC (2008) Gray's anatomy: the anatomical basis of clinical practice, 40th edn. Elsevier, London

Tanoue S, Kiyosue H, Mori H, Hori Y, Okahara M, Sagara Y (2013) Maxillary artery: functional and imaging anatomy for safe and effective transcatheter treatment. Radiographics 33(7):e209–e224

Tokugawa J, Nakao Y, Kudo K et al (2014) Posterior auricular artery-middle cerebral artery bypass: a rare superficial temporal artery variant with well-developed posterior auricular artery-case report. Neurol Med Chir (Tokyo) 54(10):841–844. https://doi.org/10.2176/nmc.cr2012-0233

Tokugawa J, Cho N, Suzuki H et al (2015) Novel classification of the posterior auricular artery based on angiographical appearance. PLoS One 10(6):e0128723. Published 2015 June 1

Torazawa S, Hasegawa H, Kin T, Sato H, Sora S (2017) Long-term patency of posterior auricular artery-middle cerebral artery bypass for adult-onset Moyamoya disease: case report and review of literature. World Neurosurg 108(994):e1–994.e5. https://doi.org/10.1016/j.wneu.2017.09.005. Epub 2017 Sep 20. PMID: 28939540

Trost O, Kadlub N, Cheynel N, Benkhadra M, Malka G, Trouilloud P (2008) Contribution de l'artèreauriculairepostérieure à la vascularisation du

nerf facial dans l'espacepréstylien : étude anatomique et conséquenceschirurgicales [Contribution of the posterior auricular artery in the vascularization of the facial nerve in prestylian space: anatomical study and surgical applications]. Morphologie 92(299):171–175. https://doi.org/10.1016/j.morpho.2008.07.016. French. Epub 2008 Oct 8. PMID: 18845467
Wang Y, Zhuang X, Jiang H et al (2008) The anatomy and application of the postauricular fascia flap in auricular reconstruction for congenital microtia. J Plast Reconstr Aesthet Surg 61(Suppl 1):S70–S76. https://doi.org/10.1016/j.bjps.2008.07.008
Wang D, Su L, Han Y, Fan X (2015) Embolization treatment of pseudoaneurysms originating from the external carotid artery. J Vasc Surg 61(4):920–926. https://doi.org/10.1016/j.jvs.2014.10.093
Zilinsky I, Erdmann D, Weissman O et al (2017) Reevaluation of the arterial blood supply of the auricle. J Anat 230(2):315–324. https://doi.org/10.1111/joa.12550
Zümre O, Salbacak A, Çiçekcibaşi AE, Tuncer I, Seker M (2005) Investigation of the bifurcation level of the common carotid artery and variations of the branches of the external carotid artery in human fetuses. Ann Anat 187(4):361–369. https://doi.org/10.1016/j.aanat.2005.03.007

Superior Thyroid Artery

7

Marc Gebara

7.1 Anatomy and Variation

The superior thyroid artery bisects the thyroid fascia and splits into anterior and posterior divisions. The anterior division supplies the thyroid's anterior surface, while the posterior division supplies the lateral and medial areas. The superior thyroid artery lies close to the external branch of the superior laryngeal nerve and is one of the main arteries supplying the thyroid gland, along with the inferior thyroid artery.

It originates from the external carotid artery inferior to the greater cornu of the hyoid bone before terminating in the upper pole of the thyroid gland. It splits into a branch that travels toward the sternocleidomastoid muscle and another that courses toward the superior laryngeal artery. Most anatomical investigations of the superior thyroid artery have focused on its potential variations in origin. Although the textbooks describe it as arising as a main branch from the external carotid artery, it can also split off the common carotid artery (Vazquez et al. 2009) or from the common carotid bifurcation (Gupta et al. 2014).

Variations of the superior thyroid artery can be rather complex. Ozgur et al. (2009) classified these variations into six types. Hu et al. (2006) proposed a similar classification. Gupta et al. (2014) conducted an angiographic study and identified three major branching types:

Type I (bifurcation): Superior thyroid artery splits into two major branches

Type II (trifurcation): It splits into three major branches in a way similar to Type I

Type III (nonbifurcation or trifurcation): Branching is random and does not resemble the Type I or II patterns

The superior thyroid artery gives off the laryngeal artery in 88% of cases (Fig. 7.1). The laryngeal artery can also branch from the external carotid artery (10%), the lingual artery (1%), or the thyrolingual/thyrolinguofacial trunk (1%). The ascending pharyngeal artery usually arises from the external carotid or common carotid artery. Power (1850) reported the artery originating from the superior thyroid artery. Lippert and Pabst (1985) found the inferior thyroid artery to be missing in 3% of cases, in which case the superior thyroid artery can supply blood to this region.

7.2 Embryology

Endocardial tubes give rise to the left and right primitive aortae, which curve in the first pharyngeal arch and around the anterior foregut. They

M. Gebara (✉)
Tulane University School of Medicine,
New Orleans, LA, USA

S. J. Bordes, Jr. et al. (eds.), *The Clinical Anatomy of the Vascular System*,
https://doi.org/10.1007/978-3-031-78326-5_7

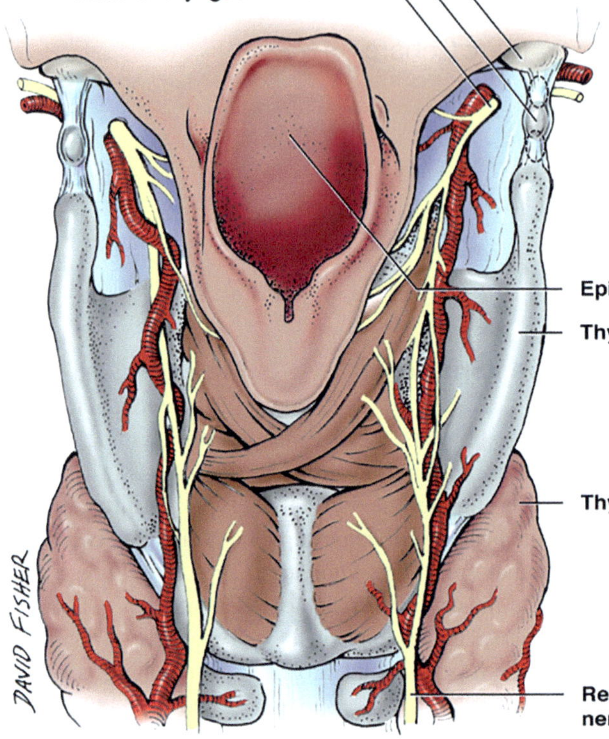

Fig. 7.1 Posterior internal view of the internal laryngeal nerve and the superior laryngeal branch of the superior thyroid artery

continue posteriorly and become two dorsal aortae, which form the aortic sac near the heart. The aortic sac and dorsal aortae are linked by six pairs of aortic arches. These arches do not all exist at the same time in the embryo, each being the origin of a critical structure in the developed human (Edwards 1948). The common carotid, proximal internal carotid, and external carotid arteries originate from the third aortic arch. The superior thyroid artery usually arises from the external carotid, as previously described.

7.3 Pathology

A case report by Hanashiro et al. (2017) described the nontraumatic rupture of a main branch of the left superior thyroid artery causing temporary extravasation. The patient was admitted to the hospital for treatment, and a follow-up enhanced CT and CT angiography were performed the following day. The rupture occurred although there were no aneurysms or dangerously thin walls, obscuring the actual point of rupture. Ten weeks later, the patient underwent left lobe hemithyroidectomy and a pathological study uncovered a parathyroid adenoma and multinodular goiter. Far fewer cases of superior thyroid artery rupture have been reported than inferior thyroid artery rupture, and nearly all have involved an abnormal superior thyroid artery or an anomalous thyroid gland. Hanashiro et al. (2017) concluded that elevated parathyroid hormone levels around the time of the hematoma suggested superior thyroid artery rupture secondary to the parathyroid adenoma versus multinodular goiter.

7.4 Clinical Significance and Complications

Toni et al. in 2003 and 2004 reported that the superior thyroid artery is critical for embolization of thyroid/parathyroid masses, and for laryngeal and thyroid transplantation. The artery can also be ligated for prophylaxis and treatment of severe hemorrhage in patients with malignant tumors (Wurm et al. 2000). It can aid with patching following carotid endarterectomy (Jenkins et al. 1997). It can further assist in esophageal reconstruction using a free jejunal graft (Kuntscher et al. 2003) and carotid axis implantation of microsurgical vascular anastomoses (Espalieu et al. 1986). It has been crucial for thyroid hemitracheal autographs in laryngotracheal reconstruction (Zur and Urken 2003). However, Salmeron et al. (1998) reported that this artery is not a good pedicle for long-segment tracheal transplants.

For thyroid resection or radical dissection of the carotid sheath, advance proximally to the common carotid bifurcation on the sheath at least 3 cm (Smith and Benton 1978). If the superior thyroid artery is ligated proximally, the internal branch of the superior laryngeal nerve can be injured during a thyroidectomy (Monfared et al. 2002). Timmermann et al. (2002) found that the external branch of the superior laryngeal nerve is at risk during superior thyroid artery ligation in thyroid surgery. Furthermore, Bellantone et al. (2001) claimed that proper distal ligation of superior thyroid artery branches is critical for preserving the external branch of the superior laryngeal nerve in thyroid procedures.

Asymmetry or absence of the superior thyroid artery can indicate a variant distribution of its associated sympathetic fibers. Toni et al. (2003) suggested that arterial variation alters the autonomic input to the thyroid gland, thereby modifying thyroid nerve regulation. Toni et al. (2004) proposed that physical injury to the cervical sympathetic chain, as in Claude Bernard-Horner Syndrome, could affect thyroid function in various ways depending on the makeup of the thyroid arterial system.

Won (2016) identified the following measurements between relevant anatomical landmarks and the superior thyroid artery:

- Within the horizontal line of the hyoid bone, the external carotid artery and mid-sagittal plane are 4.9 ± 0.5 mm apart
- The line and the superior thyroid artery are separated by a vertical distance of 0.9 ± 0.4 mm
- Within the horizontal line of the thyroid cartilage's inferior border, the superior thyroid artery and mid-sagittal plane are 3.1 ± 0.6 mm apart

The superior thyroid artery is the main artery supplying the thyroid gland and thus the major arterial supply for thyroid tumors in 80% of cases (Anagnostopoulou and Mavridis 2014). Maximum ipsilateral superior thyroid artery velocities in parathyroid adenoma were greatly increased (Varsamidis et al. 1999). Won (2016) described three complications with thyroidectomy: recurrent laryngeal nerve injury, hemorrhage, and hypoparathyroidism, found in 2% of patients (Sosa et al. 1998). Spatial relationships involving the superior thyroid artery and the thyroid must be accounted for during surgery to minimize the risk of hemorrhage and other complications.

References

Anagnostopoulou S, Mavridis I (2014) Emerging patterns of the human superior thyroid artery and review of its clinical anatomy. Surg Radiol Anat 36:33–38

Bellantone R, Boscherini M, Lombardi CP, Bossola M, Rubino F, De Crea C, Alesina P, Traini E, Cozza T, D'alatri L (2001) Is the identification of the external branch of the superior laryngeal nerve mandatory in thyroid operation? Results of a prospective randomized study. Surgery 130(6):1055–1059

Edwards JE (1948) Anomalies of the derivatives of the aortic arch system. Med Clin North Am 32:925–949

Espalieu P, Cottier M, Relave M, Youvarlakis P, Cuilleret J (1986) Radio-anatomic study of the carotid axis with regard to the implantation of microsurgical vascular anastomoses. Surg Radiol Anat 8(4):257–263

Gupta P, Bhalla AS, Thulkar S, Kumar A, Mohanti BK, Thakar A, Sharma A (2014) Variations in superior

thyroid artery: a selective angiographic study. Indian J Radiol Imaging 24:66–71
Hanashiro N, Yamashiro T, Iraha Y, Karube K, Matsumoto H, Murayama S (2017) Non-traumatic rupture of the superior thyroid artery with concomitant parathyroid adenoma and multinodular goiter. Acta Radiol Open 6(7):2058460117722813
Hu KS, Song WC, Kim SH, Choi SW, Han SH, Paik DJ et al (2006) Branching patterns of the arterial branches supplying the middle vascular pedicle of the sternocleidomastoid muscle: a topographic anatomical study with surgical applications for the use of pedicles osteomuscular flaps. Surg Radiol Anat 28:7–12
Jenkins MP, Aly S, Sarin S, Adiseshiah M (1997) Patch angioplasty following carotid endarterectomy using the ipsilateral superior thyroid artery. Eur J Vasc Endovasc Surg 14(1):60–62
Kuntscher MV, Erdmann D, Boltze WH, Germann G (2003) Use of a free jejunal graft for oesophageal reconstruction following perforation after cervical spine surgery: case report and review of the literature. Spinal Cord 41(10):543–548
Lippert H, Pabst R (1985) Arterial variations in man. Classification and frequency. J. F. Bergmann Verlag, Munich
Monfared A, Gorti G, Kim D (2002) Microsurgical anatomy of the laryngeal nerves as related to thyroid surgery. Laryngoscope 112(2):386–392
Ozgur Z, Govsa F, Celik S, Ozgur T (2009) Clinically relevant variation of the superior thyroid artery: an anatomic guide for surgical neck dissection. Surg Radiol Anat 31:151–159
Power JH (1850) Surgical anatomy of the arteries and descriptive anatomy of the heart by the Late Valentine Flood, MD. Fanninand Co, Dublin
Salmeron J, Gannon PJ, Blackwell KE, Shaari CM, Urken ML (1998) Tracheal transplantation: superior and inferior artery perfusion territory. Laryngoscope 108(6):849–853
Smith SD, Benton RS (1978) A rare origin of the superior thyroid artery. Acta Anat 101(1):91–93
Sosa JA, Bowman HM, Tielsch JM, Powe NR, Gordon TA, Udelsman R (1998) The importance of surgeon experience for clinical and economic outcomes from thyroidectomy. Ann Surg 228:320–330
Timmermann W, Hamelmann WH, Meyer T, Timm S, Schramm C, Hoppe F, Thiede A (2002) Identification and surgical anatomy of the external branch of the superior laryngeal nerve. Zentralbl Chir 127(5):425–428
Toni R, Della Casa C, Mosca S, Malaguti A, Castorina S, Roti E (2003) Anthropological variations in the anatomy of the human thyroid arteries. Thyroid 13(2):183–192
Toni R, Della Casa C, Castorina S, Malaguti A, Mosca S, Roti E, Valenti G (2004) A meta-analysis of superior thyroid artery variations in different human groups and their clinical implications. Ann Anat 186(3):255–262
Varsamidis K, Varsamidou E, Mavropoulos G (1999) Color Doppler sonography in the detection of parathyroid adenomas. Head Neck 21:648–651
Vazquez T, Cobiella R, Maranillo E, Valderrama FJ, McHanwell S, Parkin I, Sañudo JR (2009) Anatomical variations of the superior thyroid and superior laryngeal arteries. Head Neck 31:1078–1085
Won SY (2016) Anatomical considerations of the superior thyroid artery: its origins, variations, and position relative to the hyoid bone and thyroid cartilage. Anat Cell Biol 49(2):138–142
Wurm J, Gode U, Fucak A (2000) Ligature of the carotid arteries performed prophylactically as an emergency procedure in patients with malignant tumours of the head and neck. HNO 48(1):22–27
Zur KB, Urken ML (2003) Vascularized hemitracheal autograft for laryngotracheal reconstruction: a new surgical technique based on the thyroid gland as a vascular carrier. Laryngoscope 113(9):1494–1498

8 Inferior Thyroid Artery

Mia Lauter

8.1 Embryology

At the end of the third week of development, the cardiovascular system begins to form, with the pharyngeal arches developing shortly thereafter, during weeks four and five. These pharyngeal arches are penetrated by pharyngeal arteries arising from the aortic sac. The pharyngeal arch arteries convert into the following three major arteries during weeks six to eight of development: carotid, subclavian, and pulmonary. From the subclavian artery comes the thyrocervical trunk, of which the inferior thyroid artery (ITA) is a branch (Keith et al. 2020).

8.2 Anatomy

The ITA (Fig. 8.1) originates from the thyrocervical trunk, which is a branch of the subclavian artery. The path of the ITA makes an arc. From its origin, it turns medially and ascends along the medial border of the anterior scalene muscles. It passes posterior to the carotid sheath and cervical sympathetic trunk, but anterior to the vertebral artery. As it approaches the transverse process of the sixth cervical vertebra, the ITA arches downward and descends along the longus colli muscle, passing either anterior or posterior to the recurrent laryngeal nerve, until it reaches the inferior portion of the thyroid gland. Here, the ITA splits into superior and inferior branches. The superior branch forms an anastomosis with the posterior branch of the superior thyroid artery (STA) on the posterior part of the thyroid gland (Grant 1974; Stewart and Rizzolo 2007).

The ITA is considered the principal blood supply to the thyroid gland. It supplies the posterior and inferior parts of the thyroid gland as well as the four parathyroid glands. It gives rise to several branches, including the ascending cervical artery and the inferior laryngeal artery, as well as pharyngeal, tracheal, and esophageal branches (Grant 1974). These many branches go on to supply several structures, including muscles of the neck (anterior scalene, longus colli, infrahyoid, and inferior pharyngeal constrictor muscles), the laryngopharynx, esophagus, and inferior part of the larynx (Barral and Croibier 2011).

M. Lauter (✉)
Tulane University School of Medicine,
New Orleans, LA, USA
e-mail: mlauter@tulane.edu

S. J. Bordes, Jr. et al. (eds.), *The Clinical Anatomy of the Vascular System*,
https://doi.org/10.1007/978-3-031-78326-5_8

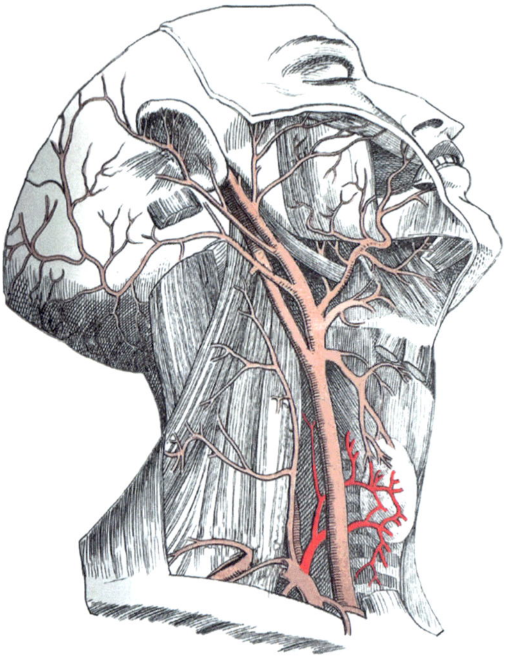

Fig. 8.1 Right neck showing both the carotid and proximal subclavian arteries. Note the inferior thyroid artery (shaded) arising from the thyrocervical trunk and sending branches to the thyroid gland. This artery has rich anastomoses with the superior thyroid artery arising from the external carotid artery. (After Allen, 1882)

8.3 Variations

Variations in the ITA are most often described in terms of its relation to the recurrent laryngeal nerve (RLN). At least 16 variations of this relationship have been described. Most often, the RLN passes posterior to the ITA, particularly on the left side. This anatomy may offer protection to the nerve during surgical procedures. On the right side of the neck, the most common variation is the anterior passage of the nerve to the artery, which is associated with more frequent injury to the RLN during thyroid gland surgery. Finally, the least common variation is when the nerve travels between branches of the artery (Noussios et al. 2020).

An aberrant origin of the ITA is also possible. Rather than deriving from the thyrocervical trunk, the ITA may arise directly from the subclavian artery, and rarely, the common carotid artery (Graves et al. 2018). It is also possible for the ITA to be absent, either unilaterally or bilaterally (Chandrakala et al. 2013).

8.4 Clinical/Surgical Significance

Partial and complete thyroidectomies are common surgical procedures. During such procedures, there is a significant risk of damaging the recurrent laryngeal nerve, which may lead to difficulty swallowing and/or hoarseness. Given the close relationship of the RLN to the ITA, the ITA serves as a useful landmark for identifying and protecting the RLN. RLN injury during thyroidectomy varies from 1.5% to 14% (Noussios et al. 2020).

Additionally, when considering the ligation of arteries that supply the thyroid and parathyroid glands, the ITA should be avoided and rather its branches ligated. Ligation of the ITA itself may contribute to devascularization of the thyroid and parathyroid glands, which can lead to symptomatic hypocalcemia in patients undergoing total thyroidectomy (Kaplan et al. 2000; Sanabria et al. 2017).

References

Barral JP, Croibier A (2011) Vessels of the thyroid. Visceral vascular manipulations. https://doi.org/10.1016/B978-0-7020-4351-2.00022-3

Chandrakala SP, Mamatha Y, Thejaswini KO (2013) Variations in the origin of inferior thyroid artery and relation of the artery with recurrent laryngeal nerve. Natl J Clin Anat 2:11–15

Grant SEK (ed) (1974) Grant's dissector, 7th edn, Baltimore Williams & Wilkins

Graves M, Henry B, Sanna B, Pękala P, Taterra D, Witczak K, Kucharska E, Barczyński M, Walocha J, Tomaszewski K (2018, 74) Anatomical variations of the inferior thyroid artery: a cadaveric examination. Przegląd lekarski

Kaplan E, Angelos P, Applewhite M et al (2000) Chapter 21 Surgery of the thyroid. [Updated 2015 Sep 25]. In: Feingold KR, Anawalt B, Boyce A et al (eds) Endotext [Internet]. MDText.com, Inc., South Dartmouth (MA). Available from: https://www.ncbi.nlm.nih.gov/books/NBK285564/

Keith L, Moore BA, Persaud TVNV, Torchia MG (2020) The developing human. In: Cardiovascular system, vol 13, pp 263–314

Noussios G, Chatzis I, Konstantinidis S et al (2020) The anatomical relationship of inferior thyroid artery and recurrent laryngeal nerve: a review of the literature and its clinical importance. J Clin Med Res 12(10):640–646

Sanabria A, Kowalski LP, Tartaglia F (2017) Inferior thyroid artery ligation increases hypocalcemia after thyroidectomy: a meta-analysis. Laryngoscope 128(2):534–541. https://doi.org/10.1002/lary.26681

Stewart WB, Rizzolo LJ (2007) Embryology and surgical anatomy of the thyroid and parathyroid glands. In: Oertli D, Udelsman R (eds) Surgery of the thyroid and parathyroid glands. Springer, Berlin, Heidelberg. https://doi.org/10.1007/978-3-540-68043-7_2

Facial Artery

9

Innokentiy Krupnov and Lisandro Montorfano

9.1 Introduction

Most of the arterial supply to the face comes from the facial artery, which branches directly from the external carotid artery (Figs. 9.1, 9.2, and 9.3). Other major external carotid artery branches include the superior thyroid, ascending pharyngeal, lingual, occipital, posterior auricular, superficial temporal, and maxillary arteries. A more minor arterial supply, mainly to the upper portion of the face, comes from the ophthalmic artery, the first branch of the internal carotid, arising just distal to the cavernous sinus. This provides the primary blood supply to the eye, extraocular muscles, lacrimal gland, upper nose, and parts of the forehead.

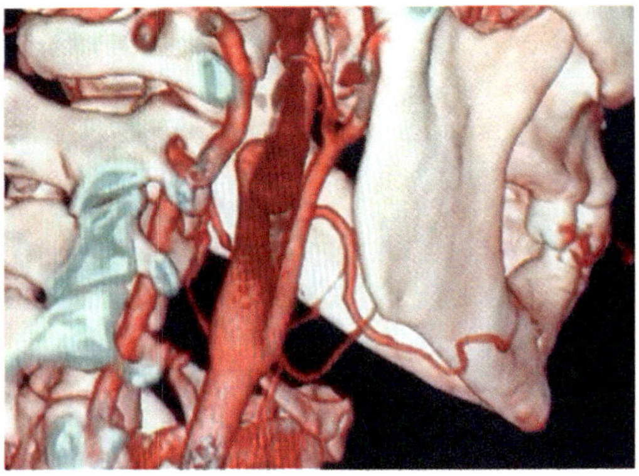

Fig. 9.1 Posterolateral view of the right facial artery traveling from its origin from the external carotid artery. Anteriorly over the mandible and onto the face

I. Krupnov
Department of Plastic and Reconstructive Surgery, Cleveland Clinic Florida, Weston, FL, USA

L. Montorfano (✉)
Department of Plastic and Reconstructive Surgery, Vanderbilt University Medical Center, Nashville, TN, USA

S. J. Bordes, Jr. et al. (eds.), *The Clinical Anatomy of the Vascular System*, https://doi.org/10.1007/978-3-031-78326-5_9

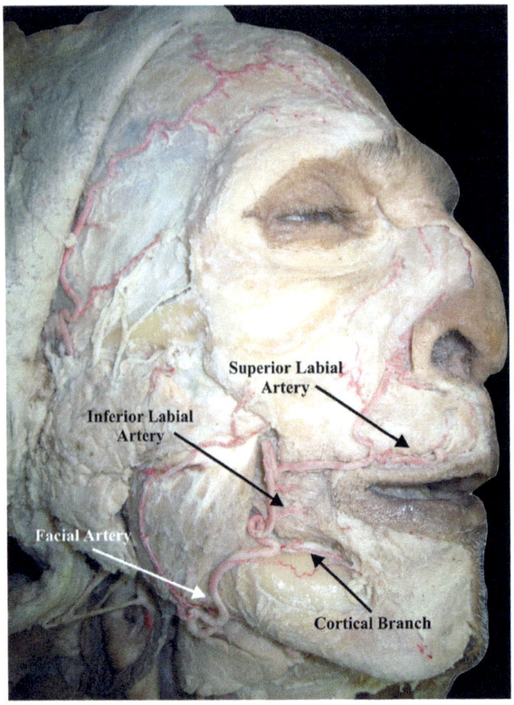

Fig. 9.2 Cadaveric dissection noting the labial branches of the facial artery. (Courtesy Loukas et al.)

9.2 Embryology

Vascular development has two main stages, vasculogenesis and angiogenesis. Vasculogenesis is the process by which hemangioblasts are differentiated into angioblasts and a primitive vascular network forms; angiogenesis refers to the formation of new capillary vessels from pre-existing blood vessels. The main mechanism of angiogenesis is sprouting, which is mainly driven by hypoxia/ischemia and related growth factors from the target tissue. As more capillaries are formed, the impedance to flow is reduced in larger arteries; this promotes flow-induced remodeling of the vessels supplying a specific parenchymal area. The development of the circulatory system supplying blood to the brain begins with the formation of the six pairs of primitive branchial arch arteries, which subsequently undergo profound modifications during development (Bertulli and Robert 2021).

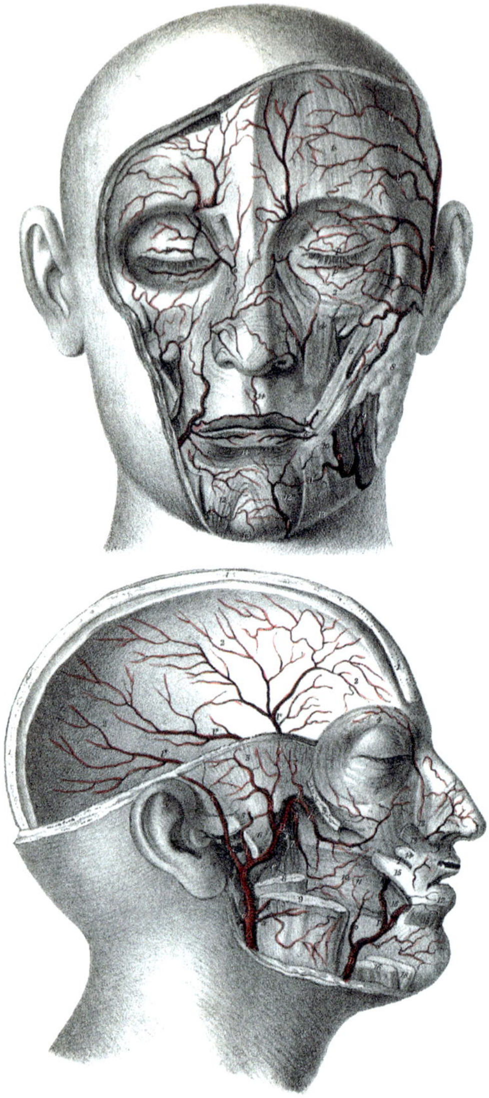

Fig. 9.3 Drawing illustrating the profuse blood supply to the face. (Quain and Wilson 1844 author's collection)

The embryological development of the carotid vascular system is very interesting. The pharyngeal arches begin to form during weeks 4–5 of gestation, and the common carotid arteries (CCAs) and proximal internal carotid arteries (ICAs) derive from the third pharyngeal arch while the distal ICAs derive from the dorsal aorta. The external carotid artery (ECA) derives from the CCA via angiogenesis.

Subsequently, between weeks four and eight, as the head and neck undergo embryological development, all the major external carotid artery branches also develop via angiogenesis. The external carotid artery gives off six branches before it divides into two terminating branches. They are, in ascending order, superior thyroid, ascending pharyngeal, lingual, facial, occipital, and posterior auricular. The two terminating branches are the maxillary and superficial temporal arteries. Occasionally, the lingual and facial arteries arise as a common trunk (Granick and Jacob 2010).

The facial artery and its branches also arise via angiogenesis as the facial features go through stages of embryological development.

9.3 Anatomy/Topography

The facial artery is a branch of the external carotid artery that supplies the anatomical structures of the superficial face. It arises from the carotid triangle, which is formed by the superior belly of the omohyoid muscle, the sternocleidomastoid, and the posterior belly of the digastric. It originates deep in the platysma and quickly becomes superficial. The tortuous course of this artery allows for more stretch during facial actions such as mastication. As it courses on the face, the vessel travels deep to the posterior belly of the digastric and stylohyoid muscles. It then continues along the posterior surface of the submandibular gland. It curves upwards over the body of the mandible and courses along the anteroinferior border of the masseter. The pulse of the facial artery is palpable as it crosses the mandible. The artery continues superiorly at an oblique angle across the cheek toward the oral commissure and then ascends along the side of the nose, terminating at the medial canthus of the eye as the angular artery.

Along its course, the facial artery gives off important cervical branches (arising in the neck):

- *Ascending Palatine Artery:* travels upwards between the styloglossus and stylopharyngeus before it splits into two branches. One of these goes to the auditory tube of the ear and the palatine tonsil; the other reaches the superior pharyngeal constrictor muscle of the pharynx.
- *Tonsillar Branch:* runs between the styloglossus and medial pterygoid muscle and pierces the superior pharyngeal constrictor before reaching the palatine tonsil. At that point, it connects with one of the branches of the ascending palatine artery.
- *Submental Artery:* travels by the underside of the chin; this is the largest branch of the facial artery. It splits off as it passes through the submandibular gland, passing over the mylohyoid muscle just behind the body of the mandible. It eventually splits further into superficial and deep branches.
- *Glandular Branches:* usually 3–4 branches that arise moving toward the submandibular salivary gland and surrounding structures.

As the facial artery continues to course upwards from the cervical to the facial region it gives rise to several additional branches:

- *Inferior Labial Artery:* emerges at the corner of the mouth before passing upwards and forwards, underneath the triangularis and the orbicularis oris muscles, and then connects with its counterpart from the other side of the head and with the mental branch.
- *Superior Labial Artery:* larger than the inferior labial artery, it runs to muscles of the upper lip, the septum, and the ala of the nose.
- *Lateral Nasal Branch:* runs along the side of the nose to supply important parts of the nose and the area around the eye.
- *Angular Artery:* this is a terminal branch of the facial artery; it supplies structures in the cheek before ultimately connecting with the

ophthalmic artery (a branch of the internal carotid). It runs upwards, reaching the medial canthus of the eye.

9.4 Imaging/Histology

Detailed examination of the anatomy of the facial artery, its course, and its variations are very important and highly beneficial for plastic surgeons in several reconstructive procedures (e.g., Abbe flap, a type of reconstructive lip flap used in rebuilding lateral upper and lower lip defects resulting from cancerous, congenital, or traumatic causes). Multiple imaging modalities, including angiography, CT, or Doppler, can be used to evaluate the anatomy of a facial artery.

Furukawa et al. (2013) used 64-slice multidetector computed tomography to examine a total of 187 facial arteries in 94 patients without facial tumors or vascular disease. They identified and classified four types of anatomical variation: type 1, a short course that terminates proximal to the superior labial artery; type 2, an intermediate course that terminates distal to the superior labial artery near the nasolabial fold; type 3, a classic course that extends to the lateral nasal ala beyond the nasolabial fold with an angular branch; and type 4, duplex with dominant lateral angular branch. The research team concluded that facial computed tomographic angiography allows the facial artery and its branches to be visualized in detail and is a potential noninvasive imaging tool for preoperative vascular mapping for facial reconstructive surgery (Furukawa et al. 2013).

9.5 Variations

The facial artery has many identified and documented anatomical variations, like most vascular structures in the human body. Examples have been described in many scholarly articles, and some will be discussed here.

The facial artery can originate as a linguofacial trunk or a thyro-linguo-facial trunk, a fairly common variation in which it branches off the lingual artery rather than directly off the external carotid. Other developmental variations have also been described such as agenesis of the facial artery, enlarged facial artery, and hypoplastic facial artery.

Bondaz et al. (2012) reported a hypoplastic facial artery as a known variation with a frequency of 0–9% according to the literature data. When the facial artery is hypoplastic, the face is supplied either by a more developed contralateral facial artery or by the transverse facial artery and the infraorbital artery on the ipsilateral side (Bondaz et al. 2012).

Mangalgiri et al. (2015) found a facial artery originating from the external carotid just below the origin of the maxillary artery within the parotid gland with an incidence of roughly 3.3%. They warned that a higher origin can be misdiagnosed as the absence of the facial artery, such as complete agenesis. Also, a facial artery with a high origin is vulnerable to injury during parotid surgeries. Control of bleeding can be difficult owing to the proximity of terminal branches of the facial nerve (Mangalgiri et al. 2015).

In rare instances, doctors have noted cases in which the facial artery never develops at all. In these cases, surrounding arteries are tasked with providing blood and oxygenation to the regions usually supplied by the facial artery. Moreover, even with a normal anatomical origin, the course and length of a facial artery and its branches within the facial part can be highly variable.

Loukas et al. (2006) examined 284 hemifaces derived from 142 formalin-fixed cadavers to examine the distribution patterns of the facial artery, paying specific attention to the superior labial artery (SLA), which is commonly used in an Abbe flap procedure. They recognized and categorized such variations into five types labeled "A" through "E." Type A (135, 47.5%): facial artery bifurcates into SLA and lateral nasal (the latter gives off inferior and superior alar and ends as angular); Type B (110, 38.7%): similar to Type A, except lateral nasal terminates as superior alar (angular artery is absent); Type C (24, 8.4%): facial artery terminates as SLA; Type D (11, 3.8%): angular artery arises directly from facial arterial trunk rather than as the termination of lateral nasal, with the facial artery ending as superior alar; and Type E (4, 1.4%): facial artery terminates as a rudimentary twig without providing significant branches. The authors' aim was to equip both anatomists and surgeons with a more thorough understanding of the vasculature of the face, and to enable plastic surgeons to have a more confident approach to reconstructive procedures in this region (Loukas et al. 2006).

9.6 Pathology

Atherosclerosis of the common carotid artery is a risk factor for embolic events leading to ischemic injury. If there is a weakness in the superficial facial musculature, there can be a vascular compromise of the facial artery. An embolus from the common carotid can travel and cause ischemic injury to a distal branch of the facial artery. However, this is a rare sequela of carotid artery atherosclerosis.

Another phenomenon documented in some patients is the development of unaccustomed headaches in chronic internal carotid artery (ICA) occlusion because collateral circulation develops. In ICA occlusion, collateral flow from branches of the external carotid artery can cause ipsilateral brisk pulsations at the angle of the jaw, brow, and cheek, the so-called ABC pulsations of Fisher. However, collateralization can predominate through the circle of Willis, so the ABC pulsations are not always brisk. Also, the diagnostic value of such pulsations is limited as they are a normal finding in 10% of adults over the age of 55 with long-standing hypertension (Thanvi and Robinson 2007).

9.7 Clinical and Surgical Significance

The facial artery arises from the external carotid and exits the neck by coursing around the inferior border of the mandible. Prior to crossing over the mandible, it gives rise to the submental artery and submental flap. The submental artery supplies the submental region, which can provide a soft tissue pedicled flap that is particularly useful in this region. It also sends branches to the inferior border of the mandible. The facial artery itself is the artery named in the facial artery musculomucosal (FAMM) flap described by Pribaz. This is an intraoral flap that has earned its place in the armamentarium of most head and neck surgeons. It incorporates both the mucosa and buccinator muscle, and the facial artery.

The facial artery is mostly used as a recipient artery for free flaps. It is usually used as a recipient vessel in the neck but can also be easily accessed in the nasolabial region. Its main advantage is its easy accessibility. This advantage is not as obvious when the patient is undergoing neck dissection than it is when a head and neck reconstruction without neck dissection is being undertaken. In this situation, the facial artery can be palpated as it traverses the border of the mandible and can easily be located through a short transverse neck incision. Because it is located under the border of the mandible, access can be difficult for microvascular anastomosis, particularly when a mandibular reconstruction is being performed. Therefore, one disadvantage of using the facial artery in this situation is that it must be dissected well back toward its origin, and sometimes it needs to be taken down below the posterior belly of the digastric muscle. Also, the nearest acceptable recipient vein can be some distance away, particularly in patients who have had a neck dissection (Neligan and Novak 2009).

9.8 Complications

The facial artery can be implicated in complications associated with various surgical procedures. Prevention of trauma to it is paramount. Thus, excessive retraction in this area should be avoided. Hemorrhage from the facial artery can also occur during submandibular gland excision. External approaches to mandible fracture repair can require it to be ligated. If this is needed, careful documentation should be made in the operative note as a reference for potential future surgeons. If there is bleeding from the facial artery, direct pressure should be applied to the angle of the mandible over the vessel until the vessel is identified and the bleeding controlled. It is also important to recognize the proximity of various nerves, including the hypoglossal nerve, to the facial artery during surgical procedures. These structures need to be identified early and preserved to minimize postoperative complications.

References

Bertulli L, Robert T (2021) Embryological development of the human cranio-facial arterial system: a pictorial review. Surg Radiol Anat 43(6):961–973. https://www.ncbi.nlm.nih.gov/pubmed/33492439. https://doi.org/10.1007/s00276-021-02684-y

Bondaz M, Ricard AS, Berge J, Caix P, Laurentjoye M (2012) Hypoplastic facial artery and allotransplantation of the lower two-thirds of the face: an anatomical study. Plast Reconstr Surg 129(2):369e–371e. https://www.ncbi.nlm.nih.gov/pubmed/22286466. https://doi.org/10.1097/PRS.0b013e31823af01d

Furukawa M, Mathes DW, Anzai Y (2013) Evaluation of the facial artery on computed tomographic angiography using 64-slice multidetector computed tomography: implications for facial reconstruction in plastic surgery. Plast Reconstr Surg 131(3):526–535. https://www.ncbi.nlm.nih.gov/pubmed/23446566. https://doi.org/10.1097/PRS.0b013e31827c6f18

Granick MS, Jacob LM (2010) Head and neck embryology and anatomy. In: Plastic surgery secrets plus. Elsevier, pp 347–354. https://doi.org/10.1016/B978-0-323-03470-8.00053-3

Loukas M, Hullett J, Louis RG, Kapos T, Knight J, Nagy R, Marycz D (2006) A detailed observation of variations of the facial artery, with emphasis on the superior labial artery. Surg Radiol Anat 28(3):316–324. https://doi.org/10.1007/s00276-006-0093-0

Mangalgiri A, Namdev LN, Mahore D, Kapre M (2015) The study of higher origin of facial artery and its surgical significance. Indian J Otolaryngol Head Neck Surg 67(1):72–74. https://www.ncbi.nlm.nih.gov/pubmed/25621238. https://doi.org/10.1007/s12070-014-0786-3

Neligan PC, Novak CB (2009) Chapter 4 – Head and neck. In: Wei F-C, Mardini S (eds) Flaps and reconstructive surgery. W.B. Saunders, pp 31–37, ISBN 9780721605197. https://doi.org/10.1016/B978-0-7216-0519-7.00004-6

Thanvi B, Robinson T (2007) Complete occlusion of extracranial internal carotid artery: clinical features, pathophysiology, diagnosis and management. Postgrad Med J 83(976):95–99. https://doi.org/10.1136/pgmj.2006.048041

10 Maxillary Artery

Nathan John, Savannah Moon, and Lisandro Montorfano

10.1 Introduction

The maxillary artery is the larger of the terminal branches of the external carotid artery (Figs. 10.1 and 10.2). It is responsible for perfusion of both the deep muscles and the bones of the face, and it contributes to the perfusion of intracranial structures such as the dura. Overall, it is a particularly important vessel for multiple surgical subspecialties including neurosurgery, otolaryngology, and oral-maxillofacial surgery.

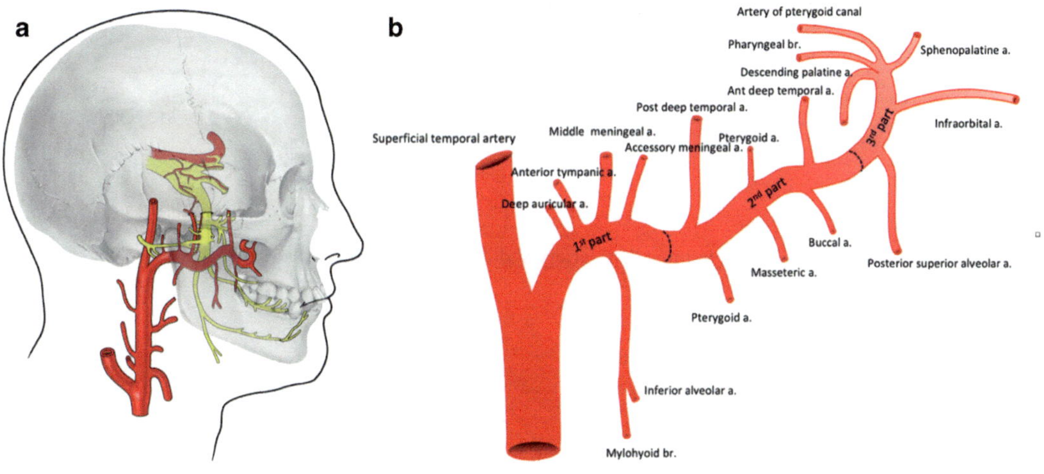

Fig. 10.1 (**a**) Highly schematic drawing of the right horizontally positioned maxillary artery and its branches. (**b**) Named branches of the maxillary artery

N. John · S. Moon
Department of Plastic and Reconstructive Surgery, Cleveland Clinic Florida, Weston, FL, USA
e-mail: johnn2@ccf.org

L. Montorfano (✉)
Department of Plastic and Reconstructive Surgery, Cleveland Clinic Florida, Weston, FL, USA

Department of Surgery, Cleveland Clinic Florida, Weston, FL, USA

Department of Plastic and Reconstructive Surgery, Vanderbilt University Medical Center, Nashville, TN, USA

S. J. Bordes, Jr. et al. (eds.), *The Clinical Anatomy of the Vascular System*,
https://doi.org/10.1007/978-3-031-78326-5_10

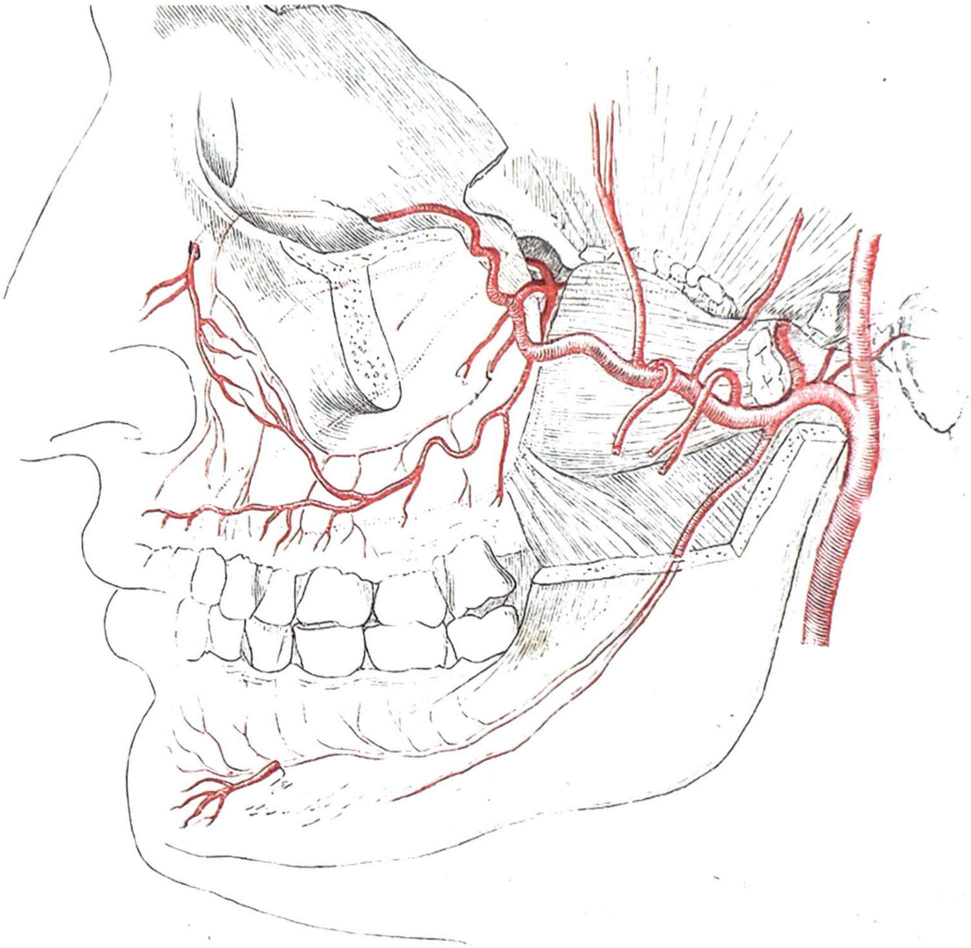

Fig. 10.2 Schematic drawing of the left maxillary artery and its branches. (After Allen, 1882)

10.2 Embryology

Blood vessels develop from embryonic mesoderm by vasculogenesis and angiogenesis. Vasculogenesis, the de novo formation of vasculature, is the source of the major vessels of the embryo such as the dorsal aorta and cardinal vessels. Angiogenesis is the growth of new vessels sprouting from the essential vascular framework produced by vasculogenesis (Risau 1997).

The primitive vasculature is initiated during the third week of development as the growing size of the embryo outpaces the ability of simple diffusion to provide adequate nutrition. During vasculogenesis, mesodermal cells are induced by FGF2 to differentiate into angioblasts in clusters described as "blood islands." These clusters elongate and coalesce into the endothelial cords that remodel into recognizable blood vessels under the influence of PDGF and TGF-β. Increasing embryonic complexity requires an increasingly complex blood supply as development continues. These new structures and organs are fed by vessels produced via angiogenesis as VEGF gradients stimulate the further proliferation and migration of endothelial cells (Leland et al. 2012a).

The heart tube develops in concert with vasculogenesis, starting during the third week with the formation of the primary heart field. The secondary heart field follows by the end of the third week, contributing to the outflow tracts growing craniocaudally. Cranial extension of the heart tube ultimately forms the aortic sac as the most distal end of the truncus arteriosus.

The six pharyngeal arches develop cranially to the heart tube throughout the fourth and fifth weeks, ultimately forming many of the structures of the face and neck. As each arch develops, each side receives its own arterial supply via an aortic arch branching from the aortic sac via angiogenesis. Each aortic arch terminates at the dorsal aorta of that laterality, remaining separate cranially and merging at their midpoints more caudally. Various sections of the arches involute as the vascular network matures. By day 27 of embryonic life, most of the first aortic arches have disappeared. The remainder persists as the maxillary artery, which remains to perfuse the first pharyngeal arch (which ultimately forms the maxilla and mandible) (Leland et al. 2012b; Tanoue et al. 2013).

10.3 Anatomy/Topography

The maxillary artery branches from the external carotid artery at the level of the neck of the mandible. It then passes through the parotid gland before traveling anteriorly between the sphenomandibular ligament and the ramus of the mandible. From there it generally travels adjacent to the lateral pterygoid muscle in a superficial course between the upper and lower heads of that muscle. Finally, it leaves the infratemporal fossa to pass obliquely through the pterygomaxillary fissure to terminate within the pterygopalatine fossa. The venous drainage for the maxillary artery's territory is the pterygoid plexus. The venous tributaries flow into the maxillary vein, which in turn drains into the retromandibular vein before returning to the internal jugular vein (von Arx et al. 2018).

The maxillary artery is divided into three segments by its relationship to the lateral pterygoid muscle. The first two divisions are considered to be within the infratemporal fossa, while the third lies within the pterygopalatine fossa.

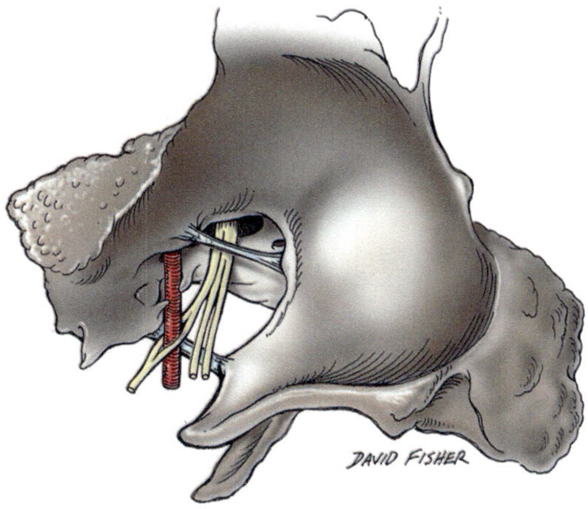

Fig. 10.3 Schematic drawings showing the middle meningeal artery ascending through the foramen spinosum and encircled by the auriculotemporal nerve

1. *The First (Mandibular) Division* is posterior to the lateral pterygoid muscle. It has five branches, all of which travel through a bony foramen:
   (a) *Deep auricular artery* enters the squamotympanic fissure and supplies the external acoustic meatus.
   (b) *Anterior tympanic artery* enters the squamotympanic fissure and supplies the deep surface of the tympanic membrane.
   (c) *Middle meningeal artery* enters the foramen spinosum into the middle cranial cavity and supplies the dura and the bone of the middle cranial fossa (Fig. 10.3).
   (d) *Accessory meningeal artery* enters the foramen ovale to complement the middle meningeal artery.
   (e) *Inferior alveolar artery* enters the mandibular foramen and supplies all the lower teeth, lower lip, and chin via the mental branch along with the mylohyoid via the mylohyoid branch.

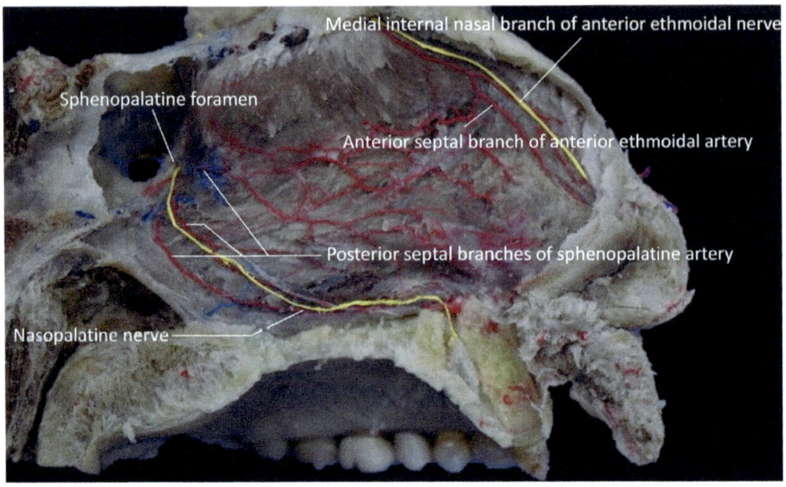

Fig. 10.4 Internal view of the nasal septum illustrating branches of the sphenopalatine artery

2. *The Second (Pterygoid) Division* is generally superficial to the lateral pterygoid muscle. It has at least five branches, often with variations:
   (a) *Anterior deep temporal artery* supplies the temporalis muscle.
   (b) *Posterior deep temporal artery* also supplies the temporalis muscle.
   (c) *Buccal artery* supplies the buccinator muscle.
   (d) *Masseteric artery* supplies the masseter muscle.
   (e) *Pterygoid arteries* are variable in number, supplying both the medial and lateral pterygoid muscles.
3. *The Third (Pterygopalatine) Section* is anterior to the lateral pterygoid muscle. It has six branches, including the terminal branch.
   (a) *Pharyngeal artery* enters the palatovaginal canal and supplies the posterior aspect of the roof of the nasal cavity, the sphenoid sinus, and the Eustachian tube.
   (b) *Greater palatine artery* enters the greater palatine foramen to supply the hard palate, while also giving off the *lesser palatine artery* to supply the soft palate.
   (c) *Artery of the pterygoid canal* enters the pterygoid canal and supplies the Eustachian tube.
   (d) *Sphenopalatine artery* enters the sphenopalatine foramen and branches into the *posterior lateral nasal arteries,* which supply the lateral wall of the nasal cavity, and the *posterior septal arteries,* which supply the nasal septum (Fig. 10.4).
   (e) *Posterior superior alveolar artery* supplies the upper molars and the surrounding gingiva (Fig. 10.5).

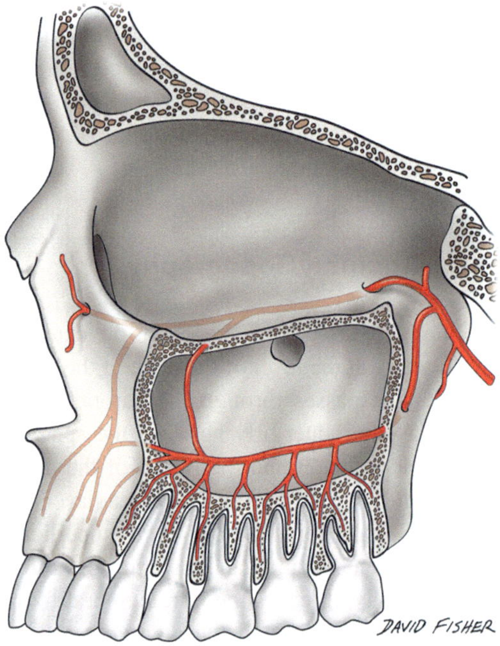

Fig. 10.5 Schematic drawing demonstrating the posterior superior alveolar artery and its branches

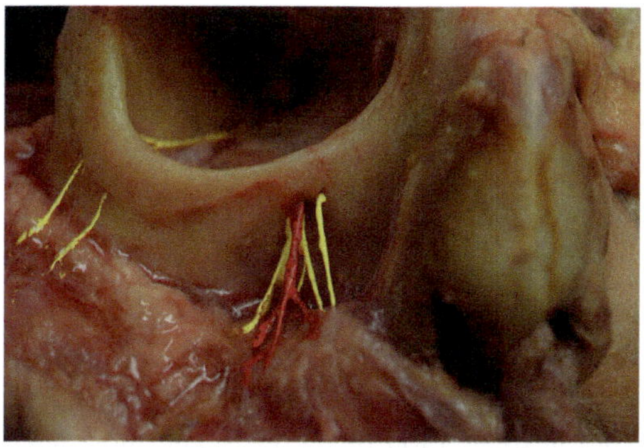

Fig. 10.6 Cadaveric view of the right infraorbital region noting the infraorbital nerve in yellow and the infraorbital artery traveling with it through the infraorbital foramen. The two nerves that are seen laterally are the zygomaticofacial nerves

(f) *Infraorbital artery* enters the inferior orbital fissure and supplies the lower eyelid and upper lip, while also giving off the *anterior superior alveolar artery* to supply the incisors and canines (Fig. 10.6).

10.4 Imaging/Histology

Imaging studies of the maxillary artery classically focus on defining its spatial relationship to the pterygoid muscles or its variability in branching pattern (discussed below). Since the maxillary artery has become increasingly popular for extracranial–intracranial bypasses and grafting procedures, the focus has shifted toward defining its relationship to the middle meningeal artery. Magnetic resonance imaging with angiography (MRA) has demonstrated that this artery receives 20.6 ± 11% of the blood flowing through the external carotid artery, second to only the facial artery (26.6 ± 10%). MRA has also pinpointed the middle meningeal artery emerging 1.37 ± 0.59 cm from the origin of the maxillary artery at an angle of 116.2 ± 35.76°. The vessel diameters at the maxillary artery origin and the middle meningeal artery origin were 2.93 ± 0.52 mm and 1.58 ± 0.43 mm, respectively, drawing a significant portion of the flow of the maxillary artery (Bettoni et al. 2018; Ji et al. 2021).

10.5 Variations

The maxillary artery was well studied in cadavers throughout the twentieth century. The courses of its first and third divisions have been generally reproducible and symmetrical. However, defining the exact position of the second division in relation to the lateral pterygoid muscle (deep vs superficial) has proven somewhat controversial. Population-based studies favor the superficial course but have shown significant variability; for example, studies of Japanese populations have demonstrated >90% superficial course, while studies of European populations have shown the superficial course in as few as 9% of specimens.

The availability and noninvasive nature of computed tomographic angiography (CTA) substantially increased the power of anatomical studies in the twenty-first century. Hussain et al. applied this novel approach to the Turkish population, evaluating 286 patients, or 572 maxillary arteries, a contrast with earlier studies that evaluated ≤20 specimens. Of these arteries, 418 (68.4%) had a course superficial to the lateral pterygoid muscle while 132 (31.6%) ran deep. This study supports the view that the superficial course is more common, although the variation seen previously across populations merits repetition beyond a single ethnic group. Furthermore, 165 (79.9%) patients were symmetrical (either both lateral or both deep) vs. 44 (21.1%) who were asymmetrical. Other groups have applied imaging to demonstrate that the asymmetry of the branching patterns on each division is ubiquitous. These studies demonstrate that asymmetry is not rare, and validate the need for preoperative imaging and for treating each laterality as a single entity during dissection (Hussain et al. 2008; Alvernia et al. 2017).

10.6 Pathology

The maxillary artery is the source of bleeding in posterior epistaxis and epidural hematomas. Epistaxis can occur spontaneously or in the setting of trauma. Although less common, posterior epistaxis is more challenging to control than anterior epistaxis because packing is often unable to reach the bleeding vessel and thus unlikely to tamponade. The pterygopalatine division of the maxillary artery is often a source of posterior epistaxis via the sphenopalatine artery, or less commonly the pharyngeal artery. If posterior epistaxis does not stop spontaneously or with a balloon tamponade, endovascular embolization of the sphenopalatine artery and anterior ethmoid artery is performed (Liao et al. 2007).

Epidural hematomas form when blunt trauma at the pterion of the lateral cranium ruptures the middle meningeal artery. Arterial flow from the maxillary artery continues through the ruptured middle meningeal between the layers of the dura. The growing hematoma forms a lens-shaped lesion trapped by the fissure lines of the dura. Subjects classically lose consciousness from the initial trauma, then awaken to a "lucid interval." As the mass effect of the growing hematoma increases intracranial pressure, the subject again loses consciousness with a high risk of mortality and morbidity. Epidural hematomas are considered neurosurgical emergencies and a craniotomy, rather than an endovascular approach, is used to relieve the intracranial pressure.

The large number of branches of the maxillary artery feed the deep muscles, soft tissues, and bones of the face and supply the dura mater and cranial nerves. Supplying so many diverse structures juxtaposes the arteriovenous territory of the maxillary artery against those of the facial artery, which supplies the superficial structures of the face, and the intracranial branches of the internal carotid artery. This allows the maxillary artery to be a major feeding artery for dural arteriovenous malformations (AVM) and irregular arterial anastomoses. In addition to causing symptoms such as debilitating headaches, these AVM can be a major source of morbidity in cases of trauma or bleeding events. Even without malformations, this overlap between arteriovenous fields allows the infection to spread intracranially from a facial nidus. For example, the lack of valves of the pterygoid plexus can allow the infection to spread from a dental source through the infratemporal sinus to the cavernous sinus (Tanoue et al. 2013).

10.7 Clinical and Surgical Significance

The maxillary artery has recently regained traction as a source for revascularization, for instance for free flap maxillary reconstruction, allotransplantation of the face, and most notably extracranial-to-intracranial diversion (EC-IC). EC-IC is historically used to bypass intracranial aneurysms to avoid their risk of rupture and as a source of emboli. The advent of endovascular techniques and stenting rendered EC-IC mostly obsolete; however, it is regaining popularity in situations where flow augmentation is ideal, such as architectural disruption due to skull base tumors, Moyamoya disease, or ischemic hypoperfusion that is otherwise refractory to medical therapy. In these cases, the maxillary artery becomes a powerful source of flow, often through the deep temporal arteries or middle meningeal arteries to the middle cerebral artery (Touré 2019; Wang et al. 2018; Nossek and Langer 2018).

10.8 Complications

Any surgical dissection into the infratemporal fossa must involve a careful identification of the maxillary artery and its many branches. Fisch type C cranial/subtemporal/infratemporal approaches to excise soft tissue or parotid gland tumors, osteotomies in mandibular surgery, or preauricular access for TMJ surgeries involve dissecting near or around the maxillary artery. Operating with such a risk for extensive intra- or postoperative hemorrhage requires careful preoperative planning with an extensive review of angiography. Even common procedures, such as the mandibular blocks routinely employed in dental practice, run the risk of intra-arterial injection, pseudoaneurysm, and hematoma formation (Hussain et al. 2008; Moore and Olsen 2007).

References

Alvernia JE, Hidalgo J, Sindou MP, Washington C, Luzardo G, Perkins E, Nader R, Mertens P (2017) The maxillary artery and its variants: an anatomical study with neurosurgical applications. Acta Neurochir 159(4):655–664. https://doi.org/10.1007/s00701-017-3092-5

Bettoni J, Pagé G, Salsac AV et al (2018) Quantitative assessment of the flow distribution in the branches of the external carotid by non-injected flow MRI. Dentomaxillofac Radiol 47(8):20180153. https://doi.org/10.1259/dmfr.20180153

Hussain A, Binahmed A, Karim A, Sándor GK (2008) Relationship of the maxillary artery and lateral pterygoid muscle in a caucasian sample. Oral Surg Oral Med Oral Pathol Oral Radiol Endod 105(1):32–36. https://doi.org/10.1016/j.tripleo.2007.04.010

Ji T, Hou K, Li C, Yu J (2021) Imaging features of internal maxillary artery and extracranial middle meningeal artery and their relationships on head CTA. Neuroradiol J 34:629. https://doi.org/10.1177/19714009211019380

Leland J et al (2012a) Chapter 6: Third to eighth weeks: the embryonic period. In: Langman's medical embryology, 12th edn. Wolters Kluwer/Lippincott Williams & Wilkins, Philadelphia

Leland J et al (2012b) Chapter 13: Cardiovascular system. In: Langman's medical embryology, 12th edn. Wolters Kluwer/Lippincott Williams & Wilkins, Philadelphia

Liao CC, Hsu YP, Chen CT, Tseng YY (2007) Transarterial embolization for intractable oronasal hemorrhage associated with craniofacial trauma: evaluation of prognostic factors. J Trauma 63(4):827–830. https://doi.org/10.1097/TA.0b013e31814b9466

Moore EJ, Olsen KD (2007) Total parotidectomy. In: Myers EN, Ferris RL (eds) Salivary gland disorders. Springer, Berlin/Heidelberg. https://doi.org/10.1007/978-3-540-47072-4_15

Nossek E, Langer DJ (2018) Internal maxillary artery to middle cerebral artery cranial bypass: the new "Work horse" for cerebral flow replacement. World Neurosurg 115:44–46. https://doi.org/10.1016/j.wneu.2018.03.214

Risau W (1997) Mechanisms of angiogenesis. Nature 386:671–674. https://doi.org/10.1038/386671a0

Tanoue S, Kiyosue H, Mori H, Hori Y, Okahara M, Sagara Y (2013) Maxillary artery: functional and imaging anatomy for safe and effective transcatheter treatment. Radiographics 33(7):e209–e224. https://doi.org/10.1148/rg.337125173

Touré G (2019) Distribution of the maxillary artery in the deep regions of the face and the maxilla: clinical applications. J Plast Reconstr Aesthet Surg 72(6):1020–1024. https://doi.org/10.1016/j.bjps.2019.02.008

von Arx T, Tamura K, Yukiya O, Lozanoff S (2018) The face – a vascular perspective. A literature review. Swiss Dent J 128(5):382–392

Wang L, Cai L, Lu S, Qian H, Lawton MT, Shi X (2018) The history and evolution of internal maxillary artery bypass. World Neurosurg 113:320–332. https://doi.org/10.1016/j.wneu.2018.02.158

Superficial Temporal Artery

11

Felice De Stefano and Lisandro Montorfano

11.1 Introduction

Knowledge of the anatomy of the arterial supply to the head and neck is imperative for surgery. This knowledge is used in a wide variety of situations, from diagnostic procedures such as a temporal artery biopsy for diagnosing temporal arteritis, to complex reconstructive surgery of the face after traumatic injuries or malformations, to oncological resections of head and neck tumors.

11.2 Anatomy/Topography

The superficial temporal artery (Figs. 11.1 and 11.2) is a terminal branch of the external carotid artery. From its origin at the level of the mandibular ramus, it travels cephalad within the parotid gland and then becomes superficial after crossing the gland. It travels between the zygomatic tubercle and the external auditory canal, superficially crosses the zygomatic arch, and reaches the temporal region; approximately 5 cm above the zygoma it bifurcates into its terminal branches, the frontal and parietal arteries (Testut and Latarjet 1988). As it courses over the zygomatic process it travels between the superficial temporal fascia, which is contiguous with the superficial musculoaponeurotic system of the face inferiorly, and the frontalis galea complex superiorly.

The artery is sometimes accompanied by the superficial temporal vein, which runs superficial to the superficial temporal fascia. Throughout its course, the artery is also close to the auriculotemporal nerve, which lies posterior to it (Shin et al. 2018).

The superficial temporal artery gives origin to several branches. Immediately after its origin, it gives off the parotid and transverse facial arteries. The former supplies the parotid tissue, while the latter provides branches supplying the parotid duct, masseter muscle, and adjacent skin. At the level of the zygomatic arch, it gives off the zygomatico-orbital and middle temporal arteries, which provide blood to the temporal and orbicularis oculi muscles. The anterior auricular branch supplies the auricle and external acoustic meatus (Barral and Croibier 2011).

F. De Stefano
Department of Surgery, Cleveland Clinic Florida, Weston, FL, USA
e-mail: desteff@ccf.org

L. Montorfano (✉)
Department of Surgery, Cleveland Clinic Florida, Weston, FL, USA

Department of Plastic and Reconstructive Surgery, Vanderbilt University, Nashville, TN, USA

S. J. Bordes, Jr. et al. (eds.), *The Clinical Anatomy of the Vascular System*,
https://doi.org/10.1007/978-3-031-78326-5_11

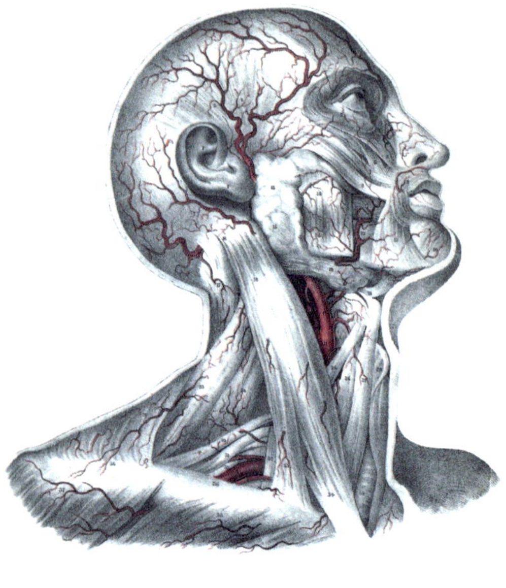

Fig. 11.1 Drawing of the superficial temporal artery (Quain and Wilson, 1844 author's collection)

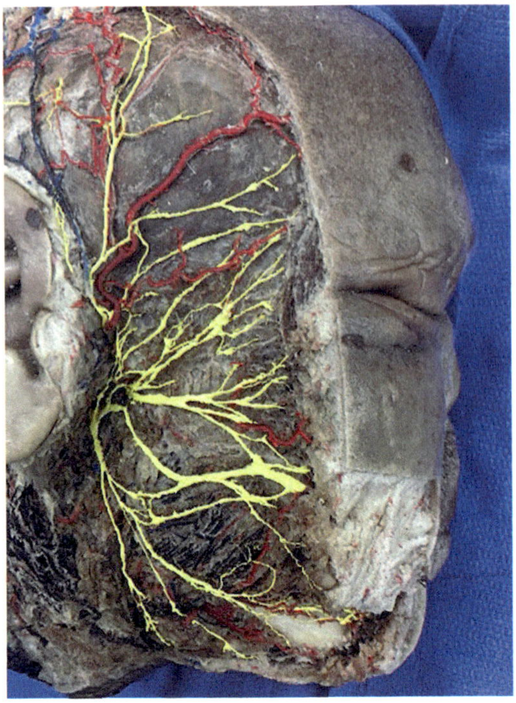

Fig. 11.2 Right-sided cadaveric dissection illustrating the superficial temporal artery and its branches injected with red latex and this relationship with the auriculotemporal and facial nerves colored yellow

11.3 Variations

The superficial temporal artery has different anatomical variations, including the following: it can be duplicated, and it can give branches that normally branch from other arteries such as the maxillary, lateral nasal, and labial arteries. Also, some of the usual branches of the superficial temporal artery can originate from adjacent arteries. For example, the transverse facial artery can originate directly from the external carotid artery. The frontal and parietal arteries can be absent unilaterally or bilaterally (Barral and Croibier 2011; Rusu et al. 2021).

11.4 Pathology: Clinical and Surgical Significance

Temporal arteritis, also known as giant cell arteritis, Horton arteritis, or cranial arteritis, is a systemic vasculitis that usually affects adults over the age of 50 and is the most common form of vasculitis in this age group in Western countries (González-Gay and García-Porrúa 1999; Kermani et al. 2010). It is characterized by inflammation of medium and large arteries that originate from the aortic arch. Histopathological changes include inflammation of the arterial wall with fragmentation and disruption of the internal elastic lamina. Multinucleated giant cells are not found in all cases and are not specific to the disease.

The condition usually presents with new onset headaches, constitutional symptoms, and visual changes among others, and can be associated with polymyalgia rheumatica. Its most feared complication is irreversible visual loss caused by ischemic optic neuropathy.

Diagnosis is based on clinical, laboratory, and pathology findings. Temporal artery biopsy is considered the gold-standard test (Niederkohr and Levin 2005). Therefore, during surgical biopsies, it is crucial to understand superficial temporal artery anatomy to prevent injury to structures that are closely anatomically related to nerves and veins.

The superficial temporal artery has also clinical significance for the management of intracranial ischemic diseases as it can be used with the middle cerebral artery for bypass. It is also considered an option for flow replacement bypass during the repair of intracranial aneurysms. In addition, it can serve as a path for intraoperative angiography and a path for endovascular treatments (Hou et al. 2019).

Owing to its superficial location, it is vulnerable to traumatic injuries. Superficial temporal artery pseudoaneurysms have been reported following head trauma (Carnevale et al. 2018; Manuel et al. 2012).

References

Barral J-P, Croibier A (2011) 20 – The superficial temporal artery. In: Barral J-P, Croibier A (eds) Visceral vascular manipulations. Churchill Livingstone

Carnevale ML, Phair J, Trestman EB, Dickinson G, Vitti MJ (2018) Superficial temporal artery pseudoaneurysm following a traumatic haircut. Int Arch Vasc Med 1:001. https://doi.org/10.23937/iavm-2017/1710001

González-Gay MA, García-Porrúa C (1999) Systemic vasculitis in adults in northwestern Spain, 1988–1997. Clin Epidemiol Aspects Med (Baltimore) 78:292

Hou K, Guo Y, Xu K, Yu J (2019) Clinical importance of the superficial temporal artery in neurovascular diseases: a PRISMA-compliant systematic review. Int J Med Sci 16(10):1377–1385. https://doi.org/10.7150/ijms.36698

Kermani TA, Schäfer VS, Crowson CS et al (2010) Increase in age at onset of giant cell arteritis: a population-based study. Ann Rheum Dis 69:780

Manuel S, Simon D, George EK, Naik RB (2012) Traumatic pseudoaneurysm of the superficial temporal artery. Clin Pract 2(2):e34. https://doi.org/10.4081/cp.2012.e34

Niederkohr RD, Levin LA (2005) Management of the patient with suspected temporal arteritis a decision-analytic approach. Ophthalmology 112:744

Rusu MC, Jianu AM, Rădoi PM (2021) Anatomic variations of the superficial temporal artery. Surg Radiol Anat 43:445–450. https://doi.org/10.1007/s00276-020-02629-x

Shin KJ, Shin HJ, Lee SH, Koh KS, Song WC (2018) Surgical anatomy of the superficial temporal artery to prevent facial nerve injury during arterial biopsy. Clin Anat 31(4):608

Testut L, Latarjet A (1988) Tratado de Anatomía Humana. Salvat Spanish Edition

Thyrocervical Trunk

12

Alex Ramos and Stephen J. Bordes, Jr.

12.1 Embryology

The proximal portion of the right subclavian artery arises from the right side of the fourth aortic arch, while the distal portion arises from the right seventh intersegmental artery. The left subclavian artery arises from the left seventh intersegmental artery. Angiogenesis, driven by hypoxia and growth factors, leads bilaterally to the formation of the thyrocervical trunks from the subclavian arteries; this process is complete by the fourth week of gestation (Sadler 2019).

12.2 Anatomy/Topography

The thyrocervical trunk originates as a branch from the first part of the subclavian artery, which is located medial to the anterior scalene muscle (Moore et al. 2018) (Fig. 12.1). It arises distal to the vertebral artery and proximal to the costocervical trunk (Netter 2019). Typically, it is a short vessel that ascends medial to the anterior scalene muscle and almost immediately gives rise to several branches. There are four major branches; these show extensive variation, as discussed below.

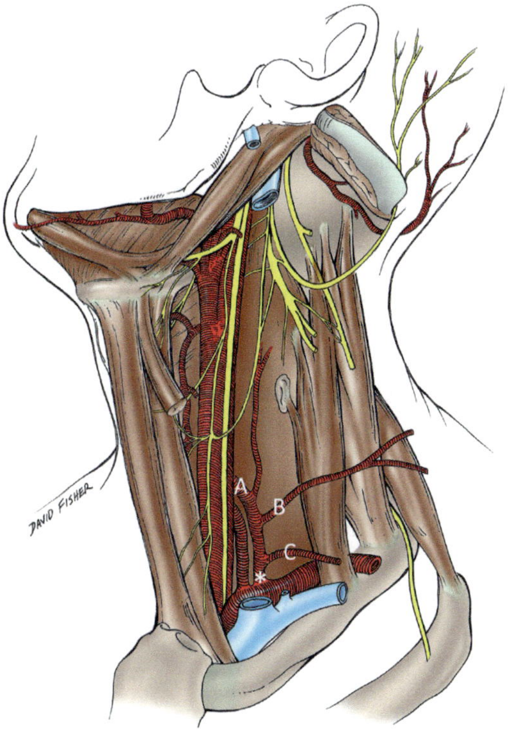

Fig. 12.1 Schematic drawing of the left-sided thyrocervical trunk (*) and its branches. (**a**) inferior thyroid artery; (**b**) transverse cervical artery; and (**c**) suprascapular artery

The most proximal of the branches arising from the thyrocervical trunk is the suprascapular artery. This courses anterior to the anterior

A. Ramos
Texas A&M University College of Medicine, Bryan, TX, USA
e-mail: aram0568@tamu.edu

S. J. Bordes, Jr. (✉)
Department of Surgery, Louisiana State University Health Sciences Center, New Orleans, LA, USA

S. J. Bordes, Jr. et al. (eds.), *The Clinical Anatomy of the Vascular System*,
https://doi.org/10.1007/978-3-031-78326-5_12

scalene muscle in an inferolateral direction to supply the infraspinatus and supraspinatus muscles, two of the rotator cuff muscles located on the posterior aspect of the scapula (Moore et al. 2018; Hacking and MacManus 2016; Wei and Mardini 2009).

The thyrocervical trunk also gives rise to the transverse cervical artery, also called the cervicodorsal trunk (Moore et al. 2018), which traverses laterally anterior to the anterior scalene muscle and superior to the suprascapular artery. The transverse cervical artery bifurcates into a superficial branch, the superficial cervical artery, and a deep branch, the dorsal scapular artery (Moore et al. 2018; Netter 2019). The superficial cervical artery courses superficially to and supplies the trapezius muscle (Markiewicz et al. 2017). The dorsal scapular artery runs medial to the scapula, contributing to the scapular anastomosis. It supplies the levator scapulae and rhomboid muscles as it passes deep into both muscles (Moore et al. 2018).

There are two terminal branches of the thyrocervical trunk, the inferior thyroid artery (Fig. 12.2) and the ascending cervical artery. The latter is the smaller of the two, ascending to supply the lateral muscles of the neck and the posterior aspect of the cervical vertebrae (Cramer and Darby 2014). The inferior thyroid artery travels medially to supply the inferior aspect of the thyroid gland at the level of C5 (Barral and Croibier 2011).

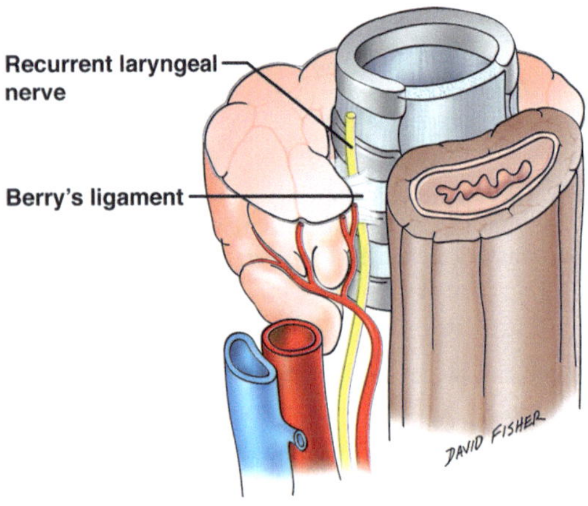

Fig. 12.2 Posterior view of the inferior thyroid artery ascending to the thyroid gland in close proximity to the left recurrent laryngeal nerve

12.3 Variations

Structurally, the thyrocervical trunk is typically depicted as a short ascending vessel that gives rise to the suprascapular artery, then the transverse cervical artery, and terminates in two branches, the ascending cervical artery laterally and the inferior thyroid artery medially (Netter 2019). There is considerable variation in this structure and in the origins of the four branches.

Three of these anatomical variations account for more than 50% of cases (Pérez-García et al. 2018). In the most common variation, the type I configuration, the thyrocervical trunk gives rise to the inferior thyroid artery and another short trunk that terminates in the suprascapular artery and the transverse cervical artery's superficial branch, the superficial cervical artery (Pérez-García et al. 2018; Muñiz et al. 2002; Jardaly and Bordoni 2021). In this variation, the deep branch of the transverse cervical artery, the dorsal scapular artery, originates directly from the subclavian artery distal to the thyrocervical trunk; and the ascending cervical artery branches off the inferior thyroid artery. In a separate study of 83 cadavers with both sides of the neck dissected, the dorsal scapular artery was a branch of the third part of the subclavian artery (lateral to the anterior scalene muscle) in 43.4%. It was a branch of the second part of the subclavian artery (posterior to the anterior scalene muscle) in 24.7%. In the same study, the dorsal scapular artery was a branch of the thyrocervical trunk in 19.9% of cases, characteristic of the type II and III configurations discussed below (Lischka et al. 1982).

In the second most prevalent variation, the type II configuration, the thyrocervical trunk branches into the inferior thyroid artery and another short trunk that gives rise to the suprascapular artery and the transverse cervical artery, which then divides into superficial and deep branches. The ascending cervical artery arises as a branch of the transverse cervical artery proximal to its terminal bifurcation (Pérez-García et al. 2018; Muñiz et al. 2002; Jardaly and Bordoni 2021).

In the third most common configuration, type III, the inferior thyroid artery branches directly from the subclavian artery proximal to the thyrocervical trunk. The thyrocervical trunk gives rise to the subscapular artery, the ascending cervical artery, and the transverse cervical artery, which bifurcates terminally into superficial and deep branches (Pérez-García et al. 2018; Jardaly and Bordoni 2021). In the aforementioned study of 83 cadavers, the inferior thyroid artery arose from the thyrocervical trunk in 87% of cases (as in the type I configuration) and from the subclavian artery in 12%; it was absent in two cases (Lischka et al. 1982).

In the less common configurations of the thyrocervical trunk, the suprascapular artery or the transverse cervical artery (and its branches) can arise directly from the subclavian artery (Pérez-García et al. 2018).

12.4 Clinical and Surgical Significance

During surgical procedures in the neck region, computed tomography angiography can be used to map the thyrocervical trunk and its branches. Mapping the vasculature associated with the thyrocervical trunk is important because of its extensive variability and potential vulnerability during surgical procedures.

Relevant Anastomoses

Branches of the thyrocervical trunk contribute to three relevant anastomoses: the scapular anastomosis, with arteries supplying the thyroid, and an anastomosis with arteries supplying the spinal cord and surrounding tissues (Pérez-García et al. 2018; Jardaly and Bordoni 2021; Perrucci et al. 2020). The scapular anastomosis supplies the muscles surrounding the scapula, ensuring sufficient circulation even if the thyrocervical trunk's relevant branches or tributaries from the axillary artery were to be obstructed.

The thyrocervical trunk contributes smaller branches, especially from the ascending cervical artery, to the anterior spinal artery to form an anastomosis that supplies the spinal cord (Jardaly and Bordoni 2021; Perrucci et al. 2020). It also has some anastomotic branches with the vertebral artery, which again supplies the spinal region. Owing to these anastomoses between the thyrocervical trunk and the arteries supplying the spinal cord, spinal infarction can result from embolization of the thyrocervical trunk or spinal artery (Pérez-García et al. 2018; Perrucci et al. 2020).

Head and Neck Cancer

The thyrocervical trunk, and more often its branches, can be sources of arterial blood to cancers of the head and neck region (Jardaly and Bordoni 2021; Lazzaro et al. 2011). The appropriate branches can be embolized to reduce the blood supply to a nearby cancer, stunting its growth.

12.5 Complications

Pseudoaneurysm

The thyrocervical trunk can develop a pseudoaneurysm, usually resulting from an iatrogenic needle puncture, or more rarely from blunt trauma to the neck region (Mehta et al. 2013; Karsonovich et al. 2019). Pseudoaneurysms of the thyrocervical trunk can be caused by puncture during an attempt to cannulate the internal jugular vein, which lies in close proximity (Mehta et al. 2013). They are generally rare in the thyrocervical trunk but can present as a slowly growing neck mass. There can be additional symptoms depending on other structures compressed by the expanding pseudoaneurysm, including Horner's syndrome. The resulting neck mass is usually pulsatile with a palpable thrill or bruit on auscultation and can be painless or painful (Mur et al. 2017). If the pseudoaneurysm were to rupture, resulting in a neck hematoma, the airway could be compromised because the hematoma could compress the trachea (Jardaly and Bordoni 2021). Diagnosis can be confirmed by magnetic resonance imaging (Mehta et al. 2013), selective angiography, or Doppler ultrasonography (Mur et al. 2017).

Pseudoaneurysms of the thyrocervical trunk can be addressed in several ways. However, it is important to be mindful of anatomical variations of the thyrocervical trunk during surgery to avoid injuring any of its branches. Open surgical repair and endovascular techniques are both viable treatment options. More novel approaches include ultrasound-guided percutaneous thrombin injection (Mehta et al. 2013), n-butyl cyanoacrylate embolization (Karsonovich et al. 2019), and coil embolization (Mur et al. 2017; Carnevale et al. 2019).

References

Barral JP, Croibier A (2011) Visceral vascular manipulations: 22 – Vessels of the thyroid. Churchill Livingstone, pp 166–181

Carnevale M, Phair J, Shariff S (2019) Post-traumatic thyrocervical trunk aneurysm following a motor vehicle accident repaired via endovascular coil embolization. In: Society for clinical vascular surgery symposium

Cramer GD, Darby SA (2014) Clinical anatomy of the spine, spinal cord, and ANS: chapter 5 – The cervical region, vol 3. Mosby, pp 135–209

Hacking C, MacManus D (2016) Thyrocervical Trunk. Radiopaedia

Jardaly AH, Bordoni B (2021) Anatomy, thorax, thyrocervical arteries. StatPearls

Karsonovich TW, Hawkins JC, Gordhan A (2019) Traumatic pseudoaneurysm of the ascending cervical artery treated with N-butyl cyanoacrylate embolization: a case report and review of the literature. Cureus 11(12):e6276

Lazzaro MA, Badruddin A, Zaidat OO, Darkhabani Z, Pandya DJ, Lynch JR (2011) Endovascular embolization of head and neck tumors. Front Neurol 2:64

Lischka MF, Krammer EB, Rath T, Riedl M, Ellböck E (1982) The human Thyrocervical trunk: configuration and variability reinvestigated. Anat Embryol 163:389–401

Markiewicz MR, Ord R, Fernandes RP (2017) Maxillofacial surgery: 43 – Local and regional flap reconstruction of maxillofacial defects, vol 1. Churchill Livingstone, pp 616–635

Mehta K, England E, Apgar J et al (2013) Post-traumatic pseudoaneurysm of the thyrocervical trunk. Skeletal Radiol 42:1169–1172

Moore KL, Dalley AF, Agur AM (2018) Clinically oriented anatomy. Wolters Kluwer, p 8

Muñiz A, Fernández A, Hernández LC, Garnacho S (2002) Variants of the thyrocervical trunk and its branches in human bodies. Eur J Anat 6(2):109–114

Mur TA, Anis MM, Soliman AM (2017) Thyrocervical trunk pseudoaneurysm presenting as a neck mass. Ear Nose Throat J 96(7):238–239

Netter FH (2019) Atlas of human anatomy, vol 7. Elsevier, pp 11–174

Pérez-García C, Malfaz C, del Valle DM et al (2018) Embolization through the thyrocervical trunk: vascular anatomy, variants, and a case series. J NeuroIntерv Surg 10:1012–1018

Perrucci L, Graziano M, Ferrante Z, Salviato E, Carnevale A, Galeotti R (2020) Pitfalls in the embolisation of a thyrocervical trunk bleeding: a case report. Patient Saf Surg 14:19

Sadler TW (2019) Landman's medical embryology, vol 14. Wolters Kluwer, pp 179–222

Wei FC, Mardini S (2009) Flaps and reconstructive surgery. Elsevier, pp 249–285

13 Ascending Cervical Artery

Lauren A. Balentine

13.1 Embryology

During the fourth week (days 28–29) of development, the first aortic arches develop from the paired dorsal aortae at the cranial ends of the primitive heart tube. Additional pairs develop in craniocaudal succession with subsequent remodeling to form the vasculature that will supply the head and neck. The dorsal aorta gives rise to dorsolateral intersegmental branches, called intersegmental arteries, which arise from the paraxial mesoderm and penetrate between the somite derivatives to supply each level (Schoenwolf et al. 2015).

The subclavian arteries develop asymmetrically. The definitive right subclavian artery is derived from (1) the right fourth aortic arch, (2) a short segment of the right dorsal aorta, and (3) the right seventh intersegmental artery. The definitive left subclavian artery is derived from the left seventh intersegmental artery.

At day 32, the vertebral arteries, the first branches of the subclavian arteries bilaterally, arise from longitudinal branches that connect the intersegmental arteries. The dorsal anastomoses contribute to the vertebral arteries, while the ventral anastomoses persist as the thyrocervical trunk (Embryology 2022). The thyrocervical trunk will give rise to multiple branches, including the inferior thyroid artery, which, before its course toward the inferior lobe of the thyroid medially, will give off the ascending cervical artery. The superior portions of the ascending cervical artery are derived from the previously mentioned cervical anastomoses of the intersegmental arteries.

L. A. Balentine (✉)
Tulane University School of Medicine,
New Orleans, LA, USA
e-mail: lbalentine@tulane.edu

13.2 Histology

The origin and longitudinal portions of the ascending cervical artery may be classified as muscular (i.e., medium or distributing) arteries (1–2 mm). Relative to larger elastic arteries (e.g., aorta), muscular arteries have a thin tunica intima with a prominent internal elastic membrane, a tunica media of predominately smooth muscle (> ten layers) with collagen fibers present, absent of fibroblasts, and sparse elastic laminae. The ascending cervical artery's smaller muscular and spinal branches may be classified as small arteries (0.1–2 mm diameter). Small arteries have a slightly thinner tunica media (8–10 smooth muscle layers) absent of elastic laminae. The tunica adventitia is thinner than the tunica media and contains connective tissue and some elastic fibers (Ross and Pawlina 2011; Arslan et al. 2018).

S. J. Bordes, Jr. et al. (eds.), *The Clinical Anatomy of the Vascular System*,
https://doi.org/10.1007/978-3-031-78326-5_13

13.3 Anatomy and Topography

The ascending cervical artery (Fig. 13.1) is typically a branch of the inferior thyroid artery that arises from the thyrocervical trunk (Daeseler 1959). The thyrocervical trunk arises from the first part of the subclavian artery as it runs posterior to the anterior scalene muscle. The inferior thyroid artery ascends along the medial border of the anterior scalene muscle, and the ascending cervical artery emerges just before the inferior thyroid artery's course toward the thyroid gland medially (Harrigan and Deveikis 2018).

The ascending cervical artery is positioned between the anterior scalene and longus capitis muscles. It ascends posterior to the internal jugular vein along the anterior tubercles of transverse processes of the cervical vertebrae (Oztürk et al. 1998). The ascending cervical artery usually runs medial to the phrenic nerve and has been noted as a useful landmark for this nerve (Canella et al. 2009). Muscular branches of the ascending cervical artery include those supplying the lateral deep muscles of the neck, including the anterior scalene and longus capitis (Zhang et al. 2016).

A cadaveric study by Arslan et al. provided scrutinous detail of the contributions of the ascending cervical artery to the anterior spine and spinal cord supply. Radicular branches include those supplying the cervical intervertebral foramina (especially C4), including branches to the foramina of C3–4, C4–5, and rarely C2–3. Concerning the C4 and C5 nerve roots, the radicular branches of the ascending cervical artery coursed anteriorly, anteroinferiorly, or anterosuperiorly and did not course into the dorsal aspect of the foramina. These branches penetrated the nerve root sleeves of C4 and C5 at the anterior aspect of the foramina. The diameter of these radicular branches of the ascending cervical artery ranged from 0.80 mm to 1.4 mm, with the C4 branch having the greatest diameter relative to the other branches in the cadaver (Arslan et al. 2018).

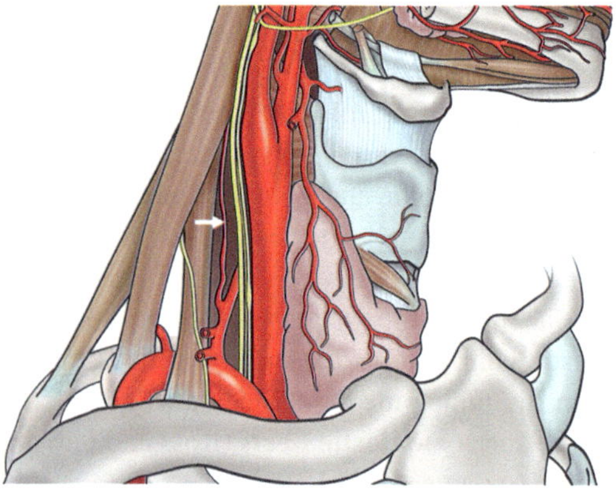

Fig. 13.1 Right ascending cervical artery (arrow) arising from the inferior thyroid artery. (After Netter, 2023)

Its ascension in the neck ends in anastomoses with branches of the ascending pharyngeal, occipital, and vertebral arteries. It is important to note that segmental narrowing and general tortuosity of the arteries increase with age. Thus, in older patients, the ascending cervical arteries may not ascend as high up in the neck as in younger patients (Kauppila and Penttilä 1994).

13.4 Variations

Usually, the ascending cervical artery is a branch of the superior portion of the inferior thyroid artery or directly from the thyrocervical trunk (Daeseler 1959). In one dissection, an anomalous ascending cervical artery was found to originate from the superior cervical artery, 1 cm lateral to the origin of the superior cervical artery from the thyrocervical trunk (Oztürk et al. 1998).

Rare variations involve the origins of ascending pharyngeal and occipital arteries arising from the ascending cervical artery instead of the external carotid artery. There is one report of a unilateral variant continuation of the ascending cervical artery as the occipital artery. Its origin from the inferior thyroid artery was enlarged, and once it continued as the occipital at the upper cervical spine, all branches usually supplied by the occipital artery were present. This was the only anomaly found in that specimen (Lasjaunias and Berenstein 1987; Tubbs et al. 2004).

An accessory ascending cervical artery was observed in a study of 87 cadavers. This accessory ascending cervical artery predominately arose independently from the subclavian artery behind the anterior scalene muscle. It supplied radicular branches to the C5–6 and C6–7 intervertebral foramina along the C6 and C7 cervical nerves, respectively (Su et al. 2000).

In another Japanese study of 67 cadavers, a small branch with a similar origin to the accessory ascending cervical artery was discovered in 77% of the bodies and was identified as the deep ascending cervical artery. Arising similarly from the subclavian artery, deep to the anterior scalene, it was concluded to have similar morphological significance as the ascending cervical artery, as it provided anastomoses of cervical intersegmental arteries and similar muscular branches (Koizumi et al. 1994).

13.5 Clinical and Surgical Significance

The ascending cervical artery provides an important supply for ventral portions of the cervical spinal cord/column, intervertebral foramina, roots of the phrenic nerve (Canella et al. 2009), and the brachial plexus (Zhong et al. 2017). This is important clinically when considering the potential for vascular compromise to the spinal cord. Consideration of the anatomical variations and topography of the thyrocervical trunk is important for many interventional vascular radiology and neuroradiology procedures (Pérez-García et al. 2018).

In a cadaveric study of the anterior cervical spine, the nutrient vessels of the sub-axial cervical spine were traced to their origins. Dissection of all ten cadavers revealed that all the anterior nutrient vessels perforating the cervical vertebrae were derived from the ascending cervical artery with a variable contribution of the transverse cervical artery (Dunbar et al. 2019).

In a study of cervical spinal cord injuries, it was noted that following the occlusion of the vertebral artery, the circulation above the C4 vertebrae was supplied by the collaterals of the ascending cervical arteries (Salkov et al. 2015). In a case of cervical canal stenosis accompanied by congenital cervical fusion and extracranial occlusion of the vertebral artery, it was found that "each collateral flow extending from the deep cervical artery and ascending cervical artery anastomosed with the proximal and distal portions of the occluded vertebral artery," further affirming the relationship between the ascending cervical artery and collateral cervical blood supply (Koizumi et al. 1994; Itoh et al. 1978). In a case of stent-assisted coiling of a ruptured vertebrobasilar junction aneurysm, the distal cervical left vertebral artery was successfully reconstituted using muscular branches of the left ascending cervical artery (Dobrocky et al. 2017). In a rare Bow Hunter syndrome case involving bilateral dynamic occlusion of the vertebral arteries, reperfusion of the vessels by the ascending cervical arteries was noted (Toluian et al. 2021).

The radicular arteries that join the anterior spinal artery are most important to consider for their potential involvement in the development of anterior spinal artery syndrome, most associated with transforaminal corticosteroid injections into the cervical foramina (Baker et al. 2003). The radicular branches of the ascending cervical artery were found to course the entire length of the intervertebral foramen, while vertebral artery branches were found within the medial portions of the foramen (Arslan et al. 2018). One dissection found that an ascending cervical artery formed a segmental medullary artery joining the anterior spinal artery. However, it was noted that it would only be injured by anterior needle misplacement during procedures (Huntoon 2005).

Endovascular catheterization of an anastomosis between the ascending cervical artery and the ipsilateral vertebral artery was utilized to perform a coil embolization of a ruptured basilar apex aneurysm. This particular patient had complete occlusion of the left vertebral artery at its origin, 80–90% stenosis of the right vertebral artery, and hypoplastic posterior communicating arteries bilaterally. The chronic stenosis had dilated collaterals sufficiently to make them suitable for catheterization, and it was decided that the ascending cervical artery was deemed most viable to access the basilar artery. The catheterization of the ascending cervical artery utilized the anastomoses with the ipsilateral vertebral artery between the C3 and C7 vertebral levels (Munger et al. 2020).

A 2021 review of cervical pseudoaneurysms with or without concomitant arteriovenous fistulas arising from the thyrocervical trunk revealed that the majority were iatrogenic (61.5%), with the remaining due to trauma (26.9%). Following a stabbing incident, a 58-year-old male patient presented with a traumatic pseudoaneurysm arising from the ascending cervical artery with a concomitant arteriovenous fistula between the ascending cervical artery and the left vertebral venous plexus. Initially, conservative approaches to preserve the ascending cervical artery were attempted due to its contributions to the spinal cord with attempts to embolize the pseudoaneurysms and ligate the arteriovenous fistula individually. However, despite the successful occlusion of a larger pseudoaneurysm, there was continued filling of an inferior pseudoaneurysm and arteriovenous fistula. Urgent treatment was indicated, and therefore, it was decided to sacrifice the ascending cervical artery via controlled injection of Onyx with great care taken so that the embolic agent did not occlude the collaterals supplying the spinal cord. This was tolerated without complication due to the presence of sufficient collateral circulation (Rallo et al. 2021).

An additional case of a blunt-traumatic pseudoaneurysm arising from the ascending cervical artery with concomitant arteriovenous fistula in an 80-year-old female was reported. Successful n-butyl cyanoacrylate embolization was utilized at the distal aspect of the right ascending cervical artery. The procedure was well tolerated, and the patient was discharged after 3 days.

The ascending cervical artery should be considered for its potential for aneurysms. A case report from 2014 stated that a saccular aneurysm was found on the C5 radiculomedullary artery, which arose from the left ascending cervical artery (Nakamura et al. 2015).

13.6 Complications

A rare complication of retropharyngeal and cervicomediastinal hematomas due to rebleeding of an ascending cervical artery following a stellate ganglion block was reported. The hematomas caused life-threatening airway obstruction, leading to respiratory arrest. The rebleed occurred 3 days following a stellate ganglion block, which is abnormal as most symptoms of retropharyngeal or cervicomediastinal hematomas occur 2+ h after the procedure (Takanami et al. 2009).

A neurofibromatosis patient was reported to develop a pseudoaneurysm arising from the ascending cervical artery following chiropractic manipulation. The patient had remarkable kyphosis and bone dysplasia consistent with neurofibromatosis. Surgery confirmed angiographical findings that the pseudoaneurysm formed within a neurofibroma, and it was speculated that cervical manipulation with torsion caused stretching of the vasculature, resulting in the aneurysm (Lennington et al. 1980).

References

Arslan M, Acar HI, Comert A, Tubbs RS (2018) The cervical arteries: an anatomical study with application to avoid the nerve root and spinal cord blood supply. Turk Neurosurg 28(2):234–240

Baker R, Dreyfuss P, Mercer S, Bogduk N (2003) Cervical transforaminal injection of corticosteroids into a radicular artery: a possible mechanism for spinal cord injury. Pain 103(1–2):211–215

Canella C, Demondion X, Delebarre A, Moraux A, Cotten H, Cotten A (2009) Anatomical study of phrenic nerve using ultrasound. Eur Radiol 20(3):659–665

Daeseler EHAB (1959) Surgical anatomy of the subclavian artery and its branches. Surg Gynecol Obstet 108:149–174

Dobrocky T, Beck J, Gralla J, Mordasini P (2017) Stent-assisted coiling of a ruptured vertebrobasilar junction aneurysm via direct vertebral artery puncture. BMJ Case Rep 2017:neurintsurg-2017-013099

Dunbar L, Vidakovic H, Löffler S et al (2019) Anterior cervical spine blood supply: a cadaveric study. Surg Radiol Anat 41(6):607–611

Embryology (2022) Aortic Arch. Ncbi.nlm.nih.gov. https://www.ncbi.nlm.nih.gov/books/NBK553173. Accessed 2 Apr 2022

Harrigan MR, Deveikis JP (2018) Handbook of cerebrovascular disease and neurointerventional technique, 3rd edn. Humana Press, Cham

Huntoon MA (2005) Anatomy of the cervical intervertebral foramina: vulnerable arteries and ischemic neurologic injuries after transforaminal epidural injections. Pain (Amsterdam) 117(1):104–111

Itoh M, Shima K, Ishikawa S, Ono Y (1978) A case of cerevical canal stenosis accompanied with congenital

cervical fusion and extracranial occlusion of vertebral artery – a clinical and embryological study (author's transl). No Shinkei Geka 6(6):591–597
Kauppila LI, Penttilä A (1994) Postmortem angiographic study of degenerative vascular changes in arteries supplying the cervicobrachial region. Ann Rheum Dis 53(2):94–99
Koizumi M, Sekiya S, Horiguchi M, Harada M (1994) A study of the deep ascending cervical artery (new definition) and its relation to some other arteries in the human cervical region. Ann Anat 176(3):233–242
Lasjaunias P, Berenstein A (1987) Surgical neuroangiography. In: Functional anatomy of craniofacial arteries, vol 1
Lennington BR, Laster DW, Moody DM, Ball MR (1980) Traumatic pseudoaneurysm of ascending cervical artery in neurofibromatosis: complication of chiropractic manipulation. AJNR Am J Neuroradiol 1(3):269–270
Munger D, O'Neill B, Priest R (2020) Embolization of basilar tip aneurysm via ascending cervical artery. World Neurosurg 140:262–266
Nakamura H, Kim P, Kanaya H, Kurokawa R, Murata H, Matsuda H (2015) Spinal subarachnoid hemorrhage caused by a mycotic aneurysm of the radiculomedullary artery: a case report and review of literature. NMC Case Rep J 2(2):49–52
Oztürk AH, Bayramoğlu A, Celik HH, Yener N (1998) An anomalous ascending cervical artery. Okajimas Folia Anat Jpn 75(5):259–261
Pérez-García C, Malfaz C, del Valle DM et al (2018) Embolization through the thyrocervical trunk: vascular anatomy, variants, and a case series. J NeuroIntervent Surg 10(10):1012
Rallo MS, Narayan V, Majmundar N et al (2021) Multimodal endovascular management of traumatic pseudoaneurysm and arteriovenous fistula of the ascending cervical artery: a technical report and review of literature. Clin Neurol Neurosurg 202:106539
Ross M, Pawlina W (2011) Histology a text and atlas with correlated cell and molecular biology, 6th edn. Lippincott Williams & Wilkins
Salkov M, Tsymbaliuk V, Dzyak L, Rodinsky A, Cherednichenko Y, Titov G (2015) New concept of pathogenesis of impaired circulation in traumatic cervical spinal cord injury and its impact on disease severity: case series of four patients. Eur Spine J 25(Suppl 1):11–18
Schoenwolf GC, Bleyl SB, Brauer PR, Francis-West PH (2015) Larsen's human embryology, 5th edn. Churchill Livingstone, Philadelphia
Su WD, Ohtsuka A, Taguchi T, Murakami T (2000) Typology of the arteries in the human scalenus region, with special reference to the accessory ascending cervical artery. Acta Med Okayama 54(6):243–252
Takanami I, Abiko T, Koizumi S (2009) Life-threatening airway obstruction due to retropharyngeal and cervicomediastinal hematomas following stellate ganglion block. Thorac Cardiovasc Surg 57(5):311–312
Toluian T, Volterra D, Gioppo A, Rigamonti P (2021) Bow Hunter's syndrome: an unusual case of bilateral dynamic occlusion of vertebral arteries. J Neurol Neurosurg Psychiatry 92(10):1131–1132
Tubbs RS, Salter G, Oakes WJ (2004) Continuation of the ascending cervical artery as the occipital artery in man. Anat Sci Int 79(1):43–45
Zhang XY, Ma TT, Liu L, Yin NB, Zhao ZM (2016) Anatomic study of the musculus longus capitis flap. Surg Radiol Anat (English Ed) 39(3):271–279
Zhong LY, Wang AP, Hong L et al (2017) Microanatomy of the brachial plexus roots and its clinical significance. Surg Radiol Anat 39(6):601–610

Internal Carotid Artery

14

Cameron Fontenot, Jonathan Richard, Stephen J. Bordes, Jr., and Alison Smith

14.1 Embryology

In human embryological development, the great arteries of the head, neck, and thorax all mature from six paired pharyngeal arches. These arches house a core of spongy, primitive mesoderm called mesenchyme. This mesenchyme present in the arches gives rise to the aortic arch arteries, embryologic precursors to the great arteries.

The carotid arterial system arises from a combination of the third aortic arches and cranial extensions of the dorsal aortae, a pair of arterial channels that fuse caudally to the aortic arch arteries while remaining separate in the head and neck region of the embryo (Frisdal and Trainor 2014). The proximal portions of the right and left internal carotid arteries (nearest to the bifurcation point of the common carotid) are an extension of the third aortic arch artery that also constitutes the common and external carotid arteries. The distal portions, however, are a derivation of segments from the dorsal aortae. During development, the dorsal aortae are connected to the six aortic arch arteries. While these connections are lost over time in an embryo, two persist in standard development: connection to the third and fourth aortic arch artery pairs. The portions of the dorsal aorta between the third and fourth connection points eventually regress, leaving behind a connection to the third arch arteries cranially. This remnant, now the distal portion of the internal carotid artery, will divide into cranial and caudal branches later in development (Dungan and Heiserman 1996).

14.2 Anatomy

The left and right internal carotid arteries (ICAs) (Fig. 14.1) originate at the bifurcation point of the common carotid arteries. In most individuals, this bifurcation point is near the superior border of the thyroid cartilage, occurring between the third and fifth cervical vertebrae. The diameter of the ICA varies among individuals due to numerous factors; however, the average, as determined by Krejza et al., was found to be approximately 4.66 mm in females and 5.11 mm in males (Krejza et al. 2006). In similar congruence to the previously mentioned study, the luminal diameter of the internal carotid arteries was found to be significantly larger in men compared to women (Limbu et al. 2006).

The ICA can be classified into four segments related to the area or anatomical structure it passes through: cervical, petrous, cavernous, and supra-

C. Fontenot · J. Richard
School of Medicine, Louisiana State University Health Sciences Center, New Orleans, LA, USA
e-mail: cfon18@lsuhsc.edu; jric29@lsuhsc.edu

S. J. Bordes, Jr. (✉) · A. Smith
Department of Surgery, Louisiana State University Health Sciences Center, New Orleans, LA, USA
e-mail: asmi60@lsuhsc.edu

S. J. Bordes, Jr. et al. (eds.), *The Clinical Anatomy of the Vascular System*,
https://doi.org/10.1007/978-3-031-78326-5_14

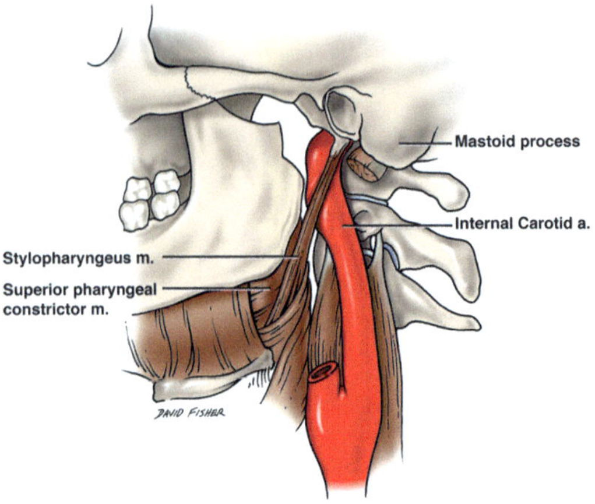

Fig. 14.1 Lateral view of the course of the internal carotid artery in the neck

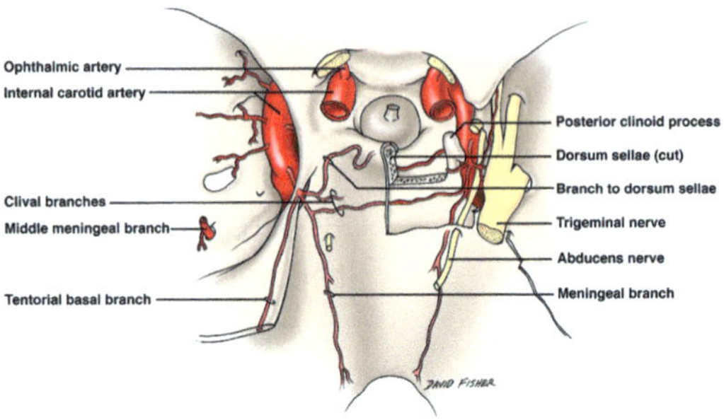

Fig. 14.2 Superior view of internal carotid artery branches at the skull base

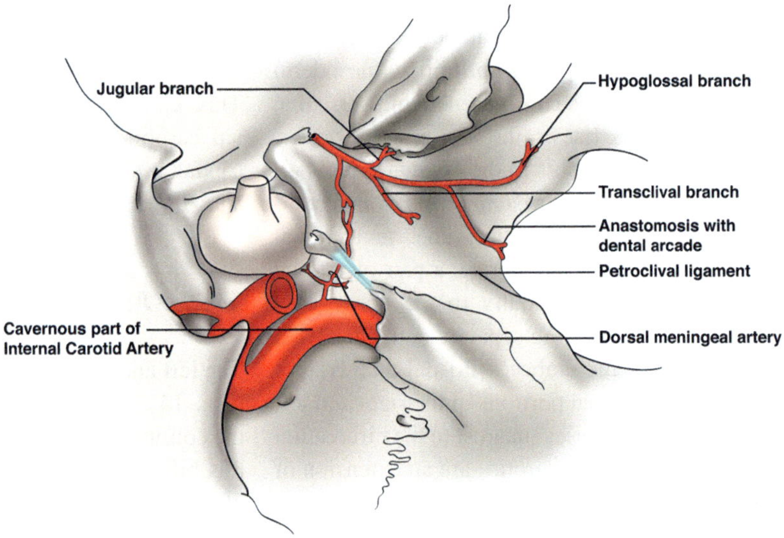

Fig. 14.3 Lateral view of internal carotid artery branches at the skull base

clinoid (Gibo et al. 1981). In total, the ICA gives off eight branches before terminally bifurcating; however, these branches are often small, inconsistent, and, in some cases, absent. Segment one, the cervical part, is the portion of the ICA in the neck compartment. This portion runs inside the fascia-enclosed compartment of the neck, the carotid sheath, the internal jugular vein, the vagus nerve, some sympathetic nerve fibers, and the deep cervical lymph nodes. It is important to note that the cervical segment does not give off any branches.

The next ICA segment, the petrous part, enters the skull through the carotid canal and travels superomedially toward the foramen lacerum and cavernous sinus. This petrous portion gives off a caroticotympanic and vidian branch while coursing through the bone. The third, cavernous part of the ICA courses toward the anterior clinoid process of the sphenoid bone, eventually exiting through the roof of the cavernous sinus (Figs. 14.2 and 14.3). The meningeal artery and inferior hypophysial artery branches originate in this segment, supplying the dura mater of the anterior cranial fossa and the posterior pituitary gland, respectively. The supraclinoid segment extends from where the ICA exits the cavernous sinus until it bifurcates into its two terminal branches contributing to the circle of Willis: the anterior and middle cerebral arteries. Before this termination, the supraclinoid segment gives off the superior hypophysial, ophthalmic, posterior communicating, and anterior choroidal arteries.

14.3 Histology and Imaging

Consistent with all other arterial vessels in the body, the ICA comprises three histologically distinct layers. The tunica intima, the innermost layer of the blood vessel, consists of endothelial cells attached to a basement membrane called the internal elastic lamina, forming the boundary between the tunica intima and media. This single layer of squamous cells lines the lumen of all blood vessels in the body and is responsible for tissue growth and repair through angiogenesis (Alberts et al. 2002). The next layer moving outward is the tunica media. This layer comprises longitudinal smooth muscle cells and elastic connective tissue fibers, with the smooth muscle layer responsible for regulating blood flow by adjusting the diameter of the vessel through smooth muscle contraction/relaxation (Charlick and Das 2021). The tunica adventitia is the outermost layer of the ICA and consists mainly of fibrous connective tissue. This layer attaches the blood vessel to the surrounding connective tissues. Of note histologically is that the adventitia does not always have a clearly defined outer border due to its intimate connection with surrounding tissues.

Various imaging techniques may be used to visualize and evaluate the ICA when diagnosing a patient. When the possibility of an ICA dissection is being explored, typically due to a partial Horner's syndrome (Borgman 2012), a conventional angiogram is the preferred form of definitive diagnostic imaging. The conventional angiogram allows for the measurement of luminal diameter, while other imaging methods, such as ultrasound and CT angiography, allow for detecting intraluminal thrombi and evaluating vessel expansion, respectively (Chengazi and Bhatt 2019).

14.4 Variations

The ICAs originate from the bifurcation point of the common carotid arteries on both sides of the body. This bifurcation point is usually found to be between the third and fourth cervical vertebrae or the fourth and fifth cervical vertebrae; however, bifurcation can also be less frequently identified at the level of or in-between the second and third cervical vertebrae or the fifth and sixth cervical vertebrae (Schulz and Rothwell 2001). Both right and left common carotid arteries tend to bifurcate at the same level in an individual (Schulz and Rothwell 2001). Cases of carotid artery nonbifurcation are rare but present with ICA origin at the level of the first or second cervical vertebrae (Michalinos et al. 2016).

One study showed a 26.2% prevalence in internal carotid morphological variations when examining tortuosity, kinking, and coiling rates (Sacco et al. 2007). Kinking of 60°–90° was found to be present in 54.0% of study subjects and was also the most common morphological variation observed bilaterally (Sacco et al. 2007). The presence of tortuosity was shown to be 44.6%, with coiling occurring much less frequently at 1.4% (Fig. 14.4) (Sacco et al. 2007). Morphological variations tended to present more proximal to the carotid artery bifurcation, and the occurrence of unilateral tortuosity and kinking were found to present on the left ICA more frequently than the right ICA (Sacco et al. 2007).

Vertebral segment hypoplasia and cellular underdevelopment were reported in 2%–6% of individuals (Chuang et al. 2012), while the presence of fenestrations was shown to be 1.3% in another study (Cooke et al. 2014). Another variation is segmental agenesis. This condition usually occurs unilaterally and is identified by the absence of any of the six segments of the ICA (Li et al. 2017).

The retropharyngeal internal carotid artery is a variation resulting from the passage of one or both ICAs through the retropharyngeal space during development or by movement over time (Gupta et al. 2013; Lukins et al. 2016). This can pose a serious risk to the patient during surgical

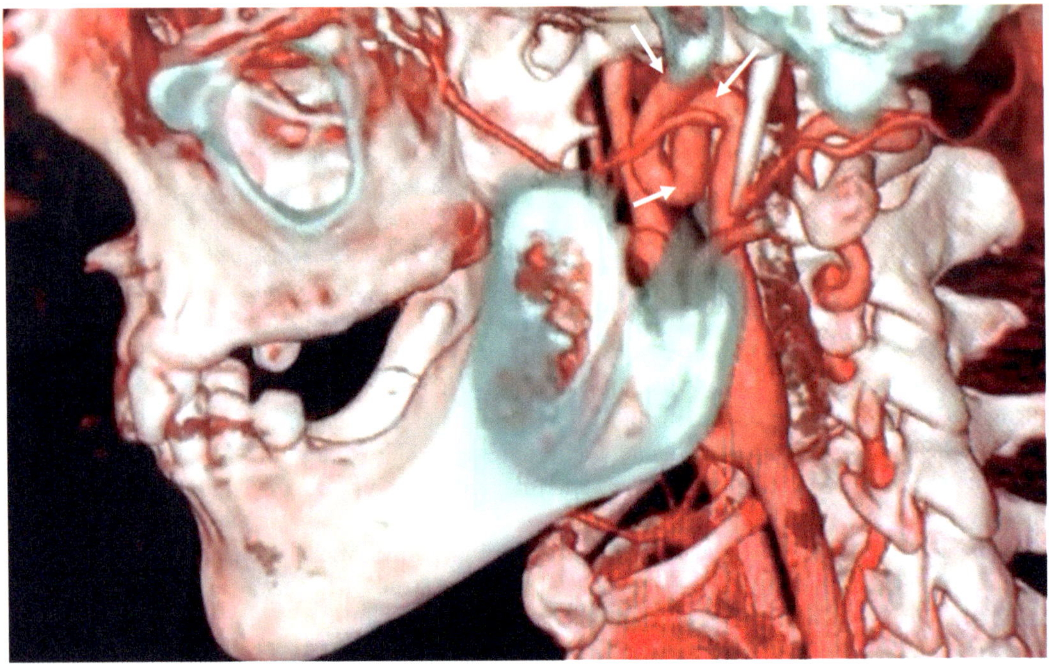

Fig. 14.4 CTA noting a very tortuous left internal carotid artery (arrows)

procedures as injury to the artery can cause severe hemorrhage and lead to stroke. According to one study, this condition is well documented and was present in 2% of the population (Kheterpal and Tremper 2009).

14.5 Pathology

The ICA is a major supplier of blood to the brain. Thus, pathologies associated with this ipsilateral artery can result in severe brain ischemic episodes. Carotid artery disease (CAD) is caused by atherosclerosis of the carotid vessels, with the ostium of the ICA being most frequently affected (Prasad 2015). Blockages of the arteries from arteriosclerosis can lead to ischemic stroke and transient ischemic attack (TIA). Risk factors for the development of CAD are hypertension, diabetes, dyslipidemia, obesity, advanced age, and history of cigarette smoking (Chengazi and Bhatt 2019). Additionally, atherosclerosis in the ICA can cause thrombosis at downstream arteries within the brain, leading to ischemic stroke (Olivato et al. 2011).

Aneurysms of the ICA are uncommon, with a vast majority presenting as pseudoaneurysms, a hematoma formed between the tunica media and tunica externa due to leakage from the lumen (Chengazi and Bhatt 2019). Pseudoaneurysms will often present asymptomatically; however, true aneurysms, a localized enlargement involving all three arterial layers, can present differently depending on the location found on the ICA. Intracranial aneurysms may present with visual defects, retro-orbital pain, and headaches (Kloss et al. 2011), while extracranial aneurysms may present with neck pain and a palpable pulsating mass (Altun et al. 2018). Causation can include trauma, hypertension, atherosclerosis, vasculitis, dissection, genetic disposition, or iatrogenic (Chengazi and Bhatt 2019). Early identification is crucial as rupture of the aneurysm can be potentially fatal.

ICA dissection is rare as it occurs in approximately every 3 per 100,000 people, according to one study (Wu et al. 2020). Arterial dissection is the tearing of the tunica intima, causing separation from the tunica media, and can occur spontaneously or as the result of trauma. This can decrease blood flow as blood exits the lumen through the tear. Symptoms of internal carotid artery dissection include "tearing" pain to the region, unilateral headache, and partial Horner syndrome (ptosis, miosis, and anisocoria) (Borgman 2012).

Fibromuscular dysplasia (FMD) of the ICA is a disease of unknown origin that affects the artery's elasticity, leading to stenosis or aneurysm in the vessel. FMD is primarily found in middle-aged women and may present as asymptomatic or with physical symptoms such as partial Horner's syndrome, dizziness, and headaches. According to one study, FMD of the ICA only affected approximately 0.3%–3% of the population (Varennes et al. 2015).

14.6 Surgical and Clinical Significance

The ICAs are of utmost clinical and surgical importance due to their location, anatomy, and ability to supply oxygenated blood to the brain. One of the most common pathologies associated with the ICA is carotid artery disease, resulting from atherosclerosis of the vessel. The presence of atherosclerosis can be assessed by examining associated risk factors such as obesity, diabetes, advanced age, history of smoking, hypertension, and dyslipidemia. At the same time, ultrasound or CT angiography can confirm the presence of cholesterol buildup within the artery (Chengazi and Bhatt 2019). Surgical intervention by carotid endarterectomy or stent placement may be warranted in cases where vessel occlusion is 50% or greater. CAD can lead to an increased risk of stroke or cerebrovascular disease due to potential occlusion of vessels or from downstream thrombosis in cerebral arteries (Olivato et al. 2011).

Aneurysms of the ICA can be life-threatening if rupture occurs due to hemorrhagic shock. Aneurysms of different segments can be assessed by examining risk factors and associated symptoms of extracranial or intracranial locations, while CT imaging is used for confirmation (Chengazi and Bhatt 2019). Current interventions include open surgery and endovascular and medical therapies, with arterial ligation reserved only for emergencies (Kraemer and Zhou 2019). The course of treatment highly varies depending on the presentation of symptoms and underlying pathology.

Dissections of the ICA are rare and usually occur spontaneously or due to trauma; however, symptoms can include neurological deficits such as partial Horner syndrome and unilateral headache with extreme pain in the area (Borgman 2012). MRI and CT angiography are most commonly used to confirm the presence of a suspected dissection (Chengazi and Bhatt 2019). Similar to aneurysms of the ICA, interventions of dissection can involve open surgery, endovascular therapy, or medical management, deepening the presentation of symptoms in the patient (Müller et al. 2000).

Other general surgical considerations include an understanding of the anatomical variations present in the population. The bifurcation point of the common carotid artery is essential to consider, as this can occur at the level of nearly any cervical vertebra and can influence surgical approaches (Schulz and Rothwell 2001; Michalinos et al. 2016). Another anatomical variation that can complicate surgery is the presence of one of the ICAs in a retropharyngeal position. This should be taken into consideration before surgery involving the retropharyngeal space, as iatrogenic complications may lead to issues such as aneurysm, dissection, or rupture.

References

Alberts B, Johnson A, Lewis J et al (2002) Molecular biology of the cell, 4th edn. Garland Science, New York. Blood Vessels and Endothelial Cells

Altun G, Pulathan Z, Hemsinli D (2018) True aneurysms of the extracranial carotid artery: an evaluation of two "Giant aneurysms" and the current literature. J Korean Neurosurg Soc 61(2):282–286. https://doi.org/10.3340/jkns.2017.0404.003

Borgman CJ (2012) Horner syndrome secondary to internal carotid artery dissection after a short-distance endurance run: a case study and review. J Optom 5:209–216. https://doi.org/10.1016/j.optom.2012.05.001

Chengazi HU, Bhatt AA (2019) Pathology of the carotid space. Insights Imaging 10(1):21. https://doi.org/10.1186/s13244-019-0704-z

Chuang YM, Chan L, Wu HM, Lee SP, Chu YT (2012) The clinical relevance of vertebral artery hypoplasia. Acta Neurol Taiwanica 21(1):1–7

Cooke DL, Stout CE, Kim WT et al (2014) Cerebral arterial fenestrations. Interv Neuroradiol 20(3):261–274. https://doi.org/10.15274/INR-2014-10027

Dungan DH, Heiserman JE (1996) The carotid artery: embryology, normal anatomy, and physiology. Neuroimaging Clin N Am 6(4):789–799

Frisdal A, Trainor PA (2014) Development and evolution of the pharyngeal apparatus. Wiley Interdiscip Rev Dev Biol 3(6):403–418. https://doi.org/10.1002/wdev.147

Gibo H, Lenkey C, Rhoton AL Jr (1981) Microsurgical anatomy of the supraclinoid portion of the internal carotid artery. J Neurosurg 55(4):560–574. https://doi.org/10.3171/jns.1981.55.4.0560

Gupta A, Shah AD, Zhang Z, Phillips CD, Young RJ (2013) Variability in the position of the retropharyngeal internal carotid artery. Laryngoscope 123(2):401–403. https://doi.org/10.1002/lary.23393

Kheterpal S, Tremper K (2009) Retropharyngeal carotid artery: an important anatomic variation for the anesthesiologist. Am Soc Anesthesiol 111:454–455

Kloss BT, Patel R, Sullivan AM (2011) Intracranial internal carotid aneurysm causing diplopia. Int J Emerg Med 4:56. https://doi.org/10.1186/1865-1380-4-56

Kraemer CJK, Zhou W (2019) Carotid aneurysm review. Int J Angiol 28(1):17–19. https://doi.org/10.1055/s-0039-1677675

Krejza J, Arkuszewski M, Kasner SE, Weigele J, Ustymowicz A, Hurst RW, Cucchiara BL, Messe SR (2006) Carotid artery diameter in men and women and the relation to body and neck size. Stroke 37(4):1103–1105. https://doi.org/10.1161/01.STR.0000206440.48756.f7

Li S, Hooda K, Gupta N, Kumar Y (2017) Internal carotid artery agenesis: a case report and review of literature. Neuroradiol J 30(2):186–191. https://doi.org/10.1177/1971400917692162

Limbu YR, Gurung G, Malla R, Rajbhandari R, Regmi SR (2006) Assessment of carotid artery dimensions by ultrasound in non-smoker healthy adults of both sexes. Nepal Med Coll J 8(3):200–203

Lukins DE, Pilati S, Escott EJ (2016) The moving carotid artery: a retrospective review of the retropharyngeal carotid artery and the incidence of positional changes on serial studies. AJNR Am J Neuroradiol 37(2):336–341. https://doi.org/10.3174/ajnr.A4533

Michalinos A, Chatzimarkos M, Arkadopoulos N, Safioleas M, Troupis T (2016) Anatomical considerations on surgical anatomy of the carotid bifurcation. Anat Res Int 2016;6907472. https://doi.org/10.1155/2016/6907472

Müller BT, Luther B, Hort W, Neumann-Haefelin T, Aulich A, Sandmann W (2000) Surgical treatment of 50 carotid dissections: indications and results. J Vasc Surg 31(5):980–988. https://doi.org/10.1067/mva.2000.104586

Olivato S, Turri G, Mazzucco S, Mansueto G (2011) Internal carotid artery thrombosis and its evolution documented by computed tomography angiography. Clin Pract 1(4):e84. https://doi.org/10.4081/cp.2011.e84

Prasad K (2015) Pathophysiology and medical treatment of carotid artery stenosis. Int J Angiol 24(3):158–172. https://doi.org/10.1055/s-0035-1554911

Sacco S, Totaro R, Baldassarre M, Carolei A (2007) Morphological variations of the internal carotid artery: prevalence, characteristics and association with cerebrovascular disease. Int J Angiol 16(2):59–61. https://doi.org/10.1055/s-0031-1278249

Schulz UG, Rothwell PM (2001) Major variation in carotid bifurcation anatomy: a possible risk factor for plaque development? Stroke 32(11):2522–2529. https://doi.org/10.1161/hs1101.097391

Varennes L, Tahon F, Kastler A et al (2015) Fibromuscular dysplasia: what the radiologist should know: a pictorial review. Insights Imaging 6(3):295–307. https://doi.org/10.1007/s13244-015-0382-4

Wu Y, Chen H, Xing S et al (2020) Predisposing factors and radiological features in patients with internal carotid artery dissection or vertebral artery dissection. BMC Neurol 20(1):445. https://doi.org/10.1186/s12883-020-02020-8

Thyroidea Ima Artery

15

Aaron Albuck and Stephen J. Bordes, Jr.

15.1 Introduction

The arterial supply to the thyroid gland is typically provided by two main arteries, the superior and inferior thyroid arteries. In a small proportion of individuals, an anomalous artery, the thyroidea ima artery, is present.

First described by Johann E. Neubauer in 1772, the thyroidea ima artery has been sporadically described as the thyroid artery of Neubauer, accessory thyroid artery, and middle thyroid artery. Typically identified when the inferior thyroid arteries are absent (Mohebati and Shaha 2012), it has routinely been detailed as either the main (Yilmaz et al. 1993), or in rare incidences even the sole (Moriggl and Sturm 1996), arterial blood supply to the thyroid. Therefore, concrete anatomical knowledge of it is of the utmost importance for head and neck surgeons to avoid iatrogenic injury.

15.2 Embryology

There is an extensive blood supply around the thyroid gland during the early stages of embryo development. Later in development, most of the arterial supply regresses, leaving only the superior and inferior thyroid arteries to supply the thyroid gland (Mariolis-Sapsakos et al. 2014). Most of the blood to the thyroid (estimated as nearly 70%) is supplied by the inferior thyroid artery (Anagnostopoulou and Mavridis 2014). In the absence of either the inferior or superior thyroid arteries, the arterial blood supply is compensated by an increase in the size or number of other accessory thyroid arteries (Cigali et al. 2008). When an inferior thyroid artery is absent, it is usually replaced by a branch from the superior thyroid artery on the ipsilateral or contralateral side (Anatomy for Surgeons 1954). However, the place of an undeveloped inferior thyroid artery is sometimes taken by the thyroidea ima artery.

A rich vascular supply is essential for proper thyroid development, as both the excessive vascularization and subsequent regression of arteries guide thyroid morphogenesis and the proper anatomical migration of the gland. With this knowledge, and the finding of an increased prevalence of thryoidea ima arteries in fetuses, Yuraskapong et al. theorized that the thyroidea ima artery is a normal structure in embryo development (Yurasakpong et al. 2022). Further studies are needed to confirm this claim, but the thyroidea ima artery is nonetheless presumed to persist throughout embryo development as a part of the branch connecting the common carotid artery to the thyroid (Won et al. 2011).

A. Albuck
Tulane University School of Medicine,
New Orleans, LA, USA

S. J. Bordes, Jr. (✉)
Department of Surgery, Louisiana State University
Health Sciences Center, New Orleans, LA, USA

S. J. Bordes, Jr. et al. (eds.), *The Clinical Anatomy of the Vascular System*,
https://doi.org/10.1007/978-3-031-78326-5_15

The prevalence of embryological conditions that allow the artery to persist into adulthood ranges in the literature from 0.4% to 12.2% (Krudy et al. 1980; Yalcin 2016). Of note, Toni et al. considered both ethnic and anthropological factors while investigating the incidence of the thyroidea ima artery. They found it to be more prevalent in individuals of Asian ancestry (10%) than European ancestry (6%) (Toni et al. 2003). Although rare, the thyroidea ima artery is a fairly consequential artery that should be considered when abnormalities in the inferior thyroid artery are observed.

15.3 Anatomy and Topography

The inferior thyroid artery is absent in about 3% of individuals, in whom either the superior thyroid artery or the thyroidea ima artery supplies the isthmus of the thymus (Yalcin 2016). The thyroidea ima artery provides an additional arterial supply to the inferior surface and isthmus of the thyroid gland. Its mediastinal origins vary, but it most commonly arises from the brachiocephalic trunk. When presented with either a hypoplastic or absent inferior thyroid artery, it courses over the anterior surface of the trachea. Once at the thyroid gland, it can terminate as a single trunk or in multiple bifurcations. Nonetheless, it has abundant collateral anastomoses on the surface of the thyroid gland and within it. These anastomoses include, but are not limited to, the superior thyroid artery arising from the common carotid artery.

The thyroidea ima artery is a relatively small arterial vessel, with an average diameter of 3–5 mm (Anatomy for Surgeons 1954). Its course is highly variable considering its origin. However, its termination is more consistent. As a terminating vessel, the thyroidea ima artery gives off no branches; however, it supplies enlarged parathyroid glands (Krudy et al. 1980), and more rarely the trachea (Kamparoudi et al. 2016). This correlates with the anatomy, as the vessel runs anteriorly to the trachea and terminates at the inferior thyroid gland close to the parathyroids.

For the surgeon, the right recurrent laryngeal nerve is of particular interest. As a branch of the vagus nerve, it innervates the larynx. Therefore, it is a vital structure to note in the surgeon's operative field during any head and neck procedure. It follows a path down the anterior surface, then up the posterior surface, of the subclavian artery, and reaches the trachea at the level of the lower pole of the thyroid gland, near the termination of the thyroidea ima (Akil et al. 2017). However, the recurrent laryngeal nerve is at very little risk when the thyroidea ima artery is present, owing to the absence of the inferior thyroid artery. The inferior thyroid artery is closely related to the nerves. When it is absent, the recurrent laryngeal nerve is at a safe and distant location from the branches of the ima artery.

15.4 Variations

The thyroidea ima artery most commonly persists into adulthood when the ipsilateral inferior thyroid artery is absent (Krudy et al. 1980). Krudy et al. (1980) first reported that if either a routine angiography is negative, or the inferior thyroid arteries are absent or small, the thyroidea ima artery can be detected at the same origin. While recent literature has shown this to be incorrect, the thyroidea ima artery has most commonly been described as arising from the brachiocephalic trunk (Yilmaz et al. 1993; Gruber 1872; Lippert and Pabst 1985). However, various literature sources challenge this, possibly in part because of conflicts about the name of the thyroidea ima and its rarity. The artery has also been noted in various case reports and meta analyses to arise from either the right (Banneheka et al. 2010) or left (Won et al. 2011; Fujimoto et al. 1974) common carotid arteries, the arch of the aorta (Tohno et al. 1989; Loukas et al. 2006), or the subclavian artery. Rarer sites of origin include the internal thoracic, subclavian, inferior thyroid, transverse scapular, pericardiophrenic, thyrocervical trunk (Toni et al. 2003), cardiophrenic, or internal mammary arteries (Lee et al. 2014). Although the thyroidea ima artery is uncommon, its numerous variations and sites of origin should not be ignored if unwarranted injury is to be prevented.

15.5 Imaging

To minimize the risk of vascular injury during surgical procedures, assessment of a patient's anatomy is of the utmost importance. Multiple imaging modalities are available to evaluate the anatomy of the thyroidea ima artery, including its course, origin, and possible deviations.

Contrast angiography, particularly digital contrast angiography (DSA) is a powerful tool for visualizing complex vasculature, and the thyroideaima artery is no exception. However, there is a growing need for less-invasive and more cost-efficient options. Most recently, Rajajee et al. found the thyroidea ima artery to be visualized best via ultrasound (Rajajee et al. 2015) for diagnostic purposes. In light of this finding, proper identification and mid-surgical assessment of the thyroidea ima artery have been accomplished via ultrasound-guided tracheostomy (Sarıtaş and Kurnaz 2019). CT angiography is another noninvasive option for visualizing this vessel (Esen et al. 2018) and has been described as the best method for demonstrating the full course of vessels. However, various case reports have used magnetic resonance angiography as an alternative noninvasive method that can provide as much information as conventional angiography for evaluation (Ozlugedik et al. 2005).

Apart from traditional imaging techniques, both the presence and origin of the thyroidea ima artery are relevant to parathyroid arteriography. The thyroidea ima supplies enlarged or inferior parathyroid glands, sometimes providing the sole arterial blood supply. A thyroidea ima artery should therefore be suspected if, upon routine arteriography, the parathyroids are negative (Krudy et al. 1980).

15.6 Clinical and Surgical Significance

While the thyroidea ima artery is present in radiographic studies, iatrogenic injury to it associated with emergent and surgical procedures can have dire consequences. The artery is of particular importance during thyroid, parathyroid, tracheal, and mediastinal surgeries. If unrecognized either before or during such procedures, injury can result in a significant source of hemorrhage.

The thyroidea ima artery is particularly susceptible to medial surgical incisions, so it should be carefully considered upon either thyroidectomies or laryngeal transplantation. Tracheostomies are the most commonly performed procedures in intensive care units worldwide (Freeman et al. 2000; Foster et al. 2020). It is estimated that there are more than 100,000 tracheostomies per year in the United States alone. Given this prevalence, tracheostomies are generally considered both routine and fairly low risk. However, this does not account for the innate risk in any surgical procedure. Although its incidence is only around 4% (Arjmand and Spector 1996), tracheal hemorrhage can occur secondary to erosion or accidental ligation of accompanying vasculature during tracheostomies. Injury to the thyroidea ima artery in the setting of tracheostomy is no exception. Therefore, the presence of this artery should be properly assessed during procedures to avoid hemorrhage (Simon et al. 2013).

15.7 Complications

As previously described, injury to the thyroideaima artery is commonly associated with midline incisions in routine head and neck procedures. Following an injury to it, excessive hemorrhage is of dire concern. If the patient is hemodynamically stable, minimally invasive ligation of the incised or ruptured artery can control the bleeding. If not, an urgent sternotomy is warranted to stop it.

Disruption of the thyroidea ima artery is not exclusive to scalpel injuries. Finger dissection of the substernal thyroid extension is a classical maneuver in a wide array of thyroid surgeries. The index finger is inserted into the mediastinum outside the thyroid capsule and is swept around until the gland is freed from the pleura or other tissue in the mediastinum (Barbetakis et al. 2014). In such cases, an ima artery could be injured or avulsed.

As thyroidea ima arteries are commonly found when inferior thyroid arteries are absent, it is a major contributor to the thyroid blood supply. Therefore, whenever a thyroidea ima artery is found, it should be preserved in order to avoid unnecessary surgical complications. Although it is a relatively well-known anomalous structure in the literature, a specific method to avoid unwarranted injury to it has yet to be established.

References

Akil F et al (2017) The anatomical relationship between recurrent laryngeal nerve and first tracheal ring in males and females. Clin Exp Otorhinolaryngol 10(1):104–108

Anagnostopoulou S, Mavridis I (2014) Emerging patterns of the human superior thyroid artery and review of its clinical anatomy. Surg Radiol Anat 36(1):33–38

Anatomy for Surgeons vol. I: the head and neck (1954) Postgrad Med J 30(349):610–610

Arjmand E, Spector J (1996) Airway control and laryngotracheal stenosis. Otorhinolaryngology: head and neck surgery. Williams & Wilkins, Baltimore, pp 468–477

Banneheka S et al (2010) Middle thymothyroid artery arising from the common carotid artery: case report of a rare variation. Anat Sci Int 85(4):241–244

Barbetakis N et al (2014) eComment. Substernal goiter. Beware of the thyroidea ima artery. Interact Cardiovasc Thorac Surg 19(1):143

Cigali BS, Ulucam E, Bozer C (2008) Accessory inferior thyroid artery and internal thoracic artery originating from the thyrocervical trunk. Anat Sci Int 83(4):283–285

Esen K et al (2018) Variations in the origins of the thyroid arteries on CT angiography. Jpn J Radiol 36(2):96–102

Foster P et al (2020) Novel approach to reduce transmission of COVID-19 during tracheostomy. J Am Coll Surg 230(6):1102

Freeman BD et al (2000) A meta-analysis of prospective trials comparing percutaneous and surgical tracheostomy in critically ill patients. Chest 118(5):1412–1418

Fujimoto Y, Suwa F, Kimura K (1974) A case of the left superior thyroid artery arising from the left common carotid artery and the A. thyroidea ima. Okajimas Folia Anat Jpn 51(5):219–230

Gruber W (1872) Ueber die Arteria thyreoidea ima. Archiv für pathologische Anatomie und Physiologie und für klinische Medicin 54(3):445–484

Kamparoudi P et al (2016) Percutaneous tracheostomy—beware of the thyroidea-ima artery. Ann Transl Med 4(22):69

Krudy AG, Doppman JL, Brennan MF (1980) The significance of the thyroidea ima artery in arteriographic localization of parathyroid adenomas. Radiology 136(1):51–45

Lee SH et al (2014) Coil embolization in ruptured inferior thyroid artery aneurysm with active bleeding. J Korean Neurosurg Soc 56(4):353–355

Lippert H, Pabst R (1985) In: Bergmen JF (ed) Arterial variations in man: classification and frequency. Verlag, Munchen

Loukas M et al (2006) A retrotracheal right subclavian artery in association with a vertebral artery and thyroidea ima. Folia Morphol (Warsz) 65(3):236–241

Mariolis-Sapsakos T et al (2014) Bilateral aberrant origin of the inferior thyroid artery from the common carotid artery. Surg Radiol Anat 36(3):295–297

Mohebati A, Shaha A (2012) Anatomy of thyroid and parathyroid glands and neurovascular relations. Clin Anat 25(1):19–31

Moriggl B, Sturm W (1996) Absence of three regular thyroid arteries replaced by an unusual lowest thyroid artery (A. thyroidea ima): a case report. Surg Radiol Anat 18(2):147–150

Ozlugedik S et al (2005) Surgical importance of highly located innominate artery in neck surgery. Am J Otolaryngol 26(5):330–332

Rajajee V, Williamson CA, West BT (2015) Impact of real-time ultrasound guidance on complications of percutaneous dilatational tracheostomy: a propensity score analysis. Crit Care 19(1):1–10

Sarıtaş A, Kurnaz MM (2019) Comparison of bronchoscopy-guided and real-time ultrasound-guided percutaneous dilatational tracheostomy: safety, complications, and effectiveness in critically ill patients. J Intensive Care Med 34(3):191–196

Simon M et al (2013) Death after percutaneous dilatational tracheostomy: a systematic review and analysis of risk factors. Crit Care 17(5):R258

Tohno S et al (1989) A case of the thyroidea ima artery arising from the aortic arch. Kaibogaku Zasshi 64(5):490–494

Toni R et al (2003) Anthropological variations in the anatomy of the human thyroid arteries. Thyroid 13(2):183–192

Won H-S et al (2011) Superior and middle thyroid arteries arising from the common carotid artery. Surg Radiol Anat 33(7):645–647

Yalcin B (2016) Thyroid gland. In: Bergman's comprehensive encyclopedia of human anatomic variation, pp 1189–1204

Yilmaz E et al (1993) Arteria thyroidea ima arising from the brachiocephalic trunk with bilateral absence of inferior thyroid arteries: a case report. Surg Radiol Anat 15(3):197–199

Yurasakpong L et al (2022) The decreasing prevalence of the thyroid ima artery: a systematic review and machine learning assisted meta-analysis. Ann Anat 239:151803

Veins of the Head and Neck: Overview

16

Samantha Morin

16.1 Introduction

Most of the veins of the face and scalp run (Fig. 16.1) concurrently with the respective arteries that supply the muscles of the head and neck region. These veins can vary in several ways but are generally divided into superficial and deep groups. The venous drainage of the deeper cranial region begins intracranially as a plexus of smaller channels that drain into larger cerebral veins and empty into the dural venous sinuses (Fig. 16.2). These sinuses comprise the superior and inferior sagittal, straight, transverse, sigmoid, and occipital sinuses; the confluence of sinuses—the point where the superior sagittal, straight, and occipital sinuses connect—comprise the cavernous, sphenoparietal, superior and inferior petrosal, and basilar sinuses (Fig. 16.3). The larger cerebral veins are joined by the diploic and emissary veins, which run through the compact bone and drain into the dural venous sinuses. The deep veins of the face constitute the pterygoid venous plexus (Marur et al. 2014; Moore et al. 2014; Onishi et al. 2017; Sinnatamby and Last 2011; Standring 2016; Watanabe et al. 2016).

Moving anteriorly to posteriorly, the superficial component consists of the supratrochlear, supraorbital, and superficial temporal veins and their branches, the posterior auricular vein, the occipital vein, and its branches (Fig. 16.4). The superficial veins drain into the common facial vein and then the internal jugular vein. The main communicating drainage veins of the face are the angular veins, considered the upper tributaries of the facial and the retromandibular veins (Figs. 16.5 and 16.6). These ultimately empty into the internal and external jugular veins and the brachiocephalic vein, leading to the superior vena cava and right atrium.

S. Morin (✉)
Tulane University School of Medicine,
New Orleans, LA, USA
e-mail: smorin1@tulane.edu

S. J. Bordes, Jr. et al. (eds.), *The Clinical Anatomy of the Vascular System*,
https://doi.org/10.1007/978-3-031-78326-5_16

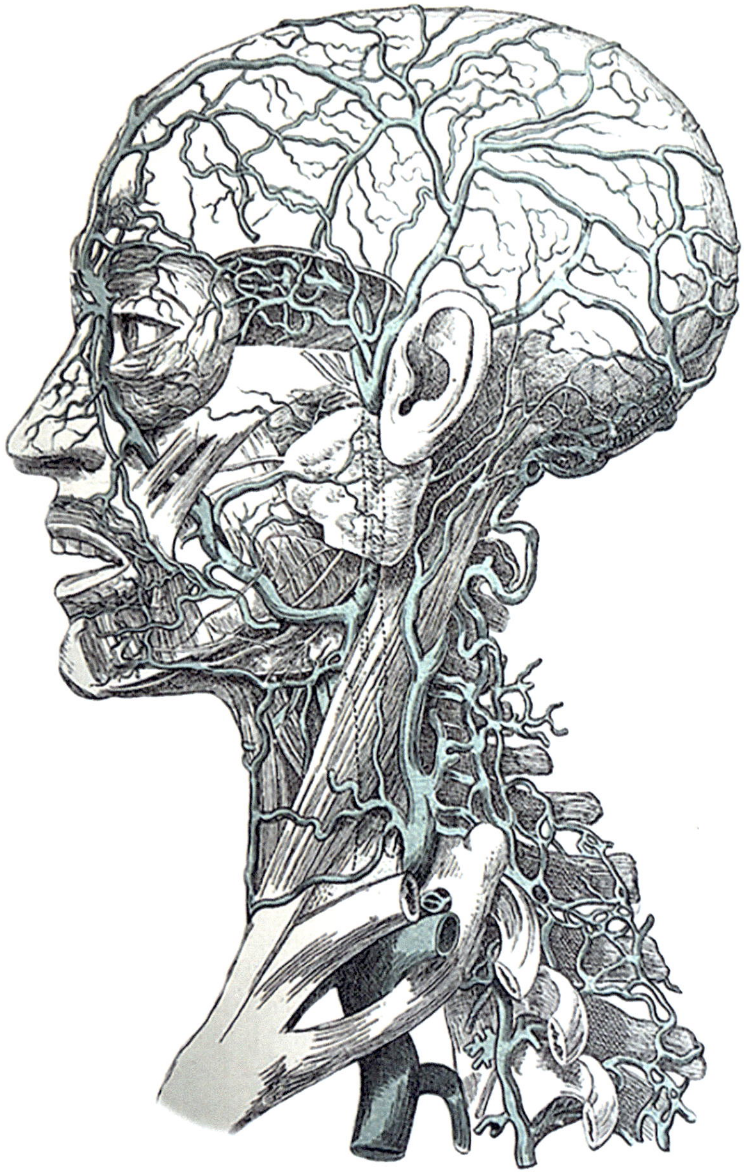

Fig. 16.1 Schematic drawing illustrating the major veins of the face, scalp, and neck. (After Allen 1882)

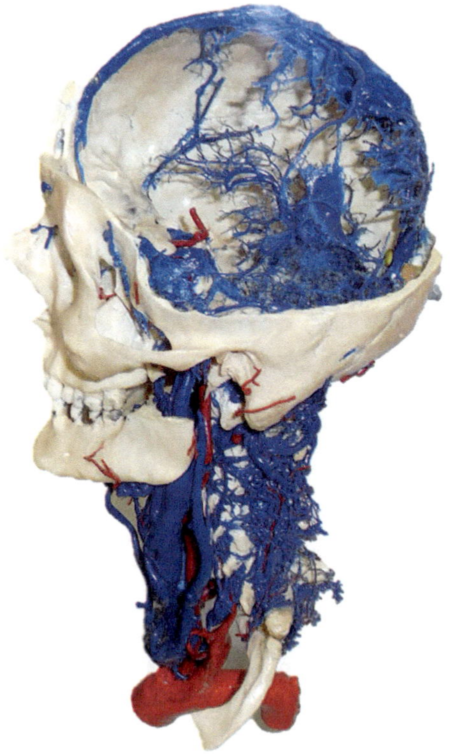

Fig. 16.2 Latex preparation showing the intracranial veins/sinuses and major venous structures of the neck

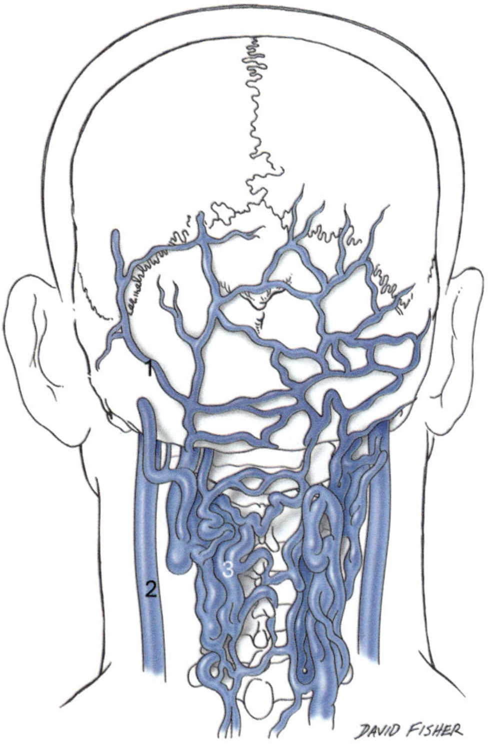

Fig. 16.4 Posterior head and neck indicating connections between the occipital veins (1), internal jugular vein (2), and veins of the deep neck (e.g., deep cervical vein), and spine (3)

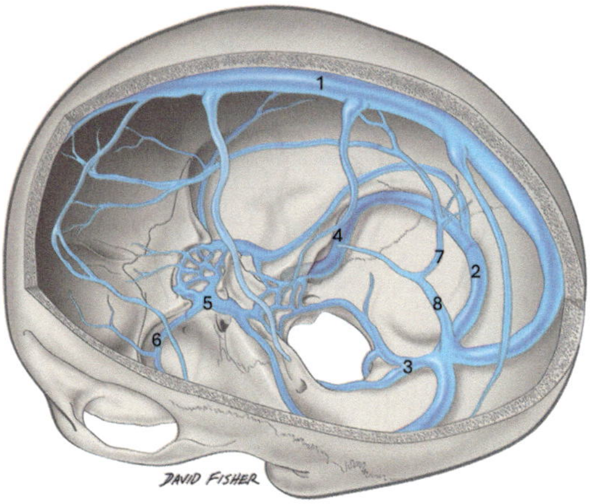

Fig. 16.3 Schematic illustrating the superior sagittal sinus (1), transverse sinus (2), occipital sinus (3), sigmoid sinus (4), cavernous sinus (5), sphenoparietal sinus (6), inferior sagittal sinus (7), and straight sinus (8)

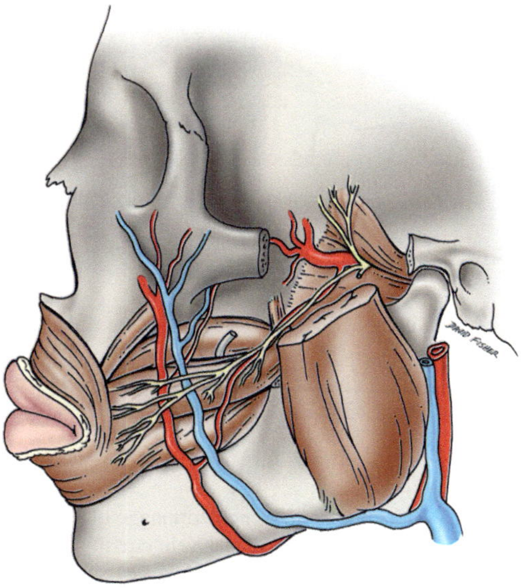

Fig. 16.5 Drawing showing the facial vein crossing the mandible just posterior to the facial artery and draining to the internal jugular vein

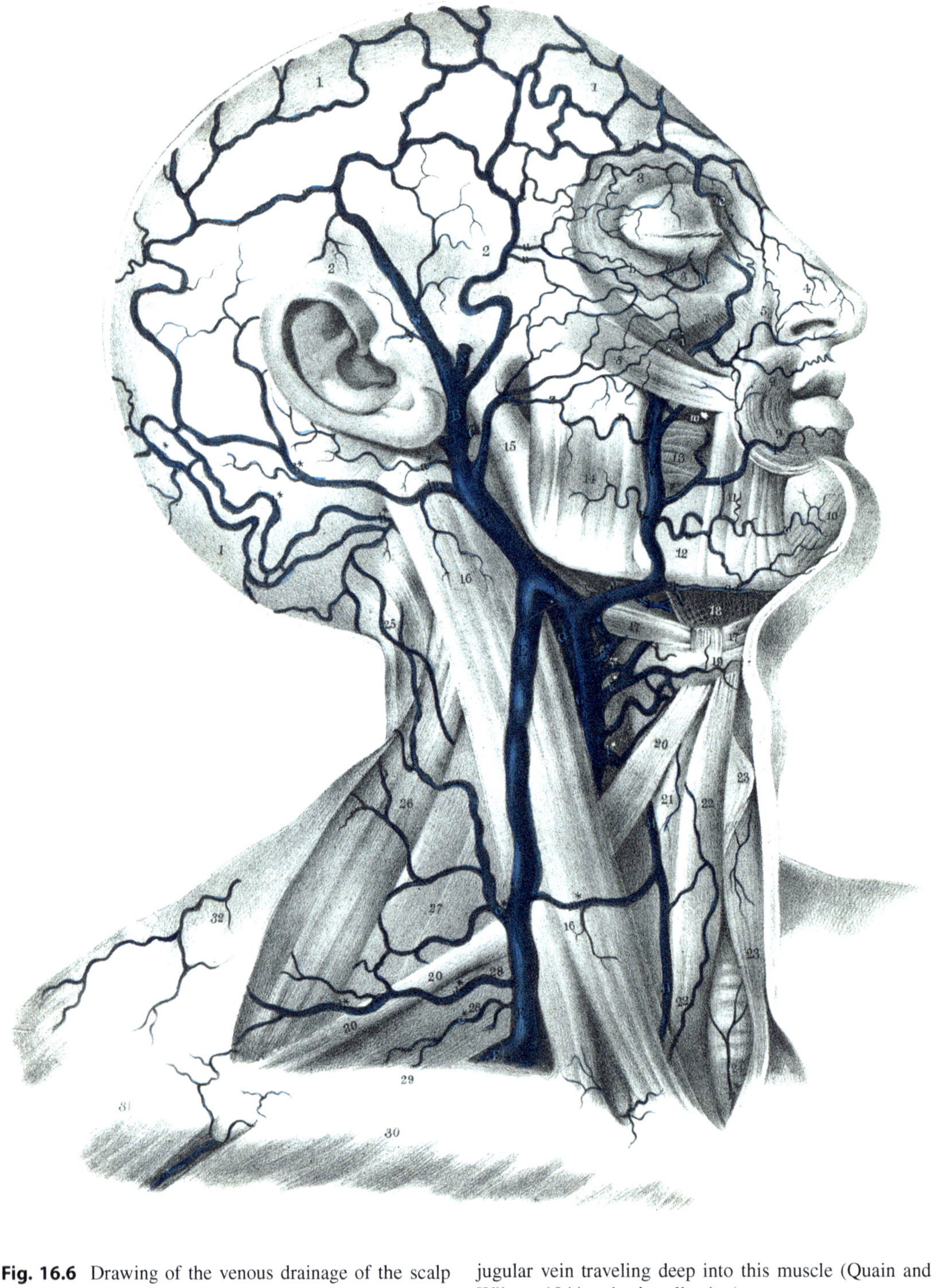

Fig. 16.6 Drawing of the venous drainage of the scalp and face. Note the external jugular vein crossing superficial to the sternocleidomastoid muscle and the internal jugular vein traveling deep into this muscle (Quain and Wilson, 1844 author's collection)

16.2 Embryology

Vascular systems develop from mesenchymal vascular networks that deepen into capillary plexuses internally and conduct early vascular flow. The neighboring mesenchyme provides protective tissue and a muscular coating for the venous structures. The first and second pharyngeal arches, formed between the fourth and seventh weeks of gestation, give rise to the scalp's veins. These veins drain the facial muscles, arising from the first and second pharyngeal arches (Adams et al. 2016); Magro et al. 2013).

The venous cardinal system, which collects blood from the head, neck, and body walls, is a component of the intra-embryonic network and comprises the ventral and dorsal cardinal veins. Within each ventral arch, five vessels gradually appear and communicate with the ventral and dorsal cardinal veins. At the heart, these cardinal veins anastomose form the common cardinal veins, which drain into the sinus venosus next to the vitelline and umbilical veins. The phrase "common cardinal veins" generally denotes embryonic veins that drain into the sinus venosus during development (Fig. 16.7). The ventral cardinal veins drain blood from the head and the neck; their cranial parts are the incipient internal jugular veins and anastomose after that with facial capillary plexuses form the external jugular veins. At the same time, the thyroid and thymic veins anastomose between the left and right ventral cardinal veins (Adams et al. 2016; Magro et al. 2013).

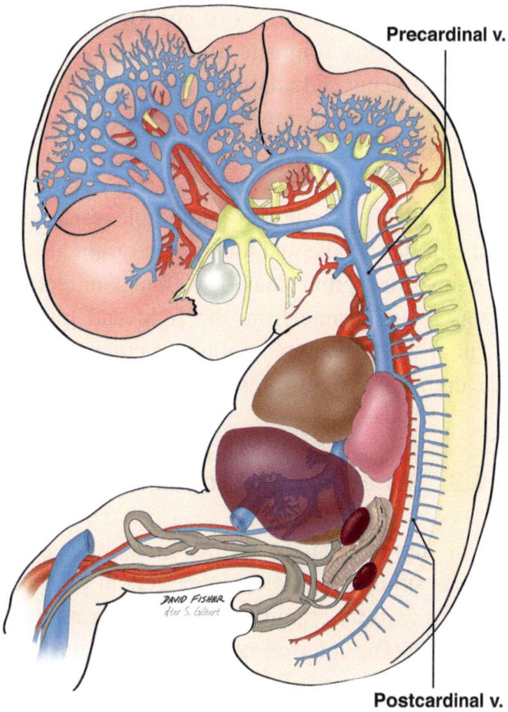

Fig. 16.7 Development of the fetal venous system highlighting the cardinal veins

The cranial part of the caval system begins to develop at week eight of gestation. The bulky anastomosis prevents blood from the left cranial cardinal vein from flowing to the right, which later becomes part of the left brachiocephalic trunk. Superior to this anastomosis, ventral cardinal veins develop into the internal jugular veins, while the mandibular ventral veins give rise to the external jugular veins. The subclavian vein develops from the upper limb venous plexuses, and the cranial vena cava finally arises from the right common cardinal vein and the proximal part of the right ventral cardinal vein.

Development of the dural venous sinuses is related to the evolution of the cerebral vesicles, the skull, and the ventral cardinal veins. Three dural venous plexuses—rostral, middle, and caudal—are formed in the early embryo within the mesenchyme of the cerebral vesicles (Fig. 16.7). These plexuses drain into the vena capitis prima or the encephalic end of the ventral cardinal vein. The rostral plexus returns venous blood from the prosencephalon and the mesencephalon, the middle plexus returns blood from part of the rhombencephalon, and the caudal plexus also drains part of the rhombencephalon.

At the end of the second month, anastomosis forms between the middle and caudal plexuses above and behind the auditory vesicles, which develop into the sigmoid part of the lateral sinus. At the same time, part of the vena capitis prima gives rise to the cavernous sinus, which contains the ophthalmic vein. The vertebral veins form from two secondary bypass pathways. One of these is the mastoid emissary vein exiting the skull via the mastoid foramen, and the other is the petrosquamous sinus. Both pathways account for the communication between the lateral sinus and the external jugular system. Afterward, the vertebral veins originate from atrophy of the emissary veins and anastomosis of a small portion of the mastoid emissary vein with the suboccipital plexus.

After the calvarial venous system communicates with the rostral and middle plexuses, they unite in the middle and the system terminates in the superior sagittal sinus. The cranial and lateral veins then separate from the right and inferior sagittal sinuses and the confluence point at which the drainage systems of the skull base meet.

16.3 Anatomy and Topography

From intracranial to extracranial, the dural venous sinuses form a complex system of drainage that exits the cranium via specific foramina in the skull. The inferior sagittal sinus drains into the straight sinus before emptying into the confluence of sinuses. In contrast, the occipital and superior sagittal sinuses drain directly into the confluence, which then drains into the transverse sinus, which in turn empties into the sigmoid sinus along with the superior petrosal sinus (Marur et al. 2014; Moore et al. 2014; Onishi et al. 2017; Sinnatamby and Last 2011; Standring 2016; Watanabe et al. 2016).

The cavernous sinus receives blood from the ophthalmic and facial veins and the sphenoparietal sinus (Fig. 16.3). Then, it gives off two small tributaries: the superior petrosal sinus, which empties into the sigmoid sinus as previously mentioned, and the inferior petrosal sinus, which drains directly into the internal jugular vein alongside the sigmoid sinus. The facial and superficial temporal veins also empty into the internal jugular veins (Fig. 16.8). Once the internal jugular veins exit the cranium via the jugular foramen, they receive tributaries from the superior and middle thyroid veins, which empty into the brachiocephalic vein and the pharyngeal veins (Figs. 16.9, 16.10, 16.11, and 16.12). (Marur et al. 2014; Moore et al. 2014; Onishi et al. 2017; Sinnatamby and Last 2011; Standring 2016; Watanabe et al. 2016).

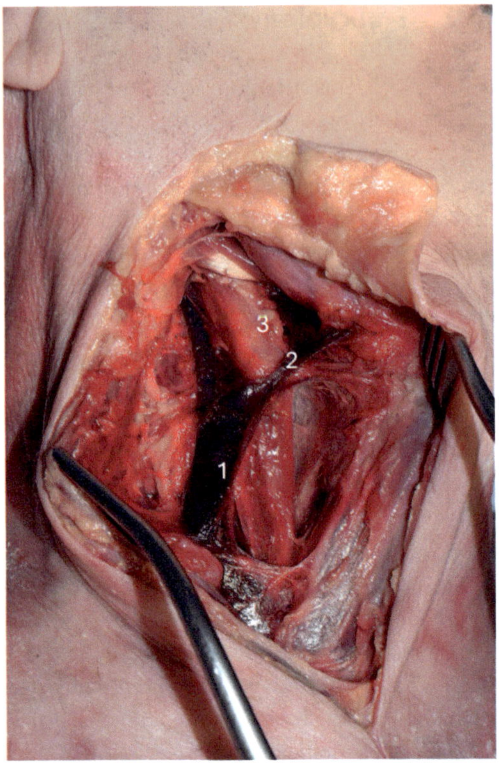

Fig. 16.8 Right-sided cadaveric dissection noting the internal jugular vein (1) and common facial vein (2). Note the medially placed carotid artery (3)

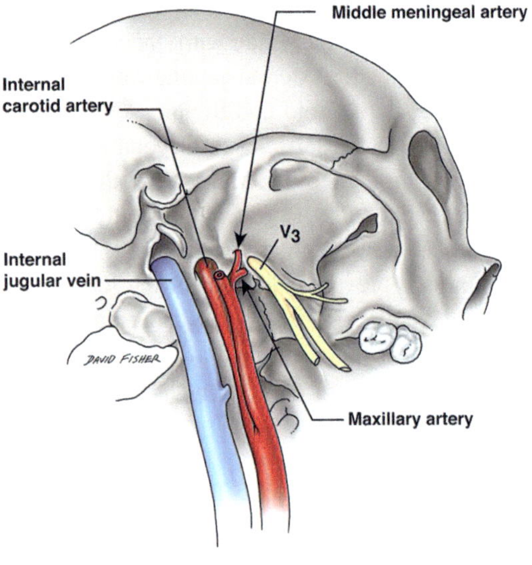

Fig. 16.9 A highly schematic drawing showing the relationship of the internal jugular vein and internal carotid artery at the skull base

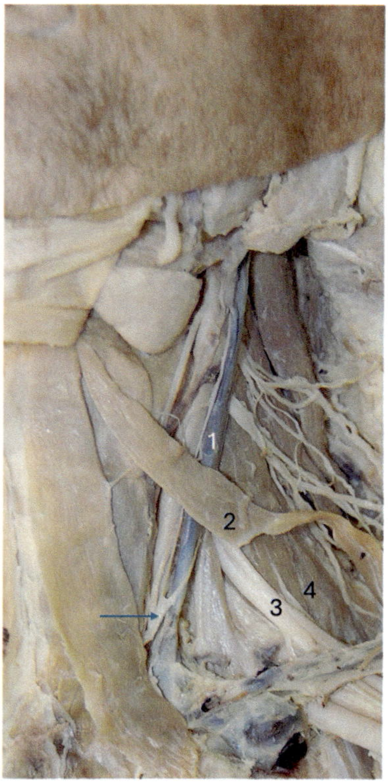

Fig. 16.10 Left-sided cadaveric section noting the internal jugular vein (1) traveling deep to the omohyoid muscle (2). For reference, note the middle scalene muscle (4) and superior trunk of the brachial plexus (3). Also, the ansa cervicalis (arrow) is seen traveling along and anterior to the internal jugular vein on its way to innervate the strap muscles of the neck

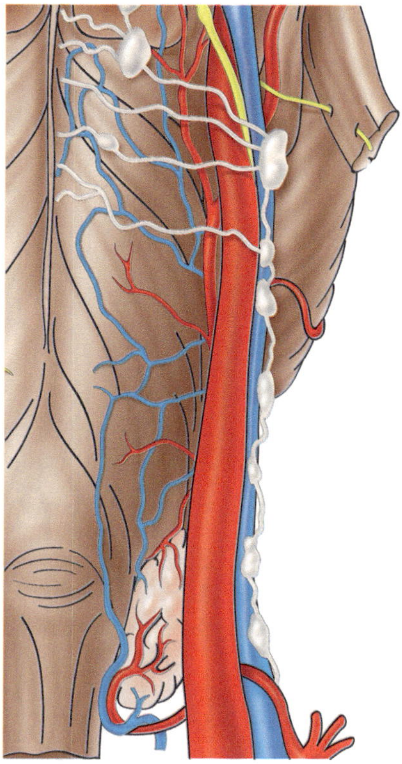

Fig. 16.11 Posterior view of the pharynx noting the multiple veins draining this region on the right side and the laterally placed right internal jugular vein. (After Netter, 2023)

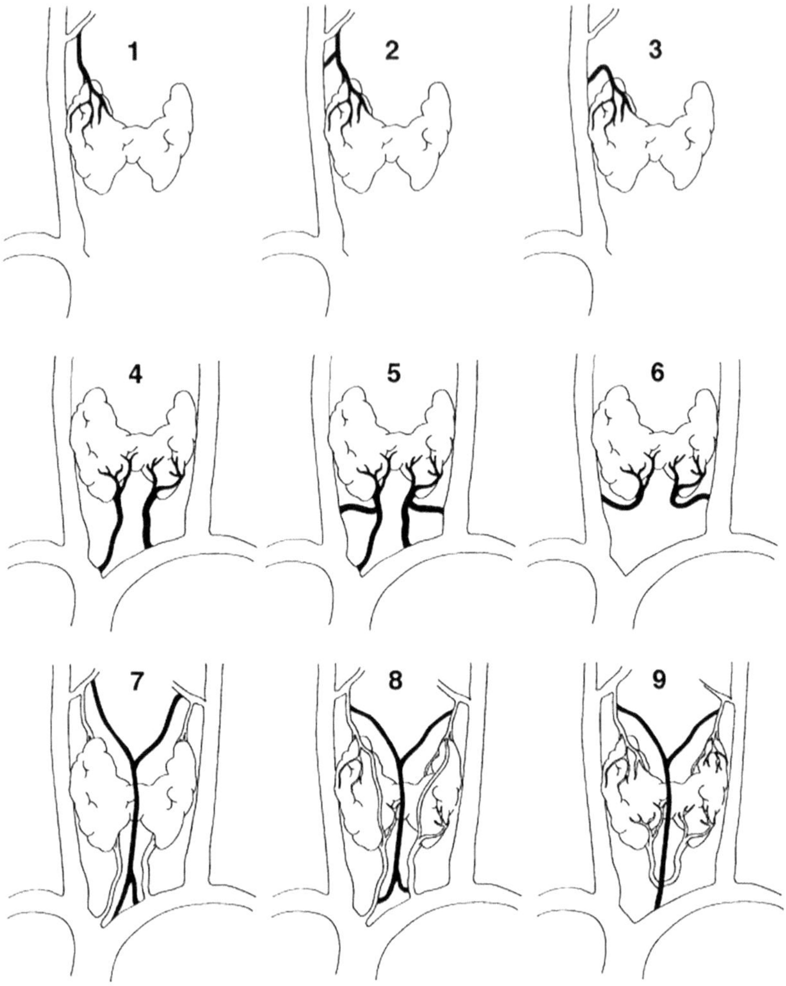

Fig. 16.12 Variations of the thyroid veins and their drainage into the internal jugular and brachiocephalic veins. (After Krmpotic-Nemanic et al. 1988)

The occipital vein drains the superficial scalp and becomes the posterior auricular vein as it travels inferiorly down the face and eventually joins the external jugular vein. The subclavian vein, a continuation of the axillary vein, receives tributaries from the external jugular and vertebral veins before joining the internal jugular vein to form the brachiocephalic vein. The right and left brachiocephalic veins fuse to form the superior vena cava, which empties blood into the right atrium of the heart (Marur et al. 2014; Moore et al. 2014; Onishi et al. 2017; Sinnatamby and Last 2011; Standring 2016; Watanabe et al. 2016).

16.4 Main Facial Veins

The veins that drain the scalp mostly follow their respective arteries. The anterior scalp, from the superciliary arches to the vertex of the head, is drained by the supratrochlear and supraorbital veins, which pass inferior to the superciliary arches and come together to form the angular vein (Figs. 16.1 and 16.13). The angular vein becomes the facial vein, the major vein draining the face as it proceeds inferiorly. The facial vein then runs across the face with the facial artery until it reaches the inferior border of the mandi-

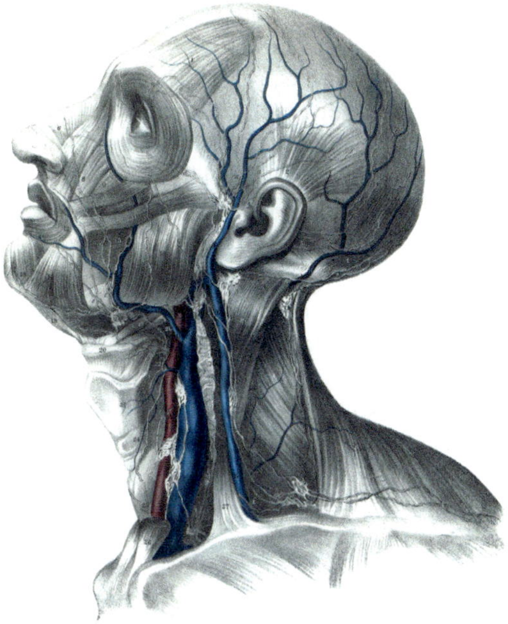

Fig. 16.13 Scalp and facial veins with drainage into the internal jugular and external jugular veins, respectively (Quain and Wilson, 1844 author's collection)

ble, where the artery and vein diverge. The vein passes superficial to the submandibular gland to empty into the internal jugular vein (Marur et al. 2014; Moore et al. 2014; Onishi et al. 2017; Sinnatamby and Last 2011; Standring 2016; Watanabe et al. 2016).

The facial vein receives tributaries from veins of the eyelids, external nose, lips, cheek, and chin as it courses with the facial artery. It also has numerous connections with venous channels that dive deep into regions of the head. Near the medial corner of the orbit, where it originates at the supratrochlear and supraorbital veins, the facial vein communicates with ophthalmic veins. In the cheek, it communicates with veins passing into the infraorbital foramen. The deep facial vein, which connects with the pterygoid plexus, also communicates with the facial vein. These venous channels connect with the intracranial cavernous sinus via emissary veins that join the intracranial and extracranial veins (Marur et al. 2014; Moore et al. 2014; Onishi et al. 2017; Sinnatamby and Last 2011; Standring 2016; Watanabe et al. 2016).

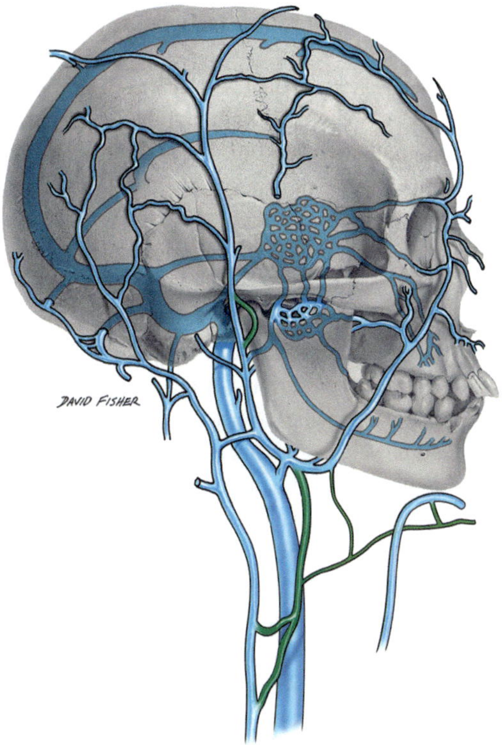

Fig. 16.14 Schematic drawing illustrating the superior and inferior ophthalmic veins and their tributaries draining internally to the cavernous sinus. Inferiorly, connections from the deep facial vein join the pterygoid venous plexus. Also note the variable connections between the facial vein and external and anterior jugular veins

The transverse facial artery is accompanied by small transverse facial veins that drain into the superficial temporal vein within the parotid gland. The lateral scalp is drained by the superficial temporal vein, which then passes inferiorly to drain into the retromandibular vein. The area of the scalp posterior to the ear is drained by the posterior auricular vein, which empties into a tributary of the retromandibular vein. The posterior aspect of the scalp, from the external occipital protuberance and superior nuchal lines to the vertex of the head, is drained by the occipital vein. This vein also passes deep to this area through the skeletal muscle of the posterior neck to form a plexus of veins in the suboccipital triangle (Marur et al. 2014; Moore et al. 2014; Onishi et al. 2017; Sinnatamby and Last 2011; Standring 2016; Watanabe et al. 2016).

16.5 Veins of the Orbit

Two main venous channels draining the orbit are the superior ophthalmic vein and the smaller inferior ophthalmic vein (Fig. 16.14). The superior ophthalmic vein originates in the anterior orbit as connecting veins between the supraorbital and angular veins. It receives tributaries from companion veins and branches of the ophthalmic artery and veins. It then drains the posterior part of the eyeball as it traverses across the superior part of the orbit, leaves the orbit posteriorly via the superior orbital fissure, and enters the cavernous sinus. The inferior ophthalmic vein begins anteriorly and passes across the inferior part of the orbit as it drains tributaries from veins from muscles of the eye and the posterior part of the eyeball. It can leave the orbit in various ways. It can join the superior ophthalmic vein, pass through the superior orbital fissure on its own to join the cavernous sinus, or pass through the inferior orbital fissure to join the pterygoid plexus of veins in the infratemporal fossa.

The eye is mostly emptied by drainage of the choroid layer through four large veins, the vortices veins. They exit through the sclera from each of the posterior quadrants of the eyeball and enter the superior and inferior ophthalmic veins. The central retinal vein also accompanies the central retinal artery.

16.6 Auricular Veins

The membranous labyrinth, the membranous ducts, and sacs of the internal ear are drained via vestibular and cochlear veins that travel with the arteries. They then combine to form a labyrinthine vein, emptying into the inferior petrosal or sigmoid sinus.

16.7 Pterygoid Venous Plexus

The network of veins between the medial and lateral pterygoid muscles and the lateral pterygoid and temporalis muscles is called the pterygoid venous plexus (Figs. 16.1 and 16.14). It is joined by the veins of the infratemporal and pterygopalatine fossae, including the tributaries that drain the nasal cavity, oral cavity, muscles of the infratemporal fossa, paranasal sinuses, and nasopharynx. Branches of the maxillary artery generally supply these areas. As mentioned earlier, the inferior ophthalmic vein can also drain into the pterygoid plexus.

The pterygoid plexus is often connected to the cavernous sinus in the cranial cavity by small emissary veins that pass through the foramen ovale, cartilage filling the foramen lacerum, and a small sphenoidal foramen on the medial side of the lateral plate of the pterygoid process at the base of the skull. It connects posteriorly with the retromandibular vein in the neck via a short maxillary vein and anteriorly with the facial vein in the face via a deep facial vein.

16.8 Retromandibular Vein

Superficial venous drainage of the neck is mainly accomplished by the external and anterior jugular veins. The retromandibular vein is formed when the superficial temporal and maxillary veins join in the parotid gland and then descend inferiorly to the angle of the mandible, where it bifurcates into anterior and posterior divisions—the anterior division drains into the facial vein and the internal jugular vein (Fig. 16.14). The posterior auricular vein, which drains the scalp above the ear, empties into a tributary of the retromandibular vein, which continues to the external jugular vein posterior to the angle of the mandible.

16.9 External Jugular Vein

The external jugular vein travels down the neck in the superficial facia and diagonally across the sternocleidomastoid muscle (Fig. 16.15). In the lower part of the neck, superior to the clavicle and posterior to the sternocleidomastoid, the cervical fascia is pierced by the external jugular vein, which passes deep to the clavicle to drain into the subclavian vein. Along the way, the external jugular vein receives tributaries from the

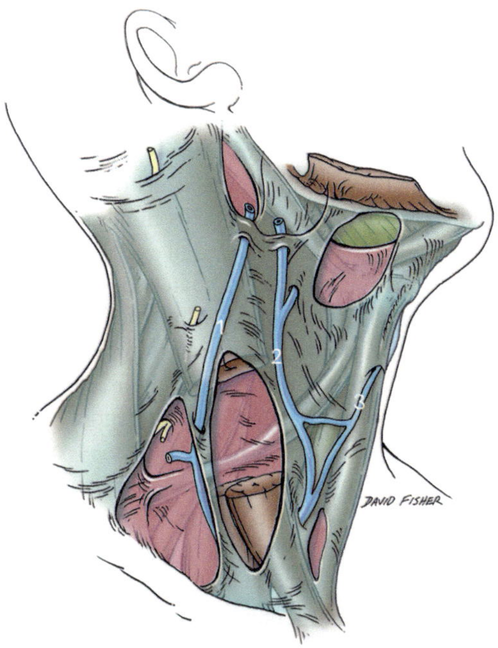

Fig. 16.15 Illustration showing the facial relationships of the external jugular (1), communicating branch (of Koch) (2), and anterior jugular veins (3) (after Morris, 1893)

posterior external jugular, transverse cervical, suprascapular, and anterior jugular veins.

16.10 Anterior Jugular Vein

Although the anterior jugular veins are quite variable, they usually drain the anterior aspect of the neck (Fig. 16.15). They form a network of small paired venous channels that come together just superior to the hyoid bone; each anterior jugular vein descends on either side of the midline of the neck, piercing the investing layer of the cervical fascia to enter the subclavian vein inferiorly near the middle attachments of the sternocleidomastoid muscle. In some cases, the anterior jugular veins enter the external jugular vein just before it enters the subclavian vein. The right and left anterior jugular veins can communicate via the jugular venous arch by the suprasternal notch. The anterior jugular is the vein most commonly encountered during procedures involving the anterior neck, such as tracheostomies.

16.11 Internal Jugular Vein

The internal jugular vein originates as a dilated continuation of the dural venous sinus, the sigmoid sinus (Fig. 16.1). The initial dilated part is known as the superior bulb of the jugular vein, and soon after it is formed, it receives the inferior petrosal sinus. It then exits the skull via the jugular foramen before entering the carotid sheath. It runs with the common carotid arteries in the carotid sheath in the anterior triangle of the neck and receives blood from nearly all structures of the head and neck. Posterior to the sternal end of the clavicle, the paired internal jugular veins join the subclavian vein to form the right and left brachiocephalic veins.

16.12 Thyroid Veins

The thyroid gland is drained by the superior thyroid vein, which mainly drains the area supplied by the superior thyroid artery, and the middle and inferior thyroid veins, which drain the remainder of the gland (Figs. 16.1 and 16.12). The superior and middle thyroid veins empty into the internal jugular, while the inferior thyroid vein drains directly into the brachiocephalic veins.

16.13 Subclavian Vein

As the subclavian vein crosses the base of the posterior triangle of the neck, it is joined by the external jugular, suprascapular, and transverse cervical veins (Fig. 16.1). It then joins the internal jugular vein to form the brachiocephalic vein near the sternoclavicular joint. The brachiocephalic veins form the superior vena cava, which carries blood to the right atrium.

16.14 Veins of the Pharynx and Larynx

The pharynx is drained superiorly by the pterygoid plexus in the infratemporal fossa and inferiorly into the facial and internal jugular veins (Fig. 16.11). The superior laryngeal veins empty into the superior thyroid veins, draining into the internal jugular. The inferior laryngeal veins drain into the inferior thyroid vein, which drains into the brachiocephalic vein.

16.15 Veins of the Nasal Cavity

The maxillary artery supplies the nasal cavity and drains into the pterygoid plexus of veins in the infratemporal fossa. Veins from the anterior regions of the nasal cavity also join the facial vein. In some individuals, an emissary vein connects a nasal vein (extracranial) with the superior sagittal sinus (intracranial). Tributaries of the superior ophthalmic vein form some of the largest emissary veins, which drain into the cavernous sinus on either side of the hypophysial fossa and accompany the anterior and posterior ethmoidal arteries.

16.16 Veins of the Oral Cavity and Related Areas

The dorsal lingual and deep lingual veins drain the tongue. The deep lingual veins, visible through the mucosa on the undersurface of the tongue, travel with the lingual arteries in the anterior parts of the tongue but separate from them posteriorly near the hyoglossus muscle. They travel on either side of the external surface of that muscle along with the hypoglossal nerve and exit through the floor of the oral cavity via an aperture formed by the edges of the mylohyoid, superior constrictor, and middle constrictor muscles. After leaving the oral cavity, the deep lingual veins empty into the internal jugular vein.

The dorsal lingual vein also follows the lingual artery between the hyoglossus and genioglossus muscles and drains into the internal jugular vein, like the deep lingual veins.

The veins that drain the parotid glands empty into the external jugular. The veins from the submandibular and sublingual glands drain into the lingual and facial veins. The veins of the palate empty into the pterygoid plexus in the infratemporal fossa or into a network of veins related to the palatine tonsil. They ultimately drain into the pharyngeal plexus or directly into the facial vein.

The inferior alveolar veins drain the lower teeth, while the superior ones drain the upper teeth. They both empty into the pterygoid plexus in the infratemporal fossa. Tributaries of the facial vein also help to drain the anterior teeth. The pterygoid plexus connects to the maxillary vein, which joins the retromandibular vein and jugular system. Vessels that pass through the mental foramen to connect with the facial vein also contribute to the venous drainage of the teeth. Veins of the gingivae drain into the facial vein or the pterygoid plexus.

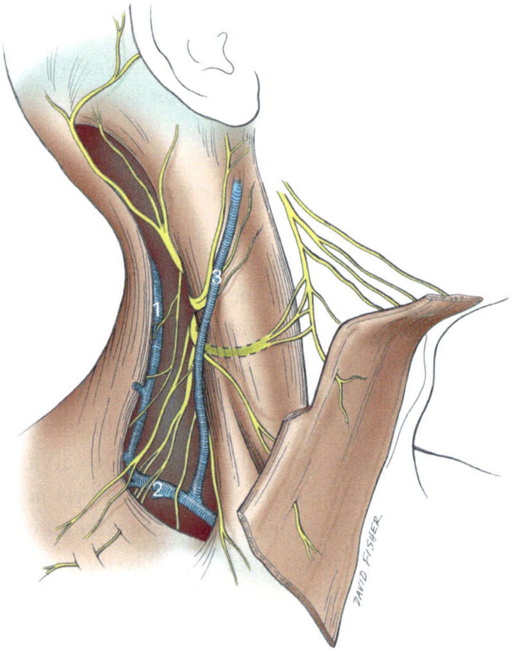

Fig. 16.16 Right-sided drawing of the neck and major nerves and veins, including the posterior jugular (1), transverse cervical (2), and external jugular veins (3). Note that this is an anatomical variation of this venous anatomy

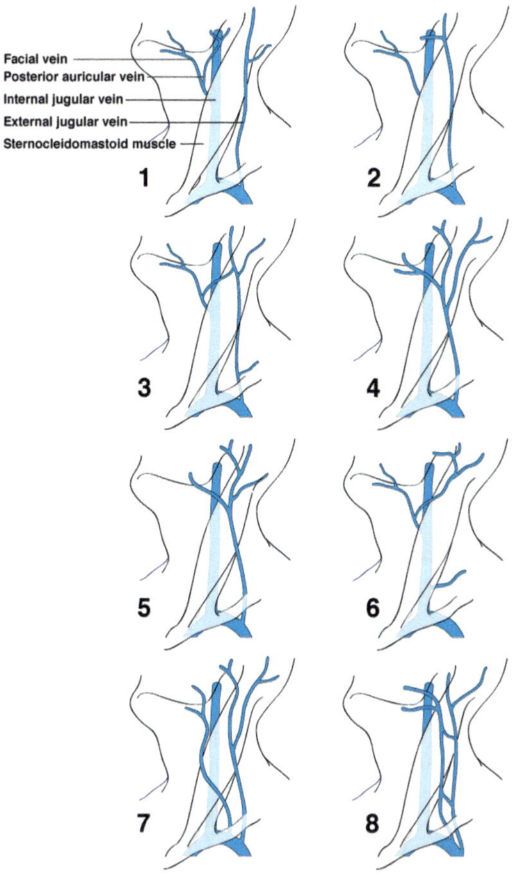

Fig. 16.17 Many variations of the internal and external jugular venous systems. (After Krmpotic-Nemanic et al. 1988)

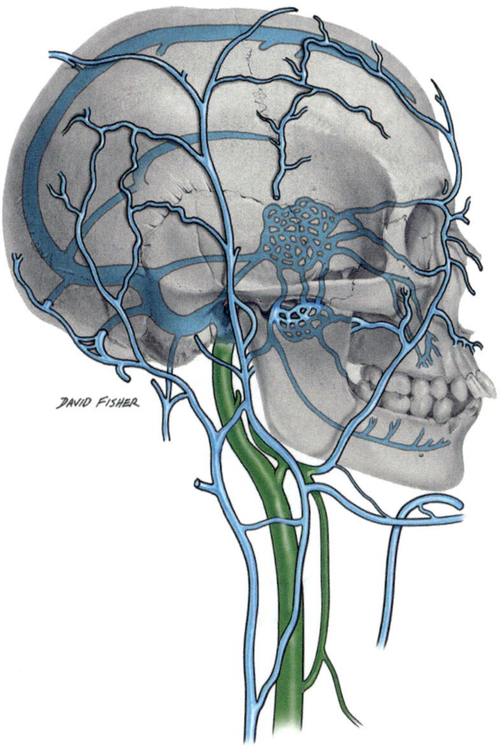

Fig. 16.18 Variable connections of the external and internal (green) venous systems

16.17 Variations

While the external jugular vein usually continues to drain into the subclavian, it is sometimes absent ipsilaterally or bilaterally. In such cases, the veins that typically drain into the external jugular drain into the internal jugular instead.

Another variation in venous drainage of the neck is the vein of Kocher or the posterior external jugular vein (Figs. 16.15, 16.16, and 16.17). This vein drains the posterosuperior superficial muscles and skin. It can be similar to or even larger than the internal jugular vein, so it is often misidentified. It originates in the occipital region as a tributary of the common facial vein, travels obliquely along the anterior border of the sternocleidomastoid muscle, and eventually drains into either the internal jugular, the middle part of the external jugular, the jugular venous arch, or the brachiocephalic vein (Dalip et al. 2018; Tubbs et al. 2016).

The anterior jugular vein can travel down the midline of the neck, in which case it is called the median cervical vein, or it can exist as two parallel anterior jugular vessels, one of which drains into the external jugular and the other into the transverse cervical vein.

The jugular venous arch is typically made by the union of the two anterior jugular veins and drains into either the subclavian or the external jugular vein. However, it can receive tributaries from small branches of the inferior thyroid vein so that it can be exposed in a low tracheostomy.

The inferior thyroid vein usually drains into the brachiocephalic vein, but alternatively, it can drain into the internal jugular or anastomose with the thymic veins (Fig. 16.11). The inferior thyroid veins on each side sometimes join the ipsilateral brachiocephalic vein separately. In other cases, they form a common trunk before joining the left brachiocephalic vein.

The superior thyroid vein sometimes terminates in the internal jugular or with other veins, such as the lingual vein. Anastomoses are commonly seen between the superior thyroid and common facial veins (Fig. 16.18) (Dalip et al. 2018; Tubbs et al. 2016).

The middle thyroid vein has variations that are important clinically for surgery involving the thyroid gland, parathyroid gland, or recurrent laryngeal nerve. Usually, it crosses anteriorly and drains into the internal jugular vein. However, it can occasionally drain directly into the brachiocephalic vein (Dalip et al. 2018).

16.18 Clinical/Surgical Significance and Complications

Scalp Lacerations and Hematoma

While lacerations of the scalp result mainly in arterial bleeds, extremely low venous pressure and the inability of vessels to close when lacerated owing to the connective tissue with which they are intermingled can contribute to excessive bleeding. Emissary veins are clinically implicated in infections as they allow for easy entry into the cranium as they are valveless.

Spread of Infection

The facial veins and other venous channels in the head lack valves, so they allow blood to move in either direction. Owing to the interconnections between the facial vein and other venous channels, infections should be addressed immediately to stop them from moving intracranially. Specifically, the area superior to the mouth is referred to as the "danger zone" as the risk of infectious spread via these valveless channels is much greater.

References

Adams A, Mankad K, Offiah C, Childs L (2016) Branchial cleft anomalies: a pictorial review of embryological development and spectrum of imaging findings. Insights Imaging 7(1):69–76. https://doi.org/10.1007/s13244-015-0454-5

Dalip D, Iwanaga J, Loukas M, Tubbs RS (2018) Review of the variations of the superficial veins of the neck. Cureus 10(6):e2826. https://doi.org/10.7759/cureus.2826

Magro E, Sénécail B, Gentric J-C, Alavi Z, Palombi O, Seizeur R (2013) Contribution of embryology in the understanding of cervical venous system anatomy within and around the transverse foramen: a review of the classical literature. Surg Radiol Anat 36:411. https://doi.org/10.1007/s00276-013-1206-1

Marur T, Tuna Y, Demirci S (2014) Facial anatomy. Clin Dermatol 32(1):14–23. https://doi.org/10.1016/j.clindermatol.2013.05.022

Moore KL, Dalley AF, Agur AMR (2014) Clinically oriented anatomy, 7th edn. Lippincott Williams & Wilkins, Philadelphia

Onishi S, Imanishi N, Yoshimura Y, Inoue Y, Sakamoto Y, Chang H, Okumoto T (2017) Venous drainage of the face. J Plast Reconstr Aesthet Surg 70(4):433–440. https://doi.org/10.1016/j.bjps.2016.11.023

Sinnatamby CS, Last RJ (2011) Last's anatomy: regional and applied, 12th edn. Churchill Livingstone, Edinburgh

Standring S (2016) Gray's anatomy, 41st edn. Elsevier Churchill Livingstone, Edinburgh

Tubbs RS, Shoja MM, Loukas M, Bergman RA (2016) Bergman's comprehensive encyclopedia of human anatomic variation. Wiley Blackwell, Hoboken

Watanabe K, Shoja MM, Loukas M, Tubbs RS (2016) Anatomy for plastic surgery of the face, head, and neck. Thieme

Dural Venous Sinuses

17

R. Shane Tubbs

The cerebral venous system can be divided into the superficial and deep systems. The superficial system comprises the sagittal sinuses and cortical veins, which drain the superficial surfaces of both cerebral hemispheres (Figs. 17.1, 17.2 and 17.3). The deep system consists of the lateral, straight, and sigmoid sinuses and the deeper draining cortical veins (Curè et al. 1994; Kiliç and Akakin 2008) (Figs. 17.4, 17.5 and 17.6). Both of these systems drain largely into the internal jugular veins (Kiliç and Akakin 2008).

The superficial cerebral veins are interlinked with the veins of Trolard and Labbé (Fig. 17.7). The superolateral surface of the hemisphere drains into the SSS, while the posteroinferior aspect drains into the transverse sinus (Kiliç and Akakin 2008).

The SSS runs in the attached convex margin of the falx cerebri along the grooves of the internal surface of the frontal bone, the adjacent margins of the two parietal bones, and the squamous part of the occipital bone (Mancall and Brock 2011) (Figs. 17.3, 17.6 and 17.7). It begins near the crista galli, a few millimeters posterior to the foramen cecum. It receives the primary tributaries from the cortical veins of the frontal lobes, the ascending frontal veins (Mancall and Brock 2011).

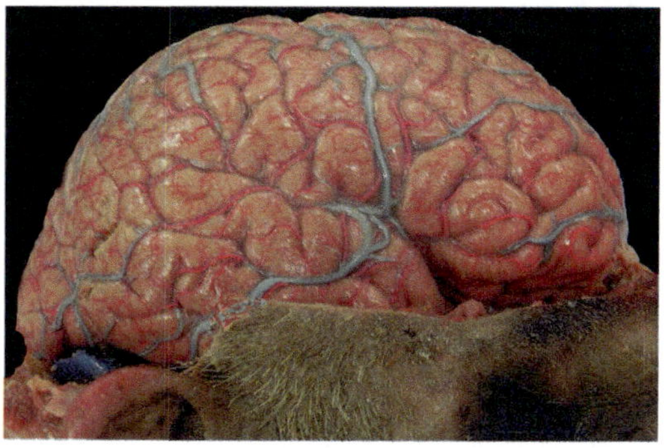

Fig. 17.1 Right side of latex injected cadaveric brain noting the superficial cortical veins. Note the right transverse sinus in the lower left corner of the photograph receiving the vein of Labbé

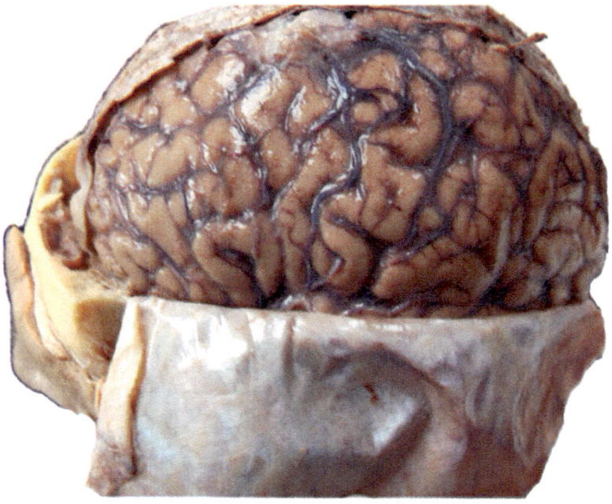

Fig. 17.2 Left-sided view of the cadaveric brain illustrating prominent superficial veins

R. S. Tubbs (✉)
Department of Neurosurgery, Tulane University School of Medicine,
New Orleans, LA, USA
e-mail: rtubbs@tulane.edu

S. J. Bordes, Jr. et al. (eds.), *The Clinical Anatomy of the Vascular System*,
https://doi.org/10.1007/978-3-031-78326-5_17

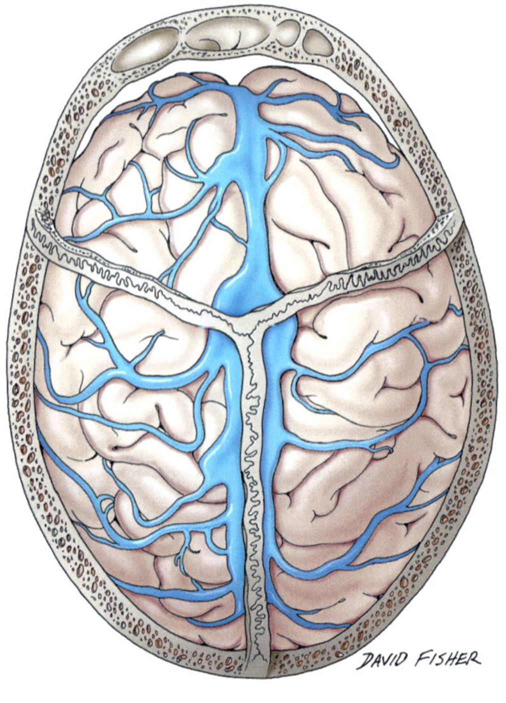

Fig. 17.3 Schematic showing the midline superior sagittal sinus collecting superficial cortical veins. The coronal and sagittal sutures are left in place to illustrate the relationship of the sinus to the midline suture

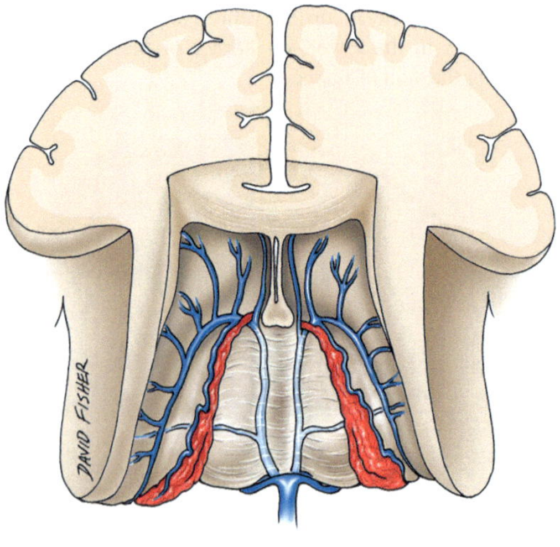

Fig. 17.4 Deep veins of the brain culminating into the great vein of Galen at 6 o'clock. (After Netter 2019)

The cortical veins of the inferior surface of the frontal lobes may drain through the anterior cortical veins into the anterior third of the superior sagittal sinus (SSS), through the anterior cerebral veins into the basal vein of Rosenthal, or the superficial middle cerebral vein (Ruíz 2006).

The SSS functions as a terminus of the superficial cortical venous system draining the anterior part of the inferior surface of the frontal lobe and the superior parts of the lateral and medial surfaces of the frontal, parietal, and occipital lobes (Brockmann et al. 2012). The more anterior veins usually enter the sinus at more or less right angles. However, the more posterior, and usually larger, veins travel in an oblique fashion anteriorly to drain into the SSS in a direction opposite the blood flow within them.

Together with the straight and occipital sinuses, the SSS empties into the transverse sinus and causes a small excavation at the internal occipital protuberance (torcular Herophili: Brockmann et al. 2012) (Fig. 17.5). While the SSS is narrow at its anterior end, the sinus runs backward and widens gradually to approximately 1 cm at its posterior end. Near the internal occipital protuberance, the SSS deviates, usually to the right (commonly three times), and continues as the transverse sinus (Mancall and Brock 2011).

In a triangular cross section, the interior of the SSS includes the openings of the superior cerebral veins and arachnoid granulation projections. Many fibrous bands traverse the SSS (see chordae Willisii below). It also communicates by small orifices with irregular venous lacunae in the dura mater near the sinus (Mancall and Brock 2011).

There usually are two or three small frontal, large parietal, or intermediate-sized occipital lacunae on each side of the sinus (Mancall and Brock 2011). In older people, the lacunae tend to become confluent, so there is one elongated lacuna on each side. Fine fibrous bands cross them, and numerous arachnoid granulations project into them (Mancall and Brock 2011).

The SSS receives the superior cerebral veins and, near the posterior end of the sagittal suture, emissary veins that pass through the parietal foramina (Mancall and Brock 2011).

The lacunae also drain the diploic veins and meningeal veins. Lateral lacunae are frequently

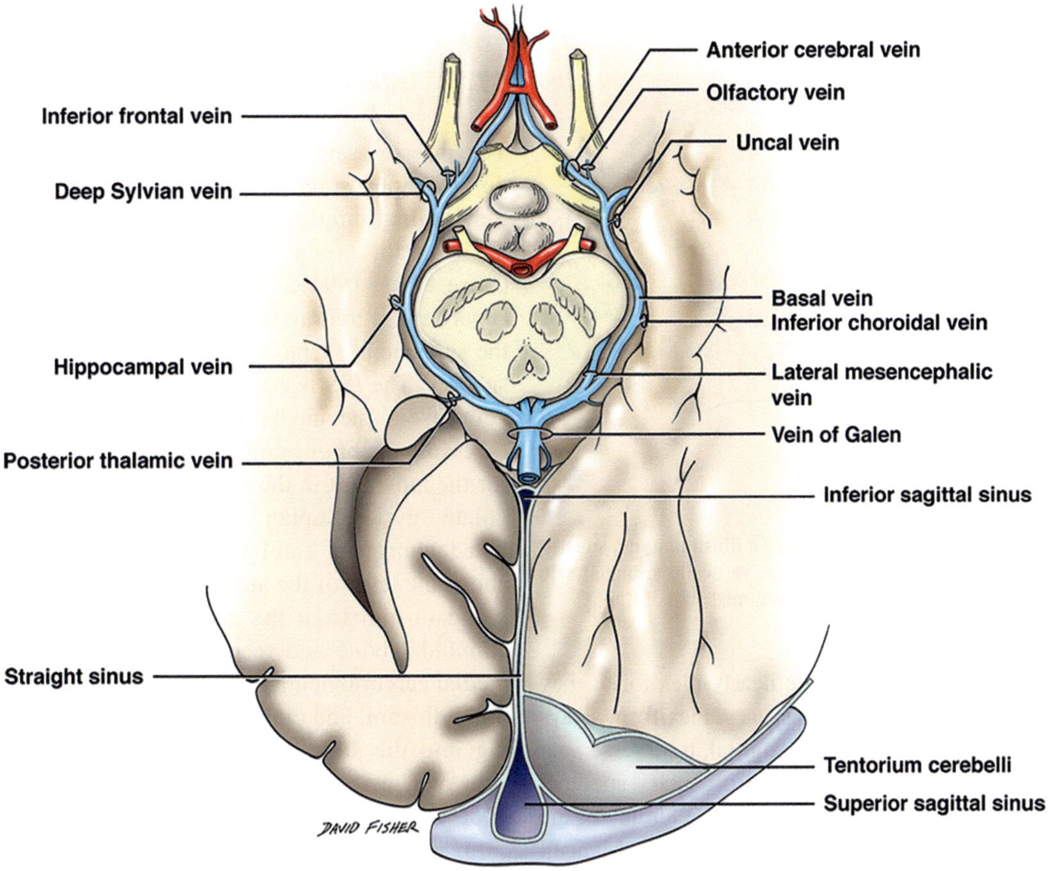

Fig. 17.5 Deep veins of the brain highlighting the tributaries of the basal veins of Rosenthal, a tributary of the vein of Galen. (After Netter 2019)

Fig. 17.6 Highly schematic drawing of the deep midline veins of the brain and their drainage posteriorly into the vein of Galen and then, straight sinus

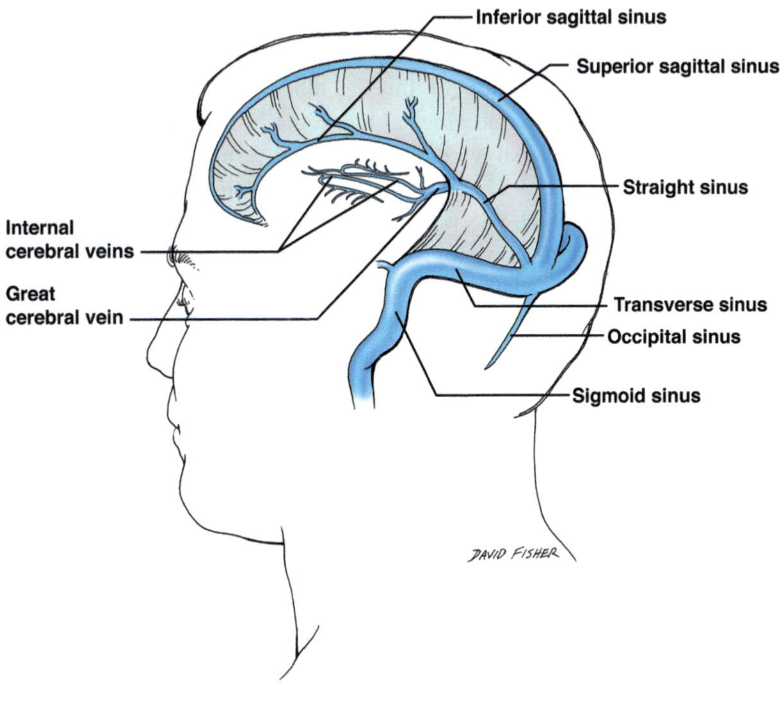

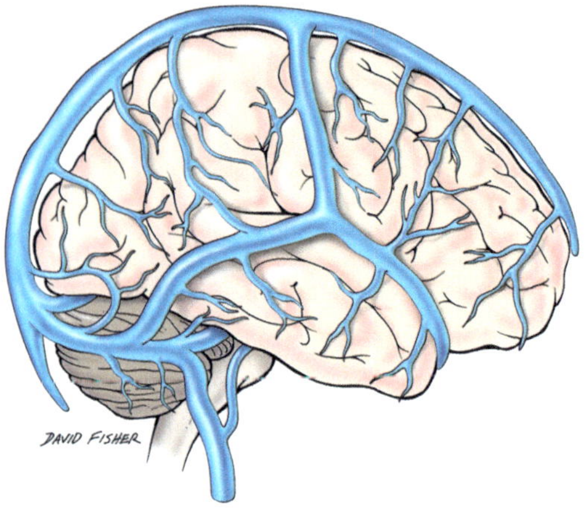

Fig. 17.7 Right side of the brain illustrating the superficial cortical veins, including the vertically arranged vein of Trolard and horizontally positioned vein of Labbé

so complex that they are nearly plexiform; they rarely are simple venous spaces. Plexiform arrays of small veins adjoin the sagittal, transverse, and straight sinuses and ridges of such "spongy" venous tissue often project into the lumina of the superior sagittal and transverse sinuses (Mancall and Brock 2011).

Variable bands and projections from its dural walls, which extend as horizontal shelves that divide its lumen into superior and inferior channels, also invade the intermediate third of the SSS. Such variable features make it impossible to provide a simple description of this or other venous sinuses, and only radiological investigations can show individual variations (Mancall and Brock 2011).

The mean length of the SSS was 25.6 + 1.6 cm in computed tomography venographies (CTV) and 19.2 + 1.4 cm in cadaveric dissections. The area of the SSS in cross section at the coronary suture measured 32.1 + 15.3 mm^2 and grows larger in the direction of flow (Brockmann et al. 2012). The mean length of the SSS ranges from 23 to 28 cm measured from the foramen cecum to the torcular herophili.

The typical shape of the SSS in a coronal view in CTV appears not to be circular, but rather triangular or polygonal (Curè et al. 1994; Brockmann et al. 2012).

17.1 Inferior Sagittal Sinus

The inferior sagittal sinus (ISS) is a paired dural venous sinus also known as the sinus longitudinal inferior (Fig. 17.6). It is a part of the superior group of dural venous sinuses (Cure et al. 1994), and as a member of the superficial cerebral venous system, it drains the superficial surfaces of both cerebral hemispheres. The ISS is established by drainage of veins from the medial aspect of the frontal lobes of the brain, collecting veins coursing for the most part in the rostral sulci of the cingulate gyri (Kaplan and Browder 1973; Swedenborg and Tafel 1887). The ISS begins above the level of the anterior body of the corpus callosum from the crista in front of the ethmoidal bone and receives blood from the medial cerebral hemispheres, falx cerebri, corpus callosum, and cingulate gyri before draining into the straight sinus together with the vein of Galen (Cure et al. 1994; Erbas et al. 2006). Most of the drainage from the corpus callosum enters the sinus at its anterior end. Small veins from the inferior sulcus of the cingulate join it in its course caudally, where it is joined frequently by branches of veins from the dorsal surface of the corpus callosum. It receives tributaries from the falx cerebri, some of which establish substantial connections between the superior sagittal sinus (SSS) and the ISS (Kaplan and Browder 1973). The ISS runs caudally in the free border of the falx cerebri to join the rostral aspect of the straight sinus and increases in size as it passes, while remaining very narrow and flat, unlike the other dural venous sinuses that tend to be triangular in shape (Swedenborg and Tafel 1887). It may lay up to 3 mm from the midline in normal patients (Cure et al. 1994). Veins typically enter the ISS just above the anterior portion of the corpus collosum, where the entrance forms an acute bend in the lateral projection, with the apex of the bend being the actual origin of the sinus itself (McCord et al. 1972). The ISS is cylindrical in form and usually increases in diameter as it passes posteriorly. In 15% of specimens, there is a bulbous termination, which continues caudally to form the rostral aspect of the straight sinus (Kaplan and Browder 1973). The inner aspect of the ISS is smooth except for an occasional slit-like orifice that joins venous channels from the falx cerebri. The ISS is absent in approximately 1% of cadavers (Kaplan and Browder 1973) and is often more pronounced in young children and infants than in adults. It may communicate with the SSS via veins that lie within the falx cerebri (Cure et al. 1994).

17.2 Transverse Sinus

The transverse sinuses (Figs. 17.8, 17.9 and 17.10) begin at the confluence of sinuses (torcular herophili). The majority of the torcular herophili and the very proximal transverse sinus are located at the internal occipital protuberance, and the majority of the inions are located inferior to the protuberance. Typically, the transverse sinus continues with the superior sagittal sinus on the right and the straight and occipital sinuses on the left. Thus, the right transverse sinus, which drains blood from the superficial part of the brain, usually is larger than the left, which drains from deep within the brain and is located inferior to the left in 53.3% of cases. Anatomical asymmetry of the transverse sinuses is common and is considered normal variation because unilateral hypoplasia or aplasia has been observed in 20% to 39% of cases. On each side, the transverse sinus passes along the posterolateral attached margins of the tentorium cerebelli, anterolaterally along the transverse sinus sulcus, reaching the posterolateral part of the petrous temporal bone, and then turns into the sigmoid sinus. The transverse sinus receives blood from the superior and inferior petrosal sinuses, cerebral, diploic, and labyrinthine veins.

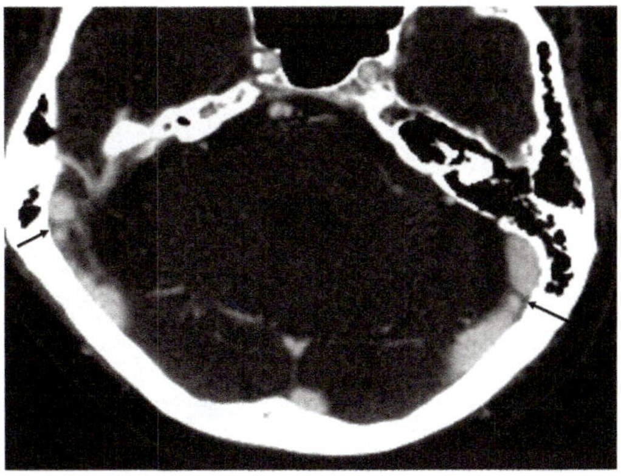

Fig. 17.8 Contrasted CT illustrating the transverse sinuses with bilateral septae (arrows)

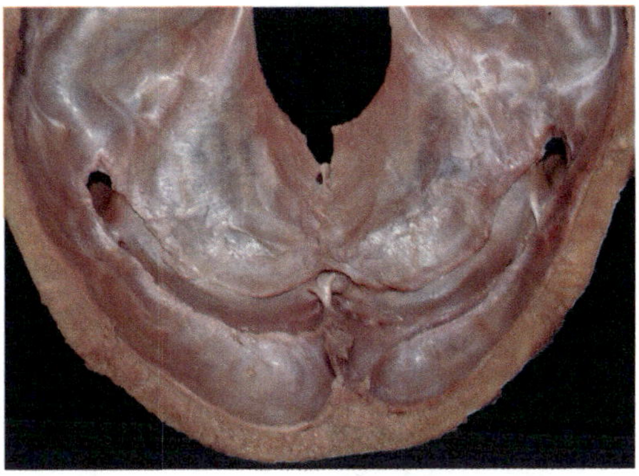

Fig. 17.9 The transverse sinuses as seen in a cadaveric dissection

Found that 53% of transverse sinuses included arachnoid granulations and trabeculae/septa, both of which could be potential venous filling defects, and are found more commonly on the right side. It is assumed that these fibrous bands or septa may act as wall reinforcements to prevent the sinus walls from dilating or collapsing. Similarly, a larger right transverse sinus needs additional reinforcement to overcome the high wall tension generated by the large amount of blood in the sinus.

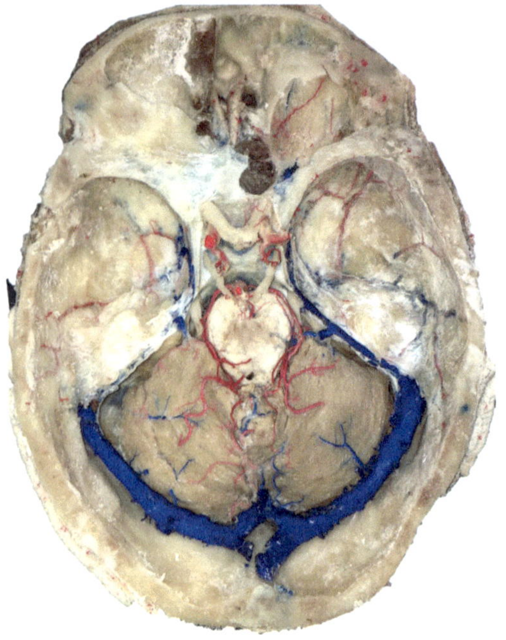

Fig. 17.10 Cadaveric dissection demonstrating the left and right transverse sinuses at the posterior border of the cerebellum

17.3 Sigmoid Sinus

The sigmoid sinus is a continuation of the transverse sinus and becomes the sigmoid sinus where the tentorium cerebelli ends (Figs. 17.11 and 17.12). It begins inferior to the temporal bone, posteromedial to the mastoid cells, and runs to the jugular foramen, extending through an S-shaped groove, where it meets the superior jugular bulb. Inconstant veins that drain into the sigmoid sinus include veins from the pons and medulla oblongata. The sigmoid sinus has anastomoses with the mastoid emissary and condylar emissary veins routinely.

The asterion is one landmark that has become a reference point used commonly. The asterion is the convergence point of the lambdoid, occipitomastoid, and parietomastoid sutures and is used because it overlays the transverse-sigmoid junction. However, many authors have found that the asterion is not a useful landmark because of its anatomical variability as well as the difficulty of identifying it. A study conducted described a new

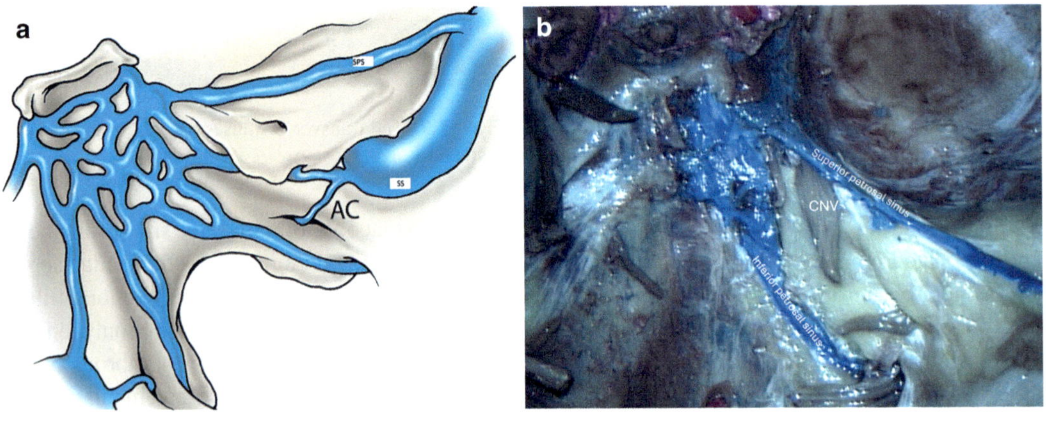

Fig. 17.11 (a) Schematic drawing illustrating venous sinuses of the skull base. Note the superior petrosal sinus (SPS) traveling along the superior aspect of the petrous part of the temporal bone, the sigmoid sinus (SS) draining into the jugular foramen, and an anterior condylar vein (AC) tributary. The basilar venous plexus and its connections to the marginal sinus are also well illustrated. (b) Cadaveric dissection noting the superior (1) and inferior petrosal sinuses (2)

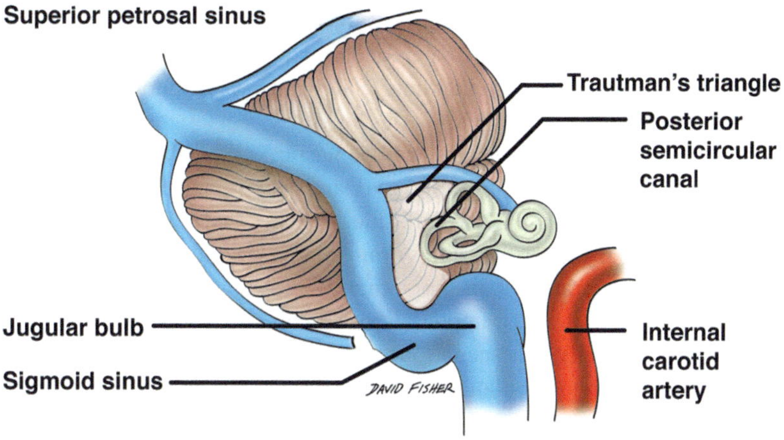

Fig. 17.12 Lateral view of the transverse, sigmoid, and superior petrosal sinuses

craniometric point to demarcate the sigmoid sinus, the digastric point, located at the top of the mastoid notch in the mastoid portion of the temporal bone. This landmark has become more useful than the asterion, because it lacks anatomic variability. The digastric point also is easy to locate in both dry skulls used in cadaveric studies and surgical procedures. In the dry skull, it is located immediately posterior to the mastoid incisures, while in dissections and surgical procedures, it is immediately above the posterior belly of the digastric muscle.

Detailed knowledge of anatomical crainometeric points is paramount. Being able to identify landmarks allows better exposure during surgical procedures and diagnostic mechanisms, all of which help reduce the likelihood of morbidity associated with any form of intervention, particularly surgical.

Variations

Although anatomical variations can lead to pathological manifestations, they should not be mistaken for pathologies in themselves. Knowledge of the anatomical variations of the sigmoid sinus is necessary not only for radiological diagnosis, but also when considering surgical or endovascular treatment of diseases at the skull base (Tanoue et al. 2010). Unlike the transverse sinus, the sigmoid sinus has fewer variations (Waltner 1944).

As stated, the sigmoid sinus occurs as a pair, one on either side. Many have had difficulty finding the correct landmarks to demarcate the sigmoid sinus because there are craniometric differences between the two sides of the same skull. It is not completely symmetrical, as one might believe. Ramos-Junior et al.'s (2014) study compared both sides of 50 dry adult skulls through the use of transillumination with LASER. 78% of the cases had a larger sigmoid sinus width at the level of the sinodural angle on the right side, thus demonstrating that the sigmoid sinus is larger on that side. Another variation discovered in the same study was that the jugular foramen width is also not symmetrical in all skulls, as it was found that the width is greater on the right side in most skulls.

In 1930, Williams (1930) found a unilateral absence of the sigmoid sinus. In 1940, he made another discovery, the complete absence of a venous sinus, which included the right sigmoid sinus.

Normally, the sigmoid receives blood from the transverse sinus, drains it into the internal jugular foramen, and straight into the internal jugular vein. However, this is not always the course in which drainage occurs. The venous drainage of the posterior cranial fossa follows asymmetric patterns with frequent anatomic variations, which helps maintain the independent development of its constituents, particularly the sigmoid and transverse sinuses. These sinuses usually exhibit lateral dominance, more so on the right side, but also can be subject to segmental variations, with greater anatomical constancy found in the sigmoid sinus (San Ruíz 2006). The drainage route also varies in cases where the sigmoid sinus is absent or hypoplastic. According to, hypoplasia of the sigmoid sinus is found most commonly on the left side. When this is the case, the transverse sinus drains into a mastoid emissary rather than the sigmoid, and the inferior petrosal sinus empties into a small jugular vein. Usually, this scenario is associated with the underdevelopment of the petrous part of the temporal bone.

In cases of transverse sinus atresia, or any form of malformation of the transverse sinus, the sigmoid sinus fills through the vein of Labbé. These are small, irregular intradural venous channels that help bypass atretic or hypoplastic segments (Curé et al. 1994). Collateral venous drainage also can occur through the emissary veins, occipital sinus, or vertebral venous system. When both the transverse and sigmoid sinuses are hypoplastic or absent, drainage occurs through the large superior petrosal sinus, which passes through the mastoid foramen (Williams 1930).

Variations in the jugular bulb include a jugular diverticulum, which appears as a fingerlike superior projection. In 1926, Rutting observed a diverticulum on the superior surface of the sigmoid sinus. These luminal out-pouches may lead to erosion into the internal auditory canal and cause tinnitus and sensorineural hearing loss. Sometimes, the sigmoid sinus can appear to be dilated aneurysmally, although this actually may be a normal variant in some people.

Conducted a study to help develop surgical classifications of the sigmoid sinus' variations in anatomy. They examined 96 temporal bones and concluded that they differed in the approximate location of the sigmoid sinus in relation to the mastoid cavity. Their classification included three types with respect to Trautmann's triangle. Trautmann's triangle is defined as the area bounded posteriorly by the anterior border of the sigmoid sinus, anteriorly by a bony labyrinth that connects the jugular bulb to the petrous apex, and superiorly by the petrous ridge. The classification is as follows:

- Type 1: Posterior sigmoid sinus causes an increase in the size of Trautmann's triangle (22.3%).
- Type 2: Anterior sigmoid sinus causes reduction of the area of Trautmann's triangle (57.5%).
- Type 3: Medial displacement of the sigmoid sinus causes a reduction of the area of Trautmann's triangle (20.2%).

A controversial issue in anatomical variation is the relation between the distance of the sigmoid sinus from the posterior border of the external auditory canal and the degree of pneumatization. The literature indicates largely that the distance is shorter in a well-pneumatized temporal bone. In these cases, the sigmoid sinus is typically situated medially. Occasionally, the vertical portion of the sigmoid sinus may follow an unusual path, coursing anteriorly and approaching the external auditory canal at its posterior wall.

According to, the sigmoid sinus in males is more posterior and lateral than in females. This is likely attributable to the difference in cranial size base between the sexes. When comparing either side, regardless of cranial width, the sigmoid sinus on the right side protrudes deeper, covering a greater cross-sectional area, and is situated more laterally. This may be correlated with the right side being the dominant side for venous sinus drainage also found that males have a right sigmoid sinus positioned more anteriorly compared to their left.

The transverse sinus is duplicated more frequently than the sigmoid sinus. This occurs through the formation of two channels separated by a bony ridge, fibrous septum, or dural sheath that causes a division of the lumen of the sinus for a shorter or longer distance (Waltner 1944). Another rare phenomenon is the contraction of the jugular foramen with the small internal jugular vein. Typically, this occurs on the left side and is associated with enlargement of the contralateral vein. The side affected is associated occasionally with an absent sigmoid sinus, transverse sinus, or both (Waltner 1944).

Occipital Sinus

The normal occipital sinus lies on the midline (Scott and Farb 2003a, b, c, d) and is positioned primarily at the base between the layers of the falx cerebelli (Fig. 17.13). It descends to the margins of the foramen magnum (Knott 1881a, b, c, d) and then extends from the foramen magnum up to drain in the torcular Herophili (Scott and Farb 2003a, b, c, d). Thus, the occipital sinus extends from the posterior margin of the foramen magnum to the confluence of sinuses, where it joins the marginal sinuses. Drainage toward the foramen magnum or into the jugular fossa or suboccipital veins also is possible (Scott and Farb 2003a, b, c, d). It also may deviate to one side, serve as a collateral channel for a small transverse sinus (Browder et al. 1976), or replace an aplastic transverse sinus (Curé et al. 1994), which is discussed further in the variations section.

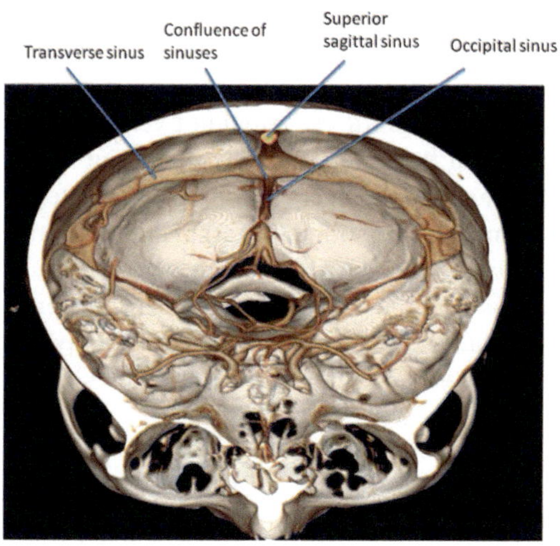

Fig. 17.13 CTA illustrating the confluence of sinuses and related sinuses

The occipital sinus is an ellipsoid structure and usually is a single channel, the size of which can range from 1 to 4 mm (Browder et al. 1976). It has been noted to be larger close to the confluence of sinuses, after which the breadth narrows as the sinus approaches the margins of the foramen magnum. Its length can vary between 15 and 70 mm, with an average length of 35 mm. The inner diameter measures between 0.33 and 7.06 mm at the midpoint, and the perimeter ranges from 0.77 to 17.61 mm.

The inner wall of the occipital sinus is tight and smooth and consists of a connection of fibrous bands, bridges, and septae that become denser as the sinus approaches the torcular Herophili. The breadth generally is wider in close proximity to the confluence of the sinuses and narrows steadily downward in the direction of the foramen magnum.

In cases in which multiple occipital sinus channels are present, they have been found to be close to each other and at the midline. The presence of multiple sinuses does not imply a larger size or diameter.

17.4 Torcular Herophili

The torcular Herophili (confluence of sinuses: CS) is a highly variable structure that constitutes a vital part of the network of dural venous sinuses. The CS is defined historically as the intersection of the superior sagittal sinus (SSS), the straight sinus (StS), the occipital sinus (OS) and the two transverse sinuses (TS) (Curé et al. 1994).

As a general rule, and with the CS as the point of reference, the SSS, OS, and the StS represent inflow tracts, while the two TS represent outflow tracts. The direction of flow observed in individual specimens depends heavily on the diameters of each of the sinuses, the specific variant of the CS present, and the presence of an intraluminal septum in the vicinity of the CS observed this intraluminal septum in 16.2% (5 of 31) of specimens and in 13% (7 of 52) of specimens.

Lamina chordae, a special type of chordae Willisii, were observed in some specimens examined, and separated the lumen into two channels.

The central aspect of the external occipital protuberance, or the inion, was believed previously to be a reliable external anatomical landmark with which to locate the CS. However, found that this marker was unreliable, and the internal occipital protuberance, which was located inferior to the inion in the majority (73.3%) of cases was a more reliable marker (Tubbs et al. 2001). The concavity at the internal occipital protuberance that houses the torcular Herophili is known as the torcular fossula or fossa of Zoja.

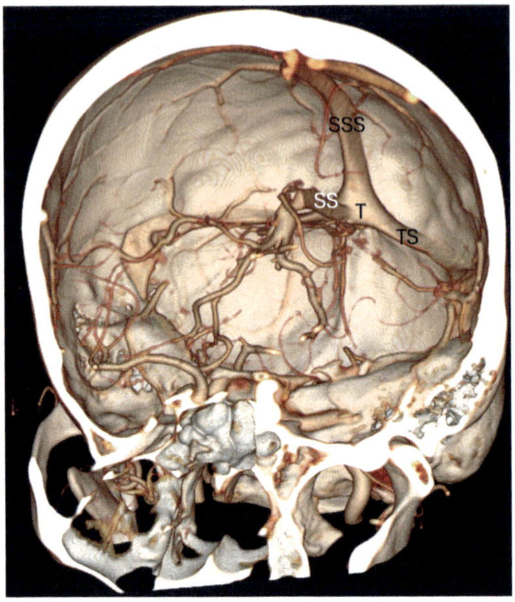

Fig. 17.14 The torcular Herlophili (T) shown on CTA. Also, note the straight sinus (SS), transverse sinus (TS), and superior sagittal sinus (SSS)

Straight Sinus

The straight sinus travels posteriorly and inferiorly beneath the splenium of the corpus callosum toward the internal occipital protuberance between the area where the falx cerebri meets the midline of the tentorium cerebelli (Figs. 17.6 and 17.14). The length of the straight sinus ranges between 4 and 7 cm, and its width at the terminal point ranges between 2 and 5 mm (Saxena et al. 1974). In 108 cadavers studied by Hasegawa et al. (1991), the length of the straight sinus was 51 ± 5.2. In a cross-sectional view, the straight sinus appears triangular (Newton and Potts 1974). The slope of the straight sinus varies with age with a normal adult tentorial angle between 27 and 52 degrees (mean 40) (Wolpert 1969).

The straight sinus will receive its major drainage from the great cerebral vein of Galen which passes caudally under the splenium of the corpus callosum to join it (Jinkins 2000a, b, c, d). The great cerebral vein of Galen joins the straight sinus at the anterior tentorial edge of the straight sinus at the level of the quadrigeminal plate. Major venous drainage also comes from the inferior sagittal sinus where it joins the anterior end after it has received its tributaries from the corpus callosum (Jinkins 2000a, b, c, d). Small contributions to the straight sinus have also been noted to come from the medial superficial cerebral veins of the occipital region, which drain centrally into the straight sinus (Jinkins 2000a, b, c, d). Although little is known about the posterior meningeal venous drainage due to its multiple variations, it has been proposed that it would logically contribute to drainage into the straight sinus as well as the tocular Herophili (Jinkins 2000a, b, c, d). The straight sinus receives a significant amount of venous drainage from the cerebellum. The superior vermian veins of the cerebellum passes anterior and medially to join the straight sinus (Jinkins 2000a, b, c, d). Arterial connections from as far as the posterior pericallosal, posterior cerebral artery, and collicular arteries have all been shown to supply the walls of straight sinus.

Variations

Although most often described as draining posteriorly toward the torcular Herophili, or commonly draining into the left transverse sinus and the superior sagittal sinus into the right transverse sinus, this arrangement can be reversed.

The straight sinus may be doubled or tripled (Saxena et al. 1973). When duplicated in a median position, one straight sinus is superior and the other inferior. The superior straight sinus represents a continuation of the inferior sagittal sinus and the inferior of the great cerebral vein. When duplicated, the two straight sinuses may either drain into the same or contralateral transverse sinuses. In paramedian position, both sinuses lie on either side of the midline. In most instances, the left represents a continuation of the great cerebral vein, and the right, the continuation of the inferior sagittal sinus. Each paramedian straight sinus drains into the ipsilateral transverse sinus (Saxena et al. 1973; Saxena et al. 1974; Browder et al. 1976; Magden 1991). Saxena et al. (1973) examined the straight sinus in 43 cadavers and found that 13.85% were doubled, with 9.3% in the median position and 4.65% in the paramedian position. In 56 human cadavers, Magden (1991) found the straight sinus in 26.7% of cadavers to be doubled, being either median in position [one superior to the other] (17.8%) or paramedian (both lying on one side of the midline at the junction of the falx cerebri with the tentorium cerebelli) (8.9%). In two cases (3.5%), the straight sinus was tripled, being either median in position or paramedian.

A triple straight sinus is rare, and when present, the sinuses are also described as median or paramedian. In both positions, a middle straight sinus is also found in addition to the superior and inferior or right and left straight sinus, which drain in either transverse sinuses (Magden 1991). Partial duplication of the straight sinus may also be found. It usually includes the anterior part of the sinus extending to the junction with the vein of Galen. Duplication of posterior and middle parts and partial tripling has also been observed (Browder et al. 1976).

The straight sinus can also be completely or partially absent. Van den Bergh et al. (1984) described a case of congenital absence of the straight sinus in a patient with Aarskog syndrome (x-linked recessive syndrome). Knott (1881a, b, c, d) also described one instance in which the sinus was completely absent. However, Hsu and Chaloupka (2012) reported a case of a 16-year-old man with a midline persistent falcine sinus (connecting the vein of Galen and the posterior one-third of the superior sagittal sinus) with the absence of only the proximal end of the straight sinus. The distal end of the straight sinus drained the cortical veins of the occipital lobe and the superior cerebellum. Kim and Lee (2010) reported two cases of persistent falcine sinus with the complete absence of the straight sinus in a 60-year-old woman and an 11-year-old girl. Ryu (2009) examined the CTA of 586 patients and found that 12 patients had a falcine sinus with one associated with a dysplastic tentorium. Three patients had an absent straight sinus, eight patients had a rudimentary straight sinus, and four patients had a normal sinus. In a case study by Desai et al. (2002), absence of the straight sinus and great vein of Galen was reported in one patient. This led to the internal vertebral veins and the basal veins of Rosenthal to drain into a large frontal interhemispheric falcine vein. In a rare case examined by, arachnoid granulations were found in the straight sinus of a 71-year-old female. A similar case was reported by MRI of arachnoid granulations of the straight sinus in a 19-year-old man who presented with dizziness (Shin et al. 2000).

In cases of absent inferior sagittal sinus, the straight sinus is formed by the continuation of the great cerebral vein. It may also communicate with the occipital sinus, the cavernous sinus through tentorial veins, or with the right transverse sinus (Saxena et al. 1974; Browder et al. 1976).

Hasegawa et al. (1991) examined the anatomic variations of the straight sinus and found that the angle of the straight sinus to the nasion-tuberculum sellae line was 51 ± 6.7 degrees; the angle to the tuberculum sellae-inion line was 44 ± 5.0 degrees, and the angle the nasion-inion line was 47 ± 5.0 degrees. The straight sinus coursed posteriorly at 1.4 ± 4.5 degrees to the bregma-inion line. The straight sinus was placed into three groups based on the relationship with the corpus callosum. They included the following:

- Type 1 (63%): The extended line of the slope of the straight sinus is tangential to the upper aspect of the splenium of the corpus callosum.
- Type 2 (26%)[high angle type]: The extended line was far above the tangent line.
- Type 3 (11%)[low angle type]: The extended line was far below the tangent line.

Multiple variations are noted in the caudal end of the straight sinus as it joins the torcular Herophili. Rare variations have even shown the inferior vermian veins to drain into the straight sinus (Lasjaunias et al. 2001).

Superior Petrosal Sinus

The superior petrosal sinus branches out of the posterosuperior aspect of the cavernous sinus and runs caudally beneath the oculomotor and trochlear nerves (Butler 1957a, b, c) and along the dorsal ridge of the petrous part of the temporal bone where the tentorium cerebelli is attached (Jinkins 2000a, b, c, d) to the petrous ridge (Rhoton Jr. 2002a, b, 2003a, b) (Figs. 17.11 and 17.12). The superior petrosal sinus also traverses the superior aspect of the trigeminal ganglion in the roof of Meckel's cave (Butler 1957a, b, c). It originates from a small cerebral vein derived from the basal surface of the hindbrain and empties into the pro-otic vein (Streeter 1915a, b, c). This long slender passage channel passes over the cochlear part of the otic capsule and empties the cavernous sinuses (Kilic and Akakin 2008) into the transverse sinus (Streeter 1915a, b, c; Butler 1957a, b, c). This passage runs in the posterosuperior margin of the pyramid of the sphenoid bone and acts as a connector between the cavernous and sigmoid sinuses (Stoltz 2006a, b), which extend to the transverse venous sinus from the cavernous sinus (Jinkins 2000a, b, c, d) from where they receive their blood supply as well as from the many veins that enter into them from the inferior and anterior parts of the cerebrum (Swedenborg 1887a, b). The anterior end of the superior petrosal sinus connects to the cavernous sinus near the proximal trigeminal nerve at the anterior aspect of the posterior clinoid process (Swedenborg 1887a, b). Conversely, stated that the medial end of the superior petrosal sinus is affixed to the posterior end of the cavernous sinus, with its lateral ends blending in with the junction of the transverse and sigmoid sinuses. The venous receptive function of the superior petrosal sinus is vast, consisting of the veins of the pons, upper medulla oblongata, cerebellum, and the inner ear (Jinkins 2000a, b, c, d). The superior petrosal sinus is responsible for draining the anterior portion of the cerebellum via the superior petrosal vein—one of the largest and most constant veins of the posterior fossa (Matsushima et al. 2015a, b) and the vein of the lateral recess or cerebellomedullary fissure (Kilic and Akakin 2008). The veins also receive the contents of the circular, cavernous, and anterior occipital sinuses via little trabeculae, which are valvular and emanate from the internal wall of the sinus. They also act in concert with the inferior petrosal sinus to drain the interior of the bones of the skull. In general, the veins that enter the superior petrosal sinus emerge from the cerebellum and brain stem (Rhoton Jr. 2002a, b, 2003a, b).

This sinus is usually larger posteriorly and becomes smaller near its apex (Tubbs et al. 2013a, b, c). It receives drainage from the cerebellum, the inferior cerebral veins, and the veins of the tympanic cavity (Tubbs et al. 2013a, b, c).

The superior petrosal sinus can be divided into three types: (1) medial if it empties into the cavernous sinus, (2) lateral if it empties into the transverse–sigmoid junction, and (3) if it empties into both (Matsushima et al. 2014, 2015a, b).

The relationship between the superior petrosal sinus and Meckel's cave also known as the porus trigeminus varies. Tubbs et al. (2013a, b, c) reported three types of sinus courses in relation to porus trigeminus, which was found to be inferior to, or around the opening of Meckel's cave, in over 30% of the specimens and 68% of specimens (Tubbs et al. 2013a, b, c). The superior petrosal sinus traverses in a somewhat triangular form via the duplication of the dura mater over the base of the cranium. This matter is impressed upon the temporal bones as they open into the cavernous sinuses and terminate in the transverse sinus.

The superior petrosal sinus can be completely absent or enlarged, passing into the mastoid foramen in order to compensate for a small or absent transverse sinus (Knott 1881a, b, c, d; Williams 1930; Waltner 1944). A communication between an ophthalmic vein and the superior petrosal sinus has been described (Knott 1881a, b, c, d). Theile also noted an aberrant termination of the superior petrosal sinus into the foramen spinosum of the middle cranial fossa. On rare occasions, the basal vein of Rosenthal may drain into the superior petrosal sinus (Curé et al. 1994). The vein of Labbé may drain into the superior petrosal sinus, which can make surgical approaches at the skull base extremely difficult or impossible. Seven percent of the 14 fixed cadaver heads examined by showed the vein of Labbé draining directly into the superior petrosal vein. The superior petrosal sinus may be absent (Knott 1881a, b, c, d; Hyrtl 1862). Matsushima et al. (2015a, b) also showed in their study of one venogram and another anatomic specimen a case where the superior petrosal vein and sinus were absent, thus necessitating that the drainage of the posterior cranial fossa be done via the vein of Galen, the lateral mesencephalic vein, and other tentorial groups of veins.

Tubbs et al. (2013a, b, c) observed different arrangements of the superior petrosal sinus from studying the relationship between the superior petrosal sinus and the porous trigeminus. In the 25 (50 sides) cadaver heads, 68% of the superior petrosal sinuses were superior to the opening of Meckel's cave, 18% were inferior, and 16% surrounded the opening of Meckel's cave. It was also observed that the superior petrosal sinuses that coursed inferiorly to the porous trigeminus were generally smaller than those that coursed superiorly or around the opening of Meckel's cave (Tubbs et al. 2013a, b, c).

Coates (1934) examined 31 cadavers (63 sinuses identified) to determine the relationship of the superior petrosal sinus to the trigeminal nerve. In eight specimens, both right and left sinuses passed over the cranial surface of the trigeminal nerve root. Both right and left sinuses passed under the trigeminal nerve root in three cases. The sinus divided on both sides in three cases with some portions passing under the trigeminal root and others passing over the trigeminal nerve root.

According to Knott, the superior petrosal sinus may occasionally communicate with the superior ophthalmic veins through an aberrant vein termed the ophthalmo-petrosal sinus in up to 9% of the cases (Bergman et al. 1988a, b). The foramen of the ophthalmo-petrosal sinus of Hyrtl is rarely seen in humans and may vary in size and be found on one or both sides (Shapiro and Robinson 1967).

Tanoue et al. (2006a, b) have classified the proximal communication of the superior petrosal sinus into three types. In type A, which is the most common, the superior petrosal sinus runs along the lesser sphenoid wing and connects with the anterior aspect of the cavernous sinus.

In type B sinuses, the superior petrosal sinus runs along the lesser sphenoid wing and connects with the foramen ovale venous plexus or the pterygoid venous plexus, in addition to the anterior aspect of the cavernous sinus.

In type C sinuses, a hypoplastic superior petrosal sinus connects with the anterior aspect of the cavernous sinus (Tanoue et al. 2006a, b). On other occasions, the superior petrosal sinus communicates via an aberrant vein with the superior ophthalmic vein (Bergman et al. 1988a, b). The superior petrosal sinus can be completely absent (Waltner 1944).

Streeter (1915a, b, c) stated that the pro-otic vein of Markowski is one in the same as the superior petrosal sinus, although there are dissenting views.

The sphenopetrosal sinus, an infrequent tributary of the superior petrosal sinus that collects drainage for the sylvian veins, may be a variation of the superior petrosal sinus.

Matsushima et al. (2015a, b) in their work on the classification of the superior petrosal sinus and veins noted that the obliteration of the superior petrosal sinus is a variation that may have a congenital or iatrogenic cause. Iatrogenically, this can result post-transpetrosal surgery. This result occurs when the superior petrosal group of veins and their tributaries are compromised, which poses an issue as these veins, most notably the veins of the cerebellopontine fissure, serve as a major drainage pathway into and from the superior petrosal sinus. This occurrence is usually seen after surgery for cerebellopontine angle meningiomas such as petroclival, tentorial, anterior petrous, and posterior petrous tumors (Watanabe et al. 2003a, b).

17.5 Inferior Petrosal Sinus

The inferior petrosal sinus can drain intra or extracranially. Intracranial drainage has been reported in roughly 20% of cases. It rests in the lower parts of the foramen and has no relationship with the jugular foramen in about 7% of people (Lang 2012). It originates from the side of the cavernous sinus projecting between the sphenoid and temporal bones (Swedenborg 1887a, b) (Fig. 17.15).

Along its course, it receives venous drainage from the internal auditory vein, medulla oblongata, pons, and inferior cerebellar surface (Mortazavi et al. 2013, 2014). It is the major drainage of the cavernous sinus, with numerous branches communicating with the anterior condylar vein (ACV), lateral condylar vein (LCV), basilar plexus, and the vertebral venous plexus (VVP) (Zhang et al. 2010a, b). The size of the IPS increases as it progresses through its path, bending enough to leave its imprints on both the temporal and sphenoid bones and entering into both the cavernous sinus and the jugular vein, where they collect the contents of the lateral sinuses (Swedenborg 1887a, b). The size ranges in diameter from 2.85 mm in its lower part to about 7 mm–10 mm through its many channels (Lang 2012).

The inferior petrosal sinus (IPS) is housed in the petrooccipital suture or groove and runs from the cavernous sinus to the internal jugular vein, close to the basilar artery (Stoltz 2006a, b; Swedenborg 1887a, b). It is posterior and superior to the internal carotid artery and carries blood to the internal jugular vein (Dolenc 2012a, b). The IPS enters the jugular foramen between CNs IX and X, while draining the basilar venous plexus. It also lies between the petrous apices and the clivus along the path of the petrooccipital fissure.

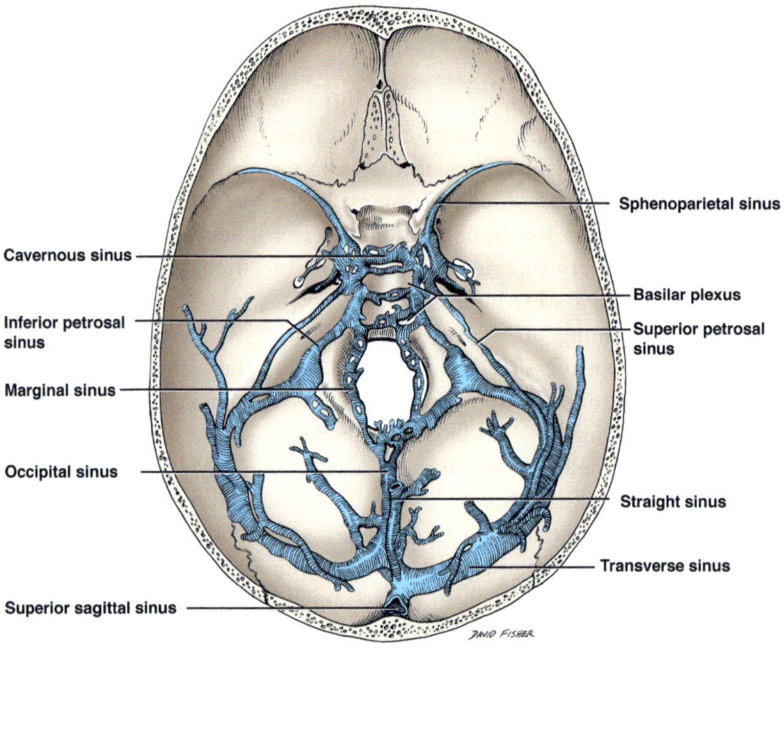

Fig. 17.15 Drawing of the dural venous sinuses of the skull base

Variations

Osborn and Jacobs (1999a, b) stated that it is common for the inferior petrosal sinus to display quite a number of variations in its appearance, ranging in size, anastomoses, and relationship to the nerves that surround it as they enter the jugular foramen (Osborn and Jacobs 1999a, b). The variations of this sinus depend on the level of dominance of anastomosing landmarks associated with the sinus. Any one of these may be more dominant, but in the majority of people, these anastomotic channels/landmarks are symmetrical. However, 1.0% of people do not have an inferior petrosal sinus that connects with the internal jugular vein (Miller et al. 1993). Despite the variance in the anatomy of the anastomoses between the inferior petrosal sinus, the internal jugular vein, and the venous plexuses at the base of the skull, it is still almost always possible to catheterize the inferior petrosal sinus.

Variations in the Level of Termination

The normal anatomy of the inferior petrosal sinus is discussed in the section of this chapter titled Anatomy. In brief, the inferior petrosal sinus forms a connection with the posteroinferior portion of the cavernous sinus. It then follows an approximately posterolateral course along the petrous ridge to drain into the internal jugular vein at the jugular bulb.

Gailloud et al. (1997) described a variant of the inferior petrosal sinus with respect to its point of termination. In the cadaver of an 85-year-old male, the IPS was noted to have its point of drainage into the IJV approximately 4.0 cm below the jugular bulb, which is its normal point of termination. This was generally the same bilaterally in this patient, with minor differences. Among these differences were the exact distance below the jugular bulb that the IPS joined the IJV, the number and size of the tributaries between the IJV and this extended version of the IPS, and finally the thickness of the connective tissue separating these two vessels.

In order to have an idea of the prevalence of this variant, Gailloud et al. (1997) observed 100 consecutive routine neuroradiological angiographies. This low-terminating version of the IPS was observed in four of these cases (the drainage pattern in the other 96 cases was not mentioned). Benndorf and Campi (2002) reported two similar cases in patients undergoing endovascular management of cavernous sinus lesions.

In the 27 cases evaluated by, three different locations of the junction of the IPS and the IJV were identified. In four cases, the junction was located within the jugular foramen. In eight cases, it was found in the extracranial opening of the jugular foramen. Its location was below the skull base in the remaining 15 cases.

The work of Knott (1881a, b, c, d) was carried out on 22 sinuses. Here, the level of termination of the IPS was near the lower margin of the jugular foramen in eight sinuses. In nine sinuses, it was slightly above the jugular foramen, and in five sinuses, it was slightly below the jugular foramen.

The level of the IPS-IJV junction was classified into six main types by Mitsuhashi et al. (2007) based on a study of 3D Rotational Venograms of 83 venous sinus sides in 63 patients. In Type A (seen in one side), the IPS drained directly into the superior part of the jugular bulb. In Type B (seen in 29 sides), the point of drainage of the IPS was at the level of the extracranial opening of the hypoglossal canal. The types where the IPS drained into the lower extracranial IJV were classified as Type C (seen in 31 sides). Type C was sub-classified into Type C-1, where the IPS had a single communication with the IJV and Type C-2 where another upper venous route existed. In Type D (seen in five sides), the IPS adopted a plexiform architecture and multiple connections between the IPS and the IJV existed around the jugular foramen. In Type E (seen on three sides), the IPS drained directly into the vertebral venous plexus without communicating with the IJV. The IPS was absent in Type F (seen in 14 sides). Types D and E correspond, respectively, to Types III and IV as mentioned by Shiu et al. (1968).

If one were to also consider the patterns of anastomosis between the IPS and the sigmoid sinus, anterior condylar vein, lateral condylar vein, and the vertebral venous plexus, then several further subdivisions of the six categories named by Mitsuhashi et al. can be obtained (Zhang et al. 2010a, b). Type A can be divided into A1, where the IPS drains directly into the jugular bulb, and Type A2, where the IPS connects with the sigmoid sinus and the jugular bulb via two branches. Type B was subdivided into two categories: B1 where the IPS drained into the IJV at the hypoglossal canal, and B2 where it also had a communication with the sigmoid sinus. To Type C, the subtype C3 was added in which one branch of the IPS drained into IJV at a lower extracranial level either alone or with a higher branch draining into the IJV or jugular bulb. Also present in C3 was a communicating branch with the sigmoid sinus. Type D was divided into two subtypes: D1 in which the plexiform IPS drained into the IJV near the jugular foramen, and D2 that was plexiform and possessed a communicating branch with the sigmoid sinus.

Variation in IPS-IJV Communication

In 1968, Shiu et al. wrote on their experience with cavernous angiography. From the venography of 53 patients, they commented on the variations of the IPS. They found four distinct patterns of drainage into the IJV as follows (Fig. 17.16):

1. Type I (45% of patients): This is the textbook version, in which the IPS drains directly into the IJV.
2. Type II (24% of patients): In these cases, the IPS drained into a communicating vein that united the internal jugular bulb with the deep cervical plexus.
3. Type III (24% of patients): In Type III variants, the IPS exists essentially as a plexus of veins. The cavernous sinus drains into various locations via this plexus.
4. Type IV (7% of patients): The IPS, in these cases, is well formed. However, instead of draining into the jugular bulb, it drains into the deep cervical plexus of veins.

The results obtained by Miller et al. (1993) had slightly different results. Type II, as defined above, was the most common type. The prevalence of Type IV in their study was only 1 in 133 subjects in their series.

Miller et al. (1993) mentioned an observed architecture described as a variant of Type II and Type III and termed as an incomplete Type IV. The anatomy of this type is characterized by a normal-sized IPS as it leaves the cavernous sinus. The channel, however, is very small from the IJV to the point where the IPS anastomoses with the vertebral venous plexus. The IPS may otherwise exhibit Type II or Type III anatomy. This incomplete Type IV was found in 2.6% of the 268 evaluated IPSs.

With regard to the clinical significance of these variants, Type I and Type II permit sufficient filling of the cavernous sinus and orbital veins during venography. In Type III, although a venous plexus now replaces the IPS, when the contrast is injected at the IPS, the cavernous sinus is usually outlined. In Type IV, investigation of the cavernous sinus via the IPS is not possible.

Variations in the IPS Course

The course of the IPS at the entrance to the jugular foramen relative to the location of the glossopharyngeal nerve may be variable. Rubinstein et al. (1995a, b) investigated these variations using 25 specimens, in which 22 of the IPS passed as inferior to the horizontal portion of the glossopharyngeal nerve at its entrance into the jugular foramen. In the remaining three specimens, the IPS passed anterior and medial or medial to the descending portion of the glossopharyngeal nerve.

Variations of the course within the jugular foramen were also examined in 20 specimens by Rubinstein et al. (1995a, b), where they noted that the IPS took a route between glossopharyngeal and vagus/accessory nerves in 16 cases. In four of the cases, IPS passed medial to all three cranial nerves.

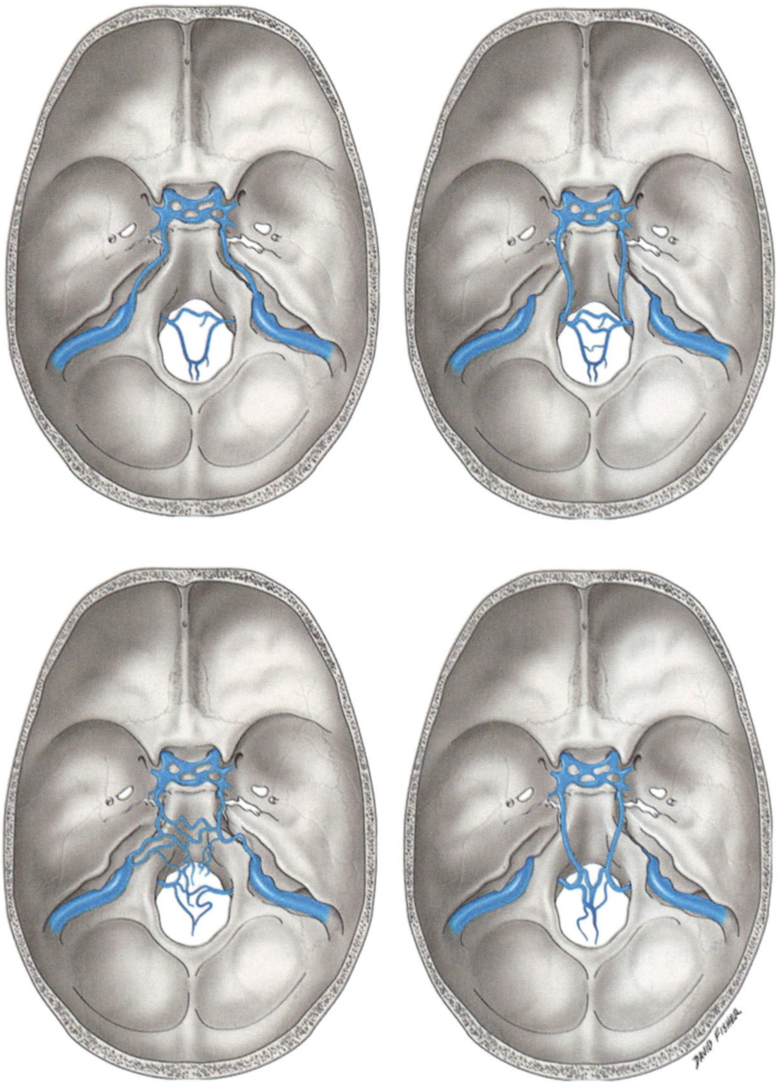

Fig. 17.16 Variations of the inferior petrosal sinuses

17.6 Basilar Venous Plexus

The basilar venous plexus consists of interconnecting channels that join both the inferior petrosal sinuses and the internal vertebral venous plexus (Krayenbühl et al. 1982; Tubbs et al. 2007a, b, c, d) (Fig. 17.17). The communicating channels lie between the dural sheets above the dorsum sellae of the sphenoid bone and the bony skull base. With the largest and most constant connection being between the cavernous sinus and the pars basilaris of the sphenoid bone. Lower portions of the plexus connect with both the inferior petrosal sinuses and with the anterior internal vertebral venous plexus (Benndorf 2010). Posteriorly, the sinus communicates with the sinus intercavernosus and sometimes with the superior petrosal and cavernous sinuses anteriorly (Curé et al. 1994). In a study by Tubbs et al. (2006a, b), using twenty latex injected cadavers, they found the plexus in all specimens and that it was always plexiform and very variable in nature. This structure was dorsal to the clivus superiorly and dorsal to the tectorial membrane inferiorly. The diameter of the channels forming the basilar plexus ranged from 0.5 mm to 1.6 mm (mean 1.1 mm). The basilar venous plexus more or less covered the entire dorsal aspect of the clivus. The

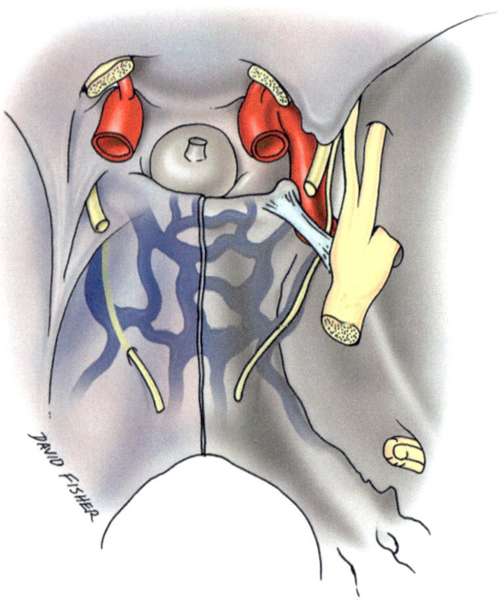

Fig. 17.17 Drawing illustrating the midline basilar venous sinus

thickness of this venous sinus was approximately 1 mm in all specimens. Communication was always found between the basilar plexus and the inferior petrosal sinuses, which were the main outflow of this system of veins. In fact, these two venous structures were more or less contiguous across the midline at multiple levels. In seven specimens (35%), the basilar plexus communicated with veins draining into the hypoglossal canal. The basilar plexus communicated with the marginal sinus in twelve specimens (60%). We found visible connections between the internal vertebral plexus and the basilar plexus in two specimens. No specimen was found to have connections with veins of the anterior brain stem or ventral surface of the clivus. No direct connections between the basilar plexus and occipital sinus were observed. The concentration of venous channels in the basilar plexus decreased as this structure descended inferior to a plane between the left and right jugular tubercles.

If the marginal sinus is enlarged, it may also communicate anteriorly with the basilar plexus (Tubbs et al. 2006a, b) or as reported, with the anterior condylar vein (venous plexus of the hypoglossal nerve) in approximately one-fourth of the studied sides. The basilar plexus may also indirectly communicate with the pterygoid plexus and facial veins through the cavernous sinus and superior ophthalmic vein, respectively, or with the suboccipital cavernous sinus (Arnautovic et al. 1997). Hollinshead and Rosse (1985) and have written that this venous sinus may also receive small veins from the brain stem or cerebellum. Additionally, Takahashi et al. (2001) reported a dural arteriovenous fistula that venography revealed communication between the right transverse pontine vein and the basilar plexus. There may also be transosseous connections between the basilar plexus and the inferior side of the clivus via the basilar canal when present. Lang (1991) in his textbook regarding the anatomy of the posterior cranial fossa also described communications of the basilar plexus with the venous network of the hypoglossal canal, condylar emissary veins, and rarely (~2%), transclival veins to the inferior aspect of the skull base.

Venous tributaries from the pons and medulla oblongata drain directly into the basilar plexus. It was found that during expiration more cranial venous blood exits the cranium via the basilar vertebral plexus when compared to the internal jugular veins (Tubbs et al. 2007a, b, c, d). This becomes even more vital when considering arteriovenous fistulas and secondary increases in venous blood pressure with pooling of blood.

Eckenhoff (1971) has pointed out that the basilar plexus is important due to its connection to the internal vertebral venous plexus of Batson, so that during expiration, more cranial venous blood exits the cranium via the basilar/vertebral plexus compared to the internal jugular veins. McDougall et al. (1997) and Ernst et al. (1999) postulated that the basilar plexus is linked to the occipital sinus via the intervening marginal sinuses encircling the foramen magnum. In an earlier study, we found that the marginal sinus communicated with the basilar venous plexus in 80% of specimens (Tubbs et al. 2006a, b) stated that the venous plexus of the hypoglossal canal communicates the basilar venous plexus with the marginal sinus depicted a basilar plexus that ran in parallel with the inferior petrosal sinus and connected the cavernous sinus and jugular bulb together.

The internal vertebral (rachidian) venous plexus in humans is already well developed at 4 months gestation (Okudera et al. 1994). This is the primary channel for cerebral venous drainage in an upright posture (Falk and Conroy 1983) and can provide for the entire drainage of venous blood from the cranial cavity described the basilar venous plexus as a continuation of the cranial part of the anterior internal vertebral venous plexus. However, found connections between these two plexuses in only 37.5% of the studied sides. We found visible connections between the internal vertebral plexus and the basilar plexus in the minority of our specimens. Padget (1956a, b, c, d) commented that the basilar plexus is a secondary anastomosis between bilateral embryonic channels near the midline. This plexus is most likely derived from the primary head vein postulated that the development of the posterior fossa dural sinuses partly occurs in the postnatal period following postural changes.

The basilar plexus is found in other species such as the baboon. In fact, the basilar plexus is more extensive in many primates and coalesces into two paired sphenoidal sinuses that drain caudally into the basisphenoid sinus of these species. Studies involving the inferior petrosal sinuses have depicted these sinuses as not communicating with the basilar plexus (Shiu et al. 1968) demonstrated that the basilar plexus receives few branches from the ipsilateral inferior petrosal sinus and connects inferiorly with the anterior spinal epidural plexus. We found that this basilar plexus communicated with the left and right inferior petrosal sinuses freely in all of our specimens and formed the main route for venous egress of the basilar plexus out of the cranium.

The basilar plexus may be visualized with dynamic computerized tomography (Bonneville et al. 1989) and is seen best on post-contrast axial and sagittal MRI. The basilar plexus can be sizable as illustrated by found ipsilateral basilar plexus enlargement in a patient with carotid-cavernous fistula. The transvenous embolization of a dural arteriovenous fistula at the right jugular bulb has been described as being approached through the basilar plexus via the left inferior petrosal sinus, in one patient, approached a carotid-cavernous fistula utilizing the basilar plexus. The basilar plexus may occasionally be a source of erroneous blood sampling of ACTH if used instead of the inferior petrosal sinus for lateralization of pituitary microadenomas. We found that the basilar plexus was more or less confluent with the inferior petrosal and cavernous sinuses in all specimens. Interestingly, Pernkopf (1963) has depicted and distinguished the more superior basilar “plexus” communicating between the inferior petrosal sinuses from the more inferior basilar “sinus” shown as a single, midline, small, vertical channel extending into the vertebral canal.

The role of the basilar plexus in providing venous collateral drainage remains to be fully understood found that after injection of contrast, the superior sagittal sinus, confluence of sinuses, and straight and petrosal sinuses are well filled but only a portion of the basilar plexus was outlined. The anastomoses between the intracranial and vertebral venous plexus, the so-called cerebrospinal venous system (CSVS), may contribute to pressure homeostasis of the intracranial venous system with changes in posture and intracranial pressure values (Arnautovic et al. 1997). Classically, postulated that these anastomoses may serve to spread tumors, infection, or emboli from the pelvis or spine to the brain. Enhanced venous outflow via the basilar plexus may also result in the formation of spinal epidural varices in such cases as internal jugular vein thrombosis/stenosis or superior vena cava obstruction. As the basilar plexus is located cephalad to the internal vertebral venous plexus and is connected to the other intracranial venous sinuses, one may hypothesize that it is, developmentally, an intermediate portion of the CSVS. The development of a vascular network is mainly the work of local hemodynamic forces that result in the selection and persistence of a particular group of channels and the formation of complex vascular variations.

We found only very small connections between the basilar plexus and internal vertebral plexus. This could be due to the very small diameter of the vessels contributing to this latter venous system.

17.7 Marginal Sinus

The marginal sinus communicates anteriorly with the basilar venous plexus on the superior surface of the inferior clivus and posteriorly with the occipital sinus (de Oliveira et al. 1985) (Fig. 17.15). The marginal sinuses may also connect the distal portions of the sigmoid sinuses (via small connections that pass across the intracranial surface) and communicate with the veins of the hypoglossal canal (anterior condylar confluence of Trolard or Luschka) via the condylar sinus of Labbé (de Oliveira et al. 1985). Distally, the marginal sinus drains freely into the vertebral venous plexus (Kimbel 1984). The occipital sinus begins at the junction of the left and right posterior parts of the marginal sinus (Go et al. 1974). However, the occipital sinus may not connect to the marginal sinus (Dora and Zileli 1980).

In 15 adult cadavers, Tubbs et al. (2006a, b) identified the marginal sinus in all.

Sinuses were found to lie between the leaves of the dura mater at the superior margin of the foramen magnum in all specimens. The maximal vertical height of the sinuses ranged from 7 to 15 mm (mean 10 mm), and in all specimens, it was located at or near the foramen magnum where the accessory nerve crossed in route to the jugular foramen. In all specimens the sinus was noted to taper to 3–5 mm as it traveled both anteriorly and posteriorly toward the basion and opisthion. The marginal sinus communicated with the basilar venous plexus in 12 of 15 specimens (80%) and with the occipital sinus in all specimens. Fourteen of 15 specimens (93%) were found to have drainage into the veins of the hypoglossal canal. The hypoglossal nerve rootlets were noted to pierce the sinus and its tributaries in 11 of 15 specimens (73%). Venous communication to the sigmoid sinus was seen in all specimens. The vertebral artery was found to course through the marginal sinus as it pierced the posterior atlantooccipital membrane on all left sides and 13 of 15 (87%) of right sides. Caruso et al. (1999) have reported that the marginal sinus had a mean diameter of 2 mm in 14 patients undergoing magnetic resonance imaging (MRI).

17.8 Cavernous Sinus

The cavernous sinuses are a pair of intradural venous compartments located centrally within the skull in the anteromedial portion of the middle cranial fossa, and are bordered medially by the pituitary gland and the body of the sphenoid bone (Fig. 17.18). The cavernous sinus is a four-walled structure with a roof, and medial, lateral, and posterior walls in the shape of a "boat" as Rhoton described. The superior wall or the roof of the sinus consists of an anterior portion that is formed by the dural lining of the clinoid process and a posterior portion formed by the "oculomotor triangle," which is a dural patch through which the oculomotor penetrates the sinus. The lateral wall is formed of two dural layers: a thick external layer and an internal, semi-transparent layer. It extends from the superior orbital fissure anteriorly to the medial border of Meckel's cave posteriorly and from the lower border of the carotid sulcus inferiorly to the anterior petroclinoid ligament superiorly. The medial wall consists of two parts, sellar and spheroidal, both of which consist of one dural layer. The borders of the medial wall extend from the superior orbital fissure anteriorly to the lateral border of the dorsum sellae posteriorly and from the lower border of the carotid sulcus inferiorly to the interclinoid ligament superiorly. The posterior wall consists of two dural layers, the periostic and meningeal. The cavernous sinus shares part of its wide posterior wall with the lateral

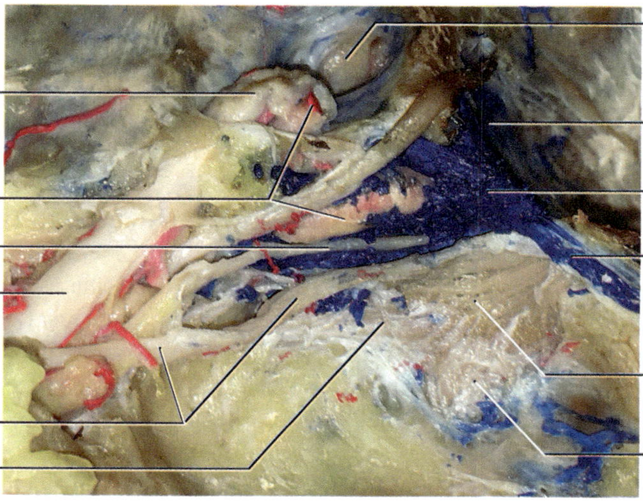

Fig. 17.18 Cadaveric dissection of the lateral view of the left cavernous sinus and its contents

border of the basilar sinus, and the remaining portion of that wall can be divided into upper, lower, medial, and lateral margins. The upper margin lays at the level of the petroclinoid ligament, while the lower is located at the upper border of the petroclival fissure above the petrous apex. The medial margin of the posterior wall of the cavernous sinus lays at the lateral aspect of the dorsum sellae, while the lateral margin is located medial to the ostium of Meckel's cave (Rhoton Jr. 2009; Campero et al. 2013).

The cavernous sinus is more than just a compartment filled with venous blood; in fact, it is a complex neurovascular structure in which fibrous and adipose tissues fill the space. Four cranial nerves are related directly to the cavernous sinus: the oculomotor (CN III), trochlear (CN IV), trigeminal (CN V), and the abducens nerve (CN VI). The lateral wall of the cavernous sinus contains CNs III and IV in its upper part, and both the ophthalmic and maxillary branches of the trigeminal (V_1 and V_2) in the lower part of the wall. The posterior wall is related directly to CN VI, while the roof of the cavernous sinus has the trigeminal triangle through which CN III pierces the dura, joins CNs IV, V_1, and VI and exits the cranium through the superior orbital fissure, which leaves V_2 as the only cranial nerve that exits the skull through the foramen rotundum (Rhoton Jr. 2009; Campero et al. 2013; Bakan et al. 2015). The internal carotid artery (ICA) is another significant component of the cavernous sinus neurovascular complex. The cavernous ICA is an S-shaped vessel that lies within the dural layers that form the sinus; its short course begins with its ascent toward the posterior clinoid process to form an upward convexity before it perforates the dural layers of the superior wall medial to the anterior clinoid process. In its short journey through the cavernous sinus, the ICA gives rise to three branches: The meningohypophysial artery that provides branches to regional cranial nerves and walls of the sinus, the inferior cavernous sinus artery, and the capsular arteries (Wallace et al. 1967).

The cavernous sinus has several venous connections that add further clinical significance to the plexus of veins in the lateral sellar compartment and can be considered the venous crossroads in the middle of the skull. The middle meningeal veins connect the cavernous sinus to the dura, while the ophthalmic veins connect it to the facial veins. The cavernous sinus is associated directly with other sinuses as well; it communicates with the transverse sinus via the superior petrosal sinus and to the internal jugular vein through the inferior petrosal sinus. The superior and inferior ophthalmic veins connect the cavernous sinus to the orbit, the middle and inferior cerebral veins connect it to the cerebral hemispheres, and the retinal veins link the sinus to the retina. The emissary veins perforate the skull to give the cavernous sinus an opportunity to communicate with the outside world by connecting it to the pterygoid venous plexus, which drains eventually into the facial vein. The small anterior and posterior intercavernous sinuses and the basal plexus veins connect both cavernous sinuses at the midline (Keller et al. 2009; Bakan et al. 2015).

Tubbs et al.'s (2008) study showed that the maxillary division of the trigeminal (V2) is not part of the cavernous sinus. Their study, which used 10 fresh cadaveric heads, found no evidence that V2 travels within the sinus itself; however, they demonstrated that the emissary veins traveling with V2 could cause that confusion.

Emanuel Swendenborg gave one of the most detailed accounts of the cavernous sinus in the nineteenth century. Swendenborg referred to the cavernous sinuses as the "receptacles of the sphenoid bone" because they filled the grooves of the sphenoid bone. The clinoid processes—the anterior and the posterior sinuses—block the sinuses on each side. The sinuses resemble the cavernous substance of the spleen and the urethra, which probably led to the name that other authors disputed later. He also mentioned the presence of CNs III, IV, and VI, as well as part of the carotid artery within the cavernous sinus. Swendenborg also outlined the accounts of several "modern" authors. Vieussens made a very important observation of the presence of a connection between the cavernous sinuses on both sides, which he discovered by injecting liquor on one side and then finding it on the other. On the other hand, Ridley did not agree with many of Vieussens' observations and suggested that the circular sinus was responsible for returning the blood to the circulation rather than the veins accompanying the carotid artery. Further, Swendenborg highlighted Winslow's comments that we mentioned earlier, in which he described the cavernous sinus as a spongy substance filled with blood.

17.9 Intercavernous Sinuses

The intercavernous sinuses connect the left and right cavernous sinuses and occupy the meningeal dural layer covering the pituitary gland and the endosteal layer covering the floor of the sella turcica (Aquini et al. 1994a, b, c; Dolenc and Rogers 2009a, b; Hong et al. 2010) (Fig. 17.19). These can be further classified as anterior, posterior, and inferior.

The anterior intercavernous sinus is said to be present in most cases and may cover the entire anterior aspect of the sella turcica (Aquini et al. 1994a, b, c; de Divitiis et al. 1981; Renn and Rhoton Jr. 1975a, b). Various reports have described it as being larger (Dolenc and Rogers 2009a, b; Kaplan et al. 1976) and longer (Krayenbuehl et al. 1982) than the posterior intercavernous sinus, which is located along the superior border of the dorsum sellae. One study found this structure in sagittal sections of all studied anatomic specimens (Aquini et al. 1994a, b, c). Some authors have suggested that the posterior intercavernous sinus is more variable and less frequently present and, under nonpathological conditions, difficult to distinguish from the basilar plexus (Renn and Rhoton Jr. 1975a, b; Schnitzlein et al. 1985; Theron et al. 1979). An inferior intercavernous sinus may be present (Kaplan et al. 1976; Tanabe et al. 1984); however, one venographic study did not identify this structure (Waga et al. 1970). The intercavernous sinuses are important clinically and surgically. Kaplan et al. (1976) emphasized the possible complications of encountering a sizable intercavernous sinus with transsellar approaches to the sella turcica. Additionally, these interconnecting sinuses may be involved with fistulas between the internal carotid artery and cavernous sinus (Kurata et al. 2011), may convey infection to the contralateral cavernous

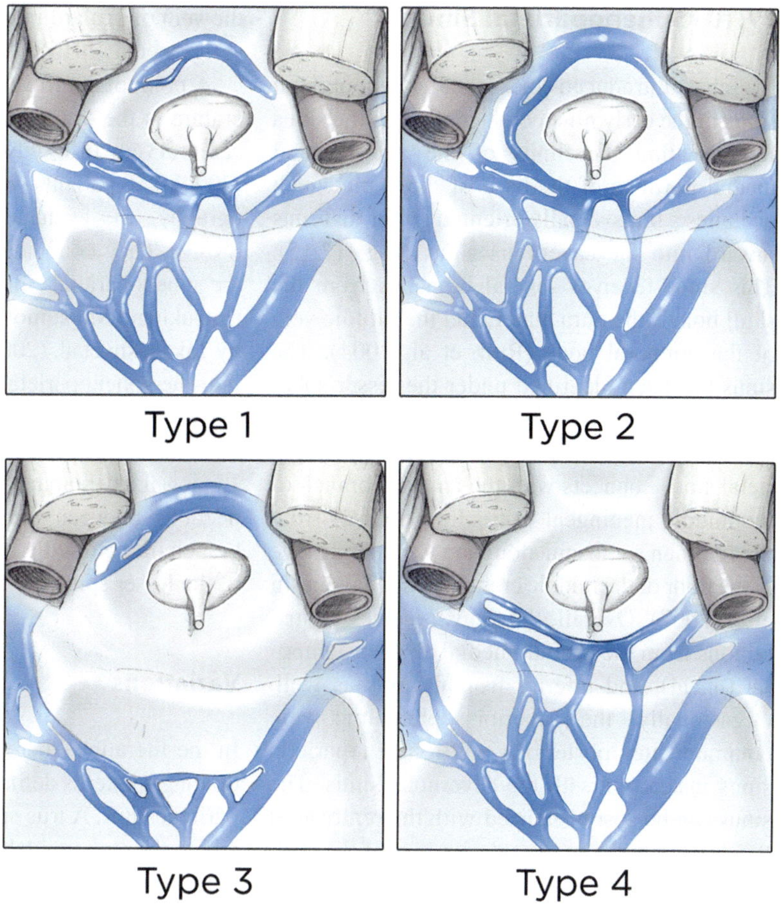

Fig. 17.19 Variations of the intercavernous sinuses

sinus (Kiddee et al. 2010), and can be cannulated so that its blood is evaluated for increased hormone levels indicative of various neurological pathologies such as tumors of the pituitary gland (Tanabe et al. 1984). We previously studied 35 cadaveric specimens and classified the different forms as types 1–4. An found a coexistent anterior and posterior intercavernous sinus 80% and classified this as a type I configuration. Of the 80% of specimens with both sinuses, a circular connection between these was found in 25% and termed a type II configuration. In six cadavers, only the anterior intercavernous sinus was found (17%) and this morphology was termed a type III configuration. In 2.9%, a posterior intercavernous sinus was found in isoloation and this was termed a type IV configuration. Knott (1881a, b, c, d) did not identify the posterior intercavernous sinus in 45.4% of cadaveric specimens. A least one of the intercavernous sinuses was present i.e., no specimen had absence of both. In most, the anterior intercavernous sinus was larger in size than the posterior intercavernous sinus. In 17% of specimens, an inferior intercavernous sinus was found. When present, these sinuses were typically much smaller than the anterior or posterior intercavernous sinuses. None of the intercavernous sinuses ran in the free edge of the diaphragma sella. The mean diameter of the anterior intercavernous sinus was 4.9 mm (range 2–6 mm), and the mean diameter of the posterior intercavernous sinus was 3 mm (range 1.5–4.8 mm). Lastly, if a large basilar plexus was present, then the intercavernous sinuses were smaller in size.

17.10 Sphenoparietal Sinus

Breschet introduced the term sphenoparietal sinus in the early nineteenth century as the sinus found within the limits of the anterior and medial portions of the base of the skull which occupies a transversally oriented gutter that runs inward into the cavernous sinus (Fig. 17.15). This sinus receives several branches from the skull bones, the dura mater, and the diploic vein of the temporal bone (Ruíz et al. 2004). The sinus has been identified under the lesser sphenoid wing which connects to the cavernous sinus medially. Near the pterion, the sphenoparietal sinus connects with the anterior branch of the middle meningeal vein which courses from the foramen spinosum along the vascular groove in the floor of the middle cranial fossa (Takahashi et al. 2007). Overall, the sinus receives contributions from the orbital, medial anterior temporal (uncal), and inferior frontal veins as well. Occasionally, the superior ophthalmic vein communicates posteriorly with sphenoparietal sinus instead of with the cavernous sinus. The sinus can be best visualized with the removal of the dura mater from the anterior part of the middle cranial fossa with an anterior viewing perspective. In a study by Tubbs et al. (2007a), 15 adult cadavers were visualized, all of which had both a left and right sphenoparietal sinus. The sinuses measured on average 2.5 and 3 mm for the left and right sides, respectively (Tubbs et al. 2007a).

Frigeri et al. (2015) described that the central lobe is drained directly into 1 of 3 sinuses: superior sagittal, transverse, and sphenoparietal. The lower one-third of the lateral surface most commonly drains to the superficial sylvian veins and from there to the transverse sinus via the vein of Labbé or anteriorly to the sphenoparietal sinus via the superficial sylvian veins. Inferior one-third of the precentral gyrus drained to the sphenoparietal sinus in 50% of their specimens, the transverse sinus 40%, and the superior sagittal sinus 10% of the time. Finally, the lower portion of the post-central drained to the sphenoparietal sinus in 35% of the specimens, the transverse sinus in 45%, and the superior sagittal sinus via the vein of Trolard in the final 20% of specimens (Frigeri et al. 2015).

A point of controversy that came up in the literature is the drainage of the superficial middle cerebral vein. In a study by Suzuki and Matsumoto (2000), 250 patients were evaluated via CT angiography to evaluate the drainage of the SMCV. They concluded that the SMCV entered the sphenoparietal sinus in 54% of their patients (Suzuki and Matsumoto 2000). In a similar study by Takahashi et al. (2007), drainage of the SMCV into the sphenoparietal sinus occurred in 67.1% of their patients. Opposition to these findings has come from Ruíz et al. (2004), who, in a study of 15 non-fixed human specimens, found no connections between the SMCV and the anterior branch of the middle meningeal vein or the sinus of the lesser sphenoid wing (Ruíz et al. 2004).

Variations

In the literature, there have been three different drainage patterns defined in regard to the sphenoparietal sinus. A true sphenoparietal sinus is only when the sphenoidal component anastomoses with the parietal component, the latter consisting of the dural veins that accompany the middle meningeal artery (Curé and Smith 1994). As mentioned previously, the medial extension of the superficial middle cerebral (Sylvian) vein can travel beneath the lesser wing and drain directly into the cavernous sinus, which is known as the cavernous type. Otherwise, it can drain into the pterygoid plexus via the basal emissary veins in what is called the paracavernous or sphenobasal pattern. Finally, it can travel posteriorly along the floor of the middle cranial fossa after rounding the tip of the temporal lobe to drain into the superior petrosal sinus or transverse sinus (Curé and Smith 1994; Browder and Kaplan 1998a, b; Ruíz et al. 2004; Tubbs et al. 2007a). Such a variant as this is probably the same as what has been described as the sinus of Kelch, ophthalmopetrous sinus of Hyrtl and accessory sinus of Verga. In a study by Takahashi et al. (2007), it was observed that the cavernous type of SMCV drainage was equal to the SMCV being completely absent in 35.4% of the specimens. The paracavernous type was present 10.4% and the laterocavernous was present in 16.7% of the specimens (Takahashi et al. 2007).

Tubbs et al. (2007b) presented a case report in which a 69-year-old male cadaver was found to have a sinus that was 5 mm in width and traveled from the sphenoparietal sinus to the veins draining into the foramen spinosum. Known as the accessory venous sinus of Hyrtl (see above), it begins near the superior orbital fissure, the sphenoparietal sinus, and travels posteriorly into the transverse sinus (Tubbs et al. 2007b).

References

Superior Sagittal Sinus

Brockmann C, Kunze SC, Schmiedek P, Groden C, Scharf J (2012) Variations of the superior sagittal sinus and bridging veins in human dissections and computed tomography venography. Clin Imaging 36:85–89
Butler H (1957a) The development of certain human dura; venous sinuses. J Anat 91:510
Curè JK, Tassel PV, Smith MT (1994) Normal and variant anatomy of the dural venous sinuses. Semin Ultrasound CT MR 15:499–519
Jinkins JR (2000a) Atlas of neuroradiologic embryology, anatomy, and variants. Lippincott Williams and Wilkins, Philadelphia, p 369
Kaplan HA, Browder J (1973) Atresia of the rostral superior sagittal sinus: substitute parasagittal venous channel. J Neurosurg 38:602–607
Kiliç T, Akakin A (2008) Anatomy of cerebral veins and sinuses. Front Neurol Neurosci 23:4–15
Mancall EL, Brock DG (2011) Gray's clinical neuroanatomy: the anatomic basis for clinical neuroscience. Elsevier Saunders, Philadelphia, pp 71–74
Scott JN, Farb RI (2003a) Imaging and anatomy of the intracranial venous systems. Neuroimag Cin N Am 13:1–12
Streeter GL (1915a) The development of the venous sinuses of the dura mater in human embryo. Am J Anat 18:2
Tubbs RS, Salter G, Elton S, Grabb PA, Oakes WJ (2001) Sagittal suture as an external landmark for the superior sagittal sinus. J Neurosurg 94:985–987

Inferior Sagittal Sinus

Cure JL, Van Tassel P, Smith MT (1994) Normal and variant anatomy of the dural venous sinuses. Semin Ultrasound CT MRI 15(6):499–519
Erbas G, Oner AY, Akpek S, Tokgoz N (2006) Corpus callosum hematoma secondary to isolated inferior sagittal sinus thrombosis. Acta Radiol 19:1085–1088
McCord GM, Goree JA, Jimenez JP (1972) Venous drainage to the inferior sagittal sinus. Radiology 105:583–589
Ruiz DSM, Fasel JHD, Gaillloud P (2012) Unilateral hypoplasia of the rostral end of the superior sagittal sinus. AJNR 33:286–291
Swedenborg E, Tafel RL (1887) The brain considered anatomically, physiologically and philosophically, vol 2. James Speirs, pp 275–276
Tubbs RS, Loukas M, Louis RG, Shoja MM, Acakpto-Satchivi L, Blount JP, Salter EG, Oakes WJ, Wellons JC III (2007a) Anatomy of the falcine venous plexus. J Neurosurg 107:155–157

Straight Sinus

Magden AO (1991) Triple straight sinus – report of 2 cases. Anat Anz 173(1):17–22
Saxena RC, Beg MA, Das AC (1973) Double straight sinus – report of six cases. J Neurosurg 39:540–542
Saxena RC, Beg MA, Das AC (1974) The straight sinus. J Neurosurg 41:724–727
Browder J, Kaplan HA, Krieger AJ (1976) Anatomical features of the straight sinus and its tributaries. J Neurosurg 44(1):55–61
Van den Bergh P, Fryns JP, Wilms G, Piot R, Dralands G, Van den Bergh R (1984) Anomalous cerebral venous drainage in Aarskog syndrome. Clin Genet 25:288–294
Knott JF (1881a) On the cerebral sinuses and their variations. J Anat Physiol 16:27–42
Hsu S, Chaloupka JC (2012) Atretic parietal cephalocele associated with sinus pericranii: embryological consideration. Brain Dev 34:325–328
Kim MS, Lee GH (2010) Two cases with persistent falcine sinus as congenital variation. J Korean Neurosurg Soc 48:82–84
Ryu CW (2009) Persistent falcine sinus: is it really rare? AJNR Am J Neuroradiol 31:367–369
Desai K, Bhayani R, Nadkarni T, Limaye U, Goel A (2002) Developmental deep venous system anomaly associated with congenital malformation of the brain. Pediatr Neurosurg 36:37–39
Lasjaunias P, Berenstein A, Brugge KT (2001) Clinical vascular anatomy and variations. Surgical neuroangiography. Springer, Berlin/Heidelberg
Shin J, Lee KH, Lee JM, Kwon US, Choi GC, Suh CD (2000) An arachnoid granulation in the straight sinus. Neuroradiology 42(10):746–748. https://doi.org/10.1007/s002340000381
Hasegawa M, Yamashita J, Yamashima T (1991) Anatomical variations of the straight sinus on magnetic resonance imaging in the infratentorial supracerebellar approach to pineal region tumors. Surg Neurol 36:354–359

Newton TH, Potts DG (1974) Radiology of the skull and brain. Mosby
Wolpert SM (1969) Dural sinus configuration: measure of congenital disease. Radiology 92(7):1511–1516. https://doi.org/10.1148/92.7.1511

Superior Petrosal Sinus

Bergman R, Thompson SA, Afifi AK, Saadeh FA (1988a) Cardiovascular system. In: Bergman R (ed) Compendium of human anatomic variation. Urban & Schwarzenberg, Baltimore/Munich, p 93
Butler H (1957b) The development of certain human dural venous sinuses, pp 511–523
Campero A, Abarca-Olivas J, Gonzalez-Lopez P, Fernandez-Cornejo V, Verdu-Martinez I, Moreno-Lopez P, Loret-Garcia J, Rhoton AL Jr (2013) Presigmoid approach. 3D Neuroanat Approac 3(3):1–10
Coates AE (1934) A note on the superior petrosal sinus and its relation to the sensory root of the trigeminal nerve. J Anat 68:428
Dolenc VV, Rogers L (2009a) Cavernous Sinus: developments and future perspectives. Springer, pp 37–41
Dolenc VV (2012a) The Cavernous Sinus: a multidisciplinary approach to vascular and tumorous lesions. Springer, pp 113–115
Jinkins JR (2000b) Atlas of neuroradiologic embryology, anatomy, and variants. Lippincott Williams & Wilkins, pp 374–381
Hyrtl J (1862) Der Sinus opththalmo-petrosus. Wien Med Wochenschr 12:290–291
Jinkins RJ (2000c) Venous drainage of the cranium. Atlas of Neuroradiologic embryology, anatomy and variants. Lippincott Williams and Wilkins, Philadelphia
Knott JF (1881b) On the cerebral sinuses and their variations. J Anat Physiol 16:27–42
Matsushima K, Matsushima T, Kuga Y, Kodama Y, Inoue K, Ohnishi H, Rhoton AL Jr (2014) Classification of the superior petrosal veins and sinus based on drainage pattern. Neurosurgery 10(Suppl 2):357–367
Matsushima K, Ribas ESC, Kiyosue H, Komune N, Koichi M, Rhoton AL (2015a) Absence of the superior petrosal veins and sinus: surgical considerations. Surg Neurol Int 6:34
Osborn AG, Jacobs JM (1999a) Diagnostic cerebral angiography, 2nd edn. Lippincott Williams &Wilkins, pp 201–207
Padget DH (1956a) The cranial venous system in man in reference to development, adult configuration and relation to the arteries. Am J Anat 98:307–355
Rhoton AL Jr (2002a) The cerebral veins [suppl]. Neurosurgery 51:159–205
Rhoton AL Jr (2003a) Rhoton's cranial anatomy and surgical approaches. Congress of Neurological Surgeons, Schaumburg
Rubinstein D, Burton B, Walker A (1995a) The anatomy of the inferior petrosal sinus, glossopharyngeal nerve, vagus nerve, and accessory nerve in the jugular foramen. Am J Neuroradiol 16:185–194
Scott JN, Farb RI (2003b) Imaging and anatomy of the intracranial venous systems. Neuroimaging Clin N Am 13:1–12
Shapiro R, Robinson F (1967) The foramina of the middle cranial fossa: a phylogenetic, anatomic and pathologic study. Am J Roentgenol Radium Therapy, Nucl Med 101:779–794
Stoltz E (2006a) Cerebral veins and sinuses. In: Baumgartner RW (ed) Handbook on neurovascular ultrasound. Front Neurosci, vol 21. Karger, Basel, pp 182–193
Streeter GL (1915b) Development of the cerebral dural sinuses and their tributaries
Tanoue S, Kiyosue H, Okahara M, Sagara Y, Hori Y, Kashiwagi J, Mori H (2006a) Para-cavernous sinus venous structures: anatomic variations and pathologic conditions evaluated on fat-suppressed 3D fast gradient-echo MR images. AJNR Am J Neuroradiol 27:1083–1089
Tanoue S, Kiyosue H, Sagara Y, Hori Y, Okahara M, Kashiwagi J, Mori H (2010) Venous structures at the craniocervical junction: anatomical variations evaluated by multidetector row CT. Br J Radiol 83:831–840
Tubbs RS, Mortazavi MM, Krishnamurthy S, Verma K, Griessenauer CJ, Cohen-Gadol AA (2013a) The relationship between the superior petrosal sinus and the porus trigeminus: an anatomical study. J Neurosurg 119(5):1221–1225
Waltner J (1944) Anatomic variations of the lateral and sigmoid sinuses. Arch Otolaryngol Head Neck Surg 39:307–312
Watanabe A, Nagaseki Y, Ohkubo S, Ohhashi Y, Horikoshi T, Nishigaya K, Nukui H (2003a) Anatomical variations of the ten triangles around the cavernous sinus. Clin Anat 16:9–14
Williams HL (1930) Apparent unilateral absence of sigmoid sinus noted at operation. Arch Otolaryng 12:339–341

Inferior Petrosal Sinuses

Benndorf G, Campi A (2002) Aberrant inferior petrosal sinus: unusual transvenous approach to the cavernous sinus. Neuroradiology 44(2):158–163
Browder J, Kaplan HA (1998a) Cerebral dural sinus and their tributaries, pp 1–12
Butler H (1957c) The development of certain human dural venous sinuses
Dolenc VV, Rogers L (2009b) Cavernous Sinus: developments and future perspectives. Springer, pp 37–41
Dolenc VV (ed) (2012b) The cavernous Sinus: a multidisciplinary Approach to vascular and tumorous lesions. Springer, pp 113–115
Gailloud P, Fasel JH, Muster M, Desarzens F, Ruefenacht DA (1997) Termination of the inferior petrosal sinus: an anatomical variant. Clin Anat 10:92–96

Jinkins JR (2000d) Atlas of neuroradiologic embryology, anatomy, and variants. Lippincott Williams & Wilkins, pp 374–381
Kilic T, Akakin A (2008) Anatomy of cerebral veins and sinuses. In: Caso V, Agnelli G, Paciaroni M (eds) Handbook on cerebral venous thrombosis. Front Neurosci, vol 23. Karger, Basel, pp 4–15
Knott JF (1881c) On the cerebral sinuses and their variations. J Anat Physiol 16:27–42
Lang J (2012) Clinical anatomy of the head: neurocranium, orbit, craniocervical regions. Springer, pp 331–334
Matsushima K, Carvalhal Ribas ES, Kiyosue H, Komune N, Miki K, Rhoton AL (2015b) Absence of the superior petrosal veins and sinus: surgical considerations. Surg Neurol Int 6:34
Miller DL, Doppman JL, Chang R (1993) Anatomy of the junction of the inferior petrosal sinus and the internal jugular vein. Am J Neuroradiol 14(5):1075–1083
Mortazavi MM, Griessenauer CJ, Krishnamurthy S, Verma K, Loukas M, Tubbs RS (2014) The inferior petrosal sinus: a comprehensive review with emphasis on clinical implications. Childs Nerv Syst 30:831–834
Mortazavi M, Denning M, Yalcin B, Shoja M, Loukas M, Tubbs RS (2013) The intracranial bridging veins: a comprehensive review of their history, anatomy, histology, pathology, and neurosurgical implications. Childs Nerv Syst 29:1073–1078
Mitsuhashi Y, Nishio A, Kawahara S, Ichinose T, Yamauchi S, Naruse H, Matsuoka Y, Ohata K, Hara M (2007) Morphologic evaluation of the caudal end of the inferior petrosal sinus using 3D rotational venography. AJNR Am J Neuroradiol 28:1179–1184
Osborn AG, Jacobs JM (1999b) Diagnostic cerebral angiography, 2nd edn. Lippincott Williams &Wilkins, pp 201–207
Padget DH (1956b) The cranial venous system in man in reference to development, adult configuration and relation to the arteries. Am J Anat 98:307–355
Rubinstein D, Burton B, Walker A (1995b) The anatomy of the inferior petrosal sinus, glossopharyngeal nerve, vagus nerve, and accessory nerve in the jugular foramen. Am J Neuroradiol 16:185–194
Scott JN, Farb RI (2003c) Imaging and anatomy of the intracranial venous systems. Neuroimaging Clin N Am 13:1–12
Shiu PC, Hanaffee WN, Wilson GH, Rand RW (1968) Cavernous sinus venography. Am J Roentgenol 104(1):57–62
Stoltz E (2006b) Cerebral veins and sinuses. In: Baumgartner RW (ed) Handbook on neurovascular ultrasound. Front Neurosci, vol 21. Karger, Basel, pp 182–193
Streeter GL (1915c) Development of the cerebral dural sinuses and their tributaries
Swedenborg E (1887a) The Pituitary gland, the cerebellum and the medulla oblongata. The brain considered anatomically, physiologically and philosophically, vol 2. James Speirs, pp 143–147
Tubbs RS, Mortazavi MM, Krishnamurthy S, Verma K, Griessenauer CJ (2013b) The relationship between the superior petrosal sinus and the porus trigeminus: an anatomical study. J Neurosurg 119(5):1221–1225
Zhang W, Ye Y, Chen J, Wang Y, Chen R, Xiong K, Li X, Zhang S (2010a) Study on inferior petrosal sinus and its confluence pattern with relevant veins by MSCT. Surg Radiol Anat 32:563–572
Zhang WG, Ye YY, Chen JH, Chen R, Kuang LQ, Li X (2010b) Imaging study of the long extracranial extension of the inferior petrosal sinus with MSCT. Clin Anat 23:160–167

Basilar Venous Plexus

Aquini M, Marrone A, Schneider F (1994a) Intercavernous venous communications in the human skull base. Skull Base Surg 4:145–150
Arnautovic KI, Al-Mefty O, Pait TG, Krisht AF, Husain MM (1997) The suboccipital cavernous sinus. J Neurosurg 86:252–262
Benndorf G (2010) Dural cavernous sinus fistulas: diagnosis and endovascular therapy. In: Baert AL, Knauth M, Sartor K (eds) Medical radiology – diagnostic imaging. Springer, Berlin, pp 42–43
Curé J, Tassel P, Smith M (1994) Normal and variant anatomy of the dural venous sinuses. Semin Ultrasound CT MRI 15:499–519
Eckenhoff JE (1971) The vertebral venous plexus. Can Anaesth Soc J 18:487–495
Ernst R, Bulas R, Tomsick T, van Loveren H, Aziz KA (1999) Three cases of dural arteriovenous fistula of the anterior condylar vein within the hypoglossal canal. AJNR 20:2016–2020
Falk D, Conroy GC (1983) The cranial venous sinus system in Australopithecus afarensis. Nature 306:779–781
Hollinshead WH, Rosse C (1985) Textbook of anatomy. Harper & Row, Philadelphia
Krayenbühl H, Yasargil M, Huber P (1982) In: Georg (ed) Cerebral angiography. Thieme Verlag, Stuttgart, pp 228–229
Lang J (1991) Floor of the posterior cranial fossa in Clinical Anatomy of the Posterior Cranial Fossa and its Foramina. Thieme, New York, pp 2–8
Padget D (1956c) The cranial venous system in man in reference to development, adult configuration and relation to the arteries. Am J Anat 98:307–355
Pernkopf E (1963) Atlas of topographical and applied human anatomy, vol 1. W. B. Saunders, Philadelphia, p 73
Ruíz D (2006) Systèmes veineux de la base du crâne et de la junction cranio-cervicale: anatomie macroscopique et radiologique et implications Cliniques. Thèse de doctorat. Univ. Genève, pp 59–76
Swedenborg E (1887b) In: Tafel RL (ed) The brain considered anatomically, physiologically and philosophically. James Speirs, London

Takahashi S, Kato K, Tomura N, Mizoi K (2001) Dural arteriovenous fistula of the cavernous sinus with cortical venous reflux of the posterior fossa via a bridging vein. Radiat Med 19:219–222

Tubbs RS, Ammar K, Liechty P, Wellons JC 3rd, Blount JP, Salter EG, Oakes WJ (2006a) The marginal sinus. J Neurosurg 104:429–431

Tubbs RS, Hansasuta A, Loukas M, Louis R, Shoja MM, Salter EG, Oakes WJ (2007b) The basilar venous plexus. Clin Anat 20:755–759

Marginal Sinus

Caruso RD, Rosenbaum AE, Chang JK, Joy SE (1999) Craniocervical junction venous anatomy on enhanced MR images: the suboccipital cavernous sinus. AJNR Am J Neuroradiol 20:1127–1131

de Oliveira E, Rhoton AL, Peace D (1985) Microsurgical anatomy of the region of the foramen magnum. Surg Neurol 24:293–352

Dora F, Zileli. (1980) Common variations of the lateral and occipital sinuses. Neuroradiology 20:23–27

Go RT, Suzuki Y, Schapiro RL, Christie JH (1974) The scintigraphic features of the occipital sinus. Radiology 112:635–638

Kimbel WH (1984) Variation in the pattern of cranial venous sinuses and hominid phylogeny. Am J Phys Anthropol 63:243–263

McDougall CG, Halbach VV, Dowd CF, Higashida RT, Larsen DW, Hieshima GB (1997) Dural arteriovenous fistulas of the marginal sinus. AJNR Am J Neuroradiol 18:1565–1572

Okudera T, Huang YP, Ohta T, Yokota A, Nakamura Y, Maehara F, Utsunomiya H, Uemura K, Fukasawa H (1994) Development of posterior fossa dural sinuses, emissary veins, and jugular bulb: morphological and radiologic study. AJNR Am J Neuroradiol 15:1871–1883

Rhoton AL Jr (2003b) The foramen magnum in Rhoton cranial anatomy and surgical approaches. Congress of Neurological Surgeons, Schaumburg, pp 587–625

Tubbs RS, Ammar K, Liechty P, Wellons JC 3rd, Blount JP, Salter EG, Oakes WJ (2006b) The marginal sinus. J Neurosurg 104:429–431

Cavernous Sinus

Aquini MG, Marrone AC, Schneider FL (1994b) Intercavernous venous communications in the human skull base. Skull Base Surg 4:145–150

Bakan AA, Alkan A, Kurtcan S et al (2015) Cavernous sinus: a comprehensive review of its anatomy, pathologic conditions, and imaging features. Clin Neuroradiol 25:109–125

Bergman RA, Thompson SA, Afifi AK et al (1988b) Compendium of human anatomic variation: catalog, atlas and world literature. Urban & Schwarzenberg, Baltimore/Munich

Bonneville JF, Cattin F, Racle A et al (1989) Dynamic CT of the laterosellar extradural venous spaces. AJNR Am J Neuroradiol 10:535–542

Keller JT, Leach JL, van Loveren HR et al (2009) Venous anatomy of the lateral sellar compartment. In: Dolenc V, Rogers L (eds) . Springer, Vienna, pp 35–51

Renn WH, Rhoton AL Jr (1975a) Microsurgical anatomy of the sellar region. J Neurosurg 43:288–298

Rhoton AL Jr (2002b) The cavernous sinus, the cavernous venous plexus, and the carotid collar. Neurosurgery 51:S375–S410

Rhoton AL Jr (2009) The middle cranial base and cavernous sinus. In: Dolenc V, Rogers L (eds) . Springer, Vienna, pp 3–25

Tanoue S, Kiyosue H, Okahara M et al (2006b) Paracavernous sinus venous structures: anatomic variations and pathologic conditions evaluated on fat-suppressed 3D fast gradient-echo MR images. Am J Neuroradiol 27:1083–1089

Tubbs RS, Griessenauer C, Loukas M et al (2013c) The circular sinus: an anatomic study with neurosurgical and neurointerventional applications. J Neurosurg 82:e475–e478

Tubbs RS, Hill M, May WR et al (2008) Does the maxillary division of the trigeminal nerve traverse the cavernous sinus? An anatomical study and review of the literature. Surg Radiol Anat 30(1):37–40

Wallace S, Goldberg HI, Leeds NE et al (1967) The cavernous branches of the internal carotid artery. Am J Roentgenol 101:34–46

Watanabe A, Nagaseki Y, Ohkubo S et al (2003b) Anatomical variations of the ten triangles around the cavernous sinus. Clin Anat New York 16:9–14

Sphenoparietal Sinus

Browder J, Kaplan H (1998b) Cerebral dural sinuses and their tributaries. Charles C Thomas, Springfield, pp 85–87

Curé Van Tassel P, Smith T (1994) Normal and variant anatomy of the dural venous sinuses. Semin Ultrasound CT MRI 15:499–519

Frigeri T, Paglioli E, Olivera E, Rhoton A (2015) Microsurgical anatomy of the central lobe. J Neurosurg 122:483–498

Padget D (1956d) The cranial venous system in man in reference to development, adult configuration and relation to the arteries. Am J Anat 98:307–355

Ruíz D, Fasel J, Rufenacht D, Gailloud P (2004) The sphenoparietal sinus of Breschet: does it exist? An anatomic study. AJNR Am J Neuroradiol 25:112–120

Scott J, Farb R (2003d) Imaging and anatomy of the normal intracranial venous system. Neuroimaging Clin N Am 13:1–12

Suzuki Y, Matsumoto K (2000) Variations of the superficial middle cerebral vein: classification using three-dimensional CT angiography. AJNR Am J Neuroradiol 21:932–938

Takahashi S, Sakuma I, Otani T, Yasuda K, Tomura N, Watarrai J, Yanagisawa T, Mizoi K (2007) Venous anatomy of the sphenoparietal sinus: evaluation by MR imaging. Interv Neuroradiol 13:84–89

Tubbs RS, Loukas M, Shoja MM, Oakes WJ (2007d) Accessory venous sinus of Hyrtl. Folia Morphol (Warsz) 66:198–199

Tubbs RS, Salter E, Wellons J III, Blount JB, Oakes WJ (2007c) The sphenoparietal sinus. Neurosurgery 60:1-5

Intercavernous Sinuses

Aquini MG, Marrone AC, Schneider FL (1994c) Intercavernous venous communications in the human skull base. Skull Base Surg 4:145–150

de Divitiis E, Spaziante R, Iaccarino V, Stella L, Genovese L (1981) Phlebography of the cavernous and intercavernous sinuses. Surg Neurol 15:306–312

Hong Y, Chen S, Guo S, Yu J, Wu Q, Zhang J (2010) A new technique for management of intercavernous sinus bleeding with titanium clips in transsphenoidal surgery. Neurol India 58:847–851

Kaplan HA, Browder J, Krieger AJ (1976) Intercavernous connections of the cavernous sinuses: the superior and inferior circular sinuses. J Neurosurg 45:166–168

Kiddee W, Preechawai P, Hirunpat S (2010) Bilateral septic cavernous sinus thrombosis following the masticator and parapharyngeal space infection from the odontogenic origin: a case report. J Med Assoc Thail 93:1107–1111

Knott JF (1881d) On the cerebral sinuses and their variations. J Anat Physiol 16:27–42

Krayenbuehl H, Yasargil MG, Huber P (1982) Cerebral angiography, 2nd edn. Thieme, New York

Kurata A, Suzuki S, Iwamoto K, Nakahara K, Sasaki M, Kijima C, Inukai M, Abe K, Niki J, Satou K, Fujii K, Kan S (2011) Dural arteriovenous fistulas in the cavernous sinus: clinical research and treatment. ISRN Neurol 453834:2011

Renn WH, Rhoton AL Jr (1975b) Microsurgical anatomy of the sellar region. J Neurosurg 43:288–298

Schnitzlein HN, Murtagh FR, Arrington JA, Parkinson D (1985) The sinus of the dorsum sellae. Anat Rec 213:587–589

Tanabe S, Sato O, Hasunuma M, Sohma H, Nakagawa T (1984) Diagnosis of pituitary microadenoma by magnification carotid angiography and intercavernous sinus venography. No Shinkei Geka 12:815–822

Theron J, Chevalier D, Delvert M, Laffont J (1979) Diagnosis of small and micro pituitary adenomas by intercavernous sinus venography: a preliminary report. Neuroradiology 18:23–30

Waga S, Kikuchi H, Handa J, Handa H (1970) Cavernous sinus venography. Am J Roentgenol Radium Therapy, Nucl Med 109:130–137

18 Emissary Veins

R. Shane Tubbs

Emissary veins connect the veins outside the cranium with the intracranial venous sinuses. These structures are more commonly found in children, and not all of the various emissary veins are identified in every individual (Eckley 1903). These veins serve an important function of equalizing intracranial pressure and can act as safety valves during cerebral congestion or in patients with lesions of the neck or skull such as obstruction of the internal jugular veins (Eckley 1903; Boyd 1930; Braun and Tournade 1977; Williams et al. 1989; Curé et al. 1994; Louis Jr. et al. 2009). The foramina for these vessels tend to be larger during childhood (Williams et al. 1989) and may communicate with the diploic veins (Curé et al. 1994). Louis Jr. et al. (2009) found that emissary veins act as the primary outflow route for venous drainage in the upright position and may serve to cool venous blood circulating through the head.

Emissary veins can often be seen on MRI, CT, and angiograms, and the various types are named according to their positions on the skull (Boyd 1930; Hacker 1974; Louis Jr. et al. 2009). Interestingly, an enlarged emissary foramen does not necessarily transmit a large emissary vein (Boyd 1930), and fossil records indicate that there is an increasing frequency of some emissary veins such as the mastoid and parietal vessels in man (Falk 1986).

R. S. Tubbs (✉)
Department of Neurosurgery, Tulane University School of Medicine,
New Orleans, LA, USA
e-mail: rtubbs@tulane.edu

18.1 Posterior Condylar Emissary Vein

The posterior condylar emissary vein connects the lower end of the sigmoid, marginal, or occipital sinuses with the internal vertebral venous plexus (Pearson 1979; Falk 1986; Williams et al. 1989; Curé et al. 1994). Irmak et al. (2004) noted that a posterior condylar emissary vein may connect the sigmoid sinus with the suboccipital plexus of veins (Curé et al. 1994). Observations have shown that the posterior condylar foramen is usually the largest emissary foramen in the retromastoid region (Boyd 1930). Lang (1985) found that this vein joins the hypoglossal plexus in 3.84%, and in another 3.84%, it joins the internal jugular vein below the jugular foramen.

18.2 Mastoid Emissary Vein

The mastoid emissary vein (Figs. 18.1, 18.2, and 18.3) passes through the mastoid canal and connects the transverse or sigmoid sinus to the posterior auricular, or occipital veins, which then join the vertebral plexus of veins (Cheatle 1925; Boyd 1930; Falk 1986; Curé et al. 1994; Irmak et al. 2004; Standring 2005). Lang Jr and Samii (1991)

S. J. Bordes, Jr. et al. (eds.), *The Clinical Anatomy of the Vascular System*,
https://doi.org/10.1007/978-3-031-78326-5_18

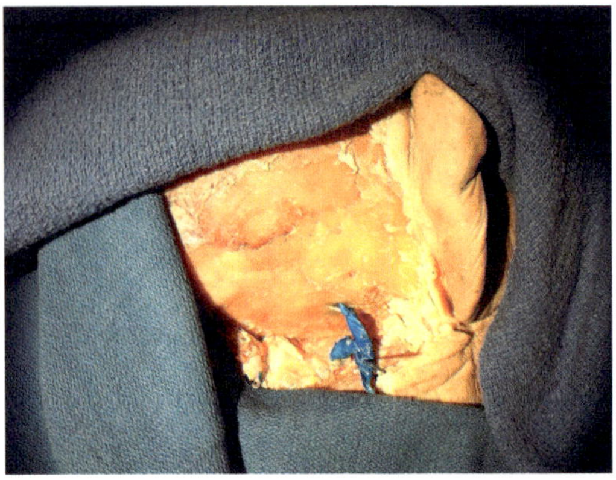

Fig. 18.1 Right-sided emissary vein draining into the skull

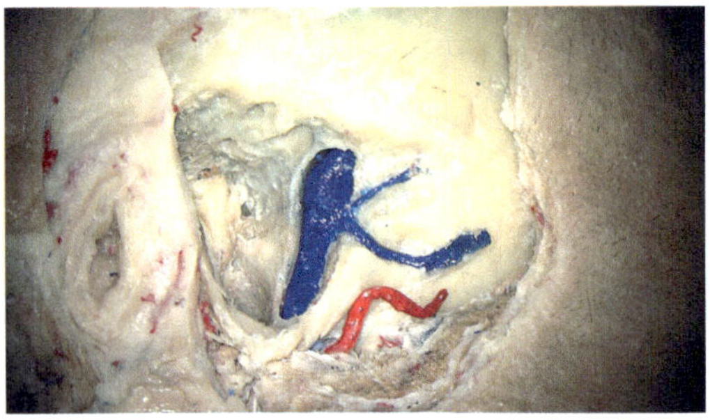

Fig. 18.2 Two mastoid emissary veins draining into the sigmoid sinus as shown after drilling away the overlying bone

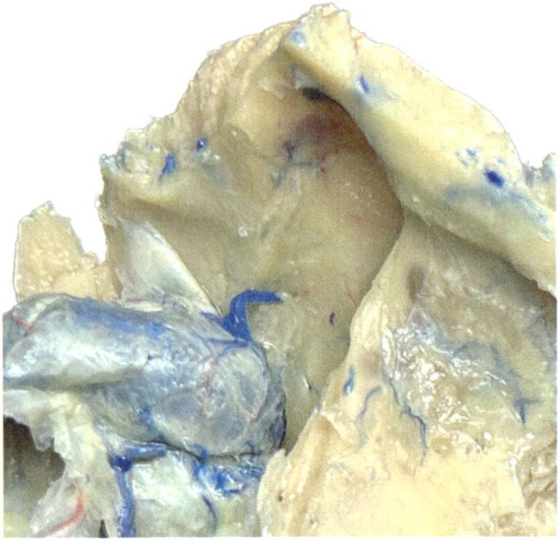

Fig. 18.3 Internal view showing the right mastoid emissary vein entering the skull and terminating in the sigmoid sinus, which has been pulled anteriorly out of its bony groove

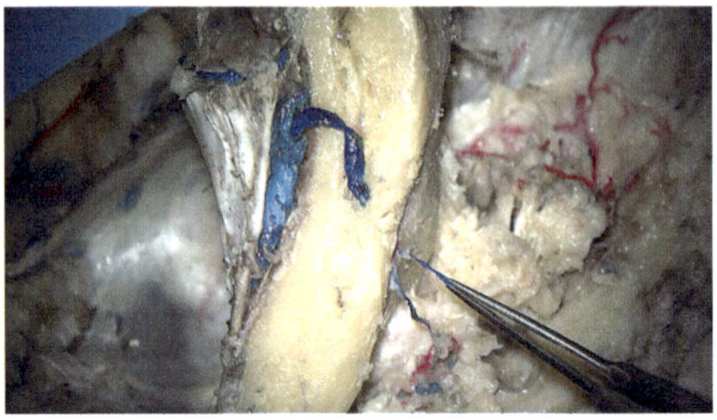

Fig. 18.4 Sagittal section through the occipital bone showing the occipital emissary vein held externally with forceps, traveling through the occipital diploic space and terminating into the confluence of sinuses

observed that the mastoid emissary foramen exists less frequently in females than males, although the reason for this is unclear. It is controversial whether mastoid foramina are more common on right versus left sides (Louis Jr. et al. 2009). Cheatle (1925) noted an infrequent association between the mastoid emissary vein and the petro-squamosal sinus. Multiple emissary veins may converge toward one mastoid foramen.

18.3 Occipital Emissary Vein

The occipital emissary vein (Fig. 18.4) connects the transverse sinus to the occipital vein, which then drains blood to the vertebral venous plexus (Falk 1986; Williams et al. 1989). On occasion, the occipital emissary vein joins the transverse-sigmoid sinus complex (Neuroangio 2013). Okudera et al. (1994) have reported that the occipital vein may drain to the confluens of sinuses. There may be several occipital veins that drain to one occipital foramen.

18.4 Parietal Emissary Vein

The parietal emissary vein connects the superior sagittal sinus with the occipital vein that then joins the vertebral venous plexus (Falk 1986; Williams et al. 1989). Parietal emissary veins may also interact with the diploic veins.

18.5 Petrosquamosal Sinus

Although it serves as the primary cerebral drainage site in many lesser primates and quadrupeds, the petrosquamosal sinus is a rare emissary vein in humans that typically disappears prior to birth in humans. The petrosquamosal sinus ranges from 2 to 4 mm in diameter and originates at the junction between the transverse and sigmoid sinuses, coursing laterally above the superior border of the temporal bone.

18.6 Ophthalmic Veins

The ophthalmic veins are considered by some to be emissary veins as they may carry out a similar function as the emissary veins by connecting the exterior with the interior skull (angular vein to cavernous sinus) and are valveless. The ophthalmic veins travel through the orbit and traverse the superior orbital fissure to reach the most anterior portion of the cavernous sinus (Dwight et al. 1913).

18.7 Sphenoidal Emissary Vein (Vesalius)

Sphenoidal emissary veins (Vesalian veins) connect the cavernous sinus with the pterygoid or pharyngeal plexuses (Dwight et al. 1913; Gray 1918; Hayashi et al. 2010). These veins are not constant and are found in only approximately one in three cases (Dwight et al. 1913). When present, the foramen for this vessel is found just medial to the foramen ovale.

18.8 Emissary Veins of the Foramen Ovale

Just like the sphenoidal emissary veins, the emissary veins of the foramen ovale connect the cavernous sinus to the pterygoid plexus (Dwight et al. 1913; Gray 1918).

18.9 Internal Carotid Venous Plexus

The internal carotid plexus of veins extends through the carotid canal and connects the cavernous sinus with the internal jugular vein (Gray 1918). This collection of veins is classified by some as emissary veins as it joins the intradural sinuses with extracranial veins (Ernst et al. 1999).

18.10 Emissary Vein of the Foramen Cecum

The single emissary vein of the foramen cecum joins the superior sagittal sinus with veins of the nasal cavity and is unpaired (Gray 1918). This vessel is larger at birth than in the adult.

18.11 Emissary Veins of the Foramen Lacerum

An inconsistent number of small veins may traverse the cartilage-filled foramen lacerum. Such emissary veins connect the cavernous sinus with the pterygoid venous plexus (Dwight et al. 1913).

18.12 Emissary Veins of the Clivus

Some have identified transclival emissary veins (Tubbs et al. 2007). If present, these structures connect the basilar venous plexus to veins on the inferior surface of the clivus adjacent to the pharynx.

18.13 Anterior Condylar Vein

The anterior condylar vein accompanies the hypoglossal nerve through the hypoglossal canal and joins the anterior condylar confluent (Chen et al. 2007).

18.14 Temporal Emissary Vein

This venous structure may connect the remnant of the petrosquamosal sinus with the deep temporal vein and likely represents the dorsal cerebral vein seen in some mammals.

18.15 Superficial Petrosal Emissary Vein

This vein connects the vessel(s) traveling along the superficial petrosal vein in the middle cranial fossa to the vein traveling with the facial nerve within the stylomastoid foramen. Such vessels may be remnants of the primary head sinus.

Hoshi et al. (2000) reported a case of removal of a hypoglossal neurinoma where the condylar emissary veins were enlarged. Coagulation of these channels resulted in dural venous sinus thrombosis and cerebellar infarction leading to the patient's death. Sphenoidal emissary veins and veins of the foramen ovale may hinder lateral extradural approaches to the skull base.

In the event of compromise of the internal jugular veins, emissary veins may increase in size and provide the primary venous outflow of the skull (Ginsberg 1994; Marsot-Dupuch et al. 2001). The main venous outflow of the posterior fossa venous sinuses may be via the occipital and mastoid emissary veins (Jeevan et al. 2008).

Tinnitus has been reported as a consequence of large posterior condylar veins (Lambert and Cantrell 1986; Forte et al. 1989; Chen et al. 2007). One case of tinnitus was alleviated after decreasing blood flow through the internal jugular vein via direct pressure on the neck, thus altering the flow within the posterior condylar emissary vein ((Lambert and Cantrell 1986). Although rare, dural arteriovenous fistulas have been identified at the anterior condylar vein. Ernst et al. (1999) reported 3 cases of dural arteriovenous fistula involving this vein within the hypoglossal canal. In a cadaver, Moore et al. identified an anastomosis between the posterior condylar vein and the vertebral venous plexus.

An enlarged anterior condylar emissary vein has also been linked to unilateral tongue atrophy (Shiozawa et al. 1996). Sinus pericranii are dilated extracranial veins found under the scalp under or inside the periosteum (Ota et al. 1975) and are characterized by communication of pericranial varicosities with an underlying dural sinus (Carpenter et al. 2004). Such vessels may be sizable and may represent ectopic diploic veins. Emissary veins such as the parietal veins may connect these structures to the diploic venous system (Fig. 18.5).

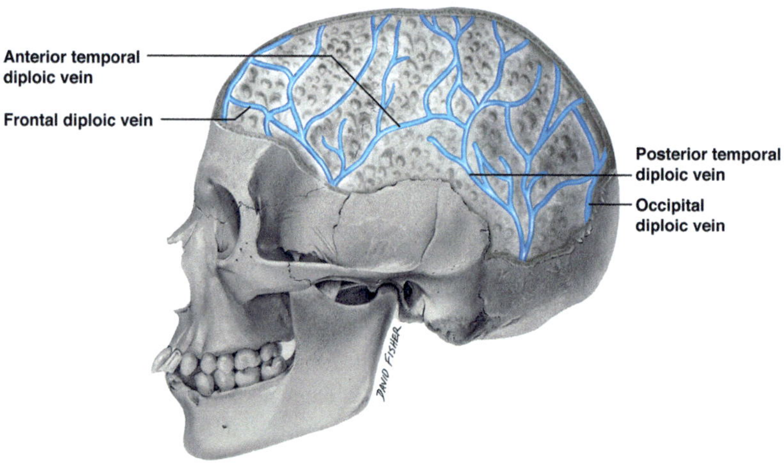

Fig. 18.5 Illustration of the major diploic veins

References

Boyd G (1930) The emissary foramina of the cranium in man and the anthropoids. J Anat 65:108–121

Braun JP, Tournade A (1977) Venous drainage in the craniocervical region. Neuroradiology 13:155–158

Carpenter JS, Rosen CL, Bailes JE, Gailloud P (2004) Sinus pericranii: clinical and imaging findings in two cases of spontaneous partial thrombosis. AJNR Am J Neuroradiol 25:121–125

Cheatle A (1925) The mastoid emissary vein and its surgical importance. Proc R Soc Med 18:29–34

Chen Z, Feng H, Zhu G, Wu N, Lin J (2007) Anomalous intracranial venous drainage associated with basal ganglia calcification. AJNR Am J Neuroradiol 28:22–24

Curé JK, Van Tassel P, Smith MT (1994) Normal and variant anatomy of the dural venous sinuses. Semin Ultrasound CT MR 15:499–519

Dwight T, Hamann C, McMurrich J, Piersol G (1913) In: Piersol G (ed) Human anatomy, vol 1. Lippincott Company, London. 879 p

Eckley W (1903) Anatomic and physiological correspondences of child and the adult. JAMA 20:1337

Ernst R, Bulas R, Tomsick T, van Loveren H, Aziz KA (1999) Three cases of dural arteriovenous fistula of the anterior condylar vein within the hypoglossal canal. AJNR Am J Neuroradiol 20:2016–2020

Falk D (1986) Evolution of cranial blood drainage in hominids: enlarged occipital/marginal sinuses and emissary foramina. Am J Phys Anthropol 70:311–324

Forte V, Turner A, Liu P (1989) Objective tinnitus associated with abnormal mastoid emissary vein. J Otolaryngol 18:232–235

Ginsberg LE (1994) The posterior condylar canal. AJNR Am J Neuroradiol 15:969–972

Gray H (1918) Anatomy of the human body. Lea & Febiger, Philadelphia. 1396 p

Hacker H (1974) Emissary veins related to the transverse sinus, sigmoid sinus, and torcular herophili. In: Newton TH, Potts DG (eds) Radiology of the skull and brain, vol 2. C.V. Mosby Company, St. Louis, pp 1852–1858

Hayashi N, Sato H, Tsuboi Y, Nagai S, Kuwayama N, Endo S (2010) Consequences of preoperative evaluation of patterns of drainage of the cavernous sinus in patients treated using the anterior transpetrosal approach. Neurol Med Chir (Tokyo) 50:373–377

Hoshi M, Yoshida K, Ogawa K, Kawase T (2000) Hypoglossal neurinoma-two case reports. Neurol Med Chir (Tokyo) 40:489–493

Irmak MK, Korkmaz A, Erogul O (2004) Selective brain cooling seems to be a mechanism leading to human craniofacial diversity observed in different geographical regions. Med Hypotheses 63:974–979

Jeevan DS, Anlsow P, Jayamohan J (2008) Abnormal venous drainage in syndromic craniosynostosis and the role of CT venography. Childs Nerv Syst 24:1413–1420

Lambert PR, Cantrell RW (1986) Objective tinnitus in association with an abnormal posterior condylar emissary vein. Am J Otol 7:204–207

Lang J (1985) Anatomy of the brainstem and the lower cranial nerves, vessels, and surrounding structures. Am J Otol Suppl:1–19

Lang J Jr, Samii A (1991) Retrosigmoidal approach to the posterior cranial fossa. An anatomical study. Acta Neurochir (Wien) 111:147–153

Louis RG Jr, Loukas M, Wartmann CT, Tubbs RS, Apaydin N, Gupta AA, Spentzouris G, Ysique JR (2009) Clinical anatomy of the mastoid and occipital emissary veins in a large series. Surg Radiol Anat 31:139–144

Marsot-Dupuch K, Gayet-Delacroix M, Elmaleh-Bergès M, Bonneville F, Lasjaunias P (2001) The petrosquamosal sinus: CT and MR findings of a rare emissary vein. AJNR Am J Neuroradiol 22:1186–1193

Neuroangio (2013) Venous anatomy of the brain. URL: http://neuroangio.org/venous-brain-anatomy/

Okudera T, Huang YP, Ohta T, Yokota A, Nakamura Y, Maehara F, Utsunoiya H, Uemura K, Fukasawa H (1994) Development of posterior fossa dural sinuses, emissary veins, and jugular bulb: morphological and radiologic study. AJNR Am J Neuroradiol 15:1871–1883

Ota T, Waga S, Handa H, Nishimura S, Mitani T (1975) Sinus pericranii. J Neurosurg 42:704–712

Pearson BW (1979) Microsurgical anatomy of the approach to the base of the skull. Clin Neurosurg 26:379–390

Shiozawa Z, Koike G, Seguchi K, Shindo K, Sugita K (1996) Unilateral tongue atrophy due to an enlarged emissary vein in the hypoglossal canal. Surg Neurol 45:477–479

Standring S (2005) Gray's anatomy, 39th edn. Elsevier Churchill Livingstone, 1600 p

Tubbs RS, Kelly DR, Humphrey ER, Chua GD, Shoja MM, Salter EG, Acakpo-Satchivi L, Wellons JC 3rd, Blount JP, Oakes WJ (2007) The tectorial membrane: anatomical, biomechanical, and histological analysis. Clin Anat 20:382–386

Williams P, Warwick R, Dyson M, Bannister L (1989) Gray's anatomy, 37th edn. Churchill Livingstone, Edinburgh

19 Ophthalmic Veins

Jenilkumar Patel

19.1 Embryology

The eye begins to form during the 22nd day of embryogenesis. According to Raybaud's developmental stages of the intracranial vascular system, the embryo is around 16–21 mm long toward the end of week seven (stage 5) and both the choroid plexus and arteries form at this time. During the seventh and eighth week of embryogenesis, the pro-otic sinus (eventually called the cavernous sinus) forms via the union of the middle dural plexus and the anterior region of the primary sinus (Raybaud 2010).

During week nine, the superior ophthalmic vein (SOV) forms from an elementary maxillary vein, which drains the contents of the optic vesicles into the pro-otic sinus (Fig. 19.1). Growth factors such as vascular endothelial growth factor (VEGF) and transforming growth factor beta (TGF-B) are vital for the differentiation of the SOV (from mesoderm) into various elements of the venous vasculature. The primary head sinus drains the embryonic dural plexus and all structures in the head region, including ocular and visceral structures, via the primitive maxillary vein into the anterior cardinal vein, which eventually becomes the internal jugular vein.

J. Patel (✉)
Tulane University School of Medicine,
New Orleans, LA, USA
e-mail: jpatel10@tulane.edu

S. J. Bordes, Jr. et al. (eds.), *The Clinical Anatomy of the Vascular System*,
https://doi.org/10.1007/978-3-031-78326-5_19

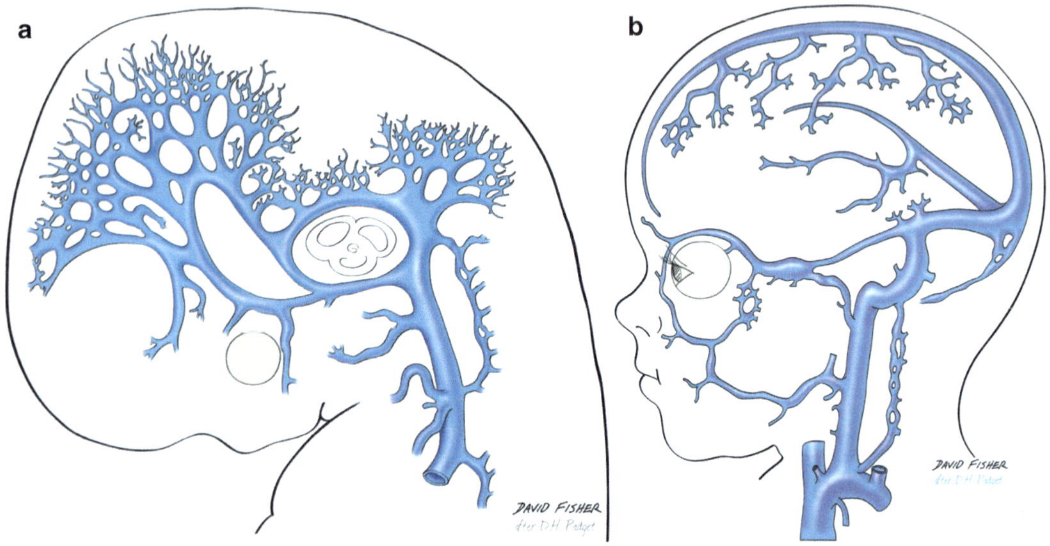

Fig. 19.1 (**a**) Schematic drawing of the development of the fetal venous system of the head and neck. Note the developing ophthalmic veins on either side of the eye (central gray circle). (**b**) In the neonate, the ophthalmic veins are seen collecting veins of the face and communicating with the deeper venous systems (e.g., cavernous sinus)

19.2 Anatomy/Topography

The venous system of the orbit is highly complex, partly because of several anastomoses and plexuses (Hayreh 2006). The ophthalmic veins drain blood from the orbit and empty into the cavernous sinus (Fig. 19.2). They also form various anastomoses with superficial facial and deep intracranial veins. The two main orbital veins are the SOV and the inferior ophthalmic vein (IOV).

The SOV emerges near the superomedial orbital rim behind the trochlea by the unification of the angular vein (a small branch of the facial vein) and supraorbital veins. It runs posteriorly between the superior rectus muscle inferiorly and the optic nerve (cranial nerve II) superiorly and leaves the orbit through superior orbital fissure at the edge of the annulus of Zinn, emptying into the cavernous sinus (CS) (Cheung and McNab 2003; Bergen 1981). In contrast, the ophthalmic artery exits the orbit through the optic canal. On its way to the CS, the SOV drains superior contents of the orbit including the eye, ethmoid air cells, and the superficial face and forehead (Cheung and McNab 2003; Bergen 1981). Some key tributaries of the SOV are the lacrimal veins, ethmoidal veins, central retinal vein, nasofrontal vein, ciliary veins, and two superior vorticose veins (Cheung and McNab 2003; Brismar 1974).

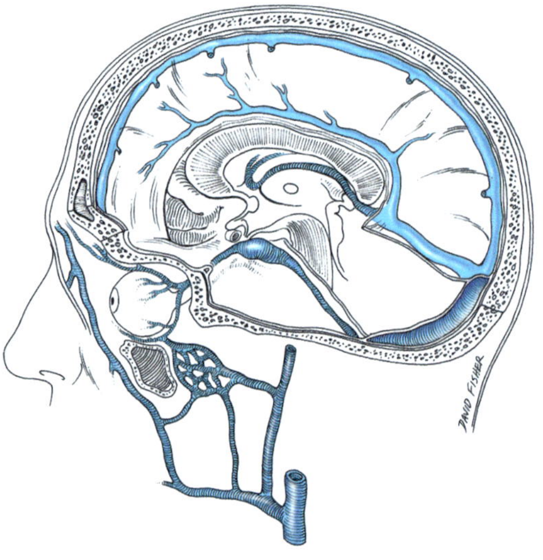

Fig. 19.2 Schematic drawing of the ophthalmic veins of the adult traveling above and below the globe

The IOV emerges near the anteromedial portion of the orbital floor as a venous plexus. It runs posteriorly below the eyeball on the surface of the inferior rectus muscle (Forrester 2016; Moore et al. 2014; Netter 2019; Snell and Lemp 1998; Standring 2016). After it exits the orbit, it either empties into the SOV or drains directly into the CS via the superior orbital fissure. Additionally, a communicating branch of the IOV exits the orbit via the inferior orbital fissure and empties into the pterygoid venous plexus (Forrester 2016; Moore et al. 2014; Netter 2019; Snell and Lemp 1998; Standring 2016). Tributaries of the IOV drain blood from the inferior rectus and inferior oblique muscles, inferior orbital contents such as the lacrimal sac, two inferior vorticose veins, and the lower eyelid (Cheung and McNab 2003; Brismar 1974). The vein also drains blood from the superficial and deep face. Blood from IOV tributaries emptying into the CS usually drains into the internal jugular vein, whereas blood emptying into the pterygoid plexus can drain into either the internal or the external jugular vein. The external jugular (EJV), internal jugular (IJV), and subclavian veins combine to form the brachiocephalic vein. The left and right brachiocephalic veins combine to form the superior vena cava, which empties directly into the right atrium of the heart.

19.3 Variations

Some consider the ophthalmic veins to be emissary veins that connect the external cranium (via angular veins) to the internal surface of skull (cavernous sinus) (Tubbs et al. 2016). Compared to the arteries of the orbit, the ophthalmic veins are highly variable and not well-defined. The diameter of the SOV ranges from 2 mm to 1 cm, and the diameter of IOV is often reported to be inconsistent (Cheung and McNab 2003). From foundational anatomical studies of the orbital venous system during the eighteenth century, two distinct perspectives on the organization of orbital veins emerged. First, the course and branching pattern of the SOV correlates closely to that of the ophthalmic artery. Second, unlike other veins in the human body, the orbital veins do not run with a corresponding ophthalmic artery. The first view has been challenged and completely opposed by many investigators; the second viewpoint has received support from many contemporary anatomists and scientists (Cheung and McNab 2003).

19.4 Pathology

The ophthalmic veins (both SOV and IOV) are considered to be types of emissary veins and they lack valves. Thus, blood from the orbit/face or from venous sinuses can flow in either direction depending on the pressure gradient. The anastomosis between the SOV and the angular vein (a branch of the facial vein) can serve as a route for the spread of infections from the superficial scalp into deep venous sinuses inside the cranial cavity (Forrester 2016; Moore et al. 2014; Netter 2019; Snell and Lemp 1998; Standring 2016). In cavernous sinus thrombosis (CST), an embolus formed in the facial vein can reach the cavernous sinus through the SOV. CST is life threatening and should be treated appropriately with antibiotics or surgical intervention as soon as it is identified (Remington and Goodwin 2022).

19.5 Clinical and Surgical Significances

Normally, the pressure in the central retinal vein is equivalent/ identical to the intraocular pressure (IOP). When the volume of blood in the central retinal vein increases, this pulsatile vein can be observed with an ophthalmoscope as it leaves the orbit through the blind spot (optic disc) (Remington and Goodwin 2022; Mathew et al. 2014). This condition is known as spontaneous venous pulsation (Remington and Goodwin 2022).

The optic nerve is considered part of the brain, and the protective sheath that enwraps it is an uninterrupted extension of the meninges that cover the brain (Remington and Goodwin 2022). Consequently, cerebrospinal fluid (CSF) bathes the optic nerve as it does throughout the cranial cavity in the subarachnoid space. Therefore, with an increase in intracranial pressure, the central retinal vein can be compressed as it leaves the meningeal sheath. The central retinal artery is not compressed during similar circumstances as its wall is thicker than that of its corresponding vein (Remington and Goodwin 2022). Compression can lead to obstruction of the retinal veins, which causes accumulation of edema within the retina, known as papilledema. Papilledema is a swelling of the optic disc secondary to increased intracranial pressure and noticeable as blurred disc margins or hemorrhages (Remington and Goodwin 2022). An increase in orbital pressure related to the central retinal vein can occlude/ compress the optic nerve and lead to blindness (Mathew et al. 2014).

A rare condition, bacterial orbital cellulitis, is an infection of the eyelid and nearby skin areas in front of the orbital septum that could result from an infectious agent or trauma. In periorbital cellulitis, the IOV can be pinched or squeezed, leading to an increase in orbital pressure and edema (Mathew et al. 2014).

Carotid-cavernous sinus fistula (CCSF) is an orphan disease that involves uncommon communication between the internal carotid artery (ICA) and the cavernous sinus (CS). It could result from a tear in the ICA wall due to trauma or a spontaneous event (Remington and Goodwin 2022). As blood from the ICA leaks into the CS, the high-pressure arterial blood travels to the ophthalmic veins, leading to dilation of those veins. Patients can subsequently present with bulging eyes (exophthalmos) or conjunctival infection (Remington and Goodwin 2022). To diagnose a CCSF, computed tomography angiography is a minimally invasive approach but appropriate for diagnosis only; the gold standard approach for both diagnosis and treatment is carotid angiography (Remington and Goodwin 2022).

Generally, endovascular embolization is regarded a primary treatment option for CCSFs. In direct CCSFs, a transarterial embolization approach is usually needed to avoid vision loss due to elevated intraocular pressure. In indirect CCSFs, a transvenous endovascular embolization approach is the preferred surgical option; it involves cannulation and ligation of the SOV or IOV under direct visualization. A recent meta-analysis and systematic review by Texakalidis et al. comparing the transarterial versus transvenous embolization approaches for direct and indirect CCSFs showed no significant differences between the two.

In the transvenous embolization approach for indirect CCSFs, direct cannulation of the SOV is considered to access the fistulas. In the SOV approach, a local anesthetic is used to numb the upper eyelid area and a small skin incision is made. The pre-aponeurotic fat is dissected until the SOV is visualized. The surgeon then isolates and exposes the vessel and it is entered with an angiocatheter. In exceptional cases, when the transvenous embolization approach using the SOV fails, an alternative approach involving percutaneous puncture of the IOV has shown promise.

References

Bergen MP (1981) A literature review of the vascular system in the human orbit. Acta Morphol Neerl Scand 19(4):273–305

Brismar J (1974) Orbital phlebography. II. Anatomy of superior ophthalmic vein and its tributaries. Acta Radiol Diagn (Stockh) 15(5):481–496

Cheung N, McNab AA (2003) Venous anatomy of the orbit. Invest Ophthalmol Vis Sci 44(3):988–995

Forrester JV (2016) The eye: basic sciences in practice, 4th edn. Elsevier, Amsterdam

Hayreh SS (2006) Orbital vascular anatomy. Eye (Lond) 20(10):1130–1144

Mathew AV, Craig E, Al-Mahmoud R, Batty R, Raghavan A, Mordekar SR, Chan J, Connolly DJ (2014) Paediatric post-septal and pre-septal cellulitis: 10 years' experience at a tertiary-level children's hospital. Br J Radiol 87(1033):20130503

Moore KL, Dalley AF, Agur AMR (2014) Clinically oriented anatomy, 7th edn. Lippincott Williams & Wilkins, Philadelphia

Netter F (2019) Atlas of human anatomy, 7th edn. Saunders, Philadelphia

Raybaud C (2010) Normal and abnormal embryology and development of the intracranial vascular system. Neurosurg Clin N Am 21(3):399–426

Remington LA, Goodwin D (2022) Orbital blood supply. In: Clinical anatomy and physiology of the visual system, 3rd edn. Essay, Elsevier, pp 212–216

Snell RS, Lemp MA (1998) Clinical anatomy of the eye, 2nd edn. Blackwell Science, Carlton

Standring S (2016) Gray's anatomy, 41st edn. Elsevier Churchill Livingstone, Edinburgh

Tubbs RS, Shoja MM, Loukas M, Bergman RA (2016) Bergman's comprehensive encyclopedia of human anatomic variation. Wiley Blackwell, Hoboken

Intracranial Lymphatics

20

Semih Topbas and Stephen J. Bordes, Jr.

20.1 Embryology, Anatomy, and Histology

The lymphatic vasculature is a vital component of the circulatory system. It mediates the drainage of interstitial fluid to maintain fluid homeostasis, transport macromolecules, and regulate immune cell trafficking, allowing for immune surveillance and response. Lymphatic vessels are composed of lymphatic endothelial cells (LECs) that differentiate from the endothelium of the cardinal vein during embryo development and commit to this lineage through vascular endothelial growth factor C (VEGF-C) signaling. Lymphatic capillaries are formed by a single layer of LECs, have discontinuous basement membranes, and lack smooth muscle cells. These initial capillaries, made highly permeable by the discontinuous button-like junctions between the LECs, begin blind-ended and drain into collecting lymphatic vessels that are larger in caliber, surrounded by smooth muscle cells, and impermeable; they have continuous tight junctions and intraluminal integrin-expressing valves for unidirectional flow, in contrast to the capillaries (Aspelund et al. 2015).

Given the vital role of the lymphatic system in immune surveillance, waste regulation, and fluid homeostasis, it seems surprising that the central nervous system (CNS) is considered to lack conventional lymphatic vasculature (Louveau et al. 2015). This has raised questions as to how cerebral interstitial fluid (ISF) is cleared of waste.

The blood-brain barrier limits the exchange of compounds; the transvascular clearance of most molecules depends on specific active transport mechanisms. In addition, the so-called glymphatic system in the brain provides a unique paravascular route of fluid exchange between the brain ISF and the CSF along glial routes without having to cross the blood-brain barrier. Downstream of this glymphatic system, most of the CSF drains into the venous circulation through arachnoid granulations. Nevertheless, several studies have shown that a proportion of the CSF is also drained into extracranial lymphatic vessels and lymph nodes, though the mechanism and routes are unclear. While it is accepted that the CNS undergoes immune surveillance within the meninges (Ransohoff and Engelhardt 2012), the entrance and exit routes and pathways of the immune cells from the CNS are not well understood. These inconsistencies led us to re-examine the role of lymphatics in the CNS and to reconsider our previously accepted dogma that the CNS lacks lymphatics. While these discoveries are quite recent, they have already added invaluable insight to our

S. Topbas
Department of Medicine, San Joaquin General Hospital, San Joaquin, CA, USA

S. J. Bordes, Jr. (✉)
Department of Surgery, Louisiana State University Health Sciences Center, New Orleans, LA, USA

S. J. Bordes, Jr. et al. (eds.), *The Clinical Anatomy of the Vascular System*,
https://doi.org/10.1007/978-3-031-78326-5_20

understanding of the complex workings of the brain and will no doubt be a key component of our goal to improve our understanding and, ultimately, treatment of many CNS disorders.

20.2 Meningeal Lymphatic Vessels

Lymphatic vessels in the brain meninges were first noted by the late eighteenth century Italian physician Paolo Mascagni (Bucchieri et al. 2015), but his claims were largely discounted for more than 200 years until two independent studies by Louveau et al. and Aspelund et al. in 2015, using modern techniques and robust methods, detailed the structural and functional characterization of a network of lymphatic vessels in the meninges (Aspelund et al. 2015; Louveau et al. 2015; Iliff and Nedergaard 2013).

Lymphatic vessels within the meninges of the CNS express the classical markers of lymphatic endothelial cells, including vascular endothelial growth factor receptor 3 (VEGFR3), LYVE-1, CD31, and Prox1, and drain both molecules and immune cells from the subarachnoid space into the deep cervical lymph nodes (Bucchieri et al. 2015). The meningeal lymphatic vessels run down toward the base of the skull along the superior sagittal sinus, transverse sinus, sigmoid sinus, and major branches of the middle and anterior meningeal arteries. A network of lymphatic vessels exits the skull along the meningeal portions of the pterygopalatine artery, a branch of the internal carotid, which gives rise to the middle meningeal artery. In the base of the skull, lymphatic vessels are observed in the distal portions of several cranial nerves (optic, trigeminal, glossopharyngeal, vagus, and accessory) and exit the skull along those nerves. Lymphatic vessels can also be seen in the dural lining of the cribriform plate, passing through the skull into the nasal mucosa. Generally, the base of the skull contains a more extensive network of lymphatics; interestingly, only these vessels contain valves like those in peripheral lymphatics (Iliff and Nedergaard 2013; Papadopoulos et al. 2020).

Lymphatic vessels, in general, can be differentiated into two structural types, initial and collecting, according to the presence or absence of surrounding smooth muscle cells, unidirectional valves, and their permeability as determined by the expression patterns of adhesion molecules. The meningeal lymphatic vessels have features characteristic of the initial lymphatic capillaries of the periphery: They are devoid of smooth muscle cells, lack integrin-expressing valves (except those in the base of the skull), and exhibit a discontinuous basement membrane. Furthermore, they maintain structural homogeneity as they remain thin and mostly unbranched, in contrast to their peripheral counterparts. The meningeal lymphatic network is markedly less complex, with significantly less tissue coverage than in the periphery (Iliff and Nedergaard 2013; Papadopoulos et al. 2020).

20.3 MRI Visualization of the Dural Lymphatic Vessels

Building on the studies of Aspelund et al. and Louveau et al., a team of researchers provided noninvasive visualization of the meningeal lymphatics in both humans and nonhuman primates (marmoset monkeys) using T2-weighted fluid-attenuation inversion recovery (T2-FLAIR) pulse sequence, commonly used to detect lesions within the brain parenchyma, and T1-weighted "black-blood" imaging sequences, which are used to measure vascular wall thickness and to detect atherosclerotic plaques (Absinta et al. 2017). These techniques revealed dural lymphatics running parallel to the dural venous sinuses and alongside branches of the middle meningeal artery, with a maximum apparent diameter of 1 mm (Absinta et al. 2017). MRI cannot visualize the tiny lymphatic capillaries revealed by immunohistochemical and immunofluorescence staining.

Furthermore, to understand the function of any vascular compartment thoroughly, the flow direction must be known. This is achieved with time-of-flight (TOF) angiography, an MRI technique often used to image arteries. Briefly, this technique exploits the key principle that blood flowing from outside to the inside of the image section appears brighter because it has a greater signal intensity than the relatively saturated protons in the imaged section (Kuo et al. 2018). Using this principle, MRI can resolve the direction of flow within vessels without using exogenous contrast agents. Interestingly, when this technique is applied to study the meningeal lymphatics running alongside the superior sagittal sinus, lymphatic flow appears posterior to anterior, running countercurrent to the direction of venous flow in the sagittal sinus. This countercurrent flow strongly supports the belief that most CNS lymphatic flow is directed toward the cribriform plate (Absinta et al. 2017; Kuo et al. 2018).

20.4 Pathology and Clinical Significance

Alzheimer's Disease

The extracellular deposition of amyloid beta (AB) aggregates constitutes one of the main pathological hallmarks of Alzheimer's disease (AD). It contributes to these patients' cognitive impairment and behavioral changes (Serrano-Pozo et al. 2011). Given the role of the meningeal lymphatics in clearing macromolecules and waste, their recent (re)discovery encourages us to reassess their role in AD pathology.

Waste is removed from the brain by a combination of transcellular transport, phagocytosis by resident microglia, and CSF/ISF efflux through the perivascular glymphatic route, to which the meningeal lymphatics are downstream. Aging, the biggest risk factor for Alzheimer's disease, has been associated with a decreased paravascular recirculation of CSF/ISF and is therefore thought to be at least partly responsible for the accumulation of AB in AD patients. Furthermore, studies have shown progressive lymphatic vessel dysfunction with aging in peripheral tissues, raising the question of whether a decline in CNS lymphatic function contributes to the amyloid pathology seen in AD.

To this end, pharmacological, surgical, and genetic mouse models of impaired meningeal lymphatic function have been studied to investigate the influence of such functional impairment on the pathology of AD. These models first demonstrated a significant decline of brain perfusion by CSF macromolecules, further establishing the functional connection between CSF influx/ISF efflux mechanisms and the meningeal lymphatics. However, more importantly for our focus on Alzheimer's pathology, the functionally impaired mice performed worse in the Morris water maze and the cued fear conditioning test, which assess learning and memory.

Interestingly, impaired brain perfusion by CSF in old mice is accompanied by a decrease in meningeal lymphatic vessel diameter and decreased drainage of CSF macromolecules into the deep cervical lymph nodes, highlighting the declining function of lymphatics with age (Zhou et al. 2020). Previous studies have underlined the role of VEGF-C in meningeal lymphatic development and differentiation. Notably, mice treated with recombinant VEGF-C showed increased meningeal lymphatic diameter and drainage of CSF into cervical lymphatics. These VEGF-C-treated elderly mice showed significant improvements in learning and memory. However, when the lymphatic afferents in the same VEGF-C-treated mice were surgically ligated, the beneficial cognitive effects were reversed, strongly supporting the hypothesis that enhanced meningeal lymphatic drainage was responsible for the cognitive improvements.

Lastly, samples from postmortem AD patients showed striking vascular amyloid pathology in the cortical leptomeninges and AB deposition in the dura mater adjacent to the superior sagittal sinus, where meningeal lymphatics are found, and in the cervical lymph nodes, to which meningeal lymphatics drain. Furthermore, AB deposition can be replicated in mouse modes after meningeal lymphatic vessel ablation. These studies further establish that AB peptides have access to locations containing meningeal lymphatics, which can then promote their drainage into the periphery (Da Mesquita et al. 2018).

Taken together, these studies strongly suggest that meningeal lymphatic dysfunction is an aggravating factor in AD pathology and opens the door for promising new therapeutic targets.

Parkinson's Disease

Our knowledge of the pathophysiology of Parkinson's disease has also been extended by recent studies highlighting the role of meningeal lymphatics in macromolecule clearance. Parkinson's disease (PD), a common neurodegenerative movement disorder, arises because of the loss of dopaminergic neurons in the substantia nigra pars compacta, the pathological hallmark being intracellular aggregates of α-synuclein (α-syn) in the form of Lewy bodies. Although the pathogenesis of PD is still not fully understood, the vast majority of PD patients show decreased removal of α-syn from the brain, leading to toxic accumulation (das Neves et al. 2021). Previous studies have reported that α-syn is degraded through various intracellular clearance mechanisms, but it was not clear whether an extracellular removal route existed. The recent rediscovery of the meningeal lymphatics and their role in macromolecule clearance raised the question of whether the abnormal lymphatic function was also involved in PD pathology.

This question was addressed using A53T transgenic mice over-expressing mutated human α-syn. Surgical ligation of the lymphatics draining into the deep cervical lymph nodes resulted in significant intracellular and extracellular exacerbation of α-syn aggregation in the substantia nigra of these mice. Furthermore, A53T transgenic mice that underwent lymphatic ligation showed significantly poorer coordination and balance than controls. The increases in α-syn after lymphatic ligation also led to higher astrocyte activation, increased levels of proinflammatory cytokines, and prominent loss of dopaminergic neurons, suggesting that decreased macromolecule clearance through the lymphatics could partly explain the secondary pathological cascades that underlie the clinical symptoms of PD (Zou et al. 2019). The contribution of meningeal lymphatic dysfunction to PD pathology was also highlighted in human studies using dynamic contrast-enhanced magnetic resonance imaging to assess meningeal lymphatic flow. In a 2021 study, patients with idiopathic PD demonstrated reduced flow through meningeal lymphatics along the superior sagittal and sigmoid sinuses and significant delays in the perfusion of the deep cervical lymph nodes by CSF tracers (Ding et al. 2021). These results support the idea that meningeal lymphatic drainage dysfunction aggravates α-syn pathology and probably contributes to the progression of PD.

References

Absinta M et al (2017) Human and nonhuman primate meninges harbor lymphatic vessels that can be visualized noninvasively by MRI. Elife 6:e29738. https://www.ncbi.nlm.nih.gov/pmc/articles/PMC5626482/

Aspelund A et al (2015) A dural lymphatic vascular system that drains brain interstitial fluid and macromolecules. J Exp Med. 212(7):991–999. https://www.ncbi.nlm.nih.gov/pmc/articles/PMC4493418/

Bucchieri F, Farina F, Zummo G, Cappello F (2015) Lymphatic vessels of the dura mater: a new discovery? J Anat 227(5):702–703. https://www.ncbi.nlm.nih.gov/pmc/articles/PMC4609205/

Da Mesquita S et al (2018) Functional aspects of meningeal lymphatics in ageing and Alzheimer's disease. Nature. 560(7717):185–191

das Neves SP, Delivanoglou N, Da Mesquita S (2021) CNS-draining meningeal lymphatic vasculature: roles, conundrums and future challenges. Front Pharmacol. 12:655052. https://www.frontiersin.org/articles/10.3389/fphar.2021.655052/full

Ding X-B et al (2021) Impaired meningeal lymphatic drainage in patients with idiopathic Parkinson's disease. Nat Med. 27(3):411–418. https://pubmed.ncbi.nlm.nih.gov/33462448/

Iliff JJ, Nedergaard M (2013) Is there a cerebral lymphatic system? Stroke. 44(601):S93–S95. https://www.ncbi.nlm.nih.gov/pmc/articles/PMC3699410/

Kuo PH, Stuehm C, Squire SW, Johnson KM (2018) Meningeal lymphatic vessel flow runs countercurrent to venous flow in the superior sagittal sinus of the human brain. Tomography 4:99–104. https://www.ncbi.nlm.nih.gov/pmc/articles/PMC6173792/

Louveau A et al (2015) Structural and functional features of central nervous system lymphatics. Nature 523(7560):337–341. https://www.ncbi.nlm.nih.gov/pmc/articles/PMC4506234/

Papadopoulos Z, Herz J, Kipnis J (2020) Meningeal lymphatics: from anatomy to central nervous system immune surveillance. J Immunol. 204(2):286–293. https://www.jimmunol.org/content/204/2/286

Ransohoff RM, Engelhardt B (2012) The anatomical and cellular basis of immune surveillance in the central nervous system. Nat Rev Immunol. 12(9):623–635. https://pubmed.ncbi.nlm.nih.gov/22903150/

Serrano-Pozo A, Frosch MP, Masliah E, Hyman BT (2011) Neuropathological alterations in Alzheimer disease. Cold Spring Harb Perspect Med. 1(1):a006189. https://www.ncbi.nlm.nih.gov/pmc/articles/PMC3234452/

Zhou Y et al (2020) Impairment of the glymphatic pathway and putative meningeal lymphatic vessels in the aging human. Ann Neurol 87(3):357–369. https://pubmed.ncbi.nlm.nih.gov/31916277/

Zou W et al (2019) Blocking meningeal lymphatic drainage aggravates Parkinson's disease-like pathology in mice overexpressing mutated α-synuclein. Transl Neurodegener. 8:7. https://pubmed.ncbi.nlm.nih.gov/30867902/

21 Lymphatics of the Face and the Scalp

Nilo Alvarez Toledo

21.1 Embryology

The lymph nodes develop during fetal life after the development of the lymphatic plexus (Alex 2003). Two models, centrifugal and centripetal, have been proposed for the embryological origin of the lymphatic system (Ribatti and Crivellato 2010). The first model proposes that the lymphatics originate from the veins by sprouting (Ranvier 1895; Sabin 1909). The second describes the origin of the lymphatics from lymphatic progenitor cells, lymphangioblasts, in the early mesenchyme in the peripheral tissues (Huntington 1908; Kampmeier 1912). Evidence for dual origin has been published, from both the veins and the peripheral lymphoblasts (Wilting et al. 2006).

21.2 Anatomy

The lymphatic vessels of the superficial tissues in the head and neck can be divided into three groups: scalp, facial, and cervical (Pan et al. 2011). This chapter focuses on the scalp and face groups (Fig. 21.1).

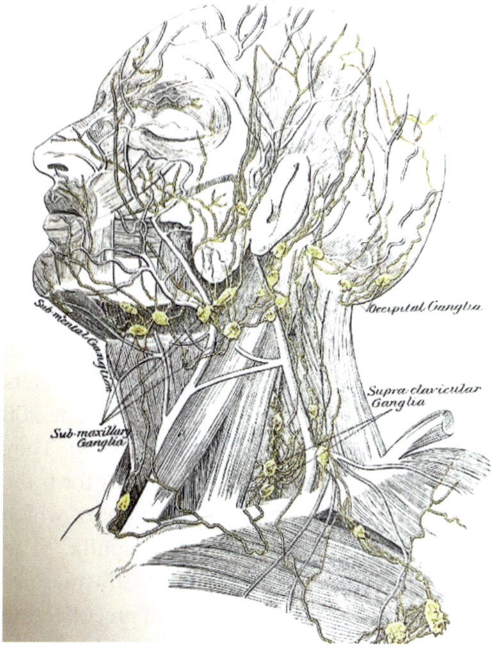

Fig. 21.1 Schematic drawing of the lymphatic drainage of the face and scalp. (After Allen 1882)

Lymphatic Drainage of the Scalp and Face

The lymph-collecting vessels are dense in the scalp but sparse in the face (Pan and Wang 2013). In the scalp, the dense array of lymph capillaries in the galea and the dermis converge in the subcutaneous tissue to form the precollecting lymph ves-

N. Alvarez Toledo (✉)
Department of Neurosurgery, Icahn School of Medicine at Mount Sinai, New York, United States
e-mail: nilo.alvareztoledo@mountsinai.org

S. J. Bordes, Jr. et al. (eds.), *The Clinical Anatomy of the Vascular System*,
https://doi.org/10.1007/978-3-031-78326-5_21

sels. These run in the subcutaneous tissue and converge at the level of the parasagittal region to form the lymph-collecting vessels, which run caudally to drain into their corresponding first-tier lymph nodes (Pan et al. 2008). The lymph-collecting vessels of the face originate from lymph capillary networks around the eyes, nose, mouth, and chin. They run obliquely, downward, and backward in the subcutaneous tissue and drain into their corresponding first-tier lymph nodes (Pan et al. 2008). In both the face and the scalp, the lymph vessels can bypass the nearest node and continue to a different one (Pan and Wang 2013). Therefore, the lymphatic drainage of a region does not always enter the nearest lymph node.

Scalp

The lymphatic drainage of the scalp is organized in a basic unit comprising the lymph capillaries, the precollecting vessels, and the lymph-collecting vessels, which drain into their corresponding first-tier lymph nodes (Pan et al. 2008). The lymph capillaries are valveless, fragile vessels that form a dense and fine polygonal network in the galea and the dermis of the scalp (Pan et al. 2008). They are less than 100 micrometers in diameter in the dermis and more than 200 micrometers in the galea (Pan et al. 2008). They converge in the subcutaneous tissue to form the precollecting vessels, the initial portions of which are often dilated into a "lymphatic ampulla." (Pan et al. 2008) Their diameters are 100–300 micrometers (Pan et al. 2008). Two types of precollecting vessels have been described, direct and indirect. The former extend from the lymph capillaries in the dermis and galea to the lymph-collecting vessels in the subcutaneous tissue. The latter extend from the lymph capillaries of the dermis, traverse the subcutaneous tissue, bypass the lymph-collecting vessels, and join the galea precollecting vessels (Pan et al. 2008).

The precollecting lymph vessels converge at the parasagittal region level, approximately two centimeters lateral to the midline, to form the lymph-collecting vessels (Pan et al. 2008; Feely et al. 2012). The lymph-collecting vessels of the scalp travel inferiorly toward their corresponding first-tier lymph nodes, running in a "snake-like" fashion through the deep subcutaneous tissue, branching, diverging, and converging along their course (Pan et al. 2011). Their diameters average approximately 200 micrometers (Pan et al. 2008). Three groups of scalp lymph-collecting vessels have been described: frontal, parietal, and occipital branches. There are 3–6 frontal, 4–12 parietal, and 4–9 occipital lymph-collecting vessels on each side (Pan et al. 2011).

Iwanaga et al. (2021) divided the scalp into frontal, temporoparietal, and occipital regions and described lymph node drainage in each of those three regions (Iwanaga et al. 2021). The submandibular nodes receive lymph from the upper nasal root area in the frontal region (Iwanaga et al. 2021). The suprafascial/superficial parotid, subfascial extraglandular, and deep intraglandular lymph nodes drain the rest of the frontal region. The suprafascial/superficial parotid and mastoid lymph nodes drain the temporoparietal regions (Iwanaga et al. 2021). The occipital and the superior deep cervical lymph nodes drain the occipital region (Iwanaga et al. 2021).

Face

The lymph capillaries in the face create networks at the levels of the upper and lower eyelids, in the outer and inner canthi, the sides of the nose and the mouth, and the chin. They converge and drain into the lymph-collecting vessels, which run radially in the subcutaneous tissue, obliquely downward and backward to their first-tier lymph nodes (Pan et al. 2011). Four groups of branches of lymph-collecting vessels in the face have been described: the eyelid, nasal, oral, and mental branches (Pan et al. 2011). The upper and lower eyelid lymph capillaries converge in the inner and outer canthi to form lymph-collecting vessels that run inferiorly (Pan et al. 2011). The lymph-collecting vessel that drains the outer canthi, called the "outer canthus branch," drains into the parotid and submandibular lymph nodes. The one that drains the inner canthi is called the "inner canthus branch" and drains into the submandibular lymph node (Pan et al. 2011). The nasal branches originate bilaterally from the lateral nose and run obliquely to their corresponding first-tier lymph nodes (Pan et al. 2011). The oral lymph-collecting vessels originate at the mouth corner and drain the lymph capillaries of the perilabial tissue. There is one oral lymph-collecting vessel on each side; typically, they drain into the buccinator or submandibular lymph nodes. In some cases, the oral lymph-collecting vessels join the inner canthus branch (Pan et al. 2011). The submental lymph-collecting vessels run in the subcutaneous tissue of the submental region and drain into the submental and submandibular lymph nodes (Pan et al. 2011). The facial lymph-collecting vessels are approximately 300 micrometers in diameter (Pan et al. 2008).

In general, the eyelid and conjunctiva lymph-collecting vessels drain into the parotid, facial, and submandibular lymph nodes (Iwanaga et al. 2021). The anterior malar region, the nasal ala, the upper lip, and the lateral lower lip drain into the submandibular and facial lymph nodes (Iwanaga et al. 2021). The central area of the lower lip and the mental region drain into the submental lymph nodes (Iwanaga et al. 2021).

Lymph Nodes and Lymphatic Drainage of the Scalp and Face

The lymph nodes of the head and neck that regularly drain the face and scalp are the occipital, mastoid, superior deep cervical, parotid, facial, submandibular, and submental (Iwanaga et al. 2021).

Occipital Lymph Nodes

The occipital lymph nodes typically lie in the angle between the origin of the trapezius muscle and the insertion of the sternocleidomastoid, in the upper part of the posterior triangle of the neck (Iwanaga et al. 2021). There are 1–3 of them and they usually occupy a subfascial plane (Iwanaga et al. 2021; Lengelé et al. 2007). They receive their afferent lymph vessels from the occipital scalp and periosteum (Iwanaga et al. 2021). Their efferent vessels run downward to join the posterior cervical accessory chain (Lengelé et al. 2007).

Mastoid Lymph Nodes

The mastoid or retroauricular lymph nodes are located in the postauricular region, lying on the mastoid process of the temporal bone (Iwanaga et al. 2021). There are 1–2 of them (Iwanaga et al. 2021; Ogassavara et al. 2016). Their afferent vessels come from the sincipital scalp, the postauricular skin, and the inferior temporal region (Iwanaga et al. 2021). Their efferent vessels join the superior deep cervical and the inferior parotid lymph nodes (Iwanaga et al. 2021).

Superior Deep Cervical Lymph Nodes

The superior deep cervical lymph nodes are located along the upper internal jugular vein and mostly medial to the sternocleidomastoid muscle (Iwanaga et al. 2021). There are 4–7 of them (Iwanaga et al. 2021), the most prominent one(s) being the sentinel or Kuttner's jugulodigastric lymph node(s), typically located in the triangle formed by the posterior belly of the digastric muscle, the internal jugular vein, and the facial vein (Iwanaga et al. 2021; Lengelé et al. 2007). Afferent lymph vessels from the occipital region join these lymph nodes (Iwanaga et al. 2021). The efferent vessels drain into the inferior deep cervical lymph nodes or directly into the jugular trunk (Iwanaga et al. 2021).

Parotid Lymph Nodes

The parotid lymph nodes comprise three groups surrounding the parotid gland: the deep intraglandular, subfascial extraglandular, and suprafascial/superficial parotid lymph nodes (Iwanaga et al. 2021).

- There are 2–4 deep intraglandular lymph nodes located underneath the parotid sheath (Iwanaga et al. 2021; Terada and Kawata 2022). Some of them lie lateral to the facial nerve (Iwanaga et al. 2021). Their afferent lymphatic vessels come from other lymph nodes, the suprafascial parotid and subfascial preauricular, and from the procephalic skin (Iwanaga et al. 2021). Their efferent vessels join the superior deep cervical and inferior parotid infra-auricular nodes (Iwanaga et al. 2021).
- The subfascial extraglandular lymph nodes are subdivided into the subfascial preauricular and inferior parotid infra-auricular groups (Iwanaga et al. 2021). One to three lymph nodes are present between the parotid sheath and the deep cervical fascia (Lengelé et al. 2007) (Iwanaga et al. 2021). They receive

afferent lymph vessels from the procephalic scalp and the eyelids (Iwanaga et al. 2021). Their efferent vessels join the deep intraglandular and the subfascial preauricular lymph nodes (Iwanaga et al. 2021).

- There are 1–2 suprafascial/superficial parotid lymph nodes, also known as subcutaneous preauricular nodes (Lengelé et al. 2007), located superficial to the superficial/investing layer of the fascia colli/deep cervical fascia (Iwanaga et al. 2021). They receive afferent lymph vessels from the procephalic region, the sincipital scalp, and the upper and lower eyelids (Iwanaga et al. 2021). Their efferent vessels reach the deep intraglandular and subfascial preauricular nodes (Iwanaga et al. 2021).

Facial Lymph Nodes

The facial lymph nodes are a collection of five groups closely related to the facial artery and vein (Iwanaga et al. 2021). They are inconsistently present (Tart et al. 1993) and are located internal to the superficial musculoaponeurotic system (Iwanaga et al. 2021). The five groups are:

- Buccinator.
- Infraorbital/nasolabial (Rouviere and Tobias 1938).
- Malar.
- Retrozygomatic.
- Mandibular.

The 2–4 buccinator nodes are present "in the fat of the buccinator space." (Tart et al. 1993) The anterior facial vein divides this group into anterior (1–2 nodes) and posterior (1–2 nodes) subgroups. The anterior subgroup is located anterior or superficial to the vein while the posterior group is located posterior or deep to it (Tart et al. 1993). Their afferent lymph vessels come from the cheek, the nose, and the lower eyelid skin and subcutaneous tissue (Tart et al. 1993). They also receive afferents from the malar, infraorbital, and retrozygomatic lymph node groups (Tart et al. 1993). Their efferent lymphatics drain into the submandibular lymph nodes via the mandibular lymph nodes (Tart et al. 1993).

The infraorbital lymph node, also known as the nasolabial lymph node (Rouviere and Tobias 1938), lies in the canine space or infraorbital space at the level of the nasolabial fold. It is typically a single lymph node closely related to the facial vessels in the area (Tart et al. 1993). It drains the nose, the nasolabial fold, and the inner canthus skin and subcutaneous tissue (Tart et al. 1993). Its efferent lymphatics travel to the buccinator and mandibular lymph node groups (Tart et al. 1993).

The malar group is located between the malar eminence and the temporozygomatic suture (Tart et al. 1993). These nodes lie between the lymph-collecting vessels that run vertically connecting the temporal region and the parotid lymph nodes (Tart et al. 1993). Their afferent lymphatics come from the skin and subcutaneous tissue of the malar, temporal, and outer canthus regions (Tart et al. 1993). Their efferents drain into the parotid lymph nodes (Tart et al. 1993).

The retrozygomatic nodes, first described by Tart et al. in 1993, are located in the infratemporal fossa, deep into the zygomatic arch. They drain deep structures of the face (Tart et al. 1993).

The 1–3 mandibular lymph nodes lie on the external surface of the mandible, anterior to the masseter muscle, and close to the facial artery (Tart et al. 1993). Also part of this group is the inferior maxillary lymph node, located in the subcutaneous tissue on the inferior border of the mandible (Tart et al. 1993). Their afferent lymph vessels come from the skin and subcutaneous tissue of the lower lip and cheek and from the buccinator and infraorbital facial groups (Tart et al. 1993). Their efferents drain into the submandibular lymph nodes (Tart et al. 1993).

In general, the afferent lymphatic vessels to the facial lymph nodes come from the palpebral region, the conjunctiva, the anteriormalar region, the nasal ala, the upper lip, and the lateral area of the lower lip (Iwanaga et al. 2021).

Submandibular Lymph Nodes

The 3–6 submandibular lymph nodes are located internally to the deep fascia in the submandibular triangle, the area bounded by the anterior and posterior bellies of the digastric muscle and the body of the mandible (Iwanaga et al. 2021; Kikuta et al. 2019). They are found in three places: anterior to the submandibular gland, around the mandibular bone close to the facial artery, and deep to the gland (Iwanaga et al.

Table 21.1 Lymph node drainage of the scalp and the face

Lymph nodes	Area of drainage
Submandibular LNs	Upper nasal root area
Suprafascial/superficial parotid LNs Mastoid LNs	Temporoparietal region
Occipital LNs Superior deep cervical LNs	Occipital region
Parotid LNs Facial LNs Submandibular LNs	Eyelid Conjunctiva
Parotid LNs Facial LNs	Posterior area of the malar region
Submandibular LNs Facial LNs	Anterior area of the malar region Nasal ala Upper lip Lateral area of the lower lip
Submental LNs	Central area of the lower lip Mental region

Adapted from Iwanaga et al. (2021)
LNs lymph nodes

2021). The afferent lymphatic vessels come from the skin of the buccal region and the lips (Iwanaga et al. 2021). Most of the facial lymph nodes also drain into the submandibular nodes (Tart et al. 1993). The efferent vessels reach the superior deep cervical lymph nodes (Iwanaga et al. 2021).

Submental Lymph Nodes

The 1–3 submental or suprahyoid lymph nodes are located in the submental triangle, the area limited by both anterior bellies of the digastric muscles and the hyoid bone, between the mylohyoid and the platysma muscles (Kikuta et al. 2019). The afferent lymphatic vessels come from the center of the mental region and the center of the lower lip. The efferent lymph vessels reach the submandibular or the superior deep cervical lymph nodes (Iwanaga et al. 2021) (Table 21.1).

21.3 Anatomical Variations

Variants of the lymph-collecting vessels of the scalp and the face are related to their corresponding lymph node drainage. The frontal lymph-collecting vessels usually drain into the parotid and submandibular lymph nodes; however, in some cases, they drain into the mastoid or retroauricular lymph nodes or into the facial lymph nodes, specifically the nasolabial and the buccinator groups (Pan et al. 2011). The lymph-collecting vessels from the temporoparietal region usually drain into the parotid lymph nodes; however, in some cases, they drain into the mastoid, the occipital, or even the lateral internal jugular lymph nodes (Pan et al. 2011). The lymph-collecting vessels from the occipital region drain in some cases into the lateral internal jugular lymph nodes (Pan et al. 2011).

The lymph-collecting vessel that drains the inner canthi, the "inner canthus branch," usually drains into the submandibular lymph node, but sometimes it drains into the parotid lymph nodes or the buccinator group of the facial lymph nodes (Pan et al. 2011).

21.4 Pathology

The lymphatic vessels of the scalp and the face can be affected by several diseases, primarily in cases such as local lymphedema or local lymph vessel malformations, or secondarily by infections and malignancies.

Lymphedema

The impairment of lymphatic drainage caused by injury to lymphatic vessels, lymphatic malformation, trauma, surgery, radiotherapy, metastasis, or mechanical obstruction produces an accumulation of protein and water in the interstitium called lymphedema (Warren et al. 2007) (Lu et al. 2009). Lymphedema can be diffuse when it involves a whole extremity or, exceptionally, can be localized to certain areas of the body (Lu et al. 2009). Localized lymphedema is extremely rare and can present as a polyp, a plaque, a pendulous swelling, or a sarcoma-like mass (Korbl et al. 2018). Localized lymphedema of the scalp (Korbl et al. 2018) and face (Lu et al. 2009; Van Duyn 1969, 1974; Warren and Slavin 2007) has been reported. Korbl et al. (2018), reported localized lymphedema of the scalp that presented with a prominent papillomatous plaque (Korbl et al. 2018). Lu et al. (2009), reported patients with local lymphedema involving the face and the eyelids (Lu et al. 2009). Warren and Slavin (2007), reported nine patients with facial scar lymphedema, a type of local lymphedema associated with microlymphatic stasis after surgery or trauma (Warren and Slavin 2007).

Lymphatic Malformations

A disorder of the embryological development of the lymphatics can produce an abnormal communication between the peripheral lymphatics and the central venous sacs, a so-called lymphatic malformation (Boardman et al. 2010). These rare lesions are benign and mostly affect the head and neck, presenting as a lymphangioma or a cystic hygroma (Hamoir et al. 2001; Chervenak et al. 1983). They are much more frequently diagnosed in children and typically manifest as an asymptomatic cystic mass secondary to lymph accumulation (Boardman et al. 2010). The parotid region in the face is one common anatomical location of lymphatic malformations (Boardman et al. 2010; Chinnakkulam Kandhasamy et al. 2018).

Lymphangioma of the scalp has been reported (Servet et al. 1992; Lee et al. 2006; Kennedy 1989; Davies 1929). Servet et al. (1992), reported a frontal cavernous lymphangioma in a 16-year-old patient (Servet et al. 1992). Lee et al. (2006), reported a lymphangioma circumscriptum of the scalp in a 12-year-old girl (Lee et al. 2006). Lymphangiomas typically present as cystic, well-circumscribed, soft tumors (Servet et al. 1992). Lymphangioma circumscriptum presents as a group of vesicle-like, fluid-filled lesions (Kennedy 1989; Whimster 1976).

Fetal cystic hygroma affects the fetus and presents as a single or multiloculated cystic mass in the occipitocervical region. It is produced by a congenital malformation of the lymphatic system (Chervenak et al. 1983; Kennedy 1989).

Lymphadenopathy and Lymphadenitis

Lymphadenopathy denotes a lymph node that has reached abnormal size and consistency owing to a disease process (Gosche and Vick 2006). Typically, the head and neck are the most frequent sites for lymphadenopathy, so they are examined regularly by the physician (Karnath 2005). The parotid, mastoid, occipital, submental, submandibular, and deep cervical lymph nodes are particularly related to lymph drainage of the face and scalp (Iwanaga et al. 2021). The causes of lymphadenopathy in this region include infection (bacterial pharyngitis, otitis, dental abscess, toxoplasmosis, tuberculosis, cytomegalovirus, adenovirus, HIV), malignancies (Hodgkin lymphoma, non-Hodgkin lymphoma, metastasis), and reactive lymphadenopathy (Screaton et al. 2002). Other rare cases such as cat-scratch disease affecting the mastoid lymph nodes (Eidlitz-Markus and Zeharia 2006), mastoid suppurative lymphadenitis (Tymofieiev et al. 2021), and post-microneedling lymphadenopathy (Elghblawi 2019) have been reported.

21.5 Clinical/Surgical Topics

Sentinel Lymph Node Biopsy

A sentinel lymph node is any lymph node that drains a tumor site directly (Uren 2004; Maeda et al. 2017; Morton et al. 2014). Biopsy of the sentinel lymph node provides the strongest prognosis for overall survival in cases of melanoma of the scalp (Cappello et al. 2013). Dominant superficial lymph drainage of the cheek and eyelids to the parotid, buccinator, and submandibular (level IB) lymph nodes enables melanomas in the cheek and eyelids to metastasize to the aforementioned lymph nodes (Hayashi et al. 2012). Melanomas of the lips can metastasize to the submental, submandibular, or buccinator lymph nodes (Hayashi et al. 2012).

Ventriculo-Subgaleal Shunt

A ventriculo-subgaleal shunt is created between the lateral ventricle and the subgaleal space. The absorptive qualities of the subgaleal space contribute to cerebrospinal fluid absorption (Eid et al. 2018). The procedure was performed for the first time in 1896 by von Mikulicz (Aschoff et al. 1999). This type of shunt has been used effectively for temporary cerebrospinal fluid diversion in post-hemorrhagic and post-infectious hydrocephalus in children (Amen et al. 2022). Other authors have reported its use in cases of posterior fossa tumors, myelomeningocele repair, shunt infections, and acute shunt blockage (Sil et al. 2021). The procedure has a much lower infection rate than external ventricular drainage (Nee et al. 2017). The lymphatics of the galea could contribute to this important absorptive property of the subgaleal space.

References

Alex JC (2003) The head and neck lymphatic system. Facial Plast Surg Clin North Am 11(1):43–47. https://doi.org/10.1016/S1064-7406(02)00050-0

Amen MM, Zaher A, Badr HI, Elshirbiny MF, Elnaggar AM, Khalil AF (2022) Ventriculosubgaleal shunt as a proposed technique for post-infectious hydrocephalus. Childs Nerv Syst 38(11):2155–2162. https://doi.org/10.1007/s00381-022-05661-x

Aschoff A, Kremer P, Hashemi B, Kunze S (1999) The scientific history of hydrocephalus and its treatment. Neurosurg Rev 22(2–3):67–95. https://doi.org/10.1007/s101430050035

Boardman SJ, Cochrane LA, Roebuck D, Elliott MJ, Hartley BE (2010) Multimodality treatment of pediatric lymphatic malformations of the head and neck using surgery and sclerotherapy. Arch Otolaryngol Head Neck Surg 136(3):270–276. https://doi.org/10.1001/archoto.2010.6

Cappello ZJ, Augenstein AC, Potts KL, McMasters KM, Bumpous JM (2013) Sentinel lymph node status is the most important prognostic factor in patients with melanoma of the scalp. Laryngoscope 123(6):1411–1415. https://doi.org/10.1002/lary.23793

Chervenak FA, Isaacson G, Blakemore KJ et al (1983) Fetal cystic hygroma. Cause and natural history. N Engl J Med 309(14):822–825. https://doi.org/10.1056/NEJM198310063091403

Chinnakkulam Kandhasamy S, Ramasamy Raju T, Sahoo AK, Gunasekaran G (2018) Adult cystic Lymphangioma of the parotid gland: an unwonted presentation. Cureus 10(5):e2644. Published 2018 May 18. https://doi.org/10.7759/cureus.2644

Davies JH (1929) Lymphangioma Circumscriptum. Proc R Soc Med 22(3):353

Eid S, Iwanaga J, Oskouian RJ, Loukas M, Jerry Oakes W, Tubbs RS (2018) Ventriculosubgaleal shunting-a comprehensive review and over two-decade surgical experience [published correction appears in Childs Nerv Syst. 2018 Nov;34(11):2339. Shane Tubbs, R [corrected to Tubbs, R Shane]]. Childs Nerv Syst 34(9):1639–1642. https://doi.org/10.1007/s00381-018-3887-6

Eidlitz-Markus T, Zeharia A (2006) Images in clinical medicine. Cat scratch disease lymphadenopathy. N Engl J Med 354(17):e17. https://doi.org/10.1056/NEJMicm050180

Elghblawi E (2019) Intense retroauricular lymphadenopathy post-microneedling. J Cosmet Dermatol 18(6):2048–2049. https://doi.org/10.1111/jocd.12947

Feely MA, Olsen KD, Gamble GL, Davis MD, Pittelkow MR (2012) Cutaneous lymphatics and chronic lymphedema of the head and neck. Clin Anat 25(1):72–85. https://doi.org/10.1002/ca.22009

Gosche JR, Vick L (2006) Acute, subacute, and chronic cervical lymphadenitis in children. Semin Pediatr Surg 15(2):99–106. https://doi.org/10.1053/j.sempedsurg.2006.02.007

Hamoir M, Plouin-Gaudon I, Rombaux P et al (2001) Lymphatic malformations of the head and neck: a retrospective review and a support for staging. Head Neck 23(4):326–337. https://doi.org/10.1002/hed.1039

Hayashi T, Furukawa H, Oyama A, Funayama E, Saito A, Yamamoto Y (2012) Dominant lymph drainage in the facial region: evaluation of lymph nodes of facial melanoma patients. Int J Clin Oncol 17(4):330–335. https://doi.org/10.1007/s10147-011-0293-4

Huntington GS (1908) Symposium on the development and structure of the lymphatic system. II. The genetic interpretation of the development of the mammalian lymphatic system. Anat Rec 2(1–2):19–45

Iwanaga J, Lofton C, He P, Dumont AS, Tubbs RS (2021) Lymphatic system of the head and neck. J Craniofac Surg 32(5):1901–1905. https://doi.org/10.1097/SCS.0000000000007230

Kampmeier OF (1912) The value of the injection method in the study of lymphatic development. Anat Rec 6(6):223

Karnath BM (2005) Approach to the patient with lymphadenopathy. Hosp Physician 41(7):29

Kennedy TL (1989) Cystic hygroma-lymphangioma: a rare and still unclear entity. Laryngoscope

99(10 Pt 2 Suppl 49):1–10. https://doi.org/10.1288/00005537-198910001-00001

Kikuta S, Iwanaga J, Kusukawa J, Tubbs RS (2019) Triangles of the neck: a review with clinical/surgical applications. Anat Cell Biol 52(2):120–127. https://doi.org/10.5115/acb.2019.52.2.120

Korbl JD, Chan J, Wood BA, Harvey NT (2018) Localised lymphoedema forming a papillated lesion on the scalp. Pathology 50(3):361–364. https://doi.org/10.1016/j.pathol.2017.09.016

Lee YH, Kim SH, Suh MK, Park SK, Kim JR (2006) Lymphangioma Circumscriptum of the scalp resembling nevus Sebaceus. Korean J Dermatol 44:1265–1267

Lengelé B, Hamoir M, Scalliet P, Grégoire V (2007) Anatomical bases for the radiological delineation of lymph node areas. Major collecting trunks, head and neck. Radiother Oncol 85(1):146–155. https://doi.org/10.1016/j.radonc.2007.02.009

Lu S, Tran TA, Jones DM et al (2009) Localized lymphedema (elephantiasis): a case series and review of the literature [published correction appears in J Cutan Pathol. 2009 Apr;36(4):502]. J Cutan Pathol 36(1):1–20. https://doi.org/10.1111/j.1600-0560.2008.00990.x

Maeda T, Yamamoto Y, Furukawa H et al (2017) Dominant lymph drainage patterns in the occipital and parietal regions: evaluation of lymph nodes in patients with skin cancer of the head. Int J Clin Oncol 22(4):774–779. https://doi.org/10.1007/s10147-017-1121-2

Morton DL, Thompson JF, Cochran AJ et al (2014) Final trial report of sentinel-node biopsy versus nodal observation in melanoma. N Engl J Med 370(7):599–609. https://doi.org/10.1056/NEJMoa1310460

Nee LS, Harun R, Sellamuthu P, Idris Z (2017) Comparison between Ventriculosubgaleal shunt and Extraventricular drainage to treat acute hydrocephalus in adults. Asian J Neurosurg 12(4):659–663. https://doi.org/10.4103/ajns.AJNS_122_16

Ogassavara B, Tucunduva Neto RR, de Souza RR, Tucunduva MJ (2016) Ultrasound evaluation of the morphometric patterns of lymph nodes of the head and neck in young and middle-aged individuals. Radiol Bras 49(4):225–228. https://doi.org/10.1590/0100-3984.2015.0002

Pan WR, Wang DG (2013) Historical review of lymphatic studies in the head and neck. J Lymphoedema 8(1):43

Pan WR, Suami H, Taylor GI (2008) Lymphatic drainage of the superficial tissues of the head and neck: anatomical study and clinical implications. Plast Reconstr Surg 121(5):1614–1624. https://doi.org/10.1097/PRS.0b013e31816aa072

Pan WR, Le Roux CM, Briggs CA (2011) Variations in the lymphatic drainage pattern of the head and neck: further anatomic studies and clinical implications. Plast Reconstr Surg 127(2):611–620. https://doi.org/10.1097/PRS.0b013e3181fed511

Ranvier L (1895) Développement des vaisseaux lymphatiques. CR Acad Sci 121:1105–1109

Ribatti D, Crivellato E (2010) The embryonic origins of lymphatic vessels: a historical review. Br J Haematol 149(5):669–674. https://doi.org/10.1111/j.1365-2141.2009.08053.x

Rouviere H, Tobias MJ (1938) Anatomy of the human lymphatic system. Edwards Brothers, Ann Arbor

Sabin FR (1909) The lymphatic system in human embryos, with a consideration of the morphology of the system as a whole. Am J Anat 9(1):43–91

Screaton NJ, Berman LH, Grant JW (2002) Head and neck lymphadenopathy: evaluation with US-guided cutting-needle biopsy. Radiology 224(1):75–81. https://doi.org/10.1148/radiol.2241010602. PMID: 12091664

Servet İNCİ, Bavbek M, Özcan OE, Behsan ÖNOL, Gedikoğlu G (1992) Lymphangioma of the scalp. Turk Neurosurg 2(4):203–206

Sil K, Ghosh SK, Chatterjee S (2021) Ventriculo-subgaleal shunts-broadening the horizons: an institutional experience. Childs Nerv Syst 37(4):1113–1119. https://doi.org/10.1007/s00381-020-04929-4

Tart RP, Mukherji SK, Avino AJ, Stringer SP, Mancuso AA (1993) Facial lymph nodes: normal and abnormal CT appearance. Radiology 188(3):695–700. https://doi.org/10.1148/radiology.188.3.8351335

Terada T, Kawata R (2022) Role of intra-parotid lymph node metastasis in primary parotid carcinoma. Life (Basel) 12(12):2053. Published 2022 Dec 7. https://doi.org/10.3390/life12122053

Tymofieiev OO, Ushko NO, Fesenko II, Tymofieiev OO, Yarifa MO, Cherniak OS (2021) Suppurative mastoid lymphadenitis mimicking mastoiditis: a case report. J Korean Assoc Oral Maxillofac Surg 47(5):398–402. https://doi.org/10.5125/jkaoms.2021.47.5.398

Uren RF (2004) Lymphatic drainage of the skin. Ann Surg Oncol 11(3 Suppl):179S–185S. https://doi.org/10.1007/BF02523625

Van Duyn J (1969) Lymphedema in face scars. South Med J 62(9):1149–1150. https://doi.org/10.1097/00007611-196909000-00026

Van Duyn J (1974) First appearance of lymphedema in facial scars. South Med J 67(12):1502–1503. https://doi.org/10.1097/00007611-197412000-00027

Warren AG, Slavin SA (2007) Scar lymphedema: fact or fiction? Ann Plast Surg 59(1):41–45. https://doi.org/10.1097/01.sap.0000258449.23979.3f

Warren AG, Brorson H, Borud LJ, Slavin SA (2007) Lymphedema: a comprehensive review. Ann Plast Surg 59(4):464–472. https://doi.org/10.1097/01.sap.0000257149.42922.7e

Whimster IW (1976) The pathology of lymphangioma circumscriptum. Br J Dermatol 94(5):473–486. https://doi.org/10.1111/j.1365-2133.1976.tb05134.x

Wilting J, Aref Y, Huang R et al (2006) Dual origin of avian lymphatics. Dev Biol 292(1):165–173. https://doi.org/10.1016/j.ydbio.2005.12.043

Lymphatics of the Neck

22

Charles J. Peper, Stephen J. Bordes, Jr., and R. Shane Tubbs

22.1 Introduction

It is estimated that up to 300 of the human body's circa 800 lymph nodes reside in the head and neck (Fig. 22.1) (Iwanaga et al. 2021). Although most lymphadenopathies result from benign viral-reactive etiologies, insidious pathologies of the head and neck often involve the lymphatic system. For instance, the mere presence of intranodal metastasis dramatically affects head and

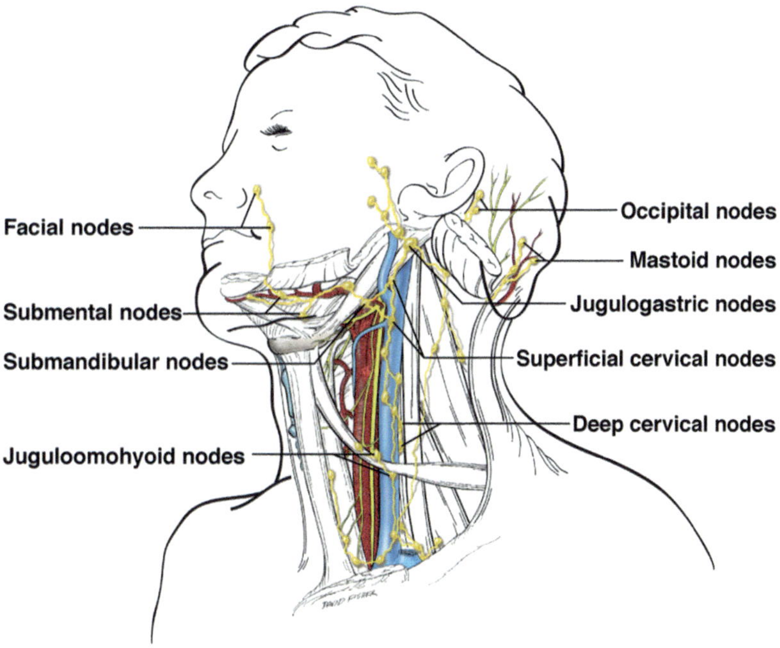

Fig. 22.1 Main lymphatic vessels and nodes in the neck

C. J. Peper
Tulane University School of Medicine,
New Orleans, LA, USA
e-mail: cpeper@tulane.edu

S. J. Bordes, Jr.
Department of Surgery, Louisiana State University Health Sciences Center, New Orleans, LA, USA

R. S. Tubbs (✉)
Department of Neurosurgery, Tulane University School of Medicine,
New Orleans, LA, USA
e-mail: rtubbs@tulane.edu

S. J. Bordes, Jr. et al. (eds.), *The Clinical Anatomy of the Vascular System*,
https://doi.org/10.1007/978-3-031-78326-5_22

neck cancer staging (Agarwal et al. 2018; Lin et al. 2006; Saindane 2013). Scientific understanding of the lymphatic system of the head and neck progressed from the foundation laid by French anatomist-physician Rouviere. Dr. George Crile described radical neck dissection, which, though it was effective for his patients, has undergone major improvements during the last 100 years (Subramanian et al. 2006). Today, advanced imaging and sentinel node biopsy (SNB) guide medical decision-making, and improvements in the surgical procedure of neck dissection have led to improved outcomes. Advances in nuclear medicine imaging allow us to anticipate yet greater accuracy in staging head and neck cancers (de Bree et al. 2015; Vidal-Sicart and Valdés Olmos 2021).

22.2 Embryology

The development of the human lymphatic system has been debated for over 100 years. For a model of lymph system development to be complete, it must explain both lymphvasculogenesis and lymphangiogenesis. Lymphvasculogenesis denotes the development of lymph sacs associated with veins. Lymphangiogenesis denotes the development of the network of vessels belonging to the system. Sabin began the discussion in 1911 by proposing a model resembling the contemporaneous "centrifugal" model as opposed to the "centripetal" model (Mayerson 1962; Lee and Suami 2020).

The centrifugal model describes the lymph sacs as arising from a cluster of endothelial cells and pinching away from their associated veins to form independent structures. Thereafter, lymphangiogenesis proceeds through "endothelial sprouting." The centripetal model describes lymph vessel development in the mesenchyme independently of lymph sac development and assumes that the connections between the vessel network and lymph sacs appear later in development (Mayerson 1962; Lee and Suami 2020).

In any case, the first lymphatic system structures to be visualized in the human embryo are lymph sacs associated with the jugular portion of the cardinal vein, bilaterally. These structures correspond to the deep cervical chain of lymph nodes in the adult (Wang et al. 2012).

22.3 Anatomy

The lateral cervical nodes are divided unequally into two groups, superficial and deep. The superficial nodes lie superficial to the sternocleidomastoid (SCM) and follow the course of the external jugular vein. The deep division of the lateral cervical nodes comprises three chains of nodes, roughly forming a triangle that can be imagined to follow the internal jugular vein, transverse cervical artery, and spinal accessory nerve (CN XI) (Som 1987; Lengelé et al. 2007).

The *internal jugular (colloquially deep cervical) chain* follows its eponymous vein. It is mostly concentrated superior to the omohyoid muscle but extends as inferiorly as the tendon of the sternal head of the SCM. Among the inferior nodes of the internal jugular chain are the nodes of Virchow, which are the only nodes of the neck lymphatic system that receive lymph from the stomach. At the superior end of this chain is the jugulodigastric node. At the crossing of the omohyoid is the juguloomohyoid node. These large nodes (jugulodigastric and juguloomohyoid) mark the superior ends of the superior and inferior groups of the internal jugular chain, respectively. Of note, although the superior and inferior internal jugular nodes are connected in a chain, they also give rise to efferent vessels that combine to form a "jugular trunk" bilaterally. The jugular trunk joins the thoracic duct on the left and the right lymphatic duct on the right.

The *spinal accessory chain* follows the course of CN XI caudally. Its superior nodes form the apex of the triangle of deep lateral cervical nodes along with the internal jugular chain. It receives lymph from the mastoid and occipital nodes and its efferent vessels carry lymph to the internal jugular and transverse cervical chains.

The *transverse cervical chain* follows the transverse cervical artery and vein and forms the base of the triangle of deep lateral cervical nodes. This group receives lymph from not only the spinal accessory chain but also the anterior chest wall, inferior and anterolateral neck, and the subclavicular nodes, while it drains into the internal jugular chain (Iwanaga et al. 2021; Som 1987; Lengelé et al. 2007).

The anterior cervical lymph nodes are found between the carotid sheaths and are divided into anterior jugular and juxtavisceral chains. The anterior jugular lies superficial to the strap muscles of the neck. Left anterior jugular nodes drain into either the thoracic duct or the anterior mediastinal nodes, while right anterior jugular nodes drain into the inferior internal jugular chain or an intrathoracic node.

The juxtavisceral chain consists of several groups of nodes, which can be found deep to the deep cervical fascia but overlie midline neck structures. Nodes over the thyrohyoid membrane and larynx drain into the deep cervical chain. The pre- and paratracheal nodes communicate with each other and drain into the jugular trunk, the inferior internal jugular nodes, or the brachiocephalic angular nodes. The most numerous nodes in this chain are paratracheal nodes found in the tracheoesophageal groove. The larger prelaryngeal node is termed the "Delphian Node." (Som 1987; Lengelé et al. 2007).

Another group of lymph nodes in the neck comprises the retropharyngeal nodes. These nodes are divided into two groups: a medial group, superior to the plane of the hyoid bone, and a lateral group, found along the length of the pharynx. "Node of Rouviere" denotes the most superior node in the lateral retropharyngeal chain bilaterally (Som 1987; Gross et al. 2004). In many adult patients, the medial group is either absent or not well visualized on standard imaging (Iwanaga et al. 2021; Gross et al. 2004).

Submental and submandibular nodes constitute two more clinically important node groups in the neck. Submental nodes are enclosed between the anterior bellies of the digastric muscles, the hyoid bone, and the mandible. They drain into either the superior internal jugular chain or the submandibular nodes. Submandibular nodes are found in the submandibular triangle, typically either in the anterior portion of the submandibular gland or near the facial artery. They drain into the superior internal jugular chain (Iwanaga et al. 2021).

In 1990, J. Shah proposed a new classification system for the deep cervical lymph nodes of the neck, which heavily influenced the 1991 commission from the AAO-HNS aimed at standardizing such systems. A revision commission was established in 2002 with the goal of revising their system. A system of six "Levels" was developed, some levels containing subgroups (Subramanian et al. 2006; Robbins et al. 2002).

Levels Ia and Ib are the submental and submandibular triangles, respectively. However, the AAO-HNS commission stated that the posterior border of Ib is more accurately described by the stylohyoid muscle than by the posterior belly of the digastric muscle (Iwanaga et al. 2021; Robbins et al. 2002; van den Brekel and Castelijns 1999).

Level IIa is bound by the plane of the body of the hyoid bone, the spinal accessory nerve, and the stylohyoid muscle. Level IIb is situated posterior to IIa so its anterior border is the spinal accessory nerve. Its upper border is the skull base, and it is also bound by the lateral border of the SCM and the stylohyoid (Saindane 2013; Robbins et al. 2002).

Levels III and IV have no sublevels. Level III nodes lie between the body of the hyoid bone and the lower margin of the cricoid cartilage. They lie lateral to the medial border of common carotid (or internal carotid) artery and anteromedial to the posterior border of the SCM (Saindane 2013; Robbins et al. 2002).

Level IV is bound by the plane of the inferior border of the cricoid cartilage, the clavicle, and the medial and lateral borders of the SCM. Of note, according to the AAO-HNS commission of 2002, the nodes between the omohyoid muscle and the interior jugular vein belong to Level III. This level typically includes the juguloomohyoid node, which marks the superior end of the inferior internal jugular chain (Saindane 2013; Som 1987; Robbins et al. 2002).

Level Va is a triangular space with an apex defined by the convergence of the SCM and the trapezius. Level Va is defined by the plane of the inferior border of cricoid cartilage, the lateral border of the SCM, and the anterior border of the trapezius muscle. Level Vb follows Level Va inferiorly. Therefore, Level Vb continues to be bordered by the lateral border of the SCM and the anterior border of the trapezius. It is contained on its superior border by the plane of the cricoid cartilage and on its inferior border by the clavicle. The intention in dividing Level V into two sublevels was for Va to correspond to the lower spinal accessory chain while Vb corresponded to the transverse cervical chain (Saindane 2013; Robbins et al. 2002).

Level VI nodes lie between the horizontal planes of the suprasternal notch and the hyoid bone. They are medial to the bilateral common carotid arteries (Saindane 2013).

Some authors and clinicians have referred to retropharyngeal nodes as "level VII," though it should be emphasized that this is not named in the 2002 revision. However, they stated that retropharyngeal nodes are found medial to the internal carotid artery and above the axial plane of the hyoid bone. All other nodes of the lower scalp and face should be referred to by their classical groupings, i.e., "occipital" or "mastoid" nodes (Robbins et al. 2002).

Of note, the sensory branches of the cervical plexus form a demarcation line equivalent to the lateral (posterior) border of the SCM. Also, the posterior border of the submandibular gland can be used as a radiographical equivalent of the stylohoid muscle dividing levels I and II (Lengelé et al. 2007; Robbins et al. 2002).

The pericervical circle: A circle of five nodal groups drains into the lymphatics of the neck. The nodal groups constituting the pericervical circle are.

1. Occipital nodes.
2. Mastoid nodes.
3. Parotid nodes.
4. Submandibular nodes.
5. Submental nodes (Som 1987; Lengelé et al. 2007).

22.4 Topography

When enlarged, lymph nodes can be palpable or even visible in the neck region. The jugulodigastric node can be felt or seen between the medial border of the SCM and the angle of the mandible. The juguloomohyoid node can be felt or seen just anterior or deep to the SCM in the plane of the cricoid cartilage. The Delphian node can be felt or seen over the thyroid cartilage of the larynx. The node or nodes of Virchow can be felt anterior or deep to the SCM and above the suprasternal notch (Som 1987) (Table 22.1).

Table 22.1 Palpation sites for sentinel nodes of the neck

Node	Palpable or visible
Jugulodigastric	Between the medial border of the SCM and the angle of the mandible
Juguloomohyoid	Anterior or deep to SCM in the plane of the cricoid cartilage
Delphian	Over the thyroid cartilage of the larynx
Virchow's	Anterior or deep to SCM and just above the suprasternal notch

22.5 Imaging

Palpation alone is insufficient for staging cancer of the neck. CT, PET/CT, and MRI have all improved sensitivity for metastatic disease, although sensitivity estimates for advanced imaging typically range between 40% and 60%. The estimated sensitivity of ultrasound-guided fine needle aspiration (USFNAC) is as high as 70–80%. These low sensitivity rates for intranodal disease are due to the dearth and ambiguity of radiological features to identify such disease. Sentinel node biopsy (SNB) has become standard of care in staging lymphatics of the neck, sensitivity estimates lying between 85 and 95% (de Bree et al. 2015). Nonetheless, for several reasons, advanced imaging remains important for treatment planning of head and neck cancer. Radiotherapy can be targeted using advanced imaging, and retropharyngeal nodes are inaccessible to sentinel node biopsy. Furthermore, SPECT/CT continues to gain importance for guiding SNB (Saindane 2013; de Bree et al. 2015; Vidal-Sicart and Valdés Olmos 2021; Castelijns and van den Brekel 2002).

Distinct radiological features signal the presence of intranodal metastasis. Irregular contrast enhancement, roundness, and size are the three primary radiological features for identifying metastases. Irregular contrast enhancement is most often due to tumor necrosis but can also be caused by keratinization or the presence of cysts. Roundness implies metastatic disease, while benign reactive lymph nodes are likely to retain their ovoid or kidney bean shape. If several nodes are contiguously grouped in the direct drainage pathway of the primary tumor, this can imply metastatic disease even without significant lymphadenopathy (Saindane 2013; Castelijns and van den Brekel 2002).

Size criteria for lymph nodes differ across institutions and training programs. There is no complete agreement that lymph nodes should be judged for suspicion on the basis of maximal or minimal diameter. Generally, the diameter threshold for suspicion is greater in levels I and II than other levels (van den Brekel and Castelijns 1999; Chung et al. 2019).

Extranodal spread is another important prognostic feature of metastatic disease, which can be depicted radiologically as irregular borders and disruption of fat planes around lymph nodes. Prognostically, histopathology informs clinicians better than radiographic findings, unless the extranodal spread is to vital structures in the neck. Radiographic evidence of invasion of vital structures is an ominous sign. It implies a complicated operation if it is operable at all. Not only vessels of the neck, but also the thoracic inlet and skull base, can be considered vital structures in this regard (Castelijns and van den Brekel 2002).

Lymphoscintigraphy is a nuclear radiography technique used to guide the SNB. It involves injecting the patient with a radiolabeled substance and using a gamma camera to record the locations of spread. Concurrent SPECT/CT imaging can improve the results further. These preoperative images can identify unexpected drainage patterns and guide SNB to locations that would not have been considered without imaging (Lin et al. 2006; Saindane 2013; Vidal-Sicart and Valdés Olmos 2021; Kim et al. 2019).

22.6 Pathology

Primary tumors can induce both angiogenesis as well as lymphangiogenesis. This is also true of head and neck squamous cell carcinomas. Lymphatic vessels grow toward the primary tumor and can enter it. Cells destined to become metastatic dissociate from the tumor and enter the interstitial space. The junctions between the lymph endothelial cells (LEC) expand because interstitial fluid pressure and volume are increased in the area of the primary tumor. Because of this increased junction size, metastatic cells can enter the lymphatic vessels after migrating through the chemokine gradient established by the LECs (Zhang et al. 2010).

22.7 Clinical and Surgical Significance

Noncancerous Causes of Lymphadenopathy

Clinicians should be aware of the many causes of reactive lymphadenopathy. Viral-reactive lymph nodes are the most common causes. Bacterial lymphadenopathy can require intervention if an abscess forms, which can be detected on CT imaging. Castleman's disease often presents with lymph node involvement localized to the neck (Som 1987; Castelijns and van den Brekel 2002).

Head and Neck Cancer and Neck Dissection

Most head and neck cancer patients die of local processes, and lymph node staging remains the single most important prognostic factor. Squamous cell carcinoma of the upper aerodigestive tract, the most common cancer of the head and neck, metastasizes preferentially through lymphatic vessels, which confers even greater importance on lymph node staging.

Neck dissection is the surgical procedure of removing lymph node levels in the neck either to treat known nodal metastasis, or electively in a cN0 (negative) neck as a prophylactic intervention for any potential occult micrometastases. Traditionally, a >20% chance of micrometastases in a cN0 neck has been considered sufficient to treat with neck dissection, but this can differ depending on institutional practice. Radical neck dissections were first described over 100 years ago, representing a major advance in treating head and neck cancers, despite the morbidity of loss of the ipsilateral SCM, internal jugular vein, and spinal accessory nerve. Modern surgeons can use advanced knowledge of lymphatic spread to guide "modified" or "selective" neck dissections. "Modified" implies preservation of at least one major structures (SCM, IJV, or CNXI), while "selective" implies preservation of at least one nodal level (Subramanian et al. 2006).

Sentinel Node Biopsy

As scientific understanding of lymphatic drainage in the head and neck progressed, so did the technology needed to inform clinicians of metastatic disease. Although micrometastases or aberrant spread can never be entirely ruled out, sentinel node biopsy allows clinicians the best chance of making accurate predictions about a patient's cancer. Best practices for sentinel node biopsies have been developed since it became evident that sensitivity for the presence of metastasis is greater than in previously used imaging methods. Typically, blue dye and technetium-99 are added to the primary tumor basin, and both gamma ray detection and visual inspection are used to assess the nodes of primary drainage. Presence or absence of metastatic disease informs the decision to treat the neck surgically. Of note, use of blue dye depends on institutional practices (de Bree et al. 2015; Vidal-Sicart and Valdés Olmos 2021; Tardelli et al. 2016; Thompson and Uren 2005).

Lymphoscintigraphy and, if possible, SPECT/CT have been performed preoperatively to indicate unanticipated spread (Lin et al. 2006; de Bree et al. 2015; Reynolds et al. 2009).

Drainage Patterns

There are four major lymphatic drainage paths through the neck:

1. Main.
2. Posterior.
3. Anterior.
4. Superficial lateral.

Prior lymphadenectomy, primary tumor removal, and former pathology in the neck can disrupt these regular flow patterns.

Main lymphatic pathway: Internal jugular chain in level IIa receives lymph from the parotid and facial nodes, which receive lymph in turn from face, forehead, and anterior scalp. Internal jugular chain also receives lymph from the submandibular nodes, which receive it from the submental, which receive it from the oral and nasal mucosa. The superior internal jugular chain then passes lymph down to the inferior jugular chain in levels III and IV and eventually to the major lymphatic ducts. Of note, the submental and submandibular nodes in the main lymphatic pathway can also receive drainage from the parotid and facial nodes (Wang et al. 2012).

Posterior lymphatic pathway: The jugulodigastric node group (forming the peak of the deep lateral cervical triangle) in level IIb receives lymph from the naso- and oro-pharynx, the tongue root, the postauricular and posterior scalp, and the suboccipital region. In this pathway, the jugulodigastric node group passes lymph through the spinal accessory chain and then further on to the transverse cervical chain in levels Va and Vb, respectively (Wang et al. 2012).

Anterior lymphatic pathway: The submental nodes in level Ia collect lymph from the tip of the tongue, the oral floor mucosa, the midline oral inferior labia, and the gingiva and buccal mucosa. The submental nodes then bypass the typical drainage that occurs when they collect lymph for the main pathway, and instead drain lymph to the internal jugular chain in levels III. The submental nodes have been shown infrequently to drain into level IV as well (Wang et al. 2012).

Superficial lateral lymphatic pathway: The superficial lateral cervical nodes drain the posterior scalp through the suboccipital and retroauricular nodes. Lymph is then passed into the internal jugular chain.

Skin

The skin of the neck generally drains into the anterior jugular chain, or the superficial lateral chain. These chains represent the superficial divisions of the anterior and lateral cervical groups, respectively. Although that pattern is more common, the skin can also drain into the juxtavisceral or deep cervical triangle chains. Of note, melanomas of the arm and upper trunk have been known to produce sentinel nodes in the neck (Wang et al. 2012; Veenstra et al. 2012).

Pharynx and Larynx

The nasopharynx primarily drains into the retropharyngeal and jugulodigastric nodes. Bilateral or contralateral nodal spread is associated with midline primary tumor locations (Wakisaka et al. 2000). It is possible for nasopharyngeal cancers to metastasize to level V nodes or to other internal jugular chain nodes besides the jugulodigastric. Nasopharyngeal cancers are notable for frequent bilateral or contralateral spread. Tongue base and oropharyngeal lymphatics typically drain to nodes in levels II-IV or to retropharyngeal nodes. This pattern is true of the lingual and palatine tonsils as well. The lymphatics of the lower pharynx drain in a similar pattern to the oropharynx and tongue base (levels II-IV and retropharyngeal) with the added drainage into level VI nodes. The lymphatic system of the larynx is notable for its complexity and its extensive midline crossings. Extensive drainage to levels II, III, IV, VI characterizes the laryngeal lymphatics, which comprise a deep and a superficial plexus. The vocal ligament and true vocal fold contain no lymphatic structures. However, cancerous involvement of the vocalis muscle can implicate the laryngeal lymphatic system.

Retropharyngeal nodes: The lateral chains of retropharyngeal nodes are most likely to drain into the superior internal jugular chain in levels II and III (Iwanaga et al. 2021; Wang et al. 2012).

Lymphatic Malformation

The neck is one of the most common regions for patients to develop lymphatic malformations, previously termed lymphangiomas. This is a disease process typically seen before two years of age, and at least 50% of cases appear to be congenital. Macro- or micro-cytic lesions (or both combined, as can happen in the head and neck) with overlying venous malformations and discoloration are typical for this diagnosis. Neck LMs have been staged according to the severity of prognosis, as depicted in Table 22.2. Although these lesions are benign, they have the propensity to grow rapidly and compress structures of the neck resulting in dysphagia, dysphonia, and difficulty breathing.

Surgical resection of these growths proves difficult because it is hard to assess the true extent of the lesion considering the irregular invasion of surrounding tissues. Deep seated lesion resections are often accompanied by morbidity and iatrogenic injury. Aside from these discouraging factors, resurgence of the disease is common (Zhou et al. 2011).

Sclerotherapy, laser therapy, and injection of cytotoxic agents are the remaining treatment modalities. Treatment should be individualized and comprehensive, while mindful of the psychosocial impact of deformity in a developing child (Zhou et al. 2011).

Table 22.2 Staging and location of nodes of the neck

Stage	Location
I	Unilateral infrahyoid
II	Unilateral suprahyoid
III	Unilateral infra- and suprahyoid
IV	Bilateral infrahyoid
V	Bilateral infra- and suprahyoid

References

Agarwal SK, Akali NR, Sarin D (2018) Prospective analysis of 231 elective neck dissections in oral squamous cell carcinoma with node negative neck—to decide the extent of neck dissection. Auris Nasus Larynx 45(1):156–161

Castelijns JA, van den Brekel MW (2002) Imaging of lymphadenopathy in the neck. Eur Radiol 12(4):727–738

Chung MS, Choi YJ, Kim SO, Lee YS, Hong JY, Lee JH et al (2019) A scoring system for prediction of cervical lymph node metastasis in patients with head and neck squamous cell carcinoma. Am J Neuroradiol 40(6):1049–1054

de Bree R, Takes RP, Castelijns JA, Medina JE, Stoeckli SJ, Mancuso AA et al (2015) Advances in diagnostic modalities to detect occult lymph node metastases in head and neck squamous cell carcinoma. Head Neck 37(12):1829–1839

Gross ND, Ellingson TW, Wax MK, Cohen JI, Andersen PE (2004) Impact of retropharyngeal lymph node metastasis in head and neck squamous cell carcinoma. Arch Otolaryngol Head Neck Surg. 130(2):169–173

Iwanaga J, Lofton C, He P, Dumont AS, Tubbs RS (2021) Lymphatic system of the head and neck. J Craniofac Surg 32(5):1901

Kim S-J, Pak K, Kim K (2019) Diagnostic accuracy of F-18 FDG PET or PET/CT for detection of lymph node metastasis in clinically node negative head and neck cancer patients; a systematic review and meta-analysis. Am J Otolaryngol 40(2):297–305

Lee BB, Suami H (2020) Chapter 2 – Ontogeny and phylogeny of lymphatics: embryological aspect. In: Gavins FNE, Alexander JS (eds) Lymphatic structure and function in health and disease. Academic Press, pp 5–17

Lengelé B, Hamoir M, Scalliet P, Grégoire V (2007) Anatomical bases for the radiological delineation of lymph node areas. Major collecting trunks, head and neck. Radiother Oncol 85(1):146–155

Lin D, Franc BL, Kashani-Sabet M, Singer MI (2006) Lymphatic drainage patterns of head and neck cutaneous melanoma observed on lymphoscintigraphy and sentinel lymph node biopsy. Head Neck 28(3):249–255

Mayerson HS (1962) Chapter XXIII – lymphatic vessels and lymph. In: Abramson DI (ed) Blood vessels and lymphatics. Academic Press, pp 701–718

Reynolds HM, Smith NP, Uren RF, Thompson JF, Dunbar PR (2009) Three-dimensional visualization of skin lymphatic drainage patterns of the head and neck. Head Neck 31(10):1316–1325

Robbins KT, Clayman G, Levine PA, Medina J, Sessions R, Shaha A et al (2002) Neck dissection classification update: revisions proposed by the American Head and Neck Society and the American Academy of Otolaryngology–Head and Neck Surgery. Arch Otolaryngol Head Neck Surg. 128(7):751–758

Saindane AM (2013) Pitfalls in the staging of cervical lymph node metastasis. Neuroimaging Clin N Am 23(1):147–166

Som PM (1987) Lymph nodes of the neck. Radiology 165(3):593–600

Subramanian S, Chiesa F, Lyubaev V, Aidarbekova A (2006) The evolution of surgery in the management of neck metastases. Acta Otorhinolaryngol Ital 26(6):309–316

Tardelli E, Mazzarri S, Rubello D, Gennaro M, Fantechi L, Duce V et al (2016) Sentinel lymph node biopsy in cutaneous melanoma: standard and new technical procedures and clinical advances. A systematic review of the literature. Clin Nucl Med 41(12):e498–e507

Thompson JF, Uren RF (2005) Lymphatic mapping in management of patients with primary cutaneous melanoma. Lancet Oncol 6(11):877–885

van den Brekel MWM, Castelijns JA (1999) Radiologic evaluation of neck metastases: The otolaryngologist's perspective. Semin Ultrasound CT MR 20(3):162–174

Veenstra HJ, Klop WMC, Speijers MJ, Lohuis PJFM, Nieweg OE, Hoekstra HJ et al (2012) Lymphatic drainage patterns from melanomas on the shoulder or upper trunk to cervical lymph nodes and implications for the extent of neck dissection. Ann Surg Oncol 19(12):3906–3912

Vidal-Sicart S, Valdés Olmos RA (2021) Sentinel lymph node biopsy in head & neck cancers. In: Reference Module in Biomedical. Elsevier, Sciences

Wakisaka M, Mori H, Fuwa N, Matsumoto A (2000) MR analysis of nasopharyngeal carcinoma: correlation of the pattern of tumor extent at the primary site with the distribution of metastasized cervical lymph nodes. Preliminary results. Eur Radiol 10(6):970

Wang Y, Ow TJ, Myers JN (2012) Pathways for cervical metastasis in malignant neoplasms of the head and neck region. Clin Anat 25(1):54–71

Zhang ZA, Helman JI, Li LJ (2010) Lymphangiogenesis, lymphatic endothelial cells and lymphatic metastasis in head and neck cancer – a review of mechanisms. Int J Oral Sci 2(1):5–14

Zhou Q, Zheng JW, Mai HM, Luo QF, Fan XD, Su LX et al (2011) Treatment guidelines of lymphatic malformations of the head and neck. Oral Oncol 47(12):1105–1109

Right Lymphatic Duct

23

Jonathan Staav, Annie Syed,
and Stephen J. Bordes, Jr.

23.1 Introduction

The small diameter and subtle appearance of lymph ducts belie their importance in collecting lymph and chyle for entry into the systemic circulation. Though not always present as a single right lymphatic duct, it is important that surgeons dissecting near the right venous angle be cautious of the variability in route and terminations of the lymphatic trunks that may come together to form the right lymphatic duct.

23.2 Anatomy/Topography/Variations

Embryologically, the lymphatics develop from and terminate into the venous system. The importance of the lymphatic system is underlined by the fact that congenital anomalies of the lymphatic system remain relatively rare (Standring 2020; Skandalakis et al. 2007; van der Putte and van Limborgh 1980).

All lymph and chyle are ordinarily returned to systemic venous circulation through either the right lymphatic duct or the thoracic duct. These lymphatic ducts are the largest lymph vessels that link the systemic lymph network with the right and left venous angles in the lower cervical region (Standring 2020).

The right lymphatic duct is a terminal lymphatic duct that is formed by the joining of the right bronchomediastinal, right jugular, and right subclavian lymphatic trunks. These three trunks serve as the emptying route for the lymph of the right half of the head and neck (right jugular), from the terminal apical axillary group of lymph nodes of the right upper limb (right subclavian), and from the right thorax including the right lung, diaphragm, and some drainage from the physiological right lobe of the liver (right bronchomediastinal). The right bronchomediastinal trunk also drains some structures positioned left of midline including the lower lobe of the left lung and most of the drainage of the heart (Standring 2020; Skandalakis et al. 2007; Tubbs et al. 2016).

There is great variability in the arrangement of the three right lymphatic trunks and how they join to form a common right lymphatic duct. In some cases, two or three of these lymphatic trunks come together and form a short right lymphatic duct that runs for 1–2 cm before emptying into the right venous angle. The trunks themselves might bifurcate or trifurcate instead of forming a right lymphatic duct and may terminate themselves in the right subclavian vein via multiple lymphovenous connections (Standring 2020; Tubbs et al. 2016).

J. Staav · A. Syed
School of Medicine, Tulane University,
New Orleans, LA, USA
e-mail: jstaav@tulane.edu; asyed1@tulane.edu

S. J. Bordes, Jr. (✉)
Department of Surgery, Louisiana State University
Health Sciences Center, New Orleans, LA, USA

S. J. Bordes, Jr. et al. (eds.), *The Clinical Anatomy of the Vascular System*,
https://doi.org/10.1007/978-3-031-78326-5_23

Where it does exist, the right lymphatic duct most commonly reaches the ventral aspect of the right venous junction by crossing the medial border of the scalenus anterior muscle at the root of the neck (Standring 2020; Skandalakis et al. 2007; Tubbs et al. 2016).

Whether through union into a right lymphatic duct, or independent drainage, the three lymphatic trunks empty into the right subclavian vein at the right venous angle where the right internal jugular vein meets the right subclavian vein. Lymphatics drain into the vein via lymphovenous portals that contain a bicuspid semilunar valve (Standring 2020; Pan 2017). The exact location of the lymphovenous portal is as variable as the arrangement of the trunks that may come together to form it (Skandalakis et al. 2007; Smith et al. 2013).

Locating the presence and location of a patient's right lymphatic duct is possible through computed tomography (CT) examinations in the head and neck region. While the right lymphatic duct has a lower detection rate than the thoracic duct, it can be visualized in a majority of instances (Kammerer et al. 2016).

23.3 Clinical Significance

While extremely uncommon, central venous catheterization of the right internal jugular vein may disrupt the right lymphatic duct (Arditis et al. 1988). Injury to the right lymphatic duct during cervical surgery is also uncommon but carries significant morbidity in the unlikely event of disruption. A leak of this vessel may result in chylothorax of the right chest. Chyle leakage can lead to hypovolemia, electrolyte imbalance, decreased lymphocytes, and hypoalbuminemia (Pan 2017; Smith et al. 2013). The mainstay of treatment in such cases is wide drainage, nil per os (NPO) status, and total parenteral nutrition (TPN). Somatostatin analogs such as octreotide may also be added. In the event the leak does not resolve over a course of days to weeks, surgical exploration may be warranted.

While exceedingly rare compared to cysts of the thoracic duct, cysts of the right lymphatic duct have been documented in the literature (Stárek et al. 2020).

Surgical interventions injuring the right lymphatic duct or its tributaries may lead to lymphedema of the head or right upper extremity. Symptoms include swelling, paresthesia, skin fibrosis, and recurrent skin infections. Point of injury is often difficult to identify and treatment involves exercise, elevation, and compression.

References

Arditis J, Giala M, Anagnostidou A (1988) Accidental puncture of the right lymphatic duct during pulmonary artery. A case reportcatheterization. *Acta anaesthesiologica Scandinavica* 32(1):67–68. https://doi.org/10.1111/j.1399-6576.1988.tb02690.x

Kammerer FJ, Schlude B, Kuefner MA, Schlechtweg P, Hammon M, Uder M, Schwab SA (2016) Morphology of the distal thoracic duct and the right lymphatic duct in different head and neck pathologies: an imaging based study. Head Face Med 12:15. https://doi.org/10.1186/s13005-016-0108-y

Pan W (2017) Atlas of lymphatic anatomy in the head, neck, chest and limbs, 1st edn. Springer, Singapore

Skandalakis JE, Skandalakis LJ, Skandalakis PN (2007) Anatomy of the lymphatics. Surg Oncol Clin N Am 16(1):1–16. https://doi.org/10.1016/j.soc.2006.10.006

Smith ME, Riffat F, Jani P (2013) The surgical anatomy and clinical relevance of the neglected right lymphatic duct: review. J Laryngol Otol 127(2):128–133. https://doi.org/10.1017/S0022215112002939

Standring S (2020) Gray's anatomy. In: The anatomical basis of clinical practice, 42nd edn. Elsevier, Amsterdam

Stárek I, Salzman R, Pokorná Z, Skanderová D (2020) Cyst of the right lymphatic duct. Eur Ann Otorhinolaryngol Head Neck Dis 137(2):123–125. https://doi.org/10.1016/j.anorl.2019.01.002

Tubbs RS, Shoja MM, Loukas M (2016) Bergman's comprehensive encyclopedia of human anatomic variation. Wiley Blackwell

van der Putte SC, van Limborgh J (1980) The embryonic development of the main lymphatics in man. Acta Morphol Neerl Scand 18(4):323–335

Part II

Spine and Spinal Cord

24 Vertebral Artery

R. Shane Tubbs

The first branch of the left and right subclavian arteries is the vertebral artery. These vessels are divided into four parts, V1–V4.

The V1 part of the vertebral artery (Fig. 24.1) ascends in the triangle of the vertebral artery, which is bordered by the scalene anterior, longus colli, and subclavian artery. In most individuals, it then enters the foramen transversarium of the sixth cervical vertebra, which is at the apex of the triangle of the vertebral artery. Branches of the V1 part of the vertebral artery include small twigs that supply the stellate ganglion. The left vertebral artery arises directly from the aortic arch in about 0.5% of cases (Fig. 24.2).

The V2 part ascends through the C6 to C1 foramina transversarium (Fig. 24.3). Along its course, it lies anterior to the anterior rami of the cervical spinal nerves. This part of the vertebral artery gives rise to the following branches:

(a) Spinal branches (radiculomedullary branches, Figs. 24.4, and 24.5) arise from the vertebral artery from C1 to C5 vertebral levels and may vary in number and side of origin. They supply the spinal cord, periosteum, and adjacent bone (Fig. 24.6).
(b) Muscular branches.

R. S. Tubbs (✉)
Department of Neurosurgery, Tulane University School of Medicine,
New Orleans, LA, USA
e-mail: rtubbs@tulane.edu

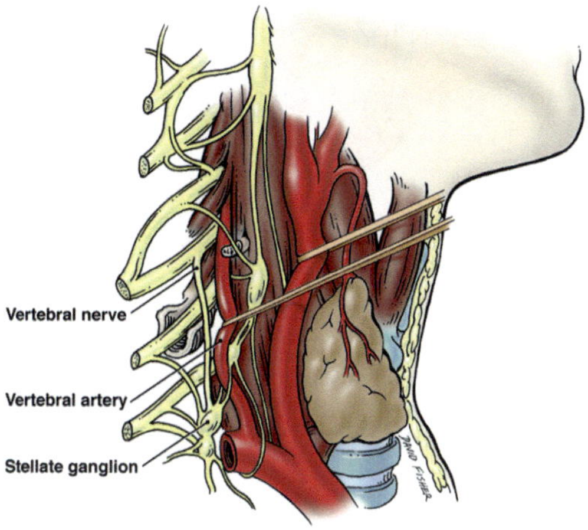

Fig. 24.1 Origin of the vertebral artery from the subclavian artery and regional muscular and nerve relationships

(c) Artery of the cervical enlargement. This artery usually arises from both vertebral arteries in the region of C4 to C6 and anastomoses with the anterior spinal artery to supply the anterior spinal cord.
(d) Anterior meningeal artery. This artery supplies the dura around the foramen magnum and ascends along the clivus.
(e) Posterior meningeal artery. Arises near the foramen magnum and supplies the adjacent dura and the falx cerebelli.

The V3 part travels along the posterior arch of the atlas in a bony groove, which is located in the

S. J. Bordes, Jr. et al. (eds.), *The Clinical Anatomy of the Vascular System*,
https://doi.org/10.1007/978-3-031-78326-5_24

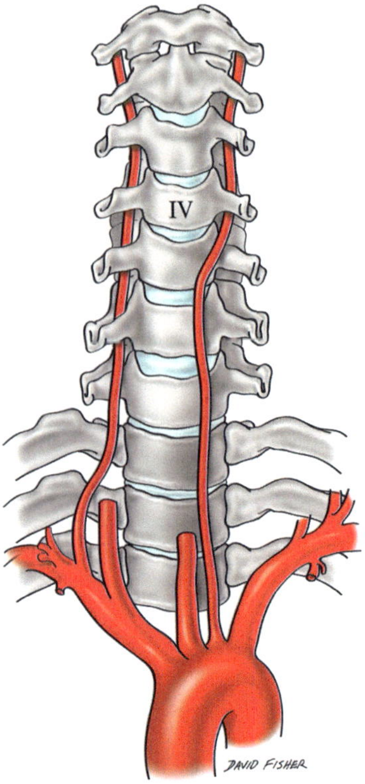

Fig. 24.2 Schematic drawing of the variant origin of the left vertebral artery from the aortic arch. (After Adachi)

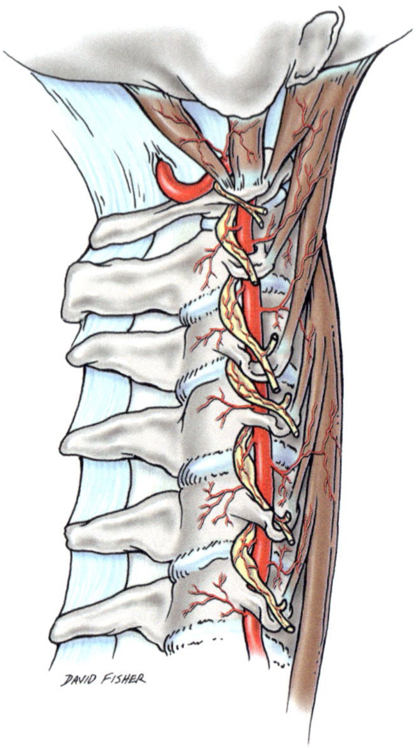

Fig. 24.3 Lateral view of the V2 segment of the vertebral artery providing multiple Bony and muscular branches. Notice the V3 segment of the vertebral artery piercing the posterior atlanto-occipital membrane

depths of the suboccipital triangle (Fig. 24.7). The suboccipital nerve lies between it and the posterior arch of the atlas. It then pierces the posterior atlanto-occipital membrane. Branches of this part of the vertebral artery include muscular branches and in some, the posterior inferior cerebellar artery (PICA).

The V4 segment is the intracranial part of the vertebral artery and ascends along the clivus to fuse with its contralateral part to form the basilar artery at the lower part of the pons. Its branches include the PICA and spinal arteries. The anterior spinal artery arises near the end of the vertebral artery, descends anterior to the medulla oblongata, and unites with a contralateral branch to travel inferiorly along the spinal cord. Branches of the anterior spinal artery are reinforced sequentially by small branches from the vertebral, ascending cervical, posterior intercostal, and first lumbar arteries. The spinal branches enter the vertebral canal via the intervertebral foramina and supply the spinal cord and its membranes. The posterior spinal artery follows a similar course but on the posterior aspect of the spinal cord. Additional details of the intracranial part of the vertebral artery are found in Sect. 1.

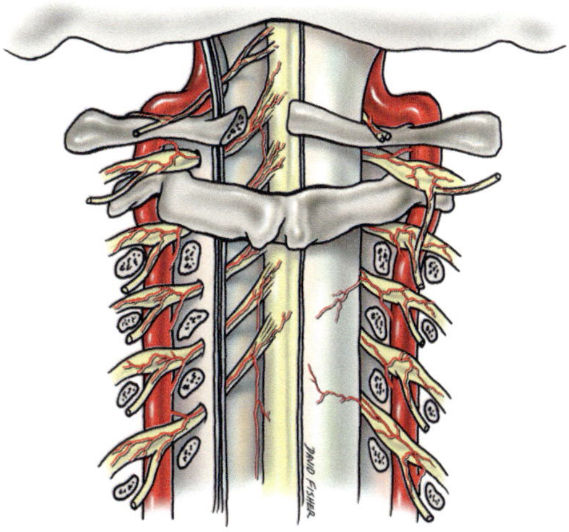

Fig. 24.4 Posterior view of the V2 and V3 segments of the vertebral artery with branches to the adjacent spinal cord and nerves

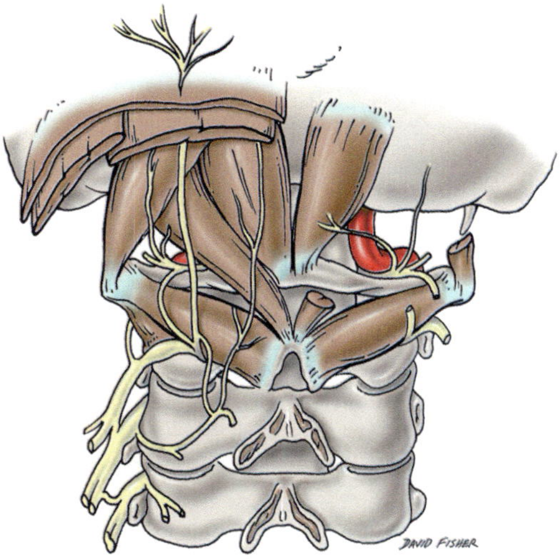

Fig. 24.7 Posterior view of the V3 segment of the vertebral artery in the depths of the suboccipital triangle

Fig. 24.5 Axial image illustrating the nerve and medullary branches supplying the spinal cord and derived from the vertebral artery

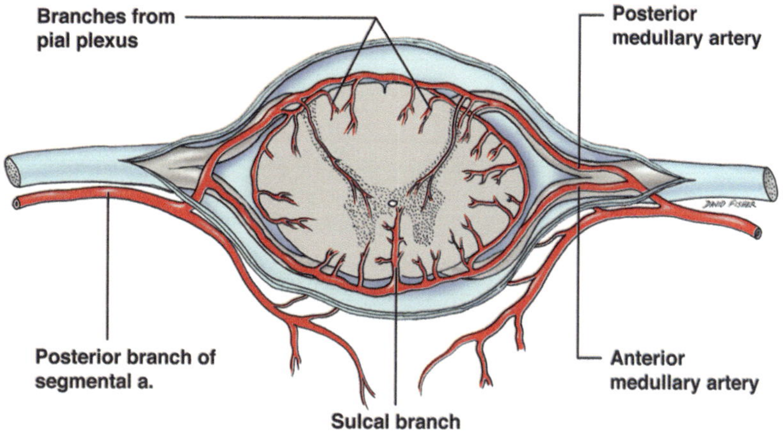

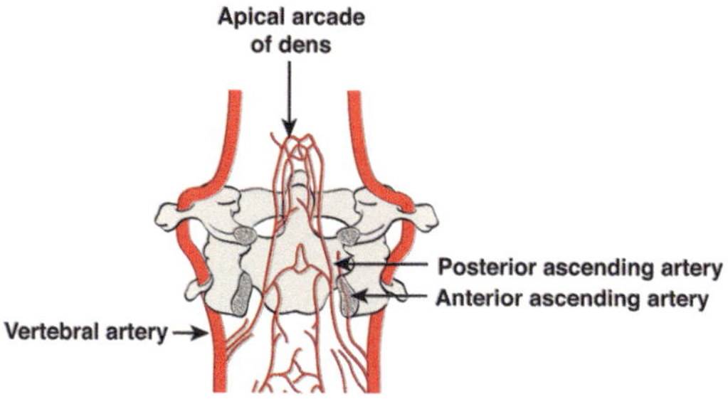

Fig. 24.6 Branches of the vertebral artery to the adjacent bone and here, contributing to the arterial arcade around the dens

Lumbar Arteries

25

Ronit Patnaik and Lisandro Montorfano

25.1 Embryology

The vascular system develops from fixed primitive angiogenic cells that are mesodermal in origin. This occurs in the middle of week 3 of fetal development. The process begins with hemangioblasts, which divide into primitive hematopoietic progenitor cells (HPCs) and endothelial precursor cells (EPCs, otherwise known as angioblasts). Hematopoiesis begins at the yolk sac of the extra-embryonic mesoderm, and flattened EPCs initiate vasculogenesis at the embryonic discs. EPCs form the vessels that eventually become the arteries and veins through which blood circulates (Ilina and Lilley 2019; Sadler 2018).

The first major arteries to appear are the right and left primitive aortae, which curve dorsally in the first pharyngeal arch, around the anterior part of the foregut, and then continue dorsally as the two dorsal aortae. The common dorsal aorta is formed by the union of the right and left dorsal aortae in the region extending from the fourth thoracic to the fourth lumbar somite segments (Ilina and Lilley 2019).

The common dorsal aorta branches into the ventral splanchnic arteries, lateral splanchnic branches, intersegmental arteries, and umbilical arteries. The intersegmental arteries later divide into the intercostal arteries in the thorax and the lumbar arteries in the abdomen.

25.2 Anatomy/Topography

The lumbar arteries supply blood to the spinal cord and the structures of the back including the vertebral bodies and surrounding muscles (Fig. 25.1). The abdominal aorta gives rise to four lumbar arteries from its dorsal surface on either side. Rarely, a fifth, smaller pair of lumbar arteries emerges from the median sacral or the iliolumbar artery, which arise from the abdominal aorta and common iliac artery, respectively (Crock and Yoshizawa 2013; Ratcliffe 1980). Lumbar arteries are direct branches of the aorta that run across the vertebral body, approximately 4 mm on average below the inferior endplate of the superior intervertebral disc space. These arteries originate near the midpoint of the vertebral body and pass posterior to the sympathetic chain and tendinous arches (Listmann and Tubbs 2020; Arslan et al. 2011). Just beyond the intervertebral foramina, each lumbar artery divides into medial, middle,

R. Patnaik
Department of Surgery, University of Texas Health, San Antonio, TX, USA

L. Montorfano (✉)
Department of Plastic and Reconstructive Surgery, Nashville, TN, USA

Department of Surgery, Cleveland Clinic Florida, Weston, FL, USA

S. J. Bordes, Jr. et al. (eds.), *The Clinical Anatomy of the Vascular System*,
https://doi.org/10.1007/978-3-031-78326-5_25

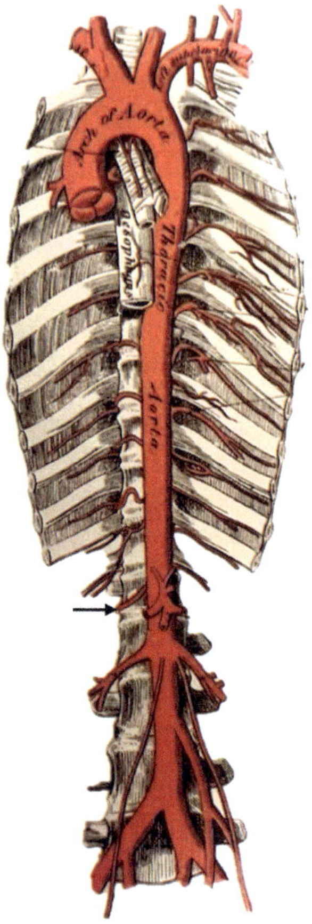

Fig. 25.1 Schematic drawing of the branches of the abdominal aorta, including a right lumbar artery (arrow). (After Allen 1882)

and lateral branches (Listmann and Tubbs 2020; Viswanathan et al. 2002).

The medial branch of the lumbar artery gives off the ganglionic and spinal branches. The ganglionic branches supply the regional sympathetic ganglia. The spinal branch, also known as the anterior segmental medullary branch, supplies the lumbar segments of the spinal cord (Fig. 25.2). The spinal branches pierce the dura in the vicinity of the dorsal root ganglia and are named according to their termination: radicular if the vessels terminate along the root, radiculopial if they terminate at the pial vessels of the spinal cord, and radiculomedullary if they anastomose with the anterior spinal artery. The artery of Adamkiewicz (the great radicular artery) is the largest radiculomedullary artery and originates between the ninth intercostal (thoracic) artery and the second lumbar artery, most commonly on the left side. Injury to the artery of Adamkiewicz can lead to spinal cord infarction with resultant neurological compromise (Biglioli et al. 2004).

The middle branch of the lumbar artery gives off anastomotic and dorsal branches. The dorsal branch takes a posterior course, passing between the transverse processes of the lumbar vertebrae, and supplies the adjacent deep muscles of the back, joints of the spine and the overlying skin of the back. The anastomotic branches form collaterals with the lumbar arteries above and below.

The lateral branches pass posterior to the psoas major and anterior to the quadratus lumborum to supply the muscles and skin of the posterior abdominal wall. They course between the internal oblique and transversus abdominis muscles. The lumbar arteries form collaterals with the lower posterior intercostal, subcostal, iliolumbar, deep circumflex iliac, and inferior epigastric arteries. This network of arteries supplies the back muscles, vertebrae and their joints, and the skin of the back (Listmann and Tubbs 2020).

The right lumbar arteries travel posterior to the inferior vena cava (IVC). The first left lumbar artery and the superior two right lumbar arteries rest behind the corresponding crus of the diaphragm. Each lumbar artery sends collaterals to anastomose with the lumbar artery one level below and above through the anastomotic branches (Arslan et al. 2011; Baniel et al. 1995).

25.3 Variations

While performing 102 retroperitoneal lymph node dissections, Baniel et al. noted that the lumbar arteries followed a fairly regular pattern. According to this team, the number of lumbar arteries exiting the aorta showed the most individual variation (range 2–4). The most common pattern of lumbar arteries (60% of cases) included three paired vessels exiting posteriorly at every third of the infrarenal aorta and coursing diagonally toward the lumbar foramina. There was more variation among the lumbar veins that drained into the inferior vena cava (Baniel et al. 1995).

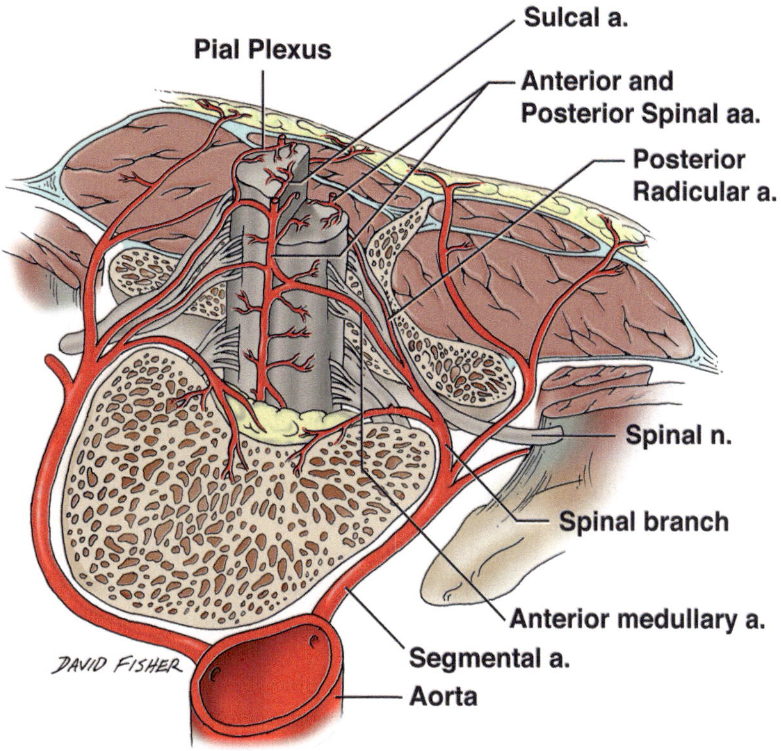

Fig. 25.2 Illustration of the spinal cord and nerve root branches of the lumbar arteries

Ratcliffe et al. described the anatomy of the fourth and fifth lumbar arteries in humans by reviewing aortographies of 100 patients (Ratcliffe 1982). They found a common artery of origin for the fourth lumbar arteries in 15 patients, and an origin from the median sacral artery for the fourth lumbar arteries in two (Ratcliffe 1982).

Karunanayake et al. dissected 109 cadavers and found the total number of lumbar artery pairs was in the range 3–5. The first three lumbar arteries were almost always present (the first two in 99% and the third in 98% of cadavers). The fourth lumbar artery was present in 83%. There was a fifth lumbar artery in one cadaver (Karunanayake and Pathmeswaran 2013). In this study, some lumbar arteries arose from a common stem from the abdominal aorta. The first and second pairs of lumbar arteries originated from a common stem in 1 and 2% of cadavers, respectively. The third and fourth pairs arose from a common stem in 3 and 11%, respectively. The fifth pair of lumbar arteries was present in only one cadaver, in which the arteries arose from a common stem.

Of note, Lo et al. have reported an unusual origin of the artery of Adamkiewicz from the fourth lumbar artery (Lo et al. 2002). Illuminati et al. stated that this artery originated from a lumbar artery between T12 and L1 in 16%, between the L1 and L2 level in 21.4%, and between L2 and L4 in 6% of individuals (Illuminati et al. 1996).

25.4 Clinical Significance and Complications

Procedures involving an anterior approach with surgical decompression and fusion of the spine (for example, an anterior lumbar interbody fusion) are performed for various pathologies of the spine including deformity, infection, metastases, trauma, and primary spinal neoplasms. The lumbar arteries can easily be injured during these approaches if care is not taken (Samudrala et al. 1999; Rajaraman et al. 1999).

Winter et al. reported a series of 1197 consecutive patients who underwent anterior approaches to the spine for correction of deformity. The exposure required ligation of multiple segmental vessels (including lumbar arteries) from T1 to L3, and there were no cases in which paraplegia resulted. According to these authors, vessel ligation is safe if the patient remains normotensive, and the ligation is unilateral and at the midportion of the vertebral body (Winter et al. 1996). Unintentional ligation of the dominant vessels of the vascular supply to the spinal cord, such as the artery of Adamkiewicz, can cause anterior spinal cord syndrome. This has devastating results: motor paralysis and loss of pain and temperature sensation below the level of the injury. Injury to a spinal branch off the medial branch of the lumbar arteries and their anastomotic network can also lead to spinal cord injury (Arslan et al. 2011; Orchowski et al. 2005; Bridwell et al. 1998).

Lumbar arteries are prone to retraction injury during mobilization of the psoas major if caution is not exercised. Arslan et al. showed that the most critical area for such injury was where the radicular vessels passed beneath the tendinous arches (Arslan et al. 2011). Traditionally, surgeons have preferred a left-sided approach because aortic repair is technically easier. However, it is important to remember that the dominant blood supply for the spinal cord, from the artery of Adamkiewicz, is on the left side.

Some case reports suggest that the lumbosacral nerve roots and plexus can suffer ischemic injury because of thrombosis in one or several lumbar arteries (Willaume et al. 2021). Willaume et al. reported the case of a 59-year-old woman who presented with spontaneous incomplete cauda equina syndrome due to diffuse lumbar nerve root infarction secondary to lumbar artery thrombosis (Willaume et al. 2021; Gloviczki et al. 1991).

Epidemiological and postmortem angiographic studies indicate that an insufficient blood supply to the lumbar spine due to atherosclerotic lesions in the abdominal aorta and the lumbar arteries (particularly stenosis of the lumbar artery ostia) could be a causative factor in disc degeneration and lower back pain (Kauppila et al. 1994, 1997; Kauppila 1995; Hadjipavlou et al. 2008).

In vascular surgery, endovascular aneurysm repairs (EVAR) have become popular for the management of abdominal aortic aneurysms. During EVARs, the ostia of the inferior mesenteric artery and many of the lumbar arteries are covered by a stent. Late complications of EVARs include endoleaks (White et al. 1997). An endoleak occurs when some part of the blood flow remains in the aneurysmal cavity despite endovascular repair. Endoleaks were divided by White et al. into four types (White et al. 1997). Type 2 endoleaks are caused by retrograde flow of blood from collateral arteries (such as lumbar arteries), which over time can result in aneurysmal sac expansion and rupture (Chung et al. 2021; Patatas et al. 2012). These can be treated by coil embolization, laparoscopy, or open surgery (Chung et al. 2021; Patatas et al. 2012).

Since the artery of Adamkiewicz usually originates from the lumbar arteries, it is also prudent to preserve blood flow from those arteries during open thoracoabdominal aortic repairs (Griepp et al. 1996). Previous abdominal aortic repair and extensive thoracic aorta exclusion by multiple stent grafts is associated with a significantly higher risk of paraplegia. After most of the intercostal and lumbar arteries have been interrupted, the residual collateral blood supply is marginal, and in some cases, the spinal cord becomes extremely prone to injury owing to arterial hypotension or low cardiac output from any cause (Ishimaru et al. 1998).

References

Arslan M et al (2011) Surgical view of the lumbar arteries and their branches: an anatomical study. Oper Neurosurg 68(suppl_1):ons16–ons22

Baniel J, Foster RS, Donohue JP (1995) Surgical anatomy of the lumbar vessels: implications for retroperitoneal surgery. J Urol 153(5):1422–1425

Biglioli P et al (2004) Upper and lower spinal cord blood supply: the continuity of the anterior spinal artery and the relevance of the lumbar arteries. J Thorac Cardiovasc Surg 127(4):1188–1192

Bridwell KH et al (1998) Major intraoperative neurologic deficits in pediatric and adult spinal deformity patients: incidence and etiology at one institution. Spine 23(3):324–331

Chung BH et al (2021) Laparoscopic lumbar artery ligation of type II endoleaks following endovascular aneurysm repair: a case report. Medicine 100(18):e25732
Crock HV, Yoshizawa H (2013) The blood supply of the vertebral column and spinal cord in man. Springer Science & Business Media
Gloviczki P et al (1991) Ischemic injury to the spinal cord or lumbosacral plexus after aorto-iliac reconstruction. Am J Surg 162(2):131–136
Griepp RB et al (1996) Looking for the artery of Adamkiewicz: a quest to minimize paraplegia after operations for aneurysms of the descending thoracic and thoracoabdominal aorta. J Thorac Cardiovasc Surg 112(5):1202–1213. discussion 1213
Hadjipavlou A et al (2008) The pathophysiology of disk degeneration: a critical review. J Bone Joint Surg Br 90(10):1261–1270
Ilina M, Lilley S (2019) Embryology of the great vessels. In: Clinical embryology. Springer, pp 275–288
Illuminati G et al. (1996) Variations in the origin of the artery of Adamkiewicz Rivista europea per le scienze mediche e farmacologiche= European review for medical and pharmacological sciences= Revue europeenne pour les sciences medicales et pharmacologiques. 18(2):61–66
Ishimaru S et al (1998) Preliminary report on prediction of spinal cord ischemia in endovascular stent graft repair of thoracic aortic aneurysm by retrievable stent graft. J Thorac Cardiovasc Surg 115(4):811–818
Karunanayake AL, Pathmeswaran A (2013) Anatomical variations of lumbar arteries and their clinical implications: a cadaveric study. Int Sch Res Notices 2013:154625
Kauppila L (1995) Can low-back pain be due to lumber-artery disease? Lancet 346(8979):888–889
Kauppila LI et al (1994) Lumbar disk degeneration and atherosclerosis of the abdominal aorta. Spine 19(8):923–929
Kauppila LI et al (1997) Disk degeneration/back pain and calcification of the abdominal aorta: a 25-year follow-up study in Framingham. Spine 22(14):1642–1647
Listmann M, Tubbs RS (2020) The abdominal aorta. Elsevier, pp 185–188
Lo D et al (2002) Unusual origin of the artery of Adamkiewicz from the fourth lumbar artery. Neuroradiology 44(2):153–157
Orchowski J, Bridwell KH, Lenke LG (2005) Neurological deficit from a purely vascular etiology after unilateral vessel ligation during anterior thoracolumbar fusion of the spine. Spine 30(4):406–410
Patatas K et al (2012) Static sac size with a type II endoleak post-endovascular abdominal aortic aneurysm repair: surveillance or embolization? Interact Cardiovasc Thorac Surg 15(3):462–466
Rajaraman V et al (1999) Visceral and vascular complications resulting from anterior lumbar interbody fusion. J Neurosurg Spine 91(1):60–64
Ratcliffe J (1980) The arterial anatomy of the adult human lumbar vertebral body: a microarteriographic study. J Anat 131(Pt 1):57
Ratcliffe J (1982) The anatomy of the fourth and fifth lumbar arteries in humans: an arteriographic study in one hundred live subjects. J Anat 135(Pt 4):753
Sadler TW (2018) Langman's medical embryology. Lippincott Williams & Wilkins
Samudrala S et al (1999) Complications during anterior surgery of the lumbar spine: an anatomically based study and review. Neurosurg Focus 7(6):E11
Viswanathan R et al (2002) Extraforaminal lumbar disk herniations: microsurgical anatomy and surgical approach. J Neurosurg Spine 96(2):206–211
White GH et al (1997) Endoleak as a complication of endoluminal grafting of abdominal aortic aneurysms: classification, incidence, diagnosis, and management. J Endovasc Ther 4(2):152–168
Willaume T et al (2021) Spontaneous infarction of lumbar roots, vertebrae and paravertebral muscles. Am J Emerg Med 46:796.e5
Winter RB et al (1996) Paraplegia resulting from vessel ligation. Spine 21(10):1232–1233

Venous Drainage of the Spinal Cord

26

Vinila Baljepally

26.1 Introduction

Venous drainage of the spinal cord is is carried out by a complex network of venous structures and in certain aspects mirrors the arterial supply of the spinal cord (Fig. 26.1). Breschet described the vertebral venous system as consisting of three interconnecting divisions: An internal division within the spinal canal, a transosseous division within the vertebrae linking the internal and external divisions, and an external division surrounding the spine (Nathoo et al. 2011). Around the same time, Cruveilhier (1836) recognized the importance of this system as collateral network with inferior vena cava occlusion. Harris credits for the first description of the concept that blood passes through the vertebral venous system in either direction due to the absence of valves even though Breschet had first described it a few decades earlier (Harris 1941). Batson (1940) made significant contributions to the understanding of the vertebral venous system. Batson's studies also showed that during a physiologic Valsalva maneuver blood flow is redirected from the body cavities into the mostly valveless vertebral venous system challenging Harvey's "paradigm of circular flow" (Harvey 1894; Herlihy 1947). The term "Batson's plexus" encompasses the extradural, internal, and external vertebral venous plexuses extending up to the intervertebral veins and associated extraspinal efferents" (Batson 1940). Anderson (1951) demonstrated shunting of blood from into the vertebral venous system in studies on living humans (Anderson 1951). More recently, proposed the term "cerebrospinal venous system" emphasizing the continuity between the cerebral and vertebral venous systems (Tobinick and Vega 2006).

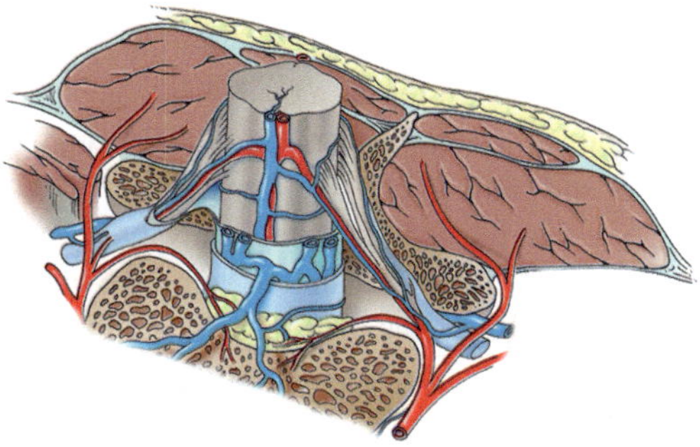

Fig. 26.1 Schematic drawing illustrating veins of the anterior spinal cord and adjacent nerve roots and epidural/spinal spaces

26.2 Embryological Aspects

The vertebral venous system begins with the development of paired longitudinal veins, the anterior and posterior cardinal vein above and below the common cardinal vein at 4 weeks. At

V. Baljepally (✉)
School of Medicine, Tulane University,
New Orleans, LA, USA
e-mail: baljepally@etsu.edu

S. J. Bordes, Jr. et al. (eds.), *The Clinical Anatomy of the Vascular System*,
https://doi.org/10.1007/978-3-031-78326-5_26

50 weeks, formation of paired sub- and supracardinal veins, parallel to the posterior cardinal veins and caudal to the common cardinal vein, is formed. These veins are connected through transverse venous channels including azygos, renal, and iliac anastomoses (Brookes and Zietman 1998; Minot et al. 1912). Transverse intersegmental veins (7 in the cervical, 12 in the thoracic, and 5 in the lumbar spine) arise with common variations from the cardinal veins bilaterally and drain the paravertebral plexus around the developing spinal elements (Sabin 1915). The right and left supracardinal veins persist as the azygos, hemiazygos, and accessory hemiazygos veins, respectively. These veins drain into the superior vena cava through the arch of the azygos vein, a remnant of the right posterior cardinal vein. The azygos and hemiazygos veins have a caudal anastomosis with the common iliac veins through the subcostal and ascending lumbar veins, the latter a remnant of the thoracolumbar chain, another longitudinal embryological structure connecting transversely oriented intersegmental veins (Brookes and Zietman 1998). With regression of the cardinal veins, the paravertebral plexus around the developing spine drains into the azygos system and ascending lumbar veins (Sabin 1915). Cranial to the common cardinal vein, a transverse branch, the brachiocephalic anastomosis, connects the anterior cardinal veins, which is the precursors of the internal and external jugular veins. The anterior cardinal vein also has transverse intersegmental branches in the cervical region and portions persist as the vertebral veins (Fig. 26.1) (Brookes and Zietman 1998; Gray 1918; Sabin 1915).

Early in development, the vascular system of the spinal cord consists of a medial spinal artery, capillary plexus, and a lateral segmental spinal vein (Sabin 1915). Zawiliński et al. described the development of the vascular supply and drainage of the developing human fetus in detail (Zawiliński et al. 2001). Most importantly, in the early fetal period, the gray matter receives a steady blood supply while the white matter appears partially avascular. Central arteries supply the capillaries of the gray matter that form "glomerular plexuses" and deliver blood to the anterior horns. Once these capillary plexuses fuse into a continuous system near the anterior columns, further vascular spread supplies the posterior horns. Further along in development, the moderately avascular white matter becomes more vascularized but never to the extent of the gray matter. With respect to location and distribution, the veins demonstrate tremendous variability as compared to the arteries. The superficial veins tend to demonstrate "high tortuosity, especially in the younger fetuses."[33] The corresponding veins usually accompany the anterior spinal and central arteries; however, no other topographical relationship with the arteries seems to exist. This high degree of individual variability is a common theme in the development of the venous system of the spinal cord. Dommisse described the veins of the spinal cord at birth as complementary to those of the adult in every aspect except for size, for which they are relatively larger than at any other point in post-natal development (Dommisse 1975).

26.3 Anatomical and Physiological Aspects

The venous system of the spine and spinal cord can be divided into intrinsic, extrinsic, and extradural components. The intrinsic system is separated into central and peripheral sections.

Intrinsic System

The intrinsic system is comprised of a network of symmetrically arranged, radially draining, venous channels with axial and longitudinal components (Fig. 26.2).

Axially Oriented Intrinsic Venous System

In the central section, capillaries from the medial parts of both sides of the spinal cord drain into axially oriented, ventral (medial and anteromedial), and dorsal (medial, and posteromedial) sulcal veins (Lasjaunias and Berenstein 1990). The ventral sulcal veins collect blood from nearby structures including the medial portion of the anterior

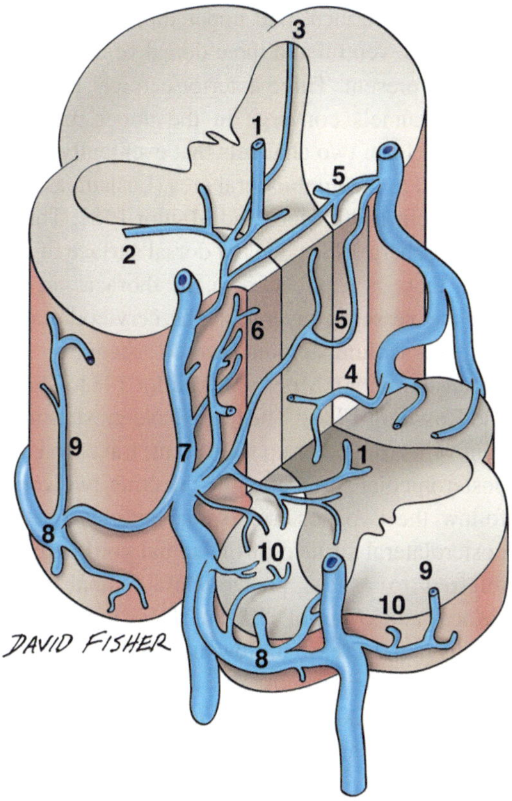

Fig. 26.2 Intrinsic and extrinsic systems after Lasjaunias and Berenstein (Lasjaunias and Berenstein 1990). Longitudinal intrinsic anastomosis (1); axial anterolateral and anteroposterior intrinsic anastomosis (2 and 3); dorsal sulcal vein (4); oblique transmedullary anastomosis (5); extrinsic longitudinal pial venous network (6); ventral median vein (7); extrinsic axial pial venous network (8); extrinsic longitudinal pial venous network and venous ring (9); radial vein (10)

horns, the anterior commissure, and white matter of the anterior funiculus (Santillan et al. 2012). The number of ventral sulcal veins varies over the length of the spinal cord. There are four to eight ventral sulcal veins per cm in the cervical and upper thoracic areas, with a diameter of 100–200 μm. In the lower third of the thoracic cord and the lumbosacral area, their number increases to 4–12 per cm (Lasjaunias and Berenstein 1990; Moes and Maillot 1981; Thron and Rossberg 1988). The dorsal sulcal veins are similar in diameter (100–300 μm), found 4–6 times per cm in the cervical cord (Moes and Maillot 1981; Thron and Rossberg 1988), and are encountered less frequently in the thoracolumbar area (1–4 times per cm) (Moes and Maillot 1981; Thron and Rossberg 1988). Thus, the ventral dominance observed in the arterial system is only partially present in the venous system of the thoracolumbar region (Lasjaunias and Berenstein 1990). In the peripheral regions, capillaries from peripheral gray matter and white matter travel toward the periphery forming radial veins (Lasjaunias and Berenstein 1990). Once these radial veins reach the spinal surface, they form a venous ring that anastomoses with the pial venous network and longitudinally oriented extrinsic venous system, which are both components of the extrinsic venous system (see below) (Santillan et al. 2012). The intrinsic system has an intricate network of anastomoses uniting the radial and sulcal veins, and central and peripheral sections. This is most prominent in the thoracic region (Thron and Rossberg 1988). Transmedullary anastomotic veins span between the ventral and dorsal surface of the cord and, when compared to sulcal veins, are larger in caliber (100–700 μm) and less concentrated (1–2 per cm), particularly in the lower thoracic and lumbar regions (Lasjaunias and Berenstein 1990; Moes and Maillot 1981; Thron and Rossberg 1988). These do not seem to collect drainage from smaller veins and travel horizontally through the midline of the spinal cord while also ascending or descending obliquely and longitudinally at the gray commissure (Lasjaunias and Berenstein 1990; Thron and Rossberg 1988). These veins may play a role in pressure equalization occurring with cerebrospinal fluid (CSF) reabsorption (Lasjaunias and Berenstein 1990).

Longitudinally Oriented Intrinsic Venous System

In addition to highly prevalent axial anastomosis, the spinal cord has longitudinal intrinsic anastomosis present in gray and white matter that link the previously described axial anastomosis (Lasjaunias and Berenstein 1990). In the periphery, radial veins have longitudinal, "intersegmental bridges" linking them. These bridges are considered part of the extrinsic venous system (Lasjaunias and Berenstein 1990).

Extrinsic System

The extrinsic system begins at the level of the spinal pia and includes the pial venous networks, the longitudinally oriented extrinsic venous system, and radicular veins (Fig. 26.2).

Pial Venous Networks

The extrinsic venous system begins with the pial network. Pial collectors receive sulcal veins while intersegmental bridges link radial collectors that eventually drain into the ventral (anterior) and dorsal (posterior) median veins and its derivatives. Such connectors are located laterally on the spinal cord and are short and less prominent. The pial network is slightly more prominent ventrally with one transverse vein occurring every 1–3 cm in the cervical and upper thoracic cord segments. The lumbosacral area is also rich in ventral pial veins. Dorsal pial veins are most frequently found in the high cervical region and at the upper and lower extents of the thoracic region. The diameter of the pial collectors is around 400–500 μm, thus comparable to the size of the ventral and dorsal median veins (Lasjaunias and Berenstein 1990; Moes and Maillot 1981; Thron and Rossberg 1988).

Longitudinally Oriented Extrinsic Venous System

While the pial venous network has some axial components, the extrinsic system tends to follow a longitudinal orientation (Santillan et al. 2012). Midline longitudinal collectors lying on the ventral and dorsal surfaces of the spinal cord represent the dominant venous channel of this system. In the dorsal cervical and lumbosacral regions, there is usually a single ventral and dorsal collector, referred to as the ventral and dorsal median veins. The diameter of this single vessel can reach 500 μm in the cervical and up to 2000 μm in the lumbosacral region (Lasjaunias and Berenstein 1990; Moes and Maillot 1981; Thron and Rossberg 1988). The anterior median vein accompanies the anterior spinal artery on the anteromedial surface of the spinal cord (Gillilan 1970; Goshgarian 2003; Lasjaunias and Berenstein 1990; Thron and Rossberg 1988). In the ventral cervical and upper thoracic regions, up to three ventral and three dorsal venous channels are present. Three anterior cervical longitudinal channels converge in the upper thoracic area and form two channels that eventually unite proximal to the lumbosacral area (Lasjaunias and Berenstein 1990; Moes and Maillot 1981; Thron and Rossberg 1988). On the dorsal surface of the cord, there is triplication in the thoracic region and the vessels converge at the cervicothoracic and thoracolumbar junctions (Lasjaunias and Berenstein 1990; Moes and Maillot 1981; Thron and Rossberg 1988). The largest and most prominent veins (diameter up to 850 μm) travels on the posteromedial surface, while the other two veins follow the two posterior spinal arteries on the posterolateral surface of the spinal cord and are referred to as the posterolateral spinal veins (Moes and Maillot 1981). This redundancy results in a decrease in flow and renders the cord susceptible for venous stasis and engorgement (Goshgarian 2003; Lasjaunias and Berenstein 1990; Moes and Maillot 1981). The anterior median vein continues along the filum terminale as a terminal vein to the end of the dural sac (Santillan et al. 2012). The posterior median vein also receives blood from radial veins and venous rings (Brockstein et al. 1994; Goshgarian 2003; Lasjaunias and Berenstein 1990; Moes and Maillot 1981; Santillan et al. 2012; Thron and Rossberg 1988). Secondary venous networks may exist that contribute to or replace these longitudinal venous channels at various segments (Santillan et al. 2012).

Radicular Veins

As mentioned previously, the anterior and posterior venous segments collect blood over an equal territory, unlike the arterial system that exhibits more anterior dominance. The exception is the thoracolumbar region where ventral sulcal veins are more prominent compared to their dorsal counterpart (Lasjaunias and Berenstein 1990). Thus, radicular veins, veins that connect the extrinsic to the extradural system, are almost equally as common on the ventral and dorsal root. These veins are analogous to the previously described radiculomedullary arteries and, to be precise, named radiculomedullary veins since they drain the spinal cord. They receive blood from the ventral or dorsal median veins (Santillan et al. 2012). There are numerous ventral (anterior) and dorsal (posterior) radiculomedullary veins and their number ranges from 8 to 20 and 5 to 10, respectively (Gillilan 1970; Lasjaunias and Berenstein 1990; Santillan et al. 2012; Thron and Rossberg 1988). One important vein is the great anterior radiculomedullary vein (GARV) with a diameter of up to 2 mm that travels with either the anterior or posterior nerve roots between T11 and L3 and it can be mistaken for the artery of Adamkiewicz (AKA) because of its size and location in the thoracolumbar region. One distinguishable characteristic is the more obtuse angle between GARV and the ventral or dorsal median veins, referred to as the "coat hook" configuration, compared to the AKA that forms a sharper, "hairpin" junction (Bowen et al. 1996). Furthermore, the AKA and GARV almost never enter or exit at the same level (Lasjaunias and Berenstein 1990). The level that these radiculomedullary veins exit through the dura, either with the nerve root (60% of cases) or a separate foramen between two spinal nerves (40% of cases) (Lasjaunias and Berenstein 1990), typically differs from the corresponding arterial supply. The quality of a venogram depends on the territory supplied by the artery and significant opacification of the arterial supply to the spinal cord allows for a high quality venograms (Lasjaunias and Berenstein 1990). These radiculomedullary veins merge and receive drainage from the spinal nerve roots via superficial nerve root veins (Jinkins 2000). Within the nerve root, longitudinal veins receive blood from the venules surrounding the neural axons and empty into draining branch veins that then exit the nerve root dura and subsequently empty into the superficial nerve root veins. This system of venous drainage of the spinal nerve root contains transverse coils in the venules, longitudinal veins, and draining branch veins that allow for compensation of extra blood volume (Jinkins 2000). The radiculomedullary veins drain into the epidural vertebral venous plexus via the spinal nerve venous channels (Lasjaunias and Berenstein 1990).

"Reflux-Regulating" Mechanisms of Radicular Veins

Groen and colleagues (Groen et al. 1997) performed a human cadaver study of the radicular veins and found no bicuspid valves. Previously, it was assumed that venous reflux within the radicular veins was prevented by bicuspid valves. They dissected spinal nerve sheaths with nerve roots and epidural and radicular veins and studied the veins macroscopically and microscopically. Macroscopic and microscopic examination found no bicuspid valves. Several other factors were discovered that appear to form a "reflux-regulating" system. The radicular veins have intravenous dural folds, undergo a very tortuous and oblique path through the dura where the lumen narrows, and contain smooth muscle fibers in the vascular walls in the intradural and extradural sections that may regulate reflux in conditions of venous hypertension (van der Kuip et al. 1999). Nevertheless, this system has limitations in the presence of spinal vascular pathologies such as arteriovenous fistulas or malformations as evidenced by the observation that exertion worsens symptoms, secondary to uncompensated pressure changes that interfere with circulation (Lasjaunias and Berenstein 1990). Interestingly, this reflux-regulating system may also play a role in heat exchange for the spinal cord (van der Kuip et al. 1999).

Extradural System

This system plays a role in draining not only the spinal cord and vertebrae but also participates in cerebrospinal fluid (CSF) reabsorption through arachnoid granulations within spinal nerve roots (Lasjaunias and Berenstein 1990). The radicular veins have a narrow lumen and take an oblique and tortuous course along the dura, allowing them to function as one-way valve. They then exit at the neural foramen after merging with venous channels of the spinal nerve and drain into the vertebral venous plexus (Lasjaunias and Berenstein 1990; Santillan et al. 2012). Commonly, there are two veins, one superior and one inferior to the spinal nerve (Lasjaunias and Berenstein 1990).

Vertebral Venous Plexus

The vertebral venous plexus, or plexus of Batson, has been divided into three intercommunicating divisions: the internal vertebral, external vertebral, and the basivertebral plexus (Dommisse 1975; Goshgarian 2003; Groen et al. 2004; Lasjaunias and Berenstein 1990; Santillan et al. 2012). Angiographically, it has a diamond-shaped appearance on an anteroposterior view (Lasjaunias and Berenstein 1990). The anterior and posterior internal vertebral venous plexuses are located epidurally within the vertebral canal and continue superiorly to communicate with the suboccipital venous system, including the basilar and hypoglossal plexuses, marginal and occipital sinuses, and vertebral veins (Tubbs et al. 2007). The internal vertebral plexus includes four longitudinal veins, two in anterior and two in the posterior internal plexus (Abrams 1957). The anterior internal vertebral plexus lies on either side of the posterior longitudinal ligament and is connected to the basivertebral plexus via transverse branches. The posterior internal vertebral plexus is anterior to the ligamentum flavum and connects to the posterior external plexus through perforating branches. The veins of the anterior internal vertebral plexus are larger than those of the posterior plexus and increase in caliber from cranial to caudal concomitant with increase in size of the vertebrae. This is not observed for the veins of the posterior internal venous plexus (Groen et al. 1997). The anterior (front of the vertebral bodies) and posterior (surrounding the posterior elements) external vertebral venous plexus surrounds the vertebral column, while the final division is the basivertebral veins, which run horizontally within the vertebral body. Along with the intrinsic and extrinsic system, the extradural vertebral venous plexus has been reported to be valveless (Tobinick and Vega 2006). A recent study, however, showed that some veins of the posterior external vertebral venous plexus possess valves directing flow toward the internal vertebral venous plexus. The implications of this finding are less clear (Stringer et al. 2012).

Extraspinal Efferents

Cervical, thoracic, lumbar, and sacral intervertebral veins connect the external and internal vertebral venous plexuses and communicate with various extraspinal longitudinal veins including the vertebral veins, azygos system, and ascending lumbar veins (Abrams 1957; Santillan et al. 2012). Drainage of these extraspinal longitudinal veins depends on the specific vertebral level (Abrams 1957). In the cervical level, drainage occurs via the vertebral, deep cervical, and jugular veins eventually leading to the superior vena cava (Lasjaunias and Berenstein 1990). Venous drainage from thoracic and lumbar intervertebral veins is accomplished through the azygos system and ascending lumbar veins. The right ascending lumbar vein continues with the azygos vein, while the left continues with the hemiazygos, which crosses the vertebral column at either T8 or T9 to join the azygos vein. The accessory hemiazygos receives upper thoracic intervertebral veins and connects to the left superior intercostal vein above and the hemiazygos below (Abrams 1957). In the thoracic region, intervertebral veins on the left drain into either the accessory hemiazygos vein from T3 to T8 or the hemiazygos vein below T8, while veins on the right drain into the azygos vein (Lasjaunias and Berenstein 1990). The azygos vein drains into the superior vena cava. In the lumbosacral level, the ascending lumbar veins communicate with the inferior vena cava via four horizontal, lumbar segmental veins. The first and second lumbar segmental veins usually converge prior to draining into the inferior vena cava (Lolis et al. 2011). Some spinal venous drainage into the left renal vein and vein of the crus of the diaphragm occurs through connections with the hemiazygos, subcostal, and ascending lumbar venous network (Gillot and Aaron 1968). The sacral veins empty into the lateral sacral vein, the sacral equivalent of the azygos and ascending lumbar veins, and into the internal iliac vein (Lasjaunias and Berenstein 1990). Numerous variations in the extraspinal venous system exist and demonstrate compensatory venous drainage through other pathways. If the inferior vena cava is occluded or congenitally absent, blood from the lower extremities enters the heart through the paravertebral and azygos systems via the superior vena cava (Dommisse 1975).

General Hemodynamics

Spinal cord venous drainage is dependent on gravity, body position, and even thoracic and abdominal cavity pressure (Lasjaunias and Berenstein 1990). The pressure in the intracranial dural venous sinuses fluctuates in the range of 10 cmH_2O when supine to negative pressures while standing. In the lower extremities, venous pressure ranges from 10 cmH_2O when supine to 100 cmH_2O while standing. Therefore, spinal venous flow characteristics are very different above or below the level of the heart. The inspiratory phase of the respiratory cycle is the main force facilitating venous drainage of the spinal cord below the level of the heart (Gillot and Aaron 1968; Lasjaunias and Berenstein 1990). In addition, the venous drainage of the spinal cord is a low-pressure system and contains many anastomoses that are capable of absorbing pressure. It is able to store blood via dilation and receive backflow from adjacent veins, redistributing the blood to other areas of its rich anastomoses. Since blood flow can be directed in either direction in the vertebral venous plexus, rapid adjustments can be made to accommodate an increase or decrease in arterial blood flow.

26.4 Pathology of Spinal Cord Venous Drainage

Spinal pathology affecting the venous system includes vascular, neoplastic, and infectious disorders. These pathologies may result in obstruction and stasis of venous flow leading to spinal cord damage and permanent myelopathy (Henderson et al. 1993).

Vascular Pathology

Classically, there are four types of spinal vascular lesions that affect the venous system: dural arteriovenous fistulas (AVF) (type I) (Lasjaunias and Berenstein 1990; Takai and Taniguchi 2012), intramedullary arteriovenous malformations (AVMs) (type II), juvenile AVMs (type III), and intradural perimedullary AVF (type IV) (Harrigan and Deveikis 2013; Takai and Taniguchi 2012).

Venous Infarction

Venous infarction has been reported even in the absence of a spinal vascular lesion (Niino et al. 1999). Procedures including surgery directly involving the spinal cord or vasculature, surgery manipulating the aorta (particularly during aortic cross clamping), and even endoscopic sclerotherapy for esophageal varices, have been implicated in obstructing venous return resulting in spinal infarction (Heller et al. 1996; Ross 1985).

Neoplastic Pathology

Batson's experiments led to the notion that aberrant malignant and pyogenic metastasis spread along the vertebral column via the venous system of the spine. At the time, he considered it a fourth pathway in addition to the pulmonary, caval, or portal system by which metastatic spread, in particular in breast and prostate cancers, could occur (Batson 1940; Maccauro et al. 2011). Increased flow during instances of elevated intra-thoracic or intra-abdominal pressure promotes metastatic spread and metastatic tumor deposits in the vertebrae (Coman and de, 1951; Geldof 1997; Maccauro et al. 2011). Under normal circumstances, 5–10% of the portal and caval blood volume is shunted into the vertebral venous system (Louis et al. 1976). The tendency of the vertebral column to be a frequent site of skeletal metastasis may be at least partially explained by the presence of Batson's plexus. Metastatic spread of other malignancies via Batson's plexus has been reported (Boaz and Natarajan 2012; Maccauro et al. 2011; Mathew et al. 2000; Onec et al. 2007).

Infectious and Inflammatory Pathology

Batson's plexus also provides a route for spread of abscesses along the spinal column (Batson 1940). Spinal epidural abscesses can cause congestion in the epidural venous plexus or basivertebral veins and produce contrast enhancement on spinal imaging. The enhancement results from direct spread of the infection along the extradural, internal vertebral plexus, as well as mechanical obstruction of venous flow (Numaguchi et al. 1993). Interestingly, dilation of the anterior internal vertebral plexus was noted in up to 20% of patients in the upper cervical region in the absence spinal canal abnormality. This observation has to be considered when images suspicious for diskitis or osteomyelitis are interpreted to avoid false diagnosis of a spinal epidural abscess. One distinguishing feature is the symmetric appearance of normally dilated veins on axial imaging compared to the venous dilation occurring in the presence of infection (Numaguchi et al. 1993).

Chronic cerebrospinal venous insufficiency secondary to internal jugular, vertebral, and azygos venous outflow obstruction has been described in patients with multiple sclerosis. Still, the role of this finding in the pathogenesis of this disease is highly controversial (Zivadinov et al. 2011).

Trauma

After acute spinal cord injury, intrathecal hypertension can contribute to compression of the spinal veins and venous stasis (Horn et al. 2008). Potential sequel include rupture of the internal vertebral plexus with subsequent formation of a spinal epidural hematoma (Al-Mutair and Bednar 2010). Interestingly, the distribution of heterotopic ossification in paraplegic patients has been found to be similar to that of Batson's plexus. A possible explanation involves immobilization producing venous stasis with resultant demineralization of the skeleton (Major et al. 1980).

Vertebro-vertrebral arteriovenous fistulas are high flow fistulas frequently traumatic in origin (Kominami et al. 1996; Madoz et al. 2006). Patients often present with cervical bruit, vertebrobasilar insufficiency, tinnitus, or rarely subarachnoid hemorrhage (Kominami et al. 1996). Imaging reveals dilation of the extradural, internal vertebral plexus in the cervical region and dilation of the extraspinal efferents. The treatment of choice is endovascular occlusion (Wang et al. 2011).

Other Pathology

Intervertebral disc herniations and degenerative spinal canal stenosis may also cause obstruction of venous drainage with venous dilation (Numaguchi et al. 1993). Lateral disc herniation can occlude radicular veins that accompany the spinal nerves, which can be visualized on CT or MRI. In rare instances, a disc herniation can result in myelopathy, not from direct compression, but from impedance of venous outflow at the level of the intervertebral foramen causing venous hypertensive myelopathy (Krishnan et al. 2004).

26.5 Conclusion

Standard anatomy and physiology of the spinal cord venous drainage system is affected by various pathological entities that result in either abnormal drainage or obstructed flow. The variability and complexity of this system can make recognizing abnormal pathology and treatment difficult, leading to devastating consequences. Previous research has laid the foundation for the understanding of normal anatomy, allowing current study to focus on pathological entities. As knowledge is gained, improvement in the diagnosis and management will reduce the morbidity associated with disease processes affecting spinal cord venous drainage.

References

Abrams H (1957) The vertebral and azygos venous systems, and some variations in systemic venous return. Radiology 69:508–526

Al-Mutair A, Bednar DA (2010) Spinal epidural hematoma. J Am Acad Orthop Surg 18:494–502

Anderson R (1951) Diodrast studies of the vertebral and cranial venous systems to show their probable role in cerebral metastases. J Neurosurg 8:411–422

Batson O (1940) The function of the vertebral veins and their role in the spread of metastases. Ann Surg 112:138–149

Boaz K, Natarajan S (2012) Have we forgotten the Batson plexus? J Oral Maxillofac Surg 70:4
Bowen B, DePrima S, Pattany P, Marcillo A, Madsen P, Quencer R (1996) MR angiography of normal intradural vessels of the thoracolumbar spine. AJNR Am J Neuroradiol 17:483–494
Breschet G (1819) Essai sur les veines du rachis. Méquignon-Marvis, Paris
Breschet G (1828–1832) Récherches Anatomiques, Physiologique et Pathologiques sur le Systéme Veineux et Spécialement sur les Cavaux Veineux des Os. Villaret et Cie, Paris
Brockstein B, Johns L, Gewertz B (1994) Blood supply to the spinal cord: anatomic and physiologic correlations. Ann Vasc Surg 8:394–399
Brookes M, Zietman A (1998) Clinical embryology: a color atlas and text. CRC Press, Boca Raton
Cruveilhier J (1836) Traite d'Anatomie Descriptive. Tome 3-ieme. Veins du Rachis. Bechet Jeune, Paris
Dommisse G (1975) The venous drainage of the spinal cord. In: The arteries and veins of the human spinal cord from birth. Churchill Livingstone, New York, pp 81–96
Geldof AA (1997) Models for cancer skeletal metastasis: a reappraisal of Batson's plexus. Anticancer Res 17:1535–1539
Gillilan L (1970) Veins of the spinal cord. Neurology 20:860
Gillot C, Aaron C (1968) On the course of the branches of the left renal vein. Arch Anat Pathol 16:A196–A202
Goshgarian H (2003) Blood supply of the spinal cord. In: Lin V, Cardenas D, Cutter N, Frost F, Hammond M, Lindblom L, Perkash I, Waters R, Woolsey R (eds) Spinal cord medicine: principles and practice. Demos Medical Publishing, New York
Gray H (1918) Anatomy of the human body. Lea & Febiger, Philadelphia
Groen R, Groenewegen H, van Alphen H, Hoogland P (1997) Morphology of the human internal vertebral venous plexus: a cadaver study after intravenous Araldite CY 221 injection. Anat Rec 249:285–294
Groen R, du Toit D, Phillips F, Hoogland P, Kuizenga K, Coppes M, Muller C, Grobbelaar M, Mattyssen J (2004) Anatomical and pathological considerations in percutaneous vertebroplasty and kyphoplasty: a reappraisal of the vertebral venous system. Spine 29:1465–1471
Harrigan MR, Deveikis JP (2013) Spinal Vascular Lesions. In: Harrigan MR, Deveikis JP (eds) Handbook of cerebrovascular disease and neurointerventional technique, 2nd edn. Humana Press, New York, pp 803–821
Harris HA (1941) A note on the clinical anatomy of the veins, with special reference to the spinal veins. Brain 64:291
Harvey W (1894) An anatomical dissertation upon the movement of the heart and blood in animals, being a statement of the discovery of the circulation of the blood: Canterbury
Heller S, Meyer J, Russell E (1996) Spinal cord venous infarction following endoscopic sclerotherapy for esophageal varices. Neurology 47:1081–1085
Henderson F, Crockard H, Stevens J (1993) Spinal cord oedema due to venous stasis. Neuroradiology 35:312–315
Herlihy WF (1947) Revision of the venous system: role of vertebral veins. Med J Aust 1:661–672
Horn E, Theodore N, Assina R, Spetzler R, Sonntag V, Preul M (2008) The effects of intrathecal hypotension on tissue perfusion and pathophysiological outcome after acute spinal cord injury. Neurosurg Focus 25
Jinkins J (2000) Spinal vasculature. In: John J, Brown B, Seigafuse S, Palumbo R (eds) Atlas of Neuroradiologic embryology, anatomy, and variants. Lippincott Williams & Wilkins, Philadelphia, pp 563–578
Kominami S, Liu Y, Alvarez H, Rodesch G, Coubes P, Lasjaunias P (1996) A case of vertebrovertebral arteriovenous fistula presenting with subarachnoid haemorrhage. A case report. Interv Neuroradiol 2:229–233
Krishnan C, Malik J, Kerr D (2004) Venous hypertensive myelopathy as a potential mimic of transverse myelitis. Spinal Cord 42:261–264
Lasjaunias P, Berenstein A (1990) Spinal and spinal cord arteries and veins. In: Surgical neuroangiography: functional vascular anatomy of brain, spinal cord, and spine. Springer-Verlag, New York
Lolis E, Panagouli E, Venieratos D (2011) Study of the ascending lumbar and iliolumbar veins: surgical anatomy, clinical implications and review of the literature. Ann Anat 193:516–529
Louis R, Ouiminga RM, Obounou D (1976) The azygos or vertebro-parietal venous anastomotic system. Bull Assoc Anat (Nancy) 60:381–397
Maccauro G, Spinelli MS, Mauro S, Perisano C, Graci C, Rosa MA (2011) Physiopathology of spine metastasis. Int J Surg Oncol 2011:107969
Madoz A, Desal H, Auffray-Calvier E, Isnard J, Liberge R, Taverneau C, De Kersaint-Gilly A (2006) Vertebrovertebral arteriovenous fistula diagnosis and treatment: report of 8 cases and review of the literature. J Neuroradiol 33:319–327
Major P, Resnick D, Greenway G (1980) Heterotopic ossification in paraplegia: a possible disturbance of the paravertebral venous plexus. Radiology 136:797–799
Mathew P, Fleming D, Adegboyega PA (2000) Myelophthisis as a solitary manifestation of failure from rectal carcinoma. A Batson phenomenon? Arch Pathol Lab Med 124:1228–1230
Minot C, Evans H, Tandler J, Sabin F (1912) The development of the blood, the vascular system, and the spleen. In: Keibel F, Mall F (eds) Manual of human embryology. J.B. Lippincott Company, Philadelphia, pp 669–687
Moes P, Maillot C (1981) Superficial veins of the human spinal cord. An attempt at classification. Arch Anat Histol Embryol 64:5–110

Nathoo N, Caris EC, Wiener JA, Mendel E (2011) History of the vertebral venous plexus and the significant contributions of Breschet and Batson. Neurosurgery 69:1007–1014. discussion 1014

Niino M, Isu T, Tashiro K (1999) Nonhemorrhagic venous infarction of the spinal cord without spinal vascular malformation. J Neurol 246:852–854

Numaguchi Y, Rigamonti D, Rothman M, Sato S, Mihara F, Sadato N (1993) Spinal epidural abscess: evaluation with gadolinium-enhanced MR imaging. Radiographics 13:545–559

Onec B, Oksuzoglu B, Hatipoglu HG, Onec K, Azak A, Zengin N (2007) Cavernous sinus syndrome caused by metastatic colon carcinoma. Clin Colorectal Cancer 6:593–596

Ross R (1985) Spinal cord infarction in disease and surgery of the aorta. Can J Neurol Sci 12:289–295

Sabin F (1915) Development of the veins in the embryo pig. In: Streeter G (ed) Contributions to embryology. Carnegie Institution of Washington, Washington, D.C., pp 12–26

Santillan A, Nacarino V, Greenberg E, Riina H, Gobin Y, Patsalides A (2012) Vascular anatomy of the spinal cord. J Neuro Intervent Surg 4:67–74

Stringer MD, Restieaux M, Fisher AL, Crosado B (2012) The vertebral venous plexuses: the internal veins are muscular and external veins have valves. Clin Anat 25:609–618

Takai K, Taniguchi M (2012) Comparative analysis of spinal extradural arteriovenous fistulas with or without intradural venous drainage: a systematic literature review. Neurosurg Focus 32:E8

Thron A, Rossberg C (1988) Vascular anatomy of the spinal cord: neuroradiological investigations and clinical syndromes. Springer, New York

Tobinick E, Vega C (2006) The cerebrospinal venous system: anatomy, physiology, and clinical implications. MedGenMed 8:53

Tubbs RS, Hansasuta A, Loukas M, Louis RG Jr, Shoja MM, Salter EG, Oakes WJ (2007) The basilar venous plexus. Clin Anat 20:755–759

van der Kuip M, Hoogland P, Groen R (1999) Human radicular veins: regulation of venous reflux in the absence of valves. Anat Rec 254:173–180

Wang Q, Song D, Chen G (2011) Endovascular treatment of high-flow cervical direct vertebro-vertebral arteriovenous fistula with detachable coils and Onyx liquid embolic agent. Acta Neurochir 153:347–352

Zawiliński J, Litwin J, Nowogrodzka-Zagórska M, Gorczyca J, Miodoński A (2001) Vascular system of the human spinal cord in the prenatal period: a dye injection and corrosion casting study. Ann Anat 183:331–340

Zivadinov R, Ramanathan M, Dolic K, Marr K, Karmon Y, Siddiqui A, Benedict R, Weinstock-Guttman B (2011) Chronic cerebrospinal venous insufficiency in multiple sclerosis: diagnostic, pathogenetic, clinical and treatment perspectives. Expert Rev Neurother 11:1277–1294

Vertebral Veins

27

Skyler Kopit and R. Shane Tubbs

27.1 Embryology

The morphology of the vertebral venous plexus (Figs. 27.1, 27.2, 27.3 and 27.4) is believed to vary with age. In 2005, Groen et al. conducted a study on human fetuses between 21 and 25 weeks gestational age. The findings confirmed previous studies showing that the posterior intravertebral venous plexus lacked transverse bridging veins in the fetus, in contrast to a mature sample. This is suspected to be a result of developmental delay, or simply a difference between prenatal and postnatal life (Groen et al. 2005).

The primitive circulatory network develops closely with the heart and the intra-embryonic, vitelline, and umbilical-allantoic networks (Magro et al. 2013). The mesenchymal and vascular networks develop together. Neighboring mesenchyme gives rise to the protective tissue and muscular coat present. The mature venous system contributes to a nutritional network that returns extra-embryonic blood (omphalomesenteric network).

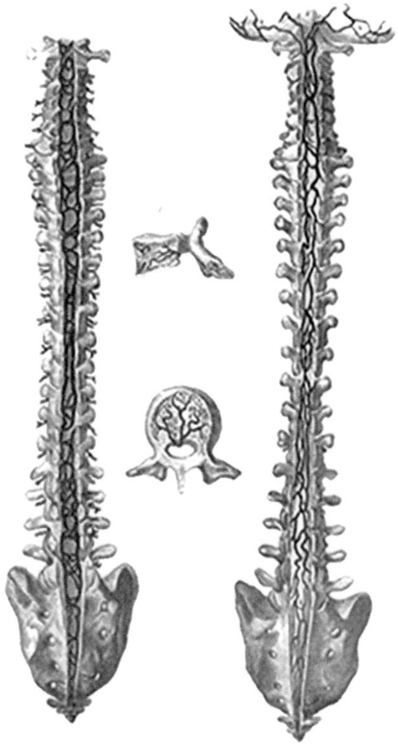

Fig. 27.1 Drawings of the vertebral venous plexus. (From Quain and Wilson 1844 author's collection)

The venous cardinal system, part of the intra-embryonic network, drains blood from the head, neck, and body walls (Larsen 2001). The vertebral veins arise from the superior intrarachidial plexuses (Poirier 1898). The vertebral venous plexus comes from the primitive branching anas-

S. Kopit
Tulane University School of Medicine, New Orleans, LA, USA
e-mail: skopit@tulane.edu

R. S. Tubbs (✉)
Department of Neurosurgery, Tulane University School of Medicine, New Orleans, LA, USA
e-mail: rtubbs@tulane.edu

S. J. Bordes, Jr. et al. (eds.), *The Clinical Anatomy of the Vascular System*,
https://doi.org/10.1007/978-3-031-78326-5_27

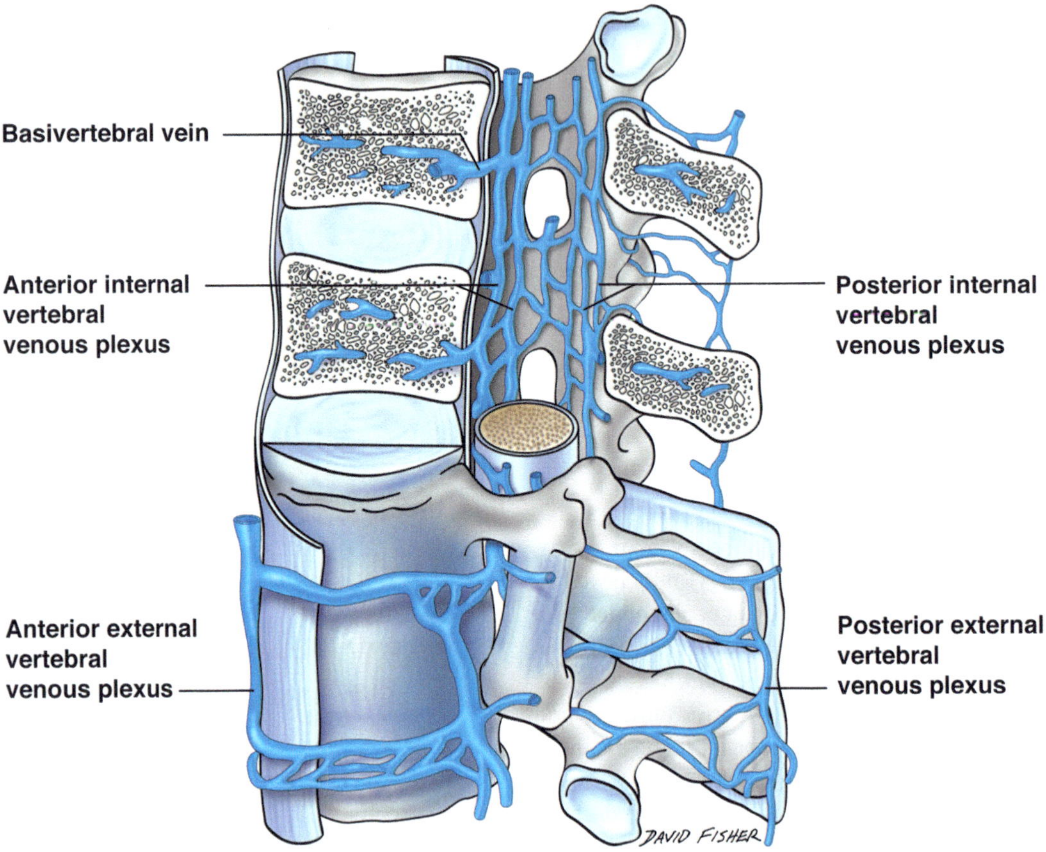

Fig. 27.2 Schematic drawing of the different parts of the vertebral venous plexus. (After Netter 2023)

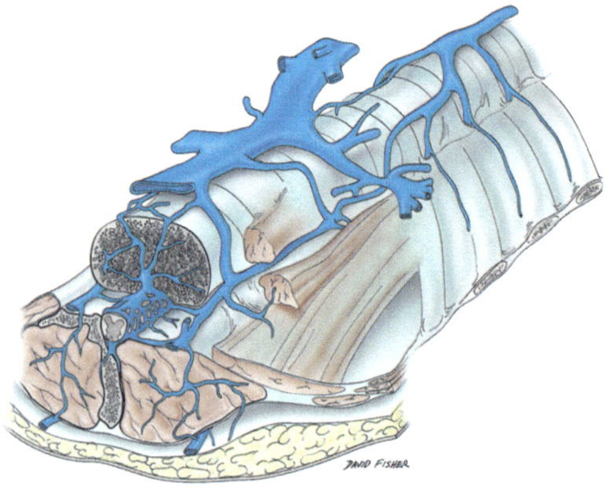

Fig. 27.3 Schematic drawing illustrating the veins surrounding the spinal cord and vertebral column and their inner communications with other regional veins. (After Woodburne)

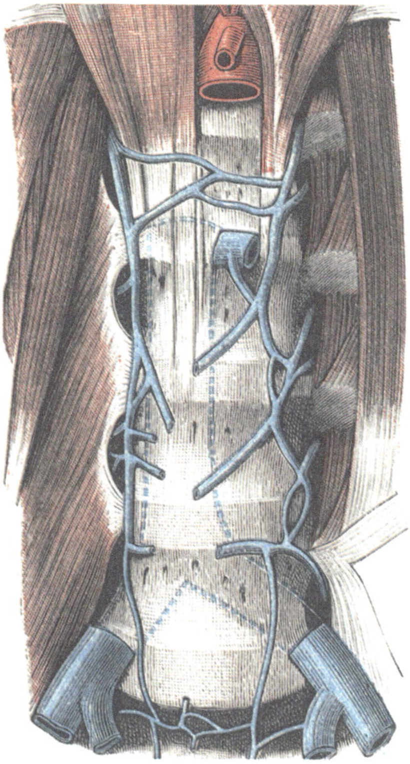

Fig. 27.4 Schematic drawing of the anterior aspect of the lumbar spine noting the inferior vena cava in outline and its lumbar vein tributaries. (After Testut)

tomosis with the occipital vein and cranial sinuses (Rouvière and Delmas 1978). At each intervertebral foramen, it anastomoses with the posterior jugular vein and with the sigmoid aspect of the lateral sinus through the mastoid emissary and external jugular veins. This is all completed through the deep tributaries of the occipital veins.

27.2 Anatomy/Topography

The external vertebral vein, which has also been termed the trachea vein of Breschet, has been considered internal in some instances. It descends dorsally to the cervical vertebrae, specifically the articular processes. The veins are united internally and externally through many anastomoses (Gérard 1921).

The first notable mention of the vertebral venous plexus was in 1819 when Gilbert Breschet provided the world's first anatomically descriptive report of the network. He noted the valveless nature of the veins and the three interconnecting divisions spanning from the cranial dural sinuses to the entire spinal column. It was not until 1940 that Oscar V. Batson detailed the cerebrospinal venous network with bidirectional flow (Nathoo et al. 2011). Although Batson's discoveries concerning the vertebral veins were significant, modern technology has enabled major advances to be made in the knowledge base regarding the plexus. One study during the late 1990s used a CY 221 mixture to differentiate the anterior and posterior parts of the plexus (Groen et al. 1997). Another study revealed venous connections between the craniocervical vertebral venous systems and the dural venous sinuses of the posterior cranial fossa using corrosion casts and Magnetic Resonance (MR) venography (San Millán Ruíz et al. 2002).

Tobinick and Vega divided the cerebrospinal venous system into the cranial venous system and the vertebral venous plexus in 2006 (Tobinick and Vega 2006). This emphasized the continuity between the two, highlighted the existence of a richly interconnected venous network, and introduced the capacity for spreading of components in either direction within the system.

Perhaps the most important clinical finding, emphasized by numerous studies, including those mentioned above, is that the vertebral veins lack valves. They also form a large-capacity network capable of bidirectional flow. They are important for regulating intracranial pressure in response to postural changes, including the Valsalva maneuver.

Batson's discoveries led to investigations of the connection between structure and function in the vertebral venous plexus. Groen's studies further revealed the complex connections between the vertebral venous plexus and other venous areas, such as the intercostal veins, intra-abdominal veins, and cranial venous sinuses.

It is important to emphasize that this system is unique. Its characteristic bidirectional flow is its critical distinction from other veins. In 1947, Herlihy challenged the idea that the vertebral venous plexus operated like any other plexus and argued that the bidirectional flow does not function in returning blood to the heart (Herlihy

1947). Batson noted this and went on to propose that its physiological role, apart from being a large-capacity fluid reserve, lay in cooling the brain and spine and in venous outflow, which is posture-dependent (Batson 1940).

Before 1966, venous outflow from the cranium was thought to be entirely due to the internal jugular vein. In 1966, Eckenhoff showed that while the internal jugular vein was significant, the vertebral venous plexus was also a major contributor. Epstein et al. studied cerebral venous outflow in rhesus monkeys using angiography (Epstein et al. 1970). By examining body position and pressure within the airway, they discovered that nearly all of the injected contrast flowed through the vertebral plexus when the monkey was erect; the internal jugular veins were barely visible. This finding challenged the idea that the internal jugular veins were the sole avenue of blood outflow from the brain. While the animal is erect, as previously stated, the vertebral venous plexus provides the major outflow pathway from the cranium. When the subject is supine, the jugular veins carry most of the outflow (Valdueza et al. 2000).

Several longitudinal channels formed by the vertebral plexus drain into the epidural space of the vertebral canal. These veins can specifically be mapped out as two pairs connecting the anterior and posterior roots to the spinal cord on each side. Additionally, two midline channels travel in parallel and pass along the anterior and posterior median fissure and sulcus. The longitudinal channels drain into a channel within the epidural space. Eventually, they connect with other major systemic veins such as the azygos system within the thorax and the intracranial veins.

The vertebral venous plexus can become significant for lower limb and pelvic drainage if the inferior vena cava becomes occluded. This collateral route through the intrapelvic vertebral venous plexus allows blood to return to the heart (Lasjaunias and Berenstein 1987). Additionally, the venous plexus arrangement within the vertebral canal has been shown to provide anatomical preconditions for spinal cord cooling through the cerebrospinal fluid. This is accomplished through the venous networks close to the skin, which anastomose externally and internally within the vertebral venous plexus (Zenker and Kubik 1996). The dura mater is highly vascularized, and it is proposed that its temperature changes could be transmitted to the cerebrospinal fluid. Warm venous blood flows from brown fat and muscle, while cool cutaneous venous blood can help to regulate the temperature within the vertebral venous plexus. The mechanisms are largely unknown, but the pathways for venous blood temperature regulation in the vertebral venous plexus, and the anastomoses involved, are being studied in cadavers (Hoogland et al. 2013).

27.3 Variation

A study in 2012 revealed that approximately one-third of patients have valves in the posterior external vertebral vein division (Clemens 1961). In 2013, the non-jugular venous system was investigated using MR images of participants between the ages of 16 and 76 (Kopelman et al. 2013), revealing age-related variations. The cross-sectional area of the vertebral venous plexus seemed to decrease with increasing age, while the internal jugular vein sections remained relatively constant. The same study showed that most of the participants had a nonsignificant asymmetry between their left and right non-jugular venous systems (including the vertebral venous plexus).

27.4 Clinical Significance

The cerebrospinal venous system is crucial for disease's metastatic and infectious spread because its mostly valveless network allows for bidirectional flow. This network has increased the incidence of cerebral metastases from 5 to 20.5%, as revealed mainly by autopsies and reports (Cushing 1932). It is important to recognize that cerebral metastatic lesions are blood-borne. As one such study illustrated, metastatic spread from lung cancers to the brain was frequent enough to establish it as a common cause of brain tumors (Richards and McKissock

1963). The severity of the cancers also warrants attention; survival rates from brain metastases scarcely approach 6 months, and such metastases are frequently associated with poor life quality (Perse 1959).

The multiplicity and widespread distribution of cerebral metastatic lesions reflect their ability to spread in a vascular fashion through the plexus (Willis 1934). A study investigating inferior vena cava agenesis, a rare anomaly with less than 1% incidence, confirmed the significance of the vertebral venous plexus. Those without an inferior vena cava are forced to develop collateral systems, one of which involves the paravertebral venous plexus, which ultimately drains into the superior vena cava (SVC) or rectal venous plexus (Singh and Bhatt 2017).

References

Batson OV (1940) The function of the vertebral veins and their role in the spread of metastases. Ann Surg 112:138–149

Clemens HJ (1961) The venous Systems of the Human Spine; morphology and functional importance. De Gruyter, Berlin

Cushing H (1932) Intracranial tumours. Notes upon a series of two thousand verified cases with surgical-mortality percentages pertaining thereto. Charles C Thomas, Springfield, p 150

Epstein HM, Linde HW, Crampton AR, Ciric IS, Eckenhoff JE (1970) The vertebral venous plexus as a major cerebral venous outflow tract. Anesthesiology 32:332–337

Gérard G (1921) Manuel d'anatomiehumaine. Masson, Paris

Groen RJ, Groenewegen HJ, van Alphen HA, Hoogland PV (1997) Morphology of the human internal vertebral venous plexus: a cadaver study after intravenous araldite CY 221 injection. Anat Rec 249: 285–294

Groen RJ, Grobbelaar M, Muller CJ, et. al. (2005) Morphology of the human internal vertebral venous plexus: a cadaver study after latex injection in the 21-25-week fetus. Clin Anat 18:397–403

Herlihy WF (1947) Revision of the venous system: the role of the vertebral veins. Med J Aust 1:661–667

Hoogland PV, Wessels Q, Vorster W, et. al. (2013) The posterior intercostal vein: a thermoregulatory gateway to the internal vertebral venous plexus. Clin Anat 26:735–740

Kopelman M, Glik A, Greenberg S, Shelef I (2013) Intracranial nonjugular venous pathways: a possible compensatory drainage mechanism. AJNR Am J Neuroradiol 34:1348–1352

Larsen WJ (2001) Human embryology. Elsevier, Amsterdam

Lasjaunias PL, Berenstein A (1987) Surgical Neuroangiography. Springer-Verlag Berlin, HeidelbergBerlin

Magro E, Sénécail B, Gentric J-C, Alavi Z, Palombi O, Seizeur R (2013) Contribution of embryology in the understanding of cervical venous system anatomy within and around the transverse foramen: a review of the classical literature. Surg Radiol Anat 36:411–418. https://doi.org/10.1007/s00276-013-1206-1

Nathoo N, Caris EC, Wiener JA, Mendel E (2011) History of the vertebral venous plexus and the significant contributions of Breschet and Batson. Neurosurgery 69:1007–1014

Netter FH.(2023) Netter atlas of human anatomy: classic regional approach (8th ed.). Elsevier.

Perse DN (1959) Diagnosis in metastatic tumors of the brain and skull. Cancer 12(609–6):13

Poirier P (1898) Traited'anatomie humaine. Masson, Paris

Richards P, McKissock W (1963) Intracranial metastases. Br Med J:1115–1118

Rouvière H, Delmas A (1978) Anatomie humaine, Tête et cou. Masson, Paris

San Millán Ruíz D, Gailloud P, Rüfenacht DA, Delavelle J, Henry F, Fasel JH (2002) The craniocervical venous system in relation to cerebral venous drainage. AJNR Am J Neuroradiol 23:1500–1508

Singh SN, Bhatt TC (2017) Inferior vena cava agenesis: a rare cause of pelvic congestion syndrome. J Clin Diagn Res 11:TD06–TD08

Tobinick E, Vega CP (2006) The cerebrospinal venous system: anatomy, physiology, and clinical implications. Med Gen Med 8:53

Valdueza JM, von Münster T, Hoffman O, Schreiber S, Einhäupl KM (2000) Postural dependency of the cerebral venous outflow. Lancet 355:200–201

Willis RA (1934) *The spread of tumours in the human body*. J. & A. Churchill, London, p 540

Zenker W, Kubik S (1996) Brain cooling in humans–anatomical considerations. Anat Embryol (Berl) 193:1–13

Part III

Thorax

28 Ascending Aorta

Caroline Csenge and Timothy Legare

28.1 Anatomy

The aorta is the major outflow vessel of the left ventricle of the heart, originating from just distal to the aortic valve and continuing through the thorax and abdomen until its termination into the left and right common iliac arteries near the pelvic inlet. The aorta is divided into three sections depending on location: the ascending aorta, the arch of the aorta, and the descending aorta. The descending aorta is further subdivided into thoracic and abdominal components. The ascending aorta is the initial portion arising distal to the aortic valve, ascending anterior to the trachea, and transitioning into the arch of the aorta at the junction with the brachiocephalic trunk, also known as the innominate artery. This junction between the ascending aorta and the arch of the aorta occurs approximately at the level of articulation between the body of the sternum and the manubrium or the sternal angle (Collins et al. 2014; Saliba et al. 2015a). The ascending aorta can be further subdivided into two sections: the aortic root and the tubular ascending aorta. The aortic root is the first portion arising from the ventricular outflow tract and extending to the sinotubular junction. The tubular ascending aorta is the portion from the sinotubular junction to the origin of the brachiocephalic trunk. The aortic arch curves to the left of the body in most individuals and gives off three main branches: the brachiocephalic trunk, the left common carotid, and the left subclavian artery. The arch then begins its descent as the descending thoracic aorta at the level of the fourth thoracic vertebral body (Collins et al. 2014).

On average, the ascending aorta measures 5 cm in length and 2.5 cm in diameter (Collins et al. 2014). Arising from the aortic root (Fig. 28.1) are the three aortic sinuses, also called the sinuses of Valsalva (Arcario et al. 2021; Loukas et al. 2014). These include the left and right coronary aortic sinuses and the noncoronary aortic sinus.

Most commonly, the coronary arteries arise from the left and right coronary aortic sinuses, respectively, at or below the level of the sinotubular junction; however, it is estimated that the coronary arteries arise above the junction in 1% of the population (Anderson 2000; Tomanek and Angelini 2019; Angelini et al. 1999). Rarely either coronary artery can arise from the noncoronary sinus (Loukas et al. 2013; Manghat et al. 2005). The coronary arteries are the only branches arising from the ascending aorta.

C. Csenge
Tulane University School of Medicine, New Orleans, LA, USA
e-mail: ccsenge@tulane.edu

T. Legare (✉)
Department of Surgery, Louisiana State University Health Sciences Center, New Orleans, LA, USA

Department of Vascular Surgery, Louisiana State University Health Sciences Center, New Orleans, LA, USA
e-mail: tlegar@lsuhsc.edu

S. J. Bordes, Jr. et al. (eds.), *The Clinical Anatomy of the Vascular System*,
https://doi.org/10.1007/978-3-031-78326-5_28

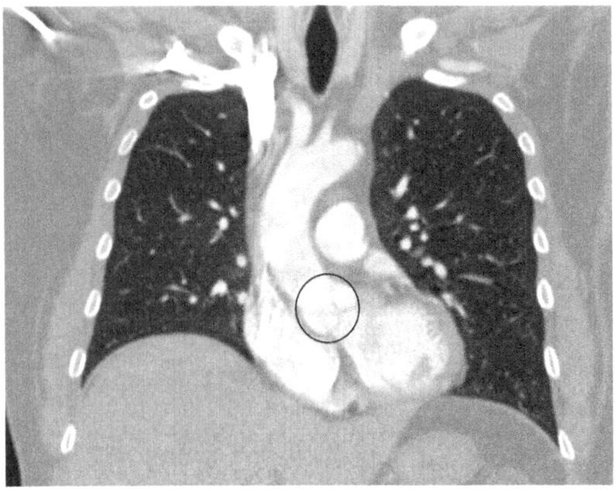

Fig. 28.1 Coronal CTA showing the aortic root (circle)

28.2 Embryology

The aorta develops during the third week of gestation due to vasculogenesis. The primitive aorta is composed of symmetrical dorsal and ventral segments with paired pharyngeal arches between these segments connecting the structures. The dorsal segments will fuse to become the midline descending aorta, while the ventral segments fuse to form the aortic sac, the superior portion of the truncus arteriosus. The truncus arteriosus is the common outflow tract from the primordial heart. During the fifth week of development, a septum is formed within the truncus arteriosus due to neural crest cell migration. The septum divides the lumen, leading to separate outflow tracts for the aorta and the pulmonary artery, with the aorta anterior and to the right of the pulmonary trunk. It is during this process that congenital malformations such as Tetralogy of Fallot and Transposition of the Great Vessels can occur (Murillo et al. 2012; Larsen 1993; Moore and Persaud 1998). The pharyngeal arches contribute to the development of various structures in the adult. Importantly, the fourth pharyngeal arch gives rise to the aortic arch (Hanneman et al. 2017).

The sinuses of Valsalva arise during the fifth week of embryogenesis. The coronary arteries follow—developing either later in the fifth week or earlier in the sixth week due to the condensation of the vascular angioblasts from within the sinuses. Disruption of this process is the basis for anomalous origins of the coronary arteries (Murillo et al. 2012; Larsen 1993; Moore and Persaud 1998).

Variations in the origin of the coronary arteries are thought to occur in 0.3%–1.6% of the population (Loukas et al. 2013). This number may, however, be underestimated since individuals are often asymptomatic, never receiving a diagnosis (Loukas et al. 2013). These variations include the origin of the coronary artery from the opposite sinus, opposite coronary artery, or pulmonary artery (Tomanek and Angelini 2019; Loukas et al. 2013; Yamanaka and Hobbs 1990). A coronary artery originating from the opposite sinus is associated with an increased risk of myocardial ischemia and sudden cardiac death (Loukas et al. 2013; Manghat et al. 2005; Zimmermann et al. 2008). Interestingly, it is thought that such variations in the origin of the coronary arteries account for a significant percentage of sudden cardiac death in young athletes (Loukas et al. 2013; Koşar et al. 2009).

28.3 Histology

The aorta is comprised of the same three histological layers that are common to many arteries in the body. The innermost layer lining the aortic lumen, the tunica intima, is composed of endothelial cells and is supported by the internal elastic lamina. The middle layer, the tunica media, consists of smooth muscle cells and elastic tissue. The external elastic lamina separates the tunica media from the outermost layer, the tunica adventitia, which is composed of fibroblasts and collagen fibers. Additionally, the vaso vasorum in the tunica adventitia provides blood to the outermost aortic wall (Collins et al. 2014). The aorta differs from most arteries of the body as it is an elastic artery. Elastic arteries contain a greater quantity of elastin fibers within the tunica media compared to muscular arteries, such as the femoral artery. It is important to note that the normal aging process leads to fragmentation of the elastin fibers—weakening the aortic wall (Collins et al. 2014).

28.4 Imaging

Three imaging modalities are commonly used to visualize the ascending aorta. These include computed tomography (CT), magnetic resonance angiography (MRA), and echocardiograph (Gudbjartsson et al. 2020; Iddriss et al. 2019). Echocardiographs can be taken using two different approaches: transthoracic (TTE) (noninvasive) or a transesophageal (TEE)(invasive) approach. TTE is widely available and easily performed at the bedside. A TTE can report cardiac function and allows for visualization of the aortic root, which is important when assessing aortic dilation. This visualization, however, is limited in obese patients, patients with chest deformities, or patients on mechanical ventilation (Gudbjartsson et al. 2020; Meredith and Masani 2009). A TTE is also limited in its ability to visualize the ascending aorta distal to the sinotubular junction, making it less sensitive for detection of aortic dissection (Gudbjartsson et al. 2020; Meredith and Masani 2009). TEE allows for a more detailed image of the ascending aorta, while still allowing for assessment of cardiac function. The TEE is limited by its invasive nature, requiring the patient to be sedated. Although sensitive, it is not commonly used in the emergency setting due to its complexity and lack of accessibility; however, TEE can be used pre- or intraoperatively for cardiac assessment and monitoring (Chong et al. 2020).

CT and MRA imaging are both well-regarded methods for measuring the ascending aorta. CT is most widely performed because of its availability, ease of use, and limited radiation exposure to the patient (Freeman et al. 2013; Frazao et al. 2020). One method used when measuring the diameter of the ascending aorta is ECG gating. This method takes the cardiac cycle into consideration to prevent overestimation of the aortic diameter by controlling for the effect of cardiac pulsation (Freeman et al. 2013). Additionally, one benefit to CT and MRA is the ability to produce 3D reconstruction of the 2D images, which can give a more complete look at complex pathologies and allow for a more thorough evaluation of vasculature (Iddriss et al. 2019).

28.5 Pathology

Cystic medial necrosis (CMN), or cystic medial degeneration, is a histological term used to describe disruption of the tunica media of large arteries, particularly the aorta, leading to the creation of cyst-like lesions. Elastin and smooth muscle fibers undergo degeneration, while collagen fibers are cross-linked, leading to fibrosis. This is commonly seen in patients with Type A aortic dissection and thoracic aortic aneurysms, especially when coexisting with connective tissue disorders like Marfan syndrome and Ehlers-Danlos (Yuan and Jing 2011). CMN is not specific to aortic aneurysms or dissection and can often be found in geriatric and hypertensive patients. CMN has also been found in individuals with no underlying disease; therefore, the role of CMN in aortic pathology remains unclear (Yuan and Jing 2011).

Aortic Aneurysm

An aortic aneurysm is an abnormal dilation of the aorta that measures at least 50% greater than expected for the individual's age and body size. An aneurysm is distinguished from ectasia, where dilation is less than 50% greater than expected (Saliba et al. 2015). Aneurysms are further classified as either true or false aneurysms. True aneurysms involve all three layers of the aortic wall. False aneurysms, or pseudoaneurysms, involve a tear through the tunica intima and the tunica media while the tunica adventitia remains intact and contains the blood. False aneurysms can be the result of iatrogenic injury from surgical procedures or a result of trauma (Takamure et al. 2007).

Dilation of the ascending aorta is associated with common cardiovascular risk factors such as hypertension, hypercholesterolemia, diabetes, and smoking (Kauhanen et al. 2020; Roman et al. 1989; Johnston et al. 1991; Rogers et al. 2013). Additionally, aneurysms of the ascending aorta are a known complication of syphilis (Yuan 2018; Goldstein et al. 2003). Thoracic aortic aneurysms are often seen in patients with connective tissue disorders, such as Marfan syndrome and Ehlers-

Danlos. Marfan syndrome is caused by a mutation in the fibrillin-1 gene, leading to defective elastin fibers, which compromises the structural integrity of the aortic wall (Freeman et al. 2013). Importantly, it is also expected that the aorta will increase in both diameter and length with age (Collins et al. 2014).

No matter the cause, an aortic aneurysm is associated with a yearly increased risk of rupture or dissection correlated with the size of the aneurysm. An aneurysm measuring 6 cm or greater carries at least a 6.9% yearly risk of rupture (Davies et al. 2002). Additionally, with an enlarged diameter of the ascending aorta, regurgitation through the aortic valve becomes a concern. Preventative surgery can be performed to mitigate these risks (Davies et al. 2002).

Aortic Dissection

Aortic dissection results from a tear in the aortic intima. An intimal flap is created which allows blood through the flap, creating a false lumen between the tunica intima and tunica media. As more blood fills the false lumen, the tear can propagate either proximal or distal to the original site. Aortic dissections are classified based on the location of the initial tear, most often described using the Stanford classification system (Gudbjartsson et al. 2020). According to this system, any dissection involving the ascending aorta, even if there is also distal involvement, is described as a Type A aortic dissection (TAAD). A Type B dissection involves the aorta distal to the left subclavian artery (Gudbjartsson et al. 2020; Nienaber 2015; Elsayed et al. 2017). The most common location of a TAAD is at the sinotubular junction (Plonek et al. 2017; Hirst Jr et al. 1958). Most TAADs occur with an aortic diameter > 40 mm (Plonek et al. 2017; Rylski et al. 2014). TAADs account for 58–62% of all aortic dissections and have an estimated mortality rate of 73% (Gudbjartsson et al. 2020; Hagan et al. 2000; Landenhed et al. 2015; Rogers et al. 2011).

Factors associated with an increased risk for TAAD include hypertension, bicuspid aortic valve, and advanced age (Gudbjartsson et al. 2020; Plonek et al. 2017; Hagan et al. 2000). Existing aortic aneurysms and conditions leading to degeneration of the aortic wall, such as chronic inflammatory conditions or connective tissue disorders, enhance the risk for aortic dissection (Gudbjartsson et al. 2020). Additionally, TAAD may be associated with trauma, most commonly involving posterior displacement of the chest wall (Boulate et al. 2018).

In the emergency setting, CT imaging visualizes the intimal flap and false lumen. This is the preferred diagnostic method because of its wide availability, ease of use, and quick results (Gudbjartsson et al. 2020). TAAD requires immediate open surgical repair. Successful surgical repair is associated with a 5-year survival rate of 84–85% (Gudbjartsson et al. 2020; Olsson et al. 2017; Van Arsdell et al. 1998; Bekkers et al. 2013; Pan et al. 2017).

Transposition of the Great Vessels (TGV)

In normal development, the right and left ventricles have separate outflow tracts. The pulmonary trunk arises from the right ventricle, and the aorta arises from the left ventricle. TGV is a congenital anomaly where the reverse is true: the pulmonary trunk arises from the left ventricle, and the aorta arises from the right ventricle. This maintains separate pulmonary and systemic circulations after birth, which are unable to mix, preventing deoxygenated blood from traveling to the lungs, and freshly oxygenated blood is unable to be delivered to tissues.

TGV accounts for 5–7% of congenital cyanotic heart defects (Cohen et al. 2010; Tynan and Anderson 2002). Surgical correction of the defect is often completed within the first 2 weeks of life and involves reconnecting the arteries to match normal anatomic positioning (Cohen et al. 2010).

28.6 Surgical Significance

Surgical repair of the ascending aorta requires more complex considerations compared to the descending aorta. The ascending aorta is a rather small section of the aorta situated between important structures like the aortic valve and the brachiocephalic trunk and contains the coronary artery ostia. The Bentall procedure, or fixation of the aortic root, performed by or with cardiac surgeons, is a complicated procedure requiring replacement of the root and aortic valve with either a mechanical or bioprosthetic valve and re-implantation of the coronary arteries (Hamandi et al. 2020). Thoracic endovascular aortic repair (TEVAR) is well established regarding the thoracic descending aorta but has limited use in the ascending aorta due to its complex anatomy (Preventza et al. 2014). More recently, there have been developments in utilizing physician-modified endografts (PMEGs), which have been altered at the back table for use in ascending arch repair. The utilization of PMEGs has given vascular surgeons another option to repair the ascending aorta (Gandet et al. 2021). That said, open repair remains the gold standard for surgical interventions of the ascending aorta (Iddriss et al. 2019). There have been several advancements in open aortic arch repair since Micheal DeBakey and Denton Cooley performed the first operation in 1955 (Cooley et al. 1955). Use of Dacron, hypothermia to reduce metabolic demand, cardiopulmonary bypass, and antegrade cerebral flow have taken this procedure with a historic major neurologic morbidity and mortality from 25% to 5% in recent case studies (Ouzounian et al. 2013). TEVAR is currently reserved for patients considered to be too high risk for open repair but who have favorable anatomy for an endovascular approach (Iddriss et al. 2019).

Surgical intervention is recommended for patients with ascending aortic diameters exceeding 5.5 cm (Leontyev et al. 2014). For patients with connective tissue disorders or other risk factors, surgery is recommended earlier in life (Leontyev et al. 2014). Management of an individual's blood pressure is essential when there is an expansion of the aortic diameter to reduce the risk of rupture (Okamoto et al. 2003).

Ultimately, surgical management of the ascending aorta is complex. Decisions are made on an individual basis and require careful consideration of anatomy and specific pathology.

References

Anderson RH (2000) Clinical anatomy of the aortic root. Heart 84(6):670–673. https://doi.org/10.1136/heart.84.6.670.5

Angelini P, Villason S, Chan AV Jr, Diez JG (1999) *Coronary artery anomalies. A comprehensive approach*. Lippincott & Wilkins, Philadelphia, pp 27–151

Arcario MJ, Lou S, Taylor P, Gregory SH (2021) Sinus of valsalva aneurysms: a review with perioperative considerations. J Cardiothorac Vasc Anesth 35(11):3340–3349. https://doi.org/10.1053/j.jvca.2020.12.016.3

Bekkers JA, Raap GB, Takkenberg JJM, Bogers AJJC (2013) Acute type A aortic dissection: long-term results and reoperations. Eur J Cardiothorac Surg 43(2):389–396. https://doi.org/10.1093/ejcts/ezs342.44

Boulate D, Fabre D, Langer NB, Fadel E (2018) Ascending aorta, aortic arch and supra-aortic vessels rupture in blunt thoracic trauma. Interact Cardiovasc Thorac Surg 27(2):304–306. https://doi.org/10.1093/icvts/ivy055.41

Chong WH, Saha BK, Wang C, Beegle S (2020) Type A aortic dissection mimicking saddle pulmonary embolism on CT imaging. J Am Coll Emerg Physicians Open 1(2):132–136. https://doi.org/10.1002/emp2.12026.20

Cohen MD, Johnson T, Ramrakhiani S (2010) MRI of surgical repair of transposition of the great vessels. Am J Roentgenol 194(1):250–260. https://doi.org/10.2214/AJR.09.3045.46

Collins JA, Munoz J-V, Patel TR, Loukas M, Shane Tubbs R (2014) The anatomy of the aging aorta: anatomy of the aging. Clin Anat 27(3):463–466. https://doi.org/10.1002/ca.22384.1

Cooley DA, Mahaffey DE, De Bakey ME (1955) Total excision of the aortic arch for aneurysm. Surg Gynecol Obstet 101(6):667–672. PMID: 1327427353

Davies RR, Goldstein LJ, Coady MA, Tittle SL, Rizzo JA, Kopf GS, Elefteriades JA (2002) Yearly rupture or dissection rates for thoracic aortic aneurysms: simple prediction based on size. Ann Thorac Surg 73(1):17–28. https://doi.org/10.1016/S0003-4975(01)03236-2.32

Elsayed RS, Cohen RG, Fleischman F, Bowdish ME (2017) Acute Type A aortic dissection. Cardiol Clin 35(3):331–345. https://doi.org/10.1016/j.ccl.2017.03.004.34

Frazao C, Tavoosi A, Wintersperger BJ, Nguyen ET, Wald RM, Ouzounian M, Hanneman K (2020) Multimodality assessment of thoracic aortic dimensions: comparison of computed tomography angiography, magnetic resonance imaging, and echocardiography measurements. J Thorac Imaging 35(6):399–406. https://doi.org/10.1097/RTI.0000000000000514.22

Freeman LA, Young PM, Foley TA, Williamson EE, Bruce CJ, Greason KL (2013) CT and MRI assessment of the aortic root and ascending aorta. Am J Roentgenol

200(6):W581–W592. https://doi.org/10.2214/AJR.12.9531.21

Gandet T, Westermann D, Conradi L, Panuccio G, Heidemann F, Rohlffs F, Kölbel T (2021) Modular endo-bentall procedure using a “Rendez-Vous Access”. J Endovasc Ther:15266028211065959. https://doi.org/10.1177/15266028211065959. Epub ahead of print. PMID: 34964371.52

Goldstein B, Carroccio A, Ellozy SH, Spielvogel D, Morrissey N, Teodorescu V, Hollier LH, Marin ML (2003) Combined open and endovascular repair of a syphilitic aortic aneurysm. J Vasc Surg 38(6):1422–1425. https://doi.org/10.1016/S0741-5214(03)00943-1

Gudbjartsson T, Ahlsson A, Geirsson A, Gunn J, Hjortdal V, Jeppsson A, Mennander A, Zindovic I, Olsson C (2020) Acute Type A aortic dissection – a review. Scand Cardiovasc J 54(1):1–13. https://doi.org/10.1080/14017431.2019.1660401.17

Hagan PG, Nienaber CA, Isselbacher EM et al (2000) The International Registry of Acute Aortic Dissection (IRAD): new insights into an old disease. JAMA 283(7):897–903. https://doi.org/10.1001/jama.283.7.897.38

Hamandi M, Nwafor CI, Baxter R, Shinn K, Wooley J, Vasudevan A, Harrington K, Schaffer J, Moore D, DiMaio JM, Ryan WH, Brinkman WT (2020) Comparison of the Bentall procedure versus valve-sparing aortic root replacement. Proc (Bayl Univ Med Cent) 33(4):524–528. https://doi.org/10.1080/08998280.2020.1771163. PMID: 33100520; PMCID: PMC7549908.51

Hanneman K, Newman B, Chan F (2017) Congenital variants and anomalies of the aortic arch. Radiographics 37(1):32–51. https://doi.org/10.1148/rg.2017160033

Hirst AE Jr, Johns JVJ, Kime JSW (1958) Dissecting aneurysm of the aorta: a review of 505 cases. Medicine (Baltimore) 37(3):217–279. https://doi.org/10.1097/00005792-195809000-00003

Iddriss A, Nakano J, Chris Malaisrie S (2019) Endovascular repair of the ascending aorta and aortic arch. Springer, Cham, pp 307–321

Johnston KW, Rutherford RB, Tilson MD, Shah DM, Hollier L, Stanley JC (1991) Suggested standards for reporting on arterial aneurysms. Subcommittee on reporting standards for arterial aneurysms, ad hoc committee on reporting standards Society for Vascular Surgery and North American Chapter, International Society for Cardiovascular Surgery. J Vasc Surg 13(3):452–458. https://doi.org/10.1067/mva.1991.26737.28

Kauhanen SP, Liimatainen T, Kariniemi E, Korhonen M, Parkkonen J, Vienonen J, Vanninen R, Hedman M (2020) A smaller heart-aorta-angle associates with ascending aortic dilatation and increases wall shear stress. Eur Radiol 30(9):5149–5157. https://doi.org/10.1007/s00330-020-06852-3.26

Koşar P, Ergun E, Oztürk C, Koşar U (2009) Anatomic variations and anomalies of the coronary arteries: 64-slice CT angiographic appearance. Diagn Interv Radiol 15(4):275–283. https://doi.org/10.4261/1305-3825.DIR.2550-09.1.16

Landenhed M, Engström G, Gottsäter A, Caulfield MP, Hedblad B, Newton-Cheh C, Melander O, Gustav Smith J (2015) Risk profiles for aortic dissection and ruptured or surgically treated aneurysms: a prospective cohort study. J Am Heart Assoc 4(1):e001513–e001513. https://doi.org/10.1161/JAHA.114.001513.39

Larsen WJ (1993) *Human embryology*. Churchill Livingstone, New York, NY, pp 111–234

Leontyev S, Misfeld M, Mohr FW (2014) Aneurysms of the ascending aorta and aortic arch. Chirurg 85(9):758–760. https://doi.org/10.1007/s00104-014-2716-z.49

Loukas M, Sharma A, Blaak C, Sorenson E, Mian A (2013) The clinical anatomy of the coronary arteries. J Cardiovasc Transl Res 6(2):197–207. https://doi.org/10.1007/s12265-013-9452-5.8

Loukas M, Bilinsky E, Bilinsky S, Christa Blaak R, Tubbs S, Anderson RH (2014) The anatomy of the aortic root. Clin Anat 27(5):748–756. https://doi.org/10.1002/ca.22295.4

Manghat NE, Morgan-Hughes GJ, Marshall AJ, Roobottom CA (2005) Multidetector row computed tomography: imaging congenital coronary artery anomalies in adults. Heart 91(12):1515–1522. https://doi.org/10.1136/hrt.2005.065979.9

Meredith EL, Masani ND (2009) Echocardiography in the emergency assessment of acute aortic syndromes. Eur J Echocardiogr 10(1):i31–i39. https://doi.org/10.1093/ejechocard/jen251

Moore KL, Persaud TV (1998) *The developing human: clinically oriented embryology*. WB Saunders, Philadelphia, pp 384–391

Murillo H, Lane MJ, Punn R, Fleischmann D, Restrepo CS (2012) Imaging of the aorta: embryology and anatomy: imaging of the aorta. Semin Ultrasound CT MRI 33(3):169–190

Nienaber CA, Clough RE (2015) Management of acute aortic dissection. Lancet (British edition) 385(9970):800–811. https://doi.org/10.1016/S0140-6736(14)61005-9.33

Okamoto RJ, Haodong X, Kouchoukos NT, Moon MR, Sundt TM (2003) The influence of mechanical properties on wall stress and distensibility of the dilated ascending aorta. J Thorac Cardiovasc Surg 126(3):842–850. https://doi.org/10.1016/S0022-5223(03)00728-1.50

Olsson C, Ahlsson A, Fuglsang S, Geirsson A, Gunn J, Hansson EC, Hjortdal V, Jarvela K, Jeppsson A, Mennander A, Nozohoor S, Wickbom A, Zindovic I, Gudbjartsson T (2017) Medium-term survival after surgery for acute Type A aortic dissection is improving. Eur J Cardiothorac Surg 52(5):852–857. https://doi.org/10.1093/ejcts/ezx302

Ouzounian M, LeMaire SA, Coselli JS (2013) Open aortic arch repair: state-of-the-art and future perspectives. Semin Thorac Cardiovasc Surg 25(2):107–115. https://doi.org/10.1053/j.semtcvs.2013.07.001. PMID: 24216527.54

Pan E, Kytö V, Savunen T, Gunn J (2017) Early and late outcomes after open ascending aortic surgery: 47-year experience in a single centre. Heart Vessel 33(4):427–433. https://doi.org/10.1007/s00380-017-1075-3.45
Plonek T, Zak M, Burzynska K, Rylski B, Gozdzik A, Kustrzycki W, Beyersdorf F, Jasinski M, Filipiak J (2017) The combined impact of mechanical factors on the wall stress of the human ascending aorta – a finite elements study. BMC Cardiovasc Disord 17(1):297–297. https://doi.org/10.1186/s12872-017-0733-9.35
Preventza O, Henry MJ, Cheong BYC, Coselli JS (2014) Endovascular repair of the ascending aorta: when and how to implement the current technology. Ann Thorac Surg 97(5):1555–1560. https://doi.org/10.1016/j.athoracsur.2013.11.066.48
Rogers IS, Banerji D, Siegel EL, Truong QA, Ghoshhajra BB, Irlbeck T, Abbara S, Gupta R, Benenstein RJ, Choy G, Avery LL, Novelline RA, Bamberg F, Brady TJ, Nagurney JT, Hoffmann U (2011) Usefulness of comprehensive cardiothoracic computed tomography in the evaluation of acute undifferentiated chest discomfort in the emergency department (CAPTURE). Am J Cardiol 107(5):643–650. https://doi.org/10.1016/j.amjcard.2010.10.039.40
Rogers IS, Massaro JM, Truong QA, Mahabadi AA, Kriegel MF, Fox CS, Thanassoulis G, Isselbacher EM, Hoffmann U, O'Donnell CJ (2013) Distribution, determinants, and normal reference values of thoracic and abdominal aortic diameters by computed tomography (from the Framingham Heart Study). Am J Cardiol 111(10):1510–1516. https://doi.org/10.1016/j.amjcard.2013.01.306.29
Roman MJ, Devereux RB, Kramer-Fox R, O'Loughlin J (1989) Two-dimensional echocardiographic aortic root dimensions in normal children and adults. Am J Cardiol 64(8):507–512. https://doi.org/10.1016/0002-9149(89)90430-X
Rylski B, Blanke P, Beyersdorf F, Desai ND, Milewski RK, Siepe M, Kari FA, Czerny M, Carrel T, Schlensak C, Krüger T, Mack MJ, Brinkman WT, Mohr FW, Etz CD, Luehr M, Bavaria JE (2014) How does the ascending aorta geometry change when it dissects? J Am Coll Cardiol 63(13):1311–1319. https://doi.org/10.1016/j.jacc.2013.12.028.37
Saliba E, Sia Y, Dore A, El Hamamsy I (2015) The ascending aortic aneurysm: When to intervene? Int J Cardiol Heart Vasc 6(C):91–100. https://doi.org/10.1016/j.ijcha.2015.01.009.24
Saliba E, Sia Y, Dore A, El Hamamsy I (2015a) The ascending aortic aneurysm: when to intervene? Int J Cardiol Heart Vasc 6(C):91–100. https://doi.org/10.1016/j.ijcha.2015.01.009.2
Takamure A, Nakagawa T, Kobayashi A, Morimoto S, Yamasaki S, Takase I, Yamamoto Y, Nishi K (2007) Traumatic intercostal artery pseudoaneurysm following a bicycle accident. Forensic Sci Med Pathol 3(3):217–220. https://doi.org/10.1007/s12024-007-0019-0.25
Tomanek R, Angelini P (2019) Embryology of coronary arteries and anatomy/pathophysiology of coronary anomalies. A comprehensive update. Int J Cardiol 281:28–34. https://doi.org/10.1016/j.ijcard.2018.11.135.6
Tynan M, Anderson RH (2002) Complete transposition. In: Anderson RH, Baker EJ, Macartney FJ, Rigby ML, Shinebourne EA, Tynan M (eds) Paediatric cardiology. Harcourt Publishers Limited, London, pp 1281–1319. https://doi.org/10.2214/AJR.09.304547
Van Arsdell GS, David TE, Butany J (1998) Autopsies in acute Type A aortic dissection. Surgical implications. Circulation 98(19 Suppl):II299-302. discussion II302-4
Yamanaka O, Hobbs RE (1990) Coronary artery anomalies in 126,595 patients undergoing coronary arteriography. Catheter Cardiovasc Diagn 21(1):28–40. https://doi.org/10.1001/ccd.1810210110
Yuan SM (2018) Syphilitic aortic aneurysm. Z Rheumatol 77(8):741–748. https://doi.org/10.1007/s00393-018-0519-1
Yuan S-M, Jing H (2011) Cystic medial necrosis: pathological findings and clinical implications. Rev Bras Cir Cardiovasc 26(1):107–115. https://doi.org/10.1590/S0102-76382011000100019.23
Zimmermann E, Schnapauff DMD, Dewey MMD (2008) Cardiac and coronary anatomy in computed tomography. Semin Ultrasound CT MRI 29(3):176–181. https://doi.org/10.1053/j.sult.2008.02.003.15

29 Thoracic Aorta

Stephen J. Bordes, Jr.

29.1 Embryology

During the third week of gestation, the aorta develops along with the endocardial tube. The very early aorta has dorsal and ventral segments that are associated with the first aortic arch. This arch later regresses. The ventral segments form the aortic sac, while the dorsal segments form the descending aorta. Six paired aortic arches form during embryogenesis. The right dorsal aortic root subsequently regresses, which leads to the formation of the physiologically normal left-sided aorta (Murillo et al. 2012).

29.2 Anatomy

The aorta is the body's largest elastic artery. It receives all of the left heart's oxygen-rich blood and terminates at its bifurcation into the common iliac arteries. The ascending aorta originates at the level of the tri-leaflet, semilunar aortic valve and extends superiorly approximately 5 cm to the origin of the brachiocephalic artery (Fig. 29.1). It gives off the coronary arteries near the aortic cusp and then continues as the aortic arch, which is defined as the portion of the aorta between the brachiocephalic trunk and the left subclavian artery. The aortic arch is typically found at the level of the T4 vertebra, also known as the plane of Ludwig. It gives off the three great vessels (brachiocephalic trunk, left common carotid artery, and left subclavian artery). It reaches its most superior position at this level before curving inferiorly and medially. The descending thoracic aorta continues inferiorly and just to the left of the vertebral column in the posterior mediastinal cavity until it reaches the aortic diaphragmatic hiatus at the T12 vertebral level. Thereafter, it becomes the abdominal aorta, which terminates at its bifurcation into the left and right common iliac arteries at the L4 vertebral level (Murillo et al. 2012; Loukas et al. 2014; Anderson 2000; White et al. 2021; Tran et al. 2021; Symbas 1974).

Branches of the aorta include:

Ascending aorta: coronary arteries.

Aortic arch (Figs. 29.2 and 29.3): brachiocephalic trunk, left common carotid artery, left subclavian artery.

Descending thoracic aorta: pericardial arteries, bronchial arteries, esophageal arteries, mediastinal arterial branches, intercostal arteries, subcostal arteries, superior phrenic arteries (Fig. 29.4).

Abdominal aorta: celiac trunk, superior mesenteric artery, inferior mesenteric artery, renal arteries, gonadal arteries, inferior phrenic arteries, suprarenal arteries, lumbar arteries, medial sacral arteries.

S. J. Bordes, Jr. (✉)
Department of Surgery, Louisiana State University Health Sciences Center, New Orleans, LA, USA

S. J. Bordes, Jr. et al. (eds.), *The Clinical Anatomy of the Vascular System*,
https://doi.org/10.1007/978-3-031-78326-5_29

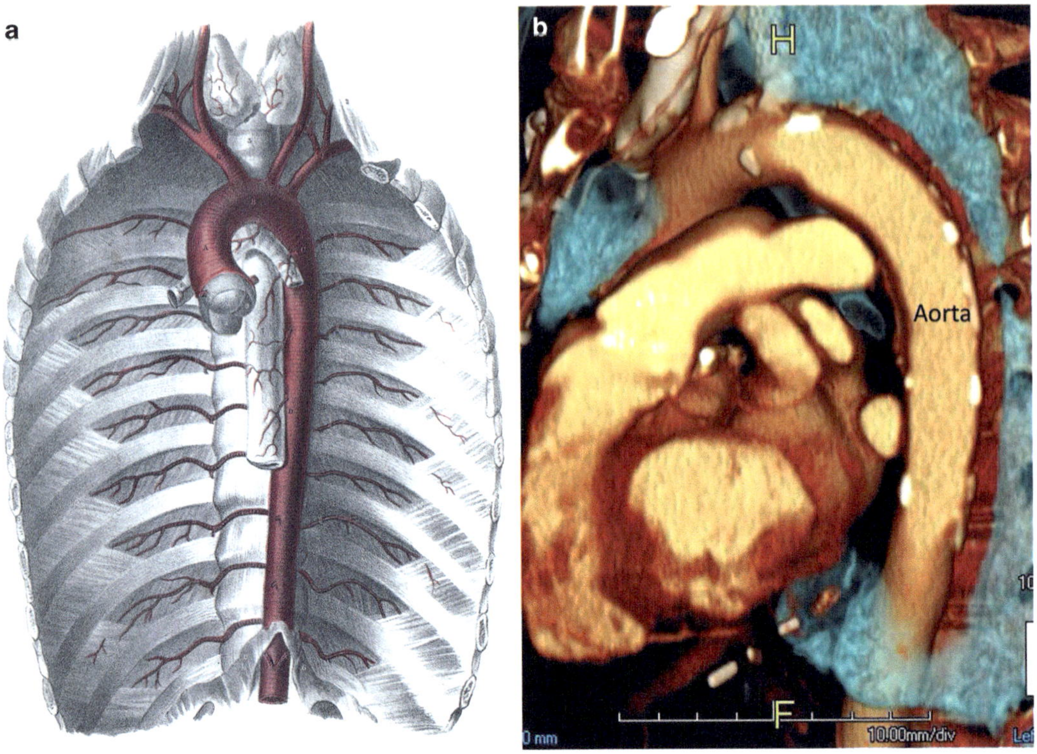

Fig. 29.1 (**a**) Drawing of the thoracic aorta and its branches (Quain and Wilson, 1844 author's collection). (**b**) CTA rendered image of the thoracic aorta

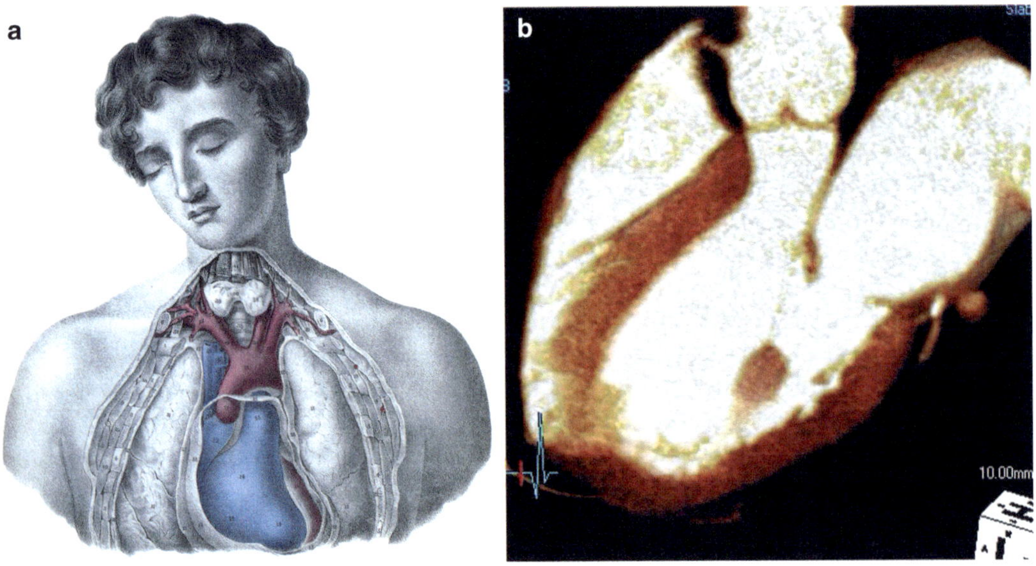

Fig. 29.2 (**a**) Drawing of the ascending and arch parts of the aorta (Quain and Wilson, 1844 author's collection). (**b**) CTA of the aortic root (AO). Also, note the left atrium (LA), left ventricle (LV), and right ventricle (RV)

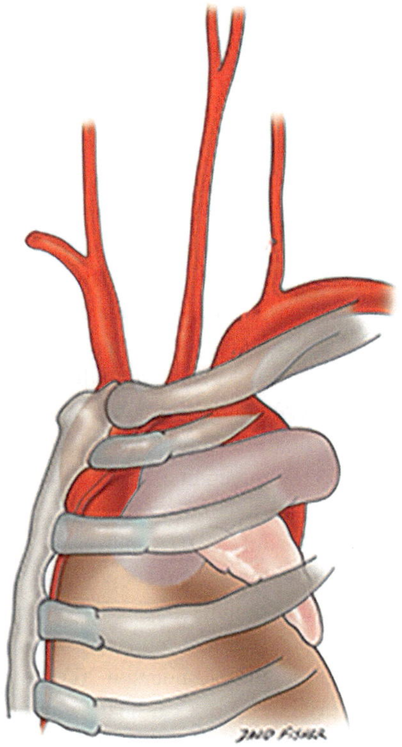

Fig. 29.3 Typical branching of the aortic arch into right brachiocephalic artery, left common carotid, and left subclavian branches

A small network of blood vessels known as the vasa vasorum supplies the aorta (Ritman and Lerman 2007; Mulligan-Kehoe 2010). This network is most abundant superior to the renal arteries; therefore, inferior to the renal arteries, the abdominal aorta is prone to oxidation, atherosclerosis, and aneurysm formation. Intricate lymphatic networks can be found along the aorta and typically drain into the cisterna chyli inferior to the diaphragm before ascending into the thorax via the thoracic duct.

Fig. 29.4 Drawing of the superior phrenic artery, a branch of the thoracic aorta or its posterior intercostal branches

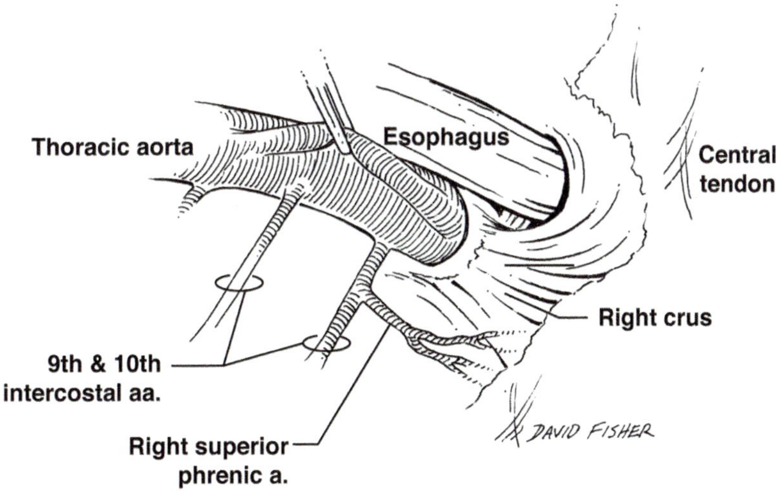

29.3 Histology

The aorta is an elastic artery. It is composed of vascular smooth muscle and elastin. Like most vascular structures, the aortic wall is made up of three layers: a thin innermost tunica intima, an elastin rich tunica media, and an outermost fibrous or adventitial layer.

29.4 Imaging

Magnetic resonance imaging (MRI) remains the gold standard imaging modality for visualization of the aorta owing to its ability to differentiate branching patterns and nearby soft tissue structures including the trachea and bronchi. However, computed tomography angiography (CTA) is more commonly used because of its cost, efficiency, and three-dimensional reconstructive capabilities, which are utilized for mapping during endovascular and open aortic procedures (Fig. 29.5). Other modalities that are often used include transthoracic or transesophageal echocardiography, chest X-ray, and invasive catheter angiography (Nagpal et al. 2015; Richter 2020).

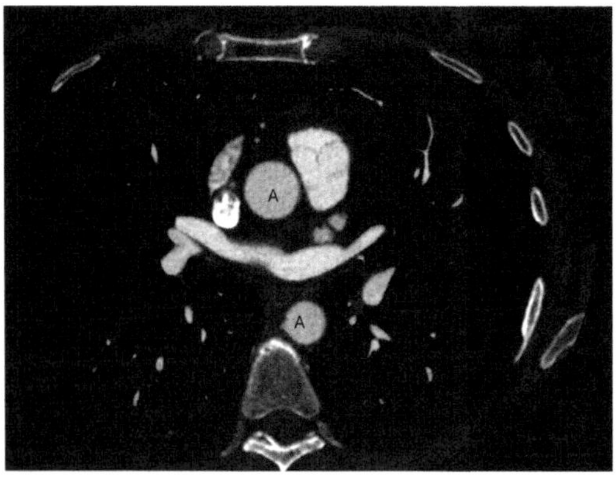

Fig. 29.5 CTA of the thorax noting the ascending and descending parts of the aorta (A)

29.5 Variations (Fig. 29.6)

Patent ductus arteriosus results when the ductus arteriosus fails to close following birth. The ductus arteriosus allows blood to flow between the aorta and pulmonary artery during intrauterine development and typically closes within a few days of delivery. A patent duct results in a right to left shunt of deoxygenated blood. Typically, a machine-like murmur can be appreciated upon auscultation. Surgical ligation or percutaneous closure is generally indicated if the duct fails to close naturally or with medical therapy such as indomethacin (Hermes-DeSantis and Clyman 2006; Pamukcu et al. 2018).

A **hypoplastic ascending aorta** can coexist with hypoplastic left heart syndrome, which is typically fatal within the first few weeks of life. This abnormality results in severely decreased left heart output and consequently diminished blood flow to the body (Anderson 2021).

Dextrocardia describes a heart with a rightward base-apex axis (dextroversion). Dextrocardia with situs inversus describes a heart anatomy that is a mirror image of the norm. Situs inversus totalis is rare; all organs are mirror images of their normal anatomies. Dextrocardia of embryonic arrest describes a heart with normal anatomy located entirely to the right of midline (dextroposition) and is typically associated with pulmonary hypoplasia. Some consider this term incorrect because the heart's base-axis still points leftward (Evans et al. 2010).

A **bovine aortic arch** is the most common variation; the innominate artery and the left common carotid arteries share a common trunk off the aortic arch. This variant has very few adverse consequences (Goldsher et al. 2020).

In an **aberrant right subclavian artery**, also known as arteria lusoria, the right subclavian artery originates from the aortic arch just distal to the left subclavian artery before crossing the

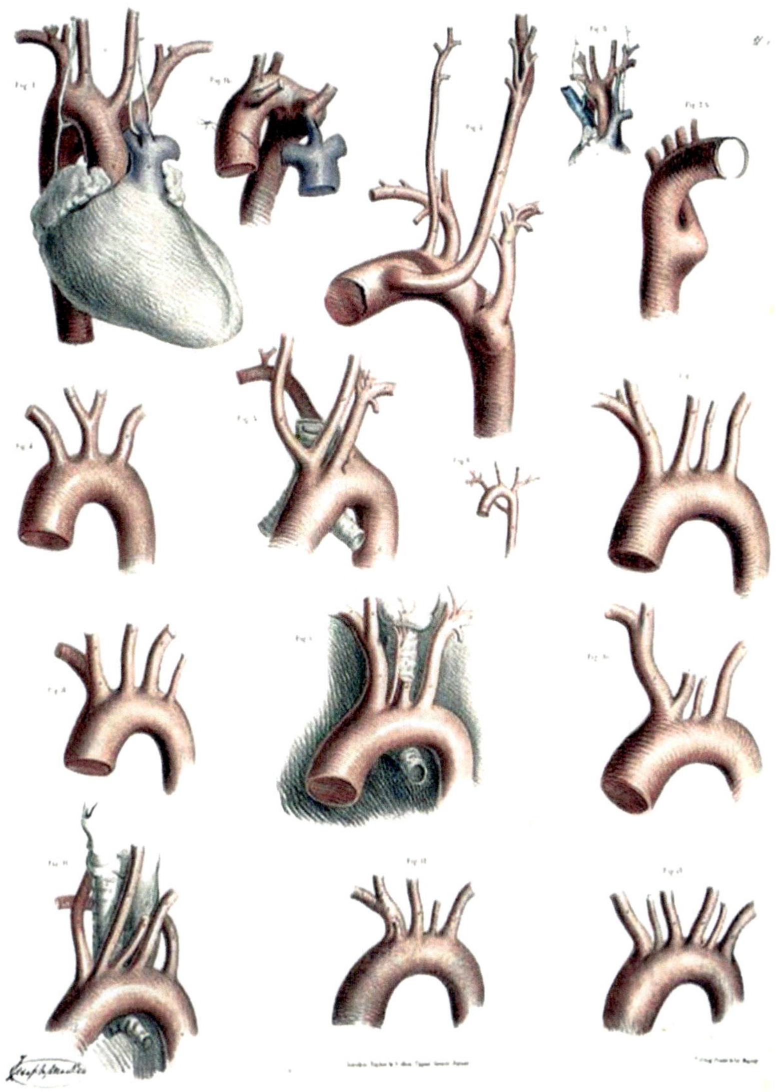

Fig. 29.6 Drawing of the many variants possible for the branching pattern of the aortic arch (Quain and Wilson, 1844 author's collection)

midline and traveling to the right upper extremity. Genetic abnormalities such as Down Syndrome, or Trisomy 21, are associated with this variation. The patient can present with dysphagia (most commonly) from esophageal compression, dyspnea, and retrosternal chest pain (Krupiński et al. 2019).

In a **right aortic arch**, the ascending aorta and arch travel posteriorly and to the right instead of the left. Furthermore, the arch crosses over the right bronchus instead of the left. The patient can be entirely asymptomatic or have an associated underlying cardiac or genetic abnormality. A right arch can be associated with an aberrant left subclavian artery traveling anterior or posterior to the trachea. The ligamentum arteriosum can form a complete vascular ring around the esophagus if the vessel travels posterior to the trachea (Prabhu et al. 2020).

A **double aortic arch** comprises a multitude of branching patterns; however, all involve a split of the ascending aorta into two vessels that typically rejoin and continue as the descending aorta. Patients with a double arch make up more than half of all patients undergoing surgery for correction of a vascular ring. They usually complain of dysphagia or dyspnea from esophageal or tracheal compression (Hanneman et al. 2017).

29.6 Pathology

While there are many infectious and developmental etiologies of aortic pathology, most aortic disease is related to modifiable risk factors including uncontrolled diabetes, tobacco use, hypertension, hyperlipidemia, and obesity, which lead to atherosclerotic deposition and oxidation of the aortic wall. The vasa vasorum supply to the infrarenal aorta is poorest, so atherosclerotic and aneurysmal disease is most common inferior to the renal arteries.

Atherosclerotic pathophysiology involves endothelial cell dysfunction, dyslipidemia, inflammatory factors, and plaque rupture. Oxidized low-density lipoprotein (LDL) is known to cause endothelial cell dysfunction leading to loss of nitric oxide-induced vasodilation. Macrophages that have taken up oxidized LDL release inflammatory cytokines and growth factors, causing plaques to form. Atherosclerosis is generally asymptomatic until the vascular lumen is 70–80% occluded. The aorta can be asymptomatic even longer owing to its large caliber diameter and ability to form significant collateral circulation over time. In symptomatic cases, atherosclerosis results in peripheral artery disease in which distal organs and extremities become intermittently ischemic, leading to claudication, rest pain, and tissue loss in severe cases. Severe aorto-iliac disease, known as Leriche syndrome, typically warrants aorto-iliac or axillo-bifemoral bypass to revascularize the distal extremities. There is no way to reverse vascular disease, but progression can be significantly slowed with lifestyle modification and medical management. Preventative measures include diet, exercise, weight loss, and smoking cessation. Statins, blood pressure medications, and aspirin are the mainstays of treatment (Perrotta and Aquila 2015).

Aortic aneurysms are true aneurysms that involve all three layers of the aortic wall and are defined as dilation greater than 50% of the normal circumference. Dilation of the aorta between 0 and 50% is termed aortic ectasia (typically a diameter > 3.0 cm in the abdominal aorta). Generally, an aneurysm of the ascending aorta is any diameter greater than 5.0 cm, while an aneurysm of the descending aorta is any diameter greater than 4.0 cm. While male sex and age are nonmodifiable factors that increase the risk for this disease, most aortic aneurysms are secondary to the modifiable factors listed above. Increased oxidation and wall stress lead to the aneurysmal dilation of the aortic wall secondary to macrophage and lymphocyte infiltration, in addition to loss of elastin and smooth muscle due to proteases. Nearly 10% of people over the age of 65 have aneurysmal dilation of the aorta. Abdominal aneurysms are much more common than thoracic aneurysms. Rupture of an aneurysm can be life-threatening. As a result, any prior smoker over the age of 65 should receive abdominal aortic aneurysm screening per the US Preventative Services Task Force. If any dilation is detected, interval screenings on a yearly basis are needed. Asymptomatic aneurysms are electively repaired when they are larger than 5.0 cm or are rapidly expanding. Symptomatic aneurysms require urgent intervention (Legg and Legg 2016; Collins et al. 2014).

Aortic dissections form false aneurysms as they involve only two layers of the aortic wall: the tunica intima and media. Usually, a tear in the intima from trauma or hypertension results in the formation of a false lumen. However, connective tissue disorders such as Marfan and Ehlers-Danlos syndrome can lead to dissection. Stanford A dissections involve the ascending aorta and/or aortic arch with or without involvement of the descending aorta. They are surgical emergencies. Stanford B dissections only involve the descending aorta and are first managed medically with blood pressure control prior to surgical or endovascular intervention. Aortic regurgitation, acute coronary syndrome, cerebrovascular accidents, and cardiac tamponade can occur in symptomatic cases (Setacci et al. 2010).

Aortic coarctation (Fig. 29.7) occurs when the lumen at the distal aortic arch or descending aorta is narrowed, generally near the ligamentum arteriosus. Coarctation is responsible for 4–6% of congenital heart defects and is associated with bicuspid aortic valve, ventricular septal defect (VSD), mitral valve defects, patent ductus arteriosus, and Turner syndrome. Severe cases can

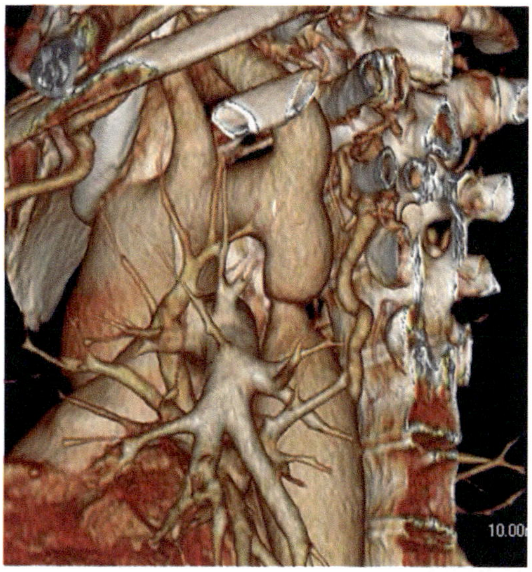

Fig. 29.7 CTA of a patient with coarctation of the aorta

lead to heart failure and death shortly after birth after the ductus arteriosus closes. Other cases can remain clinically silent for some time owing to collateralization of the intercostal arteries. Patients can be hypertensive in the upper extremities and normotensive or hypotensive in the lower extremities. Surgery is indicated in nearly all cases once discovered. Otherwise, life expectancy is 30–40 years owing to increased risks of stroke, coronary artery disease, peripheral artery disease, heart attack, and heart failure (Ho and Anderson 1979; Kim et al. 2020).

Aortitis can result from a variety of conditions that lead to aortic inflammation. Common causes include giant cell arteritis, Takayasu arteritis, anti-neutrophil cytoplasmic antibody (ANCA) vasculitis, systemic lupus erythematosus, spondyloarthropathies, sarcoidosis, syphilis, and others (Pugh et al. 2021).

Traumatic aortic rupture is the second leading cause of death in blunt trauma and typically occurs at the aortic isthmus just distal to the left subclavian artery, where the ligamentum arteriosum tethers the aorta. Ruptures are fatal in nearly all cases secondary to blood loss and hypovolemic shock. Patients who make it to the hospital in time generally undergo emergency endovascular repair (D'Alessio et al. 2020; Schachner et al. 2020).

Vascular Ring

A double aortic arch with an aberrant left subclavian artery accounts for up to 90% of symptomatic vascular rings. Aortic vascular rings can lead to complete or incomplete encompassing of respiratory or gastrointestinal structures, resulting in stridor, respiratory infections, respiratory distress, cough, vomiting, feeding difficulty, and dysphagia (Backer et al. 2016).

29.7 Surgical and Clinical Considerations

Aneurysm

A ruptured aortic aneurysm is a surgical emergency and is often fatal prior to arrival at a hospital. However, most aneurysms are screened for and followed for years prior to elective repair. An unruptured aneurysm should be deemed operative under the following conditions: (1) The aneurysm is symptomatic; (2) an asymptomatic aneurysm is rapidly growing >1 cm per year; and (3) an asymptomatic aneurysm is >5.5 cm in men or > 5.0 cm in women. Once an aneurysm is deemed operative, the surgeon must decide between open surgical repair and endovascular repair. While endovascular stent graft repairs (EVARs) are faster, entail less blood loss, incur less cardiac stress, and are associated with shorter hospital stays, the grafts often require reintervention in the long term. Therefore, open surgical repair is favored for younger patients with few comorbidities. Further, open repairs should generally be chosen for patients with aneurysms related to connective tissue disorders. EVARs are generally used in older patients with shorter life expectancies, patients with multiple comorbidities, and patients with a history of prior aortic or abdominal surgeries.

Dissection

Aortic dissections, as mentioned above, are classified as Type A (involving the aortic arch or ascending aorta) and Type B (isolated descending thoracic or abdominal aortic dissection). Type A dissections are always operative and generally require open surgical repair owing to the complex branching pattern of the aortic arch. Uncomplicated Type B dissections can be managed entirely with medical treatment. Heart rate and blood pressure control is the mainstay of treatment. Beta blockers with short half-lives and rapid clearance such as esmolol are generally preferred. Descending aortic dissections that are deemed operative are generally managed with endovascular stent grafts.

Endoleak

Endoleaks are common complications following EVARs; blood continues to fill the aneurysm sac, causing enlargement. Endoleaks are screened for following stent placement and can be classified as follows:

Type 1: Leak at either proximal (1a) or distal (1b) graft attachment sites from inadequate seal. These leaks must always be repaired.

Type 2: Back filling via branch vessel (the most common endoleak). These leaks generally resolve spontaneously over time, though a persistent leak can require embolization of the branch.

Type 3: Leak through the graft itself via fracture or other mechanical defect. These leaks should be repaired.

Type 4: Leak from graft porosity, which generally resolves within a few days of graft placement. These leaks rarely require intervention.

Type 5: Continued expansion of the aneurysm from endotension. These leaks remain poorly understood and are thought to result from the aortic pulse wave (as opposed to a true leak). While graft reinforcement can resolve them, conversion to an open repair is sometimes necessary.

Aortic Occlusion

Aorto-iliac atherosclerotic disease warrants surgical intervention if luminal narrowing is significant and patients exhibit symptoms such as claudication, rest pain, tissue loss, or, in severe cases, loss of pulses in distal extremities. Heparinization, statins, and strict blood pressure control are the initial stays of treatment while open and/or endovascular repairs are pursued. Aorto-iliac or aorto-femoral bypass can be undertaken for patients with acceptable cardiac risk stratification. Axillo-bifemoral bypass, an extra-anatomical procedure, can be preferable for patients deemed to have too high a cardiac risk because the aorta must be clamped in the aforementioned procedures. In this procedure, a subcutaneous graft is tunneled between the axillary and femoral arteries.

Trauma

Blunt traumatic injuries are commonly responsible for aortic aneurysms and dissections, which are mainly managed medically and/or endovascularly. Penetrating aortic trauma requires emergency open surgical repair to achieve proximal and distal control. If a victim of penetrating trauma loses signs of life within 15 min of hospital arrival and undergoes interim advanced cardiopulmonary resuscitation, a left-sided or bilateral anterolateral (clam shell) thoracotomy can be a life-saving temporizing measure to expose the thoracic cavity, clamp the aorta, decompress the pericardium, massage the heart, and evacuate the pleural cavities.

References

Anderson RH. Clinical anatomy of the aortic root. Heart 2000 Dec;84(6):670–673. doi: https://doi.org/10.1136/heart.84.6.670. PMID: 11083753; PMCID: PMC1729505

Anderson RH (2021 Apr 29) Guidelines to hypoplastic left heart syndrome. Eur J Cardiothorac Surg 59(4):924. https://doi.org/10.1093/ejcts/ezaa417

Backer CL, Mongé MC, Popescu AR, Eltayeb OM, Rastatter JC, Rigsby CK. Vascular rings. Semin Pediatr Surg 2016 Jun;25(3):165–175. doi: https://doi.org/10.1053/j.sempedsurg.2016.02.009. Epub 2016 Feb 22

Collins JA, Munoz JV, Patel TR, Loukas M, Tubbs RS. The anatomy of the aging aorta. Clin Anat 2014

Apr;27(3):463–466. doi: https://doi.org/10.1002/ca.22384. Epub 2014 Feb 12

D'Alessio I, Domanin M, Bissacco D, Rimoldi P, Palmieri B, Piffaretti G, Trimarchi S. Thoracic endovascular aortic repair for traumatic aortic injuries: insight from literature and practical recommendations. J Cardiovasc Surg 2020 Dec;61(6):681–696. doi: https://doi.org/10.23736/S0021-9509.20.11580-5. Epub 2020 Sep 23

Evans WN, Acherman RJ, Collazos JC, Castillo WJ, Rollins RC, Kip KT, Restrepo H. Dextrocardia: practical clinical points and comments on terminology. Pediatr Cardiol 2010 Jan;31(1):1–6. doi: https://doi.org/10.1007/s00246-009-9516-0. Epub 2009 Aug 29

Goldsher YW, Salem Y, Weisz B, Achiron R, Jacobson JM, Gindes L. Bovine aortic arch: prevalence in human fetuses. J Clin Ultrasound 2020 May;48(4):198–203. doi: https://doi.org/10.1002/jcu.22800. Epub 2019 Nov 28

Hanneman K, Newman B, Chan F (2017 January-February). Congenital variants and anomalies of the aortic arch. Radiographics 37(1):32–51. doi: https://doi.org/10.1148/rg.2017160033. Epub 2016 Nov 18

Hermes-DeSantis ER, Clyman RI. Patent ductus arteriosus: pathophysiology and management. J Perinatol 2006 May;26 Suppl 1:S14–S18; discussion S22-3. doi: https://doi.org/10.1038/sj.jp.7211465

Ho SY, Anderson RH (1979 Mar) Coarctation, tubular hypoplasia, and the ductus arteriosus. Histological study of 35 specimens. Br Heart J 41(3):268–274

Kim YY, Andrade L, Cook SC. Aortic coarctation. Cardiol Clin 2020 Aug;38(3):337–351. doi: https://doi.org/10.1016/j.ccl.2020.04.003. Epub 2020 May 31

Krupiński M, Irzyk M, Moczulski Z, Banyś R, Dwojak I, Urbańczyk-Zawadzka M. CT evaluation of aberrant right subclavian artery: anatomy and clinical implications. Cardiol Young 2019 Feb;29(2):128–132. doi: https://doi.org/10.1017/S1047951118001907. Epub 2018 Nov 23

Legg JS, Legg LM (2016 Nov) Abdominal aortic aneurysms. Radiol Technol 88(2):145–163

Loukas M, Bilinsky E, Bilinsky S, Blaak C, Tubbs RS, Anderson RH (2014 Jul) The anatomy of the aortic root. Clin Anat 27(5):748–756

Mulligan-Kehoe MJ. The vasa vasorum in diseased and nondiseased arteries. Am J Physiol Heart Circ Physiol. 2010 Feb;298(2):H295–305. doi: https://doi.org/10.1152/ajpheart.00884.2009. Epub 2009 Nov 25. PMID: 19940078; PMCID: PMC2822580

Murillo H, Lane MJ, Punn R, Fleischmann D, Restrepo CS (2012 Jun) Imaging of the aorta: embryology and anatomy. Semin Ultrasound CT MR 33(3):169–190. https://doi.org/10.1053/j.sult.2012.01.013

Nagpal P, Khandelwal A, Saboo SS, Bathla G, Steigner ML, Rybicki FJ. Modern imaging techniques: applications in the management of acute aortic pathologies. Postgrad Med J 2015 Aug;91(1078):449–462. https://doi.org/10.1136/postgradmedj-2014-133178. Epub 2015 Aug 6

Pamukcu O, Tuncay A, Narin N, Baykan A, Korkmaz L, Argun M, Ozyurt A, Sunkak S, Uzum K. Patent ductus arteriosus closure in preterms less than 2kg: Surgery versus transcatheter. Int J Cardiol 2018 Jan 1;250:110–115. doi: https://doi.org/10.1016/j.ijcard.2017.10.020. Epub 2017 Oct 6

Perrotta I, Aquila S. The role of oxidative stress and autophagy in atherosclerosis. Oxidative Med Cell Longev 2015;2015:130315. doi: https://doi.org/10.1155/2015/130315. Epub 2015 Mar 18. PMID: 25866599; PMCID: PMC4381688

Prabhu S, Mehra S, Kasturi S, Tiwari R, Joshi A, John C, Karl TR. Anatomic classification of the right aortic arch. Cardiol Young 2020 Nov;30(11):1694–1701. doi: https://doi.org/10.1017/S1047951120003601. Epub 2020 Oct 28

Pugh D, Grayson P, Basu N, Dhaun N. Aortitis: recent advances, current concepts and future possibilities. Heart 2021 Oct;107(20):1620–1629. doi: https://doi.org/10.1136/heartjnl-2020-318085. Epub 2021 Feb 16

Richter GM. State-of-the-art imaging of the aorta with respect to endoluminal intervention. Cardiovasc IntervRadiol 2020 Dec;43(12):1745–1755. doi: https://doi.org/10.1007/s00270-020-02680-6. Epub 2020 Oct 16

Ritman EL, Lerman A (2007, September 1) The dynamic vasa vasorum. Cardiovasc Res 75(4):649–658. https://doi.org/10.1016/j.cardiores.2007.06.020. Epub 2007 Jun 29. PMID: 17631284; PMCID: PMC2121590

Schachner T, Oji-Zurmeyer J, Rylski B, Kapral S, Stanger O, Widmann G, Schönhoff F, Czerny M, Jochberger S (2020) Die traumatischeAortenruptur [Traumatic aortic rupture – diagnosis and managment]. Wien Med Wochenschr. 170(7–8):178–188. https://doi.org/10.1007/s10354-019-00727-z. Epub 2019 Dec 19

Setacci F, Sirignano P, de Donato G, Chisci E, Perulli A, Setacci C (2010 Oct) Acute aortic dissection: natural history and classification. J Cardiovasc Surg 51(5):641–646

Symbas PN (1974 Dec) Surgical anatomy of the great arteries of the thorax. Surg Clin North Am 54(6):1303–1312. https://doi.org/10.1016/s0039-6109(16)40486-x

Tran CT, Wu CY, Bordes S, Lui F (2021, July 31). Anatomy, abdomen and pelvis, abdominal aorta. In: StatPearls [Internet]. Treasure Island (FL): StatPearls Publishing; 2022 January

White HJ, Bordes S, Borger J. (2021, July 31) Anatomy, abdomen and pelvis, aorta. In: StatPearls [Internet]. Treasure Island (FL): StatPearls Publishing; 2022 January

30 Brachiocephalic Trunk

Sreeganesh Manoharan and Stephen J. Bordes, Jr.

30.1 Embryology

The brachiocephalic trunk is one of the three main branches of the aortic arch. The development of the aorta begins around the third week of gestation (Kau et al. 2007). The aortic sac is considered the first portion of the aorta to form. It develops into two distinct horns, the right and left ones, giving rise to numerous important anatomical structures. The right horn specifically gives rise to the brachiocephalic trunk, while the aortic arches later develop from the aortic sac. The right horn of the aortic sac joins with the right third and right fourth aortic arch, eventually giving off the right common carotid artery and right subclavian artery, respectively. Merging the right horn with the two aortic arches forms the structure known as the brachiocephalic trunk.

30.2 Anatomy and Topography

The first branch of the aortic arch, the brachiocephalic trunk (Fig. 30.1), is directly behind the manubrium of the sternum and slightly anterior and to the right of the left common carotid and left subclavian artery (Drake et al. 2020). As the brachiocephalic trunk ascends, it begins to course posteriorly and to the right to the level of the right sternoclavicular joint; at this point, the artery bifurcates into the right common carotid and right subclavian artery. At the point of division and slightly superior to the jugular notch, there are many critical anatomical structures of the head and neck, including the sternohyoid and sternothyroid muscles.

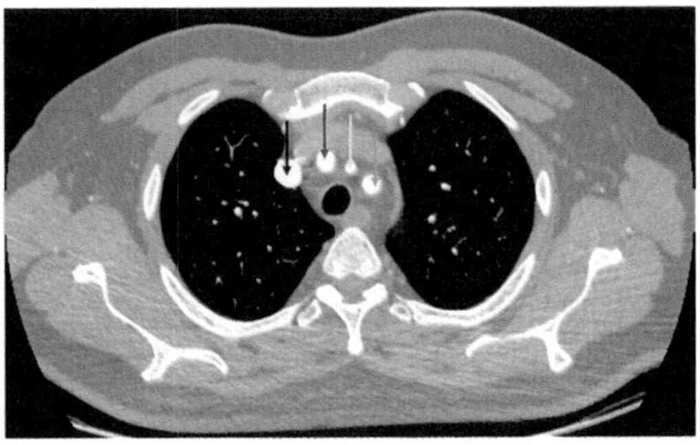

Fig. 30.1 Axial CT of the thorax noting the superior vena cava (black arrow), brachiocephalic trunk (red arrow), left common carotid artery (yellow arrow), and left subclavian artery (blue arrow)

S. Manoharan
Tulane University School of Medicine, New Orleans, LA, USA
e-mail: smanoharan@tulane.edu

S. J. Bordes, Jr. (✉)
Department of Surgery, Louisiana State University Health Sciences Center, New Orleans, LA, USA

S. J. Bordes, Jr. et al. (eds.), *The Clinical Anatomy of the Vascular System*,
https://doi.org/10.1007/978-3-031-78326-5_30

30.3 Histology

Blood vessels comprise three distinct layers: the tunica intima, the tunica media, and the tunica adventitia. The innermost layer, the tunica intima, provides a smooth luminal surrounding for blood movement. The tunica media regulates the internal diameter of the blood vessel. Finally, the tunica adventitia is very important in providing structural support and shape to the vessel. There are two main types of arteries in the body: elastic and muscular. Most of the vasculature near the heart is classified as elastic arteries. The elasticity of the brachiocephalic trunk allows it to withstand the high-pressure flow of blood from the aorta to the right side of the body. Elastic arteries are named for their relative amount of elastin and collagen contained within the tunica media, which allows for increased stretch. This allows the brachiocephalic artery to maintain a relatively constant pressure gradient despite the dramatic increases and decreases in flow through the vessel.

30.4 Variations

The complex embryological development of thoracic vasculature lends itself to numerous variants. Studies show that the natural variant is present in approximately 80% of the population (Popieluszko et al. 2018).

Bovine Arch

The bovine arch (a misnomer) is the most common variant associated with the brachiocephalic trunk. In this case, the brachiocephalic trunk shares a common origin with the left common carotid and has three branches off its base. The variant is present in ~15% of the population and is most common in individuals of African descent and least common in those of Asian descent. This is when a common single brachiocephalic trunk trifurcates into bilateral subclavian arteries and a single bicarotid trunk. Many patients with bovine arch are asymptomatic, but careful surgical considerations must be taken when approaching this area.

Aberrant Right Subclavian Artery (ARSA)

Presenting in approximately 0.5–2% of the population, the right subclavian arises as a fourth branch from the aortic arch distal to the left subclavian artery. The right subclavian then courses to the right side of the body with variable paths, most commonly posterior to the esophagus. Patients with an ARSA are usually asymptomatic, but about 10% complain of tracheoesophageal symptoms, termed dysphagia lusoria. Surgical considerations must be taken during procedures such as esophagectomy. In patients with ARSA, the right recurrent laryngeal nerve is usually non-recurrent due to the expected embryological development of the artery.

In addition to these two common natural variants of the brachiocephalic trunk, studies have observed some very rare variations in this artery, as listed below (Popieluszko et al. 2018):

- Four branches with absent brachiocephalic trunk.
- Double aortic arch.
- Left-sided arch with a left-sided brachiocephalic trunk.
- Common trunk of the brachiocephalic left common carotid and left subclavian arteries.
- Three branches: right subclavian, common carotid, and left subclavian arteries.

30.5 Pathology

Atherosclerotic disease of the brachiocephalic trunk can lead to flow-limiting obstruction and potentially distal embolization. This can result in transient ischemic attacks, stroke, upper extremity ischemia, and vertebrobasilar insufficiency. Stenosis of this artery is relatively rare but can cause dramatic changes in hemodynamics and spectral Doppler waveform patterns (Rodriguez 2016).

30.6 Clinical and Surgical Significances

The brachiocephalic trunk has very little clinical and surgical significance unless atherosclerosis leads to stenosis occlusion or trauma, resulting in direct injury to the vessel. Severe disease in this artery is considered as stenosis greater than approximately 75% of the vessel's diameter (Galland 2011). In brachiocephalic revascularizations reported by Berguer et al., there was a 40% incidence of multivessel atherosclerotic disease (Berguer et al. 1998). Once the disease is confirmed through imaging, steps can be taken for brachiocephalic trunk revascularization, although the intervention has raised questions about its necessity (de Borst 2019). Overall, we tend to see asymptomatic cases of stenosis of the brachiocephalic trunk because of the numerous collateral vessels and the low incidence of atherosclerotic disease of the upper limb (Galland 2011). General indications for revascularization include stroke or transient ischemic attacks due to atheroembolism (Galland 2011). There have been few guidelines published on decision-making when it comes to patients with asymptomatic brachiocephalic disease; some suggest intervention when there is symptomatic and presentable risk to the individual.

Dissection

Isolated brachiocephalic trunk dissection is considered rare. Supra-aortic dissection accounts for 10–25% of ischemic stroke for patients under 50 years of age, but those caused by dissection of the BCA are still under further review (Mani and Ahluwalia 2015). With this specific yet rare condition, there is a large risk of stroke caused by thromboembolism from the intimal tear rather than by hemodynamic factors (Mani and Ahluwalia 2015). Supra-aortic dissections, including the brachiocephalic trunk, can be caused by trauma, sports injuries, neck manipulation, and connective tissue or vascular disorders. Regardless of the cause, most cases are treated using a combination of anti-coagulants and endovascular techniques such as angioplasty or stenting (Mani and Ahluwalia 2015).

Tracheoinnominate Fistula

A tracheoinnominate fistula is considered highly life-threatening and described as occurring after placement of tracheostomy. These complications are associated with high mortality and massive hemorrhaging. This condition has been caused by tracheal resection or reconstruction procedures near the lower neck region. In rare cases, this complication has also been described after placement of endovascular stent grafts. When a tracheostomy is placed below the third tracheal ring, the risk of this tracheobrachiocephalic fistula vastly increases. Other known risk factors include prolonged use of steroids and immunosuppressants, radiation therapy to the neck area, and episodes of hypotension causing ischemia. Although there are many causes of a TIF, ulceration due to localized ischemia or infection ultimately progresses to the formation of a fistula between the tracheal and brachiocephalic trunk.

30.7 Complications

Severe complications arise with trauma, thrombosis, or occlusion of the brachiocephalic trunk. These complications can lead to transient ischemic attacks, stroke, upper extremity ischemia, and vertebrobasilar insufficiency (Takach et al. 2005). Most patients with asymptomatic BCA disease with proximal stenosis or occlusion of the brachiocephalic trunk see blood flow through numerous collateral vessels, minimizing the dangerous effects of typical atherosclerotic vascular disease. Still, surgical intervention may need to be pursued depending on the degree of occlusion or injury of the brachiocephalic trunk (Berguer et al. 1998; Galland 2011).

References

Berguer R et al. Transthoracic repair of innominate and common carotid artery disease: immediate and long-term outcome for 100 consecutive surgical reconstructions. J Vasc Surg 27, 1 (1998): 34–41; discussion 42. https://doi.org/10.1016/s0741-5214(98)70289-7

de Borst GJ (2019) Innominate artery stenting: the continuing Saga of 'who, when, and how'? J Endovasc Ther 26(3):391–393. https://doi.org/10.1177/1526602819839495

Drake RL, Vogl W, Mitchell AWM, Gray H (2020) Gray's Anatomy for Students. Print

Galland B (2011) Rutherford's vascular surgery, 7th ed. Ann R Coll Surg Engl 93(2):176. https://doi.org/10.1308/003588411X561062

Kau T et al (2007) Aortic development and anomalies. Semin Interv Radiol 24(2):141–152. https://doi.org/10.1055/s-2007-980040

Mani H, Ahluwalia S (2015) Isolated brachiocephalic artery dissection presenting as acute stroke. BMJ Case Rep 2015(bcr2014208815):27. https://doi.org/10.1136/bcr-2014-208815

Popieluszko P, Henry BM, Sanna B, Hsieh WC, Saganiak K, Pękala PA, Walocha JA, Tomaszewski KA (2018 Jul) A systematic review and meta-analysis of variations in branching patterns of the adult aortic arch. J Vasc Surg 68(1):298–306.e10

Rodriguez JD (2016) Brachiocephalic artery disease progression resulting in complex steal phenomena. J Diagn Med Sonogr 32(3):173–180. https://doi.org/10.1177/8756479316649950

Takach TJ et al (2005) Brachiocephalic reconstruction I: operative and long-term results for complex disease. J Vasc Surg 42(1):47–54. https://doi.org/10.1016/j.jvs.2005.03.027

31 Coronary Arteries

Baris Vefali, Huseng Vefali, and Veli Vefali

31.1 Embryology

An overview of cardiac embryology is essential for understanding the development of the coronary system. During the third to fifth weeks of development, the cardiogenic zone of the mesoderm turns into a horseshoe-shaped endocardial tube. The pericardial space is formed from the space around this tube. By the end of the third week, the two legs of the endocardial tube start to contract rhythmically and join one another to become the heart tube, known as the cor tubulare simplex. It consists of the sinus venosus, a pair of primitive atria, a primitive ventricle, and the conus arteriosus. During the fourth and fifth weeks, the heart tube morphs into an S shape known as the cor sigmoideum. The connection between the primitive atria and ventricle is modified by the atrioventricular canal. As development progresses, structures change location. The endocardial cushion forms the caps for the atrioventricular valves. By the fifth to sixth weeks, the heart has two ventricles and the pars muscularis of the septum interventriculare has formed. However, the ventricles are still connected to one another. Once the pars membranacea of the interventricular septum has completed its development during week seven, the lumina of the two ventricles are completely and irreversibly separated. The conus arteriosus twists into a spiral and the truncus pulmonalis and the aorta are formed from the saccus aorticus attached to it. The arteries to the head and neck area come off the saccus aorticus. Once the aorta forms, adult coronary artery development begins (Sharma et al. 2017).

Importantly, the cell dynamics of arterial remodeling in humans cannot be studied in fine detail owing to the sensitivity of the arteries. However, many mammalian, fish and avian yolk sac models have provided a clearer picture. On the basis of such models, we can predict that the coronary arteries are derived from plexus vessel fusion. Specifically, some fish models have shown that endothelial cells tend to migrate against the blood stream in the direction of the growing artery (Conte and Pellegrini 1984).

Our knowledge of vascular remodeling is limited because its main location, the heart, is sensitive. Like all arteries, the coronary arteries are derived from mesoderm. Two types of vascular remodeling occur over the human life span, developmental and adult. Both processes involve increased blood flow to the region, which triggers a pressure response from endothelial cells.

B. Vefali (✉)
Department of Medicine, Austin Medical Clinic, Bastrop, TX, USA

H. Vefali
Department of Interventional Cardiology, Austin Cardiology Clinic, Austin, TX, USA

V. Vefali
Department of Interventional Cardiology, Medicana Hospital, Ankara, Turkey

S. J. Bordes, Jr. et al. (eds.), *The Clinical Anatomy of the Vascular System*,
https://doi.org/10.1007/978-3-031-78326-5_31

However, the two processes differ significantly in the mechanism of this response. In developmental arterial remodeling, immature plexus vessels morph and convert into arteries covered by smooth muscle. In contrast, adult artery remodeling involves outward expansion of the arterial lumen. During embryogenesis, coronary arteries form via developmental remodeling through the development of the plexus vessels after stem formation. The resulting network of arteries then spreads and covers the myocardium. Initially the plexus is not supple, but over time the vessels spread further and anastomose with the aorta to provide proper blood flow to the myocardium (Bogers et al. 1989).

31.2 Anatomy/Topography

The heart is the only muscle that works non-stop from early in development until death. Optimal blood flow to the myocardium is necessary for adequate oxygenation. Although there are only two main coronary arteries, they have multiple essential branches, and these branches have many variations; one must be aware of such variations when dealing with cardiovascular structures. We will start this section with the normal anatomy of the coronary vasculature.

The heart is a four-chamber pump connected to the arterial system via two main arteries arising from the ascending aortic root (Fig. 31.1). These two coronary arteries come off the sinus of Valsalva. The location of the ostia allows them to be filled optimally during diastole. The left coronary artery (left main artery (LMA) originates at the left sinus of Valsalva, courses in the left anterior direction, and provides blood to the ventricular myocardium, left atrium, left atrial appendage, pulmonary artery, and aortic root via its own branches. The LMA is 1–25 mm long and 2.0–5.5 mm in diameter. It bifurcates into two arteries. The first, the (left) anterior descending (LAD) (Fig. 31.1), curves around the pulmonary artery trunk and extends to the apex through the interventricular septum. It can range from 10 to 13 cm in length and its diameter is 2.0–5.0 mm.

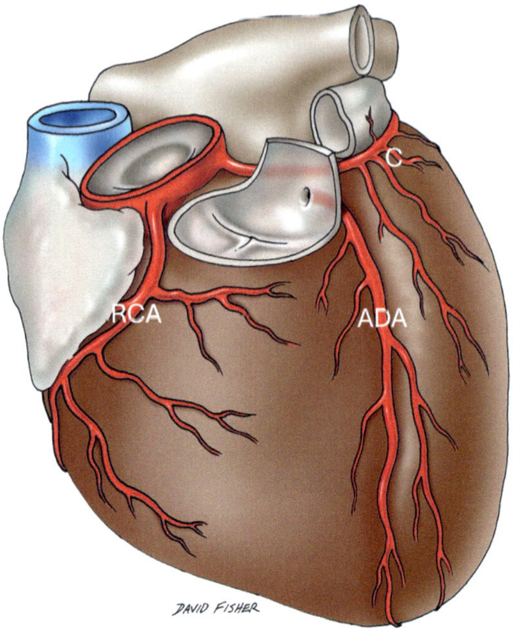

Fig. 31.1 Schematic drawing of the coronary arteries. *RCA* right coronary artery, *C* circumflex artery, and *ADA* anterior descending artery

The LAD branches further into diagonal and septal branches. Diagonals typically branch off laterally at a right angle and supply the ventricular free wall. The septal branches run medially toward the interventricular septum and supply the anterior 2/3 of the septum and the anterior papillary muscles. The second artery branching off the LMA is the left circumflex (LCx [Fig. 31.1]), which is vital for people with left dominant coronary systems (dominance is determined by the side of the coronary system that supplies the posterior interventricular septum). The LCx, about 6–8 cm long and 1.5–5.5 mm in diameter, courses along the left atrioventricular (AV) groove to form marginal branches that supply the posterolateral wall. In people with right dominant hearts (90%), the LCx terminates into the left obtuse artery and courses into the apex via the left margin of the heart. In those with left dominant hearts (10%), the LCx goes further toward the AV groove and terminates at the cardiac crux. It then branches into the posterior interventricular artery, which supplies the AV node. Although most people have only two branches arising from the LMA, some have

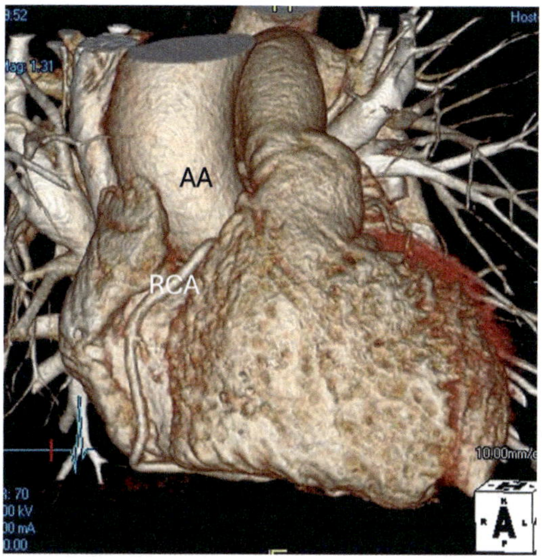

Fig. 31.2 CTA rendering of the right coronary artery (RCA) arising from the ascending aorta (AA)

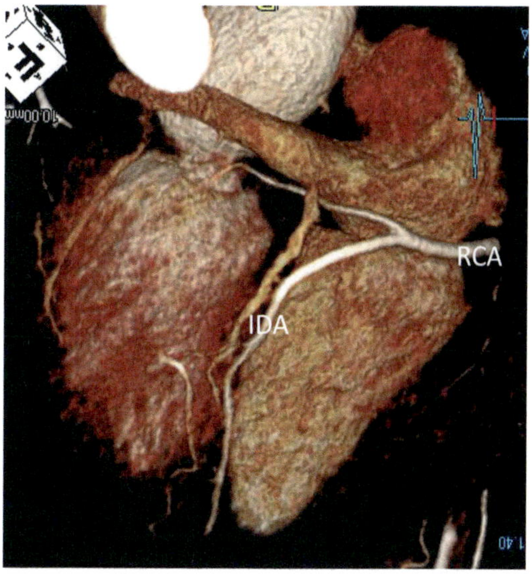

Fig. 31.3 Posterior view of CTA rendering noting the right coronary artery (RCA) and its inferior descending branch (IDA)

another artery in the left coronary system, the left intermediate artery, which provides additional blood to the left ventricle (Waller et al. 1992).

The right coronary artery (RCA) (Figs. 31.2 and 31.3), originates from the right sinus of Valsalva adjacent to the tricuspid annulus. It is about 12–14 cm long in the right dominant population and its first branch is the conus branch coming off the septate ostium of the right sinus. Although the conus artery commonly derives from the RCA, it can also arise directly from the right sinus of Valsalva. The blood supply to the SA node comes off the RCA following the conus artery. However, in patients with left dominant hearts, the sinoatrial (SA) node artery can arise from the LCx artery. The largest branch of the RCA is the right marginal artery, which supplies the free wall of the right ventricle. After all the branches have come off the RCA, it becomes the inferior interventricular artery, which extends into the base of the heart up to the apex posteriorly. It supplies the posterior wall of the heart in most people with right dominant hearts (Muresia 2016).

The coronary arteries are usually epicardial, which means they cover the surface of the epicardium and supply blood to the specific regions of the heart mentioned above. However, they are subepicardial when they run through the myocardium, in which case they are known as "mural" or "tunneled" arteries. These variants run only up to a few millimeters through the ventricular myocardium. When ventricular muscle covers the coronary artery, this is called a muscular bridge. Muscular bridges occur in up to 86% of the population. In angiography, muscular bridging can be recognized when contrast is injected into the artery and the visible narrowing and expansion of the vessel is synchronized with ventricular contractions and relaxations. Although myocardial bridging of the LAD is mostly benign, about 5% of athlete deaths can be attributed to it (Muresia 2016).

Coronary artery	Branches	Supplies	Drains
Left main artery (LMA)	LAD Left intermediate LCx	LAD Left intermediate artery (Variant) LCx	Great cardiac vein
Left anterior descending(LAD)	Diagonal arteries (D1, D2) Septals	Anterior 2/3 IVS Right ventricle Left ventricle	Great cardiac vein
Left intermediate artery	None	Left ventricle	Left marginal vein Great cardiac vein
Circumflex (LCx)	Obtuse marginal arteries (OM1, OM2)	Left atrium Left ventricle	Great cardiac vein
Right coronary artery (RCA)	AV branch Right marginal Inferior descending artery (PDA) Posterolateral arteries (PA1, PA2)	Right atrium SA and AV nodes Posterior part of interventricular septum (IVS)	Small cardiac vein Middle cardiac vein
Right marginal (Acute marginal)	None	Right ventricle Apex	Small cardiac vein Middle cardiac vein
Inferior descending artery (PDA)	None	Right ventricle Left ventricle Posterior 1/3 of IVS	Left posterior ventricular vein
Posterolateral artery	None	AV node	Small cardiac vein

31.3 Imaging/Histology

Histology

The walls of coronary arteries, like those of all arteries, consist of three layers: tunica intima, tunica media, and tunica adventitia. The tunica intima is the innermost layer, comprising an endothelium lining the lumen and a subendothelium that contains connective tissue and smooth muscle cells. The endothelium is thin enough to allow for selective diffusion. The endothelial cells are oriented longitudinally and connect to one another via occluding and gap junctions. The endothelium is also an endocrine structure. Its cells produce procyclin and PGI_2, which are antithrombotic, von Willebrand factor, which is prothrombotic, and fibrinolytic, proinflammatory, and growth factors. It is also a hub for receptors for LDL, thrombin and factor X (Waller et al. 1992).

The tunica media is about 20 μm thick. It is separated from the tunica intima by an internal elastic membrane. As arteries age, this tissue begins to lose its elasticity and stiffens. The tunica media consists of layers of smooth muscle, elastic fiber, proteoglycans, and collagen. Although most arterial walls are similar in composition, one big difference between the coronary arteries and the rest of the vasculature is the greater ratio of smooth muscle to elastic tissue in the epicardial arteries. The smooth muscles are housed in a glycoprotein mixture. The glycoprotein stains periodic acid-Schiff (PAS) positive. On the outer parts of the elastic membrane lie unmyelinated nerve axons. Medial smooth muscles receive signals from the diffusion of neurotransmitters through fenestrations in the elastic membrane (Waller et al. 1992).

The final layer, the adventitia, houses more collagen and elastic fibers, vasa vasorum (microarteries that supply the walls of larger arteries), nerves, and lymphatics. The tunica adventitia is about 400 μm thick. The collagen is arranged loosely to provide enough space for changes in coronary artery diameter (Waller et al. 1992).

Imaging

Two main types of imaging are used to visualize the coronary and cardiac systems: functional and anatomical. Functional imaging does not necessarily help us see the structures themselves, but it helps us map coronary artery disease. To achieve this, the heart needs to be stressed to induce ischemia. We use three types of functional imaging today: (1) single photon emission computed tomography (SPECT) or Positron emission tomography (PET); (2) contrast stress echocardiography; and (3) MRI. These techniques allow the perfusion and function of the heart to be assessed during rest and after stress. They allow experienced physicians and specialists to form an idea of which coronary artery could be diseased and guide treatment accordingly. The stress can be induced either through physical tasks (bicycle or treadmill) or pharmacologically (dobutamine, adenosine/dipyridamole). Physical stress increases the heart rate and myocardial oxygen demand, and areas of low perfusion lead to ischemic areas in the heart walls. Pharmacological stress using dobutamine (a beta-1 specific agonist) also increases the heart rate, contractility, and arterial blood pressure and results in increased myocardial oxygen demand and ischemia of areas of heart wall perfused by diseased coronary arteries. Dobutamine is administered at 5, 10, 20, 30, and 40 mg/kg/min doses at intervals of 5 min. If the target heart rate is not achieved, atropine (0.25–1.0 mg) can be added. If the patient deteriorates or the test needs to come to an end, beta blockers can be used as an antidote. Adenosine exerts a direct vasodilatory effect, while dipyridamole inhibits the breakdown of adenosine, causing slower vasodilation indirectly. These medications cause normal vessels to dilate further and increase perfusion 3- to five-fold, while diseased vessels show attenuated hyperemic filling. The choice between physical and pharmacological stress depends on the patient's mobility and ability to perform physical tasks. Patients unable to perform physically (obesity, inability to walk, ataxias, etc.) receive pharmaceutical stress (Schuijf 2005).

Like all functional imaging systems, SPECT compares two patient states: at rest and during physical stress. SPECT uses one of three radiopharmaceuticals—thallium-201 (most commonly used), technetium-99 m sestamibi, and technetium-99 m tetrofosmin—that emit gamma rays. These tracers are absorbed by the myocardial tissue through coronary artery perfusion. Areas with diseased coronary arteries have lower perfusion so the tracers fail to reach them. Computer-generated 3D images are obtained using CT and gamma cameras, which detect the gamma emissions. The images from rest and stress states are compared. SPECT of a patient with healthy coronary arteries should show images with complete emission of gamma rays from all myocardial tissues. However, patients with diseased arteries show either reversible or irreversible defects. Reversible defects are stress-induced and reflect ischemia; irreversible defects are apparent at rest and imply myocardial infarction (Schuijf 2005).

Echocardiography with contrast agents is another method of functional imaging for perfusion. Like SPECT, it is performed during rest and during stress on the patient's heart. Contrast is injected intravenously and microbubbles are used to visualize the microvascular circulation of the coronary arteries; they dissolve over time. The signal intensities are measured in the coronary arteries and the relative concentrations are used to assess the relative myocardial blood volumes in those regions. This enables the overall perfusion of the coronary arteries and areas of coronary artery disease (CAD) to be predicted. Resting perfusion defects suggest infarcted myocardium; stress-induced perfusion defects indicate ischemia (Schuijf 2005).

As a relatively new imaging modality, MRI pictures a number of slices through the short axis while a bolus of contrast agent (gadolinium) is injected into the patient's vein during both rest and pharmacological stress. Gadolinium changes the T1 relaxation time, which is detected as areas of high signal intensity. The more an area of myocardium is perfused, the more intense the MRI signal from that region. In contrast, areas of low perfusion, or CAD and ischemia, have lower signal intensities on the images. MRI is more sensitive than SPECT in distinguishing subendocardial from transmural perfusion defects. This allows for earlier detection of compromised blood flow (Schuijf 2005).

Anatomical imaging remains necessary for most patients. Given positive stress tests, anatomical visualization of the coronary system is still required for diagnosing CAD, assessing the extent of the disease, and guiding the treatment strategy. It is important to remember that functional imaging is not as reliable in the diabetic population as it is in healthy individuals, not least because of diffuse atherosclerosis in the coronary arteries. Anatomical imaging is therefore necessary in these patients (Schuijf 2005).

Currently, the gold standard for anatomical imaging of the coronary arteries is angiography with selective contrast injection, also known as percutaneous coronary intervention. This procedure is fairly invasive, but it is the best option for diagnosing and simultaneously treating the underlying CAD. Live images of coronary artery anatomy are obtained in a catheterization laboratory. Further detail is discussed in the Clinical and Surgical Significance section (Schuijf 2005).

Other, non-invasive, angiographic methods are being developed. These include MRI, multi-slice computed tomography (MSCT), electron beam computed tomography (EBCT), and coronary artery calcium scoring. They can all help clinicians to diagnose some patients; however, none of them is as sensitive or specific as the gold standard. Non-invasive techniques are also time-consuming and do not allow for intervention and treatment of the coronary disease. Therefore, these options are not optimal for use on patients with suspected CAD (Schuijf 2005).

31.4 Variations

As discussed above, the ostia and coronary arteries have similar locations in most people. However, there can be slight individual differences, or variations, depending on events during embryo development. Normally, the LMA arises at the level of the aortic valve cusp; however, in about 34% of cases, it branches above, and in 18% of cases, below, the sinotubular junction. The RCA usually comes off at the level of cuspal free margin, but in 19%, it arises above, and in 10%, below, the sinotubular junction (Rahalkar and Rahalkar 2009). The different locations and positions of these arteries can affect their diastolic filling potentials and result in ischemia in certain regions of perfusion.

Early branching is a common cause of variations in coronary artery anatomy. The intermediate artery (ramus intermedius) is one example. It is seen in about 30% of the population and arises between the LAD and LCx. Clinicians also refer to it as the high diagonal or high obtuse marginal artery. The inferior descending artery (PDA) can also branch early from the RCA before the crux of the heart. Another example is separate origins of the LAD and LCx arteries from the left aortic cusp instead of the LMA. Finally, arteries can be duplicated. An example seen in some patients is double PDAs arising from the coronary arteries and entering the septum (Muresia 2016).

Dominance type is also a product of variation. Patients can present with one of three types of dominance depending on the anatomical positioning and development of their coronary arteries: (1) The PDA is supplied by the RCA (right dominant); (2) the PDA is supplied by the LCx (left dominant); (3) both the LMA and RCA supply the PDA (co-dominant) (Muresia 2016).

The coronary arteries can have abnormal origins, benign in most cases. These anomalies include (1) a single coronary artery comes off the right aortic cusp and gives rise to the RCA and LCA, and the LCA courses anterior to the right ventricle; (2) a single coronary artery comes off the left aortic cusp and divides into the RCA and LCA, the RCA running posterior to the aorta; (3) the LCx comes off the right aortic cusp and runs posterior to the aorta (most common); (4) the RCA comes off the aorta superior to the cusp; and (5) finally, and very rarely, origin of coronaries from the subclavian or internal mammary arteries (Rahalkar and Rahalkar 2009).

However, not all coronary artery anomalies are benign. A malignant RCA occurs when the RCA comes off the left aortic cusp and courses between the pulmonary artery and the aorta. The counterpart of the malignant RCA is a malignant LCA. This is when the LCA arises from the right aortic cusp and courses between the pulmonary artery and aorta. In both of these situations, the coronary artery crossing the aorta and pulmonary artery is subject to compromise during high-stress events or heavy exertion; this can lead to sudden cardiac death, especially in young patients who are athletes. The origin of the LCA from the pulmonary artery (ALCAPA), also known as Bland-White-Garland syndrome, is another malignant anomaly. Although extremely rare, it has a mortality of 90% during the first year of life. The blood flows retrospectively into the pulmonary artery from the LCA owing to low pulmonary resistance. This coronary steal leaves the left myocardium hypoperfused. The patient suffers from LV dysfunction and later congestive heart failure. A variant of this anomaly is when the RCA originates from the pulmonary artery. The last and rarest of these congenital malignant anomalies is a congenital coronary artery fistula, an abnormal connection between any coronary artery and a chamber of the heart or pulmonary circulation. This abnormal connection leads to the shunting of blood from higher pressure to lower resistance areas, and the patient is liable to ischemia and heart failure. The most common form of such a fistula involves the RCA draining into the right ventricle, right atrium, or coronary sinuses (Rahalkar and Rahalkar 2009).

31.5 Pathology

Being among the most vital of arteries, coronary arteries have pathologies that it is critical to understand and treat. To achieve a deeper appreciation and better coverage of the topic, we will divide CADs into their etiologies for easy categorization. There are two main causes of CAD: non-atherosclerotic and atherosclerotic. Pathologies under these two main categories include fixed stenosis, aneurysm, dissection, thrombus, small vessel disease, and spasm. However, atherosclerosis is the leading disease in coronary artery pathology. Plaques are the root of most symptomatic coronary patients. Coronary atherosclerosis causes stable angina, which is fairly benign. More malignant symptoms such as unstable angina, myocardial infarction, and sudden cardiac death are due to plaque rupture and thrombosis superimposed on the plaque (Basso et al. 2001). In reality, the symptoms are due to decreased perfusion, increased oxygen demand, and ischemia of the heart tissue. Thus, any cascade of events that leads to a setting of increased oxygen demand or decreased perfusion will cause symptoms. Generally, the more acute the ischemic event, the more significant the symptoms. On the other hand, chronic ischemic disease processes give the heart cells time to adapt to the low oxygen state, so these patients have less severe symptoms, though the strain on the heart is greater in the long run. Data show that in patients who suffer sudden cardiac death (SCD), congenital lesions such as Kawasaki disease are the leading cause in children, while myocardial ischemia due to atherosclerosis is the main cause in the adult population (van der Wal 2007).

Non-atherosclerotic coronary artery pathologies are less common. Acquired coronary ostial pathologies are usually due to ostial narrowing of the coronary artery, which has a wide range of causes such as previous cardiac surgery, history of coronary catheterization and trauma, endocarditis in the aortic valve, vasculitis extending into the ostia, or a space-occupying lesion such as a tumor, septic embolus, or atheroma. Congenital anomalies in the coronary arteries are the most common cause of SCD in children. As discussed earlier, the highest risk anomalies are the origin of a coronary artery from an abnormal sinus of Valsalva with an interatrial course between the great arteries and the origin of one coronary artery from the pulmonary trunk. Coronary fistulas between the artery and veins, pulmonary artery, or a cardiac chamber can also decrease perfusion, so they should be considered when non-atherosclerotic causes of CAD are suspected. Coronary artery dissection, though rare, is a lethal complication that can be secondary to trauma or to intervention for a preexisting CAD with percutaneous coronary intervention (PCI) or surgical grafting. Major risk factors for dissection are connective tissue disorders and chronic hypertensive diseases. Coronary artery vasculitis is an inflammatory, non-atherosclerotic cause of CAD. Although rare, it can present either in isolation or as part of a systemic disease such as rheumatoid arthritis, SLE, Behçet syndrome, and, most importantly, in children, Kawasaki disease. Kawasaki children are the main population affected by the disease. Vasculitis in this condition has a high affinity for the coronary arteries and is the most common acquired inflammatory arterial disease in younger patients. Histology of biopsies shows immunoglobulin infiltrations in the arterial walls, implying an infectious precursor to the disease. It is important to remember that Kawasaki disease can lead to coronary artery aneurysms if left untreated; these can rupture or even result in a myocardial infarction (MI). Other examples of vasculitis that affect the coronary arteries are polyarteritis nodosa (PAN), giant cell arteritis, and isolated coronary vasculitis. A good rule of thumb is that inflammation of the coronary artery walls weakens the layers, making the artery susceptible to aneurysms (Jenette and Falk 2007).

Fibromuscular dysplasia (FMD) is neither atherosclerotic nor inflammatory. Although it most commonly affects the renal arteries, it can be seen rarely in the coronary arteries. Unlike the previously mentioned pathologies, FMD is more likely to affect intramural than epicardial arteries. In this disease, the arteries have areas of alternating stenosis and aneurysms that give a "beads on string" appearance on imaging. Patients with affected coronary arteries can display symptoms of myocardial ischemia due to the stenosis (Lee et al. 2000).

The most common pathologies affecting the coronary arteries, and the leading cause of death secondary to cardiac events, are atherosclerotic coronary diseases. In 1876, Adam Hammer described inadequate perfusion of heart muscles relative to their oxygen demand causing symptoms of angina. The pathophysiology of atherosclerotic coronary diseases is believed to be disruption of coronary plaques (a cholesterol core covered by a thin cap) and infiltration of them by macrophages (foam cells) and T cells, leading to occlusion of the coronary artery, disrupting the coronary blood supply. These acute events can lead to MIs and sudden cardiac deaths. However, as well as rupturing, vulnerable plaques can be subject to plaque erosion (thick fibrous cap or completely fibrous skeleton; generally limited inflammatory activity, but often with subendothelial rim of foam cells and T cells), protruding calcified nodules (complex fibro–lipid–calcific architecture and calcified shell with sharp edges protruding into the lumen of the artery), or intraplaque hemorrhages (complex fibro–lipid–calcific architecture and deep-seated collections of dilated thin-walled microvessels prone to leakage or disruption) (Lusis et al. 2004).

We need to classify chest pains to predict the likelihood of a cardiac etiology. The three types of chest pains are: Typical (all three need to be present—substernal, brought on by exertion, and alleviated by rest or vasodilators), atypical (2/3 of the typical criteria have to be present), and noncardiac chest pains (one or none of the typical criteria). Clinicians can use the abovementioned classification to guide their management of patients with chest pain.

Less than 15–30% of patients who present to the emergency department (ED) with nontraumatic chest pain have acute coronary syndrome. This is an umbrella term used for unstable angina, non-ST elevation myocardial infarction (NSTEMI), and ST elevation myocardial infarction (STEMI). Unstable angina and NSTEMI result from lack of perfusion in only a segment of the myocardial thickness. These patients have either ST segment depressions, deep T wave inversions with no Q waves, or no ECG changes. In a setting of STEMI the entire myocardial thickness is affected, the ECG showing ST segment elevations of at least 1 mm in two contiguous leads, at least 2 mm elevation in leads V2 and V3, or a new left bundle branch block.

Clinically, the most distressing heart diseases are due to lack of perfusion to the myocardial cells. The first stage of CAD is ischemia, the reversible loss of perfusion to areas of heart tissue. These areas have impaired function and often present with typical chest pain. If left untreated, or in a setting of plaque rupture, ischemia can lead to infarction. This is irreversible injury to myocardial cells. It is essential to reperfuse areas of low perfusion as soon as possible, regardless of the stage of disease. Even areas of a complete, irreversible injury can benefit from the reestablishment of blood flow.

For clinicians dealing with CADs, a major complication to keep in mind is coronary microembolization. Distal parts of coronary arteries with atherosclerotic disease can be subject to microembolization, especially if there is an aneurysmal component. Areas with embolization are at a higher risk of spontaneous plaque rupture followed by thrombosis, especially with interventions in which wires or stents disrupt the plaques. When PCIs are performed, the total atherothrombotic burden of the artery and the extent of intervention are major factors affecting the likelihood of plaque disruption (van der Wal 2007).

31.6 Clinical and Surgical Significance

As the major arteries supplying the heart, the coronary arteries have vital clinical and surgical significance in modern medicine. Fortunately, great strides have been made in both medical and surgical interventions in this area. Optimal medical therapy is the gold standard for all patients who present with CAD. The goal of this therapy is to reduce the progression of disease, stabilize the plaques in the arteries, prevent major adverse events, and improve life quality. However, the critical intervention in CAD is revascularization to salvage as much cardiac function as possible in the shortest possible time. The two main surgical interventions used today are PCI and coronary artery bypass grafting (CABG). Special interventions tend to be variants of one of those two procedures for special cases, depending on the disease. We briefly touched on PCI in an earlier section, so this section will focus more on the surgical procedures and compare the two procedures and their indications.

Although specialist experience and patient preference are the main considerations in choosing the method of intervention, the suggested guideline is: PCI is preferred in disease of one or two vessels not involving the proximal LAD with SYNTAX score < 23; CABG is preferred in diseases affecting more than two arteries, LMA disease and SYNTAX score > 32. (SYNTAX score is a complex tool used by cardiologists to guide treatment options. It involves dominance, the number and location of the disease and other variables used to classify disease severity) (Serruys et al. 2009).

Coronary artery bypass grafting (CABG) (Fig. 31.4) became the gold standard for treating patients with CAD. It was introduced in 1968, 9 years before the invention of PCI. As the name CABG suggests, a diseased coronary artery is bypassed using a graft to reestablish coronary perfusion. It is performed in one of three ways: traditional CABG (open heart surgery), where the anterior chest wall is opened and the aorta is

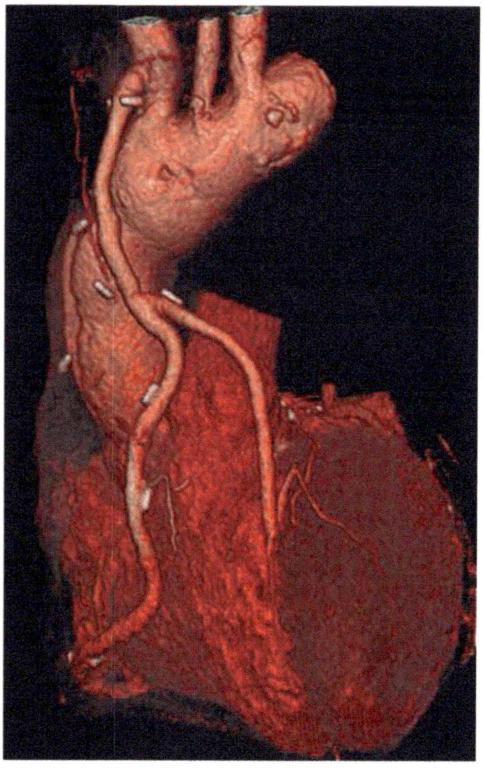

Fig. 31.4 CTA rendering illustrating a patient who has undergone a CABG procedure using the internal thoracic artery

connected to a cardiopulmonary bypass machine; off-pump CABG, where the chest wall is also open but heart continues to beat during the surgery; and minimally invasive CABG, where the surgery is performed robotically. The vessels most commonly used for grafting are the internal thoracic artery (ITA), saphenous vein (SV), radial artery (RA), right gastroepiploic artery (RGEA), and occasionally the ulnar artery (UA), splenic artery, or inferior epigastric artery (Martínez-González et al. 2017). We will briefly summarize the procedures in each CABG type below.

The most common type of CABG surgery takes 3–6 h. The patient is sedated and intubated with ventilator support throughout the procedure. Areas where the surgery will be performed are sterilized using proper technique, including the location from which the graft will be retrieved. Once sterile, the graft artery is harvested by making an incision parallel to the vessel and carefully separating it from surrounding structures. The surgeon makes a midline incision on the chest wall and then cuts through the sternum to separate the two sides of the thoracic cage and separates the ribs to expose the thoracic cavity and its contents. The pericardium is carefully incised to expose the heart for access to the coronary arteries. At this stage, the patient's heart is pharmaceutically stopped (cardioplegia) using a potassium-containing solution and the body is cooled. The patient is then placed on the cardiopulmonary bypass machine to perfuse the rest of the body and the aorta is clamped to preclude blood loss. The diseased coronary artery and the location of the disease are identified by the surgeon, and the artery is clipped to stop blood flow and prevent embolisms in that area. The graft artery is then used to connect the distal portion of the diseased artery to the aorta to create a new perfusion path. If the left or right mammary artery (LIMA or RIMA) is used, the artery is attached directly distal to the diseased artery; there is no need to harvest it owing to its location. Once bypassing is complete, the clamp is removed from the aorta, the patient is taken off the bypass machine, the body is warmed and the heart is restarted. The surgeon watches for any blood leakage and makes sure the grafts work before carefully closing the patient's chest (Diodato and Chedrawy 2014).

Off-pump coronary artery bypass (beating-heart surgery), as the name suggests, is CABG performed without using a cardiopulmonary bypass machine. Although studies show better outcomes in terms of complications and patient mortality, off-pump CABG is not as widely used as the traditional method because of surgeon comfort or lack of resources. The procedure is essentially the same as traditional CABG, but since the heart continues to beat, some modifications are needed to perform the surgery appropriately. The aorta is not clamped. The main problem with a beating heart is that the surgeon needs a stable working surface for bypass connection. Vacuum stabilizers are used to immobilize isolated areas of interest on the heart. Otherwise, the surgery is similar to the standard of care in CABG, each offering its own benefits and complications depending on patient situation, surgeon experience, and resource availability (de Jaegere 2002).

In minimally invasive CABG, instead of a single large incision, multiple small incisions are made in the chest wall for thoracoscopic/robotic access to the thoracic cavity. The patient is not placed on a cardiopulmonary bypass machine. The same logic applies as in performing traditional or off-pump CABG. Diseased arteries are bypassed using grafts but the incision sites are smaller, reducing the risk of infection. However, it is important to remember that this method is practical only for patients with single artery disease (Diodato and Chedrawy 2014).

In 1977, Andreas Gruentzig performed the first PCI in Switzerland. Also known as angioplasty, PCI has now become the gold standard for diagnosis and treatment of CAD. A number of steps are involved in safe angiography. The procedure must be performed in a cardiac catheterization lab with a team of highly trained professionals, including an interventional cardiologist, a nurse, and a technician. The patient must be connected to an ECG machine and have an IV access in place. Before the angiography begins, the site of arterial access must be sufficiently sterilized and a fully clean area must be prepared for the procedure. The site of arterial access must be sufficiently numbed with a local anesthetic. The heart is catheterized by accessing the peripheral artery of the clinician's choosing (the arteries usually preferred are the right femoral and left radial, owing to their anatomical advantage and easier access) using a hollow needle (trocar). A soft curved guide wire is then passed through the trocar and into the lumen until about a third of the wire remains outside. A hollow tube called an introducer sheath, which is a one-way valve system, is slid over the guide wire until it is inside the arterial lumen. The guide wire is pulled out with the dilatator of the introducer sheath, leaving only the sheath in the lumen. From this point on, guide wires and catheters are used in accordance with clinician preference and patient anatomy in order to access the coronary arteries. The wire is passed through the peripheral artery into the aorta, where the catheter is placed at the ostia of the coronary artery of interest. Contrast is injected selectively into the lumen, and the coronary artery is imaged at different angles to aid in optimal diagnosis of the CAD. Luminal narrowing can be measured precisely using an angiographic approach. If necessary, while the patient is catheterized, a percutaneous transluminal coronary angioplasty (PTCA) can be performed to place a stent to treat the underlying CAD. The stent is a tunnel-like device that can be passed into the location of a stenosis and opened like an umbrella using small, inflatable devices called balloons, which are expanded using normal saline to dilate these diseased locations and resolve the stenosis to reestablish good perfusion. Patients who undergo stenting must be placed on a dual antiplatelet regimen (aspirin and clopidogrel preferred), usually for 12 months.

Chronic total occlusion (CTO) of the coronary artery has been reported in nearly one in five patients with obstructive native CAD. This is when a coronary artery has been totally occluded for more than 3 months. Most such patients are reported to be symptomatic and at high risk of adverse cardiac events. However, even though perfusion is lacking in the areas affected by the CTO, studies have shown that revascularization via PCI improves patient outcomes significantly, and viable myocardial tissue has been observed in a remarkable number of patients undergoing such interventions (Vefali et al. 2014). Newer and more feasible access sites, including distal radial access, are good alternative locations for PCI interventions (Vefali 2008; Vefalı and Sarıçam 2020).

31.7 Complications

- Percutaneous coronary intervention complications
  - Dissection due to mechanical trauma
- Coronary artery bypass grafting
  - Emboli (thrombus, air bubble, foreign object)
  - Perforation
  - Infection
  - Arrhythmias
  - Myocardial infarction
  - Sudden cardiac death
  - Coronary spasms
  - Pericardial hemorrhage and tamponade
  - Renal failure
  - Reperfusion injury
  - Acute stent thrombosis
  - No reflow and slow flow
- Dissection due to mechanical trauma
- Emboli (thrombus, air bubble, foreign object)
- Perforation
- Infection
- Arrhythmias
- Myocardial infarction
- Sudden cardiac death
- Pericardial hemorrhage and tamponade
- Mediastinitis
- Non-union of the sternum
- Reperfusion injury
- Intubation complications
- Reduced output syndrome
- Renal injury
- Transfusion
- Bleeding and anemia

References

Basso C, Calabrese F, Corrado D et al (2001) Postmortem diagnosis in sudden death victims: macroscopic, microscopic and molecular findings. Cardiovasc Res 50:290–300

Bogers AJJC, Gittenberger-de Groot AC, Poelmann RE, Péault BM, Huysmans HA (1989) Development of the origin of the coronary arteries, a matter of ingrowth or outgrowth? Anat Embryol 180(5):437–441. https://doi.org/10.1007/bf00305118

Conte G, Pellegrini A (1984) On the development of the coronary arteries in human embryos, stages 14?19. Anat Embryol 169(2):209–218. https://doi.org/10.1007/bf00303151

de Jaegere PPT (2002) Off-pump coronary artery bypass surgery. Heart 88(3):313–318. https://doi.org/10.1136/heart.88.3.313

Diodato M, Chedrawy EG (2014) Coronary artery bypass graft surgery: the past, present, and future of myocardial revascularisation. Surg Res Pract 2014:1–6. https://doi.org/10.1155/2014/726158

Jenette JC, Falk RJ (2007) Nosology of primary vasculitis. Curr Opin Rheumatol 19:10–16

Lee AHS, Gray PB, Gallagher PJ (2000) Sudden death and regional ventricular fibrosis with fibromuscular dysplasia of small intramyocardial coronary arteries. Heart 83:101–102

Lusis AJ, Mar R, Pajukanta P (2004) Genetics of atherosclerosis. Annu Rev Genomics Hum Genet 5:198–218

Martínez-González B, Reyes-Hernández CG, Quiroga-Garza A, Rodríguez-Rodríguez VE, Esparza-Hernández CN, Elizondo-Omaña RE, Guzmán-López S (2017) Conduits used in coronary artery bypass grafting: A review of morphological studies. Ann Thorac Cardiovasc Surg 23(2):55–65. https://doi.org/10.5761/atcs.ra.16-00178

Muresia H (2016) Chapter 50, coronary arteries. In: Tubbs RS, Shoja MM, Loukas M (eds) Bergman's comprehensive encyclopedia of human anatomic variation. Wiley, pp 530–568. https://doi.org/10.1002/9781118430309

Rahalkar A, Rahalkar M (2009) Pictorial essay: coronary artery variants and anomalies. Indian J Radiol Imaging 19(1):49. https://doi.org/10.4103/0971-3026.45345

Schuijf JD (2005) Cardiac imaging in coronary artery disease: differing modalities. Heart 91(8):1110–1117. https://doi.org/10.1136/hrt.2005.061408

Serruys PW, Morice M-C, Kappetein AP, Colombo A, Holmes DR, Mack MJ, Ståhle E, Feldman TE, van den Brand M, Bass EJ, Van Dyck N, Leadley K, Dawkins KD, Mohr FW (2009) Percutaneous coronary intervention versus coronary-artery bypass grafting for severe coronary artery disease. N Engl J Med 360(10):961–972. https://doi.org/10.1056/nejmoa0804626

Sharma B, Chang A, Red-Horse K (2017) Coronary artery development: progenitor cells and differentiation pathways. Annu Rev Physiol 79(1):1–19. https://doi.org/10.1146/annurev-physiol-022516-033953

van der Wal AC (2007) Coronary artery pathology. Heart 93(11):1484–1489. https://doi.org/10.1136/hrt.2004.038364

Vefali V, Arslan U (2008 Apr) Our experience with transradial approach for coronary angiography. Turk Kardiyol Dern Ars 36(3):163–167

Vefalı V, Sarıçam E (2020) The comparison of traditional radial access and novel distal radial access for cardiac catheterization. Cardiovasc Revasc Med 21(4):496–500. https://doi.org/10.1016/j.carrev.2019.07.001

Vefali H, Manda Y, Shirani J (2014) Myocardial viability in coronary artery chronic Total occlusion. Curr Cardiol Rep 17(1):1. https://doi.org/10.1007/s11886-014-0552-x

Waller BF, Orr CM, Slack JD, Pinkerton CA, Van Tassel J, Peters T (1992) Anatomy, histology, and pathology of coronary arteries: a review relevant to new interventional and imaging techniques-part I and II. Clin Cardiol 15(6):451–457. https://doi.org/10.1002/clc.4960150613

Pulmonary Artery

32

Emma Lesser

32.1 Embryology

The great vessels begin to form during the third week of embryo development, deriving from both mesoderm and ectoderm. The pharyngeal mesoderm comes from mesodermal cells, while the neural crest cells originate from the ectoderm layer. Aortic arches develop from the distal-most portion of the truncus arteriosus, the aortic sac (Sadler and Langman 2010). The aortic sac and the two dorsal aortae meet at the aortic arches. The aortic arches make up six pairs of arteries and line the pharyngeal arches within the pharyngeal wall. Branches of each pharyngeal arch wrap around the primitive pharynx and terminate in the dorsal aorta.

The first and second aortic arches mostly disappear during development, but portions remain and are incorporated into the maxillary artery and stapedial artery, respectively. The third aortic arch ultimately becomes the common carotid artery and has segments that develop further into the internal and external portions of the carotid arteries. The fifth arch regresses. The sixth pharyngeal arch divides into dorsal and ventral segments. The right dorsal segment degenerates completely and ultimately disconnects from the dorsal aorta. The right ventral segment develops into the right pulmonary artery. The left dorsal branch provides the connection between the left pulmonary artery and the arch of the aorta, which is subsequently called the ductus arteriosus. Finally, the left ventral segment develops into the left pulmonary artery (Jiang et al. 2000).

32.2 Anatomy/Topography

The pulmonary artery (PA) (Figs. 32.1, 32.2 and 32.3) is in the superior part of hilum in both the right and left lungs, though each has its own characteristic arrangement (Drake et al. 2015). In the right lung, the PAs have a vertical orientation, and the openings of the two main branches are stacked on top of each other. In the left lung, the PAs follow a more circular arrangement at the superior aspect of the hilum.

The PAs originate from the pulmonary trunk, which bifurcates inferior to the T4/T5 vertebral level, allowing each PA to continue to its respective lung (Drake et al. 2015). The pulmonary trunk, encased in serous pericardium, carries deoxygenated blood that has been pumped through the pulmonary trunk from the right ventricle (Drake et al. 2015). The oxygen-poor blood then flows bilaterally through the left and right PAs to the left and right lungs, respectively.

The right and left PAs follow different characteristic furcation patterns (Kandathil and Chamarthy 2018, p. 202). The parent arterial

E. Lesser (✉)
Tulane University School of Medicine, New Orleans, LA, USA
e-mail: elesser1@tulane.edu

S. J. Bordes, Jr. et al. (eds.), *The Clinical Anatomy of the Vascular System*,
https://doi.org/10.1007/978-3-031-78326-5_32

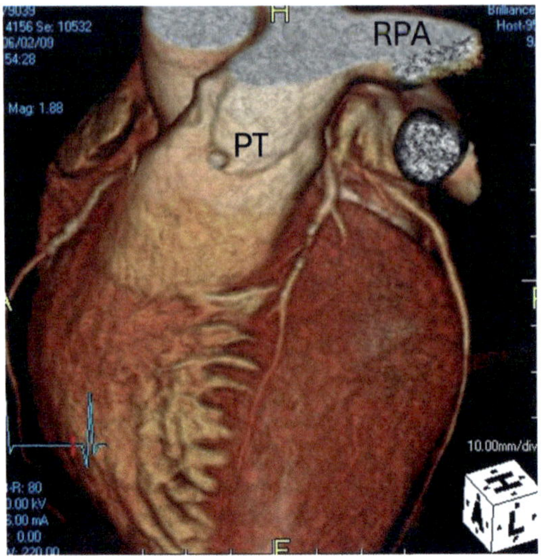

Fig. 32.1 CTA rendered image noting the pulmonary trunk and its right pulmonary artery (RPA)

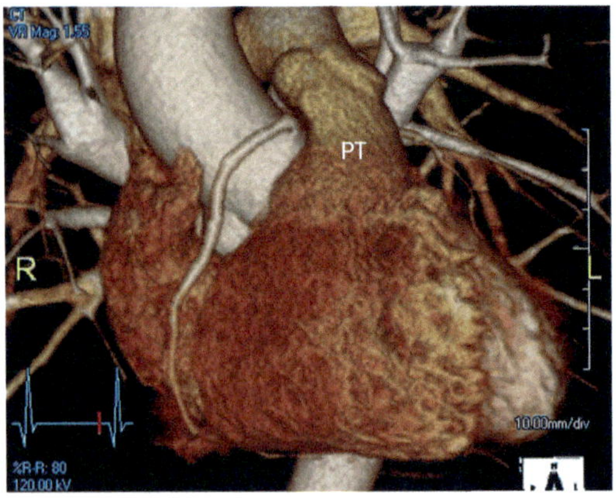

Fig. 32.2 CTA rendered imaging noting the pulmonary trunk and its relationship to the aorta seen to its right

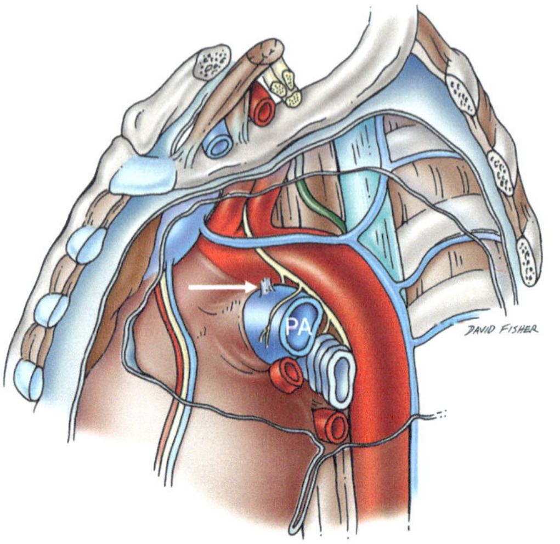

Fig. 32.3 Schematic drawing of the left mediastinum. Note the ligamentum arteriosum at the arrow connecting the left pulmonary artery (PA) to the aortic arch

segment divides into smaller offshoots (conventional branches), which originate at oblique angles and follow the branching pattern of the bronchi (Townsley 2012). Supernumerary arteries leave at 90-degree angles and are not associated with airway branches. As a general rule, the branches grow narrower as their number increases and as the vasculature approaches the periphery of the lung.

The right PA passes inferior to the aortic arch and anterior to both the tracheal bifurcation of the right main bronchus. The right pulmonary artery is posterior to the ascending aorta and superior vena cava, and the upper right pulmonary vein (Drake et al. 2015). The left PA is shorter than the right because the pulmonary trunk lies just left of the esophagus and the bronchi (Townsley 2012). Also, the pulmonary trunk travels superolaterally over the heart, from the superior end of the right ventricle to the superomedial region of the left atrium.

Upon entering the right lung, the right PA normally gives off its first branch, the truncus anterior, and continues to descend into the lung as the interlobar PA. The posterior segment splinters from the interlobar PA and extends posteriorly to supply the right upper lobe along with the apical and anterior segments from the truncus anterior. Continuing distally, the middle lobe segmental arteries arise from the anteromedial face of the interlobal artery and supply the middle lobe of the right lung. The interlobar PA has subdivisions that also extend inferiorly to supply the lower lobe of the right lung. These branches are the superior, medial basal, and anterior basal segments. Finally, the most distal portion of the interlobal PA bifurcates to form the medial basal and posterior basal segments in the right lower lobe.

The left PA extends inferiorly and posteriorly, passing deep to and over the left mainstem bronchus. It leaves the pericardium at the level of the

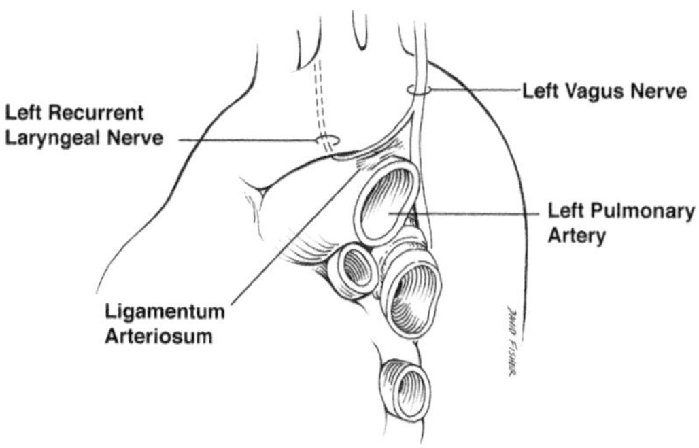

Fig. 32.4 Schematic drawing of the left pulmonary artery and its relationships. Specifically noting the left recurrent laryngeal nerve as it courses under the ligamentum arteriosum traveling up toward the larynx

ligamentum arteriosum (Fig. 32.3). At this location, the left recurrent laryngeal nerve courses under the ligamentum arteriosum to travel superiorly to the larynx (Fig. 32.4). After passing the left hilum and inside the lung, the left PA curves sharply downwards and gives two main segments, apicoposterior and anterior, which supply the left upper lobe. As the PA continues inferiorly, it is called the common basal trunk. This gives off superior, inferior, and anteromedial basal segments before bifurcating into the terminal basal segments (posterior and lateral) (Kandathil and Chamarthy 2018).

The pulmonary arterial vasculature is innervated by the vagus and cervicothoracic sympathetic nerves. The pulmonary trunk receives innervation from the right cardiac recurrent nerve, which forms an adventitial nerve plexus with its branches (Townsley 2012).

32.3 Histology

The PAs are large, elastic conducting vessels similar to the aorta. Conducting vessels carry large volumes of blood to progressively smaller arteries.

PAs have the same three layers as typical arteries: tunica intima, tunica media, and tunica adventitia. The tunica intima contains simple squamous endothelial cells, subendothelial connective tissue and an internal elastic lamina that serves as a physical barrier to the tunica media. The tunica media consists of elastic lamellae with smooth muscle fibers. Elastic arteries typically have around 50 layers of lamellae and smooth muscle fibers, facilitating even, uniform blood flow throughout the vessel. The tunica externa (or adventitia) consists largely of loose connective tissue.

When the ventricles contract, elastin is stretched to accommodate the influx of blood into the pulmonary arteries. Elastin rebounds passively (an automatic reaction) during diastole (ventricular relaxation). The collagen of the arteries regulates and prevents over-stretching.

32.4 Variations

Variations in pulmonary artery branches are more common in the left lung than the right. The number of branches supplying the left upper lobe, for example, can range from two to seven (Kandathil and Chamarthy 2018).

As with other organ systems, the pulmonary arteries are susceptible to developmental congenital disorders and variations. Some common disorders include pulmonary artery atresia, pulmonary artery sling, and persistent ductus arteriosus. In pulmonary artery atresia, the pulmonary arteries do not develop completely. The affected lung becomes hypoplastic because it is not drained adequately and is overexposed to oxygen-poor blood. The lungs are perfused by collateral flow from aortopulmonary collateral vessels. Right pulmonary artery atresia normally presents as the sole pulmonary congenital anomaly; however, left pulmonary artery atresia can occur in conjunction with other congenital cardiovascular anomalies. Other than hypoplasia of the affected lung, common symptoms of pulmonary artery atresia include thickening of the pleura surrounding the lung, chronic pulmonary infections, and respiratory difficulties.

Pulmonary artery sling occurs because the left sixth aortic arch fails to develop correctly. As a result, the left pulmonary artery arises from the right pulmonary artery, which can cause narrowing of the airways and mainstem bronchus.

Patent ductus arteriosus (PDA) occurs when the ductus arteriosus (a blood vessel allowing the fetal circulation to bypass the lungs) remains open after birth (Sadler and Langman 2010). The ductus arteriosus connects the aorta to the pulmonary trunk. Infants can be asymptomatic immediately after birth but display characteristic indications of PDA later in life, including shortness of breath, heart murmur, and other respiratory problems.

32.5 Imaging

In a clinical context, Computed Tomography Pulmonary Angiography (CTPA) is the gold standard for imaging pulmonary arterial structures (Kandathil and Chamarthy 2018). It allows us to see the divisions of the PA that supply each of the bronchopulmonary segments. The high degree of specificity and sensitivity of CTPA makes it particularly beneficial for evaluating patients with pulmonary embolism. Catheter-based pulmonary angiography is used to diagnose congenital and acquired anomalies, such as malformations and pulmonary artery stenosis (Kandathil and Chamarthy 2018).

32.6 Pathology

Tetralogy of Fallot is a relatively well-known cardiac anomaly. It frequently presents with ventricular septal defect (VSD), which allows blood to shunt from right ventricle to left ventricle, following the path of least resistance. Pulmonary arteries can also be connected directly to the aorta rather than the right ventricle (they are otherwise known as aortopulmonary collateral arteries). Often, pulmonary artery stenosis is present with Tetralogy of Fallot.

Some common acquired disorders affect the pulmonary arteries. Pulmonary arterial hypertension can be subcategorized: idiopathic, heritable, and drug and/or toxin induced. Pulmonary artery aneurysm and pseudoaneurysm usually present secondary to hypertension. Normally, the pulmonary artery narrows as it branches and travels distally into the lung. In cases of pulmonary aneurysm, imaging reveals the enlarged part(s) of the branching pattern. Pulmonary artery stenosis or obstruction can occur secondary to malignant pulmonary sarcomas, inflammatory disorders, and/or pulmonary emboli. Idiopathic dilatation of the pulmonary trunk is a congenitally dilated pulmonary trunk with normal pressures. The pulmonary diameter can be measured by cardiovascular magnetic resonance imaging.

32.7 Clinical/Surgical Significance

Appreciation of the detailed blood supply to the lungs is necessary for surgery and clinical exams. Each lung is divided into regions each supplied by a single, independent branch of the pulmonary artery; these regions are called bronchopulmonary segments (Drake et al. 2015). The bronchopulmonary segment is the "smallest functionally independent region of a lung" (Drake et al. 2015). This structuring enables surgeons to remove parts of lung without impairing function or the blood supply to the rest of the organ.

32.8 Complications

Pulmonary embolism can lead to many other complications, as mentioned. Emboli can best be visualized by spiral CTA with contrast, showing ipsilateral mediastinal deviation toward the atretic pulmonary vein. Pulmonary emboli can lead to pulmonary vein thrombosis, which can lead to respiratory compromise and chronic pneumonia.

References

Drake RL, Vogl W, Mitchell AWM, Gray H (2015) Gray's anatomy for students. Elselvier/Churchill Livingstone, Philadelphia

Jiang X, Rowitch DH, Soriano P, McMahon AP, Sucov HM (2000) Fate of the mammalian cardiac neural crest. Development 127(8):1607–1616

Kandathil A, Chamarthy M (2018) Pulmonary vascular anatomy & anatomical variants. Cardiovasc Diagn Ther *8*(3):201–207. https://doi.org/10.21037/cdt.2018.01.04

Sadler TW, Langman J (2010) Langman's medical embryology. Lippincott William & Wilkins, Philadelphia

Townsley MI (2012) Structure and composition of pulmonary arteries, capillaries, and veins. Compr Physiol. *2*(1):675–709. https://doi.org/10.1002/cphy.c100081

33 Bronchial Arteries

Jordan Lo

33.1 Introduction

The bronchial arteries (Fig. 33.1) carry oxygenated blood but do not participate in gas exchange at the level of the alveoli. Instead, they supply the parenchyma of the lungs, along with supportive structures such as the bronchial system, blood vessels, and lymph nodes (Chaudhry and Bordoni 2022). Bronchial arteries can vary widely in their origin and are the most common source of bleeding during hemoptysis.

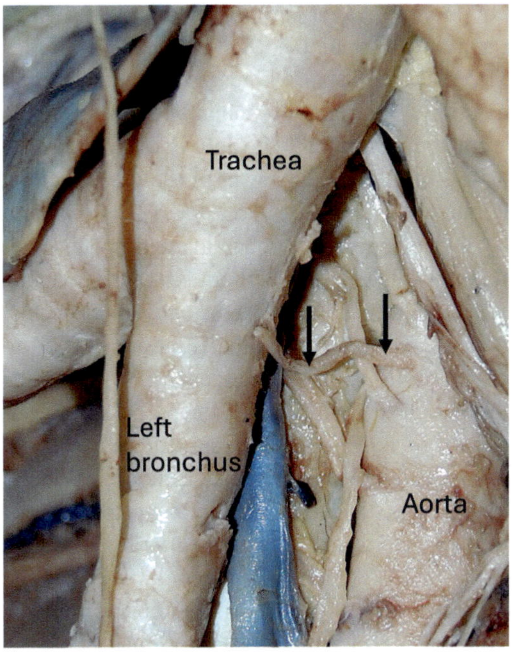

Fig. 33.1 Cadaveric dissection illustrating a left-sided bronchial artery (arrows). (Courtesy Prof. M. Loukas)

J. Lo (✉)
School of Medicine, Tulane University, New Orleans, LA, USA
e-mail: jlo2@tulane.edu

33.2 Embryology

The bronchial arteries have the potential for numerous variations, and information about the exact embryologic origins is scarce and limited. The bronchial arteries are believed to develop from the descending thoracic aorta between the ninth and 12th weeks of gestation (DeRuiter et al. 1993). During development, the descending thoracic aorta arises from the fusion of the right and left dorsal aortas distal to the aortic arches (Rosen and Bordoni 2022).

33.3 Anatomy/Topography

The bronchial arteries supply the bronchial system, lung parenchyma, and trachea. They also supply other areas in the thoracic region, such as the esophagus and posterior mediastinum (Valji 2006).

There are extreme variations in the anatomy of the bronchial arteries. In typical cases, patients develop one right bronchial artery and two left bronchial arteries. The right bronchial artery arises

S. J. Bordes, Jr. et al. (eds.), *The Clinical Anatomy of the Vascular System*,
https://doi.org/10.1007/978-3-031-78326-5_33

from the descending thoracic aorta, specifically from the right third posterior intercostal artery or intercostal-bronchial trunk (ICBT). The two left bronchial arteries typically arise from the anterior surface of the aorta just distal to the arch of the aorta. One left bronchial artery branches off, superior to the other. Both bronchial arteries run on the posterior surface of the bronchial system and terminate in the pulmonary or bronchial veins that empty into the azygous veins (Drake et al. 2005).

Because both bronchial arteries branch off the high-pressure thoracic aorta, the blood pressure in the bronchial arteries matches the high systemic pressure of the aorta. In contrast, the pulmonary arteries carry blood away from the heart at low pressure. This physiological and anatomical distinction between the blood pressures in these vessels is essential when determining the causes of hemoptysis. In 90% of hemoptysis cases, the bronchial arteries are responsible due to their highly pressurized system. In 5% of cases, the pulmonary arteries were the cause (Almeida et al. 2020).

33.4 Imaging

Bronchial arteries are small and must be visualized using higher-quality imaging techniques with nonionized contrast (Valji 2006). Contrast-enhanced computed tomography angiography (CTA) is the standard method of observing bronchial arteries under imaging (Almeida et al. 2020). When they are observed under CTA, both bronchial arteries appear tortuous as they branch from their origin and run toward the hilum of the lungs. Bronchial arteries observed under axial imaging under 2 mm in diameter are considered clinically insignificant, with arteries greater than 2 mm in diameter considered enlarged with a higher chance of rupture and bleeding.

33.5 Histology

Bronchial arteries take on the usual characteristics of a muscular artery: thin tunica intima, thick tunica media, and thin adventitia. They can easily be distinguished from pulmonary arteries by size, as they are significantly larger than bronchial ones (Mescher and Junqueira 2016).

Lack of hyaline cartilage but thick tunica media distinguishes the bronchial arteries from the neighboring bronchi. A simple stratified tunica intima distinguishes the bronchial arteries from the simple cuboidal epithelium lining bronchioles as the bronchi transition to bronchioles. Bronchial arteries are not found to accompany alveoli air spaces since bronchial arteries do not participate in gas exchange (Mescher and Junqueira 2016).

33.6 Variations

The variability of the number, origin, and course of the bronchial arteries is of importance when considering clinical intervention. There are four common types of bronchial artery variations regarding the number of right and left arteries (Table 33.1). Of all kinds of bronchial artery anatomy, the right bronchial arteries originate from the ICBT, and the left bronchial arteries arise from the anterior wall of the thoracic aorta (Chun et al. 2010).

The left and right bronchial arteries can sometimes originate from the thoracic aorta as a single stem, splitting into the right and left bronchial arteries. In other cases, the right and left bronchial arteries can branch independently from the thoracic aorta. In rare cases, the bronchial arteries can originate ectopically from different areas of the vasculature, such as the aortic arch, subclavian artery, thyrocervical trunk, internal mammillary artery, and coronary arteries (Almeida et al. 2020). The best way to identify an ectopic bronchial artery is to observe its course with the bronchi and trace it back to its origin.

A rare but substantial variation in the bronchial artery with significant clinical consequences is an artery giving spinal cord branches via the anterior spinal artery. In some patients, branches from the right bronchial artery (such as a right bronchointercostal artery) give off the great radicular artery of Adamkiewicz (Valji 2006). In 1.4–6.5% of cases involving the right bronchial artery embolization,

Table 33.1 Number of left and right bronchial arteries in the four variation types and their occurrence (Almeida et al. 2020)

Type	Left Bronchial Artery	Right Bronchial Artery	Occurrence of type
1	2	1	41%
2	1	2	21%
3	2	2	20%
4	1	2	10%

spinal ischemia in the lower two-thirds of the spinal cord has been seen (Brown and Ray 2012). Although this variation in the bronchial artery is rare, knowledge about the variant is crucial.

The right bronchial arteries tend to give off more branches than the left bronchial arteries, with increased lung volume on the right side demanding more blood flow and oxygen as an explanation for this. Sex can also play a role in the number of bronchial arteries and branches, with males showing an increased likelihood of a greater number of bronchial arteries and branches than females (Yener et al. 2015).

33.7 Pathology

Bronchial artery hypertrophy (BAH) and dilation are the most common disease states affecting the bronchial arteries, putting the patient at risk for severe hemoptysis. One of the most common congenital causes of BAH is Tetralogy of Fallot. In severe cases, blood flow from the right ventricle to the lungs via the pulmonary arteries is reduced. When the fetus is delivered, major aorticopulmonary collateral arteries (MAPCAs) persist after parturition to compensate for the reduction in blood flow in the pulmonary system. These collaterals originate from the thoracic aorta and commonly include dilated bronchial arteries (Almeida et al. 2020).

BAH also occurs in a rare anomaly in which one of the pulmonary arteries does not develop. Pulmonary artery atresia occurs most commonly on the right side during week 5 of development (Williams et al. 2019). The arterial system compensates for the absence of pulmonary arterial branches, including supplying the distal parts of the affected lung with hypertrophied bronchial arteries (Almeida et al. 2020).

Inflammatory and infectious diseases can also cause BAH through angiogenic growth factors that dilate the bronchial arteries. Bronchiectasis, cystic fibrosis, and tuberculosis are all associated with a risk of hemoptysis due to dilation of bronchial arteries in an inflammatory state. Lung abscesses, and malignancies are other causes of BAH and severe bleeding due to erosion of the bronchial arteries from persistent destruction of pulmonary tissue (Valji 2006).

33.8 Clinical and Surgical Significances

BAH and dilation increase the risk of vessel rupture and hemoptysis.

Bronchial artery embolization is typically used to treat clinically significant hemoptysis. CTA is used to identify the source of bleeding and embolize. An angiographic approach through the femoral artery is the most common in bronchial artery embolization, followed by aortography of the descending thoracic aorta to identify the bronchial arteries. The left main stem bronchus is a common landmark for bronchial artery origin. The most common agents used to embolize bronchial arteries are polyvinyl alcohol (PVA) and Gelfoam pledgets (Valji 2006).

If multiple attempts at bronchial artery embolization are unsuccessful, an open surgical approach would be warranted.

33.9 Complications

Complications mainly arise from bronchial artery embolization. Postembolization syndrome, which can be severe and last up to 1 week in patients, may result in chest pain, dysphagia, and fever. Non-target embolization is another complication that may cause bronchial or esophageal infarction and necrosis (Almeida et al. 2020).

The most severe complication of non-target embolization is spinal cord injury (transverse myelitis) due to the possibility of the bronchial artery feeding the great artery of Adamkiewicz. Vigilance on CTA and during embolization is crucial in preventing spinal cord injury. Additionally, neurological examinations should be documented in the patient's medical record (Valji 2006).

In 20–30% of cases, bronchial artery embolization fails, resulting in recurrent bleeding. Treatment involves reidentification of the bronchial arteries under imaging and repeating the procedure. Recurrent bleeding can be caused by incomplete initial embolization, collateral vessel recruitment, or disease progression. Specific risk factors place patients at a higher probability of recurrent bleeding, such as patients with cystic fibrosis or infection (Valji 2006).

References

Almeida J, Leal C, Figueiredo L (2020) Evaluation of the bronchial arteries: normal findings, hypertrophy and embolization in patients with hemoptysis. Insights Imaging 11:70. https://doi.org/10.1186/s13244-020-00877-4

Brown AC, Ray CE (2012) Anterior spinal cord infarction following bronchial artery embolization. Semin Interv Radiol 29(3):241–244. https://doi.org/10.1055/s-0032-1326936

Chaudhry R, Bordoni B (2022) Anatomy, thorax, lungs. [Updated 2021 Jul 31]. In: StatPearls [Internet]. StatPearls Publishing, Treasure Island (FL). Available from: https://www.ncbi.nlm.nih.gov/books/NBK470197/

Chun J-Y et al (2010) Radiological management of hemoptysis: a comprehensive review of diagnostic imaging and bronchial arterial embolization. Cardiovasc Intervent Radiol 33(2):240–250. https://doi.org/10.1007/s00270-009-9788-z

DeRuiter MC et al (1993) Development of the pharyngeal arch system related to the pulmonary and bronchial vessels in the avian embryo. With a concept on systemic-pulmonary collateral artery formation. Circulation 87(4):1306–1319. https://doi.org/10.1161/01.cir.87.4.1306

Drake RL, Vogl W, Mitchell AWM, Gray H (2005) Gray's anatomy for students. Elsevier/Churchill Livingstone, Philadelphia. Print

Mescher AL, Junqueira LCU (2016) Junqueira's basic histology: text and Atlas, 14th edn. McGraw-Hill Education, New York

Rosen RD, Bordoni B (2022) Embryology, Aortic Arch. [Updated 2022 Feb 10]. In: StatPearls [Internet]. StatPearls Publishing, Treasure Island. Available from: https://www.ncbi.nlm.nih.gov/books/NBK553173/

Valji K (2006) Vascular and interventional radiology, 2nd edn. W.B. Saunders, pp 347–373

Williams EA et al (2019) Proximal Interruption of the Pulmonary Artery. J Thorac Imaging 34(1):56–64. https://doi.org/10.1097/RTI.0000000000000373

Yener Ö et al (2015) The normal anatomy and variations of the bronchial arteries: evaluation with multidetector computed tomography. Can Assoc Radiol J 66(1):44–52. https://doi.org/10.1016/j.carj.2014.07.001

Esophageal Arteries

34

Marc Gebara, Joseph Asper,
and Katherine D. Bordes

34.1 Embryology

In the developing fetus, the trachea, bronchi, lungs, thyroid, and esophagus are all derived from the foregut, thus resulting in an intimate relationship in blood supply. The esophageal arteries mainly originate from the vitelline branches of the dorsal aortae (Shapiro and Robillard 1950).

34.2 Anatomy and Variation

The esophageal arterial supply is derived largely from the thoracic aorta and its major branches. Additionally, the arterial supply is categorized into cervical, thoracic, and abdominal segments of the esophagus, with numerous anastomoses (Tubbs et al. 2016). The cervical portion of the esophagus is supplied by branches of the inferior thyroid artery in most cases (68%). In addition, it may also receive small contributions from other arteries, including the subclavian, common carotid, vertebral, superior thyroid, and costocervical arteries (Shapiro and Robillard 1950; Tubbs et al. 2016; Standring 2021).

M. Gebara · K. D. Bordes (✉)
Tulane University School of Medicine,
New Orleans, LA, USA

J. Asper
Marian College of Osteopathic Medicine,
Indianapolis, IN, USA

The thoracic segment of the esophagus is supplied mainly by the bronchial arteries, as well as the esophageal branches of the descending thoracic aorta and right intercostal arteries (Shapiro and Robillard 1950; Tubbs et al. 2016; Standring 2021). At the level of the bifurcation of the trachea, the main arterial supply is derived from branches of the bronchial arteries (94%), which descend along the anterior surface of the esophagus. Some variation has been noted (6%) in which direct aortic branches were the main suppliers of this region. Below the bifurcation of the trachea, the main arterial supply is derived from two esophageal branches of the ventral aorta (Shapiro and Robillard 1950; Tubbs et al. 2016).

The distal region of the esophagus, including the lower esophageal sphincter, is supplied predominantly by esophageal branches of the left gastric artery (59%) and secondarily by the inferior phrenic artery (12.3%). These vessels run along the right anterolateral and dorsal aspect of the esophagus (Shapiro and Robillard 1950; Tubbs et al. 2016; Yazaki and Sifrim 2012).

These arterial sources come together to form a vascular chain of anastomoses obliquely down the esophagus. In the thoracic region, the esophageal branches of the bronchial artery and aorta pass from the anterior side of the esophagus to its posterior side and then give off left, right, ascending, and descending branches to anastomose the vessels supplying other regions of the esophagus.

S. J. Bordes, Jr. et al. (eds.), *The Clinical Anatomy of the Vascular System*,
https://doi.org/10.1007/978-3-031-78326-5_34

The ascending thoracic branches anastomose with the descending branch of the inferior thyroid artery of the cervical segment of the esophagus. The descending thoracic branches anastomose with the ascending branch of the left gastric and inferior phrenic arteries of the abdominal segment of the esophagus (Tubbs et al. 2016; Standring 2021).

34.3 Imaging

Unlike major arteries, esophageal arteries can be challenging to identify on axial CT. Although selective angiography can be used to visualize the esophageal artery anatomy, this method of imaging is invasive and does not provide a three-dimensional model of the vessels and their relationship with their surrounding structures (Kato et al. 2010; Morita et al. 2010).

Multi-detector helical CT can accurately provide a more detailed map of the esophageal arteries and their surrounding structures with a minimally invasive approach. A complete preoperative image allows better visualization during surgery and may reduce the risk of intraoperative hemorrhage (Kato et al. 2010; Morita et al. 2010).

34.4 Pathology

Significant esophageal hemorrhage can arise from varices secondary to portal hypertension. Other causes include esophagitis, Mallory-Weiss tears, or trauma.

Although rare, acute esophageal necrosis (AEN) may occur in patients with a history of cardiovascular disease, acute alcohol abuse, chronic obstructive pulmonary disease, renal failure, and/or diabetes mellitus. These underlying comorbidities increase the risk of damage to the esophageal circulation and mucosal defense mechanisms. Typical presentation includes hematemesis, melena, and epigastric pain, most commonly diagnosed by esophagoscopy. A case report from 2011 describes AEN secondary to a type B aortic dissection with complete thrombosis of the thoracic false lumen, which occluded the intercostal arteries, thereby restricting blood flow to the thoracic segment of the esophagus (van de Wal-Visscher et al. 2011).

34.5 Clinical Significance and Complications

There are several indications for esophageal surgery. In cancer of the esophagus and gastric cardia, esophagectomy and esophagogastrostomy are typically performed (Sweet 1945). These procedures are also required for benign tumors, extensive strictures, and lower esophageal varices (Puestow and Chess 1948; Sweet 1946; Garlock 1948a, b). Direct complications following surgical ischemia include diastasis and mediastinitis due to leakage (Shapiro and Robillard 1950). Ischemic necrosis might result in delayed stricture formation versus early perforation following gut and biliary surgery (Shapiro and Robillard 1950). This outcome can lead to progressive obstruction in those who survive the procedure for several months (Strieder 1948).

The common tracheoesophageal arteries bifurcate at the tracheoesophageal groove and go on to supply its cervical part. Blood flow could be compromised if the upper esophagus is separated too much from the trachea, as the collateral supply from the superior thyroid artery and others is insufficient (Shapiro and Robillard 1950). The Torek esophagectomy has been known to result in damage to the cervical esophagus, which could cause ischemia to the region (Sweet 1945). It is recommended that the esophagus be left in place as much as possible to keep circulation intact during these procedures (Shapiro and Robillard 1950).

The bronchial arteries are at risk of compromise if extensive mobilization occurs at the pars bifurcalis, as is the trachea-esophageal anastomotic arch connecting the inferior thyroid and bronchial arteries in several cases, according to Shapiro and Robillard. They found no significant collateral compensatory blood supply in approximately half of the cases studied. The authors believe that the high mortality of cancer operations in the mid-thoracic esophagus may be because of esophageal mobilization anterior to the arch, which compromises the blood supply in about half of the cases. Recurrent obstruction and stricture are common, but delayed complications can be linked to steady necrosis due to devascularization and obliterative fibrosis (Shapiro and Robillard 1950).

The aorta and esophagus can usually be carefully separated at the thoracic part for several centimeters without significant devascularization. Unless resection is indicated, it is imperative not to tear the two esophageal arteriae propriae in the superior one-third area. Collateral blood flow is high enough from the superior thoracic esophagus to anastomose with the arteries from the lower thoracic esophagus, should the lower region need to be displaced. The aorta should be separated from the esophagus as little as possible in preparation for an end-to-end anastomosis during infantile atresia and stricture procedures. If this cannot be done without hyper mobilization, trans-thoracic esophagogastrostomy and resection of the pertinent region may be appropriate (Shapiro and Robillard 1950).

The left gastric artery can sometimes be cut to improve gastric mobilization at the abdominal part of the esophagus, which may result in a free thoracic esophagus at the complete length and lower esophagus devascularization. As such, left gastric artery ligation should be avoided in low esophagostomy for cardiospasm or stricture. Esophageal branches originating from the celiac plexus should be spared in total abdominal gastrectomy. Historically, total gastrectomies failed when the appropriate and correct amount of esophagus was not correctly positioned; likely due to of ischemic diastasis of the collateral suture lines. Resection of the lowermost portion of the esophagus may be preferred over anastomosis in cases where the esophagus is dissected free from the hiatus (Shapiro and Robillard 1950).

References

Garlock JH (1948a) Progress in the surgical treatment of carcinoma of the esophagus and upper stomach. Surgery 23:906

Garlock JH (1948b) Resection of thoracic esophagus for carcinoma located above arch of aorta. Surgery 24(I)

Kato T, Takase K, Ichikawa H, Satomi S, Takahashi S (2010) Demonstration of the anatomy of the esophageal artery using multidetector-row helical computed tomography. J Comput Assist Tomogr 34(6). https://journals.lww.com/jcat/Fulltext/2010/11000/Demonstration_of_the_Anatomy_of_the_Esophageal.22.aspx

Morita Y, Takase K, Ichikawa H et al (2010) Bronchial artery anatomy: preoperative 3D simulation with multidetector CT. Radiology 255(3):934–943. https://doi.org/10.1148/radiol.10081220

Puestow CB, Chess SJ (1948) Resection of the esophagus for persistent stricture. Arch Surg 56:34

Shapiro AL, Robillard GL (1950) The esophageal arteries their configurational anatomy and variations in relation to surgery. Ann Surg 131(2):171–185. https://doi.org/10.1097/00000658-195002000-00004. PMID: 15402791; PMCID: PMC1616398

Standring S (ed) (2021) Gray's anatomy: the anatomical basis of clinical practice, 42nd edn. Elsevier

Strieder JW (1948) Surgical management of carcinoma of the lower two-thirds of the esophagus and cardiac end of the stomach. J Thorac Surg 17:143

Sweet RH (1945) Transthoracic resection of esophagus and stomach for carcinoma. Ann Surg 121:272

Sweet RH (1946) Subtotal esophagectomy in the treatment of extensive cicatricial obliteration of the esophagus. Surg Gynecol Obstet 83:417

Sweet RH (1948) Carcinoma of the superior mediastinal segment of the esophagus. Surgery 24:929

Tubbs RS, Shoja M, Loukas M (2016) Bergman's comprehensive encyclopedia of human anatomic variation. Wiley, Hoboken

van de Wal-Visscher E, Nieuwenhuijzen GA, van Sambeek MR, Haanschoten M, Botman KJ, de Hingh IH (2011) Type B aortic dissection resulting in acute esophageal necrosis [published correction appears in Ann Vasc Surg. 2013; 27(1):129. Nieuwenhuijsen, Grard A [corrected to Nieuwenhuijzen, Grard AP]]. Ann Vasc Surg 25(6):837.e1–837.e8373. https://doi.org/10.1016/j.avsg.2010.12.037

Yazaki E, Sifrim D (2012) Anatomy and physiology of the esophageal body. Dis Esophagus 25(4):292–298. https://doi.org/10.1111/j.1442-2050.2011.01180.x. Epub 2011 Mar 8. PMID: 21385283

Phrenic Arteries

35

Jordan Lo

35.1 Embryology

Information about the superior phrenic artery's embryological origins is limited. The inferior phrenic artery has embryological origins documented in the literature, mainly stemming from a network of ventrolateral blood vessels supplying the mesonephros around the fifth week of embryological development of the kidneys and suprarenal glands. Initially, the blood supply to the suprarenal glands develops from the superior suprarenal arteries. Once the diaphragm develops later in gestation, the inferior phrenic arteries become more prominent in their connections to the suprarenal glands (Kahn 1967).

35.2 Anatomy/Topography

The left and right superior phrenic arteries branch off from the thoracic aorta or its posterior intercostal branches (Fig. 35.1) just before it passes through the aortic hiatus at the T12 vertebral level. The superior phrenic artery runs along the superior surface of the diaphragm, giving collateral branches to the pericardiacophrenic and musculophrenic arteries to supply blood to the posterior superior surface of the diaphragm, pleura, and pericardium. The musculophrenic artery mainly supplies the anterior superior surface of the diaphragm (Loukas et al. 2006).

The right and left inferior phrenic arteries branch off from the abdominal aorta just after it passes through the aortic hiatus near the arching crura of the diaphragm. In most patients, the inferior phrenic arteries branch off the celiac trunk or immediately superior to the celiac trunk and curve upwards and laterally to supply the inferior surface of the diaphragm, as well as the left and right suprarenal glands via the superior suprarenal arteries (Kahn 1967).

The inferior phrenic arteries also provide circulation to the liver, esophagus, inferior vena cava, stomach, and retroperitoneum (Aslaner et al. 2017). In cases of hepatocellular carcinoma (discussed later in this chapter), the right inferior phrenic artery is the most common site of extrahepatic collateral circulation supplying hepatocellular carcinomas (Miyayama et al. 2010). Knowledge of this extrahepatic collateral circulation is essential when transcatheter embolization of the right inferior phrenic artery is performed for continued chemotherapeutic treatments of hepatocellular carcinoma (Loukas et al. 2005).

The right and left inferior phrenic arteries give rise to an anterior ascending branch, posterior descending branch, superior suprarenal branch, and middle suprarenal branch. The right inferior phrenic arteries also produce a branch that supplies the inferior vena cava and diaphragmatic

J. Lo (✉)
School of Medicine, Tulane University,
New Orleans, LA, USA
e-mail: jlo2@tulane.edu

S. J. Bordes, Jr. et al. (eds.), *The Clinical Anatomy of the Vascular System*,
https://doi.org/10.1007/978-3-031-78326-5_35

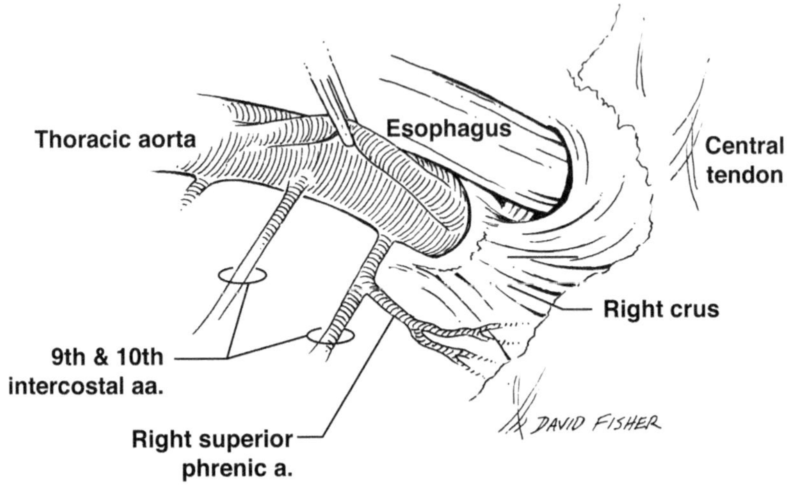

Fig. 35.1 Schematic drawing of the superior phrenic artery

branches. The left inferior phrenic artery provides accessory branches to the stomach, esophagus, and spleen (Miyayama et al. 2010).

35.3 Imaging

CT angiography is the preferred method of imaging the phrenic arteries to observe anatomical variations and clinically significant events. The inferior phrenic arteries have the most potential to show variation. By using CT angiography, clinicians can identify variations in origin, such as an inferior phrenic artery originating from the renal arteries. CT angiography before interventional procedures also eliminates the need for an aortography and saves copious amounts of time visualizing the patient's specific anatomy (Aslaner et al. 2017).

35.4 Histology

The superior and inferior phrenic arteries are smaller blood vessels and follow the histology of a typical muscular artery. Both have a tunica intima, tunica media, and adventitia in their histological appearance. Due to the smaller size of these arteries, the tunica media contains a smaller number of layers of smooth muscle, as well as a less noticeable internal elastic lamina (Mescher and Junqueira 2016).

35.5 Variations

The superior phrenic artery typically arises bilaterally from the thoracic aorta before the aortic hiatus at T12. Sometimes, the superior phrenic artery can branch from the tenth intercostal artery, either from the proximal segment or distal segment. A cadaveric study found that the left and right superior phrenic arteries branched off the thoracic aorta in most cases. The second most common site of origin was along the proximal segment of the tenth intercostal artery, with the third most common site near the distal segment of the tenth intercostal artery. When the superior phrenic arteries originated near the origin of the tenth intercostal artery, the superior phrenic arteries took a more medial to lateral course to supply the diaphragm. In cases where the superior phrenic arteries originated away from the origin of the tenth intercostal artery, the superior phrenic arteries coursed more laterally to medially (Loukas et al. 2006).

The inferior phrenic arteries show a wide variety of variations in origin. Left and right inferior phrenic arteries can originate independently or from a single stem. In most cases, both arteries arise from the celiac trunk (Type A) or immediately above the celiac trunk from the aorta (Type B), with the former being the most common. In decreasing order of incidence, the inferior phrenic arteries can arise from the right or left renal arteries (Type C), left gastric artery (Type D), or the hepatic artery proper (Loukas et al. 2005).

35.6 Clinical and Surgical Significances/Pathology

The superior phrenic artery does not commonly exhibit pathology, although it can be recruited as a collateral blood supply in cases of hepatocellular carcinoma. Tumor progression promotes the angiogenesis of nearby blood vessels to fuel the growing cancer cells. Although recruitment of the superior phrenic artery is not commonly observed, it is still within reasonable distance to the liver to have the potential to provide collateral branches to the growing tumors in the region. These superior phrenic branches may exist initially before chemotherapeutic treatment or arise post-treatment. A common technique to administer chemotherapy in unresectable cases of hepatocellular carcinoma has been through transarterial chemoembolization (TACE) through the hepatic artery. TACE is also used to embolize specific tumor-feeding branches to disrupt blood flow and the survival of the tumors. In response to the chemotherapy treatment and embolization, angiogenesis of nearby blood vessels can occur to rescue the viability of the tumor and sustain its growth. Therefore, clinicians must be aware of the possible collateral blood supply provided by the superior phrenic artery so TACE can be performed through the superior phrenic artery for continued treatment of hepatocellular carcinoma (Loukas et al. 2006).

In cases of hepatocarcinoma near the diaphragm, the right inferior phrenic artery is the most common site of extrahepatic collateral circulation. As mentioned above, TACE is performed most commonly through the hepatic artery and can cause angiogenesis of nearby blood vessels as collateral circulation to the tumor as a protective response. To counteract this occurrence, TACE must be performed through the inferior phrenic artery to sustain sufficient treatment of hepatocarcinoma and disrupt the blood supply to the tumor (Miyayama et al. 2010).

35.7 Complications

Complications can occur during the use of TACE in embolization of the inferior phrenic arteries. When performing TACE, only the tumor-feeding branches should be embolized in the inferior phrenic arteries. The gastric and esophageal branches should be avoided. Special care and a delicate approach should be observed when injecting embolic materials into the inferior phrenic arteries, especially toward the distal end of the extrahepatic collateral branches. If the embolic materials are injected with force, they can bypass the desired target and cause non-target embolization (Miyayama et al. 2010).

References

Aslaner R et al (2017) Variations in the origin of inferior phrenic arteries and their relationship to celiac axis variations on CT angiography. Korean J Radiol 18(2):336. https://doi.org/10.3348/kjr.2017.18.2.336

Kahn PC (1967) Selective angiography of the inferior phrenic arteries. Radiology 88(1):1–8. https://doi.org/10.1148/88.1.1

Loukas M, Hullett J et al (2005) Clinical anatomy of the inferior phrenic artery. Clin Anat 18(5):357–365. https://doi.org/10.1002/ca.20112

Loukas M et al (2006) Superior phrenic artery: an anatomic study. Surg Radiol Anat 29(1):97–102. https://doi.org/10.1007/s00276-006-0172-2

Mescher AL, Junqueira LCU (2016) Junqueira's basic histology: text and atlas, 14th edn. McGraw-Hill Education, New York

Miyayama S et al (2010) Inferior phrenic arteries: angiographic anatomy, variations, and catheterization techniques for transcatheter arterial chemoembolization. Jpn J Radiol 28(7):502–511. https://doi.org/10.1007/s11604-010-0456-7

36 Pericardiacophrenic Arteries

Lincoln Kavinsky and R. Shane Tubbs

36.1 Introduction

The pericardium is a fluid-filled sac that keeps the heart in its proper anatomical position, maintains stable intracardiac pressures, and shields the heart from the spread of malignancy and infection (Hayase et al. 2020; Rodriguez and Tan 2017). Knowledge of vascular supply to the pericardium is crucial for surgeons and proceduralists alike for open surgical and endovascular procedures such as percutaneous coronary interventions (PCIs), radiofrequency ablations, and embolization procedures.

36.2 Embryology

The phrenic nerve mainly determines the orientation of the pericardiacophrenic vessels (Stauffer 2021; Horton et al. 2010). The phrenic nerve moves superiorly during embryogenesis as the septum transversum moves inferiorly, separating the abdomen from the thorax. The phrenic nerve receives its arterial supply from the pericardiacophrenic artery as it passes through the superior portion of the thorax. The internal thoracic artery is derived from the eighth intersegmental somatic artery (Stauffer 2021). This artery comes from the dorsal arch superior to the seventh intersegmental artery. The pericardiacophrenic and internal thoracic arteries run at 90° angles to the intersegmental arteries and develop from anastomoses that connect the ventral portions of the intersegmental arteries.

36.3 Anatomy

The pericardiacophrenic branch (Fig. 36.1) of the internal thoracic artery supplies the pericardium (Wurtz et al. 2011). The pericardiacophrenic branches are medially the first and most superior branches of the internal thoracic arteries. They branch off the internal thoracic artery opposite the first intercostal artery and then run inferiorly toward the diaphragm (Henriquez-Pino and Prates 1992). The right pericardiacophrenic artery runs alongside the phrenic nerve on the right side of the heart between the fibrous pericardial layer and the pleura. The left pericardiacophrenic artery also passes alongside the phrenic nerve and then runs along the left obtuse margin of the heart. The left and right pericardiacophrenic arteries provide collateral blood supply to the coronary arteries (Stauffer 2021).

L. Kavinsky
Tulane University School of Medicine,
New Orleans, LA, USA

R. S. Tubbs (✉)
Department of Neurosurgery, Tulane University
School of Medicine, New Orleans, LA, USA
e-mail: rtubbs@tulane.edu

S. J. Bordes, Jr. et al. (eds.), *The Clinical Anatomy of the Vascular System*,
https://doi.org/10.1007/978-3-031-78326-5_36

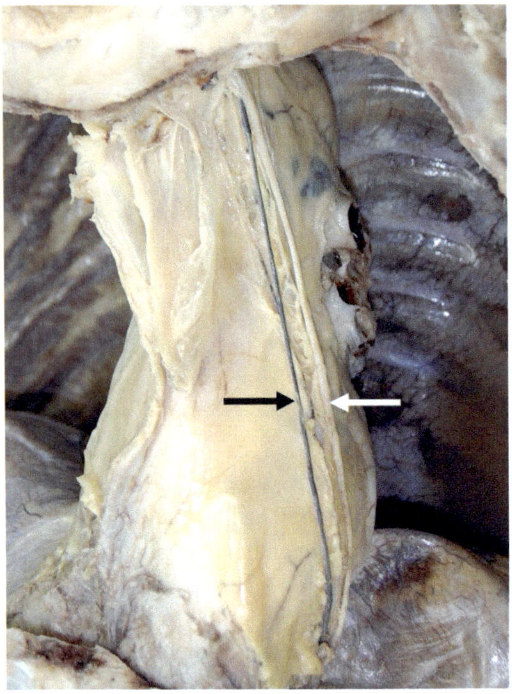

Fig. 36.1 Cadaveric view of the phrenic nerve (white arrow) and adjacent pericardiacophrenic artery (black arrow) traveling along the pericardial sac. Note the diaphragm inferiorly and that the left and right lungs have been removed. (Courtsey Prof. M Loukas)

36.4 Clinical Relevance

The pericardiacophrenic arteries have several clinical applications. First, during PCIs using transradial access, the proceduralist must be mindful of the risk of perforating the pericardiacophrenic artery during advancement of the guidewire into the aorta as the pericardiacophrenic artery and the aorta have similar orientations (Bhasin et al. 2011). Second, during radiofrequency procedures to treat atrial fibrillation, the pericardiacophrenic artery can locate the phrenic nerve since the two run together, thus avoiding injury to the phrenic nerve (Bhasin et al. 2011). In patients with massive hemoptysis, the pericardiacophrenic artery can be the source of bleeding in rare instances. This is typically caused by pulmonary tumors that have eroded anteriorly. Less stable patients may be treated with pericardiacophrenic artery embolization or open mediastinal exploration (Bhasin et al. 2011). It is important to note that the phrenic nerve may lose vascular supply should the pericardiacophrenic artery be ligated or embolized. For example, Chapman et al (2000) reported a patient with cystic fibrosis who developed paralysis of the diaphragm following embolization of the bronchial artery for hemoptysis. The authors theorized that iatrogenic obstruction of the left pericardiacophrenic artery occurred, thus devascularizing the phrenic nerve.

References

Bhasin A et al (2011) Pericardiophrenic artery embolisation for control of massive haemoptysis. Singapore J Med 52(5)

Chapman SA, Holmes MD, Taylor DJ (2000) Unilateral diaphragmatic paralysis following bronchial artery embolization for hemoptysis. Chest 118:269–270

Hayase J, Mori S, Shivkumar K, Bradfield JS (2020) Anatomy of the pericardial space. Card Electrophysiol Clin 12(3):265–270. https://doi.org/10.1016/j.ccep.2020.04.003. Epub 2020 Jun 30. PMID: 32771181

Henriquez-Pino JA, Prates JC (1992) Estudo anatômico da porção proximal da artéria torácica interna [An anatomical study of the proximal portion of the internal thoracic artery]. Arq Bras Cardiol 59(3):177–180. Portuguese. PMID: 1341168

Horton R, Di Biase L, Reddy V, Neuzil P, Mohanty P, Sanchez J, Nguyen T, Mohanty S, Gallinghouse GJ, Bailey SM, Zagrodzky JD, Burkhardt JD, Natale A (2010) Locating the right phrenic nerve by imaging the right pericardiophrenic artery with computerized tomographic angiography: implications for balloon-based procedures. Heart Rhythm 7(7):937–941. https://doi.org/10.1016/j.hrthm.2010.03.027. Epub 2010 Mar 25. PMID: 20348030

Rodriguez ER, Tan CD (2017) Structure and anatomy of the human pericardium. Prog Cardiovasc Dis 59(4):327–340. https://doi.org/10.1016/j.pcad.2016.12.010. Epub 2017 Jan 4. PMID: 28062264

Stauffer CM (2021) Anatomy, thorax, pericardiacophrenic vessels. National Center for Biotechnology Information, U.S. National Library of Medicine. https://www.ncbi.nlm.nih.gov/books/NBK559242/?report=printable

Wurtz A et al (2011) The 'thymopericardial fat flap': a versatile flap in thoracic and cardiovascular surgery. J Thorac Cardiovasc Surg 141(3):841. https://doi.org/10.1016/j.jtcvs.2010.07.005

37 Internal Thoracic Artery

Joshua Linhuber

37.1 Embryology

During the end of the 3rd week of development, vasculogenesis occurs from the posterolaterally located paraxial mesoderm. Vessels form between developing somites and join with the dorsal aorta to form intersegmental arteries. Around the 7th week, the internal thoracic artery is formed when the longitudinal branches of the intersegmental arteries from the dorsal aorta become linked via longitudinal anastomosing channels and lose their connection to the aorta (Schoenwolf et al. 2020).

37.2 Anatomy/Topography

The internal thoracic artery (Fig. 37.1) originates from the first part of the subclavian artery inferiorly and 2 cm superior to the sternal end of the clavicle. It courses anteromedially posterior to the clavicle, interior jugular vein, brachiocephalic vein, and first costal cartilage. The phrenic nerve will obliquely cross anterior to the internal thoracic artery around the entrance of the thorax. As the internal thoracic artery descends inferiorly, it is accompanied by venae comitantes. It travels behind the pectoralis major, the first six costal cartilages, the external intercostal membrane, internal intercostal muscles, and the terminal ends of the upper six intercostal nerves. The artery lies anterior to the pulmonary pleura. It is separated from the pleura by a thick layer of fascia in the first two or three spaces and by transversus thoracis for the

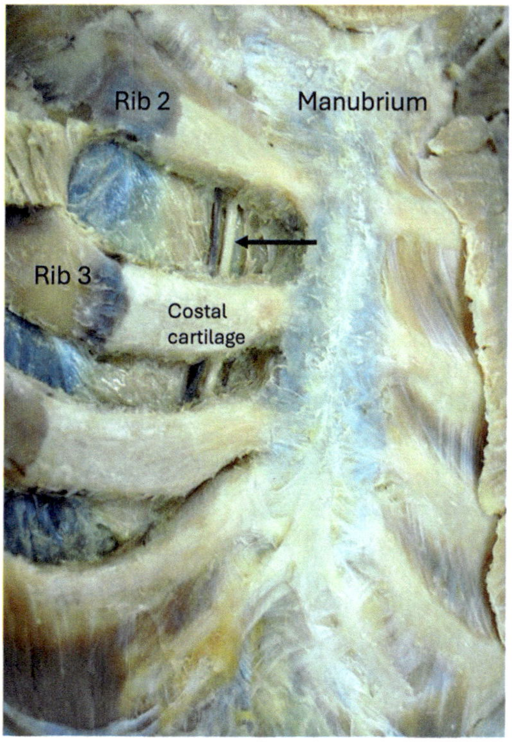

Fig. 37.1 Cadaveric dissection illustrating the right internal thoracic artery (arrow) and adjacent internal thoracic vein. (Courtsey Prof. M. Loukas)

J. Linhuber (✉)
Tulane University School of Medicine,
New Orleans, LA, USA
e-mail: jlinhuber@tulane.edu

S. J. Bordes, Jr. et al. (eds.), *The Clinical Anatomy of the Vascular System*,
https://doi.org/10.1007/978-3-031-78326-5_37

rest of its course. Along this path, the anterior intercostal artery gives off sternal, anterior intercostal, and perforating branches (Standring 2016).

The sternal branches supply the sternal red bone marrow, transversus thoracis, and the posterior sternal periosteum. Sternal branches also contribute to the subpleural mediastinal plexus by anastomosing with posterior intercostal and bronchial arteries (Standring 2016).

The anterior intercostal branches supply the intercostal muscles and have branches that supply the pectoralis major, pectoralis minor, breast, and skin. The branches travel laterally from the internal thoracic artery into the upper six intercostal spaces. They typically originate as a single vessel but immediately bifurcate into superior and inferior vessels. These vessels originate anterior to the internal intercostal muscles but eventually lie in between the intercostal muscles (Standring 2016).

The perforating branches are found in the upper five or six intercostal spaces and supply the pectoralis major, skin, and breast. They travel alongside the cutaneous branches of intercostal nerves to supply the pectoralis major and then travel laterally to supply the skin. Typically, branches 2–4 supply the breast (Standring 2016).

37.3 Imaging

CT angiography (CTA), MR angiography (MRA), conventional angiographic techniques, and ultrasonography can all visualize the internal thoracic arteries (Takehara et al. 2011). 3D CT angiography is the most commonly used method and is used for pre-operative screening for patent internal thoracic arteries to be used in coronary artery bypass grafting (CABG) (Rajiah and Schoenhagen 2013). The internal thoracic arteries are often hypertrophied in cases of aortic coarctation due to their use for collateral circulation, and this hypertrophy can be accurately visualized using 3D methods (Rowe et al. 2018).

37.4 Histology

The internal thoracic artery is a transitional artery. The tunica media has an internal layer composed of circularly arranged vascular smooth muscle cells and a few solitary elastic fibers and an external layer composed of spirally arranged elastic lamellae and smooth muscle cells. The number of elastic fibers decreased gradually as the internal thoracic artery moved further away from its origin at the subclavian artery. Most parameters in the table below appear to generally decrease with age except for the number of IEL fenestrations per 100 μm and the size of elastic lamellae fenestrations, which appear to increase with age (Borovic et al. 2010).

Factors such as increasing age, female sex, and history of myocardial infarction have been associated with a higher intimal width and intima/

Table 37.1 Age-dependent changes of the morphometric parameter of the IEL and Elastic Lamellae (Borovic et al. 2010)

	IEL thickness (μm)	Number of IEL fenestrations per 100 μm	Size of IEL fenestrations (μm)	Thickness of elastic lamellae (μm)	Number of elastic lamellae fenestrations per 100 μm	Size of elastic lamellae fenestrations
Group 1 (20–40 y/o)	2.01 ± 0.67	2.37 ± 1.88	2.05 ± 1.46	1.61 ± 0.46	2.31 ± 0.97	2.88 ± 1.42
Group 2 (41–60 y/o)	2.22 ± 0.83	3.46 ± 2.48	2.41 ± 1.55	1.51 ± 0.42	2.07 ± 1.56	2.94 ± 2.07
Group 3 (≥61 y/o)	1.60 ± 0.54	3.03 ± 2.22	2.40 ± 1.38	1.17 ± 0.29	2.00 ± 1.23	4.59 ± 1.44
Total	2.05 ± 0.75	2.79 ± 2.18	2.24 ± 1.50	1.54 ± 0.45	2.20 ± 1.25	3.00 ± 1.76

Table 37.2 Histomorphologic parameters of total population according to sex (Fonseca et al. 2019)

	Total	Females	Males
Intimal width (μm)	47.95 ± 4.62	54.33 ± 12.11	46.50 ± 5.00
Medial width (μm)	184.54 ± 9.05	163.70 ± 19.25	189.27 ± 10.17
Intima/media ratio	0.27 ± 0.02	0.33 ± 0.07	0.25 ± 0.02

media ratio. An intima/media ratio >0.25 indicates intimal hyperplasia (Fonseca et al. 2019; Kaufer et al. 1997) (Tables 37.1 and 37.2).

37.5 Variations

The internal thoracic artery typically originates from the subclavian artery. However, it may arise from the thyrocervical trunk or from a common trunk that also gives off the inferior thyroid and suprascapular arteries (8%), the suprascapular artery (4%), the transverse cervical artery (1%), the inferior thyroid artery (1%), or the suprascapular and transverse cervical arteries (1%) (Lippert and Pabst 1985).

The internal thoracic artery typically bifurcates into the musculophrenic and superior epigastric arteries at the 6th costal cartilage level but has been observed to trifurcate in multiple cadaveric studies. In one study, a trifurcation was observed in 7% of cadavers (Henriquez-Pino et al. 1997). In another study, the trifurcation, the xiphoid branch, was present in 61.3% of cadavers studied. Of these, 61.3% and 30.7% of cases were present on the right, 21% on the left, and 9.7% were present bilaterally. The xiphoid branch originated from the medial border of the internal thoracic artery at the level of the sixth costal cartilage. It traveled towards the angle between the xiphoid process and the 7th costal cartilage. It then gave off 2–3 small branches to the lower sternum. The vessel continued deep to the angle and terminated in fine branches on the anterior surface of the xiphoid process, with the terminal branches reaching towards the inferior aspect of the xiphoid cartilage (Lachman and Satyapal 1999).

The branches of the internal thoracic artery typically come off as solitary branches; however, the branches may come off from common trunks. These common trunks can arise as a combination of sternal/perforating, sternal/intercostal, perforating/intercostal, and sternal/perforating/intercostal. In one study, the left and right thoracic arteries gave off solitary branches 83.53% and 81.66% of the time, respectively. Additionally, of the total branches between the left and right internal thoracic arteries, the left gave off 53% of the total branches. In comparison, the right gave off 47% of the total branches. The left side gave off significantly more sternal (20.19%) and intercostal (19.14%) branches when compared to the right-sided sternal (16.72%) and intercostal (16.40%) branches (Pietrasik et al. 1999). A lateral costal branch was also present in 15% of cases in one study (Henriquez-Pino et al. 1997).

Extracardiac anastomoses may involve either the right or left internal thoracic arteries. The right internal thoracic artery may anastomose with the right coronary artery. The left coronary artery may anastomose with the anterior interventricular artery (Loukas et al. 2011).

A lateral internal thoracic vessel may arise from the internal thoracic artery at the entrance of the thorax or from the costocervical trunk, thyrocervical trunk, or ascending cervical artery. It typically runs about midway between the spine and the sternum, spans up to six intercostal spaces, and anastomoses with the intercostal arteries of those spaces (Bergman et al. 2000). The pericardiacophrenic artery commonly arises from the internal thoracic artery; however, it is not always present as a branch of the internal thoracic artery (Henriquez-Pino et al. 1997).

A thymic branch originating from the internal thoracic artery is common, and the thyroidea ima artery may arise from the internal thoracic artery (Ziolkowski et al. 1994).

37.6 Pathology

The internal thoracic artery is thought to be more resistant to atherosclerosis than other arteries; however, it is still susceptible to atherosclerotic changes. The proximal, more elastic part of the internal thoracic artery is more susceptible to atherosclerosis than its more muscular distal end (Sons et al. 1993).

In addition to atherosclerosis, the internal thoracic artery will typically be hypertrophied in patients with aortic coarctation and may even show signs of calcification (Chen et al. 1995; De Salazar et al. 1996).

There have been a few cases of idiopathic internal thoracic artery aneurysms (Almerey et al. 2017; Heyn et al. 2014; Okura et al. 2012). Aneurysms and pseudoaneurysms have been associated with conditions such as respiratory actinomycosis, Kawasaki disease, neurofibromatosis 1, Marfan syndrome, and Ehlers-Danlos syndrome (Rose et al. 2011; Wani et al. 2010; Ishiwata et al. 1990; Kim et al. 2005; Phan et al. 1998).

Congenital and non-congenital systemic artery-to-pulmonary artery (SA-PA) malformations are infrequent conditions; however, the internal thoracic artery has been seen to be involved with these malformations. Multiple cases have reported connections between the internal thoracic artery and the pulmonary artery and its branches (Cheng et al. 2021; Jamshidi et al. 2009). Another case reported the right internal thoracic artery anastomosing with the right pulmonary vein (Kanamori et al. 2012).

37.7 Clinical/Surgical Significance

As an artery well-protected from atherosclerosis (Kaufer et al. 1997; Mannacio et al. 2012; Otsuka et al. 2013), the internal thoracic artery is the leading vascular conduit for CABG. Bilateral internal thoracic artery grafting is the gold standard of grafting options (Kelly et al. 2012); however, it should not be blindly accepted as the only option for CABG. The condition of the internal thoracic arteries can be affected by many conditions, such as coarctation of the aorta, in which they may hypertrophy and calcify. As a result, another conduit, such as the saphenous vein, may be preferred to the internal thoracic artery (Chen et al. 1995; De Salazar et al. 1996). It is also important to note that it is typically the distal part of the internal thoracic artery used in CABG. This part is more resistant to atherosclerotic changes than the proximal portion, highlighting the importance of screening for conduit patency before any procedure.

The internal thoracic artery is commonly used as a recipient vessel for breast reconstructive surgery with free flaps (Ninkovic et al. 1995). For patients who have undergone CABG in the past and may later undergo breast reconstructive surgery, identifying the vessel used as a conduit for the CABG procedure is important before planning the procedure.

The internal thoracic artery can be injured iatrogenically during pericardiocentesis, central venous catheterization, and sternotomies (Jones et al. 2021). Pseudoaneurysms have been reported post-sternotomy (Albors et al. 2010). The artery is also commonly injured due to blunt trauma to the chest (Nomori et al. 2003).

37.8 Complications

Bleeding from the internal thoracic artery can cause a hemothorax, which can be life-threatening if not addressed quickly. Bleeding from the internal thoracic artery is typically caused by blunt trauma and is accompanied by an anterior mediastinal hematoma. Pseudoaneurysm and arteriovenous fistula have also been associated with blunt trauma to the chest (Chen et al. 2014).

References

Albors J, Bahamonde JA, Guijarro J (2010) Poststernotomy pseudoaneurysm of the internal mammary artery: endovascular treatment. J Card Surg 25(2):180–181. https://doi.org/10.1111/j.1540-8191.2009.00987.x

Almerey T, Paz-Fumagalli R, Farres H, Oldenburg WA, Hakaim AG (2017) Idiopathic internal mammary artery aneurysm in the setting of aberrant right subclavian artery. J Vasc Surg Cases Innov Tech 3(4):251–253. https://doi.org/10.1016/j.jvscit.2017.10.006

Bergman RA, Afifi AK, Miyauchi R (2000) Virtual hospital. Illustrated encyclopedia of human anatomic variation. University of Iowa. http://www.anatomyatlases.org/AnatomicVariants/AnatomyHP.shtml

Borović ML, Borović S, Perić M, Vuković P, Marinković J, Todorović V, Radak D, Lacković V (2010) The internal thoracic artery as a transitional type of artery: a morphological and morphometric study. Histol Histopathol 25(5):561–576. https://doi.org/10.14670/HH-25.561

Chen RH, Reul GJ, Cooley DA (1995) Severe internal mammary artery atherosclerosis after correction of coarctation of the aorta. Ann Thorac Surg 59(5):1228–1230. https://doi.org/10.1016/0003-4975(94)00967-c

Chen JM, Lv J, Ma K, Yan J (2014) Assessment of internal mammary artery injury after blunt chest trauma: a literature review. J Zhejiang Univ Sci B 15(10):864–869. https://doi.org/10.1631/jzus.B1400098

Cheng X, Zhang X, Gu F et al (2021) Multiple systemic arteries to pulmonary artery malformations: a case description. Quant Imaging Med Surg 11(11):4671–4675. https://doi.org/10.21037/qims-21-109

De Salazar AO, Gonzalez JA, Zuazo J (1996) Internal mammary artery atherosclerosis fifteen years after correction of aortic coarctation. Ann Thorac Surg 62(2):629

Fonseca DA, Antunes PE, Antunes MJ, Cotrim MD (2019) Histomorphometric analysis of the human internal thoracic artery and relationship with cardiovascular risk factors. PLoS One 14(1):e0211421. https://doi.org/10.1371/journal.pone.0211421

Henriquez-Pino JA, Gomes WJ, Prates JC, Buffolo E (1997) Surgical anatomy of the internal thoracic artery. Ann Thorac Surg 64(4):1041–1045. https://doi.org/10.1016/s0003-4975(97)00720-0

Heyn J, Zimmermann H, Klose A, Luchting B, Hinske C, Sadeghi-Azandaryani M (2014) Idiopathic internal mammary artery aneurysm. J Surg Case Rep 2014(12). https://doi.org/10.1093/jscr/rju125

Ishiwata S, Nishiyama S, Nakanishi S et al (1990) Coronary artery disease and internal mammary artery aneurysms in a young woman: possible sequelae of Kawasaki disease. Am Heart J 120(1):213–217. https://doi.org/10.1016/0002-8703(90)90184-y

Jamshidi P, Ghaffari S, Mahmoodi K (2009) Transpulmonary closing of left internal mammary artery to pulmonary artery fistula with polytetrafluoroethylene covered stent: a case report and review of literature. Cardiol J 16(5):469–472

Jones KA, Sadri S, Ahmad N, Weintraub JR, Reis SP (2021) Thoracic trauma, nonaortic injuries. Semin Interv Radiol 38(1):75–83. https://doi.org/10.1055/s-0041-1726005

Kanamori N, Matsumoto Y, Oribe Y, Omi W (2012) Multiple anomalous systemic-pulmonary connections and coronary-pulmonary artery fistula associated with large aneurysms in an elderly woman. J Card Surg 27(5):565–567. https://doi.org/10.1111/j.1540-8191.2012.01470.x

Kaufer E, Factor SM, Frame R, Brodman RF (1997) Pathology of the radial and internal thoracic arteries used as coronary artery bypass grafts. Ann Thorac Surg 63(4):1118–1122. https://doi.org/10.1016/s0003-4975(96)01393-8

Kelly R, Buth KJ, Legare JF (2012) Bilateral internal thoracic artery grafting is superior to other forms of multiple arterial grafting in providing survival benefit after coronary bypass surgery. J Thorac Cardiovasc Surg 144(6):1408–1415. https://doi.org/10.1016/j.jtcvs.2012.01.030

Kim SJ, Kim CW, Kim S et al (2005) Endovascular treatment of a ruptured internal thoracic artery pseudoaneurysm presenting as a massive hemothorax in a patient with type I neurofibromatosis. Cardiovasc Intervent Radiol 28(6):818–821. https://doi.org/10.1007/s00270-004-0067-8

Lachman N, Satyapal KS (1999) Origin and incidence of xiphoid branch of the internal thoracic artery. Surg Radiol Anat 21(5):351–354. https://doi.org/10.1007/BF01631338

Lippert H, Pabst R (1985) Arterial variations in man : classification and frequency. J.F. Bergmann Verlag, 121 p

Loukas M, Hanna M, Chen J, Tubbs RS, Anderson RH (2011) Extracardiac coronary arterial anastomoses. Clin Anat 24(2):137–142. https://doi.org/10.1002/ca.21088

Mannacio V, Di Tommaso L, Antignano A et al (2012) Endothelial nitric oxide synthase expression in postmenopausal women: a sex-specific risk factor in coronary surgery. Ann Thorac Surg 94(6):1934–1939. https://doi.org/10.1016/j.athoracsur.2012.06.040

Ninkovic M, Anderl H, Hefel L, Schwabegger A, Wechselberger G (1995) Internal mammary vessels: a reliable recipient system for free flaps in breast reconstruction. Br J Plast Surg 48(8):533–539. https://doi.org/10.1016/0007-1226(95)90040-3

Nomori H, Ootsuka T, Horio H, Naruke T, Suemasu K (2003) Bilateral internal thoracic artery injury induced by blunt trauma. Jpn J Thorac Cardiovasc Surg 51(5):214–216. https://doi.org/10.1007/s11748-003-0036-1

Okura Y, Kawasaki T, Hiura T, Seki H, Saito H (2012) Aneurysm of the internal mammary artery with cystic medial degeneration. Intern Med 51(17):2355–2359. https://doi.org/10.2169/internalmedicine.51.8139

Otsuka F, Yahagi K, Sakakura K, Virmani R (2013) Why is the mammary artery so special and what protects it from atherosclerosis? Ann Cardiothorac Surg 2(4):519–526. https://doi.org/10.3978/j.issn.2225-319X.2013.07.06

Phan TG, Sakulsaengprapha A, Wilson M, Wing R (1998) Ruptured internal mammary artery aneurysm presenting as massive spontaneous haemothorax in a patient with Ehlers-Danlos syndrome. Aust NZ J Med 28(2):210–211. https://doi.org/10.1111/j.1445-5994.1998.tb02972.x

Pietrasik K, Bakon L, Zdunek P, Wojda-Gradowska U, Dobosz P, Kolcsnik A (1999) Clinical anatomy of internal thoracic artery branches. Clin Anat 12(5):307–314. https://doi.org/10.1002/(SICI)1098-2353(1999)12:5<307::AID-CA1>3.0.CO;2-6

Rajiah P, Schoenhagen P (2013) The role of computed tomography in pre-procedural planning of cardiovascular surgery and intervention. Insights Imaging 4(5):671–689. https://doi.org/10.1007/s13244-013-0270-8

Rose JF, Lucas LC, Bui TD, Mills JL Sr (2011) Endovascular treatment of ruptured axillary and large internal mammary artery aneurysms in a patient with Marfan syndrome. J Vasc Surg 53(2):478–482. https://doi.org/10.1016/j.jvs.2010.08.076

Rowe SP, Johnson PT, Fishman EK (2018) Initial experience with cinematic rendering for chest cardiovascular imaging. Br J Radiol 91(1082):20170558. https://doi.org/10.1259/bjr.20170558

Schoenwolf G, Bleyl S, Brauer P, Francis-West P (2020) Larsen's human embryology, 6th edn. Elsevier, Inc

Sons HJ, Godehardt E, Kunert J, Losse B, Bircks W (1993) Internal thoracic artery: prevalence of atherosclerotic changes. J Thorac Cardiovasc Surg 106(6):1192–1195

Standring S (2016) Gray's anatomy: the anatomical basis of clinical practice, 41st edn. Elsevier Limited, xviii, 1562 p

Takehara Y, Yamashita S, Sakahara H, Masui T, Isoda H (2011) Magnetic resonance angiography of the aorta. Ann Vasc Dis 4(4):271–285. https://doi.org/10.3400/avd.di.11.00822

Wani NA, Rawa IA, Pala NA, Kosar T (2010) Pseudoaneurysm of internal mammary artery caused by pulmonary actinomycosis. Br J Radiol 83(995):e235–e238. https://doi.org/10.1259/bjr/69723351

Ziolkowski M, Bieganska-Dembowska D, Kurlej W (1994) Variations in the number and in origin of the thyroid arteries. Folia Morphol (Warsz) 53(2):105–110

38 Musculophrenic Artery

Joshua Linhuber, Sreeganesh Manoharan, and Gowri Gowda

38.1 Embryology

The musculophrenic artery is a terminal branch of the internal thoracic artery and has similar embryologic origins. During the end of the 3rd week of development, vasculogenesis occurs from the posterolaterally located paraxial mesoderm. Vessels form between developing somites and join with the dorsal aorta to form intersegmental arteries. Around the 7th week, the internal thoracic artery is formed when the longitudinal branches of the intersegmental arteries from the dorsal aorta become linked via longitudinal anastomosing channels and lose their connection to the aorta. During this process of anastomosis, the musculophrenic artery is also formed (Standring 2016).

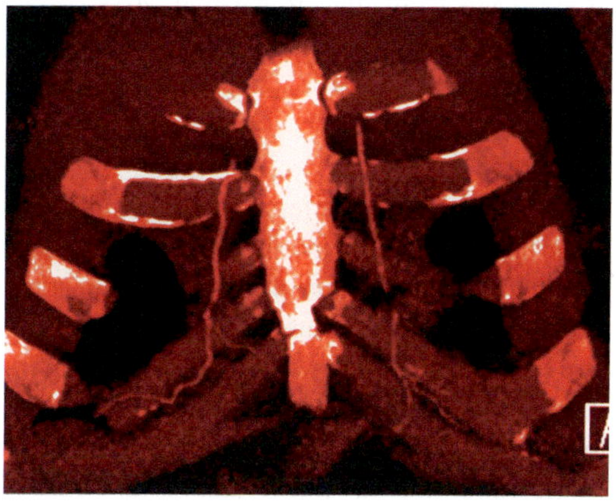

Fig. 38.1 CTA noting the internal thoracic arteries traveling lateral to the sternum and dividing inferiorly into the more medial superior epigastric and more lateral musculophrenic arteries

38.2 Anatomy/Topography

The musculophrenic artery (Fig. 38.1) originates from the internal thoracic artery, typically at the level of the sixth costal cartilage. It courses inferolaterally just posterior to the seventh and ninth costal cartilages while tracking along the anterior to the diaphragm. As it passes the 7th, 8th, and 9th intercostal spaces, the musculophrenic artery gives off anterior intercostal branches like the upper anterior intercostal arteries. The artery terminates at the costal angle. It has anastomoses with the inferior phrenic artery, the lower two posterior intercostal arteries, and the ascending branches of the deep circumflex iliac arteries. The musculophrenic artery supplies the 7th–9th intercostal spaces, diaphragm, abdominal wall, and lower part of the pericardium (Standring 2016). The 10th and 11th intercostal spaces do not receive an anterior blood supply.

J. Linhuber · S. Manoharan (✉) · G. Gowda
Tulane University School of Medicine,
New Orleans, LA, USA
e-mail: smanoharan@tulane.edu

S. J. Bordes, Jr. et al. (eds.), *The Clinical Anatomy of the Vascular System*,
https://doi.org/10.1007/978-3-031-78326-5_38

38.3 Imaging

The musculophrenic artery can be seen on CT angiography (CTA), MR angiography (MRA), conventional angiographic techniques, and ultrasonography (Takehara et al. 2011; Murray et al. 2012). CTA is the preferred modality for imaging the vessels of the thoracic wall. 3D methods for CTA, such as maximum intensity projection (MIP), volume rendering (VR), and cinematic rendering (CR), are valuable tools for accurately assessing the condition of these arteries. If CTA is not an option, 3D MIP reconstruction can be used for MRA imaging (Rowe et al. 2018).

38.4 Histology

The musculophrenic artery is a predominately muscular artery. In some musculophrenic arteries, the first 1–2 cm of the proximal portion could be described as elastomuscular or muscular with rare elastic lamellae; however, most of the artery is muscular. In portions with elastic lamellae, there is a high level of discontinuity in the circumferential internal elastic lamina. The rate of these discontinuities increases distally in the artery. The mean luminal diameter of the muscular segments of the musculophrenic artery is 0.9 mm (van Son et al. 1993).

38.5 Variations

The musculophrenic artery is typically thought to give off anterior intercostal arteries 7–9, although it may give off intercostal arteries higher than that, such as the 6th anterior intercostal artery (Sekino et al. 2005).

The variable anatomy of the musculophrenic artery has not been heavily studied; however, a 2019 study of 100 cadavers noted that the incidence of anterior intercostal arteries arising from the musculophrenic artery was 73% (Mitra and Agnihotri 2020).

38.6 Pathology

Congenital and non-congenital systemic artery-to-pulmonary artery (SA-PA) malformations are rare. On occasion, the musculophrenic artery is involved in these malformations. One case reported a connection between the musculophrenic artery and branches of the pulmonary artery (Cheng et al. 2021).

38.7 Clinical/Surgical Significance

The location and condition of the musculophrenic artery should be considered when performing invasive procedures such as pericardiocentesis and sternotomies. Damage to the musculophrenic artery or its intercostal branches could cause pseudoaneurysms and hemothorax. Of the rare injuries to the musculophrenic artery reported, two have involved pseudoaneurysm after a puncture injury, and one case caused by blunt trauma resulted in extravasation from a hematoma that eventually resulted in massive hemothorax and shock (Sekino et al. 2005; Tourmousoglou et al. 2015; Sawamoto et al. 2017).

References

Cheng X, Zhang X, Gu F et al (2021) Multiple systemic arteries to pulmonary artery malformations: a case description. Quant Imaging Med Surg 11(11):4671–4675. https://doi.org/10.21037/qims-21-109

Mitra A, Agnihotri G (2020) A morphometric study of the internal thoracic artery and its branches. Mymensingh Med J 29(3):701–708

Murray AC, Rozen WM, Alonso-Burgos A, Ashton MW, Garcia-Tutor E, Whitaker IS (2012) The anatomy and variations of the internal thoracic (internal mammary) artery and implications in autologous breast reconstruction: clinical anatomical study and literature review. Surg Radiol Anat 34(2):159–165. https://doi.org/10.1007/s00276-011-0886-7

Rowe SP, Johnson PT, Fishman EK (2018) Initial experience with cinematic rendering for chest cardiovascular imaging. Br J Radiol 91(1082):20170558. https://doi.org/10.1259/bjr.20170558

Sawamoto T, Aoki H, Otsuka H et al (2017) Successful treatment of blunt musculophrenic artery injury by transcatheter arterial embolization: a case report. Tokai J Exp Clin Med 42(3):126–129

Sekino S, Takagi H, Kubota H, Kato T, Matsuno Y, Umemoto T (2005) Intercostal artery pseudoaneurysm due to stab wound. J Vasc Surg 42(2):352–356. https://doi.org/10.1016/j.jvs.2005.03.058

Standring S (2016) Gray's anatomy: the anatomical basis of clinical practice, 41st edn. Elsevier Limited, xviii, 1562 p

Takehara Y, Yamashita S, Sakahara H, Masui T, Isoda H (2011) Magnetic resonance angiography of the aorta. Ann Vasc Dis 4(4):271–285. https://doi.org/10.3400/avd.di.11.00822

Tourmousoglou C, Zambakis P, Koletsis E, Prokakis C, Charoulis N, Dougenis D (2015) Coil embolization of traumatic pseudoaneurysm of right internal thoracic artery. Asian Cardiovasc Thorac Ann 23(8):982–984. https://doi.org/10.1177/0218492314530978

van Son JA, Smedts F, Korving J, Guyt A, de Kok LB (1993) Intercostal artery: histomorphometric study to assess its suitability as a coronary bypass graft. Ann Thorac Surg 56(5):1078–1081. https://doi.org/10.1016/0003-4975(95)90018-7

39 Intercostal Arteries

Joshua Linhuber

39.1 Embryology

During the end of the 3rd week of development, vasculogenesis occurs in the paraxial mesoderm. Vessels form between developing somites and join with the dorsal aorta to form intersegmental arteries. The ventral branches of the intersegmental vessels in the thoracic region develop into the intercostal arteries. The superior intercostal artery is derived from the 7th and 8th intersegmental arteries (Schoenwolf et al. 2020; Gailloud 2014).

39.2 Anatomy/Topography

The superior intercostal artery originates from the costocervical trunk and courses posteroinferiorly towards the 1st intercostal space, anterior to the neck of the 1st and 2nd ribs and posterior to the pulmonary pleura. It remains lateral to the stellate ganglion and medial to the 1st thoracic spinal nerve. The superior intercostal artery anastomoses with the 3rd posterior intercostal artery to provide blood to the 2nd intercostal space.

The 1st and 2nd posterior intercostal arteries originate from the superior intercostal artery anterior to the neck of the 1st and 2nd ribs, respectively. They travel posterolaterally along the thoracic wall near the inferior border of the corresponding numbered rib, sequentially passing anterior to the intercostal membrane and the internal intercostal muscle while also being situated posterior to the parietal pulmonary pleura. As they travel laterally, the 1st and 2nd intercostal arteries become located within the costal groove of the rib immediately superior while also running between the internal and innermost intercostal muscles. Around the area of the costal angle, collateral branches form that travel inferiorly towards the upper border of the rib immediately inferior and continue until they anastomose with a branch of the 1st or 2nd anterior intercostal branch of the internal thoracic artery. The 1st and 2nd posterior intercostal arteries anastomose with the 1st and 2nd anterior intercostal branches of the internal thoracic artery, respectively (Standring 2016).

Posterior intercostal arteries 3–11 originate posteriorly from the descending thoracic aorta (Fig. 39.1). Due to the descending thoracic aorta's typical location to the left of the midline, intercostal arteries destined for the right side must travel contralaterally from the left side, meaning the length of these arteries is typically longer than their left-sided counterparts. Right-sided posterior intercostal arteries must course across the anterior of the vertebral column and posterior to the esophagus, thoracic duct, and

J. Linhuber (✉)
Tulane University School of Medicine,
New Orleans, LA, USA
e-mail: jlinhuber@tulane.edu

S. J. Bordes, Jr. et al. (eds.), *The Clinical Anatomy of the Vascular System*,
https://doi.org/10.1007/978-3-031-78326-5_39

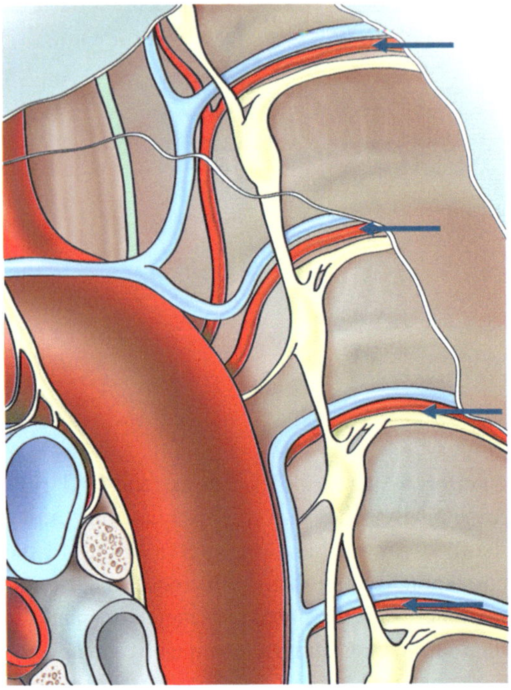

Fig. 39.1 Schematic drawing of posterior intercostal arteries (arrows)

azygous vein. The lower arteries on the left side are also anteriorly crossed by the hemiazygos veins and accessory hemiazygos veins. Both the left and right-sided arteries pass posterior to the sympathetic trunk (Standring 2016).

Posterior intercostal arteries 3–11 enter the inferior portion of the intercostal space and travel posterolaterally and superiorly until they become nestled in the costal groove (Choi et al. 2010). About 99% of intercostal arteries are shielded by the rib immediately superior at 6 cm from the corresponding spinous process (Helm et al. 2013). An intercostal vein is situated superior to lower posterior intercostal arteries, and an intercostal nerve is located inferiorly. Posterior intercostal arteries 3–6 anastomose with an anterior intercostal branch of the internal thoracic artery, while posterior intercostal arteries 7–9 anastomose with anterior intercostal branches of the musculophrenic artery (Standring 2016). Posterior intercostal arteries 10 and 11 typically extend anteromedially to the abdominal wall, where they will anastomose with the subcostal, superior epigastric, and lumbar arteries.

The dorsal intercostal arteries arise from the posterior intercostal arteries at the neck of the ribs and run dorsally, just medial to the corresponding vertebral body. It then gives off a spinal branch as it passes the intervertebral foramen and supplies the meninges, spinal cord, and vertebrae. It will also anastomose with the spinal arteries above, below, and contralaterally. Soon after the artery gives off its spinal branch, the dorsal intercostal artery will give off a medial and lateral musculocutaneous branch. The medial musculocutaneous branch will travel alongside the medial branch of the posterior rami of a thoracic spinal nerve past the transverse process to supply the posteromedial thorax. The lateral musculocutaneous branch will travel laterally alongside the lateral branch of the posterior rami of a thoracic spinal nerve just anterior to the transverse process and posterior to the inferior rib. These branches will supply the posterolateral thorax (Standring 2016).

39.3 Histology

The intercostal arteries are considered medium-sized arteries, but they have characteristics of an elastic artery in their proximal portion. Three common patterns can be used to describe an intercostal artery as having (1) a proximal elastic segment, a subsequent elastomuscular portion, and a final muscular portion; (2) an initial elastomuscular portion followed by a muscular portion; and (3) a completely muscular artery (van Son et al. 1993).

The mean luminal diameter of the 5th intercostal artery is 1.4 ± 0.3 mm at the origin and 0.9 ± 0.2 mm at 30 cm. The mean luminal diameter of all other intercostal arteries is not significantly different, measuring 1.3 ± 0.4 mm at the origin and 1.1 ± 0.3 mm at 30 cm (van Son et al. 1993) (Table 39.1).

Table 39.1 Descriptive statistics of proximal posterior intercostal artery (Mean ± SD [μm]) (Reddy et al. 2011)

	19–40 y/o	41–60 y/o	≥61 y/o	Total
Tunica intima (Ti) thickness	6.77 ± 0.87	8.21 ± 0.97	15.5 ± 8.06	10.03 ± 5.87
Tunica media (Tm) thickness	71.55 ± 10.11	75.8 ± 4.68	59.97 ± 5.94	69.2 ± 9.78
Ti/Tm ratio	0.09 ± 0.01	0.11 ± 0.02	0.27 ± 0.16	0.16 ± 0.10
Number of elastic lamina (proximal portion)	7.88 ± 0.69	6.62 ± 0.51	4.56 ± 0.82	6.42 ± 1.55

39.4 Variations

The origin of the superior intercostal artery can be quite varied. Although it most commonly arises from the costocervical trunk, the artery can also arise from the thyrocervical trunk, vertebral artery, costocervical trunk, inferior thyroid, second intercostal, and axillary artery. In rare cases, the left superior intercostal artery originated from the aortic arch or the lateral internal thoracic artery (Bergman et al. 2000).

Typically, the superior intercostal artery supplies the 1st and 2nd intercostal spaces; however, cases have been reported of the superior intercostal artery supplying only the 1st intercostal space (Bergman et al. 2000).

The 2nd posterior intercostal artery may arise from the thoracic aorta rather than the superior intercostal artery. The 1st posterior intercostal artery has also been reported as arising from the thoracic aorta; however, this is much more uncommon. The 3rd and 4th posterior intercostal arteries typically arise from the thoracic aorta; however, these can also arise from the superior intercostal artery (Bergman et al. 2000). The upper right posterior intercostal arteries may give off esophageal branches (Bergman et al. 2016). In rare cases, pairs of intercostal arteries may bifurcate at a shared common stem on the posterior median line of the aorta (Bergman et al. 2000).

39.5 Pathology

Although uncommon, the primary pathology associated with the intercostal arteries is aneurysms, typically in conjunction with diseases such as neurofibromatosis type 1, aortic coarctation, and Kawasaki disease (Batlivala and Rome 2016; Misao et al. 2012; Masuda et al. 1986). Rarer pathologies include pseudoaneurysms and intercostal artery aneurysmosis (Gutierrez Romero et al. 2014; Carr et al. 2013).

39.6 Clinical and Surgical Significances

The condition and location of the intercostal arteries should be considered when performing procedures that include the thoracic aorta, such as Thoracic Endovascular Repair (TEVAR), and pleural interventions, such as thoracocentesis, to avoid iatrogenic injury. This should especially be considered with patients who have conditions such as neurofibromatosis type 1, aortic coarctation, and Kawasaki disease.

Intercostal arteries contribute to spinal blood flow via their spinal branches and by the collateral flow of blood to the Artery of Adamkiewicz. Therefore, damage to these arteries can lead to decreased spinal blood flow and subsequent paraparesis. These injuries can be reduced by using image fusion techniques, such as combining preoperative CTA with intraoperative CT to identify the intercostal arteries more accurately intraoperatively (Koutouzi et al. 2017).

When performing thoracocenteses, the location of the posterior intercostal artery needs to be considered, as its tortuosity and position within the intercostal space can vary slightly. Due to these variations, ultrasound guidance is recommended to reduce the incidence of iatrogenic injuries such as hemothorax (Rodriguez Lima et al. 2020).

39.7 Complications

Injuries to intercostal arteries and intercostal artery bleeding can lead to significant complications for patients. Obliteration of intercostal arteries during invasive thoracic procedures can lead to spinal ischemia and subsequent paraparesis due to loss of spinal blood flow from spinal branches of the intercostal artery and collateral blood flow from the intercostal artery to the artery of Adamkiewicz. Intercostal arterial bleeding can lead to hemothorax, a potentially life-threatening condition if not addressed quickly.

References

Batlivala SP, Rome JJ (2016) Concomitant percutaneous treatment of aortic coarctation and associated intercostal aneurysms: pre-procedural recognition is key. Cardiol Young 26(2):390–395. https://doi.org/10.1017/S104795111500116X

Bergman RA, Afifi AK, Miyauchi R (2000) Virtual hospital. Illustrated encyclopedia of human anatomic variation. University of Iowa. http://www.anatomyatlases.org/AnatomicVariants/AnatomyHP.shtml

Bergman RA, Tubbs RS, Shoja MM, Loukas M (2016) Bergman's comprehensive encyclopedia of human anatomic variation. Wiley, xx, 1432 p

Carr JA, Vadlamudi V, Azher QS (2013) Intercostal artery aneurysmosis. Ann Thorac Surg 96(5):1870–1873. https://doi.org/10.1016/j.athoracsur.2013.02.055

Choi S, Trieu J, Ridley L (2010) Radiological review of intercostal artery: anatomical considerations when performing procedures via intercostal space. J Med Imaging Radiat Oncol 54(4):302–306. https://doi.org/10.1111/j.1754-9485.2010.02175.x

Gailloud P (2014) The supreme intercostal artery includes the last cervical intersegmental artery (C7) – angiographic validation of the intersegmental nomenclature proposed by Dorcas Padget in 1954. Anat Rec (Hoboken) 297(5):810–818. https://doi.org/10.1002/ar.22893

Gutierrez Romero DF, Barrufet M, Lopez-Rueda A, Burrel M (2014) Ruptured intercostal artery pseudoaneurysm in a patient with blunt thoracic trauma: diagnosis and management. BMJ Case Rep 2014. https://doi.org/10.1136/bcr-2013-202019

Helm EJ, Rahman NM, Talakoub O, Fox DL, Gleeson FV (2013) Course and variation of the intercostal artery by CT scan. Chest 143(3):634–639. https://doi.org/10.1378/chest.12-1285

Koutouzi G, Sandstrom C, Skoog P, Roos H, Falkenberg M (2017) 3D image fusion to localise intercostal arteries during TEVAR. EJVES Short Rep 35:7–10. https://doi.org/10.1016/j.ejvssr.2017.03.001

Masuda H, Shozawa T, Naoe S, Tanaka N (1986) The intercostal artery in Kawasaki disease. A pathologic study of 17 autopsy cases. Arch Pathol Lab Med 110(12):1136–1142

Misao T, Yoshikawa T, Aoe M, Ueda Y, Yodoya M, Sakurai J (2012) Recurrent rupture of intercostal artery aneurysms in neurofibromatosis type 1. Gen Thorac Cardiovasc Surg 60(3):179–182. https://doi.org/10.1007/s11748-011-0806-0

Reddy S, Kumar P, Prasad K (2011) Histomorphometric and sympathetic innervation of the human posterior intercostal artery and its clinical importance. Folia Morphol (Warsz) 70(3):161–167

Rodriguez Lima DR, Yepes AF, Birchenall Jimenez CI, Mercado Diaz MA, Pinilla Rojas DI (2020) Real-time ultrasound-guided thoracentesis in the intensive care unit: prevalence of mechanical complications. Ultrasound J 12(1):25. https://doi.org/10.1186/s13089-020-00172-9

Schoenwolf G, Bleyl S, Brauer P, Francis-West P (2020) Larsen's human embryology, 6th edn. Elsevier, Inc

Standring S (2016) Gray's anatomy: the anatomical basis of clinical practice, 41st edn. Elsevier Limited, xviii, 1562 p

van Son JA, Smedts F, Korving J, Guyt A, de Kok LB (1993) Intercostal artery: histomorphometric study to assess its suitability as a coronary bypass graft. Ann Thorac Surg 56(5):1078–1081. https://doi.org/10.1016/0003-4975(95)90018-7

40 Brachiocephalic (Innominate) Vein

Brian Le, Angel Castro, Annie Syed,
and Stephen J. Bordes, Jr.

40.1 Introduction

The brachiocephalic veins (BCV) are central veins formed by the internal jugular and subclavian veins, as they return blood from the cranial and upper limb circulations through the thoracic inlet (Fig. 40.1). The right BCV (RBCV) and left BCV (LBCV) join and form the superior vena cava (SVC). The structures bordering the BCV are the thyroid gland, common carotid arteries, trachea, phrenic nerve, and manubrium. BCV complications can arise from neck and reconstructive surgery, central venous catheterization, and congenital heart disease and malformations.

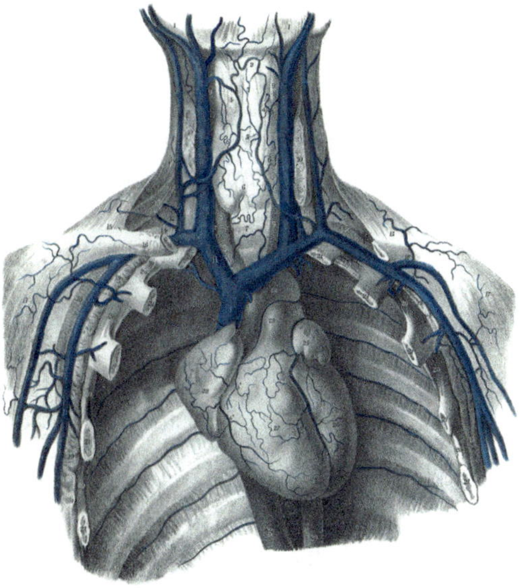

Fig. 40.1 Drawing showing the union of the internal jugular and subclavian veins forming the left and right brachiocephalic veins. (Quain and Wilson 1844 author's collection)

40.2 Embryology

The BCV arises from the primitive systemic and cardiopulmonary embryological vasculature primarily from the cardinal veins. In the 3rd–4th weeks of fetal development, segments of cardinal veins begin to form and regress with growth, forming the BCV by approximately the 8th week. The common cardinal vein splits into two during weeks 5–6 of development to eventually form the

B. Le · A. Syed
School of Medicine, Tulane University,
New Orleans, LA, USA
e-mail: ble3@tulane.edu; asyed1@tulane.edu

A. Castro
Washington University School of Medicine,
St. Louis, MO, USA
e-mail: a.h.castro@wustl.edu

S. J. Bordes, Jr. (✉)
Department of Surgery, Louisiana State University
Health Sciences Center, New Orleans, LA, USA

S. J. Bordes, Jr. et al. (eds.), *The Clinical Anatomy of the Vascular System*,
https://doi.org/10.1007/978-3-031-78326-5_40

superior vena cava and coronary sinus. Progressing through the 7th week of development, the distal segments of the anterior cardinal veins, proximal to the common cardinal veins, begin to anastomose, connecting the future left and right internal jugular veins. The anastomosis is completed by the 10th week, which gives rise to the LBCV medially. Proximal segments of the right cardinal vein form the RBCV, emptying into the superior vena cava (Carlson 2014).

40.3 Anatomy/Topography

Right Brachiocephalic Vein

The right brachiocephalic vein (RBCV) begins at the union of the internal jugular and subclavian veins as they pass through the thoracic inlet. It is approximately 3 cm in length, beginning at the sternoclavicular joint and descending to the junction joining the left brachiocephalic vein (LBCV) at the lower border of the first right costal cartilage, at which point the vessels become the superior vena cava.

The RBCV is bordered medially by the thymus and brachiocephalic artery. The inferior borders of the RBCV include the internal thoracic artery, phrenic nerve, and lung pleura. Numerous smaller tributaries join the BCV. Among them are the right vertebral vein, which joins it superiorly, and the internal thoracic vein laterally. Other contributing vessels include the anterior mediastinal, pericardial, sternal, and anterior intercostal veins (Bennett et al. 2011).

Left Brachiocephalic Vein

The left brachiocephalic vein (LBCV) is notably longer than the RBCV, owing to its development as an anastomosis connecting the left jugular and left subclavian vessels to the RBCV. The LBCV is approximately 6 cm and starts at the left sternoclavicular joint, proceeding inferomedially and posterior to the manubrium to join the RBCV at the level of the first costal cartilage.

The LBCV shares the same supply and drainage as the RBCV with a few specific additions. The thymic veins drain into the LBCV superiorly by passing just inferiorly to the thymus. The left superior intercostal vein and pericardial tributaries also join (Bennett et al. 2011).

40.4 Imaging/Histology

Diagnosing and treating vascular pathology involving the BCV relies on usual methods of vascular imaging including CT, MRI/MRA, and ultrasound. Below is an overview of imaging techniques and the relevance to the BCV.

CT/MRI (MRA)

CT is used extensively to identify and visualize anomalous brachiocephalic veins. This includes variations that can interfere with or present as surgical complications. While anomalous veins are uncommon (0.2–1% of congenital cardiovascular anomalies), understanding these variants can prevent procedural complications (Chen et al. 2005). CT/MRA are crucial in identifying other common causes of BCV related pathology including tumors, trauma, aneurysms, stenosis, thrombosis, and iatrogenic injury related to central venous catheterization (Ko et al. 2008).

Ultrasound

Ultrasound is the least invasive form of imaging and commonly used guide for central venous catheterization. While most veins in the upper limbs can be viewed easily in the transverse plane, visualization of the BCV involves using the subclavian veins as a primary landmark and indicator of patency. Doppler flow and a supraclavicular approach can visualize the medial portion of the subclavian vein and the superior aspect of the BCV. Confirmation of spontaneous, phasic, and non-pulsatile flow indicates BCV patency. Absence of these signs can be an indication of stenosis or obstruction of the BCV or the superior vena cava. Due to abundance of collateral flow, it is important to trace the internal jugular vein from the caudal aspect to the subclavian junction where the BCV begins. In the setting of some pathologies, well-developed collateral structures can be mistaken for the BCV. The BCV and other central veins are straight veins, while collateral veins tend to be serpiginous (Weber et al. 2007).

Pregnancy

The use of ultrasound has been demonstrated to be an effective means of assessment of the LBCV for the development of congenital malformations of the vessel. The LBCV is visualized in a transverse view of the upper fetal chest while measuring the diameter of the vessel with Doppler visualization. This assessment was performed mainly in the second and third trimester, but first trimester assessment was performed as well and effective at identifying congenital anomalies in utero (Sinkovskaya et al. 2012).

40.5 Variations

Congenitally Related Anomalous Brachiocephalic Veins

Brachiocephalic vein variation is relatively uncommon, but when present is associated with congenital heart disease. It has been documented in infants and adults and studied using CT/MRA techniques, often in preoperative settings in the repair of tetralogy of Fallot (TOF). The prevalence of anomalous BCV associated with congenital heart was found to be 1.7% with TOF making up 77% of the patients with BCV anomalies (Chen et al. 2003). The most common pattern of the anomalous BCV course was retroaortic crossing the midline anterior to a patent ductus arteriosus or ligamentum arteriosus (Chen et al. 2003; Kulkarni et al. 2008; Kahhouee et al. 2017). It is worth noting that although an anomalous BCV is associated with TOF, it was not seen to be associated with persistent SVC in patients with TOF, but radiologically prudent to distinguish an aberrant LBCV versus a persistent SVC as an aberrant LBCV can indicate more serious cardiac disease complications (Kahkouee et al. 2017).

Thyroid Veins

The inferior thyroid veins typically drain into the BCV and the SVC. In 41% of cases, inferior thyroid veins drain into the ipsilateral BCV, and in 35% of other cases, the thyroid veins join into a common trunk before emptying into the LBCV. It is worth noting these variations as BCV pathology can engorge tributaries like the thyroid veins which are regularly variable (Dalip et al. 2018).

40.6 Pathology

Brachiocephalic Venous Aneurysm

Brachiocephalic venous aneurysm is rare and usually diagnosed incidentally during imaging with a finding of mediastinal widening. It is most associated with congenital heart malformations and hemangiomas but can also be due to trauma and vascular wall degeneration. Presentation is usually asymptomatic but can contribute to upper extremity edema, pain, cough, as well as a mediastinal mass effect. When suspected, diagnosis can be refined through venography, ultrasound, echocardiograms, or thoracoscopy. Asymptomatic patients can be managed through conservative management, though surgical resection has a favorable success rate in symptomatic patients. The development of thrombi is associated with these aneurysms and may necessitate anticoagulation therapy (Nguyen et al. 2020).

40.7 Clinical and Surgical Significance

Brachiocephalic Vein Turn-Down

In patients who receive the Fontan procedure (treatment of hypoplastic left heart, tricuspid atresia, and double outlet right ventricle in children), complications related to systemic lymphatic hypertension are significant and include plastic bronchitis, protein-losing enteropathy, edema, and pleural effusions. Turning down the brachiocephalic veins into the Fontan atrium can lower the pressure in the thoracic duct, reducing the lymphatic overload in patients experiencing these conditions that have failed to respond to medical management. The procedure requires isolating the brachiocephalic vein without disrupting the thoracic duct and its lymphatic vessels. Depending on the length of the brachiocephalic vein, an interposition graft may be necessary to connect to the atrium. One complication of this procedure is the degree of right-to-left shunt that develops which can be modified by banding the left internal jugular vein (Hraska et al. 2020).

Central Venous Catheter Cannulation

The brachiocephalic veins are sites that can be used for central venous access in adults, children, and neonates when identified by ultrasound. When compared to the internal jugular vein, cannulation of the brachiocephalic veins presents certain advantages: ease of identification on ultrasound, low incidence of arterial overlap, and physiology conducive to stenting the lumen open. This location is also farther from possible nasal and oral contamination. Complication rates and types due to insertion in this site are similar to those seen in the internal jugular vein, with pneumothorax and hemothorax being the most common (Breschan et al. 2018).

Ligation and Division of the Left Brachiocephalic Vein

During surgery in the mediastinum, such as SVC reconstruction, mass excision, or aortic arch repair, the left brachiocephalic vein can be ligated and divided to provide more exposure to the structure requiring intervention. The most common complication is upper extremity edema, which can be managed conservatively while collateral circulation develops (McPhee et al. 2013).

Selective Venous Sampling (SVS) for Parathyroid Hormone Diseases

The central veins are the site of selective venous sampling (SVS) for the diagnosis and treatment of parathyroid diseases. Noninvasive imaging of the parathyroid can be unreliable for patients undergoing treatment for primary and secondary hyperparathyroidism. The inferior thyroid veins drain into the BCV (Dalip et al. 2018), drawing parathyroid hormone levels in the central vein assisted in the localization of parathyroid adenomas or hyperplastic parathyroid glands with higher sensitivity than both ultrasound and sestamibi scans (Reidel et al. 2006).

40.8 Complications

Thoracic Central Vein Obstruction

The most common complications involving the brachiocephalic vein relate to occlusion and stenosis, with a variety of etiologies responsible leading to development of thoracic central vein obstruction (TCVO). Damage to the vessel leads to inflammatory changes, promoting flow alterations and formation of thrombi. Thrombi can also be caused by lumen reduction from a mass effect due to an external structure. Depending on the time course of the development of the reduction in flow which can permit collaterals to form, the presentation will vary from asymptomatic in the chronic to pain and upper extremity, facial, and breast edema in more acute presentations. The most severe consequences include SVC syndrome, phlegmasia cereal dolens, and pulmonary embolism. Multimodality imaging is useful in diagnosing this condition, with doppler ultrasound permitting the assessment of vessel patency and planning for intervention, while venography remains the most useful in determining the extent and location of the impacted structure. The most common contributors to the development of TCVO in the brachiocephalic veins include central venous catheter placement and compression due to another structure. Any type of catheter intervention that terminates in the superior vena cava, from dialysis lines to pacemaker lead placement, can lead to stenosis and the formation of thrombi. This complication is common in those receiving long-term catheter placement and more likely to become symptomatic in dialysis patients, due to the increased flow when accessing AV fistulas or newly inserted lines, leading to venous and lymphatic hypertension and overload. Repeated AV fistula formation and repair can contribute to the development of TCVO. Catheter removal may help relieve symptoms, but existing TCVO complicates further central access, which requires planning for recurrence and judicious site location to maintain future vessel accessibility. Compression of the brachiocephalic veins is often related to malignancy, but other structures may contribute. These veins can be occluded and intravascularly invaded by thymomas, lymphomas, and metastatic disease, such as breast cancer. Due to the relative rapidity of TCVO in this manner, malignant lesions will cause rapid development of symptoms. Medical treatment for mediastinal tumors includes chemotherapy or radiotherapy along with steroid administration. Compression from other thoracic structures is more likely to cause chronic obstruction, such as aortosternal venous compression in which the left brachiocephalic vein's lumen is reduced due to the pressure exerted by the aorta and sternum. Definitive treatment of TCVO requires endovascular intervention. Angioplasty with stent placement is the gold standard. Unilateral stenting in bilateral disease may be preferable as it is associated with a lower rate of complications. Precise utilization of various imaging modalities is crucial to avoiding the damage of any the surrounding structures as well as the relatively high rate of vascular variability, which could lead to complications with high mortality and high morbidity. Anticoagulation may be necessary post-procedure to prevent occlusion of the stent as well as reduce the likelihood of pulmonary embolism (Fat-Ko et al. 2007; Mehta 2017; Azizi et al. 2020; Holly 2019; Dolmatch et al. 2018; Giglio et al. 2022; Otten et al. 2003; Collin et al. 2015; Mazer et al. 2018; Ma et al. 2021; Stevens et al. 2015).

Trauma

Traumatic injuries of the brachiocephalic veins can be split into iatrogenic etiologies as well as blunt and penetrating trauma. A variety of procedures can lead to perforation of these veins from open surgeries and tracheostomies to intravascular sources, such as pacemaker insertion or port placement. Central venous and hemodialysis catheters are recognized as the most common culprits. One contributing factor is the degree of variability that can be found in the arrangement of the veins. Penetrating trauma, especially to the cervicomediastinal region, is the most likely of non-iatrogenic causes while blunt injuries, associated high speed collisions, occur less often. A high index suspicion is necessary to avoid a missed brachiocephalic vein injury, especially in the context of distracting injuries. CTA with contrast can reveal an expanding mediastinal hematoma or contrast extravasation. Complications of trauma in these veins include pseudoaneurysm, mediastinal hematoma, hemothorax, and cardiac tamponade, which can require emergent surgical intervention. In the event that a line is the cause of the perforation, removal with balloon angioplasty may be enough to end the hemorrhage, though ligation of the brachiocephalic veins remains the definitive treatment. Post-surgical complications include upper extremity and facial edema but are usually transient and resolve over time with the development of collaterals.

References

Azizi AH, Shafi I, Shah N et al (2020) Superior vena cava syndrome. Cardiovasc Intervent 13(24):2896–2910

Bennett WF, Altaf F, Deslauriers J (2011) Anatomy of the superior vena cava and brachiocephalic veins. Thorac Surg Clin 21(2):197–203

Breschan C, Graf G, Jost R et al (2018) A retrospective analysis of the clinical effectiveness of supraclavicular, ultrasound-guided brachiocephalic vein cannulations in preterm infants. Anesthesiology 128(1):38–43
Carlson BM (2014) Development of the vascular system. In: Reference module in biomedical sciences. Elsevier. https://www.sciencedirect.com/science/article/pii/B9780128012383054593. https://doi.org/10.1016/B978-0-12-801238-3.05459-3
Chen S, Liu K, Chen H et al (2005) Anomalous brachiocephalic vein: CT, embryology, and clinical implications. Am J Roentgenol 184(4):1235–1240
Collin G, Jones RG, Willis AP (2015) Central venous obstruction in the thorax. Clin Radiol 70(6):654–660
Dalip D, Iwanaga J, Loukas M, Oskouian RJ, Tubbs RS (2018) Review of the variations of the superficial veins of the neck. Cureus 10(6)
Dolmatch BL, Gurley JC, Baskin KM et al (2018) Society of Interventional Radiology reporting standards for thoracic central vein obstruction: endorsed by the American Society of Diagnostic and interventional nephrology (ASDIN), British Society of Interventional Radiology (BSIR), Canadian Interventional Radiology Association (CIRA), Heart Rhythm Society (HRS), Indian Society of Vascular and Interventional Radiology (ISVIR), Vascular Access Society of the Americas (VASA), and Vascular Access Society of Britain and Ireland (VASBI). J Vasc Interv Radiol: JVIR 29(4):454–460. e3
Ellis H (2010) The great veins of the neck. Anaesth Intensive Care Med 11(1):19–20
Giglio V, Badar Z, Bhogadi Y, Van Adel B, Yip G (2022) Aortosternal venous compression: a review of two cases. Case Rep Med 2022:4591024
Holly B, Lessne ML (2019) Evaluation and management of chronic thoracic venous obstructions. Clinical presentation, anatomic patterns of TCVO, and revascularization strategies. Endovasc Today 18(5)
Hraska V, Mitchell ME, Woods RK, Hoffman GM, Kindel SJ, Ginde S (2020) Innominate vein turn-down procedure for failing Fontan circulation. Semin Thorac Cardiovasc Surg Pediatr Card Surg Annu 23:34–40
Kahkouee S, Sadr M, Pedarzadeh E et al (2017) Anomalous left brachiocephalic vein: Important vascular anomaly concomitant with congenital anomalies and heart diseases. Folia Morphol 76(1):51–57
Ko S, Huang C, Ng S et al (2008) Imaging of the brachiocephalic vein. Am J Roentgenol 191(3):897–907
Kulkarni S, Jain S, Kasar P, Garekar S, Joshi S (2008) Retroaortic left innominate vein-incidence, association with congenital heart defects, embryology, and clinical significance. Ann Pediatr Cardiol 1(2):139
Ma W, Zhao Z, Fu Q et al (2021) Comparison of management for central venous stenosis with or without previous catheter placement. Front Neurol 12:703286
Mazer BA, Hughes PG (2018) Pacemaker-associated phlegmasia cerulea dolens treated with catheter-directed thrombolysis. Clin Pract Cases Emerg Med 2(4):316
McPhee A, Shaikhrezai K, Berg G (2013) Is it safe to divide and ligate the left innominate vein in complex cardiothoracic surgeries? Interact Cardiovasc Thorac Surg 17(3):560–563
Mehta HJ (2017) Central venous stenosis: what should a nephrologist know? Indian J Nephrol 27(6):427
Nguyen Q, Olive JK, Vervoort D, Phan K, Luc JG (2020) Brachiocephalic vein aneurysm: a systematic review of the literature. J Thorac Dis 12(5):2747
Otten TR, Stein PD, Patel KC, Mustafa S, Silbergleit A (2003) Thromboembolic disease involving the superior vena cava and brachiocephalic veins. Chest 123(3):809–812
Quain and Wilson (1844) The surgical anatomy of the human body. London: Taylor & Walton
Reidel MA, Schilling T, Graf S et al (2006) Localization of hyperfunctioning parathyroid glands by selective venous sampling in reoperation for primary or secondary hyperparathyroidism. Surgery 140(6):907–913
Sinkovskaya E, Abuhamad A, Horton S, Chaoui R, Karl K (2012) Fetal left brachiocephalic vein in normal and abnormal conditions. Ultrasound Obstet Gynecol 40(5):542–548
Stevens DC, Butty S, Johnson MS (2015) Superior vena cava rupture and cardiac tamponade complicating the endovascular treatment of malignant superior vena cava syndrome: a case report and literature review. Semin Interv Radiol 32(04):439–444
Weber TM, Lockhart ME, Robbin ML (2007) Upper extremity venous Doppler ultrasound. Radiol Clin North Am 45(3):513–524. https://www.sciencedirect.com/science/article/pii/S0033838907000206. https://doi.org/10.1016/j.rcl.2007.04.005

41 Superior Vena Cava

Corey Leskanic, Ashley Yearwood, and Stephen J. Bordes, Jr.

41.1 Embryology

At 7 weeks of life, a connection develops between the left and right anterior cardinal veins. This forms the left brachiocephalic anastomosis (Ghandour et al. 2017). It leads to regression of the left anterior cardinal veins, thus allowing blood to flow from the left side of the body toward the right anterior cardinal vein (Ghandour et al. 2017). Caudally to this transverse anastomosis, near the proximal portion of the right anterior and right common cardinal veins, the superior vena cava (SVC) is derived. Cranially to the transverse anastomosis, the right brachiocephalic vein forms from the right anterior cardinal vein. During embryonic development, the cardinal veins are responsible for venous return and represent the primary drainage. The anterior cardinal veins drain the cephalic portion of the embryo, and the posterior cardinal veins drain the caudal portion. Merging of the anterior and posterior cardinal veins before the sinus venosus creates the short common cardinal veins.

C. Leskanic
Department of Radiology, HCA Florida Trinity, Trinity, USA

A. Yearwood
Department of Interventional Radiology, UCLA David Geffen School of Medicine, Los Angeles, CA, USA

S. J. Bordes, Jr. (✉)
Department of Surgery, Louisiana State University Health Sciences Center, New Orleans, LA, USA

41.2 Anatomy and Topography

The superior vena cava (SVC) (Fig. 41.1) is located behind the first right costal cartilage at the confluence of the brachiocephalic veins and extends down to the third right costal cartilage, where it drains into the right atrium (Mozes and Gloviczki 2007). Halfway along the descending course, the SVC receives the azygos vein before entering the pericardium. The azygos vein is one of the few major tributaries of the SVC. However, the esophageal, mediastinal, and pericardial veins can also drain into the SVC (Sonavane et al. 2015). The SVC has a diameter of 2.1 cm ± 0.7 and a length of 7.4 cm ± 1.4 (Sonavane et al. 2015). It is often irregular in shape on cross-sectional imaging (Sonavane et al. 2015). Its normal diameters range from 1.5 to 2.8 cm at the major axis and 1.0 to 2.4 cm at the minor axis (Sonavane et al. 2015). If the cross-sectional area of the SVC is less than 1.07 cm^2, one should consider possible obstruction or compression of the vessel (Sonavane et al. 2015). During SVC obstruction, collateral venous pathways can be identified, including the bra-

S. J. Bordes, Jr. et al. (eds.), *The Clinical Anatomy of the Vascular System*,
https://doi.org/10.1007/978-3-031-78326-5_41

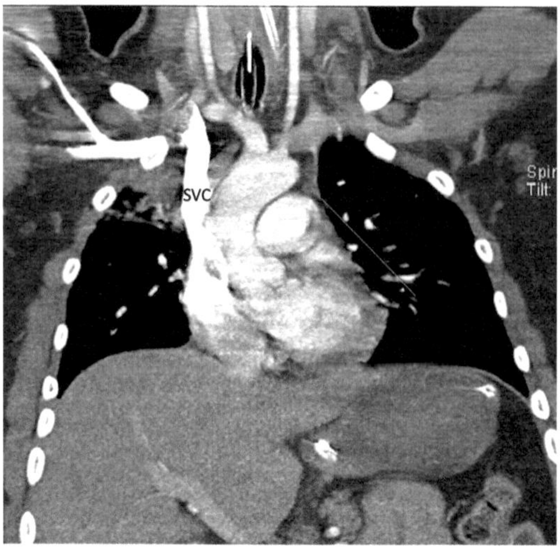

Fig. 41.1 CTA noting the superior vena cava (SVC)

chiocephalic veins, their tributaries, and the mediastinal venous plexus.

41.3 Histology

The SVC is valveless and is composed of three layers: the intima, media, and adventitia. The intima is made up of endothelial cells with underlying connective tissue. Bordering the intima is the internal elastic lamina, a layer of thick elastic fibers. The media layer contains smooth muscle but mostly comprises connective tissue fibers consisting of collagen. The adventitia consists of loose connective tissue with vasa vasorum and nerve fibers (Mozes and Gloviczki 2007). However, the adventitia is poorly differentiated from the media in larger veins such as the SVC (Mozes and Gloviczki 2007).

41.4 Variations

Variations of the SVC arise from abnormalities in embryological development. The major variations include abnormal drainage, bilateral absence of SVCs, persistent left SVC, absence of the right SVC with persistent left SVC, persistent left SVC draining into the coronary sinus with right SVC draining into the left atrium, and retroaortic innominate vein.

Abnormal drainage of the SVC is very rare and occurs when the right SVC opens into the left atrium or both atria. This develops when there is a lack of formation of a common wall that usually separates the right upper pulmonary veins from the SVC (Ghandour et al. 2017). When this common wall is not formed, the right upper pulmonary veins drain into the right SVC. This allows blood to drain into the right SVC and right atrium (Ghandour et al. 2017). Blood can then shunt into the right pulmonary vein and left atrium, resulting in bilateral drainage (Ghandour et al. 2017). Flow into the left atrium can become the preferential flow pathway if the right atrial orifice of the SVC is stenotic (Ghandour et al. 2017). Complications arise from the right-to-left shunting of blood and can include cyanosis, polycythemia, dyspnea, systemic emboli, brain abscess, and cerebrovascular complications (Ghandour et al. 2017). Surgical correction is required; interventional procedures are of minor value (Ghandour et al. 2017).

Bilateral failure of the SVC to develop is an extremely rare abnormality. It results from abnormal regression of both the right and left anterior cardinal veins (Ghandour et al. 2017). Instead, the right head and neck veins connect to the azygos vein, while the left head and neck veins drain into the hemiazygos vein. Both veins then drain into the right atrium, inferior vena cava, left renal vein, or a complex of the subdiaphragmatic plexus of veins (Ghandour et al. 2017). Bilateral absence of the SVC can be asymptomatic and require no medical treatment or surgical intervention (Derbel et al. 2019). While this anomaly is rare, it is essential to identify it and acknowledge the potential associated difficulties during cardiac surgery, such as the Glenn procedure or pacemaker placement (Ghandour et al. 2017; Derbel et al. 2019). Clamping of the inferior vena cava during other procedures can also cause death due to decreased venous return (Ghandour et al. 2017). This anomaly is associated with atrial septal defect, fetal cystic hygroma, tetralogy of Fallot, and congenital hydrothorax (Ghandour et al. 2017).

Persistence of the left SVC (PLSVC) is the most common systemic venous anomaly, affecting 0.3–0.5% of the general population (Ghandour et al. 2017). Failure of regression of the left anterior cardinal vein and its continuity with the left common cardinal vein results in PLSVC. The left SVC drains blood from the left brachiocephalic vessels into the coronary sinus and then the right atrium in 80–92% of cases (Ghandour et al. 2017). Both SVCs persist in 80–90% of cases, and in 65% of such cases the left brachiocephalic vein is absent, leading to a lack of connection between them (Ghandour et al. 2017). In 40–50% of patients with a left-sided SVC, cardiac malformations are associated with this anomaly (Bader et al. 2013). Such malformations include teratology of Fallot, atrioventricular canal defect, mitral atresia, atrial septal defect, juxtaposition of the right atrial appendage, ventricular septal defect, transposition of the great arteries, and anomalous connections of the pulmonary veins (Ghandour et al. 2017). Complications can arise when there is a right-to-left defect. The anomaly can also be associated with extracardiac malformations, including esophageal atresia (Ghandour et al. 2017). Dilation of the coronary sinus stretches the atrioventricular node and the bundle of His, potentially resulting in cardiac arrhythmias. Also, dilation of the coronary sinus can be more pronounced if the right SVC is absent (Ghandour et al. 2017).

Another PLSVC variation involves a PLSVC draining into the left atrium. This additional anomaly is due to defective portioning of the coronary sinus from the left atrium, resulting from the absence of the anterosuperior wall of the coronary sinus (Ghandour et al. 2017). In 10–20% of all PLSVC cases, the drainage of blood flows into the left atrium via its roof, or into an unroofed coronary sinus, or the left superior pulmonary vein (Povoski and Khabiri 2011). The right atrial ostium can provide interatrial communication, but stenosis of the coronary sinus orifice can necessitate an atrial septal defect for communication (Ghandour et al. 2017). Specifically, drainage into the left atrium with an unroofed coronary sinus results in termination of the left vena cava in the posterior corner of the left atrium between the left upper pulmonary vein posteriorly and the left atrial appendage anteriorly (Ghandour et al. 2017). The roof of the coronary sinus can be fenestrated, partially absent, or completely absent (Ghandour et al. 2017). The combination of a PLSVC and a defectively portioned coronary sinus from the left atrium is known as Raghib syndrome (Ghandour et al. 2017). This anomaly of a PLSVC draining into the left atrium is associated with other congenital malformations including heterotopia syndrome (Ghandour et al. 2017). Clinical manifestations can be insignificant or can include symptoms of a large atrial septal defect such as chronic hypoxia, cyanosis, polycythemia, paradoxical embolus, clubbing, brain abscess, stroke, or right heart overload (Ghandour et al. 2017). Of note, PLSVC draining into the left atrium increases the risk of subsequent paradoxical emboli owing to the presence of a right-to-left cardiac shunting (Povoski and Khabiri 2011). Surgical ligation of the left SVC is sometimes required. Other surgical interventions can also be required for the cardiac manifestations, including patch closure of the ostium of the coronary sinus, tunneling of the left SVC into the right atrium with closure of the atrial septal defect, and wide resection of the atrial septum (Ghandour et al. 2017).

Further anomalies can arise from combined etiologies of veins failing to regress while others fail to develop. Both anomalies previously discussed, including persistence of the left SVC (PLSVC) draining into the coronary sinus and the right SVC draining into the left atrium, can occur concurrently (Ghandour et al. 2017). Additionally, PLSVC can occur with the complete absence of the right SVC. In 10% of PLSVC cases, the right SVC is absent owing to complete regression of the right anterior cardinal vein and persistence of the left anterior cardinal vein (Ghandour et al. 2017). The brachiocephalic veins ultimately join the left SVC and drain into the coronary sinus (Ghandour et al. 2017). Dilation of the coronary sinus can be noted, which can also decrease the left ventricular preload (Ghandour et al. 2017). In 54% of cases this anomaly can present with a normal heart, while 46% present with congenital heart disease and 36% with rhythm disturbances (Ghandour et al. 2017).

Lastly, another combined etiology anomaly involves a retroaortic innominate vein. In this anomaly, the SVC receives drainage from the left innominate vein, which courses inferiorly and posteriorly to the aortic arch rather than superiorly and anteriorly. The left innominate vein then joins the SVC inferiorly to the azygous vein. Reduced shortening of the aortic arch during embryological development compresses the superior transverse venous plexus and prevents development (Ghandour et al. 2017). The reduced shortening of the aortic arch allows for persistence of the inferior transverse capillary plexus, which normally regresses if the aortic arch occupies this space (Ghandour et al. 2017). A high aortic arch allows the inferior transverse capillary plexus to persist, leading to the formation of an anomalous course of the innominate vein (Ghandour et al. 2017). The incidence of this anomaly is 0.2–1% of congenital heart defects (Kulkarni et al. 2008). Additionally, 80% of cases are associated with other malformations, including right ventricular outflow obstruction, total anomalous pulmonary venous connection, or a right aortic arch (Ghandour et al. 2017). An SVC receiving drainage from the left innominate vein has limited clinical significance, but there can be difficulties during pacemaker and central venous line placements (Ghandour et al. 2017). Other surgical interventions can present with difficulties such as patent ductus arteriosus ligation, exposure of pulmonary arteries during a Glenn shunt, and poor visualization of the surgical field during a Blalock Taussig shunt (Ghandour et al. 2017).

41.5 Imaging

Visualization of the SVC helps identify congenital variants and other pathological conditions of this vein (Sonavane et al. 2015). Certain congenital anomalies of the SVC are asymptomatic. Knowledge of asymptomatic SVC variants such as persistent left SVC, partial anomalous pulmonary venous return, and aneurysm helps to identify these potential incidental findings (Sonavane et al. 2015). Imaging of the SVC by radiography, computed tomography (CT), conventional venography, and magnetic resonance venography are useful for diagnosing SVC pathology accurately (Sonavane et al. 2015). Imaging is also useful for identifying acquired abnormalities, including fibrin sheath strictures, thrombus, primary neoplasm, and trauma that alter the SVC by narrowing the lumen or causing complete occlusion (Sonavane et al. 2015). The latter can lead to SVC syndrome, in which imaging is used to identify the site, extent, and cause of the occlusion (Sonavane et al. 2015). Such evaluation helps to guide the appropriate management.

41.6 Pathology

The SVC is a thin-walled, low-pressure vein that can be easily compressed. Superior vena cava syndrome (SVCS) is the mechanical obstruction of the SVC by either an intrinsic obstruction, external compression, or neoplastic invasion (Zimmerman and Davis 2018). Impairment of the blood flow results in enlargement of the SVC proximal to the site of the obstruction. Collateral pathways dilated include the azygous venous system and the inferior vena cava in later stages. These collateral pathways dilate over the course of several weeks, so upper body venous pressure can be elevated initially but decrease as time progresses (Yu et al. 2008). Elevated pressures cause interstitial edema of the head and neck. Edema of the nasal passages and larynx can require intubation. Cerebral edema can occur, leading to cerebral ischemia, herniation, and death (Yu et al. 2008). Malignancy accounts for 90% of SVCS cases, 75% being secondary to lung cancer and 15% secondary to non-Hodgkin lymphomas. Right-sided small cell lung cancer is the most common type. Additional malignancies include breast cancer, thymoma, esophageal cancer, germ cell tumors, thyroid carcinoma, and metastatic tumors (Zimmerman and Davis 2018). The severity depends on how rapidly the SVC obstruction develops and the degree of narrowing of the SVC (Yu et al. 2008).

41.7 Clinical and Surgical Significance

Superior vena cava syndrome (SVCS) can cause life-threatening cerebral edema and airway obstruction secondary to laryngeal edema (Zimmerman and Davis 2018). When a patient is stable, symptoms are improved with supplemental oxygen, elevation of the head of the bed, and parenteral steroids (Zimmerman and Davis 2018). Early anticoagulation should be initiated when SVCS is associated with a thrombus, and thrombolytic therapy can also be considered (Zimmerman and Davis 2018). As malignancy is a common cause of SVCS, radiation therapy to shrink the tumor is indicated for obstructions secondary to non-small cell carcinomas of the lung (Zimmerman and Davis 2018). Chemotherapy can also be used for chemosensitive tumors. SVCS of benign origin can benefit from endovascular therapy. Endovascular stenting, percutaneous transluminal angioplasty, and intravascular thrombolysis can be used to relieve the narrowing of the vessel (Zimmerman and Davis 2018). Endovascular stenting can also be used in malignant cases while histological diagnosis is pending or when chemotherapy and radiotherapy have failed (Zimmerman and Davis 2018). For patients presenting with severe SVCS, venous stenting has become the standard of care (Bongers et al. 2020). It has higher success rates and more rapid symptom resolution than radiation or chemotherapy. However, stenting is associated with SVC rupture and cardiac tamponade (Bongers et al. 2020). If stenting is considered as part of management, covered stenting should be used (Bongers et al. 2020).

Many variants of the SVC and their clinical and surgical significance have been discussed previously. The most common systemic venous anomaly, affecting 0.3–0.5% of the general population, is persistence of the left superior vena cava (Ghandour et al. 2017). Persistent left superior vena cava (PLSVC) results when the left superior cardinal vein fails to regress during gestation. It can present asymptomatically if it coexists with a right SVC. However, difficulties can arise during endovascular interventions including hemodialysis tunneled catheter insertion, cardio radiofrequency ablation, or implantable cardiac placement. Serious complications include arrhythmia and coronary sinus thrombosis. PLSVC should be suspected if there is failure of the guidewire or catheter to cross the midline upon imaging (Lee et al. 2017). A guidewire or catheter via internal jugular vein access will terminate along the left mediastinal border. This positioning can be falsely identified as the carotid artery, with additional high oxygen saturation characteristically leading to inadvertent carotid artery catheterization (Mackey and Dargin 2014).

References

Bader M, Bromley P, Jester I, Bennett J, Arul GS (2013) Central venous catheters in the left-sided superior vena cava: clinical implications. J Pediatr Surg 48(2):400–403. https://doi.org/10.1016/j.jpedsurg.2012.11.023

Bongers KS, Patel V, Gualano SK, Schildhouse RJ (2020) Cardiac tamponade after superior vena cava stenting. BMJ Case Rep 13(6):e234345. https://doi.org/10.1136/bcr-2020-234345

Derbel B, Krarti N, Ghedira F, Ben Mrad M, Hedhli M, Ziadi J, Denguir R (2019) Bilateral absence of the superior vena cava. Case report and literature review. JMV-J Méd Vasc. https://doi.org/10.1016/j.jdmv.2019.06.005

Ghandour A, Karuppasamy K, Rajiah P (2017) Congenital anomalies of the superior vena cava: embryological correlation, imaging perspectives, and clinical relevance. Can Assoc Radiol J 68(4):456–462. ISSN 0846-5371. https://doi.org/10.1016/j.carj.2016.11.003

Kulkarni S, Jain S, Kasar P, Garekar S, Joshi S (2008) Retroaortic left innominate vein – incidence, association with congenital heart defects, embryology, and clinical significance. Ann Pediatr Cardiol 1:139–141

Lee K-G, Tan RY, Chidambaram VA, Pang SC, Tan C-S (2017) Duplicate superior vena cava on arteriovenous fistulography. J Vasc Surg Venous Lymphat Disord 5(5):739. https://doi.org/10.1016/j.jvsv.2017.03.009

Mackey CG, Dargin JM (2014) Left-sided superior vena cava and venous hyperoxia masquerading as inadvertent carotid artery catheterization. J Emerg Med 47(2):e31–e35. https://doi.org/10.1016/j.jemermed.2014.02.004

Mozes G, Gloviczki P (2007) Venous embryology and anatomy. In: The vein book. Academic Press, pp 15–25. https://doi.org/10.1016/b978-012369515-4/50005-3

Povoski SP, Khabiri H (2011) Persistent left superior vena cava: review of the literature, clinical implications, and relevance of alterations in thoracic central

venous anatomy as pertaining to the general principles of central venous access device placement and venography in cancer patients. World J Surg Oncol 9(1):173. https://doi.org/10.1186/1477-7819-9-173
Sonavane SK, Milner DM, Singh SP, Abdel Aal AK, Shahir KS, Chaturvedi A (2015) Comprehensive imaging review of the superior vena cava. RadioGraphics 35(7):1873–1892. https://doi.org/10.1148/rg.2015150056
Yu JB, Wilson LD, Detterbeck FC (2008) Superior vena cava syndrome—a proposed classification system and algorithm for management. J Thorac Oncol 3(8):811–814. https://doi.org/10.1097/jto.0b013e3181804791
Zimmerman S, Davis M (2018) Rapid fire: superior vena cava syndrome. Emerg Med Clin North Am 36(3):577–584. https://doi.org/10.1016/j.emc.2018.04.011

42 Veins of the Mediastinum

Emma Cole

42.1 Embryology

The primitive mediastinum arises from mesenchyme during the fourth week of development. The intraembryonic coelom gives rise to the mesodermally lined pericardial cavity, peritoneal cavity, and two pericardioperitoneal canals, all of which contribute to the borders of the mediastinum (Ronson et al. 2000). Infoldings of the pericardial cavity, the pleuropericardial canals, grow medially from the lateral body wall and begin to create the division between the peritoneum and the pericardium (Schoenwolf et al. 2021). These pleuropericardial canals form the initial lateral borders of the mediastinum (Ronson et al. 2000). The mediastinal space is further defined as the bronchial lung buds extend into the pleuropericardial canals, pushing the superior and inferior surfaces of the canals away from each other (Ronson et al. 2000). The embryonic mediastinum is formed by the end of the seventh week of development when the pleuropericardial membranes growing from the lateral body wall fuse with mesoderm at the midline (Schoenwolf et al. 2021). This creates defined borders for the pericardial cavity and pleural cavities, forming a space for the mediastinum and its structures to extend from the sternum to the vertebral column and lie between the lungs (Ronson et al. 2000).

The superior vena cava (SVC) is formed from three embryological segments: the proximal right anterior cardinal vein, the right common cardinal vein, and the right horn of the sinus venosus (Van Den Bos and De Maeseneer 2012). The distal right anterior cardinal vein gives rise to the right brachiocephalic vein, while the left brachiocephalic vein arises from the anastomosis between the left and right anterior cardinal veins (Beattie 1931; Bennet et al. 2011). The right supracardinal vein gives rise to the azygos vein, while the arch of the azygos is formed from the right posterior cardinal vein. On the contralateral side, the left supracardinal vein gives rise to the hemiazygos and accessory azygos veins, which drain into the azygos and hemiazygos veins, respectively (Van Den Bos and De Maeseneer 2012).

42.2 Anatomy/Topography

The mediastinum is the central anatomical space of the thorax. It is bordered by the thoracic aperture superiorly, the diaphragm inferiorly, the sternum and costal cartilages anteriorly, and the thoracic vertebrae posteriorly (Dalley and Agur 2022). It can be subdivided into the superior, posterior, anterior, and middle mediastinum.

The veins of the superior mediastinum include the right and left brachiocephalic veins

E. Cole (✉)
Louisiana State University Health Sciences Center
School of Medicine, New Orleans, LA, USA

S. J. Bordes, Jr. et al. (eds.), *The Clinical Anatomy of the Vascular System*,
https://doi.org/10.1007/978-3-031-78326-5_42

and the SVC (Dalley and Agur 2022). Together, the brachiocephalic veins drain blood from the head, neck, and upper limbs and shunt it toward the right atrium via the SVC. The SVC then collects blood from all structures above the diaphragm, including the heart and lungs (Dalley and Agur 2022).

The right brachiocephalic vein originates from the fusion of the right internal jugular and right subclavian veins posterior to the sternoclavicular joint (Bennet et al. 2011). It courses inferiorly for 2–3 cm until it merges with the left brachiocephalic vein at the inferior border of the first right costal cartilage, where the two brachiocephalic veins join to form the SVC. Several significant anatomical relationships along the descent of the right brachiocephalic vein include the right phrenic nerve and internal thoracic artery laterally, and the thymus and brachiocephalic artery medially. Multiple tributaries feed into the right brachiocephalic vein, including the right vertebral and internal thoracic, anterior mediastinal, pericardial, sternal, and anterior intercostal veins (Bennet et al. 2011).

Similarly, the left internal jugular and the left subclavian vein merge to form the left brachiocephalic vein posterior to the level of the sternoclavicular joint (Dalley and Agur 2022). The left brachiocephalic vein is longer than the right, descending obliquely from the left to the right side for nearly 6 cm until it merges with the right vein to form the SVC (Bennet et al. 2011). Significant relationships of the left brachiocephalic vein include the brachiocephalic, left common carotid, and left subclavian arteries, and the vagus and phrenic nerves, all of which border the vein posteriorly. Many of the tributaries feeding into it are shared with the right brachiocephalic vein but also include the superior intercostal and multiple inferior thymic veins (Bennet et al. 2011).

After the two brachiocephalic veins join, the SVC is formed and descends for approximately 7 cm to the right third costal cartilage level, draining into the superior right atrium. Along the course of the SVC, the phrenic nerve is associated laterally, and the brachiocephalic artery and ascending aorta are associated medially. Additionally, the azygos vein merges with the SVC near its drainage point into the right atrium (Bennet et al. 2011).

The posterior mediastinum contains the azygos and hemiazygos veins responsible for draining the back, the thoracoabdominal wall, and the mediastinal viscera (Dalley and Agur 2022). The azygos vein arises from the roots of the inferior vena cava (IVC). It ascends along the right side of the thoracic vertebrae, draining the posterior walls of the thorax and abdomen. It continues through the aortic hiatus to the superior mediastinum, and once inside that structure, it arches over the root of the right lung to drain into the SVC (Dudiak et al. 1991). The azygos vein thus connects the SVC and IVC and helps provide critical collateral pathways if major drainage pathways are obstructed (Dudiak et al. 1991; Gray et al. 1980; Lawler et al. 2002). It also forms anastomoses with several posterior intercostal veins and the vertebral venous plexus and receives drainage from the mediastinal, esophageal, and bronchial veins (Dalley and Agur 2022).

In contrast, the hemiazygos vein originates from the roots of the renal vein and the fusion of the subcostal and ascending lumbar veins (Dalley and Agur 2022). It ascends on the left side of the thoracic vertebrae until it reaches T9, where it crosses over to the right side to join the azygos vein. The veins draining into the hemiazygos include posterior intercostal, esophageal, and various small mediastinal veins. The last major vein of the posterior mediastinum is the accessory hemiazygos, which originates at the medial side of the fourth or fifth intercostal spaces and traverses inferiorly along the left vertebral column until it crosses over to the right side to join the azygos vein. The accessory hemiazygos drains tributaries from the veins of the fourth to eighth intercostal spaces and, occasionally, the left bronchial veins (Dalley and Agur 2022).

42.3 Imaging

The SVC and azygos vein can be visualized with frontal and lateral chest radiographs (Sonavane et al. 2015). This type of imaging can detect anatomical anomalies, abnormal dilation of these veins, and mediastinal masses and is used to ensure correct placement of central venous catheters (CVCs) and peripherally inserted central catheters (PICCs). Chest radiographs can also reveal retained catheters, cardioverter-defibrillator lead fragments, fibrin sheaths, and calcified thrombi within the SVC (Sonavane et al. 2015). Similarly, nonenhanced computed tomography (CT) of the SVC can show SVC anomalies such as dilation, compression, duplication, calcifications, and CVC placement. Contrast-enhanced CT of the chest can also be used but can pose additional challenges because the dense contrast retained in the SVC leads to streak artifacts.

Stenosis or thrombus within the SVC can also be visualized using magnetic resonance (MR) venography to analyze the wall and lumen of the vessel. Conventional venography can also demonstrate venous obstruction and has the advantage of being able to proceed directly to intervention if indicated (Sonavane et al. 2015).

Doppler ultrasonography (US) and echocardiography can provide indirect evidence of brachiocephalic vein obstruction. These techniques could be better for visualizing the SVC owing to interference by the overlying costal cartilages and lungs. However, limited studies using Doppler US showed that flow velocities within the SVC were lower in normal patients and those with constrictive pericarditis than in patients with cardiac tamponade (Sonavane et al. 2015).

42.4 Variations

There are several congenital anatomical variations amongst the veins of the mediastinum. Most such anomalies are discovered incidentally, the most common variant in the thoracic venous system being a persistent left SVC (Ratliff et al. 2006). The SVC can persist on the left side if the left anterior cardinal vein, left common cardinal vein, and left horn of the sinus venosus persist during development (Mauro et al. 2001). The right SVC can be normal or otherwise and persist. Still, the obliteration of it combined with the persistence of the left SVC is associated with an increased risk for additional cardiac abnormalities such as atrial or ventricular septal defects (Sheikh and Mazhar 2014). While the right-sided SVC normally arises from the confluence of the brachiocephalic veins, a left-sided SVC would arise from the union of the left subclavian and left jugular veins (Mauro et al. 2001).

There is extensive variation in the anastomoses, origin, tributaries, and course of the azygos system of veins. Although the azygos vein normally drains into the SVC, it can also drain directly into the right brachiocephalic or right subclavian vein, the intrapericardial SVC, or even the right atrium (Dudiak et al. 1991). In some cases there is complete agenesis of the azygos vein owing to failure of the superior segment of the right supracardinal vein to develop correctly. This congenital anomaly is typically asymptomatic because the associated hemiazygos and accessory hemiazygos veins can still drain blood from the intercostal veins on both sides of the thoracic cavity (Gray et al. 1980). Another congenital variation of the azygos vein, known as an azygos continuation, occurs when the IVC is interrupted between the hepatic and prerenal segments because those segments fail to fuse during development (Fleischner and Udis 1952). As a result, the azygos and hemiazygos veins carry the shunted blood directly back to the SVC (Mandato et al. 2019). Imaging with CT can reveal the absence of the IVC below the level of the hepatic segment and dilation of the azygos and hemiazygos due to the increased flow volume (Bass et al. 2000). Although this variation only occurs in 0.6% of the population, it is significant because it has been associated with asplenia, polysplenia, and congenital heart disease (Bass et al. 2000; Fleischner and Udis 1952).

The accessory hemiazygos vein is also variable. While it usually traverses down the left side of T5–T8 before joining the azygos vein on the right side, it has also been found to join the hemiazygos vein before they both join the azygos (Dalley and Agur 2022). Additionally, the left superior intercostal vein, carrying deoxygenated blood from the first three intercostal spaces, can drain directly into the accessory hemiazygos vein rather than directly into the left brachiocephalic vein. However, in some individuals, the hemiazygos system of veins is completely absent.

42.5 Pathology

Severe obstruction of blood flow through the SVC can lead to a massive decrease in venous return from the upper extremities, head, and neck, resulting in SVC syndrome (Dahan et al. 1998; Kandpal et al. 2008; Youngson et al. 1980). Because blood flow to the right atrium via the SVC is impaired, patients with SVC syndrome often experience swelling in the upper extremities, head, and neck (Stanford et al. 1987; Sznajder et al. 1986). Intraluminal occlusion of the SVC can be caused by intrinsic factors such as stenosis, thrombus, or longstanding placement of indwelling CVCs (Dahan et al. 1998; Kandpal et al. 2008; Quaretti et al. 2014; Youngson et al. 1980). The most frequent cause of SVC stricture leading to SVC syndrome is chronic placement of CVCs in patients with end-stage renal disease requiring hemodialysis (Stanford et al. 1987; Sznajder et al. 1986). External factors can also cause compression of the SVC, narrowing its lumen and impeding blood flow to the right atrium. These factors include mediastinal tumors, fibrosing mediastinitis, lymphadenopathy, or ascending aortic aneurysm (Dahan et al. 1998; Dondelinger 2008; Kandpal et al. 2008; Rösch et al. 1992; Youngson et al. 1980). Malignancies in the mediastinum are the most common cause of SVC compression; small cell and non-small lung cell cancers, lymphoma, metastatic lymphadenopathy from thoracic malignancies, and tracheal malignancies are the most frequently observed pathologies (Dondelinger 2008; Rösch et al. 1992). The azygos system of veins normally facilitates decompression of the SVC by providing alternative routes to the right atrium through which blood can flow either anterograde or retrograde depending on the location of the obstruction (Dahan et al. 1998; Kandpal et al. 2008; Youngson et al. 1980). However, the capacity of collateral pathways to compensate for the SVC obstruction is often overwhelmed, causing characteristic swelling in the upper extremities, head, and neck (Dahan et al. 1998; Kandpal et al. 2008; Stanford et al. 1987; Sznajder et al. 1986; Youngson et al. 1980).

The azygos system offers critical collateral pathways for venous return to the heart, particularly in SVC or IVC obstruction (Dalley and Agur 2022). If the IVC is obstructed, the azygos vein can receive all the blood normally intended for it, aside from the liver, thereby draining all blood in the lower limbs below the diaphragm except for the GI tract. Similarly, if the SVC is obstructed above the level of the point where the azygos vein drains into the SVC, then blood will be shunted inferiorly to the venous system of the abdominal wall and then into the azygos system to return to the right atrium (Dalley and Agur 2022).

42.6 Clinical and Surgical Significance

Venous angioplasty and intravascular stents can be used to treat SVC stenosis or obstruction, with balloon angioplasty being the preferred method for establishing lumen patency (Dondelinger 2008; Rösch et al. 1992). These methods can help alleviate severe upper extremity and head and neck swelling caused by SVC syndrome. If the cause of SVC occlusion is malignancy, radiation therapy or chemotherapy is used with an approximately 90% success rate. Still, clinical results are not always clear for several days, and malignancy can even recur in about 20% of cases (Dondelinger 2008; Rösch et al. 1992). In contrast, intravascular SVC stents have about a 76–86% technical success rate. Still, they can effectively relieve symptoms immediately and provide sustained relief until death in patients estimated to live for about 6 months or less (Andersen and Duvnjak 2014; Sobrinho and Aguiar 2014).

Central venous catheters (CVCs) for treating critically ill patients are used to facilitate drug administration, parenteral nutrition, measurement of pulmonary arterial pressure, or short-term dialysis (Agarwal et al. 2007; Oguzkurt et al. 2005). They are commonly used for oncology patients requiring frequent chemotherapy and are often inserted into the SVC or brachiocephalic veins (Yildizeli et al. 2004).

42.7 Complications

When a stent is placed to treat SVC stenosis or occlusion, stent migration, venous perforation, and increased pulmonary edema can ensue. As the lumen of the SVC is made more patent, venous return significantly increases, potentially resulting in thrombosis or compression of the nearby recurrent laryngeal nerves (Dondelinger 2008). In addition, stent placement for treating malignant SVC compression can result in technical failure, leading to occlusion of both the SVC and the brachiocephalic veins (Andersen and Duvnjak 2014; Sobrinho and Aguiar 2014).

CVC placement can also introduce similar complications with chronic use. Prolonged use of indwelling CVCs can cause endothelial injury, thrombosis, smooth muscle proliferation, and central venous stenosis (Agarwal et al. 2007; Gonsalves et al. 2003). Sometimes, the catheter can become pinched off in the SVC or brachiocephalic veins, leading to atrial thrombosis, venous perforation, or ventricular tachycardia. However, these catheters are typically safely retrieved percutaneously (Yildizeli et al. 2004). CVC placement is further complicated by difficulty navigating the sharp turns in the large central veins within the thoracic cavity without imaging guidance in the emergency room or ICU (Wicky et al. 2002). Therefore, ultrasound-guided cannulation and fluoroscopic verification are recommended when possible.

References

Agarwal AK, Patel BM, Haddad NJ (2007) Central vein stenosis: a nephrologist's perspective. Semin Dial 20(1):53–62. https://doi.org/10.1111/j.1525-139X.2007.00242.x

Andersen PE, Duvnjak S (2014) Palliative treatment of superior vena cava syndrome with nitinol stents. Int J Angiol 23(4):255–262. https://doi.org/10.1055/s-0034-1383432

Bass JE, Redwine MD, Kramer LA, Huynh PT, Harris JH Jr (2000) Spectrum of congenital anomalies of the inferior vena cava: cross-sectional imaging findings. Radiographics 20(3):639–652. https://doi.org/10.1148/radiographics.20.3.g00ma09639

Beattie J (1931) The importance of anomalies of the superior vena cava in man. Can Med Assoc J 25(3):281–284

Bennet WF, Altaf F, Deslauriers J (2011) Anatomy of the superior vena cava and brachiocephalic veins. Thorac Surg Clin 21(2):197–203. https://doi.org/10.1016/j.thorsurg.2010.12.010

Dahan H, Arrivé L, Monnier-Cholley L, Le Hir P, Zins M, Tubiana JM (1998) Cavoportal collateralpathways in vena cava obstruction: imaging features. Am J Roentgenol 171(5):1405–1411. https://doi.org/10.2214/ajr.171.5.9798887

Dalley AF II, Agur AMR (2022) Thorax. In: Moore's clinically oriented anatomy. Wolters Kluwer Health, pp 355–409

Dondelinger R (2008) Chronic upper extremity/superior vena cava venous occlusive disease. In: Mauro MA, Murphy K, Thomson KR, Venbrux AC, Zollikofer CL (eds) Image-guided interventions. Elsevier, pp 1065–1066

Dudiak CM, Olson MC, Posniak HV (1991) CT evaluation of congenital and acquired abnormalities of the azygos system. Radiographics 11(2):233–246. https://doi.org/10.1148/radiographics.11.2.2028062

Fleischner FG, Udis SW (1952) Dilatation of the azygos vein: a roentgen sign of venous engorgement. Am J Roentgenol Radium Therapy Nucl Med 67(4):569–575

Gonsalves CF, Eschelman DJ, Sullivan KL, DuBois N, Bonn J (2003) Incidence of central vein stenosis and occlusion following upper extremity PICC and port placement. Cardiovasc Intervent Radiol 26:123–127. https://doi.org/10.1007/s00270-002-2628-z

Gray H, William PL, Warwick R (eds) (1980) Gray's anatomy of the human body, 36th edn. Churchill Livingstone, pp 622–800

Kandpal H, Sharma R, Gamangatti S, Srivastava DN, Vashisht S (2008) Imaging the inferior vena cava: a road less traveled. Radiographics 28(3):669–689. https://doi.org/10.1148/rg.283075101

Lawler LP, Corl FM, Fishman EK (2002) Multi-detector row and volume-rendered CT of the normal and accessory flow pathways of the thoracic systemic and pulmonary veins. Radiographics 22(1):45–60. https://doi.org/10.1148/radiographics.22.suppl_1.g02oc05s45

Mandato Y, Pecoraro C, Giuliano G, Tecame M (2019) Azygos and hemiazygos continuation: an occasional finding in the emergency department. Radiol Case Rep 14(1):1063–1068. https://doi.org/10.1016/j.radcr.2019.06.003

Mauro B, Guiseppe B, Lorenzo F, Gabriele B, Angelo B (2001) Left superior vena cava persistence in patients undergoing pacemaker or cardioverter-defibrillator implantation: a 10-year experience. Chest J 120(1):139–144. https://doi.org/10.1378/chest.120.1.139

Oguzkurt L, Tercan F, Yildirim S, Torun D (2005) Central venous stenosis in haemodialysis patients without a previous history of catheter placement. Eur J Radiol 55(2):237–242. https://doi.org/10.1016/j.ejrad.2004.11.006

Quaretti P, Galli F, Moramarco LP, Corti R, Leati G, Fiorina I, Maestri M (2014) Dialysis catheter-related superior vena cava syndrome with patent vena cava: long term efficacy of unilateral Viatorr stent-graft avoiding catheter manipulation. Korean J Radiol 15(3):364–369. https://doi.org/10.3348/kjr.2014.15.3.364

Ratliff HL, Yousufuddin M, Lieving WR, Watson BE, Malas A, Rosencrance G, McCowan RJ (2006) Persistent left superior vena cava: case reports and clinical implications. Int J Cardiol 113(2):242–246. https://doi.org/10.1016/j.ijcard.2005.08.067

Ronson RS, Duarte I, Miller JI (2000) Embryology and surgical anatomy of the mediastinum with clinical implications. Surg Clin N Am 80(1):157–169. https://doi.org/10.1016/s0039-6109(05)70400-x

Rösch J, Uchida BT, Hall LD, Antonovic R, Petersen BD, Ivancev K, Barton RE, Keller FS (1992) Gianturco-Rösch expandable Z-stents in the treatment of superior vena cava syndrome. Cardiovasc Intervent Radiol 15(5):319–327. https://doi.org/10.1007/BF02733957

Schoenwolf GC, Bleyl SB, Brauer PR, Francis-West PH (2021) Development of the respiratory system and body cavities. In: Larsen's human embryology, 6th edn. Elsevier, pp 245–260

Sheikh AS, Mazhar S (2014) Persistent left superior vena cava with absent right superior vena cava: review of the literature and clinical implications. Echocardiography 31(5):674–679. https://doi.org/10.1111/echo.12514

Sobrinho G, Aguiar P (2014) Stent placement for the treatment of malignant superior vena cava syndrome: a single-center series of 56 patients. Arch Bronconeumol (Engl Ed) 50(4):135–140. https://doi.org/10.1016/j.arbr.2014.03.001

Sonavane SK, Milner DM, Singh SP, Aal AKA, Shahir KS, Chaturvedi A (2015) Comprehensive imaging review of the superior vena cava. Radiographics 35(7). https://doi.org/10.1148/rg.2015150056

Stanford W, Jolles H, Ell S, Chiu LC (1987) Superior vena cava obstruction: a venographic classification. Am J Roentgenol 148(2):259–262. https://doi.org/10.2214/ajr.148.2.259

Sznajder JI, Zveibil FR, Bitterman H, Weiner P, Bursztein S (1986) Central vein catheterization: failure and complication rates by three percutaneous approaches. Arch Intern Med 146(2):259–261. https://doi.org/10.1001/archinte.1986.00360140065007

Van Den Bos RR, De Maeseneer M (2012) Endovenous thermal ablation for varicose veins: strengths and weaknesses. Phlebolymphology 19(4):163–169

Wicky S, Meuwly JY, Doenz F, Schnyder P, Denys A (2002) Life-threatening vascular complications after central venous catheter placement. Eur Radiol 12:901–907. https://doi.org/10.1007/s003300101018

Yildizeli B, Lacin T, Batirel HF, Yuksel M (2004) Complications and management of long-term central venous access catheters and ports. J Vasc Access 5(4):174–178. https://doi.org/10.1177/112972980400500407

Youngson GG, McKenzie FN, Nichol PM (1980) Superior vena cava syndrome: case report: a complication of permanent transvenous endocardial cardiac pacing requiring surgical correction. Am Heart J 99(4):503–505. https://doi.org/10.1016/0002-8703(80)90386-5

Pulmonary Veins

43

Emma Lesser

43.1 Embryology

The pulmonary veins develop from the respiratory diverticulum, which arises from the primitive foregut. The developing veins are located posteriorly to the left of the septum primum. The rudimentary pulmonary vein becomes incorporated into the atrium of the developing heart. The oblique vein (also called the vein of Marshall) is an embryological remnant that persists into adulthood in many hearts (Żabówka et al. 2020). In others, it is obliterated and becomes the ligament of Marshall. The vein of Marshall is frequently used in ablation procedures to treat atrial fibrillation as it contains myocardial tissue, making it continuous with the musculature of the coronary sinus (Fig. 43.1).

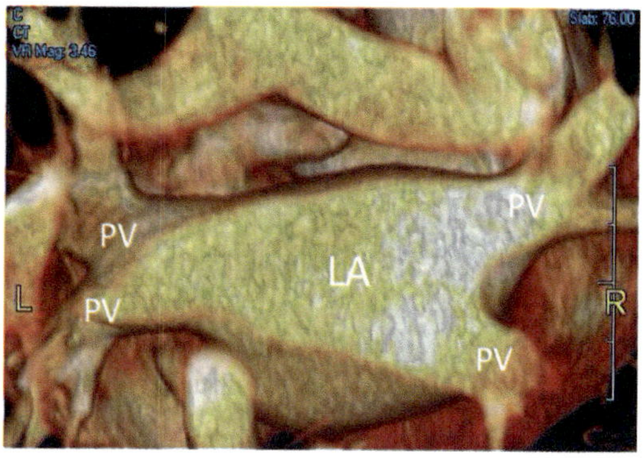

Fig. 43.1 CTA demonstrating the pulmonary veins (PV) draining into the left atrium of the heart (LA)

43.2 Anatomy/Topography

Two pulmonary veins usually feed into the left atrium of the heart on either side. The superior and inferior pulmonary veins deliver re-oxygenated blood from the lungs to the heart. They originate at the hilum, continue through the root of the lungs, and terminate in the left atrium.

The pulmonary veins ultimately drain the pulmonary capillary networks. They first join to form intra- and inter-segmental branches before forming segmental veins, which converge into the lobar veins. Typically, there are three lobar veins on the right and two on the left. They form the four pulmonary veins. The right superior pulmonary vein drains the superior lobe of the right lung, and the right inferior pulmonary vein drains

E. Lesser (✉)
Tulane University School of Medicine, New Orleans, LA, USA
e-mail: elesser1@tulane.edu

S. J. Bordes, Jr. et al. (eds.), *The Clinical Anatomy of the Vascular System*,
https://doi.org/10.1007/978-3-031-78326-5_43

both the middle and inferior lobes. The left superior pulmonary vein drains the superior lobe of the left lung, and the left inferior pulmonary vein drains the inferior lobe (Bose et al. 2021).

43.3 Histology

The venous trunks close to the heart are classified as large veins. They have robust intimal layers along with characteristic arrangements of the other internal layers of cells. Like all other vascular tissue, large veins have a tunica intima, an internal elastic lamina, a tunica media, and a tunica adventitia. The tunica intima comprises simple squamous endothelium with a thin subendothelial layer of smooth connective tissue. An internal elastic lamina layer can separate the tunica intima from the tunica media. The internal elastic lamina is a discontinuous layer of elastin that allows substances to diffuse more readily through the layers of the vein. The tunica media is thin in large veins but comprises alternate smooth muscle and connective tissue layers. It also has a helical organization of smooth muscle layers. The tunica adventitia is thicker in large veins than in medium veins and comprises connective tissue. Specifically, it is made of type I collagen, elastic fibers, and smooth muscle bundles.

43.4 Variations

According to Bose et al., 60–70% of individuals exhibit a characteristic pattern of drainage from the lobes of the lungs: the right superior pulmonary vein drains the superior lobe of the right lung, and the right inferior pulmonary vein drains its inferior and middle lobes; the left superior pulmonary vein drains the superior lobe, and lingula of the left lung and the left inferior pulmonary vein drains its inferior lobe. Suppose the upper and lower veins on one side unite to deliver oxygenated blood to the left atrium. In that case, there is only one opening into the left atrium and, therefore, only one ostium. Even though the main pulmonary veins can converge to a single vein, there are sometimes accessory veins, which are narrower than the main pulmonary veins and have their ostia in the left atrium. If more ostia are present, the left atrium is thicker. Varying degrees of atrial wall thickness can lead to many complications.

43.5 Clinical/Surgical Significance

Pulmonary vein stenosis often presents with congenital heart disease and can be diagnosed by echocardiography. It can often be repaired surgically, and the pulmonary veins can be bypassed in such cases (Vanderlaan et al. 2021).

Atrial fibrillation can also require surgical intervention. Pulmonary vein isolation ablation is a potential treatment for it. Cryoballoon ablation, as discussed by Miyazaki and Tada, is emerging as a favored technique for pulmonary vein isolation over the radiofrequency ablation approach. It allows for a more direct and structurally based strategy (Miyazaki and Tada 2019).

The oblique vein (or vein of Marshall) becomes of interest for ablations. As noted earlier, it is an embryological remnant in most hearts, though in some, it is obliterated and becomes the ligament of Marshall. The oblique vein is densely filled with myocardial tissue and thus has high electrical activity potential. These features make it a favorable site for cauterization by guidewires and catheters, as well as the placement of electrodes (Żabówka et al. 2020). In cases of pulmonary vein stenosis, there is an increased risk of bronchial congestion and hemoptysis. As a result, thoracoscopic lobectomy of the affected tissue may be necessary (Cheng et al. 2018; Nakaizumi et al. 2019). Additionally, while the cryoballoon ablation technique for pulmonary vein isolation is burgeoning, cryoballoon ablation may bring an increased risk of injury to the phrenic nerve. Right phrenic nerve injury is common and adverse outcomes of such may result from the nerve's proximity to the left atrium, the freezing procedure itself, and the size of the balloon (Miyazaki and Tada 2019).

References

Bose P, Choudhary U, Singh P, Singh B, Singh R (2021) Morphological variations of pulmonary veins draining into left atrium: a detailed cadaveric study. Natl J Clin Anatom 10(4):236–239

Cheng S, Lu X, Wang J, Liu T, Zhang X (2018) Thoracoscopic lobectomy for massive hemoptysis caused by complete pulmonary vein occlusion after radiofrequency ablation for atrial fibrillation. J Thorac Dis 10(4):E296

Miyazaki S, Tada H (2019) Complications of Cryoballoon pulmonary veinisolation. Arrhythmia Electrophysiol Rev 8(1):60–64

Nakaizumi T, Nakamura K, Nakamura K, Ishigaki M, Taniguchi H, Kajiura K (2019) Haemoptysis and bronchial congestion due to pulmonary vein stenosis after maze procedure. Respirol Case Rep 7(7):e00467

Polaczek M, Szaro P, Jakubowska L, Zych J, Religioni J, Orlowski TM (2020) Pulmonary veins variations with potential impact in thoracic surgery: a computed-tomography-based atlas. J Thorac Dis 12(3):383–393

Ramirez FD, Reddy VY, Viswanathan R, Hocini M, Jaïs P (2020) Emerging technologies for pulmonary vein isolation. Circ Res 127(1):170–183

Vanderlaan RD, Rome J, Hirsch R, Ivy D, Caldarone CA (2021) Pulmonary vein stenosis: treatment and challenges. J Thorac Cardiovasc Surg 161(6):2169–2176

Wittkampf FH, Derksen R, Wever EF, Simmers TA, Boersma LV, Vonken EP, Velthuis BK, Sreeram N, Rensing BJ, Cramer MJ (2002) Technique of pulmonary vein isolation by catheter ablation. Neth Hear J 10(5):241–244

Żabówka A, Jakiel M, Bolechała F, Jakiel R, Jasińska KA, Hołda MK (2020) Topography of the oblique vein of the left atrium (vein of Marshall). Kardiologia Polska (Polish Heart J) 78(7–8):688–693

Bronchial Veins

44

Erika Moxley and Dean John Hendrix Dante

44.1 Introduction

The bronchial circulation is a high-resistance, low-capacitance part of the systemic circulation. It is estimated to receive 1–2% of the total cardiac output and supplies oxygenated blood to the conducting airways, regional lymph nodes, vaso vasorum of the pulmonary arteries and veins, and the visceral pleura. The bronchial veins are essential in the lung, but their function, course, and regulation have yet to be extensively studied (McCullagh et al. 2010; Kumar et al. 2016).

44.2 Embryology

The bronchial circulation begins to develop during the fourth week of gestation as primitive bronchial arteries arise from the dorsal aorta. These primitive vessels eventually regress, and permanent bronchial arteries appear during weeks nine through twelve, correlating with the early canalicular stage of lung development (McCullagh et al. 2010; Sparrow and Weichselbaum 2003). By this stage, the branching of the bronchial tree is essentially complete. These vessels run adjacent to the peribronchial nerves and surrounding glia that developed during the previous pseudoglandular stage (Sparrow and Weichselbaum 2003).

44.3 Anatomy/Topography

The left and right bronchial arteries usually originate from the descending aorta, less commonly from the intercostal and internal thoracic arteries. The bronchial circulation follows the posterior aspect of the main stem bronchi, subdividing along the tracheobronchial tree and continuing distally to the level of the terminal bronchioles. They subdivide into a network of arterioles to form a submucosal plexus within the bronchial wall. This plexus further divides into capillaries and drains into submucosal and adventitial venules and, ultimately, into the bronchial veins (Kumar et al. 2016).

Venous drainage of the proximal conducting airway follows the same pattern as the arterial supply, with blood returning to the right heart via the azygos and hemiazygos veins. The more distal portions of the airway are drained by anastomoses between the bronchial and pulmonary veins, collectively referred to as bronchopulmonary circulation (McCullagh et al. 2010).

E. Moxley
Tulane University School of Medicine, New Orleans, LA, USA

D. J. H. Dante (✉)
School of Medicine, Louisiana State University Health Sciences Center, New Orleans, LA, USA
e-mail: ddante@lsuhsc.edu

S. J. Bordes, Jr. et al. (eds.), *The Clinical Anatomy of the Vascular System*,
https://doi.org/10.1007/978-3-031-78326-5_44

Marchand et al. found two types of bronchial veins, deep and superficial, communicating freely with the pulmonary veins in the aforementioned anastomotic system. The deep bronchial veins originate in the walls of the terminal bronchioles and eventually drain into the left atrium either directly or via the main pulmonary veins. The superficial veins from the right lung drain into the azygos vein, while those from the left drain into the hemiazygos, superior intercostal, or brachiocephalic veins. The deep and superficial bronchial veins are arranged so that the intrapulmonary bronchi and bronchioles drain into the deep veins and empty into the pulmonary venous system. In contrast, the subpleural and hilar regions drain into systemic veins via the superficial bronchial veins (Anderson et al. 2010; Marchand et al. 1950).

The deep bronchial veins are first apparent in the region of the terminal bronchiole. They form an intricate network surrounding the bronchiole and become more prominent as they ascend and coalesce into several larger longitudinal vessels along the outer wall of the larger bronchi. Several vessels unite near the hilum of each lung to form a single main trunk that terminates directly into the atrium or a main pulmonary vein close to the atrium. On the left, it lies anterior to the main stem bronchus just below the origin of the left upper bronchus. It typically enters the atrium below the two pulmonary veins, often near the inferior one. The termination of the main bronchial vein on the right is very similar to that of the left. Owing to the more superior position of the right upper lobe bronchus, the vein crosses the lower portion of the right main bronchus in relation to the middle lobe bronchus.

The superior bronchial veins, also termed the pleuro-hilar veins by Marchand et al., reflect the regions that drain into these vessels. They form a rich subpleural network on the anterior and posterior surfaces of the lungs, running beneath the pleura towards the lung hilum. At the hilum, the superior bronchial veins also receive tributaries from the extra-pulmonary bronchi, hilar lymph nodes, and the vasa vasorum of the pulmonary vessels. On the right, the superior bronchial veins coalesce to drain into the azygos vein before entering the superior vena cava. On the left, they drain into the hemiazygos, superior intercostal, or brachiocephalic veins.

44.4 Imaging/Histology

The bronchial veins are histologically similar to all other veins in the human body. Their walls consist of an outer tunica adventitia, a thin middle layer of tunica media, and an inner tunica intima. Bronchial veins have thinner walls than pulmonary veins and contain delicate bifoliate valves (Anderson et al. 2010).

44.5 Clinical and Surgical Significance

The pulmonary and bronchial circulations are connected via anastomoses throughout the lungs and are collectively referred to as the bronchopulmonary circulation. Approximately two-thirds of the drainage from the bronchial veins enters the pulmonary veins, the remaining one-third draining directly into the azygos and hemiazygos veins. In healthy individuals, these anastomoses constitute the normal right-to-left shunt as deoxygenated blood from the bronchial veins drains into the pulmonary vein and mixes with newly oxygenated blood.

Disease states such as lung injury and pulmonary embolism augment intrapulmonary shunting and prevent adequate systemic oxygenation. Processes causing increased back pressure from distal venous hypertension or vascular obstruction can engorge the central airway vasculature. Pulmonary venous hypertension can also result in the reversal of the blood flow from pulmonary veins into bronchial veins, leading to increased flow into the azygos system and subsequent engorgement and dilation of the venous plexus. Other etiologies include portal hypertension, fibrosing mediastinitis, congestive heart failure, and congenital cardiopulmonary defects.

Bronchoscopy presentations can range from mucosal congestion and hyperemia to vascular engorgement, development of bronchial varices, and hemoptysis. These lesions can easily bleed, so they should not be biopsied, and emphasis should be placed on treating the underlying cardiopulmonary defect (Kumar et al. 2016).

References

Anderson RH, Haworth SG, Rabinovitch M (2010) 7 Pulmonary circulation. In: Paediatric cardiology, 3rd edn. Churchill Livingstone/Elsevier, Philadelphia, pp 117–119

Kumar A, Raju S, Das A, Mehta AC (2016) Vessels of the central airways: a Bronchoscopic perspective. Chest 149(3):869–881. https://doi.org/10.1016/j.chest.2015.12.003. Epub 2015 Dec 14. PMID: 26836893

Marchand P, Gilroy JC, Wilson VH (1950) An anatomical study of the bronchial vascular system and its variations in disease. Thorax 5(3):207–221. https://doi.org/10.1136/thx.5.3.207

McCullagh A, Rosenthal M, Wanner A, Hurtado A, Padley S, Bush A (2010) The bronchial circulation--worth a closer look: a review of the relationship between the bronchial vasculature and airway inflammation. Pediatr Pulmonol 45(1):1–13. https://doi.org/10.1002/ppul.21135. PMID: 20025051

Sparrow M, Weichselbaum M (2003) The development of the bronchial circulation. Arch Physiol Biochem 111(4):353–355. https://doi.org/10.3109/13813450312331337586. PMID: 15764073

Esophageal Veins

45

Katherine D. Bordes, Annie Syed, and Stephen J. Bordes, Jr.

45.1 Anatomy

The esophagus can be divided into three parts that correlate with distinct sources of venous drainage. The inferior thyroid veins drain the cervical esophagus, the thoracic esophagus by the thoracic esophageal veins, and the abdominal esophagus, located inferior to the diaphragm, by the lower esophageal veins (Standring 2021).

Generally, blood from the esophagus drains first into the superficial submucosal plexus and then into the peri-esophageal venous plexus. The latter gives origin to the proper esophageal veins. The thoracic esophageal veins drain predominantly via the azygos venous system and, to a lesser extent, via the intercostal and bronchial veins. The lower esophageal veins drain into the left gastric vein and the portal venous system. This venous drainage includes an additional deep intrinsic venous plexus connecting the submucosal venous plexus to the peri-esophageal veins, which are absent from the thoracic portion of the esophagus. Although there are valves in these vessels, some bidirectional flow occurs, accommodating normal pressure changes during breathing and Valsalva maneuvers (Standring 2021).

45.2 Embryology

During the third week of embryogenesis, the earliest veins form from capillary plexuses derived from the mesoderm and carry blood into the primitive heart, the sinus venosus. The first veins that drain blood from each side of the embryo into the sinus venosus, the common cardinal veins, branch distally into the anterior and posterior cardinal veins. During the fifth week of development, the final portions of the posterior cardinal veins are formed, while the anterior veins change significantly to become the veins of the heart and the superior vena cava and its tributaries. On day 35 of gestation, the azygos vein and its tributaries are fully formed and functioning. The proximal esophageal veins are derived from the anterior cardinal veins, and the distal ones are derived from the posterior cardinal veins.

45.3 Histology

The esophageal veins drain deoxygenated blood from peripheral organs to the right side of the heart. It comprises three layers: the inner layer, the tunica intima; the middle muscular layer, the tunica media; and the outermost layer,

K. D. Bordes · A. Syed
Tulane University School of Medicine, New Orleans, LA, USA

S. J. Bordes, Jr. (✉)
Department of Surgery, Louisiana State University Health Sciences Center, New Orleans, LA, USA

S. J. Bordes, Jr. et al. (eds.), *The Clinical Anatomy of the Vascular System*,
https://doi.org/10.1007/978-3-031-78326-5_45

the tunica externa or adventitia. In contrast to arteries, where each layer is easily identifiable, the layers are difficult to differentiate in the walls of most veins.

The tunica intima is composed of endothelium and an internal elastic membrane. Endothelium comprises specialized simple squamous epithelial cells derived from mesenchyme. Endothelial cells act as barriers between blood and tissues and have focal regulation mechanisms that facilitate vasoconstriction, vasodilation, coagulation, and immune functions. In veins, the internal elastic membrane separating the tunicae intima and media is either absent or very thin and can be difficult to visualize on histological slides. Small to medium-sized veins also have valves formed by loose, pocket-shaped folds of the tunica intima that extend into the venous lumen. These valves promote unidirectional blood flow towards the heart. Retrograde flow is prevented by the valves pooling the blood and occluding the vein. This distinctive ability depends on the tunica media and the extent of smooth muscle contraction.

The middle layer, the tunica media, comprises a thin layer of smooth muscle cells and elastic fibers made of elastin and fibrillin. Smooth muscle cells, innervated by nerve fibers, help modulate venous blood flow and the ability of venous valves to prevent backflow. The elastic fibers provide resilience and the elasticity to accommodate changes in venous diameter if needed.

The tunica externa, or adventitia, attaches the vessel to its surrounding connective tissue. This layer prevents the overexpansion of veins and contains collagen, which anchors the veins to their surrounding structures, and a thin external elastic lamina.

45.4 Clinical Significance

Portal hypertension can result from acute or chronic liver disease such as hepatic portal thrombosis, liver cirrhosis, or pre- or post-hepatic venous thrombosis (Kovacs and Jensen 2019). Collateral circulation allows portosystemic shunting to develop between the left gastric and lower esophageal veins, forming esophageal varices, the most important clinical manifestation of this syndrome. Other clinical presentations include hemorrhoids (rectal varices), caput medusae, splenomegaly, and ascites. Esophageal varices are veins in the submucosal plexus that are abnormally dilated due to increased venous pressure. The valves become incompetent, there is retrograde blood flow, and the veins become susceptible to rupture and hemorrhage. Esophageal varices are more prevalent in males than females, and about half of all patients with this condition experience rupture and bleeding. Esophageal hemorrhage from varices has a 25% mortality rate (Standring 2021). Visible gastroesophageal vessels have a more than 50% chance of rebleeding, while ulcerations with visible clots and ulcerations over 2 cm in diameter carry a lower but still significant risk of rebleeding. All of the above warrant treatment upon endoscopic identification.

Common findings that indicate varices include hematemesis, hematochezia, and/or melena. The diagnostic screening tool used is upper GI endoscopy. Treatment involves administering vasoconstrictors and paravariceal endoscopic injection of a sclerosant, which forms a thrombus and fibrosis, to control the formation of collateral circulation and obliteration of the varices. Generally, epinephrine injection alone results in a high rebleeding rate, so it should be an adjunct to other endoscopic hemostatic agents such as clips, bands, and cautery. The transjugular intrahepatic portosystemic shunt (TIPS) procedure is commonly used if other treatments fail (Standring 2021). TIPS can worsen encephalopathy, but it serves as a bridge to liver transplant for the definitive resolution of portal venous hypertension should the problem be intrahepatic.

If there is a massive esophageal venous hemorrhage, a Sengstaken-Blakemore tube can be used as a bridge to definitive endoscopic, endovascular, or open surgical treatment. An endoscopic balloon extending from the gastric cardia to the proximal esophagus is inflated and tamponades venous bleeding. This procedure carries a very high risk of gastroesophageal ischemia and gastroesophageal perforation.

Open surgical procedures are rarely indicated to control esophageal venous hemorrhage but can be performed rapidly in an emergency. A portocaval or mesocaval shunt is created to decompress the portal venous system. For portocaval shunts, end-to-side, side-to-side (highest rate of encephalopathy), or central splenorenal shunts (distal splenic vein to left renal vein for retrograde portal flow) can be used. These shunts help to decrease ascites. Mesocaval shunts are more common and are favored because they preserve the porta hepatis for future liver transplants. Mesocaval shunts create an anastomosis between the superior mesenteric vein and the inferior vena cava (IVC) using a "C" graft. They are excellent when thrombosis does not involve the IVC, and liver function is preserved. They also treat ascites. Operative splenorenal shunts can be used in patients without splenectomy (short gastric vessels must be preserved) and have isolated gastroesophageal varices refractory to other treatments. The proximal splenic vein is ligated and anastomosed to the left renal vein. These shunts worsen ascites and are contraindicated in patients with large-volume ascites.

Isolated bleeding gastroesophageal varices due to splenic vein thrombosis can be treated with emergency splenectomy when they are refractory to other treatments.

References

Kovacs TOG, Jensen DM (2019) Varices: Esophageal, gastric, and rectal. Clin Liver Dis 23(4):625–642. https://doi.org/10.1016/j.cld.2019.07.005

Standring S (ed) (2021) Gray's anatomy: the anatomical basis of clinical practice, 42nd edn. Elsevier

46 Azygos Vein

Katherine D. Bordes

46.1 Anatomy

The azygos vein (Fig. 46.1) is usually found on the right side of the posterior mediastinum, ascending from the diaphragm and traveling along the eight inferior-most thoracic vertebral bodies, to empty into the superior vena cava (SVC). The following chapter describes the azygos vein as it is commonly seen; however, it is important to note that the origin, tributaries, anastomoses, and termination of this vein are all highly variable (Standring 2021).

The azygos vein is typically found retroperitoneally at the junction of the right ascending lumbar vein, lumbar azygos vein, and right subcostal vein. The lumbar azygos vein can stem from either the posterior aspect of the inferior vena cava (IVC) or the left renal vein. At the level of the twelfth thoracic vertebra, it is joined by the right ascending lumbar and subcostal veins to form the azygos vein. If the azygos lumbar vein is absent, the right ascending lumbar and subcostal veins alone form the azygos vein. The azygos vein enters the thoracic cavity by passing posterior to, or through, the right diaphragmatic crus, or in some cases the aortic hiatus (Standring 2021).

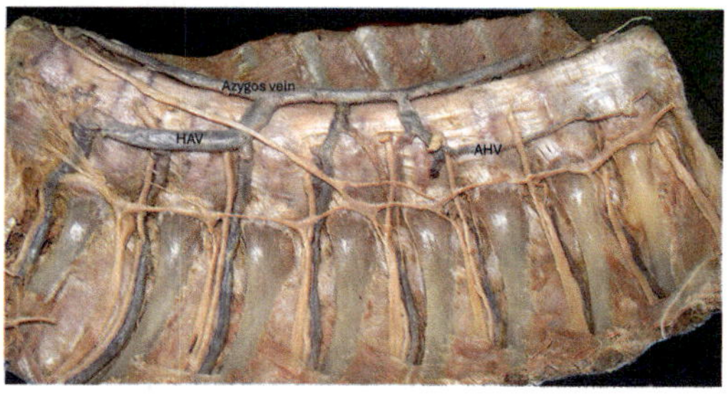

Fig. 46.1 Cadaveric dissection showing the azygos vein, accessory hemiazygos (AHV) and hemiazygos veins (HAV). (Courtsey Prof. M. Loukas)

As it ascends towards the SVC, the azygos vein lies on the spinal column, including the ventral aspect of the anterior longitudinal ligament, right posterior intercostal arteries, vertebral bodies, and intervertebral discs. To the left of the azygos vein, there is the thoracic duct, the descending thoracic aorta, esophagus, trachea, and the right vagus nerve as it arches over the hilum of the right lung. Pulsations from the descending aorta assist with venous return to the SVC owing to their close anatomical relationship. To the right of the azygos vein, there is the right thoracic sympathetic trunk, right greater thoracic splanchnic nerve, right lung, and parietal pleura. Of note, the azygos vein lies on the mediastinal surface of the right lung, forming a deep groove arching superior to the hilum and descending posterior to it. This azygos arch serves as a marker for the superior border of the lung hilum.

K. D. Bordes (✉)
Tulane University School of Medicine,
New Orleans, LA, USA

S. J. Bordes, Jr. et al. (eds.), *The Clinical Anatomy of the Vascular System*,
https://doi.org/10.1007/978-3-031-78326-5_46

After making this arch, the vein terminates in the posterior aspect of the SVC, just superior to the point where it passes into the pericardium. This anatamosis with the SVC is considered one of the anatomical landmarks for the sternal plane at the level of the inferior portion of the T5 vertebral body (Standring 2021).

Along its course, the azygos vein receives drainage primarily from the right posterior intercostal veins (PICVs). Typically, the second through fourth PICVs form a right superior intercostal vein that drains into the arch of the azygos vein, while the remaining inferior PICVs drain directly into the azygos vein. The first PICV is the only exception, emptying directly into the right brachiocephalic vein (Standring 2021).

Additional tributaries include the veins of the external vertebral plexuses, which drain partially into the azygos system and the caval system. Superficial bronchial veins, which drain the extrapulmonary bronchi, visceral pleura, and hilar lymph nodes, communicate with the pulmonary veins and drain into the azygos vein. The thoracic portion of the esophageal veins predominantly empty into the azygos vein. Distal esophageal veins drain into tributaries of the azygos or hemiazygos venous systems, in which bidirectional flow is possible, enabling pressure changes from breathing or Valsalva maneuvers to be accommodated (Standring 2021).

Numerous variable anastomoses exist between the azygos, hemiazygos, and accessory hemizygous veins, the IVC, and vertebral venous plexuses, allowing for collateral drainage pathways to the SVC. The hemiazygos vein joins the azygos vein at the level of T8 or T9, and the accessory hemiazygos joins more superiorly at the level of T7 (Standring 2021).

46.2 Embryology

In utero, the cardinal venous system is responsible for the return of blood to the developing heart. It comprises the precardinal and postcardinal veins. The precardinal veins form rostral to the heart and contribute to the formation of veins draining the head and upper extremities, including the SVC. The postcardinal veins are caudal to the heart and contribute to the formation of veins draining the lower extremities, including the IVC. As the embryo continues to grow, bilateral longitudinal channels and the anastomosing posterior cardinal system supplement the postcardinal veins to drain the developing mesonephros and gonads. More specifically, the subcardinal veins are found ventrally, and the subcentral, supracardinal, azygos line, and precostal veins are found dorsally (Standring 2021).

The proximal portion of the azygos vein, which drains into the SVC, is derived from the cranial supracardinal veins. The distal portion of the azygos vein is formed from the cranial part of the right postcardinal vein and the right azygos line (Standring 2021).

46.3 Variations

Failure of the superior segment of the right supracardinal vein to develop results in agenesis of the azygos vein; however, this is usually asymptomatic because the hemiazygos and accessory hemiazygos veins take over drainage of the right PICVs.

Agenesis of the IVC results when the right subcardinal vein fails to join with the hepatic sinusoids during embryo development. Venous blood is therefore shunted to the right supracardinal vein, which develops into the azygos system. Thus, the azygos vein is no longer a form of collateral circulation but rather the primary vessel returning venous blood from the lower extremities and abdomen to the heart. This condition occurs in 0.6–2.0% of patients with cardiovascular defects and 0.3% of otherwise healthy individuals (Standring 2021; Rivaud and Maldjian 2019).

In the case of cardinal and anterior cardinal vein malformation, the SVC fails to develop properly, and the azygos vein drains directly into the right atrium, right brachiocephalic vein, or right subclavian vein. Drainage into the right brachiocephalic or subclavian veins is thought to be caused by an elongated right common cardinal vein in utero (Standring 2021; Tubbs et al. 2016).

The hemiazygos or accessory hemiazygos veins are incompletely formed in 15% of individuals. In this case, the left PICVs drain directly into the azygos vein, which is found in a more midline position (Standring 2021; Tubbs et al. 2016).

The azygos vein arises from a combination of the lateral, intermediate, and/or medial roots. The lateral root involves the union of the ascending lumbar and subcostal veins, which occurs in 85% of cases. The intermediate root involves the lumbar azygos vein, seen in 34% of individuals. The medial root also occurs in 34% of cases; small plexiform veins closely associated with the anterior aspect of the lumbar vertebral bodies also contribute to forming the azygos vein (Tubbs et al. 2016).

As previously mentioned, the anastomotic connections between the azygos and hemiazygos veins are numerous and variable. One study of 200 cadavers revealed at least two anastomoses between these veins in 10% of the specimens, three anastomoses in 4%, and four or more in 6% (Tubbs et al. 2016).

Other tributary variations include the azygos vein receiving one or both supreme intercostal veins. The left superior intercostal vein joins the azygos vein in 57% of individuals. The left testicular, renal, and suprarenal (adrenal) veins can also terminate in the azygos vein (Tubbs et al. 2016).

One study examined variations in the course of the azygos vein with respect to the midline. Among 84 cadavers, it ascended along the right side (classical) in 5%, did not cross the midline in 30%, included part or all of its course left of the midline in 70%, and reached the extreme left side in 14% (Tubbs et al. 2016; Bales 2014).

When the azygos vein courses more laterally or superiorly along the apical region of the right lung's superior lobe, it forms an "azygos lobe." This positions the lung further into the pretracheal and retrotracheal mediastinum. Although this variation has no associated morbidity, it can be mistaken for a pulmonary nodule when found incidentally on imaging owing to the presence of a fine radiopaque line extending from the apex to the right tracheobronchial angle. This congenital anomaly is seen in 1.0% of cadaveric specimens and 0.4% of chest X-rays (Standring 2021; Rivaud and Maldjian 2019).

Typically, the azygos vein ranges from 3 mm to 7 mm in width on radiograph imaging; however, the width can reach 10 mm during pregnancy owing to increased physiological blood volumes and compensatory drainage through the azygos vein. Other conditions such as IVC thrombosis or congenital interruption of the IVC can cause the azygos vein to dilate to >10 mm, which is recognized as abnormal on imaging (Standring 2021; Rivaud and Maldjian 2019).

46.4 Clinical Significance

Conditions that increase the mean right atrial pressure, such as constrictive pericarditis, cardiac tamponade, pulmonary hypertension, and portal hypertension, can impede venous return. These hemodynamic changes can enlarge the azygos vein.

Obstruction of the SVC results in a condition called SVC syndrome. This is characterized by absent or decreased flow through the SVC because of an intraluminal blockage or external compression. As with agenesis of the SVC, this syndrome results in increased collateral blood flow through the azygos vein.

Many different conditions, such as embolism, tumors, or pregnancy, can result in occlusion or decreased blood flow through the IVC. IVC thrombosis results in increased collateral circulation through the azygos vein and also the hemiazygos and vertebral veins. Hepatic vein thrombosis, or Budd-Chiari syndrome, results in decreased outflow of blood from the liver via the IVC and thus congestion of the liver parenchyma. Consequently, there is increased venous blood flow through intrahepatic and portocaval vessels, which drain into the azygos and other systemic veins in order to decompress the liver. Most cases of IVC obstruction cause some azygos vein enlargement.

IVC interruption with azygos continuation is typically asymptomatic, as previously discussed in the context of embryology; however, it has been associated with several other cardiac and vascular abnormalities including polysplenia syndrome, dextrocardia, complete situs inversus, congenital pulmonary venolobar syndrome, pulmonary atresia, persistent left or duplicated SVC, recurrent deep vein thrombosis of the lower limbs, and sick sinus syndrome. It is also important to be aware of this variation during pediatric surgery, particularly in cases of congenital esophageal or tracheoesophageal atresia in neonates, in order to avoid ligating the primary vessel responsible for venous return to the heart. Lastly, it should be noted that on imaging, both a dilated azygos vein and continuation of the IVC with the azygos vein can mimic a mediastinal mass or lymphadenopathy (Standring 2021; Rivaud and Maldjian 2019).

Azygos vein aneurysms are rare and typically asymptomatic. Radiologically, such an aneurysm is defined as persistent dilation, either >3.75 cm at the arch or >2.5 cm in any other part of the vein; however, the aneurysm occurs almost exclusively in the arch due to the physiologic weakness in this area. Because the condition is rare and typically asymptomatic, there is no defined course of treatment. In general, conservative management is recommended unless there is symptomatic progression, in which case surgical intervention can be appropriate (Rivaud and Maldjian 2019; Wang et al. 2021).

Devices such as central venous catheters and peripherally inserted central catheters (PICCs) are placed in the SVC between the azygos arch and the atriocaval junction. Known complications of these procedures are the risk of thrombosis and fistulas. Placement of these devices can increase the risk of SVC syndrome, thrombosis, and fistula formation between the azygos vein and the airway.

The space between the esophagus and azygos vein forms the azygo-esophageal pleural recess. The appearance of this recess on imaging changes with age, looking more dextroconvex during early childhood and becoming progressively more concave by the age of twelve. However, any further displacement of this recess could signify a mass in the visceral mediastinal compartment, such as a bronchopulmonary foregut malformation, achalasia, or neoplasm (Standring 2021; Rivaud and Maldjian 2019).

References

Bales G (2014) A semi-quantitative approach to variation of the azygos vein course. Clin Anat 27(7):1030–1037. https://doi.org/10.1002/ca.22346

Rivaud Y, Maldjian PD (2019) The Azygos vein from A to Z. J Thorac Imaging 34(5) https://journals.lww.com/thoracicimaging/Fulltext/2019/09000/The_Azygos_Vein_From_A_to_Z.10.aspx

Standring S (ed) (2021) Gray's anatomy: the anatomical basis of clinical practice, 42nd edn. Elsevier

Tubbs RS, Shoja MM, Loukas M (2016) Bergman's comprehensive Encyclopedia of human anatomic variation. Wiley

Wang ZX, Yang LL, Xu ZN, Lv PY, Wang Y (2021) Surgical therapy for hemangioma of the azygos vein arch under thoracoscopy: a case report. World J Clin Cases 9(11):2655–2661. https://doi.org/10.12998/wjcc.v9.i11.2655

Accessory Hemiazygos and Hemiazygos Veins

47

Wesley Chastain

The accessory hemiazygos and hemiazygos veins (Figs. 47.1 and 47.2) are portions of the azygos venous system that drain the posterior mediastinum (Lachkar et al. 2021).

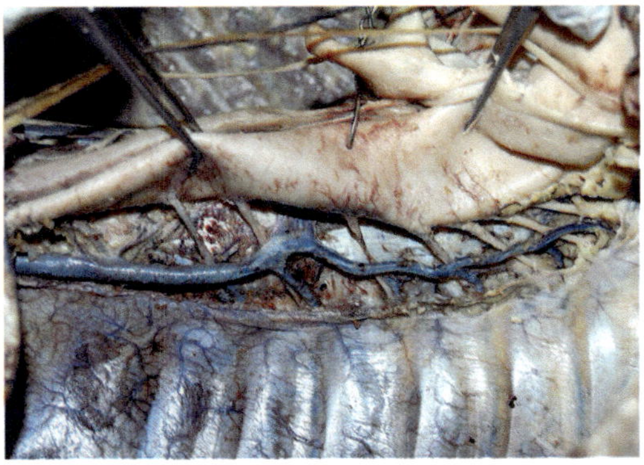

Fig. 47.1 Cadaveric view of the left posterior mediastinum noting the hemiazygos (left) and accessory hemiazygos veins (right). Note the aorta is retracted superiorly. (Courtsey Prof. M. Loukas)

47.1 Embryology

During the sixth week of development, the cardinal and subcardinal veins are united by the supracardinal veins, which are located posteriorly. The azygos venous system is formed by the cranial portion of the supracardinal veins. More specifically, the left supracardinal veins form the hemiazygos and accessory hemiazygos veins (Minniti et al. 2002; Byung-Boong 2012). The accessory hemiazygos vein courses along the left common cardinal vein and drains into the hemiazygos vein. Improper development of the superior vena cava (SVC) can influence where the hemiazygos vein is located. For example, a left-sided SVC arising from persistent left anterior and left common cardinal veins replaces the right-sided SVC; this can cause reversal of the azygos veins, where the remainder of the proximal left posterior cardinal vein, or the hemiazygos vein, can be found on the left side and drains directly into the SVC. This defect typically results from developmental arrest in the final stage of embryogenesis (Byung-Boong 2012; Lee et al. 2010). Another relatively frequent congenital anomaly is the presentation of an azygos continuation of the inferior vena cava (IVC) with a left SCV. This is described as blood flowing from right to left through an anastomosis of the azygos and hemiazygos veins, continuing to the left SVC, and finally to the right atrium of the heart (Byung-Boong 2012).

W. Chastain (✉)
Tulane School of Medicine, New Orleans, LA, USA
e-mail: wchastain@tulane.edu

S. J. Bordes, Jr. et al. (eds.), *The Clinical Anatomy of the Vascular System*,
https://doi.org/10.1007/978-3-031-78326-5_47

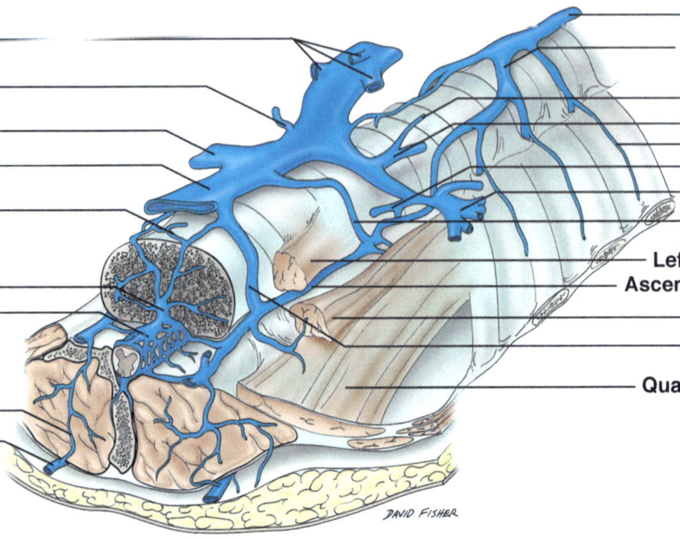

Fig. 47.2 Schematic view of the azygos and hemiazygos systems and the many interconnections. (After Woodburne)

47.2 Anatomy and Topography

In a study of 48 cadavers, the diameter of the hemiazygos vein was 3.17 ± 0.7 mm at its origin and 5.67 ± 1.17 mm at its termination. The mean diameter of the accessory hemiazygos vein at its termination was 5.65 ± 1.17 mm (Kutoglu et al. 2012).

Both the accessory hemiazygos and hemiazygos veins can vary in location and drainage.

The hemiazygos vein begins around T12-L2 and courses cranially on the left side of the vertebral column. It then crosses to the right side of the vertebral column and empties into the azygos vein around T8. Tributaries of the hemiazygos vein include the posterior left 8th–12th intercostal veins. The accessory hemiazygos vein is more variable than its counterpart. The most typical tributary of the accessory hemiazygos vein is the left superior intercostal vein, which drains the posterior 2nd–4th intercostal veins, along with the left posterior 5th–7th intercostal veins. It also courses to the left of the vertebral column but travels caudally. Around T7, the accessory hemiazygos vein empties into the azygos vein (Mozes and Gloviczki 2007).

The hemiazygos vein is located in the retrocrural space, a triangular region in the posterior inferior mediastinum that borders two crura of the diaphragm (Restrepo et al. 2008).

47.3 Variations

One study showed that 15% of 200 cadavers had incomplete accessory hemiazygos and hemiazygos veins (Celik et al. 1996). In such a case, left-sided posterior intercostal veins can drain directly into the azygos vein (Lindsay 1925). If the azygos vein is absent, both the right and left intercostal veins are drained by the accessory hemiazygos and hemiazygos veins (Minniti et al. 2002; Ozdemir et al. 2002).

Both veins usually form common channels posterior to the aorta. In rare instances (3.6%), the accessory hemiazygos and hemiazygos veins form common channels anterior to the aorta that are referred to as interazygos veins (Das and Paul 2017; Morton 1948). Another rare variation (1–2%) is the accessory hemiazygos vein draining directly into the brachiocephalic vein (Galwa et al. 2007). If the azygos vein duplicates, the hemiazygos veins are absent (Keskin et al. 2013).

47.4 Clinical/Surgical Significance

The hemiazygos veins are important for collateral blood flow if the SVC or IVC is obstructed (Mozes and Gloviczki 2007). Importantly, because of the relationship of the hemiazygos system to the vena cava, obstruction of either the SVC or IVC can shunt blood through the opposite vena cava to compensate for the return of blood to the heart. This phenomenon is associated with enlarged vessels in the hemiazygos system (Castellino et al. 1968). Collateral blood flow in this system can also be augmented in individuals who have no IVC or have an anomalous pulmonary venous return, portal hypertension, or Budd-Chiari syndrome (Castellino et al. 1968; Doyle et al. 1961; Leigh et al. 1954; Abrams 1957). Knowledge of anatomical variations in this region is important when surgical intervention is considered, especially for catheter placement (Blackmon and Franco 2011). Dilation of the hemiazygos vein should be considered as it can impede right heart catheterization using the femoral vein (van der Horst and Hastreiter 1981; Leggio et al. 2004).

When imaging is scrutinized, the hemiazygos system has the potential to be mistaken for pathology. Its presence or enlargement has been misinterpreted as irregular lymph nodes, retrocrural adenopathy, or a paratracheal mass (Blackmon and Franco 2011; Kandpal et al. 2008).

References

Abrams HL (1957) The vertebral and azygos venous systems, and some variations in systemic venous return. Radiology 69(4):508–526. https://doi.org/10.1148/69.4.508

Blackmon JM, Franco A (2011) Normal variants of the accessory hemiazygos vein. Br J Radiol 84(1003):659–660. https://doi.org/10.1259/bjr/13695502

Byung-Boong LEE (2012) Venous embryology: the key to understanding anomalous venous conditions. Damar CerDerg 19(4):170–182

Castellino RA, Blank N, Adams DF (1968) Dilated azygos and hemiazygos veins presenting as paravertebral intrathoracic masses. N Engl J Med 278(20):1087–1091. https://doi.org/10.1056/NEJM196805162782003

Celik HH, Sargon MF, Aldur MM, Cumhur M (1996) An anomalous course of the interazygos vein. Surg Radiol Anat 18(1):61–62. https://doi.org/10.1007/BF03207766

Das S, Paul S (2017) Preaortic interazygos vein: a case report. J Morphol Sci 21(3):0–0

Doyle FH, Read AE, Evans KT (1961) The mediastinum in portal hypertension. Clin Radiol 12:114–129. https://doi.org/10.1016/s0009-9260(61)80071-8

Galwa RP, Prakash M, Khandelwal N (2007) 16-MDCT depiction of accessory hemiazygos vein draining into the left brachiocephalic vein. Indian J Radiol Imaging 17(01):50–51

Kandpal H, Sharma R, Gamangatti S, Srivastava DN, Vashisht S (2008) Imaging the inferior vena cava: a road less traveled. Radiographics 28(3):669–689. https://doi.org/10.1148/rg.283075101

Keskin S, Keskin Z, Sekmenli N (2013) The independent right and left azygos veins with hemiazygos absence: a rare case presentation. Case Rep Vasc Med 2013:282416. https://doi.org/10.1155/2013/282416

Kutoglu T, Turut M, Kocabiyik N, Ozan H, Yildirim M (2012) Anatomical analysis of azygos vein system in human cadavers. Rom J MorpholEmbryol 53(4):1051–1056

Lachkar S, Iwanaga J, Newton E, Dumont AS, Tubbs RS (2021) Split Azygos vein: a case report. Cureus 13(2):e13362. Published 2021 Feb 15. https://doi.org/10.7759/cureus.13362

Lee AB, Laredo J, Neville R (2010) Embryological background of truncular venous malformation in the extracranial venous pathways as the cause of chronic cerebro spinal venous insufficiency. Int Angiol 29(2):95–108

Leggio M, Sgorbini L, Cicone E, Leggio F (2004) Aortic ectasia with hemiazygos vein dilatation: a mimicker of aortic dissection. Echocardiography 21(6):545–550. https://doi.org/10.1111/j.0742-2822.2004.03110.x

Leigh TF, Abbott OA, Rogers JV Jr, Gay BB Jr (1954) Venous aneurysms of the mediastinum. Radiology 63(5):696–705. https://doi.org/10.1148/63.5.696

Lindsay HC (1925) An Abnormal Vena Hemiazygos. J Anat 59(Pt 4):438

Minniti S, Visentini S, Procacci C (2002) Congenital anomalies of the venae cavae: embryological origin, imaging features and report of three new variants. Eur Radiol 12(8):2040–2055. https://doi.org/10.1007/s00330-001-1241-x

Morton WR (1948) Pre-aortic drainage of the hemiazygos veins; report of two cases. Anat Rec 101(2):187–191. https://doi.org/10.1002/ar.1091010205

Mozes G, Gloviczki P (2007) Venous embryology and anatomy. In: The vein book. Academic Press, pp 15–25
Ozdemir B, Aldur MM, Celik HH (2002) Multiple variations in the azygos venous system: a preaortic interazygos vein and the absence of hemiazygos vein. Surg Radiol Anat 24(1):68–70. https://doi.org/10.1007/s00276-002-0008-7
Restrepo CS, Eraso A, Ocazionez D, Lemos J, Martinez S, Lemos DF (2008) The diaphragmatic crura and retrocrural space: normal imaging appearance, variants, and pathologic conditions. Radiographics 28(5):1289–1305. https://doi.org/10.1148/rg.285075187
van der Horst RL, Hastreiter AR (1981) Congenital interruption of the inferior vena cava. Chest 80(5):638–640. https://doi.org/10.1378/chest.80.5.638

Cardiac Veins

48

Allyson Segall and Stephen J. Bordes, Jr.

48.1 Embryology

A general understanding of the development of the heart is crucial for understanding the embryology of the entire cardiac system. Heart development begins with the differentiation of cardiac progenitor cells into cardiac myoblasts just lateral to the primitive streak. "Blood islands" develop in the cardiogenic field and supply blood to the developing heart. Two cardiogenic plates of anterior splanchnic mesoderm plates fuse down the midline, forming the primitive heart tube with atrial and venous poles. The atrial pole is on the cranial aspect and shunts blood into the dorsal aorta, while the venous or ventral pole collects blood from systemic venous return. The linear heart tube then undergoes rightward looping, so it becomes S-shaped by day 28 (Gittenberger-de Groot et al. 2005). The atrial pole bends ventrocaudally and to the right while the venous pole bends dorsocranially and to the left. Between the 27th and 37th days of development, the septa form from endocardial cushions; they divide the atria from the ventricles (atrioventricular septum), the atrioventricular canals and valves, and the aortic and pulmonary channels. Around this time, the roof of the heart develops into the septum primum, the ends of which migrate to the endocardial cushions but do not fully fuse. This leaves an opening called the ostium primum. The endocardial cushions eventually fuse with the septum primum, but shortly thereafter, a perforation is produced by apoptosis of local cells, and the ostium secundum is formed, which allows blood to pass from the right to the left primitive atrium. The septum secundum forms to overlap the ostium secundum partially, but it does not fully partition the two atria, so a hole called the foramen ovale is left open. It is at this point that the adult heart begins to form.

The sinus venosus is a vein present only during embryo development. It serves to return blood to the heart before the venous system is fully developed (Chen et al. 2014). Endothelial cells from the sinus venosus migrate towards the external cardiac surface where they differentiate to form the walls of the coronary veins. During embryo development, the sinus venosus is flanked on either side by the right and left common cardiac veins. Its rightmost aspect merges with the right common cardiac vein to form the superior vena cava, while its leftmost aspect merges with the left common cardiac vein to form the coronary sinus (Saremi et al. 2012).

A. Segall
Tulane University School of Medicine,
New Orleans, LA, USA

S. J. Bordes, Jr. (✉)
Department of Surgery, Louisiana State University
Health Sciences Center, New Orleans, LA, USA

S. J. Bordes, Jr. et al. (eds.), *The Clinical Anatomy of the Vascular System*,
https://doi.org/10.1007/978-3-031-78326-5_48

48.2 Anatomy

The cardiac veins (Figs. 48.1 and 48.2) drain deoxygenated blood from the myocardium. This section covers the most common anatomy of the coronary vasculature, but it should be noted that the cardiac venous system presents many variations. It can be divided into two subsystems, the greater and the smaller cardiac venous systems (Sirajuddin et al. 2020).

The greater cardiac venous system consists of many branches and drains about 75% of the coronary blood. These veins are all tributaries of the coronary sinus, which is the largest cardiac vein. It drains most of the left ventricle (directly or indirectly) and portions of the left atrium and right ventricle. It runs on the posterior surface in the left atrioventricular groove or sulcus inferior to the pulmonary veins and superior to the circumflex branch of the left coronary artery (Tubbs et al. 2016). It returns blood into the posteromedial right atrium just inferior to the vena cava opening. At the junction of the coronary sinus and the right atrium lies the Thebesian valve; the valve of Vieussens lies at the junction of the coronary sinus and the great cardiac vein (Sirajuddin et al. 2020).

The only superior branch of the coronary sinus is the oblique vein of the left atrium, also known as the oblique vein of Marshall. This vein is 2–3 cm long and courses through the pulmonary veins and the left atrium (Sirajuddin et al. 2020). It is wrapped in the Marshall bundle, which is reported to have electrophysiological characteristics and connects the right and left atria (Han et al. 2010). As a remnant of the left superior vena cava, it can be obliterated in its superior aspect or entirely (Tubbs et al. 2016).

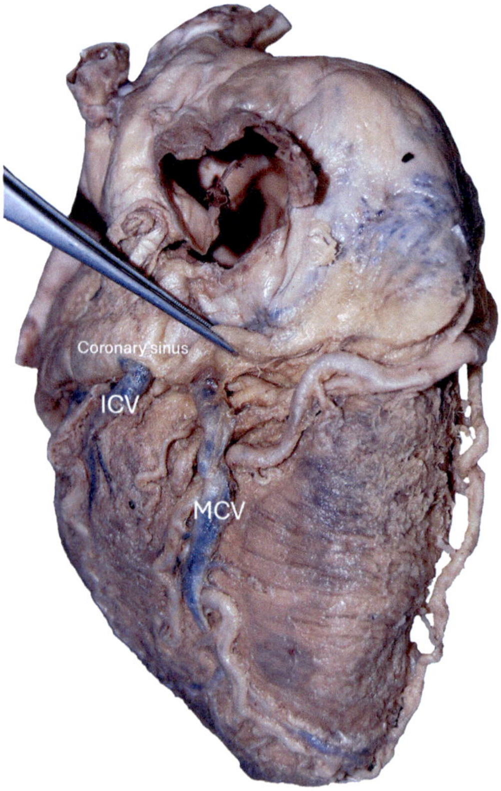

Fig. 48.1 Cadaveric heart illustrating the inferior ventricular (IVV) and middle cardiac veins (MCV) draining into the coronary sinus. (Courtsey Prof. M. Loukas)

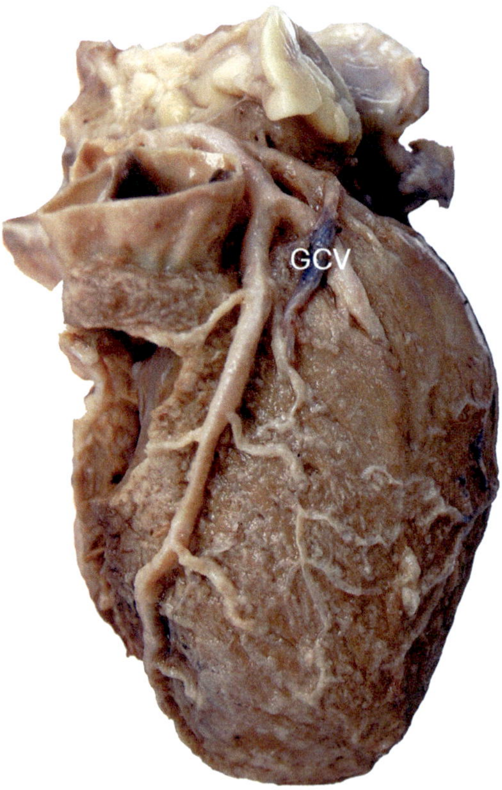

Fig. 48.2 Cadaveric heart showing the great cardiac vein (GCV) traveling with the anterior interventricular artery. (Courtsey Prof. M. Loukas)

Table 48.1 Blood drains from the left ventricle into the coronary sinus

Vein	Branches	Drains
Coronary sinus	Great cardiac vein Small cardiac vein Oblique vein of the left atrium Posterior veins of the left ventricle	Inferior portion of the left atrium Anterior wall of the right ventricle Myocardium of the left atrium Myocardium of the posterior left ventricle
Great cardiac vein	Anterior interventricular vein Left marginal vein	Anterior surface of the right and left ventricles and the interventricular septum Myocardium of the left ventricle
Small cardiac vein	Middle cardiac vein Anterior veins of the right ventricle (Right marginal vein)	Inferior walls of the left and right ventricles Myocardium of the anterior right ventricle

Most of the blood draining from the left atrium and right ventricle enters the right atrium directly, as does the blood from the right atrium itself. A small amount of blood from the left atrium and right ventricle drains into the coronary sinus before emptying into the right atrium (Tubbs et al. 2016)

As the coronary sinus passes under the pulmonary veins and continues in the anterior interventricular septum (AIVS), it becomes thinner and its name changes to the great cardiac vein. This transition is commonly demarcated by the joining of the oblique vein of the left atrium (Marshall) (Harikrishnan et al. 2005), which can occur anywhere between the distal half and the apex of the AIVS. The great cardiac vein travels in the atrioventricular groove along the inferior border of the left atrium as it wraps around from the posterior to the anterior surface. Once it reaches the apex of the left atrium (the termination of the left atrioventricular sulcus) on the anterior surface of the heart, its name changes again as its course is altered to travel inferiorly through the atrioventricular groove (Sirajuddin et al. 2020). As suggested by its location, it is now called the anterior interventricular vein, and it runs alongside the left anterior descending artery (LAD). The anterior interventricular vein forms the inferior border of the trigone of Brocq and Mouchet along with the LAD and left circumflex coronary arteries. The great cardiac vein has another tributary: the left marginal vein, also known as the obtuse marginal vein. This travels along the lateral border of the left ventricle and empties into the great cardiac vein before its transition into the coronary sinus (Tubbs et al. 2016).

Branching off the coronary sinus and traveling towards the right is the small cardiac vein, previously called the right coronary vein. This vein courses through the right atrioventricular groove onto the anterior cardiac surface. Its terminal branches are the anterior veins of the left ventricle. The main branch is termed the right marginal branch or acute marginal branch. The anterior veins can vary in number and can be absent altogether (Sirajuddin et al. 2020). Just to the left is the middle cardiac or inferior interventricular vein. This drains into the coronary sinus about 1 cm from its ostium and travels inferiorly to wrap around the apex of the heart to the anterior surface (Tubbs et al. 2016). Between the middle cardiac vein and the left marginal vein are the posterior veins of the left ventricle, which, like the anterior veins of the left ventricle, can vary in number or be absent (Table 48.1).

The smaller cardiac venous system is a deep network of sinusoids that drains the remaining 25% of deoxygenated blood from the deep one-third of the myocardium. This system is made up of Thebesian vessels, which drain more from the right side of the heart than the left. Averaging about 0.5 mm in diameter, these vessels are significantly smaller than their greater cardiac venous counterparts (Saremi et al. 2012).

48.3 Imaging

Imaging the cardiac venous system prior to cardiac surgery can be of great value. Veins of the heart are highly variable, so it is crucial to image the patient prior to an invasive procedure. Also, in a patient with heart failure, surgeons and interventionalists need to be aware of any obstruction or changes in heart anatomy prior to procedures. For example, a patient with a previous myocardial infarction can lack coronary sinus tributaries (Saremi et al. 2012). Many imaging techniques are used; a few are mentioned below.

Computed tomographic (CT) angiography allows blood flow through the cardiac veins to be mapped. An advantage of this technique is that it can show the coronary arteries, old infarctions, and the anatomy of the heart as a whole. This can provide a larger picture of potential variations (Saremi et al. 2012). It also can help to use dual-source CT, which yields a clearer image (Genc et al. 2013). However, CT has limitations, including regulations that limit radiation dosage and an inability to obtain clear images of more distal tributaries (Saremi et al. 2012).

Another common technique is retrograde venography via the coronary sinus. This is an invasive procedure as it involves inserting a catheter with an occlusive balloon through which contrast material is injected (Saremi et al. 2012). Veins can be viewed directly or indirectly, allowing for complete imaging. Direct visualization involves filling a vein with dye directly from the injection. Visualization is indirect when the veins fill with blood from communicating veins and bring contrast in the direction of blood flow (Harikrishnan et al. 2005). This technique also allows for drug administration during the procedure, which makes it advisable for mapping veins in a patient with a predetermined or chronic heart condition. One disadvantage of retrograde venography is that the balloon obscures portions of the coronary sinus (Saremi et al. 2012). Multiple images with alterations in the placement of the balloon can be used to mitigate this (Meisel et al. 2001).

An echocardiogram uses ultrasonics to track the movement of blood through the heart. A traditional echocardiogram is noninvasive. The transducer is placed on the chest of the patient, sending waves into the body that are reflected back and turned into images. A transesophageal echocardiogram (TEE) is obtained by inserting the transducer into the esophagus. It can be used to determine the diameters of blood vessels just as a CT would. It has been proven to work well on standard patients, but anatomical variations are challenging to identify (Dong et al. 2019). Another type of echocardiogram is a Doppler echocardiogram, which can be used to see blood traveling through vessels. Similarly, Doppler ultrasound shows blood flow and is a reliable technique for identifying specific developmental anomalies in a fetus. It has been used to visualize a left persistent superior vena cava, which is discussed below (Chaoui et al. 2014).

48.4 Clinical/Surgical Significance

As treatment of heart diseases continues to improve, long-term effects such as arrhythmia and heart failure are becoming increasingly common. Cardiac resynchronization therapy (CRT) is a common treatment for pacing the hearts of patients with severe heart failure. It involves inserting a biventricular pacemaker to regulate the contraction of the right and left ventricles. CRT-D, which adds a defibrillator, is recommended for patients at high risk for sudden cardiac arrest as it can administer an intense electric shock if the heartbeat becomes dangerously arrhythmic. CRT without a defibrillator is called CRT-P; it contains just the pacemaker component. This device contains three leads connecting the right atrium to both ventricles (Genc et al. 2013). The CRT-D device tends to be preferred over CRT-P but is only chosen when the patient meets particular CRT indications such as having an ejection fraction less than 35% (Kaya et al. 2017). Although the procedure is common, about one-third of patients who receive it do not respond optimally, if at all. The main reason is the lack of parameters that regulate AV and VV synchrony (Lunati et al. 2014).

Radiofrequency ablation (RFA) is another treatment option for patients with arrhythmias. RFA targets specific arrhythmias such as atrial flutter, preexcitation syndromes, and atrioventricular nodal reentry tachycardia (Fu 2015). Originally, direct current and high voltage were used for this method, but this created several other complications and required general anesthesia. Over time, it has been shown that it is safer to use a low voltage with alternating currents. This cost-effective procedure has success rates of around 90% and is safe for both children and adults (Manolis et al. 1994).

Administration of drugs through the coronary sinus is of great value for clinical treatment and procedures. Cardioplegic solutions are commonly administered during cardiac procedures in order to stop the heart temporarily. Multiple cardiothoracic surgeons have reported that compared to drug administration through the coronary arteries, administration through the coronary sinus is safe and potentially more effective (Menasché et al. 1989).

The lesser cardiac venous system is important in many heart conditions. If blood flow from the coronary arteries into the ventricles is compromised, as in coronary sinus atresia or hypoplasia, the venous system can adopt a reverse blood flow in order to compensate. This can cause the Thebesian veins to enlarge in conditions such as congenital heart disease or chronic myocardial infarction. In other clinical conditions such as congestive heart failure, the vessels of the greater cardiac system can be enlarged (Saremi et al. 2012).

In patients with a persistent left superior vena cava, the superior vena cava drains directly into the coronary sinus, causing the vessel to enlarge. The complications arising from this condition are nonspecific, making it difficult to diagnose. It can cause developmental problems with the left atrium and its relationship to the coronary sinus, called an unroofed coronary sinus. The roof of the coronary sinus can be fenestrated or completely unformed, allowing for communication between the coronary sinus and the left atrium. This left-to-right shunt can result in many drastic complications such as brain abscess or cerebral embolism (Bonardi et al. 2012). It can also lead to heart disease and is often closed with a stent.

48.5 Variations

Variations of the coronary venous system are numerous. The anatomy discussed above is based on the most common venous drainage pathways, but these pathways differ among individuals.

Beginning with the anterior/right portion of the heart, three pathways of drainage have been reported. About one-third of the population is classed as having type I drainage, whereby the small cardiac vein is non-existent and the anterior cardiac veins (including the right marginal vein) drain directly into the right atrium. In type I drainage, the anterior cardiac veins can drain individually into the right atrium, or they can drain into a common ostium in various patterns before draining into the right atrium. Type II drainage is present in about 30% of the population and is similar to type I except that there is a rudimentary small cardiac vein with a diameter of only 1–2 mm. The anterior cardiac veins can still join or drain separately into the right atrium. Type III drainage is also present in about 30% of the population. The small cardiac vein is as described in the anatomy section above with the right marginal vein. It opens into the coronary sinus but can also have openings into the right atrium along its path. The remaining anterior cardiac veins can drain individually into the right atrium, form a common ostium, or drain into the small cardiac vein. Finally, in type IV drainage, the small cardiac vein is prominent and terminates into the right atrium with no opening to the coronary sinus. In this scenario, the right marginal vein and most, if not all, of the anterior cardiac veins are tributaries of the small cardiac vein (Tubbs et al. 2016).

The posterior portion of the heart also varies, particularly in the drainage of the left ventricle. The left marginal vein, the middle cardiac vein, and the posterior veins of the left ventricle are generally responsible for this, but they can vary in size and length. Tributaries of both the middle cardiac vein and the left marginal vein can provide assistance. They vary in number, size, and location. In some cases, the posterior vein of the left ventricle is absent, so several branches can be seen entering nearby veins to drain the middle portion of the left ventricle. Sometimes, there are additional small veins that drain directly into the coronary sinus (Tubbs et al. 2016).

The middle cardiac vein can vary in number, size, course, and location. It can be duplicated, the second vein being located to the right or the left and therefore draining either the right ventricle or the left ventricle. It can differ in its course, traveling obliquely across the posterior left atrium and running in the interventricular sulcus just prior to joining the coronary sinus. Near its junction with the coronary sinus, the middle cardiac vein can become dilated (Tubbs et al. 2016).

Finally, variations in the coronary sinus are most abundant. The most drastic variation is complete absence. This is commonly found in patients with a persistent left superior vena cava; however, it can also occur without that persistent vessel (Foale et al. 1979). There can be differences in the veins that drain into the coronary sinus. One or more of them can drain directly into the right atrium via Thebesian vessels. The coronary sinus can become stenotic or enlarged for several reasons. All of these variations are relevant to catheterization and cardiac surgery and are, therefore, important to note.

References

Bonardi M, Valentini A, Camporotondo R (2012) Unroofed coronary sinus and persistent left superior vena cava: a case report. J Ultrasound 15(3):179–182. https://doi.org/10.1016/j.jus.2012.06.001

Chaoui R, Heling KS, Karl K (2014) Ultrasound of the fetal veins part 2: veins at the cardiac level. Ultraschall Med 35(4):302–321. https://doi.org/10.1055/s-0034-1366848

Chen HI, Sharma B, Akerberg BN et al (2014) The sinus venosus contributes to coronary vasculature through VEGFC-stimulated angiogenesis. Development 141(23):4500–4512. https://doi.org/10.1242/dev.113639

Dong QQ, Yang WY, Sun YP et al (2019) Comparison of transthoracic echocardiography with computed tomography in evaluation of pulmonary veins. BMC Cardiovasc Disord 19(1):315. https://doi.org/10.1186/s12872-019-01272-8

Foale RA, Baron DW, Rickards AF (1979) Isolated congenital absence of coronary sinus. Heart 42(3):355–358

Fu D (2015) Cardiac arrhythmias: diagnosis, symptoms, and treatments. Cell Biochem Biophys 73:291–296. https://doi.org/10.1007/s12013-015-0626-4

Genc B, Solak A, Sahin N, Gur S, Kalaycioglu S, Ozturk V (2013) Assessment of the coronary venous system by using cardiac CT. Diagn Interv Radiol 19(4):286–293. https://doi.org/10.5152/dir.2013.012

Gittenberger-de Groot A, Bartelings M, Deruiter M et al (2005) Basics of cardiac development for the understanding of congenital heart malformations. Pediatr Res 57:169–176. https://doi.org/10.1203/01.PDR.0000148710.69159.61

Han S, Joung B, Scanavacca M, Sosa E, Chen PS, Hwang C (2010) Electrophysiological characteristics of the Marshall bundle in humans. Heart Rhythm 7(6):786–793. https://doi.org/10.1016/j.hrthm.2010.02.028

Harikrishnan S, Nair K, Tharakan J (2005) Oblique vein of Marshall. Heart (British Cardiac Society) 91(2):e16. https://doi.org/10.1136/hrt.2004.044222

Kaya A, Tatlısu MA, Tekkesin Aİ, Alper AT (2017) CRT-D or CRT-P in CRT-indicated patients? Anatol J Cardiol 17(1):79–80. https://doi.org/10.14744/AnatolJCardiol.2017.7483

Lunati M, Magenta G, Cattafi G, Moreo A, Falaschi G, Contardi D, Locati E (2014) Clinical relevance of systematic CRT device optimization. J Atrial Fibrillation 7(2):1077. https://doi.org/10.4022/jafib.1077

Manolis AS, Wang PJ, Estes NA 3rd. (1994) Radiofrequency catheter ablation for cardiac tachyarrhythmias. Ann Intern Med 121(6):452–461. https://doi.org/10.7326/0003-4819-121-6-199409150-00010

Meisel E, Pfeiffer D, Engelmann L et al (2001) Investigation of coronary venous anatomy by retrograde venography in patients with malignant ventricular tachycardia. Circulation 104(4):442–447. https://doi.org/10.1161/hc2901.093145

Menasché P, Kucharski K, Mundler O, Veyssié L, Subayi JB, Le Pimpec F, Fauchet M, Piwnica A (1989) Adequate preservation of right ventricular function after coronary sinus cardioplegia. A clinical study. Circulation 80(5 Pt 2):III19–III24

Saremi F et al (2012) Coronary veins: comprehensive CT-anatomic classification and review of variants and clinical implications. RadioGraphics 32(1):E1. https://doi.org/10.1148/rg.321115014

Sirajuddin A, Chen MY, White CS, Arai AE (2020) Coronary venous anatomy and anomalies. J Cardiovasc Comput Tomogr 14(1):80–86. ISSN 1934-5925. https://doi.org/10.1016/j.jcct.2019.08.006

Tubbs RS, Shoja MM, Loukas M (2016) Bergman's comprehensive encyclopedia of human anatomic variation. Wiley Blackwell. Copyright@ 2016 by John Wiley & Sons, Inc

Veins of the Thoracic Wall

49

Jonathan Richard, Mallory Granier, Cameron Fontenot, Joshua Linhuber, and Stephen J. Bordes, Jr.

49.1 Embryology

Vascular formation begins in the human embryo on approximately day 18 of development (Gerecht-Nir et al. 2004). Attaining the complex vascularity of the adult involves extensive vascular regression and remodeling via vasculogenesis and angiogenesis to accommodate further growth and development. Specifically, veins of the thoracic wall undergo complex rounds of vascular modeling and remodeling. The formation of a three-membered system of paired veins consisting of the umbilical veins from the chorion, the cardinal veins from the embryo, and the vitelline veins from the yolk sac provides the framework for the eventual development of the thoracic wall veins.

The cardinal venous system, comprising the anterior and posterior pairs of cardinal veins, drains blood from the cranial and caudal regions of the developing embryo. As embryogenesis proceeds, the left anterior cardinal vein regresses, forming a shunt to the right anterior cardinal vein. This shunt eventually becomes the left brachiocephalic vein, while the right anterior cardinal vein becomes the right brachiocephalic vein. The posterior cardinal veins undergo remodeling and replacement to become the subcardinal and supracardinal veins, the latter of which drain the thoracic wall of the embryo (Yagel et al. 2010).

The internal thoracic veins form from anastomoses with the brachiocephalic veins described above. However, the supracardinal veins undergo further remodeling into the azygos and hemiazygos veins, providing initial anastomoses for the left and right intercostal veins (Yagel et al. 2010).

49.2 Anatomy

The veins of the thoracic wall follow a pathway nearly identical to that of the corresponding arteries, with very few significant variations being reported in the literature (Loukas et al. 2007; Wraight et al. 2005). Of note, the specifics of the typical anatomy of these veins are not documented in significant detail in the vast majority of sources. Most publications state only that the path of veins follows that of their arteries, and there is a significant lack of research dedicated to documenting the intricacies of these vessels further. The internal thoracic, anterior intercostal, and posterior intercostal veins are included and explored in this chapter.

J. Richard · M. Granier
School of Medicine, Louisiana State University Health Sciences Center, New Orleans, LA, USA

C. Fontenot · S. J. Bordes, Jr. (✉)
Department of Surgery, Louisiana State University Health Sciences Center, New Orleans, LA, USA

J. Linhuber
Tulane University School of Medicine, New Orleans, LA, USA

S. J. Bordes, Jr. et al. (eds.), *The Clinical Anatomy of the Vascular System*, https://doi.org/10.1007/978-3-031-78326-5_49

The internal thoracic veins (ITV), also known as the internal mammary veins, are major tributaries of the brachiocephalic veins in the neck. The ITVs begin as the venae commitantes, which travel from the inferior portions of the internal thoracic arteries and then typically form right and left common trunks at the manubriosternal joint (Loukas et al. 2007).

Following a pattern nearly identical to their corresponding arteries, the ITVs drain blood from the anterior intercostal veins, the anterior perforating branches of the ITV, and the two terminal branches, the musculophrenic and superior epigastric veins. The perforating branches, also referred to as sternal branches, run from the internal aspect of the rib cage to the external surface of the sternum to drain the corresponding arterial branches. The musculophrenic vein is formed by anastomosis of the 7th, 8th, and 9th sets of anterior intercostal veins. The superior epigastric vein receives venous blood from the superior portion of the abdominal wall and will be discussed later in this book.

The intercostal veins, comprising posterior and anterior divisions, travel alongside their corresponding arteries and nerves in the costal groove of the ribs, being the most superior of the three bundle components. The posterior and anterior intercostal veins, although they are separate veins draining as tributaries to other larger veins, anastomose at the midaxillary line, as do the nearby arteries.

The 12 pairs of posterior intercostal veins (PIVs) are distributed in a manner similar to the related arteries (Fig. 49.1). The 2nd, 3rd, and 4th PIVs on the same side anastomose together to form the left and right superior intercostal veins (Fig. 49.2), and the 12th pair of PIVs is commonly called the subcostal veins. The right and left-sided veins drain into the azygos and accessory/hemiazygos systems, large veins positioned just behind the thoracic aorta on the right and left sides of the body, respectively.

The anterior intercostal veins, again like their corresponding arteries, consist of nine sets of

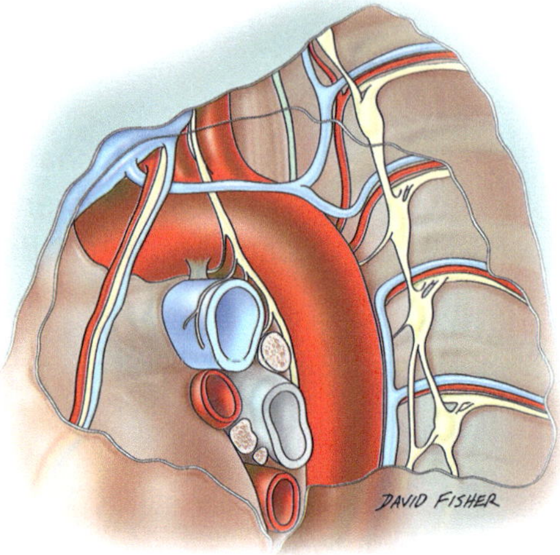

Fig. 49.1 Cadaveric dissection of the right posterior wall of the thorax noting several posterior intercostal veins (*) draining medially into the azygos vein

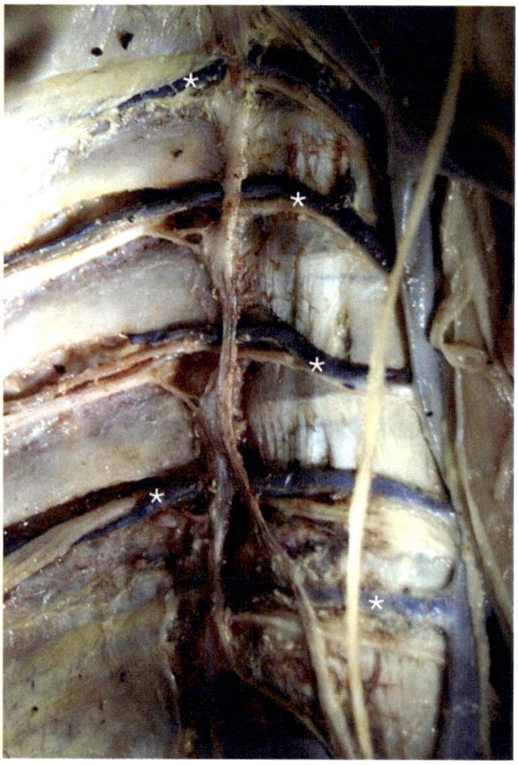

Fig. 49.2 Schematic drawing showing the left superior intercostal vein draining the upper intercostal spaces

veins traveling along the costal grooves of ribs 1–9. The first six sets drain directly into the ITVs, while sets 7–9 combine to drain into what is now called the musculophrenic vein before ultimately draining into the ITVs.

49.3 Variations

As previously stated, the veins of the thoracic wall have patterns of variation, but these are more sparsely documented than their arterial counterparts.

Loukas et al. noted significant variation amongst 100 cadavers in both the number of concomitant veins on the left and right sides and the point of bifurcation of the right or left common trunks forming the ITVs. In nearly half of the specimens examined, there were variations in the venae commitantes pattern, including two concomitant veins on the right and one on the left, two concomitant veins on the left and one on the right, and one concomitant vein on each side. Variations in the bifurcation point of the ITVs were noted, occurring at the 2nd, 3rd, or 4th ribs (Loukas et al. 2007).

Variants of the intercostal veins are not well documented, but in one example, the 10th, 11th, and 12th posterior intercostal veins form two anastomotic loops on the right side of the thoracic wall near the 10th–12th thoracic vertebrae (Badagabettu et al. 2016).

49.4 Pathology and Clinical/Surgical Significance

The thoracic wall veins are involved in many clinical and surgical diagnoses and treatments, several of which will be outlined below in view of their prevalence and documentation.

Migraine Medication-Induced Vasoconstriction

Triptans, medications commonly used in migraine treatment, have been associated with chest-related side effects. The origins of these side effects were not previously clear, but Visser et al. discovered that this class of medication can elicit strong vasoconstriction in the small veins of the thoracic wall. This vasoconstriction was found in 40–57% of patients, and 41% experienced chest symptoms when treated with triptans (Visser et al. 1996). This similarity in occurrence, according to Wackenfors et al., suggests a possible correlation with the percentage of patients treated with triptans in the general population who have chest-related symptom complaints (Wackenfors et al. 2005).

ITV Aneurysm

When diagnosing patients presenting with anterior mediastinal masses, common considerations for the origin of the mass include vascular, lymphatic, glandular (thyroid, parathyroid, thymus), and other tissue types. For vascular origins, aortic disease and pre-existing connective tissue disorders resulting in internal thoracic artery aneurysms are well documented (Common et al. 1999; Den Otter and Stam 1978; Phan et al. 1998). A case report of an anterior mediastinal mass with an ITV origin indicated that, although rare, this possible mass origin should also be considered in differential diagnoses (Almassi 2006).

Collateral Flow

The right superior intercostal vein (RSIV), formed by anastomosis of the right 2nd, 3rd, and 4th posterior intercostal veins, is important for collateral venous drainage during superior vena cava (SVC) obstruction/occlusion. The flow of blood through the RSIV in this collateral pathway depends on the level of obstruction in the vena cava. There is anterograde flow when the obstruction is parallel or superior to the point where the azygos vein drains into the SVC, and retrograde flow when it occurs anywhere inferior to this drainage point (Catalano et al. 1989).

Free-Tissue Transfers

The ITVs are ideal recipients for microsurgical reconstructions involving vascular attachments, including autologous breast reconstructions and reconstruction of chest wall defects (Catalano et al. 1989). Autologous breast reconstruction involves free-tissue transfers from, e.g., the abdomen, inner thigh, back, and buttocks. Most reconstruction techniques typically involve the ITVs as recipient veins when blood flow to the transferred tissue is re-established.

Autologous breast reconstruction, involving the ITV de facto, is used in a range of circumstances, from massive weight loss (surgical and non-surgical) to post-mastectomy reconstruction (Schwabegger et al. 2008; Yoo et al. 2022). The ITV has proven safe and reliable as a recipient vein, with very low rates of complications (Saint-Cyr et al. 2007; O'Neill et al. 2016).

References

Almassi GH (2006) Internal thoracic vein aneurysm presenting as an anterior mediastinal mass. J Thorac Cardiovasc Surg 131(2):500–501

Badagabettu SN, Shetty P, D'Souza MR (2016) A unique variation of azygos system of veins. J Cardiovasc Echogr 26(2):61–64

Catalano D, Pagliari U, Lapiccirella G (1989) The radiology of the right superior intercostal vein. RoFo: Fortschr Geb Rontgenstr Nuklearmed 151(5):582–585

Common AA, Pressacco J, Wilson JK (1999) Internal mammary artery aneurysm in Marfan syndrome: case report. Can Assoc Radiol J 50(1):47–50

Den Otter G, Stam J (1978) Aneurysm of internal mammary artery. Thorax 33(4):526–527

Gerecht-Nir S, Osenberg S, Nevo O, Ziskind A, Coleman R, Itskovitz-Eldor J (2004) Vascular development in early human embryos and in teratomas dervied from human embryonic stem cells. Biol Reprod 71(6):2029–2036

Loukas M, Tobola M, Tubbs RS, Louis R, Karapidis M, Khan I, Spentzouris G, Linganna S, Curry B (2007) The clinical anatomy of the internal thoracic veins. Folia Morphol 66:25–32

O'Neill AC, Hayward V, Zhong T, Hofer SO (2016) Usability of the internal mammary recipient vessels in microvascular breast reconstruction. J Plast Reconstr Aesthet Surg 69(7):907–911

Phan TG, Sakulsaengprapha A, Wilson M, Wing R (1998) Ruptured internal mammary artery aneurysm presenting as massive spontaneous haemothorax in a patient with Ehlers-Danlos syndrome. Aust NZ J Med 28(2):210–211

Saint-Cyr M, Youssef A, Bae HW, Robb GL, Chang DW (2007) Changing trends in recipient vessel selection for microvascular autologous breast reconstruction: an analysis of 1483 consecutive cases. Plast Reconstr Surg 119(7):1993–2000

Schwabegger AH, Ninković MM, Moriggl B, Waldenberger P, Brenner E, Wechselberger G, Anderl H (2008) Internal mammary veins: classification and surgical use in free-tissue transfer. J Reconstr Microsurg 13(1):17–23

Visser WH, de Vriend RH, Jaspers MW et al (1996) Sumatriptan in clinical practice: a 2-year review of 453 migraine patients. Neurology 47:46–51

Wackenfors A, Jarvius M, Ingemansson R, Edvinsson L, Malmsjö M (2005) Triptans induce vasoconstriction of human arteries and veins from the thoracic wall. J Cardiovasc Pharmacol 45(5):476–484

Wraight WM, Tweedie DJ, Parkin IG (2005) Neurovascular anatomy and variation in the fourth, fifth, and sixth intercostal spaces in the mid-axillary line: a cadaveric study in respect of chest drain insertion. Clin Anat 18:346–349

Yagel S, Kivilevitch Z, Cohen SM, Valsky DV, Messing B, Shen O, Achiron R (2010) The fetal venous system, Part I: normal embryology, anatomy, hemodynamics, ultrasound evaluation and Doppler investigation. Ultrasound Obstet Gynecol 35:741–750

Yoo A, Palines PA, Maier MA, Maddox SS, Hilaire HS, Stalder MW (2022) Autologous breast reconstruction with bilateral stacked free flaps in massive weight loss patients. Plast Reconstr Surg Glob Open 10(3):e4186

50 Lymphatics of the Thoracic Wall

Cameron Fontenot, Connor Plaisance, and Joshua Linhuber

50.1 Embryology

The lymphatic system begins to take form in the human embryo at approximately the sixth week of development. This primordial system appears as a collection of six lymph sacs: two jugular lymph sacs, two iliac lymph sacs, the cisterna chyli, and the retroperitoneal lymph sac. They all originate from the mesoderm and are extensions of veins, being similarly lined with endothelium. Eventually, the lymphatic sacs give rise to a network of lymph nodules interconnected via endothelium-lined channels, establishing the early lymphatic system in the human embryo.

C. Fontenot (✉)
Department of Surgery, Louisiana State University Health Sciences Center, New Orleans, LA, USA
e-mail: cfon18@lsuhsc.edu

C. Plaisance
School of Medicine, Louisiana State University Health Shreveport, Shreveport, LA, USA

J. Linhuber
Tulane University School of Medicine, New Orleans, LA, USA

50.2 Anatomy and Topography

The lymphatic system of the thoracic wall comprises networks of interconnected nodules in the axillary, intercostal, sternal, and diaphragmatic regions (Peters et al. 2009).

The parasternal or internal thoracic lymph nodes follow the internal thoracic arteries and average 4–5 nodules per individual. They collect lymphatic fluid from the breasts, deep anterior thoracic wall, supra-umbilical anterior abdominal wall, and superior surface of the liver. They typically anastomose with the paratracheal and brachiocephalic lymph nodes to form the bronchomediastinal trunk, which later drains into the internal jugular or subclavian vein, the subclavian trunk, or the thoracic duct (Standring 2020).

The intercostal lymph nodes are located in the posterior intercostal spaces near the heads of each rib and contain 1–2 nodules per space. They collect lymphatic fluid from the lateral portions of the thoracic wall, the superior nodes on the left side draining into the thoracic duct and those on the right side into the right lymphatic trunk. Inferior intercostal nodes, typically at the sixth and seventh intercostal spaces, anastomose to form a common duct and eventually drain into the thoracic duct (Standring 2020).

S. J. Bordes, Jr. et al. (eds.), *The Clinical Anatomy of the Vascular System*,
https://doi.org/10.1007/978-3-031-78326-5_50

The diaphragmatic (phrenic) lymph nodes are located on the superior surface of the diaphragm and can be divided into three distinct groups. The anterior group is located just below the inferior aspect of the sternum and consists of 2–3 nodules. These nodes collect lymphatic fluid from the anterior surface of the diaphragm and the diaphragmatic surface of the liver and drain into the parasternal lymph nodes. The lateral groups are located close to the entry point of the phrenic nerves on the diaphragm. They collect lymphatic fluid from the diaphragmatic surface of the liver and the central region of the diaphragm and drain it into the parasternal and brachiocephalic lymph nodes. The posterior group is located on the posterior surface of the diaphragm and collects lymphatic fluid from the associated area, draining into the para-esophageal lymph nodes (Standring 2020).

References

Peters AM, Fowler JC, Britton TB, Solanki CK, Ballinger JR, Ravichandran D, Mortimer PS, Purushotham AD (2009) Functional variation in lymph node arrangements within the axilla. Lymphat Res Biol 7(3):139–144

Standring SS (2020) Gray's anatomy, 42nd edn. Elsevier, Philadelphia

51 Thoracic Duct

R. Shane Tubbs

The thoracic duct typically arises from the cisterna chyli, and it carries lymph superiorly and drains it to the junction of the left internal jugular and subclavian veins (Figs. 51.1 and 51.2). While the thoracic duct is typically 45 cm in length (Smith et al. 2013). Van Pernis (1949) and Celis and Porter (1952) observed more variation, ranging from 36 to 45 cm. Jacobsson (1972) found that the diameter is usually greatest in the cervical region and smallest in the lower thoracic region. The external diameter fluctuated between 1.5 and 8 mm. Akcali et al. (2006) also measured the diameter of the thoracic duct. They found that at the levels of T5, T9, and T12, mean diameters were respectively 2.3, 2.6, and 3.3 mm. In addition, it seems to show an elastic property, being able to dilate to several times the normal width, depending on the amount of lymph flow (Smith et al. 2013).

In the retroperitoneum, the abdominal part of the thoracic duct begins at the formation of the cisterna chyli and tracks superiorly through the aortic hiatus into the thorax (Bartels 1909). As observed by Jdanov (1959), the cisterna chyli is not always present at the start of the thoracic duct.

R. S. Tubbs (✉)
Department of Neurosurgery, Tulane University School of Medicine,
New Orleans, LA, USA
e-mail: rtubbs@tulane.edu

The thoracic part of the thoracic duct is between an area where it courses after the aortic hiatus, between the aorta and azygos vein, to the superior thoracic aperture (Bartels 1909). It ascends through the posterior mediastinal cavity, anterior to the vertebral column, and varies between the right and the left from the midline. Variations in this region mainly concern the branching and doubling of the duct and how it courses up to the superior thoracic region.

Thoracic ducts in the thoracic region often show branching or doubling trunks connected by cross anastomoses that usually resolve to a single trunk. The plexus formation that was described in the abdominal part of the thoracic duct also continues through the thoracic part (Jacobsson 1972). Sometimes, the duct can communicate with the plexus of lymphatics around the aorta, at the level of the seventh and eighth thoracic vertebrae (Celis and Porter 1952). In addition, it can also connect with small plexuses of lymphatic capillaries around the aorta and esophagus (Celis and Porter 1952). Elliott and Thompson (2013) reported a thoracic duct that crossed the descending thoracic aorta in its ascent.

According to Jacobsson (1972), there are two more characteristics of the branching of the thoracic duct. One is insular doubling (Rouvière 1938), where the thoracic duct divides into two or more branches and eventually reunites. Occasionally, these two branches show different

S. J. Bordes, Jr. et al. (eds.), *The Clinical Anatomy of the Vascular System*,
https://doi.org/10.1007/978-3-031-78326-5_51

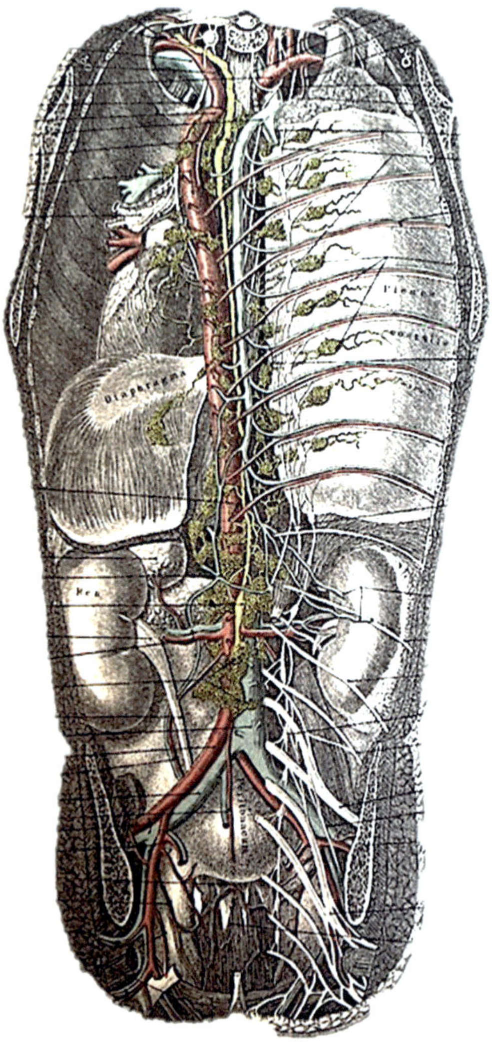

Fig. 51.1 Posterior view of the thoracic and abdominal anatomy showing the thoracic duct (yellow). (After Bock, 1890)

pathways, where one branch shows a normal course while the other shows "a most unexpected course" (Rouvière 1938). Celis and Porter (1952) examined that 4% of their total cadavers (50 total) showed division of the thoracic duct into two equal vessels, reuniting at the level of the fourth thoracic vertebra. Although Jacobsson (1972) observed more insular than plexiform configurations (24% and 8%, respectively), these duct divisions only accounted for 32% of the total specimens. In addition, the embryologic basis for the presence of insulae and plexiform divisions has been described, in which during embryologic development, numerous anastomoses between two thoracic ducts sometimes do not completely dissolve away, thus leaving behind residual communications that result in the formation of insulae or plexiform (Jacobsson 1972; Davis 1915).

The lymphatic vessels have been seen to join the thoracic duct along its thoracic part, coming from the intercostal spaces, para-aorta, and mediastinal nodes (Jacobsson 1972). Some lymph vessels can pass through the aortic hiatus from the abdominal region to join into the lower thoracic region of the duct (Jacobsson 1972). Others can discharge independently to the great cervical veins (Rouvière 1938).

As the thoracic duct ascends towards the cervical region, it starts from the right side of the median plane and crosses to the left side. Van Pernis (1949) has observed two sites of crossover from right to left. After dissecting 1081 cadavers, 44.4% (480) of the specimens crossed at the level of the fifth thoracic vertebral body, and 55.6% (601) of the specimens crossed at the level of the sixth thoracic vertebral body (Van Pernis 1949).

In rare cases of duplication of the thoracic duct, the origination of the doubling can be observed in the thoracic part of the duct (Shafiroff and Kau 1959). When the origination is below the thoracic region, in cases of complete bilateral thoracic ducts, both ducts pass through the aortic hiatus (Chen et al. 2006).

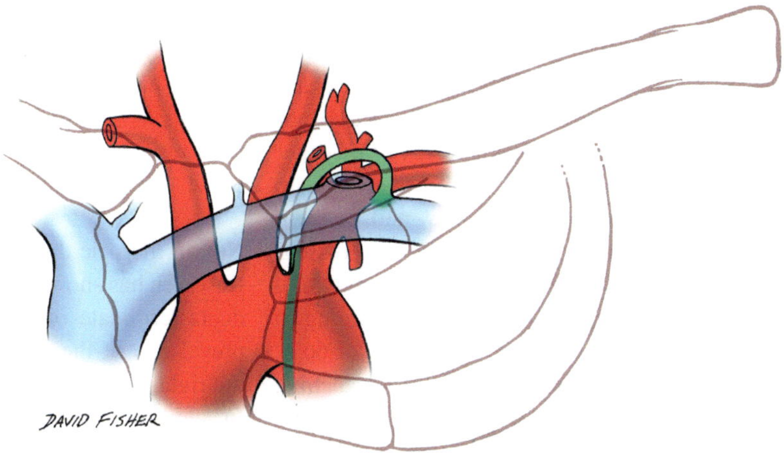

Fig. 51.2 Schematic drawing of the upper thoracic course of the thoracic duct (green) and its termination

51.1 Thoracic Duct Termination

When the thoracic duct reaches the termination site, it shows many different variation types regarding pattern and site of termination, and overall, when looking at the whole structure of the duct, the greatest number of variations is found in this region. Jacobsson (1972) identified nine different types of cervical thoracic ducts, classifying them by several branches and the presence of an island or plexus. Shimada and Sato (1997) also classified the termination of the duct by the location of the termination and the number of branches into four different types, each with multiple subtypes. Kausel et al. (1957) divided the variations into five different types, based on drainage sites and branching.

One of the most documented variations of the thoracic duct involves the location of the termination. The most frequent location of the termination is at the left internal jugular vein, followed by the junction of the left internal jugular and subclavian veins (venous angle) and the left subclavian vein. Less common and rare termination sites include the external jugular vein (Shimada and Sato 1997; Greenfield and Gottlieb 1956; Zorzetto et al. 1977; Jacobsson 1972), brachiocephalic vein (Jdanov 1959; Shafiroff and Kau 1959; Davis 1915; Rouvière 1938; Greenfield and Gottlieb 1956), transverse cervical vein (Kinnaert 1973), suprascapular vein (Zorzetto et al. 1977), and vertebral vein (Zorzetto et al. 1977).

The thoracic duct can join the venous system indirectly by terminating into the lymph nodes that are part of the cervical lymphatic chain, or cervical lymph nodes can join the terminal ampulla (Zorzetto et al. 1977). The cervical lymphatic channels have been observed to join the thoracic duct near its termination. Still, it can also drain independently or form a common trunk by joining with a horizontal lymphatic vessel (Kinnaert 1973).

The termination point within each vessel can vary from the anterior and lateral surfaces to the posterior surface. When the single duct terminates into the left internal jugular vein, it most often terminates into lateral surfaces, followed by anterior surfaces (Greenfield and Gottlieb 1956). Additionally, it terminates into the posterior surfaces (dorsal aspect) least often (Greenfield and Gottlieb 1956; Jacobsson 1972). For the venous angle and subclavian vein, termination to the frontal aspect is the least common (Jacobsson 1972). In addition, while the termination to the left subclavian vein is mainly between the internal and external jugular veins, Greenfield and Gottlieb (1956) have found one instance where the termination was lateral to the junction of the external jugular vein.

Another most documented variation involves the branching of the duct. Similar to the thoracic region, when the duct divides into multiple channels, they reunite and form a common trunk or duct (Zorzetto et al. 1977; Kinnaert 1973; Seeger et al. 2009; Rouvière 1938; Shimada and Sato 1997; Jacobsson 1972; Parsons and Sargent 1909; Archimbaud et al. 1969; Rocca-Rossetti et al. 1961). However, the incidence of reuniting into a common duct varies greatly according to different authors, ranging from 7% to 66%. Although some authors note that most branching results in merging into a common trunk, other studies have shown opposite results. For example, Zorzetto et al. (1977) found that 15.69% of the cases showed multiple channels reuniting to form a single ending branch, whereas Kinnaert's (1973) study showed a much higher rate, 65.90% in 47 studied cases. In addition, a radiologic technique using a high-resolution ultrasonographic (US) examination was performed by Seeger et al. (2009), showing that in most cases where the branching occurred, multiple channels merged to a single lumen, and only in 21 of 155 cases did channels terminate separately at different locations.

There has also been research done to show that the branches do not reunite but instead, they stay separate and terminate independently. Past research has shown that between 10% and 41% of the cadavers studied have shown multiple channels that terminate separately (Zorzetto et al. 1977; Kinnaert 1973; Parsons and Sargent 1909; Davis 1915; Greenfield and Gottlieb 1956; Shafiroff and Kau 1959; Archimbaud et al. 1969; Rocca-Rossetti et al. 1961; Jacobsson 1972; Van Pernis 1949; Correia 1926). The branches can be double, triple, or even multiple vessels, which can terminate in many different combinations of veins (Shimada and Sato 1997; Rouvière 1938; Kaur et al. 2012; Seeger et al. 2009; Tesauro 1966; Jdanov 1959). However, as the number of branches increases, the prevalence rate decreases (Parsons and Sargent 1909; Davis 1915; Jdanov 1959; Jacobsson 1972; Greenfield and Gottlieb 1956). Nevertheless, regardless of whether the branches reunite or terminate separately, the frequency of the branching is much more prevalent in the cervical region compared to the thoracic region (Jacobsson 1972).

51.2 Right-Sided Termination

Not all thoracic ducts terminate on the left side. Research has shown that some percentages of the ducts terminate on the right side. The general prevalence of the thoracic duct terminating into the right venous system is between 1% and 11%. In addition, similar to the variation on the left side, the termination endings can be single or multiple, where they terminate in the right internal jugular vein (Greenfield and Gottlieb 1956; Kausel et al. 1957; Davis 1915; Watson 1872; Peña and Zuñiga 2009), right venous angle (Van der Putte and van Limborgh 1980; Seeger et al. 2009; Nathan and Seidel 1983; Wirth and Kubik 1980; Epstein and Debord 2002; Rouvière 1938; Okumura and Fujimoto 1974; Chen et al. 2006; Nathan and Gitlin 1968), or right subclavian vein (Clark 1953; Celis and Porter 1952).

There are three main forms of variations that can terminate on the right side, which Lane and Todd (1847) developed one of the earliest classification systems:

I. Bilateral thoracic ducts may exist, with one terminating into the left and another into the right venous angle.
II. A single thoracic duct may bifurcate, with one branch terminating at the left venous angle and another branch at the right venous angle.
III. A single duct may drain into the right venous angle.

It is difficult to estimate the prevalence for each classification because the literature that has reported right-sided termination of the thoracic duct did not specify whether it was due to a bilateral duct system, bifurcation, or a single duct system. However, some findings suggest that single duct terminating at the right venous angle is less prevalent than bilateral termination (Wirth and

Kubik 1980; Seeger et al. 2009). In addition, the left-sided thoracic duct is less likely to be present if the thoracic duct is right-sided (Watson 1872).

There have been few reports where a thoracic duct terminating at the right venous angle is associated with venous malformation, such as aberrant retroesophageal right subclavian artery (RRSA) (Nathan and Gitlin 1968; Nathan and Seidel 1983; Orts Llorca 1936). When the thoracic duct is in the superior mediastinum, its normal path can be disrupted by the anomalous right subclavian artery, and instead of ascending towards the left side, it can track the artery until it terminates into the right venous angle (Nathan and Seidel 1983). In one case, the thoracic duct branched into two while tracking the inferior surface of the right subclavian artery, which reunited into one common trunk just before the venous angle termination (Nathan and Gitlin 1968). From 29 cases where RRSA was observed, thoracic ducts following the artery to the right were observed in 24 cases (Orts Llorca 1936). It is important to note, though, that not all endings at the right venous angle are accompanied by anomalous large arteries, and the extent to which an RRSA is associated with a right-sided termination is not known and requires further research (Nathan and Gitlin 1968; Nathan and Seidel 1983). In addition to the association between the RRSA and the right-sided venous angle termination, there can be, in very rare cases, a third association relating to the truncus bicaroticus, although the authors did not elaborate further (Nathan and Gitlin 1968). Lastly, Ranade et al. (2006) reported a rare case of lymphatic-venous communication, where a left branch from the duplicated thoracic duct terminated into the inferior vena cava at the level of the lower border of the L3 vertebra.

51.3 Duplication of the Thoracic Duct

As noted in previous sections, rare cases of duplication of the thoracic duct have been reported, where branches terminate on both the right and left sides (Davis 1915; Lachapèle et al. 1964; Yamada 1932; Chen et al. 2006; Shafiroff and Kau 1959; Rouvière 1938; Van Pernis 1949; Celis and Porter 1952; Ranade et al. 2006; Jacobsson 1972). The ducts can be partially duplicated, which often shows an interconnected network of anastomoses between the two partially duplicated ducts (Davis 1915). When it duplicates, the duplication site is often in the upper part of the thoracic course, which can be located in both the thorax and the neck (Shafiroff and Kau 1959; Okumura and Fujimoto 1974). Van Pernis (1949) found that the right thoracic ducts joined the left thoracic duct at the levels between the second and third thoracic vertebra. Similarly, Jacobsson (1972) noted that the branch left the main trunk just above the aortic arch to terminate into the right side of the neck. Contrastingly, a case report by Ranade et al. (2006) revealed a "Y-shaped" bifurcation at the level of the T12 vertebra without the presence of cisterna chyli. The lengths of the two branches were significantly different, measuring 8.5 and 12.5 cm for the right and left branches, respectively (Ranade et al. 2006).

In one case, Akcali et al. (2006) observed two different lymphatic drainage systems, which covered the vertebral bodies between the 7th and the 12th thoracic vertebrae. Left ascending branches formed the main duct, while inferior and right tributaries formed the right ascending trunk, eventually joining with the main duct near the 5th thoracic vertebrae (Akcali et al. 2006). In another case, the thoracic duct twice formed a ring shape by looping itself around the vertebrae, once at the upper thoracic region and another at the thoracolumbar junction (Akcali et al. 2006).

In extremely rare circumstances, completely bilateral thoracic ducts have been reported (Yamada 1932; Chen et al. 2006). In one particular case reported by Chen et al. (2006), bilateral thoracic ducts were accompanied by a persistent left superior vena cava (SVC). From both sets of cisternae chyli, the left and right thoracic ducts began at the level of 1st lumbar vertebra and entered the thoracic region via the aortic hiatus, ascending together until being separated at the level of the 5th thoracic vertebra (Chen et al. 2006). An anastomotic branch was present between the two at the level of the 10th thoracic vertebra, and the two eventually terminated at their corresponding right and left venous angles (Chen et al. 2006).

As mentioned earlier regarding the association between right-sided termination and venous malformation, the coexistence of venous and lymphatic variations is common. In addition to the association between the right-sided termination and aberrant retroesophageal right subclavian artery, Yamada (1932) reported complete double thoracic ducts with a right-sided aortic arch. In addition, the coexistence of double persistent left superior vena cava and bilateral thoracic ducts has been discovered, and the author notes that this incidence is very rare (Chen et al. 2006).

51.4 Tributaries

Usually, the cervical lymph trunks, which consist of jugular, bronchomediastinal, and subclavian lymph trunks, join the thoracic duct near the termination. However, this is not always the case. Due to the variations of the tributaries, lymph from the thoracic duct, the left upper limb, and the left side of the head and neck do not always drain together. The cervical lymph trunks can terminate into a number of vessels—including the venous angle, internal jugular vein, subclavian vein, and transverse cervical vein—independently without joining the thoracic duct (Kinnaert 1973; Davis 1915; Lane and Todd 1847). In one instance, the cervical lymph trunk joined the horizontal lymphatic vessel and formed a common trunk, which then terminated into the jugular vein (Kinnaert 1973). In addition, of the three lymph trunks, the left bronchomediastinal trunk is least likely to join the thoracic duct directly, but instead, it often joins at the venous angle (Kwon et al. 2002; Langford 2002).

51.5 Valves

Valves are present throughout the course of the thoracic duct. It can vary in size, number, type, and the characteristics of the cusps, depending on the location of the thoracic duct. Jacobsson (1972) came up with six different types of valves in the thoracic duct, where different types are location-specific. For example, type A is more present in the abdominal and cervical parts, whereas type B is more in the thoracic part of the duct (Jacobsson 1972). In addition, type D valves are most common at the junction with the venous system at the terminal endings, especially when the termination is at an oblique angle (Jacobsson 1972). Although most of the valves are bicuspid (Jacobsson 1972; Gray 1942; Seeger et al. 2009; Rouvière 1938; Bartels 1909), tricuspid valves have also been discovered (Jacobsson 1972).

The number of valves that are present throughout the course of the thoracic duct varies. The valves become more numerous as the duct reaches the termination site. On average, Jacobsson (1972) found 4.6 valves inferior to the diaphragm, 5.9 valves between the diaphragm and the aortic arch, and 11.1 valves between the aortic arch and its termination site.

51.6 Caliber and Diameter

Similar to the valves, the diameter of the thoracic duct changes depending on the location throughout the course. When the thoracic duct reaches the thoracic area, the diameter noticeably narrows, marking the smallest external diameter and gradually becoming wider as it ascends toward its termination (Jacobsson 1972). In the cervical area, the duct widens until it reaches the widest point near the termination (Jacobsson 1972). Overall, the thoracic duct is widest at the cervical region and narrowest at the thoracic region (Jacobsson 1972; Correia 1926).

The diameters measured by different authors show variable results. Van Pernis (1949) measured the external diameter to be between 2 and 4 mm. Using lymphangiography, Nasbaum et al. (1964) and Hidden and Florent (1966) measured the diameter to be between 2 and 6 mm and 0.5 and 5 mm, respectively. Shafiroff and Kau (1959) measured the widths to be 5–12 mm. Kinnaert (1973) observed that the diameters were between 2 and 5 mm when there were single duct terminations. In contrast, when the multiple channels ended in a short common duct, the diameters were between 0.5 and 3 mm. Akcali et al. (2006) showed that at T5, T9, and T12 levels, mean diameters were 2.3, 2.6, and 3.3 mm. The mean diameter of 2.5 mm was found in US imaging (Seeger et al. 2009).

A dilatation, or an ampulla, is common, especially in the terminal side of the thoracic duct (Kinnaert 1973; Correia 1926; Zorzetto et al. 1977; Seeger et al. 2009; Jacobsson 1972). Kinnaert (1973) noted that ten of the 47 patients studied showed an ampulla at the extremity of the thoracic duct, sometimes associated with lymph nodes. In addition, Jacobsson (1972) observed that ampullary dilatation is often present before the terminal valves, especially when the duct terminates into jugular, subclavian, or bronchomediastinal trunks. Zorzetto et al. (1977) observed that 20 of 51 cases showed an ampullar dilatation at the final portion of the thoracic duct.

References

Akcali O, Kiray A, Ergur I, Tetik S, Alici E (2006) Thoracic duct variations may complicate the anterior spine procedures. Eur Spine J 15:1347–1351

Archimbaud JP, Bansillon V, Bernhardt J, Bonnet P, Saubier EC (1969) Quarante dissections du canal thoracique. Sa terminaison chez l'homme vivant. Lyon Med 30:211–214

Bartels P (1909) Das Lymphgefa¨sssystem. In: Bardeleben's Handbuch der Anatomie des Menschen. Jena, Gustav Fischer, pp 85–99

Celis A, Porter JK (1952) Lymphatics of the thorax; an anatomic and radiologic study. Acta Radiol 38:461–470

Chen H, Shoumura S, Emura S (2006) Bilateral thoracic ducts with coexistent persistent left superior vena cava. Clin Anat 19:350–353

Clark ER (1953) The lymphatic system and the spleen. In: Schaeffer JP (ed) Morris' human anatomy, 11th edn. The Blakiston Division, McGraw-Hill Book Company, Inc, New York, pp 864–867

Correia M (1926) Le canal thoracique chez l'homme. Folia Anat Univ Conimbr 1:1–20

Davis HK (1915) A statistical study of the thoracic duct in man. Am J Anat 17:211–244

Elliott GE, Thompson KL (2013) A previously undocumented pre-aortic pathway of the thoracic duct. A case report. Int J Anat Var 6:221–224

Epstein DA, Debord JR (2002) Abnormalities associated with aberrant right subclavian arteries—a case report. Vasc Endovasc Surg 36:297–303

Gray H (1942) Anatomy of the human body (Gray's anatomy), 24th edn. Lea and Febiger, Philadelphia, p 692

Greenfield J, Gottlieb MI (1956) Variations in the terminal portion of the human thoracic duct. AMA Arch Surg 73:955–959

Hidden G, Florent J (1966) Radio-anatomical study of the opacified thoracic duct by segmental lymphography. J Chir 91:373

Jacobsson SI (1972) Clinical anatomy and pathology of the thoracic duct: an investigation of 122 cases. Almqvist and Wiksell, Stockholm, pp 9–103

Jdanov DA (1959) Anatomie du canal thoracique et des principaux collecteurs lymphatiques du tronc chez l'homme. Acta Anat 37:20–47

Kaur D, Jain M, Dhall U, Singh K (2012) An unusual course of the thoracic duct in relation to the vertebral vessels. Singapore Med J 1:e1–e2

Kausel HW, Reeve TS, Stein AA, Alley RD, Stranahan A (1957) Anatomic and pathologic studies of the thoracic duct. J Thorac Surg 34:631–641

Kinnaert P (1973) Anatomical variations of the cervical portion of the thoracic duct in man. J Anat 115:45–52

Kwon SS, Falk A, Mitty HA (2002) Thoracic duct injury associated with left internal jugular vein catheterization: anatomic considerations. J Vasc Interv Radiol 13:337–339

Lachapèle AP, Hughes A, Lagarde C (1964) De l'ètude anatomo-radiologique du canal thoracique d'apres 60 opacifications sur l'etre humain vivant. J Radiol Electrol Med Nucl 45:1–10

Lane S, Todd RB (eds) (1847) The cyclopaedia of anatomy and physiology. Sherwood, Gilbert and Piper, London, p 3

Langford RJ (2002) Valves in the subsidiary lymph trunks in the neck. J Craniomaxillofac Surg 30:121–124

Nasbaum M, Baum S, Hedges RC, Blakemore WS (1964) Roentgenographic and direct visualization of thoracic duct. Arch Surg 88:105–113

Nathan H, Gitlin G (1968) Thoracic duct terminating on the right side associated with aberrant retro-oesophagal right subclavian artery and truncus bicaroticus. Thorax 23:266–270

Nathan H, Seidel MR (1983) The association of a retroesophageal right subclavian artery, a right-sided terminating thoracic duct, and a left vertebral artery of aortic origin: anatomical and clinical considerations. Acta Anat 117:362–373

Okumura K, Fujimoto T (1974) A case of anomalous right subclavian artery passing behind the esophagus with a rare variation of the thoracic duct. Acta Anat Jpn 49:179–183

Orts Llorca F (1936) Über den rechtsseitigen Verlauf des Ductus thoracicus bei erhaltener rechter Aortenwurzel. Anat Anz 81:283–290

Parsons FG, Sargent PWG (1909) On the termination of the thoracic duct. Lancet 173:1773–1774

Peña E, Zuñiga J (2009) An anatomical variation: the aberrant termination of the thoracic duct [in Spanish]. Biomedica 29:204–208

Ranade AV, Rai R, Prabhu LV, Kumaran M (2006) A rare lymphatico-venous communication associated with duplication of thoracic duct. Int J Morphol 2:155–158

Rocca-Rossetti S, Aresu R, Liguori G (1961) Ricerche sulla composizione della linfa umana. I. Anatomia chirurgica del dotto toracico al collo e tecnica del cateterismo. Rass Med Sarda 63:651–656

Rouvière H (1938) Anatomie des Lymphatiques de l'Homme. Masson, Paris, pp 231–240. Translated by M.J. Tobias. Anatomy of the human lymphatic system. Ann Arbor: Edwards Brothers, pp 240–250

Seeger M, Bewig B, Günther R, Schafmayer C, Vollnberg B, Rubin D, Hoell C, Schreiber S, Fölsch UR, Hampe J (2009) Terminal part of thoracic duct: high-resolution US imaging. Radiology 252:897–904

Shafiroff BG, Kau QY (1959) Cannulation of the human thoracic lymph duct. Surgery 45:814–819

Shimada K, Sato I (1997) Morphological and histological analysis of the thoracic duct at the jugulo-subclavian junction in Japanese cadavers. Clin Anat 10:163–172

Smith ME, Riffat F, Jani P (2013) The surgical anatomy and clinical relevance of the neglected right lymphatic duct: review. J Laryngol Otol 8:1–6

Tesauro B (1966) Anatomical aspects of the thoracic duct after ligation. Bull Soc Int Chir 25:602–607

Van der Putte SC, van Limborgh J (1980) The embryonic development of the main lymphatics in man. Acta Morphol Neerl Scand 18:323–335

Van Pernis PA (1949) Variations of the thoracic duct. Surgery 26:806–809

Watson M (1872) Note on the termination of the thoracic duct on the right side. J Anat Physiol 6:427–428

Wirth W, Kubik S (1980) Lymphographic roentgen anatomy. In: Viamonte M, Rüttimann A, Agbaba M (eds) Atlas of lymphography. Thieme-Verlag, Stuttgart, pp 71–89

Yamada G (1932) A case of right sided aortic arch. Acta Anat Jpn 5:783–793

Zorzetto NL, Ripari W, De Freitas V, Seullner G (1977) Anatomical observations on the ending of the human thoracic duct. J Morphol 3:363–369

Part IV

Abdomen and Pelvis

Celiac Trunk

52

Julian Gallegos and Stephen J. Bordes, Jr.

52.1 Embryology

During the 4 mm stage of development between the 4th and 5th weeks of embryogenesis, paired dorsal aortae fuse craniocaudally to form the abdominal aorta (White et al. 2015). Simultaneously, the paired 10th vitelline segmental arteries supplying the primitive gut extend anteriorly between the level of the lower border of T12 and the upper border of L1 and fuse to form the celiac trunk (White et al. 2015). Segmental aortic fusion typically continues for approximately 1–2 cm before trifurcating to form the left gastric, common hepatic, and splenic arteries responsible for supplying blood to the primitive gut and subsequent foregut structures (Hazirolan et al. 2009; Mahajan et al. 2018).

52.2 Anatomy and Topography

Typically located anterior to the posterior fibers of the diaphragmatic crura between the T12 and L1 vertebral levels, the celiac trunk (Figs. 52.1, 52.2, 52.3, and 52.4) measures 0.8 cm in diameter and extends 1–3 cm anterolaterally from the aorta just inferior to the aortic hiatus, crossing beneath the median arcuate

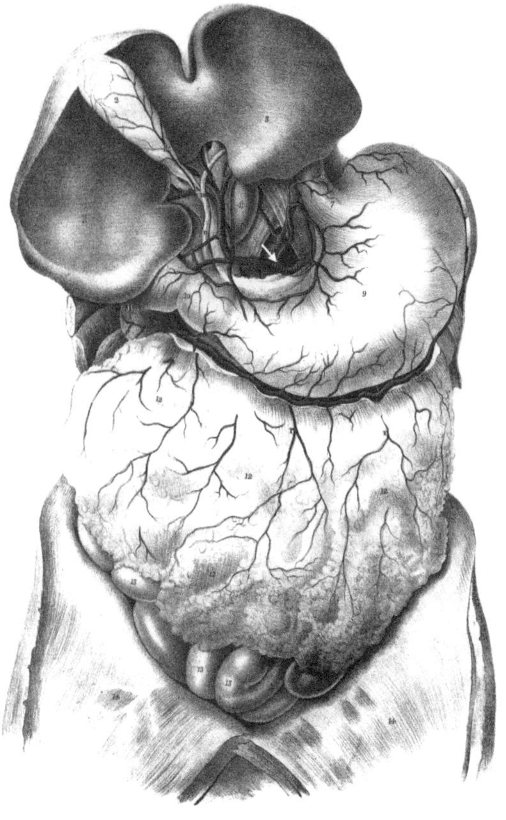

Fig. 52.1 Drawing illustrating the celiac trunk (arrow) and its branches (Quain and Wilson 1844 author's collection)

J. Gallegos
Department of Medicine, New York-Presbyterian/Weill Cornell Medicine, New York, USA

S. J. Bordes, Jr. (✉)
Department of Surgery, Louisiana State University Health Sciences Center, New Orleans, LA, USA

S. J. Bordes, Jr. et al. (eds.), *The Clinical Anatomy of the Vascular System*,
https://doi.org/10.1007/978-3-031-78326-5_52

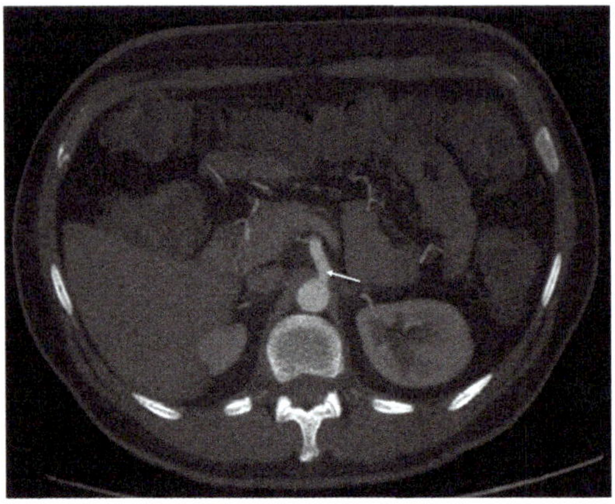

Fig. 52.2 Axial CT Illustrating the abdominal aorta and celiac trunk at the arrow

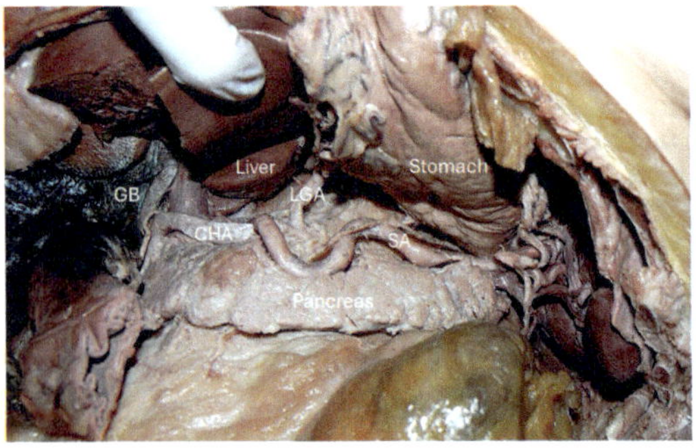

Fig. 52.3 Cadaveric dissection showing the three branches of the celiac trunk. LGA left gastric artery, SA splenic artery, CHA common hepatic artery. The gallbladder (GB) is seen for reference. (Courtsey Prof. M. Loukas)

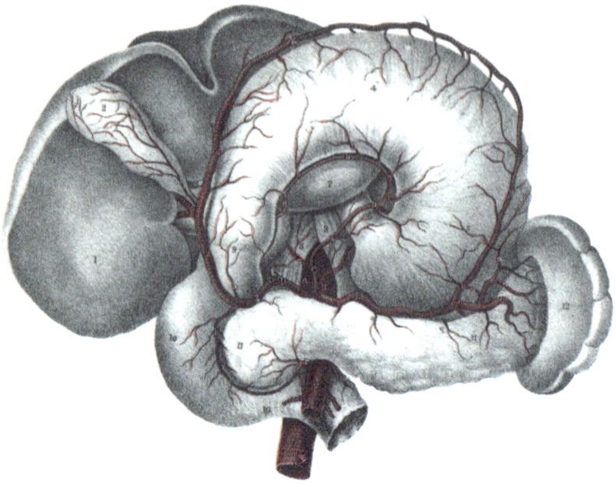

Fig. 52.4 Drawings illustrating the celiac trunk and its branches with the stomach lifted up (Quain and Wilson 1844 author's collection)

ligament and above the superior border of the body of the pancreas before trifurcating (Hazirolan et al. 2009; Silveira et al. 2009; Mishan Listmann 2020).

Most arteries are accompanied by similarly named veins; the celiac trunk is the only major artery nourishing the unpaired digestive organs (Jacob 2008). The length of the celiac trunk is accompanied by preaortic lymph contributors and nerve fibers originating within the celiac plexus. These nerve fibers receive parasympathetic input from the vagus nerve and sympathetic input from the greater and lesser splanchnic nerves (John et al. 2021). Visceral afferent fibers within the celiac trunk provide sensory input, which modulates autonomic rhythms and facilitates regulation of the associated foregut structures by providing sensory input (John et al. 2021). Owing to its anatomical location, the celiac trunk is susceptible to compression by various structures, most commonly the median arcuate ligament, pancreas, or other abdominal organs (Saleem et al. 2021). However, anastomoses between communicating celiac trunk branches can provide a vital arterial blood supply in certain disease states or during various ischemic events. The major celiac trunk anastomoses can be divided into the following three groups:

1. *Greater Curvature*
   (a) The left gastroomental (gastroepiploic) artery arises superiorly from the splenic artery.
   (b) The right gastroomental artery arises inferiorly from the gastroduodenal artery.
2. *Lesser Curvature*
   (a) The left gastric artery arises superiorly from the celiac trunk.
   (b) The right gastric artery arises inferolateral from the proper hepatic artery.
3. *Pancreaticoduodenal*
   (a) The anterior and posterior superior pancreaticoduodenal arteries arise superiorly from the gastroduodenal artery.

(b) The anterior and posterior inferior pancreaticoduodenal arteries arise inferiorly from the superior mesenteric artery (SMA).

52.3 Histology

The celiac trunk is an elastic artery branching directly from the aorta and, as such, is a major distributing artery. Elastic arteries receive their blood supply from vessels known as vasa vasorum. The tunica intima is the thinnest and innermost layer of the celiac trunk, comprising an inner layer of endothelium and subendothelial connective tissue. The middle and thickest layer of the celiac trunk is the tunica media, which is surrounded by many layers of collagen and elastin filaments that facilitate a stretch response adequate for maintaining a constant pressure gradient despite changes in pulse. To maintain a constant stretch response, muscle cells maintain regular synthesis of collagen and elastin fibers. The tunica adventitia is the outermost layer of the celiac trunk. It is composed of dense irregular connective tissue that equates to less than half the thickness of the tunica media.

52.4 Imaging

The celiac trunk is an excellent indicator for verifying the effects of our vascular manipulations, not only of the liver but also of the spleen and pancreas. In most cases, computed tomography angiography (CTA) has proved more beneficial than digital subtraction angiography for initial investigations. This method facilitates the identification of disruption to the many pathological variants within the celiac trunk caused by stenosis, occlusions, and occlusive lesions by highlighting collateral circulatory pathways. Furthermore, this method can highlight associated ischemic contributors such as ischemic bowel strictures or chronic/acute mesenteric ischemia. CTA, in conjunction with Doppler ultrasound, can provide additional information to contribute to cross-sectional imaging, which can fill gaps in information regarding bleeding or arterial dissection that cannot be identified endoscopically owing to the orientation and location of the celiac trunk (Li et al. 1995; Cademartiri et al. 2008; Upponi et al. 2013).

52.5 Variations

In 1756, Albrecht von Haller stated that trifurcation was the most common anatomical presentation of the celiac trunk; it was therefore deemed the classic presentation (Pinal-Garcia et al. 2018). Further research into anatomical variants of the celiac trunk across many studies has revealed numerous branching pattern variations. In approximately 65% of the population, a classic pattern is observed in which the left gastric artery arises from the celiac trunk proximal to terminal division into splenic and common hepatic arteries (Hazirolan et al. 2009; Olewnik et al. 2017; Song et al. 2010).

The leading cause of variation from the classic branching pattern is a disagreement in determining what constitutes a common hepatic artery and the proximity of the left gastric artery to the terminal bifurcation of the splenic and common hepatic arteries. In approximately 20% of cases, a terminal trifurcation is observed. In approximately 10% of cases, there is a quad/penta-furcation with additional branches (gastroduodenal artery, dorsal pancreatic artery, and both hepatic arteries) arising directly from the celiac trunk. In <1%, a celiacomesenteric artery has been observed where the SMA and celiac trunk originate as a single common trunk (Prakash et al. 2012).

52.6 Pathology

Ischemia

In most cases, celiac trunk ischemia is caused by an obstruction due to the development of atherosclerotic plaque within the celiac trunk (Moawad and Gewertz 1997). Ischemia has also been attributed to arterial embolism. However, arterial embolism has been found to affect the SMA more frequently than the celiac trunk owing to its short length and favorable anatomical orientation in relation to the aorta (Chang et al. 2006).

Compression

The median arcuate ligament segments meet at the anterior aspect of the aorta above the celiac trunk. However, in 10–24% of the population, the ligament crosses anterior to the celiac trunk, causing compression of nearby structures, including the celiac trunk and ganglion. Various degrees of median arcuate ligament compression are asymptomatic, with severe compression in 1% of the population (Lindner and Kemprud 1971).

Dissection

Celiac trunk dissection is rarely seen as a standalone issue; instead, it is propagated by aortic dissection. However, because of collateral anastomoses between foregut structures, it is typically asymptomatic, with few cases developing ischemic complications. Generally, celiac trunk dissection is iatrogenic, but it can present secondary to other forms of mesenteric ischemia (D'Ambrosio et al. 2007; Wang et al. 2013). In some cases, peptic ulcers in the stomach and/or the duodenum cause significant injury to the celiac trunk, leading to dissection. As the root of the celiac trunk erodes, terminal branches supplying foregut organs cannot receive oxygenated blood and nutrition, leading to various devastating effects (Daher et al. 2016; Ahluwalia et al. 2021). Splenic artery aneurysms are the most common type of visceral aneurysm within the celiac trunk associated with dissection. While they are typically asymptomatic until rupture, high paired morbidity and mortality rates make early detection of them critically important. Typical preliminary detection focuses mainly on diagnostic imaging via arteriography, while most fatal asymptomatic cases are determined following post-rupture autopsies (Shanley et al. 1996; Messina and Shanley 1997).

52.7 Clinical and Surgical Significance

Ischemia

Rather than high-risk surgical revascularization, endovascular therapy with percutaneous transluminal angioplasty or stent placement has become the gold-standard treatment option. This approach avoids complications generally associated with more invasive surgical procedures, resulting in more favorable outcomes, particularly for patients with one or more associated comorbidities (Oderich et al. 2009; Gupta et al. 2010; Upponi et al. 2013). Furthermore, celiac trunk stent placement and percutaneous transluminal angioplasty have been reported to have similar outcomes to surgical revascularization in terms of overall arterial patency (Loffroy et al. 2010; Turba et al. 2012).

Compression

Treatment options for celiac trunk compression syndrome remain a topic of debate, given that some options, such as percutaneous transluminal angioplasty or stent placement, provide symptomatic relief only in the short term without addressing the main underlying issues (Jaik et al. 2007; Duncan 2008). Other options include vasodilator therapy, which can be appropriate for patients whose symptoms do not warrant invasive or surgical intervention. The most effective treatment option is surgical transection of the median arcuate ligament or revascularization through reimplantation or surgical bypass (Grotemeyer et al. 2009; Vaziri et al. 2009).

Dissection

Owing to the contribution of the celiac trunk to many major foregut viscera, options for managing aneurysms in this structure include surgical endovascular repair, which is best done with embolization or stent grafts once the patient is hemodynamically stable (Lee et al. 1999; Vaidya and Dighe 2010; Fitzpatrick et al. 2014). While surgical intervention is the most reliable treatment for complications associated with or leading to celiac trunk dissection, alternative therapy focusing on symptom management can be appropriate for patients with limited dissection. These alternative treatment options mainly focus on preventing thromboembolic complications by using anticoagulation and blood pressure reduction therapies (Glehen et al. 2001; Fenoglio et al. 2004).

52.8 Complications

Ischemia

Untreated celiac trunk ischemia typically leads to postprandial abdominal pain. Occurring within 60 min of food ingestion and persisting for several hours, the pain associated with ischemic inflammation is caused by reduced visceral perfusion after oral food intake (Chang et al. 2006). More advanced cases can result in bowel infarction and malnutrition, with high rates of morbidity and mortality. More unusual pathological conditions include dissection, aneurysms, and vascular malformations (Moawad and Gewertz 1997).

Compression

Compression of the celiac trunk by the median arcuate ligament or other proximal viscera is typically asymptomatic owing to anastomoses between the celiac trunk and superior mesenteric artery collaterals (Loukas et al. 2007). However, a small portion of clinical cases, mainly women between the ages of 22 and 40, have been reported to develop symptoms of median arcuate ligament syndrome or Dunbar syndrome (Horton et al. 2005). Presentation varies but generally includes abdominal pain modulated by inspiration/expiration, weight loss, malnutrition, and epigastric bruit. While the exact cause of pain associated with compression of the celiac trunk is unknown, it is thought to be caused by ischemic inflammation secondary to celiac trunk compression in conjunction with direct compression of various celiac ganglia (Lindner and Kemprud 1971; Horton et al. 2005; Duncan 2008).

Dissection

Untreated acute dissection can be incredibly painful because of visceral nerve propagation associated with anastomotic contributors. Furthermore, nerve extension into the superior mesenteric artery can lead to pain associated with bowel ischemia and smooth muscle inflammation. In turn, untreated chronic celiac trunk dissection can result in mesenteric ischemia and omental inflammation, including but not limited to severe postprandial abdominal pain. Furthermore, both acute and chronic dissection can be complicated by aneurysm formation with subsequent rupture and hemorrhage, deterioration of celiac trunk integrity, and subsequent dissection being expedited by gastric or duodenal ulcers (D'Ambrosio et al. 2007; Daher et al. 2016; Ahluwalia et al. 2021).

References

Ahluwalia N, Nassereddin A, Futterman B (2021) Anatomy, abdomen and pelvis, celiac trunk. StatPearls, Treasure Island

Cademartiri F, Palumbo A, Maffei E, Martini C, Malago R, Belgrano M, La Grutta L, Bartolotta TV, Luccichenti G, Midiri M, Raaijmakers R, Mollet N, Zompatori M, Crisi G (2008) Noninvasive evaluation of the celiac

trunk and superior mesenteric artery with multislice CT in patients with chronic mesenteric ischaemia. Radiol Med 113(8):1135–1142
Chang RW, Chang JB, Longo WE (2006) Update in management of mesenteric ischemia. World J Gastroenterol 12(20):3243–3247
Daher S, Lahav Z, Rmeileh AA, Mizrahi M, Khoury T (2016) Ischemic Gastropathic ulcer mimics gastric cancer. Case Rep Gastrointest Med 2016:9745854
D'Ambrosio N, Friedman B, Siegel D, Katz D, Newatia A, Hines J (2007) Spontaneous isolated dissection of the celiac artery: CT findings in adults. AJR Am J Roentgenol 188(6):W506–W511
Duncan AA (2008) Median arcuate ligament syndrome. Curr Treat Options Cardiovasc Med 10(2):112–116
Fenoglio L, Allione A, Scalabrino E, Alberto G, Benedetti V, Pomero F, Valpreda S, Morino M, Perin PC (2004) Spontaneous dissection of the celiac artery: a pitfall in the diagnosis of acute abdominal pain. Presentation of two cases. Dig Dis Sci 49(7–8):1223–1227
Fitzpatrick J, Bhat R, Young JA (2014) Angiographic embolization is an effective treatment of severe hemorrhage in pancreatitis. Pancreas 43(3):436–439
Glehen O, Feugier P, Aleksic Y, Delannoy P, Chevalier JM (2001) Spontaneous dissection of the celiac artery. Ann Vasc Surg 15(6):687–692
Grotemeyer D, Duran M, Iskandar F, Blondin D, Nguyen K, Sandmann W (2009) Median arcuate ligament syndrome: vascular surgical therapy and follow-up of 18 patients. Langenbeck's Arch Surg 394(6):1085–1092
Gupta PK, Horan SM, Turaga KK, Miller WJ, Pipinos II (2010) Chronic mesenteric ischemia: endovascular versus open revascularization. J Endovasc Ther 17(4):540–549
Hazirolan T, Metin Y, Karaosmanoglu AD, Canyigit M, Turkbey B, Oguz BS, Ariyurek M (2009) Mesenteric arterial variations detected at MDCT angiography of abdominal aorta. AJR Am J Roentgenol 192(4):1097–1102
Horton KM, Talamini MA, Fishman EK (2005) Median arcuate ligament syndrome: evaluation with CT angiography. Radiographics 25(5):1177–1182
Jacob S (2008) Chapter 4 – Abdomen. In: Human Anatomy. Churchill Livingstone, pp 71–73
Jaik NP, Stawicki SP, Weger NS, Lukaszczyk JJ (2007) Celiac artery compression syndrome: successful utilization of robotic-assisted laparoscopic approach. J Gastrointestin Liver Dis 16(1):93–96
John RS, Dixon B, Shienbaum R (2021) Celiac Plexus Block. StatPearls, Treasure Island
Lee PC, Rhee RY, Gordon RY, Fung JJ, Webster MW (1999) Management of splenic artery aneurysms: the significance of portal and essential hypertension. J Am Coll Surg 189(5):483–490
Li KC, Hopkins KL, Dalman RL, Song CK (1995) Simultaneous measurement of flow in the superior mesenteric vein and artery with cine phase-contrast MR imaging: value in diagnosis of chronic mesenteric ischemia. Work in progress. Radiology 194(2):327–330
Lindner HH, Kemprud E (1971) A clinicoanatomical study of the arcuate ligament of the diaphragm. Arch Surg 103(5):600–605
Loffroy R, Guiu B, Cercueil JP, Krause D (2010) Chronic mesenteric ischemia: efficacy and outcome of endovascular therapy. Abdom Imaging 35(3):306–314
Loukas M, Pinyard J, Vaid S, Kinsella C, Tariq A, Tubbs RS (2007) Clinical anatomy of celiac artery compression syndrome: a review. Clin Anat 20(6):612–617
Mahajan A, Tiwari S, Mishra S (2018) A unique conglomeration of variations in the celiac, hepatic, and superior mesenteric artery: a Clinico-embryological perspective. Int J Appl Basic Med Res 8(4):256–258
Messina LM, Shanley CJ (1997) Visceral artery aneurysms. Surg Clin North Am 77(2):425–442
Mishan Listmann RST (2020) Chapter 19 - The abdominal aorta. In: Surgical anatomy of the lateral Transpsoas approach to the Lumbar spine. Elsevier, pp 185–188
Moawad J, Gewertz BL (1997) Chronic mesenteric ischemia. Clinical presentation and diagnosis. Surg Clin North Am 77(2):357–369
Oderich GS, Bower TC, Sullivan TM, Bjarnason H, Cha S, Gloviczki P (2009) Open versus endovascular revascularization for chronic mesenteric ischemia: risk-stratified outcomes. J Vasc Surg 49(6):1472–1479 e1473
Olewnik L, Wysiadecki G, Polguj M, Wasniewska A, Jankowski M, Topol M (2017) Types of coeliac trunk branching including accessory hepatic arteries: a new point of view based on cadaveric study. Folia Morphol (Warsz) 76(4):660–667
Pinal-Garcia DF, Nuno-Guzman CM, Gonzalez-Gonzalez ME, Ibarra-Hurtado TR (2018) The celiac trunk and its anatomical variations: a cadaveric study. J Clin Med Res 10(4):321–329
Prakash T, Rajini V, Mokhasi BS, Geethanjali PV, Sivacharan, Shashirekha M (2012) Coeliac trunk and its branches: anatomical variations and clinical implications. Singapore Med J 53(5):329–331
Quain and Wilson (1844) The surgical anatomy of the human body. London, Taylor & Walton
Saleem T, Katta S, Baril DT (2021) Celiac artery compression syndrome. StatPearls, Treasure Island
Shanley CJ, Shah NL, Messina LM (1996) Uncommon splanchnic artery aneurysms: pancreaticoduodenal, gastroduodenal, superior mesenteric, inferior mesenteric, and colic. Ann Vasc Surg 10(5):506–515
Silveira LA, Silveira FB, Fazan VP (2009) Arterial diameter of the celiac trunk and its branches. Anatomical study. Acta Cir Bras 24(1):43–47
Song SY, Chung JW, Yin YH, Jae HJ, Kim HC, Jeon UB, Cho BH, So YH, Park JH (2010) Celiac axis and common hepatic artery variations in 5002 patients: systematic analysis with spiral CT and DSA. Radiology 255(1):278–288

Turba UC, Saad WE, Arslan B, Sabri SS, Trotter S, Angle JF, Hagspiel KD, Kern JA, Cherry KJ, Matsumoto AH (2012) Chronic mesenteric ischaemia: 28-year experience of endovascular treatment. Eur Radiol 22(6):1372–1384

Upponi S, Harvey JJ, Uberoi R, Ganeshan A (2013) The role of radiology in the diagnosis and treatment of mesenteric ischaemia. Postgrad Med J 89(1049):165–172

Vaidya S, Dighe M (2010) Spontaneous celiac artery dissection and its management. J Radiol Case Rep 4(4):30–33

Vaziri K, Hungness ES, Pearson EG, Soper NJ (2009) Laparoscopic treatment of celiac artery compression syndrome: case series and review of current treatment modalities. J Gastrointest Surg 13(2):293–298

Wang HC, Chen JH, Hsiao CC, Jeng CM, Chen WL (2013) Spontaneous dissection of the celiac artery: a case report and literature review. Am J Emerg Med 31(6):1000 e1003–1005

White RD, Weir-McCall JR, Sullivan CM, Mustafa SA, Yeap PM, Budak MJ, Sudarshan TA, Zealley IA (2015) The celiac axis revisited: anatomic variants, pathologic features, and implications for modern endovascular management. Radiographics 35(3):879–898

Superior Mesenteric Artery

53

Lindsey Peterson and Thomas Bernik

53.1 Embryology

During fetal development, the vasculature in the abdomen starts as a dual blood supply with two aortas, one ventral and one dorsal arterial system. The vitelline arteries are a group of embryonic vessels that link the dorsal aorta to the vitelline veins within the yolk sac (Lin and Chaikof 2000). The visceral arterial vasculature develops from these primitive ventral segmental arteries, most of which regress during fetal development and are not present at birth. However, the remaining ones form a median vessel that gives rise to the three major mesenteric arteries: the celiac trunk, superior mesenteric artery (SMA), and inferior mesenteric artery (IMA). The 13th segmental artery develops into the SMA (Lin and Chaikof 2000). During embryogenesis, the midgut loops 270 degrees in total around the SMA, which acts as a central axis securing the final anatomical location of the GI system (Danowitz and Solounias 2016).

L. Peterson (✉)
Department of Surgery, Hackensack Meridian Health, Hackensack, NJ, USA

T. Bernik
Department of Vascular and Endovascular Surgery, Englewood Health, Englewood, NJ, USA

53.2 Anatomy/Topography

The SMA is the second major branch off the abdominal aorta at the level of the L1 vertebra (Figs. 53.1, 53.2, 53.3, and 53.4). It arises anteriorly approximately 1–2 cm below the celiac trunk and above the IMA. It courses anteriorly inferiorly, posterior to the pylorus of the stomach, the neck of the pancreas, and the splenic vein, and anterior to the left renal vein (Lin and Chaikof 2000).

The SMA supplies the midgut starting near the ampulla of Vater in the second part of the duodenum and continuing to the jejunum, ileum, ascending colon, and the proximal two-thirds of the transverse colon. Its first branch is the inferior pancreaticoduodenal branch traveling behind the uncinate process, supplying the head of the pancreas along with the superior pancreaticoduodenal artery from the celiac trunk (Shaikh et al. 2021). The jejunal and ileal arteries branch from the SMA to the left and to the right, the terminal branches, including the ileocolic, right colic, and middle colic arteries. Prior to entering the mesentery, the middle colic artery branches off the SMA and forms an anastomosis with the left colic artery, a branch of the IMA, to supply the transverse colon. Sometimes, the right colic artery terminates by joining the left colic artery to supply the transverse colon. This connection is

S. J. Bordes, Jr. et al. (eds.), *The Clinical Anatomy of the Vascular System*,
https://doi.org/10.1007/978-3-031-78326-5_53

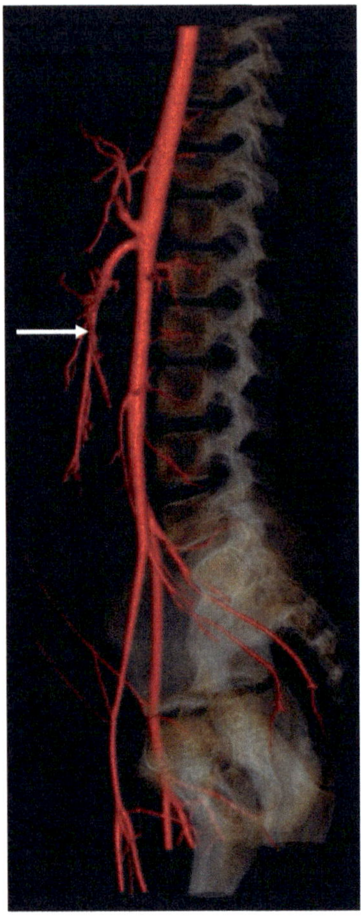

Fig. 53.1 CTA reconstruction illustrating the origin of the SMA (aorta) from the aorta in a child

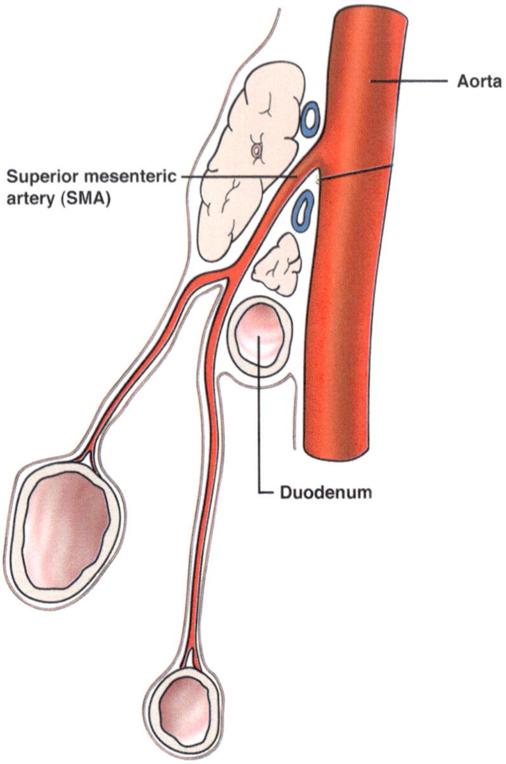

Fig. 53.2 Schematic drawing noting the structures between the SMA and aorta

known as the Riolan arcade, which anastomoses with other branches of the IMA. The Riolan arcade is not seen consistently in all people. More commonly observed is the marginal artery of Drummond, a set of vessels that forms collateral vasculature between the SMA and the IMA along the mesenteric border (Shaikh et al. 2021). This anastomosis includes the terminal branches of the right colic, ileocolic, and middle colic arteries from the SMA with the left colic and sigmoid arteries from the IMA (Shaikh et al. 2021). Lymph nodes are found near the mesenteric root and continuously along the mesentery. In the first three centimeters from the root of the SMA, the superior mesenteric vein (SMV) is always to the right (Fig. 53.4). In most cases, along the distal six centimeters, the SMV is to the left (Chou et al. 1997).

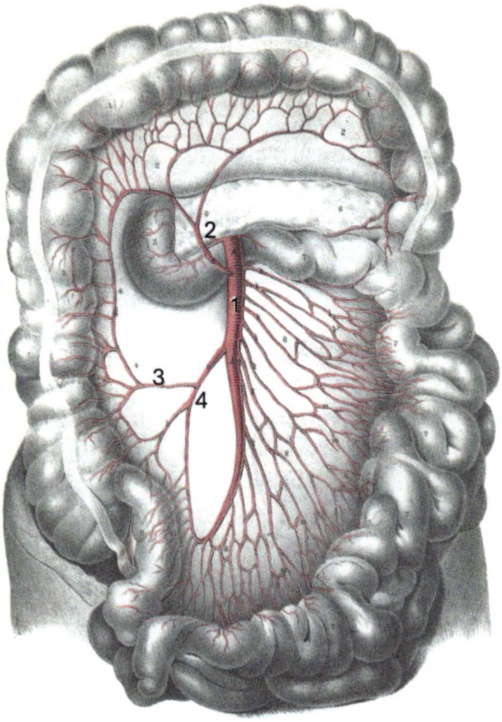

Fig. 53.3 Drawing of the major branches of the SMA (1). Note the middle colic artery (2), right colic artery (3) and ileocolic artery (4). Jejunal and ileal branches are seen arising from the left side of the artery (Quain and Wilson, 1844 author's collection)

53.3 Histology/Imaging

The SMA is an elastic artery, like the aorta and other branches from it. Histologically, there is an inner tunica intima, middle tunica media, and tunica adventitia surrounding the lumen (Munjal and Bordoni 2021). The inner layer comprises a single layer of endothelial cells supported by an underlying basement membrane. In such elastic arteries, there is a layer between the intima and the media containing fibro-elastic connective tissue called the internal elastic laminae (Munjal and Bordoni 2021). This layer allows the vessel to maintain shape during high-pressure blood flow from the heart and to rebound during diastole. The caliber of the artery determines the ratio of elastic fibers to muscle fibers within the media. In a medium-sized vessel such as the SMA, there are more elastic fibers with some interspersed muscle cells. The tunica adventitia, or outer layer, is composed of fibroblasts and myofibroblasts (Munjal and Bordoni 2021).

Generally, less invasive imaging techniques such as ultrasound are used to identify abdominal pathologies but do not always reveal vessel abnormalities. Color duplex ultrasound is often used as a screening tool; it does provide sufficient

Fig. 53.4 Cadaveric dissection showing many of the branches of the SMA. (Courtsey Prof. M. Loukas)

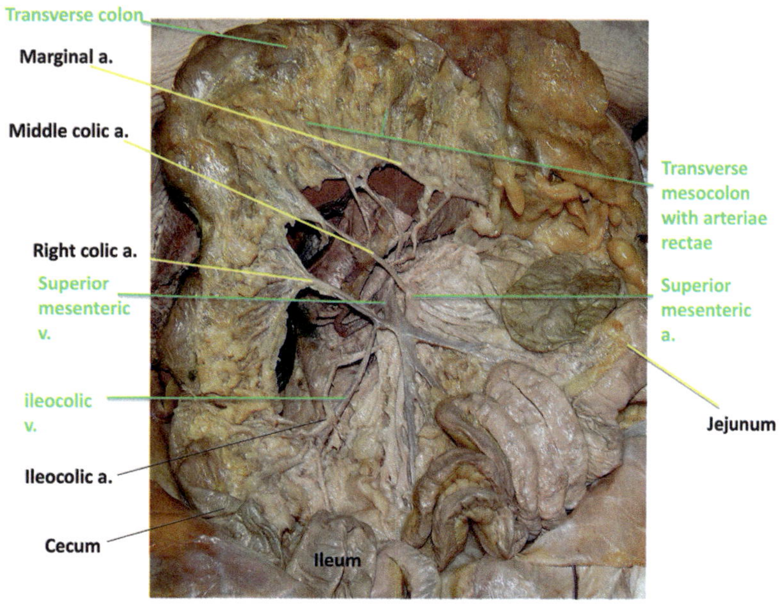

visibility of the vasculature (Zeller et al. 2010). For optimal visualization, ultrasound is done during fasting. Abnormal velocities are those greater than 275 cm/s, consistent with more than 70% stenosis.

Computed tomography angiography (CTA) is the imaging modality of choice when abdominal vasculature is suspected as a cause of symptoms. CTA is also used as a supplemental modality when ultrasound is inconclusive owing to extensive bowel gas patterns; this occurs approximately 10% of the time (Zeller et al. 2010). CTA allows for the observation of any congenital or acquired pathology, including but not limited to aneurysm, atherosclerotic disease, or stenosis (Caraiani et al. 2020). This mode of imaging allows the collateral vascularization associated with the SMA to be visualized. The ileocolic artery is most consistently observed, so it is a landmark often used in angiographical studies (Lin and Chaikof 2000).

Magnetic Resonance Imaging (MRI) can also be used to observe the abdominal vasculature, though because of its higher cost, it is not often the modality of choice.

Diagnostic angiography allows for immediate intervention if needed. X-ray with contrast material injected into the vessels is used to obtain images of the abdominal vasculature. The images appear opaque, and a continuous X-ray beam allows areas of stenosis, aneurysm, AV malformations, or other pathologies to be evaluated by showing material passing through the vessel in real-time on a screen. If there is a contrast allergy or renal insufficiency, CO_2 can be used as the contrast agent.

53.4 Variations

Different degrees of persistence of certain portions of the dual blood supply (described under the embryology section) account for the many variations in the mesenteric circulation (Walker 2009). The arterial and venous systems in the bowel are complex, allowing scope for several anatomical variations and extensive collateral pathways.

CTA and MRI angiography are used to display a personalized road map showing normal vascular anatomy, variants of the mesenteric vessels, or pathologies (Wang et al. 2014), which assists with diagnosis, treatment, and creation of a surgical plan. Angiography via a femoral or radial approach can also be used for diagnosis with the intention of treatment.

- The inferior pancreaticoduodenal artery arises as two separate branches in 60% of people or as a common single trunk with the first jejunal branch in 40% (Lin and Chaikof 2000).
- Intestinal malrotation during embryogenesis can alter the position of the SMV, which in normal anatomical position is to the right of the SMA at 3 cm and to the left at 6 cm (Chou et al. 1997).
- The ileocolic branch of the SMA is the most consistent, seen in over 95% of people with low variability. The right colic branch is the most inconsistent branch with very high variability. The middle colic is slightly less consistent than the ileocolic in approximately 95% of people with variation in origin and number (Bruzzi et al. 2019).
- The right hepatic artery arises from the SMA in 9–20% of patients and is important to document for procedure planning (Noussios et al. 2017).

53.5 Pathology

Acute Mesenteric Ischemia A rare abdominal emergency with a high mortality rate associated with 0.1% of hospital admissions leading to significant bowel necrosis, characterized by either a thromboembolic event or a non-thrombotic event. Non-thrombotic events include low-flow states caused by cardiogenic shock, hemorrhage, hypovolemia, or pancreatitis or are secondary to mechanical causes such as adhesions or strangulated hernias (Romano et al. 2011). Patients with thromboembolic events commonly present with pain out of proportion to the physical presentation and a history of cardiovascular disease such as atrial fibrillation or congestive heart failure.

The most common sources of emboli are valvular heart disease, coronary heart disease, bacterial endocarditis, or a prior myocardial infarction.

Chronic Mesenteric Ischemia Often secondary to atherosclerosis, presenting with postprandial cramping abdominal pain and ischemic gastropathy and enteropathy, nausea, vomiting, and weight loss secondary to "food fear" (Zeller et al. 2010). Diagnosis is often made late owing to the slow progression of the disease and the abundance of collateral mesenteric vessels (Hohenwalter 2009). Treatment with an endovascular stent or SMA bypass is necessary to prevent irreversible bowel necrosis.

Aneurysms SMA aneurysms can originate at the arterial axis or along its branches; this artery is the third most common site for visceral aneurysms (Lorelli et al. 2003). Generally, they arise secondary to an infectious process, atherosclerosis, or genetic processes. As the aneurysm expands, many patients present with progressive and severe abdominal pain, some with postprandial abdominal pain due to intestinal angina (Lorelli et al. 2003). Symptoms can mimic those of pancreatitis, perforated viscous, or appendicitis. Ultrasound is useful for quick and accurate diagnosis, although most cases are picked up during computed tomography or angiography. Aneurysmal dilatation was defined as greater than 50% dilation of the SMA lumen above normal for each patient as measured by CT scan (Gobble et al. 2009). Acceptable outcomes can be achieved with timely diagnosis and treatment prior to rupture.

Dissections Dissections of the SMA are a rare type of vascular disease and can present in a variety of ways, from incidental radiographic findings in an asymptomatic patient to fatal dissections. They can be isolated incidences or can be concurrent with dissection of the celiac artery. When associated with the celiac artery, dissection is believed to be secondary to hypertension; however, isolated incidences of just the SMA are hypothesized to be secondary to shear force strain on the fixed portion of the vessel (under the body of the pancreas) to a relatively unfixed (in the mesenteric root) segment (Kim 2016). Diagnosis is often made using CT, with the "double lumen" sign. Most patients present with nausea, vomiting, and epigastric abdominal pain, and pain severity correlates positively with the length of dissection of the artery.

Superior Mesenteric Artery Syndrome This condition is occasionally referred to as Wilke's syndrome and presents with epigastric pain, weight loss, and prolonged bilious vomiting. It is caused by an obstruction in the third part of the duodenum secondary to compression by the SMA (Figs. 53.1, 53.2, and 53.5). It occurs because of a reduced aorto-mesenteric angle and can sometimes be associated with anorexia nervosa (Yao et al. 2015). Computed tomography (CT) is the imaging modality of choice, showing narrowing of the 3rd part of the duodenum with a distended stomach and proximal duodenum (Shinji et al. 2015). Medical and surgical approaches to treatment are advised, often beginning with nasogastric decompression and proceeding to a laparoscopic duodenojejunostomy if there is no improvement in symptoms (Kingham et al. 2004).

Trauma Blunt or penetrating trauma is common along the intestines, causing contusion or necrosis (McMahon and Balasubramanya 2021). Both types of trauma lead to vascular trauma with equal frequency, but blurt trauma patients are more severely injured (Perkins et al. 2012). Blunt trauma occurs when the bowel is crushed by an external object compressing internal structures against the spine or is caused by acceleration and deceleration forces that stretch and tear the attachments that normally stabilize the intestines. These types of injuries often lead to the devascularization of that segment of the bowel, leading to necrosis despite the SMA and IMA collaterals (McMahon and Balasubramanya 2021).

Pancreatic Cancer Mesopancreatic dissection requires thorough knowledge of the mesenteric vasculature because central artery ligation and the SMA-first approach is the cornerstone of rad-

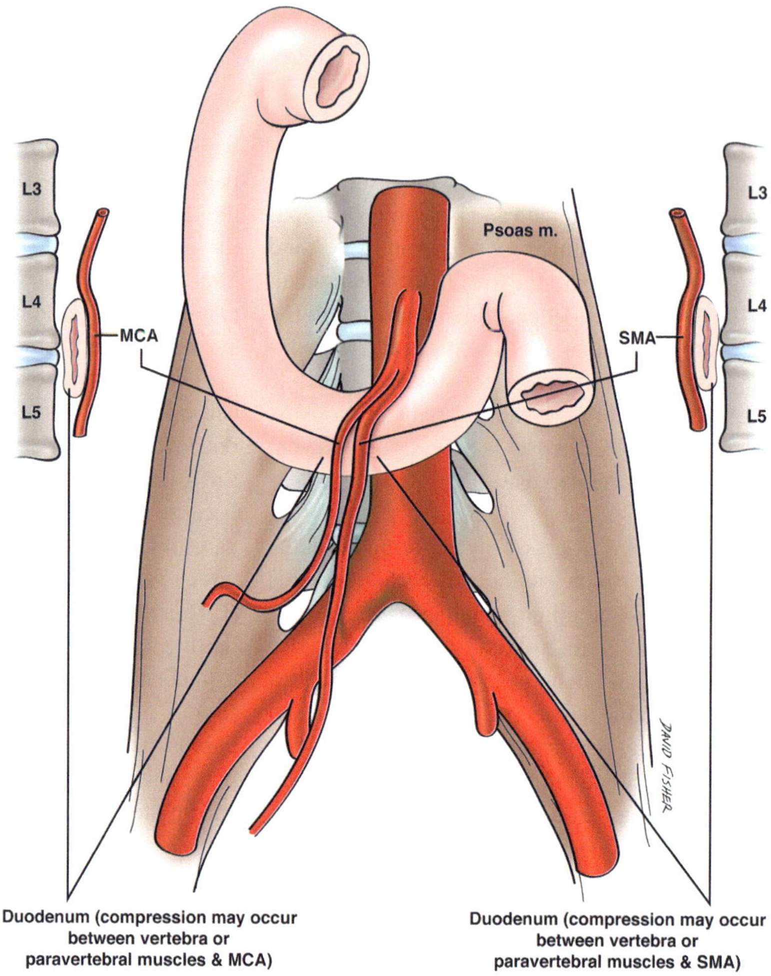

Fig. 53.5 Schematic drawing of possible compression sites involving the SMA

ical resection of the pancreatic head. The mortality rate in this type of cancer remains significantly high despite advances in medical and surgical techniques; the five-year survival rate is 5% (Negoi et al. 2018). It is often associated with a genetic abnormality. Frequently, owing to the location of the pancreas, symptoms are non-specific, and the condition is not detected or diagnosed until after metastasis to lymph nodes has occurred. Pancreatic head adenocarcinoma has a high rate of lymph node metastasis, whereas cancers in the body or tail are more likely to involve other visceral organs. The goal of treatment is to achieve an R0 resection with clear margins. For this purpose, it is crucial to determine the relationship between the mesenteric vasculature and the primary tumor using multi-detector computed tomography (MDCT) prior to surgical resection (Dolay et al. 2019). Sometimes, margins involve the SMA, requiring resection and bypass.

Proximal (Right) Colorectal Cancer The SMA supplies the ascending colon and part of the transverse colon, so any neoplasm associated with these parts needs careful consideration if resection is required. Colorectal cancer is the third most common cancer worldwide and has a high mortality rate. Cancer located in the proximal portion of the colon is either a mucinous adenocarcinoma or a serrated adenoma and is associated with mismatch repair or microsatellite genetic abnormalities. The morphology of right-sided tumors can cause symptoms to appear late, leading to more advanced disease and potential metastatic spread to the peritoneum (Baran et al. 2018). Treatment involves complete mesocolon excision (CME) using a laparoscopic technique, with the SMV and the ileocolic branch of the SMA serving as landmarks.

53.6 Clinical and Surgical Significances/Treatment

Understanding the arterial system helps treat various pathological conditions that require surgery, especially with new techniques such as minimally invasive surgery. The Riolan artery and the marginal artery of Drummond are both essential for healthy bowel because they provide collateral blood flow between the SMA and the IMA. The Riolan artery anastomosis, when present, is especially important in colorectal surgeries, vascularizing the descending colon when the IMA has been ligated. The marginal artery of Drummond, ileocolic, and middle colic vessels are consistent anatomical landmarks for surgeons once the abdominal cavity has been exposed (Xiao et al. 2019). The crossing patterns of the ileocolic and middle colic arteries in relation to the SMV provide good planes for dissection. The middle colic always crosses anteriorly after originating from the SMA, and the ileocolic usually crosses posteriorly to the SMV (Alsabilah et al. 2016). The ileocolic pedicle with the SMV allows for the accurate positioning of laparoscopic tools for a medial-to-lateral approach. Awareness of the anastomosis between the abdominal vessels is imperative because the arcades provide a collateral blood supply when necessary in ischemic scenarios.

Acute Mesenteric Ischemia Acute mesenteric ischemia (AMI) is a surgical emergency. It carries 60%–70% morbidity and mortality. Initial treatment involves correction of hydration status and electrolyte abnormalities and administration of broad-spectrum antibiotics prior to surgery (Monita and Gonzalez 2021). Emergent exploratory laparotomy or laparoscopy is required to determine the level of ischemia and bowel necrosis and to restore blood flow and resect dead bowel if necessary. AMI can later be differentiated into occlusive and non-occlusive. Acute occlusive mesenteric ischemia can have embolic or thrombotic etiologies. Embolic causes are usually treated with embolectomy or thrombolysis in rare circumstances. Thrombotic mesenteric ischemia has poor outcomes because of underlying disease. The mortality rate approaches 90%. After revascularization, within 12–48 h, it is usual to make a second inspection to reassess for overlooked or new areas of necrosis. Bowel resection follows 53% of second-look surgeries and 31% during exploratory surgery at the first attempt at revascularization (Monita and Gonzalez 2021). Open surgical repair involves transaortic endarterectomy, direct reimplantation on the aorta, and antegrade or retrograde bypass grafting. Endovascular repair using angioplasty, or a stent within the artery to restore patency, is more often used owing to the lower complication rate and decreased mortality (Monita and Gonzalez 2021).

Chronic Mesenteric Ischemia After diagnosis, prompt treatment is required to prevent complications such as continued weight loss, acute infarction, perforation, sepsis, or death (Hohenwalter 2009). The treatment of choice is endovascular repair with angioplasty or a stent. Poor surgical candidates can be managed long-term with anticoagulation such as warfarin or short-term with nitrates, although these are not curative. Open surgical repair includes transaortic endarterectomy, direct reimplantation on the aorta, and antegrade or retrograde bypass grafting, which has favorable outcomes with a 90–100% success rate (Hohenwalter 2009). Endovascular repair has a lower complication rate (9%) and a 30-day mortality rate of 3%. Treatment with stenting has a better patency rate than angioplasty alone.

Mesenteric Aneurysms The natural progression of a superior mesenteric aneurysm is expansion and eventual rupture (Lorelli et al. 2003). The Society for Vascular Surgery recommends the repair of all SMA aneurysms and pseudoaneurysms as soon as the diagnosis is made, regardless of size (Chaer et al. 2020). Therefore, repair is needed when the aneurysm is found, using ligation with or without excision to prevent a grave outcome. This technique allows the collateral blood flow to perfuse the rest of the bowel, and intestinal viability must be assessed intraoperatively by clinical evaluation, fluorescein dye, or Doppler flow (Lorelli et al. 2003). Ligation and bypass are the safer options if elective repair is planned. The endovascular technique has progressed to allow aneurysms to be treated with transcatheter coil embolization; covered stent grafts are good treatment options for poor surgical candidates. Other surgical techniques include intimectomy, transposition, resection, and bypass (Lorelli et al. 2003).

Dissections Treatment options associated with SMA dissections include expectant management, anticoagulation, open surgical repair, or endovascular repair (Gobble et al. 2009). Incidental radiographic findings are managed conservatively with bowel rest, observation, anticoagulation, and repeat CT scans to monitor the extent of dissection. The degree of additional intervention and surgical repair depends on surgeon preference, anatomical suitability, comorbidities, and symptoms. Patients experiencing an acute abdomen undergo immediate endovascular stent placement to preserve blood flow to the bowel. Open surgical repair and anticoagulation should be attempted when endovascular repair is unsuccessful (Gobble et al. 2009).

Superior Mesenteric Artery Syndrome Conservative treatment is usually the initial approach, with nasogastric decompression, prone positioning, and nutrition supplementation. Definitive treatment entails duodenojejunostomy as a laparoscopic surgery because it is safe and minimally invasive. Duodenal-jejunal resection is a better option than a bypass procedure because it avoids reversal peristalsis and blind-loop syndrome. Recently, a single incision laparoscopic-assisted umbilical mini-laparotomy was performed with a good result. This surgery has been used previously for cholecystectomy, appendectomy, and colon surgery (Shinji et al. 2015). It requires the assisted suturing device to be used under direct open supervision, which is less invasive than previous techniques and more manageable because the additional bulky equipment is eliminated (Shinji et al. 2015).

Trauma Morbidity and mortality associated with SMA trauma are due to exsanguination, especially if there is multiple vessel trauma. Successful management of patients with visceral arterial injuries depends upon rapid and adequate exposure followed by primary repair or revascularization using available surgical techniques. Failure to repair the vasculature leads to bowel ischemia and necrosis. Stable patients with blunt injury are taken to the operating room for repair if intestinal injury is noted on radiographical imaging. Diagnostic laparoscopy or laparotomy continues to be the treatment of choice for hemodynamically unstable patients (McMahon and Balasubramanya 2021). There are four stages of a trauma laparotomy: (1) control of hemorrhage, (2) control of contamination, (3) diagnosis of all injuries, and (4) reconstruction (McMahon and Balasubramanya 2021). Once homeostasis is achieved, the small bowel is run from the ligament of Treitz to the cecum to make sure there is no bowel spillage and the bowel is receiving adequate flow.

Pancreatic Cancer The location of the SMA in relation to the pancreas allows for retroperitoneal invasion of tissue when pancreatic cancer occurs, creating difficulty for radical excision of the cancer (Nagakawa et al. 1994). The artery-first approach to pancreaticoduodenectomy has been the gold standard to date. It allows the arterial planes and retroperitoneal tissue to be directly visualized and dissected to achieve a higher rate of cancer-free margins. Because the branches from the SMA and other surrounding vessels are variable, a meta-analysis has shown that the origin site of the first dissection also changes (Negoi et al. 2018). The leading cause of morbidity and mortality is secondary to iatrogenic intraoperative peripancreatic vessel injury, so identifying the SMA anatomy prior to surgery by CTA is imperative.

Right-Sided Colorectal Cancer When resection of the colon and surrounding lymph nodes is required, individualized vessel anatomy must be considered to avoid iatrogenic complications. The goal of right hemicolectomy is to eliminate cancer; the main vessels are ligated more proximally, and the lymph nodes are subjected to root dissection (Ye et al. 2017). Understanding the cross patterns of the branches is crucial during the control of bleeding to avoid injury to the SMA and SMV (Alsabilah et al. 2016). This risk is even more apparent during minimally invasive laparoscopic or robotic surgery when the threat of bleeding increases. Some individuals have a variable right colic artery, which can arise from the middle colic or ileocolic artery, potentially causing insufficient bowel vascularization depending on where the artery is ligated (Alsabilah et al. 2016). The previously mentioned Riolan and marginal artery of Drummond arcades are critical for collateral circulation.

53.7 Complications

Acute Mesenteric Ischemia A new thrombosis at the site of the arteriotomy in the SMA can occur during the first 48 h. The underlying risk factors are the main determinants of re-stenosis. Therefore, post-intervention treatment includes anticoagulation and control of co-morbidities such as atherosclerosis, atrial fibrillation, or congestive heart failure. Often, even with immediate surgery or revascularization, morbidity remains 70–90% (Zeller et al. 2010).

Chronic Mesenteric Ischemia Perioperative mortality ranges from 0 to 16% but can be as high as 50% if there is associated acute mesenteric ischemia. Morbidity rates for endovascular revascularization are 11% and for open revascularization 16%. Mortality rates are similar at 4.4% for endovascular vs. 5.6% for open (Tendler and Thomas Lamont 2021). Most late deaths in patients who have had revascularization surgery are from complications unrelated to the mesenteric vasculature, most commonly heart conditions or cancer (Tendler and Thomas Lamont 2021).

Dissection A review of 58 patients showed that those requiring intervention after failed medical management had a mortality rate of 18% (Gilyami 2020). The overall mortality rate is less than 1% in patients who present with spontaneous SMA dissection.

Superior Mesenteric Artery Syndrome Early surgical treatment shows immediate symptom relief in most patients. If there had been extensive dilation prior to surgery, duodenal atony can persist even after duodenal decompression, precipitating an intestinal motility problem.

Arteriovenous Malformation and Pseudoaneurysm This commonly occurs after blunt abdominal trauma or recent abdominal surgery. It can present as a hypervascular lymph node on CT or MRI imaging. The formulation of a fistula between the SMA and SMV is very rare; however, it is commonly associated with carcinoid tumors. The treatment of choice is endovascular repair with coils (Nasrullah et al. 2014).

References

Alsabilah J, Kim WR, Kim NK (2016) Vascular structures of the right colon: incidence and variations with their clinical implications. Scand J Surg 106(2):107–115

Baran B, MertOzupek N, YerliTetik N, Acar E, Bekcioglu O, Baskin Y (2018) Difference between left-sided and right-sided colorectal cancer: a focused review of literature. Gastroenterology Res 11(4):264–273

Bruzzi M, M'harzi L, El Batti S, Ghazaleh RA, Taieb J, Poghosyan T, Berger A, Chevallier JM, Douard R (2019) Inter-mesenteric connections between the superior and inferior mesenteric arteries for left colonic vascularization: implications for colorectal surgery. Surg Radiol Anat 41(3):255–264

Caraiani C et al (2020) Indications for abdominal imaging: when and what to choose? J Ultrasonograph 20(80):e43–e54

Chaer RA, Abularrage CJ, Coleman DM et al (2020) The Society for Vascular Surgery clinical practice guidelines on the management of visceral aneurysms. J Vasc Surg 72(1S):3S–39S
Chou C, Mak C, Hou C et al (1997) CT of the mesenteric vascular anatomy. Abdom Imaging 22:477–482
Danowitz M, Solounias N (2016) Embryology, comparative anatomy, and congenital malformations of the gastrointestinal tract. J Anatom Embryol 33:39–50
Dolay K, Malya FU, Akbulut S (2019) Management of pancreatic head adenocarcinoma: from where to where? World J Gastrointest Surg 11(3):143–154
Ganss A, Rampado S, Savarino E, Bardini R (2019 May) Superior Mesenteric Artery syndrome: a prospective study in a single institution. J Gastrointest Surg 23(5):997–1005
Gilyami R (2020, April 30) Spontaneous Mesenteric artery dissection. Retrieved From Uptodate: https://www.uptodate.com/contents/spontaneous-mesenteric-arterial-dissection#H965553988
Gobble RM, Brill ER, Rockman CB, Hecht EM, Lamparello PJ, Jacobowitz GR, Maldonado TS (2009) Endovascular treatment of spontaneous dissections of the superior mesenteric artery. J Vasc Surg 50(6):1326–1332
Hohenwalter EJ (2009) Chronic mesenteric ischemia: diagnosis and treatment. Semin Intervent Radiol 26(4):345–351
Kim YW (2016) Current understandings of spontaneous isolated superior Mesenteric Artery dissection. Vasc Special Int 32(2):37–43
Kingham TP, Shen R, Ren C (2004) Laparoscopic treatment of superior mesenteric artery syndrome. JSLS 8(4):376–379
Lin PH, Chaikof EL (2000) Embryology, anatomy, and surgical exposure of the great abdominal vessels. Surg Clin N Am 80(1):417–433
Lorelli DR, Cambria RA, Seabrook GR, Towne JB (2003) Diagnosis and management of Aneurysms involving the superior Mesenteric Artery and its branches. Vasc Endovasc Surg 37(1):59–66. https://doi.org/10.1177/153857440303700108
McMahon KR, Balasubramanya R (2021) Intestinal trauma. [Updated 2021 May 18]. In: StatPearls [Internet]. StatPearls Publishing, Treasure Island. Available from: https://www.ncbi.nlm.nih.gov/books/NBK557624/
Monita MM, Gonzalez L (2021) Acute Mesenteric ischemia. [Updated 2020 Jun 30]. In: StatPearls [Internet]. StatPearls Publishing, Treasure Island. Available from: https://www.ncbi.nlm.nih.gov/books/NBK431068/
Munjal A, Bordoni B (2021) Histology, vascular. [Updated 2021 Feb 7]. In: StatPearls [Internet]. StatPearls Publishing, Treasure Island. Available from: https://www.ncbi.nlm.nih.gov/books/NBK554407/
Nagakawa T, Mori K, Kayahara M, Ohta T, Ueno K, Sanada H, Miyazaki I (1994) Three-dimensional studies on the structure of the tissue surrounding the superior mesenteric artery. Int J Pancreatol 15(2):129–188
Nasrullah A, Bell JK, Reddy S, Ray A, Rudralingam V (2014) Looking beyond lymph nodes: beware of the lurking pseudoaneurysm! Gastrointest Interv 3(2):112–115., ISSN 2213-1795. https://doi.org/10.1016/j.gii.2014.10.001
Negoi I, Beuran M, Hostiuc S, Negoi RI, Inoue Y (2018) Surgical anatomy of the superior Mesenteric vessels related to Pancreaticoduodenectomy: a systematic review and meta-analysis. J Gastrointest Surg 22(5):802–817
Noussios G, Dimitriou I, Chatzis I, Katsourakis A (2017) The main anatomic variations of the hepatic artery and their importance in surgical practice: review of the literature. J Clin Med Res 9(4):248–252
Perkins ZB, De'Ath HD, Aylwin C et al (2012) Epidemiology and outcome of vascular trauma at a British Major Trauma Centre. Eur J Vasc Endovasc Surg 44(2):203–209
Quain and Wilson (1844) The surgical anatomy of the human body, London. Taylor & Walton
Romano N, Prosperi V, Basili G et al (2011) Acute thrombosis of the superior mesenteric artery in a 39-year-old woman with protein-S deficiency: a case report. J Med Case Rep 5:17
Shaikh H, Wehrle CJ, Khorasani-Zadeh A (2021) Anatomy, abdomen and pelvis, superior mesenteric artery. [Updated 2020 Aug 15]. In: StatPearls [Internet]. StatPearls Publishing, Treasure Island. Available from: https://www.ncbi.nlm.nih.gov/books/NBK519560/
Shinji S, Matsumoto S, Kan H, Fujita I, Kanazawa Y, Yamada T, Hagiwara N, Koizumi M, Onodera H, Ko K, Machida T, Uchida E (2015) Superior mesenteric artery syndrome treated with single-incision laparoscopy-assisted duodenojejunostomy. Asian J Endosc Surg 8(1):67–70
Tendler DA, Thomas Lamont M (2021, Jan 07) Chronic Mesenteric Ischemia. Retrieved from Uptodate: https://www.uptodate.com/contents/chronic-mesentericischemia#H23494069
Walker TG (2009) Mesenteric vasculature and collateral pathways. Semin Interv Radiol 26:167–174
Wang Y, Cheng C, Wang L, Li R, Chen J, Gong S (2014) Anatomical variations in the origins of the celiac axis and the superior mesenteric artery: MDCT angiographic findings and their probable embryological mechanisms. Eur Radiol 24(8):1777–1784
Xiao Y, Lu JY, Xu L, Zhang GN (2019) Clinical anatomy study of superior mesenteric vessels and its branches. Zhonghua Wai Ke Za Zhi 57(9):673–680. Chinese
Yao SY, Mikami R, Mikami S (2015) Minimally invasive surgery for superior mesenteric artery syndrome: a case report. World J Gastroenterol 21(45):12970–12975
Ye K, Lin J, Sun Y, Wu Y, Xu J, He S (2018) Variation and treatment of vessels in laparoscopic right hemicolectomy. Surg Endosc 32(3):1583–1584
Zeller T, Rastan A, Sixt S (2010) Chronic atherosclerotic mesenteric ischemia (CMI). Vasc Med 15(4):333–338

Renal Artery

54

Corey Leskanic and Stephen J. Bordes, Jr.

54.1 Embryology

During the development of the kidneys, the renal blood supply changes (Moore and Persaud 1998). Initially, it comes from branches of the common iliac arteries; later, it comes from the distal end of the aorta (Moore and Persaud 1998). As the kidneys ascend during development, newer branches arise from a higher level of the aorta (Moore and Persaud 1998). The caudal branches normally involute (Moore and Persaud 1998). By the ninth week of development, the kidneys meet the suprarenal glands and become fixed in their position (Moore and Persaud 1998). The most cranial branches become the permanent renal arteries (Moore and Persaud 1998).

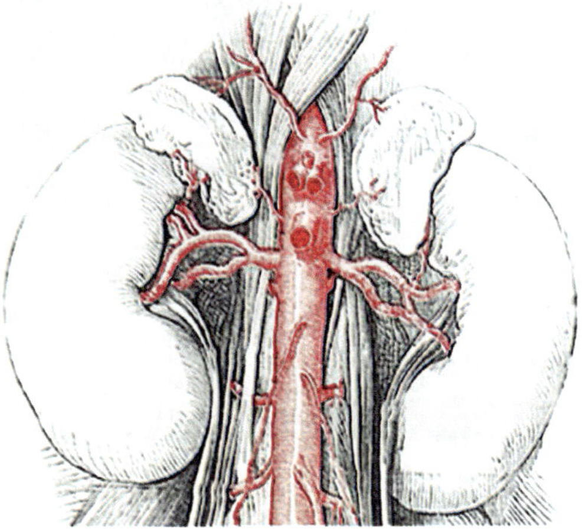

Fig. 54.1 Schematic drawing of the left and right renal arteries, which are bifurcated in this image. (After Allen 1892)

The renal arterial supply is most commonly a single artery from the abdominal aorta (Murillo et al. 2012) (Figs. 54.1 and 54.2). However, accessory arteries (Figs. 54.3 and 54.4) can arise from incomplete regression of arteries formed earlier, transient renal arteries, or additional arteries that do not perfuse the kidney entirely owing to its final position (Murillo et al. 2012). Two to four renal arteries are found in approximately 25% of adult kidneys (Moore and Persaud 1998). The term

C. Leskanic
Department of Radiology, HCA Florida Trinity, Trinity, USA

S. J. Bordes, Jr. (✉)
Department of Surgery, Louisiana State University Health Sciences Center, New Orleans, LA, USA

S. J. Bordes, Jr. et al. (eds.), *The Clinical Anatomy of the Vascular System*,
https://doi.org/10.1007/978-3-031-78326-5_54

"accessory artery" can denote any artery other than the main renal artery arising from the abdominal aorta and supplying the kidney (Ramulu and Prasanna 2016). It can enter a pole directly or pass through the renal hilum (Ramulu and Prasanna 2016). Accessory arteries can present superior or inferior to the main renal artery (Moore and Persaud 1998).

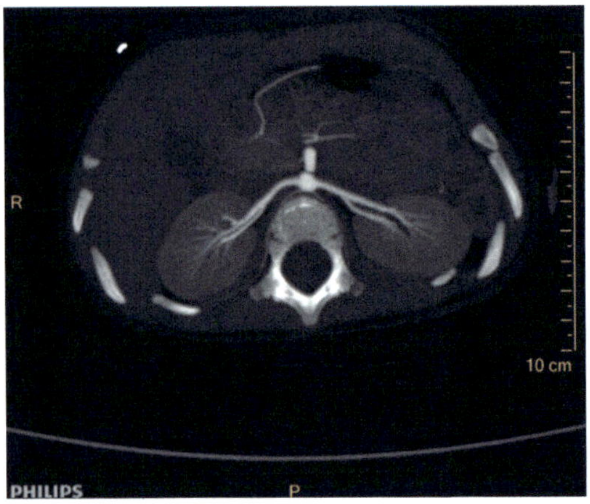

Fig. 54.2 Axial CT with contrast illustrating the left and right renal arteries arising from the abdominal aorta and supplying the kidneys

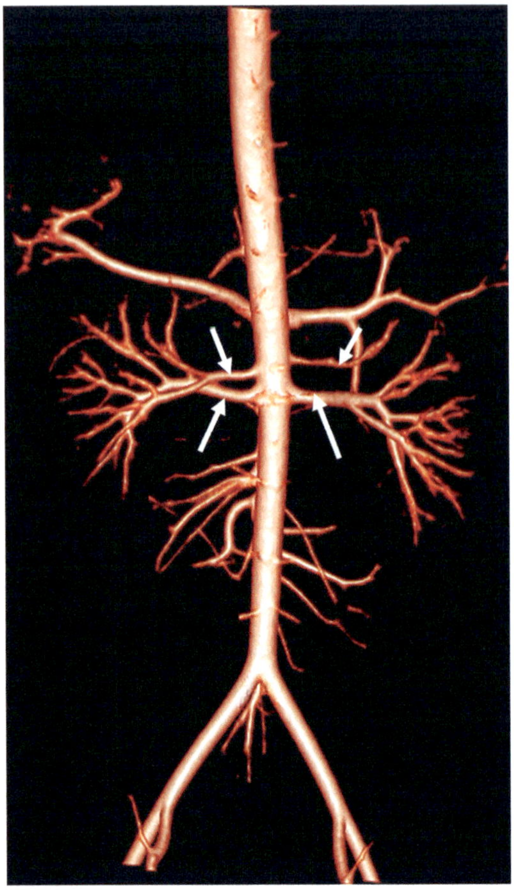

Fig. 54.4 Volume rendered image of the patient shown in Fig. 54.3 noting the accessory renal arteries

Fig. 54.3 CTA of the abdomen illustrating accessory renal arteries

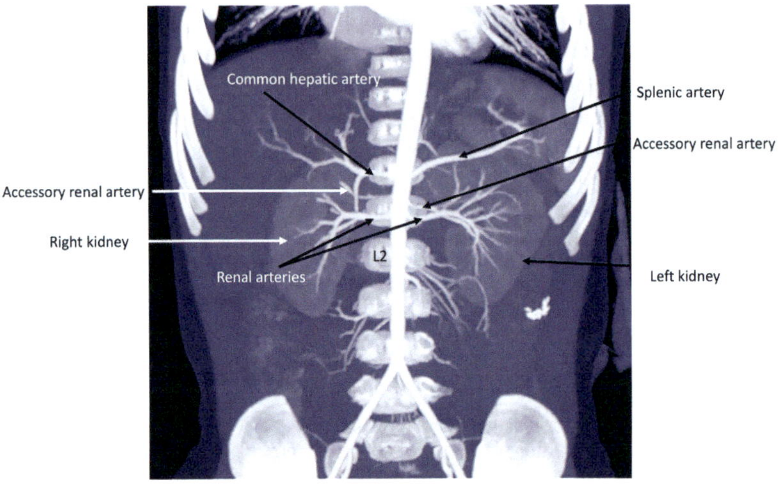

54.2 Anatomy

The renal parenchyma can be subdivided into five segments using the classic Graves descriptions: the middle, lower, posterior, apical, and upper segments. Each segment is supplied by an arterial branch from the main renal artery (Macchi et al. 2017), which forms two divisions, anterior and posterior (Trivedi et al. 2012). The anterior division divides into four segmental arteries (Trivedi et al. 2012): the lower, middle anterior, superior anterior, and apical segmental arteries (Trivedi et al. 2012). The posterior division does not divide and continues solely as the posterior segmental artery (Trivedi et al. 2012). Recent radiological studies have suggested that "pre-segmental arteries" can also arise between the main artery and the segmental branches (Macchi et al. 2017).

The renal arteries use approximately 20% of the cardiac output to supply the kidneys (Al-Katib et al. 2017). Usually, the main renal arteries arise at the L2 vertebral body level of the abdominal aorta, below the superior mesenteric artery (Al-Katib et al. 2017). The right main renal artery is slightly superior to and longer than the left (Al-Katib et al. 2017). A distinctive characteristic of the right main renal artery is that it is a significant vessel located posteriorly to the inferior vena cava (Al-Katib et al. 2017). The renal arteries commonly measure 5–6 mm in diameter and 4–6 cm in length (Al-Katib et al. 2017).

Accessory renal arteries are found in approximately 30% of individuals, and 10% are present bilaterally (Al-Katib et al. 2017). They can arise from anywhere between the T11 and L4 vertebral levels, originating from the aorta or iliac arteries (Al-Katib et al. 2017). In rare cases, an accessory renal artery arises from the lower thoracic aorta (Al-Katib et al. 2017).

54.3 Histology

The renal vessel walls comprise three layers: endothelium, muscular media, and adventitia (Mohamed and Sequeira-Lopez 2018). The muscular media is more prominent in the renal arteries than the renal veins (Mohamed and Sequeira-Lopez 2018). The adventitia comprises connective tissue, vasa vasorum, and autonomic nerves (Mohamed and Sequeira-Lopez 2018).

54.4 Imaging

The renal vasculature can be visualized via ultrasonography (US), computed tomographic (CT) angiography, and magnetic resonance (MR) angiography (Bedayat et al. 2020). US is useful for real-time qualitative and quantitative analyses of the renal artery (Bedayat et al. 2020). A Doppler US can reveal physiological data such as angle-corrected peak systolic velocity (PSV) (Bedayat et al. 2020). The PSV of the main renal artery can range from 60 to 100 cm/s (Bedayat et al. 2020). The acceleration time, the time from the start of systole to the systolic peak, can also be measured (Bedayat et al. 2020). Usually, the acceleration time for the main renal artery is less than 70 ms (Bedayat et al. 2020). These measurements are significant when investigating renal artery stenosis (RAS) (Bedayat et al. 2020). RAS can result in an increased PSV in the stenotic segment and an increased acceleration time in the segment distal to the stenosis (Bedayat et al. 2020). While the US can be a valuable real-time tool, it has limitations (Bedayat et al. 2020). Large body habitus, bowel gas, and anatomical variants make accurate US imaging difficult (Bedayat et al. 2020).

CT and MRI angiography offer anatomical information (Bedayat et al. 2020). CT angiography helps detect and classify renal abnormalities (Bedayat et al. 2020). A scanning delay optimized for arterial enhancement is required to identify renal arteries, specifically after a rapid contrast injection rate of 5 mL/s (Bedayat et al. 2020). Patients with endovascular renal stents are commonly evaluated with CT angiography because MR imaging gives artifacts (Bedayat et al. 2020). Additional advantages of CT imaging include fast data acquisition and the availability of volumetric data sets for image processing (Bedayat et al. 2020). Three-dimensional maximum intensity projection (3D MIP) is helpful for small vessel pathology, including aneurysms and pseudoaneurysms of the renal arteries (Bedayat et al. 2020). Curved planar reformations can classify stenotic diseases such as tortuous renal arteries (Bedayat et al. 2020).

MRI imaging is an option for patients if their renal dysfunction contraindicates iodinated contrast administration (Bedayat et al. 2020). When patients can receive gadolinium-based contrast, MRI angiography with subtraction imaging helps visualize vascular structures (Bedayat et al. 2020). Gadolinium contrast agents do not cause contrast-induced nephropathy at standard doses (Bedayat et al. 2020). To ensure proper timing and renal artery visualization, a 1 mL test bolus is used to time the arrival of contrast at the level of the renal arteries (Bedayat et al. 2020). Time-resolved imaging can be beneficial by visualizing real-time blood flow and detecting arterial stenoses, malformations, or arteriovenous fistulas (Bedayat et al. 2020). MR angiography at 3 T is the preferred MRI option for renal artery imaging (Bedayat et al. 2020). In addition, parallel imaging can decrease the required length of breath-holding (Bedayat et al. 2020).

54.5 Variations

Six variations in the anterior division of segmental branches of the main renal artery have been classified (Trivedi et al. 2012). In Type 1, the anterior division divides into upper and lower presegmental branches (Trivedi et al. 2012). The lower branch is divided into the middle and lower segmental arteries, while the upper one is divided into the apical and superior segmental arteries (Trivedi et al. 2012). Type 1 accounts for 29.16% of variants (Trivedi et al. 2012). In Type 2, the anterior division divides at one common point into four segmental arteries; this accounts for 20.83% of variants (Trivedi et al. 2012). In Type 3, the anterior division divides into presegmental and segmental branches (Trivedi et al. 2012). The presegmental branch tracks to the lower pole, creating superior, middle, and lower segmental arteries (Trivedi et al. 2012), while the segmental branch tracks to the superior pole and becomes the apical segmental branch (Trivedi et al. 2012). Type 3 accounts for 16.67% of variants (Trivedi et al. 2012). In Type 4, the anterior division first branches to produce the apical and lower segmental arteries (Trivedi et al. 2012) and then continues to track as the middle presegmental branch (Trivedi et al. 2012). Later, it divides further into the superior and middle segmental arteries (Trivedi et al. 2012). Type 4 accounts for 15.28% of variants (Trivedi et al. 2012). Type 5 involves the anterior division continuing to the lower pole while producing the superior, middle, and lower segmental arteries (Trivedi et al. 2012); the apical artery originates from the renal artery or aorta. Type 5 accounts for 9.72% of variants (Trivedi et al. 2012). In Type 6, the anterior division divides into the apical, superior, and middle segmental arteries (Trivedi et al. 2012). The lower segmental artery originates directly from the renal artery or aorta (Trivedi et al. 2012). Type 6 accounts for 8.33% of variants (Trivedi et al. 2012).

54.6 Pathology

Renal Artery Stenosis

Renal artery stenosis accounts for 1–5% of patients suffering from hypertension and is the leading cause of secondary hypertension (Bedayat et al. 2020). The most common cause includes atherosclerosis, typically found in the proximal 2 cm of the renal artery (Bedayat et al. 2020). An increase in blood pressure via activation of the renin-angiotensin cascade results from decreased kidney perfusion (Bedayat et al. 2020). Renal artery stenosis presents bilaterally in about 30% of affected patients (Bedayat et al. 2020). Additional causes include dissection, embolic disease, fibromuscular dysplasia, vasculitis, and external compression (Bedayat et al. 2020). Classifying the degree of renal stenosis is controversial (Bedayat et al. 2020). While more than 50% stenosis is generally considered significant, some patients only present symptomatically once the stenosis exceeds 70% (Bedayat et al. 2020). A meta-analysis of 947 patients showed no differences between renal stenting and medical therapy for managing renal artery stenosis (Bedayat et al. 2020). Thus, understanding the cause and degree of stenosis can help to guide management (Bedayat et al. 2020). Renal Doppler US is commonly used as first-line imaging for evaluating renal artery stenosis (Bedayat et al. 2020). Stenosis is significant when peak systolic velocity (PSV) is greater than 200 cm/s (Bedayat et al. 2020). In addition, if the renal artery PSV/aorta ratio exceeds 3.5:1, the stenosis is significant (Bedayat et al. 2020). Another option for imaging is a CTA, which provides anatomical evaluation (Bedayat et al. 2020). The sensitivity and specificity for diagnosing renal artery stenosis are 88–96% and 77–98%, respectively (Bedayat et al. 2020). MRA evaluation has sensitivity and specificity for diagnosing renal artery stenosis of 88–100% and 71–100%, respectively (Bedayat et al. 2020).

Fibromuscular Dysplasia

Fibromuscular dysplasia (FMD) is considered a nonatherosclerotic, noninflammatory, and idiopathic disease involving the muscular layer of the arterial walls (Bedayat et al. 2020). It is more common in young females and presents most frequently as hypertension (Bedayat et al. 2020). In the Assessment of Renal and Cervical Artery Dysplasia (ARCADIA) study, 469 patients with FMD were screened, and 77.4% were hypertensive (Bedayat et al. 2020). FMD can lead to segmental stenosis of small and medium-sized arteries (Bedayat et al. 2020). Atherosclerotic renal artery stenosis typically occurs in the proximal artery, while FMD predominantly affects the middle to distal portion of the renal artery (Al-Katib et al. 2017). Different types of FMD can coexist (Bedayat et al. 2020). The medial type is the most common, accounting for 60–70% of FMD cases (Bedayat et al. 2020). It is characterized by fibrous tissue replacing the media smooth muscle cells (Bedayat et al. 2020). Imaging reveals multifocal stenosis alternating with dilation of the renal artery, a string-of-beads appearance (Bedayat et al. 2020). The medial type is also considered the most common type in women aged 30–50 years (Bedayat et al. 2020). The second most common type is perimedial or subadventitial, which accounts for 10–20% of cases (Bedayat et al. 2020). This type is characterized by the involvement of the external layer of the media (Bedayat et al. 2020); stenoses tend to be more tubular, and interval segments do not exceed the average diameter of the renal artery (Bedayat et al. 2020). Finally, the intimal type is described as intimal fibrosis and subendothelial proliferation (Bedayat et al. 2020). This type is mainly seen in male children (Bedayat et al. 2020).

Renal Artery Occlusion

The main renal artery can be occluded by thrombosis or thromboembolism (Al-Katib et al. 2017). The kidney can become infarcted within 60 minutes of onset without collateral blood flow (Al-Katib et al. 2017). Typically, acute thrombosis cases have underlying atherosclerotic disease (Al-Katib et al. 2017). Renal artery embolism is relatively uncommon and accounts for 2% of cases (Al-Katib et al. 2017). Emboli commonly originate from the heart if there is a history of valvular heart disease, atrial fibrillation, or recent myocardial infarction (Al-Katib et al. 2017). Presenting symptoms can include hematuria, nausea, vomiting, or flank pain (Al-Katib et al. 2017). The most sensitive serological marker for renal infarction is lactate dehydrogenase (Al-Katib et al. 2017). The Doppler signal is absent in US Doppler (Al-Katib et al. 2017). CT with contrast can show a wedge-shaped nonenhancing region in the renal parenchyma or global nonenhancement (Al-Katib et al. 2017). When blood flow from the capsular perforating vessels is maintained, a 1–3 mm rim of subcortical enhancement can be visualized (Al-Katib et al. 2017). An advantage of CT is that it can also reveal infarcts of other organs, which typically occur in 34% of patients (Al-Katib et al. 2017). Bilateral renal infarction can also occur and accounts for approximately 16% of cases, typically with a cause secondary to atrial fibrillation or coagulopathy (Al-Katib et al. 2017). Contrast-enhanced MR imaging can also be used (Al-Katib et al. 2017). Catheter angiography can detect renal artery occlusion and help to classify its degree (Al-Katib et al. 2017). Renal artery occlusion is typically treated with anticoagulants to control hypertension (Al-Katib et al. 2017). Thrombolytic therapy can preserve renal function when the renal parenchyma is viable (Al-Katib et al. 2017). Although revascularization tends to have a low success rate, patients can benefit from endovascular revascularization when renal function deteriorates, and the patient has presented within two days of the infarct (Al-Katib et al. 2017).

Renal Artery Aneurysm

Renal artery aneurysms (RAA) are rare, with a prevalence of around 0.1%, and in approximately 73% of cases are found in systemic hypertensive patients (Al-Katib et al. 2017; Bedayat et al. 2020). In addition, 34% of patients have underlying FMD (Al-Katib et al. 2017). RAAs are typically asymptomatic and are diagnosed as incidental findings during imaging (Bedayat et al. 2020). Symptomatic RAAs are associated with renal infarction, pain, and hematuria (Bedayat et al. 2020). Complications include thrombosis, embolism, and rupture (Bedayat et al. 2020). RAAs are most common at the level of renal bifurcation but can occur in either the main right or left renal artery (Bedayat et al. 2020). CT angiography is typically used to assess RAAs (Bedayat et al. 2020). Treatment depends on comorbidities and aneurysms (Bedayat et al. 2020). Underlying conditions that require treatment for RAAs owing to the increased risk of rupture include pregnancy, inflammatory diseases such as polyarteritis nodosa, and uncontrolled systemic hypertension (Bedayat et al. 2020). Aneurysms larger than 1.5 cm have a 10% mortality rate and require treatment (Bedayat et al. 2020). Smaller aneurysms with no additional risk factors can be observed and reevaluated by MRA and CTA after 1–2 years (Bedayat et al. 2020).

Renal Artery Dissection

Renal artery dissections can extend from the aorta or be limited to the renal artery (Bedayat et al. 2020). Symptoms include decreased renal function, increased blood pressure, or acute flank pain (Bedayat et al. 2020). Iatrogenic causes include complications from endovascular interventions (Bedayat et al. 2020). Blunt abdominal traumas can cause dissection of the renal artery (Bedayat et al. 2020). Rare cases of spontaneous dissection typically involve the distal portions of the renal arteries in patients with underlying comorbidities, including FMD, Ehlers-Danlos, Marfan syndrome, cystic medial necrosis, or a history of cocaine use (Al-Katib et al. 2017; Bedayat et al. 2020). Contrast-enhanced CTA is the imaging method of choice for evaluating renal artery dissection (Bedayat et al. 2020). However, small flaps and thromboses of small vessels can go unrecognized (Bedayat et al. 2020). In these cases, it is helpful to recognize secondary signs, including focal vessel cutoff, uniform lumen narrowing, or distal ischemic changes (Al-Katib et al. 2017). Recommended treatments for stable patients include observation and anticoagulation. Unstable patients undergo endovascular stent repair as the treatment of choice (Bedayat et al. 2020).

Renal Artery Entrapment

Renal artery entrapment typically occurs in young patients with hypertension (Bedayat et al. 2020). Compression of the proximal portion of the renal artery by the psoas or crus of the diaphragm is a common underlying etiology (Bedayat et al. 2020). The left renal artery is often involved (Bedayat et al. 2020). The imaging method of choice is CT angiography (Bedayat et al. 2020). Initial treatment includes stenting the affected artery (Al-Katib et al. 2017; Bedayat et al. 2020). However, surgical repair is sometimes indicated because of the mechanical forces exerted by muscle contractions and diaphragm motion (Al-Katib et al. 2017; Bedayat et al. 2020).

Pseudoaneurysm

Pseudoaneurysms of the renal artery are commonly saccular (Al-Katib et al. 2017; Bedayat et al. 2020). They are often a consequence of direct penetration trauma to the arterial wall (Bedayat et al. 2020). Extravasation leads to a collection of blood contained in the adventitia or neighboring tissue (Bedayat et al. 2020). CT angiography is used to visualize the pseudoaneurysm, and MR angiography without contrast or with non-gadolinium contrast can be used for patients with functional renal impairment (Bedayat et al. 2020). Pseudoaneurysms are considered high risk when larger than 2 cm (Bedayat et al. 2020). High-risk pseudoaneurysms in the emergency setting are typically treated with transcatheter embolization, with a success rate of 93% for preserving parenchyma and controlling hemorrhage (Al-Katib et al. 2017). Smaller pseudoaneurysms of the renal artery tend to resolve spontaneously (Bedayat et al. 2020).

Mycotic Renal Artery Aneurysm

Mycotic renal artery aneurysm is a subtype of renal pseudoaneurysm that results from arterial wall degeneration and disruption from infectious arteritis (Al-Katib et al. 2017; Bedayat et al. 2020). It develops in patients with renal abscesses, septic emboli, or pyelonephritis (Bedayat et al. 2020). It is associated with wall thickening and perivascular fluid or fat stranding (Bedayat et al. 2020). Initial treatment includes intravenous antibiotics, but sizes greater than 2 cm or symptomatic presentations require urgent endovascular or surgical repair (Bedayat et al. 2020).

Arteriovenous Communications

Arteriovenous communications (AVMs) are abnormal connections between arteries and veins (Bedayat et al. 2020). They can occur as a fistula involving a single artery and a dilated vein. Alternatively, they can present as complex nidi with tortuous channels between single or multiple arteries and veins referred to as cirsoids (Al-Katib et al. 2017; Bedayat et al. 2020). When AVMs involve the renal vasculature, patients can present with hematuria, hypertension, flank pain, or high-output heart failure (Bedayat et al. 2020). About 75% of AVMs result from trauma due to renal biopsy or neoplasms (Bedayat et al. 2020). AVMs are typically three times more common in women than men (Bedayat et al. 2020). The right kidney is often affected (Bedayat et al. 2020). CT angiography is used to identify an AVM and can detect complications such as subcapsular and perinephric hematomas (Bedayat et al. 2020). MRA can show a flow void on a T2-weighted sequence (Bedayat et al. 2020). In simple AVMs, percutaneous embolization is the treatment of choice (Bedayat et al. 2020). However, in complex AVMs, partial or total nephrectomy could be indicated (Bedayat et al. 2020).

Systemic Disorders Affecting the Renal Arteries

Renal artery vasculitis is underlying inflammation of the vessel wall, which typically causes stenosis and thickening (Bedayat et al. 2020). In some instances, reduced blood flow and the potential for arterial occlusion (Bedayat et al. 2020) result in kidney ischemia (Bedayat et al. 2020). More recently, vasculitis classification has focused on disease manifestation and size (Bedayat et al. 2020).

Small renal vasculitis includes granulomatosis with polyangiitis, Henoch-Schonlein, microscopic polyarteritis, and cryoglobulinemic vasculitis (Bedayat et al. 2020). Medium and large vasculitis, most commonly polyarteritis nodosa, usually affects the main renal artery (Bedayat et al. 2020). Polyarteritis nodosa creates necrotizing inflammatory lesions, leading to systemic hypertension in about 90% of cases (Al-Katib et al. 2017; Bedayat et al. 2020). CT typically shows diffuse enlargement and hypoattenuation of the kidneys, and T2-weighted MR images show a diffuse high-intensity signal (Al-Katib et al. 2017). Less common forms of medium or larger vessel vasculitis involving the renal arteries include Takayasu, giant cell arteritis, and Kawasaki disease (Bedayat et al. 2020).

Secondary vasculitis is commonly caused by systemic lupus erythematosus, amphetamines, cocaine, or neurofibromatosis (Bedayat et al. 2020). Malignancies can also cause renal artery vasculitis via a paraneoplastic reaction or as a consequence of radiation or chemotherapy (Bedayat et al. 2020). CT and MR angiography are typically used to evaluate vasculitis of the renal artery as they can visualize vessel wall thickness, stenosis, edema, and overall renal parenchymal changes (Bedayat et al. 2020). Catheter angiography is limited because it visualizes only thrombus formation and luminal abnormalities (Bedayat et al. 2020). Vasculitis affecting the renal artery is typically treated medically, but endovascular treatments can be used if complications arise, such as aneurysms or significant arterial stenosis (Bedayat et al. 2020).

54.7 Clinical and Surgical Significance

Knowledge of the classical anatomy of the renal arterial vasculature is essential for excision and reconstructive partial nephrectomy outcomes (Macchi et al. 2017). Clamping the main renal artery helps maintain a bloodless field during surgical procedures; however, prolonged clamping and ischemia can impair renal function in the long term (Macchi et al. 2017). To preserve renal function, identifying and clamping only the specific segmental arteries that supply the defect or site of interest helps to preserve uninvolved renal parenchyma (Macchi et al. 2017). However, selective clamping could be less valuable than supposed because segmental arterial branches can supply more than a single renal segment (Macchi et al. 2017).

The renal artery divides into anterior and posterior parts (Bedayat et al. 2020). The junction between these divisions is known as the avascular plane of Brodel (Bedayat et al. 2020). It is relatively avascular, and access to it can preclude bleeding complications during percutaneous nephrostomy (Bedayat et al. 2020).

Clinical manifestations of renal artery pathology, such as renal artery stenosis (RAS), can present with hypertension (HTN), renal dysfunction, or incidental findings (Safian 2021). Several HTN syndromes are associated with RAS (Safian 2021). Patients can present with essential HTN with coexisting atherosclerotic RAS, the RAS being an incidental finding (Safian 2021). Another scenario includes the new onset of severe HTN, which, depending on the patient's risk factors, can lead to suspicion of FMD or atherosclerotic RAS (Safian 2021). Patients with severe bilateral atherosclerotic RAS and uncontrolled or resistant HTN could have an underlying renin-dependent HTN superimposed on essential HTN (Safian 2021). In these cases, renal revascularization can reduce the HTN and requirement for medication, but patients typically still need medical management (Safian 2021). Severe forms of RAS can also present with malignant HTN, including such symptoms as acute pulmonary edema, papilledema, acute aortic syndrome, cerebral ischemia, or acute encephalopathy (Safian 2021).

Renal artery disease can present clinically as renal excretory dysfunction (Safian 2021). In atherosclerotic RAS or FMD, an acute rise in creatinine can follow the administration of angiotensin-converting enzyme (ACE) inhibitors, angiotensin receptor blockers (ARBs), or other diuretics (Safian 2021). Patients with atherosclerotic RAS tend to have unremarkable urinary sediment and absent proteinuria (Safian 2021). In elderly cases with atherosclerotic RAS, an unexplained acute or chronic rise in creatinine can indicate renal excretory dysfunction (Safian 2021).

Evaluation of renal excretory function in renal artery disease includes serum creatinine measurements, blood urea nitrogen, and urinalysis to investigate proteinuria (Safian 2021). Renal US Doppler and renal resistive indices (RRI) can also help the assessment. RRI is measured by a renal US Doppler and is calculated as (peak systolic velocity—end-diastolic velocity)/peak systolic velocity (Safian 2021). An RRI greater than 0.8 indicates intrarenal fibrosis and high intrarenal resistance (Safian

2021). An RRI of less than 0.6 indicates average intrarenal resistance but does not exclude abnormalities of the parenchyma (Safian 2021). When renal excretory dysfunction resulting from nephropathy is compared with impaired blood flow through the renal artery, creatinine measurement is not very sensitive (Safian 2021). For example, kidney donors tend to have Cr < 1 mg/dL, unilateral RAS typically have Cr > 2.0 mg/dL, and severe bilateral RAS can have markedly high creatinine levels but normal kidneys (Safian 2021). Overreliance on any single parameter of renal function can overestimate or underestimate the effect of renal artery disease on renal function (Safian 2021).

References

Al-Katib S, Shetty M, Jafri SMA, Jafri SZH (2017) Radiologic assessment of native renal vasculature: a multimodality review. Radiographics 37(1):136–156. https://doi.org/10.1148/rg.2017160060

Bedayat A, Hassani C, Prosper AE, Chalian H, Khoshpouri P, Ruehm SG (2020) Recent innovations in renal vascular imaging. Radiol Clin North Am 58(4):781–796. https://doi.org/10.1016/j.rcl.2020.02.010

Macchi V, Crestani A, Porzionato A, Sfriso MM, Morra A, Rossanese M et al (2017) Anatomical study of renal arterial vasculature and its potential impact on partial nephrectomy. BJU Int 120(1):83–91. https://doi.org/10.1111/bju.13788

Mohamed T, Sequeira-Lopez MLS (2018) Development of the renal vasculature. Semin Cell Dev Biol. https://doi.org/10.1016/j.semcdb.2018.06.0

Moore KL, Persaud TV (1998) The developing human: clinically oriented embryology. WB Saunders, Philadelphia, pp 384–391

Murillo H, Lane MJ, Punn R, Fleischmann D, Restrepo CS (2012) Imaging of the aorta: embryology and anatomy. Semin Ultrasound CT MRI 33(3):169–190. https://doi.org/10.1053/j.sult.2012.01.013

Netter FH (2023) Netter Atlas of Human Anatomy(8th ed.). Philadelphia: Elsevier. ISBN: 9780323680424

Ramulu MV, Prasanna LC (2016) Accessory renal arteries—anatomical details with surgical perceptions. J Anat Soc India 65:S55–S57. https://doi.org/10.1016/j.jasi.2016.06.003

Safian RD (2021) Renal artery stenosis. Prog Cardiovasc Dis 65:60–70. https://doi.org/10.1016/j.pcad.2021.03.003

Trivedi S, Dixit A, Kotgiriwar S, Athavale S, Gupta V, Deopujari R (2012) Variations in the branching pattern of renal artery and its clinical implications. J Anatom Soc India 61(2):229–233. https://doi.org/10.1016/s0003-2778(12)80036-0

55 Suprarenal Arteries

Logan Gold and Stephen J. Bordes, Jr.

55.1 Embryology

The suprarenal arteries start to form during the fifth week of development. The paired dorsal aortae fuse below the level of the fourth thoracic segment and then subdivide into three branching segments: ventral, lateral, and dorsolateral intersegmental. The suprarenal arteries develop from the lateral branches of the descending aorta arising from the upper lumbar level. The suprarenal gland receives blood from branches of the renal and inferior phrenic arteries, but the descending aorta is the predominant source (Schoenwolf et al. 2009). The suprarenal gland vasculature is fully established by the eighth week of gestation (Ishimoto and Jaffe 2011). It has been proposed that the wide variations in origin of the suprarenal arteries are linked to the persistence of anomalous branches from the lateral segments of the dorsal aorta, which normally disappear in the caudal direction. Other possible causes of such variation include growth factors, hemodynamic patterns, genetics, and environmental agents such as teratogens (Sushma et al. 2014).

L. Gold
School of Medicine, Louisiana State University Health Sciences Center, New Orleans, LA, USA

S. J. Bordes, Jr. (✉)
Department of Surgery, Louisiana State University Health Sciences Center, New Orleans, LA, USA

55.2 Anatomy/Topography

The suprarenal arteries supply blood to the suprarenal gland: the superior, middle, and inferior suprarenal arteries. The suprarenal glands receive more blood per unit weight than any other organ in the human body, presumably because they secrete hormones that mediate a range of vital systemic processes (Venkateshwara Reddy et al. 2014). Blood flows through them from the outer cortex to the inner medulla and drains via a single suprarenal vein on each side (Megha et al. 2021). Specifically, the suprarenal arteries branch into a subcapsular arteriolar plexus in the gland's capsule before giving rise to a capillary network that is continuous within the thin-walled sinusoids of the three cortical zones (Ishimoto and Jaffe 2011). One study describes this transition from arteriolar plexus to capillary network as particularly abrupt, naming it a "vascular dam" that can predispose the gland to hemorrhagic necrosis in certain instances (Vella et al. 2001).

The superior suprarenal artery can be single or multiple and arises almost exclusively from the inferior phrenic artery (Venkateshwara Reddy et al. 2014; Megha et al. 2021; Dutta 2010). It supplies the superomedial aspect of the gland (Fowler et al. 2013). In most cases, the middle suprarenal artery is a single artery arising directly from the aorta between the inferior phrenic and renal arteries (Venkateshwara

S. J. Bordes, Jr. et al. (eds.), *The Clinical Anatomy of the Vascular System*,
https://doi.org/10.1007/978-3-031-78326-5_55

Reddy et al. 2014; Megha et al. 2021; Dutta 2010). It supplies the anteromedial aspect of the organ and can also be multiple or completely absent (Fowler et al. 2013). The inferior suprarenal artery is often referred to as the most important vessel supplying the suprarenal gland. It can be single or multiple and usually arises from the renal artery (Venkateshwara Reddy et al. 2014; Megha et al. 2021; Dutta 2010). It supplies the posterior and inferolateral aspects of the gland (Fowler et al. 2013).

Arising from each of the suprarenal arteries are multiple capillary branches that enter the gland, sequentially penetrating the zona glomerulosa, zona fasciculata, zona reticularis, and finally the medulla before draining into the venous system. Each of these sections of the suprarenal gland preferentially secretes specific hormones: aldosterone, cortisol, androgens, and epinephrine, respectively (Pignatti et al. 2017). Blood flow through the suprarenal glands is therefore essential for maintaining homeostasis via such mechanisms as water and electrolyte balance, blood pressure control, and the body's adaptive stress response (Molina 2010).

55.3 Imaging

The suprarenal arteries are small-caliber vessels and can usually be visualized with catheter aortography in patients without evidence of suprarenal disease (Toni et al. 1988). They can also be seen on abdominal CT angiography (Fowler et al. 2013; Oğuzkurt et al. 1997).

55.4 Variations

The precise anatomy of the suprarenal arteries is highly variable. Knowledge of their common branching patterns is necessary for abdominal surgery or radiological intervention (Sushma et al. 2014; Dutta 2010). The superior suprarenal artery seems the least variable of the vessels, arising from the inferior phrenic artery in more than 90% of documented cases (Dutta 2010). It has also been documented as arising from the abdominal aorta, the splenic artery, and, more rarely, the celiac trunk or an intercostal artery (Venkateshwara Reddy et al. 2014; Vella et al. 2001; Dutta 2010). The middle suprarenal artery is the most variable. Reddy et al. found that when it was not present as a single entity arising from the aorta, it could be multiple or absent altogether and could originate from the inferior phrenic, renal, or superior polar artery (Venkateshwara Reddy et al. 2014). Another study confirmed these findings, some specimens having a middle suprarenal artery arising from the celiac trunk (Vella et al. 2001). Aside from its usual origin from the renal artery, the inferior suprarenal artery has been shown to arise from the aorta or from an accessory renal, inferior phrenic, or gonadal artery (Venkateshwara Reddy et al. 2014; Megha et al. 2021; Dutta 2010).

55.5 Pathology

While pathology of the suprarenal arteries is rare, the suprarenal gland itself is subject to a plethora of disease states including Cushing's syndrome (hypercortisolism), Addison's disease (hypocortisolism), and medullary diseases such as pheochromocytoma (Gorman 2013). Because each of these abnormalities involves dysregulation of hormone secretion, the role of the suprarenal vessels in the systemic dissemination of the hormones should be appreciated.

Aneurysm

There have been limited reports in the literature of aneurysms of the suprarenal arteries. Most documented cases have occurred among patients with elevated blood flow, such as pregnant women or patients with uncontrolled hypertension (Fowler et al. 2013). One case has been reported of an inferior suprarenal artery aneurysm in a 16-year-old male with polyarteritis nodosa. It has been speculated that hemorrhages of unknown origin in polyarteritis nodosa patients could be due to undiagnosed microaneurysm ruptures of the suprarenal arteries. Angiography is the most useful tool for diagnosing visceral vessel aneurysms (Oğuzkurt et al. 1997).

55.6 Surgical Significances and Complications

Laparoscopic Adrenalectomy

Laparoscopic adrenalectomy is widely considered the gold standard for treating many suprarenal gland tumors and other pathologies (Alemanno et al. 2017). However, it is important to note that laparoscopic procedures entail challenges that are not encountered during open surgical procedures. Adrenalectomies performed laparoscopically eliminate the surgeon's ability to palpate blood vessels that are prone to massive bleeding if not avoided. Achieving control of the suprarenal veins is necessary for laparoscopic adrenalectomy, but there is little data to establish whether the suprarenal arteries should be similarly addressed during the surgery (Joel et al. 2005). During an extensive study of laparoscopic adrenalectomies, Joel et al. found that while control of the suprarenal arteries was less crucial than control of the suprarenal veins, the inferior suprarenal artery definitively required formal ligation in a number of procedures (Joel et al. 2005).

Suprarenal Artery Embolization

Owing to their deep retroperitoneal position in the body, the suprarenal glands are relatively protected from traumatic injury, though it occurs occasionally. Such an injury can lead to suprarenal hemorrhage, which can be severe and life-threatening (Joel et al. 2005). Adrenal hemorrhage can occur in a wide variety of clinical settings, and one proposed mechanism for its manifestation is increased blood flow to the gland through the suprarenal arteries (Vella et al. 2001; Tan and Sutherland 2016). Adrenal artery embolization is a minimally invasive procedure often indicated for treating suprarenal hemorrhage. For this procedure, patients are moderately sedated, and a local anesthetic is administered. Access to the suprarenal arteries is usually achieved through the common femoral artery. The choice of embolic agent depends on physician preference, cost, and whether permanent or temporary arterial occlusion is desired. Other indications for suprarenal artery embolization include palliative pain relief and debulking in patients with suprarenal tumors, suprarenal hormone suppression in high-risk patients who cannot safely undergo adrenalectomy, and, in particularly rare cases, suprarenal artery aneurysms (Megha et al. 2021; Joel et al. 2005).

References

Alemanno G, Bergamini C, Prosperi P, Valeri A (2017) Adrenalectomy: indications and options for treatment. Updat Surg 69(2):119–125

Dutta S (2010) Suprarenal gland-arterial supply: an embryological basis and applied importance. Romanian J Morphol Embryol 51(1):137–140

Fowler AM, Burda JF, Kim SK (2013) Adrenal artery embolization: anatomy, indications, and technical considerations. AJR Am J Roentgenol 201(1):190–201

Gorman LS (2013) The adrenal gland: common disease states and suspected new applications. Clin Lab Sci 26(2):118–125

Ishimoto H, Jaffe RB (2011) Development and function of the human fetal adrenal cortex: a key component in the feto-placental unit. Endocr Rev 32(3):317–355

Joel AB, Rubenstein JN, Arredondo S, Meng MV, Duh QY, Stoller ML (2005) Laparoscopic appreciation of the adrenal artery: fact or fiction? J Endourol 19(7):793–796

Megha R, Wehrle CJ, Kashyap S, Leslie SW (2021) Anatomy, abdomen and pelvis, adrenal glands (suprarenal glands). In: StatPearls. StatPearls Publishing

Molina PE (2010) Endocrine physiology. McGraw-Hill Medical, New York, pp 131–133

Oğuzkurt L, Çekirge S, Balkancı F (1997) Inferior suprarenal artery aneurysm in polyarteritis nodosa. Pediatr Radiol 27(3):234–235

Pignatti E, Leng S, Carlone DL, Breault DT (2017) Regulation of zonation and homeostasis in the adrenal cortex. Mol Cell Endocrinol 441:146–155

Schoenwolf GC, Bleyl SB, Brauer PR, Francis-West PH (2009) Larsen's human embryology, 4th edn. Churchill Livingstone, p 410

Sushma RK, Dhoot M, Harode HA, D'souza AS, Mamatha H (2014) Anatomical variations in the arterial supply of the suprarenal gland. Int J Health Sci Res 4:31–36

Tan GX, Sutherland T (2016) Adrenal congestion preceding adrenal hemorrhage on CT imaging: a case series. Abdom Radiol (New York) 41(2):303–310

Toni R, Mosca S, Favero L, Ricci S, Roversi R, Toni G, Vezzadini P (1988) Clinical anatomy of the suprarenal arteries: a quantitative approach by aortography. Surg Radiol Anat 10(4):297–302

Vella A, Nippoldt TB, Morris JC 3rd (2001) Adrenal hemorrhage: a 25-year experience at the Mayo Clinic. Mayo Clin Proc 76(2):161–168

Venkateshwara Reddy M, Ahmed MM, Praveen Kumar P, Raghuramaiah G (2014) A study on vasculature of the suprarenal glands. Int J Anat Res 2(1):195–101

Gonadal Arteries

56

Bryant McLafferty

56.1 Embryology

The origin of the gonadal arteries is embryologically complex. They arise from lateral splanchnic branches of the aorta that enter the primitive kidney system, known as the mesonephros. The mesonephric arteries supply the developing gonads, suprarenal glands, diaphragm, and kidneys (Kayalvizhi et al. 2017). The gonadal arteries arise from one of these nine lateral mesonephric arteries (Hinata et al. 2015). This is inferred from Felix's "ladder theory," which proposes that all segmental mesonephric arteries persist as the definitive phrenic, adrenal, renal, accessory renal, gonadal, and accessory gonadal arteries (Felix 1912). The nine mesonephric arteries can be divided into cranial, middle, and caudal groups. The gonadal arteries develop from those branches, usually from one of the caudal mesonephric arteries; the others atrophy. The gonadal arteries lengthen as the sex glands descend and the kidneys ascend (Gupta et al. 2011). Anatomical variations in the gonadal arteries can result from variations in origin between the three groups of mesonephric arteries (Hinata et al. 2015).

B. McLafferty (✉)
Tulane University School of Medicine,
New Orleans, LA, USA
e-mail: bmclafferty@tulane.edu

56.2 Anatomy/Topography

The gonadal arteries (Figs. 56.1 and 56.2) are paired vessels that primarily supply oxygenated blood to the testes or ovaries. In both sexes, they arise from the anterolateral abdominal aorta, just inferior to the origin of the ipsilateral renal artery at the level of the second lumbar vertebra (Shoja et al. 2007). They seldom give off any branches during their course.

In males, the gonadal arteries are called the testicular arteries. From their origin, they traverse inferolaterally under the posterior parietal perineum along the psoas major muscle (Standring 2020). Before entering the pelvis, the right gonadal artery crosses over the inferior vena cava and lies posterior to the middle colic and ileocolic arteries and the terminal ileum (Shoja et al. 2007). The left pre-pelvic gonadal artery lies posterior to the inferior mesenteric vein, left colic artery, and descending colon (Standring 2020). As both arteries enter the pelvis, they cross anteriorly over the genitofemoral nerves, ureters, and external iliac arteries (Standring 2020). They then travel through the deep inguinal ring and within the ipsilateral spermatic cord until they reach the scrotum (Standring 2020).

The proximal portions of the testicular arteries supply the upper and middle parts of the ureter (Standring 2020). The distal portions give off one or more testicular arteries, an inferior testicular

S. J. Bordes, Jr. et al. (eds.), *The Clinical Anatomy of the Vascular System*,
https://doi.org/10.1007/978-3-031-78326-5_56

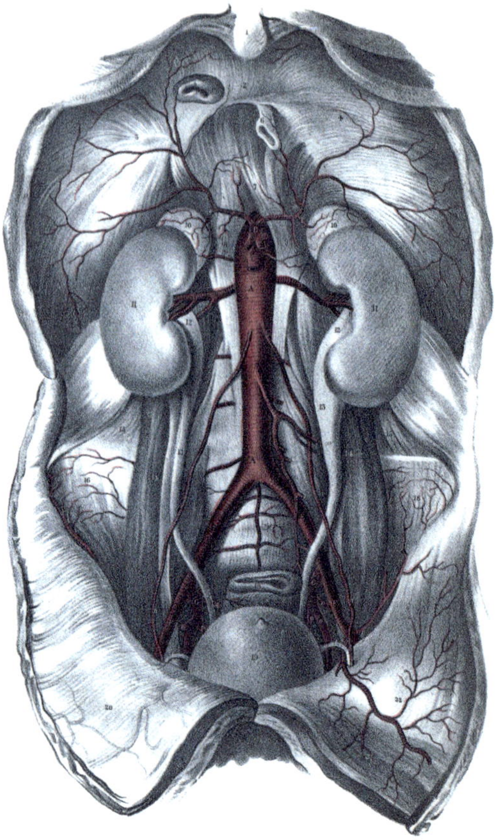

Fig. 56.1 Drawing illustrating the gonadal arteries in the male arising between the renal and inferior mesenteric arteries (Quain and Wilson, 1844 author's collection)

artery, and branches that supply the head, body, and tail of the epididymis (Standring 2020). Further distally, the arteries furcate to supply the anterior, medial, and lateral parts of the inferior pole of the testis and the anterior segment of the superior pole (Standring 2020).

In females, the gonadal artery is called the ovarian artery. It follows a similar anatomical path to the testicular artery as it exits the aorta and transverses inferolaterally until it reaches the pelvis. As it enters the pelvis, it crosses anteriorly over the common iliac artery and vein. Then, it turns medially to split into branches that supply the ovary broad ligament and the uterine tube (Standring 2020). Small branches of the ovarian artery travel through the inguinal canal to supply blood to the skin of the labium majus and the inguinal region (Standring 2020). The primary

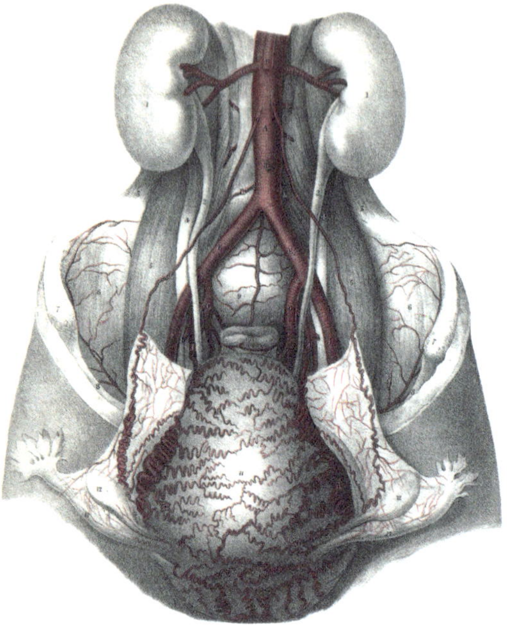

Fig. 56.2 Drawing illustrating the gonadal arteries in the female (Quain and Wilson, 1844 author's collection)

branch of the ovarian artery continues lateral to the uterus to anastomose with the uterine artery (Standring 2020).

56.3 Histology

The gonadal artery is a muscular, also known as a distributing artery. Muscular arteries arise from elastic arteries and feed into small arteries and arterioles. A prominent internal elastic lamina within the tunica intima allows muscular arteries to be identified quickly. They also have a thick layer of smooth muscle within the tunica media with fewer elastic fibers than elastic arteries. Their tunica externa is more prominent than that in elastic arteries and arterioles.

56.4 Variations

Variations in gonadal arteries are common and can involve their origin, number, and course. In both sexes, there are similar variations in the origin of the gonadal artery from the aorta. Accessory gonadal arteries have been observed in males and are classified as either medial or lateral (Kayalvizhi et al. 2017). The most common variations in origin are gonadal arteries that arise from the renal arteries, arch over the renal vein, or pass through a hiatus in the renal vein (Gupta et al. 2011). The arched gonadal artery over the renal artery is thought to result from the gonadal artery persisting from one of the cranial mesonephric arteries during embryo development.

Interestingly, this variation is mostly found on the left, possibly because the left kidney has to travel further than the right as it ascends. A gonadal artery arising from a renal artery (called an aberrant gonadal artery) indicates that it has persisted from a middle mesonephric artery. Anomalous origins of gonadal arteries are most common at the renal level of the abdominal aorta from associated anomalous renal vessels (Kayalvizhi et al. 2017). A high origin of the gonadal arteries is associated with anatomical variations and pathological findings in the kidneys (Ozan et al. 1995).

56.5 Pathology

The gonadal arteries are paired muscular arteries that can become occluded through atherosclerosis. The pathogenesis of this condition has been well established in the literature. Injury to the endothelial wall can result in the development of fibrous plaques. If severe enough, these plaques can cause claudication and even tissue loss. Risk factors for atherosclerosis include hypertension, smoking, diabetes, and obesity.

Sánchez-Sierra et al. (2019) published the first case of testicular artery occlusion, which was caused by a nematode parasite, *Angiostrongylus costaricensis*. After the orchiectomy, the right testicle showed signs of necrosis. Histological examination of the testicular artery revealed the worm (Sánchez-Sierra et al. 2019).

56.6 Clinical and Surgical Significance

The gonadal arteries are paired vessels that branch off the abdominal aorta and course into the pelvis. Knowledge of their anatomy and its variations is important for clinicians planning and performing endovascular and open surgeries within the abdomen and pelvis. Aneurysm or even rupture of the gonadal arteries is very rare. If it occurs, an endovascular technique can access and embolize the vessel(s) (García-Alva et al. 2018).

In boys with cryptorchidism, orchidopexy is sometimes needed to move the undescended testicle(s) into the scrotum. If the intra-abdominal testis is too high to grasp and pull down through the inguinal ring, the Fowler-Stephens orchidopexy technique can be used. This is a two-stage procedure. The first stage mobilizes the colon to expose the gonadal vessels proximally and then ligates the testicular artery and internal spermatic vein as far away from the testis as possible to preserve collateral circulation. The surgeon then waits 6 months for the arteries of collateral circulation to enlarge. Stage two is completed by dividing the peritoneum around the testis and vas deferens to mobilize the structures so they can be pulled down through the inguinal canal (Hutson 2019).

Orchiectomy is performed to remove a cancerous lesion or a necrotic testis if there is torsion. Typically, an inguinal incision is made. After the testes are dissected and divided from the surrounding fascia, the spermatic cord and its contents are ligated separately using sutures to ensure hemostasis (Owyong and Ramasamy 2021).

When performing an oophorectomy (one or both ovaries), the surgeon must carefully identify and ligate the ovarian artery, which lies within the suspensory ligament of the ovary. In patients with a history of breast cancer, it is essential to ligate the ovarian artery and vein 2 cm proximal to the ovary after meticulous dissection of the vessels. This technique ensures that all cancerous tissue of the ovary is removed from the patient.

Ovarian artery ligation is occasionally used to control hemorrhage from various causes in damage-control situations. These causes include hysterectomy and carcinoma of gynecological structures (Thavarasah et al. 1989). Postpartum hemorrhage is a severe complication in which a patient suffers from heavy bleeding immediately after giving birth. Laceration of the ovarian and/ or uterine vessels is one cause of this. Ligation of the uterine artery or uterine-ovarian branches is the preferred approach to control. If bleeding continues, the vessels of the uterine-ovarian arcade are ligated.

56.7 Complications

During orchiopexy and orchiectomy, meticulous care must be taken to secure control of the gonadal vessels to prevent retraction into the retroperitoneum and subsequent hemorrhage (Hutson 2019). Accidental ligation of the testicular artery has been reported during microsurgical varicocelectomy; however, it is exceedingly rare (Thavarasah et al. 1989). Similar care must be taken for the ovarian arteries during oophorectomy. Lack of proper ligation can lead to serious hemorrhaging.

References

Felix W (1912) Mesonephric arteries (aa. mesonephricae). In: Keibel F, Mall FP (eds) Manual of human embryology, vol 2. Lippincott, Philadelphia/London, pp 820–825

García-Alva R, Guerrero-Hernández M, Anaya-Ayala JE et al (2018) Successful embolization of a ruptured ovarian artery aneurysm in a postmenopausal woman: case report and literature review of gonadal artery aneurysms. Vasc Endovasc Surg 52(2):159–163

Gupta A, Singal R, Singh D (2011) Variations of gonadal artery: embryological basis and clinical significance. Int J Biol Med Res 2(4):1006–1010

Hinata N, Suzuki R, Ishizawa A et al (2015) Fetal development of the mesonephric artery in humans with reference to replacement by the adrenal and renal arteries. Ann Anat 202:8–17

Hutson JM (2019) Orchidopexy. In: Puri P, Höllwarth ME (eds) Pediatric surgery. Springer, pp 569–578

Kayalvizhi I, Narayan RK, Kumar P (2017) Anatomical variations of testicular artery: a review. Folia Morphol (Warsz) 76(4):10

Owyong M, Ramasamy R (2021) Surgical anatomy: orchiectomy and fertility preservation options. In: Nikolavsky D, Blakely SA (eds) Urological care for the transgender patient: a comprehensive guide. Springer International Publishing, pp 47–59

Ozan H, Gümüşalan Y, Onderoğlu S, Simşek C (1995) High origin of gonadal arteries associated with other variations. Ann Anat 177(2):156–160

Sánchez-Sierra LE, Martínez-Quiroz RA, Antúnez HS, Cabrera-Interiano H, Barrientos-Melara FJ (2019) Right testicular artery occlusion and acute appendicitis by Angiostrongylus costaricensis. Case Rep Surg 2019:5670802

Shoja MM, Tubbs RS, Shakeri AB, Oakes WJ (2007) Origins of the gonadal artery: embryologic implications. Clin Anat 20(4):428–432

Standring S (2020) Male reproductive system. In: Gray's anatomy: the anatomical basis for clinical practice, 42nd edn. Elsevier

Thavarasah AS, Sivalingam N, Almohdzar SA (1989) Internal iliac and ovarian artery ligation in the control of pelvic Haemorrhage. Aust N Z J Obstet Gynaecol 29(1):22–25

57 Inferior Mesenteric Artery

R. Shane Tubbs

The inferior mesenteric artery (IMA) (Figs. 57.1 and 57.2) is the final ventral unpaired branch of the abdominal aorta, originating at the L3 vertebra. It is relatively short, irrigating a large area from the splenic flexure of the colon to the mid-rectum. Due to this extensive coverage, its branches fan out like a short string over a long bowel segment (Skandalakis et al. 2004). The IMA initially gives rise to the left colic artery, which splits into superior and inferior divisions, followed by two or three sigmoid arteries, and continues as the superior (ano)rectal artery. Although traditionally described as following a standard pattern, this is observed in only between 15–16% and 30–40% of cases (Michels et al. 1965; Nelson et al. 1988; Goligher 1949).

57.1 Variations in Origin

The IMA typically arises from the ventral aspect of the aorta. Benton and Cotter (1963) documented the first case of duplicated IMA, with two separate origins from the aorta. Another case with trunks 2.6 cm apart was reported by IMA absence is exceedingly rare but has been documented (Adachi 1928; Mori et al. 1960). Variations include a common trunk with other vessels, such as celiacomesenteric trunks and hepato-mesenteric trunks (Osawa et al. 2004). The IMA may also originate from the superior mesenteric artery (SMA) in less than 0.1% of cases (Lippert and Pabst 1985).

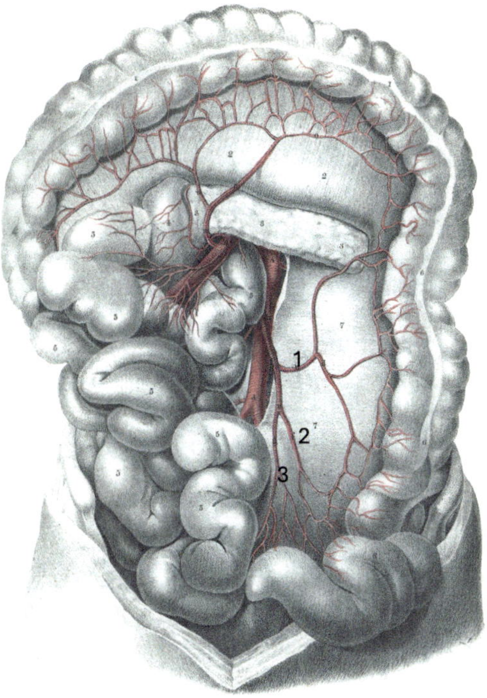

Fig. 57.1 Branches of the inferior mesenteric artery; 1 = left colic artery, 2 = sigmoid artery, 3 = superior (ano) rectal artery (Quain and Wilson, 1844 author's collection)

R. S. Tubbs (✉)
Department of Neurosurgery, Tulane University School of Medicine,
New Orleans, LA, USA
e-mail: rtubbs@tulane.edu

S. J. Bordes, Jr. et al. (eds.), *The Clinical Anatomy of the Vascular System*,
https://doi.org/10.1007/978-3-031-78326-5_57

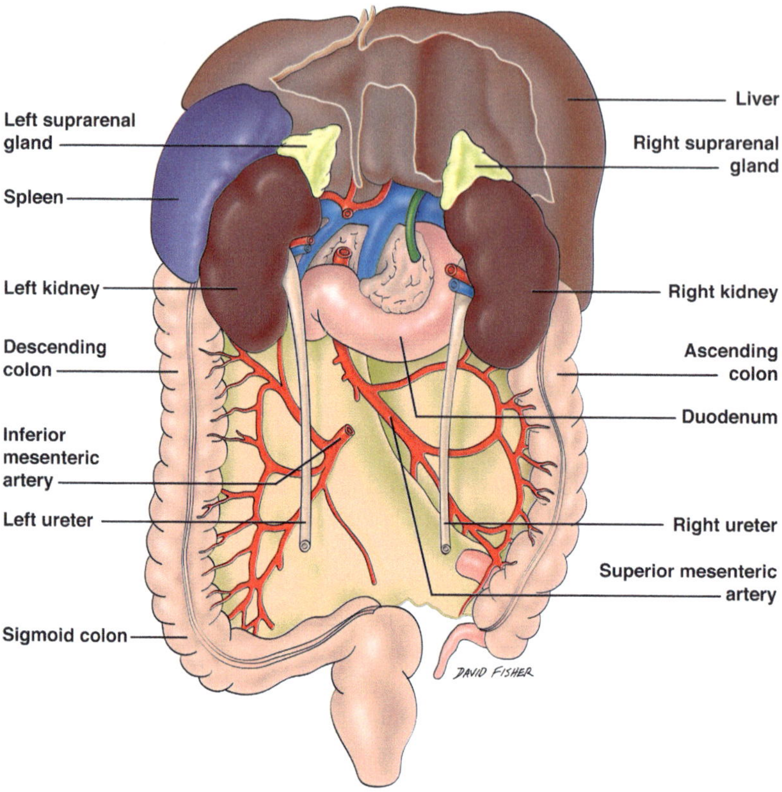

Fig. 57.2 Posterior view of the inferior mesenteric artery and its branches

57.2 Vertebral Level of Origin

The IMA predominantly originates at the L3 vertebra (Michels et al. 1965). However, its origin can vary, typically within one vertebral level (George 1935). Zebrowski et al. (1971) found that the IMA originates at L3 in 67% of cases, between L3 and L4 in 20%, at L4 in 8%, between L2 and L3 in 4%, and between L4 and L5 in 1%. Distal shifts in the origin with age have been noted (George 1935). The distance between the SMA and IMA origins is 5–7 cm (George 1935; Michels et al. 1965; Yahel and Arensburg 1998).

57.3 Branches

Left Colic Artery (LCA)

The LCA typically arises directly from the IMA in about 30–60% of cases (Goligher 1949; Nelson et al. 1988; Niculescu et al. 2005). In 53–56% of cases, it originates via a common trunk with the sigmoid artery (Nelson et al. 1988; Michels et al. 1965). Absence of the LCA is rare, reported in 0.7–12% of cases (Sonneland et al. 1958).

Sigmoid Arteries

The sigmoid arteries, ranging from one to seven in number, arise from the lateral aspect of the IMA and fan out toward the sigmoid colon (Michels et al. 1965; Nelson et al. 1988). Variations in their origins include the LCA, IMA trunk, or the left middle colic artery.

Superior Anorectal Artery

The superior anorectal artery is the continuation of the IMA trunk beyond the left common iliac vessels and is considered the most significant end branch. It typically bifurcates into two main branches at S2 or S3 and supplies the rectum (Michels et al. 1965; Vandamme et al. 1982).

Sudeck's and Griffith's Points

Sudeck's point, the area of the last sigmoid artery's origin from the IMA, has been debated for its anatomical and surgical significance. It is generally considered less relevant today due to advances in surgical techniques (Vandamme et al. 1982).

Griffith's point at the splenic flexure marks the critical point for marginal artery function, which is crucial for collateral circulation. The anastomosis at this point varies significantly, influencing the arterial supply to the descending colon and splenic flexure (Meyers 1976; Kahn and Abrams 1964).

57.4 Communications Between SMA and IMA

Marginal Artery

The marginal artery, running alongside the mesenteric border of the colon, forms an essential anastomosis between the SMA and IMA. It is a constant structure in most specimens, failing only in rare cases (Steward and Rankin 1933).

Arc of Riolan and Villemin Arcade

The Arc of Riolan is an anatomical communication between the SMA and IMA (Gourley and Gering 2005). It varies in presence and configuration, with some interpretations considering it synonymous with the marginal artery.

The Villemin arcade, present in 12–18% of cases, represents another communication point between these arteries (Vandamme and Bonte 1990).

Aberrant Branches

The IMA can give rise to various aberrant branches, including the middle colic artery, accessory middle colic artery, accessory right hepatic artery, accessory renal artery, or a common artery for both umbilical arteries.

References

Adachi B (1928) Anatomie der Japaner I, Das Arteriensystem der Japaner. Band II. Verlag der Kaiserlich-Japanischen Universitat zu Kyoto, Kyoto

Adachi B (1930) Das Fehlen der A. mesenterica inferior bei einem Japaner. Anat Anz 69:431–433

Amonoo-Kuofi HS, el-Badawi MG, el-Naggar ME (1995) Anomalous origins of colic arteries. Clin Anat 8:288–293

Armstrong BG, Hunt TH, Price CW, Resnick MI (1979) Common origin of inferior mesenteric and accessory renal artery. Urology 14:298–299

Ault GW, Castro AF, Smith RS (1952) Clinical study of ligation of the inferior mesenteric artery in left colon resections. Surg Gynecol Obstet 94:223–228

Benton RS, Cotter WB (1963) A hitherto undocumented variation of the inferior mesenteric artery in man. Anat Rec 145:171–173

Bertelli L, Lorenzini L, Bertelli E (1996) The arterial vascularization of the large intestine: anatomical and radiological study. Surg Radiol Anat 18:51–59

Cauldwell EW, Anson BJ (1943) The visceral branches of the abdominal aorta: topographical relationships. Am J Anat 73:27–57

Covey AM, Brody LA, Maluccio MA, Getrajdman GI, Brown KT (2002) Variant hepatic arterial anatomy revisited: Digital subtraction angiography performed in 600 patients. Radiology 224:542–547

Drummond H (1913) The arterial supply of the rectum and pelvic colon. Brit J Surg 1:677–685

Drummond H (1914) Some points relating to the surgical anatomy of the arterial supply of the large intestine. Proc R Soc M London 7:185–193

Feller I, Woodburne RT (1961) Surgical anatomy of the abdominal aorta. Ann Surg 154:239–252

Forgue E, Milhaud (1923) La circulation du segment sigmoido-rectal. La valeur reelle du point de sudeck. Contribution a la technique de I'extirpation abdomino-perineale du rectum. Rev Chir 61:61

Garti I, Nissenkorn I, Lerner M (1986) Common origin of inferior mesenteric and main renal artery. Eur Urol 12:215–216

George R (1935) Topography of the unpaired visceral branches of the abdominal aorta. J Anat 69:196–205

Gerard G (1921) Manuel d'anatomie humaine, 2nd edn. Masson e Cie Éditeurs, Paris

Goligher JC (1949) The blood supply to the sigmoid colon and rectum with reference to the technique of rectal resection with restoration of continuity. Brit J Surg 37:157–162

Gourley EJ, Gering SA (2005) The meandering mesenteric artery: a historic review and surgical implications. Dis Colon Rectum 48:996–1000

Gwyn DG, Skilton JS (1966) A rare variation of the inferior mesenteric artery in man. Anat Rec 156:235–238

Haller Von A (1745) Arteriae celiacae. In: Haller Von A (ed) Iconum anatomicarum, vol 2. Vandenhoeck A, Gottingae, pp 1–50

Hollinshead WH (1956) Anatomy for surgeons: Volume 2, The thorax, abdomen and pelvis. In: Vessels and nerves associated with the posterior abdominal wall. Paul. B. Hoeber Inc, pp 589–595

Kadir S (1991) In: Kadir S (ed) Atlas of normal and variant angiographic anatomy. W.B. Saunders Company, Philadelphia, pp 387–429

Kahn P, Abrams HL (1964) Inferior mesenteric arterial patterns: angiographic study. Radiology 82:429–441

Kaneko O (1990) The arterial distribution to the abdominal digestive organs in human fetuses. J Nippon Med Sch 57:448–464

Kitamura S, Nishiguchi T, Sakai A, Kumamoto K (1987) Rare case of the inferior mesenteric artery arising from the superior mesenteric artery. Anat Rec 217:99–102

Koizumi M, Horiguchi M (1990) Accessory arteries supplying the human transverse colon. Acta Ansat (Basel) 137:246–251

Larobina M, Nottle PD (2005) Extrahepatic biliary anatomy at laparoscopic cholecystectomy: is aberrant anatomy important? ANZ J Surg 75:392–395

Latarjet M, Ruiz-Liard A (1989) Anatomia humana, 2nd edn. Medicina Panamericana Editora do Brasil Ltda, Sao Paulo

Lippert H, Pabst R (1985) Arterial variations in man: classification and frequency. JF Bergmann Verlag, Munchen, pp 34–39. 23–73

Lochhart-Mummery P (1923) Diseases of rectum and colon and their surgical treatment. Bailliere, Tindall and Cox, London

Mayo CW (1955) Blood supply of the colon: surgical considerations. Surg Clin North Am 35:1117–1121

McGowan JR (1955) Collateral circulation in the sigmoid colon. Arch Surg 71:531–537

McNulty JG, Hickey N, Khosa F, O'Brien P, Callaghan JP (2001) Surgical and radiological significance of variants of Buhler's anastomotic artery: a report of three cases. Surg Radiol Anat 23:277–280

Meyers MA (1976) Griffiths' Point: critical anastomosis at the splenic flexure* significance in ischemia of the colon. Am J Roentgenol 126:77–94

Michels NA, Siddharth P, Kornblith PL, Parke WW (1963) The variant blood supply to the small and large intestines: its importance in regional resections. A new anatomic study based on four hundred dissections with a complete review of the literature. J Int Coll Surg 39:127–170

Michels NA, Siddharth P, Kornblith PL, Parke WW (1965) The variant blood supply to the descending colon, rectosigmoid and rectum based on 400 dissections. Its importance in regional resections: a review of medical literature. Dis Col Rect 8:251–278

Michiya K, Satoshi M, Takehiro O, Kana M, Ken O, Tsutomu N, Yasuhiro O, Keijiro A (2006) Preoperative assessment of vascular anatomy of inferior mesenteric artery by volume- rendered 3D-CT for laparoscopic lymph node dissection with left colic artery preservation in lower sigmoid and rectal cancer. World J Gastroenterol 12:1442–1446

Mori Y, Ito I, Hatashita S, Yoshikawa K (1960) An anomalous case of the mesenteric arteries. Absence of a mesenterica inferior. J Osaka Med Coll 20:77–79

Nassar L, Atweh LA, Jurjus A, Kutoubi AA (2012) Unusual arterial pattern of the gastrointestinal tract: inferior mesenteric artery arising from the iliac artery and corkscrew external lilac. Vasc Endovasc Surg 46:418–421

Nelson TM, Pollak R, Jonasson O, Abcarian H (1988) Anatomic variants of the celiac, superior mesenteric and inferior mesenteric arteries and their clinical relevance. Clin Anat 1:75–91

Niculescu MC, Niculescu V, Ciobanu IC, Daescu E, Jianu A, Sisu AM, Petrescu CI, Motoc A (2005) Correlations between the colic branches of the mesenteric arteries and the vascular territories of the colon. Romanian J Morphol Embryol 46:193–197

Osawa T, Feng XY, Sasaki N, Nagato S, Matsumoto Y, Onodera M, Nara E, Fujimura A, Nozaka Y (2004) Rare case of the inferior mesenteric artery and the common hepatic artery arising from the superior mesenteric artery. Clin Anat 17:518–521

Poireir P, Charpy A (1902) Traite'd' anatomie humaine, vol II, 2nd edn part 2

Pope CE, Buie LA (1929) A description of the arterial blood supply of the pelvic colon. Trans Am Proct Soc 30:78

Pope CE, Judd ES (1929) The arterial blood supply of the sigmoid, rectosigmoid and rectum. Surg Clin North Am 9:957

Prakash, Mokhasi V, Rajini T, Shashirekha M (2011) The abdominal aorta and its branches: anatomical variations and clinical implications. Folia Morphol (Warsz) 70:282–286

Quain R (1844) Anatomy of the arteries of the human body, London

Salve VM, Ratanprabha C (2011) Multiple variations of branches of abdominal aorta. Kathmandu Univ Med J 33(1):72–76

Shivarama CH, Shivarama B, Shetty RK, Avadhani R (2012) Multiple variations of branches of abdominal aorta: a case study. NUJHS 2:48–50

Shuang-Qin LJY, Terayama H, Naito M, Iimura A, Itoh M (2008) A rare case of inferior mesenteric artery arising from the superior mesenteric artery, with a

review of the review of the literature. Surg Radio Anat 30:159–165
Skandalakis LJ, Skandalakis PN, Mirilas PS (2004) Surgical anatomy: the embryonic and anatomic basis of modern surgery, vol 2. PMP, Thessaloniki, pp 1098–1102
Songur A, Toktas M, Alkoc O, Acar T, Uzun I, Bas O, Ozen AO (2010) Abdominal aorta and its branches: morphometry- variations in autopsy cases. Eur J Gen Med 7:321–325
Sonneland J, Anson BJ, Beaton. (1958) Surgical anatomy of the arterial supply of the colon from the superior mesenteric artery based upon a study of 600 specimens. Surg Gynecol Obstet 106:385–398
Ssoson-Jaroschewitsch AJ (1926) Z ges Anat Abt 1, Bd. LXXIX, p 44
Standring S, Borley NR, Collins P, Crossman AR, Gatzoulis MA, Healy JC et al (2008) Gray's anatomy. The anatomical basis of clinical practice. In: Posterior abdominal wall and retroperitoneum, 40th edn. Churchill Livingstone, NewYork, p 1074
Steward JA, Rankin FW (1933) Blood supply of the large intestine its surgical considerations. Arch Surg 26:843–891
Sunderland S (1942) Blood supply of the distal colon. Aus New Zeal J Surg 11:253–263
Uysal LL, Cicekcibasi AE, Yilmaz MT, Seker M, Sanli O (2010) Multiple variations of the abdominal aorta in a single cadaver. Singapore Med J 51:e94–e97
Vandamme JP, Bonte J (1990) Vascular anatomy in abdominal surgery. Georg Thieme, Stuttgart, pp 4–42
Vandamme JP, Bone J, Schueren G (1982) Re-evaluation of the colic irrigation from the inferior mesenteric artery. Acta Anat (Basel) 112:18–30
Von Bardeleben K (1906) Lehrbuch der systematischen Anatomie des Menschen, Berlin
Von Haller A (1966) The large intestine. In: Cullen WW (ed) First lines of physiology. A reprint of the 1786 ed (Sources of science 32). Johnson Reprint Corporation, New York/London, pp 139–140
Williams P, Dyson M, Dussak JE, Bannister LH, Berry MM, Collins P, Ferguson MWJ (2000) Gray's anatomy, 38th edn. ELBS with Churchill Livingstone, Edinburg/London, pp 1553–1620
Wu Y, Peng W, Wu H, Chen G, Zhu J, Xing C (2013) Absence of the superior mesenteric artery in an adult and a new classification method for superior–inferior mesenteric arterial variations. Surg Radiol Anat. https://doi.org/10.1007/s00276-013-1183-4
Yahel J, Arensburg B (1998) The Topographic relationships of the unpaired visceral branches of the aorta. Clin Anat 11:304–309
Yamasaki M, Nakao T, Ishizawa A, Ogawa R (1990) A rare case of the inferior mesenteric artery and some colic arteries in man. Anat Anz 171:343–349
Yi SQ, Li J, Terayama H, Naito M, Iimura A, Itoh M (2008) A rare case of inferior mesenteric artery arising from the superior mesenteric artery, with a review of the review of the literature. Surg Radiol Anat 30:159–165
Yoo SJ, Ku MJ, Cho SS, Yoon SP (2011) A case of inferior mesenteric artery arising from the superior mesenteric artery in a Korean woman. J Korean Med Sci 26:1382–1385
Yoshida T, Suzuki S, Sato T (1993) Middle mesenteric artery; an anomalous origin of a middle colic artery. Surg Radiol Anat 15:361–363
Zebrowski W, Augustyniak E, Zajac S (1971) Variation of origin and branches of the inferior mesenteric artery and its anastomoses. Folia Morphol (Warsz) 30:510–517

Common Iliac Artery

58

Andrea Yazbeck and Stephen J. Bordes, Jr.

58.1 Embryology

In the middle of the third week of embryo development, the primordial heart and vascular system begin to develop (Ozgüner and Sulak 2011). The arterial system arises from both ectoderm- and mesoderm-derived angiogenic cells. The aortic arches and the right and left dorsal aorta unite during development to form the descending aorta, which begins to form posterolateral intersegmental branches throughout the thorax (Ozgüner and Sulak 2011). As the dorsal aorta descends through the abdomen, branches to the gastrointestinal and renal systems form, including the intersegmental arteries (Ozgüner and Sulak 2011). The umbilical arteries also arise from the dorsal aorta and disconnect from it after anastomosing with the intersegmental arteries. The fifth intersegmental artery gives rise to the external iliac artery (while the umbilical artery is attached distally). Later, the umbilical artery contributes to the formation of the internal iliac artery. The common iliac arteries then arise from the proximal portion of the patent umbilical arteries (Szpinda et al. 2012).

A. Yazbeck
Department of Anatomical Sciences, Tulane University School of Medicine, New Orleans, LA, USA

S. J. Bordes, Jr. (✉)
Department of Surgery, Louisiana State University Health Sciences Center, New Orleans, LA, USA

58.2 Anatomy/Topography

The abdominal aorta is a retroperitoneal vessel lying anterior to the posterior abdominal wall and vertebral column (Selçuk et al. 2018). As it descends through the abdominal cavity toward the pelvis and lower limbs, it bifurcates into the right and left common iliac arteries (RCIA and LCIA, respectively) (Figs. 58.1 and 58.2). Typically, this bifurcation is located approximately at the level of the fourth or fifth lumbar vertebra, though there are variants (Tubbs et al. 2016). The lateral angle of the bifurcation from the midline is reported to average 60 degrees in males and 68 degrees in females (Anson 1966).

According to the literature, the RCIA is slightly longer than the LCIA because the abdominal aorta lies to the left of the midline (Tubbs et al. 2016). Among the several structures surrounding the RCIA are:

- *Anteriorly: peritoneum, small intestines, sympathetic nerve branches, right ureter.*
- *Posteriorly: termination of the inferior vena cava (IVC), right and left common iliac veins (RCIV and LCIV).*
- *Laterally: IVC and RCIV, psoas major muscle.*
- *Medially: LCIV.*

S. J. Bordes, Jr. et al. (eds.), *The Clinical Anatomy of the Vascular System*,
https://doi.org/10.1007/978-3-031-78326-5_58

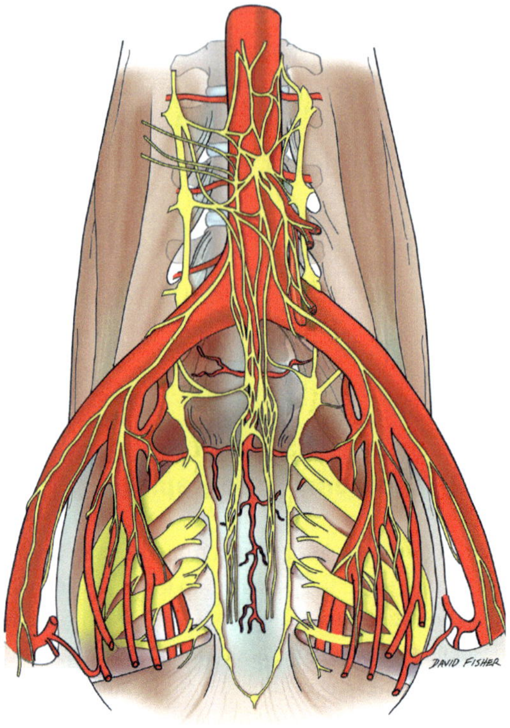

Fig. 58.1 Schematic drawing illustrating the left and right common iliac arteries and their terminal branches, and the internal and external iliac arteries. Note the relationship of these vessels to the surrounding neurovascular anatomy

Similarly, the several structures surrounding the LCIA include:

- *Anteriorly: peritoneum, small intestines, sympathetic nerve branches, superior rectal artery, left ureter.*
- *Posteriorly: lumbar vertebrae, LCIV.*
- *Laterally: psoas major muscle.*
- *Medially: LCIV.*

The diameter of the RCIA is typically in the range 7–14 mm. That of the LCIA is typically 6–16 mm (Tubbs et al. 2016). Both CIAs run parallel to the medial border of the psoas muscles and travel inferiorly to the pelvic brim (Anson 1966; Moore et al. 2014). They terminate opposite the lumbosacral articulation (Anson 1966; Moore et al. 2014).

At their termination, they bifurcate into the internal and external iliac arteries. These give rise to several branches that travel into the pelvic region and the lower extremities.

Fig. 58.2 Cadaveric dissection illustrating the common iliac artery (*) on the left side

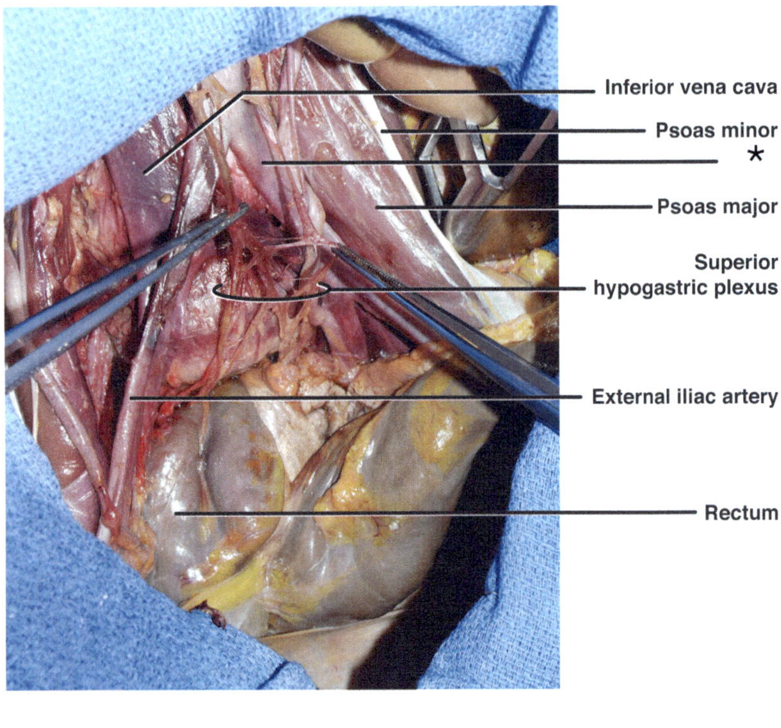

58.3 Histology

The CIAs are classed as elastic (large) arteries because of several properties, most distinctively the greater proportion of elastic tissue in their walls than in muscular (distributing) arteries. This enables them to make an important contribution to maintaining arterial blood pressure throughout the systemic and pulmonary circulations (Tucker et al. 2021).

The tunica intima of an elastic artery comprises endothelium, subendothelial connective tissue, and a non-prominent internal elastic membrane. The subendothelial layer consists of collagen fibers, elastic fibers, and smooth muscle cells. Surrounding the tunica intima, the tunica media is the thickest layer of the vessel. It comprises circumferential layers of vascular smooth muscle cells separated by elastic lamellae (fenestrated sheets). The elastin enables the blood vessel diameter to accommodate to pressure changes between systole and diastole (Tucker et al. 2021).

The tunica adventitia is the outermost layer of this vessel. In elastic arteries it is relatively thin and contains collagen fibers, elastic fibers, fibroblasts, and macrophages. It also contains small networks of blood vessels ("vasa vasorum") and vasoconstrictor nerves ("nervi vasorum").

Elastic arteries are on average >10 mm in diameter. The typical diameter of the CIAs is in the range 9.7–12.3 mm, consistent with their classification as elastic arteries (Richards et al. 2009).

58.4 Imaging

The CIAs can be visualized through several imaging modalities. Imaging is necessary for confirming diagnoses, finding details of aorto-iliac anatomy for pre-procedural surgical planning, and monitoring patient progress through follow-up.

An accepted non-invasive technique for visualizing the CIAs is ultrasound of the abdomen and pelvis. This is commonly used for surveillance of conditions such as CIA aneurysms (Dhanji et al. 2020). However, overlying bowel gas and variations in the iliac vessels themselves can limit its value (Nelson 1980; Päivänsalo et al. 1987; Parry et al. 2001).

Computed tomography (CT) angiography imaging enables the aorto-iliac vasculature to be visualized in greater detail (Baliyan et al. 2019). In most cases, iodinated contrast material is administered intravenously to enhance the quality of imaging (Baliyan et al. 2019). Magnetic resonance (MR) imaging and MR angiography are appropriate alternatives to CT angiography in respect of the quality of imaging and anatomical detail.

In one study, MR angiography, CT angiography, and digital angiography were used to evaluate the anatomy of the internal iliac artery (a branch of the CIA) in 21 males (mean age 73.2 years) before selective embolization of the pelvic arteries (Bilhim et al. 2011). The results showed that CT angiography provided more detailed anatomy than MR angiography, and digital angiography was considered the "gold standard" (Bilhim et al. 2011).

58.5 Variations

The CIAs can vary in several different ways. They can be between 1.2 and 11 cm long (Tubbs et al. 2016), though most are between 3.7 and 7.5 cm (Tubbs et al. 2016). In one study, the average length of the RCIA was 8.7 cm (range 4.5–16.9 cm) and that of the LCIA 8.6 cm (range 4.0–16.9 cm) (Tubbs et al. 2016). These differences in length could depend upon whether the aorta bifurcates below or above the typical bifurcation point (at the L4 vertebra level) (Khamanarong et al. 2009).

The level of aortic bifurcation can vary significantly. This can be caused by lumbosacral anomalies such as sacralization of the L5 vertebra and lumbarization of S1 (Chithriki et al. 2002). In one prospective study, 441 cadavers with mean age of 73 ± 10 years at death were examined (Chithriki et al. 2002). In 67%, the aorta bifurcated at the L4 level (Chithriki et al. 2002). If L5 was sacralized, the aortic bifurcation was at the L3 level in 59% (Chithriki et al. 2002). If S1 was lumbarized, the aorta bifurcated at the L4 level in 40% and at the L4/5 disc space in 33% (Chithriki et al. 2002). There were no significant sex or age differences in the aortic bifurcation level (Chithriki et al. 2002). However, a lower bifurcation is more common than a higher one (Tubbs et al. 2016).

In rare instances, the CIAs are absent bilaterally (Tubbs et al. 2016). In this situation, the external and internal iliac arteries arise directly from the abdominal aorta (Bergman et al. 1988; Mansfield and Howard 1964; Shetty et al. 2013) at the L4/5 disc level. Unilateral absence of the RCIA or LCIA has also been reported (Dabydeen et al. 2008; Dumanian et al. 1965; Llauger et al. 1995).

While the main branches of the CIAs are the internal and external iliac arteries, there are also several small distributing branches to surrounding structures. In 2% of cases, an artery to the neighboring psoas major stems from the CIA bilaterally (Tubbs et al. 2016). The CIA also provides an iliolumbar artery in 3.7–4.8% (Tubbs et al. 2016). On occasion, a lateral or median sacral, a lumbar, or an accessory or replaced renal artery can arise from the CIA (Tubbs et al. 2016).

58.6 Pathology

Clinical and Surgical Significance

Aorto-iliac peripheral artery disease occurs when there is a build-up of fat or calcium in peripheral arteries, more commonly referred to as plaque deposits. Over time, this continued build-up leads to atherosclerosis and can cause narrowing/obstruction of the artery and reduced blood flow to the distal organs. In mild to moderate cases, patients can be asymptomatic or experience only moderate pain and numbness in the lower legs (Javed and Hawkins 2021). As the disease becomes more severe, patients can experience a larger range of symptoms including claudication and limb-threatening ischemia (Javed and Hawkins 2021). Claudication is identified by a sensation of cramping pain during or after exercise, relieved by resting.

For milder cases, treatment initially includes counseling the patient on lifestyle modifications and prescribing anticoagulants, antihypertensives, and cholesterol-reducing medication. However, endovascular or surgical intervention is typically considered for more severe cases (Javed and Hawkins 2021). Non-invasive endovascular procedures such as angioplasty and stenting are considered first. If the situation is more complex, an endarterectomy can be advised—the obstruction in the artery is surgically removed—or a bypass, in which a healthy vein is used to create an alternative blood flow path bypassing the blocked artery.

Iliac artery aneurysm (IAA) is defined as a focal dilation of the CIA to a diameter over 50% greater than normal (Johnston et al. 1991). The CIA diameter normally averages 1.2 ± 0.2 cm in men and 1.0 ± 0.2 cm in women (Johnston et al. 1991). Aneurysms in the CIA typically measure >1.85 cm in males and >1.5 cm in females (Bacharach and Slovut 2008). This condition develops when the vessel wall loses mechanical integrity. Inflammation, atherosclerotic changes, and proteolytic degradation of arterial wall tissue have all been reported as contributing factors (Wilderman and Sanchez 2011). Clinically, patients with an IAA can be asymptomatic depending on its diameter and location. Larger aneurysms are classed as those with a diameter of >6 cm and are more likely to be associated with compression, thrombosis, thromboembolism, or iliac artery rupture.

As in treatment for aorto-iliac peripheral artery disease, endovascular and surgical repairs are considered for treating IAA. Typically, endovascular approaches such as stent graft placement are preferred because the evidence indicates significantly lower perioperative morbidity. This is important because most patients with this diagnosis are older and frequently have comorbidities. In a systematic review of 13 observational studies of isolated IAAs, most patients (181/202) underwent endovascular repair (Perini et al. 2021). Perioperative mortality was lower for endovascular repairs than open surgical repairs (2.8% versus 8.3%) (Chaer et al. 2008). To determine the appropriate treatment, healthcare professionals must consider several factors including the length of normal artery proximal and distal to the aneurysm (proximal/distal landing zones), involvement of the internal iliac artery, bilateral versus unilateral aneurysmal disease, and presence or absence of a coexisting abdominal aortic aneurysm (AAA) (Uberoi et al. 2011). In patients with bilateral CIA aneurysms, regardless of a coexisting AAA, aorto-bi-iliac or aorto-bi-femoral graft placement is considered (Bacharach and Slovut 2008). The origin of the graft is placed just below the renal arteries owing to the increased risk of future aortic arterial wall degeneration due to aneurysms. IAAs associated with compressive symptoms are repaired directly by endoaneurysmorrhaphy to decompress the aneurysm sac rather than performing ligation and bypass.

Lastly, May Thurner's syndrome (MTS) is another common pathology associated with the common iliac vessels. This condition is defined by compression of the veins by the arterial system against bony structures in the iliocaval territory. Variants of this condition involve several different structures. However, most commonly, the left common iliac vein is compressed between the right common iliac artery and the fifth lumbar vertebra. This compression is associated with intimal hyperplasia, which increases the risk of venous stasis and subsequent thrombosis (Peters et al. 2012).

58.7 Complications

Endovascular procedures involving CIA aneurysms often have technical complications associated with the endograft placement. These include vascular injury during access or device deployment; endoleak from inadequate fixation, sealing of the graft to the vessel wall, or breakdown of the graft material; stent fractures; component separations; and endograft collapse. Complications during open surgical repair include lower extremity ischemia due to graft thrombosis or distal embolization during intraoperative manipulation, pelvic ischemia from disruption of internal iliac artery blood flow, refilling of the aneurysm sac from retrograde collaterals resulting in rupture, arterio-enteric fistula, graft infection, and ureteral or iliac vein injury.

CIA aneurysms have been reported following hysterectomies (Cooper and Hester Jr. 1975).

Iatrogenic iliac artery stenosis has been reported as a complication following renal transplantation, and can lead to hypertension and renal dysfunction (Merkus et al. 1993).

References

Anson BJ (1966) Morris' human anatomy. In: the cardiovascular system: a complete systematic treatise, 12th edn. The Blakiston Division, McGraw-Hill Book Co., New York

Bacharach JM, Slovut DP (2008) State of the art: management of iliac artery aneurysmal disease. Catheter Cardiovasc Interv 71(5):708–714. https://doi.org/10.1002/ccd.21507

Baliyan V, Shaqdan K, Hedgire S, Ghoshhajra B (2019) Vascular computed tomography angiography technique and indications. Cardiovasc Diagn Ther 9(Suppl 1):S14–S27. https://doi.org/10.21037/cdt.2019.07.04

Bergman RA, Thompson SA, Afifi AK, Saadeh FA (1988) Compendium of human anatomic variation: text, atlas, and world literature. In: Cardiovascular system. Urban and Schwarzenber, Baltimore/Munich

Bilhim T, Casal D, Furtado A, Pais D, O'Neill JE, Pisco JM (2011) Branching patterns of the male internal iliac artery: imaging findings. Surg Radiol Anat 33(2):151–159. https://doi.org/10.1007/s00276-010-0716-3

Chaer RA, Barbato JE, Lin SC, Zenati M, Kent KC, McKinsey JF (2008) Isolated iliac artery aneurysms: a contemporary comparison of endovascular and open repair. J Vasc Surg 47(4):708–713. https://doi.org/10.1016/j.jvs.2007.11.008

Chithriki M, Jaibaji M, Steele RD (2002) The anatomical relationship of the aortic bifurcation to the lumbar vertebrae: a MRI study. Surg Radiol Anat 24(5):308–312. https://doi.org/10.1007/s00276-002-0036-3

Cooper JM, Hester LL Jr (1975) Iliac artery aneurysm presenting as a complication of hysterectomy. Obstet Gynecol 45(6):689–690. https://doi.org/10.1097/00006250-197506000-00020

Dabydeen D, Shabashov A, Shaffer K (2008) Congenital absence of the right common iliac artery. Radiol Case Rep 3:47

Dhanji A, Murray HE, Downing R (2020) Ultrasound surveillance of common iliac artery aneurysms. Ann Vasc Surg 65:166–173. https://doi.org/10.1016/j.avsg.2019.10.063

Dumanian AV, Frahm CJ, Benchik FA, Wooden TF (1965) Intermittent claudication secondary to congenital absence of iliac arteries. Arch Surg 91:604–606

Javed IN, Hawkins BM (2021) Aorto-iliac peripheral artery disease. Prog Cardiovasc Dis 65:9–14. https://doi.org/10.1016/j.pcad.2021.02.010

Johnston KW, Rutherford RB, Tilson MD, Shah DM, Hollier L, Stanley JC (1991) Suggested standards for reporting on arterial aneurysms. Subcommittee on Reporting Standards for Arterial Aneurysms, Ad Hoc Committee on Reporting Standards, Society for Vascular Surgery and North American Chapter, International Society for Cardiovascular Surgery. J Vasc Surg 13(3):452–458. https://doi.org/10.1067/mva.1991.26737

Khamanarong K, Sae-Jung S, Supa-Adirek C, Teerakul S, Prachaney P (2009) Aortic bifurcation: a cadaveric study of its relationship to the spine. J Med Assoc Thail 92(1):47–49

Llauger J, Sabate JM, Guardia E, Escudero J (1995) Congenital absence of the right common iliac artery: CT and angiographic demonstration. Eur J Radiol 21:128–130

Mansfield AO, Howard JM (1964) Absence of both common iliac arteries. A case report. Anat Rec 150:363–364

Merkus JW, van Asten WN, Hoitsma AJ, Buskens FG, Koene RA, Skotnicki SH (1993) Iliac artery stenosis after kidney transplantation. Acta Chir Belg 93(5):242–248

Moore KL, Agur AMR, Dalley AF (2014) Clinically oriented anatomy, 7th edn. Wolters Kluwer/Lippincott Williams & Wilkins Health, Philadelphia. Print

Nelson RP (1980) Isolated internal iliac artery aneurysms and their urological manifestations. J Urol 124(2):300–303. https://doi.org/10.1016/s0022-5347(17)55418-x

Ozgüner G, Sulak O (2011) Development of the abdominal aorta and iliac arteries during the fetal period: a morphometric study. Surg Radiol Anat 33(1):35–43. https://doi.org/10.1007/s00276-010-0696-3

Päivänsalo M, Lähde S, Leinonen A (1987) Bilateral isolated internal iliac aneurysms presenting as pelvic tumours. Rontgenblatter 40(11):360–362

Parry DJ, Kessel D, Scott DJ (2001) Simplifying the internal iliac artery aneurysm. Ann R Coll Surg Engl 83(5):302–308

Perini P, Mariani E, Fanelli M et al (2021) Surgical and endovascular management of isolated internal iliac artery aneurysms: a systematic review and meta-analysis. Vasc Endovasc Surg 55(3):254–264. https://doi.org/10.1177/1538574420981812

Peters M, Syed RK, Katz M et al (2012) May-Thurner syndrome: a not so uncommon cause of a common condition. Proc (Bayl Univ Med Cent) 25(3):231–233. https://doi.org/10.1080/08998280.2012.11928834

Richards T, Dharmadasa A, Davies R, Murphy M, Perera R, Walton J (2009) Natural history of the common iliac artery in the presence of an abdominal aortic aneurysm. J Vasc Surg 49(4):881–885. https://doi.org/10.1016/j.jvs.2008.11.025

Selçuk İ, Ersak B, Tatar İ, Gungör T, Huri E (2018) Basic clinical retroperitoneal anatomy for pelvic surgeons. Turk J Obstet Gynecol 15(4):259–269. https://doi.org/10.4274/tjod.88614

Shetty S, Kantha L, Sheshgiri C (2013) Bilateral absence of common iliac artery– a cadaveric observation. Int J Anat Var 6:7–8

Szpinda M, Szpinda A, Woźniak A, Daroszewski M, Mila-Kierzenkowska C (2012) The normal growth of the common iliac arteries in human fetuses – an anatomical, digital and statistical study. Med Sci Monit 18(3):BR109–BR116. https://doi.org/10.12659/msm.882516

Tubbs S, Shoja M, Loukas M (2016) Bergman's comprehensive encyclopedia of human anatomic variation. Wiley Blackwell. Copyright@ 2016 by John Wiley & Sons, Inc

Tucker WD, Arora Y, Mahajan K (2021) Anatomy, blood vessels. In: StatPearls. StatPearls Publishing, Treasure Island. August 11, 2021

Uberoi R, Tsetis D, Shrivastava V, Morgan R, Belli AM (2011) Subcommittee on Reporting Standards for Arterial Aneurysms of The Society for Vascular Surgery. Standard of practice for the interventional management of isolated iliac artery aneurysms. Cardiovasc Intervent Radiol 34(1):3–13. https://doi.org/10.1007/s00270-010-0055-0

Wilderman M, Sanchez L (2011) Chapter 46 – iliac artery aneurysms. In: Moore WS, Ahn SS (eds) Endovascular surgery. Elsevier/Saunders, Philadelphia, pp 487–495

59 External Iliac Artery

Annie Syed

59.1 Embryology

During week three of development, the arterial and venous systems are established via differentiation of the mesoderm. The arterial system is comprised of paired dorsal aortae, which branches to form two umbilical arteries and a dorsal intersegmental artery during the beginning of the fourth week of development (Tamisier et al. 1990). Shortly after, the umbilical arteries anastomose with the dorsal intersegmental artery to form the dominant placental-aortic connection, which becomes the common and internal iliac arteries after birth. The sciatic and external iliac arteries branch from this umbilical-dorsal anastomosis at the end of week 4 of fetal development. Between weeks 4 and 6, the sciatic artery is the primary blood supply to the developing lower limbs (Tamisier et al. 1990). After week 6, the external iliac artery and its branches fully develop and become the main blood supply to the lower limbs, causing the sciatic artery to regress.

59.2 Anatomy/Topography

The external iliac artery originates from the bifurcation of the common iliac artery at the level of the pelvic brim, anterior to the sacroiliac joint at the level of the L1 or L2 vertebra (Fig. 59.1). It extends laterally along the medial border of the psoas major muscle and then travels under the inguinal ligament between the pubic symphysis

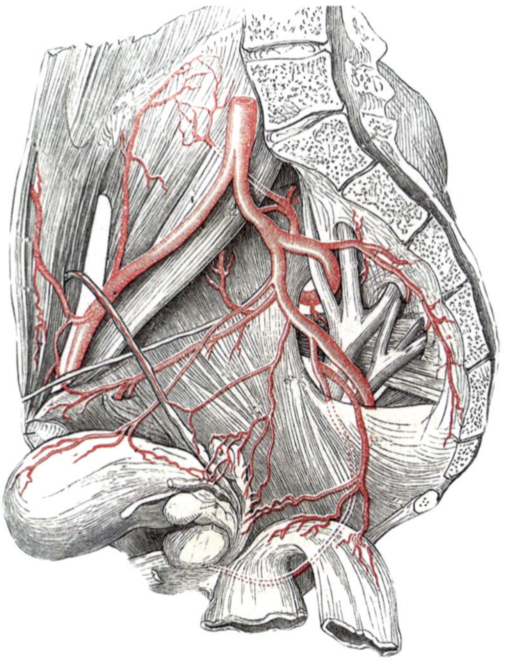

Fig. 59.1 Drawing of the internal and external iliac areries and their branches. (After Allen 1892)

A. Syed (✉)
Department of Surgery, Louisiana State University Health Sciences Center, New Orleans, LA, USA
e-mail: asyed1@tulane.edu

S. J. Bordes, Jr. et al. (eds.), *The Clinical Anatomy of the Vascular System*,
https://doi.org/10.1007/978-3-031-78326-5_59

and the anterior superior iliac spine. Once the vessel passes the inguinal ligament and enters the thigh, it becomes the femoral artery.

The external iliac artery indirectly supplies muscles of the anterolateral abdominal wall and the lower limbs. It mainly acts as a conduit between smaller and more functional arteries derived from its branches. The inferior epigastric branch of the external iliac artery supplies the rectus abdominis, while the deep circumflex branch supplies the transversus abdominis and internal oblique muscles. Its continuation past the inguinal ligament, the femoral artery, provides the main arterial supply to the entirety of the lower limb.

59.3 Histology

The external iliac artery is a muscular vessel that provides oxygenated blood to the lower limbs. It comprises three layers: the inner layer, also known as the tunica intima; the middle muscular layer, or the tunica media; and the outermost layer, known as the tunica externa.

The tunica intima's innermost layer comprises endothelium and an internal elastic membrane. The endothelium is a specialized form of mesenchymal-epithelial simple squamous cells. Endothelial cells act as a barrier between blood and tissues and have focal regulation mechanisms that facilitate vasoconstriction, vasodilation, clotting, and immune functions. An elastic membrane allows the entire vessel to separate the tunica intima and media. The middle layer, the tunica media, comprises smooth muscle cells and elastic fibers of elastin and fibrillin. Smooth muscle cells are innervated by nerve fibers, which help modulate blood flow and pressure regulation through vasoconstriction or vasodilation. The elastic fibers provide resilience and elasticity to allow for changes in arterial diameter. The media is the thickest layer of this particular vessel.

The outermost layer, the tunica externa or adventitia, attaches the vessel to its surrounding connective tissue. This layer prevents the overexpansion of arteries and contains collagen, which anchors the artery to the surrounding structures and an external elastic lamina.

59.4 Variations

The corona mortis is one of the most clinically significant variations involving the external iliac artery. A vascular connection between the external iliac artery and the obturator artery occurs in an estimated 9.8% of patients. It is located superiorly or behind the superior pubic ramus, creating a pelvic anastomosis (Darmanis et al. 2007).

Another significant anomaly is external iliac artery agenesis or hypoplasia associated with a persistent sciatic artery. This anomaly is categorized into three groups: anomalies due to the artery's course or origin, hypoplasia or atresia with the sciatic artery, and isolated hypoplasia of the external iliac artery. In most cases, if the external iliac artery is absent or hypoplastic, the sciatic artery persists, and branches of the internal iliac arteries create a collateral circulation to supply the lower limbs (Tamisier et al. 1990).

59.5 Clinical and Surgical Significance

A persistent sciatic artery with external iliac artery agenesis can present with physical exam findings that may confuse clinicians. This may result in the absence of a femoral artery pulse; however, collateral blood flow to the lower limb creates a palpable popliteal pulse (van Hooft et al. 2009).

Surgical procedures involving the transfemoral approach to endovascular catheterization depend on a widely patent external iliac artery. Many cardiac and aortic endovascular procedures involve the insertion of a catheter into the common femoral artery that then travels superiorly through the iliac arteries, eventually reaching the aorta and left side of the heart (Anjum et al. 2017). This common approach replaces aortic valves, repairs abdominal aortic aneurysms, and treats occluded vessels distal or proximal to the external iliac artery via angioplasty and stent deployment.

The external iliac artery also plays a role in kidney transplantation. Due to its ease of access and low complication rate, it is the most common site of donor kidney implantation (Marimuthu et al. 2022).

59.6 Complications

Persistent Sciatic Artery

In patients with a persistent sciatic artery and a hypoplastic external iliac artery, there is an increased risk of arterial thrombosis and aneurysm formation. Due to the external iliac artery's hypoplastic nature and its communication with the sciatic artery, the hypoplastic artery can form a thrombus within its compromised structure, which may lead to the formation of an aneurysm, vessel occlusion, or emboli (van Hooft et al. 2009).

Corona Mortis

Surgeons must be aware of this vascular anomaly as it is appropriately named the "crown of death." The location of the corona mortis behind the superior pubic ramus makes it vulnerable to injury during open pelvic procedures. If the external iliac or obturator arteries are lacerated, severe hemorrhage may ensue with little ability to obtain vascular control (Darmanis et al. 2007). However, it has been advised that surgeons should not change their surgical approach due to the extremely low incidence of such a complication.

Atherosclerosis

The most common complication involving the external iliac artery is atherosclerosis. Atherosclerosis in the external iliac artery can cause arterial stenosis, which may be flow-limiting and obstruct vascular access during cannulation. These complications can decrease the efficacy of endovascular abdominal aorta aneurysm repair, heart valve replacements, and peripheral angioplasty. Chronic atherosclerosis also leads to aortoiliac occlusive disease and peripheral vascular disease, which commonly presents with cramping leg pain that is reproducible and worse with exercise (claudication) and erectile dysfunction (Leriche Syndrome). In cases of severe occlusion, vascular compromise may lead to rest pain, tissue loss, and limb loss.

Transfemoral Catheterization

Though most complications involving the transfemoral approach occur at the access site, there is still a risk for external iliac artery damage in a post-operative setting. Damage to the external iliac artery can be caused by improper catheter insertion and maneuvering techniques, leading to arterial dissection and aneurysm. Fortunately, these dissections are uncommon and easily repaired with endovascular stent implantations, decreasing the incidence of aneurysms. Other complications of transfemoral interventions result from the perforation of the external iliac artery, which can cause retroperitoneal hemorrhage or the formation of an arterial thrombus. Though both complications are rare, they can quickly lead to death if not immediately treated.

References

Anjum I, Khan MA, Aadil M, Faraz A, Farooqui M, Hashmi A (2017) Transradial vs. Transfemoral approach in cardiac catheterization: a literature review. Cureus 9(6):e1309. Published 2017 Jun 3. https://doi.org/10.7759/cureus.1309

Darmanis S, Lewis A, Mansoor A, Bircher M (2007) Corona mortis: an anatomical study with clinical implications in approaches to the pelvis and acetabulum. Clin Anat 20(4):433–439. https://doi.org/10.1002/ca.20390

Marimuthu V, Khamitkar Shankar Rao S, Jadhav S, Nayak MH, Alur N (2022) Management of iatrogenic external iliac artery perforation with a self-made covered stent. Cureus 14(8):e28460. Published 2022 Aug 27. https://doi.org/10.7759/cureus.28460

Tamisier D, Melki JP, Cormier JM (1990) Congenital anomalies of the external iliac artery: case report and literature review. Ann Vasc Surg 4(5):510–514. https://doi.org/10.1016/S0890-5096(07)60081-8

van Hooft IM, Zeebregts CJ, van Sterkenburg SM, de Vries WR, Reijnen MM (2009) The persistent sciatic artery. Eur J Vasc Endovasc Surg 37(5):585–591. https://doi.org/10.1016/j.ejvs.2009.01.014

Deep Circumflex Iliac Artery

60

Logan Roberts, Bretton Urban, Cameron Fontenot, and Stephen J. Bordes, Jr.

60.1 Embryology

The vasculature begins to form during the third week of human embryo development, around day 18, when mesenchymal cells differentiate into angioblasts. These angioblasts cluster together, flatten, and fuse with other clusters to begin forming primitive endothelium-lined channels. The endothelium then differentiates to form distinct layers while new vessels begin to develop by budding from a preexisting vessel. During the fourth week, the umbilical arteries anastomose with the dorsal intersegmental arteries to form an umbilical artery, precursor of the common and internal iliac arteries. During the fifth week, the external iliac and sciatic arteries arise from the same umbilical artery root and interconnect, the sciatic arteries being destined eventually to regress. By the third month of development, the external iliac artery forms the major blood supply to the lower extremities through its branches, including the deep circumflex iliac artery and its continuation as the femoral artery (Hammond et al. 2021; Greebe 1977; Tamisier et al. 1990; Senior 1919; Evans 1909).

L. Roberts · B. Urban
School of Medicine, Louisiana State University Health Sciences Center, New Orleans, LA, USA

C. Fontenot · S. J. Bordes, Jr. (✉)
Department of Surgery, Louisiana State University Health Sciences Center, New Orleans, LA, USA

60.2 Anatomy/Topography

The deep circumflex iliac artery (DCIA) (Fig. 60.1) is a bilateral artery of the pelvic region that averages 2 mm in diameter and gives rise to multiple perforating branches that supply blood to the iliac fossa and crest, anterolateral abdominal wall muscles, and the skin over the iliac crest. This artery can become a vital vascular detour in patients with congenital or acquired aortic constriction because of its many anastomoses (Taylor et al. 1979a, b; Shin et al. 2018). In most cases, it branches off the external iliac artery opposite the inferior epigastric artery branch and just deep to the inguinal ligament. Its origin is deep to the transversalis fascia, and it ascends along the medial aspect of the iliac bone in a superolateral direction to the anterior superior iliac spine (ASIS). It gives rise first to an ascending branch 1 cm medially to the ASIS. This branch averages 1 mm in diameter and supplies the internal oblique and transversus abdominis muscles. It anastomoses with the lumbar, inferior epigastric, and lateral circumflex arteries (Hanasono and Butler 2021; Kim et al. 2013; Ogami et al. 2017; Bergeron et al. 2007). A transverse branch can arise from the same trunk as the ascending branch and continue posteriorly along the inner surface of the iliac crest, producing several additional branches (Shin et al. 2018). After giving rise to the ascending branch, the DCIA travels through a fibro-osseus tunnel along the superior rim of the

S. J. Bordes, Jr. et al. (eds.), *The Clinical Anatomy of the Vascular System*,
https://doi.org/10.1007/978-3-031-78326-5_60

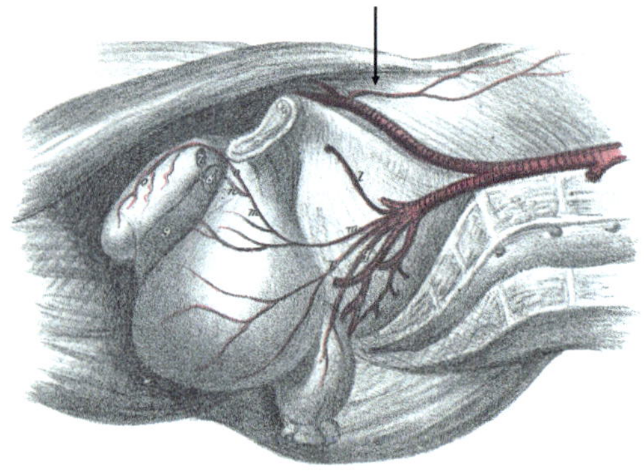

Fig. 60.1 Schematic drawing illustrating the deep circumflex iliac artery (arrow) arising from the external iliac artery (Quain and Wilson, 1844 author's collection)

iliac crest, where it produces nutrient branches to the iliac bone and perforators that travel through the abdominal wall muscles to the overlying skin. The DCIA pierces the transversus abdominis muscle around the mid-axillary line and forms anastomoses with the iliolumbar and superior gluteal arteries between the internal oblique and transversus abdominus muscles (Taylor et al. 1979a, b; Hanasono and Butler 2021). An anastomotic branch can also emerge from its deep aspect, penetrating the transversus abdominis muscle, and proceed to anastomose with adjacent arteries such as the ascending branch of the lateral femoral circumflex (Shin et al. 2018). A groove created by the iliacus and transversalis facia is prominent in recognizing the pathway of the DCIA, which is consistently external and adjacent to this groove (Taylor et al. 1979a). Many perforators from the DCIA and other arteries are found around the richly vascular region superior to the iliac crest (Bergeron et al. 2007).

60.3 Variations

Amuti et al. found that each of the 52 cadavers they studied had bilaterally symmetrical DCIAs (Sarna et al. 2020). The variations in the artery involved differences in origin, length, and branching patterns. In all the cadavers, the DCIA originated from the external iliac artery (EIA) posterior to the inguinal ligament (IL). Similarly, Vasanthakumar et al. indicated that out of 50 cadavers, 36 (72%) followed this trend of arising at or posterior to the level of the IL, while 14 (28%) arose from the femoral artery below the level of the IL (Vasanthakumar and Priyadarisini 2017). Kim et al. found that when the DCIA arose below the IL, there was a point where it crossed the IL (Rendenbach et al. 2018). The mean distance between this point and the ASIS was 3.99 cm. The study also showed that the highest level at which the DCIA originated was 1.48 cm above the IL. According to Sarna et al., the average distance from the origin to the ASIS was 7.28 ± 0.99 cm and that from the origin to the pubic tubercle was 5.91 ± 1.03 cm, the overall length of the DCIA ranging from 3.7 cm to 9.5 cm. There were many possible branching patterns: bifurcation into horizontal and ascending branches (78%), trifurcation (6%), or an arborization pattern (4%). The point of branching occurred at an average distance of 4.5 cm from the ASIS and 8.26 cm from the pubic tubercle (Sarna et al. 2020).

Owing to the variations in origin of the DCIA relative to the IL, care must be taken to prevent iatrogenic lacerations during femoral triangle incisions. This can best be achieved by being aware of the artery relative to the inguinal ligament and ASIS and by keeping the lacerations consistent. Because the human thumb is similar in length to the distance from origin to the ASIS (6 cm) and similar in width to the height of the origin above the IL (2 cm), it has conventionally been used to mark a safe incision line (Vasanthakumar and Priyadarisini 2017). Even though some origins lie below the IL, its crossing can be detected along the incision line.

60.4 Clinical/Surgical Significance

Owing to its robust blood supply and optimal contour, the osteocutaneous flap using the DCIA and iliac crest can be used in maxillary or mandibular reconstruction (Hanasono and Butler 2021). While the DCIA and associated soft tissue are being harvested, it is important to recognize variations in origin and perforator numbers, alternative iliac blood supply, and common anatomical encounters. Since the DCIA can originate from either the EIA or the femoral artery, it can often be mistaken for its ascending branch. This misinterpretation can be avoided by making medial to lateral incisions while noting that the ascending branch travels superiorly and the DCIA superolaterally. Other perforators can originate from the lumbar, iliolumbar, superior circumflex iliac, or intercostal arteries. These usually have accompanying nerves and are in posterior orientation. Nerves that can be encountered in this area are the iliohypogastric, iliolingual, and lateral femoral cutaneous nerves. Common complications are related to mechanical and functional deficits such as reduced hip mobility, ability to stand up from seated position, venous congestion to the skin paddle, and sensory deficits. However, overall long-term morbidity is expected to be low (Rendenbach et al. 2018; Wu et al. 2022). Using the DCIA flap along with venous superdrainage from the superficial circumflex iliac vein in order to reduce venous congestion has also reduced venous insufficiency (Wu et al. 2022).

The common femoral is the most widely used artery for inflow in peripheral artery disease revascularization surgeries, but this is not always feasible if the artery is diseased by calcification, atherosclerosis, or wall abnormalities. The DCIA has been used as a viable alternative route of inflow in those with limb-threatening ischemia. In a case report by Haga, the DCIA used for revascularization was 3 mm in diameter, pulsating well, and had developed as a collateral artery. Postoperative limb and skin perfusion and ischemic pain all improved following revascularization (Haga et al. 2019). The size of the vessel should be confirmed with Computed Tomography (CT) angiograms preoperatively to confirm its patency.

60.5 Pathology/Surgical Complications

Isolated cases of pathology intrinsic to the DCIA are rare. However, spontaneous hemorrhage of the artery has been reported (Katsumori and Nakajima 1998). This is generally considered to result from non-traumatic injury to the vessel secondary to increased muscle contraction, weakness in the vasculature walls, or bleeding tendencies. Typical presentation is sudden lower abdominal pain with an acute swelling due to hematoma. Imaging studies of DCIA hemorrhage typically show arterial extravasation and the condition can be successfully treated with transcatheter arterial embolization.

Rarely, the DCIA can be injured following surgeries (Neo et al. 2000; Kim et al. 2016), abdominal trauma, surgical drain placements (Park et al. 2009), and abdominal paracentesis (Kalantari et al. 2019). The possible surgical complications include pseudoaneurysm, hematoma, hemoperitoneum and hemorrhage, and careful perioperative monitoring should be used to recognize these scenarios. Some symptoms that indicate the need for immediate assessment are continuous abdominal tenderness, abdominal ecchymosis, decreasing hemoglobin concentration, hemodynamic instability, and signs of abdominal wall malperfusion and paresthesia. A duplex ultrasound with B mode imaging, color flow imaging, and pulse wave analysis color is a quick and noninvasive tool for diagnosing pseudoaneurysms. A "to-and-fro" waveform indicates an arterial wall deficiency and resulting bidirectional blood flow through the neck of the pseudoaneurysm. The "to" represents arterial blood entering the pseudoaneurysmal sac during systole, while the "fro" represents blood exiting during diastole (Mahmoud 2015). A Computed Tomography Arteriogram can be used if obesity obscures imaging or if active extravasation is suspected.

Once the diagnosis has been confirmed, an efficient treatment strategy should be implemented that does not subject the patient to overtreating with invasive measures, but does not underestimate the severity of the situation. For pseudoaneurysms <2 cm and non-enlarging, a conservative observational approach has been shown to result in spontaneous closure. Embolization in these cases entails a higher risk of unintended arterial occlusion and ischemia. For larger pseudoaneurysms, embolization with ultrasound-guided thrombin injection is recommended. In those that fail to respond to this injection, or initially present with signs of hemodynamic instability or active extravasation, open surgical repair can be considered. Hemorrhages carry the highest risk of mortality and morbidity, so appropriate fluid or blood infusions and hemorrhage ligation should be achieved in order to maintain adequate perfusion (Aumiller and Motaganahalli 2014).

References

Aumiller BJ, Motaganahalli RL (2014) Arterial pseudoaneurysms. In: Dean SM, Satiani B, Abraham WT (eds) Color atlas and synopsis of vascular diseases. McGraw Hill. Accessed March 27, 2022. https://accesssurgery-mhmedical-com.ezproxy.lsuhsc.edu/content.aspx?bookid=1201&sectionid=71013737

Bergeron L, Tang M, Morris SF (2007) The anatomical basis of the deep circumflex iliac artery perforator flap with iliac crest. Plast Reconstr Surg 120(1):252–258. https://doi.org/10.1097/01.prs.0000264392.42396.a3

Evans HM (1909) On the development of the aorta, cardinal and umbilical veins and the other blood vessels of vertebral embryos from capillaries. Anat Rec 3:498–518

Greebe J (1977) Congenital anomalies of the iliofemoral artery. J Cardiovasc Surg 18(3):317–323

Haga M, Motohashi S, Inoue H, Akasaka J, Shindo S (2019) Deep circumflex iliac artery as an inflow alternative for surgical revascularization: a case report. Vasc Endovasc Surg 54(1):85–88. https://doi.org/10.1177/1538574419877730

Hammond E, Nassereddin A, Costanza M (2021) Anatomy, abdomen and pelvis, external iliac arteries. In: StatPearls. StatPearls Publishing, Treasure Island

Hanasono MM, Butler CE (2021) Handbook of reconstructive flaps. Thieme. Accessed March 24, 2022. https://search.ebscohost.com/login.aspx?direct=true&db=e000xna&AN=2568364&site=eds-live&scope=site

Kalantari J, Nashed MH, Smith JC (2019) Post paracentesis deep circumflex iliac artery injury identified at angiography, an underreported complication. CVIR Endovasc 2(1):24. https://doi.org/10.1186/s42155-019-0068-y

Katsumori T, Nakajima K (1998) A case of spontaneous hemorrhage of the abdominal wall caused by rupture of a deep iliac circumflex artery treated by transcatheter arterial embolization. Eur Radiol 8(4):550–552. https://doi.org/10.1007/s003300050432

Kim HS, Kim BC, Kim HJ, Kim HJ (2013) Anatomical basis of the deep circumflex iliac artery flap. J Craniofac Surg 24:605–609

Kim H-S, Kim J-Y, Hur H, Nam W (2016) Herniation after deep circumflex iliac artery flap: two cases of rare complication. Maxillofac Plast Reconstr Surg 38(1):10. https://doi.org/10.1186/s40902-016-0055-3

Mahmoud MZ (2015) "To-and-fro" waveform in the diagnosis of arterial pseudoaneurysms. World J Radiol 7(5):89. https://doi.org/10.4329/wjr.v7.i5.89

Neo M, Matsushita M, Morita T, Nakamura T (2000) Pseudoaneurysm of the deep circumflex iliac artery. Spine 25(14):1848–1851. https://doi.org/10.1097/00007632-200007150-00018

Ogami K, Murata H, Sakai A et al (2017) Deep and superficial circumflex iliac arteries and their relationship to the ultrasound-guided femoral nerve block procedure: a cadaver study. Clin Anat 30:413–420

Park SW, Chang S-H, Yun IJ, Lee HW (2009) Deep circumflex iliac artery-related hemoperitoneum formation after surgical drain placement: successful transcatheter embolization. Cardiovasc Intervent Radiol 33(2):402–405. https://doi.org/10.1007/s00270-009-9571-1

Rendenbach C, Goehler F, Hansen L et al (2018) Evaluation of long-term functional donor-site morbidity after deep circumflex iliac crest artery bone flap harvest. Microsurgery 39(4):304–309. https://doi.org/10.1002/micr.30358

Sarna K, Amuti T, Butt F, Kamau M, Muriithi A (2020) Variations of the anatomy and bony landmarks of deep circumflex iliac artery in a select Kenyan population. Craniomaxillofac Trauma Reconstr 13(4):300–304. https://doi.org/10.1177/1943387520958333

Senior HD (1919) The development of the arteries of the human lower extremity. Am J Anat 25:55–95

Shin K-J, Lee S-H, Koh K-S, Song W-C (2018) Anatomical consideration for the safe elevation of the deep circumflex iliac artery in flap surgery. Plast Reconstr Surg 142(1):193–201. https://doi.org/10.1097/PRS.0000000000004514

Tamisier D, Melki JP, Cormier JM (1990) Congenital anomalies of the external iliac artery: case report and review of the literature. Ann Vasc Surg 4(5):510–514. https://doi.org/10.1016/S0890-5096(07)60081-8

Taylor GI, Townsend P, Corlett R (1979a) Superiority of the deep circumflex iliac vessels as the supply for free groin flaps. Experimental work. Plast Reconstr Surg 64(5):595–604

Taylor GI, Townsend P, Corlett R (1979b) Superiority of the deep circumflex iliac vessels as the supply for free groin flaps. Clinical work. Plast Reconstr Surg 64(6):745–759. https://doi.org/10.1097/00006534-197912000-00001

Vasanthakumar T, Priyadarisini SE (2017) A study on the variations in the origin of deep circumflex iliac artery. Int J Anat Res 5(4.1):4451–4453. https://doi.org/10.16965/ijar.2017.367

Wu W-J, Zheng L, Zhang J et al (2022) Venous superdrainage using superficial circumflex iliac vein in deep circumflex iliac artery perforator flap with iliac crest for Oromandibular reconstruction. Int J Oral Maxillofac Surg 51(1):38–43. https://doi.org/10.1016/j.ijom.2021.03.014

Inferior Epigastric Artery 61

Kanav Chhabra, George Tadros, Savannah Moon, and Lisandro Montorfano

61.1 Embryology

Vasculogenesis, the formation of blood vessels, begins with hemangioblasts in the yolk sac, which differentiate into two different cell types: angioblasts and hematopoietic progenitor cells (Aghi and Chiocca 2006). Angioblasts migrate from the mesoderm to specific portions of the embryo, where they form vascular plexuses of veins and arteries (Ribatti et al. 2015).

Arteries develop when angiogenic cells from the neural crest and the pharyngeal mesoderm coalesce during the third week of development. A primitive heart tube forms during angiogenesis and then transforms into the dorsal aorta, which eventually splits into ventral, lateral, and postero-lateral branches. The embryonic dorsal arterial system gives rise to the systemic-systemic collateral pathways, starting as somatic branches from the dorsal aorta and then dividing into the intercostal and lumbar arteries. These vessels branch further and anastomose to form the internal mammary artery superiorly and the inferior epigastric artery inferiorly (Hardman et al. 2011).

K. Chhabra
Department of Surgery, University of Texas Southwestern Medical Center, Dallas, TX, USA
e-mail: Kanav.Chhabra@UTSouthwestern.edu

G. Tadros · S. Moon
Department of Surgery, Cleveland Clinic Florida, Weston, FL, USA

L. Montorfano (✉)
Department of Plastic and Reconstructive Surgery, Vanderbilt University Medical Center, Nashville, TN, USA

61.2 Anatomy/Topography

The inferior epigastric artery supplies the muscles of the anterior abdominal wall and typically originates from the external iliac artery (Figs. 61.1, 61.2, and 61.3). It then ascends under the peritoneum and on the anterior abdominal wall medial to the deep inguinal ring, behind the spermatic cord, to form the lateral portion of the Hesselbach's triangle. The medial and inferior borders of this triangle consist of the rectus abdominis muscle and the inguinal ligament. Now, within the lateral umbilical fold, the inferior epigastric artery ascends obliquely and medially, piercing the transversalis fascia and crossing the arcuate line. The inferior epigastric artery advances superiorly until it pierces the rectus abdominis muscle. The artery has two main intramuscular divisions: the lateral branch anastomoses with the lower four intercostal arteries and the medial branch with the superior epigastric artery (Joy et al. 2017; Offman et al. 2005). The inferior epigastric artery also branches into the cremasteric artery, which supplies the spermatic cord and the cremaster muscle (Lin and Troyer 2014). The inferior epigastric vein accompanies the artery and drains into the external iliac vein.

S. J. Bordes, Jr. et al. (eds.), *The Clinical Anatomy of the Vascular System*,
https://doi.org/10.1007/978-3-031-78326-5_61

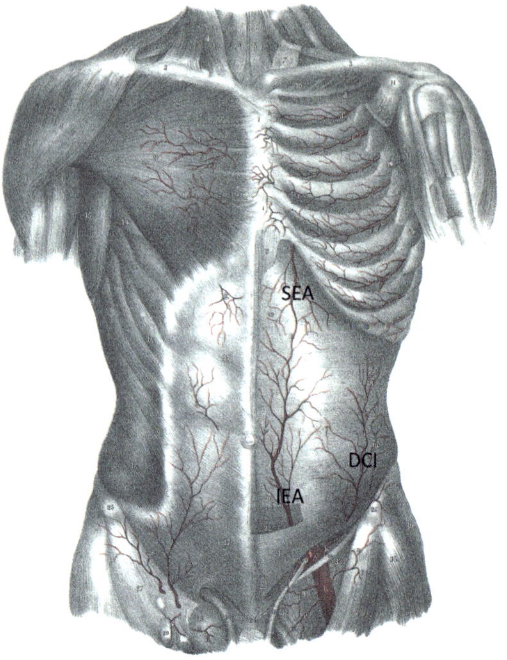

Fig. 61.1 Arteries of the anterolateral abdominal wall including the inferior epigastric artery (IEA). Also note the superior epigastric artery (SEA) and deep circumflex iliac artery (DCI). (After Quain and Wilson)

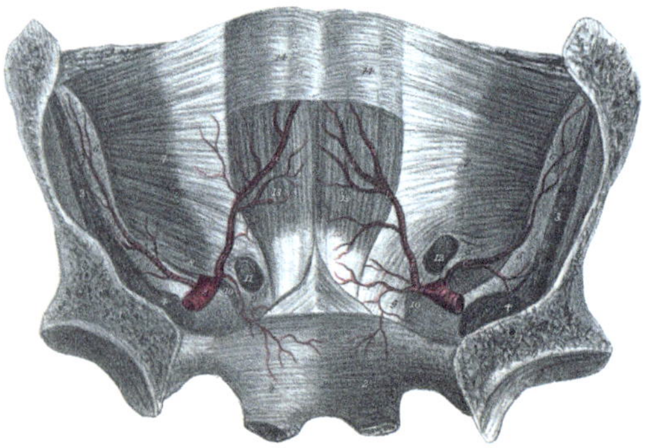

Fig. 61.2 Posterior view of the inferior epigastric artery. (After Quain and Wilson)

The inferior epigastric artery gives off 5 ± 2 perforators focused in the peri-umbilical region. It anastomoses with the contralateral inferior epigastric artery and with portions of the superficial inferior epigastric, superior epigastric, and lateral branches of the posterior intercostal arteries. The superficial inferior epigastric artery branches from the medial side of the femoral artery and supplies the subcutaneous tissues of the abdominal wall (Offman et al. 2005).

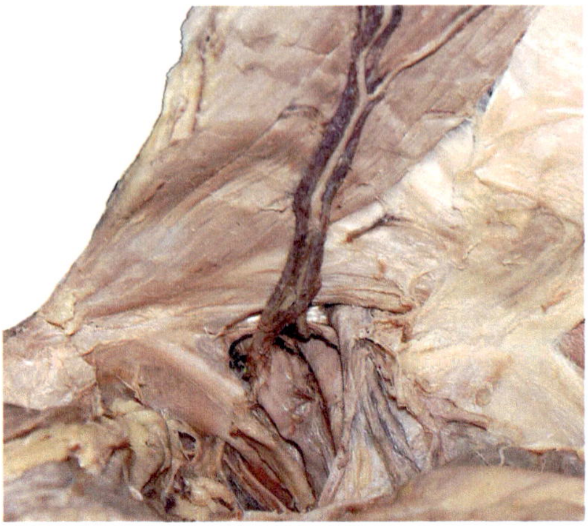

Fig. 61.3 Cadaveric dissection noting the inferior epigastric artery and vein traveling on the posterior aspect of the anterior abdominal wall. (Courtesy Prof. M. Loukas)

61.3 Imaging/Histology

The inferior epigastric artery is a muscular artery with three layers: the tunica intima, media, and adventitia. The tunica intima is composed of the endothelium, which is layered over the internal elastic lamina and projects into the lumen. The intima has minimal subendothelial connective tissue. It is surrounded by the tunica media, which predominantly comprises smooth muscle along with elastic tissue. The media is then surrounded by the tunica adventitia, the outermost layer, which contains bundles of longitudinal smooth muscle cells and dense irregular connective tissue (LabudovicBorovic et al. 2015; Cisneros et al. 2002).

For preoperative mapping of the inferior epigastric artery, CT angiography is the gold standard. Doppler ultrasonography is typically used for vascular mapping, but because most perforating arteries of the inferior epigastric artery have diameters less than 15 mm, the threshold for reliable visualization with Doppler ultrasonography, CT angiography must be used (Karunanithy et al. 2011).

61.4 Variations

The inferior epigastric artery is referenced as the deep inferior epigastric artery when a superficial inferior epigastric artery is present and branching from the femoral artery. A superficial inferior epigastric artery was found in 65% of patients studied (Taylor and Daniel 1975).

Jakubowicz et al. dissected 75 lower limbs and found different sites of origin for the inferior epigastric artery. In 76% of cases, the vessel arose from the external iliac artery above the inguinal ligament; in 12%, it originated from the external iliac artery below the inguinal ligament in the vascular lacuna; in 8%, it arose from the femoral artery; and in 4%, it originated with the common trunk with the obturator artery (Jakubowicz and Czarniawska-Grzesiñska 1996).

The branching of the inferior epigastric artery has many variations. Moon et al. found three distinct branching patterns of the deep inferior epigastric artery above the arcuate line in 64 cadavers (The Vascular Anatomy of Rectus Abdominis Musculocutaneous n.d.):

I. Type I (29% of patients): the vessel ascended as a single intramuscular artery.
II. Type II (57%): the vessel divided at the arcuate line into two major intramuscular vessels.
III. Type III (14%): the vessel divided into three branches at the arcuate line.

Corona Mortis is a variation in which a branch of the deep inferior epigastric artery connects to the external iliac and obturator arteries. Pungpapong et al. found Corona Mortis in 77.27% of the patients studied (Pungpapong and Thum-umnauysuk 2005). This is medically significant because the Corona Mortis connects to major arteries, and if it is transected, it can result in major hemorrhagic shock (Al Talalwah 2016).

Gençhellaç et al. studied CT images of 200 patients to measure the distances between the inferior epigastric artery and the midline. The mean distances on the right and left sides were 4.01 and 4.47 cm at the umbilical level, 3.81 and 4.26 cm at the mid-level, and 5.62 and 5.51 cm at the origin level (Gençhellaç et al. 2014; Rao et al. 2013).

The average diameter of the inferior epigastric artery is approximately 3.5 mm and its average length is 7 cm (Castro Ferreira et al. 2009).

On the CT scans of 100 patients, Saber et al. found the epigastric arteries to be between 4 and 8 cm from the midline. The safety zone of entry into the abdominal wall therefore lies outside this area both medially and laterally and should be considered during laparoscopic or open surgical procedures along with image-guided biopsy and paracentesis. Patients who are 68 years old or more and those with a BMI of at least 26.3 typically show greater displacement of the inferior epigastric artery from the midline than those who are younger and have lower BMIs (Saber et al. 2004).

Rao et al. studied 50 cadavers and found significant variations in the length of the inferior epigastric artery in the anterior abdominal wall: on the left side it was as short as 3.5 cm, while on the right it was as long as 7 cm. The authors assert that given the wide variation in the course, length, and distance from the midline of the inferior epigastric arteries, IEA mapping should be considered as part of the preoperative evaluation protocol for laparoscopic surgeries (Rao et al. 2013).

61.5 Pathology

Occlusions within the inferior epigastric artery are often caused by atherosclerosis. Cisneros et al (2002) found them in 28% of specimens. In atherosclerosis, the endothelial wall of the artery is injured and then becomes permeable to leukocytes and monocytes that form an initial soft lesion. Smooth muscle cells enter the intima from the media and begin to proliferate, increasing the size of the plaque. The plaque can restrict blood flow, leading to major tissue loss in severe cases (Rafieian-Kopaei et al. 2014).

61.6 Clinical and Surgical Significance

Because the inferior epigastric artery has a low rate of occlusion due to atherosclerosis, is easy to harvest, and has useful average length and diameter, it is a good candidate for coronary revascularization (Cisneros et al. 2002).

In cases of external iliac blockage, the Winslow pathway, a connection between the internal thoracic artery and the inferior epigastric artery, serves as a useful collateral pathway (Tada et al. 2022). This pathway becomes important in patients with aortoiliac occlusion, and its contribution should be considered especially when coronary revascularization is necessary, given that interruption of the pathway can cause severe leg ischemia (Yurdakul et al. 2006). The Winslow pathway also becomes relevant in cases of Takayasu arteritis with stenosis of the abdominal aorta (Babu et al. 2019).

The deep inferior epigastric artery perforator flap (DIEP) includes skin and fat from the lower abdominal wall. It is often used in autologous tissue breast reconstruction. The procedure has low morbidity; however, several studies have been conducted to reduce abdominal wall morbidity further by minimizing dissection through the rectus abdominis muscle by mapping the deep inferior epigastric artery perforator system, which is highly variable (Ireton et al. 2014).

Another option for autologous tissue breast reconstruction is to use the superficial inferior epigastric artery flap (SIEA). This procedure is advantageous because it does not require the rectus abdominis muscle to be incised, leading to a shorter hospital stay than the DIEP flap. Disadvantages of the SIEA flap are its smaller pedicle diameter and shorter pedicle length than the DIEP flap (Chevray 2004).

Injury to the inferior epigastric artery has been reported after surgical incisions in the lower abdominal wall, abdominal peritoneocentesis, and trocar placements at laparoscopic port sites, often resulting in abdominal wall hematomas. Therefore, it is important to determine safety zones. After studying 30 adult cadavers, Praisy et al. concluded that to avoid injury to the inferior epigastric artery, trocars can be safely inserted 5.5 cm away from the midline or slightly more than one-third of the distance between the midline and a sagittal plane running through the anterior superior iliac spine (Joy et al. 2017).

61.7 Complications

Inferior epigastric artery (IEA) false aneurysms, although rare, are likely to complicate abdominal wall surgery or trauma. In a literature review, Georgiadis et al. found 16 cases of IEA false aneurysm, 13 of which were iatrogenic and followed abdominal wall procedures. Treatment options are open surgery, percutaneous coil embolization, or ultrasound-guided thrombin injection (Georgiadis et al. 2007).

Retroperitoneal bleeding from unintentional puncture of the inferior epigastric artery during cardiac catheterization is another possible complication. Quick diagnosis via femoral angiography provides information to guide treatment options such as local external compression, balloon tamponade, coil or gel-foam embolization, thrombin injection, covered stent placement, or surgery (Kawamura et al. 2006).

References

Aghi M, Chiocca E (2006) Contribution of bone marrow-derived cells to blood vessels in ischemic tissues and tumors. Mol Ther 12:994–1005. https://doi.org/10.1016/j.ymthe.2005.07.693

Al Talalwah W (2016) A new concept and classification of corona mortis and its clinical significance. Chin J Traumatol 19(5):251–254. https://doi.org/10.1016/j.cjtee.2016.06.004

Babu SNM, Chacko S, Irodi A, Joseph E, Joseph G (2019) Winslow pathway collaterals in Takayasu arteritis with middle aortic syndrome. IHJ Cardiovasc Case Rep (CVCR) 3(1):39–43. https://doi.org/10.1016/j.ihjccr.2019.01.001

Castro Ferreira M, Henrique Ishida L, Munhoz A (2009) Chapter 19—Rectus flap. In: Wei F-C, Mardini S (eds) Flaps and reconstructive surgery. W.B. Saunders, pp 207–223. https://doi.org/10.1016/B978-0-7216-0519-7.00019-8

Chevray PM (2004) Breast reconstruction with superficial inferior epigastric artery flaps: a prospective comparison with TRAM and DIEP flaps. Plast Reconstr Surg 114(5):1077–1083. https://doi.org/10.1097/01.PRS.0000135328.88101.53

Cisneros AI, Whyte JR, Torres A, Alconchel MD, Sarrat R (2002) Intimal thickenings in the inferior epigastric artery and coronary bypass. Eur J Morphol 40(5):303–308. https://doi.org/10.1076/ejom.40.5.303.28904

Gençhellaç H, Dursun M, Temizöz O, Cağlı B, Demir MK (2014) Two-detector computed tomography map of the inferior epigastric vessels for percutaneous transabdominal intervention procedures. Balkan Med J 31(1):72–76. https://doi.org/10.5152/balkanmedj.2014.13002

Georgiadis GS, Souftas VD, Papas TT, Lazarides MK, Prassopoulos P (2007) Inferior epigastric artery false aneurysms: review of the literature and case report. Eur J Vasc Endovasc Surg 33(2):182–186. https://doi.org/10.1016/j.ejvs.2006.08.006

Hardman RL, Lopera JE, Cardan RA, Trimmer CK, Josephs SC (2011) Common and rare collateral pathways in Aortoiliac occlusive disease: a pictorial essay. Am J Roentgenol 197(3):W519–W524. https://doi.org/10.2214/AJR.10.5896

Ireton JE, Lakhiani C, Saint-Cyr M (2014) Vascular anatomy of the deep inferior epigastric artery perforator flap: a systematic review. Plast Reconstr Surg 134(5):810e. https://doi.org/10.1097/PRS.0000000000000625

Jakubowicz M, Czarniawska-Grzesiñska M (1996) Variability in origin and topography of the inferior epigastric and obturator arteries. Folia Morphol (Warsz) 55(2):121–126

Joy P, Prithishkumar IJ, Isaac B (2017) Clinical anatomy of the inferior epigastric artery with special relevance to invasive procedures of the anterior abdominal wall. J Minim Access Surg 13(1):18–21. https://doi.org/10.4103/0972-9941.181331

Karunanithy N, Rose V, Lim AKP, Mitchell A (2011) CT angiography of inferior epigastric and gluteal perforating arteries before Free flap breast reconstruction. Radiographics 31(5):1307–1319. https://doi.org/10.1148/rg.315105089

Kawamura A, Piemonte TC, Nesto RW, Gossman DE (2006) Retroperitoneal hemorrhage from inferior epigastric artery: value of femoral angiography for detection and management. Catheter Cardiovasc Interv 68(2):267–270. https://doi.org/10.1002/ccd.20835

LabudovicBorovic M, Rakocevic J, Borovic S, Zaletel I, Dzelebdzic S, Puskas N, Trifunović Z (2015) Structural features of arterial grafts important for surgical myocardial revascularization: Part II – histology of the radial, inferior epigastric, and right gastroepiploic arteries. Vojnosanitetski Pregled 72:1111–1117. https://doi.org/10.2298/VSP140516080L

Lin JB, Troyer D (2014) Testicular anatomy and physiology. In: McManus LM, Mitchell RN (eds) Pathobiology of human disease. Academic Press, pp 2464–2475. https://doi.org/10.1016/B978-0-12-386456-7.05102-9

Offman SL, Geddes CR, Tang M, Morris SF (2005) The vascular basis of perforator flaps based on the source arteries of the lateral lumbar region. Plast Reconstr Surg 115(6):1651–1659. https://doi.org/10.1097/01.prs.0000161464.11134.c1

Pungpapong S, Thum-umnauysuk S (2005) Incidence of corona mortis; preperitoneal anatomy for laparoscopic hernia repair. J Med Assoc Thai 88 Suppl 4:S51–S53

Rafieian-Kopaei M, Setorki M, Doudi M, Baradaran A, Nasri H (2014) Atherosclerosis: process, indicators, risk factors and new hopes. Int J Prev Med 5(8):927–946

Rao MP, Swamy V, Arole V, Mishra P (2013) Study of the course of inferior epigastric artery with reference to laparoscopic portal. J Minim Access Surg 9(4):154–158. https://doi.org/10.4103/0972-9941.118826

Ribatti D, Nico B, Crivellato E (2015) The development of the vascular system: a historical overview. In: Ribatti D (ed) Vascular morphogenesis, vol 1214. Springer, New York, pp 1–14. https://doi.org/10.1007/978-1-4939-1462-3_1

Saber AA, Meslemani AM, Davis R, Pimentel R (2004) Safety zones for anterior abdominal wall entry during laparoscopy. Ann Surg 239(2):182–185. https://doi.org/10.1097/01.sla.0000109151.53296.07

Tada K, Murai A, Nakamura Y, Suganuma S, Tsuchiya H (2022) Deep inferior epigastric artery as a collateral pathway to the lower extremities: a case report. JPRAS Open 32:13–17. https://doi.org/10.1016/j.jpra.2022.01.003

Taylor GI, Daniel RK (1975) The anatomy of several free flap donor sites. Plast Reconstr Surg 56(3):243–253

The Vascular Anatomy of Rectus Abdominis Musculocutaneous Fl... : Plastic and Reconstructive Surgery (n.d.) Retrieved June 16, 2022, from https://journals.lww.com/plasreconsurg/abstract/1988/11000/the_vascular_anatomy_of_rectus_abdominis.14.aspx

Yurdakul M, Tola M, Özdemir E, Bayazit M, Cumhur T (2006) Internal thoracic artery-inferior epigastric artery as a collateral pathway in aortoiliac occlusive disease. J Vasc Surg 43(4):707–713. https://doi.org/10.1016/j.jvs.2005.12.042

62 Superior Epigastric Artery

Ronit Patnaik and Lisandro Montorfano

62.1 Embryology

The vascular system develops from fixed primitive angiogenic cells that are mesodermal in origin; this occurs in the middle of week 3 of fetal development. The process initiates with hemangioblasts, which divide into two categories: primitive hematopoietic progenitor cells (HPCs) and endothelial precursor cells (EPCs, otherwise known as angioblasts). Hematopoiesis begins in the yolk sac of the extraembryonic mesoderm, and flattened EPCs initiate vasculogenesis at the embryonic discs. EPCs form the vessels that eventually become arteries and veins through which blood will circulate.

The first major arteries to appear are the right and left primitive aortae. The primitive aortae curve dorsally in the first pharyngeal arch and around the anterior part of the foregut and then continue dorsally as the two dorsal aortae. The common dorsal aorta is formed by the union of the right and left dorsal aortae in the region extending from the fourth thoracic to the fourth lumbar somite segments.

Systemic-systemic collateral pathways arise from connections in the embryonic dorsal arterial system. These arteries start as somatic branches from the dorsal aorta. The common dorsal artery branches into the ventral splanchnic arteries, lateral splanchnic branches, intersegmental arteries and umbilical arteries. The intersegmental arteries later divide into the intercostal arteries in the thorax and the lumbar arteries in the abdomen. They anastomose both ventrally and dorsally, and their ventral anastomoses produce the internal thoracic artery superiorly, which terminates as the superior epigastric artery (SEA) and anastomoses with the inferior epigastric artery inferiorly. The embryological development of the abdominal and lower extremity arteries allows collateral flow to develop by linking the subclavian and external iliac arteries. These collaterals are clinically significant, as will be discussed in a later section (Ilina and Lilley 2019).

62.2 Anatomy/Topography

The epigastric arteries provide the main arterial blood flow to the abdominal wall. The three epigastric arteries supplying the abdominal wall are the superficial, inferior (IEA), and superior (SEA) epigastric arteries. While the superficial

R. Patnaik
Department of Surgery, University of Texas Health, San Antonio, TX, USA

L. Montorfano (✉)
Department of Surgery, Cleveland Clinic Florida, Weston, FL, USA

Department of Plastic and Reconstructive Surgery, Vanderbilt University, Nashville, TN, USA

S. J. Bordes, Jr. et al. (eds.), *The Clinical Anatomy of the Vascular System*,
https://doi.org/10.1007/978-3-031-78326-5_62

epigastric artery supplies the superficial tissue of the lower abdominal wall, the IEA and SEA anastomose with each other and contribute to the system of perforators that perfuse the anterior abdominal musculature bilaterally (Netter 2014).

The SEA is one of the two terminal branches of the internal mammary (internal thoracic) artery, which runs just posterior and lateral to the sternum. The internal mammary artery bifurcates into terminal branches at the level of the sixth or seventh costal cartilage. One of these is the musculophrenic artery, which supplies the muscles of the diaphragm. The other is the SEA, which descends caudally behind the costal cartilage, leaving the thoracic cavity between the xiphoid and costal origins of the diaphragm. Before entering the rectus sheath, it divides into superficial and deep branches. The superficial branch pierces the rectus fascia directly under the xiphoid. The deep branch (DSEA) enters the rectus muscle on its deep surface, approximately 7 cm below the costal margin. When it enters the rectus sheath, it runs behind the rectus abdominis muscle. It then descends on the abdominal wall, between the posterior aspect of the rectus sheath and the anterior aspect of the transversus abdominis muscle. The SEA provides the arterial supply for the rectus abdominis musculature until it anastomoses with the IEA, approximately above the level of the umbilicus. Along its course, the SEA is accompanied by the superior epigastric vein, which drains into the internal thoracic vein. This anatomy is mirrored bilaterally.

The SEA gives off a mean of 5 ± 1 perforators and supplies 150 ± 40 cm^2 of abdominal skin surface area (Offman et al. 2005). This territory extends from the inferior origins of the pectoralis major to just beyond the termination of its deep course in most cases. Its medial and lateral borders are the midline and linea semilunaris, respectively (Offman et al. 2005). The SEA sends a superficial branch (diameter >1 mm) through the origin of the rectus abdominis muscle to supply the overlying skin. This branch was originally described by Manchot, and then Boyd et al. coined the term “superficial superior epigastric artery” during their extensive cadaver dissections (Manchot 1889; Boyd et al. 1984).

62.3 Variations

In their study of 17 cadavers, Bhatti et al. reported that the SEA lies at a mean of 4.3 cm from the midline, with standard deviations of ±0.2 cm. They outlined the “danger zones” and “safe zones” on the anterior abdominal wall where the SEA is most easily injured (Bhatti et al. 2008). According to them, the “danger zones” were strips starting from 3 to 8.5 cm from the midline on either side of the abdominal wall.

Saber et al. reviewed 100 CT scans and recorded the locations of the epigastric vessels (SEA and IEA) at different levels on the abdominal wall. They found they are usually located in the area between 4 cm (SEA) and 8 cm (IEA) from the midline. Staying away from this area, either medially or laterally, will determine the safe zone of entry into the abdominal wall without risk of injury to these vessels. To avoid such injury, these anatomical landmarks should be considered whenever the anterior abdominal wall is violated, not only during laparoscopic or open surgical procedures but also during image-guided biopsy, endoscopic procedures, and paracentesis. Saber et al. also noted that the average distance of the epigastric vessels from the midline bilaterally is consistently higher for those with BMI ≥26.3. Epigastric vessels in older patients with high BMI are further from the midline, on average, than those of younger patients (Saber et al. 2004).

62.4 Clinical Significance

The SEA can be injured during laparoscopic trocar placement, percutaneous endoscopic gastrostomy, drain insertion in cardiothoracic surgery, or any procedure in the vicinity such as paracentesis. This could lead to rectus abdominis hematoma formation and severe blood loss and increase the chance of surgical site infection (Immè and Cardì 2006; Ubogu and Zaidat 2002).

In fact, injury to the epigastric arteries is the most common complication during laparoscopic entry into the abdomen, occurring 0.2–2% of the time (Lin and Grow 1999).

Some authors have recommended transillumination from the light of either the endoscope, for PEG tube insertion, or the laparoscope prior to port placement (Hurd et al. 2003). This can be problematic in cases of dark skinned patients and patients with a high BMI (Saber et al. 2004).

Schematic of systemic-systemic pathways. Systemic collateral pathways arise from the embryological dorsal artery. The dorsal artery begins as somatic branches from the dorsal aorta. Arteries later divide into intercostal, lumbar, and internal mammary arteries, SEA and IEA. This common embryological origin allows collateralization to occur. There are collateral pathways between the intercostal, lumbar, and iliolumbar arteries and inferior epigastric and deep circumflex arteries (Hardman et al. 2011; Krupski et al. 1984).

Coarctation of the aorta is a congenital narrowing of the aorta distal to the origin of the left subclavian artery. It can be pre-ductal or post-ductal. In the pre-ductal type, the narrowing is proximal to the level of the ductus arteriosus. The ductus remains patent to maintain circulation to the lower body. With post-ductal coarctation, the narrowing of the aorta is distal to the level of the ductus arteriosus, which closes to form the ligamentum arteriosum. In post-ductal coarctation, the circulation to the lower part of the body is supported by collateral circulation, which includes engorgement of collateral flow to the intercostal arteries and to the SEA and IEA through the internal thoracic (mammary) artery (Ilina and Lilley 2019; Edwards and Clagett 1948).

Aorto-iliac occlusive disease is atherosclerotic disease of the distal aorta or proximal common iliac, which leads to ischemic symptoms. It can present with symptoms of Leriche syndrome: the triad of claudication, absence of femoral pulses, and impotence. The SEA, IEA, internal thoracic arteries, and subclavian arteries are collateral vessels to the thoracic aorta and abdominal aorta. In aortoiliac occlusive disease, as in coarctation of the aorta, various collateral pathways maintain blood flow to the lower extremities. The internal thoracic artery to IEA pathway is one of these collaterals; it can easily be overlooked owing to its distance from the aortoiliac segment (Yurdakul et al. 2006).

62.5 Use in Surgical Procedures

Several authors have described the use of the SEA for plastic surgery and reconstructive flaps. Vertical and TRAM (transverse rectus abdominis myocutaneous) flaps were the original myocutaneous flaps using the SEA as the blood supply (Moon and Taylor 1988; Miller et al. 1988; Hartrampf et al. 1982). The superior epigastric artery perforator (SEAP) flap, a perforator flap, is now also available for soft-tissue reconstructions because of its easy, quick elevation (Uemura 2007; Hamdi et al. 2009; Hamdi et al. 2014). Hallock et al. were the first to describe a perforator flap based on the SEA (Hallock 2005). The SEAP flap has been used by multiple teams for reconstruction of breast, chest, sternal, and abdominal wall defects without disrupting the muscular continuity of the abdominal wall (Tai and Hasegawa 1974; Lejour and De Mey 1983; Mah et al. 2009; Oni et al. 2011; Woo et al. 2010; Ziegler et al. 2011)

62.6 Complications

As previously noted, the SEA can be lacerated during procedures in its anatomical vicinity. These procedures include laparoscopic trocar placement, percutaneous endoscopic gastrostomy, drain insertion in cardiothoracic surgery, paracentesis, parenchymal liver biopsies, etc. (Vijayaraghavan et al. 2011). Abdominal wall bleeding/hematomas after laceration of a SEA can be controlled by a variety of techniques. These include application of direct pressure, full thickness abdominal wall suture ligation or tamponade with a Foley catheter ball through the trocar site (Aharoni et al. 1997). If these maneuvers fail, exploration with direct suture ligation of the SEA can be necessary (Saber et al. 2004).

Injury to the SEA as described above could also lead to inability to use a SEA-based flap in a future reconstructive surgery.

In diseases such as coarctation of the aorta and aorto-iliac occlusive disease, transection of the epigastric arteries can lead to more severe bleeding and potential critical limb ischemia owing to the disruption of collateral flow.

References

Aharoni A, Condrea A, Leibovitz Z, Paz B, Levitan Z (1997) A comparative study of Foley catheter and suturing to control trocar-induced abdominal wall haemorrhage. Gynaecol Endosc 6(1):31–32

Bhatti A, Iqbal S, Lee T (2008) Variation in surface marking of superior epigastric vessels. A guide to safe laparoscopic port insertion. Surgeon 6(1):50–52

Boyd JB, Taylor GI, Corlett R (1984) The vascular territories of the superior epigastric and the deep inferior epigastric systems. Plast Reconstr Surg 73(1):1–16

Edwards J, Clagett O (1948) The collateral circulation in coarctation of the aorta. Paper presented at: Proceedings of the staff meetings. Mayo Clinic

Hallock GG (2005) The Superior EpigastricRECTUS ABDOMINIS Muscle Perforator Flap. Ann Plast Surg 55(4):430–432

Hamdi M, Van Landuyt K, Ulens S, Van Hedent E, Roche N, Monstrey S (2009) Clinical applications of the superior epigastric artery perforator (SEAP) flap: anatomical studies and preoperative perforator mapping with multidetector CT. J Plast Reconstr Aesthet Surg 62(9):1127–1134

Hamdi M, Craggs B, Stoel A-M, Hendrickx B, Zeltzer A (2014) Superior epigastric artery perforator flap: anatomy, clinical applications, and review of literature. J Reconstr Microsurg 30(07):475–482

Hardman RL, Lopera JE, Cardan RA, Trimmer CK, Josephs SC (2011) Common and rare collateral pathways in aortoiliac occlusive disease: a pictorial essay. Am J Roentgenol 197(3):W519–W524

Hartrampf C, Scheflan M, Black PW (1982) Breast reconstruction with a transverse abdominal Island flap. Plast Reconstr Surg 69(2):216–225

Hurd WW, Amesse LS, Gruber JS, Horowitz GM, Cha GM, Hurteau JA (2003) Visualization of the epigastric vessels and bladder before laparoscopic trocar placement. Fertil Steril 80(1):209–212

Ilina M, Lilley S (2019) Embryology of the great vessels. In: Clinical embryology. Springer, pp 275–288

Immè A, Cardì F (2006) Incisional hernia at the trocar site in laparoscopic surgery. Chir Ital 58(5): 605–609

Krupski WC, Sumchai A, Effeney DJ, Ehrenfeld WK (1984) The importance of abdominal wall collateral blood vessels: planning incisions and obtaining arteriography. Arch Surg 119(7):854–857

Lejour M, De Mey A (1983) Experience with 33 epigastric rectus flaps in breast reconstruction. Handchir Mikrochir Plast Chir 15(4):257–260

Lin P, Grow DR (1999) Complications of laparoscopy: strategies for prevention and cure. Obstet Gynecol Clin N Am 26(1):23–38

Mah E, Rozen WM, Ashton MW, Flood S (2009) Deep superior epigastric artery perforators: anatomical study and clinical application in sternal reconstruction. Plast Reconstr Surg 123(6):1719–1723

Manchot C (1889) Die Hautarterien des menschlichen Körpers. Vogel

Miller LB, Bostwick J 3rd, Hartrampf CR Jr, Hester TR Jr, Nahai F (1988) The superiorly based rectus abdominis flap: predicting and enhancing its blood supply based on an anatomic and clinical study. Plast Reconstr Surg 81(5):713–724

Moon HK, Taylor GI (1988) The vascular anatomy of rectus abdominis musculocutaneous flaps based on the deep superior epigastric system. Plast Reconstr Surg 82(5):815–832

Netter FH (2014) Atlas of human anatomy, Professional Edition E-Book: including NetterReference.com Access with full downloadable image Bank. Elsevier Health Sciences

Offman SL, Geddes CR, Tang M, Morris SF (2005) The vascular basis of perforator flaps based on the source arteries of the lateral lumbar region. Plast Reconstr Surg 115(6):1651–1659

Oni G, Sharma R, Rao R, Unger J, Saint-Cyr M (2011) Bilateral superior epigastric pedicle perforator flaps for total chest wall coverage. J Plast Reconstr Aesthet Surg 64(8):1104–1107

Saber AA, Meslemani AM, Davis R, Pimentel R (2004) Safety zones for anterior abdominal wall entry during laparoscopy: a CT scan mapping of epigastric vessels. Ann Surg 239(2):182

Tai Y, Hasegawa H (1974) A transverse abdominal flap for reconstruction after radical operations for recurrent breast cancer. Plast Reconstr Surg 53(1):52–54

Ubogu EE, Zaidat OO (2002) Rectus sheath hematoma complicating percutaneous endoscopic gastrostomy. Surg Laparosc Endosc Percutan Tech 12(6):430–432

Uemura T (2007) Superior epigastric artery perforator flap: preliminary report. Plast Reconstr Surg 120(1):1e–5e

Vijayaraghavan G, Sheehan D, Zheng L, Hussain S, Ferrucci J (2011) Unusual complication after left-lobe liver biopsy for diffuse liver disease: severe bleeding from the superior epigastric artery. Am J Roentgenol 197(6):W1135–W1139

Woo K-J, Pyon J-K, Lim S-Y, Mun G-H, Bang S-I, Oh K-S (2010) Deep superior epigastric artery perforator 'propeller' flap for abdominal wall reconstruction: A case report. J Plast Reconstr Aesthet Surg 63(7):1223–1226

Yurdakul M, Tola M, Özdemir E, Bayazit M, Cumhur T (2006) Internal thoracic artery-inferior epigastric artery as a collateral pathway in aortoiliac occlusive disease. J Vasc Surg 43(4):707–713

Ziegler K, Schmidt M, Huemer GM (2011) A pedicled superior epigastric artery perforator (SEAP-) propeller flap for 2-cavity reconstruction after oncologic rib resection. Microsurgery 4(31):335–336

Internal Iliac Artery 63

Mahant Malempati and Vinila Baljepally

63.1 Embryology

The vessels of the vascular system come from mesoderm- and ectoderm-derived angiogenic cells. First, the right and left primitive aortae develop as the endocardial cardiac tubes lengthen. The aortae curve posteriorly into the first pharyngeal arch and go around the anterior foregut to continue posteriorly as the dorsal aortae. Next these two dorsal aortae are dorsal to the primitive gut and fuse caudally at the distal ends to form the caudal aorta while the cranial ends stay separate (Khalid and Bordoni 2022). The right and left dorsal aortae combine at the fourth thoracic to fourth lumbar somite segments (Bhatia et al. 2021).

The dorsal aortae are the embryological equivalents of the descending aorta. They give off multiple branches: ventral (visceral segmental branches), lateral (visceral), segmental, dorsolateral (parietal), and intersegmental. The branches relevant to the development of the internal iliac arteries are the dorsolateral intersegmental and the ventral segmental branches. From the ventral segmental branches, vessels arise from the umbilical vesicle (celiac trunk, superior mesenteric artery, and inferior mesenteric artery) and from the allantois. The relevant vessels that arise from the allantois are the umbilical arteries (Accueil n.d.; Khalid and Bordoni 2022).

The umbilical arteries arise from a vessel plexus in the lower dorsal aorta at the allantois level. They arise as ventral segmental arteries, which anastomose with the intersegmental fifth lumbar artery. The cranial ends of the umbilical arteries atrophy, removing connections with the dorsal aorta. The fifth intersegmental lumbar artery serves as the new attachment point for the umbilical artery. This process is called the caudal displacement of the umbilical artery. The fifth intersegmental lumbar artery, unlike the umbilical artery, comes from the dorsolateral intersegmental branches of the dorsal aortae (Accueil n.d.; Hegazy 2016; Khalid and Bordoni 2022).

Both the fifth lumbar intersegmental lumbar artery and the umbilical artery give rise to the iliac arteries. The fifth lumbar intersegmental artery becomes the external iliac artery while the umbilical artery, which displaces and attaches to this branch, becomes the internal iliac artery. The proximal portions of these arteries become the common iliac arteries (Accueil n.d.; Khalid and Bordoni 2022).

63.2 Anatomy

The abdominal aorta divides into right and left common iliac branches at the L4/L5 vertebral level (Fig. 63.1). At the level of the sacroiliac

M. Malempati (✉) · V. Baljepally
School of Medicine, Tulane University, New Orleans, LA, USA
e-mail: mmalempati@tulane.edu

S. J. Bordes, Jr. et al. (eds.), *The Clinical Anatomy of the Vascular System*,
https://doi.org/10.1007/978-3-031-78326-5_63

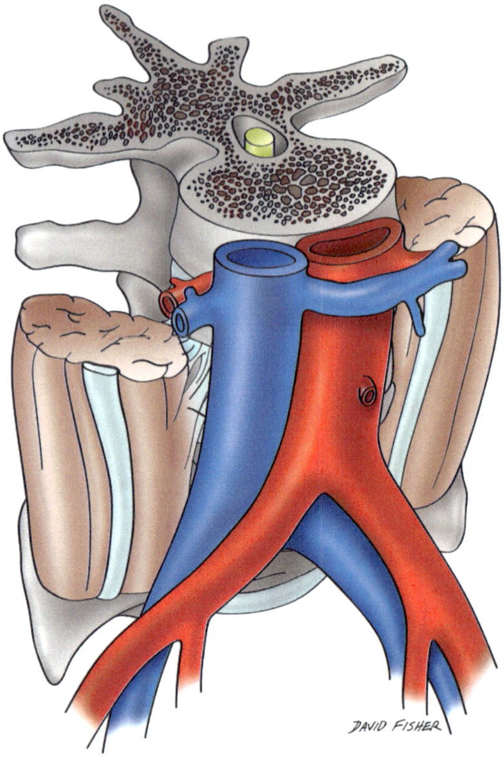

Fig. 63.1 Schematic drawing showing the bifurcation of the abdominal aorta into left and right common iliac arteries, which then bifurcate into the more medially placed internal iliac artery and more laterally placed external iliac artery. (After Netter, 2023)

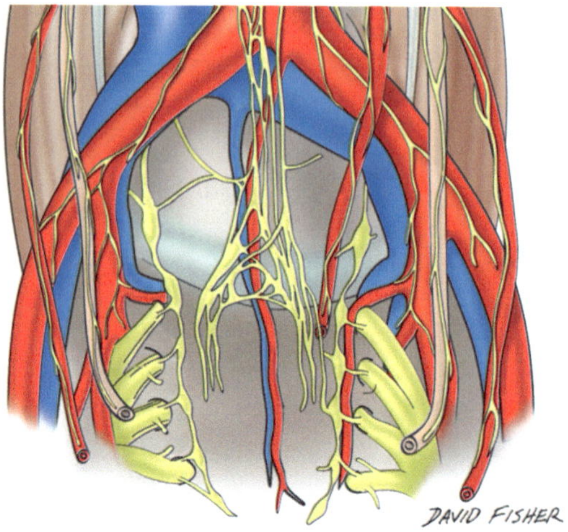

Fig. 63.2 Schematic drawing illustrating the relationship of the internal and external iliac arteries and surrounding nerves

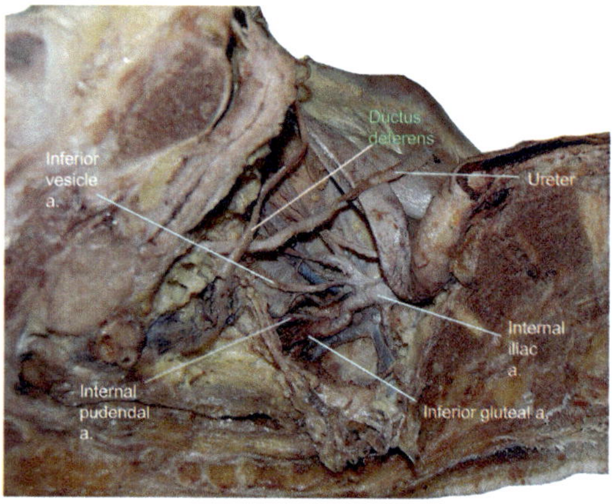

Fig. 63.3 Cadaveric dissection noting many branches of the internal iliac artery in a male specimen. (Courtesy Prof. M. Loukas)

joint, both common iliac arteries bifurcate into external and internal iliac arteries (Fig. 63.2). The external iliac artery continues under the inguinal ligament as the femoral artery while the internal iliac artery provides the main blood supply to the pelvis (Standring 2016). The internal iliac branch runs over the pelvic brim and divides into anterior and posterior aspects at the level of the greater sciatic foramen (Selçuk et al. 2018).

The posterior portion of the internal iliac artery gives rise to:

- The iliolumbar artery: Anastomoses with superior gluteal and circumflex iliac arteries (Selçuk et al. 2018).
- The lateral sacral artery: Anastomoses with median sacral artery (Selçuk et al. 2018).
- The superior gluteal artery (Fig. 63.4): Anastomoses with lateral sacral, inferior gluteal, and internal pudendal arteries. It is also the main portion and continuation of the posterior trunk, which usually runs between the lumbosacral trunk and the first sacral nerve (Agur and Dalley 2016; Baggish and Karram 2016; Selçuk et al. 2018).

The anterior portion of the internal iliac artery often gives rise to:

- The obliterated umbilical artery: The superior vesical artery arises from its proximal part and

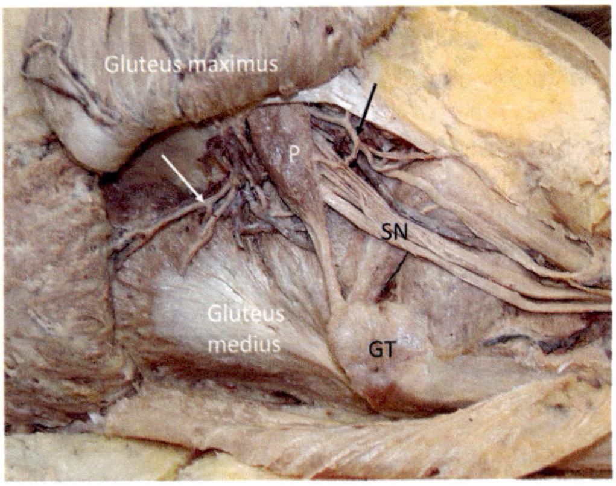

Fig. 63.4 Left gluteal region in a cadaver noting the superior and inferior gluteal arteries. For reference, the greater trochanter (GT), sciatic nerve (SN), and piriformis muscles (P) are shown. (Courtesy Prof. M. Loukas)

supplies the superior bladder; the inferior vesical artery (vaginal artery in females) supplies the base of the bladder; and the umbilical artery itself becomes the median umbilical ligament as it approaches the anterior abdominal wall (Selçuk et al. 2018).

- The uterine artery: Attaches antero-medially to the lateral part of the uterine cervix below the isthmus. It runs between the anterior and posterior layers of the broad ligament and above the cardinal ligament. It anastomoses with the ovarian artery and gives off a vaginal branch (Selçuk et al. 2018).
- The obturator artery: Runs antero-infero-laterally on the obturator fascia, exiting the pelvis through the obturator foramen. It usually arises near the umbilical artery and follows the obturator nerve. There are anastomoses with the inferior epigastric or external iliac artery via a pubic branch called the corona mortis, which is important for surgeons to identify because severe bleeding ensues if it is injured (Selçuk et al. 2018).
- The internal pudendal artery (Fig. 63.3): Runs infero-laterally, anterior to the piriformis muscle and sacral plexus. Travels through the greater sciatic foramen posterior to the ischial spine and the sacrospinous ligament before entering the ischio-anal fossa through the lesser sciatic foramen. Gives rise to many branches in the perineum (Selçuk et al. 2018).
- The inferior gluteal artery (Figs. 63.3 and 63.4): Passes between the S1 and S2 sacral nerves but also S2 and S3 in some cases. Travels through the greater sciatic foramen below the piriformis before exiting through the lesser sciatic foramen (Selçuk et al. 2018).
- The superior vesicle artery: Supplies the distal ureter and bladder (Selçuk et al. 2018).

63.3 Topography

The mean diameter of the internal iliac artery is 7.32 ± 1.60 mm and its mean length is 36.97 ± 14.12 mm according to one study (Ongidi et al. 2021). In general, the length is commonly expressed as 4 cm (Ozgüner and Sulak 2011).

63.4 Histology/Imaging

The iliac arteries are elastic arteries. Arterial walls are composed of three layers of tissue. The adventitia is the outer layer supporting the tissue; the tunica media constitutes the middle layer; and the inner layer is called the tunica intima. Elastic arteries have less smooth muscle and more elastic tissue in the tunica media than muscular arteries (Uflacker 2007).

MR angiography, CT angiography, and digital angiography can detect most branches of the internal iliac artery (Naguib et al. 2008). MR and CT angiography are minimally invasive techniques that are usually pursued before more invasive angiography. MR angiography detects most of the branches of the internal iliac artery (Naguib et al. 2008). However, CT angiography showed smaller branches better and allowed the researchers to see blood vessels within the context of bony landmarks if they chose to do so. CT angiography is not suggested if iodinated contrast injections are contraindicated for a patient. In that event, the MR angiography can be conducted. The gold standard, although more invasive, is angiography proper.

63.5 Variations

In a study of 50 bisected human pelvises, the origin of the internal iliac artery was found at the L5/S1 vertebral level in 72% of cases, the S1 level in 24%, and the L5 level in 4% (Mamatha et al. 2015). The distance between the origin of the internal iliac artery and the greater sciatic foramen also varied. The distances are summarized in the table below but ranged from less than 1 cm to over 3.5 cm (Mamatha et al. 2015).

Finally, this study also observed common variations in branching from the internal iliac artery. Overall, 60% of cadavers had some variation in branching. Some common branching patterns included the iliolumbar arising from the trunk of internal iliac artery (6%), the obturator artery arising from its posterior division (14%), the obturator artery branching from the inferior vesicle artery (8%), and the obturator artery branching from the inferior epigastric artery (12%) (Mamatha et al. 2015).

Another study of 645 pelvic cadaver halves distinguished five types of internal iliac artery branching based on Adachi's classification (Adachi 1928; Yamaki et al. 1998). In Type I, the superior gluteal artery arises first, while the inferior gluteal and internal pudendal arteries arise from a second common trunk. Type I branching was seen in 58% of cadavers. In Type II, the superior and inferior gluteal arteries arise from the same trunk while the internal pudendal artery originates from a second branch (13.6%). Type III shows all three arteries arising independently, the internal pudendal artery serving as the final branch (22.8%). Type IVa has all three arteries arising from the same trunk, the superior gluteal branching off first, the trunk then dividing to give off the other two arteries. Type IVb has the internal pudendal branches stemming first before the trunk divides into the other two arteries; 5.4% of cadavers were type IV. Type V (0.2%) has the interior pudendal and superior gluteal arteries stemming from a common trunk while the inferior gluteal artery arises independently (Yamaki et al. 1998).

63.6 Pathology, Clinical, Surgical, and Complications

Internal iliac stenosis is associated with atherosclerotic plaque formation and smoking. This condition can be difficult to diagnose since the pain presentation in the lower back, hip, buttock, or thigh is vague. However, such symptoms can suggest claudication or pseudo-claudication (Jaquinandi et al. 2007). Claudication refers to vasculogenic pain, while pseudo-claudication suggests lumbar spine stenosis, hip osteoarthritis, venous congestion, bone metastasis, or sciatica (Katz and Harris 2008; Lane 2007; White 2007).

The incidence of internal iliac stenosis is low, but it is often associated with common iliac artery stenosis. The main symptoms include lower back, hip, buttock, or thigh claudication. Pain occurs during exercise in muscle groups supplied by the internal iliac artery branches, followed by a relief period with rest. When severe, pain during rest lasts longer before dissipating. Internal iliac stenosis can lead to ambulatory impairment and sexual dysfunction (Rayt et al. 2008). On average, delay to diagnosis of this condition was estimated at 2 years (Naguib et al. 2008).

Imaging studies were discussed above for anatomical investigation. In terms of clinically significant studies, duplex ultrasound is the cheapest and least invasive procedure. Compared to arteriography, ultrasound has sensitivity and specificity of 91 and 93% for diagnosing aorto-iliac disease (Currie et al. 1995).

Proposed functional assessments for diagnosing pelvic ischemia include exercise transcutaneous oxygen pressure ($TcPO_2$) (reflects tissue oxygenation) and exercise near-infrared spectroscopy (NIRS) (reflects muscle oxygenation). NIRS showed 55% accuracy and Exercise-$TcPO_2$ 82% accuracy in detecting arteriographically proven lesions (Bouye et al. 2005).

Treatments for internal iliac stenosis include daily aspirin. This can be taken in conjunction with clopidogrel in certain cases (CAPRIE Steering Committee 1996). Statin drugs lowered the risk of major vascular events by 22% compared to placebo in patients with peripheral artery disease (Heart Protection Study Collaborative Group 2007). ACE inhibitors reduced the risk of adverse cardiovascular events by 25% in this patient population (Yusuf et al. 2000). Cilostazol and pentoxifylline can also be used to improve walking distances while mitigating the symptoms of claudication (Selçuk et al. 2019).

Revascularization is a possible treatment for patients with severe symptoms who show minimal improvement following maximum medical treatment and exercise therapy. For isolated internal iliac stenosis, endovascular treatment is more commonly performed since it is less risky and less demanding than open revascularization. The two most common endovascular treatments are percutaneous transluminal angioplasty (PTA) and primary stent placement. Both procedures have been shown to be safe with minimal complications, to alleviate symptoms of claudication, and to have high technical success rates (Donas et al. 2009; Thompson et al. 2010). One study of 34 patients showed partial or complete relief of symptoms in 79% of patients (Prince et al. 2013).

In the event of uncontrollable hemorrhage, the internal iliac artery can be ligated to establish control. Indications for ligation include obstetric factors (uterine atony, uterine laceration or rupture, abruptio placenta, placental adhesion disorder, placenta accrete, increta, percreta, and cervical pregnancy) and gynecological factors (gynecological oncological procedures, mass/tumor in deeper pelvis, exenteration surgery, radical hysterectomy, extended radical operations) (Selçuk et al. 2018). Some common operative complications during ligation include injury to the external or the internal iliac vein, ligation or laceration of the external iliac artery, injury to the internal iliac artery, and injury to the ureter (Selçuk et al. 2018).

References

Accueil, Universities of Fribourg, Lausanne and Bern (Switzerland), https://embryology.ch/

Adachi B (1928) Das Arteriensystem der Japaner. Bd. 1. Die KaiserlichJapanischeUniversitatzu Kyoto, Kyoto

Agur AMR, Dalley AF (2016) Grant's atlas of Anatomy. Wolters Kluwer Health

Baggish M, Karram M (2016) Atlas of pelvic Anatomy and gynecologic surgery, 4th edn. Elsevier

Bhatia A, Shatanof RA, Bordoni B (2021) Embryology, gastrointestinal. In: StatPearls [Internet]. StatPearls Publishing, Treasure Island

Bouye P, Jacquinandi V, Picquet J, Thouveny F, Liagre J, Leftheriotis G et al (2005) Near-infrared spectroscopy and transcutaneous oxygen pressure during exercise to detect arterial ischemia at the buttock level: comparison with arteriography. J Vasc Surg 41:994–999

CAPRIE Steering Committee (1996) A randomised, blinded, trial of clopidogrel versus aspirin in patients at risk of ischaemic events (CAPRIE). CAPRIE steering committee. Lancet 348:1329–1339

Currie IC, Jones AJ, Wakeley CJ, Tennant WG, Wilson YG, Baird RN et al (1995) Non-invasive aortoiliac assessment. Eur J VascEndovasc Surg 9:24–28

Donas KP, Schwindt A, Pitoulias GA, Schonefeld T, Basner C, Torsello G (2009) Endovascular treatment of internal iliac artery obstructive disease. J Vasc Surg 49:1447–1451

Heart Protection Study Collaborative Group (2007) Randomized trial of the effects of cholesterol-lowering with simvastatin on peripheral vascular and other major vascular outcomes in 20,536 people with peripheral arterial disease and other high-risk conditions. J Vasc Surg 45:645–654

Hegazy AA (2016) Anatomy and embryology of umbilicus in newborns: a review and clinical correlations. Front Med 10(3):271–277

Jaquinandi V, Abraham P, Picquet J, Paisant-Thouveny F, Leftheriotis G, Saumet JL (2007) Estimation of the functional role of arterial pathways to the buttock circulation during treadmill walking in patients with claudication. J Appl Physiol 102:1105–1112

Katz JN, Harris MB (2008) Clinical practice. Lumbar spinal stenosis. N Engl J Med 358:818–825

Khalid N, Bordoni B (2022) Embryology, great vessel. [Updated 2021 May 8]. In: StatPearls [Internet]. StatPearls Publishing, Treasure Island. 2022 Jan-. Available from: https://www.ncbi.nlm.nih.gov/books/NBK545254/

Lane NE (2007) Clinical practice. Osteoarthritis of the hip. N Engl J Med 357:1413

Mamatha H, Hemalatha B, Vinodini P, Souza AS, Suhani S (2015) Anatomical study on the variations in the branching pattern of internal iliac artery. Indian J Surg 77(Suppl 2):248–252

Naguib NN, Nour-Eldin NE, Hammerstingl RM, Lehnert T, Floeter J, Zangos S, Vogl TJ (2008) Three-dimensional reconstructed contrast-enhanced MR angiography for internal iliac artery branch visualization before uterine artery embolization. J Vasc Interv Radiol 19:1569–1575

Ongidi IH et al (2021) Variability in morphology and branching of the internal iliac artery: implications for pelvic surgery. Transl Res Anat 22:100097
Ozgüner G, Sulak O (2011) Development of the abdominal aorta and iliac arteries during the fetal period: a morphometric study. Surg Radiol Anat 33(1):35–43
Prince JF, Smits ML, van Herwaarden JA, Arntz MJ, Vonken EJ, van den Bosch MA et al (2013) Endovascular treatment of internal iliac artery stenosis in patients with buttock claudication. PLoS One 8:e73331. https://doi.org/10.1371/journal.pone.0073331
Rayt HS, Bown MJ, Lambert KV, Fishwick NG, McCarthy MJ, London NJ et al (2008) Buttock claudication and erectile dysfunction after internal iliac artery embolization in patients prior to endovascular aortic aneurysm repair. Cardiovasc Intervent Radiol 31:728–734
Selçuk İ, Yassa M, Tatar İ, Huri E (2018) Anatomic structure of the internal iliac artery and its educative dissection for peripartum and pelvic hemorrhage. Turk J Obstet Gynecol 15(2):126–129
Selçuk İ, Uzuner B, Boduç E, Baykuş Y, Akar B, Güngör T (2019) Step-by-step ligation of the internal iliac artery. J Turk Ger Gynecol Assoc 20(2):123–128
Standring S (2016) Gray's anatomy. The anatomical basis of clinical practice, 40th edn. Elsevier
Thompson K, Cook P, Dilley R, Saeed M, Knowles H, Terramani T et al (2010) Internal iliac artery angioplasty and stenting: an underutilized therapy. Ann Vasc Surg 24:23–27
Uflacker R (2007) Atlas of vascular anatomy, an angiographic approach. ISBN:078176081X
White C (2007) Clinical practice. Intermittent claudication. N Engl J Med 356:1241–1250
Yamaki K, Saga T, Doi Y, Aida K, Yoshizuka M (1998) A statistical study of the branching of the human internal iliac artery. Kurume Med J 45(4):333–340
Yusuf S, Sleight P, Pogue J, Bosch J, Davies R, Dagenais G (2000) Effects of an angiotensin-converting-enzyme inhibitor, ramipril, on cardiovascular events in high-risk patients. The heart outcomes prevention evaluation study investigators. N Engl J Med 342:145–153

Internal Pudendal Artery

64

Julisa Nuñez, Brian Le, Timothy Acheson, and Bryan S. Torres

64.1 Embryology

Vessels begin to appear in the splanchnic mesoderm of the embryo around gestational day 18 (Streeter 1919). As the embryo grows, small branches appear at fairly regular intervals around the primitive aorta (DeSesso 2017). Around the 28th day of gestation, the hind limb buds appear and mark the formation of the intersegmental arteries, which later form the common iliac and external iliac arteries as the first vessels to penetrate the nascent hind limb buds (DeSesso 2012). Primarily, the dorsal intersegmental arteries coalesce with the ventral segmental arteries to form a single central artery, the common iliac artery (CIA) (DeSesso 2017). The CIA serves as the stem for the external iliac and the umbilical artery (presumptively the internal iliac artery) (DeSesso 2017). When the embryo is less than 1.5 mm long, the umbilical artery arises at about the level where the fourth cervical mesodermal somite and its portion ventral to the external iliac artery come together to form the origin of the internal iliac artery (Sakthivelavan et al. 2010; DeSesso 2017). The internal iliac artery (IIA) is the major artery of the pelvis. It divides into anterior and posterior divisions, the internal pudendal artery (Figs. 64.1, 64.2, and 64.3) (IPA) being one of the parietal branches of the anterior division (Sakthivelavan et al. 2010). However, given the variability of fetal branching patterns, four types of arrangements of branches of the fetal IIA have been distinguished. The IPA is one of the three largest branches, along with the

J. Nuñez
Georgetown University School of Medicine, Washington, DC, USA

B. Le · B. S. Torres (✉)
Tulane University School of Medicine, New Orleans, LA, USA
e-mail: btorres@tulane.edu

T. Acheson
University of Southern California School of Medicine, Los Angeles, CA, USA

S. J. Bordes, Jr. et al. (eds.), *The Clinical Anatomy of the Vascular System*,
https://doi.org/10.1007/978-3-031-78326-5_64

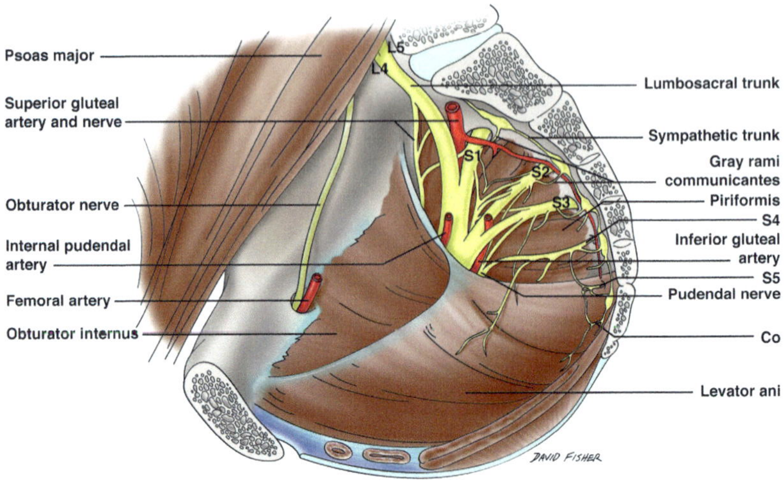

Fig. 64.1 Medial view of the internal pudendal artery and its relationship to the sacral plexus and pelvic diaphragm

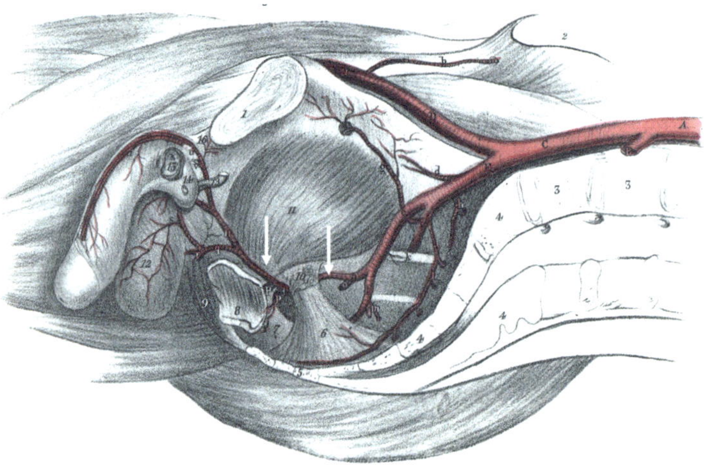

Fig. 64.2 Medial view of the internal and external iliac arteries. Note the course of the internal pudendal artery (arrows) proximal and distal to the ischial spine (Quain and Wilson, 1844 author's collection)

superior and inferior gluteal arteries, used for classification into types. Type 1 is defined as the IPA and inferior gluteal artery arising from the anterior trunk of the IIA (Sakthivelavan et al. 2010). In Types 2 and 4, the three arteries arise independently from the IIA, or from a common stem, respectively. In Type 3, the IPA remains distinct while the superior and inferior gluteal arteries arise from the common trunk (Sakthivelavan et al. 2010).

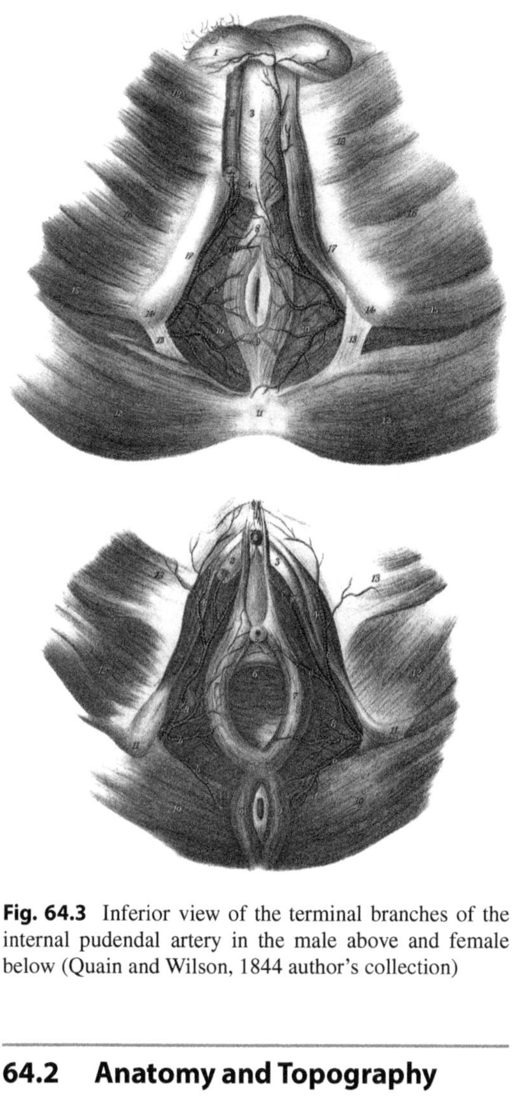

Fig. 64.3 Inferior view of the terminal branches of the internal pudendal artery in the male above and female below (Quain and Wilson, 1844 author's collection)

64.2 Anatomy and Topography

The IPA arises from the anterior division of the IIA within the pelvis. It is the principal artery of the perineum and external genital organs. It passes through the greater sciatic foramen inferior to the piriformis muscle, accompanied by the pudendal nerve. Together, the artery and nerve hook over the ischial spine and over the sacrospinous ligament travel under the sacrotuberous ligament and exit the pelvis. They then enter the perineum via the lesser sciatic foramen and then enter the pudendal canal (Alcock's canal), which is located deep to the obturator internus fascia on the lateral wall of the ischio-anal fossa (Colebunders et al. 2021). Similar to the pudendal nerve, the branches of the IPA include the inferior rectal artery, the perineal artery, and branches to the erectile tissues of the penis or clitoris (Drake et al. 2009).

The proximal branch of the inferior rectal artery arises within the pudendal canal as the IPA passes through that canal and crosses the ischio-anal fossa medially toward the anal canal, to branch and supply the inferior portion of the rectum and related skin. It supplies the anal canal inferior to the pectinate line, the internal and external anal sphincters, and perianal skin (Edozien et al. 2016). The inferior rectal artery anastomoses with the middle and superior rectal arteries, branches of the IIA, and the inferior mesenteric artery, respectively, to supply the entire rectum and anal canal superior to the pectinate line (Drake et al. 2009).

The perineal artery is a distal branch of the IPA, branching near the end of the pudendal canal as the vessel crosses over the superficial transverse perineal muscle (Drake et al. 2009). Upon exiting the pudendal canal, the artery travels anteriorly and gives rise to the posterior labial artery in females, which supplies the skin adjacent to the labia majora.

The terminal part of the IPA travels alongside the dorsal nerve of the penis or clitoris into the deep perineal pouch to supply the tissues of the deep perineal pouch and erectile tissue. It gives rise to the artery of the bulb of the penis (male) or vestibule (female), the urethral artery, a deep artery of the penis/clitoris, and the dorsal artery of the penis/clitoris. The artery of the bulb of the penis/vestibule of the vagina pierces the perineal membrane to supply the bulb of the penis in males, including the bulbourethral gland and bulbar urethra, and the bulb of the vestibule and the greater vestibular gland in females. The deep artery of the penis pierces the perineal membrane to supply the crura of the corpora cavernosa of the penis/clitoris, supplying most erectile tissue. The final branch, the dorsal artery of the penis/clitoris, pierces the perineal membrane and traverses the suspensory ligament of the penis/clitoris to run dorsally on the penis/clitoris to the glans. It supplies the deep perineal pouch, skin of the penis, fascia of penis/clitoris, and distal to the corpus spongiosum, including the spongy urethra and glans of penis/clitoris (Edozien et al. 2016).

64.3 Histology

The IPA is a distributing muscular artery as it distributes blood to the perineum and helps to regulate blood pressure by smooth muscle action in its tunica media, which comprises a thin subendothelial layer and a prominent internal elastic lamina. This is surrounded by a tunica media, which can contain up to 40 layers of larger smooth muscle cells and variable elastic lamellae. An external elastic lamina can be present if there is variant muscular tissue in the IPA. The outermost layer is the tunica adventitia, which comprises organized, dense, and irregular connective tissue (Junqueira et al. 2018).

64.4 Variations

Kawanishi et al. (2008) described five types of variation in the origin of the IPA:

Type 1. The IPA originates from the anterior trunk of the IIA at the level of the linea terminalis and the major ischial notch.
Type 2. The IPA originates from the anterior trunk at the level of the major ischial notch or distal to it.
Type 3. The IPA originates from the IIA more proximally than the level of the linea terminalis.
Type 4. The IPA branches at the same level as the inferior and superior gluteal arteries.
Type 5. The penile blood supply in males comes from other arteries such as the obturator artery.

Henry et al. described variations in supply to the tissues that the IPA usually provides. Type 1 is the normal supply to the external genitalia, rectum, and anus exclusively by the IPA. A type 2 variation involves blood being supplied by the IPA and the accessory pudendal artery, a branch of the inferior vesical artery. In a type 3 variation, blood is supplied by the (IPA) only. These variations were seen 61.9%, 32.8%, and 5.4% of the time, respectively (Henry et al. 2017).

Among the branches of the IPA, the inferior rectal artery shows four different branching patterns according to Fronhöfer et al.:

- Type 1. There is one inferior rectal artery running toward the anal canal on either side (43%) (Fronhöfer et al. 1996).
- Type 2. There are two inferior rectal arteries (31%) (Fronhöfer et al. 1996):
  - 2a. Both vessels branch from the IPA (21%).
  - 2b. The proximal branch comes from the IPA and the distal branch arises from the transverse perineal artery (10%).
- Type 3. Three inferior rectal arteries on one side (4%) (Fronhöfer et al. 1996).
- Type 4. Two or more arteries present on both sides (22%):
  - 4a. Both distal branches arise from the IPA (14%).
  - 4b. Both distal branches branch from the transverse perineal artery (8%).

64.5 Imaging

The imaging modalities used to evaluate and study the male pelvic arteries (e.g., the IPA) are CT angiography (CTA) and selective pelvic digital subtraction angiography (DSA) (Bilhim et al. 2014). Magnetic resonance (MR) angiography is another alternative; however, the lack of a bony landmark can limit the resolution when smaller vessels inside the pelvis are studied and DSA findings are compared (Bilhim et al. 2014). Depending on the clinical presentation and comorbidities, CTA is the gold standard for diagnosing cases such as arteriogenic erectile.

dysfunction (ED) (Pomposelli et al. 2010; Bilhim et al. 2014). The use of DSA for the male pelvic arteries can present challenges as many IIAs overlap, hindering the best projection for depicting each artery (Bilhim et al. 2014). However, an ipsilateral anterior oblique (steep

35–45 degrees) projection with caudal-cranial angulation (−10 degrees) can make it easier to identify the major branches of the IIA (Bilhim et al. 2014). For studying the IPA, intraarterial nitroglycerin (50–200 μg) with or without papaverine (30–100 mg) can be used to avoid spasms and improve identification (Bilhim et al. 2014).

64.6 Pathology

Multiple pathologies involve the IPA, including atherosclerosis. Atherosclerosis involving the IPA is common in patients with hypercholesteremia and/or diabetes mellitus. In patients with diabetes mellitus or coronary artery disease (CAD) there is an increased risk for accelerated atherosclerosis because of elevated blood glucose, oxidative stress, and increased inflammation. All these factors combine to increase the amount of atherosclerosis and stenosis. With increased stenosis, there is a high incidence of compromise of the pudendal artery and its branches (Poznyak et al. 2020).

64.7 Clinical and Surgical Significance

There is well-documented evidence that the pathologies described above can compromise blood flow, which eventually leads to ED. ED is, by definition, an inability to attain or maintain a penile erection sufficient for satisfactory sexual performance. The causes of the onset of ED can be categorized broadly into organic, psychogenic, or a mixture of both; organic etiologies, which include vasculogenic, neurogenic, hormonal, or drug-induced, make up 80% of all cases. Among these organic causes, vasculogenic accounts for most ED owing to abnormalities in arterial inflow or venous outflow (Ostfeld et al. 2021).

Park et al. showed that most patients with CAD have ED (88/91, 96.7%). In addition, the extent of stenosis in the IPA increased with age and was due to repeated exposures to risk factors that caused CAD, atherosclerosis, and stenosis. This decrease in blood flow led to decreased blood flow to the penis and, therefore, to ED (Park et al. 2019).

The IPA can be useful for reconstructive flaps in the perineal region. Owing to the nature of the perineal region, adequate vasculature is needed to supply the perineum. This artery has been used successfully as a flap following abdominoperineal excision of the rectum, pelvic exenteration, or excision of pilonidal disease—all procedures that result in large pelvic cavities and pelvic dead space. Flap reconstruction with the IPA is straightforward as it does not require Doppler visualization, has a reliable vasculature, and is relatively simple to perform. This reconstruction allows for a deep three-dimensional blood supply following perineal excisions, making the IPA an excellent option for these patients (Nassar et al. 2021).

In pudendal entrapment syndrome, the IPA and accompanying nerve most commonly become entrapped as they traverse the sacrospinous and sacrotuberous ligaments. This can also occur at the level of Alcock's canal and less commonly at the level below the piriformis muscle. Consequently, patients can present with urinary, sexual, and anorectal symptoms (Bollens et al. 2021).

64.8 Complications

Iatrogenic surgical injury is known to occur on occasion. Injuries can be repaired primarily or endovascularly with embolization. Surgeons need comprehensive knowledge of pelvic anatomy to avoid such complications.

References

Bilhim T, Pereira JA, Fernandes L, Tinto HR, Pisco JM (2014) Angiographic anatomy of the male pelvic arteries. Am J Roentgenol 203(4):W373–W382

Bollens R, Mjaess G, Sarkis J, Chemaly AK, Nemr E, Daher K et al (2021) Laparoscopic transperitoneal pudendal nerve and artery release for pudendal entrapment syndrome. Surg Endosc 35(11):6031–6038

Colebunders R, Njamnshi AK, Menon S, Newton CR, Hotterbeekx A, Preux P.-M, Hopkins A, Vaillant M, Siewe Fodjo JN (2021) Onchocerca volvulus and epilepsy: A comprehensive review using the Bradford Hill criteria for causation. PLoS Neglected Tropical Diseases, 15(1), e0008965. PubMed+1PubMed+1

DeSesso JM (2012) Comparative gestational milestones in vertebrate development. In: Developmental and reproductive toxicology: a practical approach. Informa Healthcare, London, pp 93–138

DeSesso JM (2017) Vascular ontogeny within selected thoracoabdominal organs and the limbs. Reprod Toxicol 70:3–20

Drake R, Vogl AW, Mitchell AW (2009) Gray's anatomy for students E-book. Elsevier Health Sciences

Edozien LC, O'Brien PM, Mackenzie IZ (2016) Uterine artery embolization versus surgery for symptomatic uterine fibroids. The Cochrane Database of Systematic Reviews, (12), CD005073.

Fronhöfer 1G, Niernberger T, Anderhuber F (1996) Variations in the branching of the internal pudendal artery in the ischioanal fossa. Ann Anat;178(2):133–5. PMID: 8638766. https://doi.org/10.1016/S0940-9602(96)80030-6.

Henry BM, Pękala PA, Vikse J, Sanna B, Skinningsrud B, Saganiak K, Walocha JA, Tomaszewski KA (2017) Variations in the Arterial Blood Supply to the Penis and the Accessory Pudendal Artery: A Meta-Analysis and Review of Implications in Radical Prostatectomy. Tomaszewski J Urol. 198:345–353. PMID: 28202357; https://doi.org/10.1016/j.juro.2017.01.080.

Junqueira LC, Carneiro J (2018) Basic Histology: Text and Atlas (14th ed.). McGraw-Hill Education.

Nassar MK, Jordan DJ, Quaba O (2021) The internal pudendal artery turnover (IPAT) flap: a new, simple and reliable technique for perineal reconstruction. J Plast Reconstr Aesthet Surg 74(9):2104–2109

Ostfeld RJ, Allen KE, Aspry K, Brandt EJ, Spitz A, Liberman J et al (2021) Vasculogenic erectile dysfunction: the impact of diet and lifestyle. Am J Med 134(3):310–316

Park H, Her S, Park B, Han D, Yuk SM, Kim D et al (2019) Correlation between internal pudendal artery stenosis and erectile dysfunction in patients with suspected coronary artery disease. PLoS One 14(11):e0225179

Pomposelli FB, Kansal N, Hamdan AD, Belfield A, Sheahan M (2010) A decade of experience with balloon-expandable stent grafts for the treatment of thoracic aortic disease. Journal of Vascular Surgery, 52(3),606–613.

Poznyak A, Grechko AV, Poggio P, Myasoedova VA, Alfieri V, Orekhov AN (2020) The diabetes mellitus–atherosclerosis connection: the role of lipid and glucose metabolism and chronic inflammation. Int J Mol Sci 21(5):1835

Sakthivelavan S, Sendiladibban SD, Felicia C (2010) Branching pattern of fetal internal iliac artery. patrón de ramificación de la arteria ilíaca interna fetal. Revista Argentina De AnatomíaClínica 2(3):100–105

Streeter GL (1919) A human embryo (mateer) of the presomite period. Carnegie Institution of Washington

Superior Vesical Artery 65

Justin David and Stephen J. Bordes, Jr.

65.1 Embryology

The umbilical cord forms during the fourth through eighth weeks of gestation (Spurway et al. 2012). Blood flow through the two umbilical arteries and the umbilical vein, which originate from mesodermal and ectodermal angiogenic cells, begins by the fifth week of gestation (Balcerzak et al. 2022). The two umbilical arteries and common iliac arteries arise from the dorsal aorta (Spurway et al. 2012; Balcerzak et al. 2022). Their proximal portions develop into the internal iliac and superior vesical arteries inside the pelvis, and their distal portions obliterate to form the medial umbilical ligaments coursing on the deep side of the anterior abdominal wall (Spurway et al. 2012).

Meanwhile, the cloaca splits into the urogenital sinus and the anal canal. The superior portion of the urogenital sinus becomes the urinary bladder, a continuation of the allantois, which resides in the umbilical cord along with the arteries and vein. The allantois obliterates to form the urachus, which connects the bladder to the umbilicus as the median umbilical ligament (Sadler 2019).

J. David
School of Medicine, Louisiana State University Health Sciences Center, New Orleans, LA, USA

S. J. Bordes, Jr. (✉)
Department of Surgery, Louisiana State University Health Sciences Center, New Orleans, LA, USA

65.2 Anatomy/Topography

At the level of the sacroiliac joint, the common iliac artery splits into the internal and external iliac arteries. The external iliac artery descends into the lower limb as the femoral artery, while the internal iliac artery remains in the pelvis; it then divides into posterior and anterior divisions. The posterior division gives off the iliolumbar, lateral sacral, and superior gluteal arteries. The most distal portion of the anterior division is the obliterated umbilical artery, which remains as the medial umbilical fold. Just proximally to it, the superior vesical artery arises, along with the obturator, middle rectal, inferior vesical, inferior gluteal, and internal pudendal arteries (Mamatha et al. 2015).

The superior vesical artery divides into several branches to supply the fundus of the bladder, the distal portion of the ureter, the seminal glands, and the ductus deferens (Standring 2020; Shermadou et al. 2021). The branch that supplies the bladder and seminal glands is sometimes referred to as the middle vesical artery. In males, the branch of the superior vesical artery that forms the artery of the ductus deferens runs through the spermatic cord with the ductus defer-

S. J. Bordes, Jr. et al. (eds.), *The Clinical Anatomy of the Vascular System*,
https://doi.org/10.1007/978-3-031-78326-5_65

ens until it anastomoses with the testicular artery (Standring 2020; Sjostrand 1965).

65.3 Imaging/Histology

The superior vesical artery is difficult to identify on computed tomography (CT), compared with most of the branches of the internal iliac artery. Occasionally, it can be seen lateral to the bladder. If it is injured, contrast can be seen extravasating into the bladder wall on contrast-enhanced CT (Yoon et al. 2004).

Magnetic resonance angiography (MRA) can detect most branches of the internal iliac artery. In a case series, MRA detected the superior and inferior gluteal arteries and the lateral sacral, iliolumbar, obturator, internal pudendal, and uterine arteries in more than 83% of patients, but the superior vesical artery in only 21% (Naguib et al. 2008).

65.4 Variations

The superior vesical artery most commonly arises from the umbilical artery, but occasionally from the uterine artery in females, the artery to the ductus deferens in males, or the obturator artery. There is typically, but not always, an anastomosis between the superior and inferior vesical arteries. It is rarely seen in common with the middle rectal artery or the uterine artery prior to its descent superior to the ureter (Selçuk et al. 2018).

The number of superior vesical arteries can vary. One to five arteries have been visualized. Two is the most common number, occurring in more than 70% of cases. One superior vesical artery is about as frequent as three. Four and five arteries are the most infrequent variations.

65.5 Clinical and Surgical Significances

Recurrent gross hematuria is a common complication in patients with a bladder cancer that is not suitable for curative treatment. Historically, transurethral flushes of hemostatic radiotherapy, urinary diversions, or palliative cystectomy have been performed in these cases. A more recent, less invasive, approach is percutaneous arterial embolization of the anterior division of the internal iliac or superior vesical artery. This allows the desired vessel to be closed selectively to prevent recurrent hematuria (Saadi et al. 2017). Embolization of the superior vesical artery has been shown to be safe and effective for controlling recurrent bladder hemorrhages, but it is best used for severe cases with poor surgical candidacy (Cho et al. 2008). In addition to recurrent gross hematuria in the setting of bladder cancer, the superior vesical artery can be selectively embolized in cases of hemorrhage after pelvic surgery, including radical prostatectomy (Bonne et al. 2017).

The superior vesical artery is relevant in pelvic surgery. During a cystoprostatectomy, it is typically used as the lateral pedicle, while the inferior vesical artery is used as the posterior pedicle. It is found by bluntly dissecting in the Denonvilliers space posterior to the bladder and anterior to the rectum. Once the artery is identified, it is ligated and divided bilaterally.

In addition to cystoprostatectomy, it can be important to identify the superior vesical artery during prostatic artery embolization, a procedure for treating symptomatic benign prostatic hypertrophy. The origin of the prostatic artery can vary, but it sometimes arises from the superior vesical artery. Identification of these arteries is important when the prostatic artery is to be embolized (Bilhim et al. 2014).

Because the superior vesical artery supplies the distal ureter and fundus of the bladder, care is needed to preserve it during ureteral reconstruction. A method for repairing long ureteral defects is to use a spiral bladder muscle flap. During this procedure, the superior vesical arteries are dissected, and a cut is made on the bladder along the route of the artery. The bladder flap is then brought up, and a psoas hitch is performed to reduce anastomotic tension. Maintenance of the superior vesical artery as the vascular pedicle helps maintain blood flow to the distal ureter and bladder muscle flap (Li et al. 2014).

65.6 Complications

Complications involving the superior vesical artery typically result from trauma or surgery. Owing to the small size and location of this artery, isolated injuries to it from trauma are rare. A case report described pelvic hematomas due to bleeding from the superior vesical artery after a low-energy impact had caused a minimally displaced right superior pubic ramus fracture. Pubic rami injuries rarely affect the branches of the internal iliac artery, but CT imaging and angiography is recommended for pelvic fracture patients who develop hypotension (Wee et al. 2013).

Blood supply from the superior vesical artery is frequently disrupted during distal ureteral injury, whether from trauma or iatrogenic injury during pelvic surgery. These injuries often require ureteroneocystostomy, the reinsertion of the ureter into the bladder. Besides injury to the ureter, other indications for this procedure include vesicoureteral reflux in children, distal ureteral cancers, other pelvic malignancies that involve the ureter, and renal transplantation.

References

Balcerzak A, Hajdys J, Tubbs RS, Karauda P, Georgiev GP, Olewnik Ł (2022) Clinical importance of variability in the branching pattern of the internal iliac artery – an updated and comprehensive review with a new classification proposal. Ann Anat 239:151837. https://doi.org/10.1016/j.aanat.2021.151837

Bilhim T, Pereira JA, Fernandes L, Rio Tinto H, Pisco JM (2014) Angiographic anatomy of the male pelvic arteries. AJR Am J Roentgenol 203(4):W373–W382. https://doi.org/10.2214/AJR.13.11687

Bonne L, Gillardin P, De Wever L, Vanhoutte E, Joniau S, Oyen R, Maleux G (2017) Endovascular management of severe arterial haemorrhage after radical prostatectomy: a case series. Cardiovasc Intervent Radiol 40(11):1698–1705. https://doi.org/10.1007/s00270-017-1715-0. Epub 2017 Jun 7

Cho CL, Lai MH, So HS, Kwok KK, Chan JC, Velayudhan V (2008 Dec) Superselective embolisation of bilateral superior vesical arteries for management of haemorrhagic cystitis. Hong Kong Med J 14(6):485–488

Li Y, Li C, Yang S, Song C, Liao W, Xiong Y (2014 May) Reconstructing full-length ureteral defects using a spiral bladder muscle flap with vascular pedicles. Urology 83(5):1199–1204. https://doi.org/10.1016/j.urology.2014.01.027

Mamatha H, Hemalatha B, Vinodini P, Souza AS, Suhani S (2015) Anatomical study on the variations in the branching pattern of internal iliac artery. Indian J Surg 77(Suppl 2):248–252. https://doi.org/10.1007/s12262-012-0785-0

Naguib NN, Nour-Eldin NE, Hammerstingl RM, Lehnert T, Floeter J, Zangos S, Vogl TJ (2008) Three-dimensional reconstructed contrast-enhanced MR angiography for internal iliac artery branch visualization before uterine artery embolization. J Vasc Interv Radiol 19(11):1569–1575. https://doi.org/10.1016/j.jvir.2008.08.012

Saadi A, Bouzouita A, Rebai MH et al (2017) Superselective embolisation of bilateral superior vesical arteries for management of intractable hematuria in context of metastatic bladder cancer. Asian J Urol 4(2):131–134. https://doi.org/10.1016/j.ajur.2016.03.003

Sadler TW (2019) Langman's medical embryology. Wolters Kluwer

Selçuk İ, Yassa M, Tatar İ, Huri E (2018) Anatomic structure of the internal iliac artery and its educative dissection for peripartum and pelvic hemorrhage. Turk J Obstet Gynecol 15(2):126–129. https://doi.org/10.4274/tjod.23245. Epub 2018 Jun 21

Shermadou ES, Rahman S, Leslie SW (2021) Anatomy, abdomen and pelvis, bladder. [Updated 2021 Jul 31]. In: StatPearls [Internet]. StatPearls Publishing, Treasure Island. 2021 Jan-

Sjostrand NO (1965) The adrenergic innervation of the vas deferens and the accessory male genital glands. Acta Physiol Scand 65:7–82

Spurway J, Logan P, Pak S (2012) The development, structure and blood flow within the umbilical cord with particular reference to the venous system. Australas J Ultrasound Med 15(3):97–102. https://doi.org/10.1002/j.2205-0140.2012.tb00013.x

Standring S (2020) Gray's anatomy: the anatomical basis of clinical anatomy, 42nd edn. Elsevier

Wee J, Lua W, Louange DT (2013) Massive haemorrhage from the internal iliac artery following a low energy superior pubic ramus fracture in a 73-year-old man. J Clin Orthop Trauma 4(2):98–101. https://doi.org/10.1016/j.jcot.2013.03.004

Yoon W, Kim JK, Jeong YY, Seo JJ, Park JG, Kang HK (2004) Pelvic arterial hemorrhage in patients with pelvic fractures: detection with contrast-enhanced CT. Radiographics 24(6):1591–1605.; discussion 1605–1606. https://doi.org/10.1148/rg.246045028

Inferior Vesical/Vaginal Artery

66

Andrea Yazbeck

66.1 Embryology

In the middle of the third week of embryonic development, the primordial heart and vascular system begin to develop (Ozgüner and Sulak 2011). The arterial system arises from both ectoderm- and mesoderm-derived angiogenic cells. The aortic arches and right and left dorsal aortae unite during development to form the descending aorta, generating posterolateral intersegmental branches throughout the thorax (Ozgüner and Sulak 2011). As the dorsal aorta descends through the abdomen, branches to the gastrointestinal and renal systems are formed, including the intersegmental arteries (Ozgüner and Sulak 2011). The umbilical arteries arise from the dorsal aorta and disconnect from it after anastomosing with the intersegmental arteries. The common iliac arteries then arise from the proximal portions of the patent umbilical arteries (Szpinda et al. 2012). The external iliac artery arises from the common iliac arteries (Bergman et al. 2016). The fifth intersegmental artery (fifth dorsal branch of the aorta) anastomoses with the umbilical artery to produce the internal iliac artery (Bergman et al. 2016), which develops anterior and posterior divisions that subsequently give off further branches to the pelvic area. This region is initially vascularized through a capillary plexus (Bergman et al. 2016), which progresses into an arterial pattern thanks to the persistence of necessary vascular channels and the regression of others, forming the branches of the internal iliac artery; these include the inferior vesical/vaginal artery (Bergman et al. 2016). The variability in branching often seen within the iliac vessels can be attributed to genetics and the vascular demands of subsequent tissues (Balcerzak et al. 2022).

66.2 Anatomy/Topography

The abdominal aorta is a retroperitoneal vessel that is superficial to the posterior abdominal wall and anterior to the vertebral column (Selçuk et al. 2018). As it descends through the abdominal cavity toward the pelvis and lower extremities, it bifurcates into the right and left common iliac arteries (RCIA and LCIA, respectively). Typically, this bifurcation is located approximately at the level of the fourth or fifth lumbar vertebrae, although there are variations (Bergman et al. 2016). At the level of the intervertebral disc between the L5 and S1 vertebrae, the common iliac arteries subsequently bifurcate into the internal and external iliac arteries (IIA and EIA).

A. Yazbeck (✉)
Department of Anatomical Sciences,
Tulane University School of Medicine,
New Orleans, LA, USA
e-mail: ayazbeck@tulane.edu

S. J. Bordes, Jr. et al. (eds.), *The Clinical Anatomy of the Vascular System*,
https://doi.org/10.1007/978-3-031-78326-5_66

The external iliac artery continues inferiorly and enters the lower limb. As it passes the inguinal ligament it becomes the femoral artery. The internal iliac artery continues to the margin of the greater sciatic foramen. The origin of the bifurcation of the internal iliac artery can be considered as approximately 4–4.5 cm from either the sacral promontory or the common iliac bifurcation. This can also be considered as 3.5 cm from the pelvic midline (Bergman et al. 2016). At this point, the internal iliac artery forms anterior and posterior divisions. The anterior division continues toward the ischial spine. The posterior trunk travels toward the greater sciatic foramen. The anterior division gives off the inferior vesical (male) or vaginal (female) artery, along with several other branches. The branching pattern of the internal iliac artery is significantly variable; however, the branches are typically divided as listed below.

Anterior Division of IIA:

- Obturator A.
- Middle Rectal A.
- Umbilical A.
- Uterine A. (female).
- *Inferior Vesical A. (male).*
- *Vaginal A. (female).*
- Inferior Gluteal A.
- Internal Pudendal A.

Posterior Division of IIA:

- Iliolumbar A.
- Lateral Sacral A.
- Superior Gluteal A.

The inferior vesical artery (IVA) is most commonly found in males, although some studies have shown that it can also arise in females (discussed in more detail in the "variations" section) (de Treigny et al. 2017). The origin of the IVA is significantly variable. In one study, researchers studied 66 cadaveric pelvic halves (Bergman et al. 2016) and found that in most cases, the inferior and middle vesicle arteries were derived from the pudendal, inferior gluteal, or middle rectal arteries (listed in order of frequency) (Bergman et al. 2016). In males, this artery typically supplies blood to the lower parts of the bladder, bladder neck, ureter, seminal glands, and prostate (de Treigny et al. 2017).

The vaginal artery (VA) is widely considered the female homologue of the inferior vesical artery in males (Vasantha Lakshmi et al. 2019). It typically arises superior and posterior to the origin of the uterine artery (which also arises from the anterior division of the IIA), although there are variations (Vasantha Lakshmi et al. 2019). The VA descends inferiorly to supply the fundus of the urinary bladder, the upper portion of the vagina, and the adjacent portion of the rectum (Vasantha Lakshmi et al. 2019). It also has branches that anastomose the uterine artery and help to supply the cervix as well as support the vasculature of the upper portion of the vagina (Vasantha Lakshmi et al. 2019).

66.3 Imaging

The internal iliac arteries and subsequent branches, including the inferior vesical artery/vaginal artery, can be visualized through several imaging modalities. Imaging is essential for finding appropriate diagnoses, providing details of pelvic vasculature for pre-procedural surgical planning, and monitoring patient progress through follow-up imaging. It is important to note that variability in branching can make the imaging more difficult to analyze and interpret (Balcerzak et al. 2022).

Computed tomography (CT) angiography imaging enables the pelvic vasculature to be visualized in greater detail (Baliyan et al. 2019). In most cases, iodinated contrast material is administered intravenously to enhance the imaging quality (Baliyan et al. 2019). Magnetic resonance (MR) imaging and MR angiography are appropriate alternatives to CT angiography with respect to the quality of imaging and anatomical detail (Pavone et al. 1990).

In one study, MR angiography, CT angiography, and digital angiography images from 21 males (mean age 73.2 years) were evaluated to define the anatomy of the internal iliac artery before selective embolization of the pelvic arteries (Bilhim et al. 2011). The final results showed that CT angiography provided more detailed anatomy than MR angiography, and digital angiography was considered the "gold standard." (Bilhim et al. 2011).

66.4 Variation

Inferior Vesical Artery

As stated previously, the inferior vesical artery is commonly seen in the male population (de Treigny et al. 2017). However, some studies indicate that it can be found in the female population as well (de Treigny et al. 2017). In one study, 50 topographies of bladder vascularization were visualized (de Treigny et al. 2017). The IVA was observed in 92% of the male and 47.4% of the female subjects (de Treigny et al. 2017). This finding is significant as these variations must be considered to prevent unintentional bladder devascularization (de Treigny et al. 2017).

The IVA is significantly variable in origin, as is the surrounding vasculature in this region. In a literature review, the IVA commonly arises as a single branch with the middle rectal artery, from a common trunk of the internal pudendal and superior gluteal arteries, or as a branch of the internal pudendal artery (Bergman et al. 2016). One annual research report (Parsons and Keith, 1897) included 58 observations of the inferior vesical artery (Bergman et al. 2016; Parsons and Keith 1896). In 68.9% of cases, the IVA arose independently from the hypogastric trunk, in 22.4% from the anterior division of the internal iliac artery, in 3.5% from the internal iliac artery, in 3.5% with the middle rectal artery, and in 1.7% with the superior vesical artery (Bergman et al. 2016; Parsons and Keith 1896). For clarification, the term hypogastric trunk has been used to denote a trunk that arises from the internal iliac artery and gives rise to subsequent branches, including the vaginal, middle rectal, and superior/inferior vesical arteries (Bergman et al. 2016). The researchers also noted that if the IVA arose from the middle rectal artery, a duplicate IVA was present (Bergman et al. 2016; Parsons and Keith 1896).

In a study of 150 cadaveric specimens, Shafiroff et al. reported the points of origin of all branches of the iliac system (Bergman et al. 2016; Shafiroff et al. 1959). The inferior vesical artery was found in 78% of specimens (Bergman et al. 2016; Shafiroff et al. 1959). In more than 10%, it originated from the anterior division of the IIA, the umbilical artery, or the vaginal artery (Bergman et al. 2016; Shafiroff et al. 1959). In fewer than 10%, it originated from the vesicodeferential artery or the internal pudendal artery (Bergman et al. 2016; Shafiroff et al. 1959).

In a study of the hypogastric artery and its branches, Lipshutz (1918) reported that the IVA arose in 34 subjects as a separate branch from the prostatic artery and arose three times as a branch of the vaginal, eight times as a branch of the umbilical artery distal to the uterine artery, nine times as a branch of the vesicodeferential artery, and five times as a branch of the internal pudendal artery" (Bergman et al. 2016; Lipshutz 1918).

Some studies have even reported the IVA originating from the obturator artery, which is typically derived from the posterior branch of the IIA (Bergman et al. 2016). Anastomoses between the IVA and neighboring arteries are also common anastomoses with the uterine arteries were found in 60–70% of subjects (Bergman et al. 2016). In some cases, the IVA is absent; the regional vascularization in such cases is typically supplied by the superior vesical artery (Bergman et al. 2016; Badagabettu et al. 2014).

The IVA is typically 0.09 ± 0.05 cm in diameter and 0.35 to 2.28 cm in length (Dzmitry and Aksana 2019). There is limited reporting of possible differences in length and diameter between the right and left sides.

Vaginal Artery

The vaginal artery (VA) is often considered the female homologue of the IVA in males (Vasantha Lakshmi et al. 2019). It typically originates from the uterine artery and can present as a single branch or multiple branches (Bergman et al. 2016). It has also been seen to originate from the internal iliac, middle rectal, superior vesical, inferior vesical, inferior gluteal, or obturator artery, the ischiopudendal trunk, or via a common trunk with the uterine artery (Bergman et al. 2016). Lipshutz (1918) reported that 40% of VAs are duplicated (Bergman et al. 2016; Lipshutz 1918). In these instances, the proximal branch originated from the anterior division of the internal iliac, uterine, or internal pudendal artery and the more distal branch from the internal pudendal, obturator, or inferior gluteal artery, the ischiopudendal trunk, or as a common trunk with the middle rectal artery (Bergman et al. 2016; Lipshutz 1918).

In a study of 150 cadaveric specimens, Shafiroff et al. reported the points of origin for all branches of the iliac systems (Bergman et al. 2016; Shafiroff et al. 1959). They found the VA to originate from the uterine artery, anterior division of IIA, ischiopudendal trunk, or inferior gluteal artery in more than 10% of cases (Bergman et al. 2016; Shafiroff et al. 1959). Less than 10% of the time, the VA originated from the obturator or internal pudendal artery (Bergman et al. 2016; Shafiroff et al. 1959).

In another cadaveric study, the VA was reported to be more variable than the uterine artery in its site of origin and size (Bergman et al. 2016; Roberts and Krishingner 1967). In one study that included 197 patients and extensive angiogram imaging (approximately 375 angiograms), the VA arose below the uterine artery in 50% of cases, specifically as a single branch of the anterior division of the IIA (Bergman et al. 2016; Pelage et al. 1999). In this same study, 9% had a VA that shared a common trunk with the uterine artery (Bergman et al. 2016; Pelage et al. 1999).

The average length of the VA is approximately 6.57 ± 0.15 cm (range: 6.39–6.87 cm) on the left side and 6.76 ± 0.18 cm (range: 6.44–6.76 cm) on the right (no significant difference) (Vasantha Lakshmi et al. 2019). The diameter of the VA was 1.33 ± 0.54 mm (range: 0.76–2.13 mm) on the left side and 1.48 ± 0.57 mm (range: 0.82–2.28 mm) on the right (Vasantha Lakshmi et al. 2019).

66.5 Pathology, Clinical, and Surgical Significance Complications

Adequate preoperative and interopertative reviewing of the branching variations in each specific patient is crucial for physicians and healthcare professionals to reduce unnecessary complications (Balcerzak et al. 2022).

Internal iliac artery aneurysms (IIAA) are rare, asymptomatic anomalies typically found during imaging examinations for other conditions (Machado et al. 2016). IIAAs account for approximately 0.3% of all aortoiliac aneurysms (Machado et al. 2016). Surgical or endovascular repair of them is indicated under certain circumstances, including but not limited to rupture, symptomatic, or rapidly expanding aneurysms. Endovascular repair is typically preferred over open surgical repair because perioperative morbidity is significantly lower, length of stay is shorter, and operative time and blood loss are decreased (Perini et al. 2021). Open operative treatment can include endoaneurysmorrhaphy with graft interposition (opening the aneurysm sac and placing a graft), aneurysm resection, or proximal and distal internal iliac artery ligation (Dix et al. 2005; Nakajima et al. 2001).

Pelvic fractures or blunt trauma to the pelvic region can lead to moderate or severe hemorrhaging. In most cases, severe bleeding is venous in origin (Wijffels et al. 2019). However, angioembolization can be considered to control pelvic arterial hemorrhage, typically in combination with open surgical exploration and pelvic packing (Vaidya et al. 2018; Gourlay et al. 2005). In one study, pelvic angiograms were collected from 556 patients in order to assess the presentation of pelvic arterial hemorrhage associated with pelvic fractures (Gourlay et al. 2005). Among all the patients studied, 42 (7.5%) underwent a second angiogram for suspected recurrent hemorrhage (Gourlay et al. 2005). In the repeat angiograms, the origin of bleeding was a new bleeding site in 68% (compared to the initial angiogram), at a previously embolized site in 18%, and both in 14% (Gourlay et al. 2005). Selective arterial embolization can minimize tissue ischemia or necrosis in the pelvic region (Biffl et al. 2001; Rossaint et al. 2007).

Uterine fibroids (leiomyomas) or benign tumors of the uterus are clinically apparent in approximately 25% of females (Stewart 2001). While some myomas cause no symptoms, many women have significant symptoms caused by uterine fibroids, including significantly prolonged or heavy menstrual bleeding (menorrhagia or hypermenorrhea), pelvic pressure or pain, and, in some cases, reproductive dysfunction (Stewart 2001). As this is a common gynecological condition, there are several options for treatment including but not limited to hormonal therapy, hysteroscopic or abdominal myomectomy, and hysterectomy (Stewart 2001; Van der Kooij and Hehenkamp 2022). Uterine artery embolization is a minimally invasive technique for treating myomas based on the hypothesis that control of the arterial blood flow will decrease the size of the fibroids and the symptoms associated with them (de Bruijn et al. 2016; Kim 2019). As the VA is often a branch of the uterine artery, it can be affected during this procedure. In a hysterectomy, the uterine vessels can be ligated to establish hemostasis and operation duration (Hebisch 2002; Sinha et al. 2008).

Postpartum hemorrhage (PPH) is a common and potentially fatal complication of childbirth. PPH is characterized by excessive blood loss during delivery. Per the American College of Obstetrics and Gynecology in 2017, the current definition of postpartum hemorrhage is cumulative blood loss over 1000 mL with signs and symptoms of hypovolemia within 24 h of delivery. This characterization is considered regardless of route of delivery (vaginal, cesarian, etc.) (Committee on Practice Bulletins-Obstetrics 2017). PPH can be defined as "primary" if the bleeding is within the first 24 h after delivery and "secondary" when it occurs 24 h to 12 weeks postpartum (Dossou et al. 2015). Treatment of this condition is typically cause-specific and focuses on maintaining hemodynamic stability (Maswime and Buchmann 2017). In some cases, internal iliac ligation or uterine artery embolization is necessary (Singhal et al. 2015).

References

Badagabettu S, Kumar N, Shetty S, Sirasanagandla S (2014) Variant branching pattern of the right internal iliac vessels in a male – a case report. Arch Clin Exp Surg ACES 3(3):197

Balcerzak A, Hajdys J, Tubbs RS, Karauda P, Georgiev GP, Olewnik Ł (2022) Clinical importance of variability in the branching pattern of the internal iliac artery – an updated and comprehensive review with a new classification proposal. Ann Anat 239:151837

Baliyan V, Shaqdan K, Hedgire S, Ghoshhajra B (2019) Vascular computed tomography angiography technique and indications. Cardiovasc Diagn Ther 9(Suppl 1):S14–S27

Bergman RA, Tubbs RS, Shoja MM, Loukas M (eds) (2016) Bergman's comprehensive encyclopedia of human anatomic variation. Wiley

Biffl WL, Smith WR, Moore EE et al (2001) Evolution of a multidisciplinary clinical pathway for the management of unstable patients with pelvic fractures. Ann Surg 233(6):843–850

Bilhim T, Casal D, Furtado A, Pais D, O'Neill JEG, Pisco JM (2011) Branching patterns of the male internal iliac artery: imaging findings. Surg Radiol Anat SRA. 33(2):151–159

Committee on Practice Bulletins-Obstetrics (2017) Practice bulletin no. 183: postpartum hemorrhage. Obstet Gynecol 130(4):e168–e186

de Bruijn AM, Ankum WM, Reekers JA et al (2016) Uterine artery embolization vs hysterectomy in the treatment of symptomatic uterine fibroids: 10-year outcomes from the randomized EMMY trial. Am J Obstet Gynecol 215(6):745.e1–745.e12

de Treigny OM, Roumiguie M, Deudon R et al (2017) Anatomical study of the inferior vesical artery: is it specific to the male sex? Surg Radiol Anat 39(9):961–965

Dix FP, Titi M, Al-Khaffaf H (2005) The isolated internal iliac artery aneurysm–a review. Eur J Vasc Endovasc Surg Off J Eur Soc Vasc Surg 30(2):119–129

Dossou M, Debost-Legrand A, Déchelotte P, Lémery D, Vendittelli F (2015) Severe secondary postpartum hemorrhage: a historical cohort. Birth Berkeley Calif 42(2):149–155

Dzmitry V, Aksana V (2019) Anatomy of the visceral branches of the iliac arteries in newborns. MOJ Anat Physiol. 6(2):68–72

Gourlay D, Hoffer E, Routt M, Bulger E (2005) Pelvic angiography for recurrent traumatic pelvic arterial hemorrhage. J Trauma Inj Infect Crit Care 59(5):1168–1173. Published online November 2005:1168–1174

Hebisch G (2002) Vaginal uterine artery ligation avoids high blood loss and puerperal hysterectomy in postpartum hemorrhage. Obstet Gynecol 100(3):574–578

Kim MD (2019) Uterine artery embolization for Leiomyomas and Adenomyosis: a pictorial essay based on our experience from 1300 cases. Korean J Radiol 20(10):1462. https://doi.org/10.3348/kjr.2019.0205

Lipshutz B (1918) A composite study of the hypogastric artery and its branches. Ann Surg 67(5):584–608
Machado RM, Rego DNC, de Oliveira PNFP, de Almeida RMGF (2016) Endovascular treatment of internal iliac artery aneurysms – single center experience. Braz J Cardiovasc Surg. 31(2):127–131. Published online 2016
Maswime S, Buchmann E (2017) A systematic review of maternal near miss and mortality due to postpartum hemorrhage. Int J Gynaecol Obstet Off Organ Int Fed Gynaecol Obstet 137(1):1–7
Nakajima T, Kawazoe K, Komoda K et al (2001) Failure of exclusion of internal iliac artery aneurysms. J Vasc Surg 33(3):476–480
Ozgüner G, Sulak O (2011) Development of the abdominal aorta and iliac arteries during the fetal period: a morphometric study. Surg Radiol Anat SRA 33(1):35–43
Parsons FG, Keith A (1896) Sixth annual report of the Committee of Collective Investigation of the Anatomical Society of Great Britain and Ireland, 1895–96. J Anat Physiol 31(Pt 1):31–44
Pavone P, Di Cesare E, Di Renzi P et al (1990) Abdominal aortic aneurysm evaluation: comparison of US, CT, MRI, and angiography. Magn Reson Imaging 8(3):199–204
Pelage JP, Le Dref O, Soyer P et al (1999) Arterial anatomy of the female genital tract: variations and relevance to transcatheter embolization of the uterus. Am J Roentgenol 172(4):989–994
Perini P, Mariani E, Fanelli M et al (2021) Surgical and endovascular Management of Isolated Internal Iliac Artery Aneurysms: a systematic review and meta-analysis. Vasc Endovasc Surg 55(3):254–264
Roberts WH, Krishingner GL (1967) Comparative study of human internal iliac artery based on adachi classification. Anat Rec 158(2):191–196
Rossaint R, Duranteau J, Stahel PF, Spahn DR (2007) Nonsurgical treatment of major bleeding. Anesthesiol Clin 25(1):35–48
Selçuk İ, Ersak B, Tatar İ, Güngör T, Huri E (2018) Basic clinical retroperitoneal anatomy for pelvic surgeons. Turk J Obstet Gynecol 15(4):259–269
Shafiroff BGP, Grillo EB, Baron H (1959) Bilateral ligation of the hypogastric arteries. Am J Surg 98(1):34–40
Singhal M, Gupta P, Sikka P, Khandelwal N (2015) Uterine artery embolization following internal iliac arteries ligation in a case of post-partum hemorrhage: a technical challenge. J Obstet Gynecol India 65(3):202–205
Sinha R, Sundaram M, Nikam YA, Hegde A, Mahajan C (2008) Total laparoscopic hysterectomy with earlier uterine artery ligation. J Minim Invasive Gynecol 15(3):355–359
Stewart EA (2001) Uterine fibroids. Lancet 357(9252):293–298
Szpinda M, Szpinda A, WoŸniak A, Daroszewski M, Mila-Kierzenkowska C (2012) The normal growth of the common iliac arteries in human fetuses – an anatomical, digital and statistical study. Med Sci Monit Int Med J Exp Clin Res 18(3):BR109–BR116
Vaidya R, Waldron J, Scott A, Nasr K (2018) Angiography and embolization in the management of bleeding pelvic fractures. J Am Acad Orthop Surg 26(4):e68–e76
Van der Kooij SM, Hehenkamp WJK (2022) Uterine fibroids (leiomyomas): treatment with uterine artery embolization. Published online March 17, 2022. https://www.uptodate.com/contents/uterine-fibroids-leiomyomas-treatment-with-uterine-artery-embolization
Vasantha Lakshmi A, Mustafa VS, Venugopala RB, Nirmala P, Joydeep DC (2019) A detailed study of the course, branching pattern and dimensions of the uterine and vaginal arteries and their branches. MOJ Anat Physiol 6(6):218–224
Wijffels DJ, Verbeek DO, Ponsen KJ, Carel Goslings J, van Delden OM (2019) Imaging and endovascular treatment of bleeding pelvic fractures: review article. Cardiovasc Intervent Radiol 42(1):10–18

67 Uterine Artery

Andrea Yazbeck

67.1 Embryology

The primordial heart and vascular system begin to develop in the middle of the third week of embryo development (Ozgüner and Sulak 2011). The arterial system arises from both ectoderm- and mesoderm-derived angiogenic cells. The aortic arches and right and left dorsal aortae unite to form the descending aorta, forming posterolateral intersegmental branches throughout the thorax (Ozgüner and Sulak 2011). As the dorsal aorta descends through the abdomen, branches are formed to the gastrointestinal and renal systems, including the intersegmental arteries (Ozgüner and Sulak 2011). The umbilical arteries also arise from the dorsal aorta but disconnect from it after anastomosing with the intersegmental arteries. The common iliac arteries then arise from the proximal portions of the patent umbilical arteries (Szpinda et al. 2012), and give rise in turn to the external iliac artery (Bergman et al. 2016). The fifth intersegmental artery (fifth dorsal branch of the aorta) anastomoses with the umbilical artery to develop the internal iliac artery (Bergman et al. 2016), which develops anterior and posterior divisions that subsequently give off further branches to the pelvic area. The vasculature of this region is initially formed through a capillary plexus (Bergman et al. 2016), which progresses into an arterial pattern thanks to the persistence of necessary vascular channels and the regression of other channels, forming the branches of the internal iliac artery, including the uterine artery (Bergman et al. 2016). The frequent variability in branching seen within the iliac vessels can be attributed to genetics and the vascular demands of tissues during later stages of development (Balcerzak et al. 2022).

67.2 Anatomy/Topography

The abdominal aorta is a retroperitoneal vessel lying superficial to the posterior abdominal wall and anterior to the vertebral column (Selçuk et al. 2018). As it descends through the abdominal cavity toward the pelvis and lower limbs, it bifurcates into the right and left common iliac arteries (RCIA and LCIA, respectively). Typically, this bifurcation is at the level of the fourth or fifth lumbar vertebra, although there are variations (Bergman et al. 2016). At the level of the intervertebral disc between L5 and S1, the common iliac arteries bifurcate into internal and external iliac arteries (IIA and EIA).

The EIA continues inferiorly and enters the lower limb. As it passes the inguinal ligament it

A. Yazbeck (✉)
Department of Anatomical Sciences,
Tulane University School of Medicine,
New Orleans, LA, USA
e-mail: ayazbeck@tulane.edu

S. J. Bordes, Jr. et al. (eds.), *The Clinical Anatomy of the Vascular System*,
https://doi.org/10.1007/978-3-031-78326-5_67

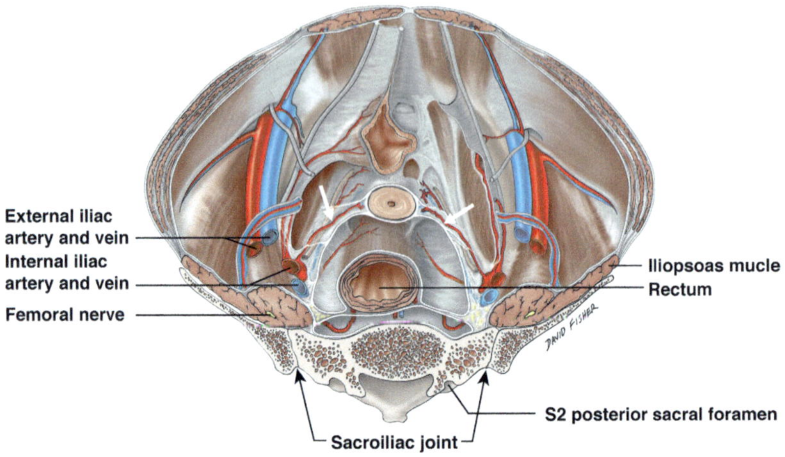

Fig. 67.1 Schematic showing the pelvis cut at the level of the cervix of the uterus, which is supplied by the uterine arteries (arrows). (After Netters, 2023)

becomes the femoral artery. The IIA continues to the margin of the greater sciatic foramen. According to authors cited by Bergman, its bifurcation originates approximately 4–4.5 cm from either the sacral promontory or the common iliac bifurcation. This can also be considered as 3.5 cm from the pelvic midline (Bergman et al. 2016). At this point, the IIA forms anterior and posterior divisions. The anterior division continues toward the ischial spine. The posterior trunk passes toward the greater sciatic foramen. The anterior division gives off the uterine artery (Fig. 67.1) and several other branches. The branching pattern of the IIA is significantly variable, but the standard divisions are:

Anterior Division of IIA:

- Obturator A.
- Middle Rectal A.
- Umbilical A.
- Uterine A. (female).
- Inferior Vesical A. (male).
- Vaginal A. (female).
- Inferior Gluteal A.
- Internal Pudendal A.

Posterior Division of IIA:

- Iliolumbar A.
- Lateral Sacral A.
- Superior Gluteal A.

The uterine artery (UA) is specific to females. Its exact point of origin from the IIA varies, as will be discussed in the "variations" section (Bergman et al. 2016). It travels medially toward the uterus, coursing through the broad ligament of the uterus, also referred to as the "cardinal ligament." At the level of the ischial spine, it passes anterior to the ureter, which descends from the kidneys to the bladder. This relationship is clinically important and is often described by the phrase "water (ureter) under the bridge (uterine artery)." The UA then divides into ascending and descending branches, typically at the level of the uterine isthmus.

The ascending (marginal) branch travels superiorly, running alongside the lateral boundary of the uterus, eventually creating an anastomoses with the ovarian artery, which arises directly from the abdominal aorta. The descending (cervicovaginal) branch runs inferiorly toward the cervix and vagina (Moore et al. 2014). The UA gives off subsequent branches within the myometrium of the uterus including the arcuate, radial, spiral, and basal arteries. The arcuate arteries supply blood to the walls of the uterus and terminate as spiral arteries. Spiral arteries are functionally critical as they supply the endometrium during menstruation and the decidua and placenta during pregnancy. Basal arteries assist in regenerating the basal zone of the endometrium throughout the menstrual cycle.

The UA is critical for supplying and maintaining blood to the uterus throughout all physiological processes. Developmentally, it is often considered the homolog of the artery supplying the ductus deferens in males (Moore et al. 2014).

67.3 Imaging

The IIA and its subsequent branches, including the UA, can be visualized by several imaging modalities. Imaging is essential for confirming a diagnosis, providing details of the pelvic vasculature for pre-procedural surgical planning, and monitoring patient progress through follow-up.

Computed tomography (CT) angiography imaging enables the pelvic vasculature to be visualized in greater detail (Baliyan et al. 2019). In most cases, iodinated contrast material is administered intravenously to enhance the quality of imaging (Baliyan et al. 2019). Magnetic resonance (MR) imaging and MR angiography are appropriate alternatives to CT angiography in respect of the quality of imaging and anatomical detail (Pavone et al. 1990).

In one study, 21 males (mean age 73.2 years) were evaluated by MR angiography, CT angiography, and digital angiography imaging to define the IIA anatomy before selective embolization of the pelvic arteries (Bilhim et al. 2011). CT angiography showed more anatomical detail than MR angiography but digital angiography was considered the gold standard (Bilhim et al. 2011).

In some pregnancy-related cases such as fetal growth delay, diabetes, hypertensive disorders, and preeclampsia, blood flow to the fetus and placenta is interrupted. Doppler ultrasound can be used throughout pregnancy to determine the severity of UA obstruction, if any. In conjunction with color imaging, these studies can help to determine flow, mean arterial pressure, and resistance within the UA.

67.4 Variations

The UA is variable in origin and branching, as is the surrounding vasculature. It can arise from the IIA completely separately, but it can also branch together with other vessels such as the internal pudendal, vaginal, middle rectal, and superior/inferior vesical arteries (Bergman et al. 2016). In one case, the UA was reported to share a common origin with the obturator artery (Lipshutz 1918). Parsons and Keith reported that it arose directly from the hypogastric trunk in 50%, from a separate branch of the anterior division of the IIA in 16.9%, from the inferior vesical artery in 11%, and from the middle rectal artery in 5.5% (Bergman et al. 2016; Parsons and Keith 1896).

From the arteriograms of 257 patients (i.e., 514 UAs were identified), the UA was described as following a characteristic curve with a lateral descending segment, a medial transverse segment, and an ascending segment along the side of the uterus (Gomez-Jorge et al. 2003). Four types of origin were distinguished. In Type 1, the most common (45% of the cases studied), the UA arose as the first branch of the inferior gluteal artery (Gomez-Jorge et al. 2003). The second most common origin (43%) was Type 3, a trifurcation with the superior and inferior gluteal arteries (Gomez-Jorge et al. 2003). In Type 2, the UA arose as the second or third branch of the anterior division if the IIA (Gomez-Jorge et al. 2003). Lastly, in Type 4, the UA was the first branch of the internal iliac trunk (Bergman et al. 2016; Obimbo et al. 2010).

In a more recent study of 53 Kenyan cadavers, 70% of UAs followed a Type 2 origin pattern (Obimbo et al. 2010). This study also provided data about the ascending branch of the UA. Obimbo et al. found that in the population studied, the UA ascended as a single branch in 76.4% of cases, bifurcated in 17.1% and trifurcated in 6.7% (Obimbo et al. 2010). This is clinically important because bifurcation and trifurcation of the UA are associated with spontaneous abortion and reduced gestational size (Bergman et al. 2016; Obimbo et al. 2010).

The study also confirmed that in most cases (96.2% of cadavers studied), the UA passes anteriorly to the ureter, as discussed previously (Bergman et al. 2016; Obimbo et al. 2010).

Dubreuil-Chambardel documented several branching patterns of the UA (Dubreuil-Chambardel 1925):

- Type I—Primary UA crosses the ureter and gives rise to ascending and descending branches.
- Type II—Follows a similar branching pattern as Type I, but branches at the ureter rather than distal to it.
- Type III—Primary UA gives rise to an ascending branch and separate cervical and vaginal branches.
- Type IV—Primary UA follows a branching pattern similar to Type III, but branches proximal to the ureter.
- Type V—IIA gives rise to ascending, cervical, and vaginal branches (no primary UA present).

In some instances, the UA is absent or replaced by alternative vasculature. In an angiographic study including 43 patients, 2.3% had no right UA (Baba et al. 2012). Saraiya et al. reported an absent left UA, which was replaced by an enlarged artery originating from the inferior epigastric artery, which in turn arose from the EIA (Saraiya et al. 2002).

The UA anastomoses with several arteries within the pelvic region including the ovarian, vaginal, and cervical arteries (Bergman et al. 2016). These anastomoses vary significantly. Dubreuil-Chambardel reported five types of anastomoses between the ovarian and uterine arteries (Dubreuil-Chambardel 1925). Functionally, anastomoses provide alternative pathways for blood flow in the event of vascular compromise (Bergman et al. 2016). One angiographic study of 43 patients demonstrated significant anastomoses between the uterine and ovarian arteries (Baba et al. 2012). Among these patients, 23 had uterine fibroids, 58% of which had significant anastomoses (Baba et al. 2012). Moreover, 20 patients presented with hemorrhage, 47.4% of whom had significant anastomoses (Baba et al. 2012).

67.5 Pathology, Clinical, and Surgical Significance Complications

Owing to its origin from the IIA, the UA is often affected in clinical cases in which blood flow in the IIA and the UA itself is disrupted. It is important to note that there is collateral circulation in the pelvic region, which provides some degree of continuing blood flow in the event of compromise.

Beginning with conditions associated with the IIA: IIA aneurysms (IIAA) are rare (approximately 0.3% of all aortoiliac aneurysms) and asymptomatic and are typically found during imaging examinations for other conditions (Machado et al. 2016). Surgical or endovascular repair for IIAA is indicated under certain circumstances, including but not limited to rupture, rapid growth, or other symptoms. Endovascular repair is typically preferred over open surgical repair because perioperative morbidity is significantly less, length of stay is shorter, and operative time and blood loss are decreased (Perini et al. 2021). Operative treatment can include endoaneurysmorrhaphy with graft interposition (opening the aneurysm sac and placing a graft), aneurysmectomy (aneurysm resection), or proximal and distal IIA ligation (Dix et al. 2005; Nakajima et al. 2001).

Uterine fibroids (leiomyomas) or benign tumors of the uterus are clinically apparent in approximately 25% of females (Stewart 2001). While some myomas cause no symptoms, uterine fibroids can cause significant symptoms in many women, including significantly prolonged or heavy menstrual bleeding (menorrhagia or hypermenorrhea), pelvic pressure or pain, and, in some cases, reproductive dysfunction (Stewart 2001). As this is a common gynecological condition, there are several options for treatment, including but not limited to hormonal therapy, hysteroscopic or abdominal myomectomy, and hysterectomy (Stewart 2001; Van der

Kooij and Hehenkamp 2022). UA embolization is a minimally invasive technique for treating myomas based on the hypothesis that control of arterial blood flow will control the symptoms associated with these fibroids (de Bruijn et al. 2016; Kim 2019). In a hysterectomy, the uterine vessels can be ligated to reduce excess blood loss and operative time (Hebisch 2002; Sinha et al. 2008). Throughout this procedure, it must be recognized that the vaginal artery can be either a branch of the UA or a separate vessel. An error about this can result in internal bleeding (Balcerzak et al. 2022). Care must also be taken to avoid injuring the ureter, which as stated previously, has an important relationship with the UA.

Postpartum hemorrhage (PPH), excessive blood loss during delivery, is a common and potentially fatal complication of childbirth. Per the American College of Obstetrics and Gynecology in 2017, the current definition of PPH is cumulative blood loss over 1000 mL with signs and symptoms of hypovolemia within 24 h of the birth process. This definition is presumed regardless of the route of delivery (vaginal, cesarian, etc.) (Committee on Practice Bulletins-Obstetrics 2017). PPH is defined as "primary" if the bleeding occurs during the first 24 h after delivery and "secondary" when it occurs between 24 h and 12 weeks postpartum (Dossou et al. 2015). Treatment is typically cause-specific and focuses on maintaining hemodynamic stability (Maswime and Buchmann 2017). In some cases, internal iliac ligation or UA embolization is considered (Singhal et al. 2015).

References

Baba Y, Hayashi S, Ikeda S, Nakajo M, Yoshinaga M, Douchi T (2012) Visualization of utero-ovarian anastomoses: comparison between patients with and without uterine fibroid. J Biomed Sci Eng 05(12):767–770. https://doi.org/10.4236/jbise.2012.512A096

Balcerzak A, Hajdys J, Shane Tubbs R, Karauda P, Georgiev GP, Olewnik Ł (2022) Clinical importance of variability in the branching pattern of the internal iliac artery– An updated and comprehensive review with a new classification proposal. Ann Anat 239:151837. https://doi.org/10.1016/j.aanat.2021.151837

Baliyan V, Shaqdan K, Hedgire S, Ghoshhajra B (2019) Vascular computed tomography angiography technique and indications. Cardiovasc Diagn Ther 9(Suppl 1):S14–S27. https://doi.org/10.21037/cdt.2019.07.04

Bergman RA, Tubbs RS, Shoja MM, Loukas M (eds) (2016) Bergman's comprehensive encyclopedia of human anatomic variation. Wiley

Bilhim T, Casal D, Furtado A, Pais D, O'Neill JEG, Pisco JM (2011) Branching patterns of the male internal iliac artery: imaging findings. Surg Radiol Anat SRA 33(2):151–159. https://doi.org/10.1007/s00276-010-0716-3

Committee on Practice Bulletins-Obstetrics (2017) Practice bulletin no. 183: postpartum hemorrhage. Obstet Gynecol 130(4):e168–e186. https://doi.org/10.1097/AOG.0000000000002351

de Bruijn AM, Ankum WM, Reekers JA et al (2016) Uterine artery embolization vs hysterectomy in the treatment of symptomatic uterine fibroids: 10-year outcomes from the randomized EMMY trial. Am J Obstet Gynecol 215(6):745.e1–745.e12. https://doi.org/10.1016/j.ajog.2016.06.051

Dix FP, Titi M, Al-Khaffaf H (2005) The isolated internal iliac artery aneurysm–a review. Eur J Vasc Endovasc Surg Off J Eur Soc Vasc Surg 30(2):119–129. https://doi.org/10.1016/j.ejvs.2005.04.035

Dossou M, Debost-Legrand A, Déchelotte P, Lémery D, Vendittelli F (2015) Severe secondary postpartum hemorrhage: a historical cohort. Birth Berkeley Calif 42(2):149–155. https://doi.org/10.1111/birt.12164

Dubreuil-Chambardel L (1925) Traité des variations du système artériel. Variations des artères du pelvis et du membre inférieur. Masson et cies

Gomez-Jorge J, Keyoung A, Levy EB, Spies JB (2003) Uterine artery anatomy relevant to uterine leiomyomata embolization. Cardiovasc Intervent Radiol 26(6):522–527. https://doi.org/10.1007/s00270-003-2652-7

Hebisch G (2002) Vaginal uterine artery ligation avoids high blood loss and puerperal hysterectomy in postpartum hemorrhage. Obstet Gynecol 100(3):574–578. https://doi.org/10.1016/S0029-7844(02)02121-X

Kim MD (2019) Uterine artery embolization for leiomyomas and adenomyosis: a pictorial essay based on our experience from 1300 cases. Korean J Radiol 20(10):1462. https://doi.org/10.3348/kjr.2019.0205

Lipshutz B (1918) A composite study of the hypogastric artery and its branches. Ann Surg 67(5):584–608. https://doi.org/10.1097/00000658-191805000-00012

Machado RM, Rego DNC, de Oliveira PNFP, de Almeida RMGF (2016) Endovascular treatment of internal iliac artery aneurysms – single center experience. Braz J Cardiovasc Surg. https://doi.org/10.5935/1678-9741.20160023. Published online 2016

Maswime S, Buchmann E (2017) A systematic review of maternal near miss and mortality due to postpartum hemorrhage. Int J Gynaecol Obstet Off Organ Int Fed Gynaecol Obstet 137(1):1–7. https://doi.org/10.1002/ijgo.12096

Moore KL, Dalley AF, Agur AMR (2014) Clinically oriented anatomy, 7th edn. Wolters Kluwer/Lippincott Williams & Wilkins Health

Nakajima T, Kawazoe K, Komoda K et al (2001) Failure of exclusion of internal iliac artery aneurysms. J Vasc Surg 33(3):476–480. https://doi.org/10.1067/mva.2001.111975

Obimbo MM, Ogeng'o JA, Saidi H (2010) Variant anatomy of the uterine artery in a Kenyan population. Int J Gynaecol Obstet Off Organ Int Fed Gynaecol Obstet 111(1). https://doi.org/10.1016/j.ijgo.2010.04.033

Ozgüner G, Sulak O (2011) Development of the abdominal aorta and iliac arteries during the fetal period: a morphometric study. Surg Radiol Anat SRA 33(1):35–43. https://doi.org/10.1007/s00276-010-0696-3

Parsons FG, Keith A (1896) Sixth Annual Report of the Committee of Collective Investigation of the Anatomical Society of Great Britain and Ireland, 1895–96. J Anat Physiol 31(Pt 1):31–44

Pavone P, Di Cesare E, Di Renzi P et al (1990) Abdominal aortic aneurysm evaluation: comparison of US, CT, MRI, and angiography. Magn Reson Imaging 8(3):199–204. https://doi.org/10.1016/0730-725X(90)90089-K

Perini P, Mariani E, Fanelli M et al (2021) Surgical and endovascular management of isolated internal iliac artery aneurysms: a systematic review and meta-analysis. Vasc Endovasc Surg 55(3):254–264. https://doi.org/10.1177/1538574420981812

Saraiya PV, Chang TC, Pelage JP, Spies JB (2002) Uterine artery replacement by the round ligament artery: an anatomic variant discovered during uterine artery embolization for leiomyomata. J Vasc Interv Radiol JVIR 13(9 Pt 1):939–941. https://doi.org/10.1016/s1051-0443(07)61779-5

Selçuk İ, Ersak B, Tatar İ, Güngör T, Huri E (2018) Basic clinical retroperitoneal anatomy for pelvic surgeons. Turk J Obstet Gynecol 15(4):259–269. https://doi.org/10.4274/tjod.88614

Singhal M, Gupta P, Sikka P, Khandelwal N (2015) Uterine artery embolization following internal iliac arteries ligation in a case of post-partum hemorrhage: a technical challenge. J Obstet Gynecol India 65(3):202–205. https://doi.org/10.1007/s13224-014-0595-3

Sinha R, Sundaram M, Nikam YA, Hegde A, Mahajan C (2008) Total laparoscopic hysterectomy with earlier uterine artery ligation. J Minim Invasive Gynecol 15(3):355–359. https://doi.org/10.1016/j.jmig.2008.01.012

Stewart EA (2001) Uterine fibroids. Lancet 357(9252):293–298. https://doi.org/10.1016/S0140-6736(00)03622-9

Szpinda M, Szpinda A, WoŸniak A, Daroszewski M, Mila-Kierzenkowska C (2012) The normal growth of the common iliac arteries in human fetuses – an anatomical, digital and statistical study. Med Sci Monit Int Med J Exp Clin Res 18(3):BR109–BR116. https://doi.org/10.12659/MSM.882516

Van der Kooij SM, Hehenkamp WJK (2022) Uterine fibroids (leiomyomas): treatment with uterine artery embolization. Published online March 17, 2022. https://www.uptodate.com/contents/uterine-fibroids-leiomyomas-treatment-with-uterine-artery-embolization

Gluteal Arteries

68

Nicole Levine

68.1 Introduction

Arterial blood is supplied to the gluteal region by the superior and inferior gluteal arteries. The branching pattern of these arteries varies among individuals, but their ultimate destination remains consistent. They have proved clinically advantageous, especially through their use in breast reconstruction surgeries. This chapter will explore the development, pathological findings, histology, imaging, and clinical significance of these arteries.

68.2 Embryology

The formation of the superior and inferior gluteal arteries during development is important for understanding their origin, supply, and variations in anatomical location. The vascular system begins to form during the third week of gestation (Ilina and Lilley 2019). The early development of this system is essential for supplying oxygen to the growing cells. It arises from angiogenic cells, similar to monocytes, and promotes endothelial growth (Shepherd et al. 2006). Growth takes place near the boundary between the endoderm and mesoderm (Noden 1989).

The two umbilical arteries merge at the 5 mm stage of fetal development. They develop through anastomosis with branches of the primitive dorsal aorta. As the embryo develops, they move caudally and eventually away from the primitive dorsal aorta. After birth, part of them becomes obliterated and forms the medial umbilical ligament, while another portion gives rise to the superior vesical artery. During fetal life, the common iliac artery is a branch of the umbilical artery posterior to the origin of the external iliac artery. The internal iliac artery is another branch of the umbilical artery anterior to the origin of the external iliac. However, there is a rearrangement after birth; the common iliac artery bifurcates into the external and internal iliac arteries. This bifurcation appears at S1 during the first trimester of fetal development, but at birth, it is located near L5. During adult life, the bifurcation occurs between L5 and S1, proving that it begins at a more inferior position during fetal development and moves superiorly throughout life. The internal iliac artery forms smaller branches through two divisions: anterior and posterior. The anterior division gives rise to the umbilical artery (which gives off the superior vesicle artery), the middle rectal, inferior vesicle (vaginal in women), obturator, uterine (in women), internal pudendal, and inferior gluteal arteries. The posterior division branches into the

N. Levine (✉)
Tulane University School of Medicine,
New Orleans, LA, USA
e-mail: nlevine@tulane.edu

S. J. Bordes, Jr. et al. (eds.), *The Clinical Anatomy of the Vascular System*,
https://doi.org/10.1007/978-3-031-78326-5_68

superior gluteal, iliolumbar, and lateral sacral arteries. The superior and inferior gluteal arteries are formed at birth, but their anatomical orientation is not always consistent. Different individuals have different branching patterns, so the anterior and posterior divisions of the internal iliac artery are not always perfectly aligned. For example, Sakthivelavan and his team found that in certain subjects, the superior or inferior gluteal artery arose directly from the internal iliac artery, while in others, the superior gluteal artery arose from the anterior instead of the posterior division of the internal iliac artery. There are many variations, but they are not due to problems during embryo development.

68.3 Anatomy and Topography

The superior and inferior gluteal arteries are branches of the internal iliac artery (Fig. 68.1). The superior gluteal artery is a part of the posterior trunk of the internal iliac artery runs above the piriformis muscle into the greater sciatic foramen. Sometimes, the vessel branches off the internal pudendal artery or another source. The inferior gluteal artery, in contrast, is a branch of

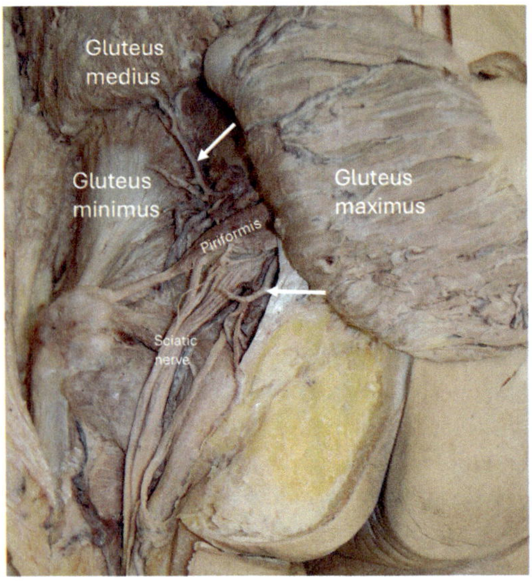

Fig. 68.1 Cadaveric dissection of the left gluteal region noting the superior (upper arrow) and inferior gluteal arteries (lower arrow). (Courtesy Prof. M. Loukas)

the anterior trunk. It also runs through the greater sciatic foramen but below the piriformis muscle. It enters the thigh and travels near the posterior cutaneous and sciatic nerves. The gluteus maximus receives vasculature from both the inferior and superior gluteal arteries through the center of the muscle. The inferior gluteal artery proceeds over the greater trochanter of the femur alongside the sciatic nerve.

Although the anatomical positioning and branching of the superior and inferior gluteal arteries are not constant throughout the population, the ultimate destination of each artery is. Kieser et al. explored the positioning of these arteries during a posterior sacrectomy approach. They found that the anatomical landmarks for both gluteal arteries were easily locatable. The superior gluteal artery's relationship to the posterior-inferior aspect of the sacroiliac joint (PISIJ) is fairly consistent amongst individuals because a posterior branch of the artery travels inferiorly to the PISIJ. The inferior gluteal artery's position is similarly constant, but its path is below the PISIJ and the midpoint of the sacrococcygeal joint (Kieser et al. 2017).

The superior gluteal artery eventually supplies the gluteus maximus, the skin of the gluteus maximus, the gluteus medius, the gluteus minimus, and the tensor fascia lata. The inferior gluteal artery supplies the gluteus maximus and the skin of the posterior thigh.

68.4 Histology

The superior and inferior gluteal arteries are classified as muscular as opposed to elastic arteries. Histologically, elastic arteries have a prominent media with an elastic lamina of 40–70 layers. Muscular arteries have a prominent internal elastic lamina, media, and smooth muscle up to 40 layers thick. This structure contrasts with veins: veins have valves, lack an internal and external elastic lamina, contain thin tunicae mediae, and have a thicker adventitia. Muscular arteries are continuations of elastic arteries with more musculature so that blood can be properly distributed throughout the body. In the superior and inferior gluteal arteries, the musculature is necessary for supplying the aforementioned muscles.

68.5 Pathology

The superior and inferior gluteal arteries can be injured during pelvic trauma and fractures, which can lead to pseudoaneurysms. Schorn et al. explored the case of a man with a left superior gluteal artery aneurysm. The patient had sciatic pain with no relief from a transcatheter embolization. A surgical team divided the gluteus maximus muscle from the left iliac crest to locate the aneurysm in the pelvic outlet. An endoaneurysmorrhaphy was performed. The left iliac artery was occluded with a balloon catheter to control bleeding. After the procedure, the patient was completely relieved of his pain (Schorn et al. 1995). This technique proved that gluteal artery aneurysms can be repaired satisfactorily.

Patients experiencing buttock claudication may have involvement of their superior gluteal artery. The artery can be stenosed even without atheroma of other arteries, suggesting anatomical compression. Bouillanne et al. found compression of the superior gluteal artery at the suprapiriformis canal. Angiographic and CT imaging demonstrated that the anatomical relationship between the superior gluteal artery, the greater sciatic notch, and the piriformis allows entrapping of the artery. Additionally, pathomorphological studies revealed histological changes showing indentation of the tunica media due to repeated trauma to this artery (Bouillanne et al. 2000).

68.6 Clinical/Surgical Significance

The superior and inferior gluteal arteries have proved advantageous in numerous surgical procedures over the last 30 years. The inferior gluteal artery was helpful in repairing a sciatic wound. The superior gluteal artery has been used to repair sacral bedsores, breasts, and spinal meningoceles (Zheng et al. 2019). The superior gluteal artery perforator (SGAP) flap allows surgeons to use gluteal tissue for breast reconstruction. This technique has become widely accepted and encouraged, especially when the use of a muscle is not required. The SGAP flap is an updated version of the gluteus maximus myocutaneous flap. Its location and abundance allow for optimal transfer in most situations, especially because most patients have an adequate area from which to take tissue. The abdomen used to be the preferred source of tissue because of its vasculature. Still, this approach occasionally faced resistance owing to excessive scarring or low BMI patients with little tissue available (Hunter et al. 2016; Gagnon and Blondeel 2006).

The gluteal myocutaneous flap was first initiated for breast reconstruction by Fujino and partners in 1975. Although it initially failed to fulfill its purpose, Koshima et al. continued with the study in 1993. They were followed by Allen and Tucker and by Blondeel in 1999. The SGAP took time, but it finally allowed for proper mobility and led to fewer cases of morbidity. When this area of the body is used, patients are less affected by the operation since the muscle can be sacrificed with little loss of function. Although the deep inferior epigastric artery perforator (DIEAP) flap is often the preferred choice, the SGAP is a close second for many surgeons (Gagnon and Blondeel 2006).

Before flap harvest for breast reconstruction, it is of paramount importance to assess the gluteal vasculature. This is important for determining whether there is proper blood flow to the gluteal region and whether there are any alterations or abnormalities. Doppler is often used to locate important perforators. This device is easy for surgeons to use, inexpensive, and mobile. Although it has advantages, it provides only approximate locations and vessel sizes rather than accurate estimates. As a result, false positives and false negatives can arise. Some surgeons, therefore, started to use multidetector computed tomography scans. These provide more accurate and detailed depictions of the flaps and the vessels being used so a more accurate plan can be devised before an operation. Surgeons try to locate the perforator farthest from the superior gluteal artery's exit from the pelvis. This allows for a larger pedicle and minimizes further complications (Gagnon and Blondeel 2006).

For a typical patient, four or five different perforators are located near the superior gluteal artery. When designing the superior gluteal perforator flap before surgery, a surgeon marks the connection between the posterior superior iliac spine (PSIS) and the greater trochanter. A line is also drawn between the PSIS and the coccyx. The piriformis is then located by joining the middle of the PSIS-coccyx line to the greater trochanter. Perforators superior to the piriformis muscle and lateral to the point at which the superior gluteal artery exits the pelvis have potential for use during the surgery (Gagnon and Blondeel 2006).

The flap design is then mapped out above the piriformis muscle. Surgeons often use a shape similar to a bikini line during breast reconstruction. The oblique incision is usually ideal for patients because the scarring is often covered, and there are minimal complications afterward. The flap size is at most 30 cm long and 13 cm wide. If a larger flap is needed, surgeons usually take tissue from the medial rather than the lateral side to avoid unusual incisions and scarring (Gagnon and Blondeel 2006).

When the flap is harvested, an anesthetic is injected, and some subcutaneous fat is removed. Sometimes, nerves are dissected with the flap, and they can be anastomosed to the anterior ramus of the fourth intercostal nerve once implanted in the breast region. This provides the patient with sensation in the reconstructed breast area. The gluteus maximus fascia is opened while the patient is paralyzed, and complete hemostasis is required. Next the desired perforator is located, and the surgeon ensures that a vein wider than 1 mm is present. The muscle is split along the fibers in order to reveal the three-dimensional perforator size. If the perforator is sufficient, a complete dissection is performed and the deepest part of the gluteus maximus with anterior fascia is found. A subgluteal fat pad appears with sufficient vasculature. The surgeon must be very careful to maintain hemostasis. With breast reconstruction, it is usually important to have a longer pedicle so that the flap insertion runs smoothly. Once the pedicle is dissected, the flap is detached further from the muscle (Gagnon and Blondeel 2006).

Usually, the breast is reconstructed on the ipsilateral side as the gluteal region donor site. The flap is transferred, microanastamosis develops, and symmetry is ensured. The internal mammary artery and vein are targeted in the breast region. The third rib is usually used because it has fewer venous branches than the fourth. The flap can be repositioned in many ways before the orientation providing the best result is decided. The pedicle is protected to ensure it does not coil. The flap attaches to the pectoral fascia and is sutured in place (Gagnon and Blondeel 2006).

The SGAP is the most common clinical use of the superior gluteal artery; however, there are other options. Herskowitz et al. proved that the superior gluteal artery can be used to access an internal iliac artery aneurysm. This approach was used to preclude expansion, especially since the aneurysm in this case had been treated previously (Herskowitz et al. 2014).

The superior gluteal artery is also at risk during S2 sacroiliac screw placement, so surgeons must be aware of the vessel location before surgery. The deep and superficial branches of the artery are often close to the screw site. In some situations, the screw is placed between the deep and superficial branches; in others, the gluteal arteries are spared because they are relatively distant from the screw placement area (Zhao et al. 2021). In order to avoid complications, it is necessary to avoid these branches and locate the vessels properly beforehand.

References

Bouillanne PJ, Batt M, Hassen-Khodja R, Declemy S, Pittaluga P (2000) Anatomic bases for compression of the superior gluteal artery at the level of the gluteal canal: SRA. Surg Radiol Anat 21(5):309–312. https://doi.org/10.1007/s00276-999-0309-1

Gagnon AR, Blondeel PN (2006) Superior gluteal artery perforator flap. Semin Plast Surg 20(2):79–88. https://doi.org/10.1055/s-2006-941714

Herskowitz MM, Walsh J, Jacobs DT (2014) Direct sonographic-guided superior gluteal artery access for treatment of a previously treated expanding internal iliac artery aneurysm. J Vasc Surg 59(1):235–237. https://doi.org/10.1016/j.jvs.2013.03.001

Hunter C, Moody L, Luan A, Nazerali R, Lee G (2016) Superior gluteal artery perforator flap: the beauty of the buttock. Ann Plast Surg 76:S191–S195. https://doi.org/10.1097/SAP.0000000000000723

Ilina M, Lilley S (2019) Embryology of the great vessels. In: Clinical embryology. Springer, p 275

Kieser DC, Coudert P, Cawley DT et al (2017) Identifying the superior and inferior gluteal arteries during a sacrectomy via a posterior approach. J Spine Surg 3(4):624–629. https://doi.org/10.21037/jss.2017.12.06

Noden DM (1989 Oct) Embryonic origins and assembly of blood vessels. Am Rev Respir Dis 140(4):1097–1103. https://doi.org/10.1164/ajrccm/140.4.1097

Schorn B, Reitmeier F, Falk V, Oestmann JW, Dalichau H, Mohr FW (1995) True aneurysm of the superior gluteal artery: case report and review of the literature. J Vasc Surg 21(5):851–854. https://doi.org/10.1016/s0741-5214(05)80017-5

Shepherd RM, Capoccia BJ, Devine SM et al (2006) Angiogenic cells can be rapidly mobilized and efficiently harvested from the blood following treatment with AMD3100. Blood 108(12):3662–3667. https://doi.org/10.1182/blood-2006-06-030577

Zhao Y, Li W, Lian W et al (2021) Anatomic relationship between S2 sacroiliac screws' entry points and pelvic external branches of superior gluteal artery. Eur J Trauma Emerg Surg 48(2):857–862. https://doi.org/10.1007/s00068-021-01622-2

Zheng HP, Xu YQ, Lin J, Hu DQ (2019) Superior and inferior gluteal artery perforator flap. In: Atlas of perforator flap and wound healing. Springer, Singapore. https://doi.org/10.1007/978-981-13-1553-4_13

69 The Obturator Artery

Ramesses Akamefula, Brianna Hines, Annie Syed, and Stephen J. Bordes, Jr.

69.1 Embryology

Compared to other vessels during embryological development, the obturator artery arises relatively late as a supplier to a primary capillary plexus (Arey 1963). The obturator artery is subsequently accompanied by the lower limb's sciatic artery, which travels and develops alongside the sciatic nerve (Sanduno et al. 1993). As the embryo's development progresses, the obturator artery and other essential vessels in the caudal region enlarge while other primitive vessels regress (Fitzerald 1978).

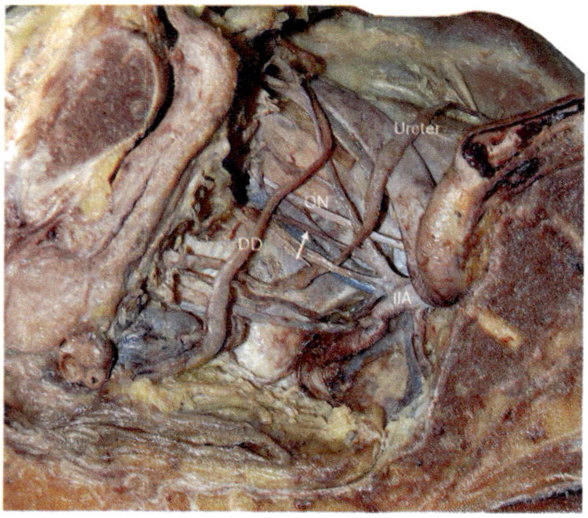

Fig. 69.1 Medial cadaveric view noting the internal iliac artery (IIA) giving off the obturator artery (arrow) traveling with the obturator nerve (ON). Note the ureter and ductus deferens (DD). (Courtesy Prof. M. Loukas)

69.2 Anatomy and Topography

After descending posteriorly through the greater sciatic foramen, the internal iliac artery bifurcates into anterior and posterior divisions (Williams et al. 1995; Snell 2000). Usually, the anterior division of the internal iliac artery gives rise to the superior and inferior vesicle arteries, the middle rectal artery, the vaginal artery (in females), the obturator artery (Fig. 69.1), the internal pudendal artery, and the inferior gluteal artery (Snell 2000; Sinnatamby 2000). Anatomically, the obturator artery travels inferiorly and anteriorly along the lateral pelvic wall with the obturator fascia. Medial to the obturator artery, other vital landmarks and structures such as the ductus deferens and ureter may also be found within the pelvic cavity (Netter 2019; Standring 2016).

Before leaving the pelvic cavity, the obturator artery gives rise to multiple branches. Ilial branches (intrapelvic) give rise to vessels that

R. Akamefula · B. Hines · A. Syed
Tulane University School of Medicine,
New Orleans, LA, USA
e-mail: rakamefula@tulane.edu; bhines@tulane.edu; asyed1@tulane.edu

S. J. Bordes, Jr. (✉)
Department of Surgery, Louisiana State University Health Sciences Center, New Orleans, LA, USA

S. J. Bordes, Jr. et al. (eds.), *The Clinical Anatomy of the Vascular System*,
https://doi.org/10.1007/978-3-031-78326-5_69

perfuse the iliac bone and muscle. Moreover, these branches have been seen to anastomose with the iliolumbar artery. A pubic branch (intrapelvic) may also arise via the obturator artery near the obturator canal. This branch courses alongside the pubic bone and has been seen to anastomose with a pubic branch of the inferior epigastric artery (Netter 2019; Standring 2016).

As the obturator artery courses through the obturator canal and leaves the pelvis, it further divides into an anterior and posterior branch as it travels adjacent to the upper margin of the obturator foramen (Netter 2019; Standring 2016).

Anterior Branch (Extrapelvic)

This branch typically runs superficially to the obturator membrane. It then deviates inferiorly and travels adjacently along the anterior margin of the obturator foramen. This branch typically supplies blood to muscles such as the pectineus, gracilis, obturator externus, and other hip adductors. This branch has been seen to anastomose with the medial femoral circumflex artery and the posterior division of the obturator artery.

Posterior Branch (Extrapelvic)

This branch courses alongside the posterior margin of the obturator canal. It then deviates as it reaches the inferior ramus of the ischium to anastomose with the anterior division of the obturator artery. Although this branch typically is not the main blood supply for any dedicated musculature, the posterior branch of the obturator artery has been seen to partially perfuse muscles that attach to the ischial tuberosity, such as the ischiocavernosus and the ligamentum teres. This branch has been seen as anastomosing with the inferior gluteal artery.

Acetabular Branch (Extrapelvic)

This branch typically travels alongside the ligamentum teres and supplies blood to the femoral head. This branch is generally derived from the posterior division of the obturator artery.

69.3 Histology

The histology of the obturator artery follows most arteries in the human body. It is a distributing artery comprised of three layers. The innermost layer closest to the lumen is the tunica intima. Furthest from the lumen, the outermost layer is the tunica adventitia, and between these two layers is the tunica media.

69.4 Imaging

Multiple imaging modalities may be used when evaluating arteries in the pelvic region. These methods include imaging techniques such as Computer Tomography Angiography (CTA), Digital Subtraction Angiography (DSA), and Magnetic Resonance Angiography (MRA). Different imaging modalities may be indicated, each technique providing unique information to the viewer, depending on specific needs (Bilhim et al. 2011; Uflacker 2006; Riche et al. 1981). CTA may be performed before DSA to view pelvic vascular anatomy and help guide subsequent DSA. MRA may also be used to study the obturator artery and related structures, although some surgeons report not having enough resolution when viewing smaller vessels of the pelvis. Moreover, comparison to DSA may be difficult when utilizing MRA due to a lack of available landmarks (Bilhim et al. 2014).

It should be noted that delineating the obturator and its branches using DSA may be difficult. As branches of the internal iliac artery can overlap, it is advised that physicians first establish an optimal projection to define all vascular structures and branches correctly. Bilhim et al. reported that the ipsilateral anterior oblique projection (steep oblique 35–40 degrees) with a caudal-to-cranial angulation of −10 degrees lets viewers fully delineate the internal iliac artery and its many branches (Bilhim et al. 2014).

Contrast-enhanced Computed Tomography Angiography (CTA) may be indicated in the case of potential iatrogenic injury. Practitioners have used this modality to illustrate an extravasion or hematoma associated with the obturator internus muscle or sidewall of the pelvis (Yoon et al. 2004). Furthermore, it should be noted that interventional radiology is the suggested first step in diagnosing and treating using transarterial embolization (Daeubler et al. 2003).

69.5 Variations

Variation in the obturator artery is not uncommon. Therefore, it is advised that surgeons take time to understand the complex anatomy of retropubic vasculature. Knowledge of different variations in the origin of the obturator artery is critical when intervening surgically and performing procedures such as extraperitoneal inguinal hernioplasty, laparoscopic herniorrhaphy, and other procedures involving precise dissection and exposure of the femoral and inguinal rings (Lau and Lee 2003; Bendavid 1992).

The origination of the obturator artery from multiple locations other than the anterior division of the internal iliac artery has been described in the literature. Missankov et al. reported that origination directly from the external iliac artery may occur up to 25% of the time (Missankov et al. 1996), Bergman et al. found this variation to be as low as 1.1% (Bergman et al. 1988). Bergman et al. further illustrated that a variant seen in up to 20–30% of cases was one in which the inferior epigastric artery and obturator artery had a common origin point of the external iliac artery (Bergman et al. 1988). Pai et al. described several other variants of the obturator artery (Pai et al. 2009). These include distinct origins from the internal iliac artery, the posterior division of the internal iliac artery, and the external iliac artery. Regarding the posterior division of the internal iliac artery, the obturator artery has been seen to have common points with the superior gluteal artery and/or the iliolumbar artery. Regarding the external iliac artery, the obturator artery has been seen sharing an originating point with the inferior epigastric artery (Pai et al. 2009).

Corona Mortis

In some occurrences, the obturator artery has given rise to an accessory branch that anastomoses with the external iliac artery and may be located along the superior pubic ramus. This branch is more commonly called the Corona Mortis (crown of death artery) and has considerable implications regarding vascular surgery and surgical complications (described later in this chapter) (Darmanis et al. 2007a).

69.6 Surgical Considerations

The many variations of the obturator artery, along with its proximity to other structures of the pelvis, make it highly prone to iatrogenic injury during procedures such as prostatectomy and inguinal hernioplasty (Negura et al. 1989). It should be noted that iatrogenic injury due to transection or pseudoaneurysm may result in severe hemorrhage. Particular attention should be devoted to identifying the obturator artery to decrease the chance of adverse outcomes.

Corona Mortis

An anastomosis between the obturator and inferior epigastric artery may occur and travel across the superior pubic ramus (Corona Mortis). When it is suspected that injury to the Corona Mortis has occurred, embolization of the inferior epigastric and obturator branches may be indicated to control hemorrhage (Suzuki et al. 2022). It is strongly encouraged to utilize angiographic imaging to help guide surgeons when intervening in this region. Surgeons performing an anterior approach to access the acetabulum should proceed cautiously when dissecting near the superior pubic ramus due to this possible variant (Darmanis et al. 2007b).

Surgical Ligation

Ligation of the internal iliac artery to control hemorrhage during pelvic surgery is not uncommon. Due to the possibility of necrosis and claudication of the buttocks, ligation of the internal iliac artery should be performed distal to the posterior division (Iliopoulos et al. 1987). Due to the many variants of the obturator artery and the possibility of the obturator artery originating from the posterior division of the internal iliac artery, ligation of the internal iliac artery may be more complicated. Thus, a strong understanding of the vascular anatomy of both the anterior and posterior divisions is necessary when attempting to control hemorrhage.

It should also be noted that the prostate gland is supplied via the inferior vesicle artery, which arises from the anterior division of the internal iliac artery (Williams et al. 1995). Sometimes, the inferior vesicle artery may originate from the obturator artery. Regarding prostatectomy or related surgery, it should be noted that proper knowledge of the obturator artery, its branches, and variations are needed to ligate the inferior vesicle artery.

69.7 Clinical Significance

A comprehensive understanding of the vascular anatomy of the pelvis is needed when looking for successful outcomes during lower abdominal and pelvic surgery (Brick et al. 1995). Complex surgeries such as pelvic organ or tumor removal may also require ligation of vessels such as the obturator artery (Dzsinich 2004). A possible Corona Mortis that courses alongside the superior pubic ramus may require special attention due to its susceptibility when dissecting the preperitoneal space or placing mesh during hernia repair (Lau and Lee 2003). In the case of hemorrhage, surgeons must be familiar with standard and variant anatomy and pelvic blood supply to decrease the risk of venostasis or ischemia (Dzsinich 2004; Rajive and Pillay 2015).

References

Arey LB (1963) The development of peripheral blood vessels. In: Orbison JL, Smith DE (eds) The peripheral blood vessels. Williams and Wilkins, Baltimore, pp 1–16

Bendavid R (1992) The space of Bogros and the deep inguinal venous circulation. Surg Gynecol Obstet 174:355–358

Bergman RA, Thompson SA, Afifi AK, Saadeh FA (1988) Compendium of human anatomic variation: catalog, atlas and world literature. Urban and Schwazenberg, Baltimore and Munich

Bilhim T, Casal D, Furtado A et al (2011) Branching patterns of the male internal iliac artery: imaging findings. Surg Radiol Anat 33:151–159

Bilhim T, Pereira JA, Fernandes L, Rio Tinto H, Pisco JM (2014) Angiographic anatomy of the male pelvic arteries. AJR Am J Roentgenol 203(4):W373–W382

Brick WG, Colborn GL, Gadacz TR, Skandalakis JE (1995) Crucial anatomic lessons for laparoscopic herniorrhaphy. Am Surg 61:172–177

Daeubler B, Anderson SE, Leunig M, Triller J (2003) Hemorrhage secondary to pelvic fracture: coil embolization of an aberrant obturator artery. J Endovasc Ther 10:676–680

Darmanis S, Lewis A, Mansoor A, Bircher M (2007a) Corona mortis: an anatomical study with clinical implications in approaches to the pelvis and acetabulum. Clin Anat 20:433–439

Darmanis S, Lewis A, Mansoor A, Bircher M (2007b) Corona mortis: an anatomical study with clinical implications in approaches to the pelvis and acetabulum. Clin Anat 20(4):433–439

Dzsinich C (2004) Ligation of the large vessels in the pelvis. CME J Gynecologic Oncol 9:38–40

Fitzerald MJT (1978) Human embryology. Harper International, New York, pp 38–56

Iliopoulos JI, Howanitz PE, Pierce GE, Kueshkerian SM, Thomas JH, Hermreck AS (1987) The critical hypogastric circulation. Am J Surg 154(6):671–675

Lau H, Lee F (2003) A prospective endoscopic study of retropubic vascular anatomy in 121 patients undergoing endoscopic extraperitoneal inguinal hernioplasty. Surg Endosc 17:1376–1379

Missankov AA, Asvat R, Maoba KI (1996) Variations of the pubic vascular anastomoses in black south Africans. Acta Anat 155:212–214

Negura A, Marderos GG, Marderos GH, Zanoschi C (1989) Anatomo-surgical study of hemorrhagic risks in urethrovesical Burch-type suspension. Rev Fr Gynecol Obstet 84:941–943

Netter F (2019) Atlas of human Anatomy, 7th edn. Saunders, Philadelphia

Pai MM, Krishnamurthy A, Prabhu LV, Pai MV, Kumar SA, Hadimani GA (2009) Variability in the origin of the obturator artery. Clinics 64(9):897–901

Rajive AV, Pillay M (2015) A study of variations in the origin of obturator artery and its clinical significance. J Clin Diagn Res 9(8):AC12–AC15
Riche MC, Thiebot J, Tubiana JM, Merland JJ, Chiras J (1981) Arteriography of the pelvis: diagnostic and therapeutic procedures. Springer, New York
Sanduno JR, Roig M, Rodriguez A, Ferreira B, Domenech JM (1993) Rare origin of the OBA, inferior epigatric and femoral arteries from a common trunk. J Anat 183:161–163
Sinnatamby CS (2000) Last's Anatomy. Regional and applied, vol 148, 10th edn. Churchill Livingstone, Edinburgh, p 291
Snell RS (2000) Clinical anatomy for students, 6th edn. Lippincott, William & Wilkins, Philadelphia, pp 292–293
Standring S (2016) Gray's Anatomy, 41st edn. Elsevier Churchill Livingstone, Edinburgh
Suzuki M, Kato M, Nakamura R, Uehara K, Sugiki D, Matsushima H (2022) Post-surgical pseudoaneurysm of the corona mortis artery treated by arterial embolization from two arteries: a case report. Radiol Case Rep 17(4):1132–1135. https://doi.org/10.1016/j.radcr.2022.01.032
Uflacker R (2006) Atlas of vascular anatomy: an angiographic approach. Lippincott Williams & Wilkins, Philadelphia
Williams PL, Bannister LH, Berry MM, Collins P, Dyson M, Dussek JE, Ferguson MWJ (1995) Gray's Anatomy. The anatomical basis of medicine and surgery, 38th edn. Churchill Livingstone, Edinburgh
Yoon W, Kim JK, Jeong YY, Seo JJ, Park JG, Kang HK (2004) Pelvic arterial hemorrhage in patients with pelvic fractures: detection with contrast-enhanced CT. Radiographics 24:1591–1605

70 Lateral Sacral Artery

Yousef Raslan Hakim and Yonis Hakim

70.1 Embryology

In the first few weeks of gestational development, 31 somites develop in a craniocaudal direction, and in the third week of gestation, the dorsal aortae form (Lasjaunias et al. 2001; DeAnda et al. 2022). The dorsal aorta is formed by vasculogenesis from extraembryonic mesoderm that differentiates into angioblasts (Sadler 2019; Moore et al. 2020). The angioblasts form angiogenic clusters, in which inner angioblasts undergo apoptosis, leading to the formation of blood islands (Moore et al. 2020). Some angioblasts flatten to form the endothelial lining of the blood islands (Moore et al. 2020). These blood islands join to form the longitudinal dorsal aorta (Sadler 2019). The dorsal aorta gives rise to paired segmental arteries, and each pair supplies one of the 31 somites (Lasjaunias et al. 2001). The segmental arteries, via their branches, supply the derivatives of the somites, which are bone and muscle derivatives as well as neural derivatives. Outside the spinal canal, longitudinal intersegmental anastomoses develop (Lasjaunias et al. 2001). In the sacral region, these longitudinal intersegmental arteries form the lateral sacral arteries. The lateral sacral arteries arise from the internal iliac arteries, the persisted proximal portions of the umbilical arteries (Lasjaunias et al. 2001; Sadler 2019). After month 4 of gestation, maturation of the formed vessels and increased tortuosity occur; the arteries fully develop from month 6 onward (Bosmia et al. 2015).

70.2 Anatomy/Topography

The lateral sacral arteries (Fig. 70.1) are posterior branches of the internal iliac artery (Standring 2021). They are accompanied by the lateral sacral veins in their course and are usually doubled (Standring 2021). If there is a single lateral sacral artery, the artery will divide into a superior and inferior branch. The superior branch medially approaches the anterior sacral foramina, and its spinal branches enter the first or second foramen, supply the sacrum and the contents of the sacral canal, including the meninges surrounding the sacral nerves' roots, and may pass posteriorly through the posterior sacral foramen to supply the erector spinae and the skin overlying the sacrum (Standring 2021; Dalley II and Agur 2022). The inferior branch passes obliquely anterior to the piriformis muscle, supplies the piriformis muscle, and descends lateral to the

Y. R. Hakim
Geisinger Commonwealth School of Medicine,
Scranton, PA, USA
e-mail: YHakim@som.geisinger.edu

Y. Hakim (✉)
Tulane University School of Medicine,
New Orleans, LA, USA
e-mail: yhakim@tulane.edu

S. J. Bordes, Jr. et al. (eds.), *The Clinical Anatomy of the Vascular System*,
https://doi.org/10.1007/978-3-031-78326-5_70

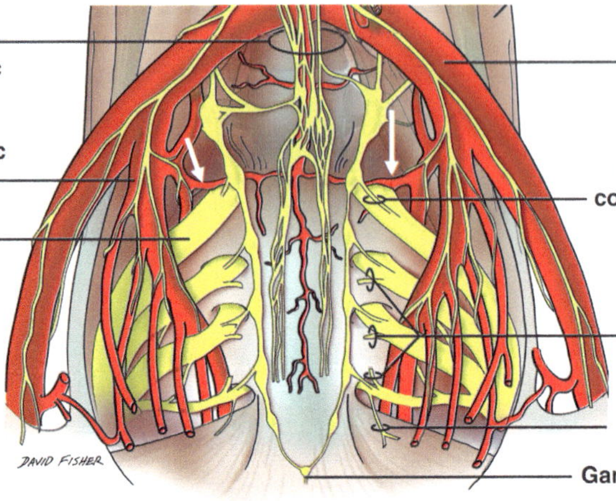

Fig. 70.1 Schematic drawing of the internal iliac artery and its branches noting the lateral sacral arteries (arrows)

sympathetic trunk (Standring 2021). The spinal branches of the inferior branch enter the anterior sacral foramina, supply the same structures as the spinal branches of the superior branch, and may pass posteriorly to supply the erector spinae and the skin (Dalley II and Agur 2022; Standring 2021).

As the lateral sacral arteries pass obliquely lateral to the sympathetic trunk, the distance between them and the sacral midline on the right has been reported as 10.6 ± 3.5 mm and on the left as 14.9 ± 2.1 mm (Güvençer et al. 2009).

The superior branches of the lateral sacral arteries anastomose with the inferior branches and the small parietal branches of the median sacral artery anterior to the coccyx (Standring 2021; Dalley II and Agur 2022). The posterior branches of the lateral sacral arteries, the branches that pass through the posterior sacral foramina and supply the erector spinae and skin, anastomose with the superficial branch of the superior gluteal artery posterior to the sacrum.

70.3 Variations

Origin

The lateral sacral arteries or the superior branch of the lateral sacral arteries can arise from the median sacral artery (Lasjaunias et al. 2001). Also, the lateral sacral arteries and the other arteries of the posterior division of the internal iliac artery can arise from a common trunk from the common iliac artery bifurcation or independently from the internal iliac artery (Bleich et al. 2007).

Branches

The lateral sacral arteries and the other arteries of the posterior division of the internal iliac artery can give rise to the obturator artery, which usually arises from the anterior division of the internal iliac artery (Mamatha et al. 2015). The middle rectal artery, which originates from the anterior division of the internal iliac artery, can arise from the lateral sacral artery (Honma and Sonomura 2018).

Course

The spinal branches of the lateral sacral arteries can enter the anterior sacral foramina in its upper medial quadrant or its inferior lateral quadrant (Liguoro et al. 1999). If the median sacral artery is absent, branches of the lateral sacral arteries will supply the anterior sacral regions that are normally supplied by the median sacral artery (Bono et al. 2018).

Development

The posterior division of the internal iliac artery, including the lateral sacral arteries, has been reported to be missing unilaterally on the left side (Fänder et al. 2018).

70.4 Histology

The lateral sacral arteries are muscular, medium-sized arteries. This was deduced based on the diameter of the median sacral artery, with which the lateral sacral arteries anastomose and which has a similar size to the lateral sacral arteries (Standring 2021). The median sacral artery has a mean diameter of 1.98 ± 0.12 mm, making it a muscular, medium-sized artery since the diameter is between 1 and 10 mm (Dimov et al. 2021; The circulatory system 2021). As muscular, medium-sized arteries, the lateral sacral arteries have an intima with prominent internal elastic lamina, media with multiple layers of smooth muscles and less elastic material, and adventitia with lymphatic capillaries, nerves, and less connective tissue than the media (The circulatory system 2021).

References

Bleich AT, Rahn DD, Wieslander CK, Wai CY, Roshanravan SM, Corton MM (2007) Posterior division of the internal iliac artery: anatomic variations and clinical applications. Am J Obstet Gynecol 197(6):658.e1–658.e6585. https://doi.org/10.1016/j.ajog.2007.08.063

Bono CM, Bhalla A, Garfin SR (2018) Applied anatomy of the spine. In: Eismont FJ, Bell GR, Fischgrund JS, Bono CM (eds) Rothman-Simeone and Herkowitz's the spine. Elsevier Inc, pp 17–55

Bosmia AN, Hogan E, Loukas M, Tubbs RS, Cohen-Gadol AA (2015) Blood supply to the human spinal cord: part I. Anatomy and hemodynamics. Clin Anat (New York, N.Y.) 28(1):52–64. https://doi.org/10.1002/ca.22281

Dalley AF II, Agur AM (2022) Pelvis and perineum. Moore's clinically oriented anatomy. Wolters Kluwer, pp 559–671

DeAnda A, Worsham J, Mell M (2022) The aorta. In: Townsend M, Beauchamp RD, Evers BM, Mattox KL (eds) Sabiston textbook of surgery. Elsevier Inc, pp 1744–1766

Dimov I, Sobczak S, Grabs D, Nayouf A (2021) The median sacral artery in rectal blood supply: a cadaveric study. Clin Anat (New York, N.Y.) 34(3):342–347. https://doi-org.gcsom.idm.oclc.org/10.1002/ca.23611

Fänder J, Büttner M, Kielstein H, Jasinski-Bergner S (2018) A missing posterior division of the internal iliac artery. EJVES Short Rep 40:18–20. https://doi.org/10.1016/j.ejvssr.2018.08.004

Güvençer M, Dalbayrak S, Tayefi H, Tetik S, Yilmaz M, Erginoğlu U, Baskan Ö, Güran S, Naderi S (2009) Surgical anatomy of the presacral area. Surg Radiol Anat: SRA 31(4):251–257. https://doi.org/10.1007/s00276-008-0435-1

Honma S, Sonomura T (2018) The middle rectal artery arising from the lateral sacral artery. Anat Sci Int 93(4):559–562. https://doi.org/10.1007/s12565018-0430-5

Lasjaunias P, Berenstein A, Ter Brugge KG (2001) Spinal and spinal cord arteries and veins. In: Surgical neuroangiography. Springer-Verlag, Berlin Heidelberg, pp 73–164

Liguoro D, Viejo-Fuertes D, Midy D, Guerin J (1999) The posterior sacral foramina: an anatomical study. J Anat 195(2):301–304. https://doi.org/10.1046/j.1469-7580.1999.19520301.x

Mamatha H, Hemalatha B, Vinodini P, Souza AS, Suhani S (2015) Anatomical study on the variations in the branching pattern of internal iliac artery. Indian J Surg 77(Suppl 2):248–252. https://doi.org/10.1007/s12262-012-0785-0

The circulatory system (2021) Mescher AL (Ed) Junqueira's basic histology text and atlas, 16e. McGraw Hill. https://accessmedicinemhmedicalcom.gcsom.idm.oclc.org/content.aspx?bookid=3047&sectionid=255121291

Moore, K. L., Persaud, T. V., & Torchia, M. G. (2020). Cardiovascular system. Before we are born (pp. 175–208). Elsevier Inc.

Sadler TW (2019) Cardiovascular system. In: Langman's medical embryology. Wolters Kluwer, pp 179–222

Standring S (2021) Lesser pelvis and perineum. In: Gray's anatomy. Elsevier Inc., pp 1244–1258

71 Middle Rectal Artery

Yonis Hakim and Yousef Raslan Hakim

71.1 Anatomy/Topography

The middle anoanorectal artery, a branch of the anterior division of the internal iliac artery, descends through the pelvis and passes through the mesorectum's lateral fascia to supply the rectum (Standring 2021; Dalley II and Agur 2022). The middle anorectal artery's branches pass within the lateral rectal ligament (Boxall et al. 1963; Jorge and Habr-Gama 2011; Church et al. 1987). The middle anorectal artery terminates at the rectal ampulla's lateral and inferior walls, and it anastomoses with the superior and inferior anorectal arteries and inferior mesenteric artery (Honma and Sonomura 2018; Dalley II and Agur 2022; Bilhim et al. 2013; Dnyandeo Sonje et al. 2020). If the middle anorectal artery is bilateral, the right and left middle rectal arteries anastomose within the rectal walls (Patrício et al. 1988).

The middle anorectal artery supplies the inferior portion of the rectum, the vagina in females, and the prostate, prostatic urethra, and seminal vesicles in males (Dalley II and Agur 2022). Some studies note that blood supply to the rectum is important (Boxall et al. 1963). On the other hand, some studies suggest that the middle anorectal artery does not significantly contribute to the rectal vasculature (Goligher 1949). The middle anorectal artery contributes to the blood supply of the anal canal, supplying it below the pectinate line (Dalley II and Agur 2022; Sundel et al. 2019).

Course

The middle anorectal artery can pass into the rectum laterally, anterolaterally, or posterolaterally, or it can pass instead to the prostate, bladder, or vagina (Jones et al. 1999; Sato and Sato 1991). One study reported that only 47.4% of the middle rectal arteries passed to the rectum, and the remainder passed to the prostate, urinary bladder, and vagina (Sato and Sato 1991). Another study found that the middle anorectal artery passed into the rectum from the lateral side in 43% of cadavers, the anterolateral side in 11% of cadavers, and the posterolateral side in 7% of cadavers (Jones et al. 1999). The middle anorectal artery's passage from the anterolateral side joins the prostatic artery or inferior vesical artery to form a common trunk (Kiyomatsu et al. 2017). This variation is characterized by a short path and a small diameter vessel (Kiyomatsu et al. 2017). Regarding the anterior and posterior rectal walls, one study reported an anastomosis in the anterior

Y. Hakim (✉)
Tulane University School of Medicine,
New Orleans, LA, USA
e-mail: yhakim@tulane.edu

Y. R. Hakim
Geisinger Commonwealth School of Medicine,
Scranton, PA, USA
e-mail: YHakim@som.geisinger.edu

S. J. Bordes, Jr. et al. (eds.), *The Clinical Anatomy of the Vascular System*,
https://doi.org/10.1007/978-3-031-78326-5_71

and posterior rectal walls between the branches of the right and left middle rectal arteries in the case of bilateral middle rectal arteries (Patrício et al. 1988).

Diameter

The middle anorectal artery's diameter ranges from 0.5 mm to 2.5 mm (Jones et al. 1999; Kiyomatsu et al. 2017). The average diameters are 1.27 mm, 1.6 mm, and 1.7 mm (Lin et al. 2010; Sato and Sato 1991; DiDio et al. 1986).

71.2 Embryology

The middle anorectal artery originates mainly from the branches of the internal iliac artery or the artery itself (Sadler 2018; Zaunbrecher and Samra 2022). Embryologically, the internal iliac artery originates from the umbilical arteries (Sadler 2018; Zaunbrecher and Samra 2022). The umbilical arteries are initially ventral branches of the dorsal aorta that join the common iliac artery and lose their origin during the fourth week of gestation (Sadler 2018). The internal iliac arteries and the superior vesical arteries originate from the proximal portions of the umbilical arteries (Sadler 2018; Zaunbrecher and Samra 2022). On the other hand, the distal portions of the umbilical arteries eventually obliterate and form the medial umbilical ligaments (Sadler 2018; Zaunbrecher and Samra 2022).

71.3 Histology

The average diameter of the middle anorectal artery has been reported to be 1.27 mm, 1.6 mm, and 1.7 mm. It is classified as a muscular artery (Lin et al. 2010; Sato and Sato 1991; DiDio et al. 1986; The circulatory system 2021). As such, the middle anorectal artery has a prominent internal elastic lamina in the intima, multiple layers of smooth muscles in the media, nerves, lymphatics, and less connective tissue/adventitia (The circulatory system 2021).

71.4 Variations

Frequency and Number

The middle anorectal artery has high variability (Ayoub 1978; Kiyomatsu et al. 2017; Boxall et al. 1963; Jones et al. 1999; Iwasa et al. 2021; Shafik and Mostafa 1996; DiDio et al. 1986; Sato and Sato 1991; Bilhim et al. 2013; Lin et al. 2010). The middle anorectal artery can be unilateral, bilateral, or present as two middle rectal arteries (Sato and Sato 1991; Kiyomatsu et al. 2017). In one study, all middle rectal arteries identified were unilateral (Jones et al. 1999). In another, most of the identified middle rectal arteries were unilateral, except for one-third, which was bilateral (Bilhim et al. 2013). Lin et al. (2010) found the frequency of unilateral middle rectal arteries (31.3%) to be a little more than double the frequency of bilateral middle rectal arteries (12.5%) (Lin et al. 2010). Iwasa et al. (2021) found the frequency of unilateral middle rectal arteries (34.3%) to be nearly equal to the frequency of bilateral middle rectal arteries (31.4%). DiDio et al. (1986) noted that the middle anorectal artery had a frequency of 56.7%, more of which were bilateral (36.7%).

Origin

The middle anorectal artery can originate from the inferior gluteal artery, internal pudendal artery, obturator artery, or common gluteal-pudendal trunk (Bilhim et al. 2011). The middle anorectal artery can also originate from the posterior division of the internal iliac artery and then form a common trunk with the lateral sacral artery or from the lateral sacral artery (Honma and Sonomura 2018). The middle anorectal artery may share the exact origin of the inferior vesical artery (Bilhim et al. 2011; Brennan et al. 2019).

Studies found that the majority of middle rectal arteries originate from the internal pudendal artery, followed by the inferior gluteal artery, then the internal iliac artery (DiDio et al. 1986; Sato and Sato 1991; Bilhim et al. 2013; Lin et al.

2010). A study found that 67%, 22%, and 11% of the identified middle rectal arteries originated from the internal pudendal (DiDio et al. 1986). The study reported middle rectal arteries originating from the internal pudendal artery with a frequency of 40% among identified middle rectal arteries, followed by the inferior gluteal artery and internal iliac artery, respectively (Lin et al. 2010). Similarly, another study reported that 63%, 21%, and 16% of the middle rectal arteries originated from the arteries above, respectively (Sato and Sato 1991). Two studies reported similar results but with slight differences (Bilhim et al. 2013; DiDio et al. 1986). One of these studies reported identical frequencies but reported the middle anorectal artery originating from the common gluteal pudendal trunk and obturator artery (Bilhim et al. 2013). The authors reported that among the middle rectal arteries, most originated from the internal pudendal artery, followed by the inferior gluteal artery, common gluteal-pudendal trunk, and obturator artery, with frequencies of 60%, 21.3%, 16.2%, and 2.5%, respectively (Bilhim et al. 2013). The second study found a similar trend regarding the origin of the middle anorectal artery but reported a lesser frequency for origination from the internal pudendal artery and the internal iliac artery, with frequencies of 26.7% and 16.8% (DiDio et al. 1986).

In juxtaposition to the inferior vesical artery, there is a similarity in the origin of the middle anorectal artery (Bilhim et al. 2011; Bilhim et al. 2013). One study found that 70% of the middle rectal arteries had a common origin with inferior vesical arteries (Bilhim et al. 2013). Another study found that the inferior vesical artery originated from the same vessels as the middle anorectal artery. The authors reported that the inferior vesical artery originated from the internal pudendal artery, followed by the common gluteal-pudendal trunk, obturator artery, and the inferior gluteal artery in 56%, 28%, 6%, and 4% of the cases, respectively (Bilhim et al. 2011). In contrast to the middle anorectal artery, the inferior vesical artery had different frequencies for its origin from the inferior gluteal artery, common gluteal-pudendal trunk, and obturator artery. The prostatic artery originates from the inferior gluteal artery with lesser frequency, 4%, in contrast to 21–27% (Bilhim et al. 2011, 2013; DiDio et al. 1986; Sato and Sato 1991; Lin et al. 2010). The artery may also originate from the common gluteal-pudendal trunk and obturator artery with higher frequencies of 28% and 6%, in contrast to 16.2% and 2.5% for middle rectal arteries, respectively (Bilhim et al. 2011, 2013).

71.5 Imaging

The middle anorectal artery and pelvic arteries can be mapped and visualized by three-dimensional (3D) contrast-enhanced magnetic resonance (MR) angiography (Naguib et al. 2008). However, detecting the middle anorectal artery may be difficult due to the artery's small diameter (Naguib et al. 2008).

The middle anorectal artery can also be visualized using contrast-enhanced magnetic resonance imaging (MRI), which is more practical than angiography (Iwasa et al. 2021). Contrast-enhanced MRI also allowed the detection of lateral lymph node metastasis (LLNM), the incidence of which was higher when the middle anorectal artery was present than when absent and when bilateral than unilateral (Iwasa et al. 2021).

71.6 Pathology

Hemorrhoids

Hemorrhoids cause anorectal bleeding and occur as a result of dilation of the sinusoids that are supplied by the superior anorectal artery and, to a lesser extent, by the middle and inferior anorectal arteries (Makris et al. 2021; Lunsford et al. 2022; Rakinic and Poola 2014). The dilation of the sinusoids is associated with increased pressure in the pelvic floor, which results in the breakdown of the surrounding connective tissue and prolapse of the anal cushions (Lunsford et al. 2022; Felt-Bersma and Bartelsman 2009). Predisposing factors include chronic diarrhea, anal intercourse, constipation with straining, pregnancy, and connective tissue disorders (Lunsford et al. 2022; Felt-Bersma and Bartelsman 2009).

For the treatment of hemorrhoids, if conservative therapy fails, hemorrhoidectomy is the gold standard surgical procedure in which hemorrhoids are excised (Makris et al. 2021). There are less invasive procedures, such as the Doppler-Guided Hemorrhoidal Arteries Ligation (DG-HAL), but they involve anal canal trauma (Makris et al. 2021). Endovascular embolization of the superior anorectal artery, Emborrhoid, has been used recently (Talaie et al. 2022). It avoids anorectal trauma and allows for the occlusion of the anastomosis with the middle anorectal artery (Talaie et al. 2022; Ferrer Puchol et al. 2020). In Emborrhoid, the superior anorectal artery is occluded with coils and the middle anorectal artery with particles (Ferrer Puchol et al. 2020). Hemorrhoidal blood supply restoration occurs mainly through the middle anorectal artery, and in these cases, the middle anorectal artery has to be occluded with coils (Ferrer Puchol et al. 2020).

Rectal Cancer

Rectal cancers can spread longitudinally or laterally, approaching the middle anorectal artery's region (Gulevski et al. 2019). Gulevski et al. (2019) found that 90% of lateral lymph node metastasis (LLNM) in low rectal cancer occurred in the regions of the middle rectal and obturator arteries. Some of the risk factors for LLNM are female sex, greater than 6.8 nm/mL serum carcinoembryonic antigen levels after chemoradiotherapy and preoperatively, and low rectal tumors, classified as being less than or equal to 5 cm from the anal verge (Gulevski et al. 2019).

In the surgical treatment of rectal cancers, the rectum can be mobilized in most cases without the identification of the middle anorectal artery, given its varied prevalence (Chen and Glasgow 2019). However, coagulation should be minimized if the middle anorectal artery is present, and the artery can be divided by electrocautery (Pugliese et al. 2008; Fung et al. 2021). Also, given LLNM surrounding the middle anorectal artery, the artery can be followed, and surrounding lymphatic tissue can be removed (Elhusseini and Aly 2020).

71.7 Surgical and Clinical Considerations

Cervical Cancer

Cervical cancer is treated with radical hysterectomy as the gold standard (Panici et al. 2011). Radical hysterectomy is performed through laparotomy as a standard approach (Panici et al. 2011). A total extraperitoneal radical hysterectomy is also possible (Panici et al. 2011). In this surgical procedure, the middle anorectal artery is isolated, clipped, and ligated close to its origin (Panici et al. 2011). In another approach, the class III nerve-sparing radical hysterectomy (NSRH), the middle anorectal artery is used as a landmark to distinguish between vascular and neural tissue (Rob et al. 2010).

Uterus Organ Harvest

In organ harvest surgery for uterine transplant, the internal iliac artery distal to the uterine artery's origin is clipped and ligated. Still, the middle anorectal artery is preserved as it is proximal to the uterine artery's origin (Puntambekar et al. 2019).

Benign Prostatic Hyperplasia (BPH)

For the diagnosis of BPH, transrectal ultrasound (TRUS)-guided needle biopsy can be used (Kilciler et al. 2008). However, care should be taken not to injure the middle anorectal artery and other vessels passing through the rectum when obtaining tissue samples (Kilciler et al. 2008).

A new treatment method called prostatic artery embolization (PAE) can be used. One study performed bilateral embolization of the prostatic arteries, unless there was an anastomosis between the prostatic artery and the middle anorectal artery or arteriosclerosis (Kisilevzky and Faintuch 2016). To reduce the risk of nontarget embolization during PAE, protective embolization of nontarget arteries, including the middle anorectal artery, can be performed (Pilan et al. 2022).

Rectal Bleeding

Rectal bleeding involves bleeding from multiple foci, exceeding the foci in intestinal bleeding (Park et al. 2020). Hence, a complete angiographic evaluation of the internal iliac and inferior mesenteric arteries is important at the beginning of the assessment because a bleeding focus may be missed if it is from the middle or inferior rectal arteries (Park et al. 2020).

Diverticulitis and Intestinal Ischemia

Given that the superior anorectal artery is a branch of the inferior mesenteric artery and that it anastomoses with the middle anorectal artery, the inferior mesenteric artery is not ligated during surgical treatment of diverticulitis to avoid reduction in the blood supply of the rectum, especially in older patients, who may have compromised ability to compensate for the blood supply's changes (Prevot et al. 2016; Hawkins et al. 2020). Also, due to the anastomoses and collaterals in the rectum, the rectum is spared when treating intestinal ischemia (Gore et al. 2008).

References

Ayoub SF (1978) Arterial supply to the human rectum. Acta Anat 100(3):317–327. https://doi.org/10.1159/000144913

Bilhim T, Pisco JM, Furtado A, Casal D, Pais D, Pinheiro LC, O'Neill JE (2011) Prostatic arterial supply: demonstration by multirow detector angio CT and catheter angiography. Eur Radiol 21(5):1119–1126. https://doi.org/10.1007/s00330-010-2015-0

Bilhim T, Pereira JA, Tinto HR, Fernandes L, Duarte M, O'Neill JE, Pisco JM (2013) Middle rectal artery: myth or reality? Retrospective study with CT angiography and digital subtraction angiography. Surg Radiol Anat 35(6):517–522. https://doi.org/10.1007/s00276-012-1068-y

Boxall TA, Smart PJ, Griffiths JD (1963) The blood-supply of the distal segment of the rectum in anterior resection. Br J Surg 50:399–404. https://doi.org/10.1002/bjs.18005022209

Brennan P, Standring S, Wiseman S (eds) (2019) Gray's surgical anatomy e-book. Elsevier Health Sciences

Chen I, Glasgow SC (2019) The technical aspects of rectal cancer surgery. Semin Colon Rectal Surg 30(2):68–74

Church JM, Raudkivi PJ, Hill GL (1987) The surgical anatomy of the rectum—a review with particular relevance to the hazards of rectal mobilisation. Int J Color Dis 2(3):158–166. https://doi.org/10.1007/BF01648000

Dalley AF II, Agur AM (2022) Pelvis and perineum. Moore's Clinically oriented anatomy. Wolters Kluwer, pp 559–671

DiDio LJ, Diaz-Franco C, Schemainda R, Bezerra AJ (1986) Morphology of the middle rectal arteries. A study of 30 cadaveric dissections. Surg Radiol Anat 8(4):229–236. https://doi.org/10.1007/BF02425072

Dnyandeo Sonje P, Kanaskar N, Vatsalaswamy P (2020) Investigating the middle rectal artery and its clinical significance. Anat Sci J 17(1):37–42

Elhusseini M, Aly EH (2020) Lateral pelvic lymph node dissection in the management of locally advanced low rectal cancer: summary of the current evidence. Surg Oncol 35:418–425. https://doi.org/10.1016/j.suronc.2020.09.021

Felt-Bersma RJ, Bartelsman JF (2009) Haemorrhoids, rectal prolapse, anal fissure, peri-anal fistulae and sexually transmitted diseases. Best Pract Res Clin Gastroenterol 23(4):575–592. https://doi.org/10.1016/j.bpg.2009.04.010

Ferrer Puchol MD, Esteban Hernández E, Blanco González FJ, Ramiro Gandia R, Solaz Solaz J, Pacheco Usmayo A (2020) Selective intra-arterial embolization to treat hemorrhoids. Embolización intraarterial selectiva como tratamiento de la patología hemorroidal. Radiologia 62(4):313–319. https://doi.org/10.1016/j.rx.2019.12.004

Fung T, Tsukada Y, Ito M (2021) Essential anatomy for total mesorectal excision and lateral lymph node dissection, in both trans-abdominal and trans-anal perspective. Surgeon 19(6):e462–e474. https://doi.org/10.1016/j.surge.2020.09.011

Goligher JC (1949) The blood-supply to the sigmoid colon and rectum with reference to the technique of rectal resection with restoration of continuity. Br J Surg 37(146):157–162. https://doi.org/10.1002/bjs.18003714604

Gore RM, Thakrar KH, Mehta UK, Berlin J, Yaghmai V, Newmark GM (2008) Imaging in intestinal ischemic disorders. Clin Gastroenterol Hepatol 6(8):849–858. https://doi.org/10.1016/j.cgh.2008.05.007

Gulevski S, Russell MM, Buchwald P (2019) Significance, diagnosis and treatment of lateral lymph nodes in rectal cancer: a systematic review. Int J Surg Open 21:73–82

Hawkins AT, Wise PE, Chan T, Lee JT, Glyn T, Wood V, Eglinton T, Frizelle F, Khan A, Hall J, Ilyas M, Michailidou M, Nfonsam VN, Cowan ML, Williams J, Steele SR, Alavi K, Ellis CT, Collins D, Winter DC et al (2020) Diverticulitis: an update from the age old paradigm. Curr Probl Surg 57(10):100862. https://doi.org/10.1016/j.cpsurg.2020.100862

Honma S, Sonomura T (2018) The middle rectal artery arising from the lateral sacral artery. Anat Sci Int 93(4):559–562. https://doi.org/10.1007/s12565-018-0430-5

Iwasa Y, Koyama F, Marugami N, Kuge H, Nakamoto T, Obara S, Nishiwada S, Takei T, Sadamitsu T, Yamauchi S, Kichikawa K, Sho M (2021) The middle rectal artery detected by contrast-enhanced magnetic resonance imaging predicts lateral lymph node metastasis in lower rectal cancer. Int J Color Dis 36(8):1677–1684. https://doi.org/10.1007/s00384-021-03887-w

Jones OM, Smeulders N, Wiseman O, Miller R (1999) Lateral ligaments of the rectum: an anatomical study. Br J Surg 86(4):487–489. https://doi.org/10.1046/j.1365-2168.1999.01080.x

Jorge JMN, Habr-Gama A (2011) Anatomy and embryology. In: The ASCRS textbook of colon and rectal surgery. Springer, New York, NY, pp 1–22

Kilciler M, Erdemir F, Demir E, Güven O, Avci A (2008) The effect of rectal Foley catheterization on rectal bleeding rates after transrectal ultrasound-guided prostate biopsy. J Vasc Interv Radiol 19(9):1344–1346. https://doi.org/10.1016/j.jvir.2008.06.005

Kisilevzky N, Faintuch S (2016) MRI assessment of prostatic ischaemia: best predictor of clinical success after prostatic artery embolisation for benign prostatic hyperplasia. Clin Radiol 71(9):876–882. https://doi.org/10.1016/j.crad.2016.05.003

Kiyomatsu T, Ishihara S, Murono K, Otani K, Yasuda K, Nishikawa T, Tanaka T, Hata K, Kawai K, Nozawa H, Yamaguchi H, Watanabe T (2017) Anatomy of the middle rectal artery: a review of the historical literature. Surg Today 47(1):14–19. https://doi.org/10.1007/s00595-016-1359-8

Lin M, Chen W, Huang L, Ni J, Yin L (2010) The anatomy of lateral ligament of the rectum and its role in total mesorectal excision. World J Surg 34(3):594–598. https://doi.org/10.1007/s00268-009-0371-1

Lunsford TN, Atia MA, Kagbo-Kue S, Harris LA (2022) A pain in the butt: hemorrhoids, fissures, fistulas, and other anorectal syndromes. Gastroenterol Clin N Am 51(1):123–144. https://doi.org/10.1016/j.gtc.2021.10.008

Makris GC, Thulasidasan N, Malietzis G, Kontovounisios C, Saibudeen A, Uberoi R, Diamantopoulos A, Sapoval M, Vidal V (2021) Catheter-directed hemorrhoidal dearterialization technique for the management of hemorrhoids: a meta-analysis of the clinical evidence. J Vasc Interv Radiol 32(8):1119–1127. https://doi.org/10.1016/j.jvir.2021.03.548

Naguib NN, Nour-Eldin NE, Hammerstingl RM, Lehnert T, Floeter J, Zangos S, Vogl TJ (2008) Three-dimensional reconstructed contrast-enhanced MR angiography for internal iliac artery branch visualization before uterine artery embolization. J Vasc Interv Radiol 19(11):1569–1575. https://doi.org/10.1016/j.jvir.2008.08.012

Panici PB, Di Donato V, Plotti F, Musella A, Sansone M, Angioli R, Perniola G, Bellati F (2011) Feasibility and safety of type C2 total extraperitoneal abdominal radical hysterectomy (TEARH) for locally advanced cervical cancer. Gynecol Oncol 120(3):423–429. https://doi.org/10.1016/j.ygyno.2010.12.348

Park S, Kim Y, Shin JH, Yang WJ, Noh SY, Chu HH, Chen CS, Yoon HK (2020) Outcome of rectal arterial embolization for rectal bleeding in 34 patients: a single-center retrospective study over 20 years. J vasc Interv Radiol 31(4):576–583. https://doi.org/10.1016/j.jvir.2019.05.013

Patrício J, Bernades A, Nuno D, Falcão F, Silveira L (1988) Surgical anatomy of the arterial blood-supply of the human rectum. Surg Radiol Anat 10(1):71–75. https://doi.org/10.1007/BF02094074

Pilan BF, de Assis AM, Moreira AM, Rodrigues V, Carnevale FC (2022) Protection of non-target structures in prostatic artery embolization. Radiol Bras 55(1):6–12. https://doi.org/10.1590/0100-3984.2021.0021

Prevot F, Sabbagh C, Mauvais F, Regimbeau JM (2016) Colectomy in patients with previous colectomy or occlusive vascular diseases: Pitfalls and precautions. J Visc Surg 153(2):113–119. https://doi.org/10.1016/j.jviscsurg.2016.02.002

Pugliese R, Di Lernia S, Sansonna F, Scandroglio I, Maggioni D, Ferrari GC, Costanzi A, Magistro C, De Carli S (2008) Results of laparoscopic anterior resection for rectal adenocarcinoma: retrospective analysis of 157 cases. Am J Surg 195(2):233–238. https://doi.org/10.1016/j.amjsurg.2007.02.020

Puntambekar S, Puntambekar S, Telang M, Kulkarni P, Date S, Panse M, Sathe R, Agarkhedkar N, Warty N, Kade S, Manchekar M, Chitale M, Parekh H, Parikh K, Mehta M, Kinholkar B, Jana JS, Pare A, Kanade S, Sadre A et al (2019) Novel anastomotic technique for uterine transplant using utero-ovarian veins for venous drainage and internal iliac arteries for perfusion in two laparoscopically harvested uteri. J Minim Invasive Gynecol 26(4):628–635. https://doi.org/10.1016/j.jmig.2018.11.021

Rakinic J, Poola VP (2014) Hemorrhoids and fistulas: new solutions to old problems. Curr Probl Surg 51(3):98–137. https://doi.org/10.1067/j.cpsurg.2013.11.002

Rob L, Halaska M, Robova H (2010) Nerve-sparing and individually tailored surgery for cervical cancer. Lancet Oncol 11(3):292–301. https://doi.org/10.1016/S1470-2045(09)70191-3

Sadler TW (2018) Langman's medical embryology. Lippincott Williams & Wilkins

Sato K, Sato T (1991) The vascular and neuronal composition of the lateral ligament of the rectum and the rectosacral fascia. Surg Radiol Anat 13(1):17–22. https://doi.org/10.1007/BF01623135

Shafik A, Mostafa H (1996) Study of the arterial pattern of the rectum and its clinical application. Acta Anat 157(1):80–86. https://doi.org/10.1159/000147869

Standring S (2021) Lesser pelvis and perineum. In: Gray's anatomy. Elsevier Inc., pp 1244–1258

Sundel MH, Voltaggio L, Leeds IL, Fang SH (2019) Anatomy and histology of the anus. In: Anal cancer. Springer, Cham, pp 1–9

Talaie R, Torkian P, Moghadam AD, Tradi F, Vidal V, Sapoval M, Golzarian J (2022) Hemorrhoid embolization: a review of current evidences. Diagn Interv Imaging 103(1):3–11. https://doi.org/10.1016/j.diii.2021.07.001

The circulatory system (2021) In: Mescher AL (ed) Junqueira's basic histology text and atlas, 16e. McGraw Hill. https://accessmedicine-mhmedical-com.gcsom.idm.oclc.org/content.aspx?bookid=3047&sectionid=255121291

Zaunbrecher N, Samra NS (2022) Anatomy, abdomen and pelvis, internal iliac arteries. StatPearls Publishing, StatPearls

Iliolumbar Artery

72

Ryan Schroeder, Brendan Tate,
and Evan Sinnathamby

72.1 Embryology

During the third week of development, a primitive heart and vasculature form. The branches of the lower abdominal aorta begin to form at the 6 mm stage embryo (Qazi et al. 2022). In the fourth week of fetal development, the umbilical arteries anastomose with the fifth lumbar intersegmental artery branches to become the dominant placental-aortic connection (Qazi et al. 2022). The umbilical arteries arise as a branch off the dorsal aorta, but once they anastomose and communicate with the intersegmental artery, they disconnect from the dorsal aorta (Hegazy 2016). The fifth lumbar intersegmental artery gives rise to the external iliac artery, while the umbilical artery is attached to its distal end, which later forms the internal iliac artery (Greebe 1977). The proximal part of the internal iliac artery will persist as the distal end obliterates (Zaunbrecher and Samra 2022). The internal iliac artery bifurcates into an anterior and posterior division as development continues. Branches of the anterior division include the superior and inferior vesical arteries, middle rectal artery, uterine arteries, and vaginal artery. The first branch of the posterior division to develop is the iliolumbar artery. The lateral sacral and superior gluteal arteries also develop as branches of the posterior division of the internal iliac artery.

R. Schroeder · B. Tate · E. Sinnathamby (✉)
School of Medicine, Louisiana State University
Health Sciences Center, New Orleans, LA, USA
e-mail: esinna@lsuhsc.edu

72.2 Anatomy/Topography

The iliolumbar artery classically originates from the posterior trunk of the internal iliac artery. The mean diameter of the iliolumbar artery at its origin was 3.5 ± 0.5 mm (Teli et al. 2013). It extends obliquely from the posterior division of the internal iliac artery in a superior and lateral fashion anterior to the sacroiliac joint and lumbosacral trunk (Teli et al. 2013). It is important to note that the iliolumbar artery is the only artery that ascends from the pelvic cavity (Talalwah et al. 2014). It runs deep to the external iliac vessels between the obturator nerve and the lumbosacral plexus and then crosses deep to the psoas major muscle. When it reaches the lateral border of the psoas major muscle, the iliolumbar artery will run between the iliac fascia and the iliacus muscle toward the iliac crest (Quain 1840).

The standard branches leaving the iliolumbar artery are the lumbar, iliac, and spinal branches. The lumbar branch runs posteriorly and leads into and supplies the psoas and quadratus lumborum muscles. It also forms anastomoses with the lumbar artery (Quain 1840). The iliac branch travels laterally into the iliac fossa to supply the iliacus muscle and the ilium. This iliolumbar iliac branch anastomoses with nearby branches of the

S. J. Bordes, Jr. et al. (eds.), *The Clinical Anatomy of the Vascular System*,
https://doi.org/10.1007/978-3-031-78326-5_72

deep circumflex iliac artery, the lateral circumflex femoral artery, the obturator artery, and the superior gluteal artery (Quain 1840). These anastomoses allow the iliolumbar artery to contribute to the supply of the abdominal wall and gluteal muscles. The small spinal branch will pass between the first sacral and fifth lumbar vertebrae to enter the vertebral canal and supply the cauda equina. The iliolumbar artery branches again to supply the ilium and is its largest nutrient-supplying vessel (Rusu et al. 2010a).

72.3 Histology

Histologically, the iliolumbar artery is considered a muscular artery. Its tunica intima consists of a layer of endothelium with a subendothelial connective tissue layer. Like most muscular arteries, it has a well-defined internal elastic lamina. The intima is surrounded by the tunica media, composed of smooth muscle and elastic fibers. The tunica adventitia, the outermost layer, mainly consists of connective tissue.

72.4 Variation

The iliolumbar artery commonly branches from the posterior division of the internal iliac artery. However, it is common to encounter the iliolumbar artery originating from the trunk of the internal iliac artery before its bifurcation into anterior and posterior branches. Several cadaveric studies have found the iliolumbar artery arising from the posterior division of the internal iliac in approximately 52% of cases and from the internal iliac artery before its bifurcation in 47% of cases (Sankaranarayanan and Rajasekhar 2019). The distance between the bifurcation of the common iliac artery and the iliolumbar artery is 11–20 mm (Sankaranarayanan and Rajasekhar 2019). The iliolumbar artery passes laterally between the obturator anteriorly and the lumbosacral trunk posteriorly, with this relationship observed in several studies (Kiray et al. 2010; Sankaranarayanan and Rajasekhar 2019; Koç et al. 2016). In a minority of cases, the iliolumbar artery will pass between the L4 and L5 roots of the lumbosacral trunk (Rusu et al. 2010b).

The length of the iliolumbar artery varies between 10 and 14 mm depending on its origin. The artery is longer when it originates from the common trunk of the internal iliac artery, as it must traverse a greater distance before bifurcating (Kiray et al. 2010; Sankaranarayanan and Rajasekhar 2019; Souza et al. 2015). The iliolumbar artery has diameters ranging from 1.2 to 3.7 mm. The iliolumbar artery anastomoses with other significant arteries within the pelvis, namely the lower lumbar, middle sacral, and lateral sacral arteries (Koç et al. 2016). As mentioned earlier, the anastomosis is formed via the lumbar branch of the iliolumbar artery. Together, these arteries supply the lower lumbar vertebrae (Tezuka et al. 2016). The iliolumbar artery supplies the psoas major and quadratus lumborum muscles through its lumbar branch and the iliacus and ilium via its iliacus branch.

72.5 Clinical and Surgical Significance

Due to its proximity to the obturator nerve, lumbosacral trunk, and other adjacent structures, it is important to consider variations in the course and branching pattern of the iliolumbar artery during surgical intervention, particularly in orthopedic procedures involving the lumbosacral or sacroiliac joint (Kiray et al. 2010). Examples of procedures in which the location of the iliolumbar artery is of particular importance include L5-S1 disc excision and lumbar spinal fixation (Kiray et al. 2010; Koç et al. 2016).

The iliolumbar artery is particularly useful in reconstructive surgery because of its appropriate size, location, and ease of access. It can be used as a nutrient pedicle in an iliac crest flap in lumbar spine reconstruction (Winters et al. 2002). It has also been used for maxillofacial reconstruction procedures and to facilitate healing in severely displaced femoral neck fractures (Singh and Andi 2017; Chen et al. 1999).

The iliac vessels are particularly susceptible to damage from penetrating wounds to the lower abdomen, which can lead to hemorrhage, shock, and death if not treated promptly (Ksycki et al. 2012). Because of its relationship with the sacroiliac joint, the iliolumbar artery can also be damaged in open book and shearing fractures of the pelvis brought on by blunt force trauma (Kiray et al. 2010). One must consider damage to the iliolumbar artery in patients with injuries of this nature.

Occlusion of the vasculature in any area proximal to the iliolumbar artery's point of origin will lead to decreased blood flow to the structures supplied by this artery. Stenosis of the internal iliac artery can cause reduced blood flow to the legs and buttocks, leading to claudication and other neurologic deficits (Prince et al. 2013; Huétink et al. 2008). Aortoiliac occlusive disease, also known as Leriche syndrome, is caused by atherosclerotic changes of the aortic bifurcation and can lead to claudication, impotence, and a reduced pulse in the lower extremities (Guan et al. 2021). Surgical intervention may be required in these cases if symptoms are not improved with conservative management and lifestyle modification.

References

Chen RS, Liu YX, Liu CB et al (1999) Anatomic basis of iliac crest flap pedicled on the iliolumbar artery. Surg Radiol Anat 21(2):103–107

Greebe J (1977) Congenital anomalies of the iliofemoral artery. J Cardiovasc Surg 18(3):317–323

Guan J, Peng Y, Ouyang L et al (2021) Evaluation of therapeutic effects on collateral circulation in patients with chronic Leriche syndrome: a case-control study of intraluminal stent implantation and surgical bypass grafting. Cardiovasc Intervent Radiol 44(4):550–557

Hegazy AA (2016) Anatomy and embryology of umbilicus in newborns: a review and clinical correlations. Front Med 10(3):271–277

Huétink K, Steijling JJF, Mali WPTM (2008) Endovascular treatment of the internal iliac artery in peripheral arterial disease. Cardiovasc Intervent Radiol 31(2):391–393

Kiray A, Akçalı O, Tayefi H, Koşay C, Ergür I (2010) Anatomical variations of iliolumbar artery and its relation with surgical landmarks. Acta Orthop Traumatol Turc 44(6):464–468

Koç T, Gilan İY, Aktekin M, Kurtoğlu Z, Dağtekin A, Aytaç G, Coşgun E (2016) Evaluation of the origin and branching patterns of the iliolumbar artery and its implications on pelvic and vertebral surgery. Saudi Med J 37(4):457–460

Ksycki M, Ruiz G, Perez-Alonso A et al (2012) Iliac vessel injuries: difficult injuries and difficult management problems. Eur J Trauma Emerg Surg 38(4):347–357

Prince JF, Smits MLJ, van Herwaarden JA et al (2013) Endovascular treatment of internal iliac artery stenosis in patients with buttock claudication. PLoS One 8(8):1

Qazi E, Wilting J, Patel NR et al (2022) Arteries of the lower limb—embryology, variations, and clinical significance. Can Assoc Radiol J 73(1):259–270

Quain R (1840) The anatomy of the arteries of the human body, with its applications to pathology and operative surgery, in lithographic drawings, with practical commentaries. Prov Med Surg J 1(12):203

Rusu MC, Cergan R, Dermengiu D, Curcă GC, Folescu R, Motoc AG, Jianu AM (2010a) The iliolumbar artery-anatomic considerations and details on the common iliac artery trifurcation. Clin Anat 23(1):93–100

Rusu MC, Cergan R, Demengiu D, Curcă GC, Folescu R, Motoc AG et al (2010b) The iliolumbar artery-anatomic considerations and details on the common iliac trifurcation. Clin Anat 23:93–100

Sankaranarayanan G, Rajasekhar SSSN (2019) Anatomical variations of the principal nutrient pedicle for iliac crest graft: the ilio-lumbar artery. Surg Radiol Anat 41(1):125–132

Singh T, Andi K (2017) Iliolumbar artery: a useful pedicle for the iliac crest free flap in maxillofacial reconstruction. Br J Oral Maxillofac Surg 55(3):270–273

Souza AD, Ankolekar VH, Manjappa AB, Souza ASD, Hosapatna M (2015) Variations and morphometry of iliolumbar artery and its relations with surgical landmarks. Acta Biomed Sci 2(212):214

Talalwah A, Waseem, Al Dorazi S, Soames R (2014) The origin variability of the iliolumbar artery and iatrogenic sciatica. Eur J Orthop Surg Traumatol 25:199–204

Teli CG, Kate NN, Kothandaraman U (2013) Morphometry of the iliolumbar artery and the iliolumbar veins and their correlations with the lumbosacral trunk and the obturator nerve. J Clin Diagn Res 7(3):422–426

Tezuka F, Sakai T, Nishisho T, Takata Y, Higashino K, Takao S, Harada M, Sairyo K (2016) Variations in arterial supply to the lower lumbar spine. Eur Spine J 25(12):4181–4187

Winters HAH, van Harten SM, van Royen BJ (2002) The iliolumbar artery as the nutrient pedicle for an iliac crest graft: a new technique in reconstruction of the lumbar spine. Plast Reconstr Surg 109(1):249–252

Zaunbrecher N, Samra NS (2022) Anatomy, abdomen and pelvis, internal iliac arteries. [Updated 2022 Jan 26]. In: StatPearls [Internet]. StatPearls Publishing, Treasure Island

Veins of the Abdominal Wall

73

Ronit Patnaik and Lisandro Montorfano

73.1 Embryology

The embryo has three broad types of veins: cardinal, umbilical, and vitelline (Ilina and Lilley 2019). The veins of the systemic circulation are derived from the cardinal veins; a portion of the inferior vena cava (IVC), along with the portal venous system, is derived from the vitelline veins of the abdominal wall, which drain into the systemic veins.

Initially, the left and right umbilical veins are symmetrical and connected to the sinus venosus. With the expansion and dominance of the right vitelline veins, the right umbilical vein regresses entirely. Its remnants, along with the right umbilical artery, form the right medial umbilical ligament, which is found in the anterior abdominal wall in the adult (Sadler 2018).

As the liver develops and impinges on the remaining left umbilical vein, a channel in the liver called the ductus venosus develops that allows blood to flow from the distal portion of the left umbilical vein through the liver and into the sinus venosus. This bypasses the proximal portion of the left umbilical vein, which then regresses. The net result of this remodeling is that oxygenated blood flows from the placenta into the left umbilical vein, through the ductus venosus in the liver, and then into the inferior vena cava and the right atrium of the heart. After birth, the left umbilical vein regresses along with the left umbilical artery to form the left medial umbilical ligament on the anterior abdominal wall and the round ligament of the liver (ligamentum teres hepatis), where the left umbilical vein was connected to the liver. The ductus venosus within the liver regresses to form the ligamentum venosum (Sadler 2018).

73.2 Anatomy/Topography

The venous drainage of the abdominal wall consists of an elaborate subcutaneous venous plexus that is more variable than the arterial pathways (Reardon et al. 2004). In general, however, the veins follow their namesake arteries; they can present as individual veins or as venae comitantes, accompanying veins (Rozen et al. 2009).

The skin of the anterior abdominal wall is drained by branches of the superior and inferior epigastric veins. The skin of the flanks is drained by branches of the intercostal, lumbar, and deep circumflex iliac veins. The inguinal region is drained by the superficial epigastric, the superficial circumflex iliac, and the superficial external

R. Patnaik
Department of Surgery, University of Texas Health, San Antonio, TX, USA

L. Montorfano (✉)
Department of Surgery, Cleveland Clinic Florida, Weston, FL, USA

Department of Plastic and Reconstructive Surgery, Vanderbilt University, Nashville, TN, USA

S. J. Bordes, Jr. et al. (eds.), *The Clinical Anatomy of the Vascular System*,
https://doi.org/10.1007/978-3-031-78326-5_73

pudendal veins, which drain into the femoral vein.

This plexus drains superiorly and laterally to the lateral thoracic veins, which drain into the axillary vein. The thoracoepigastric vein runs along the lateral aspect of the trunk between the superficial epigastric vein below and the lateral thoracic vein above, establishing a communication between the femoral and axillary veins. It is unique because it drains to both the superior and inferior vena cava. Hence, it serves as a link between the two caval systems. Furthermore, the thoracoepigasatric vein is connected to the portal vein via the paraumbilical veins, so it also serves as a portocaval anastomosis (Netter 2014).

The venous plexus drains superiorly and medially to the internal thoracic (mammary) vein, which drains into the brachiocephalic vein bilaterally. The superior epigastric vein (medial) and the musculophrenic vein (lateral) are the two terminal tributaries to the internal thoracic vein. The superior epigastric vein anastomoses with the inferior epigastric vein at the umbilicus, where it also interacts with the paraumbilical veins. These veins provide outflow/drainage for the upper abdominal wall.

The venous plexus drains inferiorly to the superficial and inferior epigastric veins. The inferior epigastric vein drains into the external iliac vein, just above the inguinal ligament. It runs upward, passes medial to the deep inguinal ring, pierces the transversalis fascia at the lateral border of the rectus abdominis, and enters the rectus sheath. Within the rectus sheath, it drains the rectus muscle and anastomoses with the superior epigastric vein around the umbilicus. The superficial epigastric vein drains into the greater saphenous vein at the fossa ovalis. It travels alongside the superficial epigastric artery and drains the superficial lower abdominal wall. In addition to the inferior epigastric vein, the deep circumflex iliac vein also drains the lower abdominal wall musculature. The flank muscles receive venous drainage from small tributaries of the intercostal, subcostal, and lumbar veins, which accompany their corresponding arteries.

The paraumbilical veins are small veins around the falciform ligament that drain venous blood from the anterior part of the abdominal wall and diaphragm directly into the liver, communicating with other anterior abdominal wall veins. They are also known as the superior vein of Sappey, the inferior vein of Sappey, and the vein of Burow (Yoshimitsu et al. 2001).

The superior vein of Sappey drains the upper portion of the falciform ligament and the medial part of the diaphragm and enters peripheral portal branches of the left hepatic lobe; it also communicates with branches of the superior epigastric veins and the internal thoracic vein (Yoshimitsu et al. 2001; Martin and Tudor 1980). Similarly, the inferior vein of Sappey drains the lower portion of the falciform ligament and enters peripheral portal branches of the left hepatic lobe; it descends along the round ligament and communicates with branches of the inferior epigastric veins around the umbilicus. The vein of Burow is the third vein around the falciform ligament, but it does not enter the liver directly. Instead, it terminates in the middle portion of the collapsed umbilical vein, although there are small communicating branches between it and the inferior vein of Sappey (Yoshimitsu et al. 2001; Martin and Tudor 1980).

73.3 Variation

The venous anatomy of the abdominal wall is more variable than the arterial anatomy. For example, Reardon et al. studied the superficial inferior epigastric vessels anatomically in 22 cadavers. They identified the superficial inferior epigastric vein (SIEV) in 21 of these dissections. In 12 cases, the venous drainage was through an individual vein; in eight, there was a pair of venae comitantes; and in one case, both patterns were observed. There was no consistency in its relationship to the superficial inferior epigastric artery (SIEA). In fact, the SIEV was found lateral to the SIEA as often as medial.

While the venous drainage of the anterior abdominal wall tends to be more variable than the arterial pathways, veins typically follow the course of arteries (Costa et al. 1987; Taylor 2003). As Rozen et al. and others have indicated, a better understanding of anterior abdominal wall venous drainage systems is needed to better manage abdominal wall flaps (Rozen et al. 2009; Blondeel et al. 2000).

73.4 Clinical Significance

The venous drainage of the abdominal wall is significant in several aspects of medical and surgical diagnosis and treatments.

Collateral flow: Abdominal wall veins can be dilated in patients with portal hypertension. They can also be engorged in patients with large pelvic masses, obstruction of the large abdominal veins such as the superior vena cava (SVC), inferior vena cava (IVC), azygos, or iliac veins (Ex. May-Thurner syndrome) (Bashist et al. 1996; Mousa and AbuRahma 2013).

Collateral veins on the abdominal wall or across the pubis in patients with complaints of the lower limb have an excellent positive predictive value for deep venous obstructive disease at the level of the groin or higher. Such collateral veins should not be removed since they can contribute to outflow from the leg. Symptomatic patients should be offered further diagnostics and treatment of their obstruction (Kurstjens et al. 2016).

Four main collateral routes are commonly reported in caval (SVC or IVC) obstruction:

1. The azygos, hemi-azygos route—Includes the azygos, hemiazygos, intercostal, and lumbar veins.
2. The vertebral venous plexus route—Includes the innominate, vertebral, intercostal, lumbar, and sacral veins to collateralize to the azygos and internal thoracic pathways.
3. Two routes on the anterior abdominal wall.
   (a) The internal thoracic vein route—Includes the internal thoracic, superior epigastric, and inferior epigastric veins along with the superficial veins of the thorax.
   (b) The lateral thoracic (thoracoepigastric) vein route—includes the lateral thoracic, thoracoepigastric, superficial circumflex, great saphenous, and femoral veins to collateralize to the IVC (Bashist et al. 1996).

Another unusual but important collateral pathway is the cavoportal collateral pathway, whereby the flow is directed from the vena cava (SVC or IVC) to the portal vein (Dahan et al. 1998; Kapur et al. 2010). Four cavoportal pathways have been identified (Kapur et al. 2010):

1. The caval-superficial-umbilical portal pathway seen in SVC or IVC obstruction: The superficial, superior, and inferior epigastric veins anastomose with the internal mammary veins. These superficial collateral veins then communicate with a recanalized/patent paraumbilical vein, draining into the left portal vein.
2. Caval-mammary-phrenic-hepatic capsule: Portal pathways are seen with SVC obstruction. In these pathways, blood flows from the internal thoracic vein to the inferior phrenic vein, which then communicates with the hepatic capsular veins, draining into the intrahepatic portal tributaries.
3. Caval-renal portal and caval-retroperitoneal portal pathways: Mainly reported with mid-IVC occlusion, they can result in duodenal varices.
4. Caval-mesenteric-portal pathway: Can lead to portal hypertension and engorgement of the hemorrhoidal plexus.

In summary, engorgement of abdominal wall venous collaterals is a clinical sign of central venous obstruction and indicates the need for further workup. It is inappropriate to treat it as a cosmetic abnormality (Kurstjens et al. 2016).

Caput medusa: When a patient experiences portal hypertension, there can be congestion (backup) of blood entering the caval system via the thoracoepigastric vein. When this occurs, there can be externally visible dilation of the paraumbilical veins, leading to the appearance of "Caput Medusae." A caput medusa is characterized by the appearance of distended veins emanating from the umbilicus. The shape of these veins and their arrangement around the umbilicus resembles the snake-like hair of the mythological Greek monster, Medusa, and is a harbinger of portal hypertension. There can be many causes of portal hypertension, hepatic cirrhosis being the most common (Soni et al. 2021; Yang and Chen 2005).

73.5 Surgical Considerations

Engorged abdominal wall collateral veins can be access points for interventional venous radiology procedures (Ono et al. 2007). For example, bleeding from mesenteric varices associated with portal hypertension is occasionally life-threatening. These varices can be addressed by a minimally invasive approach via interventional radiology, using the dilated abdominal wall veins for access. The literature has well reported this (Ono et al. 2007; Ikeda et al. 2014).

The venous anatomy of the abdominal wall is critical for reconstructive flaps because they have an arterial inflow and a venous outflow. Venous congestion is one of the most common causes of flap loss (Blondeel et al. 2000). For example, Rozen et al. and other authors have explored venous anatomical features associated with congestion in DIEP (deep inferior epigastric artery perforator) flaps (Rozen et al. 2009; Blondeel et al. 2000; Nahabedian et al. 2002; Utley et al. 1998).

Transplantation of an abdominal wall composite allograft can be used to reconstruct and close the abdominal compartment in patients with complex abdominal wall defects (Levi et al. 2003). Like a kidney allograft, the arteries and veins of the abdominal wall graft are implanted into the recipient's common iliac artery and vein.

Preoperative optimization is essential for patients with cirrhosis and portal hypertension. The abdominal wall collateral veins can be dilated in such patients. In such scenarios, it is best to optimize the patient medically before elective operative intervention. Emergent operative intervention before optimization can result in high morbidity and mortality (Abbas et al. 2017).

References

Abbas N, Makker J, Abbas H, Balar B (2017) Perioperative care of patients with liver cirrhosis: a review. Health Serv Insights 10:1178632917691270

Bashist B, Parisi A, Frager DH, Suster B (1996) Abdominal CT findings when the superior vena cava, brachiocephalic vein, or subclavian vein is obstructed. AJR Am J Roentgenol 167:1457–1463

Blondeel PN, Arnstein M, Verstraete K, Depuydt K, Van Landuyt KH, Monstrey SJ, Kroll SS (2000) Venous congestion and blood flow in free transverse rectus abdominis myocutaneous and deep inferior epigastric perforator flaps. Plast Reconstr Surg 106:1295–1299

Dahan H, Arrive L, Monnier-Cholley L, Le Hir P, Zins M, Tubiana J (1998) Cavoportal collateral pathways in vena cava obstruction: imaging features. AJR Am J Roentgenol 171:1405–1411

e Costa C, Carriquiry C, Vasconez LO, Grotting JC, Herrera RH, Windle BH (1987) An anatomic study of the venous drainage of the transverse rectus abdominis musculocutaneous flap. Plast Reconstr Surg 79:208–217

Ikeda O, Nakasone Y, Yokoyama K, Inoue S, Takamori H, Baba H, Yamashita Y (2014) Successful treatment of mesenteric varices by retrograde transvenous obliteration by the delivery of N-butyl-2-cyanoacrylate via an abdominal wall vein. Cardiovasc Intervent Radiol 37:533–536

Ilina M, Lilley S (2019) Embryology of the great vessels. In: Clinical embryology. Springer, pp 275–288

Kapur S, Paik E, Rezaei A, Vu DN (2010) Where there is blood, there is a way: unusual collateral vessels in superior and inferior vena cava obstruction. Radiographics 30:67–78

Kurstjens RL, van Vuuren TM, de Wolf MA, de Graaf R, Arnoldussen CW, Wittens CH (2016) Abdominal and pubic collateral veins as indicators of deep venous obstruction. J Vasc Surg Venous Lymphat Disord 4:426–433

Levi DM, Tzakis AG, Kato T, Madariaga J, Mittal NK, Nery J, Nishida S, Ruiz P (2003) Transplantation of the abdominal wall. Lancet 361:2173–2176

Martin B, Tudor R (1980) The umbilical and paraumbilical veins of man. J Anat 130:305

Mousa AY, AbuRahma AF (2013) May–Thurner syndrome: update and review. Ann Vasc Surg 27:984–995

Nahabedian MY, Momen B, Galdino G, Manson PN, Namnoum JD (2002) Breast reconstruction with the

free TRAM or DIEP flap: patient selection, choice of flap, and outcome. Plast Reconstr Surg 110:466–475
Netter FH (2014) Atlas of human anatomy, Elsevier
Ono S, Irie T, Kuramochi M, Kamoshida T, Hirai S, Oka Y (2007) Successful treatment of mesenteric varices with balloon-occluded retrograde transvenous obliteration via an abdominal wall vein. J Vasc Interv Radiol 18:1033–1035
Reardon C, O'Ceallaigh S, O'Sullivan S (2004) An anatomical study of the superficial inferior epigastric vessels in humans. Br J Plast Surg 57:515–519
Rozen WM, Pan W-R, Le Roux CM, Taylor GI, Ashton MW (2009) The venous anatomy of the anterior abdominal wall: an anatomical and clinical study. Plast Reconstr Surg 124:848–853
Sadler TW (2018) Langman's medical embryology. Lippincott Williams & Wilkins
Soni N, Israrahmed A, Singh P, Lal H (2021) Caput medusa: a sign of portal hypertension in case of chronic Budd-Chiari syndrome. BMJ Case Rep 14:e242076
Taylor GI (2003) The angiosomes of the body and their supply to perforator flaps. Clin Plast Surg 30:331–342
Utley DS, Koch RJ, Goode RL (1998) The failing flap in facial plastic and reconstructive surgery: role of the medicinal leech. Laryngoscope 108:1129–1135
Yang P-M, Chen D-S (2005) Caput medusae. N Engl J Med 353:e19
Yoshimitsu K, Honda H, Kuroiwa T, Irie H, Aibe H, Shinozaki K, Masuda K (2001) Unusual hemodynamics and pseudolesions of the noncirrhotic liver at CT. Radiographics 21:S81–S96

Hepatic Portal Vein

74

Annie Syed and Stephen J. Bordes, Jr.

74.1 Embryology

Development of the portal venous systems begins during the fourth week of gestation and ends during the twelfth week. During the fourth week, the vitelline veins form the inferior and middle intervitelline anastomoses and the superior or subhepatic anastomosis, while the umbilical veins divide into two branches, one ending in the primordial liver parenchyma and the other opening into the sinus venosus, the developing heart. By week six, the umbilical circulation is transferred to the liver; all oxygenated blood from the umbilical vein passes through the fetal liver before reaching the heart. Meanwhile, the inferior and middle vitelline vein anastomoses and segments of the left and right vitelline veins begin to form the origin of the mature hepatic portal vein. By the end of week six, the remaining portion of the subhepatic anastomosis enlarges to form the portal sinus, which connects the umbilical circulation to the portal venous circulation, establishing the intrauterine pattern of the hepatic portal system. The umbilical circulation begins to collapse immediately after birth, starting with the umbilical arteries. The fully closed umbilical arteries induce a reversal of blood flow toward the portal sinus and into the portal system. By this stage, the hepatic portal vein and the portal circulation are fully functional (Collardeau-Frachon and Scoazec 2008).

74.2 Anatomy/Topography

The hepatic portal vein is formed from the union of the superior mesenteric vein and the splenic vein at the L1/L2 level, posterior to the neck of the pancreas. It travels obliquely within the free edge of the hepatoduodenal ligament of the lesser omentum, which lies to the right of the porta hepatis (Fig. 74.1). It bifurcates into the left and right portal veins at the porta hepatis level.

The hepatic portal vein drains nutrient-rich, deoxygenated blood from the gastrointestinal tract, spleen, pancreas, and gallbladder into the liver. The portal circulation is essential for filtering toxins from the blood, including pharmaceutical agents and fats in the liver sinusoids, before they reach the right atrium (Kadel and Pandit 2020).

74.3 Histology

The hepatic portal vein brings deoxygenated yet nutrient-rich blood to the liver sinusoids. Its wall comprises three layers: the inner layer, the tunica intima; the middle muscular layer, the tunica media; and the outermost layer, the tunica

A. Syed
Department of Surgery, Louisiana State University Health Sciences Center, New Orleans, LA, USA

S. J. Bordes, Jr. (✉)
Department of Surgery, Louisiana State University Health Sciences Center, New Orleans, LA, USA

S. J. Bordes, Jr. et al. (eds.), *The Clinical Anatomy of the Vascular System*,
https://doi.org/10.1007/978-3-031-78326-5_74

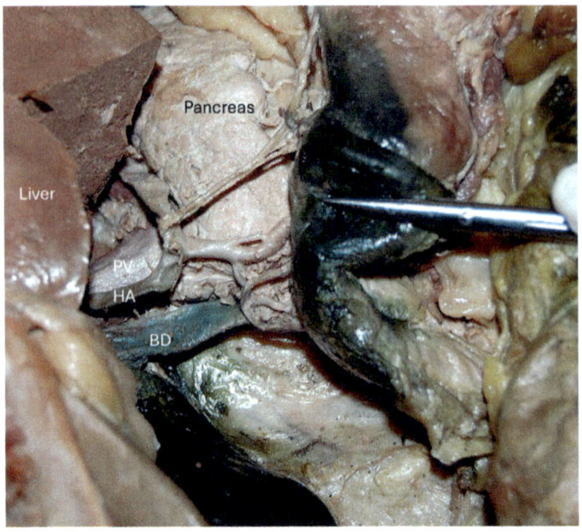

Fig. 74.1 Cadaveric dissection noting the hepatic portal vein (PV), proper hepatic artery (HA), and bile duct (BD). (Courtesy Prof. M. Loukas)

externa. In contrast to arterial walls, where each layer is easily identifiable, vein wall layers are mostly difficult to differentiate.

The innermost layer, the tunica intima, is composed of endothelium and an internal elastic membrane. The cells of the endothelium are specialized mesenchymal simple squamous epithelial cells. They form a barrier between blood and tissues and have focal regulation mechanisms that facilitate vasoconstriction, vasodilation, coagulation, and immune functions. In veins, the internal elastic membrane between the tunica intima and media is either absent or very thin and is not normally visible on histological slides.

The middle layer, the tunica media, is composed of a thin layer of smooth muscle cells and elastic fibers of elastin and fibrillin. The nerve fibers supplying the smooth muscle help to modulate venous blood flow. The elastic fibers provide resilience and elasticity to allow for changes in venous diameter if needed.

The outermost layer, the tunica externa or adventitia, attaches the vessel to its surrounding connective tissue and prevents vein overexpansion. It contains collagen, which anchors the vein to its surrounding structures, and a thin external elastic lamina.

74.4 Variations

The hepatic portal vein forms differently in approximately one-third of individuals, making preoperative CT and Doppler ultrasound essential for procedural planning. The most common anomaly is a hepatic portal vein formed from the fusion of the superior mesenteric, inferior mesenteric, and splenic veins (Imamura et al. 2004).

In another anomaly, the right anterior branch of the portal vein originates from the left portal vein instead of the main hepatic portal vein.

There are also congenital anomalies, the most common being agenesis of the left or right portal vein originating from the hepatic portal vein. This can cause collateral vessels to be used.

74.5 Clinical and Surgical Significance

The most clinically significant disease involving the hepatic portal vein is portal hypertension resulting from cirrhosis. Cirrhosis, or fibrosis of the liver parenchyma, increases the blood pressure in the portal veins, which can increase the risk of portal vein thrombosis. Budd-Chiari is an important surgical diagnosis. It is defined as a portal venous outflow obstruction that sometimes also involves the post-hepatic inferior vena cava. Anticoagulation can be attempted, but transplantation is often definitive.

Because of variations in the branching patterns of the hepatic portal vein, knowledge of its anatomy is important for liver and pancreatic surgeries. The hepatic portal vein is important in liver transplantation (Kadel and Pandit 2020). In patients awaiting liver transplantation for hepatic diseases such as hepatocellular carcinoma, cirrhosis, hepatitis, nonalcoholic hepatic steatosis, nonalcoholic fatty liver disease, Budd-Chiari, and others, a portosystemic shunt can be required for temporization, especially in cases of refractory bleeding varices. In such cases, surgical interventions such as portocaval, mesocaval, or mesoatrial shunts decompress the portal system. Mesocaval shunts are preferred as they preserve the porta hepatis for transplantation; however, the inferior vena cava must be disease-free (as in Budd-Chiari cases). Mesoatrial shunts are reserved for cases of caval obstruction. Portocaval shunts can be performed in an end-to-side or side-to-side fashion, but they alter the portal anatomy. Endovascular TIPS (transjugular intrahepatic portosystemic shunt) procedures create an intrahepatic shunt between the portal and hepatic veins to decompress the portal venous system. However, they increase the risk of encephalopathy.

The inferior vena cava must be clamped pre- and post-hepatically to ensure total vascular exclusion of the liver. The Pringle Maneuver must also be performed to occlude the portal vein and hepatic artery; a clamp is placed across the hepatoduodenal ligament, which contains the hepatic artery, portal vein, and bile duct (Imamura et al. 2004).

Branches of the hepatic portal vein can be important targets of embolization in cases of liver partition and hepatectomy. The extent of the functional liver remnant can be increased by reducing the blood flow to the diseased portion of the liver and increasing the flow to the healthy future remnant. Generally, a liver remnant should be at least 30% of the original liver volume, but 40% is needed in cases of cirrhosis or other intrahepatic diseases.

74.6 Complications

Portal Hypertension

The progression of portal hypertension can lead to many complications, including esophageal and gastric varices, ascites, and hepatorenal syndrome. As the hepatic portal vein dilates because of the increased pressure, it increases the risk of variceal channel formation and, eventually, variceal hemorrhages, which contribute to the high mortality rate from portal hypertension. Cirrhosis-related portal hypertension also increases the risk of portosystemic venous thromboses, which can lead to portal system occlusion, causing a significant decrease in blood flow and hemodynamic instability (Kadel and Pandit 2020).

Transplantation Failures

Inadequate analysis of donor and recipient vessel anatomy and technical errors during surgery can lead to liver transplantation failure. These failures can be acute (within days of transplantation, usually secondary to a technical failure with the anastomosis, such as stenosis) or chronic (weeks to months), generally leading to portal vein thrombosis. Portal venous ultrasounds on postoperative days one and seven are used to diagnose this issue. If a problem is found, the patient requires an emergency take-back to revise the anastomosis and salvage the graft.

References

Collardeau-Frachon S, Scoazec J-Y (2008) Vascular development and differentiation during human liver organogenesis. Anat Rec 291(6):614–627

Imamura H, Kokudo N, Sugawara Y et al (2004) Pringle's maneuver and selective inflow occlusion in living donor liver hepatectomy. Liver Transpl 10(6):771–778

Kadel M, Pandit TK (2020) Variations in the formation of hepatic portal vein: a cadaveric study. J Nepal Health Res Counc 18(2):224–227. Published 2020 Sep 7

Superior Mesenteric Vein 75

Hannah Kulba and Annie Syed

75.1 Embryology

It is well documented that the embryonic vitelline vein contributes to the portal venous system, developing at 5 weeks gestation. It is suggested that after its partial regression, the left vitelline vein ultimately remains as the superior mesenteric vein (SMV) (Franceschi et al. 2022; Marks 1969). However, a 2017 study of 5- to 6-week-old embryos suggested that the SMV develops instead from a tissue cleft formed by the regression of the left vitelline vein in a peritoneal fold of the midgut mesentery (Abe et al. 2017).

75.2 Anatomy/Topography

As part of the hepatic portal system, the SMV (Fig. 75.1) joins the splenic vein at the level of the second lumbar vertebra (L2) to form the hepatic portal vein (HPV) (Marks 1969).

Running anterolaterally to the right of the superior mesenteric artery (SMA), the SMV has a highly variable diameter with a mean value of 11.3 mm (Katz et al. 2008; Negoi et al. 2018). It lies anterior to the uncinate process of the head of the pancreas, the third part of the duodenum, and

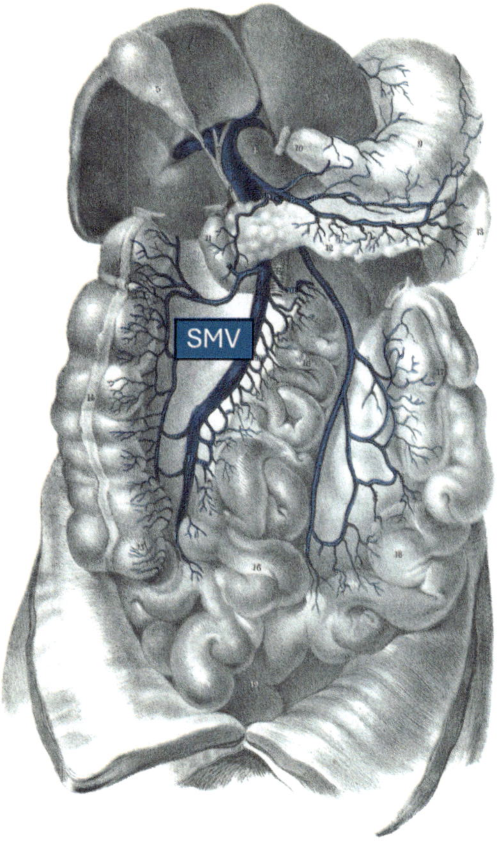

Fig. 75.1 Drawing showing the superior mesenteric vein (SMV) and many of its tributaries (Quain and Wilson 1844 author's collection)

H. Kulba · A. Syed (✉)
Tulane University School of Medicine, New Orleans, LA, USA

Department of Surgery, Louisiana State University Health Sciences Center, New Orleans, USA
e-mail: hkulba@tulane.edu; asyed1@tulane.edu

S. J. Bordes, Jr. et al. (eds.), *The Clinical Anatomy of the Vascular System*,
https://doi.org/10.1007/978-3-031-78326-5_75

the inferior vena cava (IVC), and posterior to the neck of the pancreas (Marks 1969).

The primary tributaries of the SMV drain the small intestine and include the ileal and jejunal branches. In most cases (incidence 71.8%), the first jejunal vein approaches the SMV posteriorly to the SMA (Negoi et al. 2018). Following the anatomy of the branches of the SMA, the SMV receives blood from the anterior and posterior inferior pancreaticoduodenal veins, the ileal and ileocolic veins, and the middle colic and right colic veins. Lastly, it helps to drain the stomach's greater curvature via the right gastroepiploic vein.

75.3 Variations

An inverted SMV-SMA relationship in which the SMV is located to the left of the SMA occurs in roughly 4.1% of cases without intestinal malrotation (Negoi et al. 2018).

In roughly 7.5% of cases, the common SMV trunk is entirely lacking (Negoi et al. 2018), and the splenic vein is joined by the jejunal and ileal branches directly at its junction with the portal vein. This phenomenon can also be referred to as "double trunks." This complicates resection procedures because venous reconstruction will require an alternative approach (Papavasiliou et al. 2012; Sakaguchi et al. 2010).

75.4 Histology

The SMV wall comprises an outer tunica adventitia, a middle tunica media, and an innermost tunica intima. As a tributary to the hepatic portal vein, it has no valves. Instead, the deoxygenated blood is propelled upwards by contraction of distal circular muscular layers and increased distal venous pressure (Pernambuco et al. 1983).

75.5 Imaging

CT angiography preoperatively reveals anatomical variations of the SMV and its tributaries so that surgery can progress as quickly and smoothly as possible.

The diagnosis of SMV pathologies commonly begins with an abdominal contrast-enhanced CT scan. This strategy has the added benefit of revealing an ischemic bowel if present (Essrani et al. 2020). Magnetic resonance imaging (MRI) provides similar rates of detection of SMV thrombosis and can be used in place of contrast-enhanced CT if there is renal insufficiency or hypersensitivity to the contrast dye (Choudhary et al. 2000).

75.6 Pathology/Clinical/Surgical Significance

Pancreatic Cancer

Owing to the SMV's location posterior to the neck of the pancreas and anterior to the uncinate process of the neck of the pancreas, it is especially vulnerable to thrombosis in cancers involving the pancreatic head or neck. Further, the primary intestinal tributaries can be affected by pancreatic head cancer without the involvement of the common trunk of the SMV (Katz et al. 2008; Papavasiliou et al. 2012).

The procedure for treating pancreatic cancer around the head/neck is often pancreaticoduodenectomy (Whipple procedure). It is critically important for the physicians involved in this procedure to be aware of a patient's anatomical variations, as the leading cause of morbidity/mortality is iatrogenic injury to the major vessels around the pancreas. As such, an artery-first approach is often taken to minimize vascular injury (Negoi et al. 2018).

It has been noted that in pancreatic head cancer that affects one of the two first-order intestinal branches, if venous return is preserved in either the jejunal or ileal branch, segmental resection of the other branch can be performed safely and leads to successful pancreatoduodenectomy in combination with preoperative CT imaging (Katz et al. 2008; Negoi et al. 2018).

Mesenteric Vein Thrombosis (MVT)

The SMV is involved in 95% of cases of mesenteric vein thrombosis (MVT), which is characterized by a diffuse, colicky pain in the abdomen that is much greater than physical findings suggest. There is no sensitive or specific laboratory test for MVT, so it cannot be diagnosed reliably without advanced imaging techniques. Contrast-enhanced CT scanning is the standard diagnostic method. Anticoagulation therapy with nasogastric decompression treatment and IV fluid administration is sufficient for MVT in many cases unless there is bowel infarction, in which case surgical resection is essential. A high level of suspicion leading to early diagnosis is most beneficial for treating MVT (Choudhary et al. 2000; Grendell and Ockner 1982).

References

Abe H, Yamamoto M, Yanagisawa N, Morimoto R, Murakami G, Rodríguez-Vázquez JF, Abe S (2017) Regressing vitelline vein and the initial development of the superior mesenteric vein in human embryos. Okajimas Folia Anat Jpn 94(3):87–92

Choudhary AM, Grayer D, Nelson A, Roberts I (2000) Mesenteric venous thrombosis: a diagnosis not to be missed! J Clin Gastroenterol 31(2):179–182

Essrani R, Mehershahi S, Essrani R, Sudhakaran AM, Mehmood A (2020) Isolated thrombosis of the superior mesenteric vein. Cureus 12(3):e7477

Franceschi CL, Pombar X, Cameron J, Simpson M, Pillai S, Seske LM (2022) Persistent aneurysmal vitelline vein with thrombus successfully managed with anticoagulation. J Pediatr Surg Case Rep 78:102179

Grendell JH, Ockner RK (1982) Mesenteric venous thrombosis. Gastroenterology 82:358–372

Katz MHG, Fleming JB, Pisters PWT, Lee JE, Evans DB (2008) Anatomy of the superior mesenteric vein with special reference to the surgical management of first-order branch involvement at pancreatoduodenectomy. Ann Surg 248(6):1098–1102

Marks C (1969) Developmental basis of the portal venous system. Am J Surg 117(5):671–681

Negoi I, Beuran M, Hostiuc S, Negoi RI, Inoue Y (2018) Surgical anatomy of the superior mesenteric vessels related to pancreaticoduodenectomy: a systematic review and meta-analysis. J Gastrointest Surg 22:802–817

Papavasiliou P, Arrangoiz R, Zhu F, Shin Chun Y, Edwards K, Hoffman JP (2012) The anatomic course of the first jejunal branch of the superior mesenteric vein in relation to the superior mesenteric artery. Int J Surg Oncol 2012:538769. 4 pages

Quain and Wilson, The Surgical Anatomy of the Human Body, 1844, London: Taylor & Walton

Pernambuco JRB, de Carvalho CAF, de Souza RR, Rodrigues AJ Jr (1983) Functional architecture of the human superior mesenteric vein. Okajimas Folia Anat Jpn 59(6):351–361

Sakaguchi T, Suzuki S, Morita Y, Oishi K, Suzuki A, Fukumoto K, Inaba K, Kamiya K, Ota M, Setoguchi T, Takehara Y, Nasu H, Nakamura S, Konno H (2010) Analysis of anatomic variants of mesenteric veins by 3-dimensional portography using multidetector-row computed tomography. Am J Surg 200(1):15–22

Splenic Vein

76

Yousef Raslan Hakim, Yonis Hakim, and Stephen J. Bordes, Jr.

76.1 Anatomy/Topography

Three to six veins drain blood from the spleen and form the splenic vein, which lies at the T12 to L1 vertebral level (Sharma et al. 2012; Seidelmann et al. 1977). The caliber of the splenic vein is normally under 10 mm. A wider diameter can indicate an underlying pathology such as portal hypertension (Carneiro et al. 2019). The splenic vein runs from the splenic hilum in the splenorenal ligament medially from the spleen and posterior to the pancreas (Sharma et al. 2012). It passes along a groove on the posterosuperior surface of the pancreas, lying anterior to the left kidney and abdominal aorta and inferior to the splenic artery. The superior mesenteric artery separates it from the abdominal aorta. After running medially from the spleen, the splenic vein joins the superior mesenteric vein to form the portal vein in the right paraspinal area (Seidelmann et al. 1977). Specifically, this confluence occurs behind the neck of the pancreas (Kadel and Pandit 2020; Corness et al. 2006). Forty percent of the blood that drains into the portal vein flows from the spleen and pancreas through the splenic and other veins (Brennan et al. 2019).

Multiple veins drain into the splenic vein throughout its course. These include the short gastric, posterior gastric, left gastric, left gastroepiploic, pancreatic, inferior mesenteric, and retroperitoneal veins. The latter veins are also referred to as Retzius's retroperitoneal venous plexus (Brennan et al. 2019).

The splenic artery is more tortuous than the splenic vein, which allows the two splenic vessels to be distinguished (Badagabettu et al. 2014).

76.2 Embryology

During weeks 4–12 of gestation, the portal venous system that includes the splenic vein develops. Around weeks 4–5, the right and left vitelline veins form three anastomoses (Lee et al. 2011). These vitelline veins drain blood from the yolk sac into the sinus venosus (Corness et al. 2006). The three anastomoses are the caudal-ventral and the cranial-ventral anastomoses, separated by the dorsal anastomosis, which lies between them. They lie around the developing duodenum and drain into the sinus venosus of the primordial heart after entering the septum transversum of the primordial liver, forming sinusoids

Y. R. Hakim
Geisinger Commonwealth School of Medicine, Scranton, PA, USA

Y. Hakim
Tulane University School of Medicine, New Orleans, LA, USA

S. J. Bordes, Jr. (✉)
Department of Surgery, Louisiana State University Health Sciences Center, New Orleans, LA, USA

S. J. Bordes, Jr. et al. (eds.), *The Clinical Anatomy of the Vascular System*,
https://doi.org/10.1007/978-3-031-78326-5_76

within it. Later, the caudal-ventral anastomosis, the caudal part of the right vitelline vein, and the cranial part of the left vitelline vein regress (Lee et al. 2011). The left portal vein develops from the cranial-ventral anastomosis, the common portal vein from the dorsal anastomosis, and the superior mesenteric vein from the right vitelline vein (Lee et al. 2011; Sadler 2018). The splenic vein also develops from these anastomoses (Lee et al. 2011).

76.3 Variations

Inferior Mesenteric Vein

The inferior mesenteric vein drains into the splenic vein (referred to as Type 1 herein) in approximately 50% of cases, into the superior mesenteric vein (referred to as Type 2) in 30%, and into the splenomesenteric confluence (the confluence of the splenic vein and the superior mesenteric vein, referred to as Type 3) in the remaining 20% (Brennan et al. 2019; Carneiro et al. 2019; Graf et al. 1997; Mourad et al. 1994; Krumm et al. 2011; Arimoto et al. 2019; Kadel and Pandit 2020).

Portal Vein Formation

In normal anatomy, the splenic vein joins the superior mesenteric vein to form the portal vein posterior to the neck of the pancreas (Kadel and Pandit 2020). However, variations in the formation of the portal vein have been reported. In 5% of the cadavers studied, the portal vein was formed by the confluence of the splenic, superior mesenteric, inferior mesenteric, and first jejunal veins (Kadel and Pandit 2020). Another variation, reported in around 4% of cadavers, was a confluence between the splenic, superior mesenteric, and colic veins (Kanasker et al. 2019). The splenic and superior mesenteric veins can also join within the liver. In this case, a small main portal vein is formed from the splenomesenteric confluence (Yang et al. 2021).

Jejunal Veins, Omental Vein, Middle Colic Vein, and Splenic Flexure Vein

In certain cases, the splenic vein can receive the jejunal, omental, and middle colic veins, or the splenic flexure vein (Nayak et al. 2013; Gerber et al. 1951; Chrysikos et al. 2018; Arimoto et al. 2019). A variant of the omental vein joins the left gastroepiploic vein to form a common trunk before draining into the spleen, or it drains into the splenic vein separately (Gerber et al. 1951). The middle colic vein empties into the splenic vein 1–3% of the time (Ogino et al. 2014; Yamaguchi et al. 2002), but drains into the superior mesenteric vein in most cases (68%). It also drains into the gastrocolic trunk, jejunal vein, or inferior mesenteric vein, in decreasing order of prevalence (Ogino et al. 2014).

The splenic flexure typically empties into the inferior mesenteric vein (93.9%) but joins the splenic vein directly in 3% of cases. This flexure can also be drained by the middle colic vein (3%) (Arimoto et al. 2019).

Left Gastric (Coronary Vein)

In most patients (59%), the left gastric vein drains into the splenomesenteric confluence. In 24%, it drains into the portal or splenic vein (Little and Moossa 1981; Gerber et al. 1951).

Splenic Hilar Branches

There can be 2–6 hilar branches, which join together to form a splenic vein (Gerber et al. 1951).

Absence of the Portal Vein

Another variation includes congenital absence of the portal vein. In this case, there is a portosystemic shunt. The splenic and superior mesenteric veins drain into a systemic vein (inferior vena cava or right internal iliac vein), completely bypassing the liver and forming a total shunt or end/side shunt (Corness et al. 2006; Badagabettu et al. 2014). Those two veins either join or fail to join before draining into the systemic vein (Corness et al. 2006). If they do join, they form a common vein.

Alternatively, if the portal vein is absent, there can be a partial shunt; a fraction of the portal venous blood drains into the systemic vein and the rest into the liver (Corness et al. 2006).

Treating Portosystemic Shunt Obstructions

When the portal vein is absent, nearly all of the portal venous blood drains into a systemic vein, completely bypassing the liver. Obstruction of this shunt by, e.g., a thrombus or an atherosclerotic plaque is life-threatening. However, outcomes from a partial shunt are more promising and ligation or banding can be used to treat this variant (Corness et al. 2006).

Relationship to the Pancreas

The splenic vein can run anteriorly to the pancreas, in which case the splenomesenteric confluence can be located anterior to the neck of the pancreas. The splenic vein can also enter the pancreas. A case report described it traveling anteriorly in a groove on the pancreatic body, inserting into the pancreatic body, and joining the superior mesenteric vein (Badagabettu et al. 2014). Another case report described the splenic vein entering a pancreas with a solitary fibrous tumor (Li et al. 2020).

Relationship to the Stomach and Hepatoduodenal Ligament

A case report described two splenic veins, one normal and one anomalous. The anomalous splenic vein ran posterior to the gastric fundus, anterior to the gastroesophageal junction, along the gastric lesser curvature, into the hepatoduodenal ligament, finally draining into the portal vein (Lau et al. 2005).

Left Pancreas

Regarding drainage of the left pancreas (including the neck, body, and tail), anatomical variations involving the splenic vein have been reported. Collateral branches drain blood from the left pancreas mainly into the splenic vein. One variation involves drainage of these collateral branches, which are short veins, into the superior or inferior margins of the splenic vein. They can also drain into other veins, including the left and right gastric, posterior gastric, right gastroepiploic, and inferior mesenteric veins, and the first jejunal branch and the portal trunk. Among those veins, the one that typically receives blood from these collateral branches is the left gastric vein; the others were observed in only one or two cases. The number of inferior branches is in the range 3–11, and the number of superior branches is in the range 2–6 (Mourad et al. 1994).

76.4 Histology

Upon entering the spleen, the splenic artery branches into trabecular arteries, which branch into central arterioles. The central arterioles pass through the periarteriolar lymphoid sheath (PALS), composed mainly of T cells. After passing through the PALS, they branch into penicillar arterioles. These either enter the splenic sinusoids directly (closed circulation) or terminate before doing so. If the penicillar arterioles terminate before entering the sinusoids, the blood they carry flows into splenic cords and then into the sinusoids (open circulation). The sinusoids drain into trabecular veins, which join to form the splenic vein ("The immune system" 2021).

76.5 Imaging

Radial endosonography, ultrasonography (US), magnetic resonance (MR), conventional angiography, and computerized tomography (CT) can be used to image the splenic vein (Sharma et al. 2012; Lee et al. 2011; Graf et al. 1997). More invasive imaging techniques such as splenoportography are used if the imaging is inconclusive or the treatment involves a percutaneous intervention (Lee et al. 2011).

CT imaging modalities include conventional CT, CT angiography, helical CT venography, multidetector CT (MDCT), 3D CT, and positron emission CT (Seidelmann et al. 1977; Graf et al. 1997; Badagabettu et al. 2014; Ogino et al. 2014). CT angiography entails less patient morbidity and is less expensive than conventional angiography (Graf et al. 1997). MDCT with 3D reconstruction allows for high resolution imaging of vessels because it uses isotropic acquisition (Lee et al. 2011). It allows vascular variations to be visualized and tumor growth and size to be assessed (Badagabettu et al. 2014) with a short acquisition time (Lee et al. 2011; Badagabettu et al. 2014; Lee et al. 2011). However, it uses ionizing radiation, does not provide qualitative information about blood flow dynamics, and can cause nephrotoxicity owing to the contrast substance used (Lee et al. 2011).

In contrast to MDCT, Doppler and Gray-scale ultrasonography (US) provide information about blood flow dynamics and do not use ionizing radiation. Doppler and Gray-scale US allow for real-time imaging. However, their acoustic window is small, and it is difficult to identify vascular variants and distinguish among the large vessels using them (Lee et al. 2011).

Magnetic resonance (MR) allows soft tissues to be visualized and does not employ ionizing radiation. However, in contrast to MDCT, it increases the risk of nephrogenic systemic fibrosis if a gadolinium-based contrast agent is used. It is also difficult to identify vascular calcification and stent-containing vessels by MR imaging, its acquisition speed is lower, and its resolution is lower than that of MDCT (Lee et al. 2011). The signal can also be lost at the portal confluence, where the splenic and superior mesenteric veins join at a right angle and the blood from both vessels mixes because flow areas form (Kadel and Pandit 2020). MR has lower resolution than US, and US provides more representative data (Lee et al. 2011).

US outweighs MR imaging and CT outweighs US. Hence, CT is more often used to image the portal venous system, including the splenic vein; 3D contrast-enhanced MR imaging can be used when CT is not feasible owing to a patient's clinical symptoms. Unenhanced MR imaging that employs flow-sensitive techniques can be used if administering gadolinium-based contrast agents is not recommended.

76.6 Pathology

Splenic Vein Thrombosis

Splenic vein thrombosis (SVT) is related to cirrhosis, cancer, and diseases of the pancreas, the latter being the most common cause because of the location of the splenic vein in relation to the pancreas (Valeriani et al. 2019; Simpson et al. 1990). Among pancreatic diseases, pancreatitis is the most common culprit (Butler et al. 2011). Because of its local, inflammatory, and prothrombotic mechanisms, pancreatitis can cause SVT. The splenic vein can also be compressed by pseudocysts (Agarwal et al. 2008). Neoplastic cells can directly or indirectly activate the coagulation cascade, injure the endothelium, adhere to the subendothelium, and cause venous stasis, all of which can encourage thrombosis (Piccioli et al. 1996). In cirrhosis, reduced blood flow in the portal venous system and increased production of thrombin can lead to SVT (Valeriani et al. 2019).

SVT increases the pressure distal to the obstruction, resulting in left-sided portal hypertension and increased pressure in collateral pathways (Simpson et al. 1990). The increased pressure increases blood flow in the short and posterior gastric veins and the gastroepiploic vein (Köklü et al. 2007; Chen et al. 2013). The veins dilate and gastric varices are formed (Köklü et al. 2007) and can bleed (Chen et al. 2013). SVT can also cause splenomegaly, which can progress to hypersplenism (Simpson et al. 1990; Freiberg et al. 2020).

Splenic Vein Occlusion by Tumor

The splenic vein can be occluded by a tumor thrombus or by the tumor itself. Tumor thrombus has been reported in the splenic vein in association with gastric malignancy (Ahmad et al. 2021). The thrombus travels through the gastric veins and occludes the splenic vein (Ahmad et al. 2021). Metabolic activity can be used to differentiate the tumor thrombus from thromboembolism prior to starting a treatment regimen (Ahmad et al. 2021). Splenic vein occlusion due to splenic B-cell lymphoma has been reported (Chen et al. 2013).

Splenic Vein Aneurysm

A splenic vein aneurysm can result from congenital or developmental weaknesses in the walls of the vessel or from mechanical forces that disrupt blood flow, causing venous stasis (Kocakoc et al. 2005). Splenic vein aneurysms have been associated with splenomegaly and portal vein hypertension, which disrupt blood flow in the vasculature (Kocakoc et al. 2005). Complications include thrombosis, rupture, and compression of other vessels and organs (Gharabaghi et al. 2012).

76.7 Clinical and Surgical Considerations

Consideration of Anatomical Variations

Variations in the portal vein and its tributaries are important in surgical procedures including interventional catheterization, management of traumatic mesenteric rupture, and hepatic, pancreatic, and duodenal surgeries such as hepatectomy and liver transplantation. In pancreaticoduodenectomy (Whipple), injury to the portal vein can cause postoperative hemorrhage if anatomical variations are not considered (Kadel and Pandit 2020; Kanasker et al. 2019). Such injury can occur at the confluence of the splenic n, superior mesenteric, and portal veins and can be resected (Badagabettu et al. 2014).

During the treatment of colon cancer by complete mesocolic excision, life-threatening hemorrhage can result if anatomical variations of the middle colic vein emptying into the splenic vein have not been investigated preoperatively (Ogino et al. 2014; Chrysikos et al. 2018).

Shunting Procedures Involving the Splenic Vein

Surgical shunting procedures can be used to manage esophageal varices and portal hypertension. Two of these procedures, the central or proximal splenorenal shunt and the distal splenorenal shunt (Warren shunt), involve the splenic vein (Soares and Jarnagin 2020). The central splenorenal shunt connects the proximal segment of the splenic vein to the left renal vein, which allows the portal vein to be decompressed (Soares and Jarnagin 2020). This shunt can be used when resection of the spleen is needed as it does not involve the distal segment of the splenic vein ("Shunting procedures" 2022).

The distal splenorenal shunt joins the distal segment of the splenic vein and the left renal vein, which decompresses the gastroepiploic collaterals (Soares and Jarnagin 2020). This shunt is used in the management of patients with good liver function (Child-Pugh A or B classification) when octreotide and endoscopic sclerotherapy or banding fail (Jenkins et al. 1999). However, it can lead to complications during future liver transplantation surgery (Jenkins et al. 1999). Distal splenorenal shunts are effective and have rebleeding rates between 5% and 8% (Henderson et al. 2006). However, the 3–5-year survival rate is between 75% and 80%, which is why this method is used after failure of octreotide, sclerotherapy, or banding, and in patients with good liver function (Henderson et al. 2006).

Pancreatic Cancer

Occlusion of the splenic vein by thromboembolism or by tumor infiltration of the vessel entails a severe risk for pancreatic cancer patients. Splenic vein thrombosis has been associated with intraoperative blood loss and reduced survival rates (Hicks et al. 2018). Similarly, infiltration of pancreatic cancer into the splenic vein is associated with poor survival even if the cancer is resectable (Crippa et al. 2018). Hence, radiographic diagnosis is essential for managing these patients.

References

Agarwal AK, Raj Kumar K, Agarwal S, Singh S (2008) Significance of splenic vein thrombosis in chronic pancreatitis. Am J Surg 196(2):149–154. https://doi.org/10.1016/j.amjsurg.2007.07.039

Ahmad W, Asghar N, Zafar T, Bashir H (2021) Complete splenic vein tumour thrombus in primary gastric malignancy on F18-fluorodeoxyglucose positron emission tomography-computed tomography. J Pak Med Assoc 71(1(A)):175–176

Arimoto A, Matsuda T, Hasegawa H, Yamashita K, Nakamura T, Sumi Y, Suzuki S, Kakeji Y (2019) Evaluation of the venous drainage pattern of the splenic flexure by preoperative three-dimensional computed tomography. Asian J Endosc Surg 12(4):412–416. https://doi.org/10.1111/ases.12657

Badagabettu SN, Shetty SD, Regunathan D, George BM, Mishra S, Sirasanagandla SR (2014) Dangerous course and atypical termination of the splenic vein in relation to pancreas–a case report. JOP 15(6):609–610. https://doi.org/10.6092/1590-8577/2693

Brennan P, Standring S, Wiseman S (eds) (2019) Gray's surgical anatomy E-Book. Elsevier Health Sciences

Butler JR, Eckert GJ, Zyromski NJ, Leonardi MJ, Lillemoe KD, Howard TJ (2011) Natural history of pancreatitis-induced splenic vein thrombosis: a systematic review and meta-analysis of its incidence and rate of gastrointestinal bleeding. HPB 13(12):839–845. https://doi.org/10.1111/j.1477-2574.2011.00375.x

Carneiro C, Brito J, Bilreiro C, Barros M, Bahia C, Santiago I, Caseiro-Alves F (2019) All about portal vein: a pictorial display to anatomy, variants and physiopathology. Insights Imaging 10(1):38. https://doi.org/10.1186/s13244-019-0716-8

Chen BC, Wang HH, Lin YC, Shih YL, Chang WK, Hsieh TY (2013) Isolated gastric variceal bleeding caused by splenic lymphoma-associated splenic vein occlusion. World J Gastroenterol 19(40):6939–6942. https://doi.org/10.3748/wjg.v19.i40.6939

Chrysikos D, Sgantzos M, Tsiaoussis J, Piperos T, Varlatzidou A, Bonatsos V, Theodoropoulos P, Noussios G, Troupis T, Papapanagiotou I, Mariolis-Sapsakos T (2018) Middle colic vein draining to splenic vein: a rare anatomic variation encountered during a right hemicolectomy. J Surg Case Rep 2018(8):rjy220

Corness JA, McHugh K, Roebuck DJ, Taylor AM (2006) The portal vein in children: radiological review of congenital anomalies and acquired abnormalities. Pediatr Radiol 36(2):87–171. https://doi.org/10.1007/s00247-005-0010-4

Crippa S, Cirocchi R, Maisonneuve P, Partelli S, Pergolini I, Tamburrino D, Aleotti F, Reni M, Falconi M (2018) Systematic review and meta-analysis of prognostic role of splenic vessels infiltration in resectable pancreatic cancer. Eur J Surg Oncol 44(1):24–30. https://doi.org/10.1016/j.ejso.2017.10.217

Freiberg B, Emre S, Morotti R, Dillon B, Koral A, Hattangadi SM, Valentino PL (2020) Progressive splenomegaly and hypersplenism: an unusual case of splenic vein stenosis with histologic findings of hepatoportal sclerosis. J Pediatr 218:222–227.e1. https://doi.org/10.1016/j.jpeds.2019.11.018

Gerber AB, Lev M, Goldberg SL (1951) The surgical anatomy of the splenic vein. Am J Surg 82(3):339–343. https://doi.org/10.1016/0002-9610(51)90323-6

Gharabaghi MA, Yazdi NA, Abrishami Z, Amini S (2012) Splenic vein aneurysm. BMJ Case Rep 2012:bcr1020115007. https://doi.org/10.1136/bcr.10.2011.5007

Graf O, Boland GW, Kaufman JA, Warshaw AL, Fernandez del Castillo C, Mueller PR (1997) Anatomic variants of mesenteric veins: depiction with helical CT venography. AJR Am J Roentgenol 168(5):1209–1213. https://doi.org/10.2214/ajr.168.5.9129413

Henderson JM, Boyer TD, Kutner MH, Galloway JR, Rikkers LF, Jeffers LJ, Abu-Elmagd K, Connor J, DIVERT Study Group (2006) Distal splenorenal shunt versus transjugular intrahepatic portal systematic shunt for variceal bleeding: a randomized trial. Gastroenterology 130(6):1643–1651. https://doi.org/10.1053/j.gastro.2006.02.008

Hicks AM, DeRosa A, Raj M, Do R, Yu KH, Lowery MA, Varghese A, O'Reilly EM (2018) Visceral thromboses in pancreas adenocarcinoma: systematic review. Clin Colorectal Cancer 17(2):e207–e216. https://doi.org/10.1016/j.clcc.2017.12.001

Jenkins RL, Gedaly R, Pomposelli JJ, Pomfret EA, Gordon F, Lewis WD (1999) Distal splenorenal shunt: role, indications, and utility in the era of liver transplantation. Arch Surg 134(4):416–420. https://doi.org/10.1001/archsurg.134.4.416

Kadel M, Pandit TK (2020) Variations in the formation of hepatic portal vein: a cadaveric study. J Nepal Health Res Counc 18(2):224–227. https://doi.org/10.33314/jnhrc.v18i2.2421

Kanasker N, Sonje P, Bankar J (2019) Cadaveric study on variations in the formation of extrahepatic portion of portal vein. Int J Anat Radiol Surg 8(3):AO13–AO16. https://doi.org/10.7860/IJARS/2019/39644:2494

Kocakoc E, Kiris A, Bozgeyik Z, Uysal H, Artas H (2005) Splenic vein aneurysm with calcification of splenic and portal veins. J Clin Ultrasound 33(5):251–253. https://doi.org/10.1002/jcu.20123

Köklü S, Coban S, Yüksel O, Arhan M (2007) Left-sided portal hypertension. Dig Dis Sci 52(5):1141–1149. https://doi.org/10.1007/s10620-006-9307-x

Krumm P, Schraml C, Bretschneider C, Seeger A, Klumpp B, Kramer U, Claussen CD, Miller S (2011) Depiction of variants of the portal confluence venous system using multidetector row CT: analysis of 916 cases. RoFo: Fortschritte auf dem Gebiete der Rontgenstrahlen und der Nuklearmedizin 183(12):1123–1129. https://doi.org/10.1055/s-0031-1281745

Lau ST, Kim SS, Lee SL, Ledbetter DJ (2005) The anomalous splenic vein: a case report and review of the literature. J Pediatr Surg 40(9):1492–1494. https://doi.org/10.1016/j.jpedsurg.2005.05.060

Lee WK, Chang SD, Duddalwar VA, Comin JM, Perera W, Lau WF, Bekhit EK, Hennessy OF (2011) Imaging assessment of congenital and acquired abnormalities of the portal venous system. Radiographics 31(4):905–926. https://doi.org/10.1148/rg.314105104

Li J, Li J, Xiong Y, Xu T, Xu J, Li Q, Yang G (2020) Atypical/malignant solitary fibrous tumor of the pancreas with spleen vein invasion: case report and literature review. Medicine 99(17):e19783. https://doi.org/10.1097/MD.0000000000019783

Little AG, Moossa AR (1981) Gastrointestinal hemorrhage from left-sided portal hypertension. An unappreciated complication of pancreatitis. Am J Surg 141(1):153–158. https://doi.org/10.1016/0002-9610(81)90029-5

Mourad N, Zhang J, Rath AM, Chevrel JP (1994) The venous drainage of the pancreas. Surg Radiol Anat 16(1):37–45. https://doi.org/10.1007/BF01627919

Nayak SB, Aithal AP, Melanie RD, Guru A, Kumar N (2013) Unusual jejunal tributaries of the splenic vein and their surgical importance: a case report. OA Case Rep 2(4):36–37

Ogino T, Takemasa I, Horitsugi G, Furuyashiki M, Ohta K, Uemura M, Nishimura J, Hata T, Mizushima T, Yamamoto H, Doki Y, Mori M (2014) Preoperative evaluation of venous anatomy in laparoscopic complete mesocolic excision for right colon cancer. Ann Surg Oncol 21(Suppl 3):S429–S435. https://doi.org/10.1245/s10434-014-3572-2

Piccioli A, Prandoni P, Ewenstein BM, Goldhaber SZ (1996) Cancer and venous thromboembolism. Am Heart J 132(4):850–855. https://doi.org/10.1016/s0002-8703(96)90321-x

Sadler TW (2018) Langman's medical embryology. Lippincott Williams & Wilkins

Seidelmann FE, Cohen WN, Bryan PJ, Brown J (1977) CT demonstration of the splenic vein-pancreatic relationship: the pseudodilated pancreatic duct. AJR Am J Roentgenol 129(1):17–21. https://doi.org/10.2214/ajr.129.1.17

Sharma M, Babu CS, Garg S, Rai P (2012) Portal venous system and its tributaries: a radial endosonographic assessment. Endosc Ultrasound 1(2):96–107. https://doi.org/10.7178/eus.02.008

Shunting procedures for portal hypertension (2022) In: Ellison EC, Zollinger RM Jr, Pawlik TM, Vaccaro PS, Bitans M, Baker AS (eds) Zollinger's atlas of surgical operations, 11th edn. McGraw Hill. https://accesssurgery-mhmedical-com.gcsom.idm.oclc.org/content.aspx?bookid=3101&sectionid=260096364

Simpson WG, Schwartz RW, Strodel WE (1990) Splenic vein thrombosis. South Med J 83(4):417–421. https://doi.org/10.1097/00007611-199004000-00014

Soares KC, Jarnagin WR (2020) Liver & portal venous system. In: Doherty GM (ed) Current diagnosis & treatment: surgery, 15th edn. McGraw Hill. https://accesssurgery-mhmedical-com.gcsom.idm.oclc.org/content.aspx?bookid=2859&sectionid=242158357

The immune system & lymphoid organs (2021) In: Mescher AL (ed) Junqueira's basic histology text and Atlas, 16th edn. McGraw Hill. https://accessmedicine-mhmedical-com.gcsom.idm.oclc.org/content.aspx?bookid=3047&sectionid=255121636

Valeriani E, Riva N, Di Nisio M, Ageno W (2019) Splanchnic vein thrombosis: current perspectives. Vasc Health Risk Manag 15:449–461. https://doi.org/10.2147/VHRM.S197732

Yamaguchi S, Kuroyanagi H, Milsom JW, Sim R, Shimada H (2002) Venous anatomy of the right colon: precise structure of the major veins and gastrocolic trunk in 58 cadavers. Dis Colon Rectum 45(10):1337–1340. https://doi.org/10.1097/01.DCR.0000027284.76452.84

Yang Q, Li J, Wang H, Wang S (2021) A rare variation of duplicated portal vein: left branch derived from splenic vein mimicking cavernous transformation. BMC Gastroenterol 21(1):404. https://doi.org/10.1186/s12876-021-01970-8

The Inferior Mesenteric Vein 77

Joseph Mason and Stephen J. Bordes, Jr.

77.1 Embryology

The portal venous system develops between the 4th and 10th weeks of gestation. During the 4th week, two vitelline veins emerge from the anterior portion of the yolk sac, connecting to and draining into the sinus venosus (Papamichail et al. 2018). Within the septum transversum, vitelline plexuses develop that connect to the vitelline veins, creating a network of veins that surround the primitive liver, later becoming hepatic sinusoids (Schoenwolf et al. 2020). By the end of the third month of development the left vitelline vein has regressed, causing blood from the left side of the abdominal viscera to drain to the right vitelline vein via transverse anastomoses (Schoenwolf et al. 2020). The portion of the right vitelline vein that remains, just caudal to the liver, becomes the portal vein and superior mesenteric vein (Schoenwolf et al. 2020). Left-to-right remodeling of the vitelline anastomoses leads to the formation of the inferior mesenteric vein (IMV), draining the embryo's hindgut (Schoenwolf et al. 2020).

The development of the IMV varies because it can drain into four different locations (Nepal et al. 2021). Malformations in the embryo can cause it to drain into the systemic system via the inferior vena cava instead of the portal system, forming a portosystemic shunt. Such shunts are thought to develop because of incomplete involution of the vitelline venous system in response to the formation of hepatic sinusoids (Papamichail et al. 2018). Statistically, the IMV is involved in extrahepatic portosystemic shunts in 12.5% of reported cases (Takeuchi et al. 2014).

77.2 Anatomy and Topography

The IMV (Fig. 77.1) is responsible for the venous drainage of the descending colon, sigmoid colon, and rectum. Its major tributaries include the left

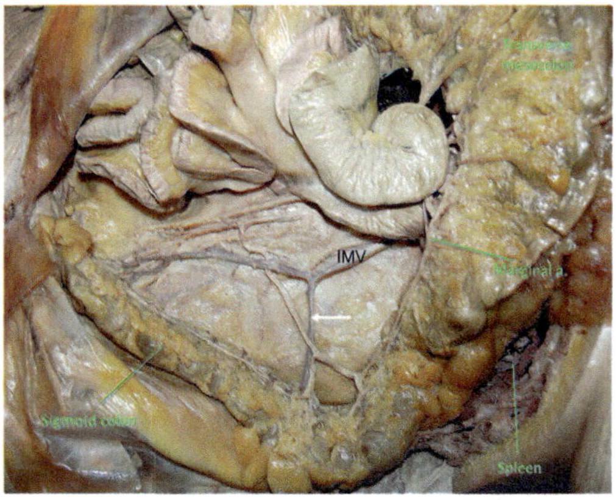

Fig. 77.1 Cadaveric dissection of the inferior mesenteric vein (IMV), noting the left colic vein (arrow). (Courtesy Prof. M. Loukas)

J. Mason
School of Medicine, Louisiana State University Health Sciences Center, New Orleans, LA, USA

S. J. Bordes, Jr. (✉)
Department of Surgery, Louisiana State University Health Sciences Center, New Orleans, LA, USA

S. J. Bordes, Jr. et al. (eds.), *The Clinical Anatomy of the Vascular System*,
https://doi.org/10.1007/978-3-031-78326-5_77

colic vein, the sigmoid veins, and the superior rectal veins. The latter emerges from the rectal (hemorrhoidal) plexus, composed of the superior, middle, and inferior rectal veins. This venous plexus is one of the locations where the portal venous system anastomoses with the systemic venous system. The superior rectal vein joins the IMV as it crosses over the left common iliac vessels (Akpinar et al. 2008). The IMV continues superiorly on the anterior surface of the psoas major, lateral to the inferior mesenteric artery (Akpinar et al. 2008). During its course, it can travel over or medial to the ovarian or testicular arteries (Akpinar et al. 2008). In most patients, it empties into the splenic vein above the duodenojejunal juncture, medial to the body of the pancreas (Akpinar et al. 2008; Nishino et al. 2022). In others, it has been observed to empty into the superior mesenteric vein or the junction of the superior mesenteric and splenic veins (Nishino et al. 2022). The IMV has a mean diameter of 4.1 ± 0.6 mm, the normal upper limit being 5.1 mm (Zhang et al. 2007). It dilates in disease states such as rectal neoplasms (Ivan et al. 2017).

77.3 Histology

The IMV is valveless and its wall consists of three layers: the tunica intima, media, and adventitia. The tunica intima, the innermost layer, comprises an endothelium surrounded by loose connective tissue and sparse smooth muscle. The intimal layer is surrounded by an incomplete internal elastic lamina, a layer of thick elastic fibers (Bergan and Bunke-Paquette 2013). The tunica media comprises mainly smooth muscle and collagen fibers (Bergan and Bunke-Paquette 2013). The outermost layer, the tunica adventitia, is poorly differentiated from the tunica media and consists of loose connective tissue (Bergan and Bunke-Paquette 2013).

77.4 Imaging

Imaging of the IMV can be useful for determining the drainage patterns of the portal venous system in a patient and for detecting anastomotic flow within the intestines when surgical removal of neoplastic tissue is considered. In most cases, three-dimensional computed tomography angiography (CTA) is used to determine the drainage pattern of the IMV. Consideration of the location at which the IMV joins the rest of the portal system points to four main drainage classifications (see *Variations*) (Nepal et al. 2021). Three-dimensional CTA can also be used to examine the relationship of the inferior mesenteric artery (IMA) and left colic artery (LCA) to the IMV, which can prove useful in laparoscopic procedures on the left colon and rectum (Murono et al. 2015).

77.5 Variations

There are four known drainage classifications of the IMV (Nepal et al. 2021). In Type I, the IMV drains into the confluence of the superior mesenteric vein (SMV) and the splenic vein (SV). In Type II, it drains into the SMV. In Type III, it drains into the SV (Nepal et al. 2021). A newly recognized Type IV classification describes the IMV draining into the left colic vein, which then drains into the SMV (Nepal et al. 2021). In patients with Type IV, the distance between the IMV and the inferior mesenteric artery (IMA) was over 50 mm (Nepal et al. 2021). In this scenario, it has been hypothesized that the IMV could be a marginal vein, and that special care must be taken when its ligation is considered (Nepal et al. 2021).

77.6 Clinical and Surgical Significance

Patients presenting with mesenteric venous thrombosis (MVT) within the IMV who show signs of peritoneal involvement or transmural infarction are candidates for surgical intervention (Singal et al. 2013). Doppler ultrasonography or infusion with fluorescein is used to determine the viability of the bowel. On average, 50–60 cm of the bowel is resected, more commonly with thrombi limited to small veins or mesenteric vein involvement than within the combined portal-mesenteric venous system (Singal et al. 2013).

The IMV can be used in a portal vein recanalization-transjugular intrahepatic porto-systemic shunt (PVR-TIPS) procedure when the trans-splenic approach is not viable (Entezari et al. 2020). This procedure can relieve portal hypertension while patients await a liver transplant (Entezari et al. 2020).

In patients undergoing a pancreaticoduodenectomy (Whipple) for the resection of pancreatic cancer, particularly within the head of the pancreas, the splenic vein-IMV confluence is often disrupted when the neoplastic tissue is removed. This disruption of venous drainage can lead to sinistral portal hypertension (SPH), a long-term complication of pancreaticoduodenectomies (Ono et al. 2021). The IMV can be used in the anastomosis of the splenic vein to the colonic marginal vein (Ono et al. 2021). While determined not to be critical for preventing SPH, the collateral flow that the IMV provides should be maintained in the event that collateral flow between the arc of Barkow and the colonic marginal vein fails to form postoperatively (Ono et al. 2021).

The IMV is often ligated to attain better mobilization when appreciating the anatomy of the lower border of the pancreas (Al-Asari et al. 2013). Special care must be taken when the IMV is ligated owing to the proximity of a collateral blood vessel known as the meandering mesenteric artery, or arc of Riolan, in 7–10% of patients (Al-Asari et al. 2013). This artery communicates between the middle and left colic arteries and provides collateral blood flow to the splenic flexure (Al-Asari et al. 2013). The proximity of these vessels can be found between the IMV, inferior border of the pancreas, and left colic artery, forming a triangular region known as the "critical zone of the IMV" (Al-Asari et al. 2013).

When considering a patient with an aneurysm in the IMV, care must be taken to determine whether it is to be managed conservatively or surgically, smaller or asymptomatic aneurysms being treated conservatively (Gorolay et al. 2021). The risk of forming a thrombus or rupture could be a reason to consider surgical intervention. An aneurysm can be treated surgically either endovascularly or using an open repair approach, via aneurysmoplasty or aneurysmectomy (Gorolay et al. 2021; Guirgis et al. 2018).

77.7 Complications

Thrombosis can occur in the IMV because of inflammatory processes such as diverticulitis or neoplasm, but IMV thromboses (only 4–11% of reported cases) are rarer than SMV thromboses (Medina-Perez et al. 2021). Acute mesenteric ischemia (AMI) caused by IMV thrombosis has a mortality risk ranging from 15% to 23% (Medina-Perez et al. 2021). Mesenteric venous thrombosis (MVT) can present in either an acute, subacute, or a chronic timeframe. Patients with an acute or subacute mesenteric thrombosis usually present with abdominal pain, whereas a patient with chronic MVT shows signs of portal hypertension (Singal et al. 2013).

Portal hypertension occurs in patients with cirrhosis of the liver, leading to increased pressure within the IMV and backflow of venous blood into the IMV away from the liver, known as hepatofugal flow. This increase in flow and volume is correlated with the formation of rectal varices, created at the anastomosis of the superior rectal vein with the inferior and middle rectal veins, which drain into the iliac venous system (Maruyama and Shiina 2021). This increase in pressure within the IMV can manifest as internal hemorrhoids in the rectum, where the superior rectal veins anastomose with the systemic venous system via the middle and inferior rectal veins. Hepatofugal flow within the IMV is associated with more severe ascites, more severe Child-Pugh classification, and a higher incidence of decompensated cirrhosis (Maruyama et al. 2013).

Aneurysms are another potential complication for the IMV. Classified as extrahepatic portal vein aneurysms (PVA), aneurysms of the IMV are rarer than those of the splenic or portal vein, accounting for only 0.5% of all visceral venous aneurysms (Guirgis et al. 2018). They can be classed as acquired or congenital, an acquired aneurysm being a complication secondary to other comorbidities such as portal hypertension and liver cirrhosis (Guirgis et al. 2018). Congenital causes of IMV aneurysms are indicated by the presence of congenital extrahepatic portosystemic shunts, also known as Abernethy malformations (Guirgis et al. 2018). IMV aneurysms have been noted in Type II Abernethy malformations, where the portal vein and inferior vena cava are joined side-to-side (Guirgis et al. 2018). Complications caused by the aneurysm include thrombosis, gastrointestinal bleeding, jaundice, and rupture (Gorolay et al. 2021).

References

Akpinar E, Turkbey B, Karcaaltincaba M, Karaosmanoglu D, Akata D (2008) MDCT of inferior mesenteric vein: normal anatomy and pathology. Clin Radiol 63(7):819–823. https://doi.org/10.1016/j.crad.2007.09.001

Al-Asari SF, Lim D, Min BS, Kim NK (2013) The relation between inferior mesenteric vein ligation and collateral vessels to splenic flexure: anatomical landmarks, technical precautions and clinical significance. Yonsei Med J 54(6):1484–1490. https://doi.org/10.3349/ymj.2013.54.6.1484

Bergan JJ, Bunke-Paquette N (2013) Venous embryology and anatomy. In: The vein book, 2nd edn. Oxford University Press, pp 17–25. Accessed 24 Jan 2022. https://web-s-ebscohost-com.ezproxy.lsuhsc.edu/ehost/ebookviewer/ebook/ZTAwMHhuYV9fNjY5OTU5X19BTg2?sid=3264f2df-7f69-4d2b-97df-c8ecf4354324@redis&vid=0&format=EB&lpid=lp_17&rid=0

Entezari P, Riaz A, Thornburg B, Salem R (2020) Percutaneous ultrasound-guided superior and inferior mesenteric vein access for portal vein recanalization–Transjugular intrahepatic portosystemic shunt: a case series. CardioVasc Interv Radiol 23:1–4. https://doi.org/10.1007/s00270-020-02713-0

Gorolay V, Nguyen D, Samra J, Neale M (2021) Asymptomatic thrombosis of extrahepatic portal vein aneurysm necessitating hybrid operative repair. Vascular 29(5):762–766. https://doi.org/10.1177/1708538120976977

Guirgis M, El-Khoury H, Sieunarine K (2018) Management of an enormous inferior mesenteric vein aneurysm. J Vasc Surg Venous Lymphat Disord 6(2):241–243. https://doi.org/10.1016/j.jvsv.2017.11.002

Ivan CV, Mullineux JH, Verma R et al (2017) Assessment of the inferior mesenteric vein diameter as a surrogate marker to evaluate response to neoadjuvant chemoradiotherapy for locally advanced rectal adenocarcinoma. Color Dis 19(12):1076–1080. https://doi.org/10.1111/codi.13811

Maruyama H, Shiina S (2021) Collaterals in portal hypertension: anatomy and clinical relevance. Quant Imaging Med Surg 11(8):3867881–3863881. https://doi.org/10.21037/qims-20-1328

Maruyama H, Kamezaki H, Kondo T et al (2013) Effects of inferior mesenteric vein flow in patients with cirrhosis. Clin Gastroenterol Hepatol 11(12):1648–1654. https://doi.org/10.1016/j.cgh.2013.06.026

Medina-Perez R, Campbell DJ, Acosta Rullan JM, Gonzalez S (2021) Partial thrombosis of inferior mesenteric vein with thrombophlebitis. Cureus 13(8):e16900. https://doi.org/10.7759/cureus.16900

Murono K, Kawai K, Kazama S et al (2015) Anatomy of the inferior mesenteric artery evaluated using 3-dimensional CT angiography. Dis Colon Rectum 58(2):214–219. https://doi.org/10.1097/DCR.0000000000000285

Nepal P, Mori S, Kita Y et al (2021) Anatomical study of the inferior mesenteric vein using three-dimensional computed tomography angiography in laparoscopy-assisted surgery for left-sided colorectal cancer. Surg Today 51(10):1665–1670. https://doi.org/10.1007/s00595-021-02292-8

Nishino H, Zimmitti G, Ohtsuka T et al (2022) Precision vascular anatomy for minimally invasive distal pancreatectomy: a systematic review. J Hepato-Biliary-Pancreat Sci 29:136. https://doi.org/10.1002/jhbp.903

Ono Y, Inoue Y, Kato T et al (2021) Sinistral portal hypertension after pancreaticoduodenectomy with splenic vein resection: pathogenesis and its prevention. Cancers 13(21):5334. https://doi.org/10.3390/cancers13215334

Papamichail M, Pizanias M, Heaton N (2018) Congenital portosystemic venous shunt. Eur J Pediatr 177(3):285–294. https://doi.org/10.1007/s00431-017-3058-x

Schoenwolf G, Bleyl S, Brauer P, Francis-West P (2020) Development of the vasculature. In: Larsen's human embryology, 6th edn. Elsevier, pp 298–334

Singal AK, Kamath PS, Tefferi A (2013) Mesenteric venous thrombosis. Mayo Clin Proc 88(3):285–294. https://doi.org/10.1016/j.mayocp.2013.01.012

Takeuchi H, Takeda Y, Takahashi M, Hayashi S, Fukuzawa Y, Nakano T (2014) A rare congenital extrahepatic portosystemic shunt affecting the inferior mesenteric vein, inferior vena cava, and left ovarian vein. Surg Radiol Anat 36(7):729–732. https://doi.org/10.1007/s00276-013-1230-1

Zhang XM, Zhong TL, Zhai ZH, Zeng NL (2007) MR venography of the inferior mesentery vein. Eur J Radiol 64(1):147–151. https://doi.org/10.1016/j.ejrad.2007.02.017

Renal Veins

78

Vinila Baljepally

A vast amount of variation is reported in the renal venous complex. Many communications exist between the subcardinal, posterior cardinal, and supracardinal veins at the level of the kidney prior to the 7th week of development, thus leading to the possibility of variations in venous drainage. Many have reported variations in the gross appearance of the suprarenal veins and their accompanying renal veins (Fig. 78.1). It is, therefore, impossible to describe the variations in the venous drainage of the suprarenal glands independently.

Anson and McVay illustrated numerous variations based on the suprarenal and renal vein configurations and combinations. The left suprarenal vein can enter the renal vein without receiving the inferior phrenic. This variant was first described by Cade but has also been seen in studies by Loukas et al. and Scholten et al. The "normal" configuration of the left suprarenal vein, where the inferior phrenic vein joins it before entering the renal, can also be seen in combination with a gonadal vein draining directly opposite, instead of lateral, to this common trunk into the renal, or may drain into one of two renal veins. This "normal" configuration may be somewhat modified by an additional inferior phrenic that enters the gland before joining the suprarenal vein, a variant also noted by Cade. The left suprarenal can be bifid and join the renal vein independently to the inferior phrenic. Double suprarenal veins, with the inferior phrenic contributing to only one, draining into a single renal vein, observed by Scholten et al., or one of the double renal veins, were described. The suprarenal vein may be joined by one of doubled gonadal veins and join the renal vein independently from the inferior phrenic. Stack et al. reported the left suprarenal vein draining directly into the IVC. Jyothsna et al. reported a case where the left gonadal vein and the second left lumbar vein joined the left suprarenal vein to form a trunk, which together entered the left renal vein. Johnstone performed a cadaveric study to examine the anatomy of the suprarenal veins. All

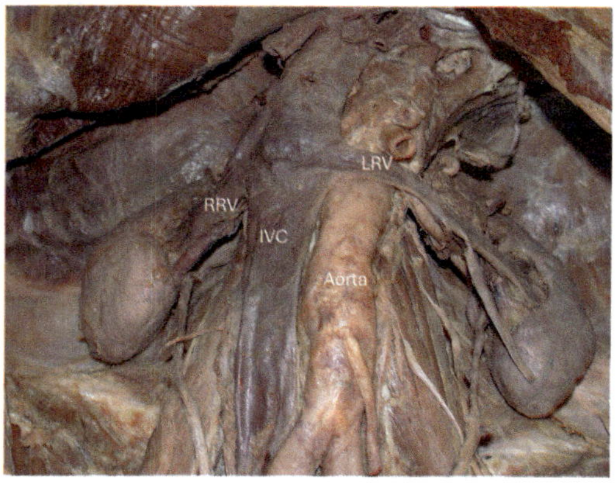

Fig. 78.1 Cadaveric view of the left (LRV) and right renal veins (RRV) and their connections to the inferior vena cava (IVC). (Courtesy Prof. M. Loukas)

V. Baljepally (✉)
School of Medicine, Tulane University,
New Orleans, LA, USA
e-mail: baljepally@etsu.edu

S. J. Bordes, Jr. et al. (eds.), *The Clinical Anatomy of the Vascular System*,
https://doi.org/10.1007/978-3-031-78326-5_78

ten subjects had a single left suprarenal vein. One subject had two left inferior phrenic veins that merged with the left suprarenal vein. Another subject had a small collateral venule from the spleen joining the left inferior phrenic vein. Another subject had no inferior phrenic vein but a retrocaval venule instead.

Pearl discussed communication between the right suprarenal vein and small hepatic veins. Matsuura et al. observed this communication in 8% of their 104 patients undergoing multidetector computed tomography for vascular disease. A study by Johnstone showed a single right suprarenal vein in eight out of ten subjects, from the eight, half drained into the IVC and half into a right hepatic vein. There were two right suprarenal veins in one subject and three in another; both variations were observed by Scholten et al. in their laparoscopic study of 546 patients. In a study of 142 laparoscopic adrenalectomy patients, Parnaby et al. discovered two patients with two right suprarenal veins. In the first, both suprarenal veins directly drained into the IVC. In the second, the IVC and the right renal vein received a tributary. Another variation about the right suprarenal-renal venous complex was noted in cadaveric dissection by the authors. In one of the cadavers, it was noted that the right suprarenal vein drained directly into the IVC, while the renal vein was found to be doubled. The inferior renal vein was noted to receive the right gonadal vein.

78.1 Renal Veins

The renal veins show less variation than the renal arteries. One study found multiple renal veins rare on the left side (1%) and common on the right side (28%). In another study, the right renal vein was found to have one to three accessories (18% of 203 cases) and the left renal vein to have one or two accessories (9% of 203 cases). Poynter pointed out that the renal veins tend to form two or more trunks as the arteries do, but in the veins the frequency is only about 7%. In about 2% of cases, the left renal vein may be retroaortic or even circumaortic; either single vessel or doubled with one limb anterior and the other posterior to the aorta reported frequency 1.5–4.4% of all abdominal CT scans). In some cases, it may form a circumaortic renal venous collar (reported frequency 1.5–8.7%). Dill et al. (2012), in an MRI study, found a total of left renal vein variations, retroaortic left renal vein, and circumaortic left renal veins in 71/2644 (2.68%), 44/2644 (1.66%), and 27/2644 (1.02%), respectively. In cases with a retroaortic left renal vein, the numbers of males and females with their corresponding percentages were 19/44 (43.2%) and 25/44 (56.8%), respectively. In cases with a circumaortic left renal vein, the numbers of males and females with their corresponding percentages were 13/27 (48.1%) and 14/27 (51.9%), respectively.

The right renal vein may be doubled, even though the left renal vein is usually single. The left renal vein may anastomose with the splenic vein and also receive lumbar veins. When the left inferior vena cava persists, the renal and testicular veins show many irregularities.

Communication between the vena cava and accessory renal veins is neither unusual nor an anastomosis between the splenic and left renal veins. Lejars found the left azygos (hemiazygos) vein communicating with the left renal vein in 88% of subjects studied.

Li et al. (2011) described the renal vein variations seen in 61 cases of left donor nephrectomy. Seven types of posterior lumbar tributaries were seen including a reno-hemi-azygo-lumbar trunk in 16.4% (10/61 cases), no posterior lumbar tributaries in 16.4% (10/61 cases), one lumbar vein in 47.5% (29/61 cases), two lumbar veins in 32.8% (20/61 cases), and three lumbar veins in 3.3% (2/61 cases).

References

Dill A, Ayaz UY, Karabacak OR, Tatar IG, Hekimoglu B (2012) Study of the left renal variations by means of magnetic resonance imaging. Surg Radiol Anat 34:267–270

Li G, Dong J, Lu JS, Zu Q, Yang SX, Li HZ, Ma X, Zhang X (2011) Anatomical variation of the posterior lumbar tributaries of the left renal vein in retroperitoneoscopic left living donor nephrectomy. Int J Urol 18:503–509

Wrisberg E (1777) De vena azyga duplici. Novi commentari Soc. R. Scientiarium Gottingensis Obs III

Suprarenal Veins

79

MaKelle Peña

79.1 Embryology

The suprarenal veins originate from the caudal subcardinal veins: The left and right suprarenal veins develop from the left and right subcardinal veins, respectively (Cesmebasi). The left suprarenal vein can be 30 mm long, while the right is generally much shorter at 6 mm (Avisse).

79.2 Anatomy/Topography

The suprarenal venous drainage pathway varies among individuals. The suprarenal vein on the right side of the body drains from the right suprarenal gland directly into the inferior vena cava (Rishi). This vessel is 3–5 cm higher than the right renal vein when it enters the inferior vena cava posterolaterally (Cesmebasi). The left suprarenal vein drains from the left suprarenal gland and enters the superior aspect of the left renal vein (Rishi) 3–5 cm away from its junction with the inferior vena cava (Cesmebasi) (Fig. 79.1).

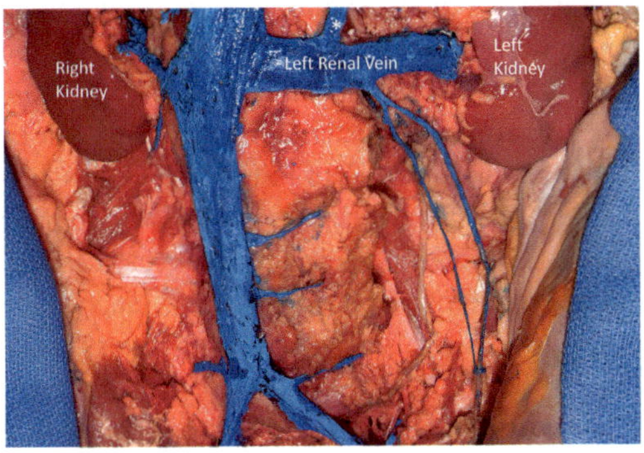

Fig. 79.1 Cadaveric dissection noting the inferior vena cava and left renal vein. The left suprarenal vein is shown at the * and note the gonadal vein draining into the left renal vein from below

79.3 Variations

Both the left and right suprarenal veins vary independently. The left suprarenal vein has many known variants. Most commonly, it joins the renal vein alone. However, it can receive drainage from the inferior phrenic vein before emptying into the renal vein. Bifid suprarenal veins can drain independently into the left renal vein. In another case, the inferior phrenic vein joined one of the doubled suprarenal veins before emptying into the left renal vein. The left suprarenal vein can receive the left gonadal vein or drain directly into the inferior vena cava. The left suprarenal vein can receive a lumbar vein with or without the left gonadal vein before joining the left renal vein.

M. Peña (✉)
Tulane University School of Medicine, New Orleans, LA, USA
e-mail: mpena4@tulane.edu

S. J. Bordes, Jr. et al. (eds.), *The Clinical Anatomy of the Vascular System*,
https://doi.org/10.1007/978-3-031-78326-5_79

Regarding variants of the right suprarenal vein, the posterior hepatic veins sometimes join before its junction with the inferior vena cava. In others, two or three right suprarenal veins join the inferior vena cava. Two right suprarenal veins have been seen, one emptying into the inferior vena cava and the other joining the right renal vein. In rare instances, the right suprarenal vein also receives the right gonadal vein (Cesmebasi).

79.4 Pathology

Suprarenal venous sampling can differentiate unilateral from bilateral aldosterone production from the suprarenal glands. It can be used as a test for primary aldosteronism. Cannulation of the right suprarenal vein before the left is recommended since the right is more likely to have anatomical variations, so this reduces the time between samples collected from each side (Quencer). It is also necessary to have knowledge of the variations of these vessels for laparoscopic adrenalectomy because an essential step in this procedure is to correctly ligate the associated suprarenal vein(s) (Parnaby, Deinum).

79.5 Complications

Complications involving the suprarenal veins are not common. Most are related to iatrogenic injury during procedures such as suprarenal vein sampling, during which dissection or suprarenal gland hemorrhage can occur. As mentioned above, the right suprarenal vein is more frequently injured owing to its greater variation and difficulty with cannulation (Monticone).

References

Avisse C, Marcus C, Patey M, Ladam-Marcus V, Delattre JF, Flament JB (2000) Surgical anatomy and embryology of the adrenal glands. Surg Clin North Am 80(1):403–415. https://doi.org/10.1016/s0039-6109(05)70412-6

Cesmebasi A, Du Plessis M, Iannatuono M, Shah S, Tubbs RS, Loukas M (2014) A review of the anatomy and clinical significance of adrenal veins. Clin Anat 27(8):1253–1263. https://doi.org/10.1002/ca.22374. Epub 2014 Apr 15

Daunt N et al (2005) Adrenal vein sampling: how to make it quick, easy, and successful. Radiographics 25:S143. https://doi.org/10.1148/rg.25si055514

Deinum J, Prejbisz A, Lenders JWM, van der Wilt GJ (2018) Adrenal vein sampling is the preferred method to select patients with primary aldosteronism for adrenalectomy: con side of the argument. Hypertension 71(1):10–14. https://doi.org/10.1161/HYPERTENSIONAHA.117.09294

Monticone S et al (2015) Clinical management and outcomes of adrenal hemorrhage following adrenal vein sampling in primary aldosteronism. Hypertension 67:146. https://www.ahajournals.org/doi/10.1161/hypertensionaha.115.06305

Parnaby CN, Galbraith N, O'Dwyer PJ (2008) Experience in identifying the venous drainage of the adrenal gland during laparoscopic adrenalectomy. Clin Anat 21(7):660–665. https://doi.org/10.1002/ca.20706

Quencer KB (2021) Adrenal vein sampling: technique and protocol, a systematic review. CVIR Endovasc 4(1):38. https://doi.org/10.1186/s42155-021-00220-y. PMID: 33939038; PMCID: PMC8093361

Gonadal Veins 80

Cameron Belding, Davis Martin,
and Jonathan Richard

80.1 Embryology

The early embryologic cardinal, vitelline, and umbilical veins initially develop from capillary plexuses and ultimately drain into the sinus venosus. The anterior and posterior cardinal veins run cranially and caudally to the heart, later developing into the superior vena cava (SVC) and inferior vena cava (IVC), respectively. The posterior cardinal veins regress, and the subcardinal and supracardinal veins arise. The right supracardinal vein drains the lower abdomen, including what will ultimately become the renal and gonadal veins (Figs. 80.1 and 80.2) in gestational weeks 5–7, into the iliac anastomosis (Mozes and Gloviczki 2007; Nallikuzhy et al. 2018; Kara et al. 2012).

At four to six months gestational age, the left gonadal vein is found to join the left renal artery at an approximately 60° angle, while the right gonadal vein enters directly into the IVC at an approximately 20° angle. These angles increase to approximately 90° on the left and approximately 40° on the right in adulthood, primarily due to elongation of the trunk, cranial migration of the kidney, and gonadal descent. The left gonadal vein is consistently longer than the right, although the ratio of their lengths gradually approaches one throughout development (Szpinda et al. 2005a).

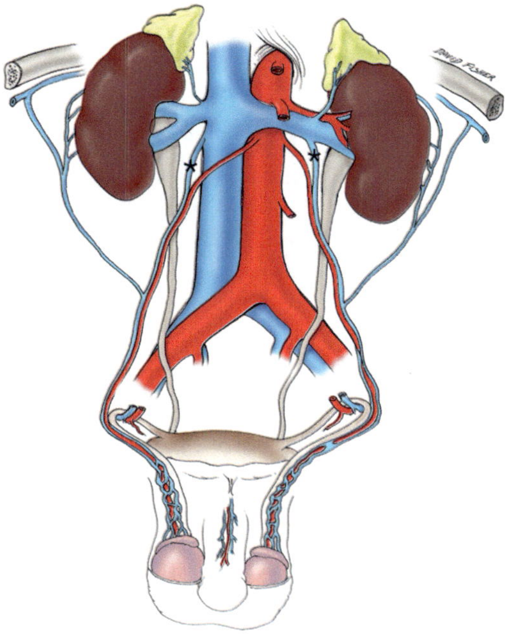

Fig. 80.1 Schematic drawing of the testicular veins (*)

At four to five months gestational age, both testicular veins originate at the level of the

C. Belding
Tulane University School of Medicine,
New Orleans, LA, USA

D. Martin
School of Medicine, Louisiana State University
Health Sciences Center, New Orleans, LA, USA

J. Richard (✉)
Department of Surgery, Louisiana State University
Health Sciences Center, New Orleans, LA, USA

S. J. Bordes, Jr. et al. (eds.), *The Clinical Anatomy of the Vascular System*,
https://doi.org/10.1007/978-3-031-78326-5_80

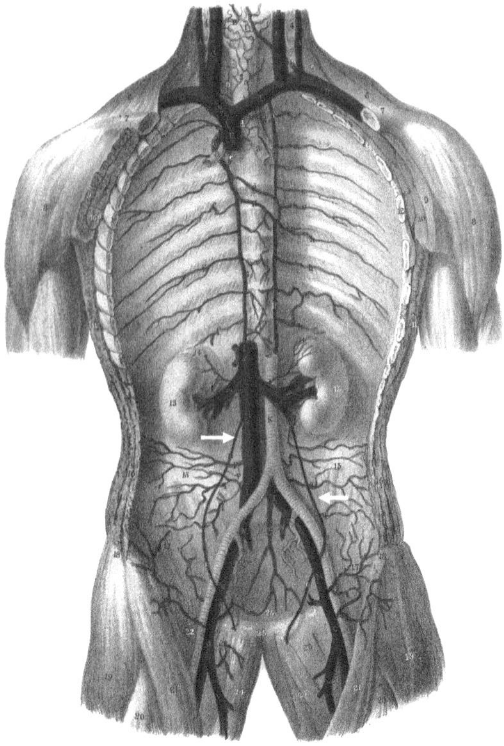

Fig. 80.2 Drawing illustrating the testicular veins (arrows) (Quain and Wilson, 1844 author's collection)

sacral apex, descending inferior to the vertebral column by month six. The right testicular vein consistently terminates at the IVC at the level of L2 throughout this stage of embryogenesis. In contrast, the left testicular vein's termination gradually descends from the level of T12 to L2 secondary to migration of the left renal vein. The origins of both ovarian veins project onto S1 at month four and descent to S3 by month six. The ovarian veins terminate at the same spinal levels as their respective testicular veins (Szpinda et al. 2005b).

80.2 Anatomy/Topography

Multiple veins emerge from the ovary to form the pampiniform plexus within the mesovarium and infundibulopelvic ligament (suspensory ligament of the ovary) (Standring 2021; Sharma and Abdi 2012). The plexus then coalesces into two ovarian veins that ascend alongside the ovarian artery, coursing retroperitoneally, anterior to the psoas and ureter (Standring 2021; Ghosh and Chaudhury 2019; Sharma and Abdi 2012). These veins coalesce into a single ovarian vein before termination (Standring 2021).

Draining the testis and the epididymis, multiple testicular veins initially emerge and immediately anastomose into the pampiniform plexus. Unlike the pampiniform plexus of the ovarian vein, this pampiniform plexus surrounds the testicular artery throughout the length of the spermatic cord anterior to the ductus deferens. The web-like structure of the pampiniform plexus coalesces into three or four veins within the inguinal canal and ascends into the abdomen via the deep inguinal ring. Within the abdomen, these veins course anterior to the psoas and ureter, then join to form the testicular vein until termination (Standring 2021; Kara et al. 2012; Sharp et al. 2013).

In males and females, the left gonadal vein lies anterior to the left ureter and runs cephalad to its termination at a right angle with the left renal vein. It lies posterior to the left colic vein and inferior mesenteric vein (Standring 2021; Meyers et al. 2010). The right gonadal vein lies posterior to the ileocolic vessels and lateral to the right ureter, coursing cephalad to the IVC (Meyers et al. 2010). Its termination occurs at an acute angle approximately 2–4 cm inferior to the termination of the left renal vein at the IVC (Standring 2021; Kara et al. 2012). The gonadal veins have minimal adventitial sheaths, predisposing them to collapse in the case of any external compression (Dunphy and Tang 2021).

80.3 Variation

A study of 47 cadavers showed that 3 of the 94 ovarian veins (3.19%) demonstrated some anatomic variation (Ghosh and Chaudhury 2019). One right ovarian vein drained into the right renal vein 19 mm from the IVC. One cadaver exhibited double right renal veins, both draining into the IVC. One cadaver exhibited a bifurcation of the left ovarian vein 40 mm distal to its termination in the left renal vein (Ghosh and Chaudhury 2019).

A study of 324 CT scans found that 32 (9.9%) of right ovarian veins drained into the right renal vein, with no other right ovarian vein variations identified (Koc et al. 2006).

The exact course of the ovarian veins may vary based on the location of the ovaries and their attachment to the broad ligament (Meyers et al. 2010).

A meta-analysis of testicular vein anatomy found that variations of the left testicular vein were found in 24.61% of cases. In comparison, the right testicular vein variations were found in 18.4% of cases. They proposed the following classification system for variations in the testicular veins (Nallikuzhy et al. 2018).

I. *Number of Veins (N)*
   (a) N1—single testicular vein (<80% of left, >12% of right).
   (b) N2—multiple testicular veins (>20% of left, <8% of right). These cases were further stratified into double (16.61% of L, 4.12% of R), triple (2.15% of L, 0.48% of R), and quadruple (0.12% of L) testicular veins, as well as cases that reported multiple testicular veins without a specific number (3.39% of R). As many as six testicular veins have been reported on one side. The mechanism of this variation is believed to be that the three or four testicular veins arising from the pampiniform plexus may not coalesce fully and can remain multiple veins until termination (Standring 2021). This failure of the testicular veins to coalesce is attributed to failure of part of the intersubcardinal anastomosis to regress in utero (Kara et al. 2012). Incidences of multiple testicular veins can be further classified as follows:
      (i) N2a—multiple veins throughout the course of the testicular vein.
      (ii) N2b—multiple veins joining to form a single testicular vein.
      (iii) N2c—single testicular vein dividing into multiple veins before drainage.

II. *Site of Drainage (D)*. 10.41% of cases demonstrated variable right testicular venous drainage sites, including the right renal vein (7.63%) and the confluence of the IVC and the right renal vein (2.78%) (Standring 2021; Nallikuzhy et al. 2018). Only 0.59% of cases demonstrated variable left testicular vein drainage directly into the IVC. Incidences of variable drainage sites can be further classified as follows:
   (a) D1—drainage into the IVC on the right side and the renal vein on the left.
   (b) D2—drainage into the renal vein on the right side and the IVC on the left.
   (c) D3—drainage into the confluence of the IVC and the renal vein on both sides.
   (d) D4—drainage into other veins.

Other variations in testicular vein anatomy include anastomoses with the cremasteric vein, external pudendal vein, or veins of the ductus deferens (Standring 2021). Exceedingly rare incidences of bilateral drainage into subcostal veins (Tubbs et al. 2005) and the left renal vein (Daniel and Havergal 2016) have been reported.

80.4 Pathology

The significant pathologies of the gonadal veins are further discussed below.

Pelvic Congestion Syndrome Pelvic congestion syndrome (PCS) may be present in up to 30% of women with chronic pelvic pain. Defined as pelvic pain, perineal discomfort, or incontinence for greater than six months, pelvic congestion syndrome is believed to be caused by ovarian vein reflux or obstruction (Chaar 2017). A study of 324 women found that 18.2% of subjects exhibited pelvic varices (58/59 in the left ovarian vein) and 17.6% exhibited ovarian vein reflux (Koc et al. 2006). PCS is also associated with an ovarian vein diameter of >6 mm, further strengthened by a mean ovarian vein diameter of 6.2–6.3 mm in patients with pelvic varices vs. 3.1–3.3 mm in patients without varices (Koc et al. 2006; Chaar 2017). Traditional treatment methods include ovarian suppression, dihydroergotamine, uterine ventrosuspension, hysterectomy, ovarian vein ligation, and excision (Chaar 2017).

Adnexal Torsion/Ovarian Torsion Adnexal torsion is a unilateral twisting of the fallopian tube and ovary, while ovarian torsion is an isolated twisting of the ovary without the involvement of the fallopian tube (Childress and Dietrich 2017). Adnexal torsion is significantly more common (approximately 2/3 of torsion cases), but both pathologies involve occlusion of the ovarian vein (Childress and Dietrich 2017). The ovarian vein is easily collapsed in torsion, leading to the classic appearance of an engorged, discolored, edematous ovary (Childress and Dietrich 2017). Ovarian artery occlusion is a later complication that ultimately leads to ischemia and necrosis of the ovary (Childress and Dietrich 2017). Patients typically present with unilateral, sudden onset pelvic or abdominal pain, as well as a constellation of other nonspecific symptoms, including tachycardia, fever, and nausea (Childress and Dietrich 2017). The diagnosis of torsion is made based on clinical presentation and Doppler ultrasonography, which visualizes ovarian vein compression in approximately 93% of cases (Childress and Dietrich 2017). Torsion is treated by surgical detorsion, with oophorectomy being found necessary in approximately 60% of cases due to necrosis, hemorrhage, or malignancy (Childress and Dietrich 2017). Follow-up Doppler ultrasonography after detorsion should demonstrate restored flow in the ovarian vein and normal follicular development (Childress and Dietrich 2017). Unlike testicular torsion, in which bilateral orchiopexy is considered necessary to prevent recurrence, oophoropexy remains controversial due to the potential loss of fertility, distortion of adnexal anatomy, and a lack of established supporting evidence (Sharp et al. 2013; Childress and Dietrich 2017).

Ovarian Vein Thrombosis Typically occurring in the postpartum period, ovarian vein thrombosis (OVT) occurs after 0.05–0.18% of pregnancies and 1–2% of cesarean sections (Sharma and Abdi 2012; Dunphy and Tang 2021). Any hypercoagulable state can predispose a patient to OVT; pregnancy is by far the most common risk factor, but other associations include systemic lupus erythematosus, factor V Leiden, antiphospholipid syndrome, and nephrotic syndrome, among others (Sharma and Abdi 2012; Dunphy and Tang 2021). The right ovarian vein is involved in 70–90% of cases, likely due to its length, fewer competent veins, and susceptibility to compression due to dextrorotation of the uterus in pregnancy (Sharma and Abdi 2012; Dunphy and Tang 2021). Patients classically present 2–15 days postpartum with fever, leukocytosis, and unilateral lower abdominal pain radiating to the flank, upper abdomen, or groin (Sharma and Abdi 2012; Dunphy and Tang 2021). Although these symptoms are mainly nonspecific, diagnosis can be confirmed with imaging; ultrasound is first- line, but MRI and CT are more sensitive (sensitivities: ultrasound 52%, MRI 92%, CT 100%) (Sharma and Abdi 2012). Patients should be promptly treated with anticoagulation, mainly because as many as 33% of women with OVT develop pulmonary embolism (Sharma and Abdi 2012). Other complications include extension of the thrombus to the site of ovarian vein termination and septic thrombophlebitis (Dunphy and Tang 2021).

Ovarian Vein Syndrome Though the exact mechanism remains controversial, ovarian vein syndrome (OVS) is known to occur due to compression of the ureter by a dilated ovarian vein (Bhutta et al. 2009). OVS is more common in multiparous women and can present either acutely or chronically (Bhutta et al. 2009). Symptoms are typically nonspecific (abdominal pain, urinary irregularity) but are characteristically most severe just before menstruation (Bhutta et al. 2009). Diagnosis can be made using an abdominal ultrasound demonstrating a dilated ovarian vein ($\geq$6 mm) and IV urography demonstrating hydroureter with a well-demarcated transverse defect at the level of L3-L4. Secondary imaging modalities such as retrograde pyelography, percutaneous ovarian venography (gold standard), CT, and MRI may help confirm the diagnosis (Bhutta et al. 2009).

While embolization and laparoscopic ligation of the ovarian vein are well-established treatments for OVS, more recent chemical sclerotherapy methods have demonstrated 98–100% success rates with a mean pain score reduction of 65% (Bhutta et al. 2009).

Testicular Vein Syndrome Though similar to OVS in pathology, testicular vein syndrome (TVS) has only been recorded in 6 total cases as of 2011, with no reported cases since (Arvind et al. 2011). In TVS, only 3 of the 6 reported cases involved changes in the offending testicular vein (Arvind et al. 2011). Treatment varies in each case, with the 2011 case demonstrating the first recorded successful laparoscopic excision of a testicular vein and ureteroureterostomy for TVS (Arvind et al. 2011).

Varicocele Varicoceles are characterized by dilation of the pampiniform plexus within the scrotum and are found in 15–20% of the general population (Masson and Brannigan 2014). Varicoceles have a well-established association with male infertility, and 35–40% of infertile men have evidence of varicoceles (Masson and Brannigan 2014). The left testicular vein is longer than the right, and this, combined with its right-angle termination at the left renal vein, likely contributes to the increased incidence of varicocele on the left (Standring 2021; Masson and Brannigan 2014). Absence or incompetence of valves and inheritance have also been suggested as risk factors (Masson and Brannigan 2014). Collateral flow to the testicular veins from the abnormal anastomoses discussed in "Variation" was found in 48% of right gonadal veins and 52% of left gonadal veins (Lechter et al. 1991), likely contributing to the recurrence of varicoceles even with proper ablation of the testicular vein (Standring 2021; Lechter et al. 1991; Masson and Brannigan 2014). These collaterals are typically not appreciated on CT imaging and may only be found during surgery (Kara et al. 2012). Diagnosis of varicocele is normally performed via palpation of a mass above the testis that resembles a "bag of worms," reduces while lying supine, and engorges with Valsalva. Scrotal ultrasound (93% sensitivity) can further investigate inconclusive physical exams (Masson and Brannigan 2014). Indications for treatment include infertility, hypogonadism, and pain (Masson and Brannigan 2014). Treatment has generally been found to result in significantly improved semen parameters, although data regarding pregnancy after varicocelectomy remain variable with mixed results as to significant increases in pregnancy (Masson and Brannigan 2014).

Testicular Torsion Testicular torsion is a twisting and occlusion of the spermatic cord that is considered a surgical emergency due to threatened loss of fertility and the involved testicle (Sharp et al. 2013). Testicular torsion has an annual incidence of 3.8 per 100,000 male children, accounting for approximately 10–15% of childhood scrotal disease (Sharp et al. 2013). Although torsion is typically considered a pathology of arterial occlusion and ischemia, the testicular veins are the first to occlude. This results in increased venous pressure and congestion, decreased arterial flow, and ischemia (Sharp et al. 2013). Testicular torsion is a clinical diagnosis based on the presence of exquisite unilateral testicular pain, nausea, high-riding testicles, and other nonspecific symptoms (Sharp et al. 2013). Treatment involves either manual or surgical detorsion followed by bilateral orchiopexy to prevent recurrence (Sharp et al. 2013). Orchiectomy is necessary in 39–71% of testicular torsion cases, especially when care is delayed by more than 8 hours, and the testicle appears grossly necrotic (Sharp et al. 2013). While Doppler ultrasonography should not be used for diagnosis due to possible delay of emergent surgical exploration, it should be used to document successful treatment with restored arterial and venous flow (Sharp et al. 2013).

80.5 Surgical Considerations

Pelvic Congestion Syndrome Embolization methods have shown a 70–90% improvement in symptoms (Chaar 2017). Cyanoacrylates are a productive and safe method of intervention for endovenous sealing. This nonthermal, nontumescent intervention is potentially faster and more comfortable and produces less adverse events than traditional endothermal ablation with similar efficacy (Chaar 2017). Sclerosing agents, namely, sodium tetradecyl sulfate, can also be used to occlude the ovarian veins with a relatively low incidence of thrombotic complications (Chaar 2017).

Nutcracker Syndrome Patients with nutcracker syndrome (compression of the left renal vein between the superior mesenteric artery and the aorta) may experience left ovarian vein reflux symptoms, including pelvic pain, dyspareunia, and worsening menses (White and Ryjewski 2021). In these patients, transposition of the left ovarian vein to the IVC or the left iliac vein, as described by White and Ryjewski (2021), is very effective in relieving pelvic symptoms (White and Ryjewski 2021).

Varicocele The definitive treatment for varicocele is surgical disruption of the offending testicular vein (Masson and Brannigan 2014). The two most common methods for dissection are the inguinal and subinguinal microsurgical approaches described by Masson et al., with the subinguinal approach generally being preferred due to its preservation of the external oblique fascia and decreased postoperative discomfort (Masson and Brannigan 2014). Laparoscopic approaches demonstrate ample safety and efficacy but present a higher incidence of postoperative complications such as hydrocele and varicocele (Masson and Brannigan 2014). Percutaneous embolization is a promising method with faster recovery time and less postoperative pain. Still, it is less successful than microsurgical approaches and may pose a further risk to fertility due to radiation exposure (Masson and Brannigan 2014).

Renal Transplantation In procuring a donor kidney from a living donor, it is essential to identify and ligate the left gonadal vein while leaving adequate length for clipping the vein stump if necessary (Englesbe and Mulholland 2015). Recent methods, such as those described by Tomizawa et al. and Feng et al., have shown that a graft from the gonadal vein is both easily accessible and productive for use in extending the renal artery (Tomizawa et al. 2020; Feng et al. 2012). These methods are still novel and have limited recorded evidence to support them, so they should be thoroughly studied before attempting.

Variations in gonadal vein anatomy should be considered and identified before any surgery involving the renal pedicles or gonads (e.g., renal transplantation, tumor embolization, varicocelectomy, orchiopexy, etc.) (Nallikuzhy et al. 2018). Varicocelectomy is especially prone to failure or recurrence in cases of multiple testicular veins or abnormal anastomoses, as discussed above (Standring 2021; Kara et al. 2012; Lechter et al. 1991; Masson and Brannigan 2014).

References

Arvind NK, Singh O, Gupta SS (2011) Testicular vein syndrome and its treatment with a laparoscopic approach. JSLS. 15(4):580–584. https://doi.org/10.4293/108680811X13176785204797. PMID: 22643523; PMCID: PMC3340977

Bhutta HY, Walsh SR, Tang TY, Walsh CA, Clarke JM (2009) Ovarian vein syndrome: a review. Int J Surg 7(6):516–520. https://doi.org/10.1016/j.ijsu.2009.09.008. Epub 2009 Oct 8

Chaar CIO (2017) Current management of venous diseases. Springer International Publishing AG. ProQuest Ebook Central

Childress KJ, Dietrich JE (2017) Pediatric ovarian torsion. Surg Clin North Am 97(1):209–221. https://doi.org/10.1016/j.suc.2016.08.008. PMID: 27894428

Daniel DK, Havergal KER (2016) Right testicular vein variations and its implications: a case report. Int J Anat Res 4:2634–2636

Dunphy L, Tang AW (2021) Ovarian vein thrombosis in the postnatal period. BMJ Case Rep 14(6):e243872. Web

Englesbe MJ, Mulholland MW (2015) Operative techniques in transplantation surgery. Wolters Kluwer, Philadelphia. Print

Feng J-Y et al (2012) Renal vein lengthening using gonadal vein reduces surgical difficulty in living-donor kidney transplantation. World J Surg 36(2):468–472. Web

Ghosh A, Chaudhury S (2019) A cadaveric study of ovarian veins: variations, measurements and clinical significance. Anat Cell Biol 52(4):385–389. Web

Kara T, Younes M, Erol B, Karcaaltincaba M (2012) Evaluation of testicular vein anatomy with multidetector computed tomography. Surg Radiol Anat 34(4):341–345. https://doi.org/10.1007/s00276-011-0898-3. Epub 2011 Nov 22

Koc Z, Ulusan S, Oguzkurt L (2006) Right ovarian vein drainage variant: is there a relationship with pelvic varices? Eur J Radiol 59(3):465–471. Web

Lechter A, Lopez G, Martinez C, Camacho J (1991) Anatomy of the gonadal veins: a reappraisal. Surgery 109:735–739

Masson P, Brannigan RE (2014 Feb) The varicocele. Urol Clin North Am 41(1):129–144. https://doi.org/10.1016/j.ucl.2013.08.001

Meyers MA et al (2010) Meyers' dynamic radiology of the abdomen: normal and pathologic anatomy. Springer, New York

Mozes G, Gloviczki P (2007) Chapter 2—venous embryology and anatomy. In: The vein book. Academic, Oxford, pp 15–25

Nallikuzhy TJ, Rajasekhar SSSN, Malik S, Tamgire DW, Johnson P, Aravindhan K (2018) Variations of the testicular artery and vein: a meta-analysis with proposed classification. Clin Anat 31(6):854–869. https://doi.org/10.1002/ca.23204. Epub 2018 Jul 3

Sharma P, Abdi S (2012) Ovarian vein thrombosis. Clin Radiol 67(9):893–898. Web

Sharp VJ, Kieran K, Arlen AM (2013) Testicular torsion: diagnosis, evaluation, and management. Am Fam Physician 88(12):835–840

Standring S (2021) Gray's anatomy: the anatomical basis of clinical practice, 42nd edn. Elsevier, New York

Szpinda M, Elminowska-Wenda G, Wiśniewski M, Frąckiewicz P (2005a) Morphometric analysis of the gonadal veins in human foetuses. Ann Anat 187(4):399–403

Szpinda M, Frąckiewicz P, Brazis P, Wiśniewski M (2005b) Skeletopic analysis of the gonadal veins in human foetuses. Folia Morphol 64(2):84–88

Tomizawa M, Hori S, Nishimura N et al (2020) Arterial reconstruction using the donor's gonadal vein in living renal transplantation with multiple renal arteries: a case report and a literature review. BMC Nephrol. 21(1):190. https://doi.org/10.1186/s12882-020-01848-z. Published 2020 May 20

Tubbs RS, Salter EG, Oakes WJ (2005) Unusual drainage of the testicular veins. Clin Anat 18:536–539

White JV, Ryjewski C (2021) A simplified surgical approach for left ovarian vein transposition for the treatment of pelvic venous disease from nutcracker syndrome. J Vasc Surg Cases Innov Tech 7(3):411–414. Published 2021 May 21. https://doi.org/10.1016/j.jvscit.2021.05.008

Lymphatics of the Abdominopelvic Region

81

Gilad Hampel

81.1 Embryology

Embryo development of the lymphatic system is less well documented than that of the vasculature. The role of the lymphatic system in disseminating pathogens and cancers has sparked interest in it and its development. The identification of pathogen-specific markers has provided new information about the latter (Oliver 2004).

One model of lymphatic system development envisages a four-stage process: lymphatic endothelial cell (LEC) competence, bias, specification, and vessel differentiation/maturation (Oliver 2004). First, venous endothelial cells gain the ability to respond to inductive signals. The mechanisms of competence are not fully understood, but they are thought to involve hyaluronan signaling (Knudson and Knudson 1993). Only the cells that receive the inductive signals retain the ability to be biased to LEC determination; the other venous endothelial cells do not. The Prox1 gene is thought to function in this step; it was shown to be necessary for lymphatic development in a murine model (Wigle and Oliver 1999). Prox1 knockouts developed a vascular system but not a lymphatic system. It was also shown that Prox1 is required for the acquisition of the LEC phenotype (Wigle et al. 2002). Petrova et al. showed that Prox1 could reprogram vascular endothelial cells for that phenotype (Petrova et al. 2002). Lymphatic markers (VEGFR3, T1α/podoplanin, Nrp2, etc.) become increasingly expressed, while blood vascular markers such as CD34 and laminin are downregulated (Wigle et al. 2002; Chen et al. 1997; Kaipainen et al. 1995; Schacht et al. 2003). T1α/podoplanin is especially important since its deficiency results in abnormal lymphatic vasculature development and lymphedema (Schacht et al. 2003). Biased venous epithelial cells then differentiate separately from the region. The early lymphatic system develops by budding throughout the embryo to yield primary lymph sacs; the LECs originally bud from cardinal veins. Lastly, lymphatic vessels differentiate and mature stepwise from the lymph sacs during terminal differentiation (Oliver and Harvey 2002). The blood and lymphatic vasculatures are separated by SLP-76 and Syk signaling (Farhad et al. 2003). Mice with mutations or knockouts in these proteins have "abnormal lymphatic connections, embryonic hemorrhages, and arteriovenous shunts." (Oliver 2004)

Anatomy and Topography

Thoracic Duct Origin and Abdominal Lymph Trunks

The thoracic duct's abdominal origin (Fig. 81.1) typically resides near the midline at the first or second lumbar vertebrae, sometimes between

G. Hampel (✉)
Tulane University School of Medicine, New Orleans, LA, USA
e-mail: ghampel@tulane.edu

S. J. Bordes, Jr. et al. (eds.), *The Clinical Anatomy of the Vascular System*,
https://doi.org/10.1007/978-3-031-78326-5_81

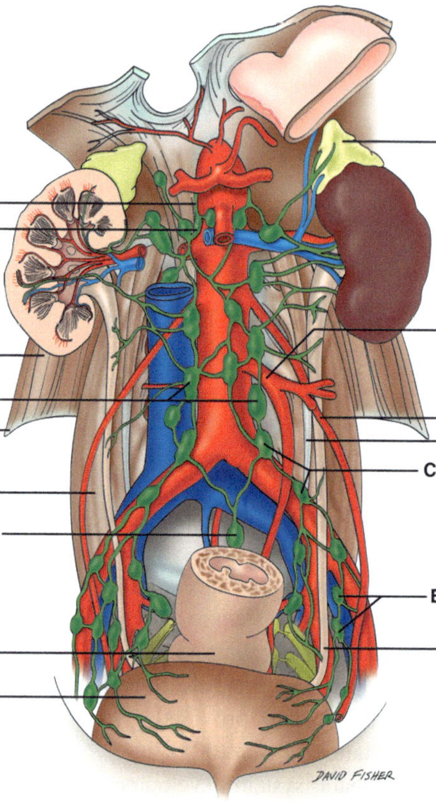

Fig. 81.1 Regional lymphatic structures of the abdominopelvic region. (After Netter 2023)

them. It collects almost all lymph from below the diaphragm via the cisterna chyli, a saccular or fusiform lymphatic enlargement generally formed by the merging of the intestinal trunk and the left lumbar trunk. The structure of the origin of the thoracic duct (cisterna chyli) can vary and may even be absent. Typically, the cisterna chyli is found posterior to the right crus of the diaphragm, and the right of the abdominal aorta, although it can be located at more proximal vertebral levels. The most common formation, seen in about two-thirds of people, is a single structure formed by the union of the intestinal trunk and the left (or less commonly, the right) lumbar trunk. The paired lumbar trunks are formed by lymphatic vessels draining the lateral aortic nodes. The intestinal trunk, which may be double, collects vessels draining the celiac, superior, and inferior mesenteric nodes, which collectively drain the entire abdominal gastrointestinal tract down to the anal canal.

Retroperitoneal Nodes

The lymphatic drainage of the rectum, colon, stomach, pancreas, esophagus, and other organs is often described in terms of lymph node stations and dissection levels (Fig. 81.1). Retroperitoneal lymph nodes around the abdominal aorta are categorized into preaortic, lateral, and retro-aortic nodes.

Preaortic Nodes

Preaortic nodes drain the gastrointestinal viscera, including the pancreas, liver, and spleen. They are located in front of the aorta, clustered around its anterior unpaired visceral branches, and can be subdivided into celiac, superior, and inferior mesenteric nodes. Their efferent lymphatic vessels contribute to the formation of the intestinal lymph trunk.

Celiac Nodes

Celiac nodes drain lymph from nodes around the stomach, splenic hilum, porta hepatis, cystic duct, lesser omentum, portacaval nodes, pancreatic nodes, and pancreaticoduodenal nodes. They also receive lymph from the superior and inferior mesenteric nodes, and their efferent lymphatic vessels drain into the intestinal trunks.

Superior Mesenteric Nodes

Superior mesenteric nodes overlap significantly with the drainage area of celiac nodes and include nodal groups such as pancreatic, pancreaticoduodenal, portacaval, ileocolic, mesenteric, and inferior mesenteric nodes. Their efferent vessels drain into the intestinal trunks directly or via the celiac nodes.

Inferior Mesenteric Nodes

Inferior mesenteric nodes, like the superior rectal nodes, drain lymph from hindgut structures, including the distal transverse, descending, and sigmoid colons and the rectum.

Lateral Aortic Nodes

Lateral aortic nodes (Fig. 81.1) are situated on either side of the abdominal aorta and inferior vena cava, in front of the medial margins of the psoas major, the crura of the diaphragm, and the sympathetic trunks. Recognized nodal groups include retrocrural (behind the diaphragmatic crura at the aortic hiatus), left and right hilar (kidney), aortocaval, lateral caval, retrocaval, and precaval groups. Retro-aortic nodes are sometimes included in this group. These nodes drain into the paired intestinal trunks, which terminate directly or indirectly in the cisterna chyli and thoracic duct. There are lymphatic connections between lateral aortic, preaortic, retro-aortic, and contralateral lateral aortic nodes. The lateral aortic nodes drain the deep layers of the body wall, retroperitoneal paired viscera, and tributaries from the iliac nodes. Lymph from the right ovary or testis drains via lymphatic vessels traveling with the gonadal vessels to the right lateral aortic and preaortic nodes at the level of the second lumbar vertebra, while lymph from the left ovary or testis drains to the left lateral aortic nodes just inferior to the left renal vein.

Iliac Nodes

Iliac nodes are distributed around the common, external, and internal iliac arteries and veins. These include the common iliac, external iliac, internal iliac, and obturator nodes (Fig. 81.1). The obturator nodes, near the obturator foramen, along with the iliac nodes, are common sites for lymph node metastasis in prostate cancer. The iliac nodes drain the pelvic viscera and walls, except for the ovaries and parts of the rectum, which are drained by superior anorectal nodes. They also collect lymph from the inguinal nodes and lower limbs.

The reproductive organs bring their vessels, nerves, and lymphatics with them during development and descent (Drake et al. 2005). For example, the ovarian lymphatics pass with the ovarian vessels and nerves through the suspensory ligament of the ovary to the superior pole. The pelvic viscera primarily drains into a series of lymph nodes (LNs) along the internal and external iliac arteries and branches. The external iliac LNs lie above the pelvic inlet and receive lymph from inguinal LNs and the superior pelvic viscera (especially the anterior and middle parts of the pelvis) (Moore and Dalley 2018). They are arranged in three chains: an external chain of 3–4 nodes lateral to the external iliac artery, a middle chain of 2–3 nodes on the anterior surface of the external iliac vein, and an internal chain of 3–4 nodes below the external iliac vein along the pelvic wall (Herman et al. 1963). The hypogastric group of LNs lies near the origin of the hypogastric artery (and the junction of the external and common iliac groups of LNs) and contains 4–8 nodes. The most posterior chain, the lateral sacral nodes, lies along the lateral sacral artery internal to the 2nd/3rd sacral foramen (Herman et al. 1963). The internal iliac LNs are found around the anterior and posterior divisions of the internal iliac artery. They drain the inferior pelvic viscera, the deep perineum, and the gluteal region. The external iliac, internal iliac, and hypogastric LNs drain into the common iliac group and eventually into the lateral aortic or lumbar nodes. The common iliac LNs form three groups: the external, middle, and internal chains. The channels of the two common iliac groups meet to form the nodes of the promontory in front of the sacral promontory in 50% of individuals (Herman et al. 1963). The common iliac groups of LNs are considered a continuation of the three external iliac chains (Herman et al. 1963). The three channels of the common iliac groups unite to yield the single aortic chain at the lowermost point of the aorta. The lateral aortic and lumbar nodes drain into the lumbar trunks, which unite to form the origin of the thoracic duct at approximately the T12 vertebral level. Sacral LNs are located anterior to the sacrum in the space formed by its concavity. They drain the posterior-inferior pelvic viscera and empty into the internal or common iliac LNs. Lymphatics from the female reproductive organs exit the pelvic cavity superiorly along the ovarian arteries to drain directly into the lateral aortic or lumbar LNs. In some variations, they drain directly into preaortic nodes anterior to the aorta. Lymph flow is dictated by external forces such as skeletal muscle contraction; valves prevent backflow (Hsu and Itkin 2016).

Histology/Imaging

Extracellular fluid enters blind-ended lymphatic capillaries that feed into the lymphatic vessels. The lymph passes through LN chains and larger lymphatic vessels before reentering the venous circulation via the thoracic duct and the right lymphatic ducts at the base of the neck (Gartner and Hiatt 2012). On a histological slide, lymphatic vessels can appear either collapsed or full of lymph. If collapsed, they are not easily discernible. Full lymphatic vessels appear like veins with thin walls and an endothelial lining. Lymphocytes can be seen within them. The endothelium is relatively thin and contains openings held in place by anchoring filaments through which lymph enters the vessels (Junqueira and Carneiro 2005). Lymphatic vessels contain valves formed by infoldings of endothelial cells not anchored to the basal lamina. The valves prevent backflow. The larger lymphatic vessels contain more connective tissue and smooth muscle arranged in indistinct tunicas (Junqueira and Carneiro 2005).

Lymphangiography can image the abdominopelvic lymphatics. This technique involves injecting oily contrast into lymphatic vessels via a cannula to amplify inguinal, pelvic, perineal, and retroperitoneal lymphatics (Tanagho and Mcaninch 2007). Nowadays, lymphangiography is seldom used since CT imaging of enlarged lymph nodes has become more common and efficient (Tanagho and Mcaninch 2007; Herring and Richens 2006; Mabey and Peeling 2002).

Clinical and Surgical Significance

Cancers typically spread to nearby LNs; therefore, it is common surgical practice to remove associated LNs (lymphadenectomy) along with the malignancies. The aforementioned main groups of LNs are the primary pathways of cancer spread in the pelvis and perineum. Perineal tumors, such as tumors of the vulva and lower vagina in women and the scrotum and penis in men, typically spread via the superficial inguinal pathway. Pathologies of the upper vagina and cervix typically spread retroperitoneally (Mabey and Peeling 2002). The inguinal and subinguinal node groups can also receive spread (Tanagho and Mcaninch 2007). The first nodal junction is the saphenofemoral node, followed by metastatic spread to the deep inguinal and external iliac groups (Paño et al. 2015). Cancers of the testis and prostate typically spread to the internal iliac, external iliac, and preaortic LNs (Tanagho and Mcaninch 2007). Midline masses of the upper abdomen in young males suggest testicular cancer. There are reports of metastases from male and female urogenital cancers spreading along abdominal lymphatic pathways instead of traditionally via the superficial inguinal pathway. Typically, a modified LN dissection technique is used to spare the lateral half of the iliac arterial lymphatics to reduce the risk of complications such as lower extremity edema (Tanagho and Mcaninch 2007).

Complications

Lymphocele is a known surgical complication of lymphadenectomy for treating pelvic and perineal cancers. Treatment can include drainage by the creation of a window into the peritoneal cavity. Lymphocele formation is also a known complication of kidney transplantation. Reports of incidence range from 6 to 18% (Tanagho and Mcaninch 2007). Most post-transplant lymphoceles are asymptomatic and resolve spontaneously (Khauli et al. 1993). Those that require treatment are usually managed by marsupialization and drainage to the peritoneal cavity.

References

Chen H, Chédotal A, He Z, Goodman CS, Tessier-Lavigne M (1997) Neuropilin-2, a novel member of the Neuropilin family, is a high affinity receptor for the Semaphorins Sema E and Sema IV but not Sema III. Neuron 19(3):547–559. https://doi.org/10.1016/S0896-6273(00)80371-2

Drake RL, Gray H, Vogl W, Mitchell AW (2005) Gray's anatomy for students. Elsevier Health Sciences TW, Toronto

Farhad A, Anastasia G, Eric S, Min-Min L, Rong Z, Attila M, Myers Erin E, Bin H, Jackson David G, Ferrari Victor A, Victor T, Lowell Clifford A, Lepore John J, Koretzky Gary A, Kahn Mark L (2003) Regulation of blood and lymphatic vascular separation by signaling proteins SLP-76 and Syk. Science 299(5604):247–251. https://doi.org/10.1126/science.1079477

Gartner LP, Hiatt JL (2012) Color atlas and text of histology. Lippincott Williams & Wilkins

Herman PG, Benninghoff DL, Nelson JH, Mellins HZ (1963) Roentgen anatomy of the Ilio-pelvic-aortic lymphatic system. Radiology 80(2):182–193. https://doi.org/10.1148/80.2.182

Herring A, Richens J (2006) Lymphogranuloma Venereum. Sex Transm Infect 82(suppl 4):iv23–iv25

Hsu MC, Itkin M (2016) Lymphatic anatomy. Tech Vasc Interv Radiol 19(4):247–254. https://doi.org/10.1053/j.tvir.2016.10.003

Junqueira LC, Carneiro J (2005) Basic histology: Text & Atlas. McGraw-Hill Professional

Kaipainen A, Korhonen J, Mustonen T, Van Hinsbergh VW, Fang G-H, Dumont D, Breitman M, Alitalo K (1995) Expression of the Fms-like tyrosine kinase 4 gene becomes restricted to lymphatic endothelium during development. Proc Natl Acad Sci 92(8):3566–3570

Khauli RB, Stoff JS, Lovewell T, Ghavamian R, Baker S (1993) Post-transplant lymphoceles: a critical look into the risk factors, pathophysiology and management. J Urol 150(1):22–26

Knudson CB, Knudson W (1993) Hyaluronan-binding proteins in development, tissue homeostasis, and disease. FASEB J 7(13):1233–1241

Mabey D, Peeling RW (2002) Lymphogranuloma Venereum. Sex Transm Infect 78(2):90–92

Moore KL, Dalley AF (2018) Clinically oriented anatomy. Wolters kluwer india Pvt Ltd

Oliver G (2004) Lymphatic vasculature development. Nat Rev Immunol 4(1):35–45

Oliver G, Harvey N (2002) A stepwise model of the development of lymphatic vasculature. Ann N Y Acad Sci 979(1):159–165. https://doi.org/10.1111/j.1749-6632.2002.tb04876.x

Paño B, Sebastià C, Ripoll E, Paredes P, Salvador R, Buñesch L, Nicolau C (2015) Pathways of lymphatic spread in gynecologic malignancies. Radiographics 35(3):916–945

Petrova TV, Mäkinen T, Mäkelä TP, Saarela J, Virtanen I, Ferrell RE, Finegold DN, Kerjaschki D, YlaÈ-Herttuala S, Alitalo K (2002) Lymphatic endothelial reprogramming of vascular endothelial cells by the Prox-1 Homeobox transcription factor. EMBO J 21(17):4593–4599

Schacht V, Ramirez MI, Hong Y-K, Hirakawa S, Feng D, Harvey N, Williams M, Dvorak AM, Dvorak HF, Oliver G (2003) T1α/Podoplanin deficiency disrupts Normal lymphatic vasculature formation and causes lymphedema. EMBO J 22(14):3546–3556

Tanagho E, Mcaninch J (2007) Smith's general urology. McGraw-Hill Prof Med/Tech

Wigle JT, Oliver G (1999) Prox1 function is required for the development of the murine lymphatic system. Cell 98(6):769–778. https://doi.org/10.1016/S0092-8674(00)81511-1

Wigle JT, Harvey N, Detmar M, Lagutina I, Grosveld G, Gunn MD, Jackson DG, Oliver G (2002) An essential role for Prox1 in the induction of the lymphatic endothelial cell phenotype. EMBO J 21(7):1505–1513

82 Lymphatics of the Abdominal Wall

Caitlin Li and Lisandro Montorfano

82.1 Embryology

Lymph nodes form during the 11th week of gestation from a condensation of capillaries and mesenchymal cells, which then bulge into a sac. Granulopoiesis and erythropoiesis within this sac during the 12th to 14th weeks of gestation lead to the formation of lymphoid cells, including T and B cells (Markgraf et al. 1982). Lymphatic vessels were thought to originate from adjacent veins. However, recent studies have questioned this paradigm; some mesenteric lymph channels arise separately from the mesenteric vein (Semo et al. 2016). Regardless their precise origin, lymphatic vessels can be categorized on the basis of size, as arteries or veins can. Lymph capillaries are lined by a single-cell layer of endothelium linked by button-like junctions, which open in response to increased interstitial fluid pressure (Baluk et al. 2007). The larger pre-collecting and collecting vessels contain valves and smooth muscle cells, preventing backflow (Butler et al. 2009).

C. Li
Department of Surgery, Cleveland Clinic Florida, Weston, FL, USA
e-mail: lic9@ccf.org

L. Montorfano (✉)
Department of Plastic and Reconstructive Surgery, Vanderbilt University, Nashville, TN, USA

82.2 Anatomy

The lymphatic drainage of the abdominal wall comprises a superficial and a deep system. The superficial system drains the skin and subcutaneous tissue while the deep drains the musculature and bone. The superficial system consists of submillimeter blind-ended lymph capillaries, located superficial to Scarpa's fascia and deep to the subdermal venules. These drain into larger lymph vessels within the subcutaneous fat, which themselves drain into one of several lymph node basins, depending upon their origin (Tourani et al. 2013).

The superficial lymphatic drainage of the anterior abdominal wall is divided into two regions, superior or inferior to the level of the umbilicus (Suami and Scaglioni 2017). Superiorly, the superficial lymphatic channels drain into the anterior or pectoral axillary nodes; inferiorly, they drain into the superficial inguinal nodes. As in other parts of the body, the deep lymphatic system of the anterior abdominal wall mirrors the vascular supply. It consists of three major pathways that follow three major blood vessels: the superior epigastric pathway, which follows the superior epigastric artery and drains into the parasternal nodes; the inferior epigastric pathway, which follows the inferior epigastric artery and drains into the external iliac nodes; and the intercostal pathway, which follows the intercostal and subcostal arteries and drains into the posterior mediastinal nodes.

S. J. Bordes, Jr. et al. (eds.), *The Clinical Anatomy of the Vascular System*,
https://doi.org/10.1007/978-3-031-78326-5_82

Similarly, superficial lymphatic drainage of the posterior abdominal wall is divided between regions superior or inferior to the level of the iliac crest. Superiorly, the superficial posterior lymphatic channels drain into the posterior axillary nodes; inferiorly, they drain into the superficial inguinal nodes. The deep system of the posterior abdominal wall follows two major pathways: the lumbar pathway, which follows the lumbar arteries and drains into the lateral aortic nodes; and the iliac pathway, which follows the deep circumflex iliac artery and drains into the external iliac nodes (Mikhael and Khan 2021).

82.3 Clinical Significance

Lymphatic disruption or obstruction can lead to failed drainage and subsequent lymphedema. The superficial inguinal nodes drain not only the abdominal skin and subcutaneous tissue below the umbilicus but also the anal canal, scrotum, vulva, and lower extremity. Disruption or excision of these nodes and their associated lymph vessels can, therefore, lead to lower extremity lymphedema. In developing countries, certain pathogens, notably *Wuchereria bancrofti*, can cause lower limb lymphedema. In developed countries, the most common cause is the excision and/or radiation of lymph nodes to treat malignancy. Inguinal lymphadenectomy is most often performed for malignant skin tumors in the anogenital region, abdominal wall, and lower extremities, including melanoma and squamous cell carcinoma (Swan et al. 2004). Numerous techniques have been described to minimize postoperative morbidity from this procedure, including saphenous vein preservation, fascia lata preservation, omental flap pediculoplasty, and myocutaneous flap closure, descriptions of which are beyond the scope of this chapter (Gupta et al. 2017). The advent of new procedures to restore lymphatic drainage, such as lymphovenous anastomosis (LVA) and vascularized lymph node transfer (VLNT), has renewed interest in defining the abdominal wall lymphatic drainage further. Care must be taken to preserve the lymphatic drainage of the lower extremity when the inguinal region is used as a donor site for VLNT. The superficial lymphovenous collecting vessels must be dissected and identified in order to perform a LVA. This is most often done with the aid of contrast agents such as indocyanine green and surgical microscopy (Schaverien and Coroneos 2019).

References

Markgraf R, von Gaudecker B, Müller-Hermelink HK (1982) The development of the human lymph node. Cell Tissue Res 225:387–413. https://doi.org/10.1007/BF00214691

Semo J, Nicenboim J, Yaniv K (2016 Mar 15) Development of the lymphatic system: new questions and paradigms. Development 143(6):924–935. https://doi.org/10.1242/dev.132431

Baluk P, Fuxe J, Hashizume H, Romano T, Lashnits E, Butz S, Vestweber D, Corada M, Molendini C, Dejana E, McDonald DM (2007 Oct 1) Functionally specialized junctions between endothelial cells of lymphatic vessels. J Exp Med 204(10):2349–2362

Butler MG, Isogai S, Weinstein BM (2009) Lymphatic development. Birth Defects Res C Embryo Today 87(3):222–231. https://doi.org/10.1002/bdrc.20155

Tourani SS, Taylor GI, Ashton MW (2013) Anatomy of the superficial lymphatics of the abdominal wall and the upper thigh and its implications in lymphatic microsurgery. J Plast Reconstr Aesthet Surg 66(10):1390–1395. https://doi.org/10.1016/j.bjps.2013.05.030

Suami H, Scaglioni MF (2017) Lymphatic territories (Lymphosomes) in the rat: an anatomical study for future lymphatic research. Plast Reconstr Surg 140(5):945–951. https://doi.org/10.1097/PRS.0000000000003776

Mikhael M, Khan Y, (2021) Anatomy, abdomen and pelvis, lymphatic drainage. [Online] Ncbi.nlm.nih.gov. Available at: https://www.ncbi.nlm.nih.gov/books/NBK557720/. Accessed 29 Aug 2021

Swan MC, Furniss D, Cassell OC (2004) Surgical management of metastatic inguinal lymphadenopathy. BMJ 329(7477):1272–1276. https://doi.org/10.1136/bmj.329.7477.1272

Gupta MK, Patel AP, Master VA (2017) Technical considerations to minimize complications of inguinal lymph node dissection. Transl Androl Urol 6(5):820–825. https://doi.org/10.21037/tau.2017.06.06

Schaverien MV, Coroneos CJ (2019 Sep) Surgical treatment of lymphedema. Plast Reconstr Surg 144(3):738–758. https://doi.org/10.1097/PRS.0000000000005993

83 Inferior Vena Cava

Kathryn Ruckstuhl

83.1 Embryology

The development of the inferior vena cava (IVC) involves three paired embryonic veins and takes place during weeks 6–10 of gestation. The posterior cardinal veins are the first to develop posteriorly and mostly regress, except for their distal aspect, which leads to the formation of the iliac bifurcation (Ureyen et al. 2013). The suprarenal IVC forms from an anastomosis between the right subcardinal and vitelline veins (Berman and Minocha 2020). Next, the subcardinal veins develop anteromedially to the posterior cardinal veins. The right subcardinal vein persists to the right of the IVC throughout normal development, whereas the left subcardinal vein completely regresses (Ureyen et al. 2013). The subcardinal veins are responsible for forming most of the infrarenal IVC (Berman and Minocha 2020). Finally, the paired supracardinal veins form dorsally to the subcardinal veins. As with the subcardinal veins, the left supracardinal vein regresses while the right persists and eventually forms the infrarenal portion of the IVC (Ureyen et al. 2013).

83.2 Anatomy and Topography

The IVC is the largest vein in the human body (Fig. 83.1). It is responsible for draining the lower limbs, pelvis, and abdominal region directly to the heart's right atrium. The right and left common iliac veins join posteriorly to the common iliac veins to form this structure at the level of the fifth lumbar vertebra. To put the IVC into anatomical perspective, it lies to the right of the vertebral column. It sits behind major structures such as the duodenum, portal vein, and liver (Schainfeld and Rajachandran 2011). Leaving the abdominal cavity, it pierces the diaphragm's central tendon to continue intrathoracically to the right atrium (Fig. 83.2). Notable tributaries of the IVC include the third and fourth lumbar veins, right gonadal vein, renal veins, right suprarenal vein, inferior phrenic veins, and finally the hepatic veins. These tributaries vary significantly among individuals. The superior vena cava and the IVC are connected collaterally through the azygous venous system (Tucker et al. 2021).

K. Ruckstuhl (✉)
Louisiana State University Health Sciences Center,
New Orleans, LA, USA
e-mail: kruckstuhl1@lsuhsc.edu

S. J. Bordes, Jr. et al. (eds.), *The Clinical Anatomy of the Vascular System*,
https://doi.org/10.1007/978-3-031-78326-5_83

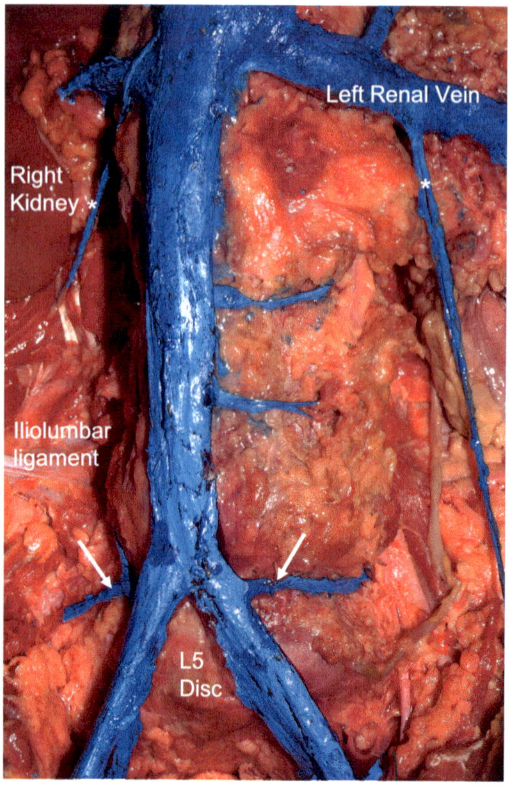

Fig. 83.1 Cadaveric dissection illustrating the inferior vena cava. The left and right iliolumbar veins are seen at the arrows and the left gonadal vein (*) is shown draining into the left renal vein

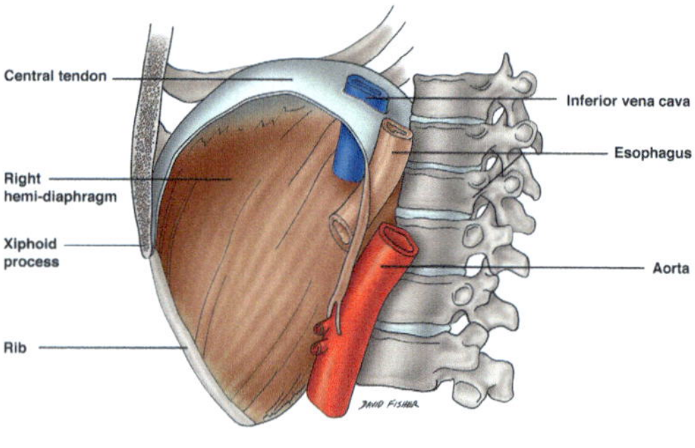

Fig. 83.2 Schematic drawing showing the course of the inferior vena cava, esophagus, and aorta through the diaphragm. (After Netter 2023)

83.3 Histology

Histologically, the wall of the IVC consists of three layers, the intima, media, and adventitia. The tunica intima is the innermost layer and consists of endothelial cells and a thin layer of connective tissue. The bicuspid valves of the IVC are made of endothelium. The tunica media can be identified histologically by its abundance of collagen and smooth muscle cells. The outermost anchoring layer, the adventitia, consists of loose connective tissue made primarily of type I collagen and elastic fibers (Schainfeld and Rajachandran 2011).

83.4 Imaging

Proper visualization of the IVC is essential for identifying congenital variants along with pathological conditions involving the vein. The mean diameter of the IVC is between 20 and 23 mm, but this can vary significantly among individuals. Patients with an IVC diameter greater than 28 mm are considered to have megacava. The threshold for IVC flattening is 9–13 mm (Shin et al. 2019). It is important to note that there is currently no universally accepted standard for acquiring this measurement (Berman and Minocha 2020).

83.5 Variations

Most variations of the IVC present as duplications or sometimes as a left-sided IVC. In general, anomalies of the IVC are infrequent, duplicate IVCs having a 0.2–3% prevalence rate (Shin et al. 2019).

A duplicated IVC is an embryological anomaly caused by abnormal persistence of both the left and right supracardinal veins. Most cases of this nature are clinically insignificant and asymptomatic and are therefore discovered incidentally through unrelated imaging. These findings on imaging are commonly misdiagnosed as pathologies such as lymphadenopathy or left pyeloureteric dilation (Ureyen et al. 2013). A double IVC, if poorly understood, can lead to significant complications such as accidental ligation in retroperitoneal surgeries. There are also several documented cases of thromboembolic complications associated with a double IVC, putting such patients at greater risk for thrombosis (Ureyen et al. 2013).

Left-sided IVC variations originate embryonically from the same persistent left supracardinal vein as seen in duplication variants, but the right supracardinal vein typically regresses, leaving the patient with a left IVC. This is the second most common IVC variation, with a 0.2–0.5% prevalence rate (Shin et al. 2019). Like the duplicated IVC, this anomaly is typically asymptomatic but must be carefully assessed before a surgical intervention to avoid potential complications. Left-sided IVC variants can be encountered during organ procurement procedures, which leave the surgeon with little time to assess their patients preoperatively (Ramia et al. 2022). If this anomaly is misidentified and injured, it can lead to bleeding and/or graft loss (Ramia et al. 2022).

Another variation of the IVC involves a circumcaval or retrocaval ureter, which occurs when the infrarenal portion of the IVC is formed from a persistent right posterior cardinal vein while the right supracardinal vein regresses. The IVC entraps the proximal portion of the ureter posteriorly and medially. Patients with this variation typically present with symptoms of urinary reflux or obstructions leading to frequent urinary tract infections. Surgical reimplantation of the ureter is the preferred treatment option for patients with severe presentations (Malaki et al. 2012).

One of the rarest IVC variations is an absent infrarenal IVC, which can be either complete or incomplete. It arises embryologically from failure of the three paired venous systems to develop. Complete absence of the infrarenal IVC typically results from failure of the posterior cardinal and supracardinal veins to form. This variation can bring about symptoms that resemble those of lower limb venous insufficiency or of deep vein thrombosis (DVT) (Petik 2015). As expected, this structural absence leads to insufficient blood flow, resulting in many cases in bilateral stasis and varices (Malaki et al. 2012).

An isolated accessory continuous hemiazygos vein is another rare IVC variation. It is caused by failure of the right subcardinal vein to join the vitelline vein during development, resulting in failure of an intrahepatic IVC to form, causing this azygos vein to compensate (Petik 2015).

83.6 Pathology

Primary or secondary malignancy can involve the IVC. Usually, IVC masses are intraluminal (Schainfeld and Rajachandran 2011). If the condition is symptomatic, these patients most commonly present with abdominal pain and edema of the lower extremity.

Primary IVC leiomyosarcoma is the most common malignancy to arise directly from the IVC. These masses originate from mesenchymal smooth muscle cells in the vein walls and commonly involve central necrosis (Shin et al. 2019). Leiomyosarcoma is a very rare type of tumor that accounts for around 5–7% of all soft-tissue sarcomas. Currently, the only suitable treatment option is surgical resection; unfortunately, many diagnoses are made at advanced stages owing to a lack of identifiable risk factors (Gaignard et al. 2020).

IVC thrombosis is associated with significant morbidity and mortality (Teter et al. 2019). Presentation ranges widely from asymptomatic discoveries on unrelated imaging to symptoms of venous insufficiency such as lower extremity edema and pain in the lower back and buttock region. In extreme cases, patients present with acute and severe cardiopulmonary symptoms. Congenital and acquired etiologies can account for this pathology, but congenital manifestations are relatively rare and require collateral involvement to produce symptoms. Patients can develop acquired IVC thrombosis from various causes such as external IVC compression, IVC malignancies, deep vein thrombosis from iliac veins, and endothelial injury. Ninety percent of IVC thrombosis patients who go untreated develop post-thrombotic syndrome (Shin et al. 2019).

Chronic inferior vena cava syndrome is caused by a chronic obstruction leading to severe and debilitating abdominal distension and edema of the lower torso, perineum, and legs, causing compromised mobility (Shin et al. 2019). Inferior vena cava syndrome has a very similar pathophysiology. Its etiology depends on the site of altered blood flow and can be classified as extrinsic compression or intrinsic stenosis. Treatment options are severely lacking, and an early and accurate diagnosis is critical for proper management.

Aortocaval fistulas form because of surgical complications from abdominal aortic aneurysm rupture/repair and occasionally trauma. Abdominal aortic aneurysms can erode into the wall of the IVC thrombosis, which accounts for the fistula formation (Patelis et al. 2018). This specific fistula has inconsistent symptoms but can present as a pulsating abdominal mass with symptoms including abdominal or back pain, abdominal bruit, pelvic and lower limb hypertension, shock, or congestive heart failure. Diagnosis can be made through CT imaging with contrast, and management includes surgical or endovascular repair (Shin et al. 2019).

Finally, blunt and penetrating traumas can injure the IVC, incurring extremely high mortality rates. The IVC has a very low intraluminal pressure. Tamponade is key until definitive vascular repair can be carried out (Shin et al. 2019).

83.7 Clinical and Surgical Significance

IVC filter placement is a relatively low-risk and increasingly common preventative procedure for pulmonary embolism. Individuals undergoing such intervention generally present with lower extremity or pelvic deep vein thromboses and are not candidates for anticoagulation owing to either active bleeding or high risk for future bleeding. Between 12 and 17% of all venous thromboembolism patients receive an IVC filter placement as prophylactic treatment (Grewal et al. 2016). Understanding potential IVC anomalies is imperative for proper IVC filter implantation, as in cases of duplicated IVCs.

IVC stent placement can reduce edema, ascites, and azotemia in patients with IVC obstruction. In rare cases, stent migration can cause cardiopulmonary complications (Shin et al. 2019).

References

Berman Z, Minocha J (2020) Inferior vena cava filter placement: anatomical evaluation and approach to variant anatomy. In: Placement and retrieval of inferior vena cava filters, pp 21–31. https://doi.org/10.1007/978-3-030-45150-9_3

Gaignard E et al (2020) Inferior Vena Cava Leiomyosarcoma: what method of reconstruction for which type of resection? World J Surg 44(10):3537–3544. https://doi.org/10.1007/s00268-020-05602-2

Grewal S et al (2016) Complications of inferior vena cava filters. Cardiovasc Diagn Ther 6(6):632–641. https://doi.org/10.21037/cdt.2016.09.08

Malaki M et al (2012) Congenital anomalies of the inferior vena cava. Clin Radiol 67(2):165–171. https://doi.org/10.1016/j.crad.2011.08.006

Netter FH (2023) Netter Atlas of Human Anatomy (8th ed.). Elsevier

Patelis N et al (2018) Aortocaval Fistula: an unusual complication of ruptured abdominal aortic aneurysm. BMJ Case Rep 2018:bcr2018224998. https://doi.org/10.1136/bcr-2018-224998

Petik B (2015) Inferior vena cava anomalies and variations: imaging and rare clinical findings. Insights Imaging 6(6):631–639. https://doi.org/10.1007/s13244-015-0431-z

Ramia JM et al (2022) Left-sided inferior vena cava during organ procurement. ANZ J Surg 93:363. https://doi.org/10.1111/ans.17766

Schainfeld RM, Rajachandran M (2011) Inferior vena cava. In: Dieter RS et al (eds) Venous and lymphatic diseases. McGraw Hill. https://accesscardiology.mhmedical.com/content.aspx?bookid=1782&sectionid=121398880

Shin DS et al (2019) The inferior vena cava: a pictorial review of embryology, anatomy, pathology, and interventions. Abdom Radiol 44(7):2511–2527. https://doi.org/10.1007/s00261-019-01988-3

Teter K et al (2019) Presentation and management of inferior vena cava thrombosis. Ann Vasc Surg 56:17–23. https://doi.org/10.1016/j.avsg.2018.08.082

Tucker WD et al (2021) Anatomy, Abdomen and Pelvis, Inferior Vena Cava. National Center for Biotechnology Information, U.S. National Library of Medicine. https://pubmed.ncbi.nlm.nih.gov/29493975/

Ureyen I et al (2013) Double vena cava inferior: a report of three cases. Asian Pac J Reprod 2(3):248–250. https://doi.org/10.1016/s2305-0500(13)60156-6

84 Internal Iliac Veins

Graham Blair and Katherine D. Bordes

84.1 Embryology

In the developing vascular system, the sinus venosus constitutes the inflow end of the developing heart, the anterior cardinal veins drain the cranial part of the embryo, and the posterior cardinal veins drain caudally to the heart. Unlike its anterior counterpart, the posterior cardinal veins almost entirely regress except for the iliac anastomosis, supracardinal, and subcardinal sections. The iliac anastomosis gives rise to the common, external, and internal iliac veins and the median sacral veins. Most anatomic variations of the external, internal, and/or common iliac veins result from deficits in embryogenesis (Mozes and Gloviczki 2014; Mehta et al. 2019).

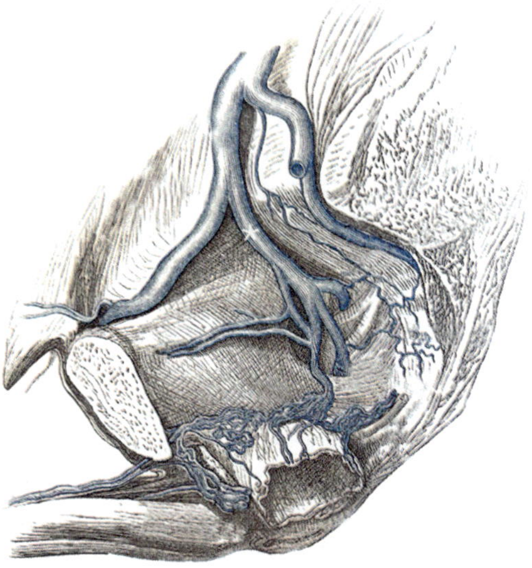

Fig. 84.1 Schematic drawing of the iliac veins of the pelvis. Note the internal iliac vein (*). (After Allen 1882)

84.2 Anatomy/Topography

The internal iliac vein (IIV) (Fig. 84.1), sometimes called the hypogastric vein, is bilaterally superior to the greater sciatic foramen in the retroperitoneum and runs posteromedial to the internal iliac arteries. The internal iliac vein is mainly responsible for the drainage of the pelvis and viscera, as well as the perineum, external genitalia, gluteal region, and the medial region of the thigh. Its short trunk is formed by the confluence of extra and intrapelvic tributaries, which are more variable than their arterial counterparts but drain the same territories. From the intrapelvic venous system, the IIV receives the lateral sacral vein and visceral veins (middle rectal, vesical, uterine, and vaginal), which drain the presacral and pelvic venous plexuses, including the rectal, vesical, uterine, vaginal, and prostatic plexuses. Its contributing vessels outside the pelvis include the superior gluteal, inferior gluteal, internal pudendal, and obturator veins, which

G. Blair · K. D. Bordes (✉)
Tulane University School of Medicine,
New Orleans, LA, USA

S. J. Bordes, Jr. et al. (eds.), *The Clinical Anatomy of the Vascular System*,
https://doi.org/10.1007/978-3-031-78326-5_84

drain the pelvis and perineum. The internal iliac vein ascends posteromedially to join with the external iliac vein (EIV) on the medial border of the psoas major, thus forming the common iliac vein (CIV) at the linea terminalis, anterior to the lower part of the sacroiliac joint. Of note, these veins, the IIV, EIV, and CIV, are valveless (Mozes and Gloviczki 2014; Standring 2021; Moore et al. 2014; Sinnatamby 2011; Brennan et al. 2020; Tubbs et al. 2016).

84.3 Variations

In most cases, approximately 70–80% of the internal iliac vein anatomy follows the standard course, as described above, in which each internal iliac vein joins with the ipsilateral external iliac vein to form a common iliac. Then, the two common iliac veins form the inferior vena cava (IVC) (Cardinot et al. 2006; Chong et al. 2015; Tubbs et al. 2016).

However, maldevelopment of the posterior cardinal veins during early embryogenesis can result in variations, including branching, junction, location, or number differences. Anomalies are more common in females at a ratio of 1.7:1. These variations may be isolated or associated with other venous malformations, such as in Klippel-Trenaunay syndrome. Recognizing variants can aid in identifying possible disease states, such as supine hypotensive syndrome, in which variants can compromise collateral venous circulation. In addition, recognizing variants can also ensure treatment efficacy, such as in the placement of a venous valve implant to treat chronic venous insufficiency. Due to the significant variability of the IIV, special anatomical considerations must be taken while performing pelvic surgeries to avoid iatrogenic injury (Tubbs et al. 2016; Mehta et al. 2019).

The location of the internal iliac veins in relation to the artery is variable. It was found by Bleich et al. in 2007 that the left IIV lies lateral to the artery about 70% of the time, but the right IIV lies lateral to the artery 93% of the time (Bleich et al. 2007). The point at which the EIV and IIV join together to form the common iliac vein is also variable. In approximately 8% of cases, the left common iliac vein forms at a higher level, thus resulting in a shorter overall length (Tubbs et al. 2016).

Several variations in drainage patterns of the EIV, IIV, and common iliac veins exist. Hamabe et al. detailed several patterns of anatomical variants of intrapelvic vessels. Type I involves each internal iliac vein draining into the ipsilateral common iliac vein in both halves of the pelvis, which occurs 47.5% of the time. In type II, each internal iliac vein drained into the ipsilateral common iliac vein in one or both halves of the pelvis, occurring 40% of the time. Three subtypes include types IIa, IIb, and IIc. Type IIa is defined by two internal iliac veins on the left half of the pelvis, type IIb involves two internal iliac veins on the right half of the pelvis, and type IIc involves four total internal iliac veins, with two on each half of the pelvis which only occurs about 1.7% of the time and may involve interconnecting channels (Cardinot et al. 2006; Hamabe et al. 2020; Tubbs et al. 2016). Type III occurs in about 10% of cases and involves one of the internal iliac veins draining into the contralateral common iliac vein while the other drains into the ipsilateral common iliac vein. There are two subtypes, IIIa and IIIb. Type IIIa involves the right internal iliac vein draining in the left common iliac vein, which is more common. Type IIIb involves the left internal iliac vein draining in the right common iliac vein (Hamabe et al. 2020). In this type of variation, there may also be an anastomotic vein connecting the right and left common iliac veins (Tubbs et al. 2016; Mehta et al. 2019).

Other variations in drainage patterns include the following: the right and left internal iliac veins join together to form a common internal iliac vein, which then meets the two external iliac veins to form the IVC (Cardinot et al. 2006; Tubbs et al. 2016; Mehta et al. 2019). There may be two anomalous channels between the left and right IIVs, creating a common trunk that drains into the IVC and the common iliac veins (Tubbs et al. 2016). The left common iliac vein may join with the right internal iliac vein, which then joins the isolated right external iliac vein to form the

IVC. Meanwhile, the right common iliac vein joins with the left internal iliac vein, which joins the isolated left external iliac vein to form the IVC (Cardinot et al. 2006).

Accessory internal iliac veins are also possible. In 1.6% of cases, the right separated internal iliac vein can drain into a central right common iliac vein. In 4.8% of cases, the left internal iliac vein will drain into a left common iliac vein. A separated bilateral internal iliac vein drains into a central left common iliac vein in 4.8% of cases. More commonly, in 11.1% of cases, the right separated internal iliac vein will drain into a central left common iliac vein (Tubbs et al. 2016).

Aplasia of the common iliac vein may be unilateral or bilateral, which significantly affects the drainage of the internal iliac vein. In the case of unilateral aplasia of the common iliac vein, three main methods exist to accommodate this malformation. The EIV and IIV may drain directly into the origin of the IVC. Alternatively, the EIV and IIV may join the opposite common iliac vein to form the IVC. Lastly, the EIV can drain into the IVC while the IIV drains into the opposite common iliac vein. In the case of bilateral aplasia of the common iliac vein, there may be the absence of the EIV, IIV, and part of the femoral vein on one side, while the EIV and IIV on the contralateral side drain into the retroperitoneal collateral veins (Tubbs et al. 2016; Mehta et al. 2019).

84.4 Clinical and Surgical Significance

The internal iliac vein can be injured during trauma of surrounding areas such as the sacrum, lumbar spine, and pelvis. Contrary to the thick-walled internal iliac artery, the IIV has weaker walls that are prone to tearing, potentially leading to hemorrhagic shock. In the instance of pelvic bone fractures, the presence of many venous plexuses and large draining veins in the retroperitoneal space can result in severe retroperitoneal hemorrhage. During surgery, isolation and vascular exposure of the IIV and EIV are crucial procedures to identify and gain control of the hemorrhage. Venography can help identify these injuries (Kataoka et al. 2005; Sinnatamby 2011; Brennan et al. 2020).

Pelvic congestion syndrome (PCS), also called female varicocele, is characterized by chronic pelvic pain in the setting of pelvic venous varicosities. The majority of cases are related to severe enlargement of ovarian veins, especially during and after pregnancy, but internal iliac vein reflux is also a known cause. PCS is diagnosed with venography and possibly transvaginal ultrasound, which reveals excessive pelvic varicose veins in the broad ligament. In cases of significant IIV reflux, endovascular embolization may be used for treatment if varicosities are demonstrated, but there are substantial risks (Mozes and Gloviczki 2014; Durham and Machan 2013).

Thrombosis of the external iliac veins or inferior vena cava can block venous drainage of the leg. In this case, the internal iliac tributaries, specifically the gluteal veins, can enlarge and offer collateral circulation between the femoral and internal iliac veins. Therefore, iatrogenic damage or trauma to the IIV tributaries can jeopardize collateral venous drainage in one or both lower limbs (Standring 2021).

The lateral sacral vein provides a pathway for thromboembolism or metastasis. In the case of cancer of the pelvic organs, malignant cells can spread from one of the valveless pelvic venous plexuses to the vertebral venous plexus, allowing metastatic cells to deposit in the vertebrae. This may result from a sudden increase in abdominal pressure, leading to reflux from the internal vertebral plexus (Sinnatamby 2011; Brennan et al. 2020).

Case reports have described free-floating thrombi moving from the IIV into IVC as rare causes of pulmonary embolisms, even when compression ultrasonography of lower limbs does not reveal any deep vein thrombosis. Despite their low prevalence, free-floating thrombi of the IIV can be life-threatening (Brodmann et al. 2012). Embolisms can also occur secondary to ballistic fragments entering the IIV and then migrating to distant sites such as the heart, lungs, or other vessels. Diagnosis can be easily missed unless the index of suspicion is high or radiographs confirm all entry and exit sites (Kalimi et al. 2002). Iliac vein aneurysms can also occur due to previous arteriovenous fistulas and/or prior trauma but are extremely rare (Zarrintan et al. 2019).

A more recent case report described a circumstance in which a patient diagnosed with COVID-19 had coexisting and related IIV thrombosis and cerebral venous thrombosis (Beretta et al. 2021).

References

Mozes G, Gloviczki P (2014) Venous embryology and anatomy. In: The vein book, 2nd edn. Oxford University Press, pp 17–26

Standring S (ed) (2021) Lesser pelvis and perineum. In: Gray's anatomy: the anatomical basis of clinical practice. 42nd edn. Elsevier Limited

Moore KL, Dalley AF, Agur AMR (2014) Pelvis and perineum. In: Clinically oriented anatomy, 7th edn. Wolters Kluwer – Lippincott Williams & Wilkins, p 760

Cardinot TM, Aragão AH, Babinski MA, Favorito LA (2006) Rare variation in course and affluence of internal iliac vein due to its anatomical and surgical significance. Surg Radiol Anat 28(4):422–425. https://doi.org/10.1007/s00276-006-0110-3

Bleich AT, Rahn DD, Wieslander CK, Wai CY, Roshanravan SM, Corton MM (2007) Posterior division of the internal iliac artery: anatomic variations and clinical applications. Am J Obstet Gynecol 197(6):658.e1–658.e6585. https://doi.org/10.1016/j.ajog.2007.08.063

Hamabe A, Harino T, Ogino T et al (2020) Analysis of anatomical variations of intrapelvic vessels for advanced pelvic surgery. BMC Surg 20(1):47. https://doi.org/10.1186/s12893-020-00711-0

Chong GO, Lee YH, Hong DG, Cho YL, Lee YS (2015) Anatomical variations of the internal iliac veins in the presacral area: clinical implications during sacral colpopepxy or extended pelvic lymphadenectomy. Clin Anat 28(5):661–664. https://doi.org/10.1002/ca.22482

Kataoka Y, Maekawa K, Nishimaki H, Yamamoto S, Soma K (2005) Iliac vein injuries in hemodynamically unstable patients with pelvic fracture caused by blunt trauma. J Trauma 58(4):704–710. https://doi.org/10.1097/01.ta.0000159346.62183.8f

Kalimi R, Angus LD, Gerold T, DiGiacomo JC, Weltman D (2002) Bullet embolization from the left internal iliac vein to the right ventricle. J Trauma 52(4):772–774. https://doi.org/10.1097/00005373-200204000-00030

Zarrintan S, Tadayon N, Kalantar-Motamedi SMR (2019) Iliac vein aneurysms: a comprehensive review. J Cardiovasc Thorac Res 11(1):1–7. https://doi.org/10.15171/jcvtr.2019.01

Sinnatamby CS (2011) Pelvic vessels and nerves. In: Last's anatomy – regional and applied, 12th edn. Elsevier Limited

Brodmann M, Gary T, Hafner F, Tiesenhausen K, Deutschmann H, Pilger E (2012) Massive pulmonary embolism caused by internal iliac vein thrombosis with free-floating thrombus formation in the inferior vena cava. Ann Vasc Surg 26(3):420.e5. https://doi.org/10.1016/j.avsg.2011.07.016

Beretta S, Da Re F, Francioni V et al (2021) Case report: concomitant massive cerebral venous thrombosis and internal iliac vein thrombosis related to Paucisymptomatic COVID-19 infection. Front Neurol 12:622130. Published 2021 Feb 10. https://doi.org/10.3389/fneur.2021.622130

Brennan PA, Standring SM, Wiseman SM (eds) (2020) Gray's surgical anatomy, 1st edn. Elsevier Limited

Tubbs RS, Shoja M, Loukas M (2016) Bergman's comprehensive encyclopedia of human anatomic variation. Wiley, Hoboken

Mehta K, Iwanaga J, Tubbs RS (2019) Absence of the right common iliac vein with the right internal iliac vein arising from the left common iliac vein: case report. Cureus 11(4):e4575. Published 2019 Apr 30. https://doi.org/10.7759/cureus.4575

Durham JD, Machan L (2013) Pelvic congestion syndrome. Semin Intervent Radiol 30(4):372–380. https://doi.org/10.1055/s-0033-1359731

Part V

Upper Limb

Subclavian Artery

85

Adalgisa Varuolo and Stephen J. Bordes, Jr.

85.1 Embryology

The aortic arch and its branches are formed during the fourth to sixth weeks of development from six pairs of embryological aortic arches (Bordes et al. 2021). The aortic sac develops as the first portion of the aorta, superior to the truncus arteriosus. The aortic sac then advances into two horns, right and left. The right horn gives rise to the brachiocephalic trunk, while the left horn forms the aortic arch proximal to the brachiocephalic trunk. The subclavian arteries are derived from the fourth aortic arch. As the aortic arches continue to develop asymmetrically, the origins of the right and left subclavian arteries differ (Rosen and Bordoni 2021). The right fourth aortic arch contributes to the right proximal subclavian artery, and the left gives rise to the medial segment of the arch of the aorta. The dorsal aorta gives rise to the intersegmental arteries. The cervical intersegmental arteries merge to form the vertebral arteries, except the left seventh intersegmental artery, which fuses with the aortic arch and exclusively forms the left subclavian artery. On the right side, the seventh intersegmental artery gives rise to the distal right subclavian artery (Rosen and Bordoni 2021). To summarize: the left subclavian artery is derived exclusively from the left seventh intersegmental artery, while the right subclavian artery is derived from the right horn of the aortic sac, the fourth aortic arch proximally and the right seventh intersegmental artery distally.

A. Varuolo
Department of Surgery, Valley Health System, Las Vegas, NV, USA

S. J. Bordes, Jr. (✉)
Department of Surgery, Louisiana State University Health Sciences Center, New Orleans, LA, USA

85.2 Anatomy and Topography

As embryonic development continues, the right and left subclavian arteries progress posterior to the sternoclavicular joints and ascend through the thoracic outlet, which connects the thorax with the root of the neck. At this point, the subclavian artery can be divided into three parts in relation to the anterior scalene muscle.

The first part of the subclavian artery lies medial to the anterior scalene muscle (Figs. 85.1 and 85.2). This first portion has three branches: the vertebral, internal thoracic (Fig. 85.3), and thyrocervical arteries. The second portion lies posterior to the anterior scalene muscle and has only one branch, the costocervical trunk (Figs. 85.3 and 85.4). The third portion lies lateral to the anterior scalene muscle and consists of one branch, the dorsal scapular artery. A common mnemonic used to remember the branches of the subclavian artery is VIT C&D.

The right subclavian artery originates from the brachiocephalic trunk. It continues to course

S. J. Bordes, Jr. et al. (eds.), *The Clinical Anatomy of the Vascular System*,
https://doi.org/10.1007/978-3-031-78326-5_85

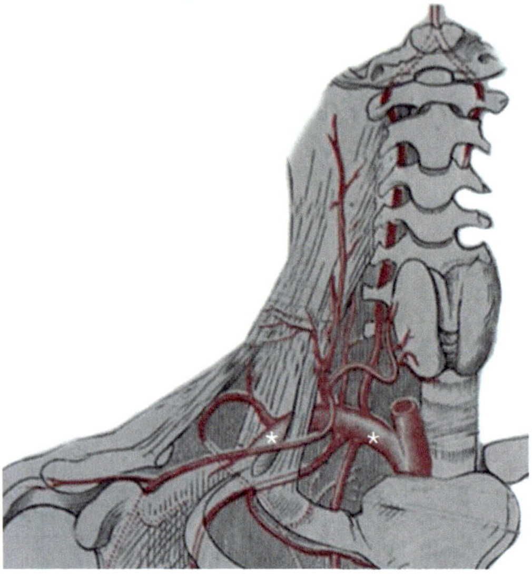

Fig. 85.1 Schematic drawing of the right subclavian artery arising from the brachiocephalic trunk and traveling posterior to the anterior scalene muscle. (After Allen 1892)

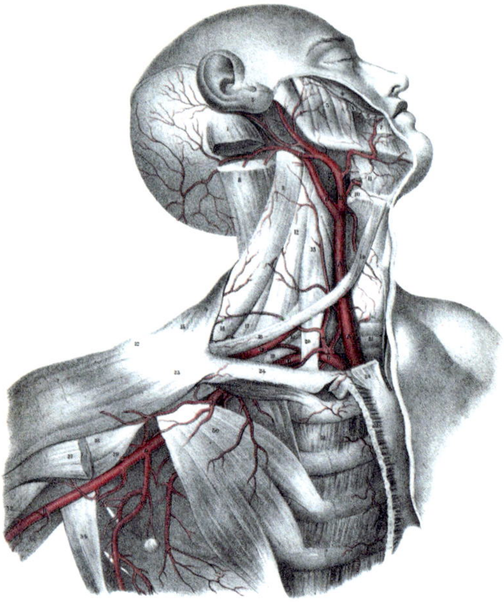

Fig. 85.2 Schematic drawing of arteries of the upper limb, thorax, and neck and face. Note the subclavian artery traveling proximal, posterior to, and distal to the anterior scalene muscle (20). (Quain and Wilson 1844 author's collection)

superior to the clavicle and then posterior to the anterior scalene muscle to the lateral border of the first rib. The lateral border of the first rib is the landmark to identify the transition of the subclavian to the axillary artery (Aghoghovwia 2020).

The left subclavian artery has an entirely different anatomy from the right. In contrast to the right subclavian, which originates from the brachiocephalic trunk, it arises directly from the aortic arch, independent of the left brachiocephalic trunk. It branches off the aorta after the brachiocephalic trunk and the left common carotid artery. It then ascends towards the medial border of the left anterior scalene muscle, continuing behind the muscle and ending at the lateral border of the first rib, where it becomes the axillary artery (Aghoghovwia 2020).

The structures anterior and posterior to the subclavian arteries can be divided to three parts described above. Notice the differences in structure between the left and right subclavian arteries, as each artery takes a different course.

First part of the subclavian artery (Jones 2021).

Structures that lie anterior:

1. Common carotid artery.
2. Internal jugular vein.
3. Vertebral vein.
4. Sternocleidomastoid, sternohyoid, and sternothyroid muscles.
5. Thoracic duct (left only).
6. Vagus and phrenic nerves (left only).
7. Sympathetic trunk.
8. Ansa cervicalis.

Structures that lie posterior:

1. Apex of the lung.
2. Brachial plexus (lower branches).
3. Middle scalene muscle.

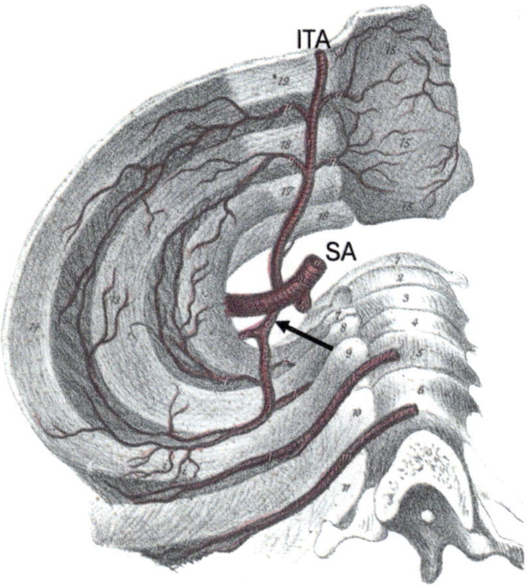

Fig. 85.3 Drawing of the subclavian artery (SA) giving rise to the internal thoracic artery (ITA) and the costocervical trunk (arrow), which will divide into a superior traveling deep cervical artery and an inferiorly traveling supreme intercostal artery. (Quain and Wilson 1844 author's collection)

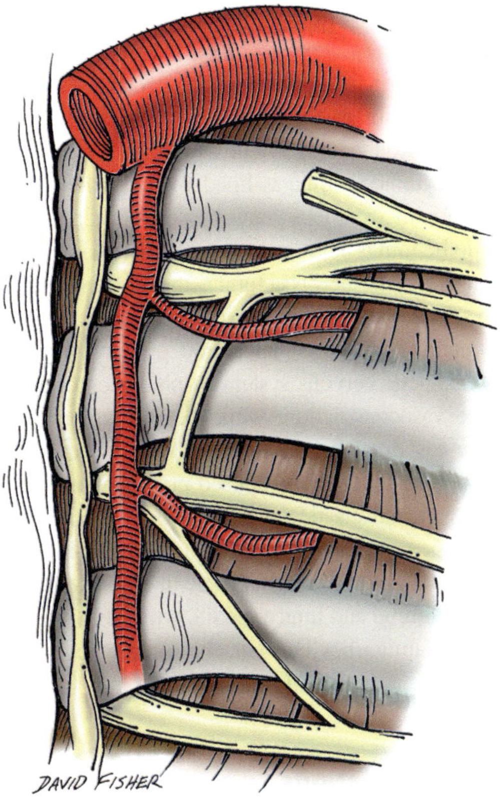

Fig. 85.4 Schematic drawing of the subclavian artery (top) giving rise to the supreme intercostal artery, in this case arising as a separate branch of the subclavian artery and not from the costocervical trunk. It supplies the upper two posterior intercostal spaces traveling with the intercostal nerves. Also, note its relationship to the sympathetic trunk. (Vertical yellow band medial to the arteries)

Second part of the subclavian artery (Jones 2021).

Structures that lie anterior:

1. Anterior scalene muscle.
2. Phrenic nerve (right only).
3. Sternocleidomastoid muscle.

Structures that lie posterior:

1. Apex of the lung.
2. Brachial plexus (lower branches).
3. Middle scalene muscle.

Third part of the subclavian artery (Jones 2021).

Structures that lie anterior:

1. Suprascapular and transverse cervical arteries and veins.
2. Subclavian and anterior jugular vein.

Structures that lie posterior:

1. Apex of the lung.
2. Brachial plexus (lower branches).
3. Middle scalene muscle.

85.3 Histology

The subclavian arteries belong to a group of large vessels termed elastic arteries. Other elastic arteries include the aorta, pulmonary trunk, common carotid artery, vertebral arteries, and common iliac arteries. Elastic arteries are characterized by the abundance of elastic fibers in the arterial wall. Arteries and veins share three common vessel wall layers: the tunica intima, tunica media, and tunica adventitia/externa. Elastic arteries contain two additional layers, the internal and external elastic laminae, to accommodate for the high pressures on leaving the heart (Crumbie 2020).

The tunica intima is lined by an endothelium comprising simple squamous cells. Deep to the endothelium is a basement membrane, followed by connective tissue. The tunica media is a middle layer of smooth muscle containing alpha- and beta-adrenergic receptors innervated primarily by the sympathetic nervous system. Alpha receptors produce vasoconstriction, while beta receptors produce vasodilation. The tunica adventitia/externa is the outermost layer, composed primarily of type 1 collagen and elastic connective tissue. This provides an anchor to neighboring organs. The internal elastic lamina is a band of elastic fibers between the tunica intima and tunica media, while the external elastic lamina lies between the tunica media and the tunica adventitia (Crumbie 2020).

85.4 Variations

The normal anatomy of the aortic arch is the branching of the brachiocephalic trunk (innominate artery), followed by the left common carotid and the left subclavian arteries. This normal anatomy is seen in 70–80% of the population (Hanneman et al. 2017). Of note, a few anatomical variations occur in <0.1% of the population; both carotid arteries can form a trunk (bicarotid trunk), and the right and left subclavian arteries can emerge from the aortic arch; the right subclavian artery can arise directly from the bicarotid trunk or originate from it (Tubbs et al. 2016). In this section, the variations are divided into those on the left and those on the right.

Variations in the Left Subclavian Artery

These variations are based on a double aortic arch system. Typically, the right aortic arch is associated with the aberrant left subclavian artery, and the left aortic arch is associated with the aberrant right subclavian artery.

- Right aortic arch with aberrant left subclavian artery: This anomaly occurs when there is a regression of the left fourth aortic arch between the left common carotid and the left subclavian artery. This artery has a retroesophageal course from right caudal to left cranial and can present with or without a retroesophageal diverticulum, depending on the regression of the left sixth aortic arch. If the left sixth arch does not regress, it will contribute to the left dorsal aorta and form a retroesophageal diverticulum. This is the more common presentation of a right aortic arch with an aberrant left subclavian artery (Hanneman et al. 2017).
- Right aortic arch with isolation of the left subclavian artery: This anomaly is rare and is only found with two ipsilateral breaks in the arch system. In this case, isolation means the vessel arises from the pulmonary artery via the ductus arteriosus or ligamentum arteriosum with no direct connection to the aorta. The variant results from regression of the left aortic arch at two segments: between the left common carotid artery and the left subclavian artery and between the segment distal to the ductus arteriosus and the left subclavian artery. This is most commonly associated with Tetralogy of Fallot, although it has been described in over half of the cases of congenital heart disease (Hanneman et al. 2017).

Variations in the Right Subclavian Artery

- Left aortic arch with aberrant right subclavian artery: This results from regression of the right arch between the right common carotid artery and right subclavian artery. Usually, the right fourth arch becomes the proximal subclavian artery. In this case, the distal right dorsal aorta becomes the proximal subclavian artery, forming the retroesophageal portion. The vessel usually takes a retroesophageal path, although it has also been documented to pass anterior to the trachea or between the two structures (Leite et al. 2017). This anomaly, arteria lusoria, is most often isolated and represents the most common congenital anomaly of the aortic arch, occurring in 0.5–2.5% of cases (Leite et al.

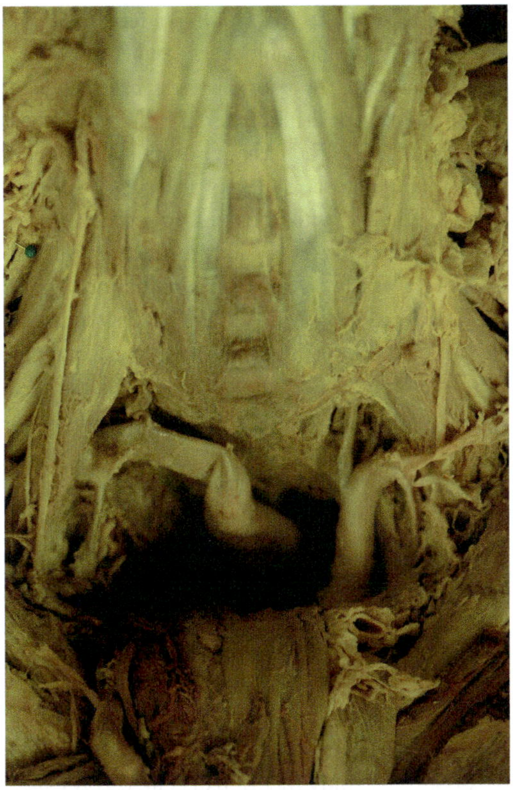

Fig. 85.5 Cadaveric example of a right retroesophageal artery. The esophagus is folded anteriorly

2017). This presentation is generally asymptomatic, but occasionally results in compression of the esophagus due to the retroesophageal course of the aberrant right subclavian artery (Fig. 85.2). This is referred to as dysphagia lusoria (Hanneman et al. 2017).

Furthermore, four different types of the right subclavian retroesophageal artery have been distinguished by the Adachi-Williams classification: G, CG, H, and N. Type G occurs where the right subclavian artery arises from the distal portion of the aortic arch as the last branch, while the other aortic arch branches follow the normal anatomical trend (deAraújo et al. 2015). Type CG occurs when the right subclavian artery is anomalous, as in type G, and the left vertebral artery originates directly from the aortic arch (deAraújo et al. 2015). Type H is seen with a right subclavian artery as in type G and the common right and left carotid arteries arising from a bicarotid trunk. Type N mirrors an anomalous right subclavian artery as in Type G; a right aortic arch and the left subclavian artery origin succeeds both carotid arteries (deAraújo et al. 2015) (Fig. 85.5).

85.5 Imaging

The noninvasive imaging modality of choice is Duplex ultrasound with color flow. Significant obstruction can be seen through waveform dampening or monophasic changes, color aliasing (an artifact that arises when the Doppler shift of the moving blood is higher than half of the pulse repetition frequency), and increased blood flow velocities at the suspected site of stenosis. Ultrasound testing can also detect the reversal of flow seen in subclavian steal syndrome. More invasive tests include computed tomography angiography (CTA) and magnetic resonance angiography (MRA). CTA and MRA provide an excellent anatomical resolution; however, angiography remains the gold standard (Caesar-Peterson et al. 2021).

85.6 Pathology

The subclavian arteries can be subject to congenital and idiopathic pathologies with intrinsic and extrinsic manifestations. The congenital pathologies associated with aberrant variations are discussed above and will not be considered further in this section.

85.7 Thoracic Outlet Syndrome

One example of an extrinsic pathology is compression of the subclavian arteries resulting from thoracic outlet syndrome. Thoracic outlet syndrome can be caused by hypertrophied musculature, an extracervical rib, or a previously fractured clavicle narrowing the thoracic outlet. It presents as numbness and tingling down the affected arm due to nerve compression and possibly diminished pulse on the ipsilateral side. Adson's maneuver is used to diagnose thoracic outlet syndrome by having the patient extend their neck, turn their head to the ipsilateral side being tested, and hold their breath. The radial pulse is then palpated; there is a vascular compromise if a diminished pulse is palpated. The subclavian artery can also be auscultated during Adson's maneuver and can reveal a bruit in cases of thoracic outlet syndrome (Rahimi and Geiger 2021).

85.8 Takayasu's Arteritis

Takayasu's arteritis is an inflammatory condition that can result in subclavian narrowing and subsequent aortic arch syndrome. A large vessel vasculitis commonly affects the aorta and its branches, namely the subclavian arteries. It primarily affects women less than 40 years old and is more common in those of Asian descent. The inflammatory trigger for this disease is unknown, although it is thought to be autoimmune. Long-term inflammation in the arterial wall leads to narrowing and reduction of flow. Diagnosis can generally be made with angiography, showing stenosis of the aorta or its branches. Treatment of Takayasu remains challenging, with corticosteroids or immune modulators to suppress inflammation. If medical management fails, surgical options such as bypass surgery or percutaneous angioplasty can be employed.

85.9 Subclavian Steal Syndrome

Subclavian steal syndrome occurs with proximal subclavian stenosis unilaterally and a distal drop in pressure leading to retrograde flow in the ipsilateral vertebral artery. This redirection of blood flow away from the Circle of Willis and brainstem can manifest with stroke-like symptoms, dysarthria, diplopia, and vertigo, among other consequences. There can also be ipsilateral arm claudication; this can be diagnosed by a > 10 mmHg blood pressure discrepancy between the two arms. Continuous wave Doppler, duplex ultrasonography, and angiography can be used to diagnose this syndrome. Patients with incidental stenosis who are asymptomatic can be managed without revascularization. Aspirin, ACE inhibitors, beta-blockers, and statins have reduced long-term mortality in patients with lower extremity peripheral artery disease and are recommended as the treatment for asymptomatic subclavian stenosis. There is an exception, however: Patients with planned coronary artery bypass graft (CABG) and ipsilateral internal mammary artery (IMA) graft must undergo preemptive revascularization of the subclavian artery (Potter and Pinto 2014). Symptomatic patients warrant surgical or percutaneous revascularization. Proximal stenosis can be managed percutaneously, while distal stenosis is better managed surgically.

85.10 Clinical and Surgical Significance

Although a definitive diagnosis of subclavian stenosis is based on imaging, a thorough physical exam can indicate the presumed diagnosis. In an observational study by English et al. (2001), noninvasive blood pressure measurements of 492 patients undergoing cardiac catheterization revealed a 99% negative predictive value using a > 10 mmHg pressure difference as a cutoff (English et al. 2001). Similar studies using systolic blood pressure cutoffs of 15 and 20 mmHg gave positive predictive values of 100% and 67%, respectively, and 100% negative predictive values for both. Given these results, a systolic cuff pressure difference of >10 mmHg can be used to justify a more invasive assessment.

The subclavian artery has little clinical significance unless it is affected by one of the aforementioned pathologies causing stenosis. Most patients with subclavian artery stenosis will remain asymptomatic owing to the slow progression of disease. Symptoms appear when the vessel lumen has narrowed by approximately 50% (Caesar-Peterson et al. 2021). Symptoms can range from weak ipsilateral pulses, 10 mmHg pressure difference between the ipsilateral and contralateral upper extremities, and digital ischemia. Once imaging is obtained and subclavian artery occlusion is confirmed in the setting of such symptoms, an endovascular approach is attempted before proceeding to open subclavian artery revascularization.

Angioplasty and stenting are first-line interventions for symptomatic subclavian stenosis. The long-term outcomes of endovascular intervention showed high primary success but only satisfactory outcomes after 10 years. Stone et al. (2010) compared endovascular treatment and open surgery for subclavian stenosis. Their study showed that endovascular repair is effective and safe, but open surgery carries better long-term durability and should be the preferred approach for low-risk patients. Balloon angioplasty or stenting can be performed in >90% of patients with 5-year patency rates of 85% (Potter and Pinto 2014). If endovascular methods fail and restenosis occurs, surgical revascularization becomes necessary.

Open surgical revascularization consists of bypasses in the form of carotid-subclavian bypass, carotid transposition, or axilla-axillary bypass surgeries with patency rates greater than 70% at 5 years. However, the success rate is >80% if the common carotid is used in the bypass. Axillo-axillary bypass, in contrast, has significantly lower patency rates (46%) and is therefore typically reserved for patients with elevated perioperative risk (Potter and Pinto 2014).

85.11 Complications

Severe complications arise with thrombosis or occlusion of the subclavian artery. This can lead to upper limb claudication, upper extremity pain, vertebrobasilar insufficiency due to subclavian steal syndrome, and upper extremity ischemia. Prognosis is good with early intervention before the development of symptoms. However, once a patient becomes symptomatic, surgical revascularization is usually warranted, and 10% of patients can experience restenosis. Such patients must be medically managed with aspirin, ACE inhibitors, beta-blockers, and statins (Rosen and Bordoni 2021).

References

Aghoghovwia B (2020, October 29) Subclavian artery and its branches. Kenhub. https://www.kenhub.com/en/library/anatomy/the-subclavian-artery-and-its-branches. Accessed 11 May 2021

Bordes SJ, Iwanaga J, Zarrintan S et al (2021, February 19) Accessory vertebral artery: an embryological review with translation from Adachi. Cureus 13(2):e13448. https://doi.org/10.7759/cureus.13448

Caesar-Peterson S, Bishop MA, Qaja E (2021 January) Subclavian artery stenosis. [Updated 2021 Mar 3]. In: StatPearls [Internet]. Treasure Island: StatPearls Publishing. Available from: https://www.ncbi.nlm.nih.gov/books/NBK470221

Crumbie L (2020, November 13) Blood vessel histology. Kenhub. https://www.kenhub.com/en/library/anatomy/histology-of-the-vascular-network. Published. Accessed 11 May 2021

deAraújo G, JunqueiraBizzi JW, Muller J, Cavazzola LT (2015) Dysphagialusoria—right subclavianretroesophagealarterycausingintermittentesophagealcompressionandeventualdysphagia—a case report and literature review. Int J Surg Case Rep 10:32–34

English JA, Carell ES, Guidera SA, Tripp HF (2001, September) Angiographic prevalence and clinical predictors of left subclavian stenosis in patients undergoing diagnostic cardiac catheterization. Catheter Cardiovasc Interv 54(1):8–11. https://doi.org/10.1002/ccd.1230. PMID: 11553941

Hanneman K, Newman B, Chan F (2017) Congenital variants and anomalies of the aortic arch. Radiographics 37(1):32–51

Jones J. Subclavian artery: radiology reference article. Radiopaedia Blog RSS. https://radiopaedia.org/articles/subclavian-artery?lang=us. Accessed 11 May 2021

Leite T, Pires L, Cisne R, Babinski M, Chagas CA (2017) Clinical discussion of the arteria lusoria: a case report. J Vasc Bras 16:339–342. https://doi.org/10.1590/1677-5449.007617

Netter FH (2023) Atlas of human anatomy (8th ed.). Elsevier. ISBN: 9780323793735

Potter BJ, Pinto DS (2014) Subclavian steal syndrome. Circulation 129(22):2320–2323. https://doi.org/10.1161/circulationaha.113.006653

Rahimi O, Geiger Z (2021 January). Anatomy, Thorax, Subclavian Arteries. [Updated 2020 Jul 27]. In: StatPearls [Internet]. StatPearls Publishing, Treasure Island. Available from: https://www.ncbi.nlm.nih.gov/books/NBK539736/

Rosen RD, Bordoni B (2021 January). Embryology, Aortic Arch. [Updated 2021 Feb 11]. In: StatPearls [Internet]. StatPearls Publishing, Treasure Island. Available from: https://www.ncbi.nlm.nih.gov/books/NBK553173/

Stone PA, Srivastiva M, Campbell JE, Mousa AY (2010) Diagnosis and treatment of subclavian artery occlusive disease. Expert Rev Cardiovasc Ther 8(9):1275–1282

Tubbs RS, Shoja M, Loukas M (2016) Bergman's comprehensive encyclopedia of human anatomic variation. Wiley, Hoboken

86 Transverse Cervical Artery

Arianna Lehuquet and R. Shane Tubbs

86.1 Embryology

The intersegmental arteries are a collection of 30 arteries from the dorsal aorta, each supplying one somite and its derivatives. The left subclavian artery arises from the seventh intersegmental artery; the right subclavian artery develops in several segments. Its proximal segment originates from the fourth aortic arch, also known as the systemic arch, its medial segment from the dorsal aorta, and its distal segment from the seventh intersegmental artery (Kau et al. 2007; Gailloud 2014).

Beginning around day 32 of embryo development, the vertebral artery, the first branch of the subclavian artery, is formed by an anastomosis that joins the cervical intersegmental arteries. These arteries regress, but the dorsal intersegmental anastomoses remain part of the vertebral arteries. The ventral anastomoses continue as the thyrocervical trunk, which gives rise to four branches: the inferior thyroid, suprascapular, ascending cervical, and transverse cervical arteries.

A. Lehuquet
Tulane University School of Medicine, New Orleans, LA, USA
e-mail: alehuquet@tulane.edu

R. S. Tubbs (✉)
Department of Neurosurgery, Tulane University School of Medicine, New Orleans, LA, USA
e-mail: rtubbs@tulane.edu

86.2 Anatomy and Topography

The transverse cervical artery (Fig. 86.1) arises from the thyrocervical trunk, the second ascending branch of the subclavian artery (Rosen and Bordoni 2022). It courses laterally and slightly posteriorly across the base of the posterior triangle of the neck, 2–3 cm superior to the clavicle (Kaufman et al. 2012). During its course, it passes anteriorly to the anterior scalene muscle, the phrenic nerve and the brachial plexus trunks, and posterior to the inferior belly of the omohyoid muscle (Tessler et al. 2017). Its origin and the beginning of its course lie in a well-defined "anatomical triangle," the medial border formed by the posterior edge of the sternocleidomastoid muscle and the lateral border by the external jugular vein. In contrast, the superior edge of the clavicle forms the inferior border (Weiglein et al. 2005).

When the transverse cervical artery reaches the deep surface of the trapezius muscle, it divides into superficial and deep branches. The superficial branch continues on the deep surface of the trapezius muscle, and the deep branch continues deep to the rhomboid major and rhomboid minor muscles on the medial border of the scapula. The deep branch, also known as the dorsal scapular artery, supplies the rhomboid major and rhomboid minor muscles and the levator scapulae. It continues inferiorly along the medial border of the scapula to its inferior angle and anastomoses with the subscapular artery (Stern 2009).

S. J. Bordes, Jr. et al. (eds.), *The Clinical Anatomy of the Vascular System*,
https://doi.org/10.1007/978-3-031-78326-5_86

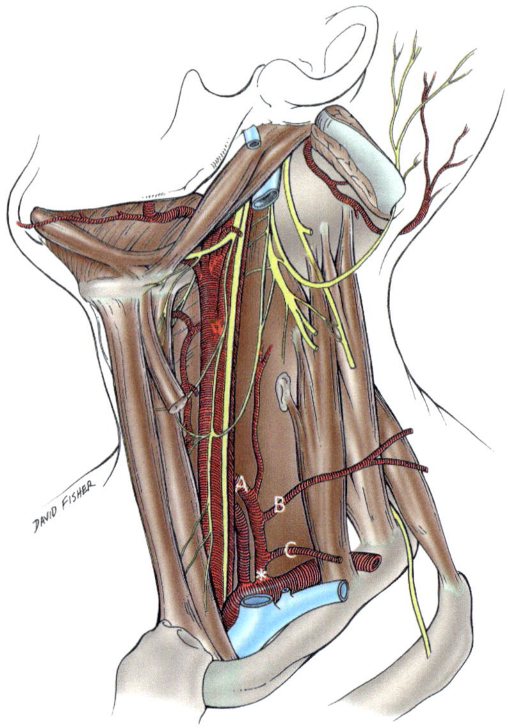

Fig. 86.1 Schematic drawing of the transverse cervical artery (B). The origin of this artery is from the thyrocervical trunk (*). Also, note the suprascapular (C) and inferior thyroid arteries (A). The ascending cervical artery (unlabeled) is shown arising from the inferior thyroid artery

86.3 Anatomical Variations

The transverse cervical artery can arise as a branch of the thyrocervical trunk, which Chen et al. reported as occurring in 54.2% of cases, or as a branch of the subclavian artery, which can occur in other cases (Chen et al. 2010). In both events, it can run either anteriorly or posteriorly to the trunks or divisions of the brachial plexus, but it has also been reported to course between those trunks (Aggarwal et al. 2010).

The dorsal scapular artery (DSA) is anomalous and can arise from the thyrocervical trunk, more specifically as a continuation of the deep branch of the transverse cervical artery. However, it has also been reported to branch directly from the second or third part of the subclavian artery (Chaijaroonkhanarak et al. 2014). When it branches from the transverse cervical artery, it courses anteriorly or posteriorly to the brachial plexus trunks. A DSA with subclavian artery origins is more likely to pass between the upper and middle trunks of the brachial plexus (Chaijaroonkhanarak et al. 2014). In a case study by Veranna et al., variation in the course of the DSA occasionally resulted in compression of the C8 and T1 anterior rami of the brachial plexus, which are often responsible for thoracic outlet symptoms (Verenna et al. 2016). Although the more common cause of thoracic outlet syndrome is mechanical compression of the brachial plexus by bone or muscle, this case study emphasized the possibility of vascular compression of the nerve roots (Verenna et al. 2016).

86.4 Histology and Pathology

The transverse cervical artery is a muscular artery with higher smooth muscle than elastin content (Leloup et al. 2015). The vessel wall contains three tissue layers: an inner tunica intima, a layer of endothelial cells lining the lumen; a tunica media, which comprises smooth muscle and elastic fibers; and an outer tunica adventitia, which consists mainly of connective tissue. Arterial dissections can occur when the tunica intima is disrupted, creating a false lumen. Arterial dissection is more common in cervical arteries or the aorta.

Another histopathology common among arteries is atherosclerotic plaque, a risk factor for embolism or stenosis, ultimately resulting in tissue ischemia. Atherosclerosis results from a buildup of plaque that occludes the lumen of the arteries. In a cadaveric study by Tessler et al., the transverse cervical artery had less atherosclerotic plaque on average than the degree of stenosis within the external carotid artery (Tessler et al. 2017). Any shoulder weakness or pain in the musculature supplied by the transverse cervical artery can indicate vascular compromise.

86.5 Clinical and Surgical Significance

A supraclavicular block is a regional anesthetic technique that anesthetizes the upper limb and shoulder. Typically, ultrasound precedes needle insertion to avoid large vessels such as the dorsal scapular artery, transverse cervical artery, and suprascapular artery coursing through the area. Anatomical variations in the course of the transverse cervical and dorsal scapular arteries have clinical implications in regional anesthetic techniques and should be considered. All branches should be carefully identified using color Doppler, which can help reduce the risk of vascular injury (Murata et al. 2012).

In a cadaveric study by Heo et al., four relationships were observed between the transverse cervical artery and the accessory nerve. In 38% of cadavers, the accessory nerve ran parallel and medial to the transverse cervical artery; in 38%, the nerve ran parallel and lateral to the artery; in 19%, the accessory nerve and transverse cervical artery crossed one another during their course; and in 5%, the nerve and artery twisted (Heo et al. 2020). An accessory nerve block from a posterior approach provides a safer alternative to injecting near the styloid process of the temporal bone, which can stimulate the vagus nerve, leading to asystole (Scott 1996). It is safer than injecting through the posterior triangle of the neck given the extensive innervation in that location (Townsley et al. 2011). With a posterior approach, the anatomical course of the accessory nerve and its relationship with the transverse cervical artery must be confirmed first to preclude arterial injection and injury (Heo et al. 2020).

Kaufman et al. demonstrated that an adherent or tortuous transverse cervical artery can cause compression of the phrenic nerve, ultimately leading to hemidiaphragm paralysis, described as "Red Cross syndrome." (Kaufman et al. 2012) This term came about as the transverse cervical artery crosses anteriorly to the phrenic nerve at an almost perfect 90° angle, which can only be diagnosed intraoperatively (Kaufman et al. 2012).

Head and neck flaps are used to reconstruct defects in the cervicofacial region after burns or trauma. Other well-known applications for supraclavicular flaps are reconstruction of the pharyngeal wall after tumor removal or pharyngeocutaneous fistula repair (Luca et al. 2017). An anterior cutaneous branch arises from the transverse cervical artery and has been used successfully in reconstructive flaps as it can be rotated and advanced more easily than other flaps. A pre-expanded transverse cervical artery flap is preferable for reconstructing deformities in the cervical region, lower and middle face, and the lateral and posterior neck and shoulder (Song and Chai 2017). The possibility of harvesting a perforator version of this flap was proposed by Ogawa et al., who found that the transverse cervical perforator was large enough for use as a vascular pedicled flap for such reconstructions (Ogawa et al. 2006). Luca et al. noted that the only contraindication for using this type of flap is a prior surgical history in the supraclavicular region, compromising the region's vascular integrity and making the flap's outcome more difficult to predict (Luca et al. 2017).

86.6 Complications

The transverse cervical artery can be implicated in complications arising from numerous surgical procedures. Pseudoaneurysm formation is a rare but documented complication of radical neck dissection. This complication was described in a case study by Singh et al. and was believed to result from thermal injury caused by the cautery used during the dissection, which weakened the vessel wall. Pseudoaneurysm of the transverse cervical artery was treated with coil embolization in this case (Singh et al. 2019).

References

Aggarwal A, Puri N, Aggarwal AK, Harjeet K, Sahni D (2010) Anatomical variation in the formation of brachial plexus and its branching. Surg Radiol Anat 32(9):891–894. https://doi.org/10.1007/s00276-010-0683-8

Chaijaroonkhanarak W, Kunatippapong N, Ratanasuwan S et al (2014) Origin of the dorsal scapular artery and its relation to the brachial plexus in Thais. Anat Sci Int 89(2):65–70. https://doi.org/10.1007/s12565-013-0200-3

Chen WL, Da-ming Z, Zhao-hui Y et al (2010) Extended supraclavicular fasciocutaneous Island Flap based on the transverse cervical artery for head and neck reconstruction after cancer ablation. J Oral Maxillofac Surg 68(10):2422–2430. https://doi.org/10.1016/j.joms.2010.01.015

Gailloud P (2014) The supreme intercostal artery includes the last cervical intersegmental artery (C7) – angiographic validation of the intersegmental nomenclature proposed by Dorcas Padget in 1954: the supreme intercostal artery redefined. Anat Rec 297(5):810–818. https://doi.org/10.1002/ar.22893

Heo Y, Cho N, Cho H, Won HS, Yang M, Kim YD (2020) New insights into pathways of the accessory nerve and transverse cervical artery for distal selective accessory nerve blockade. Korean J Pain 33(1):48–53. https://doi.org/10.3344/kjp.2020.33.1.48

Kau T, Sinzig M, Gasser J et al (2007) Aortic development and anomalies. Semin Interv Radiol 24(2):141–152. https://doi.org/10.1055/s-2007-980040

Kaufman MR, Willekes LJ, Elkwood AI et al (2012) Diaphragm paralysis caused by transverse cervical artery compression of the phrenic nerve: The Red Cross syndrome. Clin Neurol Neurosurg 114(5):502–505. https://doi.org/10.1016/j.clineuro.2012.01.048

Leloup AJ, Van Hove CE, Heykers A, Schrijvers DM, De Meyer GR, Fransen P (2015) Elastic and muscular arteries differ in structure, basal NO production and voltage-gated Ca(2+)-Channels. Front Physiol 6:375. Published 2015 Dec 15. https://doi.org/10.3389/fphys.2015.00375

Luca N, Santana MJ, Festa BM, Collurà F, Righini S (2017) Transverse cervical artery perforator flap. Ann Plast Surg 79(6):577–582. https://doi.org/10.1097/SAP.0000000000001144

Murata H, Sakai A, Hadzic A, Sumikawa K (2012) The presence of transverse cervical and dorsal scapular arteries at three ultrasound probe positions commonly used in supraclavicular brachial plexus blockade. Anesth Analg 115(2):470–473. https://doi.org/10.1213/ANE.0b013e3182568557

Ogawa R, Murakami M, Vinh VQ, Hyakusoku H (2006) Clinical and anatomical study of superficial cervical artery flaps: retrospective study of reconstructions with 41 flaps and the feasibility of harvesting them as perforator flaps. Plast Reconstr Surg 118(1):95–101

Rosen R, Bordoni B (2022) Embryology, Aortic Arch. Ncbi.nlm.nih.gov. https://www.ncbi.nlm.nih.gov/books/NBK553173/. Accessed 1 Apr 2022

Scott PV (1996) Asystole from tetanic stimulation of the accessory nerve. Anaesthesia 51(12):1148–1150. https://doi.org/10.1111/j.1365-2044.1996.tb15053.x

Singh A, Kumar R, Bhalla AS, Goyal A, Sagar P (2019) Transverse cervical artery pseudoaneurysm: an unusual delayed complication of radical neck dissection. Indian J Otolaryngol Head Neck Surg 71(S1):348–351. https://doi.org/10.1007/s12070-018-1305-8

Song H, Chai J (2017) Pre-expanded transverse cervical artery perforator flap. Clin Plast Surg 44(1):41–47. https://doi.org/10.1016/j.cps.2016.08.002

Stern J (2009) Clinically oriented anatomy by Keith L. Moore, Arthur F. Dalley, and Anne M. Agur Clin Anat 22(5):635–635. https://doi.org/10.1002/ca.20801

Tessler O, Gilardino M, Bartow M et al (2017) Transverse cervical artery: consistent anatomical landmarks and clinical experience with its use as a recipient artery in complex head and neck reconstruction. Plast Reconstr Surg 139(3):745e–751e. https://doi.org/10.1097/PRS.0000000000003085

Townsley P, Ravenscroft A, Bedforth N (2011) Ultrasound-guided spinal accessory nerve blockade in the diagnosis and management of trapezius muscle-related myofascial pain. Anaesthesia 66(5):386–389. https://doi.org/10.1111/j.1365-2044.2011.06691.x

Verenna AM, Alexandru D, Karimi A et al (2016) Dorsal scapular artery variations and relationship to the brachial plexus, and a related thoracic outlet syndrome case. J Brachial Plex Peripher Nerve Inj 11(01):e21–e28. https://doi.org/10.1055/s-0036-1583756

Weiglein AH, Moriggl B, Schalk C, Künzel KH, Müller U (2005) Arteries in the posterior cervical triangle in man. Clin Anat N Y N 18(8):553–557. https://doi.org/10.1002/ca.20143

Suprascapular Artery

87

Dean John Hendrix Dante and Aaron Yu

87.1 Introduction

The suprascapular artery (SSA) supplies the scapular region and also gives off branches to the ipsilateral regions of the chest and shoulder. Still, most of its content is delivered to the posterior surface of the scapula and nearby tissues (Figs. 87.1 and 87.2). Its role in collateral circulation confers broad clinical significance on it. Furthermore, the SSA, which travels closely with the suprascapular nerve, is associated with two important anatomical landmarks along its course to the scapula: the superior transverse scapular ligament and the spinoglenoid notch. Knowledge of these associations, especially its relationship to the suprascapular nerve, has critical applications in surgery.

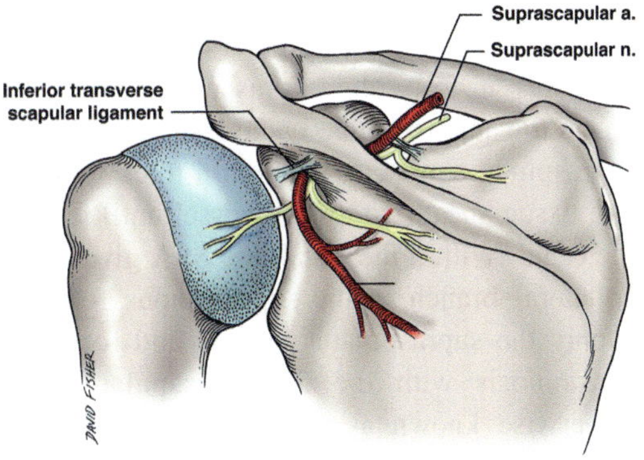

Fig. 87.1 Schematic drawing of the suprascapular artery from a posterior view and noting its relationship to the suprascapular nerve

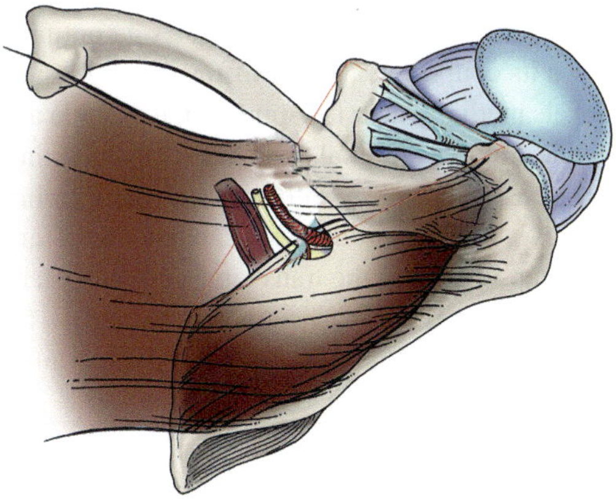

Fig. 87.2 Schematic view of the suprascapular artery from a posterior superior view

D. J. H. Dante
LSUHSC School of Medicine,
New Orleans, LA, USA
e-mail: ddante@lsuhsc.edu

A. Yu (✉)
Tulane School of Medicine, New Orleans, LA, USA
e-mail: ayu2@tulane.edu

S. J. Bordes, Jr. et al. (eds.), *The Clinical Anatomy of the Vascular System*,
https://doi.org/10.1007/978-3-031-78326-5_87

87.2 Anatomy and Topography

The SSA usually arises as the most inferior branch of the thyrocervical trunk near the medial border of the anterior scalene muscle, deep to the sternocleidomastoid muscle, and immediately superior to the clavicle. From the level of its origin, it descends laterally to the upper border of the scapula. It crosses anterior to several structures, including the brachial plexus and the third part of the subclavian artery (Pyrgakis et al. 2013). Immediately before the SSA reaches the superior border of the scapula through the suprascapular notch, it crosses over the superior transverse scapular ligament, which separates it from the suprascapular nerve. It enters the suprascapular canal lying deep to the supraspinatus muscle, wherein a thick connective tissue sheath wraps the vessel until it exits the canal. Within the canal, the SSA gives off a collateral branch in the supraspinous fossa to supply the supraspinatus muscle and forms an anastomosis with the dorsal scapular artery (otherwise known as the deep branch of the transverse cervical artery).

The SSA then loops around the lateral end of the spine of the scapula, passing through the spinoglenoid notch below the spinoglenoid ligament (also known as the inferior transverse scapular ligament) (Yepes et al. 2007). As it continues down its course towards the infraspinous fossa, it travels medially near the spine of the scapula and divides into its terminal branches. These terminal branches anastomose with some of the terminal branches of the circumflex scapular artery and the deep branch of the transverse cervical artery. Altogether, these anastomoses supply the infraspinous fossa and its related muscles: infraspinatus, teres major, and teres minor.

87.3 Variations

Various cadaveric studies and surgical settings have shown that the SSA varies anatomically in its positional relationship with other neighboring structures, its anastomotic pattern, its level of origin, and its persistence. Polguj et al. classified the positioning of the suprascapular neurovasculature into four types (Polguj et al. 2015) in terms of its relationship to the superior transverse scapular ligament (STSL) and other suprascapular vessels:

- **Type 1**: An SSA runs above the STSL, while the vein and nerve run below (61.3%).
- **Type 2**: Both suprascapular vessels run above the STSL, while the nerve is underneath (17%).
- **Type 3**: All suprascapular vessels and nerves are below the STSL (12.3%).
- **Type 4**: There are variations of the previous three types (9.4%) and accessory suprascapular veins (2.4%) (Polguj et al. 2015).

Yang et al. described the relationship between the suprascapular vessels and nerve differently:

- **Type I**: All suprascapular vessels run over the STSL (59.4%).
- **Type II**: The SSA and vein cross over and under the STSL (29.7%); over half of these (15.5%) contained an accessory suprascapular vein.
- **Type III**: All suprascapular vessels run under the STSL (10.9%) (Yang et al. 2012).

Naidoo et al. similarly classified the positional variants of the SSA in relation to the STSL: 80% ran over the ligament and 20% underneath (Naidoo et al. 2014). However, the most common positional variant described in the literature is Polguj et al.'s Type 1 (half of the shoulders classified as Type II by Yang et al.).

Although the textbook anatomy of the SSA typically involves its origin from the thyrocervical trunk, in rare cases, it arises from a segment of the subclavian or axillary artery (10%). It occasionally (30%) originates from a common trunk with the transverse cervical artery (Pyrgakis et al. 2013). In some cases, a SSA arises from the axillary or different segments of the subclavian artery (10%). More infrequently, it originates from the internal thoracic artery, costocervical trunk, or dorsal scapular artery. In a sample of 31 cadavers, Pyrgakis et al. found that the rarest anomalous origin (1.6%) was from the third segment of the subclavian artery (Pyrgakis et al. 2013).

The SSA has also been categorized by its anastomotic pattern with the circumflex scapular artery. The Shin et al. (2016) classification distinguishes two types with two corresponding subtypes:

- **Type Ia** (dual supply with anastomosis): The SSA and circumflex scapular artery supply the infraspinous fossa and anastomose (20.3%) (Shin et al. 2016).
- **Type Ib** (dual supply without anastomosis): Both the SSA and circumflex scapular artery supply the infraspinous fossa without anastomosing (except at some of the terminal branches) (42.4%) (Shin et al. 2016).
- **Type IIa** (one supply with anastomosis): The circumflex scapular artery supplies the infraspinous fossa, while the SSA anastomoses with the circumflex scapular artery without giving any terminal branches (28.8%) (Shin et al. 2016).
- **Type IIb** (one supply without anastomosis): The circumflex scapular artery supplies the infraspinous fossa without anastomosing with the SSA, which terminates at the spinoglenoid notch (8.5%) (Shin et al. 2016).

Duplication of the SSA is an extremely rare variation and should be distinguished from the aforementioned entities (Pyrgakis et al. 2013). The arterial vessels are duplicated bilaterally, one from the thyrocervical trunk (coursing normally) and the other from the third segment of the subclavian artery (traversing below the transverse scapular ligament). Hypoplastic or completely absent (3%) SSAs are also rare occurrences (Pyrgakis et al. 2013).

87.4 Pathology and Imaging

Aneurysms and pseudoaneurysms of the thyrocervical trunk and, by extension, its branches are rare and debilitating pathological manifestations. Among the few reported cases of such a vascular abnormality involving the thyrocervical system, the following represent the presumed etiologies: direct trauma, central venous cannulation, thoracic outlet syndrome, and repetitive overhead arm motions (Summers et al. 2020). Summers et al. described one of the few known cases of a SSA aneurysm. Extensive osteoarthritic changes to the glenoid are assumed to cause arterial injury and eventual aneurysm formation (Summers et al. 2020). Prater et al. described a case of pseudoaneurysm arising from direct trauma, previously undiscovered, during the repair of a clavicular fracture from an automobile accident that is also suspected to have caused the suprascapular arterial trauma (Prater et al. 2010). Coil embolization of the pseudoaneurysm here was indicated as the most effective treatment option, ultimately resulting in a successful recovery.

Zardi et al. described the only reported true aneurysm of the SSA with anomalous origin from the subclavian artery. CTA and selective angiography confirmed the aneurysm and the SSA's abnormal origin from the third segment of the right subclavian artery. Such a case constitutes a surgical emergency, given the risk of rupture and the large size of the aneurysm. To avoid complications due to a trifurcation near the origin of the aneurysm, transcatheter embolization with a liquid embolic agent (Onyx) was performed to occlude both the collaterals and the true aneurysm. Post-embolization digital subtraction angiography (DSA) confirmed successful intraoperative results (Zardi et al. 2009).

Another rare pathological presentation of the SSA involves an arteriovenous malformation (AVM). AVMs are developmental vascular anomalies, normally present at birth, that can occur anywhere in the body. Some remain quiescent, but others manifest later in adulthood, their size increasing secondary to an infection, trauma, or hormonal influence (Rajesh et al. 2006). Mendoca & Miyamoto confirmed a large axillary AVM nourished primarily by the SSA and branches of the axillary artery by magnetic resonance angiography (MRA). The SSA was embolized with platinum coils (Mendonça and Miyamotto 2008).

Given the potential severity of pathologies involving the SSA, adequate implementation and interpretation of radiological imaging can guide safe interventional approaches to its vascular territory. Several imaging techniques can be used to diagnose pathologies of the SSA reliably. In a report by Summers et al., a CT scan revealed multiloculated effusion within the left glenohumeral joint (Summers et al. 2020). Prater et al. described the combined preoperative use of CT and selective DSA to visualize the pseudoaneurysm of interest arising from the SSA (Prater et al. 2010). Other noninvasive to minimally invasive imaging modalities, such as Duplex ultrasound, computerized tomography angiography (CTA), and MRA, can also be used to identify pseudoaneurysms (Prater et al. 2010).

87.5 Clinical and Surgical Significance

As previously mentioned, the SSA courses through the suprascapular notch, usually above the superior transverse scapular ligament, as it enters the supraspinous fossa. Closely related to this anatomical landmark is the suprascapular nerve. Irritation or injury to this nerve causes suprascapular neuropathy or suprascapular nerve entrapment. This syndrome can cause severe shoulder pain associated with weakness and atrophy of the supraspinatus and infraspinatus muscles (Arce et al. 2021). A surgical procedure is often necessary to relieve the patient's pain and prevent further degeneration of the associated muscles. This suprascapular nerve release incises the suprascapular ligament. The surgeon must be aware of the location of the SSA, given its direct association with the suprascapular ligament and the significant positional variations of the suprascapular vessels (Arce et al. 2021).

Furthermore, in other shoulder arthroscopies involving the spinoglenoid notch, the surgeon must recognize the risk of injury to the SSA, which loops around the notch (Yepes et al. 2007). Iatrogenic injury to the SSA can also be caused by scapular surgeries such as open reduction and internal fixation of a scapular fracture (Shin et al. 2016). Other surgical operations, such as radical and modified neck dissection to mitigate the spread of head and neck cancers via the lymphatics, can necessitate ligation of the SSA (Pyrgakis et al. 2013). Therefore, adequate familiarity with its anatomical structure and variations is imperative to facilitate safe and effective interventional approaches to the SSA.

References

Arce G et al. (2021) Suprascapular nerve release: technique based on anatomic landmarks. Arthrosc Tech, Elsevier. https://www.sciencedirect.com/science/article/pii/S2212628720303042

Atsas S, et al (2011) The rare origin of the Suprascapular artery arising off the internal thoracic artery in the presence of the Thyrocervical trunk: clinical and surgical implications. Pulsus. Internation Variations of Anatomical Variations. https://www.pulsus.com/scholarly-articles/the-rare-origin-of-the-suprascapular-artery-arising-off-the-internal-thoracic-artery-in-the-presence-of-the-thyrocervica.pdf

Darev P (2012) Novel ultrasound-guided suprascapular nerve block. Reg Anesth Pain Med. U.S. National Library of Medicine. https://pubmed.ncbi.nlm.nih.gov/23086351/

Mendonça CT, Miyamotto M (2008) Multidisciplinary approach in the management of a giant axillary arteriovenous malformation. J Vasc Surg 47(3):657. https://doi.org/10.1016/j.jvs.2007.02.056

Naidoo N et al (2014) The variant course of the suprascapular artery. Folia Morphol (Warsz). https://journals.viamedica.pl/folia_morphologica/article/view/35840

Polguj M et al (2015) The variable morphology of suprascapular nerve and vessels at suprascapular notch: a proposal for classification and its potential clinical implications. In: Knee surgery, sports traumatology, arthroscopy. Springer, Berlin Heidelberg. https://www.ncbi.nlm.nih.gov/pmc/articles/PMC4555201/

Prater S et al (2010) Endovascular management of suprascapular artery pseudoaneurysm. Taylor & Francis, Baylor Scott & White Health. https://www.tandfonline.com/doi/abs/10.1080/08998280.2010.11928576

Pyrgakis P et al (2013) Anomalous origin and course of the Suprascapular artery .... *PubMed.* N Am J Med Sci, https://www.najms.org/article.asp?issn=1947-2714;year=2013;volume=5;issue=2;spage=129;epage=133;aulast=Pyrgakis;type=0

Rajani S (2018) Variations in the origin and course of the suprascapular artery: case report and literature review. J Vasc Brasileiro. U.S. National Library of Medicine, https://pubmed.ncbi.nlm.nih.gov/29930683/

Rajesh et al (2006) Arteriovenous malformation and color Doppler: posterior auricular artery. Indian J

Otolaryngol Head Neck Surg 58(3):318–320. https://doi.org/10.1007/BF03050860

Reineck JR, Krishnan SG (2009) Subligamentous suprascapular artery encountered during arthroscopic suprascapular nerve release: a report of three cases. J Shoulder Elbow Surg 18:e1. US National Library of Medicine. https://www.ncbi.nlm.nih.gov/pubmed/19046638

Falconer RW, Munro R, Jamieson EN (2019) Suprascapular artery pseudoaneurysm. ANZ J Surg. US National Library of Medicine. https://pubmed.ncbi.nlm.nih.gov/28695991/

Shin K-J et al (2016) Arterial supply and anastomotic pattern of the infraspinous fossa focusing on the surgical significance. J Plast Reconstr Aesthet Surg. Churchill Livingstone. https://www.sciencedirect.com/science/article/pii/S1748681515006051

Summers T et al (2020) Suprascapular artery aneurysm secondary to severe shoulder joint osteoarthritis. Karger. Karger Publishers. https://www.researchgate.net/publication/341174607_Suprascapular_Artery_Aneurysm_Secondary_to_Severe_Shoulder_Joint_Osteoarthritis/fulltext/5ee4291e458515814a5b76c2/Suprascapular-Artery-Aneurysm-Secondary-to-Severe-Shoulder-Joint-Osteoarthritis.pdf

Tountas CP. Chapter 3 Arteries. Anat Var Upper Extrem. Churchill Livingstone, pp 190–192

Tubbs RS, et al (2003) Anomalous traversement of the suprascapular artery through the suprascapular notch: a possible mechanism for undiagnosed shoulder pain?. Med Sci Monit. U.S. National Library of Medicine. https://www.ncbi.nlm.nih.gov/pubmed/12640333

Yang H-J et al (2012) Topographical anatomy of the suprascapular nerve and vessels at the suprascapular notch. Wiley Online Library. Wiley. https://onlinelibrary.wiley.com/doi/10.1002/ca.21248

Yepes H, et al (2007) Vascular anatomy of the subacromial space: a map of bleeding points for the arthroscopic surgeon. Arthroscopy. WB Saunders. https://www.sciencedirect.com/science/article/pii/S0749806307004318

Zardi EM et al (2009) Anomalous origin and aneurysm of the suprascapular artery: the first case observed. BMJ Case Rep BMJ Publishing Group. https://casereports.bmj.com/content/2009/bcr.09.2008.0869

88 Dorsal Scapular Artery

Nicholas O. Gerard III

88.1 Embryology

During embryological development, the primordial circulatory system begins to oxygenate tissue around week 3. The first structure to arise from the heart tube is the aortic sac, which feeds into the pharyngeal arches. By week 4, the fourth pharyngeal arch has begun to develop; this is the precursor for the arch of the aorta and the subclavian arteries. The final branch of the subclavian artery will later give rise to the dorsal scapular artery, oxygenating the musculature of the superficial back (Moore et al. 2010). A recognizable shoulder develops around week 10. The complex glenohumeral joint can be observed along with noted osteogenesis of the humeral head (week 10) and scapula (week 11) (Hita-Contreras et al. 2018). Its blood supply includes the axillary artery and scapular anastomoses including the dorsal scapular, suprascapular, and circumflex scapular arteries.

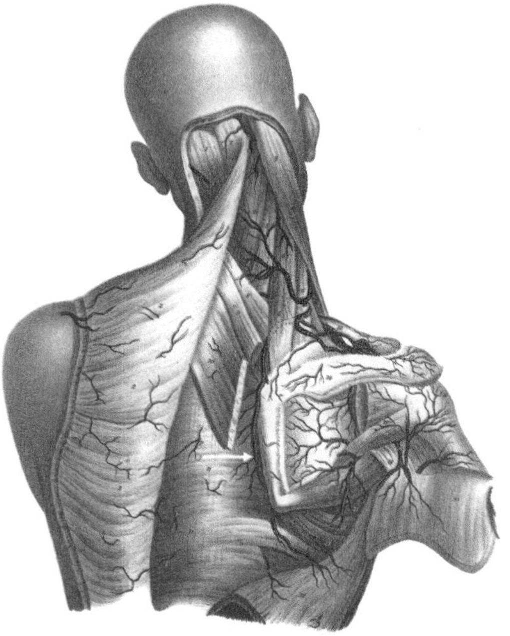

Fig. 88.1 Posterior drawing of the scapular region noting the dorsal scapular artery. (Quain and Wilson, 1844 author's collection)

88.2 Anatomy/Topography

The dorsal scapular artery (Fig. 88.1) is typically the final branch of the subclavian artery supplying three of the five superficial back muscles that work to retract and elevate the scapula: the rhomboid major, rhomboid minor, and levator scapulae. The dorsal scapular artery arises from the subclavian artery prior to its transition to the axillary artery as it crosses the first rib. It typically originates lateral to the anterior scalene muscle before diving deep to the trapezius muscle, running along the anterior borders of the muscles it supplies. When it passes lateral to the anterior

N. O. Gerard III (✉)
Tulane University School of Medicine, New Orleans, LA, USA
e-mail: ngerard@tulane.edu

S. J. Bordes, Jr. et al. (eds.), *The Clinical Anatomy of the Vascular System*,
https://doi.org/10.1007/978-3-031-78326-5_88

scalene, it approaches the brachial plexus. In 90% of people, the dorsal scapular artery courses either between the superior and middle trunks or between the middle and lower trunks of the brachial plexus (Nambyiah et al. 2011). When it passes the brachial plexus, it can give off small arterial branches supplying distal trunks and proximal cords (Gao et al. 2018; Levy et al. 2003). These direct branches communicate with branches from the subclavian and axillary arteries that also supply the brachial plexus.

The dorsal scapular artery continues posteriorly, running with the dorsal scapular nerve (C4, C5 nerve roots), which also innervates the rhomboid major, rhomboid minor, and levator scapulae (Tubbs et al. 2005). It contributes to the scapular anastomosis via the suprascapular artery superiorly and the circumflex scapular artery inferiorly. This scapular anastomosis joins the subclavian artery proximally and the axillary artery distally and can provide collateral circulation if the axillary artery is occluded (Moore et al. 2010). In 2003, Franz Haas found in a study of 124 cadaveric dissections that the dorsal scapular artery averaged 2.7 mm in diameter at its origin and 1.95 mm at the levator scapulae (Haas et al. 2004).

88.3 Imaging

The dorsal scapular artery can be visualized via ultrasound and radiographically via computed angiography and MRI (Gorospe et al. 2016; Verenna et al. 2016; Kinjo and Frankel 2012). Anatomical studies have used traditional dissections coupled with arteriography to evaluate variations (Ikka et al. 2016). Methylene blue can also be injected into the vasculature of fresh cadavers to trace its terminal branches (Tan and Tan 2000). To examine the contribution of the dorsal scapular artery to the brachial plexus, researchers have used gelatin–lead oxide microvascular radiography and macro–micro anatomical methods (Gao et al. 2018; Levy et al. 2003).

The use of ultrasound imaging to identify neurovascular structures, including the dorsal scapular artery, is essential during a supraclavicular nerve block (Kohli et al. 2014). It is important to protect these structures to ensure both the safety and efficacy of such blocks (Kinjo and Frankel 2012). Furthermore, proper visualization of neurovascular structures helps physicians to decide among various local anesthetic approaches (Hong et al. 2021). A 2012 study examined 53 patients with guided ultrasound in the supraclavicular region. They determined that the dorsal scapular artery could be visualized 34% of the time when the probe was placed in one of the following three positions (Murata et al. 2012):

- Position A: the brachial plexus and the subclavian artery both lying on the first rib.
- Position B: the brachial plexus on the first rib; the artery on the pleura.
- Position C: the brachial plexus between the anterior and middle scalene muscles.

88.4 Variation

The dorsal scapular artery was first named in 1958 (Huelke 1958). Previously, it was called the descending scapular artery or the deep branch of the transverse cervical artery (Moore et al. 2010). The former described the artery as it arose directly from the subclavian artery, while the latter described it as it arose from the transverse cervical artery, which stems from the thyrocervical trunk. In 1962, Huelke showed that the dorsal scapular artery arose from the subclavian artery in about 70% of 943 cadaveric halves and from the transverse cervical artery in about 30%. Thus, he argued the artery should have a single name regardless of variations in its origin. He proposed that it should take the name of the nerve with which it is intertwined, the dorsal scapular nerve. This naming was more consistent with current standards of describing anatomical structures (Huelke 1962).

In 1996, Reiner and Kasser reexamined the frequency of the origin of the dorsal scapular artery in 157 cadaveric halves. They determined that it arose from the subclavian artery about 75% of the time and from the transverse cervical artery in about 25%. Interesting, they noted instances in which it arose from the subclavian artery while the rhomboids received arterial contributions from perforating branches of the transverse cervical artery (Reiner and Kasser 1996). In view of the inconsistency of terminology in the literature regarding the dorsal scapular artery and the vasculature of the posterior cervical triangle, Weiglein and colleagues examined 498 cadaveric sides and categorized variations of the dorsal scapular artery. These variations are relevant to reconstructive surgery of the anterior, lateral, and posterior neck. They found that the dorsal scapular artery arose directly from the subclavian artery (42%), the cervico-dorsal trunk (30%), the cervico-dorso-scapular trunk (24%), and the dorso-scaplar trunk (4%) (Weiglein et al. 2005). These data demonstrate the high variability of the dorsal scapular artery.

In 2016, 16 cadaveric sides were dissected in individuals of Asian background. In no specimen did the dorsal scapular artery arise from the subclavian (Yang et al. 2016). This suggests possible ethnic differences that have not been thoroughly explored.

88.5 Pathology

Owing to its proximity, the dorsal scapular artery is thought to be implicated in thoracic outlet syndrome and brachial plexus compression syndromes. Verenna et al. reported surgical decompression of a 17-year-old male overhead-throwing athlete, revealing a fibrous dorsal scapular artery (2 mm in diameter) compressing the C8 ventral ramus superiorly as it coursed posteriorly. This compression proved to be the cause of physical therapy-resistant paresthesia in the patient's dominant arm. After surgery, the patient returned to the sport with a full range of motion (Verenna et al. 2016). Compression of the T1 ventral ramus of the brachial plexus by the dorsal scapular artery has also been reported in a cadaveric dissection (Tubbs et al. 2006). Lastly, an interesting case of a delayed pseudoaneurysm involving the dorsal scapular artery was reported after blunt trauma to the anterior chest wall. The bleed was embolized uneventfully, improving both pain and hematoma in an 83-year-old female patient (Gorospe et al. 2016).

88.6 Surgical Significance

The dorsal scapular artery is an important consideration for numerous surgeries of the head and neck. Portions of it can be transplanted in free flap procedures, or it can be preserved for flaps repositioned through a pedicled approach. Less commonly, it can be a recipient vessel pedicle in a vessel-depleted neck (Rosko et al. 2017). The tissue atop the dorsal scapular artery is compliant and offers rotational coverage of nearly the entire head and neck. Surgical dissections to harvest or isolate the dorsal scapular artery most often involve placing the patient in the lateral decubitus position, which can lead to complications. However, owing to its conspicuous location and minimal functional and esthetic deficits, grafts involving the dorsal scapular artery are an option in salvage head and neck reconstruction procedures (Yang et al. 2016).

It should be noted that because the head and neck have a rich blood supply, non-dorsal scapular artery flaps are often preferred. For example, the *superior flap* uses the blood supply from the occipital artery and cervical paraspinal perforators. The *lateral island trapezius myocutaneous flap*, sometimes referred to as the *superior trapezius flap*, is based on the transverse cervical artery (Yang et al. 2016). Grafting from the dorsal scapular artery often involves more complicated disease states, which could explain why the success rates of these procedures have been called into question. Flap procedures involving the dorsal scapular artery include:

1. The lower trapezius island musculocutaneous flap (LTIMF).

- First described by Baek in 1980, the LTIMF is a standard in head and neck reconstruction. It provides a long paddle of thin, pliant, hairless skin and muscle that can be rotated as far as the face and scalp in a single stage (Baek et al. 1980). A 1991 review examined 10 studies spanning 100 cases and found a success rate of 73% using LTIMF (Urken et al. 1991). A 2013 study examined 32 cases spanning 20 years and found a 93.75% success rate (Ou et al. 2013). This flap is indicated for patients who have received local radiation, chemotherapy, or previous surgeries involving the head and neck.

2. The extended lower trapezius island myocutaneous flap.
   - First described by Tan in 2000, this flap offers a wider arc of rotation of soft tissue and can be advantageous over the standard superior trapezius flap. Dissection beyond the inferior portion of the trapezius, around the level of the T10 vertebra, involves terminal perforators of the dorsal scapular artery (Tan and Tan 2000). Surgeons have also used this flap to treat postburn contractures in the axilla, demonstrating its versatility (Elshaer 2004). A 2002 study reviewed 32 cases in which a version of this procedure had a 91% flap survival rate (Angrigiani et al. 2003).
3. The dorsal scapular artery perforator flap.
   - First described by Marr in 2021, this flap is deemed a complete muscle-sparing technique. Doppler ultrasound was used to identify perforators of the dorsal scapular artery and the pedicled fasciocutaneous flap was passed superiorly, below the trapezius, to preserve complete muscle function and minimize cosmetic deformity. The procedure was pioneered to meet the needs of a 61-year-old male patient who did not meet the criteria for other surgeries (Marr et al. 2021).
4. Osteomuscular dorsal scapular flap (OMDS flap).

- First described by Vacher in 2008, this pedicled flap is indicated when mandibular reconstruction calls for bone and soft tissue but the fibula-free flap is contraindicated. Typically, mandibular autografts are taken from the ilium or fibula. However, three cases used the medial border of the scapula for mandibular reconstruction through the OMDS flap following excision of a lateral perilingual epidermoid carcinoma (Vacher 2008).

References

Angrigiani C, Grilli D, Karanas YL, Longaker MT, Sharma S (2003 Jan) The dorsal scapular Island flap: an alternative for head, neck, and chest reconstruction. Plast Reconstr Surg. 111(1):67–78. https://doi.org/10.1097/01.PRS.0000037682.59058.6B

Baek SM, Biller HF, Krespi YP, Lawson W (1980 Aug) The lower trapezius Island myocutaneous flap. Ann Plast Surg 5(2):108–114. https://doi.org/10.1097/00000637-198008000-00004

Elshaer WM (2004) Extended lower trapezius Island myocutaneous flap in the repair of postburn axillary contracture. Plast Reconstr Surg. 113(7):2076–2081; discussion 2082-4. https://doi.org/10.1097/01.prs.0000121187.58166.62

Gao AB, Lv YC, Wang AP, Zhong LY, Tang ML, Thomas BP, Peng TH (2018) The zonal pattern of arterial supply to the brachial plexus and its clinical significance. Surg Radiol Anat 40(7):815–822. https://doi.org/10.1007/s00276-018-2024-2. Epub 2018 May 8

Gorospe L, Palomera-Rico A, María Muñoz-Molina G (2016) Delayed dorsal scapular artery pseudoaneurysm rupture after blunt chest injury. Asian Cardiovasc Thorac Ann 24(3):293–294. https://doi.org/10.1177/0218492314558520. Epub 2014 Nov 4

Haas F, Weiglein A, Schwarzl F, Scharnagl E (2004 May) The lower trapezius musculocutaneous flap from pedicled to free flap: anatomical basis and clinical applications based on the dorsal scapular artery. Plast Reconstr Surg. 113(6):1580–1590. https://doi.org/10.1097/01.prs.0000117188.03152.10

Hita-Contreras F, Sánchez-Montesinos I, Martínez-Amat A, Cruz-Díaz D, Barranco RJ, Roda O (2018) Development of the human shoulder joint during the embryonic and early fetal stages: anatomical considerations for clinical practice. J Anat. 232(3):422–430. https://doi.org/10.1111/joa.12753. Epub 2017 Nov 28. PMID: 29193070; PMCID: PMC5807935

Hong B, Oh C, Jo Y, Chung W (2021) Unexpected visualization of the dorsal scapular artery during supraclavicular block. Korean J Anesthesiol 74(6):561–563. https://doi.org/10.4097/kja.21193. Epub 2021 Jun 9. PMID: 34107208; PMCID: PMC8648505

Huelke DF (1958 Nov) A study of the transverse cervical and dorsal scapular arteries. Anat Rec 132(3):233–245. https://doi.org/10.1002/ar.1091320302

Huelke DF (1962 Jan) The dorsal scapular artery–a proposed term for the artery to the rhomboid muscles. Anat Rec 142:57–61. https://doi.org/10.1002/ar.1091420109

Ikka L, Mihalea C, Achour NB, Khalek HA, Vacher C (2016) The origin of the dorsal scapular artery: anatomic variations and surgical applications. Surg Radiol Anat 38(9):1021–1027. https://doi.org/10.1007/s00276-016-1656-3. Epub 2016 Feb 26

Kinjo S, Frankel A (2012) Failure of supraclavicular block under ultrasound guidance: clinical relevance of anatomical variation of cervical vessels. J Anesth

26(1):100–102. https://doi.org/10.1007/s00540-011-1252-y. Epub 2011 Oct 16

Kohli S, Yadav N, Prasad A, Banerjee SS (2014) Anatomic variation of subclavian artery visualized on ultrasound-guided supraclavicular brachial plexus block. Case Rep Med 2014:394920. https://doi.org/10.1155/2014/394920. Epub 2014 Jul 20. PMID: 25143765; PMCID: PMC4124782

Levy SM, Taylor GI, Baudet J, Guérin J, Casoli V, Pan WR, Houseman ND (2003 Dec) Angiosomes of the brachial plexus: an anatomical study. Plast Reconstr Surg 112(7):1799–1806. https://doi.org/10.1097/01.PRS.0000091162.25751.6B

Marr KL, Varga A, Dumestre DO (2021) The dorsal scapular artery perforator flap: a complete muscle-sparing technique. Plast Reconstr Surg Glob Open 9(11):e3933. https://doi.org/10.1097/GOX.0000000000003933. PMID: 34840920; PMCID: PMC8613356

Moore KL, Agur AMR, Dalley AF (2010) Clinically oriented anatomy. LWW ISBN:1451119453

Murata H, Sakai A, Hadzic A, Sumikawa K (2012) The presence of transverse cervical and dorsal scapular arteries at three ultrasound probe positions commonly used in supraclavicular brachial plexus blockade. Anesth Analg. 115(2):470–473. https://doi.org/10.1213/ANE.0b013e3182568557. Epub 2012 Apr 20. Erratum in: AnesthAnalg. 2012 Dec;115(6):1470

Nambyiah P, Umbarje K, Amir R, Parikh M, Oosthuysen SA (2011) Sonographic assessment of arterial frequency and distribution within the brachial plexus: a comparison with the cadaveric record. Anaesthesia 66(10):931–935. https://doi.org/10.1111/j.1365-2044.2011.06832.x.Epub

Ou KL, Dai YH, Wang HJ, Chen TM, Dai NT, Chang SC, Deng SC, Tzeng YS, Wang CH, Chen SG (2013 Dec) The lower trapezius musculocutaneous flap for head and neck reconstruction: two decades of clinical experience. Ann Plast Surg 71(Suppl 1):S48–S54. https://doi.org/10.1097/SAP.0000000000000036

Reiner A, Kasser R (1996 Feb) Relative frequency of a subclavian vs. a transverse cervical origin for the dorsal scapular artery in humans. Anat Rec 244(2):265–268. https://doi.org/10.1002/(SICI)1097-0185(199602)244:2<265::AID-AR14>3.0.CO;2-N

Rosko AJ, Ryan JT, Wizauer EJ, Ibrahim M, Srinivasan A, Stucken CL, Spector ME (2017) Dorsal scapular artery as a recipient vessel in the vessel-depleted neck during free tissue transfer in head and neck reconstruction. Head Neck 39(7):E72-E76. https://doi.org/10.1002/hed.24785. Epub 2017 Apr 28

Tan KC, Tan BK (2000 Apr) Extended lower trapezius Island myocutaneous flap: a fasciomyocutaneous flap based on the dorsal scapular artery. Plast Reconstr Surg. 105(5):1758–1763. https://doi.org/10.1097/00006534-200004050-00025

Tubbs RS, Tyler-Kabara EC, Aikens AC, Martin JP, Weed LL, Salter EG, Oakes WJ (2005 May) Surgical anatomy of the dorsal scapular nerve. J Neurosurg 102(5):910–911. https://doi.org/10.3171/jns.2005.102.5.0910

Tubbs RS, Tyler-Kabara EC, Salter EG, Sheetz J, Zehren SJ, Oakes WJ (2006) Additional vascular compression of the brachial plexus in a cadaver with a cervical rib: case illustration. Surg Radiol Anat 28(1):112–113. https://doi.org/10.1007/s00276-005-0048-x. Epub 2005 Oct 15

Urken ML, Naidu RK, Lawson W, Biller HF (1991 May) The lower trapezius Island musculocutaneous flap revisited. Report of 45 cases and a unifying concept of the vascular supply. Arch Otolaryngol Head Neck Surg 117(5):502–511. https://doi.org/10.1001/archotol.1991.01870170048012

Vacher C (2008) The osteo-muscular dorsal scapular (OMDS) flap. Anatomic basis of a new pedicled flap for mandibular reconstruction. Surg Radiol Anat 30(3):233–238. https://doi.org/10.1007/s00276-008-0316-7. Epub 2008 Feb 19

Verenna AA, Alexandru D, Karimi A, Brown JM, Bove GM, Daly FJ, Pastore AM, Pearson HE, Barbe MF. Dorsal scapular artery variations and relationship to the brachial plexus, and a related thoracic outlet syndrome case. J Brachial Plex Peripher Nerve Inj 2016 May 10;11(1):e21-e28. https://doi.org/10.1055/s-0036-1583756. PMID: 28077957; PMCID: PMC5152701

Weiglein AH, Moriggl B, Schalk C, Künzel KH, Müller U (2005 Nov) Arteries in the posterior cervical triangle in man. Clin Anat 18(8):553–557. https://doi.org/10.1002/ca.20143

Yang HJ, Lee DH, Kim YW, Lee SG, Cheon YW (2016) The trapezius muscle flap: a viable alternative for posterior scalp and neck reconstruction. Arch Plast Surg 43(6):529–535. https://doi.org/10.5999/aps.2016.43.6.529. Epub 2016 Nov 18. PMID: 27896183; PMCID: PMC5122541

89 The Axillary Artery

Ramesses Akamefula and Stephen J. Bordes, Jr.

89.1 Embryology

During fetal development, the seventh intersegmental artery gives rise to the axial artery (Moore and Persaud 1998). As the axial artery continues to grow distally, it gives rise to the subclavian, axillary, and anterior interosseous arteries before terminating in the hand (Khalid and Bordoni 2021). Around 33 days of development, as the axial trunk gives rise to the subclavian and axillary arteries and extends into the limb bud, it extends past the neural plate and develops into a network of capillaries. The subclavian artery forms ventral to the neural plate as the limb bud develops. The neural plate divides into anterior and posterior divisions around 37 days of development, the forming axillary artery coursing through the anterior division. As the axillary artery becomes better defined, around 41 days of embryo development, it continues distally as the brachial artery, extending to the elbow (Rodríguez-Niedenführ et al. 2001).

On the right side, normal embryogenesis of the cardiovascular system causes the distal segment of the right distal aorta to curve. Development of the aorta allows both the fourth vascular arch and the seventh intersegmental artery to give rise to the brachiocephalic trunk and subclavian artery. Only the subclavian artery develops as a product of the seventh intersegmental artery on the left side. As the lateral branch of the seventh intersegmental artery develops within the upper limb bud, this artery enlarges and forms much of the axial artery, which originates from its proximal portion (Moore and Persaud 1998).

89.2 Anatomy and Topography

The subclavian artery courses through the axilla and becomes the axillary artery once it reaches the lateral margin of the first rib (Fig. 89.1) (Uflacker and Guimaraes 2021). The axillary artery runs superomedially to inferolaterally through the axilla. Prior to its termination, marked by the inferior margin of the teres major muscle, it divides into three distinct parts, each giving rise to specific branches (Drake et al. 2020).

Part 1: The most proximal portion of the axillary artery is defined as the segment between the lateral margin of the first rib and the proximal edge of the pectoralis minor (Wind and James Valentine 2013). This portion is enclosed within a fibrous axillary sheath that extends from a prevertebral layer of cervical fascia and

R. Akamefula
Tulane University School of Medicine,
New Orleans, LA, USA
e-mail: rakamefula@tulane.edu

S. J. Bordes, Jr. (✉)
Department of Surgery, Louisiana State University
Health Sciences Center, New Orleans, LA, USA

S. J. Bordes, Jr. et al. (eds.), *The Clinical Anatomy of the Vascular System*,
https://doi.org/10.1007/978-3-031-78326-5_89

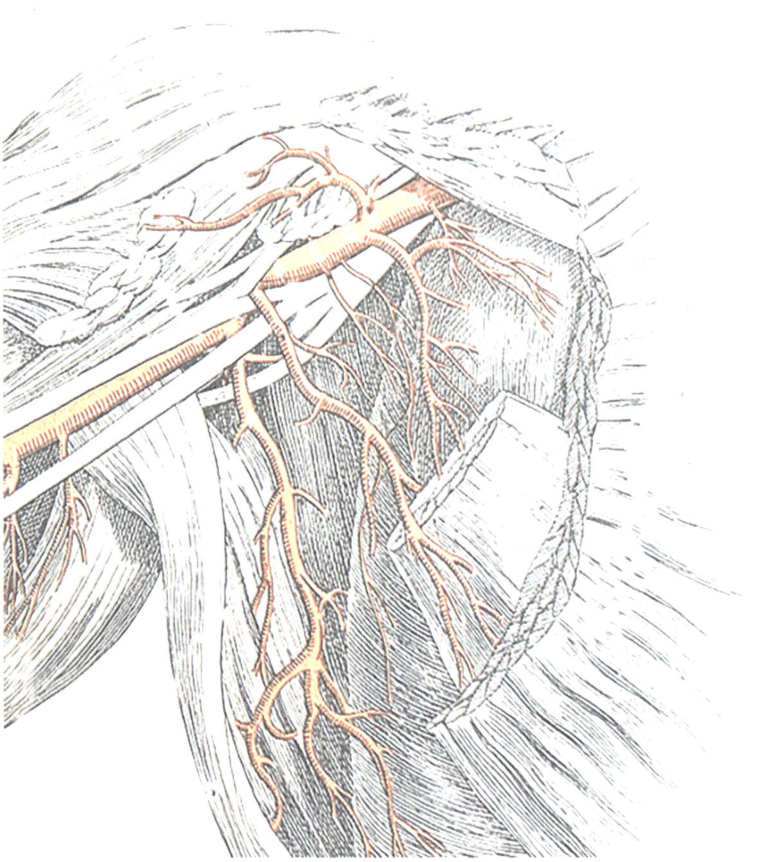

Fig. 89.1 Drawing illustrating the axillary artery and its branches. (After Allen 1882)

travels along with the axillary vein and the brachial plexus (Gray 1995). Branches of this portion include the superior thoracic artery.

Part 2: The second portion of the axillary artery lies directly posterior to the pectoralis major. Branches of this portion include the thoracoacromial artery and lateral thoracic artery.

Part 3: The most lateral portion is marked by the lateral margin of the pectoralis minor and the inferior margin of the teres major muscle. Branches of this portion include the subscapular, anterior circumflex humeral, and posterior circumflex humeral arteries.

89.3 Histology

Like other human arteries and veins, the axillary artery is considered to be a muscular distributing artery and comprises three layers. The layer closest to the lumen is the *tunica intima*, the layer furthest from the lumen is the *tunica adventitia*, and between those lies the *tunica media*.

89.4 Imaging

Imaging the axillary artery and its branches is needed to diagnose and plan treatment for patients experiencing vascular pathology. To evaluate the axillary artery accurately, a multimodality imaging approach is usual. Traditionally, digital subtraction angiography (DSA) has been the gold standard for imaging the axillary artery owing to its high spatial recognition and superior resolution, though other noninvasive forms of imaging such as ultrasound, color Doppler, CT angiography (CTA), MR angiography (MRA), and 3D CT are becoming more common for assessing vascular pathology. Because of technological improvements in CTA, MRA, and 3D CT, DSA is now used less frequently as a diagnostic imaging technique (Dave and Fleischmann 2016; Bozlar et al. 2013).

Deciding which imaging modality should be used for evaluating the axillary artery depends on variables such as urgency. CTA is typically indicated in a traumatic or emergency setting or in the event of an embolism. In nonemergency settings, MRA is preferred over CTA. An adequate combination of clinically related history and physical exam can also help to determine the best imaging technique for examining the axillary artery (Bannas et al. 2015; Hellinger et al. 2010).

Contraindications for using CTA to evaluate the axillary artery include increased exposure to iodinated contrast and radiation, leading some clinicians to lean towards MRA when CTA is not explicitly indicated or if the setting is semi-urgent (Ghouri et al. 2019).

89.5 Variations

Because of its involvement in complex arterial development, anastomosis, and periarticular networking, variations in the anatomy of the axillary artery are common (Ozan et al. 1994; Huelke 1959; Poynter 1920). Therefore, a strong understanding of deviations from the traditional native anatomy of this artery is crucial for clinicians, anatomists, and surgeons. These deviations include variations or duplication of the branching pattern and agenesis of any of the branches or even the axillary artery itself (Cavdar et al. 2000; Karlsson and Niechajev 1982; Kogan and Lewinson 1998). Moreover, variations in axillary artery development can include the persistence of vessels that usually obliterate or form through variant pathways (Collins 1995; Arey 1966).

Yang et al. examined the relationships between the origins of axillary artery branches and the branching patterns. Among the nine types of variant branching patterns, three main ones were observed. *Type A* included a common branch for the lateral thoracic and subscapular arteries; *Type B* included a common branching point for the anterior and posterior circumflex humeral arteries; and *Type C* included a common branching point for the thoracoacromial and lateral thoracic arteries. As the origins of the axillary artery branches were not consistent, Yang et al. proposed that their findings indicated simultaneous development with nonrandom branching patterns (Yang et al. 2021). Rarer variant branching and development patterns also include the axillary artery giving rise to the radial artery with a more proximal origin than in normal anatomy. Sometimes, the ulnar artery also derives directly from the axillary artery (Gupta et al. 2012; Chokkappan and Srinivasan 2015; Haładaj et al. 2018).

89.6 Pathology

It is essential to rule out vascular trauma whenever a patient presents with injury in the periclavicular or axial region. Hard indicators of vascular injury can include but are not limited to hematomas, absence of peripheral pulse, hemorrhage, and bruits upon auscultation (Demetriades and Asensio 2001). If severe enough, major injury to the axillary artery can result in limb loss owing to vessel rupture and subsequent thrombosis. Examples of injuries that can lead to severe axillary artery injury include shoulder dislocations, lacerations, and humeral neck fractures (Qi et al. 2014; Rosseykin et al. 2018).

Axillary Artery Aneurysm

Aneurysms of the axillary artery commonly arise from blunt trauma or penetration into the vessel itself. In nonacute settings, axillary artery aneurysms can evolve from atherosclerotic plaques over time or improper use of crutches. Such aneurysms can present with signs of neuropathology of the upper limb because of the close relationship between the axillary artery and the brachial plexus, occlusion of distal arteries such as the radial or ulnar arteries by emboli, and ischemic fingers (Thiel et al. 2021).

Axillary Artery Compression

In rare cases, overhead throwing athletes are prone to compression of the third part of the axillary artery from anterior displacement by the humerus. Constant pressure can lead to vascular insufficiency, potentially causing thrombosis or an aneurysm either at the site of the axillary artery or downstream. Axillary artery compression can present as numbness, unusual arm fatigue, ischemic pain, or cold intolerance following activity. Clinicians should also note that these symptoms do not always present immediately after the initial compression but in a chronic setting (Thiel et al. 2021).

89.7 Clinical Significance

Although injury to the axillary artery during surgical intervention is rare, such injuries can lead to significant complications such as ischemia, limb loss, and even death in the most extreme cases (Mazzini et al. 2011a). During surgical interventions, risk factors for axillary artery injury include atherosclerosis, male sex, fractures to the humerus, ribs, or scapula, shoulder dislocation, and brachial plexus injuries (Menendez et al. 2015).

Axillary artery injuries are most common during proximal humerus fractures (Menendez et al. 2015). Iatrogenic injuries to the axillary artery are also possible, especially during open shoulder surgery (Abbenhaus and Kamineni 2020). Other examples of iatrogenic axillary artery injury include acute thrombosis during humeral resurfacing or shoulder arthroplasty, avulsion of the artery during shoulder arthroplasty, delayed occlusion following shoulder arthroplasty, and direct trauma during drilling or dissection (Mehta et al. 2018; Wallace et al. 2011; Wilkerson et al. 2019). Special care should be taken of the axillary artery when the axilla is involved in a surgical setting; if an injury occurs, vascular surgery consultation is advised (Stone et al. 2021).

89.8 Surgical Considerations

It is vital to consider all of the related structures in the axilla when axillary artery injuries are evaluated. For example, the brachial plexus should be considered with both injury and surgical exposure of the axillary artery, as 33% of patients with axillary artery injuries also have associated brachial plexus injuries. In addition, it is essential to rule out injury to the thorax, such as pneumothorax, when patients present with axillary injuries (Makins 1919).

It is also paramount to assess the severity of axillary artery injuries to determine which diagnostic tests are indicated. Angiography is indicated if a patient with an axillary artery injury presents with "soft" injury signs such as slow bleeding, nerve injuries, diminished pulses, and stable hematomas. Imaging is not necessarily warranted in the unstable patient unless the origin of injury cannot be determined by any other diagnostic tool (Prichayudh et al. 2009; Degiannis et al. 1995). For example, these patients can present with multiple gunshot wounds or fractures with significant comminution resulting in severe bleeding, requiring urgent exploration.

Exposure

Regardless of the approach, exposure of the axillary artery requires careful dissection of the pectoralis major muscle and retraction of the pectoralis minor after detaching it from its insertion on the coracoid process and reflecting the muscle belly inferiomedially. Following exposure of the artery, both proximal and distal control should be established, and evaluation of the artery should commence. If it is injured, it is not uncommon for surgeons to debride and resect the injured area prior to surgical repair.

Because the axillary artery is anatomically situated posterior to the clavicle, direct compression of it to mitigate bleeding can be difficult. The use of a balloon tamponade with a Foley catheter can be beneficial in such cases. Excessive abduction of the arm during surgical exploration of the axillary artery should be avoided as it can result in a more complex exposure (Mazzini et al. 2011b).

Axillary Artery Cannulation

The right axillary artery has been used as a cannulation site in cardiac surgery. This is typically indicated during surgical procedures involving

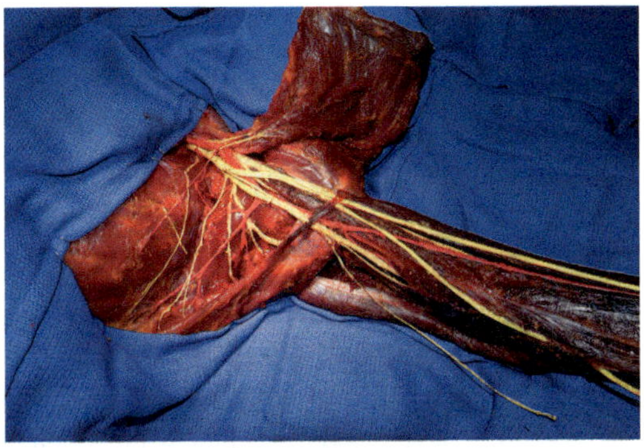

Fig. 89.2 Cadaveric dissection following reflection of the pectoralis major and minor and showing branches of the axillary artery and branches of the brachial plexus

the aortic root and aortic arch when standard ascending aorta cannulation is insufficient. Axillary artery cannulation promotes systemic blood flow from the heart to the limbs (Calvaruso et al. 2012). It also entails less risk of cerebral embolization than femoral cannulation (Neri et al. 1999) (Fig. 89.2).

Calvaruso et al. previously described indications and contraindications for axillary artery cannulation, which include (but are not limited to):

Indications

- Type A aortic dissection.
- Isolated aortic arch dissection.
- Circulatory arrest.
- Porcelain aorta.
- Ascending aortic aneurysm.

Contraindications

- Extension of aortic dissection into the subclavian artery.
- Axillary artery stenosis.
- Subclavian artery stenosis.
- High patient BMI making exposure more complicated.
- Atherosclerosis.

Axillary Artery Stenting

Although open surgical resection as a surgical modality in response to axillary artery aneurysm is often considered the typical approach, endovascular stenting has gained popularity as a feasible option. It has promising long-term outcomes and acceptable mortality rates (He et al. 2013; Epstein et al. 2017; Fokou et al. 2012). This rise in popularity could be attributed to the less invasive nature of endovascular stenting while allowing pathologies such as aneurysms to be repaired. Moreover, a less invasive approach can allow for more rapid recovery from surgical wounds than open resection (Gupta et al. 2005; Park et al. 2014; Schultz 2014).

Deciding whether to treat an axillary artery aneurysm via open repair or endovascular stenting depends on multiple factors, which can be either patient-focused or surgically focused (Esteves et al. 2013). Patient-focused considerations include surgical preference, associated comorbidities, and operative risk (Mees et al. 2013). Endovascular approaches are favored for elderly and high-risk patients; however, open approaches have better long-term outcomes and are usually preferred for younger, lower-risk patients. Surgical considerations include aneurysm size, variation, location, and surgeon familiarity with the operative technique. Depending on these considerations and the severity of the injury, endovascular stenting is not always sufficient to reduce the axillary artery and associated structures that require a longer stent, leading to stent migration or collapse (Mohan et al. 2016). Surgeons should also consider pathology related to a mobile shoulder joint. Common pathologies that put the axillary artery at risk include shoulder dislocation, shoulder separation, fractures, and compression of structures near or at the site of the thoracic outlet (Vijayvergiya et al. 2005).

Axillary Artery Bypass

Axillary-axillary or carotid-axillary bypass can be required in subclavian occlusion or steal syndrome cases. Axillofemoral bypass, an extra-anatomical repair, can also be considered in patients with distal aorto-iliac disease who are deemed too high risk for open aorto-femoral repair.

References

Abbenhaus E, Kamineni S (2020) Shoulder arthroplasty death with axillary artery and brachial 298 plexus damage: lessons from a tragedy. JSES Int 4:133–137. https://doi.org/10.1016/j.jses.2019.09.001

Arey LB (1966) Development of the arteries. In: Developmental anatomy, 7th edn. WB Saunders, Philadelphia \London, pp 371–374

Bannas P, Francois CJ, Reeder SB (2015) Magnetic resonance angiography of the upper extremity. Magn Reson Imaging Clin N Am 23:479–493. https://doi.org/10.1016/j.mric.2015.04.004

Bozlar U, Ogur T, Norton PT et al (2013) CT angiography of the upper extremity arterial system: part 1-anatomy, technique, and use in trauma patients. AJR Am J Roentgenol 201:745–752

Calvaruso D, Voisine P, Mohammadi S, Dumont E, Dagenais F (2012) Axillary artery cannulation. Multimedia manual of cardiothoracic surgery: MMCTS, 2012, mms004

Cavdar S, Zeybek A, Bayramicli M (2000) Rare variation of the axillary artery. Clin Anat 13:66–68

Chokkappan K, Srinivasan S (2015) Superficial brachioradial artery (arising from axillary artery) – a variant that can cause failure of percutaneous coronary intervention. J Vasc Access 16(5):e95–e96

Collins P (1995) Embryonic circulation. In: Bannister LH, Berry MM, Collins P, Dyson M, Dussek JE, Ferguson MWJ (eds) Gray's Anatomy, 38th edn. Churchill Livingstone, Edinburgh, pp 318–319

Dave RB, Fleischmann D (2016) Computed tomography angiography of the upper extremities. Radiol Clin North Am 54:101–114. https://doi.org/10.1016/j.rcl.2015.08.008

Degiannis E, Levy RD, Potokar T, Saadia R (1995) Penetrating injuries of the axillary artery. Aust NZ J Surg 65(5):327–330

Demetriades D, Asensio JA (2001) Subclavian and axillary vascular injuries. Surg Clin North Am 81:1357–1373. xiii

Drake RL et al (2020) Gray's anatomy for students, 4th edn. Elsevier, Philadelphia. Print

Epstein DA, Ibrahim MT, Franz RW (2017) Endovascular treatment of massive axillary artery aneurysm for palliation. Vasc Endovasc Surg 51:188e90

Esteves FP, Ferreira AV, Santos VP et al (2013) Subclavian and axillary arterial aneurysms: two case reports. J Vasc Brasileiro 12:329e34

Fokou M, Eyenga VC, Guifo ML et al (2012) Giant pseudoaneurysm of the left axillary artery following a stab wound: case report-online article. Cardiovasc. J Afr 23:4e6

Ghouri MA, Gupta N, Bhat AP, Thimmappa ND, Saboo SS, Khandelwal A, Nagpal P (2019) CT and MR imaging of the upper extremity vasculature: pearls, pitfalls, and challenges. Cardiovasc Diagn Ther 9(Suppl 1):S152–S173

Gray H (1995) Axillary artery. In: Bannister LH, Berry MM, Collins P, Dyson M, Dussek JE, Ferguson MWJ (eds) Gray's anatomy, 38th edn. Churchill Livingstone, London, pp 1537–1538

Gupta S, Lee DC, Goldstein RS et al (2005) Axillary artery aneurysm. J Emerg Med 28:215e6

Gupta G, Singh K, Chhabra S, Gupta V, Kayalvizhi I (2012) Bilateral superficial ulnar artery with high origin from the axillary artery: its anatomy and clinical significance. Folia Morphol (Warsz) 71(1):48–51

Haładaj R, Wysiadecki G, Dudkiewicz Z, Polguj M, Topol M (2018) The high origin of the radial artery (brachioradial artery): its anatomical variations, clinical significance, and contribution to the blood supply of the hand. Biomed Res Int 2018:11

He C, Wu X, Cao J et al (2013) Endovascular management of spontaneous axillary artery aneurysm: a case report and review of the literature. J Med Case Rep 7:140

Hellinger JC, Epelman M, Rubin GD (2010) Upper extremity computed tomographic angiography: state of the art technique and applications in 2010. Radiol Clin North Am 48:397–421, ix. https://doi.org/10.1016/j.rcl.2010.02.022

Huelke DF (1959) Variation in the origins of the branches of the axillary artery. Anat Rec 35:33–41

Karlsson S, Niechajev IA (1982) Arterial anatomy of the upper extremity. Acta Radiol Diagn (Stockh) 23:115–121

Khalid N, Bordoni B (2021) Embryology, great vessel. [Updated 2021 May 8]. In: StatPearls [Internet]. StatPearls Publishing, Treasure Island (FL)

Kogan I, Lewinson D (1998) Variation in the branching of the axillary artery: a description of a rare case. Acta Anat (Basel) 162:238–240

Makins GW (1919) Gunshot injuries to the blood vessels. John Wright & Sons, Bristol

Mazzini FN, Vu T, Prichayudh S, Sciarretta JD, Chandler J, Lieberman H et al (2011a) Operative 329 exposure and management of axillary vessel injuries. Eur J Trauma Emerg Surg 37:451. https://doi.org/10.1007/s00068-011-0134-1

Mazzini FN, Vu T, Prichayudh S, Sciarretta JD, Chandler J, Lieberman H, Marini C, Asensio JA (2011b) Operative exposure and management of axillary vessel injuries. Eur J Trauma Emerg Surg 37(5):451

Mees B, Robinson D, Verhagen H et al (2013) Non-aortic aneurysms: natural history and recommendations for referral and treatment. Aust Fam Physician 42:370
Mehta VM, Mandala CL, Shriver RJ, Shah R (2018) Proximal humeral fracture-dislocation with 335 axillary artery involvement treated with reverse shoulder arthroplasty. In JSES Open Access 336(2):100–103
Menendez ME, Ring D, Heng M (2015) Proximal humerus fracture with injury to the axillary 339 artery: a population-based study. Injury 46:1367–1371. https://doi.org/10.1016/j.injury.2015.04.026
Mohan S, Schanzer A, Robinson WP et al (2016) Endovascular management of radiation-induced subclavian and axillary artery aneurysms. J Vasc Surg 64:1135e7
Moore KL, Persaud TVN (1998) The cardiovascular system. In: The developing human, clinically oriented embryology, 6th edn. WB Saunders, Philadelphia, pp 335–341
Neri E, Massetti M, Capannini G, Carone E, Tucci E, Diciolla F et al (1999) Axillary artery cannulation in type a aortic dissection operations. J Thorac Cardiovasc Surg 118:324–329
Ozan H, Simsek C, Onderoglu S, Kirici Y, Basar R (1994) High division of the axillary artery: a rare case of superficial ulnar artery. Acta Anat (Basel) 151:68–70
Park SK, Hwang JK, Park SC et al (2014) Endovascular treatment of a spontaneous aneurysm in the axillary artery. Interact Cardiovasc Thorac Surg 20:140e2
Poynter CWM (1920) Congenital anomalies of the arteries and veins of the human body with bibliography. Univ Stud Univ Nebraska 22:1–106
Prichayudh S, Verananvattna A, Sriussadaporn S, Kritayakirana K, Pak-art R, Capin A et al (2009) Management of upper extremity vascular injury: outcome related to the mangled extremity severity score. World J Surg 33(4):857–863
Qi F, Li J, Qi X, Xiao LW (2014) Judgment of defect length of extremities artery trauma and reconstruction. Zhongguo Gu Shang 27(3):199–202
Rodríguez-Niedenführ M, Burton GJ, Deu J, Sañudo JR (2001) Development of the arterial pattern in the upper limb of staged human embryos: normal development and anatomic variations. J Anat 199(Pt 4):407–417. https://doi.org/10.1046/j.1469-7580.2001.19940407.x
Rosseykin E, Kobzev E, Bazylev V (2018 Sep) One more method for arterial cannulation in aortic arch surgery ("Penza cannulation"). Asian Cardiovasc Thorac Ann 26(7):584–586
Schultz T (2014) Axillary artery aneurysm. Sonography 1:49e51
Stone MA, Ihn HE, Gipsman AM, Iglesias B, Minneti M, Noorzad AS, Omid R (2021) Surgical anatomy of the axillary artery: clinical implications for open shoulder surgery. J Shoulder Elb Surg 30(6):1266–1272
Thiel R, Munjal A, Daly DT (2021) Anatomy, shoulder and upper limb, axillary artery. [Updated 2021 Jul 26]. In: StatPearls [Internet]. StatPearls Publishing, Treasure Island (FL)
Uflacker A, Guimaraes M (2021) Uflacker's atlas of vascular anatomy : an angiographic approach, 3rd edn. Wolters Kluwer, Philadelphia. Print
Vijayvergiya R, Kumar RM, Ranjit A et al (2005) Endovascular management of isolated axillary artery aneurysm: a case report. Vasc Endovasc Surg 39:199e201
Wallace WA, Kalogrianitis S, Manning P, Clark D, McSweeney S (2011) The elderly fragile axillary artery – management lessons learned from four serious iatrogenic injuries. Orthop Proc 93-B:295–295. https://doi.org/10.1302/0301-620X.93BSUPP_III.0930295a
Wilkerson J, Napierala M, Shalhub S, Warme WJ (2019) Axillary artery intimal dissection with thrombosis and brachial plexus injury after reverse total shoulder arthroplasty. J Shoulder Elbow Surg 28:e393–e397. https://doi.org/10.1016/j.jse.2019.07.044
Wind GG, James Valentine R (2013) Anatomic exposures in vascular surgery, 3rd edn. Wolters Kluwer Health/Lippincott Williams & Wilkins, Philadelphia. Print
Yang K, Lee H, Choi IJ, Jeong W, Kim HT, Wei Q, Lee JH (2021) Topography and anatomical variations of the axillary artery. Biomed Res Int 2021:6393780

Superior Thoracic Artery 90

Ramesses Akamefula and Stephen J. Bordes, Jr.

90.1 Embryology

Around the fourth week of fetal development, the upper limb bud arises from a communication between the lateral plate mesoderm and ectoderm. One of the structures that develops from this interaction is the axial trunk, followed by the axillary artery. Like the other branches of the axillary artery, the superior thoracic artery (also known as the *supreme thoracic artery* and *highest thoracic artery*) arises from the vascular plexus that is subsequently derived from the developing upper limb bud. The axillary artery and its branches are typically formed by day 41 of fetal development (Rodríguez-Niedenführ et al. 2001; Arora and Sapra 2022).

Hox gene expression has been associated with the positioning of limb buds, including the upper limb bud. Irregular expression of Hox can alter the development of the upper limb bud and, subsequently, any structure-derived downstream, including vascular plexuses and the superior thoracic artery. This can lead to anatomical variations (Iimura and Pourquié 2007).

R. Akamefula
Tulane University School of Medicine, New Orleans, LA, USA
e-mail: rakamefula@tulane.edu

S. J. Bordes, Jr. (✉)
Department of Surgery, Louisiana State University Health Sciences Center, New Orleans, LA, USA

90.2 Anatomy and Topography

The axillary artery comprises three distinct portions. The superior thoracic artery is the only branch to arise from the first portion (Fig. 90.1). The landmarks signifying the boundaries of this

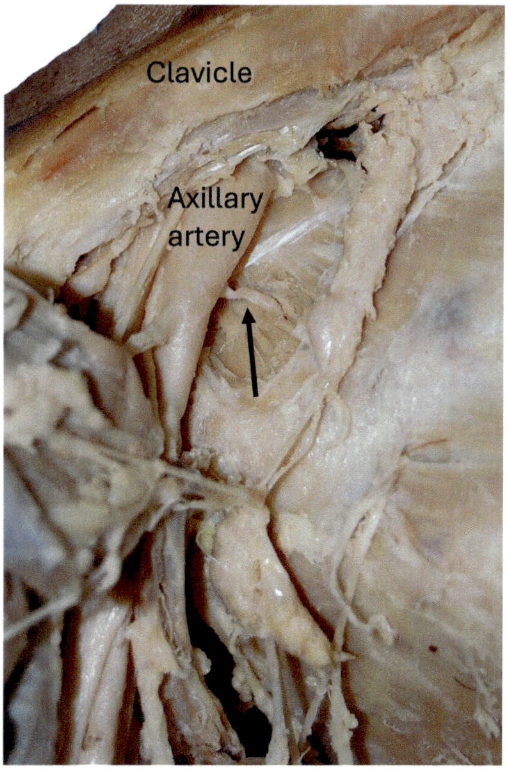

Fig. 90.1 Cadaveric dissection of the right axillary artery noting its first branch, the superior thoracic artery (arrow)

S. J. Bordes, Jr. et al. (eds.), *The Clinical Anatomy of the Vascular System*,
https://doi.org/10.1007/978-3-031-78326-5_90

portion of the axillary artery are the distal portion of the first rib (medial border) and the proximal portion of the pectoralis minor (lateral border). The superior thoracic artery travels anteromedially along the superior border of the pectoralis minor and inferior to the subclavius muscle before anastomosing with the anterior intercostal arteries and the internal thoracic artery (Wind and James Valentine 2013).

The superior thoracic artery supplies blood to the first and second intercostal spaces and, therefore, to the associated intercostal muscles. In some individuals, it also supplies blood to the pectoralis major and pectoralis minor. Terminal branches of this artery have also been seen to supply blood to the superior portion of the serratus anterior muscle. Through its anastomoses, it communicates directly and indirectly with the internal thoracic artery and the anterior intercostal arteries of the first and second intercostal spaces (Shahoud et al. 2021).

90.3 Histology

Like the axillary artery from which it arises, the superior thoracic artery comprises three layers. The innermost layer closest to the lumen is the *tunica intima*; the outermost layer, furthest from the lumen, is the *tunica adventitia*; between those two layers is the *tunica media*.

90.4 Imaging

Proper imaging is needed to correctly identify injury or pathology of the superior thoracic artery and plan treatment. Although it is routine to obtain chest radiographs when evaluating the thorax, the current gold standard of imaging the superior thoracic artery is CT angiography (CTA). Clinically, involvement of the superior thoracic artery is best indicated using CT with contrast, which can illustrate a hematoma posterior to the clavicle (Wong et al. 2013). It should also be noted that considerable technological advances have been made with ultrasound, color Doppler, MR angiography (MRA), and 3D CT. Thanks to these improvements, a multimodality imaging approach for evaluating vascular pathology has become common (Dave and Fleischmann 2016; Bozlar et al. 2013).

Urgency can influence the choice of imaging technique. In many emergency settings, such as severe trauma or embolism, CTA is typically indicated. When the setting is nonurgent, MRA can be preferred over CTA. Supplemental information from a physical exam and a detailed patient history can help clinicians decide the most appropriate imaging technique for evaluating the superior thoracic artery (Bannas et al. 2015; Hellinger et al. 2010).

90.5 Variations

Many anatomical variations can arise during the development of the superior thoracic artery. One of the most common variants includes this artery branching off the second portion of the axillary artery instead of the first. Specifically, this variation consists of the superior thoracic artery branching off the thoracoacromial artery (Aastha et al. 2015). Also, variations in the order of axillary artery branches are not uncommon (Bagoji et al. 2013). In a case report, Yursel et al. mentioned the discovery of the lateral thoracic artery, which typically branches distally from the axillary artery relative to the superior thoracic artery, branching before the latter. Moreover, it is common for other arteries to branch off the first portion of the axillary artery in conjunction with the superior thoracic artery (Yuksel and Yuksel 1997).

Another variation includes a trajectory by which the superior thoracic artery courses through the brachial plexus. Aastha et al. illustrated this variation in a case report, specifically reporting the superior thoracic artery coursing through the divisions of the lateral cord and then towards the first and second intercostal spaces (Aastha et al. 2015).

Functionally, variations of the superior thoracic artery do not affect the blood supply to the intercostal muscles, the pectoralis major or minor, or the serratus anterior. Variations, therefore, tend to be discovered during the dissection of cadaveric specimens (Arora and Sapra 2022). However, a comprehensive understanding of superior thoracic artery variation is advised in the event of injury, especially in cases of iatrogenic injury, penetrating trauma, or shoulder dislocation.

90.6 Clinical Significance

Injury to the superior thoracic artery and related structures is rare. If it does occur, it can have devastating results and has been linked to increased mortality and morbidity. Patients can present with pleuritic chest pain, tachycardia, hypotension, tachypnea, or hemothorax/pneumothorax.

Injury to the superior thoracic artery is commonly caused by penetrating trauma to the neck, chest, or axillary region. Fourteen percent of cases involving superior thoracic injury involve concomitant injury to the clavicle or the first rib (Wong et al. 2013). This suggests that injury to the superior thoracic artery is not uncommon when there is trauma to the ribs or clavicles and should be evaluated in such cases. Associated structures should also be considered when injury to this artery is evaluated. These include the axillary artery and brachial plexus. Anterior shoulder dislocations can also put the superior thoracic artery at risk for injury.

Endovascular Intervention

Endovascular interventions can also place the superior thoracic artery at risk for injury. This is especially emphasized during processes such as sheath insertion and stent placement; injury from this procedure has been noted in the literature (Uchikado et al. 2017).

Coronary Artery Bypass

Coronary bypass surgery and other open-heart procedures can also put the superior thoracic artery at risk. It is not uncommon to extract the internal thoracic artery as a graft in bypass surgery. In this setting, it is crucial to clip the anastomosing arteries of the internal thoracic artery properly, including the superior thoracic artery. If it is clipped improperly, postoperative complications such as hemothorax or steal syndrome can result can result (Dar et al. 2017).

90.7 Surgical Considerations

Because the superior thoracic artery is small, iatrogenic injuries during surgical intervention in areas close to it are rare. However, special care should be taken during surgical interventions in the neck, chest, and axilla to avoid superior thoracic artery injury.

References

Aastha, Jain A, Kumar MS (2015) An unusual variation of axillary artery: a case report. J Clin Diagn Res 9(1):AD05–AD7. https://doi.org/10.7860/JCDR/2015/11680.5477

Arora Y, Sapra A (2022) Anatomy, thorax, superior thoracic arteries. [Updated 2021 Jul 31]. In: StatPearls [Internet]. StatPearls Publishing, Treasure Island (FL)

Bagoji IB, Hadimani GA, Bannur BM, Patil BG, Bharatha A (2013 Dec) A unique branching pattern of the axillary artery: a case report. J Clin Diagn Res 7(12):2939–2940

Bannas P, Francois CJ, Reeder SB (2015) Magnetic resonance angiography of the upper extremity. Magn Reson Imaging Clin N Am 23:479–493. https://doi.org/10.1016/j.mric.2015.04.004

Bozlar U, Ogur T, Norton PT et al (2013) CT angiography of the upper extremity arterial system: part 1-anatomy, technique, and use in trauma patients. AJR Am J Roentgenol 201:745–752

Dar MI, Dar AH, Bilal M, Ahmad M, Haseeb A (2017 Aug 20) Association of Internal Mammary Artery Flow with different comorbidities and post-coronary artery bypass graft complications. Cureus 9(8):e1584

Dave RB, Fleischmann D (2016) Computed tomography angiography of the upper extremities. Radiol Clin North Am 54:101–114. https://doi.org/10.1016/j.rcl.2015.08.008

Hellinger JC, Epelman M, Rubin GD (2010) Upper extremity computed tomographic angiography: state of the art technique and applications in 2010. Radiol Clin North Am 48(397–421):ix. https://doi.org/10.1016/j.rcl.2010.02.022

Iimura T, Pourquié O (2007 May) Hox genes in time and space during vertebrate body formation. Develop Growth Differ 49(4):265–275

Rodríguez-Niedenführ M, Burton GJ, Deu J, Sañudo JR (2001 Oct) Development of the arterial pattern in the upper limb of staged human embryos: normal development and anatomic variations. J Anat 199(Pt 4):407–417

Shahoud JS, Kerndt CC, Burns B (2021) Anatomy, thorax, internal mammary (internal thoracic) arteries. StatPearls Publishing

Uchikado Y, Yoshino S, Ohmure K, Tabata H, Tateishi S, Ohishi M (2017 Nov) Angle-tipped guidewire-induced vascular perforation at branch of superior thoracic artery during sheath insertion: case report. J Cardiol Cases 16(5):158–161

Wind GG, James Valentine R (2013) Anatomic exposures in vascular surgery, 3rd edn. Wolters Kluwer/Lippincott Williams & Wilkins, Philadelphia. Print

Wong VW, Gordy SD, Schreiber M, Tieu BH (2013) Penetrating neck injury to the superior thoracic artery managed by video-assisted thoracoscopic surgery. Case Rep Surg 2013:413462

Yuksel M, Yuksel E (1997) Anomalous branching order of the superior and lateral thoracic arteries. Clin Anat 10(6):394–396

91 Thoracoacromial Artery

Brian Le and Angel Castro

91.1 Embryology

The axillary arteries begin to form around the fourth week of embryo development in the early upper limb paddles. Starting from a primitive capillary, a network enlarges into a central axial artery and later contributes to the three axillary artery segments (Carlson 2014). In the second segment, the thoracoacromial artery forms proximally to the lateral thoracic artery. It arises anteriorly at the superior border of the pectoralis minor. Variations of branching are thought to result from inherited variability or de novo mutations during embryo development. Overall, the causes are numerous, and our understanding of them is limited (Carlson 2014).

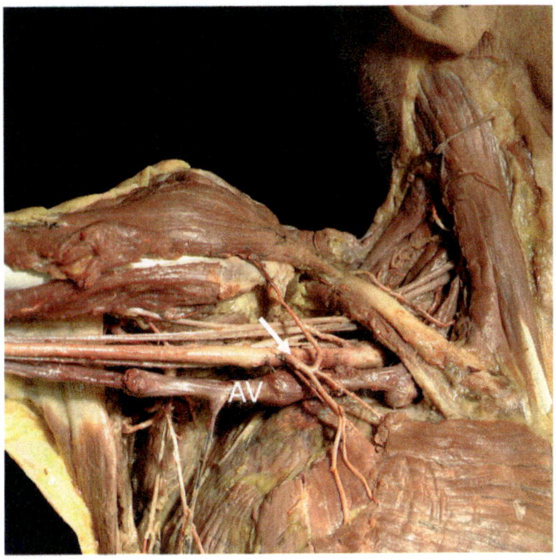

Fig. 91.1 Cadaveric dissection of the right thoracoacromial artery (arrow). Note the axillary vein (AV)

91.2 Anatomy/Topography

The thoracoacromial artery (Fig. 91.1) branches off the second segment of the axillary artery, posterior to the superior border of the pectoralis minor at the level of the suprascapular notch. It courses inferiorly in the subclavicular region before it divides into four terminal branches: pectoral, deltoid, acromial, and clavicular.

The pectoral branch travels medially and inferiorly, descending between the pectoralis major and minor. It bifurcates and supplies the pectoralis minor on the anterior surface while piercing the posterior surface of the sternoclavicular region of the pectoralis major. Its anterior portion divides further into three secondary branches: The lateral branch anastomoses with the lateral thoracic artery, and the medial and caudal branches medially with the internal thoracic artery. Finally, the pectoral branch gives off per-

B. Le (✉)
Tulane School of Medicine, New Orleans, LA, USA
e-mail: ble3@tulane.edu

A. Castro
Washington University School of Medicine,
St. Louis, MO, USA
e-mail: a.h.castro@wustl.edu

S. J. Bordes, Jr. et al. (eds.), *The Clinical Anatomy of the Vascular System*,
https://doi.org/10.1007/978-3-031-78326-5_91

forating musculocutaneous branches that supply the breast skin and lateral region of the nipples.

The deltoid branch courses laterally from the thoracoacromial artery over the pectoralis minor and supplies both the pectoralis major and deltoid muscles. It also typically gives rise to the acromial branch of the thoracoacromial artery.

The acromial branch courses superiorly over the coracoid process of the scapula to supply the deltoid muscle. When it emerges from the deltoid, it forms an anastomosis with the posterior circumflex humeral artery and the deltoid branch of the thoracoacromial artery.

The clavicular branch terminates superiorly to supply the subscapular muscle and the sternoclavicular joint.

91.3 Imaging/Histology

The thoracoacromial artery and its branches are often visualized for preoperative planning in cases such as perforator flaps for reconstructive surgery. A high-frequency linear ultrasound probe or contrast-enhanced magnetic resonance angiography (CE-MRA) can be used for such visualization.

- Ultrasound.
  - The perforator vessels are visualized using an ultrasound Doppler approach. A linear high-frequency ultrasound probe is ideal for noninvasive vascular imaging. Vessel size and flow rates can be approximated by applying continuous color Doppler flow detection to the vessels of interest. Color Doppler flow can also locate the perforator vessels, which are used as thin flaps for face and neck reconstruction (Portenard et al. 2019). The perforator vessels used for pectoral branch flap pedicles are around 6-8 cm long in cadavers (Zhang et al. 2013), similar to the findings from ultrasound studies of live patients (Kompatscher et al. 2005).
- CE-MRA.
  - Magnetic resonance imaging is performed on the desired vessel with parametric contrast (typically gadolinium). Imaging depends on capturing the first pass of the contrast through the artery for optimal visualization of the artery. Therefore, cardiac output and fluid volume status are major determinants of contrast circulation and image quality. Upper limb visualization with CE-MRA is more challenging than lower limb imaging because the venous circulation of contrast is faster (arterial-venous time window). Resolution can be improved by optimizing the variables of IV contrast delivery, such as flow rate, contrast dose, and saline flush volume. Injection rates of 2–3 ml/s are sufficient for first pass visualization; the greatest signal intensity is obtained at faster injection rates of 5 ml/s, but there is signal degradation at rates above 6 ml/s. It is worth noting that the subclavian region is susceptible to MRA artifacts, resulting in an artifactual finding of subclavian artery stenosis.

91.4 Variations

Least Variable: Thoracoacromial Trunk, Pectoral and Deltoid Branches

The thoracoacromial trunk is absent in a small number of cases (8.8%), the branches arising from the axillary artery (Astik and Dave 2012). Although the thoracoacromial artery is universally characterized by its four terminal branches, only the pectoral and deltoid are consistently found in most patients. The pectoral branch is the most commonly seen and least variable (Kompatscher et al. 2005). The second most prevalent is the deltoid branch, which is known to give rise to the acromial branch of the thoracoacromial artery in some cases. On occasion, the artery branches inversely from the acromial branch.

Most Variable: Acromial and Clavicular Branches

There is great variability in the thoracoacromial artery's acromial and clavicular branches. The clavicular branch is the smallest and the most frequently absent (Astik and Dave 2012; Nyemb et al. 2020). The acromial branch typically arises from the deltoid branch but is known to give rise, inversely, to the deltoid branch. Phillippe also noted that it was absent in half of their dissections. This contradicts a study of 13 cases where the acromial branch had variable origin but was consistently present with robust blood flow for perforator flaps (Portenard et al. 2019).

Nonclassical Branches

The axillary artery is highly variable owing to a host of developmental factors that promote frequent nonclassical anastomoses of the common regional arteries. In several large studies in India, Korea, and the US, the lateral thoracic artery was found to be a branch of the thoracoacromial artery. This was the most prevalent morphological variant, found in 67.7% of cases (Kocsis et al. 2020). In other cases, radial arteries branch off the thoracoacromial artery (Loukas et al. 2005; Lee et al. 2015).

91.5 Pathology

The thoracoacromial artery has rarely been studied as a primary shoulder pathology source. Because of its robust blood supply and anastomoses with other vessels including the internal thoracic, lateral thoracic, posterior humeral circumflex, and internal mammary arteries, and their perforators, it functions as part of a network of robust collateral blood supply that supports the tissues of the shoulder and pectoral region.

91.6 Clinical and Surgical Significance

Since 1979, the branches of the thoracoacromial artery have been studied and used in pectoralis major myocutaneous flaps for reconstructive head and neck surgeries. These flaps have been used to reconstruct areas in which patients had either recurrent or double primary cancers of the head and neck and for radical mastectomies.

The intercostal perforator of the internal mammary gland is the most commonly used musculocutaneous flap, but the thoracoacromial artery is often studied as an alternative. It has a smaller diameter than the internal mammary artery with a more restricted blood flow, which can lead to complications if it is used as a flap, such as vessel size mismatch and thrombotic events. The thoracoacromial artery flap is a desirable alternative if a thinner flap is required, the pectoral flap is needed for another use, there is medial chest scarring, or the internal mammary artery is absent (Xin; Zhang et al. 2013).

Breast Reconstruction

The thoracoacromial artery can serve as a recipient vessel for microsurgeons performing breast reconstruction using a deep inferior epigastric perforator flap. It was found to have outcomes comparable to the internal mammary artery flap, with less pain in recovery.

Head, Neck, and Shoulder Reconstruction

The thoracoacromial artery has also been used as a propeller flap in reconstructing a high-voltage burn wound with an exposed shoulder.

One drawback of using the thoracoacromial artery flap is that it takes longer to isolate its pedicle vessels than the internal mammary flap. The variability of the perforating branches of the thoracoacromial artery occasionally necessitates isolation of the smaller vessel proximal towards the trunk during flap preparation, which can prove more tedious than the larger internal mammary artery (Iida and Yoshimatsu 2019; Zhang et al. 2013; Onoda et al. 2011).

91.7 Complications

Thrombosis is a known complication when the thoracoacromial artery flap is used as an alternative to the internal mammary artery. The perforating vessels of the thoracoacromial artery are thinner than the internal mammary artery, so blood flow is slower, and the risk of postoperative thrombosis is higher. In Onoda's study, two out of eight patients developed microvascular thrombosis from risk factors that were atypical of head and neck surgeries. Both patients had histories of radiation therapy and prior surgeries. Still, Onoda emphasized that education of postoperative staff concerning the applied medical devices and pressure on the anastomosed vessels over the clavicle should be reviewed (Onoda et al. 2011).

References

Astik R, Dave U (2012) Variations in the branching pattern of the axillary artery: a study in 40 human cadavers. J Vasc Bras 11(1):12–17. https://doi.org/10.1590/S1677-54492012000100003

Carlson BM (2014) Human embryology and developmental biology, 5th edn. Elsevier/Saunders, Philadelphia

Iida T, Yoshimatsu H (2019) Anatomical study and clinical application of free thoracoacromial artery true-perforator flap for reconstruction of the face. J Craniofac Surg 30(1):205–207. https://doi.org/10.1097/SCS.0000000000004968

Kocsis L, Harşa M, Dénes L, Pap Z (2020) An anatomical variation of terminal branches of the thoracoacromial artery–case report. J Interdiscip Med 5(3):110–113. https://doi.org/10.2478/jim-2020-0018

Kompatscher P, Manestar M, Schuster A, Lang A, Beer GM (2005) The thoracoacromial vessels as recipient vessels in microsurgery and supermicrosurgery: An anatomical and sonographic study. Plast Reconstr Surg 115(1):77–83. https://doi.org/10.1097/01.PRS.0000146034.16752.68

Lee H, Moon Y, Park HS, Kim H, Choi I (2015) A radial artery originating from the thoracoacromial artery. Surg Radiol Anat 38(5):631–633. https://doi.org/10.1007/s00276-015-1576-7

Loukas M, Louis RG Jr, Almond J, Armstrong T (2005) A case of an anomalous radial artery arising from the thoracoacromial trunk. Surg Radiol Anat 27(5):463–466. https://doi.org/10.1007/s00276-005-0028-1

Nyemb PMM, Fontaine C, Duquennoy-Martinot V, Demondion X (2020) Anatomical study of the acromial branch of the thoracoacromial artery summary. MOJ Anat Physiol 7:43–49

Onoda S, Sakuraba M, Asano T et al (2011) Thoracoacromial vessels as recipients for head and neck reconstruction and cause of vascular complications. Microsurgery 31(8):628–631. https://doi.org/10.1002/micr.20947

Portenard A, Auquit-Auckbur I, Gardeil L, Elbaz B, Carricaburu A, Duparc F (2019) Anatomical study of the perforator flap based on the acromial branch of the thoraco-acromial artery (abTAA flap): a cadaveric study. Surg Radiol Anat 41(11):1361–1367. https://doi.org/10.1007/s00276-019-02322-8

Wright AE, Wall M, Slaney P, Downing R (2014) An unusual presentation of thoracoacromial artery pseudoaneurysm following shoulder arthroplasty. BMJ Case Rep 2014:bcr2014204625. https://doi.org/10.1136/bcr-2014-204625

Zhang YX, Yongjie H, Messmer C et al (2013) Thoracoacromial artery perforator flap: Anatomical basis and clinical applications. Plast Reconstr Surg. 131(5):759e–770e. https://doi.org/10.1097/PRS.0b013e3182865bf5

Lateral Thoracic Artery

92

Davis Martin, Brandon Carroll, Jonathan Legier, and Stephen J. Bordes, Jr.

92.1 Embryology

The embryology and development of the arteries in the thorax and upper limb regions are quite complex but well understood. The lateral thoracic artery (LTA), the focus of this chapter, is no exception.

The left and right subclavian arteries arise from the primordial left seventh and right seventh intersegmental arteries, respectively (Stojanovska et al. 2012). Once the subclavian arteries are developed, they continue, now named the axillary arteries as they cross the lateral margin of the first rib on either side of the body. As outlined in many anatomy textbooks throughout the past two centuries, the LTA (or long thoracic artery) develops as a branch from the second part of the axillary artery, which is designated as just medial to the border of the pectoralis minor (Abrahams et al. 2008; Agur and Lee 1999).

D. Martin · B. Carroll · J. Legier
School of Medicine, Louisiana State University Health Sciences Center, New Orleans, LA, USA
e-mail: dmar19@lsuhsc.edu; bcarr8@lsuhsc.edu; jlegie@lsuhsc.edu

S. J. Bordes, Jr. (✉)
Department of Surgery, Louisiana State University Health Sciences Center, New Orleans, LA, USA

92.2 Anatomy

The LTA (Fig. 92.1) typically arises from the second part of the axillary artery just medial to the pectoralis minor. Specifically, Tinhofer et al. found that on average, its origin lies 1–3 cm medial to the pectoralis minor, about halfway between the acromion and the jugular

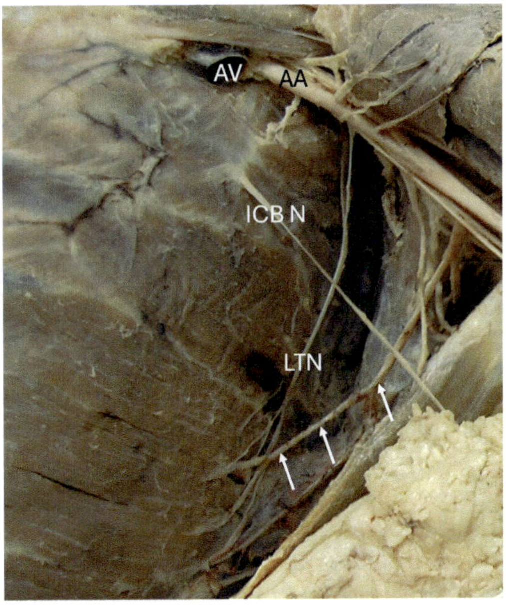

Fig. 92.1 Cadaveric dissection of the left axilla. The lateral thoracic artery is seen at the arrows. Note the axillary artery (AA), axillary vein (AV), intercostobrachial nerve (ICBN), and long thoracic nerve (LTN). (Courtesy Prof. M. Loukas)

S. J. Bordes, Jr. et al. (eds.), *The Clinical Anatomy of the Vascular System*,
https://doi.org/10.1007/978-3-031-78326-5_92

notch. However, recent studies have suggested extensive variation in its origin, with the possibility that in most individuals, it originates from the thoracoacromial artery (Loukas et al. 2014).

The LTA traverses the lateral border of the pectoralis minor in a ventromedial direction deep to the pectoralis major muscle and on top of the serratus fascia (Fong et al. 2016). It courses the serratus anterior fascia, but its course is unrelated to the long thoracic nerve (Tinhofer et al. 2017). This path continues, giving off multiple unnamed branches, until it terminates around the fifth intercostal space at the inferolateral border of pectoralis major (Fong et al. 2016; De Garis and Swartley 1928). Typically, the LTA gives off both cutaneous branches that wrap around the pectoralis major to supply the skin of the breast region and deep branches supplying the lymph nodes, the serratus anterior, pectoralis major and minor, and the subscapularis (Tinhofer et al. 2017).

The muscular or deep branches are given off in the first 3–4 cm of the artery's course before it continues down the lateral aspect of the pectoralis minor (McCulley et al. 2015). It also gives off perforating cutaneous branches just above the level of the nipple, relevant in many surgical flaps (McCulley et al. 2015). Finally, in females, the LTA supplies the superolateral breast parenchyma by giving off lateral mammary branches that wrap around the lateral portion of the pectoralis major, anastomosing with branches from the internal thoracic artery (Jesinger 2014). Because it supplies the breast parenchyma and is a major contributor to the blood supply to the nipple-areolar complex, it is often much larger in females than males (Loukas et al. 2014); its average mean external diameter is roughly 2 ± 0.50 mm (Fong et al. 2016; Tinhofer et al. 2017).

Although the exact origin of the LTA varies significantly, as discussed in detail in the following section, its overall path and the structures it supplies remain fairly constant in the vast majority of individuals (Loukas et al. 2014).

92.3 Variation

Although the most common origin of the LTA is the second part of the axillary artery, numerous patterns of variation have been described in the literature. Some sources state that an individual can either lack an LTA altogether or have multiple LTAs branching from the axillary artery, but the occurrence rates are less than 5% (Loukas et al. 2014; De Garis and Swartley 1928; Huelke 1959). Other studies have described the LTA arising from either the first or third part of the axillary artery; however, this pattern is comparatively rare (Huelke 1959; Daimi et al. 2010; Bergman et al. 1988). Most important are those patterns of variation concerning an alternative branch point within the second part of the axillary artery. While most sources agree on the most common branching variations, they largely disagree about their prevalence in the general population. The most common branching patterns will be discussed below in order of their relative frequency.

The most common variant is an LTA originating from the thoracoacromial artery. The prevalence of this branching pattern ranged from 67.2% (Stojanovska et al. 2012) in one study to 29.3% (right side)—39.8% (left side) (Pandey and Shukla 2004) in another. DeGaris and Swartley found the LTA to arise from the thoracoacromial artery in 43% of their subjects (De Garis and Swartley 1928). Interestingly, DeGaris and Swartley also compared the relative branching patterns between black and white individuals and found that an LTA branching from the thoracoacromial artery was the most common branching pattern in African Americans, whereas branching directly from the axillary artery was most common in white individuals (De Garis and Swartley 1928). Other studies found this branching pattern in less than 15% of their samples, demonstrating the extensive variability within this one branching pattern (P'an 1940; Pellegrini 1906; Trotter et al. 1930). Given the large number of individuals with this branching pattern, further research is needed to determine the most common origin of the LTA (Loukas et al. 2014).

The next most common branching pattern is the LTA arising from the subscapular or the thoracodorsal artery. The incidence of this branching pattern ranges from 14% to 25%, depending on source (Huelke 1959; P'an 1940; Trotter et al. 1930). Other studies found lower prevalence rates, in the range 5–10% (Loukas et al. 2014; Olinger and Benninger 2010). Overall, this branching pattern appears to be much less prevalent than the LTA originating from the thoracoacromial artery.

While these three branching patterns are the most commonly cited, many others with far lower prevalence rates have been observed. DeGeris and Swartley described over 15 unique patterns among the 500 cadavers they examined, showing the true extent of variability in the LTA's origin from the axillary artery (De Garis and Swartley 1928). Overall, the most common origin appears to be either from the second part of the axillary artery or the thoracoacromial artery. Regardless of the origin, more than 90% of individuals had LTAs that supplied the same structures and maintained a consistent course through the body (Loukas et al. 2014). The only exceptions to this rule are the rare cases in which individuals presented with no LTA or multiple LTAs.

92.4 Pathology

The LTA delivers blood from the axillary artery to several muscles in the thoracic cage as it arborizes to send perforating branches supplying the superolateral portion of the breast parenchyma. The vessel can be occluded by atherosclerotic calcification, including Monckeberg medial calcific sclerosis (Jesinger et al. 2011). Risk factors for this condition include aging, chronic hypertension, diabetes, and hyperlipidemia. Severe blockage of the artery can prevent perfusion to the lateral thoracic cage (Jesinger et al. 2011; Kemmeren et al. 1998). This condition can be revealed during mammography as parallel opaque lines following the vessel and is more common in postmenopausal women. Breast atherosclerosis can also be associated with increased risk of coronary artery disease (Moshyedi et al. 1995).

Necrosis is another possible pathology due to occlusion of the LTA. The artery can be blocked by trauma, radiation therapy, embolus, or malignant mass expansion. Examples of mass expansions include fibroadenomas, lactating adenomas, or hamartomas (Jesinger et al. 2011). Infarction of the LTA can also lead to necrosis of the breast, nipple-areolar complex, or fat in the area.

True aneurysms involving the LTA are rare diagnoses that usually result from prior trauma. They present as slowly enlarging pulsatile masses that can be viewed as lumps in the breast tissue. Pseudoaneurysms due to spontaneous hemorrhage can occur in the acute trauma setting. Hemorrhage of the LTA can be caused by coagulopathies, hypertension, or breast malignancy. The diagnoses of both aneurysms and pseudoaneurysms are confirmed by Doppler ultrasound (Jesinger et al. 2011; Dehn and Lee 1986).

92.5 Clinical and Surgical Significances/Complications

The LTA has been extensively documented for its use in plastic surgery and reconstructive flaps. Initially described by Ariyan as supplying additional circulation to the pectoralis major myocutaneous flap (PMMF) supplementing the thoracoacromial artery, the LTA's importance in perfusing the medial and lateral portions of the muscle has been highlighted in later studies (Ariyan 1979; Moloy and Gonzales 1986). The LTA's significance should be noted as a substantial number of pectoralis major muscular and myocutaneous flaps develop complications, though complete flap loss is rare (Pellegrini 1906). The PMMF is used primarily in head and neck reconstruction, especially involving cancer treatment, and remains relevant as an alternative to free flap surgery in those areas when applicable (Dedivitis and Guimarães 2002; Aničin et al. 2015; Liu et al. 2015; Meier et al. 2021; Sen et al. 2019; Pradhan et al. 2019).

The LTA is significant as part of the lateral thoracic perforator flap (LTPF), which can serve as an alternative to the latissimus dorsi perforator flap and the thoracodorsal perforator flap (Kim

2005; Kim et al. 2011). The LTPF is an important option for axillary reconstructions in patients with burns and hidradenitis suppurativa (Dietrich et al. 2021; Baghaki et al. 2015). It has also been used as a potential perforator flap in breast reconstructions following breast cancer (McCulley et al. 2015; Mangialardi et al. 2021). The LTA is an important potential supplier of the nipple-areola complex (NAC) (van Deventer 2004). Therefore, it is notable in various breast reduction and mastopexy procedures as part of a pedicle to prevent nipple necrosis (Başaran et al. 2011; van Deventer and Graewe 2010).

Because of the aforementioned high likelihood of variation concerning the LTA, caution is required during operations to ensure the integrity of the vessel.

References

Abrahams PH, Boon JM, Spratt JD (2008) McMinn's clinical atlas of human anatomy, 6th edn. Elsevier, p 145

Agur AMR, Lee MJ (1999) Grant's atlas of anatomy, 10th edn. Lippincott Williams and Wilkins, p 6

Aničin A, Šifrer R, Strojan P (2015) Pectoralis Major Myocutaneous Flap in Primary and Salvage Head and Neck Cancer Surgery. J Oral Maxillofac Surg 73(10):2057–2064

Ariyan S (1979) The pectoralis major myocutaneous flap. A versatile flap for reconstruction in the head and neck. Plast Reconstr Surg 63(1):73–81

Baghaki S, Cevirme M, Diyarbakirli M, Tatar C, Aydin Y (2015) Locoregional use of lateral thoracic artery perforator flap as a propeller flap. Ann Plast Surg 74(5):532–535

Başaran K, Ucar A, Guven E, Arinci A, Yazar M, Kuvat SV (2011) Ultrasonographicallydetermined pedicled breast reduction in severe gigantomastia. Plast Reconstr Surg 128(4):252e–259e

Bergman R, Thompson S, Afifi A, Saadeh F (1988) Compendium of human anatomic variation. Urban and Schwarzenberg. Inc., Baltimore, p 9

Daimi SR, Siddiqui AU, Wabale RN (2010) Variations in the branching pattern of axillary artery with high origin of radial artery. Int J Anat Var 3:76–77

De Garis CF, Swartley WB (1928) The axillary artery in white and Negro stocks. Am J Anat 41(2):353–397

Dedivitis RA, Guimarães AV (2002) Pectoralis major musculocutaneous flap in head and neck cancer reconstruction. World J Surg 26(1):67–71

Dehn TC, Lee EC (1986) Aneurysm presenting as a breast mass. Br Med J (Clin Res Ed) 292(6530):1240

Dietrich S, Reumuth G, Kuentscher M, Panchulidze I (2021) Perforator based flaps of lateral thoracic artery (LTAP) and thoracodorsal artery (TDAP) for axillary reconstruction in patients with hidradenitis suppurativa. Handchir Mikrochir Plast Chir 53(4):370–375

Fong HC, Goh T, Chui CHK, Rasheed MZ, Ong YS (2016) The lateral thoracic vessels: a useful recipient in immediate breast reconstruction. J Reconstr Microsurg Open 1(01):002–007

Huelke DF (1959) Variation in the origins of the branches of the axillary artery. Anat Rec 135:33

Jesinger RA (2014) Breast anatomy for the interventionalist. Tech Vasc Interv Radiol 17(1):3–9

Jesinger RA, Lattin GE Jr, Ballard EA, Zelasko SM, Glassman LM (2011) Vascular abnormalities of the breast: arterial and venous disorders, vascular masses, and mimic lesions with radiologic-pathologic correlation. Radiographics 31(7):E117–E136

Kemmeren JM, van Noord PA, Beijerinck D, Fracheboud J, Banga JD, van der Graaf Y (1998) Arterial calcification found on breast cancer screening mammograms and cardiovascular mortality in women: The DOM project. Doorlopend Onderzoek Morbiditeit en Mortaliteit. Am J Epidemiol 147(4):333–341

Kim JT (2005) Two options for perforator flaps in the flank donor site: latissimus dorsi and thoracodorsal perforator flaps. Plast Reconstr Surg 115(3):755–763

Kim JT, Ng SW, Naidu S, Kim JD, Kim YH (2011) Lateral thoracic perforator flap: additional perforator flap option from the lateral thoracic region. J Plast Reconstr Aesthet Surg 64(12):1596–1602

Liu T, Zhang H, Xiong J, Yi S, Gu L, Zhou M (2015) Inhibition of MDM2 homodimerization by XIAP IRES stabilizes MDM2, influencing cancer cell survival. Mol Cancer 14(1):65

Loukas M, du Plessis M, Owens DG, Kinsella CR Jr, Litchfield CR, Nacar A, Lu O, Tubbs RS (2014) The lateral thoracic artery revisited. Surg Radiol Anat 36(6):543–549

Mangialardi ML, Baldelli I, Salgarello M, Raposio E (2021) Breast Reconstruction Using the Lateral Thoracic, Thoracodorsal, and Intercostal Arteries Perforator Flaps. Plast Reconstr Surg Glob Open 9(1):e 3334

McCulley SJ, Schaverien MV, Tan VKM, Macmillan RD (2015) Lateral thoracic artery perforator (LTAP) flap in partial breast reconstruction. J Plast Reconstr Aesthet Surg 68(5):686–691

Meier JK, Spoerl S, Spanier G, Wunschel M, Gottsauner MJ, Schuderer J, Reichert TE, Ettl T (2021) Alternatives to free flap surgery for maxillofacial reconstruction: focus on the submental Island flap and the pectoralis major myocutaneous flap. BMC Oral Health 21(1):198

Moloy PJ, Gonzales FE (1986) Vascular anatomy of the pectoralis major myocutaneous flap. Arch Otolaryngol Head Neck Surg 112(1):66–69

Moshyedi AC, Puthawala AH, Kurland RJ, O'Leary DH (1995) Breast arterial calcification: association with coronary artery disease. Work in progress. Radiology 194(1):181–183

Olinger A, Benninger B (2010) Branching patterns of the lateral thoracic, subscapular, and posterior circumflex humeral arteries and their relationship to the posterior cord of the brachial plexus. Clin Anat 23(4):407–412

P'an M-T (1940) The origin of branches of the axillary artery in the Chinese. Am J Phys Anthrop 27:269–279

Pandey SK, Shukla VK (2004) Anatomical variation in origin and course of the thoracoacromial trunk and its branches. Nepal Med Coll J 6(2):88–91

Pellegrini A (1906) Le arteriae subclavia e axillaris nell'uomo studiate col metodo statistico. Arch Ital Anat Embriol 5(205):466

Pradhan P, Samal S, Preetam C (2019) Pectoralis Major Myocutaneous Flap for the Reconstruction of the Palatal Defect. Indian J Otolaryngol Head Neck Surg 71(1):132–135

Sen S, Gajagowni JG, Pandey JK, Dasgupta P, Sahni A, Gupta S, Mp S, Ravi B (2019) Effectiveness of pectoralis major myocutaneous flap in the surgical management of oral cancer: a retrospective study. J Stomatol Oral Maxillofac Surg 120(1):21–27

Stojanovska J, Cascade PN, Chong S, Quint LE, Sundaram B (2012) Embryology and imaging review of aortic arch anomalies. J Thorac Imaging 27(2):73–84

Tinhofer IE, Meng S, Steinbacher J, Roka-Palkovits J, Györi E, Reissig LF, Cheng MH, Weninger WJ, Tzou CH (2017) The surgical anatomy of the vascularized lateral thoracic artery lymph node flap—A cadaver study. J Surg Oncol 116(8):1062–1068

Trotter M, Henderson J, Gass H, Brua R, Weisman S, Agress H, Curtis G, Westbrook E (1930) The origins of branches of the axillary artery in whites and in American negroes. Anat Rec 46(2):133–137

van Deventer PV (2004) The blood supply to the nipple-areola complex of the human mammary gland. Aesth Plast Surg 28(6):393–398

van Deventer PV, Graewe FR (2010) Enhancing pedicle safety in mastopexy and breast reduction procedures: the posteroinferomedial pedicle, retaining the medial vertical ligament of Würinger. Plast Reconstr Surg 126(3):786–793

Subscapular Artery 93

Jeremy Lalla

93.1 Embryology

The peripheral vasculature starts to develop in the embryo around day 26. A capillary plexus forms dorsal to the aorta as the primitive limb bud develops. Around day 28, this plexus differentiates into the early proximal parts of the upper limb vasculature. By day 33, the subclavian and axillary arteries start to form. At this time, the nerves of the upper limb enter the limb bud. Around day 37, the initial nerve plate divides; the anterior division later gives rise to the musculocutaneous, ulnar, and median nerves, while the posterior division becomes the radial nerve. This time period determines the topography of the branches of the axillary artery relative to the brachial plexus surrounding it. By day 41, the axillary artery has continued distally and become the brachial artery.

93.2 Anatomy, Topography, and Histology

The subscapular artery (Fig. 93.1) branches from the third part of the axillary artery near the lateral border of the pectoralis minor and the inferior border of the subscapularis muscle. As a muscular artery, it has a well-developed tunica media with smooth muscles and limited elastic fibers, but histological examination nevertheless reveals a prominent internal elastic lamina. In one study (Jesus et al. 2008), the subscapular artery was consistent in 96.7% of the cadavers examined. It is generally described as the common trunk of the circumflex scapular and thoracodorsal arteries. This arrangement was found 82.6% of the time in a cadaver study. It has the largest lumen of any branch of the axillary artery with an average caliber of ~5.0 mm and an average length of ~18.0 mm (The subscapular artery and the thoracodorsal branch: an anatomical study). The third part of the axillary artery lies most commonly between the lateral and medial cords of the brachial plexus while remaining anterior to the posterior cord. The subscapular artery, which branches from the inferior part of the axillary artery, crosses anterior to the medial cutaneous nerve of the arm and posterior to the ulnar nerve.

The subscapular artery often gives off a collateral subscapular branch to the subscapularis muscle at its lateral border before dividing into its terminal branches inferiorly. Its larger branch, the thoracodorsal artery, continues inferiorly, giving off collateral branches to the subscapularis and serratus anterior muscles and terminal branches to the latissimus dorsi and serratus anterior muscles. It commonly runs with the thoracodorsal nerve along the anterolateral border of the scapula inferiorly towards the latissimus dorsi.

J. Lalla (✉)
Tulane University School of Medicine,
New Orleans, LA, USA
e-mail: jlalla@tulane.edu

S. J. Bordes, Jr. et al. (eds.), *The Clinical Anatomy of the Vascular System*,
https://doi.org/10.1007/978-3-031-78326-5_93

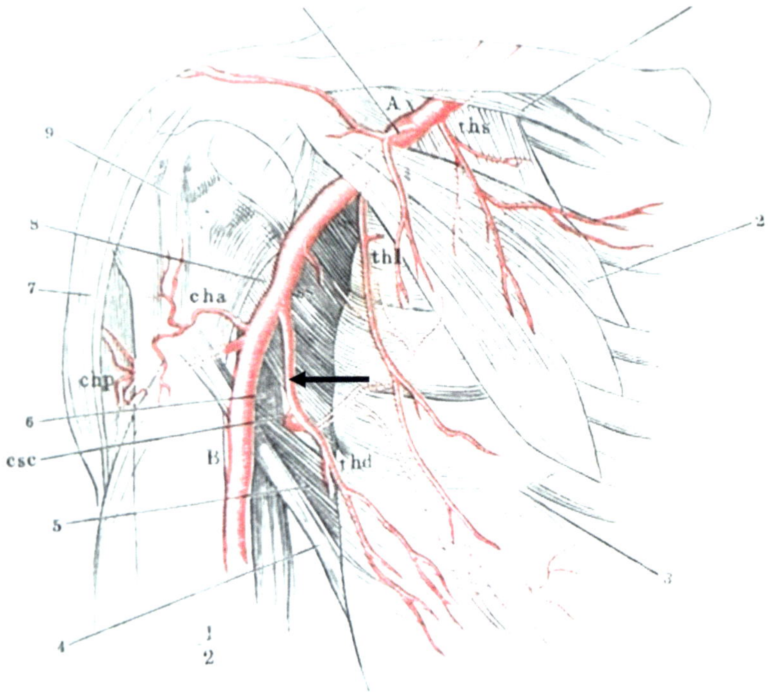

Fig. 93.1 Drawing of the axillary artery and its branches including the subscapular artery (arrow). (After Allen 1892)

The smaller of the two branches, the circumflex scapular artery, travels more laterally. It has an average caliber of 1.5–3.0 mm the artery has an important landmark; it passes anteriorly through a triangular space bounded by the teres minor superiorly, the teres major and subscapularis muscles inferiorly, and the long head of the triceps laterally. As the circumflex scapular artery emerges posteriorly from the triangular space, it curves around the lateral portion of the scapula. The vessel then gives small branches to supply the posterior portion of the deltoid muscle, part of the long head of the triceps brachii, and the teres minor. The circumflex scapular artery and the posterior circumflex humeral artery often supply an outer portion of the glenoid labrum.

93.3 Variations

The subscapular artery is a highly variable branch of the axillary artery. Standard anatomy describes it as a common trunk that gives rise to the thoracodorsal and circumflex scapular arteries. Yamada described two variations of the subscapular artery (Aizawa et al. 1995). One is a common trunk arising from the axillary artery proximal to the point at which the axillary artery dives between the lateral and medial cords of the brachial plexus. In this case, the subscapular artery crosses superiorly over the medial cord, arising from a common trunk with the lateral thoracic artery. Yamada termed this variation "A. subscapularis superficialis." In Yamada's second variation, the subscapular artery follows the course mentioned in the above topography section, arising distally to the point where the axillary artery dives between the lateral and medial cords.2 This is referred to as a deep subscapular artery.

More recent cadaveric dissections led to three different classifications based on the branching of the subscapular artery in the deep axial region: S-Type, I-Type, and P-type. The S-Type classification is almost identical to the "A. subscapularis superficialis" described by Yamada (Aizawa et al. 1995); the subscapular artery arises from a common trunk containing the lateral thoracic artery proximal to the point at which the axillary artery dives between the lateral and medial cords of the brachial plexus. The I-Type follows the second variation described by Yamada, the subscapular trunk arising distal to that point. In this classification, the subscapular artery often gives rise to the circumflex scapular artery. The thoracodorsal artery can be its own branch of the axillary or can be associated with the lateral thoracic artery. The circumflex humeral can also be a branch of the circumflex scapular in the I-type arrangement. The P-type also follows the second variation described by Yamada, but the branching arises more distally than in the I-type and is more closely associated with the circumflex humeral arteries.

One study of 83 cadavers (166 axillae) quantified some of the more common branching patterns (Olinger and Benninger 2010). It showed the thoracodorsal artery arising from the lateral thoracic artery in 7.2% of cadavers. A lateral thoracic artery gave off a subscapular trunk of the thoracodorsal and circumflex scapular arteries in 5.4%. A subscapular artery gave rise to the lateral thoracic in 4.2%. A subscapular artery gave off the circumflex humeral artery in 12%.

93.4 Clinical and Surgical Significance

There are multiple clinically significant aspects of the subscapular artery and its branches, a major one being an anastomosis between a branch of the axillary artery and a branch of the subclavian artery via the scapular anastomosis. This anastomosis is formed on the posterior side of the scapula between the circumflex scapular artery on the inferolateral border of the scapula, the suprascapular artery just inferior to the spinous process, and the dorsal scapular artery (deep branches of the transverse cervical when dorsal scapular is not present) at the medial border of the scapula. It creates a path through which the upper limb can remain perfused if the distal subclavian or proximal axillary artery is occluded. Another important clinical feature is the collateral supply to the muscles perfused by the subscapular artery, which makes the subscapular artery and/or its branches prime targets for surgical flaps. Both the circumflex scapular artery and thoracodorsal arteries can be used for pedicle flaps. The circumflex scapular artery pedicle flap is commonly used in breast reconstruction following mastectomy.

The thoracodorsal artery is commonly used in free flap surgeries in which the perforating arteries and thoracodorsal artery are removed as a segment with the myocutaneous region they supply. Because of the thoracodorsal artery length and the size of the latissimus dorsi, this particular flap can vary greatly and is a versatile and favorable flap target. If the underlying muscle is not required, a fasciocutaneous flap can be harvested using the perforating arteries of the thoracodorsal artery, the deep fascia, and overlying tissue. This flap is widely used because of its thinness and large area. Both the circumflex scapular and thoracodorsal arteries can be harvested in an osseocutaneous flap, which contains bone as well as soft tissue. This flap is often used in facial reconstructions involving the mandible.

References

Aizawa Y, Ohtsuka K, Kumaki K (1995) Examination of the courses of the arteries in the axillary region I. The course of the subscapular artery system, especially the relationships between the arteries and the posterior cord of the brachial plexus. Kaibogaku Zasshi 70:554–568

Jesus RC, Lopes MCH, Demarchi GTS et al (2008) The subscapular artery and the thoracodorsal branch: an anatomical study. Folia Morphol (Warsz) 67:58–62

Olinger A, Benninger B (2010) Branching patterns of the lateral thoracic, subscapular, and posterior circumflex humeral arteries and their relationship to the posterior cord of the brachial plexus. Clin Anat 23:407–412

94 Thoracodorsal Artery

Jessica Dang

94.1 Embryology

The vascular system starts to develop during the fourth to fifth week of gestation, when six pairs of primitive pharyngeal arch arteries form. The fourth right pharyngeal arch generates the proximal part of the right subclavian artery, while the seventh intersegmental artery forms its distal part. The fourth left pharyngeal arch contributes to the entire left subclavian artery. The seventh intersegmental artery, derived from the dorsal aorta, contributes to the formation of the axial arteries bilaterally, which then contribute to the subclavian artery. Owing to its complex development, the axillary artery can have various origins. A few of the named vessels can arise as common trunks, or their branches can branch separately from the axillary artery. The embryological variants of the axillary artery and subsequently the thoracodorsal artery become significant as these vessels have implications for reconstructive surgery that will be explored later in this chapter (Kumar et al. 2018).

J. Dang (✉)
Tulane University School of Medicine,
New Orleans, LA, USA
e-mail: jdang3@tulane.edu

94.2 Anatomy and Topography

The thoracodorsal artery (Fig. 94.1) is one of two terminal branches of the subscapular artery, which arises from the third part of the axillary artery. Its length averages 8.5 cm but can range from 6.5 to 12 cm, and its average diameter is 3 mm with a range of 2–4 mm. As it arises from the subscapular artery and courses from the axilla, the thoracodorsal artery travels inferiorly along the serratus anterior muscle and posterior to the lateral thoracic artery. It then travels along the anterior border of the latissimus dorsi muscle, entering this muscle from beneath. At a clearly defined point 2.5 cm (range 1–4 cm) medial to the lateral edge of the latissimus dorsi muscle and approximately 4 cm (range 3–6 cm) distal to the inferior scapular border, it commonly divides into two major muscular branches at the undersurface of the latissimus dorsi: the transverse or horizontal branch, and the lateral or vertical branch (Heitmann et al. 2003a). These branches travel from the deep surface of the muscle and become intramuscular.

The transverse branch usually runs parallel to the free upper border of the latissimus dorsi, 3.5 cm from its edge, and the vertical branch usually runs towards the iliac crest, 2.1 cm from the lateral edge of the muscle (O'Connell et al. 2017). Both branches distribute several cutaneous perforator arteries (Kumar et al. 2018). Most of the thoracodorsal cutaneous perforators originate within 8 cm distal to the branching

S. J. Bordes, Jr. et al. (eds.), *The Clinical Anatomy of the Vascular System*,
https://doi.org/10.1007/978-3-031-78326-5_94

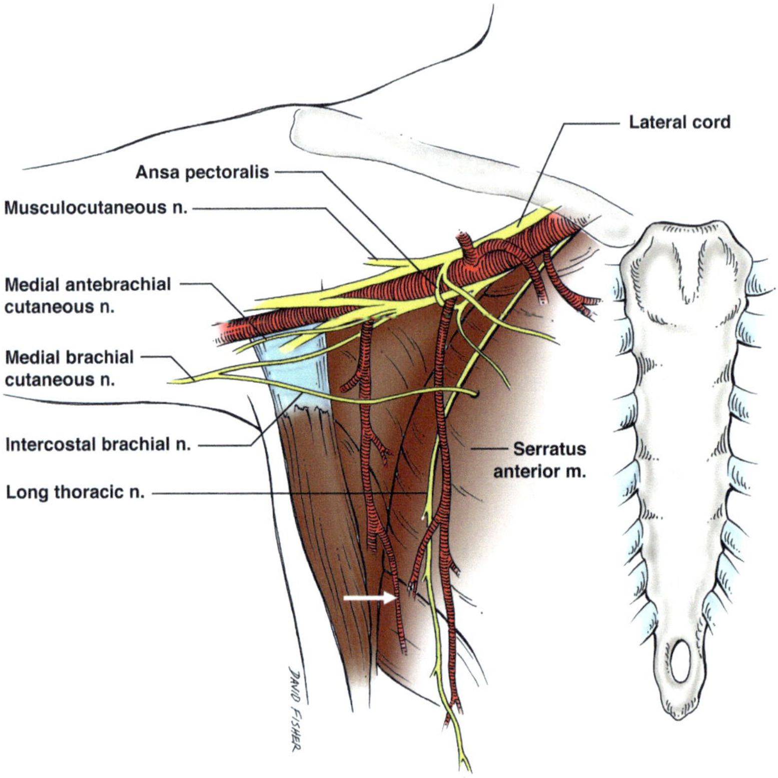

Fig. 94.1 Schematic drawing of the axillary artery and its branches and relations to the branches of the brachial plexus. Note the thoracodorsal artery (arrow) branching from the subscaular artery as its terminal branch

point and have clinical implications for thoracodorsal artery perforator flap reconstructive surgeries (Heitmann et al. 2003a).

The thoracodorsal artery provides the main arterial blood supply to the latissimus dorsi muscle; however, along its course, it also supplies minor branches to the axillary skin and to the serratus anterior, the subscapularis, and the teres major muscles (Germann and Reichenberger 2009). It gives off 1–3 branches of about 1 mm diameter to the serratus anterior, penetrating the muscle at its center along the course of the long thoracic nerve. The thoracodorsal artery usually terminates around the inferior angle of the scapula by dividing into 2–4 branches that supply a large portion of the latissimus dorsi (Germann and Reichenberger 2009).

94.3 Imaging

Knowledge of the anatomy of the thoracodorsal artery, its course, and its variations is beneficial in providing the most detailed illustrations for reconstructive procedures, namely the thoracodorsal artery perforator (TDAP) flap, a type of pedicled flap frequently used in breast reconstruction post-radical mastectomies (Sjøberg et al. 2020). Different imaging modalities have been used to evaluate the anatomy of this artery, such as Doppler, MR angiography, and multidetector computed tomography angiographies (MD-CTA) (Xie et al. 2018; Şentürk et al. 2015). Unwanted vascular injuries can be avoided during surgical and other clinical procedures if any vascular anomaly is identified in advance. Preoperative imaging can also provide information about the location and quality of perforators. Such information can simplify the surgical procedure and reduce the operating time.

Anatomical techniques, including a combination of vascular injection studies and three-dimensional (3D) computerized angiography, have provided the most detailed vascular anatomy of the upper extremity (Morris et al. 2010).

94.4 Variations

While variants of the thoracodorsal artery are rare overall, the most common ones involve it (1) as a branch of the lateral thoracic artery or (2) arising from a common trunk of the axillary artery along with the lateral thoracic, circumflex scapular, and occasionally circumflex humeral arteries (Germann and Reichenberger 2009; Rowsell et al. 1984). More commonly, the lateral thoracic artery arises from the thoracodorsal artery. Occasionally, the thoracodorsal artery originates directly from the axillary artery. Variations in the branching of the axillary artery, mainly involving its lateral thoracic, circumflex humeral, and subscapular branches, are frequently encountered and reported.

Rowsell et al. examined 100 dissections of the subscapular thoracodorsal arterial system in 58 cadavers, 10 fresh and 48 preserved. The thoracodorsal artery proved to be a branch of the subscapular artery in 94% of the dissections and a direct branch of the axillary artery in 5%. In four of these direct branches, three originated from the second part and one from the third portion of the axillary artery (Rowsell et al. 1984). The thoracodorsal artery also gives a cutaneous branch in about 75% of cases before it penetrates and supplies the latissimus dorsi. The cutaneous branch arises 0.5–2 cm beyond the bifurcation of the subscapular artery (Germann and Reichenberger 2009; Heitmann et al. 2003b).

In more than 90% of cases, the thoracodorsal artery bifurcates into the medial and the slightly larger vertical branch. The vertical branch characteristically parallels the lateral border of the thoracodorsal muscle, running approximately 2.5 cm from its edge. Within the muscle, both branches split into lesser branches that run medially and then anastomose with intercostal and lumbar arteries perforators. A significant feature of these anastomoses is their large diameter, important for supplying robust blood flow in surgical procedures. Thirty-six perforators (56%) originated from the lateral branch and 28 (44%) from the horizontal branch. There was an average of 1.8 perforators (range 1–4) from the vertical branch and of 1.4 perforators (range 1–3) from the horizontal branch (Heitmann et al. 2003a).

94.5 Histology

The thoracodorsal artery is a muscular artery. Typical features of a muscular artery are:

- The tunica intima, the innermost and thinnest layer, comprises an inner layer of endothelium surrounded by subendothelial connective tissue and a prominent internal elastic lamina.
- The tunica media surrounds the tunica intima with mostly circumferentially arranged smooth muscle, minimal elastin, and a thick external elastic lamina.
- The most external layer (outermost wall) of the artery, the tunica adventitia, consists of organized, dense, irregular connective tissue.

94.6 Clinical and Surgical Significance

Owing to its large diameter and the infrequency of its anatomical variations, the thoracodorsal artery provides a reliable blood supply for an abundance of muscular flap reconstruction surgeries. As the latissimus dorsi is one of the largest muscles in the human body, it is one of the most versatile and widely used tissue transfers in reconstructive surgery (Heitmann et al. 2003a). The thoracodorsal artery supplies the latissimus dorsi, and its versatility is implicated in reconstructions of the breast, axilla, neck, chest wall, torso, upper and lower limbs, and maxillofacial region.

The dominant pedicle of the latissimus dorsi is the thoracodorsal artery, but the muscle also receives segmental circulation from perforators off the posterior intercostal and lumbar arteries. The thoracodorsal artery has numerous musculocutaneous perforators that allow for a versatile skin island design (Sood et al. 2018).

The advantages of the pedicled latissimus dorsi flap include a bigger surface area and a less obvious scar than with other pedicled regional flaps such as the pectoralis major (Sood et al. 2018). Disadvantages include a requirement to turn the patient (precluding simultaneous tumor ablation and flap raising), prolonged wound drainage and seroma formation, and risk of injury to the brachial plexus (Barton Jr et al. 1983). Another shortcoming, and a criticism of the traditional latissimus dorsi flap, is that the skin island overlying the flap is difficult to orient properly for appropriate breast volume replacement while providing sufficient coverage for the latissimus dorsi muscle.

More recently, the thoracodorsal artery perforator (TDAP) has been used with and without the latissimus dorsi for volume replacement in either partial or total breast tissue defects (Sood et al. 2018; Hamdi et al. 2008). The TDAP flap relies on perforators arising from the thoracodorsal artery and uses the overlying skin while avoiding the underlying muscle. It has been used as a pedicled flap for reconstructing defects that require a large amount of cutaneous tissue, including the trunk, axilla, and breast, or as a free flap in reconstructing various distant tissue defects such as those in the face, elbow, forearm, and lower limb. Its advantages include simultaneous flap elevation, easy dissection, minimal donor-site morbidity, and constant vascular anatomy with a long pedicle (Sever et al. 2012; Mun et al. 2008). A single perforator is sufficient to supply a skin paddle of up to 15 cm by 25 cm (Angrigiani and Siebert 1995). Perforators originating in the vertical branch have a shorter intramuscular course and are therefore preferred over those derived from the transverse branch. Abandoning the muscle entails a less robust blood flow; however, preoperative perforator mapping to locate the dominant perforators and optimize the flap design ensures adequate tissue perfusion.

Owing to the high success rate of microvascular free tissue transfer and more versatile flap design, the latissimus dorsi is now more commonly used as a free rather than a pedicled flap. Microvascular free tissue transfer involves harvesting the patient's soft tissues and/or bone from anatomically remote sites. These flaps, along with their defined vasculature, are then used to reconstruct defects (Futran and Alsarraf 2000). Microvascular free tissue transfer is considered the gold standard in complex cases of head and neck reconstructive surgeries when patients present with extensive primary or recurrent tumors. A failure rate of 0–8% has been recorded for the free latissimus dorsi flap used in head and neck reconstruction (O'Connell et al. 2017).

Complications

The thoracodorsal artery serves as the posterior boundary during axillary lymph node dissection. Therefore, any significant variation in its normal course can compromise surgical results in such cases. If there are multiple thoracodorsal arteries, isolation of the vascular pedicle for free flap transfer becomes risky and can result in postoperative complications or in failure of the flap itself (Sophia et al. 2021).

References

Angrigiani C, Grilli D, Siebert J (1995) Latissimus Dorsi musculocutaneous flap without muscle. Plast Reconstr Surg 96(7):1608–1614. https://doi.org/10.1097/00006534-199512000-00014

Barton FE Jr, Spicer TE, Byrd SH (1983) Head and neck reconstruction with the Latissimus Dorsi Myocutaneous flap: anatomic observations and report of 60 cases. Plast Reconstr Surg 71(2):199–204

Futran ND, Alsarraf R (2000) Microvascular free-flap reconstruction in the head and neck. JAMA 284(14):1761–1763. https://doi.org/10.1001/jama.284.14.1761

Germann G, Reichenberger M (2009) 23 - Latissimus Dorsi Flap. In: Flaps and reconstructive surgery. W.B. Saunders, pp 287–303

Hamdi M, Van Landuyt K, Hijjawi JB, Roche N, Blondeel P, Monstrey S (2008) Surgical technique in Pedicled thoracodorsal artery perforator flaps: a clinical experience with 99 patients. Plast Reconstr Surg 121(5):1632–1641. https://doi.org/10.1097/PRS.0b013e31816c3bfa

Heitmann C, Guerra A, Metzinger SW, Levin LS, Allen RJ (2003a) The thoracodorsal artery perforator flap: anatomic basis and clinical application. Ann Plast Surg 51(1):23–29. https://doi.org/10.1097/01.SAP.0000054189.14799.F3

Heitmann C, Pelzer M, Kuentscher M, Menke H, Germann G (2003b) The extended Latissimus Dorsi flap revisited. Plast Reconstr Surg 111(5):1697–1701. https://doi.org/10.1097/01.PRS.0000055444.84307.75

Kumar N, Aithal AP, Nayak SB, Bhaskar R (2018) A rare case of atypical thoracodorsal artery: a challenge for flap reconstruction. Surg Radiol Anat 40(8):899–902. https://doi.org/10.1007/s00276-018-2009-1

Morris SF, Tang M, Almutari K, Geddes C, Yang D (2010) The anatomic basis of perforator flaps. Clin Plast Surg 37(4):553–570. https://doi.org/10.1016/j.cps.2010.06.006

Mun GH, Lee SJ, Jeon BJ (2008) Perforator topography of the thoracodorsal artery Perforator flap. Plast Reconstr Surg 121(2):497–504. https://doi.org/10.1097/01.prs.0000299184.57535.81

O'Connell JE, Bajwa MS, Schache AG, Shaw RJ (2017) Head and neck reconstruction with free flaps based on the thoracodorsal system. Oral Oncol 75:46–53. https://doi.org/10.1016/j.oraloncology.2017.10.019

Rowsell AR, Davies DM, Eizenberg N, Ian TG (1984) The anatomy of the subscapular-thoracodorsal arterial system: study of 100 cadaver dissections. Br J Plast Surg 37(4):574–576. https://doi.org/10.1016/0007-1226(84)90152-8

Şentürk S, Efe D, Özkan A, Göncü RG, Zümrüt M (2015) Multidetector computed tomography angiography to evaluate the subscapular arterial tree: MDCTA to evaluate the subscapular arterial tree. Microsurgery 35(8):640–644. https://doi.org/10.1002/micr.22499

Sever C, Uygur F, Kulahci Y, Karagoz H, Sahin C (2012) Thoracodorsal artery perforator fasciocutaneous flap: a versatile alternative for coverage of various soft tissue defects. Indian J Plast Surg 45(3):478–484. https://doi.org/10.4103/0970-0358.105956

Sjøberg T, Mercer JB, Weum S, de Weerd L (2020) The value of dynamic infrared thermography in Pedicled thoracodorsal artery perforator flap surgery. Plast Reconstr Surg Glob Open 8(7):e2799. https://doi.org/10.1097/GOX.0000000000002799

Sood R, Easow JM, Konopka G, Panthaki ZJ (2018) Latissimus Dorsi flap in breast reconstruction. Cancer Control J Moffitt Cancer Cent 25(1):1073274817744638. https://doi.org/10.1177/1073274817744638

Sophia L, Singh D, Xalxo N et al (2021) Upper limb arterial pattern: clinical correlation and embryological perspective. J Vasc Bras 20:e20210008. https://doi.org/10.1590/1677-5449.210008

Xie Q, Lei X-Y, Zheng X-Y (2018) Analysis of high frequency ultrasound in subscapular region. Eur Rev Med Pharmacol Sci 22(23):8098–8103. https://doi.org/10.26355/eurrev_201812_16500

Brachial Artery

95

Briggs S. Ham, Katherine D. Bordes,
and Stephen J. Bordes, Jr.

95.1 Embryology

At the 8 mm crown-rump length (CRL) stage, the subclavian artery, a continuation of the seventh intersegmental artery, crosses the lateral margin of the first rib and continues as the axillary artery (Rodríguez-Niedenführ et al. 2001a, b). The axillary artery extends through the posterior axillary fold and continues as the brachial artery after passing the inferior border of the teres major tendon in the 11 mm CRL stage (Rodríguez-Niedenführ et al. 2001a, b). From the 11 to 14 mm CRL stages, the brachial artery descends the upper limb, pierces the entepicondylar foramen, and continues as the interosseous artery, supplying the embryonic forearm before regressing (Rodríguez-Niedenführ et al. 2001a, b). A well-developed brachial artery can be observed from the 14 to 17 mm CRL stage, and defined structures corresponding to the ulnar, median, and interosseous arteries can be identified (Rodríguez-Niedenführ et al. 2001a, b). By the 23 mm CRL stage, the interosseous and median arteries have regressed. The definitive adult arterial morphology can now be visualized, in which the ulnar and newly formed radial artery branches descend the forearm to supply the hand.

95.2 Anatomy and Topography

The brachial artery (Fig. 95.1) is the main arterial supply of the upper limb and is the continuation of the axillary artery at the inferior border of the teres major tendon. The length of the brachial artery is 23.0 ± 8.64 cm in males and 22.7 ± 0.77 cm in females. This vessel descends

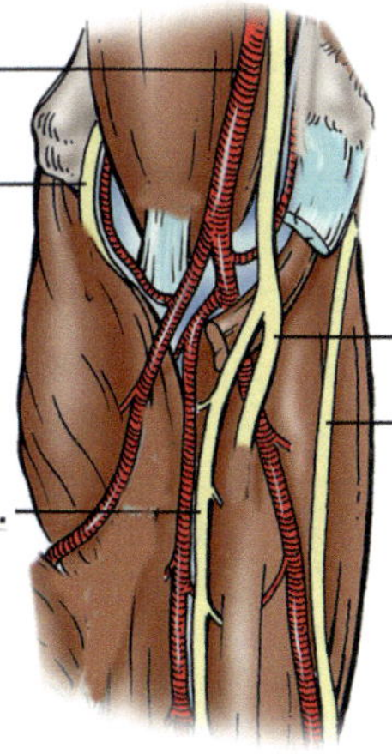

Fig. 95.1 Schematic drawing illustrating the brachial artery and its branches. (After Netter, 2023)

B. S. Ham
School of Medicine, University of Queensland, Brisbane, QLD, Australia

K. D. Bordes
Tulane University School of Medicine, New Orleans, LA, USA

S. J. Bordes, Jr. (✉)
Department of Surgery, Louisiana State University Health Sciences Center, New Orleans, LA, USA

S. J. Bordes, Jr. et al. (eds.), *The Clinical Anatomy of the Vascular System*,
https://doi.org/10.1007/978-3-031-78326-5_95

the anterior surface of the upper limb in the medial intermuscular septum and travels laterally to assume a position midway between the lateral and medial epicondyle of the humerus (Epperson and Varacallo 2021). The brachial artery (Fig. 95.2) passes deep to the bicipital aponeurosis toward the cubital fossa. Proximally, the median nerve courses along the lateral aspect of the brachial artery (Chakravarthi et al. 2014). At the midpoint of the upper limb, the median nerve crosses the brachial artery and continues medially. The vessel terminally bifurcates at the level of the radial neck, approximately 1 cm proximal to the cubital fossa, but variant bifurcation is commonly encountered (Tubbs et al. 2016). The brachial artery and its branches supply the humerus, biceps brachii, coracobrachialis, and brachialis muscles, elbow, and forearm (Chakravarthi et al. 2014).

In the normal arterial pattern of the upper limb, the first branch of the brachial artery is the deep brachial artery (Fig. 95.3). This vessel courses with the radial nerve. It supplies the posterior compartment of the upper limb (Epperson and Varacallo 2021). The second and third branches are the superior (Fig. 95.4) and inferior ulnar collateral arteries, respectively, both arising from the medial surface of the brachial artery and course distally. The humeral nutrient artery passes through a foramen in the anteromedial surface of the humeral shaft at the level at which the coracobrachialis muscle attaches dis-

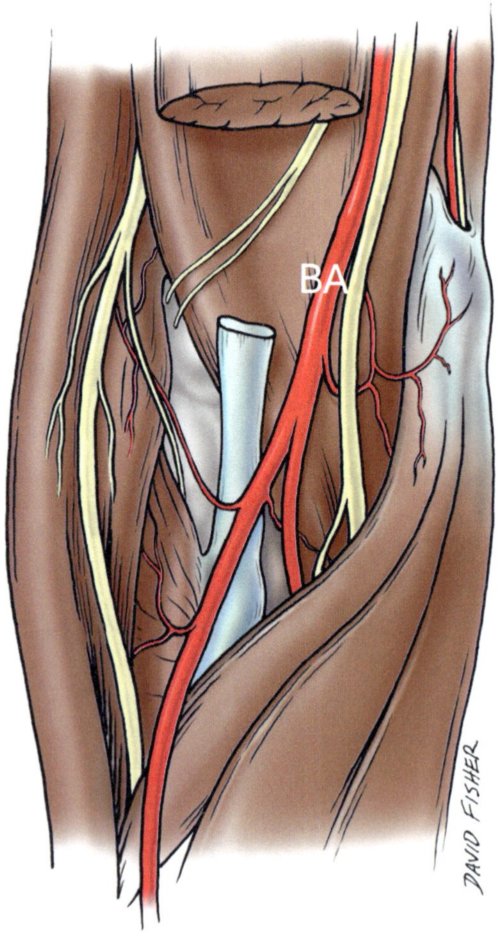

Fig. 95.2 The brachial artery (BA) near the cubital fossa. (After Netter 2023)

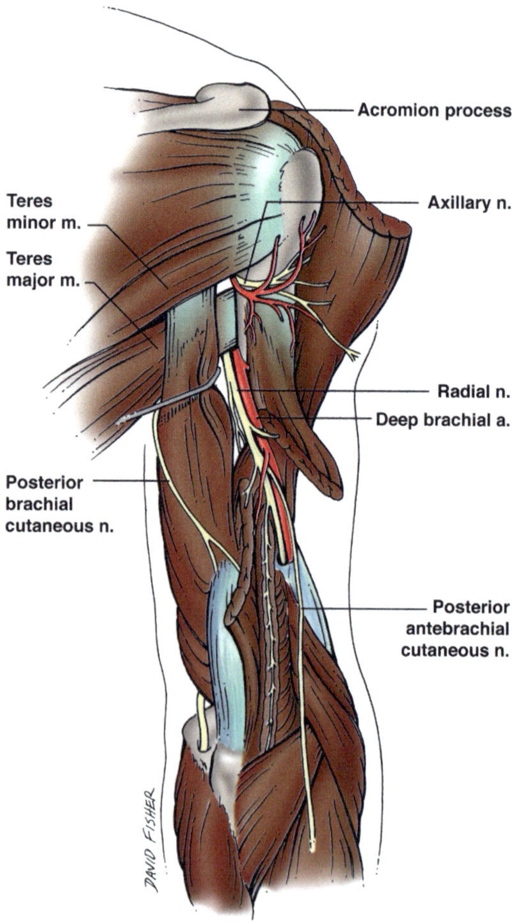

Fig. 95.3 Posterior view of the proximal upper limb showing the deep brachial artery traveling with the radial nerve in the triangular interval. (After Netter 2023)

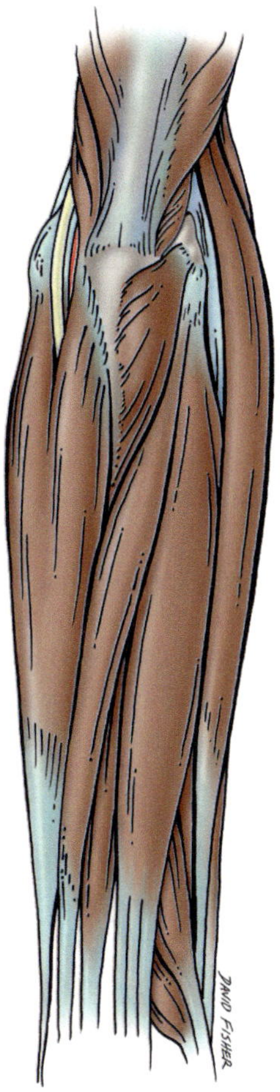

Fig. 95.4 Posterior view of the elbow illustrating the ulnar nerve and superior ulnar collateral artery entering the cubital tunnel posterior to the medial epicondyle of the humerus. (After Netter 2023)

tally (Chakravarthi et al. 2014). The coracobrachialis, bicep brachii, and brachialis muscles are supplied by muscular branches of the brachial artery (Epperson and Varacallo 2021). The terminal bifurcation of this vessel at the level of the radial neck gives rise to the radial and ulnar arteries, which continue distally through the forearm to the hand.

95.3 Histology

The brachial artery is classified as a muscular, distributing artery. Like other major muscular arteries, it has a characteristic thick layer of smooth muscle in the tunica media with a very thin subendothelial layer. The tunica media layer regulates the internal diameter of the vessel and, thus, blood pressure (Tucker et al. 2022).

95.4 Variations

Variation of the brachial artery is encountered in 25% of upper limbs and can be explained by the complex embryological development of this extremity (Tsoucalas et al. 2020). These variations can be divided into superficial coursing and variant bifurcation. Below are the results of a study that analyzed the right and left upper limbs of 192 adult cadavers (91 males and 101 females) aged 56 to 103 years at the time of death.

Type 1 Superficial course: a singular brachial artery of normal origin that courses anterior to the median nerve instead of posterior. This artery will bifurcate classically into the radial and ulnar arteries.

(a) Adopts its superficial course proximal to the median nerve root (1.8%).
(b) Adopts its superficial course distal to the median nerve root (2.9%).

Type 2 Premature bifurcation of the brachial artery occurs in approximately 20% of upper limbs, most commonly in the proximal third of the brachial artery (Tsoucalas et al. 2020). The two branches of premature bifurcation are delineated superficial and deep and can reunite or descend the upper limb in parallel, with or without communication. Premature bifurcation has also been observed proximal to the inferior border of the teres major tendon, which results in similar morphologies. Listed below are the possible outcomes of premature bifurcation of the brachial artery and their incidence in individual upper limbs.

(a) The superficial branch courses anterior to the medial nerve and reunites with the deep branch before bifurcating normally into the radial and ulnar arteries. The superficial branch is commonly called an accessory brachial artery (0.26%).

(b) The superficial branch courses parallel to the brachial artery laterally and passes deep, or more commonly superficial, to the bicipital aponeurosis before adopting radial artery morphology in the forearm. The superficial branch is frequently referred to as a brachioradial artery. The deep branch will divide into the ulnar artery and the common interosseous trunk (13.8%).
(c) The superficial branch courses parallel to the brachial artery medially and passes superficial, or more commonly deep, to the bicipital aponeurosis before adopting ulnar artery morphology in the forearm. The superficial branch is frequently referred to as the brachioulnar artery. The deep branch will divide into the radial artery and common interosseous trunk (4.5%).
(d) The superficial branch descends the arm, passes deep to the bicipital aponeurosis, and adopts the morphology of a common interosseous or persistent median artery in the forearm- the latter being more frequent. The deep branch usually divides into the radial and ulnar arteries (0.5%).

95.5 Imaging

Digital subtraction angiography has historically been the preferred technique for visualizing the brachial artery (Ghouri et al. 2019). More recently, however, vascular ultrasound, magnetic resonance angiography, and computed tomography angiography have emerged as more viable methods with specific indications (Ghouri et al. 2019). Doppler ultrasound is an efficient, noninvasive method of evaluating endothelial function, but low spatial resolution and restricted windows, angles, and depth of access limit its diagnostic usefulness. Additionally, it has shown comparable specificity and sensitivity to angiography (Ekim and Tuncer 2009). Magnetic resonance angiography is a high contrast technique for assessing vascular pathologies and characterizing vascular and soft tissue anatomy. This method has a higher degree of accuracy and predictability than Doppler ultrasound but is costly, slow, and of limited availability (Ghouri et al. 2019). Computed tomography angiography is the preferred method for evaluating acute trauma or acute occlusion due to faster scan times, higher availability, and improved ability to concurrently visualize vascular and soft tissue compared to Doppler ultrasound (Ghouri et al. 2019). Patient history and physical examination are critical in determining the most appropriate imaging technique.

95.6 Pathology

Brachial artery aneurysms: Aneurysms are divided into two categories: true aneurysms and pseudoaneurysms. True aneurysms display dilation of all three layers of a vessel's wall, while pseudoaneurysms commonly involve only the tunica media and adventitia. The brachial artery is located at 0.5% of peripheral aneurysms, 0.17% of true aneurysms, and 0.33% of pseudoaneurysms (Heydari et al. 2015). True brachial artery aneurysms are often the result of blunt trauma, infection, atherosclerosis, and Kawasaki disease (Ko et al. 2011). Common brachial artery pseudoaneurysm etiologies include catheterization complications, intravenous drug use, connective tissue pathologies, and penetrating trauma, in addition to infection and atherosclerosis (Deşer 2017). The primary complications of brachial artery aneurysms include occlusion, rupture, thrombosis, and embolic events, which can cause critical limb ischemia if undiagnosed or untreated (Schunn and Sullivan 2002). Ligation and excision are the most appropriate management of both true aneurysm and pseudoaneurysm in the brachial artery (Schunn and Sullivan 2002; Deşer 2017).

Ischemic Compartment Syndrome: In cases of severe upper limb injury, swelling of soft tissue can raise intercompartmental pressure and compress the muscles and nerves surrounding the brachial artery. This process results in ischemia. Necrotic muscle tissue is replaced by fibrotic scar tissue (Epperson and Varacallo 2021). Scarring can lead to contracture of compartmental muscles, causing pain, paralysis, and paresthesia in the associated areas of the upper limb.

95.7 Clinical and Surgical Significance

Ankle-Brachial Index The ankle-brachial index test is a noninvasive measurement commonly used to assess peripheral vascular disease. The ankle-brachial index is the ratio of the highest blood pressure measured at the ankle to the highest blood pressure measured in the upper arm when the patient is at rest. Physiologically, the blood pressure in the ankle should be greater than the blood pressure in the brachial artery, and ratios below the acceptable level of 0.9 can indicate blood supply occlusion in the lower limbs (McDermott et al. 2000). Values above 1.3 can also suggest peripheral vascular disease, which may indicate vessel wall calcification (McDermott et al. 2000). This test has a sensitivity of 90% and specificity of 98% but is limited in populations with relatively incompressible vessels, such as diabetics, the elderly, and patients with renal disease (McDermott et al. 2000).

Blood Pressure The brachial artery is superficial and the most common location for measuring blood pressure with a sphygmomanometer.

Brachial Artery Catheterization The brachial artery can be a vascular access site of the upper extremity. Both open and percutaneous approaches have clinically been proven safe (Epperson and Varacallo 2021). Vascular access from the brachial artery allows for complex endovascular interventions. Still, the risk of thrombosis must be considered as vessel occlusion, which is a well-documented complication of brachial artery catheterization (Lavingia et al. 2018). This phenomenon can be managed with surgical thromboembolectomy or balloon angioplasty, although more conservative techniques are often indicated (Siddiqui et al. 2013). It is essential to be familiar with the anatomical patterns of the upper limb, for vulnerable structures such as the median nerve lie in close proximity to the brachial artery and have severe complications if injured.

Supracondylar Humerus Fractures This fracture is often the result of forced hyperextension of the elbow when breaking a fall with an outstretched hand. This motion can lead to posterior displacement of a distal humeral fragment, and the proximal humeral fragment may injure the brachial artery (Vineet and Ajai 2016). The supracondylar area undergoes remodeling between 6 and 7 years of age, making the area thin and prone to fracture before maturing into its adult morphology (Vineet and Ajai 2016).

Volkmann's Ischemic Contracture This is a permanent flexion contracture of the hand at the wrist and can manifest in a claw deformity of the fingers and hand (Epperson and Varacallo 2021). The causes of this can vary, but damage or occlusion of the brachial artery is the critical component. This condition is often the result of prolonged ischemic compartment syndrome but can also be caused by the improper use of tourniquets and plaster casts (Epperson and Varacallo 2021).

AV Fistula
The brachial artery is a common vessel used to create an arteriovenous (AV) fistula when creating long-term dialysis access in individuals with end-stage renal disease. A vein of the upper extremity, such as the cephalic, may be ligated distally and attached end-to-side to the brachial artery. The newly created fistula allows for the maturation of vein over 6–8 weeks to provide reliable dialysis access. Typically, the nondominant arm is chosen. Distal arterial and venous targets are a surgeon's first choice, followed by more proximal vessels. In the event a good target is not available in the upper extremity, a graft may be placed.

Trauma and Surgical Considerations
In cases of trauma, the brachial artery is the most commonly injured artery of the upper extremity and accounts for 28% of all vascular injuries. Blunt and penetrating injuries are the most common causes of brachial artery injury. Supracondylar fractures or humeral dislocation may result in brachial artery injury. The median, radial, and ulnar nerves are also subject to damage due to their anatomical proximity to the brachial artery (Epperson and Varacallo 2021).

Collateral circulation systems exist between the deep brachial and radial recurrent arteries and between the superior ulnar collateral, inferior ulnar collateral, anterior ulnar recurrent, and posterior ulnar recurrent arteries. This network maintains limb viability in the case of brachial artery injury. Ischemia may result when the injury occurs proximal to the deep brachial branch—the first branch of the collateral circulation network (Levin et al. 1971).

Prior to the First World War, amputation was the primary means of managing upper extremity injuries. However, due to advances in vascular repair techniques and an improved understanding of the extensive collateral circulation of the upper extremity, amputation in the context of brachial artery injury is rare and only indicated in the case of obvious necrotic changes (Zellweger et al. 2004; Ekim and Tuncer 2009).

There are a variety of techniques for surgical repair of brachial artery injuries, including lateral repair, primary repair, resection with end-to-end anastomosis, reverse saphenous vein interposition graft, endovascular repair, and ligation. The resection with end-to-end anastomosis technique is most commonly used and is typically favored if it can be done successfully without tension or damage to significant collateral vessels. The reverse saphenous vein interposition graft is another favorable option and has been shown to have better patency rates and resistance to infection compared with synthetic grafts. Endovascular techniques are used more frequently for managing penetrating and blunt axillary artery injuries. As with any endovascular approach, there is a risk of thrombosis, which may be managed conservatively or with surgical thrombo-embolectomy or balloon angioplasty. Functional disability following repair depends on the extent of nerve damage, major venous injuries, fractures, and surrounding tissue damage (Ekim and Tuncer 2009; Epperson and Varacallo 2021).

References

Chakravarthi KK, Ks S, Venumadhav N, Sharma A, Kumar N (2014) Anatomical variations of brachial artery – its morphology, embryogenesis and clinical implications. J Clin Diagn Res 8(12):17–20

Deşer SB (2017) Management of iatrogenic brachial artery pseudoaneurysm: surgical treatment of iatrogenic brachial artery pseudoaneurysm. Int J Cardiovasc Acad 3(1–2):9–10

Ekim H, Tuncer M (2009) Management of traumatic brachial artery injuries: a report on 49 patients. Ann Saudi Med 29(2):105–109. https://doi.org/10.4103/0256-4947.51797

Epperson T, Varacallo M (2021) Anatomy, shoulder and upper limb, brachial artery. StatPearls Publishing. [Online]

Ghouri M, Gupta N, Bhat A, Thimmappa N, Saboo S, Khandelwal A, Nagpal P (2019) CT and MR imaging of the upper extremity vasculature: pearls, pitfalls, and challenges. Cardiovasc Diagn Ther 9(1):152–173

Heydari F, Taheri M, M. Esmailian M. (2015) Brachial artery aneurysm as a limb threatening condition: a case report. Emergency 3(2):75–77

Ko S, Han IY, Cho KH, Lee YH, Park KT, Kang MS (2011) Recurrent true brachial artery aneurysm. Korean J Thorac Cardiovasc Surg 44(5):364–367

Lavingia KS, Dua A, Stern JR (2018) Upper extremity access options for complex endovascular aortic interventions. J Cardiovasc Surg 59(3):360–367

Levin MPM, Rich LNM, Hutton MJE Jr (1971) Collateral circulation in arterial injuries. Arch Surg 102(4):392–399. https://doi.org/10.1001/archsurg.1971.01350040154030

McDermott M, Criqui M, Liu K, Guralnik J, Greenland P, Martin GJ, Pearce W (2000) Lower ankle/brachial index, as calculated by averaging the dorsalis pedis and posterior tibial arterial pressures, and association with leg functioning in peripheral arterial disease. J Vasc Surg 32(6):1164–1171

Rodríguez-Niedenführ M, Burton G, Deu J, Sanudo JR (2001a) Development of the arterial pattern in the upper limb of staged human embryos: normal development and anatomic variations. J Anat 199(4):407–417

Rodríguez-Niedenführ M, Vazquez T, Nearn L, Ferreira B, Parkin I, Sanudo J (2001b) Variations of the arterial pattern in the upper limb revisited: a morphological and statistical study, with a review of the literature. J Anat 199:547–566

Schunn CD, Sullivan TM (2002) Brachial arteriomegaly and true aneurysmal degeneration: case report and literature review. Vasc Med 7(1):25–27

Siddiqui MU, Khurram D, Elder M (2013) Management of brachial artery thrombosis post catheterization. J Invasive Cardiol 25(3):60–62
Tsoucalas G, Eleftheriou A, Panagouli E (2020) High bifurcation of the brachial artery: an embryological overview. Cureus 12(2):e7097
Tubbs RS, Mohammadali M, Loukas M (2016) Upper limb arteries. In: Bergman's comprehensive encyclopedia of human anatomic variation. Wiley, pp 583–618
Tucker WD, Arora Y, Mahajan K (2022) Anatomy, blood vessels [Updated 2021 Aug 11]. In: StatPearls [Internet]. StatPearls Publishing, Treasure Island. 2022 Jan-. Available from: https://www.ncbi.nlm.nih.gov/books/NBK470401/
Vineet K, Ajai S (2016) Fracture supracondylar humerus: a review. J Clin Diagn Res 10(12):1–6
Zellweger R, Hess F, Nicol A, Omoshoro-Jones J, Kahn D, Navsaria P (2004) An analysis of 124 surgically managed brachial artery injuries. Am J Surg 188(3):240–245. https://doi.org/10.1016/j.amjsurg.2004.02.005

Circumflex Humeral Arteries

96

Katherine D. Bordes and Stephen J. Bordes, Jr.

96.1 Embryology

At the end of the fourth week of the embryonic period, the development of the upper limb arteries, along with other parts of the upper limb, begins with the activation of mesenchymal cells from the lateral mesoderm. At this stage, upper limb buds begin to form from a mass of mesenchymal cells covered by ectoderm. Later in its development, this mass differentiates into bone, cartilage, and blood vessels. Specifically, the anterior circumflex humeral artery (ACHA) and posterior circumflex humeral artery (PCHA) develop from branches of the axis artery, which is a lateral branch of the seventh intersegmental artery (Gilbert and Nelson 2021; Tang and Varacallo 2021).

96.2 Anatomy

The anterior circumflex humeral artery (AHCA) originates as a proximal, lateral branch of the third part of the axillary artery. It lies distal to the lower border of the pectoralis minor muscle and travels behind the coracobrachialis muscle toward the bicipital groove. It then terminates in an ascending branch and a main branch that anastomoses with the posterior circumflex humeral artery and other arterial branches, including the deep brachial and thoracoacromial arteries (Hapugoda 2021a; Gilbert and Nelson 2021).

The posterior circumflex humeral artery (PHCA) (Fig. 96.1) is larger than its anterior counterpart and arises medially as a branch of the third part of the axillary artery. It then travels posteriorly and, within the quadrangular space, divides into its own respective anterior and posterior branches, which wrap around the surgical neck of the humerus. The posterior branch continues distally and terminates in anastomoses with the ascending branch of the deep brachial artery. The anterior branch anastomoses with the anterior circumflex humeral artery (Hapugoda 2021b).

The anastomosis between the anterior and posterior circumflex humeral arteries is responsible for much of the blood supply to the humeral head. Both of these arteries supply blood to the glenohumeral joint, teres major and minor, and deltoid (Mostafa et al. 2021).

96.3 Variations

Variations of the axillary artery are relatively common and are typically described as either variations of branching patterns of the whole ves-

K. D. Bordes
Tulane University School of Medicine, New Orleans, LA, USA

S. J. Bordes, Jr. (✉)
Department of Surgery, Louisiana State University Health Sciences Center, New Orleans, LA, USA

S. J. Bordes, Jr. et al. (eds.), *The Clinical Anatomy of the Vascular System*,
https://doi.org/10.1007/978-3-031-78326-5_96

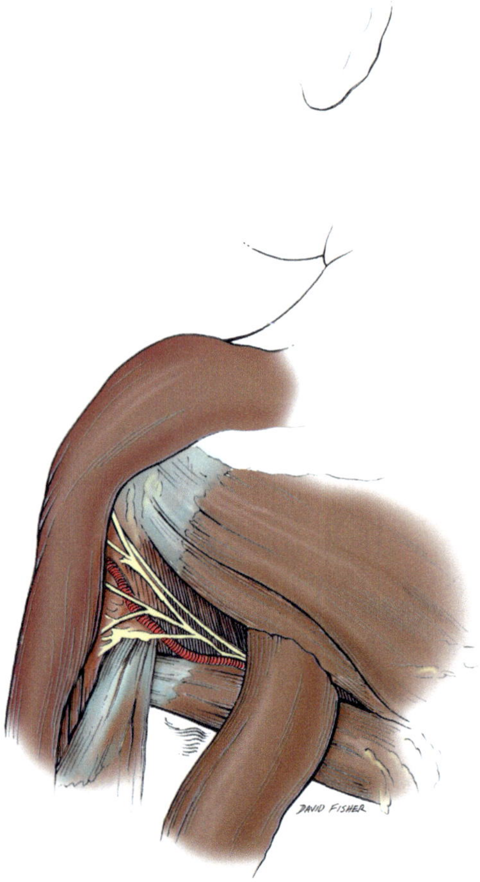

Fig. 96.1 Schematic drawing showing the posterior circumflex humeral artery and axillary nerve exiting the quadrangular space to supply the deltoid muscle

sel or the frequencies with which specific branches originate from specific locations.

The anterior and posterior circumflex humeral arteries originate as a common trunk in about 20% of cases. On occasion, when the anterior and posterior circumflex arteries originate as a common trunk, the lateral thoracic artery also originates with the thoracoacromial trunk.

In about 2%, the lateral thoracic, thoracoacromial, anterior, and posterior circumflex humeral and subscapular arteries originate as a common trunk. In about 20%, the PHCA originates from the subscapular artery rather than the axillary artery. In about 5%, the deep brachial artery arises from the PHCA (Tubbs et al. 2016).

96.4 Clinical/Surgical Significance

Traditionally, a deltopectoral and transdeltoid approach has been used in open reduction and internal fixation of proximal humerus fractures; however, injuries to the anterior circumflex humeral artery and axillary nerve are common complications. As a result, studies have been conducted to identify the mean distances between the anterior circumflex humeral artery and other anatomical landmarks including the infraglenoid tubercule, the coracoid process, the midclavicular line, and the inferior border of the medial acromion (Gilbert and Nelson 2021; Chen et al. 2012).

Among the four rotator cuff muscles, the anterior circumflex humeral artery is the salient blood supply to the teres minor. Rotator cuff injuries are common owing to degeneration, overuse, and trauma, which often result in chronic pain, specifically night pain. A study using Doppler ultrasonography demonstrated that this night pain could be related to the hemodynamics of the anterior circumflex humeral artery (Gilbert and Nelson 2021; Terabayashi et al. 2014).

Poor posture can also cause significant pain. One study showed that anatomical differences in the anterior circumflex humeral artery determined the degree of movement during scapular motion. During internal rotation of the scapula, the anterior branch proved more limited and the posterior branch more freely mobile, thus decreasing the blood flow to the anterior circumflex humeral artery (Gilbert and Nelson 2021; Hagiwara et al. 2015).

Proximal humerus fractures can result from trauma, pathological conditions, or osteoporosis. The anterior circumflex humeral artery was originally thought to be the primary blood supply to the humerus; however, the posterior circumflex humeral artery predominates. This could explain the relatively low rate of osteonecrosis in displaced fractures of the proximal humerus (Hettrich et al. 2010).

References

Chen Y-F, Zhu N-F, Zhang C-Q, Wang L, Wei H-F, Lu Y (2012) The relevance of the anatomical basis of fracture for the subsequent treatment of the anterior circumflex humeral artery and the axillary nerve. Int Orthop 36(4):783–787. https://doi.org/10.1007/s00264-011-1394-4

Gilbert GM, Nelson R (2021) Anatomy, shoulder and upper limb, anterior circumflex humeral artery. In: StatPearls. StatPearls Publishing. Accessed July 20, 2021. http://www.ncbi.nlm.nih.gov/books/NBK557892/

Hagiwara Y, Kanazawa K, Ando A et al (2015) Blood flow changes of the anterior circumflex humeral artery decrease with the scapula in internal rotation. Knee Surg Sports Traumatol Arthrosc 23(5):1467–1472. https://doi.org/10.1007/s00167-013-2823-2

Hapugoda S. Anterior circumflex humeral artery | Radiology Reference Article | Radiopaedia.org. Radiopaedia. Accessed July 20, 2021a. https://radiopaedia.org/articles/anterior-humeral-circumflex-artery?lang=us

Hapugoda S. Posterior circumflex humeral artery | Radiology Reference Article | Radiopaedia.org. Radiopaedia. Accessed July 20, 2021b. https://radiopaedia.org/articles/posterior-humeral-circumflex-artery?lang=us

Hettrich CM, Boraiah S, Dyke JP, Neviaser A, Helfet DL, Lorich DG (2010) Quantitative assessment of the vascularity of the proximal part of the humerus. J Bone Joint Surg Am 92(4):943–948. https://doi.org/10.2106/JBJS.H.01144. PMID: 20360519

Mostafa E, Imonugo O, Varacallo M (2021) Anatomy, shoulder and upper limb, Humerus. [Updated 2021 Aug 11]. In: StatPearls [Internet]. StatPearls Publishing, Treasure Island. Available from: https://www.ncbi.nlm.nih.gov/books/NBK534821/

Tang A, Varacallo M (2021) Anatomy, head and neck, posterior circumflex humeral artery. [Updated 2021 Aug 11]. In: StatPearls [Internet]. StatPearls Publishing, Treasure Island. Available from: https://www.ncbi.nlm.nih.gov/books/NBK538283/

Terabayashi N, Watanabe T, Matsumoto K et al (2014) Increased blood flow in the anterior circumflex humeral artery correlates with night pain in patients with rotator cuff tear. J Orthop Sci 19(5):744–749. https://doi.org/10.1007/s00776-014-0604-5

Tubbs RS, Shoja M, Loukas M (2016) Bergman's comprehensive Encyclopedia of human anatomic variation. Wiley, Hoboken

97 Radial Artery

Baris Vefali, Annie Syed, Panida Charnvitayapong, Huseng Vefali, and Veli Vefali

97.1 Embryology

Sprouting theory suggests that arteries of a growing limb arise as a bud of the higher artery trunk. In the upper limb, subclavian arteries give off the axillary, brachial, and interosseous arteries, while the radial and ulnar arteries develop in later stages (Standring 2008).

All arteries start as networks of capillaries that then develop and differentiate into major vessels. The first bud is the axial trunk that comes off the dorsal aorta. By day 26 of development, the upper limb bud has started to form, and a diverse yet undifferentiated network of capillaries is present. Nerve trunks are present but have not yet entered the axial bud. By day 28, the axial trunk has become larger but remains undifferentiated at its terminal branches. The terminal branches are still capillaries, with no skeletal, muscular, or neural elements. By day 32, the subclavian artery has branched off the axial trunk, giving off further undifferentiated capillaries. On day 33, the axial artery has given rise to subclavian and axillary arteries with capillary branches and neurons. By day 37, neural and musculocutaneous structures begin to differentiate, with neural structures integrating into surrounding structures. On day 41, the budding of the brachial artery is observed from the axillary artery, extending to the elbow and forming into capillaries. By day 44, the ulna and radius have become more defined, and thus, vascular patterns start to include the ulnar, median, and interosseous arteries, as well as the proximal radial artery. However, the distal part of the radial artery remains as a capillary net at this stage. On day 47, the bony structures are defined to the hands with clearly defined arteries and easily recognizable nerves running down to the fingers. By day 52–56, all arteries are clearly defined. The fully developed distal parts of the radial artery now give rise to the palmar arches. Once all nerves innervate the surrounding structures, they are fully developed (Rodríguez-Niedenführ et al. 2001).

97.2 Anatomy/Topography

The arm is divided into two compartments, anterior (flexor) and posterior (extensor), by the intermuscular septum. The radial artery splits into

B. Vefali (✉)
Department of Medicine, Austin Medical Clinic, Bastrop, TX, USA

A. Syed
Department of Surgery, Louisiana State University Health Sciences Center, New Orleans, LA, USA

P. Charnvitayapong
St. George's University School of Medicine, St. George's, Grenada

H. Vefali
Department of Interventional Cardiology, Austin Cardiology Clinic, Austin, TX, USA

V. Vefali
Department of Interventional Cardiology, Medicana Hospital, Ankara, Turkey

S. J. Bordes, Jr. et al. (eds.), *The Clinical Anatomy of the Vascular System*,
https://doi.org/10.1007/978-3-031-78326-5_97

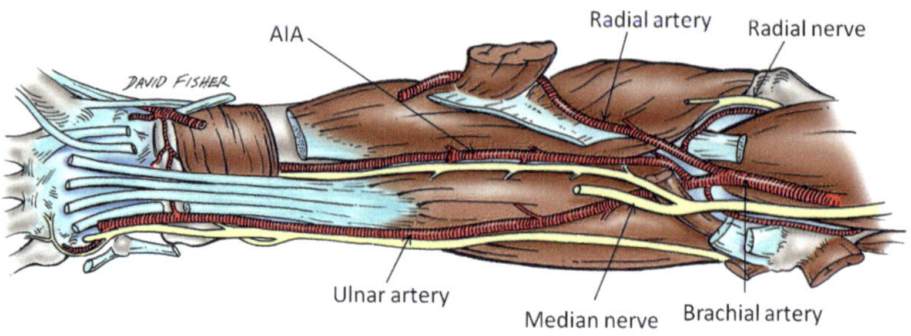

Fig. 97.1 Drawing of the radial and ulnar arteries at the wrist and hand. (After Allen 1892)

proximal, middle, and distal sections. The proximal portion lies under the brachioradialis muscle, while the middle third lies near the superficial branch of the radial nerve. The distal third of the radial artery becomes superficial and lies anterior to the radius and pronator quadratus muscle between the brachioradialis and flexor carpi radialis tendons.

The radial artery supplies the posterior or extensor aspect of the upper limbs. At the level of the cubital fossa, the brachial artery bifurcates into the radial and ulnar arteries. The radial artery, a classically smaller but more direct route of the two terminal branches of the brachial artery, runs on the lateral side of the forearm and supplies the forearm and hand. As it courses into the hand, it passes over the radius, producing the palmar carpal branch and palmar branch that runs through the floor of the anatomical snuffbox between the first and second metacarpal bones. Eventually, the radial artery anastomoses with the deep branch of the ulnar artery within the hand and completes the deep palmar arch (Fig. 97.1). While there are variations in which the radial artery travels into the hand, the artery will ultimately supply blood to the radial nerve, carpal bones and joints, thumb, and index finger (Sgueglia et al. 2018).

97.3 Imaging/Histology

Like most other arteries, the radial artery is a muscular vessel that provides oxygenated blood to the upper extremities. It is composed of three layers: an inner layer known as the tunica intima, a middle muscular layer, or the tunica media, and an outermost layer, known as the tunica externa.

The tunica intima is composed of endothelium and an internal elastic membrane. The endothelium is a thin, specialized form of mesenchymal-epithelial tissue made of simple squamous cells. All vessels, including the circulatory and lymphatic systems, are lined by an endothelium. Endothelial cells not only act as a barrier between blood and tissues but are specialized in that they release substances that control vasoconstriction and vasodilation, as well as clotting and immune function. The elastic membrane is a flexible barrier between the tunica intima and tunica media.

The tunica media is composed of smooth muscle cells and elastic fibers. Smooth muscle cells are innervated by nerve fibers, which help modulate the constriction or dilation of the vessels and thus regulate blood flow and blood pressure. Elastic fibers are made of elastin and fibrillin. These are structural and mechanical units of arteries that provide resilience and elasticity to allow for changes in arterial diameter and vascular compliance.

The tunica externa and tunica adventitia attach the vessel to the surrounding tissue. This layer of the artery prevents the overexpansion of arteries and contains collagen, which anchors the artery to the surrounding structures, as well as the external elastic lamina (Tennant and McGeachie 1990).

97.4 Variations

Branching variation of the radial artery is among the most common vascular variations seen in the upper limb and is usually determined by the development of the vessel itself. Many other investigators have previously reported different anatomical variations. Awareness of these distinct branching patterns guides clinical decision-making during preoperative evaluations that determine the best postoperative outcomes in diagnostic, therapeutic, and surgical interventions (Nasr 2012).

Classic branching, in which the radial, ulnar, and common interosseous arteries all arise directly from the brachial artery, occurs about 80% of the time. The high origin of the radial artery is the most common variation and can either arise from the brachial artery (~8.6%) or the axillary artery (~1.2%) proximal to the antecubital fossa. It can occur either unilaterally or bilaterally. Occasionally, the radial artery arises from the deep brachial artery.

Other variations include:

1. Radial and ulnar arteries arising from the superficial brachial artery (~6% of the time).
2. Radial artery arising from the superficial brachial artery while the ulnar and common interosseous arteries branch off the brachial artery (~5% of the time).
3. Radial artery arises from the superficial brachial artery and brachial artery, while the ulnar and common interosseous arteries arise from the brachial artery. An anastomosis typically forms between the radial and brachial artery (~3% of the time).
4. Radial artery doubles in the arm, one traveling in its standard anatomical course and the other running more superficially towards the hand (~1% of the time). The palmar branch may come off the superficial radial artery and operate as the typical radial artery.
5. Cases in which the radial artery is absent have occurred with an incidence of 0.03%, making it extremely rare. The anterior interosseous artery would supply the forearm and hand (Tubbs et al. 2016).

97.5 Clinical and Surgical Significances

Access via the radial artery in various procedures has been widespread for many reasons. Its superficial course allows for better patient outcomes than the transfemoral approach in certain situations. Interventional specialists prefer a trans-radial approach over a femoral approach for patients undergoing percutaneous coronary interventions and peripheral diagnostic and revascularization procedures. This is due to its easier access for interventionalists, decreased recovery time, easy compressibility to control hemostasis, and reduced post-procedural complications. Utilization of the radial artery also lowers risk of injury to nearby structures, such as nerves and other vasculature, compared to cannulation of the femoral artery (Bhat et al. 2017).

In addition to its use in percutaneous coronary interventions and revascularization procedures, the radial artery may also be used for coronary artery bypass grafting and as a fistula site for hemodialysis access. Fortunately, anomalies in the origin of the radial artery do not typically affect its use as a graft but can change the operative approach.

Therefore, accurate knowledge of the anatomical variations within the radial artery is increasingly important. Knowledge of radial variation decreases the risk of iatrogenic injury, including thrombosis and gangrene, which may require amputation of the extremity in extreme situations (Gupta et al. 2012).

97.6 Complications

Procedure Failure

Procedural failures of the trans-radial approach can happen due to anomalies in radial artery anatomy. Anatomic variations may interfere with surgical, diagnostic, and therapeutic procedures that utilize the trans-radial approach and are more likely to occur in the right arm (<8% of cases). However, the radial artery can still guide clinical procedures when other access or repeat access is needed. Some studies discuss the disadvantages of the trans-radial approach, including a more significant learning curve, longer procedure time, increased access failure, and a greater chance of site crossover (Bhat et al. 2017).

Radial Artery Occlusion

Despite the reduced complication rates of trans-radial access, there is still about a 30% chance of radial artery occlusion. It remains to be the most common complication of trans-radial interventions. However, occlusion of the artery is often asymptomatic due to the dual blood supply to the hand. The rich collateral network to the radial and ulnar arteries ensures sufficient blood supply and function to the hand and fingers despite acute or persistent radial artery occlusion. Possible symptoms include paresthesia, pain at the occlusion site, loss of hand function, ischemia distally, and potential loss of access to the radial artery in future procedures (Brzezinski et al. 2009).

Patients with smaller post-procedural radial artery diameters were at higher risk for radial artery occlusion in later stages of recovery. Some studies have shown that simultaneous ulnar and radial compression (SURC) may help decrease the incidence of radial artery occlusion compared to conventional hemostasis techniques (Ahmed et al. 2022).

Atherosclerosis

The rising popularity of the radial artery in various clinical procedures has led to an increasing assessment of the artery and the incidence of atherosclerotic disease. Risk factors include coronary artery disease, diabetes, peripheral vascular disease, smoking, and advanced age.

Radial Artery Catheterization

Complications of indwelling radial artery catheters include intimal damage and proliferation due to endothelial injury. Studies have shown that 6 hours of catheterization can lead to increased scarring within the arterial wall, potentially reducing the diameter of the artery, stenosis, or occlusion (Brzezinski et al. 2009).

References

Ahmed TAN, Abbas E, Bakr AH, Demitry SR, Algowhary MI (2022) Prevention of radial artery occlusion by simultaneous ulnar and radial compression (PRO-SURC). A randomized duplex ultrasound follow-up study. Int J Cardiol 363:23

Bhat FA, Changal KH, Raina H, Tramboo NA, Rather HA (2017) Transradial versus transfemoral approach for coronary angiography and angioplasty - a prospective, randomized comparison. BMC Cardiovasc Disord 17(1):23

Brzezinski M, Luisetti T, London MJ (2009) Radial artery cannulation: a comprehensive review of recent anatomic and physiologic investigations. Anesth Analg 109(6):1763–1781

Gupta C, Ray B, Dsouza AS, Nair N, Pai SR, Manju M (2012) A morphological study of variations in the branching pattern and termination of the radial artery. Singapore Med J 53(3):208–211

Nasr AY (2012) The radial artery and its variations: anatomical study and clinical implications. Folia Morphol (Warsz) 71(4):252–262

Rodríguez-Niedenführ M, Burton GJ, Deu J, Sañudo JR (2001) Development of the arterial pattern in the upper limb of staged human embryos: normal development and anatomic variations. J Anat 199:407–417

Sgueglia G, Di Giorgio A, Gaspardone A et al (2018) Anatomic basis and physiological rationale of distal radial artery access for percutaneous coronary and endovascular procedures. J Am Coll Cardiol Intv 11(20):2113–2119

Standring S (2008) Grays anatomy:the anatomical basis of clinical practice. In: Gray's anatomy: the anatomical basis of clinical practice, 40th edn. Churchill Livingstone, London, pp 905–906

Tennant M, McGeachie JK (1990) Blood vessel structure and function: a brief update on recent advances. Aust N Z J Surg 60(10):747–753

Tubbs RS, Shoja MM, Loukas M (2016) Bergman comprehensive Encyclopedia of human anatomic variation. Wiley

98 Radial Recurrent Artery

Briggs S. Ham and Stephen J. Bordes, Jr.

98.1 Anatomy and Topography

The radial recurrent artery (RRA) (Fig. 98.1) branches from the radial, ulnar, or brachial artery, traveling radially and oriented transversely. In upper limbs with normal arterial axes, the most common morphology features the RRA as the largest lateral branch of the radial artery. The vessel supplies blood to the brachioradialis, brachialis, extensor carpi radialis longus, and brevis and supinator muscles (Honma et al. 2008). In this pattern, the RRA branches from the lateral aspect of the radial artery in the cubital fossa, just distal to the transepicondylar line, and ascends the arm superiorly through the branches of the radial nerve (Honma et al. 2008). It courses proximally on the surface of the supinator muscle, then deep between the brachioradialis and brachialis muscles (Gupta et al. 2012). It terminates by anastomosing with the anterior branch of the deep brachial artery, the radial collateral artery, just proximal to the distal humerus. The average diameter of the RRA at its origin is 2.67 ± 0.60 mm in males and 2.35 ± 0.49 mm in females (Hamahata et al. 2012). Standard arterial patterns have shown that the RRA branches from the radial artery from 1.3 to 6.2 cm below the transepicondylar line (Yamaguchi et al. 1997). Arterial branches of the RRA cross the posterior interosseous nerve and enter the Arcade of Frohse to form what is commonly known as the Leash of Henry.

Vazquez et al. (2013) found a secondary or accessory RRA in 102 of 332 (30.8%) cadaver limbs sampled. In 100% of these cases, the second RRA originated from the brachial artery and coursed behind the bicipital tendon, supplying the brachioradialis, brachialis, and biceps brachii muscles.

B. S. Ham
Department of Anatomical Sciences, University of Queensland School of Medicine, Brisbane, QLD, Australia

S. J. Bordes, Jr. (✉)
Department of Surgery, Louisiana State University Health Sciences Center, New Orleans, LA, USA

S. J. Bordes, Jr. et al. (eds.), *The Clinical Anatomy of the Vascular System*,
https://doi.org/10.1007/978-3-031-78326-5_98

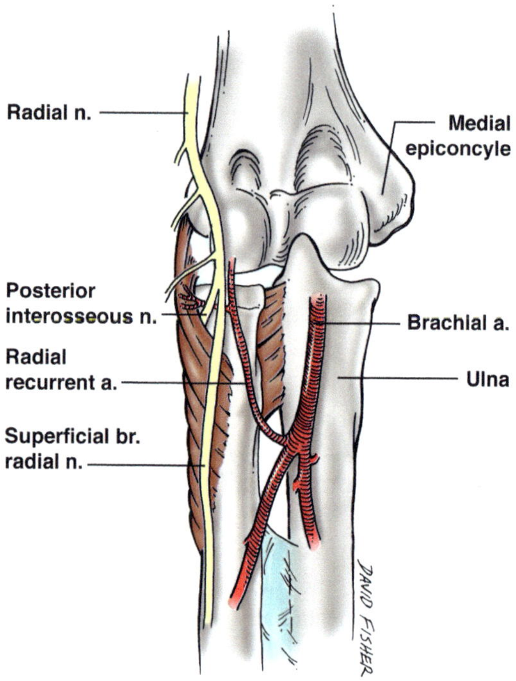

Fig. 98.1 Schematic drawing of the radial recurrent artery and related anatomy

98.2 Embryology

Researchers in the Autonomous University of Barcelona School of Medicine analyzed the development of arterial patterns in both upper limbs of 112 serially sectioned human embryos. Crown-rump length (CRL) in millimeters was used as the metric for the progression of embryo development.

At the 11 mm CRL, the axillary artery continues as the brachial artery and extends to the elbow before ramifying into the capillaries of the forearm and hand (Rodriguez-Niedenfuhr et al. 2001). The radial artery cannot be identified at this point, and the RRA has yet to form. At the 14 mm CRL, the humerus, ulna, and radius have chondrified, and the radial and ulnar arteries have bifurcated from the brachial artery below the elbow (Rodriguez-Niedenfuhr et al. 2001). The initial part of the radial artery can be identified using three-dimensional computer-aided reconstruction software, but its distal portion remains a capillary net. The RRA can first be identified, branching laterally from the radial artery, when the embryo CRL reaches 17 mm (Rodriguez-Niedenfuhr et al. 2001). The distal portion of the radial artery has yet to mature and remains a capillary plexus at this point in development. At a CRL of 21 mm, the RRA has coursed superiorly, crossing the elbow and anastomosing with the radial collateral artery (Rodriguez-Niedenfuhr et al. 2001). Excluding the distal portion of the radial artery (which remains in a capillary state), the arterial pattern of the arm has achieved its definitive adult morphology. From CRLs of 24–31 mm, the arterial wall of the RRA continues to develop, at which point the arterial pattern of the entire arm has achieved its mature configuration (Rodriguez-Niedenfuhr et al. 2001).

98.3 Variations

Vazquez et al. (2013) analyzed the RRAs in the right and left limbs of 166 adult cadavers (79 males and 87 females) aged 57–101. The RRAs in 215 of the 332 (64.8%) limbs originated from the radial artery, 210 coursing anterior and five posterior to the biceps brachii tendon (BBT). Thirty of the 332 (9.0%) RRAs originated from the posterior radioulnar division, 25 coursing anterior and five posterior to the BBT. Twenty-six of the 332 (7.8%) originated from the brachioradial artery (BR), all coursing anterior to the BBT. Twenty-four (7.2%) originated from the brachial artery, 12 coursing anterior and 12 posterior to the BBT. Eighteen (5.4%) originated from the anterior radioulnar division, all but one traveling anterior to the BBT. The remaining RRAs branched (in order of frequency) from the ulno-interosseous trunk, the anastomosis of the BR and brachial artery, the posterior radial-interosseous division, the anterior radio-interosseous division, and the interosseous trunk.

In the Vasquez et al. study, 66 of the 322 cadaver limbs examined had variations of the major arterial trunks in the upper limb. In nine of the 332, there was a superficial brachial artery. The RRAs in six originated from the radial artery, two from the posterior radioulnar division, and one from the ulno-interosseous trunk. In 44 of the 332 limbs, there was a brachioradial artery. The RRAs in 26 originated from the brachioradial artery, 12 from the brachial artery, and six from the anastomosis of the brachioradial and brachial arteries. In 10 of the 332, there was a superficial brachioulnar artery. The RRAs in seven of these originated from the radial artery, two from the posterior radial-interosseous division, and one from the anterior radio-interosseous division. In three of the 332 total limbs, there was a superficial brachioulnoradial artery. The RRAs in two of these originated from the radial artery and one from the interosseous trunk.

The variability of the muscular blood supply territory in the elbow requires surgeons to have a detailed understanding of the morphology of the blood vessels in this area. Without such knowledge, there is a risk of bleeding and vascular complications in operations in which the RRA is closely linked (Zeltser and Strauch 2015). A well-studied example of such operations is a distal bicep tendon repair. In the Vazquez et al. study, all 332 limbs contained an RRA that coursed anterior or posterior across the bicep brachii tendon. During surgery in the deep proximal forearm such as a repair of this tendon, the surgeon must be aware of the multitude of origins and locations and the possibility of an accessory RRA to achieve safe surgical exposure.

98.4 Clinical and Surgical Significance

The free radial forearm flap is often used in the primary reconstruction of defects involving the tongue, intraoral soft tissue, oropharynx, soft palate, hypopharynx, cervical esophagus, selected cutaneous defects, and more recently in the reconstructions of extended hemilaryngeal defects (Sanger et al. 1994). Hamahata et al. (2012) showed that this flap is also effective in salvage operations following head and neck cancer resections because of its long pedicle. However, as Hamahata et al. identified, pedicle length deficiency and size discrepancy among the anastomotic vessels used in these procedures still occur and can hinder the flap's success. When anastomosis is performed with a shortened pedicle length and a mismatch in vessel size, there is a greater chance of anastomosis failure and loss of the flap. When the pedicle of the radial artery is too short, Hamahata et al. suggest using the RRA as an alternative anastomotic vessel for effective transfer of the free radial artery flap during salvage operations. In this study, the authors reviewed salvage cases of four Japanese patients (three males and one female, aged 51–71, average 60.8 years), in which the RRA was used as an anastomotic vessel in place of the radial artery when the radial forearm flap was transferred. The patient in each case had a different primary disease (recurrence of tongue cancer, recurrence of nasopharyngeal cancer, maxillary cancer, floor-of-mouth cancer), and three different recipient arteries (facial, superior thyroid, and lingual) were used in the surgery. Shima et al. stated that the average facial artery diameter was 1.8 ± 0.3 mm, the average lingual artery diameter was 1.6 ± 0.3 mm, and the average superior thyroid artery diameter was 1.7 ± 0.7 mm (Shima et al. 1998). These diameters are significantly closer to that of the RRA at the 20 mm point past bifurcation (1.84 ± 0.59 mm) than the average radial artery diameter of 3.14 ± 0.64 mm (Hamahata et al. 2012). The pedicle length and diameter discrepancies were solved in each of these four cases. Because no postoperative events occurred, Hamahata et al. concluded that anastomosis using the RRA should be one of the strategies considered when the radial forearm flap is transferred during salvage operations.

References

Gupta C, Ray B, Dsouza AS, Nair N, Pai SR, Manju M (2012) A morphological study of variations in the branching pattern and termination of the radial artery. Singapore Med J 53:208–211

Hamahata A, Nakazawa H, Takeuchi M, Sakurai H (2012) Usefulness of radial recurrent artery in transplant of radial forearm flap: an anatomical and clinical study. J Reconstr Microsurg 28:195–198

Honma S, Tokiyoshi A, Kawai K, Koizumi M, Kodama K (2008) Radial artery running beneath the biceps tendon and its interrelation between the radial recurrent arteries. Anat Sci Int 83:232–238

Rodriguez-Niedenfuhr M, Burton G, Deu J, Sanudo JR (2001) Development of the arterial pattern in the upper limb of staged human embryos: normal development and anatomic variations. J Anat 199(4):407–417

Sanger JR, Ye Z, Yousif NJ, Matloub HS (1994) The brachioradialis forearm flap: anatomy and clinical application. Plastic Reconstr Surg 94:667–674

Shima H, von Luedinghausen M, Ohno K, Michi K (1998) Anatomy of microvascular anastomosis in the neck. Plast Reconstr Surg 101(1):33–41

Vazquez T, Sañudo JR, Carretero J, Parkin I, Rodríguez-Niedenführ M (2013) Variations of the radial recurrent artery of clinical interest. Surg Radiol Anat 35(8):689–694

Yamaguchi K, Sweet F, Bindra R, Morrey BF, Gelberman RH (1997) The extraosseous and intraosseous arterial anatomy of the adult elbow. J Bone Joint Surg 79-A:1653–1662

Zeltser DW, Strauch RJ (2015) Vascular anatomy relevant to distal biceps tendon repair. J Shoulder Elb Surg 25(2):283–288

Ulnar Artery

99

Corey Leskanic and Stephen J. Bordes, Jr.

99.1 Embryology

Around the 26th day of embryo development, mesoderm cells replicate rapidly under the influence of morphogens, leading to the development of the upper limb (Guéro 2018). Initially, the upper limb contains only a capillary network, which develops into the overall vasculature (Al-Qattan et al. 2009). The development of the ulnar artery is preceded by a border vein that appears on day 14 (Al-Qattan et al. 2009). Subclavian–axillary–brachial axis arteries appear between days 33 and 36 of development, the brachial artery eventually giving rise to the median artery (Al-Qattan et al. 2009). Around 41 days of development, the ulnar artery appears, replacing part of the median artery (Al-Qattan et al. 2009).

99.2 Anatomy

The ulnar artery (Fig. 99.1) arises from the brachial artery at the proximal forearm and the inferior aspect of the cubital fossa. It is the larger of the two terminal arteries of the brachial artery. It gives rise to the ulnar recurrent arteries and the common interosseous artery (Fig. 99.2). It also gives rise to the palmar and dorsal carpal branches. (Reddy and Vollala 2007)

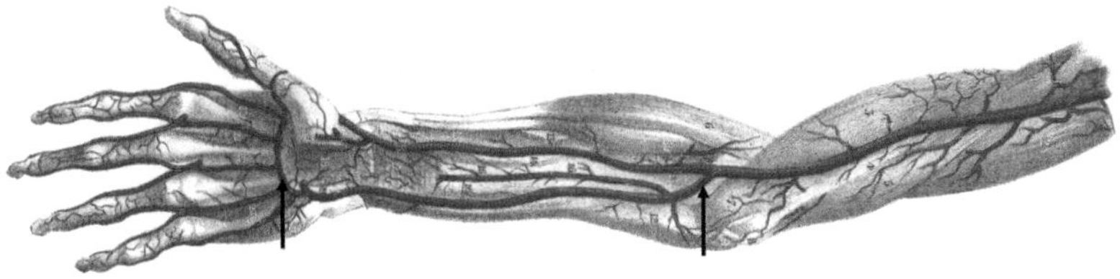

Fig. 99.1 Drawing of the ulnar artery arising (right arrow) from the brachial artery and continuing into the hand as the superficial palmar arch (left arrow) (Quain and Wilson, 1844 author's collection)

C. Leskanic
Department of Radiology, HCA Florida Trinity, Trinity, USA

S. J. Bordes, Jr. (✉)
Department of Surgery, Louisiana State University Health Sciences Center, New Orleans, LA, USA

S. J. Bordes, Jr. et al. (eds.), *The Clinical Anatomy of the Vascular System*,
https://doi.org/10.1007/978-3-031-78326-5_99

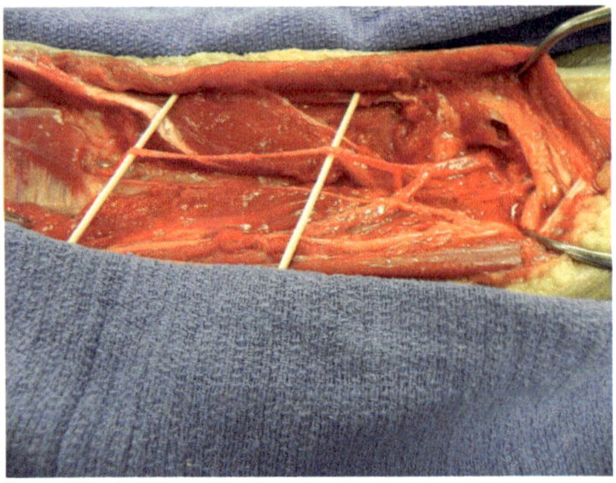

Fig. 99.2 Cadaveric dissection of the forearm demonstrating the anterior interosseous branch (traveling over the wooden sticks) of the common interosseous artery from the ulnar artery

The ulnar artery lies deep to the flexor muscles of the anterior compartment of the forearm; however, it runs superficial to the flexor retinaculum in the wrist (Reddy and Vollala 2007). It is lateral to the ulnar nerve when it enters the wrist and is deep and lateral to the flexor carpi ulnaris tendon, eventually terminating into the arches. (Gonzalez et al. 2005) The pronator teres muscle, flexor carpi radialis, and flexor digitorum superficialis encase it (Ginzburg et al. 2008).

99.3 Histology

The ulnar artery is a muscular artery with the primary role of maintaining blood pressure in the arterioles of the hand (Barry et al. 2007; Robertson and Watton 2013). The layers of its wall follow the usual pattern of a muscular artery (Robertson and Watton 2013). The innermost layer, tunica intima, includes the subendothelium and internal elastic lamina (Robertson and Watton 2013). The next layer is the tunica media, which includes circumferentially aligned layers of smooth muscle with interspersed collagen and a few elastin fibers (Robertson and Watton 2013). The outermost layer is the tunica adventitia, composed of collagen and some elastin (Robertson and Watton 2013). In muscular arteries such as the ulnar artery, the tunica adventitia occupies a larger percentage of the wall thickness (Robertson and Watton 2013). One study indicates that the adventitia of the ulnar artery contains sympathetic and parasympathetic nerves (Barry et al. 2007), which could be associated with spasms of this vessel (Barry et al. 2007).

99.4 Imaging

The ulnar vessel wall, lumen, and flow dynamics can all be assessed cost-effectively at the bedside (Ghouri et al. 2019). The ulnar artery can be visualized upon initial screening with color Doppler and vascular ultrasound (Ghouri et al. 2019).

In traumatic or occlusive cases, computed tomography angiography (CTA) is the most common imaging modality for evaluating the upper extremity (Ghouri et al. 2019). First, patients with contraindications to IV iodinated contrast should be identified (Ghouri et al. 2019). When the upper limb is to be imaged, the patient should ideally be positioned supine with the arm raised above the head, palm oriented ventrally, and fingers extended (Ghouri et al. 2019). Prone positioning is an alternative option, with the upper limbs raised and the palms facing the table (Ghouri et al. 2019). In traumatic scenarios, the upper extremities can be placed by the patient's side and supported using wedge pillows (Ghouri et al. 2019).

Magnetic resonance angiography (MRA) is helpful for disorders including hypothenar hammer syndrome, thromboangiitis obliterans, Raynaud's syndrome, and other disorders requiring multiple imaging evaluations (Ghouri et al. 2019). MRA can be performed with or without IV gadolinium-based contrast (Ghouri et al. 2019). A disadvantage of non-contrast MRA is limited visualization of the peripheral vasculature and overestimation of stenosis of the blood flow (Ghouri et al. 2019). Time of flight (TOF) and phase contrast (PC) are non-contrast MRAs that can be used to determine dynamic flow characteristics (Ghouri et al. 2019).

99.5 Variations

The anatomy of the ulnar artery can vary in several ways, including complete absence in some cases. The classical branching patterns of the radial and ulnar arteries occurred 84% of the time, according to Lippert and Pabst (Lippert and Pabst 1985; Tubbs et al. 2016). These authors also reported a superficial ulnar artery joining and replacing the normal one (Lippert and Pabst 1985; Tubbs et al. 2016). Another review revealed that the ulnar, radial, and common interosseous arteries branched directly from the brachial artery in 70% of cases (Tubbs et al. 2016).

A superficial ulnar artery is a rare variation, occurring in 0.67% to 9.38% of cases (Panagouli et al. 2009). In 0.17% to 2%, this artery arises from the axillary artery (Panagouli et al. 2009). A superficial ulnar artery can also arise from abnormal bifurcation of the brachial artery into a common radial-interosseous trunk and the superficial ulnar artery (Narayanan and Murugan 2018). In one reported case, a superficial ulnar artery gave off only minute muscular branches in the forearm (Quadros et al. 2015); the usual main branches of the ulnar artery arose instead from the radial artery (Quadros et al. 2015).

The ulnar artery can also arise from the axillary artery. In one reported case, it arose from the axillary artery at the level of the two median nerve roots (Natsis et al. 2006). In a cadaver in which a superficial ulnar artery originated from the axillary artery, it crossed the palmaris longus tendon from the radial to the ulnar side (Sato et al. 2018). The normal branching can vary so that the brachial artery bifurcates at a much higher point, leading to a longer ulnar artery or one that bifurcates earlier (Aharinejad et al. 1997). The ulnar artery's pattern in Guyon's canal can also vary considerably. This, in turn, can result in variation of the branching arteries that form the deep palmar arches (Murata et al. 2006).

99.6 Pathology

Ulnar Artery Occlusion

Injury to the ulnar artery resulting from thrombosis or laceration typically presents minimal symptoms owing to collateral circulation (Netscher and Janz 2008). The prevalence of asymptomatic occlusion is unknown (Netscher and Janz 2008). However, one study showed a higher prevalence of ulnar artery occlusion in the elderly male population (Carpentier et al. 2009). Among the symptomatic population, 12% reported unusual sensitivity of the fingertips to cold, and only 25% experienced blanching or cyanosis (Netscher and Janz 2008). Unilateral Raynaud's phenomenon can imply occlusive disease, but nonhealing ulcers indicate more severe vascular compromise from embolism or thrombosis (Netscher and Janz 2008). An association with Raynaud's phenomenon was found in 36% of patients in one study (Carpentier et al. 2009). Because there is an abundant collateral arterial system in the forearm, wrist, and hand, a single occlusion of the ulnar artery rarely leads to a critical vascular event in the vast majority of patients (Netscher and Janz 2008). Ischemic pain and ulceration typically present when there are two or more levels of arterial occlusion, such as occlusion of both the ulnar and digital arteries (Netscher and Janz 2008).

Posttraumatic Pseudoaneurysm

Upper limb arterial pseudoaneurysms are uncommon, but subfascial deep arteries are more likely to be affected than subcutaneous and perifascial arteries (Maertens et al. 2020). Deep upper extremity arteries such as the ulnar artery are more susceptible to pseudoaneurysms owing to their larger lumen diameters and lower pressure requirements for aneurysm formation, according to Laplace's Law (Maertens et al. 2020). Pseudoaneurysms typically result from direct penetrating or nonpenetrating trauma (Maertens et al. 2020). One study suggests that vasculitis, such as allergic granulomatous angiitis, can also be associated with the formation of ulnar artery pseudoaneurysms after eosinophil infiltration results in vascular injury (Sekino et al. 2005). A false saccular aneurysm can ensue from partial vessel wall injury or hematoma (Maertens et al. 2020). Repeated microtraumas can cause a dysplastic lesion of the vessel wall, which can evolve later into a fusiform aneurysm (Maertens et al. 2020). This is usually the main mechanism in hypothenar hammer syndrome (Maertens et al. 2020). The typical clinical presentation of a pseudoaneurysm is a painful and pulsatile mass (Maertens et al. 2020). Confirmational imaging can include US Doppler, MRA, CTA, or angiography (Maertens et al. 2020). Vascular complications are more significant in arterial ulnar pseudoaneurysms when the anastomotic network between the radial artery and the superficial palmar arch is either stenotic or absent (Maertens et al. 2020). Further complications can include thromboembolism with a risk of ischemia, rupture, or hemorrhage (Maertens et al. 2020).

Ulnar Artery Aneurysm

Ulnar artery aneurysms are rare but are usually associated with repetitive hand trauma believed to disrupt the internal elastic lamellae. They present as a pulsatile mass (Goodenough et al. 2020). The etiology can involve arteriosclerosis as well as trauma (Conkbayir et al. 2017). Untreated aneurysms are at risk of leading to arterial occlusions with distal ischemia, mural thrombi, or distal embolic events (Goodenough et al. 2020). In contrast to pseudoaneurysms, true aneurysms contain all layers of the arterial vessel (Conkbayir et al. 2017). Ulnar artery aneurysms can be visualized under Doppler ultrasound and definitively by angiography (Conkbayir et al. 2017). Treatment typically involves surgical resection of the affected arterial segment and potential reconstruction (Goodenough et al. 2020).

Hypothenar Hammer Syndrome

Hypothenar hammer syndrome (HHS) occurs secondary to repetitive trauma to the hypothenar eminence of the hand (Carr et al. 2019). As the ulnar artery exits Guyon's canal and branches into the deep palmar artery and superficial arch, it is susceptible to injury owing to its superficial location crossing the hamate bone (Carr et al. 2019). Classically, HHS appears in individuals repetitively using the hypothenar aspect of their palm as a hammer (Carr et al. 2019). The repetitive trauma causes intimal and medial wall degeneration and thrombosis, which can (though not necessarily) simultaneously cause an aneurysm (Carr et al. 2019). Symptomatic ischemia can result from subsequent embolic occlusion of the digital arteries (Carr et al. 2019). Individuals typically at risk include those who hammer objects with the hypothenar aspect of their palms such as carpenters, mechanics, and butchers, and athletes who sustain force to this area of the palm (Carr et al. 2019; Modarai and McIrvine 2008). However, repetitive trauma is not the only factor (Carr et al. 2019). Underlying fibromuscular dysplasia and smoking history can predispose to intimal and medial wall abnormalities (Carr et al. 2019). Symptoms of HHS include cold intolerance, numbness, and pain over the hypothenar eminence (Carr et al. 2019). In addition, cyanosis, ulceration, and ischemia of the middle and index fingers can occur (Carr et al. 2019). The severity of symptoms varies depending on the extent of arterial injury, the superficial palmar arch branching pattern, and additional collaterals (Carr et al. 2019). The ulnar artery injury of HHS can be visualized under ultrasound, digital subtraction arteriogram, CTA, or MRA (Carr et al. 2019). One study defined HHS as a segmental palmar ulnar artery occlusion or a patent palmar ulnar artery with elongation and a corkscrew appearance, both of which are associated with symptomatic embolic occlusions (Ferris et al. 2000). Conservative treatment is warranted initially; this includes avoidance of trauma to the hypothenar eminence, smoking cessation, antiplatelet therapy, vasodilator therapy, and palmar padding (Carr et al. 2019). If the symptoms are severe, including digital ischemia, or if a large aneurysm is noted upon imaging, then surgical intervention is indicated (Carr et al. 2019). In the initial setting of moderate symptoms, surgical intervention can prevent the complications of HHS that eventually occur in most cases when aneurysms are treated conservatively, such as Raynaud's phenomenon, ulnar nerve compression, or digital gangrene (Carr et al. 2019). Interventional options include sympathectomy, segmental excision of thrombus or aneurysm, or placement of an autologous venous or arterial graft (Carr et al. 2019).

References

Aharinejad S, Nourani F, Hollensteiner H (1997) Rare case of high origin of the ulnar artery from the brachial artery. Clin Anat 10(4):253–258. https://doi.org/10.1002/(SICI)1098-2353(1997)10:4<253::AID--CA7>3.0.CO;2-T

Al-Qattan MM, Yang Y, Kozin SH (2009) Embryology of the upper limb. J Hand Surg Am 34(7):1340–1350. https://doi.org/10.1016/j.jhsa.2009.06.013

Barry M, Touati G, Chardon K, Laude M, Libert JP, Sevestre H (2007) Histologic study of coronary, radial, ulnar, epigastric and internal thoracic arteries: application to coronary artery bypass grafts. Surg Radiol Anat 29(4):297–302. https://doi.org/10.1007/s00276-007-0214-4

Carpentier PH, Biro C, Jiguet M, Maricq HR (2009) Prevalence, risk factors, and clinical correlates of ulnar artery occlusion in the general population. J Vasc Surg 50(6):1333–1339. https://doi.org/10.1016/j.jvs.2009.07.076

Carr MP, Becker GW, Taljanovic MS, McCurdy WE (2019) Hypothenar hammer syndrome: case report and literature review. Radiol Case Rep 14(7):868–871. https://doi.org/10.1016/j.radcr.2019.04.015

Conkbayir C, Emiroğlu O, Birtan H, Cerit L (2017) Ulnar artery aneurysm – a rare cause of mass on wrist. Cor Vasa 59(5):e496–e498. https://doi.org/10.1016/j.crvasa.2016.11.0

Ferris BL, Taylor LM, Oyama K, McLafferty RB, Edwards JM, Moneta GL, Porter JM (2000) Hypothenar hammer syndrome: proposed etiology. J Vasc Surg 31(1):104–113. https://doi.org/10.1016/s0741-5214(00)70072-3

Ghouri MA, Gupta N, Bhat AP, Thimmappa ND, Saboo SS, Khandelwal A, Nagpal P (2019) CT and MR imaging of the upper extremity vasculature: pearls, pitfalls, and challenges. Cardiovasc Diagn Ther 9(S1):S152–S173. https://doi.org/10.21037/cdt.2018.09.15

Ginzburg E, Chong CK, Rich NM (2008) Chapter 62 – vascular anatomy of the extremities. In: Asensio JA, Trunkey DD (eds) Current therapy of trauma and surgical critical care. Mosby, Philadelphia, pp 467–472

Gonzalez MH, Weinzweig N, Graf CN, Elhassan B (2005) Chapter 6 – management of the ulnar mutilating injury. In: Weinzweig N, Weinzweig J (eds) The mutilated hand. Mosby, pp 87–95

Goodenough CJ, Afifi RO, Prakash SK, Buja LM, Lamaris GA (2020) Ulnar artery aneurysm as a late sequela of Marfan syndrome. J Hand Surg Am 45:1090.e1. https://doi.org/10.1016/j.jhsa.2020.01.010

Guéro S (2018) Developmental biology of the upper limb. Hand Surg Rehabil 37(5):265–274. https://doi.org/10.1016/j.hansur.2018.03.007

Lippert H, Pabst R (1985) Arterial variations in man: classification and frequency. Bergmann Verlag, Munich

Maertens A, Ritz FJT, Poumellec MA, Camuzard O, Balaguer T (2020) Posttraumatic pseudoaneurysm of a superficial branch of the ulnar artery: a case report. Int J Surg Case Rep 75:317–321., ISSN 2210-2612. https://doi.org/10.1016/j.ijscr.2020.09.029

Modarai B, McIrvine A (2008) The hypothenar hammer syndrome. J Vasc Surg 47(6):1350. https://doi.org/10.1016/j.jvs.2007.04.078

Murata K, Tamai M, Gupta A (2006) Anatomic study of arborization patterns of the ulnar artery in Guyon's canal. J Hand Surg Am 31(2):258–263. https://doi.org/10.1016/j.jhsa.2005.11.004

Narayanan S, Murugan S (2018) Bifurcation of brachial artery into a common radial-interosseous trunk and superficial ulnar artery: a rare variation. Anat Sci Int 93(3):400–403. https://doi.org/10.1007/s12565-017-0427-5

Natsis K, Papadopoulou AL, Paraskevas G, Totlis T, Tsikaras P (2006) High origin of a superficial ulnar artery arising from the axillary artery: anatomy, embryology, clinical significance and a review of the literature. Folia Morphol (Warsz) 65(4):400–405

Netscher DT, Janz B (2008) Treatment of symptomatic ulnar artery occlusion. J Hand Surg Am 33(9):1628–1631. https://doi.org/10.1016/j.jhsa.2008.08.013

Panagouli E, Tsaraklis A, Gazouli I, Anagnostopoulou S, Venieratos D (2009) A rare variation of the axillary artery combined contralaterally with an unusual high origin of a superficial ulnar artery: description, review of the literature and embryological analysis. Ital J AnatEmbryol 114(4):145–156

Quadros LS, Bhat N, D'Souza AS (2015) Superficial ulnar artery: a case report of its unusual course. Malays J Med Sci 22(3):65–67

Reddy S, Vollala VR (2007) The superficial ulnar artery: development and clinical significance. J Vasc Bras 6(3):285–288. SociedadeBrasileira de Angiologia e de Cirurgia Vascular São Paulo, Brasil

Robertson AM, Watton PN (2013) Mechanobiology of the arterial wall. In: Transport in biological media. Elsevier, Amsterdam, pp 275–347. https://doi.org/10.1016/b978-0-12-415824-5.00008-4

Sato K, Murakami K, Mimata Y, Kikuchi Y, Oikawa R, Doita M (2018) Superficial ulnar artery crossing over the palmaris longus tendon at the wrist in a cadaver: a case report. J Hand Surg Asian Pac 23(1):137–139. https://doi.org/10.1142/S2424835518720074

Sekino S, Takagi H, Kato T, Matsuno Y, Sekido Y, Umemoto T (2005) Nontraumatic pseudoaneurysm of the proximal ulnar artery with eosinophilia. J Vasc Surg 42(6):1233–1235. https://doi.org/10.1016/j.jvs.2005.08.019

Tubbs RS, Shoja MM, Loukas M (2016) Bergman's comprehensive encyclopedia of human anatomic variation. Wiley

Ulnar Recurrent Arteries 100

Katherine D. Bordes
and Stephen J. Bordes, Jr.

100.1 Embryology

During the fourth week of development, the upper limb bud forms from a mass of mesenchymal cells from the lateral mesoderm, covered by ectoderm, which later differentiates into bone, cartilage, and blood vessels. There are alternative models of arterial development of the arm, including the proposal that the forearm arterial supply is derived from one single axis artery from the seventh segmental artery. However, according to the leading model, the arterial supply begins as a gradually developing capillary network via angiogenesis and in situ differentiation of mesoderm tissue into endothelial cells, which differentiates proximally and distally to form the axillary, brachial, and interosseous arteries. After the brachial artery has developed, the structures distal to the elbow develop, including the ulnar, median, interosseous, and radial arteries (Berezovsky and Bordoni 2021; Rodríguez-Niedenführ et al. 2001).

K. D. Bordes
Tulane University School of Medicine,
New Orleans, LA, USA

S. J. Bordes, Jr. (✉)
Department of Surgery, Louisiana State University
Health Sciences Center, New Orleans, LA, USA

100.2 Anatomy

The ulnar artery contributes extensively to the vascular supply of the upper limb and hand. Its two main proximal branches are the anterior and posterior ulnar recurrent arteries (Marques and Bordoni 2021).

The anterior ulnar recurrent artery is the first branch. It splits off the ulnar artery just distal to the elbow joint and medial to the cubital fossa. It then ascends anterior to the medial epicondyle of the humerus, passing between the brachialis and pronator teres muscles and supplying both. At the distal end of the humerus, it anastomoses with the inferior ulnar collateral artery, a branch of the branchial artery, and contributes to the arterial anastomosis of the elbow (Hacking 2021a; Berezovsky and Bordoni 2021; Moore et al. 2014).

The posterior ulnar recurrent artery branches laterally from the ulnar artery just distal to its anterior counterpart and ascends posterior to the medial epicondyle of the humerus. It then courses between the flexor digitorum profundus and the flexor digitorum superficialis and runs deep to the flexor carpi ulnaris. It anastomoses with the superior ulnar collateral artery, thereby contributing to the arterial anastomosis of the elbow, like its anterior counterpart (Hacking 2021b; Berezovsky and Bordoni 2021; Moore et al. 2014).

S. J. Bordes, Jr. et al. (eds.), *The Clinical Anatomy of the Vascular System*,
https://doi.org/10.1007/978-3-031-78326-5_100

100.3 Clinical/Surgical Significance

Reconstructive surgery involving the use of perforator flaps has proved less invasive and more successful in preserving muscle function than other methods such as the myocutaneous flap. The perforator flap reconstruction involves transferring the skin flap of a vessel that perforates a corresponding envelope of target tissue. Ulnar artery perforator flaps can be classified on the basis of the branch's origin, including proximal, middle, and distal flaps. Although there have been relatively few cases, anterior and posterior recurrent arteries have been used successfully as proximal ulnar perforator flaps for elbow and forearm reconstruction following burns or traumatic wounds. In one analysis of posterior ulnar recurrent artery perforator (PURAP) flap reconstruction by Mateev et al., 95.5% of cases were judged excellent or good and required no further operations (Yamamoto et al. 2019; Mateev et al. 2009).

References

Berezovsky DR, Bordoni B (2021) Anatomy, shoulder and upper limb, forearm arteries. In: StatPearls. StatPearls Publishing. http://www.ncbi.nlm.nih.gov/books/NBK545155/. Accessed 27 Dec 2021

Hacking C (2021a) Anterior ulnar recurrent artery | Radiology Reference Article | Radiopaedia.org. Radiopaedia. https://radiopaedia.org/articles/anterior-ulnar-recurrent-artery?lang=us. Accessed 27 Dec 2021

Hacking C (2021b) Posterior ulnar recurrent artery | Radiology Reference Article | Radiopaedia.org. Radiopaedia. https://radiopaedia.org/articles/posterior-ulnar-recurrent-artery?lang=us. Accessed 27 Dec 2021

Marques E, Bordoni B (2021) Anatomy, shoulder and upper limb, ulnar artery. In: StatPearls. StatPearls Publishing. http://www.ncbi.nlm.nih.gov/books/NBK546619/. Accessed 7 Jan 2022

Mateev MA, Trunov L, Hyakusoku H, Ogawa R (2009) Analysis of 22 posterior ulnar recurrent artery perforator flaps: a type of proximal ulnar perforator flap. Eplasty 10:e2

Moore KL, Dalley AF, Agur AMR (2014) Upper limb. In: Clinically oriented anatomy, 7th edn. Lippincott Williams & Wilkins

Rodríguez-Niedenführ M, Burton GJ, Deu J, Sañudo JR (2001) Development of the arterial pattern in the upper limb of staged human embryos: normal development and anatomic variations. J Anat 199(Pt 4):407–417

Yamamoto T, Yamamoto N, Kageyama T et al (2019) Definition of perforator flap: what does a "perforator" perforate? Glob Health Med 1(2):114–116. https://doi.org/10.35772/ghm.2019.01009

Interosseous Arteries

101

Katherine D. Bordes and Stephen J. Bordes, Jr.

101.1 Embryology

During the fourth week of development, the upper limb forms from a mass of mesenchymal cells from lateral mesoderm, covered by ectoderm, which later differentiates into bone, cartilage, and blood vessels. Although there are alternative theories regarding the arterial development of the arm, the arterial supply likely initiates as a gradually developing capillary network via angiogenesis and in situ differentiation of mesoderm tissue into endothelial cells, which differentiates in a proximal to distal fashion to form axillary, brachial, and interosseous arteries. It has also been considered that the forearm arterial supply is derived from one single axis artery from the seventh segmental artery (Berezovsky and Bordoni 2021; Rodríguez-Niedenführ et al. 2001).

101.2 Anatomy

The common interosseous artery (Fig. 101.1) is a short branch off the proximal part of the ulnar artery, found at the distal end of the cubital fossa and at the level of the pronator teres. Then, it almost immediately divides into the anterior and posterior interosseous branches (Common interosseous artery 2021; Moore et al. 2014).

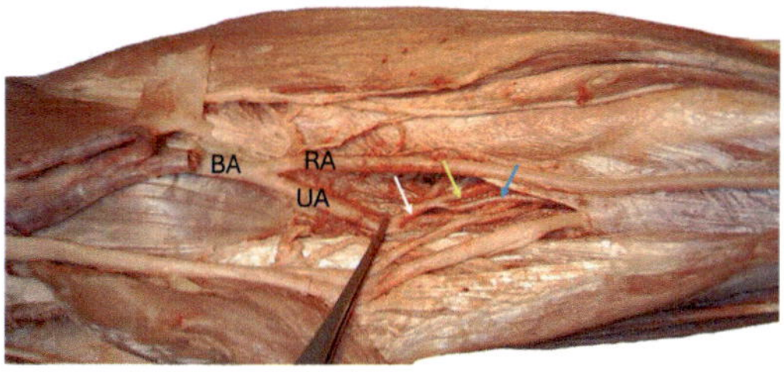

Fig. 101.1 Cadaveric dissection of the interosseous arteries. BA = brachial artery, RA = radial artery, UA = ulnar artery, common interosseous artery (white arrow), posterior interosseous artery (yellow arrow), anterior interosseous artery (blue arrow). (Courtesy Prof. M. Loukas)

The anterior interosseous artery (Fig. 101.1) passes proximal to the border of the pronator quadratus, running with the anterior interosseous nerve along the anterior aspect of the interosseous membrane, between the flexor digitorum profundus and the flexor pollicis longus. It then bifurcates at the level of the pronator quadratus, and one branch continues distally to anastomose with the palmar carpal arch, and another branch pierces the interosseous membrane to anastomose with the posterior interosseous artery (Hacking 2021a; Moore et al. 2014; Drake et al. 2015).

The posterior interosseous artery (Fig. 101.1), also a branch of the common interosseous artery, runs alongside the posterior interosseous nerve

K. D. Bordes
Tulane University School of Medicine, New Orleans, LA, USA

S. J. Bordes, Jr. (✉)
Department of Surgery, Louisiana State University Health Sciences Center, New Orleans, LA, USA

S. J. Bordes, Jr. et al. (eds.), *The Clinical Anatomy of the Vascular System*,
https://doi.org/10.1007/978-3-031-78326-5_101

between the radial and ulnar arteries. After piercing the interosseous membrane, it courses distally between the superficial and deep layers of the extensor compartment. Specifically, it lies deep to the supinator and superficial to the abductor pollicis longus, supplying both. It terminates by anastomosing with a branch of the anterior interosseous artery, contributing to the dorsal carpal arch of the wrist (Hacking 2021b; Moore et al. 2014; Drake et al. 2015).

The interosseous recurrent artery is a proximal branch of the posterior interosseous artery that passes between the lateral epicondyle of the humerus and the olecranon of the ulna. It also courses between the supinator and anconeus muscles but may pierce the former to supply the superficial extensor muscles. It contributes to the arterial anastomosis of the elbow via the middle collateral artery (Hacking 2021c; Moore et al. 2014; Drake et al. 2015).

101.3 Variations

Variations in arterial branching have been reported in up to 20% of human adults, even more so if the median artery is considered (Rodríguez-Niedenführ et al. 2001). Many variations of forearm arterial anatomy are suspected to be either due to improper involution or branching during embryologic development or possible changes in capillary differentiation pattern (Berezovsky and Bordoni 2021). Variations of forearm arteries are only detectable approximately 7 weeks into embryologic development (Rodríguez-Niedenführ et al. 2001).

The classic branching pattern of the radial and ulnar arteries occurs 84% of the time; however, the radial, ulnar, and common interosseous arteries all arise directly from the brachial artery in 70% of cases. In about 3% of cases, the radial artery arises from the superficial brachial artery, which gives rise to the ulnar and common interosseous arteries. In this case, there is an extensive anastomosis between the radial and brachial arteries (Tubbs et al. 2016).

Ulnar arterial branch variations include anterior and posterior interosseous arteries arising separately from the ulnar artery, common interosseous arising from the axillary artery, common interosseous arising from the high radial artery, and interosseous recurrent originating from the common interosseous rather than the posterior interosseous artery (Tubbs et al. 2016).

In about 32% of cases, there is a congenital absence of the common interosseous artery. Regarding the anterior and posterior interosseous branches, there is a congenital absence in 1.5% and 7.1% of cases, respectively (Al Talalwah et al. 2015).

In relation to the common interosseous artery, it is found that in about 3% of cases, the median artery emerges from the ulnar artery distal to the origin of the common interosseous artery, 2% of cases originate proximal to its origin, 2% of cases emerge directly from its origin, and 1% branch from the anterior interosseous artery (Tubbs et al. 2016).

The radial, ulnar, and interosseous arteries construct collateral pathways for the arterial supply of the wrist and hand via palmar arches. In 50% of cases, a dorsal carpal branch of the radial artery and the anterior interosseous artery form the dorsal carpal arch. In 30% of cases, the anterior interosseous and dorsal carpal branches from the radial and ulnar arteries contribute to the dorsal carpal arch. Finally, in 3% of cases, the dorsal carpal arch is formed by the anterior interosseous artery and a dorsal carpal branch from the ulnar artery, with no contribution from the radial artery (Tubbs et al. 2016). It is also possible that the anterior interosseous artery and median artery supply the superficial palmar arterial arch (Tubbs et al. 2016). One case, reported by, describes a complete unilateral absence of the radial artery in which the anterior interosseous artery compensates for the supply of the forearm and hand, with no deep palmar arterial arch present in this subject (Tubbs et al. 2016).

Of note, there is an incidence of 16% of Down's syndrome patients possessing an anomalous radial arterial pattern. In this case, eight of the 77 subjects possessed an enlarged anterior interosseous artery, of which seven had no actual radial artery (Tubbs et al. 2016).

101.4 Clinical/Surgical Significance

Due to collateral arterial supply to the wrist and hand via radial, ulnar, and interosseous arteries, occlusion or harvest of the radial or ulnar arteries does not result in significant complications such as ischemia.

Additionally, in the reconstructive use of forearm flaps, the tissue and arterial supply of the forearm are harvested together to heal the damaged area receiving the graft. Although ulnar and radial forearm free flaps are most common, posterior ulnar recurrent artery, posterior interosseous artery, and anterior interosseous artery and their perforating branches have also been reported (Berezovsky and Bordoni 2021).

References

Al Talalwah W, Regassa D, Soames R (2015) The morphology of common interosseous artery and its clinical significance. Sch J Appl Med Sci 3:1126–1131

Berezovsky DR, Bordoni B (2021) Anatomy, shoulder and upper limb, forearm arteries. In: StatPearls. StatPearls Publishing. http://www.ncbi.nlm.nih.gov/books/NBK545155/. Accessed 23 Dec 2021

Common interosseous artery | Radiology Reference Article | Radiopaedia.org. https://radiopaedia.org/articles/common-interosseous-artery?lang=us. Accessed 19 Dec 2021

Drake RL, Vogl AW, Mitchell AWM (2015) Regional anatomy – posterior compartment of the forearm. In: Gray's anatomy for students, 3rd edn. Churchill Livingstone

Hacking C (2021a) Anterior interosseous artery | Radiology Reference Article | Radiopaedia.org. Radiopaedia. https://radiopaedia.org/articles/anterior-interosseous-artery?lang=us. Accessed 19 Dec 2021

Hacking C (2021b) Posterior interosseous artery | Radiology Reference Article | Radiopaedia.org. Radiopaedia. https://radiopaedia.org/articles/posterior-interosseous-artery?lang=us. Accessed 19 Dec 2021

Hacking C (2021c) Interosseous recurrent artery | Radiology Reference Article | Radiopaedia.org. Radiopaedia. https://radiopaedia.org/articles/interosseous-recurrent-artery?lang=us. Accessed 19 Dec 2021

Moore KL, Dalley AF, Agur AMR (2014) Upper limb. In: Clinically oriented anatomy, 7th edn. Lippincott Williams & Wilkins

Rodríguez-Niedenführ M, Burton GJ, Deu J, Sañudo JR (2001) Development of the arterial pattern in the upper limb of staged human embryos: normal development and anatomic variations. J Anat 199(Pt 4):407–417. https://doi.org/10.1046/j.1469-7580.2001.19940407

Tubbs RS, Shoja M, Loukas M (2016) Bergman's comprehensive encyclopedia of human anatomic variation. Wiley, Hoboken

Arteries of the Hand and Digits

102

Katherine D. Bordes, Lisandro Montorfano, and Stephen J. Bordes, Jr.

102.1 Anatomy

The ulnar and the radial arteries course distally down the forearm and are responsible for all the blood supply to the hand. The palmar carpal branch of the ulnar artery passes laterally across the anterior aspect of the distal ulna and deep to the tendons of the flexor digitorum profundus. The corresponding palmar carpal branch of the radial artery stems just distal to the inferior edge of the pronator quadratus, then continues inferomedially along the anterior aspect of the distal radius, also deep to the flexor muscles. These two vessels anastomose centrally to form the palmar carpal arch, which is responsible for the vascular supply of the anterior wrist (Gray 1918).

Superficial Palm

The superficial palmar branch of the radial artery arises just proximal and superficial to the flexor retinaculum and then passes into the deep plane of the palm, over the floor of the anatomical snuffbox. From there, it runs between the heads of the first dorsal interosseous and the adductor pollicis muscles, supplying the thenar muscles. In rare instances, it passes superficial to the thenar muscles (Drake et al. 2015; Singer et al. 2018).

The superficial palmar arch can be formed by anastomosis of the terminal branch of the ulnar artery and the superficial branch of the radial or the radialis indicis artery. Of note, the arch remains incomplete in almost two-thirds of the population. Approximately 1–2 cm distal to the edge of the transverse carpal ligament, this vascular network crosses the palm between the flexor tendons and the palmar aponeurosis (Gray 1918; Tan and Lahiri 2020; Kong and Varacallo 2022).

Three common palmar digital arteries course from the superficial palmar arch toward the second, third, and fourth lumbrical muscles. They supply the second, third, and fourth digits correspondingly. The thumb receives some of its blood supply from the radial and ulnar pollicis arteries, which are branches of the first common digital artery (*see deep palm section*). The fifth digit is supplied by the digiti minimi ulnaris artery, a branch of the fourth common digital artery (Gray 1918; Tan and Lahiri 2020; Kong and Varacallo 2022).

At the webs of the fingers, the common palmar digital arteries anastomose with their corresponding palmar metacarpal arteries from

K. D. Bordes
Tulane University School of Medicine, New Orleans, LA, USA

L. Montorfano
Department of Surgery, Cleveland Clinic, Weston, FL, USA

S. J. Bordes, Jr. (✉)
Department of Surgery, Louisiana State University Health Sciences Center, New Orleans, LA, USA

S. J. Bordes, Jr. et al. (eds.), *The Clinical Anatomy of the Vascular System*,
https://doi.org/10.1007/978-3-031-78326-5_102

the deep palmar arch, and then divide into a pair of proper palmar digital arteries. These run along the contiguous sides of digits two, three, four, and the lateral side of the little finger and pass lateral to the corresponding digital nerve. Within the subcutaneous tissue near the interphalangeal joints, they anastomose with branches from the contralateral digital artery to form transverse arches. These arteries can also give off dorsal branches at the distal interphalangeal joint, which anastomose with the dorsal digital arteries to supply the second and third phalanges, including the matrix of the fingernail (Gray 1918; Tan and Lahiri 2020; Kong and Varacallo 2022).

Deep Palm

The deep palmar branch of the ulnar artery originates near the base of the 5th metacarpal and then passes distally through, and supplies, the abductor digiti minimi and flexor digiti minimi brevis, and through the origin of the opponens digit minimi. The dorsal radial artery courses through the anatomical snuffbox and enters the palmar aspect of the hand through the two heads of the first dorsal interosseous muscle. It then turns medially and anastomoses with the deep palmar branch of the ulnar artery to form the deep palmar arch. In contrast to the superficial palmar arch, the deep palmar arch is typically complete. It lies superficial to the metacarpal bones and deep to the flexor tendons and is located approximately 1 cm proximal to the superficial palmar arch. The deep branch of the ulnar nerve runs alongside it toward the radial side of the hand. This arch gives rise to the palmar metacarpal arteries and can also give rise to the princeps pollicis artery to the thumb (Gray 1918; Kong and Varacallo 2022).

There are three or four slender palmar metacarpal arteries, which run distally and lie superficial to the interossei muscles. They anastomose with the previously mentioned common palmar digital arteries to form the proper palmar digital arteries. They also perforate the interosseous spaces to anastomose with the dorsal metacarpal arteries (Gray 1918).

As the radial artery turns on to the palmar aspect of the hand, it gives off the princeps pollicis artery (*first palmar metacarpal artery*). This courses distally along the anteromedial aspect of the first metacarpal, running deep to the flexor pollicis longus tendon, and passes between the first dorsal interosseous muscle and the oblique head of the adductor pollicis muscle. At the base of the first phalanx, it divides into two branches, which form an arch on the palmar surface of the last phalanx, supplying the skin and subcutaneous tissue of the thumb (Gray 1918). Both the ulnar and radial arteries provide branches for the palmar digital arteries of the thumb. In most cases, both contributions arise directly from the princeps pollicis (Tan and Lahiri 2020; Kong and Varacallo 2022).

The radialis indicis artery usually arises from the radial artery, although it can originate instead from the princeps pollicis artery or the first common palmar artery. After passing between the first dorsal interossei muscle and adductor pollicis transversus muscle, it runs along and supplies the lateral side of the index finger. It then anastomoses with the corresponding proper digital artery, which is responsible for supplying the medial side of the digit. Depending on its origin, this artery can anastomose with the princeps pollicis artery and/or contribute a communicating branch to the superficial palmar arch (Gray 1918; Kong and Varacallo 2022).

Dorsal Hand (Fig. 102.1)

The dorsal carpal branch of the radial artery is a small vessel that transverses medially by crossing the scaphoid and trapezium. It passes through the floor of the anatomical snuffbox, between the two heads of the first dorsal interosseous muscle, and beneath the extensor pollicis tendons. Before passing through the first dorsal interosseous muscle, it gives rise to the first dorsal metacarpal artery and a transverse branch. The corresponding dorsal carpal branch of the ulnar artery arises above the pisiform bone and passes deep to the tendon of the flexor carpi ulnaris and the extensor tendons. This artery gives off a small branch that

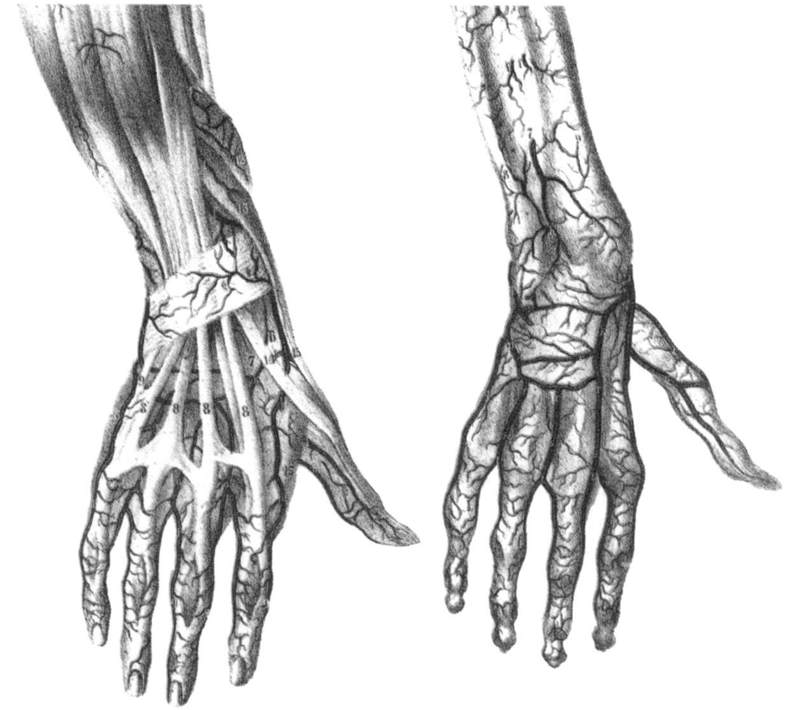

Fig. 102.1 Arteries on the dorsum of the hand (Quain and Wilson, 1844 author's collection)

runs along the fifth metacarpal and supplies the ulnar side of the dorsal surface of the fifth digit. The dorsal carpal branches along with the anterior and posterior interosseous arteries anastomose to form the dorsal carpal arch (Gray 1918; Tan and Lahiri 2020; Tubbs et al. 2016).

Three dorsal metacarpal arteries branch off from the dorsal carpal arch, running along the second, third, and fourth dorsal interossei muscles. Proximal perforating branches of the dorsal metacarpal arteries contribute to the deep palmar arch by anastomosing either with the dorsum of the deep palmar arch or the palmar metacarpal branches (Olave et al. 1998). Distal perforating branches of the dorsal metacarpal arteries connect each dorsal metacarpal artery with a common or proper palmar digital branch (Edwards 1960). Recurrent branches arise from the deep palmar arch, supply intercarpal articulations, and end in the palmar carpal network (Gray 1918).

These metacarpal branches continue, forming the proper dorsal digital arteries, which travel distally over the middle phalanges to communicate with their corresponding common palmar digital arteries of the superficial palmar arch at the level of the proximal interphalangeal joints, forming a superficial arcade (Gray 1918; Tan and Lahiri 2020; Kong and Varacallo 2022).

The first dorsal metacarpal artery (FDMCA) almost always arises from the dorsal carpal branch of the radial artery or the terminal radial branch of the dorsal carpal arch. It runs in the first intermetacarpal space, just distal to the extensor pollicis longus tendon, and passes between the two heads of the first interosseous dorsalis muscle. It then divides into a radial branch to the thumb, an intermediate branch to the first web space, and an ulnar branch to the second finger. This artery is responsible for supplying the skin of the proximal interphalangeal joint, the first interosseous dorsalis muscle, and the second metacarpal (Gray 1918; Tan and Lahiri 2020; Edwards 1960; Muyldermans and Hierner 2009).

The dorsal arteries of the thumb are more variable than their palmar counterparts. They typically originate from the first dorsal metacarpal artery, but they can also arise from the ulnar or radial arteries. The palmar or dorsal arteries of the thumb are both capable of supplying this digit independently should the other vascular supply become compromised (Tan and Lahiri 2020).

102.2 Variations

The arteries of the hand can vary and frequently anastomose with each other, forming superficial and deep arches as well as connections between the palmar and dorsal portions of the digits, resulting in a rich vascular network. In about half the population, the branching patterns differ between the hands. The superficial or deep palmar arches can also be duplicated. As a result, it is important to be aware that the "classic" descriptions of arterial supply can be less common than the variations (Tubbs et al. 2016).

The superficial palmar arch is either complete, meaning there is an anastomosis, or incomplete/open, meaning no anastomosis. Reports on the incidence of an incomplete arch range widely from 3.6% to 58% of cases. With an incomplete arch, there are at least five variations in which the main arteries take responsibility for blood supply to the digits. In 37% of cases, the ulnar artery gives rise to the radialis indicis artery and the three palmar metacarpal arteries. In 13%, the radial artery gives rise to the radialis indicis artery, and the palmar metacarpal arteries come from the ulnar artery. In 4%, the radialis indicis and first palmar metacarpal arteries stem from the median artery, while the medial two palmar metacarpal arteries are derived from the ulnar artery. In 3%, the radialis indicis and first palmar metacarpal arteries come from the radial artery and the medial two palmar metacarpal arteries from the ulnar artery. In 1%, the radialis indicis artery stems from the radial artery, the first metacarpal artery from the medial artery, and the medial two palmar metacarpal arteries from the ulnar artery. With a complete arch, there are also several variations in composition including the ulnar artery comprising the entire arch, or an anastomosis between the median and ulnar arteries (4%), or an anastomosis of the radial, median, and ulnar arteries (1%), or an anastomosis with the deep palmar arch and an ulnar artery-initiated superficial arch (2%) (Tan and Lahiri 2020; Tubbs et al. 2016).

There are four primary variations of the deep palmar arch. The first is the most common; in 79% of cases, the deep branch of the ulnar artery joins the deep palmar arch, which then gives off the princeps pollicis as well as three palmar metacarpal arteries. In 13%, there are two deep branches of the ulnar artery, which contribute to the deep palmar arch. In 5%, there are contributions from the anterior interosseous, radial, and ulnar arteries. In 3%, the deep palmar arch is incomplete. It is rare (<1% of cases) for the radial artery to be the only contributor to the deep palmar arch (Tubbs et al. 2016).

The common palmar digital arteries are classically described as being formed by the convergence of palmar metacarpal arteries from the superficial and deep palmar arterial arches. However, in practice, this only occurs in one third of cases. In 60% the palmar metacarpal arteries of the deep palmar arch provide no functional anastomosis and therefore do not contribute to the blood supply to the digits. In 10%, the palmar metacarpal arteries anastomose directly with the proper palmar digital arteries (Tubbs et al. 2016).

With regard to the two palmar digital arteries to the thumb, in 90% of cases the radial branch originates from the principal artery of the thumb. In the other 10% it comes directly from the superficial palmar branch of the radial artery. The ulnar counterpart is more variable; in 50–70% of the population it stems from the principal artery of the thumb, but in the other 50% it originates at the superficial palmar arch, the superficial branch of the radial artery, or the first dorsal metacarpal artery (Tan and Lahiri 2020).

In 10% of cases, the princeps pollicis artery arises from the deep palmar arch, while the radialis indicis and three palmar metacarpal arteries arise from the superficial palmar arch. In about 2% of the population, the princeps pollicis artery is completely absent (Tubbs et al. 2016).

In 77% of cases, the radialis indicis artery and all three palmar metacarpal arteries arise solely from the superficial palmar arch. In 11% it arises from the deep palmar arch, while the three palmar metacarpal arteries arise from the superficial palmar arch. In only 2% of cases, the radialis indicis artery and the first palmar metacarpal artery branch from the deep palmar arterial arch, while the remaining two palmar metacarpal arteries arise from the superficial palmar arch (Tubbs et al. 2016).

Although less clinically relevant, the dorsal arterial supply also has many variations. The dorsal carpal arch is present in approximately 80% of the population and includes a contribution from the anterior interosseous artery in 50–66%. In one third of cases, an anastomosis between the anterior interosseous and the dorsal carpal branches of the radial and ulnar arteries forms the dorsal carpal arch. In 8%, the arch is formed entirely by the dorsal carpal branch of the radial artery. The classic description of the dorsal carpal arch, in which the dorsal carpal branch of the ulnar artery joins the corresponding branch of the radial artery, actually only occurs in 5% of the population. In 4%, the dorsal carpal arch and dorsal metacarpal arteries are completely absent. Even less common, in 3% of cases, the arch is formed by the anterior interosseous artery and the dorsal carpal branch of the ulnar artery (Tan and Lahiri 2020; Tubbs et al. 2016).

The terminal branches of the first dorsal metacarpal artery can vary; however, the main branch itself is quite constant. The ulnar and radial dorsal arteries of the thumb typically originate from the first dorsal metacarpal artery, but in a few cases, they originate from the princeps pollicis artery. They are absent altogether in about 30% of the population (Tan and Lahiri 2020; Couceiro et al. 2018).

102.3 Clinical Significance

Kaplan's cardinal line (KCL) is a notable surgical landmark for identifying the superficial palmar arch on the surface of the palm. It was originally described by E.B. Kaplan in 1953 as "drawn from the apex of the interdigital fold between the thumb and the index finger toward the ulnar side of the hand, parallel with the middle crease of the hand." It is commonly used to avoid iatrogenic arterial injuries, especially in carpal tunnel releases (Kaplan 1953; Panchal and Trzeciak 2010).

The radial artery is often used in the clinical setting for drawing blood, cannulation for arterial line placement, catheterization, or as a graft. It is therefore important, before any procedures, to assess the extent to which this artery is responsible for the hand's blood supply. The Allen test can be used to determine easily whether there is collateral circulation. The use of radial or ulnar artery flaps in reconstructive surgery needs to be more carefully planned in patients in whom a dynamic change is detected (Kong and Varacallo 2022). This is equally important in the context of vascular and trauma surgery as the radial artery can safely be sacrificed, while control of hemostasis is established during damage-control exploratory procedures in the vast majority of individuals, the ulnar supply to the palmar arch being generally dominant. Furthermore, cardiothoracic and vascular surgeons can consider the radial or ulnar arteries for graft/harvest for arterial bypass (Tan and Lahiri 2020). The radial artery can be harvested or used as a graft without adverse effects to the palmar arch in most individuals.

A first dorsal metacarpal artery flap is a distant pedicled flap that can be used in cases of distal soft tissue defects and thumb reconstruction. It is harvested from the dorsal index finger, including the first dorsal metacarpal artery, a corresponding vein, and a branch of the superficial radial nerve. It can also be used in combination with the second dorsal metacarpal artery. With this procedure, there is minimal donor site morbidity and successful postoperative range of motion, similar to that of the contralateral thumb (Couceiro et al. 2018; Prabhu et al. 2013).

The dorsal metacarpal artery perforator flap, also called the "Quaba flap," is a vascular island flap used for soft tissue defects on the dorsal aspect of the fingers. It is usually applied to a partial defect proximal to the fingertip and typically entails few morbidities apart from decreased sensation. The transverse communications between the proper digital arteries allow for successful vascularization in the use of distally based vascular island flaps (Tan and Lahiri 2020; Aizawa et al. 2018).

References

Aizawa T, Inada C, Kubo S, Aoki S, Azuma R, Kiyosawa T (2018) Quaba flap for reconstruction after cancer resection of an amputated finger stump. J Hand Microsurg 10(2):119–122. https://doi.org/10.1055/s-0038-1630145

Couceiro J, de Prado M, Menendez G, Manteiga Z (2018) The first dorsal metacarpal artery flap family: a review. Surg J (N Y). 4(4):e215–e219. https://doi.org/10.1055/s-0038-1675369. Published 2018 Oct 29

Drake RL, Vogl AW, Mitchell AWM (2015) Gray's anatomy for students, 3rd edn. Chuchill Livingstone

Edwards EA (1960) Organization of the small arteries of the hand and digits. Am J Surg 99(6):837–846. https://doi.org/10.1016/0002-9610(60)90472-4

Gray H (1918) Anatomy of the human body. Lea & Febiger, Philadelphia. Bartleby.com, 2000. www.bartleby.com/107/

Kaplan EB (1953) Surface anatomy of the hand and wrist. In: Spinner E (ed) Functional and surgical anatomy of the hand. J.B. Lippincott Co, Philadelphia, pp 227–231

Kong AC, Varacallo M (2022) Anatomy, shoulder and upper limb, hand volar arch arteries. In: StatPearls. StatPearls Publishing. http://www.ncbi.nlm.nih.gov/books/NBK545377/. Accessed 21 Jan 2022

Muyldermans T, Hierner R (2009) First dorsal metacarpal artery flap for thumb reconstruction: a retrospective clinical study. Strategies Trauma Limb Reconstr 4(1):27–33. https://doi.org/10.1007/s11751-009-0056-1

Olave E, Carlos Prates J, Gabrielli C, Mandiola E (1998) Perforating branches: important contribution to the formation of the dorsal metacarpal arteries. Scand J Plast Reconstr Surg Hand Surg 32(2):221–228. https://doi.org/10.1080/02844319850158859

Panchal AP, Trzeciak MA (2010) The clinical application of Kaplan's cardinal line as a surface marker for the superficial palmar arch. Hand (N Y) 5(2):155–159. https://doi.org/10.1007/s11552-009-9229-0

Prabhu M, Powar R, Sulhyan SR (2013) FDMA flap: a versatile technique to reconstruct the thumb. IJPMBS 2(4):7

Singer G, Marterer R, Till H, Schmidt B (2018) A rare anatomic variation of the superficial palmar branch of the radial artery causing pain. Surg Radiol Anat 40(3):349–352. https://doi.org/10.1007/s00276-017-1936-6

Tan RES, Lahiri A (2020) Vascular anatomy of the hand in relation to flaps. Hand Clin 36(1):1–8. https://doi.org/10.1016/j.hcl.2019.08.001

Tubbs RS, Shoja M, Loukas M (2016) Bergman's comprehensive encyclopedia of human anatomic variation. Wiley, Hoboken

Palmar Arches

103

George Tadros and Lisandro Montorfano

103.1 Embryology

The human hand serves sophisticated functions and is one of the most complex structures in the body. It has a rich blood supply and extensive collaterals to sustain its high metabolic demand.

The blood supply to the hand is derived from the radial and ulnar arteries. As these arteries enter the hand they curve, divide, and anastomose to form two arches known as the superficial palmar arch and the deep palmar arch. These arches give off palmar and dorsal digital arteries and form an extensive network of blood vessels that supply oxygen and nutrients to all musculoskeletal components of the hand and fingers.

The current and most compelling suggestion is that development involves a two-step process. At Carnegie stage 12, the dorsal aorta gives rise to an initial capillary network that spreads out into the growing limb bud at a similar rate. Concurrently, the proximal parts of this network start to expand, and the arterial wall begins to characterize by stage 13. From stages 13 to 21, differentiation proceeds in a proximal to a distal pattern, the subclavian and axillary arteries being differentiated by stage 15, and the brachial artery by stage 17; the forearm arteries including the ulnar, interosseous, and radial arteries mature by stage 18. By stage 21, the definitive arterial pattern of the hand with the palmar arches has been established.

G. Tadros
Department of Surgery, Cleveland Clinic Florida, Weston, FL, USA

L. Montorfano (✉)
Department of Surgery, Cleveland Clinic Florida, Weston, FL, USA

Department of Plastic and Reconstructive Surgery, Vanderbilt University Medical Center, Nashville, TN, USA

103.2 Anatomy/Topography

The hand is supplied by two arterial arches, the superficial palmar arch and the deep palmar arch, which are considered the continuations of the ulnar and radial arteries, respectively.

The superficial palmar arch is considered the dominant vascular structure of the palm. It is formed mainly by the arch of the superficial division of the ulnar artery and is completed by the superficial palmar branch of the radial artery.

The ulnar artery enters the hand by crossing superficial to the medial part of the flexor retinaculum where it divides into superficial and deep divisions. The deep division, also known as the deep palmar branch of the ulnar artery, dips backward between the abductor digiti minimi and flexor digiti minimi brevis muscles, then passes through the opponens digiti minimi muscle to anastomose with the termination of the radial artery and form the deep palmar arch, discussed below. The superficial division, which is

S. J. Bordes, Jr. et al. (eds.), *The Clinical Anatomy of the Vascular System*,
https://doi.org/10.1007/978-3-031-78326-5_103

considered the continuation of the ulnar artery, runs distally just medial to the hook of hamate, then curves laterally across the palm to anastomose with the superficial palmar branch of the radial artery and form the superficial palmar arch.

The superficial palmar arch lies immediately deep to the palmar aponeurosis and superficial to the digital branches of the median nerves and the long flexor tendons, together with the common synovial sheath that surrounds them. It extends more distally than the deep palmar arch. It gives a palmar digital artery to the medial side of the little finger, and three common palmar digital arteries that run in the second, third, and fourth intermetacarpal spaces to reach the clefts between the medial four fingers. Each artery is joined by a palmar metacarpal artery from the deep palmar arch, and then it divides into two digital arteries, which run along the adjacent sides of two fingers.

The deep palmar arch is formed predominantly by the terminal part of the radial artery in the palm with a contribution from the deep division of the ulnar artery. It lies in front of the metacarpal bones just distal to their bases, which are concave proximally.

At the wrist, before it reaches the flexor retinaculum, the radial artery gives off its superficial branch and then continues as the dorsal radial artery. The superficial branch of the radial artery penetrates the thenar muscle to anastomose with the ulnar artery in the palm and complete the superficial palmar arch.

The dorsal radial artery curves backward below the styloid process of the radius, entering the anatomical snuff box to reach the proximal end of the first dorsal interosseus space. Then it travels forward between the two heads of the first dorsal interosseus muscle to reach the palm of the hand. It appears in the palm between the two heads of the adductor pollicis muscle and then it curves medially to anastomose with the deep branch of the ulnar artery opposite the base of the fifth metacarpal bone to form the deep palmar arch.

The surface anatomy of the deep palmar arch can be identified by drawing an arch one inch proximal to the superficial palmar arch and parallel to it.

The deep palmar arch gives rise to three palmar metacarpal arteries, which pass into the second, third, and fourth intermetacarpal spaces to anastomose with the palmar digital branches of the superficial palmar arch at the level of the metacarpophalangeal joints. It also supplies blood to the thumb and the lateral side of the index finger via the princeps pollicis and radialis indicis arteries. The former of these arises immediately after the radial artery enters the palm and divides into two digital branches to supply the thumb, while the latter runs along the lateral side of the index finger. Other branches of the deep palmar arch include the perforating arteries, which pass backward, perforating the second, third, and fourth interosseus muscles to join the dorsal metacarpal arteries on the dorsum of the hand, and the recurrent branch, which ascends in front of the wrist to join the palmar carpal arch.

103.3 Imaging/Histology

It is imperative to evaluate the collateral circulation of the hand before any vascular intervention on the hand or the radial and ulnar artery. This is achieved by obtaining a thorough general and local physical exam, supplemented by imaging and vascular studies. Allen's test is valuable for assessing palmar arch collaterals. A variety of noninvasive methods are available to support the physical examination including digital pressures, plethysmography, pulse oximetry, and Doppler ultrasonography (Habib et al. 2012).

Pencil Doppler is commonly used in practice along with the physical exam and the combination yields more information than the physical exam alone. After the superficial palmar arch has been located with the Doppler, it is assessed for character and flow while the radial and ulnar arteries are compressed sequentially. The operator can move the Doppler over the arch and finger pulps and listen for a change in the signal. This technique can almost always diagnose the site of an arterial occlusion if performed carefully.

The digital–brachial index (DBI) is the ratio of the blood pressure measured in the fingers to that measured in the brachial artery. As in the ankle–brachial index (ABI), a DBI value below 0.7 implies a significant occlusive lesion somewhere in the forearm or hand.

Laser Doppler flowmetry, which measures blood flow in the capillary bed, can quantify blood flow at the skin level. It is beneficial for evaluating the results of treatment, particularly for vasospastic problems; however, the equipment is quite expensive and is not available to most practitioners.

Nonetheless, for planning a surgical procedure, a higher level of detail should be obtained. Conventional angiography, computerized tomographic angiography (CTA), and magnetic resonance angiography (MRA) are valid options. Traditionally, digital subtraction angiography (DSA) of vascular arterial abnormalities of the upper extremity has been the primary imaging modality for vascular assessments because of its higher spatial resolution. It still remains the gold standard. However, with the technical advances of CTA and MRA and 3D reconstructions, use of these imaging modalities for assessing vascular pathologies has increased and they are replacing DSA as a pure diagnostic imaging modality (Rodríguez-Niedenführ et al. 2001).

A number of other types of study can aid in the diagnosis of vascular problems in the hand. Color-flow ultrasonography shows real-time flow and can be used to evaluate an anastomosis, but it is most useful for the venous system. Vasospasm is commonly evaluated by cold-stress testing, which gives a measure of the vascular system's ability to recover from exposure to cold. In this test, the digital temperature is measured at baseline and the hand is then placed in cold water (5–8 °C) for five minutes. Thermistor probes or laser Doppler are used to measure the time required for digital rewarming to baseline. A prolonged rewarming response is often seen in women and can be diagnostic of Raynaud's. Smokers often have a delayed rewarming response.

103.4 Variations

A deep understanding of the anatomical variations of the palmar arches is crucial for diagnosing different diseases of the hand. Improved knowledge of the possible variants has also led to enhanced surgical techniques especially in microsurgery. Because of the highly branching nature of the hand's circulation, there are numerous anatomical variations of the arterial patterns. These variants generally demonstrate either complete or incomplete superficial and deep palmar arches. A complete arch implies an anastomosis between the vessels; an incomplete arch is present when there is no such anastomosis. A recent meta-analysis by Zarzecki et al. reported a complete superficial volar arch in 81.3% of cases (Zarzecki et al. 2018). Research has reported seven different types of complete superficial palmar arch, the most common being a complete arch formed by anastomosis between the UA and RA (72%); a complete arch formed by the UA alone is the second most common (21%). Coleman and Anson examined 670 superficial palmar arches (Coleman and Anson 1961). They found that the dominant type for the complete SPA was an arch formed by the UA only, constituting 37% of their sample.

On the other hand, only five subcategories of the incomplete superficial palmar arches have been described in the literature, the most common variant being the UA supplying the third finger from both the radial and ulnar sides and the radial artery supplying the fourth and fifth fingers. The meta-analysis by Zarzecki et al. reported that a complete deep volar arch was found more consistently (95.2% of cases) (Zarzecki et al. 2018).

One explanation for the deviation from normal hand vascular anatomy was offered by (Arey 1957). They suggested that in some individuals, vessels that normally regress during fetal life persist and contribute to the vascular supply of the hand, generating a variant of the normal arterial pattern. An example is the median artery (MA), which predominates as the axis artery of the upper limb during embryogenesis and normally degenerates when the UA and RA are ready to supply blood to the hand.

103.5 Pathology

As discussed above, the radial and ulnar arteries anastomose and supply the major portion of the blood that flows into the hand. The anterior and posterior interosseous vessels can form collaterals and contribute to the arterial supply to the hand when there are chronic vascular problems; however, these two vessels are not big enough to provide adequate perfusion if there is acute injury to both the radial and ulnar vessels. Studies have shown that the median artery, when present, can provide significant blood flow to the hand, particularly in the event of injury or harvesting of the radial artery.

Hand ischemia can result from iatrogenic injury, trauma, or vascular disease. Iatrogenic injury is discussed below in the complications section.

Trauma is a common cause of hand ischemia. It can be either an open or a closed injury to the vessels of the forearm or hand. Open injuries are particularly common in the forearm and hand. They account for most traumatic injuries (around 79%) and most are the result of lacerations. Injury to both the radial and ulnar arteries is required to generate acute ischemia of the hand. However, a single vessel injury can lead to aneurysm formation and potentially to distal embolization, which can be devastating.

Closed injuries are less common than open injuries. A common type of closed trauma is ulnar artery thrombosis in the palm, also known as hypothenar hammer syndrome. It is seen in patients who suffer a blow to the hypothenar area. The ulnar artery is susceptible to injury in the Guyon canal owing to its relatively superficial position. Different pathologies can ensue, including thickening of the vessel wall (with chronic low-grade trauma), disruption of the intima and thrombosis, or even aneurysm formation. Regardless of the eventual pathology, the patient develops distal ischemia, primarily because a thrombus embolizes from the site of injury.

Hand ischemia can also be caused by intrinsic vascular disease, which can be either arteriosclerotic or vasospastic. Nowadays, a combination of diabetes and renal insufficiency is a major cause of arteriosclerotic disease in the upper limb. This combination of disease processes leads to accelerated occlusive vascular disease in the upper and lower extremities and can result in critical ischemia of the hand in severe cases.

Likewise, patients with a collagen vascular disease (primarily scleroderma) suffer from vasculitis and vasospastic disease, which can lead to vasoocclusive disease. These patients can present difficult management issues; surgical intervention, while helpful, often fails to provide a long-term solution.

Buerger disease, also called thromboangiitis obliterans (TAO), is a nonatherosclerotic, inflammatory disease that affects small to medium-sized vessels of the extremities in a segmental pattern. It is seen predominantly in young smokers, who present with ischemic ulcers or gangrene of their digits.

103.6 Clinical And Surgical Significance

Palpation of the radial and ulnar arterial pulses is a fundamental part of the general and local exam. Radial artery palpation is used to count the pulse and assess for asymmetry between the right and left upper extremities, which occurs in patients with acute limb ischemia, and between the upper and lower extremities as in coarctation of the aorta or Buerger disease.

The radial pulse can be useful for diagnosing thoracic outlet syndrome with Adson's test. Adson's test is done by abducting the arm while palpating the radial pulse, then extending and rotating the head to the ipsilateral side as the radial pulse to determine whether the blood supply is affected. A positive test is a decrease in pulse strength or complete absence of the pulse, indicating thoracic outlet syndrome (Ghouri et al. 2019).

Allen's test assesses collateral blood flow in the hand before arterial blood gas sampling or arterial line placement in the radial artery. It is performed by occluding both the radial and ulnar arteries and then releasing the ulnar artery to see if the hand turns pink again from collateral blood flow; this indicates that it is safe to sample the blood in the radial artery or to place an arterial line in it. Allen's test can also be used to evaluate the collateral blood flow from the radial artery by releasing the radial artery while still occluding the ulnar artery (Mills and Porter 1993).

The blood supply to the scaphoid bone arises solely from a small branch off the radial artery, which supplies the scaphoid bone in a retrograde fashion. Patients with a fractured scaphoid bone can suffer avascular necrosis and nonunion if the injury affects the blood supply.

The radial artery is one of the most commonly used arteries for numerous interventions. It is often punctured to obtain an arterial blood gas. It is also a common site for inserting an arterial line, e.g., for blood pressure monitoring in an intensive care unit. The increasing trend toward management of vascular problems with endovascular techniques has led to increased interest in the radial artery, which offers a relatively easy way into the arterial system with minimal complications. The radial artery is commonly used as a harvested graft for patients undergoing coronary revascularization and for radial forearm flap reconstruction.

103.7 Complications

Fortunately, iatrogenic injury is unusual; however, there is a potential for affecting flow to the hand negatively in many procedures. This happens in patients with anatomical variations, where collateral blood flow is not formed. Therefore, the surgical considerations for the arteries of the hand are to ensure that the structures will have proper blood flow before and after any intervention. An Allen's test should be performed routinely prior to any intervention involving the radial artery.

As mentioned above, the radial artery is punctured or harvested for many clinical uses. Although isolated radial artery puncture is exceedingly unlikely to cause ischemia in the hand, indwelling catheters are known to cause permanent ischemic complications in approximately 0.09% of patients. These patients often have severe vascular disease in the radial artery, which is exacerbated by the effects of distal embolization from the site of injury.

Another iatrogenic cause of ischemia is vascular access in the upper extremity for hemodialysis. This happens in 2–5% after a vascular access procedure and can be due to distal embolization or a "steal" phenomenon; in this case, most of the arterial flow is diverted across the fistula, leading to ischemia in the distal hand. Often the fistula has to be banded or even ligated to decrease the shunting of blood from the hand; however, vascular surgeons are usually unwilling to ligate these because of the difficulties of long-term access in dialysis patients.

Another option is the so-called distal revascularization interval ligation, or DRIL, procedure. In this operation, a vein graft is sewn into the artery above the site of the fistula and the bypass is anastomosed to the distal artery at a level beyond it. The native artery is then ligated just distal to the fistula. In principle, this should lead to the hand receiving most of the flow rather than the shunt, and it is effective in many patients. Diabetic patients on dialysis present other difficulties in that they often develop worsening arteriosclerosis of the upper extremity's vessels and ongoing ischemia.

References

Arey LB (1957) Developmental anatomy. In: Development of the arteries, 6th edn. W.B. Saunders Co, Philadelphia, pp 375–377

Coleman SS, Anson BJ (1961) Arterial patterns in the hand based upon a study of 650 specimens. Surg Gynecol Obstet 113:409–424

Ghouri MA, Gupta N, Bhat AP et al (2019) CT and MR imaging of the upper extremity vasculature: pearls, pitfalls, and challenges. Cardiovasc Diagn Ther 9(Suppl 1):S152–S173. https://doi.org/10.21037/cdt.2018.09.15. (CTA vs MRA article)

Habib J, Baetz L, Satiani B (2012) Assessment of collateral circulation to the hand prior to radial artery harvest. Vasc Med 17(5):352–361

Mills JL, Porter JM (1993) Buerger's disease: a review and update. Semin Vasc Surg 6:14

Rodríguez-Niedenführ M, Burton GJ, Deu J, Sañudo JR (2001) Development of the arterial pattern in the upper limb of staged human embryos: normal development and anatomic variations. J Anat 199(Pt 4):407–417

Zarzecki MP, Popieluszko P, Zayachkowski A, Pękala PA, Henry BM, Tomaszewski KA (2018) The surgical anatomy of the superficial and deep palmar arches: a meta-analysis. J Plast Reconstr Aesthet Surg 71(11):1577–1592

104 Veins of the Shoulder and Arm

Alex Ramos

104.1 Embryology

By the fourth week of development, the embryo has upper limb outpouchings, which later develop into limb buds. By the eighth week, it has visible limbs with fingers and toes (Sadler 2019). Both lateral plate and somite-derived mesoderm, or myotome, migrate to the upper limb buds to form the vasculature and musculature of the upper limb, respectively (Dudek 2014). As the limb buds elongate, the lateral plate mesoderm tissue develops into veins that elongate with them. Initially, the limb buds are drained by the preaxial and postaxial marginal veins (Cotter et al. 2014). Eventually, the preaxial vein becomes the cephalic vein and the postaxial vein the basilic vein (Cotter et al. 2014).

104.2 Anatomy and Topography

The major superficial veins of the anterior shoulder and arm are the basilic vein (Fig. 104.1), which runs medially, the cephalic vein (Figs. 104.1 and 104.2), which runs more laterally, and the medial cubital vein, which connects the basilic and cephalic veins in an anastomosis in the cubital fossa. The veins deep to the invest-

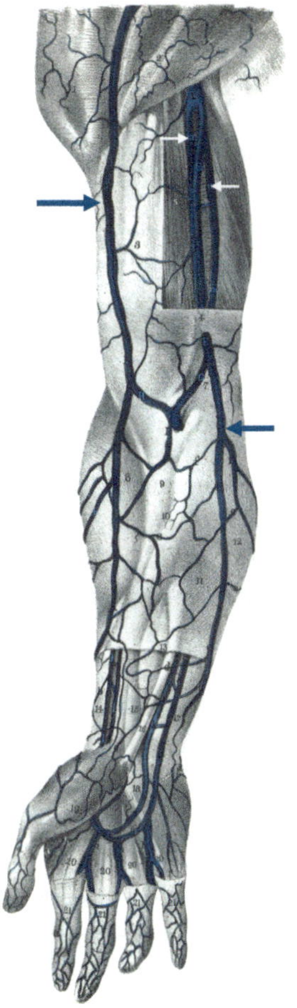

Fig. 104.1 Drawing of the upper limb highlighting the superficial veins. Note the cephalic (left blue arrow) and basilic (right blue arrow) veins as well as the brachial veins (white arrows) (Quain and Wilson, 1844 author's collection)

A. Ramos (✉)
College of Medicine, Texas A&M University, Houston Methodist Hospital, Houston, TX, USA
e-mail: aram0568@tamu.edu

S. J. Bordes, Jr. et al. (eds.), *The Clinical Anatomy of the Vascular System*,
https://doi.org/10.1007/978-3-031-78326-5_104

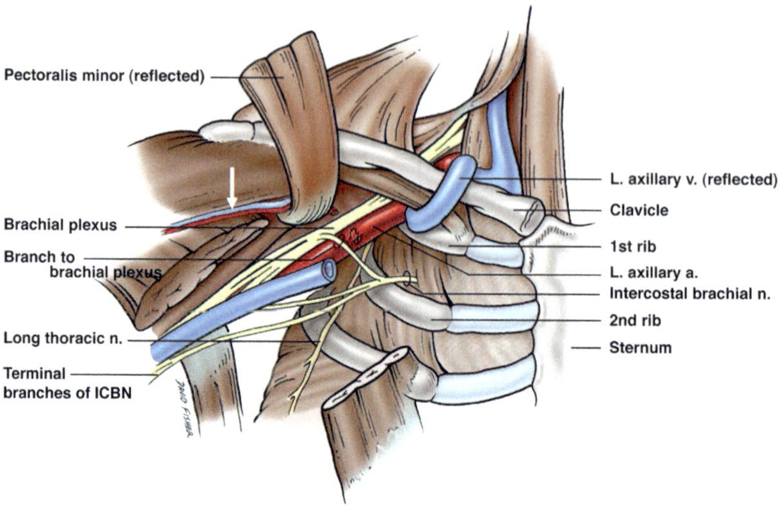

Fig. 104.2 Schematic drawing of the axilla noting the cephalic vein (arrow) traveling in the deltopectoral triangle

ing fascia of the shoulder and arm include the superior-most axillary vein, its distal continuation, the subclavian vein, and the brachial veins (Fig. 104.1) (Netter 2019).

The cephalic vein (Fig. 104.1) ascends laterally from the cubital fossa, running just lateral to the biceps brachii muscles. It then runs superficially in the delto-pectoral groove, defined medially by the pectoralis major muscle, laterally by the deltoid muscle, superiorly by the clavicle, and inferiorly by the biceps brachii. Just inferior to the clavicle and costocoracoid ligament, it penetrates through the costocoracoid membrane in the deltopectoral groove to join the axillary vein (Netter 2019; Loukas et al. 2008). Once the axillary vein passes the first rib superiorly, it is referred to as the subclavian vein.

The basilic vein courses medially in the arm, running on the medial border of the biceps brachii muscle. After it ascends in the arm, it eventually pierces through the deep fascia at the basilic hiatus. At the inferior border of the teres major muscle, the basilic vein and deeper brachial veins (Fig. 104.1) join to form the axillary vein (Netter 2019; Bajzer 2004) (Fig. 104.3).

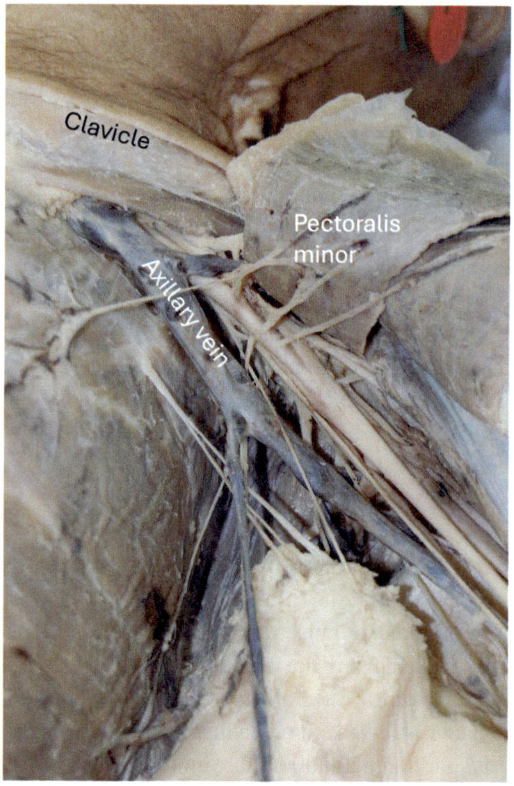

Fig. 104.3 Cadaveric dissection demonstrating the axillary vein. (Courtesy Prof. M. Loukas)

The brachial veins are formed by the union of the major veins that drain the forearm, the ulnar, radial, anterior, and posterior interosseous veins. Usually ascending as a pair, the two brachial veins lie on either side of the brachial artery and course deep to the median cubital vein in the antecubital fossa, ascend the upper arm, and terminate by joining the basilic vein to form the axillary vein (Netter 2019).

The median cubital vein is a superficial anastomosing connection between the cephalic and basilic veins in the antecubital fossa. Typically, it originates from the more lateral cephalic vein and courses in a superomedial direction to join the more medial basilic vein. The median cubital vein is short and its diameter is sufficient for venipuncture, which is its primary clinical significance (Mikuni et al. 2013). Perforating veins in the upper arm connect the superficial and deep veins, but they are highly variable in location and course.

Many of the smaller veins of the arm and shoulder course with their arterial counterparts, including the anterior and posterior circumflex humeral veins and the subscapular vein, all of which drain into the axillary vein. The scapular region is similarly drained by veins that course with their counterpart arteries, including the subscapular, suprascapular, circumflex scapular, transverse cervical, and dorsal scapular veins.

104.3 Variations

Variations of the cephalic vein in the deltopectoral triangle were evaluated in 100 cadavers. Rarely, the cephalic veins were absent; in most cases they were symmetrical (Loukas et al. 2008; Lee et al. 2015). They received one or two tributaries in the deltopectoral triangle, more commonly two. In 20% of cases, the cephalic vein never emerged from the investing fascia of the deltopectoral triangle. Three quarters of the specimens had accessory cephalic veins (Loukas et al. 2008).

While the connection between the cephalic vein and the basilic vein is conventionally the median cubital vein of the cubital fossa, there is considerable variation (Sadeghi et al. 2017). Just over 50% of 200,000 subjects investigated had this anatomy, but variations included the cephalic and basilic veins merging into a median antebrachial vein, lack of a median cubital vein, or lack of a cephalic vein and bifurcation of the basilic vein in the cubital fossa (Mikuni et al. 2013; Lee et al. 2015; Sadeghi et al. 2017).

The brachial veins typically ascend the arm as a pair on either side of the brachial artery. In some cases, there is a single brachial vein rather than a pair, the basilic vein branching off of the unpaired brachial vein in the mid-arm region (Kumar et al. 2012; Anaya-Ayala et al. 2011). In another variation, the traditionally paired brachial veins with the basilic vein branch mid-arm, distal to the usual branch point (Anaya-Ayala et al. 2011).

The venous drainage of the scapular region is highly variable, with many anastomosing vessels that ultimately drain into the axillary vein. In many cases, the veins that drain the shoulder, especially the thoracoacromial vein, drain directly into the cephalic vein (Lee et al. 2019; Podgórski et al. 2014). This pattern was also seen in the deltoid, clavicular, acromial, and pectoral veins, which drained directly into the cephalic vein (Lee et al. 2019). The other veins of the shoulder are also variable in their course, sometimes with variable unpaired veins contributing to shoulder drainage, including a superficial vein of the axillary base (Lee et al. 2019) or a suprascapular notch vein (Podgórski et al. 2014).

104.4 Clinical and Surgical Significances

The veins of the arms and shoulders are susceptible to the same vascular pathologies as other vessels, including thrombosis in the deep veins or phlebitis in the superficial veins, varicose veins (although these are more common in the lower extremities), venous occlusion, vasculitis, and venous insufficiency. Deep vein thrombosis, although more common in the lower extremities, can occur in the deep veins of the upper extremities, most often the axillary or subclavian veins (Engelberger and Kucher 2012). Upper extremity venous occlusion, especially in the axillary region, can occur if the muscles of that region enlarge, press against the vessel walls, and narrow the vein.

An interesting, but rarer, pathology of the upper extremity venous system is venous thoracic outlet syndrome. In this case, the subclavian vein is compressed between the first rib inferiorly and the clavicle superiorly because of a cervical rib, hypertrophied scalene muscle, or other abnormality (Modi et al. 2021). Venous thoracic outlet syndrome is increasingly seen in young athletes, particularly those who repeatedly abduct their arms or undergo cervical extension as part of their training, resulting in injury to the entrapped vasculature. If this is left untreated, thrombosis or Paget-Schroetter syndrome can follow, with more urgent need to manage a potential venous thromboembolism (Modi et al. 2021; Moore and Wei 2015).

Perhaps the most clinically relevant vein for venipuncture is the median cubital vein, which is sufficiently wide and superficial for this purpose. It is commonly used for both routine blood draws and intravenous access or cannulation. Understanding the anatomy and possible variations of the median cubital vein is especially important for venipuncture, although the cephalic vein can be used as an alternative. Another use of the cephalic vein is for central venous access, especially if alternative veins are incompetent or narrowed. Central access through the cephalic vein, or cephalic vein cutdown, reduces the risk of preoperative complications (Loukas et al. 2008; Jung and Moon 2014). The cephalic vein can also be used as an alternative to the great saphenous vein for carotid artery grafts, percutaneous pacemaker implantation, long-term venous catheters, and defibrillator implantation (Loukas et al. 2008). Generally, for transvenous pacemaker or defibrillator lead insertion, access is gained via cephalic cutdown or venipuncture of the subclavian or axillary veins (Al-Hadithi et al. 2018).

The basilic vein can also be used for venipuncture, and more distally, in the forearm, as an arteriovenous fistula for hemodialysis access (Lee et al. 2015). Like other superficial veins, it is adequate for transfusion, use as a bypass graft, and cardiac catheterization, although it is less often used for catheterization than the femoral or radial arteries (Sadeghi et al. 2017).

References

Al-Hadithi A, Do D, Boyle N (2018) Vein management for cardiac device implantation. Card Electrophysiol Clin 10(4):561–571

Anaya-Ayala JE, Younes HK, Kaiser CL et al (2011) Prevalence of variant brachial-Basilic vein anatomy and implications for vascular access planning. J Vasc Surg 53(3):720–724

Bajzer CT (2004) Guide to peripheral and cerebrovascular intervention: venous drainage of the upper extremities. Remedica, London

Cotter M, Davis E, Krabbenhoft K, Traynor S (2014) Mind and motion anatomy: limb development. University of Wisconsin – Madison, School of Medicine and Public Health, pp 301–327

Dudek RW (2014) BRS embryology. Wolters Kluwer 6:226–232

Engelberger RP, Kucher N (2012) Management of Deep Vein thrombosis of the upper extremity. Circulation 126:768–773

Jung KH, Moon SB (2014) Cephalic vein cutdown for totally implantable central venous port in children: a retrospective analysis of prospectively collected data. Can J Surg 57(1):21–25

Kumar N, Aithal AP, Rao MK, Nayak SB (2012) The venous chiasma between the Basilic vein and the brachial vein: a case report. J Clin Diagn Res 6(9):1539–1540

Lee H, Lee SH, Kim SJ, Choi WI, Lee JH, Choi IJ (2015) Variations of the cubital superficial vein investigated by using the intravenous illuminator. Anat Cell Biol 48(1):62–65

Lee H, Bang J, Kim S, Yang H (2019) The axillary vein and its tributaries are not in the Mirror image of the axillary artery and its branches. PLoS One 14(1):e0210464

Loukas M, Myers CS, Wartmann CT, Tubbs RS, Judge T, Curry B, Jordan R (2008) The clinical anatomy of the cephalic vein in the Deltopectoral triangle. Via Medica 67(1):72–77

Mikuni Y, Chiba S, Tonosaki Y (2013) Topographical anatomy of superficial veins, cutaneous nerves, and arteries at venipuncture sites in the cubital fossa. Anat Sci Int 88:46–57

Modi BP, Chewning R, Kumar R (2021) Venous thoracic outlet syndrome and Paget-Schroetter syndrome. Semin Pediatr Surg 30(6):151125

Moore R, Wei LY (2015) Venous thoracic outlet syndrome. Vasc Med 20(2):182–189

Netter FH (2019) Atlas of human anatomy. Elsevier 7:443–519

Podgórski M, Sibiński M, Major A, Stefańczyk L, Topol M, Polguj M (2014) The Suprascapular vein: a possible etiology for Suprascapular nerve entrapment and risk of complication during procedures around the Suprascapular foramen region. Orthop Traumatol Surg Res 100(5):515–519

Sadeghi A, SetayeshMehr M, Esfandiari E, Mohammadi S, Baharmian H (2017) Variation of the cephalic and Basilic veins: a case report. J Cardiovasc Thorac Res 9(4):232–234

Sadler TW (2019) Landman's medical embryology. Wolters Kluwer 14:167–178

Veins of the Forearm

105

Cole Fiedler

105.1 Embryology

Veins, like arteries, are derived from mesenchymal tissue originating from the mesodermal germ layer of the developing embryo. During development, the superficial veins of the upper extremity develop from a plexus of capillaries, which fuse and enlarge. The ulnar capillary plexus develops into the basilic, axillary, and subclavian veins. The radial capillary plexus atrophies; the cephalic vein develops from an anastomotic channel of the external jugular vein (Araujo et al. 2018).

As in most systemic circulation, the veins of the upper extremity derive from the embryonic cardinal veins, specifically the anterior cardinal veins (Araujo et al. 2018). The paired anterior cardinal veins, located cranial to the heart, join the paired posterior cardinal veins, located caudal to the heart, to form the common cardinal veins. These empty into the sinus venosus of the primitive heart. As development continues, the anterior cardinal veins form the superior vena cava and its tributaries, receiving blood from the head and upper extremities (Sidawy and Perler 2023).

C. Fiedler (✉)
University of South Florida Morsani College of Medicine, Tampa, FL, USA
e-mail: colefiedler@usf.edu

105.2 Anatomy/Topography

The venous system of the forearm comprises superficial and deep veins. The superficial venous system drains blood from the skin and superficial fascia, while the deep venous system drains blood from deeper tissue, muscle, and bone. The venous drainage in the forearm receives deoxygenated blood from the digits and hands and ultimately empties into the right atrium of the heart via the superior vena cava.

The major superficial veins of the forearm include the basilic and cephalic veins (Figs. 105.1 and 105.2). They arise from the dorsal venous arch of the hand. The basilic vein ascends posteromedially and drains the ulnar aspect of the forearm. The cephalic vein, arising in the anatomical snuff box, ascends anterolaterally and drains the radial side. It receives additional drainage from the accessory cephalic vein just distal to the elbow. As these two veins approach the cubital fossa, they form a connection called the median cubital vein. Anatomical variations in this region are discussed below. The palmar venous plexus of the hand drains into the median antebrachial vein, which ascends anteriorly in the forearm and connects with the median cubital vein or basilic vein (Yammine and Eric 2017). The basilic vein continues up the proximal arm to empty into the brachial or axillary vein, while the cephalic vein drains into the subclavian vein.

S. J. Bordes, Jr. et al. (eds.), *The Clinical Anatomy of the Vascular System*,
https://doi.org/10.1007/978-3-031-78326-5_105

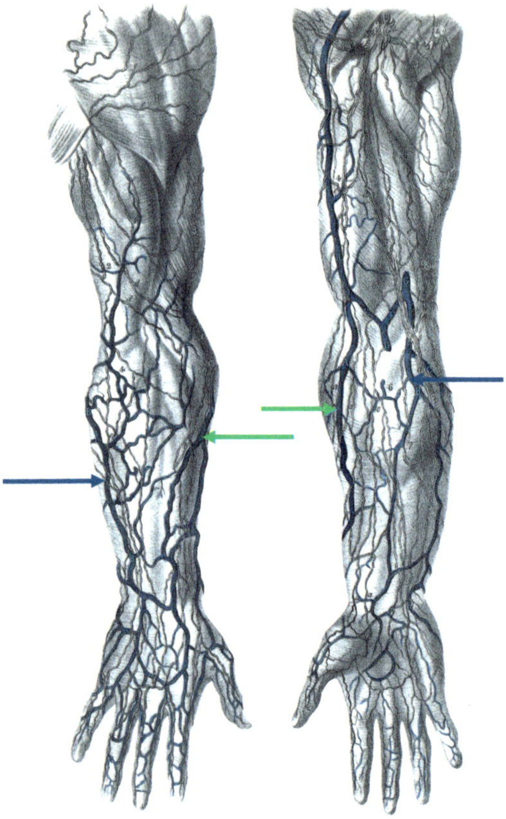

Fig. 105.1 Drawings of the superficial veins of the forearm, posterior (left) and anterior (right) surfaces. The cephalic veins are noted at the green arrows and the basilic veins at the blue arrows. (After Quain and Wilson)

The deep venous system of the forearm is more closely related to its arterial counterparts than the superficial venous system. From the deep and superficial palmar arches of the hand arise the ulnar and radial veins, which provide the main drainage of the forearm. Each of these veins forms pairs around its corresponding artery, called venae comitantes, which lie close to the artery, so arterial pulsations facilitate venous return. At the cubital fossa, the ulnar and radial veins merge to form the brachial veins, which become the axillary vein at the axilla. The anterior and posterior interosseous veins arise from the palmar and dorsal veins of the hand. These veins travel within the interosseous space between the radius and ulna, anastomose, and empty into the ulnar vein (Moore et al. 2014).

The superficial and deep veins connect via a system of perforating veins, which provide collateral venous drainage. These connections are highly variable.

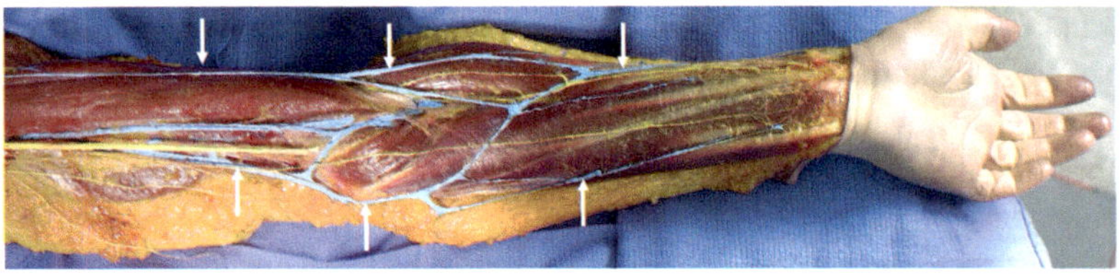

Fig. 105.2 Cadaveric dissection noting the cephalic (upper arrows) and basilic veins (lower arrows)

105.3 Variations

The venous system has a high potential for anatomical variations because it arises from various vascular channels that can develop into multiple configurations (Sidawy and Perler 2023).

One such site of variation is the cubital fossa, which usually includes the basilic, cephalic, median cubital, and median antebrachial veins. According to a meta-analysis by Yammine and Eric, there are eight reported patterns of superficial veins in the cubital fossa (Yammine and Eric 2017). In type 1, also called the "M" or "Y"-shaped arrangement, the median antebrachial vein splits into two terminal branches, the median basilic and median cephalic veins, which join the basilic and cephalic veins, respectively. In type 2 ("N" or "H"-shaped arrangement), the median antebrachial vein drains into the median cubital vein, which connects the basilic and cephalic veins in the cubital fossa. Other patterns are noted, such as redundant median cubital veins or complete absence of communicating veins; however, types 1 and 2 are the most common. Interestingly, the researchers found that the "M" arrangement (type 1) is more prevalent in males, and the "N" arrangement (type 2) has a higher frequency in Indian and Japanese populations (Yammine and Eric 2017). There was no significant difference in laterality. These findings suggest that the anatomical variations in the venous system of the cubital region have a genetic component and warrant consideration when venous access is performed in different patient populations.

The size of the superficial veins also varies. Typically, the basilic vein is larger than the cephalic vein (Shima et al. 1996). The drainage of the basilic vein also varies; it can empty into either the brachial or the axillary vein.

105.4 Clinical Significance

The cubital fossa, specifically the median cubital vein, is one of the most clinically significant sites in the forearm. It is regularly used for venipuncture, transfusion, infusion, dialysis access, or cardiac catheterization. It is suitable for use given its superficial location and ease of access; the veins in this region are not paired with arteries as they are in most of the systemic circulation throughout the body (Yammine and Eric 2017). Furthermore, the median cubital vein is typically the most superficial in the body, sometimes still visible or palpable in cases of obesity or shock. Another important clinical consideration is the proximity of the medial and lateral antebrachial cutaneous nerves, which lie superficial to the veins of the cubital fossa and can be injured during phlebotomy (Yammine and Eric 2017).

Thrombophlebitis is another important clinical consideration, given the frequency of venous access to the veins of the forearm. Symptoms can include tenderness, erythema, pain, and swelling. Factors affecting the occurrence of thrombophlebitis include catheter size, frequency of access, and insertion site. Catheter placement location can affect its incidence; an observational study by Cicolini et al. showed a higher risk of developing thrombophlebitis when the dorsal veins of the hand were used. They concluded that the cubital fossa veins (median cubital vein) should be preferred for venous access and catheter placement (Cicolini et al. 2009).

Deep venous thrombosis (DVT), while much more common in the veins of the lower extremity, can also develop in the veins of the upper extremity and forearm. The ulnar and radial veins are potential sites of DVT formation, and complications can include postthrombotic syndrome, venous insufficiency, thrombophlebitis, and, rarely, pulmonary embolism. DVTs of the upper extremity often present asymptomatically, and they are considered less severe than those of the lower extremity (Heil et al. 2017).

105.5 Surgical Considerations

The radial forearm free flap is a graft procedure used for reconstructive surgery, often in the head and neck (Imanishi et al. 2000; Ichinose et al. 2003). In this procedure, skin from the forearm is removed along with the radial artery and venous drainage; if necessary, the palmaris longus tendon can also be harvested. The venous drainage used in this procedure is not as clear-cut. The radial veins (venae comitantes) and cephalic vein are both surgical considerations and sometimes both venous systems can be used for microanastomoses (Imanishi et al. 2000; Ichinose et al. 2003).

Veins of the forearm can also be used in arteriovenous fistula formation for hemodialysis. Historically, the upper extremity cephalic vein has been used to create an arteriovenous fistula; however, Uzun et al. found the basilic vein system in the medial forearm to be a suitable alternative (Uzun 2019). In general, arteriovenous fistulas should be formed using distal cephalic vessels in the nondominant hand. Contraindications include small caliber vessels and preexisting intravenous catheters proximally, such as PICC lines. Surgeons then move proximally from the forearm to the arm until they find a suitable cephalic or basilic vein. A prosthetic graft can be used if a suitable vein cannot be identified in either upper extremity.

Varicosities can also affect the veins of the upper extremity and forearm. These engorged veins often result from venous insufficiency, but congenital varicose veins of the forearm have been reported (Jebens 1938). Typically, varicose veins are benign, and removal is often performed for cosmetic reasons.

References

Araujo RC, Pires LAS, Andrade ML, Perez MC, Filho CSL, Babinski MA (2018) Embryological and comparative description of the cephalic vein joining the external jugular vein: a case report. Morphologie 102(336):44–47. https://doi.org/10.1016/j.morpho.2017.10.002

Sidawy AN, Perler BA (2023) Rutherford's vascular surgery and endovascular therapy. Elsevier

Yammine K, Eric M (2017) Patterns of the superficial veins of the cubital fossa: a meta-analysis. Phlebology 32(6):403–414. https://doi.org/10.1177/0268355516655670

Moore KL, Dalley AF, AMR A (2014) Clinically oriented anatomy, 7th edn. Wolters Kluwer Health/Lippincott Williams & Wilkins Wolters Kluwer/Lippincott Williams & Wilkins Health

Shima H, Ohno K, Michi K, Egawa K, Takiguchi R (1996) An anatomical study on the forearm vascular system. J Craniomaxillofac Surg 24(5):293–299. https://doi.org/10.1016/s1010-5182(96)80062-x

Cicolini G, Bonghi AP, Di Labio L, Di Mascio R (2009) Position of peripheral venous cannulae and the incidence of thrombophlebitis: an observational study. J Adv Nurs 65(6):1268–1273. https://doi.org/10.1111/j.1365-2648.2009.04980.x

Heil J, Miesbach W, Vogl T, Bechstein WO, Reinisch A (2017) Deep vein thrombosis of the upper extremity. Dtsch Arztebl Int 114:244. https://doi.org/10.3238/arztebl.2017.0244

Imanishi N, Nakajima H, Aiso S (2000) Anatomic study of the venous drainage architecture of the forearm skin and subcutaneous tissue. Plast Reconstr Surg 106(6):1287–1294. https://doi.org/10.1097/00006534-200011000-00008

Ichinose A, Tahara S, Terashi H et al (2003) Importance of the deep vein in the drainage of a radial forearm flap: a haemodynamic study. Scand J Plast Reconstr Surg Hand Surg 37(3):145–149. https://doi.org/10.1080/28443103100007746

Uzun HA (2019) Transposition of basilic vein in forearm for arteriovenous fistula creation: our mid-term results. The Turk J Thorac Cardiovasc Surg 27(4):508–511. https://doi.org/10.5606/tgkdc.dergisi.2019.17600

Jebens EH (1938) Congenital varicose veins of forearm. Proc R Soc Med 31(5):545–546

Veins of the Hand

106

Bora Kahramangil and Lisandro Montorfano

106.1 Embryology

The upper limbs develop mostly during the first eight weeks of gestation, arising from the limb buds, which are mesodermal outgrowths into the overlying ectoderm. Two types of mesodermal cells migrate into the limb buds: the somatic and lateral plate mesoderm (Kozin 2017). The vascular structures of the upper extremity originate from the somatic mesoderm, along with the neural and muscular structures. The lateral plate mesoderm gives rise to the bony, cartilaginous, and tendinous structures (Kozin 2017).

106.2 Anatomy/Topography

The venous system of the hand collects deoxygenated blood from the hand and the digits and connects proximally with the veins of the forearm. The hand veins can be grouped into superficial and deep systems on the basis of their anatomy. The superficial venous system is the dominant drainage route. It can be divided into the superficial dorsal and palmar venous systems. Most of the superficial palmar veins connect to the dorsal veins, so the superficial dorsal system collects most of the venous blood from the hand. From a mechanical standpoint, the blood return from the hand is facilitated by flexion-extension of the fingers, which compresses the veins (Standring 2020).

The superficial dorsal venous system is formed most distally by the two dorsal digital veins that course along the radial and ulnar sides of each finger. The oblique venous branches of the fingers also drain into these superficial dorsal digital veins. As they course more proximally, these veins unite to form the three dorsal metacarpal veins over the joining surfaces of metacarpals 2–5. The dorsal metacarpal veins, in return, create the dorsal venous network over the metacarpal bones. On the radial side, the dorsal venous network is joined by the dorsal digital veins of the thumb and by the dorsal digital vein from the radial side of the index finger. More proximally, these veins drain into the cephalic vein (Fig. 106.1). On the ulnar side, the dorsal venous network is joined by the dorsal digital vein from the ulnar side of the little finger. More proximally, these veins drain into the basilic vein (Standring 2020).

The superficial palmar venous system is less prominent than its dorsal counterpart. The palmar digital veins connect to the dorsal digital veins

B. Kahramangil
Department of Surgery, Cleveland Clinic Florida, Weston, FL, USA

L. Montorfano (✉)
Department of Surgery, Cleveland Clinic Florida, Weston, FL, USA

Department of Plastic and Reconstructive Surgery, Vanderbilt University Medical Center, Nashville, TN, USA

S. J. Bordes, Jr. et al. (eds.), *The Clinical Anatomy of the Vascular System*,
https://doi.org/10.1007/978-3-031-78326-5_106

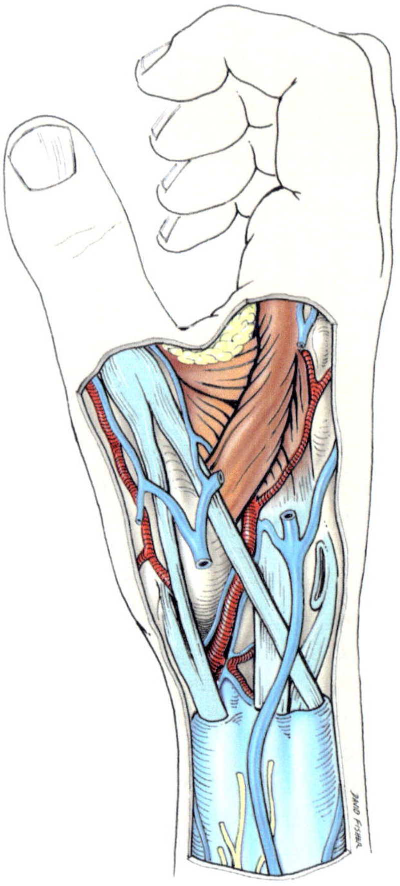

Fig. 106.1 Schematic drawing of the dorsal veins of the hand draining toward the cephalic vein over the anatomical snuffbox

via the oblique veins between the metacarpal heads. These veins communicate more proximally with the superficial vascular plexuses overlying the thenar and hypothenar eminences (Standring 2020).

The deep venous system of the hand is less developed than the superficial venous system. Although venae comitantes generally accompany the named hand arteries, those accompanying the proper and common digital arteries are absent in some individuals. Like their arterial counterparts, the venae comitantes of the common palmar digital arteries drain into the superficial palmar venous arch. In contrast, the palmar metacarpal arteries connect with the deep palmar venous arch. The venae comitantes of the palmar metacarpal arteries also communicate with the dorsal metacarpal veins through perforating branches. The dorsal metacarpal veins, in return, connect with the radial veins and the dorsal superficial venous network (Standring 2020).

106.3 Variations

The veins of the hand show considerable variation. The dorsal superficial venous network is the dominant route for drainage from the hand and fingers. The small, deep veins accompanying the named arteries are not constant and are absent in some individuals (Standring 2020).

106.4 Clinical and Surgical Significance

The superficial dorsal venous network's veins are frequently punctured to gain intravenous access. In addition, prominent and bulging hand veins have been recognized as causing cosmetic dissatisfaction in some patients. Although this is not a true disease process, these veins can be obliterated by various procedures, including phlebectomy, endovenous ablation, and sclerotherapy, to improve cosmesis (Shamma 2007; Tremaine 2014).

References

Kozin SH (2017) Embryology of the upper extremity. In: Wolfe SW (ed) Green's operative hand surgery, 7th edn. Elsevier, Philadelphia

Shamma AR (2007) Laser ablation of unwanted hand veins. Plast Reconstr Surg 120(7):2017–2024

Standring S (2020) Gray's anatomy, 42nd edn. Elsevier, Philadelphia

Tremaine AM (2014) Foam sclerotherapy for reticular veins of the dorsal hands: a retrospective review. Dermatologic Surg 40(8):892–898

107 Lymphatics of the Shoulder and Arm

Nicholas O. Gerard III and Charles J. Peper

107.1 Embryology

The development of the lymphatic system is closely associated with the circulatory system. The primordial circulatory system begins oxygenating tissues at week 3, while the initial lymphatic sacs begin communicating with venous structures at week 5. By the end of week 8, there are six primary lymphatic sacs in the developing fetus including the cisterna chyli, two jugular lymph sacs, two iliac lymph sacs, and a retroperitoneal lymph sac. Vessels begin to connect the lymphatics sacs, forming the early lymphatic system.

Embryonic development from conception to the end of week 8 can be subclassified into 23 stages known as the Carnegie classification system. The upper extremity limb bud begins to form at Carnegie stages 15–16, the primordial humerus and scapula can be identified at Carnegie stages 17–18, the glenohumeral joint begins to take shape in Carnegie stages 19–20, and a clearly identifiable articular joint space is seen at Carnegie 23. By week 13–14, the fetus will have a high degree of differentiation and osteogenesis of the upper extremity (Hita-Contreras et al. 2018).

N. O. Gerard III (✉) · C. J. Peper
Tulane University School of Medicine,
New Orleans, LA, USA
e-mail: ngerard@tulane.edu; cpeper@tulane.edu

107.2 Anatomy/Topography

The lymphatic vessels of the upper extremity are fed by an extensive cutaneous lymphatic plexus at the level of capillaries in the dermis. The lymphatic vessels of the upper extremity originate from the pulp of the fingertips and course proximally towards the axilla. Lymphatic vessels are thin, porous, valve containing tubes and, unlike vascular structures of the upper extremity, do not increase in diameter as they progress proximally towards the trunk (Suami et al. 2005). The function of these vessels is to return extravasated fluid to circulation, transport fat, and provide immune responses for the body (Loukas et al. 2011).

There are superficial subcutaneous lymphatic vessels that are closely associated with subcutaneous veins, specifically the cephalic and basilic veins (Suami et al. 2007). These lymphatic vessels run along the basilic vein to the elbow communicating with the cubital, or epitrochlear lymph nodes, terminating in the lateral axillary lymph nodes. The lymphatic vessels running along the cephalic vein communicate directly with the axillary lymph nodes (Drake et al. 2015).

Since the lymphatic system was first described, there have been various advancements in mapping the complex network of lymphatics throughout the body (Loukas et al. 2011). Recently, a new postmortem method of mapping the deep lymphatics of the upper extremity has been described by Ma et al. This method involves

S. J. Bordes, Jr. et al. (eds.), *The Clinical Anatomy of the Vascular System*,
https://doi.org/10.1007/978-3-031-78326-5_107

injecting hydrogen peroxide adjacent to lymphatic structures to inflate them, and then canulating those structures with a radiopaque substance to later visualize (Suami et al. 2007). Through this process, authors dissected 6 arms and were the first to describe the deep lymph vessels of the arm. The deep lymph vessels of the arm include the radial, ulnar, and brachial lymphatic vessels, which have an average diameter of 0.4 mm, 0.35 mm, and 0.4 mm, respectively. The deeper lymphatic vessels of the arm include the deep brachial, anterior, and posterior interosseous lymph vessels, which all have an average diameter of 0.3 mm (Ma et al. 2019). All the aforementioned structures run along corresponding vasculature and drain the subfascial compartment of the upper extremity. Ma et al. also applied this technique in the hand. The lymphatics of the hand consist of the superficial palmar arch lymph vessel, the deep palmar arch lymph vessel, the thenar lymph vessel, the hypothenar lymph vessel, and the deep finger web lymph vessel, all of which drain to superficial subcutaneous lymphatic structures at the wrist (Ma et al. 2018).

The lymph nodes of the upper limb can be grouped into two large clusters, the epitrochlear (interval) and the axillary (regional) clusters. The epitrochlear nodes drain into the axillary lymph nodes which are made up of the lateral, posterior, anterior, central, and apical nodes. The lateral or humeral nodes (Suami et al. 2005, 2007; Loukas et al. 2011) drain the lymphatic vessels of the arm and communicate with the central nodes. The posterior or subscapular nodes (Suami et al. 2007; Drake et al. 2015) drain the subscapular region as well as the posterior thoracic wall and communicate with the central nodes. The anterior or pectoral nodes (Hita-Contreras et al. 2018; Suami et al. 2005) drain the pectoral region, breast, and communicate with the central nods. Thus, the lateral, posterior, and anterior nodes communicate with the central nodes which lie deep in adipose tissue in the axilla. Finally, lymph is drained from these lymph node clusters to the apical lymph nodes (Suami et al. 2007; Drake et al. 2015; Ma et al. 2018, 2019; Suami and Scaglioni 2018; Kinmonth et al. 1957; Grada and Phillips 2017) which is the final step of lymph filtration prior to reentering vascular circulation via the subclavian vein. Collectively, there are approximately 20–40 axillary lymph nodes draining the upper extremity (Suami and Scaglioni 2018).

107.3 Pathology

Many pathologies of the upper extremity are associated with the lymphatic system and manifest due to its dysfunction. Lymphatic dysfunction may result in lymphedema which is the accumulation of interstitial fluid and fibroadipose tissue, when severe, in the extremities. According to Kinmonth's classification described in 1957, lymphedema can be classified as primary or secondary lymphedema (Kinmonth et al. 1957). Primary lymphedema is congenital while secondary can be caused by various forms of injury to the lymphatic system including malignancy, iatrogenic injury, and infection.

Primary lymphedema occurs in approximately 1 in every 100,000 people (Grada and Phillips 2017). The cause of primary lymphedema is generally idiopathic. However, there is clearly a multifactorial genetic component with ongoing research (Michelini et al. 2018). Primary lymphedema of the upper extremity was studied in 27 patients presenting with unilateral primary arm lymphedema via indocyanine green (ICG) injections and fluorescent imaging at 30- and 120-minute timepoints. Liu et al. found 25 of the 27 had no identifiable ICG uptake in the superficial lymphatic vessels of the affected arm. In this subset of patients, the most common symptom was swelling and fibrosis in the dorsum of the hand and fingers. The authors postulate that lymphedema arose from a defective dermal lymphatic plexus rather than absent lymphatic structures in the arm (Halsted 1921).

Secondary lymphedema of the arm and shoulder is well understood in the context of breast cancer. Breast cancer can commonly metastasize to the axillary lymph nodes. During mastectomy, surgeons resect these nodes and associated lymphatic vessels, disrupting the flow of lymph in the upper extremity. This results in significant edema in the arm causing gross disfigurement first described by Haslted in 1921 (Halsted 1921). Historically, the more aggressive radical mastectomy left women with arms referred to as lymphostatic "elephantiasis" derived from the Greek *elphas* or "elephant" (Routh and Bhowmik 1993). Almost all patients with upper extremity lymphedema have a history of lymphadenectomy due to breast cancer, as approximately 20% of women will have lymphedema postoperatively (Petrek and Heelan 1998; DiSipio et al. 2013). There are emerging risk factors that increase a woman's likelihood of developing lymphedema following axillary lymph node dissections. These include the intensity of the treatment, both surgically and medically, as well as the patient's preoperative body mass index (DiSipio et al. 2013).

Filariasis infection can cause lymphedema. The word "filiariasis" is derived from the Latin filum or "thread" caused by the parasites Wuscheria bancrofti and Brugia malayi/timori. Filiariasis is transmitted by a mosquito vector and is the most common cause of peripheral lymphedema in the developing world. These nematodes infiltrate the lymphatic vessels where they sexually reproduce, releasing thousands of larvae into the lymphatics. It is estimated that 51 million people are infected with these parasites worldwide, with 15 million people having lower extremity lymphedema. Drug regiments including ivermectin, albendazole, and diethylcarbamazine effectively kill the parasites (Pfarr et al. 2009). Unfortunately, the death of the adult worm triggers acute and subsequent chronic inflammation, further driving the pathophysiological response worsening symptoms (Jungmann et al. 1991). There are case reports of filiariasis in the upper extremity causing swelling of the arm, but it is exceedingly rare (Phukan et al. 2012; Maldjian et al. 2014; Sharma et al. 2018). There are other causes of peripheral lymphedema including environmental exposures (Price 1972) and obesity (Greene et al. 2015). However, these more commonly manifest in the lower extremity.

107.4 Imaging

Findings on advanced imaging that can imply metastatic disease include irregular contrast enhancement, size, and appearance (van den Brekel and Castelijns 1999). Other prognostic factors can be ascertained including extra-nodal spread and invasion of nearby anatomy. The first line test for staging breast cancer and upper extremity neoplasm is lymphoscintigraphy-assisted sentinel node biopsy (SNB), although advanced imaging can still aid clinical decision making (Wang et al. 2007). Lymphoscintigraphy is traditionally a nuclear radiography technique involving injecting a blue dye as well as the radiolabeled substance, but some clinicians may prefer hemosiderin for visualization. SPECT/CT can be performed perioperatively to assist in the evaluation of draining nodes which are difficult to visualize during SNB (Vermeeren et al. 2010; Miura et al. 2014). After treatment of breast cancer, lymphedema commonly presents as sequelae. When this develops, indocyanine green lymphography can help determine the compromised drainage pathways (Narushima et al. 2016). This imaging modality is also used for lymphaticovenular anastomosis procedures as well. Lymphoscintigraphy can be used for this purpose, although ICG or MRI studies are preferred (Mihara et al. 2012; Chao et al. 2021).

Breast implant rupture has been known to affect the lymphatic system including the nodal groups in the axilla. Ultrasound has been used to identify breast implant rupture through "snowstorm" appearance of silicone in axillary lymph nodes. Sensitivity of this modality is lower than MRI but likely better than mammography and can be a cost-effective option for patients (Samreen et al. 2018).

107.5 Surgical Significance

Lymphedema of the upper extremity can be classified according to the International Society of Lymphedema staging system 0-III. Stage 0 is characterized by lymphatic accumulation that is not visually noticeable, but may cause a patient to have discomfort and/or dysfunction of their extremity. Stage I is characterized by mild pitting edema with some relief derived from elevation and compression therapy. Stage II is characterized as initial tissue fibrosis and swelling not relived by compression therapy. Finally, stage III is characterized by severe edema and skin changes including acanthosis, severe fibroadipose deposition, lichenification, papillomas, and superimposed cutaneous infections like erysipelas and cellulitis (International Society of Lymphology 2013). Stages II-III may require surgical intervention if patients are unresponsive to conservative complete decongestive compression therapy. The two main approaches for surgical intervention include ablative and physiologic surgery. Ablative procedures involve the removal of dysfunctional, edematous, or fibrotic tissue, while physiologic therapies are aimed at rerouting or restoring lymphatic function to an affected area (Kareh and Xu 2020; Granzow et al. 2014; Carl et al. 2017).

Ablative Surgeries

Suction-assisted lipectomy, or liposuction, is a form of ablative therapy that can be effective in managing the symptoms of lymphedema. Liposuction was first described as a therapy for lymphedema by O'Brien in 1989. Eighty-four percent of patients experienced subjective improvement, while 90% had a measurable reduction in edema (O'Brien et al. 1989). Liposuction has been shown to be most effective when coupled with proper compression therapy postoperatively as the removal of fatty edematous tissue and fluid does not address the underlying pathophysiology. Furthermore, liposuction has been shown to be effective and does not further exacerbate the problem (Brorson et al. 1998). Schaverien et al. found that across 4 studies with 210 patients, liposuction effectively reduced mean limb volume, reduced the chance of infection, and improved appearance and function of the upper extremity in patients with lymphedema (Schaverien and Coroneos 2019). Brorson explains that in stage II lymphedema, when pitting is no longer present, excess fibroadipose tissue is the cause of swelling. Thus, removal of this tissue is appropriate and showed results in 116 patients followed for 15 years after undergoing liposuction and compression therapy (Brorson 2016).

Excision procedures for lymphedema therapy are aimed at physically removing edematous tissue and grafting healthy tissue to the affected area. These ablative procedures take careful planning and are often preformed in stages to ensure success. However, they have fallen out of favor due to the effectiveness of minimally invasive liposuction procedures but can be used in more severe cases. The Charles procedure was first described in 1912 (Charles 1912), and the Homan procedure in 1936 (Homans 1936). Both involve radical circumferential subcutaneous skin resections and full thickness skin grafts (Park et al. 2020). The Charles and Homan procedures have been described when treating lower limb lymphedema (Schaverien and Coroneos 2019). Radical reduction of lymphedema with preservation of perforators (RRPP) is another excisional procedure. This technique is possible due to advancements in microsurgery and preoperative imaging. This allows surgeons to visualize the minute blood supply in three-dimensional sections of tissue. Salgado et al. applied this method to 11 patients with stage II lymphedema of the upper extremity following treatment for breast cancer. They reported a significant decrease in circumference between the affected and normal limbs (Salgado et al. 2009).

Physiologic Surgeries

Lymphovenous anastomosis (LVA), or lymphovenous bypass, is a physiologic therapy for lymphedema. In LVA procedures, surgeons connect lymphatic vessels to veins in either an end-to-end, end-to-side, or side-to-side fashion. The concept of draining excess lymph by rerouting it into the venous system was first described in 1977 by O'Brien et al. (O'Brien et al. 1977). In 2000, Koshima et al. applied this concept to the upper extremity (Koshima et al. 2000). There have since been various techniques of LVA described including microsurgical lymphaticovenous implantation (Yamamoto and Sugihara 1998), intravascular stenting (Narushima et al. 2010), and lymphovenous bypass with sleeve-in anastomosis (Chung et al. 2019). In an extensive 2020 review, Forte et al. examined 22 studies evaluating LVA procedures. These studies included both retrospective and prospective analyses with 494 total patients. Forte et al. reported that 21 of the 22 studies reported positive outcomes with objective volume reduction and subjective symptom relief following LVA in the upper extremity (Forte et al. 2020). A 2021 systematic review of the literature found similar results examining 16 studies with 349 patients. Gupta et al. further confirmed the safety and efficacy of LVA for upper extremity lymphedema and noted that these procedures are more successful in the earlier stages of lymphedema (Gupta et al. 2021).

Vascularized lymph node transfer (VLNT) is an emerging physiologic therapy for upper extremity lymphedema. The concept of transferring functional lymph nodes via a pedicled flap to restore lymphatic function was first described by Clodius et al. in 1982 (Clodius et al. 1982). It is thought that transferring lymph nodes to an edematous area will promote lymphangiogenisis and reestablish

the "lymphatic pump" or the "lymphatic wick," but our understanding of these mechanisms is limited (Tourani et al. 2016). This concept was first applied to the breast and arm by Becker et al. in 2006 (Becker et al. 2006).

Ozturk et al. published a systematic review in 2016 examining various postsurgical outcomes of the VLN transfer when applied as a monotherapy. The analysis included 18 studies with 191 patients treated with VLN transfer for secondary lymphedema related to breast cancer. The techniques applied across these studies included the VLNT with a skin island and VLNT with an autologous flap. The authors concluded that there is exploratory evidence that VLNT provides improvement for patients with lymphedema (Ozturk et al. 2016). Further studies are needed to determine whether these surgical procedures would be superior if used in combination with one another.

References

Becker C, Assouad J, Riquet M, Hidden G (2006) Postmastectomy lymphedema: long-term results following microsurgical lymph node transplantation. Ann Surg 243(3):313–315

Brorson H (2016) Liposuction in lymphedema treatment. J Reconstr Microsurg 32(1):56–65

Brorson H, Svensson H, Norrgren K, Thorsson O (1998) Liposuction reduces arm lymphedema without significantly altering the already impaired lymph transport. Lymphology 31(4):156–172

Carl HM, Walia G, Bello R, Clarke-Pearson E, Hassanein AH, Cho B et al (2017) Systematic review of the surgical treatment of extremity lymphedema. J Reconstr Microsurg 33(6):412–425

Chao AH, Schulz SA, Povoski SP (2021) The application of indocyanine green (ICG) and near-infrared (NIR) fluorescence imaging for assessment of the lymphatic system in reconstructive lymphaticovenular anastomosis surgery. Expert Rev Med Devices 18(4):367–374

Charles R (1912) Elephantiasis scroti. Churchill, London, p 8

Chung JH, Baek SO, Park HJ, Lee BI, Park SH, Yoon ES (2019) Efficacy and patient satisfaction regarding lymphovenous bypass with sleeve-in anastomosis for extremity lymphedema. Arch Plast Surg 46(1):46–56

Clodius L, Smith PJ, Bruna J, Serafin D (1982) The lymphatics of the groin flap. Ann Plast Surg 9(6):447–458

DiSipio T, Rye S, Newman B, Hayes S (2013) Incidence of unilateral arm lymphoedema after breast cancer: a systematic review and meta-analysis. Lancet Oncol 14(6):500–515

Drake RL, Vogl W, Mitchell AWM, Gray H (2015) Gray's anatomy for students, 3rd edn. Churchill Livingstone/ Elsevier, Philadelphia

Forte AJ, Sisti A, Huayllani MT, Boczar D, Cinotto G, Ciudad P et al (2020) Lymphaticovenular anastomosis for breast cancer-related upper extremity lymphedema: a literature review. Gland Surg 9(2):539–544

Grada AA, Phillips TJ (2017) Lymphedema: pathophysiology and clinical manifestations. J Am Acad Dermatol 77(6):1009–1020

Granzow JW, Soderberg JM, Kaji AH, Dauphine C (2014) Review of current surgical treatments for lymphedema. Ann Surg Oncol 21(4):1195–1201

Greene AK, Grant FD, Slavin SA, Maclellan RA (2015) Obesity-induced lymphedema: clinical and lymphoscintigraphic features. Plast Reconstr Surg 135(6):1715–1719

Gupta N, Verhey EM, Torres-Guzman RA, Avila FR, Jorge Forte A, Rebecca AM et al (2021) Outcomes of lymphovenous anastomosis for upper extremity lymphedema: a systematic review. Plast Reconstr Surg Glob Open 9(8):e3770

Halsted WS (1921) The swelling of the arm after operations for cancer of the breast—elephantiasis chirurgica—its cause and prevention. Bull Johns Hopkins Hosp 32:309–313

Hita-Contreras F, Sánchez-Montesinos I, Martínez-Amat A, Cruz-Díaz D, Barranco RJ, Roda O (2018) Development of the human shoulder joint during the embryonic and early fetal stages: anatomical considerations for clinical practice. J Anat 232(3):422–430

Homans J (1936) The treatment of elephantiasis of the legs. N Engl J Med 215(24):1099–1104

International Society of Lymphology (2013) The diagnosis and treatment of peripheral lymphedema: 2013 Consensus Document of the International Society of Lymphology. Lymphology 46(1):1–11
Jungmann P, Figueredo-Silva J, Dreyer G (1991) Bancroftian lymphadenopathy: a histopathologic study of fifty-eight cases from northeastern Brazil. Am J Trop Med Hyg 45(3):325–331
Kareh AM, Xu KY (2020) Surgical management of lymphedema. Mo Med 117(2):143–148
Kinmonth JB, Taylor GW, Tracy GD, Marsh JD (1957) Primary lymphoedema; clinical and lymphangiographic studies of a series of 107 patients in which the lower limbs were affected. Br J Surg 45(189):1–9
Koshima I, Inagawa K, Urushibara K, Moriguchi T (2000) Supermicrosurgical lymphaticovenular anastomosis for the treatment of lymphedema in the upper extremities. J Reconstr Microsurg 16(6):437–442
Loukas M, Bellary SS, Kuklinski M, Ferrauiola J, Yadav A, Shoja MM et al (2011) The lymphatic system: a historical perspective. Clin Anat 24(7):807–816
Ma CX, Pan WR, Liu ZA, Zeng FQ, Qiu ZQ (2018) The deep lymphatic anatomy of the hand. Ann Anat 218:105–109
Ma CX, Pan WR, Liu ZA, Zeng FQ, Qiu ZQ, Liu MY (2019) Deep lymphatic anatomy of the upper limb: an anatomical study and clinical implications. Ann Anat 223:32–42
Maldjian C, Khanna V, Tandon B, Then M, Yassin M, Adam R et al (2014) Lymphatic filariasis disseminating to the upper extremity. Case Rep Radiol 2014:985680
Michelini S, Paolacci S, Manara E, Eretta C, Mattassi R, Lee BB et al (2018) Genetic tests in lymphatic vascular malformations and lymphedema. J Med Genet 55(4):222–232
Mihara M, Hara H, Araki J, Kikuchi K, Narushima M, Yamamoto T et al (2012) Indocyanine green (ICG) lymphography is superior to lymphoscintigraphy for diagnostic imaging of early lymphedema of the upper limbs. PLoS One 7(6):e38182
Miura H, Ono S, Shibutani K, Seino H, Tsushima F, Kakehata S et al (2014) Contribution of dynamic sentinel lymphoscintigraphy images to the diagnosis of patients with malignant skin neoplasms in the upper and lower extremities. Springerplus 3:625
Narushima M, Mihara M, Yamamoto Y, Iida T, Koshima I, Mundinger GS (2010) The intravascular stenting method for treatment of extremity lymphedema with multiconfiguration lymphaticovenous anastomoses. Plast Reconstr Surg 125(3):935–943
Narushima M, Yamamoto T, Ogata F, Yoshimatsu H, Mihara M, Koshima I (2016) Indocyanine green lymphography findings in limb lymphedema. J Reconstr Microsurg 32(1):72–79
O'Brien BM, Khazanchi RK, Kumar PA, Dvir E, Pederson WC (1989) Liposuction in the treatment of lymphoedema; a preliminary report. Br J Plast Surg 42(5):530–533
O'Brien BM, Sykes P, Threlfall GN, Browning FS (1977) Microlymphaticovenous anastomoses for obstructive lymphedema. Plast Reconstr Surg 60(2):197–211
Ozturk CN, Ozturk C, Glasgow M, Platek M, Ashary Z, Kuhn J et al (2016) Free vascularized lymph node transfer for treatment of lymphedema: a systematic evidence based review. J Plast Reconstr Aesthet Surg 69(9):1234–1247
Park KE, Allam O, Chandler L, Mozzafari MA, Ly C, Lu X et al (2020) Surgical management of lymphedema: a review of current literature. Gland Surg 9(2):503–511
Petrek JA, Heelan MC (1998) Incidence of breast carcinoma-related lymphedema. Cancer 83(12 Suppl American):2776–2781
Pfarr KM, Debrah AY, Specht S, Hoerauf A (2009) Filariasis and lymphoedema. Parasite Immunol 31(11):664–672
Phukan JP, Sinha A, Sengupta S, Bose K (2012) Cytodiagnosis of filariasis from a swelling of arm. Trop Parasitol 2(1):77–79
Price EW (1972) The pathology of non-filarial elephantiasis of the lower legs. Trans R Soc Trop Med Hyg 66(1):150–159
Routh HB, Bhowmik KR (1993) History of elephantiasis. Int J Dermatol 32(12):913–916
Salgado CJ, Sassu P, Gharb BB, Spanio di Spilimbergo S, Mardini S, Chen HC (2009) Radical reduction of upper extremity lymphedema with preservation of perforators. Ann Plast Surg 63(3):302–306
Samreen N, Glazebrook KN, Bhatt A, Venkatesh SK, McMenomy BP, Chandra A et al (2018) Imaging findings of mammary and systemic silicone deposition secondary to breast implants. Br J Radiol 91(1089):20180098
Schaverien MV, Coroneos CJ (2019) Surgical treatment of lymphedema. Plast Reconstr Surg 144(3):738–758
Sharma GM, Bhardwaj AR, Relwani NR, Dubey S (2018) Lymphatic filariasis presenting as a swelling over the upper arm: a case report. Malays Orthop J 12(1):63–65
Suami H, Scaglioni MF (2018) Anatomy of the lymphatic system and the lymphosome concept with reference to lymphedema. Semin Plast Surg 32(1):5–11
Suami H, Taylor GI, Pan WR (2005) A new radiographic cadaver injection technique for investigating the lymphatic system. Plast Reconstr Surg 115(7):2007–2013
Suami H, Taylor GI, Pan WR (2007) The lymphatic territories of the upper limb: anatomical study and clinical implications. Plast Reconstr Surg 119(6):1813–1822
Tourani SS, Taylor GI, Ashton MW (2016) Vascularized lymph node transfer: a review of the current evidence. Plast Reconstr Surg 137(3):985–993
van den Brekel MW, Castelijns JA (1999) Radiologic evaluation of neck metastases: the otolaryngologist's perspective. Semin Ultrasound CT MR 20(3):162–174

Vermeeren L, van der Ploeg IM, Olmos RA, Meinhardt W, Klop WM, Kroon BB et al (2010) SPECT/CT for preoperative sentinel node localization. J Surg Oncol 101(2):184–190

Wang L, Yu JM, Wang YS, Zuo WS, Gao Y, Fan J et al (2007) Preoperative lymphoscintigraphy predicts the successful identification but is not necessary in sentinel lymph nodes biopsy in breast cancer. Ann Surg Oncol 14(8):2215–2220

Yamamoto Y, Sugihara T (1998) Microsurgical lymphaticovenous implantation for the treatment of chronic lymphedema. Plast Reconstr Surg 101(1):157–161

Forearm Lymphatics 108

Katherine D. Bordes

108.1 Anatomy

Throughout much of the twentieth century, most knowledge of the lymphatic system was based on studies that used mercury injection to map the channels. Given mercury's toxic effects, safer techniques have since been established, mainly using lymphangiography, to better our understanding of the anatomy of lymphatic channels (Suami et al. 2007; Kinmonth 1952).

The lymphatic system of the forearm consists of superficial and deep channels. The superficial and deep lymphatic vessels form separate but complex networks with rich anastomoses. A few perforating vessels proximal to the wrist and in the cubital fossa connect the superficial and deep lymphatic system, allowing the lymph to pass in either direction (Ma et al. 2019; Gaisne et al. 1994). Lymphovenous anastomoses may also occur inconsistently in the upper arm but are more common in the forearm. These anastomoses may exist at the lymph nodes or along the lymphatic vessels (Gaisne et al. 1994). The lymphatic vessels of the forearm and hand do not vary significantly in diameter along their course and ultimately drain into the axillary group of lymph nodes (Suami et al. 2007).

K. D. Bordes (✉)
Tulane University School of Medicine,
New Orleans, LA, USA

The superficial lymphatic vessels of the forearm primarily accompany the basilic and cephalic veins and are vastly interconnected (Suami et al. 2007; Ma et al. 2019). They can be divided into anterior radial and ulnar networks and posterior radial and ulnar networks. As the posterior channels travel proximally, they dive deeper toward the biceps brachii tendon and the cubital fossa. The superficial lymph nodes that are found along the cubital and basilic vessels are few in number and inconsistent. By contrast, the supratrochlear lymph nodes, which are located superficial to the medial epicondyle of the humerus, are relatively consistent and easily identifiable (Gaisne et al. 1994).

The deep lymphatic vessels of the forearm typically have a tortuous course and can be differentiated into radial, ulnar, anterior interosseous, and posterior interosseous lymphatic vessels, according to the major neurovascular bundle with which they run. The radial and ulnar lymphatic vessels (RLV and ULV, respectively) are formed by the deep lymphatic vessels of the palm and eventually converge within the cubital fossa and continue proximally toward the axillary lymph nodes as the brachial lymphatic vessels. The anterior interosseous lymphatic vessel (AILV) arises deep to the proximal margin of the pronator quadratus, travels alongside the anterior interosseous artery, and merges with the ULV just distal to the cubital fossa. The posterior interosseous lymphatic vessel (PILV) emerges between the superficial and deep extensor muscles, travels

S. J. Bordes, Jr. et al. (eds.), *The Clinical Anatomy of the Vascular System*,
https://doi.org/10.1007/978-3-031-78326-5_108

alongside the posterior interosseous artery, and then also converges with the ULV (Ma et al. 2019). Along the deep lymphatic vessels are small and inconsistent lymph nodes, except those found at the root of the ulnar artery and the humeral vessels (Gaisne et al. 1994).

108.2 Variations

In some cases, the ulnar lymphatic vessel may not converge with the radial lymphatic vessel and continue through the subcutaneous tissue of the arm and ultimately enter the axillary lymph node (Ma et al. 2019).

108.3 Clinical/Surgical Significance

Lymphatic channels and nodes of the forearm are clinically relevant to the metastatic spread of cancer, the spread of infection, and primary and secondary lymphedema.

Although less common than cancer metastasis from the trunk of the body, there have been cases of cancers spreading distally from the forearm to the more proximal axillary lymph nodes, including melanoma, leiomyosarcoma, and eccrine porocarcinoma (Gupta et al. 2012; Ma et al. 2021; Hochwald et al. 2011).

Secondary lymphedema is prevalent following surgical ablation or radiation of axillary lymph nodes in cancer treatment, especially breast cancer. Breast cancer-related lymphedema is relatively common, largely irreversible, and remains difficult to treat. It is likely that the injury to the axilla results in increased resistance and pressure gradient needed for lymphatic drainage of the forearm. This leads to impaired lymph drainage in the subfascial and epifascial compartments of the forearm (Stanton et al. 2003).

It should also be noted that the lymphatic channels have been known to regenerate following trauma or damage, forming new anastomoses with surrounding vessels (Gaisne et al. 1994).

References

Gaisne E, Chaise F, Bellemère P, Friol JP (1994) Anatomie du système lymphatique du membre supérieur [Anatomy of the lymphatic system of the arm]. Ann Chir Main Memb Super 13(3):172–178. https://doi.org/10.1016/s0753-9053(94)80043-x

Gupta S, Husain N, Kumar A, Sunder S (2012) Adult primary pleomorphic leiomyosarcoma of forearm with axillary lymph node metastasis: a case report and literature review. J Cancer Res Ther 8(3):448–450. https://doi.org/10.4103/0973-1482.103532

Hochwald SN, Kissane N, Grobmyer SR, Lopes J (2011) Epitrochlear lymph node dissection. Ann Surg Oncol 18(2):505. https://doi.org/10.1245/s10434-010-1227-5

Kinmonth JB (1952) Lymphangiography in man; a method of outlining lymphatic trunks at operation. Clin Sci 11(1):13–20

Ma CX, Pan WR, Liu ZA, Zeng FQ, Qiu ZQ, Liu MY (2019) Deep lymphatic anatomy of the upper limb: an anatomical study and clinical implications. Ann Anat 223:32–42. https://doi.org/10.1016/j.aanat.2019.01.005

Ma C, Huang K, Liu M et al (2021) Primary eccrine Porocarcinoma of the thumb with metastasis: a rare case report and literature review. Am J Dermatopathol 43(12):e285–e289. https://doi.org/10.1097/DAD.0000000000002051

Stanton AW, Mellor RH, Cook GJ et al (2003) Impairment of lymph drainage in subfascial compartment of forearm in breast cancer-related lymphedema. Lymphat Res Biol 1(2):121–132. https://doi.org/10.1089/153968503321642615

Suami H, Taylor GI, Pan WR (2007) The lymphatic territories of the upper limb: anatomical study and clinical implications. Plast Reconstr Surg 119(6):1813–1822. https://doi.org/10.1097/01.prs.0000246516.64780.61

Lymphatics of the Hand

109

Luis Felipe Okida and Lisandro Montorfano

109.1 Introduction

The lymphatic system is distributed throughout the human body, with a few exceptions, such as the bone marrow and cartilage. The primary function of this system is to return fluid extravasated from blood capillaries into the interstitial region. Still, it also has a crucial role in carrying pathogens to lymph nodes, initiating an immunological chain reaction. Regarding the lymphatic system of the hand, Bonamy et al. described the deep lymphatic vessels of the upper limb in 1847 (Ma et al. 2019). However, the limitations of the mercury infusion technique precluded a detailed anatomical description at that time. As methods became more sophisticated over the decades, better assessment and understanding of the lymphatics of the upper limb became possible.

L. F. Okida
Department of Surgery, Cleveland Clinic Florida, Weston, FL, USA
e-mail: okidal@ccf.org

L. Montorfano (✉)
Department of Surgery, Cleveland Clinic Florida, Weston, FL, USA

Department of Plastic and Reconstructive Surgery, Vanderbilt University Medical Center, Nashville, FL, USA

109.2 Embryology

The lymphatic system starts to develop at approximately the end of the fifth week of embryogenesis, roughly one month after the appearance of the first blood vessel (Witte et al. 2006). The lymphatic vessels are believed to develop from clefts in the mesenchyme linked to the venous system. The cells lining these mesenchymal clefts assume an endothelial shape, and subsequent sprouting of this layer results in the fusion of the clefts and the formation of the lymphatic channels (Rockson 2018).

109.3 Anatomy

To elucidate the lymphatic system of the hand, the first requirement is to understand the types of structures involved in the circuit. In general, lymphatic vessels are divided into three categories based on size and anatomy: lymph capillaries, pre-collectors, and lymph-collecting vessels (Suami et al. 2008).

- Lymph capillaries (20–70 μm): These are located immediately underneath the epidermis and have no valves. A fibrous anchoring structure connects the endothelial cells to the surrounding tissue. When interstitial fluid levels increase, these anchoring filaments pull the endothelial cells outward to open the junctions

S. J. Bordes, Jr. et al. (eds.), *The Clinical Anatomy of the Vascular System*,
https://doi.org/10.1007/978-3-031-78326-5_109

between them and drain the excess fluid into the capillary lumen.
- Pre-collectors (70–150 μm): These capillaries are connected to the pre-collector system in the deep dermis. These small vessels have valves, which ensure unidirectional lymph flow from superficial to deep layers. They converge within the dermis to form larger vessels, which become efferent pre-collectors.
- Lymph-collecting vessels (150–500 μm): The pre-collectors are connected to the lymph-collecting vessels (also called collectors) in the subcutaneous adipose layer. These collecting vessels have a three-layered wall of endothelial cells, smooth muscle cells, collagen fibers, and fibroblasts. This composition enables them to contract and regulate the lymph flow (Suami and Scaglioni 2018). It is essential to understand that the collector vessels are divided into two subcategories, superficial and deep, according to their anatomical relationships to the deep fascia. The hand's superficial and deep lymphatic systems will be elucidated further in this chapter.

In contrast to the vascular system, it was initially believed that no lymphatic vessels connected the superficial and deep lymph-collecting systems directly under normal conditions. Another type of lymphatic vessel that runs in the subcutaneous tissue joins the deep lymph-collecting vessels, but was originally thought not to form any connection between the superficial and deep systems. These "perforating" lymph vessels run along the perforating artery, which goes into the deep fascia. However, more recent studies on the deep lymphatic system of the upper limb have suggested that the perforating vessels do actually establish a connection between the superficial and deep lymphatic circuits, as in the venous system (Ma et al. 2019). This has proved vitally important for the development of the lymphaticovenous anastomosis technique in patients with lymphedema of upper limbs, which will be explained in detail later in this chapter.

109.4 Superficial Lymphatic System of the Hand

As described above, the superficial lymphatic system of the hand starts with the lymph capillaries and pre-collector vessels in the skin of the fingers, palm, and dorsum. The lymphatic fluid then drains into lymph-collecting vessels in the subcutaneous tissue. One to three superficial lymph-collecting vessels run along both sides of the fingers in the neurovascular bundle and then run axially in the dorsal hand in a layer slightly deeper than the cutaneous veins. They are evenly spaced around the wrist and gradually change course around the elbow, running toward the anteromedial aspect of the forearm and then parallel to the basilic vein in the upper arm. They finally drain into one or two dominant lymph nodes in the lateral axillary region (Suami and Scaglioni 2018).

109.5 Deep Lymphatic System of the Hand

The deep lymphatic system of the hand was first assessed by Bonamy using mercury infusion, but became better understood after more recent injection techniques were introduced. The deep lymphatic system contains fewer vessels and accompanies the major deep veins in the hand and upper limb (Moore et al. 2017). With numerous valves in their lumina, the deep lymphatic vessels of the hand lie along the major neurovascular bundles, traveling tortuously either alongside adjacent arteries, veins, and nerves or sometimes between those structures.

A lymph-collecting vessel runs subcutaneously in the three web spaces from the index to the little finger. These vessels are called deep finger web lymph vessels. The deep lymphatic system of the palm consists of deep palmar arch, thenar, and hypothenar lymph vessels. On the lateral aspect of the hand, the deep palmar arch and thenar deep lymphatics drain into one or two radial lymph vessels, which travel beside the radial neurovascular bundle towards the cubital fossa. On the medial aspect, the hypothenar lymph vessel drains into one or two ulnar lymph vessels, which travel beside the ulnar vascular bundle. The radial and ulnar lymph vessels travel along the forearm and converge in the cubital fossa.

Besides the lymphatic vessels of the hand, the distal upper limb has two other groups of lymphatic vessels: anterior interosseous and posterior interosseous. Arising from the deep side of the proximal margin of the pronator quadratus muscle, one or two anterior interosseous lymph vessels travel with the anterior interosseous vascular bundle and merge with the ulnar lymph vessel near the cubital fossa. The posterior interosseous lymph vessel arises in the intermuscular septum of the superficial and deep extensor muscles in the distal forearm, travels with the posterior interosseous vascular bundle, and merges with the ulnar lymph vessel near the cubital fossa (Ma et al. 2019).

109.6 Clinical Significance

Knowing the normal anatomy of the lymphatic system of the upper limb is essential for predicting which lymph node groups are likely to be sites of metastatic disease from a primary tumor and understanding the pathological changes seen during lymphedema. If there are tumors in the hand or forearm, the cubital lymph node can be the sentinel lymph node, which can be detected with lymphoscintigraphy (Reynolds et al. 2007). Pan et al. demonstrated that lymph vessels from either the superficial or the deep tissue of the hand and forearm can drain into superficial or deep cubital lymph nodes via perforating lymph vessels (Pan 2017). This information helps search for sentinel lymph nodes in the superficial or deep layers of the cubital fossa during surgical harvesting.

During lymph node biopsies or mastectomies, the structure of the lymphatic system changes significantly. The affected lymphatic vessels regenerate from a stump and attempt to recanalize (Gray 1940). However, the superficial lymph-collecting vessels often become obstructed in the proximal aspect after such surgical intervention, resulting in changes in the distal regions of the lymphatic system. Since there is an obstruction or disruption, the lymph flows backward from the superficial lymph-collecting vessels to the precollectors and lymph capillaries in the skin of the hand. This phenomenon is called "dermal backflow" and can be detected by intradermal injection of dye, lymphoscintigraphy, or fluorescent lymphangiography (Stanton et al. 2006). Lymphedema is the accumulation of excess lymphatic fluid in the skin and subcutaneous adipose layer, and "dermal backflow" is a specific feature of this condition. In advanced stages of lymphedema, the superficial lymph-collecting vessels can deteriorate and disappear.

Among the treatment options for lymphedema of the upper limb, lymphaticovenous anastomosis is now a surgical alternative (Neligan et al. 2016). A thorough understanding of the lymphatic system of the upper limb is essential for preoperative planning and intraoperative feasibility, affecting postoperative outcomes. Studies have shown the importance of creating as many anastomoses as possible to obtain better results. Usually, most of the anastomoses are made in the superficial layer. As previously mentioned, some authors have demonstrated that the deep lymphatic system can be drained by the superficial lymphatic vessels via perforating vessels so that superficial lymphaticovenous anastomoses can drain both systems (Poumellec et al. 2017). Wolfs et al. studied the patency of the lymphaticovenous anastomoses in patients with breast cancer-related lymphedema during a 12-month follow-up (Wolfs et al. 2020). They showed that the level of patency was directly linked to the number of anastomoses created, and patients with three lymphaticovenous anastomoses had up to 100% patency after one year of surgery. The ideal number of anastomoses is still debated in the literature, but many authors maintain that several anastomoses are needed for best outcomes (Chang et al. 2013). The procedure is usually performed under local anesthesia and the lymph-collecting vessels can be initially located by methods such as indocyanine green lymphography. A small incision is made in the forearm to search for subdermal venules (0.5 mm caliber) and lymph-collecting vessels with substantial drainage (0.3–0.5 mm caliber). Subsequently, an end-to-end or end-to-side lymphaticovenous anastomosis is performed using 11–0 nylon sutures under ×20 to ×30 magnification (Wolfs et al. 2020; Koshima et al. 2000).

109.7 Conclusion

Technical difficulties in identifying the lymphatic system during cadaveric dissection have restricted the anatomical description and study of this system for almost a century. As different dye injections and imaging exams emerged, a more detailed analysis of the lymphatic anatomy of the hand became possible. Knowing the hand's superficial and deep lymph systems is the first step in clinical management and developing further techniques for treating patients with lymph-related disorders, such as lymphedema or tumors of the upper limbs.

References

Chang DW, Suami H, Skoracki R (2013) A prospective analysis of 100 consecutive lymphovenous bypass cases for treatment of extremity lymphedema. Plast Reconstr Surg 132:1305–1314. https://doi.org/10.1097/prs.0b013e3182a4d626

Gray JH (1940) Studies of the regeneration of lymphatic vessels. J Anat 74(Pt 3):309–335

Koshima I, Inagawa K, Urushibara K, Moriguchi T (2000) Supermicrosurgical lymphaticovenular anastomosis for the treatment of lymphedema in the upper extremities. J Reconstr Microsurg 16(6):437–442

Ma C-X, Pan W-R, Liu Z-A, Zeng F-Q, Qiu Z-Q, Liu M-Y (2019) Deep lymphatic anatomy of the upper limb: an anatomical study and clinical implications. Ann Anat 223:32–42

Moore KL, Dalley AF II, Agur AMR (2017) Clinically oriented anatomy. Wolters Kluwer, p 1153

Neligan PC, Masia J, Piller NB (2016) Lymphedema: complete medical and surgical management. Thieme

Pan W-R (2017) Atlas of lymphatic anatomy in the head, neck. Chest and Limbs [Internet]. https://doi.org/10.1007/978-981-10-3749-8

Poumellec M-A, Foissac R, Cegarra-Escolano M, Barranger E, Ihrai T (2017) Surgical treatment of secondary lymphedema of the upper limb by stepped microsurgical lymphaticovenous anastomoses. Breast Cancer Res Treat 162(2):219–224

Reynolds HM, Dunbar PR, Uren RF, Blackett SA, Thompson JF, Smith NP (2007) Three-dimensional visualisation of lymphatic drainage patterns in patients with cutaneous melanoma. Lancet Oncol 8(9):806–812

Rockson SG (2018) Embryology of the lymphatic system and lymphangiogenesis. In: Lee B-B, Rockson SG, Bergan J (eds) Lymphedema: a concise compendium of theory and practice. Springer, Cham, pp 47–55

Stanton AWB, Modi S, Mellor RH, Peters AM, Svensson WE, Levick JR et al (2006) A quantitative lymphoscintigraphic evaluation of lymphatic function in the swollen hands of women with lymphoedema following breast cancer treatment. Clin Sci 110(5):553–561

Suami H, Scaglioni MF (2018) Anatomy of the lymphatic system and the lymphosome concept with reference to lymphedema. Semin Plast Surg 32(1):5–11

Suami H, Shinaoka A (2019) The methodology of lymphatic anatomy studies in a cadaver model: an overview. Plast Aesthet Res 2019. Available from: https://parjournal.net/article/view/3298

Suami H, Pan W-R, Mann GB, Taylor GI (2008) The lymphatic anatomy of the breast and its implications for sentinel lymph node biopsy: a human cadaver study. Ann Surg Oncol 15(3):863–871

Witte MH, Jones K, Wilting J, Dictor M, Selg M, McHale N et al (2006) Structure function relationships in the lymphatic system and implications for cancer biology. Cancer Metastasis Rev 25(2):159–184

Wolfs JAGN, Wolfs JAG, de Joode LGEH, van der Hulst RRWJ, Qiu SS (2020) Correlation between patency and clinical improvement after lymphaticovenous anastomosis (LVA) in breast cancer-related lymphedema: 12-month follow-up [Internet]. Breast Cancer Res Treat 179:131–138. https://doi.org/10.1007/s10549-019-05450-2

Part VI

Lower Limb

Femoral Artery

110

James Michael, Adalgisa Varuolo, and Stephen J. Bordes, Jr.

110.1 Embryology

The femoral artery undergoes many morphological angiogenic changes before its final development. Three regions of the embryo are involved in arterial development: the anterior, posterior, and middle (Lederle et al. 2019). The anterior region is the most complex, with six aortic arches; the middle region undergoes fusion of the two aortae to form the final thoracic and abdominal aorta; the posterior region is the least complex, and the femoral artery develops here (Lederle et al. 2019).

A pair of umbilical arteries arise from the dorsal aorta at the posterior end of the ventral segmental arteries.

The umbilical arteries (arising from the paired fourth ventral branches) remain paired in utero and migrate caudally during embryogenesis, giving off small external branches to the lower limb (Lederle et al. 2019). The aorta caudal to the fifth dorsal branch regresses to the median sacral artery. The lower limb bud receives blood from the dorsal-segmental arterial system via a capillary plexus. After that, a single dorsal axial/sciatic/ischiadic artery is formed, and the sciatic artery is the main supply (Tubbs et al. 2016). The dorsal root of the umbilical artery gives rise to the axial artery, which supplies blood to the lower limb. This occurs when the umbilical artery reaches a diameter of 6 mm (Swift and Bordoni 2021). A femoral artery entering the lower limb is distinct at 14 mm, communicating to the sciatic and axial arteries (Swift and Bordoni 2021). Once the femoral artery reaches approximately 22 mm, the sciatic artery regresses, and the femoral artery becomes the primary vascular supply to the lower limb (Swift and Bordoni 2021).

Nevertheless, these anatomical arterial phases have established an adult pattern by the eighth week of development. The placental circulation is interrupted at birth, and the umbilical arteries begin to obliterate. However, the proximal portion develops into the primitive iliac, hypogastric, and superior vesicle arteries (Kau et al. 2007). Eventually, the common iliac artery gives rise to the external iliac artery, which then gives rise to the femoral artery and supersedes the sciatic artery.

110.2 Anatomy and Topography

Nomenclature defines the femoral artery as originating immediately after the external iliac artery traverses deep to the inguinal ligament. The neurovascular bundle, femoral artery (Fig. 110.1),

J. Michael
Department of Surgery, Rutgers New Jersey Medical School, Newark, NJ, USA

A. Varuolo
Department of Surgery, Hackensack Meridian Health, Hackensack, NJ, USA

S. J. Bordes, Jr. (✉)
Department of Surgery, Louisiana State University Health Sciences Center, New Orleans, LA, USA

S. J. Bordes, Jr. et al. (eds.), *The Clinical Anatomy of the Vascular System*,
https://doi.org/10.1007/978-3-031-78326-5_110

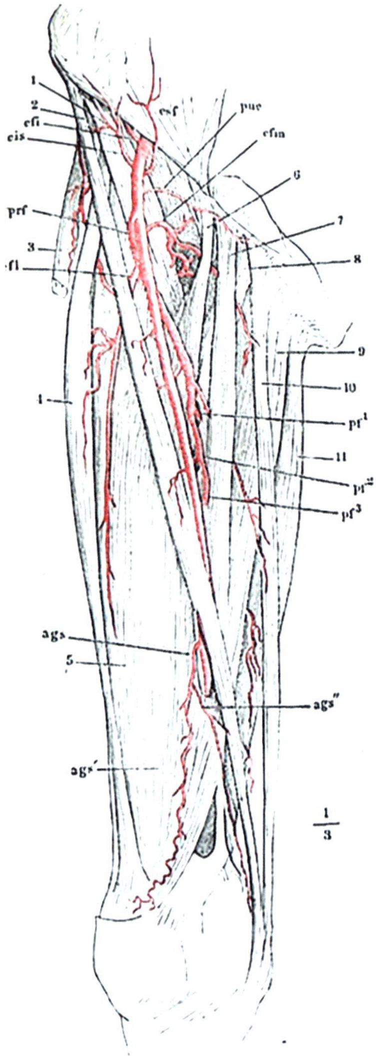

Fig. 110.1 Schematic drawing noting the femoral artery and its branches. (After Allen 1892)

femoral vein, and femoral nerve exit the femoral triangle, which is immediately inferior/caudal to the inguinal ligament, a conduit for structures to travel to and from the anterior thigh. The femoral triangle is defined by the following borders:

1. Roof, or superficial border: tensor fasciae latae.
2. Floor, or deepest border: iliopsoas, pectineus, and adductor longus muscles.
3. Superior border: inguinal ligament, which courses from the anterior superior iliac spine to the pubic tubercle.
4. Lateral border: medial border of sartorius.
5. Medial border: medial border of adductor longus muscle.

Delineating the borders of the femoral triangle and understanding the spatial anatomy carries many important clinical implications, which will be discussed later in this chapter. The femoral artery, vein, and canal (lymph nodes and vessels) are contained within the femoral sheath.

After the femoral artery exits the femoral triangle via the femoral sheath, it runs into the iliopectineal groove to travel along the anteromedial thigh before addressing the adductor hiatus, which has clinically important borders. The adductor hiatus is a narrow conical tunnel allowing vessels (the femoral artery and vein—the femoral vein runs posteriorly to the artery) to travel from the anterior to the posterior aspect of the thigh prior to entering the popliteal fossa. The hiatus is an aponeurotic canal in the middle third of the thigh, approximately 8–13.5 cm superior to the adductor tubercle (Kale et al. 2012). The following anatomical boundaries make up the adductor hiatus:

1. Anterior and lateral borders: vastus medialis oblique (VMO).
2. Posterior border: adductor longus and adductor magnus.
3. Roof and medial border: sartorius.

When the femoral artery and vein exit the canal, they become the popliteal artery and vein.

110.3 Histology

The femoral artery is a muscular as opposed to an elastic artery. Muscular arteries are named by their anatomical location (brachial artery, radial artery, etc.) (Taylor and Bordoni 2021). Muscular arteries continue from elastic arteries and histologically contain more muscle layers to control blood flow distribution throughout the body (Taylor and Bordoni 2021). Histologically, muscular arteries are distinguished by an internal elastic lamina, thick, rippled, and best visualized under the microscope with Verhoeff's stain (Taylor and Bordoni 2021).

110.4 Imaging

Evaluation of the femoral artery is essential if disease is suspected and before potential intervention. There are three main modalities for assessing femoral artery function: clinical, noninvasive diagnostics, and invasive diagnostics. Clinically, we can inspect, palpate, auscultate, and measure the ankle-brachial index (ABI). Issues with a patent femoral artery are indicated by dermal changes and irregular trophic hair distribution upon inspection. Thrills can be palpated, and bruits heard upon auscultation. The ABI is the best initial peripheral artery disease (PAD) screening test. It is the ratio of the highest systolic pressure of the ankle over the brachium, measured with a handheld Doppler and a blood pressure cuff (Kau et al. 2007). The ABI gives us a more objective measure of the severity of femoral artery disease; a value less than 0.9 is considered abnormal (0.4–0.7 is moderate, and 0–0.4 is severe) (Kau et al. 2007). Although studies have demonstrated different sensitivities and specificities of the ABI for diagnosing PAD, the consensus values are approximately 79–95% sensitivity and 96–100% specificity, with a low intra-observer variability and reproducibility of 0.10 (Kau et al. 2007).

Nonetheless, the value of the ABI is severely limited in patients with significant atherosclerosis or arteriosclerosis as the artery becomes noncompressible. These patients include those with longstanding diabetes, PAD, advanced age, end-stage renal disease on dialysis, and Monckeberg medial calcific sclerosis (Kau et al. 2007). The ABI in this patient population is greater than 1.4.

Noninvasive diagnostic evaluations include pulse-volume recordings (PVR) and continuous-wave Doppler ultrasound. Although less commonly used, PVR measures the pulsatile changes in the lower limb, extrapolating to the volumetric changes with each cardiac cycle. These changes can then be documented and graphed along the lower limb using plesythmography. Continuous-wave Doppler is used with PVR to obtain segmental velocity waveforms and systolic BP measurements in both the upper and lower limbs. These techniques are applied to patients with noncompressible arteries for the initial diagnosis of PAD. They can also be used to assess the severity of the disease and as a surrogate for postoperative follow-up (Kau et al. 2007).

Duplex ultrasound offers vascular imaging and flow velocity with sensitivity and specificity above 90%. However, recent diagnostic advances offer CTA (CT angiography) and MRA (MR angiography). New multi-detector CTA allows for fast diagnosis of stenosis. In one breath, CTA can image the abdomen and lower limb in slices of 1 mm or less and offer a 3-D reconstruction with sensitivity and specificity of over 90% for stenoses around 50% (Kau et al. 2007). MRA has better eccentric resolution; however, CTA is much less expensive, is not hindered by retained metallic foreign bodies or prostheses, and can provide images of arterial calcification.

Nevertheless, CTA risks nephrotoxicity due to the contrast and ionizing radiation (Kau et al. 2007). MRA has sensitivity and specificity of 90–100% for stenoses of 50%, which is similar to the invasive diagnostic technique of catheter angiography. Some studies suggest that MRA is superior to angiography in detecting outflow vessels that can be used for bypass in limb ischemia (Kau et al. 2007). MRA is used for preoperative planning, defining a roadmap for revascularization procedures, and for follow-up postoperatively (Kau et al. 2007). Unfortunately, it is limited in patients with retained metal foreign bodies and prostheses. It can also overestimate the degree of stenosis owing to turbulent flow. More recently, nephrogenic systemic fibrosis due to gadolinium has been reported in patients with elevated creatinine (Kau et al. 2007). Although the procedure carries more risk, diagnostic catheter angiograms are the gold standard method for evaluating patients with PAD and suspected need for revascularization (Kau et al. 2007).

110.5 Variations

The femoral artery (FA) shows wide-ranging variation. From superior/cephalad to inferior/caudal, it typically gives rise to the superficial epigastric artery, superficial circumflex iliac artery, superficial external pudendal artery, deep external pudendal artery, deep femoral artery deep to the adductor longus, and descending geniculate branch.

The FA represents the major source of arterial blood to the lower limb except in cases where the sciatic (axial) artery fails to involute and persists. This variant is rare and is reported at 0.25–0.04%; fewer than 100 case reports (Taylor and Bordoni 2021). The FA can arise as a variant from the inferior gluteal or internal iliac arteries, leaving the pelvis to travel posteriorly along with the sciatic nerve and inferior gluteal, deep femoral, lateral circumflex, or any other branch of the femoral artery (Tubbs et al. 2016). There are other case reports of bifurcated or trifurcated FAs, and Patterson (1909) reported a case with no recognizable left common, external, or internal iliac artery branching from the aorta. Therefore, tributary vessels from the aorta fed both the femoral artery and the distal part of the internal iliac artery (Tubbs et al. 2016).

110.6 Open and Endovascular Approaches to the Femoral Artery

Before 1953, the mainstay of aorto-iliofemoral pathology was open surgical correction. An interventional radiologist, Dr. Sven Idar Seldinger, described the Seldinger technique in 1953, an endovascular approach that has become popular during the last 60 years for treating many arterial and cardiac-related conditions (Ciau-Uitz et al. 2016). Guidelines and reporting standards indicated by the Society of Vascular Surgery for endovascular use include (among others) chronic mesenteric ischemia (CMI), thoracic aneurysm endovascular repair (TEVAR), endovascular aortic repair of aneurysms involving the renal-mesenteric arteries (FEVAR), management of visceral artery aneurysms, STS (Society of Thoracic Surgeons) reporting standards for type b aortic dissections, chronic limb-threatening ischemia, follow-up after vascular surgery arterial procedures, abdominal aortic aneurysm, transcarotid artery revascularization (TCAR), inferior vena cava (IVC) filter placement, management of the diabetic foot, peripheral artery disease, and venous leg ulcers.

Lederle et al. randomly assigned 881 patients with asymptomatic abdominal aortic aneurysms to either endovascular or open repair; 444 were assigned to endovascular and 437 to open repair (Lederle et al. 2019). Primary outcome was all-cause mortality; secondary outcome was all-cause mortality in prespecified subgroups and secondary therapeutic procedures that resulted directly or indirectly from the initial procedure. The study demonstrated no difference in all-cause mortality. However, long-term survival was higher in younger patients who underwent an endovascular approach and slightly lower among older patients who underwent endovascular as opposed to an open repair (Lederle et al. 2019).

Kotsis et al. performed a retrospective study on twelve patients with aortoiliac disease. Three patients underwent only endovascular repair, five underwent open repair, and four underwent a combination (Kotsis et al. 2014). The study demonstrated 100% technical success with 100% patency at 26.3 months follow-up (Kotsis et al. 2014). Finally, Rocha-Neves et al. compared technical, clinical, and economic outcomes between endovascular and open surgical approaches in patients with TASC Type D aortoiliac occlusive disease (Rocha-Neves et al. 2020). Twenty-seven patients were assigned to an endovascular approach and 32 to an open repair. The study demonstrated that the endovascular approach is at least equivalent to the open surgery approach and is more cost-efficient (Rocha-Neves et al. 2020).

110.7 Pathology

Pathologies associated with the femoral artery can be categorized as idiopathic, iatrogenic, and traumatic etiologies. Etiologies of idiopathic pathology include (among others) atherosclerosis, arteriosclerosis, thromboses, and Monkeberg's medial calcinosis. Iatrogenic etiologies include septic vasculitis, pseudoaneurysm, arteriovenous (AV) fistula, aneurysms, hematomas, and thromboses. Traumatic etiologies include arterial tears due to fractures and dislocations.

Atherosclerosis is a systemic disease with segmental distribution and predilection. The femoral artery is at significant risk of atherosclerosis. Suurkula et al. demonstrated ultrasound diagnostic appreciation of large plaques in 62% of men in a high-risk group (associated comorbidities with hypertension, diabetes, smoking, etc.) (Suurkula et al. 1996). Trans-Atlantic Inter-Society Consensus (TASC II) classified atherosclerotic lesions for therapeutic and prognostic applications.

Graham et al. studied 172 aneurysms in 100 male patients ranging from 47 to 91 years old and demonstrated bilateral lesions in 72%. Forty of the 172 patients were asymptomatic (Graham et al. 1980).

110.8 Clinical and Surgical Significances

The anatomical location of the femoral artery allows for many important clinical and interventional indications. It is superficially located within the femoral triangle, allowing for a relatively less complicated access point for coronary angiography and endovascular procedures. The anatomical course of the femoral artery places the femoral head directly beneath it along the inguinal fold, allowing pressure hemostasis in the event of trauma. Therefore, it is beneficial to access the artery at the level of the femoral head rather than superior or inferior to that level, making hemostasis more difficult to sustain (Swift and Bordoni 2021). Chronic occlusion of the femoral artery instigates the formation of extensive collaterals, thereby increasing distal blood flow. Therefore, comprehensive clinical and diagnostic evaluation is needed prior to an open or endovascular procedure in a patient with associated risk factors (Swift and Bordoni 2021). Huseyin et al. reported results on patients who underwent operations for true and false aneurysms following a catheterization procedure. They found 55 aneurysms among 12,261 patients who had undergone catheterization from January 2003 to January 2013 and determined that pseudoaneurysms of 2 cm or less should be managed conservatively with regular ultrasonographic follow-up, while pseudoaneurysms larger than 2 cm required primary repair (Huseyin et al. 2013).

The FA is the most commonly accessed vessel for interventional diagnostic and therapeutic vascular procedures (Ciau-Uitz et al. 2016). Although very common, vascular access carries many associated risks, including bleeding. The mainstay for managing arterial bleeds is manual compression. However, this is tedious and time-consuming; 20–30 min of manual compression with postoperative modifications (e.g., bed rest) diminishes the feasibility and efficiency of a percutaneous endovascular technique (Noori and Eldrup-Jørgensen 2018). Ortiz et al. (2014) reported a complication rate of 3.5%, which has decreased since the inception of the Seldinger technique (Ortiz et al. 2014). Although three-quarters of that percentage was stratified as minor, the remaining 25% of patients had moderate or severe complications, requiring blood transfusion, thrombin injection, or surgery in severe cases (Noori and Eldrup-Jørgensen 2018). In recent years, vascular closure devices (VCDs) have reduced arterial bleeding time and associated morbidity. VCDs can be categorized as active or passive approximators. Examples of active approximator VCDs are StarClose SE, PrecloseProglide, and Prostar XL. In brief, these are three FDA-approved active VCDs that approximate the arterial wall edges with a clip or suture. In trained hands, these devices should stop bleeding immediately. There are six FDA-approved passive VCDs that seal the arteriotomy with a plug, sealant, or gel: Angio-Seal, Vascade Vascular Closure System, Mynx Ace and Mynx Grip, Exoseal, FISH (Femoral Introducer Sheath and Hemostasis) CombiClose and Control Close, and Closure Vascular Sealing System (Noori and Eldrup-Jørgensen 2018). Finally, there are external hemostatic VCDs that Noori et al. studied, including FemoStop, CompressAR, Strong Arm and Super Comfort, and QuicKlamp. These devices do not shorten the time of arteriotomy closure, but alleviate the personal manual compression (Noori and Eldrup-Jørgensen 2018).

In most categories, the VCDs were superior. Nelson et al. (2014) demonstrated more favorable outcomes with PrecloseProGlide (PG) and 10F Prostar (PS) than with traditional Femoral Exposure (FE) concerning shorter times to hemostasis and procedure completion, groin pain, and overall quality of life (Nelson et al. 2014). Cox et al. (2015) published a systematic review of 34 randomized controlled trials assessing quality of life and primary outcomes (hemostasis, time to ambulation, time to discharge, bleeding, hematoma, pseudoaneurysm, and limb ischemia) in comparing femoral cutdown with VCDs (Cox et al. 2015). VCDs proved superior in all regards (Cox et al. 2015).

110.9 Complications

Accessing the femoral artery carries many risks, adverse effects, and complications. As discussed above regarding angiography, there are inherent risks of bleeding, infection, atheroembolism, dissection, vessel disruption, perforation, hematoma, pseudoaneurysm, and AV fistulas occurring in anywhere from 1.9 to 2.9% (percentages vary according to operator experience) (Kasapis and Gurm 2009). Other complications with femoral artery access, as discussed earlier, are pseudoaneurysms, which are generally caused by puncturing distal or inferior to the femoral artery bifurcation. In contrast, retroperitoneal hemorrhage is usually caused by puncturing proximal or superior to the bifurcation (Rashid and Hughes 2016). Techniques to reduce complications include using ultrasonography- and fluoroscopy-guided femoral punctures.

References

Ciau-Uitz A, Patient R, Medvinsky A (2016) Encyclopedia of immunobiology. Ontogeny Hematopoietic Syst 1:1–14

Cox T, Blair L, Huntington C, Lincourt A, Sing R, Heniford BT (2015) Systematic review of randomized controlled trials comparing manual compression to vascular closure devices for diagnostic and therapeutic arterial procedures. Surg Technol Int 27:32–44

Graham LM, Zelenock GB, Whitehouse WM et al (1980) Clinical significance of arteriosclerotic femoral artery aneurysms. Arch Surg 115(4):502–507. https://doi.org/10.1001/archsurg.1980.01380040124022

Hapugoda S, Hsu CC-T, Kwan GNC (2016) Duplication of the superficial femoral artery: comprehensive review of imaging literature and insight into embryology. Acta Radiologica Open 5(7)

Huseyin S, Yuksel V, Sivri N et al (2013) Surgical management of iatrogenic femoral artery pseudoaneurysms: a 10-year experience. Hippokratia 17(4):332–336

Kale A, Gayretli O, Oztürk A et al (2012) Classification and localization of the adductor hiatus: a cadaver study. Balkan Med J 29(4):395–400. https://doi.org/10.5152/balkanmedj.2012.030

Kasapis C, Gurm HS (2009) Current approach to the diagnosis and treatment of femoral-popliteal arterial disease. A systematic review. Curr Cardiol Rev 5(4):296–311. https://doi.org/10.2174/157340309789317823

Kau T, Sinzig M, Gasser J et al (2007) Aortic development and anomalies. Semin Intervent Radiol 24(2):141–152. https://doi.org/10.1055/s-2007-980040

Kotsis T, Louizos LA, Pappas E, Theodoraki K (2014) Complex common and internal iliac or aortoiliac aneurysms and current approach: individualised open-endovascular or combined procedures. Int J Vasc Med 2014:178610. https://doi.org/10.1155/2014/178610

Lederle F, Kyriakides A, Tassos C, Stroupe K, Freischlag T, Julie A, Padberg F, Matsumura T, Jon S, Huo Z, Johnson GR (2019) Open versus endovascular repair of abdominal aortic aneurysm. N Engl J Med 380:2126–2135. https://doi.org/10.1056/NEJMoal715955

Nelson PR, Kracjer Z, Kansal N, Rao V, Bianchi C, Hashemi H, Jones P, Bacharach JM (2014) A multicenter, randomized, controlled trial of totally percutaneous access versus open femoral exposure for endovascular aortic aneurysm repair (the PEVAR trial). J Vasc Surg 59(5):1181–1193. https://doi.org/10.1016/j.jvs.2013.10.101

Noori VJ, Eldrup-Jørgensen J (2018) A systematic review of vascular closure devices for femoral artery puncture sites. J Vasc Surg 68(3):887–899. https://doi.org/10.1016/j.jvs.2018.05.019

Ortiz D, Jahangir A, Singh M, Alaqaband S, Bajwa TK, Mewissen MW (2014) Access site complications following peripheral vascular interventions: incidence, predictors and outcomes. Circ Cardiovasc Interv 7:821–828

Pansky B (n.d.) Embryonic development and stem cell compendium. The Aortic Arches, 124. https://discovery.lifemapsc.com/library/review-of-medical-embryology/chapter-123-development-of-the-arterial-system

Rashid S, Hughes S (2016) Risk factors for femoral arterial complications and management. Br J Cardiol 23:155–158. https://doi.org/10.5837/bjc.2016.040

Rocha-Neves J, Ferreira A, Sousa J, Pereira-Neves A, Vidoedo J, Alves H, Teixeira J, Azevedo A (2020) Endovascular approach versus aortobifemoral bypass grafting: outcomes in extensive aortoiliac occlusive disease. Vasc Endovasc Surg 54(2):102–110. https://doi.org/10.1177/1538574419888815

Suurkula M, Fagerberg B, Wendelhag I, Agewall S, Wikstrand J, On Behalf of the Risk Intervention Study (RIS) Group (1996) Atherosclerotic disease in the femoral artery in hypertensive patients at high cardiovascular risk: the value of ultrasonographic assessment of intima-media thickness and plaque occurrence. Arterioscler Thromb Vasc Biol 16:971–977. https://doi.org/10.1161/01.ATV.16.8.971

Swift H, Bordoni B (2021) Anatomy, bony pelvis and lower limb, femoral artery [Updated 2021 Feb 7]. In: StatPearls [Internet]. StatPearls Publishing, Treasure Island. Available from: https://www.ncbi.nlm.nih.gov/books/NBK538262/

Taylor AM, Bordoni B (2021) Histology, blood vascular system [Updated 2021 Feb 7]. In: StatPearls [Internet]. StatPearls Publishing, Treasure Island. Available from: https://www.ncbi.nlm.nih.gov/books/NBK553217/

Tubbs S, Shoja M, Loukas M (2016) Bergman's comprehensive encyclopedia of human anatomic variation. Wiley Blackwell. Copyright@ 2016 by John Wiley & Sons, Inc

Deep Femoral Artery 111

Preston Terle and Stephen J. Bordes, Jr.

111.1 Embryology

Owing to the lateral rotation of the upper limb and medial rotation of the lower limb, the appendices stand separate from one another, causing flexion and extension to occur in opposite orientations (Kawashima and Sasaki 2010). In rare cases, the deep femoral artery does not develop, and the branches usually associated with it arise directly from the femoral artery instead (John Wiley & Sons Inc 2016). Typically, the deep femoral artery arises from the posterolateral portion of the femoral artery (51%). Still, in other cases, it can arise as follows concerning the femoral artery: posteriorly (29%), laterally (11%), and posteromedially (10%). Medial and anteromedial origins of the deep femoral artery are uncommon, occurring in less than 4% and 2%, respectively. These averages take account of possible anatomical differences between the right and left legs (Tzouma et al. 2020).

P. Terle
School of Medicine, Tulane University,
New Orleans, LA, USA
e-mail: pterle@tulane.edu

S. J. Bordes, Jr. (✉)
Department of Surgery, Louisiana State University
Health Sciences Center, New Orleans, LA, USA

111.2 Anatomy and Topography

The deep femoral artery (profunda femoris artery) is the largest branch of the femoral artery. It supplies blood to the proximal femur, extensors, flexors, adductors, and skin of the medial thigh. It originates in the upper third of the thigh, specifically 3.29–4.4 cm from the inguinal ligament, with a diameter of 6.9 ± 1.6 mm (Swift and Bordoni 2021; Siddharth et al. 1985; Claassen et al. 2021; Sur et al. 2016). It typically arises from the posterolateral aspect of the femoral artery. Three branches from it can be found: the lateral circumflex femoral, medial circumflex femoral, and perforating femoral arteries (Tzouma et al. 2020). The lateral circumflex femoral artery wraps anteriorly to the femoral head and forms transverse ascending and descending branches that help to supply the head of the femur, quadriceps femoris, and the most proximal skin of the thigh. The medial circumflex femoral artery wraps posteriorly to the femoral head. It bifurcates into transverse and ascending branches supplying the adductors of the thigh: gracilis, obturator externus, adductor brevis, adductor longus, and adductor magnus. The lateral and medial circumflex femoral arteries not only anastomose with each other to supply the head of the femur, but also form connections with the inferior gluteal artery, creating the cruciate anastomosis. Distally, the deep femoral artery runs inferiorly and parallel to the superficial

S. J. Bordes, Jr. et al. (eds.), *The Clinical Anatomy of the Vascular System*,
https://doi.org/10.1007/978-3-031-78326-5_111

femoral artery along the linea alba, traveling between the pectineus and adductor longus muscles and then between the adductor longus and brevis, piercing the adductor magnus and terminating with an anastomosis with the popliteal artery in some cases (Sur et al. 2016).

The perforating branches (Fig. 111.1) given off by the deep femoral artery supply the posterior thigh muscles. In relation to the vastus ridge, the first through third branches can be found distally as follows: 8.1 ± 1.5 cm, 13.7 ± 1.8 cm, and 20.4 ± 3.3 cm. These three branches often give rise to subbranches, many of which pierce their surrounding muscle tissue but can ascend or descend to anastomose with their surrounding branches. Each branch contributes to the blood flow supplying the sciatic nerve and the perineural vascular plexus (Sur et al. 2016).

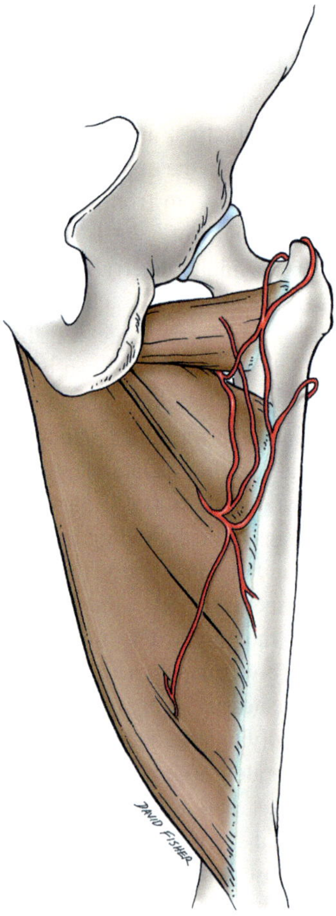

Fig. 111.1 Schematic drawing of the perforating branches of the deep femoral artery in the posterior compartment of the thigh, specifically around and through the adductor magnus and quadratus femoris

111.3 Histology

The deep femoral artery is a muscular artery that distributes blood to the surrounding bone, soft tissue, and musculature. Its wall comprises three layers or tunics: intima, media, and adventitia. The interior layer is the intima, which is composed of simple squamous endothelium followed by an internal elastic lamina that separates it from the tunica media. The tunica media of the deep femoral artery contains many layers of smooth muscle and elastic fibers that contribute to its ability to maintain compliance under systemic arterial pressure (Mescher 2013; Barallobre-Barreiro et al. 2020). External to the tunica media is the external elastic lamina, which separates the media from the tunica adventitia. The adventitia contains vasa vasorum, small arteries, and veins that supply the outer walls of the artery with oxygen and nutrients. Also present in the adventitia of the deep femoral artery are sympathetic nerve endings that contribute to the ability of the deep femoral artery to vasoconstrict (Barallobre-Barreiro et al. 2020).

111.4 Imaging

Imaging is essential for diagnosing and treating pathologies involving the deep femoral artery. The most common types of imaging for this purpose are computerized tomography (CT), duplex ultrasound, and angiography. These methods are typically used when surgery is imminent. Careful examination of any imaging is necessary to ensure that blood flow, diameter, length, and location are accurately recorded before intervention. Some studies recommend duplex scanning followed by angiography to obtain the best mapping outcomes (Yahel and Witz 1996).

111.5 Variations

As the largest branch of the femoral artery, the deep femoral artery vascularizes much of the thigh. Variations in it are of great importance to both surgeons and interventional radiologists. The most common site of origin of the deep femoral artery (51% of cases) is the posterolateral femoral artery (Tzouma et al. 2020). Other variations have been found where the deep femoral artery arises from the external iliac (less than 2%). The medial and lateral circumflex arteries originate from the deep femoral artery in 63% and 74% of cases, respectively, the second most likely site of origin being the femoral artery (Fig. 111.2). Other variations in the medial circumflex femoral artery have been observed, including a common medial circumflex femoral artery and deep femoral artery originating from the femoral artery. However, this is rare, occurring in less than 3% of cases (Tzouma et al. 2020). The descending branch of the lateral circumflex femoral artery originated from the lateral circumflex femoral artery in approximately 83% of cases (Tzouma et al. 2020). The origin and trajectory of these branches around the femoral neck and head are pertinent to orthopedic surgeons in intertrochanter and trochanter osteotomies. Preventing injury to the anastomoses created by these arteries is critical for precluding necrosis of the femoral head.

111.6 Pathology

Aneurysm

True aneurysms of the deep femoral artery are uncommon and account for less than 1% of all peripheral aneurysms (Pappas et al. 1964; Dent et al. 1972; Valiulis and Johnston 1980). This is partly because of the anatomical position of the deep femoral artery between the adductor muscles, preventing symptomatic recognition of the aneurysm and leading to late diagnosis. Deep femoral artery aneurysms are asymptomatically diagnosed in nearly 73% of cases (Kuzuya et al. 2005). Asymptomatic diagnosis typically occurs because over 60% of deep femoral artery aneurysms present simultaneously with other peripheral aneurysms (Posner et al. 2004; Kibrik et al. 2020; Harbuzariu et al. 2008). They most commonly present in older men with atherosclerosis (Posner et al. 2004; Kibrik et al. 2020). Symptomatic presentation usually takes the form of lower leg edema, pain, and or a palpable mass (Posner et al. 2004; Kibrik et al. 2020). Deep femoral artery aneurysms typically rupture when their diameters exceed 35 mm, larger aneurysms having a greater risk of rupture (Levi and Schroeder 1999). Early surgical intervention is critical for deep femoral artery aneurysms owing to their late diagnosis and high rate of rupture (Jang and Park 2017).

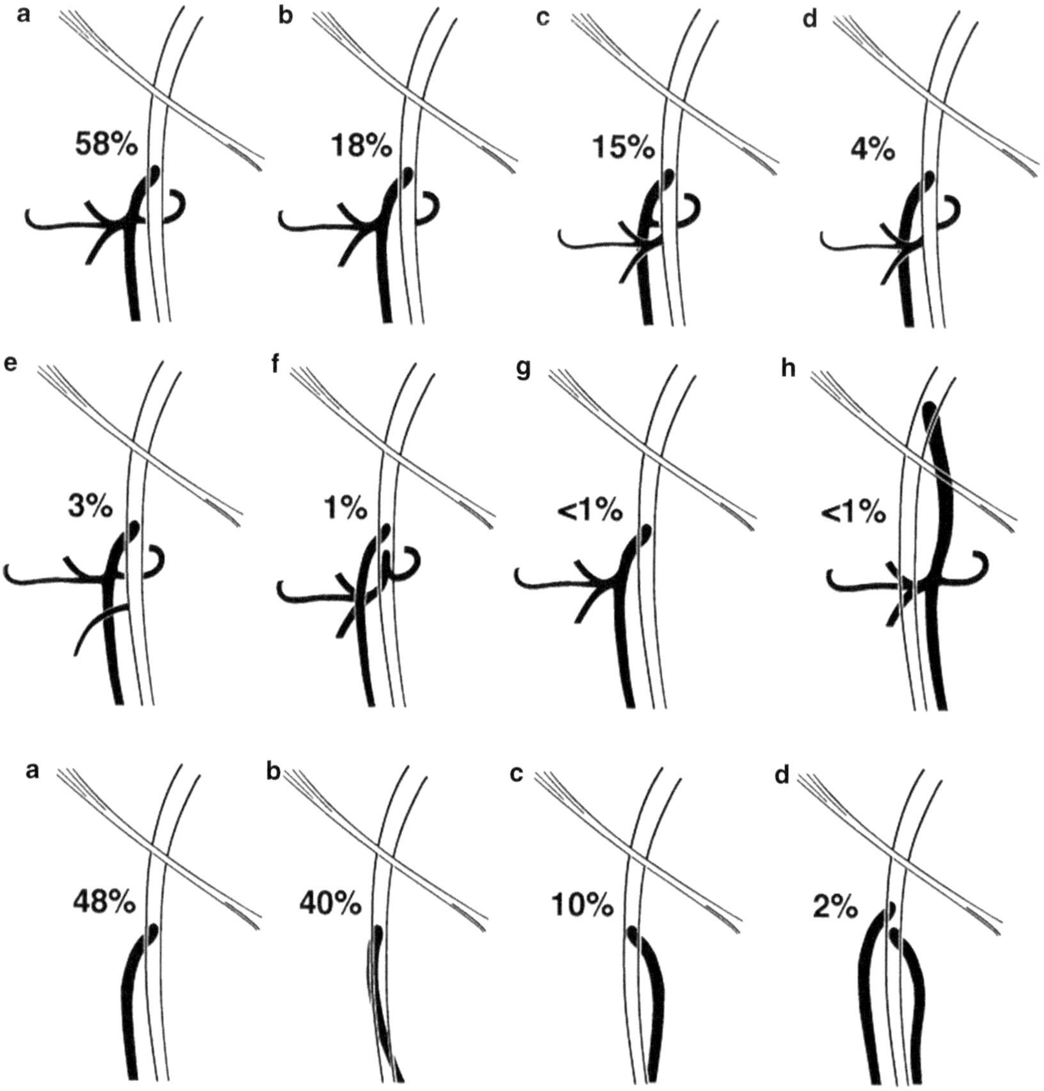

Fig. 111.2 Rows 1 and 2 illustrate the major variations of the deep femoral and its branches. (a) The circumflex and deep femoral arteries all arise from a common trunk; (b) the femoral artery gives rise to the medial circumflex femoral artery; (c) the femoral artery gives rise to the lateral circumflex femoral artery; (d) the femoral artery gives rise to both circumflex arteries; (e) similar to a or b but a large branch of the lateral circumflex artery arises from the femoral artery; (f) both circumflex arteries arise from a common trunk separate from the deep femoral artery; (g) absence of the medial circumflex artery with this territory supplied by a branch of the obturator artery; (h) the external iliac artery gives rise to the deep femoral artery and sometimes, together with the inferior epigastric artery. Row 3 illustrates the major variations of the course of the deep femoral artery. (a) Lateral or dorsolateral position of the deep femoral artery in relation to the femoral artery; (b) a dorsal position of the deep femoral artery to the femoral artery; (c) the main part of the deep femoral artery is dorsomedial or medial to the femoral artery; (d) branches of the deep femoral are located medial and lateral to the femoral artery. (After Lippert H and Pabst R Arterial Variations in Man, 1985)

Peripheral Artery Disease

Peripheral artery disease refers to the narrowing or obstruction of vessels perfusing the extremities. Typically, this is because of atherosclerotic plaques (Chan and Junia 2020). Peripheral artery disease affects nearly 120 million people worldwide (GBD 2016 Disease and Injury Incidence and Prevalence Collaborators 2017). Current treatment options are set to provide symptom relief and aid in limb salvage via medication or revascularization of the affected limb. Peripheral artery disease typically presents in more than one vascular bed (Chan and Junia 2020; Bauersachs et al. 2019). These compounding factors lead to an increased risk of myocardial infarction (Chan and Junia 2020; Cooke and Meng 2020). Peripheral artery disease can lead to reduced flow to the lower extremities, leading to intermittent claudication and general weakness of adduction, flexion, and extension of the thigh (Hamburg and Creager 2017).

111.7 Clinical and Surgical Significance

Anterolateral Flap

The anterolateral thigh flap (ALT flap) is a septocutaneous flap supplied by the descending branch of the lateral circumflex femoral artery and was first described in 1984 (Choi et al. 2007). It is of great interest to plastic surgeons for myocutaneous procedures because of its low rate of necrosis (<4%). Furthermore, it can be harvested and used as a myocutaneous or fasciocutaneous flap (Wong and Wei 2010; Colborn et al. 1995; Agostini et al. 2014). The ALT flap is most frequently used in head and neck cancer (Smith et al. 2017). Its use has presented difficulties because the vascular anatomy that supplies the anterolateral thigh is variable, particularly the musculocutaneous perforators that supply the lateral flap in 87% of cases (Wong and Wei 2010; Hsu et al. 2021). The ALT flap is typically harvested based on the descending branch of the lateral circumflex femoral artery, traveling between the rectus femoris and vastus lateralis before entering the vastus lateralis distally. This branch also gives off a large branch proximal to its origin from the lateral circumflex femoral artery, a main contributor to the rectus femoris; hence, preservation of this vessel is critical. In roughly 34% of patients, an oblique branch of the lateral circumflex femoral artery runs alongside the descending branch. When the oblique is present, either it or the descending branch can be used as the pedicle for the ALT procedure (Wong and Wei 2010).

Percutaneous Transluminal Angioplasty

In patients who present with atherosclerotic blockage in the superficial femoral artery, the deep femoral artery is critical for maintaining leg perfusion (Dick et al. 2006). Owing to the diffuse nature of atherosclerosis, it is unlikely that the deep femoral artery would be plaque-free. Profundaplasty or widening of the junction of the femoral artery and deep femoral artery can serve to revascularize the lower limb (Dick et al. 2006). Percutaneous transluminal angioplasty is another method to increase flow in the deep femoral artery and prevent ischemia (Dick et al. 2006). This procedure, with the addition of an endovascular stent, alleviates symptoms of claudication caused by peripheral artery disease (Yellen 2007). Technical success (residual stenosis <30%) exists in 85% of cases. During follow-up with a median of 13 months, reintervention-free rates and patency were observed at 61% and 71%, respectively, though sustained clinical improvement was seen in only 29% of patients (Dick et al. 2006).

Orthopedic Injuries and Trauma

Femoral artery pseudoaneurysms are a complication associated with trauma to the thigh due to orthopedic procedures, among other causes (Pecoraro et al. 2017). Treatment has involved embolization or ligation of the deep femoral artery in 47% and 38% of cases, respectively.

References

Agostini T, Lazzeri D, Spinelli G (2014) Anterolateral thigh flap thinning: techniques and complications. Ann Plast Surg 72(2):246–252. https://doi.org/10.1097/SAP.0b013e31825b3d3a

Barallobre-Barreiro J, Loeys B, Mayr M, Rienks M, Verstraeten A, Kovacic JC (2020) Extracellular matrix in vascular disease. Part 2/4: JACC focus seminar. J Am Coll Cardiol 75(17):2189–2203. https://doi.org/10.1016/j.jacc.2020.03.018

Bauersachs R, Zeymer U, Brière JB, Marre C, Bowrin K, Huelsebeck M (2019) Burden of coronary artery disease and peripheral artery disease: a literature review. Cardiovasc Ther 2019:8295054. https://doi.org/10.1155/2019/8295054

Chan KA, Junia A (2020) Lower extremity peripheral artery disease: a basic approach. Br J Hosp Med (Lond) 81(3):1–9. https://doi.org/10.12968/hmed.2019.0263

Choi SW, Park JY, Hur MS, Park HD, Kang HJ, Hu KS, Kim HJ (2007) An anatomic assessment on perforators of the lateral circumflex femoral artery for anterolateral thigh flap. J Craniofac Surg 18(4):866–871. https://doi.org/10.1097/scs.0b013e3180a03304

Claassen H, Schmitt O, Schulze M, Wree A (2021) Deep femoral artery: a new point of view based on cadaveric study. Ann Anat 237:151730. https://doi.org/10.1016/j.aanat.2021.151730

Colborn GL, Mattar SG, Taylor B, Skandalakis JE, Lumsden AB (1995) The surgical anatomy of the deep femoral artery. Am Surg 61(4):336–346

Cooke JP, Meng S (2020) Vascular regeneration in peripheral artery disease. Arterioscler Thromb Vasc Biol 40(7):1627–1634. https://doi.org/10.1161/ATVBAHA.120.312862

Dent TL, Lindenauer SM, Ernst CB, Fry WJ (1972) Multiple arteriosclerotic arterial aneurysms. Arch Surg 105(2):338–344. https://doi.org/10.1001/archsurg.1972.04180080184031

Dick P, Mlekusch W, Sabeti S, Amighi J, Schlager O, Haumer M, Minar E, Schillinger M (2006) Outcome after endovascular treatment of deep femoral artery stenosis: results in a consecutive patient series and systematic review of the literature. J Endovasc Ther 13(2):221–228. https://doi.org/10.1583/05-1766R.1

GBD 2016 Disease and Injury Incidence and Prevalence Collaborators (2017) Global, regional, and national incidence, prevalence, and years lived with disability for 328 diseases and injuries for 195 countries, 1990–2016: a systematic analysis for the Global Burden of Disease Study 2016. Lancet 390(10100):1211–1259. https://doi.org/10.1016/S0140-6736(17)32154-2. Erratum in: Lancet. 390(10106):e38, 2017

Hamburg NM, Creager MA (2017) Pathophysiology of intermittent claudication in peripheral artery disease. Circ J 81(3):281–289. https://doi.org/10.1253/circj.CJ-16-1286

Harbuzariu C, Duncan AA, Bower TC, Kalra M, Gloviczki P (2008) Profunda femoris artery aneurysms: association with aneurysmal disease and limb ischemia. J Vasc Surg 47(1):31–34; discussion 34–5. https://doi.org/10.1016/j.jvs.2007.09.040

Hsu CC, Loh CYY, Wei FC (2021) The anterolateral thigh perforator flap: its expanding role in lower extremity reconstruction. Clin Plast Surg 48(2):235–248. https://doi.org/10.1016/j.cps.2020.12.008

Jang LC, Park SS (2017) The true deep femoral artery aneurysm: a case report. Vasc Spec Int 33(1):40–42. https://doi.org/10.5758/vsi.2017.33.1.40

Kawashima T, Sasaki H (2010) Reasonable classical concepts in human lower limb anatomy from the viewpoint of the primitive persistent sciatic artery and twisting human lower limb. Okajimas Folia Anat Jpn 87(3):141–149. https://doi.org/10.2535/ofaj.87.141. Erratum in: Okajimas Folia Anat Jpn. 87(4):195, 2011

Kibrik P, Arustamyan M, Stern JR, Dua A (2020) A systematic review of the diagnosis, management, and outcomes of true profunda femoris artery aneurysm. J Vasc Surg 71(6):2145–2151. https://doi.org/10.1016/j.jvs.2019.10.086

Kuzuya A, Mizuno K, Miyake H et al (2005) Atherosclerotic aneurysms of the profunda femoris arteries – case report and literature review. Int J Angiol 14:99–101. https://doi.org/10.1007/s00547-004-1077-8

Levi N, Schroeder TV (1999) True and anastomotic femoral artery aneurysms: is the risk of rupture and thrombosis related to the size of the aneurysms? Eur J Vasc Endovasc Surg 18(2):111–113. https://doi.org/10.1053/ejvs.1999.0842

Mescher AL (2013) Junqueira's basic histology: text and atlas

Pappas G, Janes JM, Bernatz PE, Schirger A (1964) Femoral aneurysms. Review of surgical management. JAMA 190:489–493

Pecoraro F, Dinoto E, Bracale UM, Badalamenti G, Farina A, Bajardi G (2017) Symptomatic deep femoral artery pseudoaneurysm endovascular exclusion. Case report and literature review. Ann Vasc Surg 42:303.e5–303.e9. https://doi.org/10.1016/j.avsg.2016.11.026

Posner SR, Wilensky J, Dimick J, Henke PK (2004) A true aneurysm of the profunda femoris artery: a case report and review of the English language literature. Ann Vasc Surg 18(6):740–746. https://doi.org/10.1007/s10016-004-0116-4

Recorded Books Inc (2016) Bergman's comprehensive encyclopedia of human anatomic variation. Wiley-Blackwell, Malden

Siddharth P, Smith NL, Mason RA, Giron F (1985) Variational anatomy of the deep femoral artery. Anat Rec 212(2):206–209. https://doi.org/10.1002/ar.1092120216

Smith RK, Wykes J, Martin DT, Niles N (2017) Perforator variability in the anterolateral thigh free flap: a systematic review. Surg Radiol Anat 39(7):779–789. https://doi.org/10.1007/s00276-016-1802-y

Sur YJ, Morsy M, Mohan AT, Zhu L, Lachman N, Saint-Cyr M (2016) The first perforating branch of the deep femoral artery: a reliable recipient vessel for vascularized fibular grafts: an anatomical study. J Plast Reconstr Aesthet Surg 69(3):351–358. https://doi.org/10.1016/j.bjps.2015.10.024

Swift H, Bordoni B (2021) Anatomy, bony pelvis and lower limb, femoral artery. In: StatPearls [Internet]. StatPearls Publishing, Treasure Island

Tzouma G, Kopanakis NA, Tsakotos G, Skandalakis PN, Filippou D (2020) Anatomic variations of the deep femoral artery and its branches: clinical implications on anterolateral thigh harvesting. Cureus 12(4):e7867. https://doi.org/10.7759/cureus.7867

Valiulis AP, Johnston KW (1980) Isolated arteriosclerotic aneurysm of the profunda femoris artery. J Cardiovasc Surg 21(4):498–500

Wong CH, Wei FC (2010) Anterolateral thigh flap. Head Neck 32(4):529–540. https://doi.org/10.1002/hed.21204

Yahel J, Witz M (1996) Isolated true atherosclerotic aneurysms of the deep femoral artery. Case report and literature review. J Cardiovasc Surg 37(1):17–20

Yellen EA (2007) Synergy and peripheral percutaneous transluminal angioplasty. J Vasc Nurs 25(1):7–11. https://doi.org/10.1016/j.jvn.2006.10.001

112 Popliteal Artery

Skyler Jenkins and Stephen J. Bordes, Jr.

112.1 Embryology

In 1919, Senior detailed the embryological origin of the lower limb (Senior 1919). During the 9 mm stage of embryonic development, a primitive iliac artery arises from the umbilical artery and courses caudally to supply the lower limb. This artery bifurcates into a dorsal internal iliac axis, which forms the axial artery, and a ventral external iliac artery, which becomes the femoral artery at the 12 mm stage (Rosset et al. 1995). At the 12 mm stage, the walls of the femoral artery consist only of a thin wall of endothelium, resembling the associated femoral vein (Senior 1919). During the 14 mm stage, the axial artery gives off three segments: the more proximal ischiatic artery, the intermediate deep popliteal artery, and the distalmost interosseous artery. At this stage, the deep popliteal artery lies between the popliteus muscle and the posterior margin of the tibia. Just proximal to the popliteus muscle, the deep popliteal artery gives off two branches: the superficial posterior tibial artery and the superficial posterior fibular artery (Senior 1919). These two branches will later fuse to form the superficial popliteal artery. At the 22 mm stage, the ischiatic artery involutes, and the only remaining supply to the leg is the femoral artery. Distal to the popliteus muscle, a medial communicating branch arises and connects the superficial popliteal artery to the most distal extremity of the deep popliteal artery. Subsequently, this communicating branch will become the proximal segment of the anterior tibial artery. In contrast, the portion of the superficial popliteal artery above the communicating branch will become the tibiofibular trunk (Rosset et al. 1995). Finally, at the 24.5 mm stage, the distal part of the deep popliteal artery, deep to the popliteus muscle, involutes. The popliteal artery is then made up of a small, proximal segment of the deep and superficial popliteal artery (Senior 1919).

112.2 Anatomy and Topography

The adult popliteal artery (Fig. 112.1) is a continuation of the superficial femoral artery. It most commonly begins at the distal edge of the adductor hiatus, an opening in the tendinous slip of the great adductor muscle of the thigh. It is 12–18 cm in length with a diameter of 5–9 mm (Kropman 2007). It then courses through the popliteal fossa and terminates at the distal border of the popliteus muscle by bifurcating into the anterior and posterior tibial arteries (Moore 1985; Gray 1985). It is divided into three segments: P1, P2, and P3 (Kropman 2007). Segment P1 is the section

S. Jenkins
Department of Family Medicine, University of Arkansas for Medical Sciences, Jonesboro, USA

S. J. Bordes, Jr. (✉)
Department of Surgery, Louisiana State University Health Sciences Center, New Orleans, LA, USA

S. J. Bordes, Jr. et al. (eds.), *The Clinical Anatomy of the Vascular System*,
https://doi.org/10.1007/978-3-031-78326-5_112

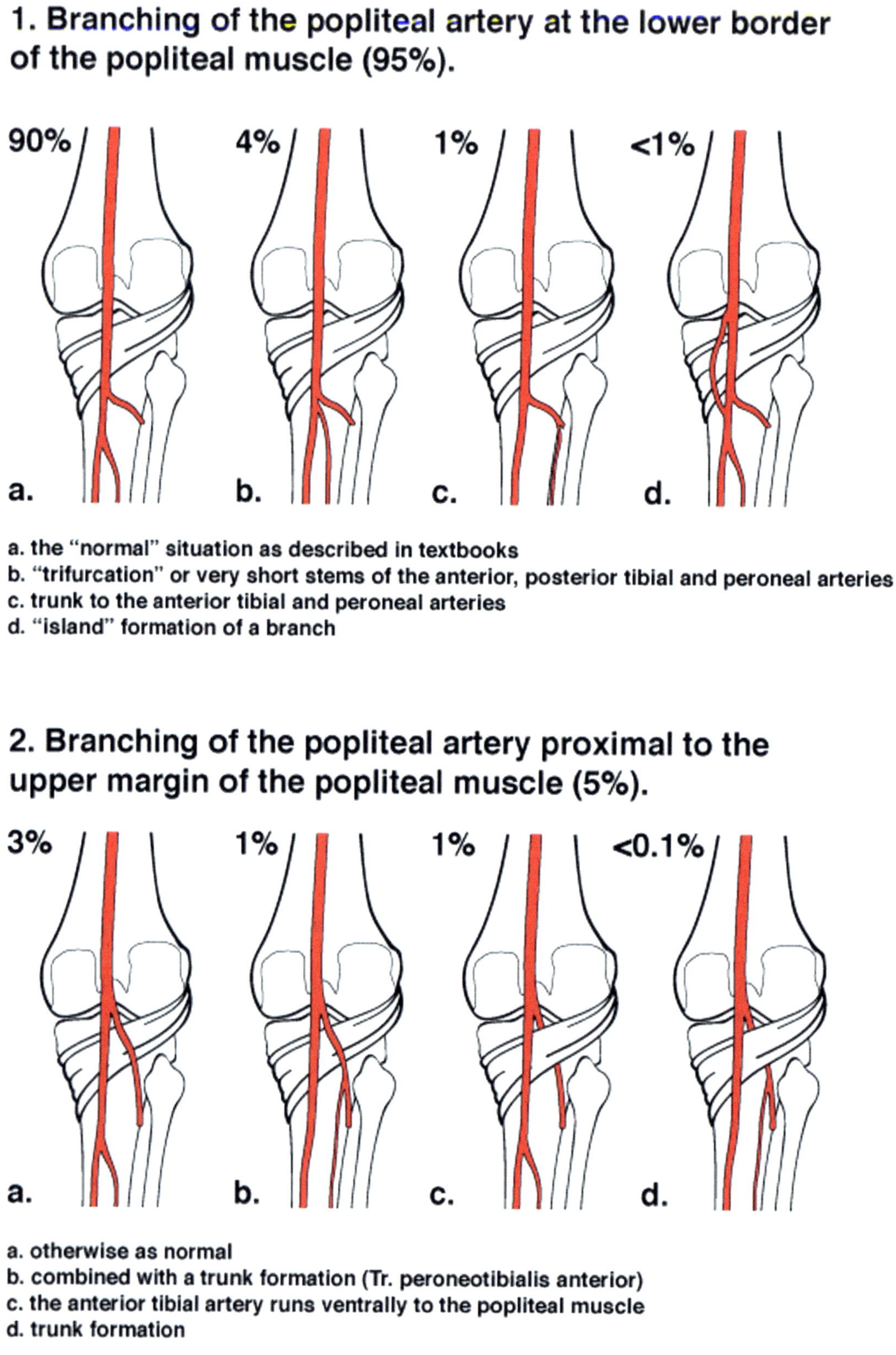

Fig. 112.1 Schematic drawing indicating the more common variations of the popliteal artery and its branches. (After Lippert H and Pabst R Arterial Variations in Man, 1985)

between the adductor hiatus and the top of the kneecap (Kropman 2007). Segment P2 continues as far as the split of the knee articulation. Segment P3 is the length up to the division into the crural arteries (Kropman 2007). The bending point of the knee is between P1 and P2. As the popliteal artery enters the popliteal fossa of the knee, it is the most ventral structure of the neurovascular bundle. It is predicted to be 8.66 ± 2.17 mm dorsal to the joint capsule and 45.58 ± 3.27 mm lateral to the medial epicondyle at the level of the medial epicondyle in full extension (Sora et al. 2015). The popliteal artery is the deepest and most medial structure within the neurovascular bundle. The popliteal vein is more superficial and lateral, while the tibial nerve is the most superficial and lateral (Sora et al. 2015). Through the entire popliteal fossa, the neurovascular bundle lies lateral to the lateral border of the medial gastrocnemius muscle and medial to the medial border of the lateral gastrocnemius. It has been shown that when the knee is flexed to 90 degrees, the neurovascular bundle moves posterolaterally from the posterior bony aspect (Keyurapan et al. 2016). Within the region of the knee, the popliteal artery gives off genicular (superior, middle, and inferior medial and lateral) branches to the knee and muscular branches to the hamstring, gastrocnemius, soleus, and plantaris (Moore and Agur 2002).

112.3 Histology

The popliteal artery is a muscular (distributing) artery. Its tunica intima comprises an inner layer of endothelium surrounded by subendothelial connective tissue and a prominent internal elastic lamina. The intima is surrounded by tunica media, mostly circumferentially arranged smooth muscle, minimal elastin, and a thick external elastic lamina. The tunica adventitia is the outermost wall of organized, dense, irregular connective tissue with multiple layers of elastic tissue.

The popliteal vein is a medium vein with less distinct layers. It has a similar intima media that contains few layers of smooth muscle and outer adventitia.

112.4 Variations

The variations of the popliteal branching pattern arise from various embryological abnormalities. In 1985, Lippert and Pabst created the first classification system, modified by Kim et al. in 1989, to give us the system most commonly used today (Lippert and Pabst 1985; Kim et al. 1989). Kim et al. described three types of branching variations and various subtypes:

Type I. Normal level of branching proximal to the lower border of the popliteus muscle (96.1% in 27 studies) (Tomaszewski et al. 2017).

- A. The anterior tibial artery is the first branch, followed by the tibioperoneal artery, which bifurcates into the fibular and posterior tibial arteries (92.6%).
- B. The anterior, posterior, and fibular arteries arise within 0.5 cm of each other without a tibioperoneal trunk (2.4%).
- C. The posterior tibial is the first branch, followed by the tibioperoneal trunk, which bifurcates into the fibular and anterior tibial arteries (1.1%).

Type II. High division of the popliteal artery (3.9% in 27 studies) (Tomaszewski et al. 2017).

- A. The anterior tibial artery begins at or above the knee joint (2.5%).
  - (a) Normal course of proximal anterior tibial (2.2%).
  - (b) Medial initial curve in anterior tibial (0.3%).
- B. The posterior tibial begins at or above the knee joint. Common trunk for fibular and anterior tibial (1.2%).
- C. The fibular arises at or above the knee joint. Common trunk for anterior and posterior tibial (0.2%).

Type III. Hypoplastic or aplastic branching with altered distal supply (3.0% in 17 studies) (Tomaszewski et al. 2017).

- A. Hypoplastic-aplastic posterior tibial. Distal posterior tibial replaced by fibular (1.7%).

B. Hypoplastic-aplastic posterior and anterior tibial. Dorsalis pedis replaced by fibular (1.2%).
C. Hypoplastic-aplastic posterior and anterior tibial. Posterior tibial and dorsalis pedis replaced by fibular. Termed the peronea arteria magna (0.1%).

Anand et al. and Longo et al. suggested modifying the Kim classification, adding another class (IV) or subclass (IIID) that includes aberrant or fibular artery insufficiency (Anand et al. 2006; Longo et al. 2014). In a unique case study by Slaba et al., the anterior tibial artery had a high origin at the femorotibial space and a tibioperoneal trunk greater than 5 cm with early bifurcation of the fibular artery into the anterior and posterior branches (Slaba et al. 2007).

Another series of studies revealed that a patient with a variation in one leg had a 19–33% chance of having a variation in the other (Keen 1961; Bardsley and Staple 1970; Kil and Jung 2009).

112.5 Imaging

Imaging is vital for evaluating popliteal artery pathology. Noninvasive studies are typically the first line after the physical exam. Color duplex ultrasonography (US) is most commonly used. It aids in diagnosing occlusion, stenosis, aneurysm, or popliteal mass (Baker cyst) (Wright et al. 2004). An ankle-brachial index (ABI) with segmental lower-extremity arterial pressure readings or pulse volume recordings helps to diagnose and assess the degree of peripheral artery disease (Wright et al. 2004). Computed tomography (CT) angiography can also be used. These studies can be supplemented by magnetic resonance (MR) angiography using either standard contrast or gadolinium-enhanced 3D time-of-flight angiography and/or stress angiography (Wright et al. 2004). Angiography is typically performed when endovascular or surgical intervention is necessary. MR imaging can also be used to determine the artery-muscle relationship, assessing the patency of the lumen in relation to surrounding structures.

112.6 Pathology

Atherosclerosis

The most common cause of popliteal artery occlusion or stenosis is atherosclerosis, and the pathogenesis of atherosclerosis has been well established. The endothelial wall of the artery undergoes injury, followed by the overproduction of cellular mediators, leading to the formation of a fibrous plaque. This plaque can then restrict blood flow, with consequences ranging from intermittent claudication to major tissue loss.

Popliteal Artery Aneurysms

Popliteal artery aneurysms (PAAs) can be classified as either true (where all arterial wall layers are dilated) or false (due to a defect in the wall, typically from trauma or infection). The popliteal artery is considered dilated if its diameter exceeds 0.7 cm or is more than 1.5 times the normal diameter (Wright et al. 2004; Kropman 2007). True PAAs are the most common form of peripheral arterial aneurysms, more prevalent in males (95–99% of cases), and most likely due to arteriosclerosis (Loukas et al. 2007). The plaques of arteriosclerosis can cause turbulent blood flow distal to the lesion, resulting in dilation of the vessel and often compounded by cardiovascular comorbidities such as hypertension (Galland 2005; Hamish et al. 2006). Other causes of PAAs can be related to infections of the popliteal fossa or a genetic disease such as Marfan syndrome or Klippel-Trenaunay syndrome (Akagi et al. 2006; Hamish et al. 2006). Pseudoaneurysms are typically caused by femoral exostosis, orthopedic surgery, or direct trauma to the knee (Raherinantenaina et al. 2016).

Popliteal Artery Entrapment Syndrome

Popliteal entrapment is due to an abnormality in the embryological development of the popliteal fossa that causes the popliteal artery to be compressed by the gastrocnemius muscle or from functional hypertrophy of the gastrocnemius. This syndrome can be caused by an aberrant course of the popliteal artery or gastrocnemius muscle, hypertrophy of the gastrocnemius muscle, or the development of a fibrous band that crosses the artery (Qazi et al. 2021). According to Wady et al., the overall incidence ranges from 0.17% to 3.5% of the general population, 85% male predominance, and approximately 60% of cases occurring in young athletes during the third decade of life (Wady et al. 2018). Whelan proposed a classification system that is most commonly used for this syndrome (Whelan and Haimovici 1984):

Type I: The popliteal artery has an aberrant medial course around the medial head of the gastrocnemius, which has a normal insertion above the femoral condyle.
Type II: The medial head of gastrocnemius inserts more laterally on the distal femur, but the popliteal artery is not displaced.
Type III: An aberrant accessory slip of the medial head of gastrocnemius slings around and surrounds the popliteal artery.
Type IV: The popliteal artery is located deep in the popliteal fossa and entrapped by the popliteus muscle or fibrous band.
Type V: Entrapment of both popliteal artery and vein.

Cystic Adventitial Disease of the Popliteal Artery

Cystic adventitial disease (CAD) typically comprises an intramural cyst containing gelatinous material between the tunica media and adventitia. However, there are cases in which it lies entirely within the media (Yoshino et al. 2019). These cysts compress the arterial lumen either eccentrically or concentrically, forming an "hourglass or scimitar" sign (Peterson et al. 2003). The etiology of CAD of the popliteal artery is still debated, but trauma, ganglion dysfunction, systemic disorder, embryonic developmental abnormality, or combinations thereof are all possible (Kawarai et al. 2012; Levien and Been 1998). The trauma hypothesis states that repetitive trauma causes degeneration of the adventitia, allowing for cystic change; the ganglion hypothesis proposes that synovial cysts make their way from nearby joints to the adventitia. CAD could also be a part of a global connective tissue disorder or incorrect migration of mesenchymal cells during embryological development (Tsilimparis et al. 2007).

Trauma

Penetrating injuries, traumatic and periprosthetic fractures, knee dislocations, and iatrogenic injury are the most frequent causes of trauma to the popliteal artery. In 28–46% of cases of blunt trauma to the lower extremity, the popliteal artery experiences trauma in the form of transection, occlusion, laceration, perforation, arteriovenous fistula, or intimal injury (Imerci et al. 2014). The popliteal artery is susceptible to injury because of its ligamentous fixation and relationship to surrounding bony structures and knee joint. These injuries typically include knee dislocations, complex tibial plateau fractures, or supracondylar femur fractures (Kim et al. 2010).

Arteriovenous Fistula (AVF) of the Popliteal Artery and Vein

An AVF is an abnormal communication between an artery and vein that pressurizes the venous system and causes dilation of both vessels. AVFs are mainly caused by trauma or are iatrogenic, but they can occur due to Marfan disease, human immunodeficiency virus (HIV) arteritis, or unknown causes (Ying et al. 2020).

112.7 Clinical and Surgical Significance

Popliteal Artery Aneurysm

Patients with PAAs (or pseudoaneurysms) typically present during the 6th–seventh decade of life on a spectrum from an asymptomatic pulsatile mass to severe lower limb ischemia. About 42% of patients present asymptomatically (Kropman 2007). Depending on the size of the aneurysm, the patient can also have secondary nerve or vein compression. Treatment of the aneurysm is based on its size and the presence of symptoms. The risk of amputation in an asymptomatic patient who undergoes treatment is roughly 0%; however, in an emergency or a symptomatic patient, the risk is much greater. PAAs are also associated with abdominal aortic aneurysms (AAA). An AAA is present in 30–50% of patients with a PAA (Wright et al. 2004). In addition, PAAs are present bilaterally in 50–70% of cases (Wright et al. 2004).

Recommendations for surgical intervention require the aneurysm to be greater than 2.0 cm with a mural thrombus if the patient is asymptomatic or in any symptomatic patient (Dawson et al. 1994; Lowell et al. 1994; Aulivola et al. 2004).

Intervention Options for the PAA

Nonsurgical Interventions

- Intra-arterial thrombolysis: In selected patients with acute limb ischemia due to aneurysmal thrombosis, intra-arterial thrombolysis can allow for partial or complete recanalization of the thrombosed PAAs (67% of cases) (Dorigo et al. 2002).

Open Surgery

- The gold standard of treatment with excellent long-term patency rates (76% at 5 years) (Pulli et al. 2006; Huang et al. 2007).
- Medial approach: Supragenicular and infragenicular incisions are made in the medial aspect of the leg, exposing the popliteal artery. The artery is ligated proximal and distal to the aneurysm, and a graft (autologous venous or polytetrafluoroethylene) bypass is performed.
- Posterior approach: An S-shaped incision is made in the posterior aspect of the knee with dissection between the two heads of the gastrocnemius muscle. The popliteal artery is clamped proximal and distal to the aneurysm, the aneurysmal sac is opened, and the thrombus is removed. The aneurysm is then bypassed with an end-to-end anastomosis.
- Endovascular repair: The aneurysm is accessed via an ipsilateral antegrade common femoral artery puncture. The appropriate stent graft is then determined from measurements of landing zone diameters and lengths (Kropman 2007).

In the largest analysis of PAA management to date, endovascular repair had a lower wound complication rate and shorter length of hospital stay than open repair (Leake et al. 2017). However, the open repair option allows for a greater primary patency rate at 3 years postoperation (Leake et al. 2017).

Popliteal Artery Entrapment Syndrome

The typical clinical presentation is a young male with symptoms of intermittent claudication, leg swelling, aching pain, rest pain, and calf cramping. Angiography is used for screening and diagnosis with the ankle plantar-flexed; this can be coupled with transverse T1-weighted MR imaging to evaluate aberrant muscular anatomy or popliteal artery deviation (Tercan et al. 2005).

Treatment depends on the severity of entrapment at the time of diagnosis. In cases of aberrant anatomy, surgical correction should be performed to avoid vascular injury or occlusion over time (Grimm et al. 2020). Surgical correction is typically done from a medial approach if more proximal or from a posterior approach if distal. Surgery can include myotomy or resection of the fibrous band, with or without vascular rerouting of the popliteal artery, depending on the pathology (Grimm et al. 2020). Results for myotomy are excellent, with 100% patency at one and ten years postoperatively (Lejay et al. 2016). If symptoms have been chronic, leading to severe stenosis or complete occlusion, a saphenous vein graft is commonly used for repair or revascularization (Grimm et al. 2020).

For functional entrapment, clinical trials of botulinum toxin A are being explored as an alternative to surgery.

Cystic Adventitial Disease

Although it is a rare cause of calf claudication (1:1200), 80% of individuals who present with CAD are men between 40 and 50 (Kawarai et al. 2012). Management is related to the condition of the popliteal artery. Aspiration, under ultrasound or, CT guidance, or endovascular treatment of the cyst, is minimally invasive but leads to a high recurrence rate (Belmir et al. 2020). Surgical cyst excision is another option, but recurrence is common if enucleation is incomplete (Belmir et al. 2020). Finally, the most supported treatment, surgical excision of the cyst coupled with popliteal bypass reconstruction with autologous saphenous vein graft, gives the lowest rate of recurrence and complications (1%) (Hernandez Mateo et al. 2014).

Trauma

Because the risk of amputation following traumatic popliteal injury is so great, a preoperative screening tool, the popliteal scoring assessment for vascular extremity injuries in trauma (POPSAVEIT), could be helpful in stratifying patients into low and high-risk categories (O'Banion et al. 2021). Plain radiographs can be taken following a well-documented pulse examination and ankle-brachial index (ABI) with neurological evaluation. If the ABI is <0.9, heparin is administered before CT angiography to determine the location of occlusion (Fox and Moore 2020). The surgical approach to be used can then be decided.

Orthopedic Trauma

With the increased rate of total knee arthroplasties (TKA), the reported prevalence of vascular injury has ranged from 17 to 230 injuries per 100,000 TKAs. However, Sundaram et al. found that the prevalence of significant vascular injury related to TKA was 54/100,000 (Sundaram et al. 2020). The popliteal artery was injured in 0.07% of cases (Sundaram et al. 2020). Therefore, the "danger zone" during a TKA is between 11 and 3 o'clock if the popliteal vein is at 12 o'clock and the anteriormost position of the tibia is 6 o'clock (Rubash et al. 1993). There have also been cases, although rare, of popliteal artery lesions related to high tibial osteotomy (Gerich et al. 2014). In addition, though this is extremely rare, the literature describes a popliteal pseudoaneurysm and an arteriovenous fistula forming post-arthroscopy (Alserr et al. 2013).

Arteriovenous Fistula

Surgical options for the treatment of an AVF include open repairs, endovascular repairs, and combined techniques.

112.8 Complications

Popliteal Artery Aneurysm

Patients with untreated, asymptomatic PAAs have been reported to have cumulative complication rates as high as 24% at 1 year, increasing to 68% at 5 years post-diagnosis (Dawson et al. 1991). The main complications of PAAs are thrombosis, distal embolization, and rupture. Thrombosis can result from multiple embolic events occluding the crural arteries, precipitating thrombosis of a PAA (Kropman 2007). Thrombosis of a PAA leads to acute limb ischemia at a rate of 7–68% (Kropman 2007). The risk of amputation greatly increases with acute, severe limb ischemia due to a thrombosed aneurysm (16–43%) (Halliday et al. 1991).

Popliteal Artery Entrapment Syndrome

Extreme cases that are left untreated can progress to chronic lower extremity ischemia, which can lead to amputation (Lejay et al. 2016).

Trauma

Traumatic vascular injury that remains untreated for as little as 6–8 h can have devastating consequences for the patient. Severe ischemia can lead to long-term morbidity or amputation of the affected limb in as many as 30–60% of patients (Imerci et al. 2014; Subasi et al. 2001).

Arteriovenous Fistula

Complications, if left unrepaired, include venous congestion with subsequent ulceration, venous thromboembolism, arterial steal, and high-output cardiac failure (Ying et al. 2020).

References

Akagi D, Ishii S, Kitagawa T, Nagawa H, Miyata T (2006) Popliteal arterial aneurysm associated with Klippel-Trenaunay Syndrome: case report and literature review. J Vasc Surg 43:1287–1289

Alserr AHK, Antonopoulos CN, Papapetrou A, Kakisis JD, Brountzos E, Liapis CD (2013) Endovascular repair of popliteal artery pseudoaneurysm with arteriovenous fistula after knee arthroscopy. Vasc Endovasc Surg 48(2):166–170

Anand R, Mourouzis C, Wilbourn M, Tilley E, Brennan PA (2006) An unreported variation of the course of peroneal artery during fibula flap harvest. Br J Oral Maxillofac Surg 45(7):588–589. https://doi.org/10.1016/j.bjoms.2006.08.018

Aulivola B, Hamdan AD, Hile CN, Sheahan MG, Skillman JJ, Campbell DR, Scovell SD, LoGerfo FW, Poposelli FB Jr (2004) Popliteal artery aneurysms: a comparison of outcomes in elective versus emergent repair. J Vasc Surg 39:1171–1177

Bardsley JL, Staple TW (1970) Variations in branching of the popliteal artery. Radiology 94(3):581–587

Belmir H, Tijani Y, El Khloufi S, Sefiani Y, El Mesnaoui A, Lekehel B (2020) Cystic adventitial disease of the popliteal artery. J Vasc Surg Cases Innov Tech 6:344–345

Dawson I, van Bockel JH, Brand R, Terpstra JL (1991) Popliteal artery aneurysms. long-term follow-up of aneurysmal disease and results of surgical treatment. J Vasc Surg 13:398–407

Dawson I, Sie R, van Baalen JM, van Bockel JH (1994) Asymptomatic popliteal aneurysm: elective operation versus conservative follow-up. Br J Surg 81:1504–1507

Dorigo W, Pulli R, Turini F, Pratesi G, Credi G, Innocenti AA et al (2002) Acute leg ischaemia from thrombosed popliteal artery aneurysms: role of preoperative thrombolysis. Eur J Vasc Endovasc Surg 23:251–254

Fox CJ, Moore EE (2020) The posterior approach to the mid-popliteal vessels is the preferred approach for an isolated knee injury. J Trauma Acute Care Surg 89:e101

Galland RB (2005) Popliteal aneurysms: controversies in their management. Am J Surg 190:314–318

Gerich T, Lens V, Seil R, Pape D (2014) Open wedge osteotomy of the tibial head. management of vascular complications. Orthopade 43(11):1008–1015

Gray H (1985) Anatomy of the human body. CD Clemente, 30th edn. Lea and Febiger, Philadelphia

Grimm NL, Danilkowicz R, Shortell C, Toth AP (2020) Popliteal artery entrapment syndrome. J Bone Joint Surg 8(1):e0035

Halliday AW, Taylor PR, Wolfe JH, Mansfield AO (1991) The management of popliteal aneurysm: the importance of early surgical repair. Ann R Coll Surg Engl 73:253–257

Hamish M, Lockwood A, Cosgrove C, Walker AJ, Wilkins D, Ashley S (2006) Management of popliteal artery aneurysms. ANZ J Surg 76:912–915

Hernandez Mateo MM, Serrano Hernando FJ, Martinez Lopez I et al (2014) Cystic adventitial degeneration of the popliteal artery: report on 3 cases and review of the literature. Ann Vasc Surg 28:1062e9

Huang Y, Gloviczky P, Noel AA, Sullivan TM, Kalra M, Gullerud RE et al (2007) Early complications and long-term outcome after open surgical treatment of popliteal artery aneurysms: is exclusion with saphenous vein bypass still the gold standard? J Vasc Surg 45:706–713

Imerci A, Özaksar K, Gürbüz Y, Sadik Sügün T, Canbek U, Savran A (2014) Popliteal artery injury associated with blunt trauma to the knee without fracture or dislocation. West J Emerg Med 15(2):145–148

Kawarai S, Fukasawa M, Kawahara Y (2012) Adventitial cystic disease of the popliteal artery. Ann Vasc Dis 5(2):190–193. https://doi.org/10.3400/avd.cr.11.00069

Keen JA (1961) A study of the arterial variations in the limbs, with special reference to symmetry of vascular patterns. Am J Anat 108:245–261

Keyurapan E, Phoemphunkunarak W, Lektrakool N (2016) Location of the neurovascular bundle of the knee during flexed and extended position: an MRI study. J Med Assoc Thail 99(10):1102–1109

Kil SW, Jung GS (2009) Anatomical variations of the popliteal artery and its tibial branches: analysis in 1242 extremities. Cardiovasc Intervent Radiol 32(2):233–240

Kim D, Orron DE, Skillman JJ (1989) Surgical significance of popliteal arterial variants. A unified angiographic classification. Ann Surg 210:776–781

Kim JW, Sung CM, Cho SH et al (2010) Vascular injury associated with blunt trauma without dislocation of the knee. Yonsei Med J 51:790–792

Kropman R (2007) Surgical and endovascular treatment of atherosclerotic popliteal artery aneurysms. J Cardiovasc Surg 48:281–288
Leake AE, Segal MA, Chaer RA, Eslami MH, Al-Khoury G, Makaroun MS, Avgerinos ED (2017) Meta-analysis of open and endovascular repair of popliteal artery aneurysms. J Vasc Surg 65(1):246–256
Lejay A, Delay C, Georg Y, Gaetner S, Ohana M, Thaveau F, Lee JT, Geny B, Chakfe N (2016) Five year outcomes of surgical treatment for popliteal artery entrapment syndrome. Eur J Vasc Endovasc Surg 51(4):557–564
Levien LJ, Been CA (1998) Adventitial cystic disease: a unifying hypothesis. J Vasc Surg 28:193–205
Lippert H, Pabst R (1985) Arterial variations in man: classification and frequency. J.F. Bergmann-Verlag, Munchen
Longo B, Sorotos M, Nicolotti M, Santanelli di Pompeo F (2014) Retrospective analysis of incidence of peroneal artery hypoplasia in 101 free fibula transfers and new classification of popliteal branch. Injury 45:394–398. https://doi.org/10.1016/j.injury.2013.09.027
Loukas M, Klaassen Z, Tubbs RS, Apaydin N (2007) Popliteal artery aneurysms: a review. Folia Morphol (Warsz) 66(4):272–276
Lowell RC, Gloviczki P, Hallett JW Jr, Naessens JM, Maus TP, Cherry KJ Jr, Bower TC, Pairolero PC (1994) Popliteal artery aneurysms: the risk of nonoperative management. Ann Vasc Surg 8:14–23
Moore KL (1985) Clinically oriented anatomy, 2nd edn. WB Saunders, Baltimore
Moore KL, Agur AMR (2002) Chapter 3: Abdomen. In: Essential clinical anatomy, 2nd edn. Lippincott Williams & Williams, Baltimore
O'Banion LA, Dirks R, Farooqui E, Saldana-Ruiz N, Yoon WJ et al (2021) Popliteal scoring assessment for vascular extremity injuries in trauma study. J Vasc Surg 74:804
Peterson JJ, Kransdorf MJ, Bancroft LW et al (2003) Imaging characteristics of cystic adventitial disease of the peripheral arteries: presentation as soft-tissue masses. AJR Am J Roentgenol 180:621–625
Pulli R, Dorigo W, Troisi N, Innocenti AA, Pratesi G, Azas L et al (2006) Surgical management of popliteal artery aneurysms: which factors affect outcomes? J Vasc Surg 43:481–487
Qazi E, Hilting J, Patel NR, Alenezi AO, Kennedy SA, Tan KT, Jaberi A, Mafeld S (2021) Arteries of the lower limb – embryology, variations, and clinical significance. Can Assoc Radiol J 73:259
Raherinantenaina F, Rajaonanahary TMA, Rakoto Ratsimba HN (2016) Prise En Charge Des Faux Anevrismes de l'artere Poplitee Associes Aux Osteochondromes et Ceux Dus Aux Traumatismes Des Membres Ou Secondaires a La Chirurgie Orthopedique. Annales de Cardiologie et d'angeiologie 65(4):265–274
Rosset E, Hartung O, Brunet C, Roche PH, Magnan PE, Mathieu JP, Branchereau A, Farisse J (1995) Popliteal artery entrapment syndrome: anatomic and embryologic bases, diagnostic and therapeutic considerations following a series of 15 cases with a review of the literature. Surg Radiol Anat 17:161–169
Rubash HE, Berger RA, Britton CA, Nettrour WS, Seel MJ (1993) Avoiding neurologic and vascular injuries with screw fixation of the tibial component in total knee arthroplasty. Clin Orthop Relat Res 286:56–63
Senior HD (1919) The development of the arteries of the human lower extremity. Am J Anat 25(1):54–95. https://doi.org/10.1002/aja.1000250105
Slaba S, Khalil SA, Younan T, Nassar-Slaba J, Khoury S, Kheir C (2007) Unusual variation of popliteal arterial branches: 4 axes by early division of the peroneal artery. J Mal Vasc 32(4–5):212–215. https://doi.org/10.1016/j.jmv.2007.07.002
Sora MC, Dresenkamp J, Gabriel A, Matusz P, Wengert GJ, Bartl R (2015) The relationship of neurovascular structures to the posterior medial aspect of the knee. An anatomic study using plastinated cross-sections. Romanian J Morphol Embryol 56(3):1035–1041
Subasi M, Cakir O, Kesemenli C et al (2001) Popliteal artery injuries associated with fractures and dislocations about the knee. Acta Orthop Belg 67:259–266
Sundaram K, Udo-Inyang I, Mont M, Molloy R, Higuera-Rueda C, Piuzzi NS (2020) Vascular injuries in total knee arthoplasty: a systematic review and meta-analysis. JBJS 8(1):e0051
Tercan F, Oguzkurt L, Kizilkilic O, Yeniocak A, Gülcan Ö (2005) Popliteal artery entrapment syndrome. Diagn Interv Radiol 11(4):222–224
Tomaszewski KA, Popieluszko P, Graves MJ, Pekala PA, Henry BM, Roy J, Hsieh WC, Walocha JA (2017) The evidence-based surgical anatomy of the popliteal artery and the variations in its branching patterns. J Vasc Surg 65(2):521–529. https://doi.org/10.1016/j.jvs.2016.01.043
Tsilimparis N, Hanack U, Yousefi S et al (2007) Cystic adventital disease of the popliteal artery: an argument for the development theory. J Vasc 45:1249–1252
Wady H, Badar Z, Farooq Z, Shaw K, Kobayashi P (2018) Avoiding the trap of misdiagnosis: valuable teaching points derived from a case of longstanding popliteal artery entrapment syndrome. Case Rep Med 2018:1. https://doi.org/10.1155/2018/3214561
Whelan TJ, Haimovici H (1984) Vascular surgery: principles and techniques. McGraw-Hill, New York
Wright LB, Matchett WJ, Cruz CP, James CA, Culp WC, Eidt JF, McCowan TC (2004) Popliteal artery disease: diagnosis and treatment. Radiographics 24:467–479
Ying A, Al-Asady R, Vicaretti M (2020) Endovascular treatment of a large iatrogenic popliteal arteriovenous fistula. J Vasc Surg Cases Innov Tech 6(1):129–132
Yoshino S, Inoue K, Yoshiya K, Kurose S, Nakayama K, Morisaki K, Furuyama T, Matsumoto T, Oda Y, Mori M (2019) Cystic arterial disease locate only in the media of the popliteal artery: a case report. Ann Vasc Dis 12(4):530–533

113 Anterior Tibial Artery

Skyler Jenkins and Stephen J. Bordes, Jr.

113.1 Embryology

During the 14 mm stage of embryonic development, a branch arises off the distal bend of the axial artery and passes through the proximal end of the tibiofibular interspace; the proximal part, termed the ramus perforans cruris, participates in the formation of the anterior tibial artery (Senior 1919). At the 17.8 mm stage, the anterior tibial artery pars distalis arises from the ramus perforans cruris and travels to the rete dorsalis (Senior 1919). Prior to this stage, the developing anterior tibial artery has received its blood supply from the deep popliteal artery and ramus perforans cruris; however, a connection is now formed between the developing fibular artery and the distal part of the deep popliteal artery (Senior 1919). The ramus communicans medius will become the definitive proximal end of the adult anterior tibial artery (Senior 1919). Therefore, the adult anterior tibial artery comprises four segments from embryonic development. The first segment extends from the origin of the artery to the root of the recurrent posterior tibial artery and corresponds to the whole medial ramus communicans. The second, near the origin of the recurrent posterior tibial artery, represents a persisting portion of the embryonic deep popliteal artery. The third extends from the recurrent posterior tibial artery to the root of the recurrent anterior tibial artery. It is derived from the proximal part of the embryonic ramus perforans cruris. The final section extends from the end of the third section to the proximal end of the dorsalis pedis artery, representing the entire embryonic anterior tibial artery pars distalis.

113.2 Anatomy and Topography

The anterior tibial artery (Fig. 113.1) typically arises from the popliteal artery at the lower border of the popliteus muscle. In the posterior compartment, it gives off the branches of the posterior tibial recurrent and circumflex fibular arteries. It runs dorsal to the popliteus muscle and passes between the heads of the tibialis posterior. It then passes anteriorly into the anterior compartment of the leg between the tibia and fibula through the upper opening of the interosseus membrane below the ligament of Barkow. May et al. found that it passes through the interosseus membrane an average of 50.9 ± 6.9 mm distal to the tibial plateau joint line (May et al. 2020). The anterior tibial artery courses through the membrane an average of 36.5 ± 6.0 mm distal to the fibular head (May et al. 2020). As it travels distally in the

S. Jenkins
Department of Family Medicine, University of Arkansas for Medical Sciences, Jonesboro, USA

S. J. Bordes, Jr. (✉)
Department of Surgery, Louisiana State University Health Sciences Center, New Orleans, LA, USA

S. J. Bordes, Jr. et al. (eds.), *The Clinical Anatomy of the Vascular System*,
https://doi.org/10.1007/978-3-031-78326-5_113

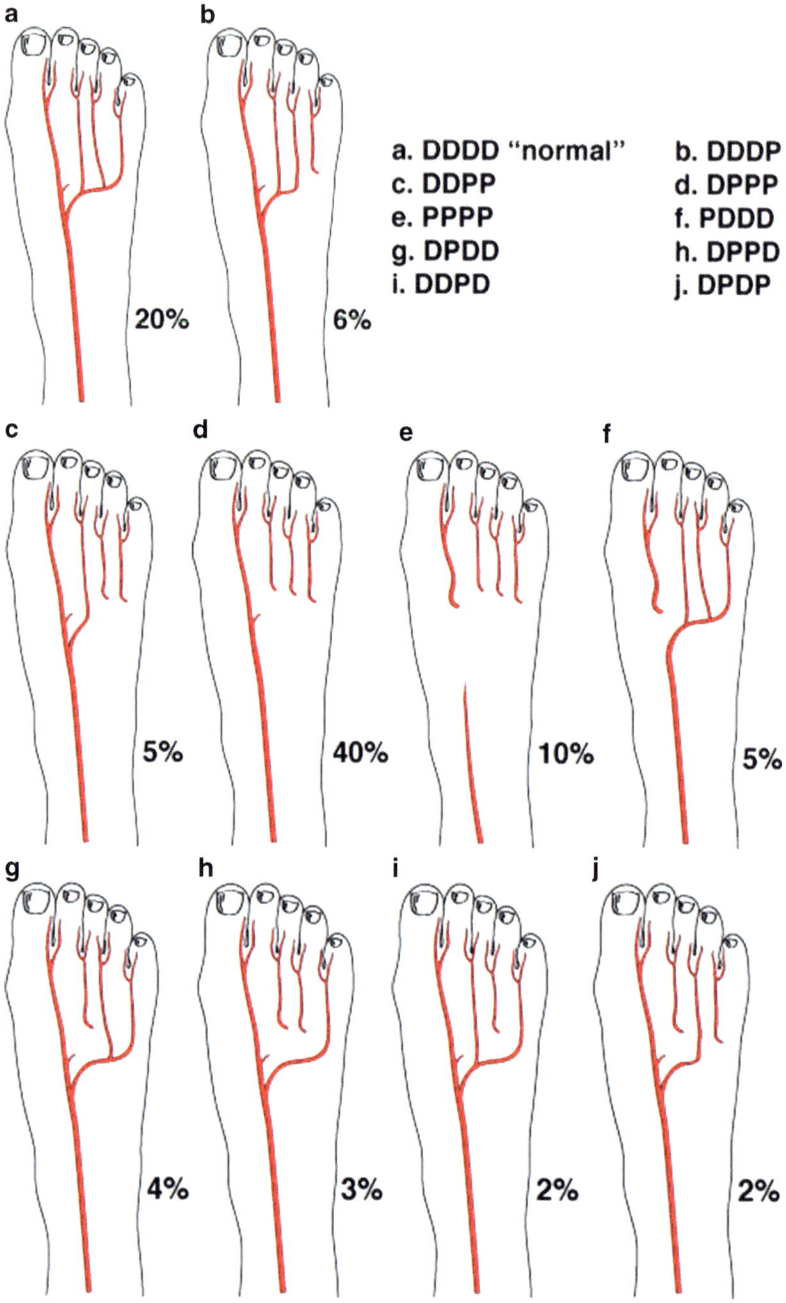

Fig. 113.1 Schematic drawing showing variant anatomy of the distal anterior tibial artery. Dorsal foot: a–j represents the more common anatomical variations with a more dorsal (D) or plantar (P) origin of the arteries on the dorsal side of the foot. (After Lippert H and Pabst R Arterial Variations in Man, 1985)

proximal two-thirds, it lies between the tibialis anterior and extensor digitorum longus muscles and along the interosseous membrane with the deep fibular nerve. It is relatively fixed to the interosseus membrane by deep fascia and the anterior and posterior recurrent tibial and fibular arteries (Ebraheim et al. 1998). Its course runs from the lateral to the anterior surface of the tibia and enters the ankle between the tibialis anterior and extensor hallucis longus muscles, and subsequently it becomes the dorsalis pedis artery (Ebraheim et al. 1998). Ebraheim et al. found that the artery intersects the anterior longitudinal axis of the tibia

5.3 ± 0.4 cm above the intermalleolar midpoint (Ebraheim et al. 1998). Branches of the anterior tibial artery include the anterior tibial recurrent artery, which courses rostrally in the tibialis anterior muscle to anastomose with the lateral genicular branches, the muscular branches supplying the muscles of the anterior compartment, the perforating branches passing behind the extensor digitorum longus to pierce the deep fascia and supply the skin of the anterior leg, the anterior medial malleolar artery, and the anterior lateral malleolar artery.

113.3 Histology

The anterior tibial artery is a muscular (distributing) artery. Its tunica intima comprises an inner layer of endothelium surrounded by subendothelial connective tissue and a prominent internal elastic lamina. The intima is surrounded by tunica media, mostly circumferentially arranged smooth muscle, minimal elastin, and a thick external elastic lamina. The tunica adventitia, the outermost wall of organized, dense, irregular connective tissue, contains multiple layers of elastic tissue.

The anterior tibial vein is a medium vein with less distinct layers. It has a similar intima, media containing few layers of smooth muscle, and outer adventitia.

113.4 Variations

Kim et al. described three types of branching variations and various subtypes (Kim et al. 1989):

Type I. Normal level of branching proximal to the lower border of the popliteus muscle (96.1% in 27 studies) (Tomaszewski et al. 2017).

- A. The anterior tibial artery is the first branch, followed by the tibiofibular artery, which bifurcates into the fibular and posterior tibial arteries (92.6%).
- B. The anterior, posterior, and fibular arteries arise within 0.5 cm of each other without a tibiofibular trunk (2.4%).
- C. The posterior tibial is the first branch, followed by the tibiofibular trunk, which bifurcates into the fibular and anterior tibial arteries (1.1%).

Type II. High division of the popliteal artery (3.9% in 27 studies) (Tomaszewski et al. 2017).

- A. The anterior tibial artery begins at or above the knee joint (2.5%).
  - (a) Normal course of proximal anterior tibial (2.2%).
  - (b) Medial initial curve in anterior tibial (0.3%).
- B. The posterior tibial begins at or above the knee joint. Common trunk for fibular and anterior tibial (1.2%).
- C. The fibular arises at or above the knee joint. Common trunk for anterior and posterior tibial (0.2%).

Type III. Hypoplastic or aplastic branching with altered distal supply (3.0% in 17 studies) (Tomaszewski et al. 2017).

- A. Hypoplastic-aplastic posterior tibial. Distal posterior tibial replaced by fibular (1.7%).
- B. Hypoplastic-aplastic posterior and anterior tibial. Dorsalis pedis replaced by fibular (1.2%).
- C. Hypoplastic-aplastic posterior and anterior tibial. Posterior tibial and dorsalis pedis replaced by fibular; termed the peronea arteria magna (0.1%).

Anand et al. and Longo et al. each suggested a modification of the Kim classification, adding another class (IV) or subclass (IIID) that includes aberrant or fibular artery insufficiency (Anand et al. 2006; Longo et al. 2014). In a unique case study by Slaba et al., the anterior tibial artery had a high origin at the femorotibial space, and there was a tibiofibular trunk greater than 5 cm with early bifurcation of the fibular artery into anterior and posterior branches (Slaba et al. 2007). Olewnik et al. described a further Type I subtype characterized by the diameter of the anterior tibial artery versus the tibiofibular trunk (Olewnik et al. 2019).

Even though the course of the anterior tibial artery around the ankle is usually predictable, it varies in up to 12% of cases (Fanter et al. 2017). Three variants have been described in the literature. The first is a hypoplastic or absent anterior tibial artery; an enlarged perforating branch of the fibular artery communicates with the dorsalis pedis artery (Vazquez et al. 2006). In the second, the anterior tibial artery provides a branch that travels laterally to supply the lateral aspect of the ankle (Vazquez et al. 2006). In the third, the anterior tibial artery runs laterally deep to the extensor digitorum longus and peroneus tertius to overlie the lateral malleolus prior to regaining its usual orientation in the foot (Vazquez et al. 2006).

113.5 Imaging

The anterior tibial artery and vessels of the lower extremity have been evaluated by contrast arteriography, specifically digital subtraction angiography (DSA), which has been considered the gold standard for many years (Pomposelli 2010). However, more recently, magnetic resonance angiography (MRA) and computed tomography angiography (CTA) have become viable alternatives as they are less invasive and provide image quality similar to that of DSA (Pomposelli 2010). All three imaging modalities have advantages and disadvantages regarding patient clinical presentation and comorbidities (Pomposelli 2010).

Vascular ultrasound of the lower extremity is a noninvasive diagnostic modality using sound waves to visualize blood flow through vessels. It is a mainstay of vascular imaging, including B-mode, continuous wave, pulsed wave Doppler, and duplex ultrasound (Varaki et al. 2018). In addition to vascular ultrasound, the ankle-brachial index (ABI) is a simple, noninvasive method for assessing arterial perfusion. This test has a high sensitivity and specificity. Still, its accuracy is questionable in specific patient populations such as the elderly, diabetics, or patients with renal disease because the vessels are largely incompressible (Varaki et al. 2018). Other similar methods are segmental blood pressure measurements to locate stenosis, occlusion, and toe brachial index (Varaki et al. 2018).

113.6 Pathology

Atherosclerosis is often responsible for occlusion or stenosis within the anterior tibial artery. Risk factors include high cholesterol, hypertension, smoking, diabetes, and obesity. The pathogenesis of atherosclerosis has been well established. The endothelial wall of the artery is injured, followed by the overproduction of cellular mediators, so a fibrous plaque forms. This plaque can then restrict blood flow, leading to intermittent claudication in moderate disease and major tissue loss in severe disease.

Thromboangiitis obliterans is typically a vasculitis that affects the small and medium vessels of the upper and lower extremities. It is primarily found in young smokers and is characterized by peripheral artery occlusion in the extremities. Histopathologically, this disease follows three phases: acute, intermediate, and end stage. In the acute phase, the lesions include an occlusive, highly cellular, inflammatory thrombus with the possible presence of polymorphonuclear leukocytes, microabscesses, and multinucleated giant cells (Arkkila 2006). The intermediate stage is marked by a progression to loosely organized thrombi in arteries and veins (Arkkila 2006). Finally, those thrombi become more organized, and fibrosis is found in the end-stage vessels (Arkkila 2006).

Eosinophilic vasculitis is characterized by eosinophilic infiltration with intra- and extravascular granuloma formation associated with a centralized zone of eosinophilic necrosis, histiocytes, multinucleated giant cells, and neutrophils (Khoury et al. 2014).

Vasculitis can co-occur with antiphospholipid syndrome, but there is no causal connection (Lally and Sammaritano 2015). Typically, they both occur in patients with systemic lupus erythematosus (SLE) or other connective tissue disease, with 80% of cases involving small vessels (Lally and Sammaritano 2015). Owing to the inflammatory nature of the disease, histopathology presents with endothelial cell activation and production of inflammatory cytokines with the recruitment of neutrophils and monocytes to the vascular lumen (Lally and Sammaritano 2015).

Mikulicz disease is a systemic inflammatory disease characterized by fibrosclerosis, lymphocytic infiltration, and IgG4-positive plasma cells, so it is considered IgG4-related (Ikeoka et al. 2019). It is typically involved in large and medium vessel arterial inflammation, but rare cases of small vessel peripheral occlusive disease have been reported (Ikeoka et al. 2019). Histopathology includes intimal hyperplasia and infiltration of IgG4-expressing plasma cells and eosinophils into the vascular wall (Ikeoka et al. 2019).

113.7 Clinical and Surgical Significance

The anterior tibial artery can undergo atherosclerotic changes, leading to peripheral arterial disease in which the lumen becomes occluded, resulting in decreased blood flow and poor perfusion. Treatment is initially aimed at lifestyle modifications and medications; however, surgical intervention with angioplasty, clot removal, and bypass surgery is sometimes required.

The leg contains four compartments divided by a strong, inflexible fascial intermuscular septum. The anterior tibial artery supplies blood to the anterior compartment, which contains the tibialis anterior, extensor digitorum longus, extensor hallucis longus, and fibularis tertius muscles. The anterior compartment is the most common site for compartment syndrome, either acutely following trauma (fractures, crush injuries, contusions, gunshot wounds) or chronically as an exertional syndrome commonly seen in athletes (Mabvuure et al. 2012). Fractures are the most common cause, accounting for 65% of cases (Taylor et al. 2012). Other causes include tight casts or dressings, extravasation of intravenous infusion, circumferential burns, postischemic swelling, crush injuries, bleeding disorders, and arterial injury (Mabvuure et al. 2012). Clinical presentation typically includes pain on passive stretch (the most sensitive finding), pallor, paresthesia, pulselessness, poikilothermia, and paralysis. Resting compartment pressure greater than 30 mmHg or "Delta P" less than 300 mmHg confirms the diagnosis. The gold standard of time to treatment is within six hours of ischemia, after which necrosis develops. Acute compartment syndrome is a medical emergency and is treated with fasciotomy. Treatment of chronic compartment syndrome can include activity modification, medications, stretching, orthotics, and physical therapy.

During orthopedic procedures, iatrogenic injury of the anterior tibial artery has been reported following high tibial osteotomy, external fixator application, locking screw insertion during intramedullary nail fixation, and during placement of the wires for an Ilizarov ring fixator (Ebraheim et al. 1998). The locations for tibial transfixation most likely to preclude iatrogenic injury are the anterior and middle thirds of the proximal tibia, anterior third of the mid-tibia, and posterior third of the distal tibia (Ebraheim et al. 1998). Within the interosseus foramen, the artery is liable to compression post-valgus osteotomy and stretching during this procedure (Steel et al. 1971; Ebraheim et al. 1998).

In patients with thromboangiitis obliterans, the anterior tibial artery is most commonly affected. This is associated with ischemia, leading to claudication, ulceration, and gangrene of the digits (Sasaki et al. 2000). Treatment of this disease is focused on complete smoking cessation, as one or two cigarettes or smokeless tobacco can keep the disease active (Arkkila 2006). Medical treatment can include Iloprost, aspirin, or intra-arterial thrombolytic therapy (Arkkila 2006). Surgical interventions such as arterial revascularization and bypass surgery are also possible options but have varied results (Arkkila 2006). Spinal cord stimulators and vascular endothelial growth factor therapy are other alternatives (Swigris et al. 1999; Isner et al. 1998).

The literature has also described the anterior tibial artery as being affected in other forms of vasculitis such as eosinophilic vasculitis, antiphospholipid syndrome, and Mikulicz's disease (Kim et al. 2005; Goldberger et al. 1992; Ikeoka et al. 2019).

The anterior tibial vein is the least common site of deep vein thrombosis. It is extremely rare, with only an isolated case, 0–0.3% (Yao et al. 2015). Treatment includes anticoagulation, thrombolytic therapy, vena cava filters, and compression socks.

There have been very few cases of ankle injuries causing isolated or combined injury to the anterior, posterior, or fibular arteries (Christofilopoulos et al. 2008). Fatality due to extravasation after open ankle dislocation with anterior and posterior tibial artery tears is even rarer, with one case reported to date (De-Giorgio et al. 2017).

113.8 Complications

Orthopedic Surgery

During the placement of distal locking bolts during tibial intramedullary nail fixation, the normal anatomical course of the anterior tibial artery and its variants could potentially be injured, leading to bleeding, hematoma, wound healing issues, or pseudoaneurysm formation (Fanter et al. 2017; Bono et al. 2003; Roberts et al. 1999). The literature has shown a roughly 0.7–30% neurovascular complication rate associated with tibial intramedullary nails. Knowledge of the neurovascular anatomy of the lower leg could help prevent such complications (Fanter et al. 2017).

References

Anand R, Mourouzis C, Wilbourn M, Tilley E, Brennan PA (2006) An unreported variation of the course of peroneal artery during fibula flap harvest. Br J Oral Maxillofac Surg 45(7):588–589

Arkkila P (2006) Thromboangiitis obliterans (Buerger's disease). Orphanet J Rare Dis 1:14. https://doi.org/10.1186/1750-1172-1-14

Bono CM, Sirkin M, Sabatino CT et al (2003) Neurovascular and tendinous damage with placement of anteroposterior distal locking bolts in the tibia. J Ortho Trauma 17:677–682

Christofilopoulos P, Panos A, Masterson K, Abrassart S, Assal M (2008) Pseudoaneurysm of the anterior tibial artery following an ankle sprain: a case report of an uncommon ankle trauma with review of the literature. Foot Ankle Surg 14:40–42

De-Giorgio F, Marchese L, Masullo M, D'Aloja E, Izzo V, Cerciello S (2017) Fatal complication following pure ankle dislocation in a domestic setting: case report and review of the literature. Forensic Sci Med Pathol 13(2):209–212

Ebraheim NA, Lu J, Hao Y, Biyani A, Yeasting RA (1998) Anterior tibial artery and its actual projection on the lateral aspect of the tibia: a cadaveric study. Surg Radiol Anat 20:259–262

Fanter NJ, Inouye SE, Beiser CW (2017) The risk of iatrogenic injury to anterior tibial artery variations during tibial nail distal interlocking: a cadaveric study. J Orthop Surg 25(1):1–7. https://doi.org/10.1177/2309499016684471

Goldberger E, Elder R, Schwartz R, Phillips P (1992) Vasculitis in the antiphospholipid syndrome. A cause of ischemia responding to corticosteroids. Arthritis Rheumatol 35(5):569–572

Ikeoka K, Watanabe T, Ohkawa T, Tsuzuki T, Ikeda Y, Takeda M, Hoshida S (2019) IgG4-related small-sized occlusive vasculitis in Mikulicz's disease. J Vasc Surg Cases Innov Tech 5(3):289–292

Isner JM, Baumgartner I, Rauh G, Schainfeld R, Blair R, Manor O, Razvi S, Symes JF (1998) Treatment of thromboangiitis obliterans (Buerger's disease) by intramuscular gene transfer of vascular enothelial growth factor: preliminary clinical results. J Vasc Surg 28:964–973

Khoury P, Grayson P, Klion A (2014) Eosinophils in vasculitis: characteristic and roles in pathogenesis. Nat Rev Rheumatol 10(8):474–483

Kim D, Orron DE, Skillman JJ (1989) Surgical significance of popliteal arterial variants. A unified angiographic classification. Ann Surg 210:776–781

Kim SH, Kim T, Yun Y, Shin J, Oh I, Sir J, Kim K et al (2005) Hypereosinophilia presenting as eosinophilic vasculitis and peripheral artery occlusions without organ involvement. J Korean Med Sci 20:677–679

Lally L, Sammaritano L (2015) Vasculitis in antiphospholipid syndrome. Rheum Dis Clin N Am 41:109–123

Longo B, Sorotos M, Nicolotti M, Santanelli di Pompeo F (2014) Retrospective analysis of incidence of peroneal artery hypoplasia in 101 free fibula transfers and new classification of popliteal branch. Injury 45:394–398

Mabvuure NT, Malahias M, Hindocha S, Khan W, Juma A (2012) Acute compartment syndrome of the limbs: current concepts and management. Open Orthop J 6:535–543

May H, Kastan O, Yunus Emre T, Cetin M, Unal M, Kose O (2020) Anterior tibial artery and its clinical importance in the posterolateral approach to the Tibial plateau: an angiographic study on 219 lower limbs. J Knee Surg 35:725. https://doi.org/10.1055/s-0040-1716849

Olewnik L, Labetowicz P, Podgorski M, Polguj M, Ruzik K, Topol M (2019) Variations in the terminal branches of the popliteal artery: a cadaveric study. Surg Radiol Anat 41(12):1473–1482. https://doi.org/10.1007/s00276-019-02262-3

Pomposelli F (2010) Arterial imaging in patients with lower extremity ischemia and diabetes mellitus. J Vasc Surg 52:81S–91S

Roberts C, King D, Wang M et al (1999) Should distal interlocking or tibial nails be performed from a medial or lateral direction? J Ortho Trauma 13:27–32

Sasaki S, Sakuma M, Kunihara T, Yasuda K (2000) Distribution of arterial involvement in thromboangiitis obliterans (Buerger's disease): results of a study conducted by the Intractable Vasculitits Syndromes Research Group in Japan. Surg Today 30:600–605

Senior HD (1919) The development of Teh arteries of the human lower extremity. Am J Anat 25(1):54–95. https://doi.org/10.1002/aja.1000250105

Slaba S, Khalil SA, Younan T, Nassar-Slaba J, Khoury S, Kheir C (2007) Unusual variation of popliteal arterial branches: 4 axes by early division of the peroneal artery. J Mal Vasc 32(4–5):212–215. https://doi.org/10.1016/j.jmv.2007.07.002

Steel HH, Sandrow RE, Sullivan PD (1971) Complicationsof tibial osteotomy in Childrenfor Genu Varum or Valgum. J Bone Joint Surg 53A:1629–1635

Swigris JJ, Olin JW, Mekhail NA (1999) Implantable spinal cord stimulator to treat the ischemic manifestations of thromboangiitis obliterans (Buerger's disease). J Vasc Surg 29:928–935. https://doi.org/10.1016/S0741-5214(99)70221-1

Taylor RM, Sullivan MP, Mehta S (2012) Acute compartment syndrome: obtaining diagnosis, providing treatment, and minimizing medicolegal risk. Curr Rev Musculoskelet Med 5(3):206–213

Tomaszewski KA, Popieluszko P, Graves MJ, Pekala PA, Henry BM, Roy J, Hsieh WC, Walocha JA (2017) The evidence-based surgica anatomy of the popliteal artery and the variations in its branching patterns. J Vasc Surg 65(2):521–529

Varaki ES, Gargiulo GD, Penkala S, Breen PP (2018) Peripheral vascular disease assessment in the lower limb: a review of current and emerging non-invasive diagnositc methods. Biomed Eng Online 17:61

Vazquez T, Rodriguez-Niedenfuhr M, Parkin I et al (2006) Anatomic study of blood supply of the dorsum of the foot and ankle. Arthroscopy 22:287–290

Yao Y, Yang X, Dai X, Xu X, Rong Z, Song K, Shi D et al (2015) Why are there so few cases of anterior tibial vein thrombosis? A case report after Total joint arthroplasty and literature review. Int J Clin Exp Med 9(11):22590–22594

Posterior Tibial Artery

114

Skyler Jenkins and Stephen J. Bordes, Jr.

114.1 Embryology

During the 14 mm stage of embryonic development, one of the branches of the deep popliteal artery descends through the posterior crural region, resembling the adult posterior tibial artery (Senior 1919). The other branch arises more distally and is the posterior superficial fibular artery. The former passes backward along the medial side of the tibialis nerve to reach the posterior aspect of the leg and then descends along the posterior aspect of the nerve until it enters the sole of the foot, becoming the medial plantar artery in the adult (Senior 1919). The latter passes from the medial to lateral aspect of the tibial nerve, traveling anterior and lateral to the nerve until it forms branches that end blindly (Senior 1919). At the 17.8 mm stage, the portion of the posterior superficial tibial artery between the popliteal artery and the origin of the lateral plantar artery becomes the adult posterior tibial artery. During the 18 mm stage, behind the medial malleolus, the posterior superficial tibial artery finds its relationship between the tibial artery and the flexor hallucis longus (Senior 1919). At the 24.5 mm stage, the proximal portion of this artery blends with the corresponding section of the posterior superficial fibular artery as far as the termination of the popliteal artery, forming the adult posterior tibial artery (Senior 1919).

S. Jenkins
Department of Family Medicine, University of Arkansas for Medical Sciences, Jonesboro, USA

S. J. Bordes, Jr. (✉)
Department of Surgery, Louisiana State University Health Sciences Center, New Orleans, LA, USA

114.2 Anatomy and Topography

The adult leg is divided into four compartments by deep, fibrous, intermuscular septae: the anterior, lateral, and deep and superficial posterior. The superficial posterior compartment contains the gastrocnemius, soleus, and plantaris muscles, while the deep posterior compartment contains the tibialis posterior, flexor digitorum longus, and flexor hallucis longus muscles. The superficial and deep posterior compartments receive blood supply and innervation from the posterior tibial artery (Figs. 114.1 and 114.2) and tibial nerve, respectively, lying within the deep posterior compartment.

Within the popliteal fossa, the posterior tibial artery arises off the tibiofibular trunk, one of two bifurcating branches of the popliteal artery. It becomes the main continuation of the popliteal artery and provides most of the blood supply to the leg (Chmielewski et al. 2016). Typically arising at the inferior border of the popliteal muscle, the tibiofibular trunk continues for approximately 2.5 cm before dividing into the posterior tibial and fibular arteries. The tendinous arch, created by the soleal line of the

S. J. Bordes, Jr. et al. (eds.), *The Clinical Anatomy of the Vascular System*,
https://doi.org/10.1007/978-3-031-78326-5_114

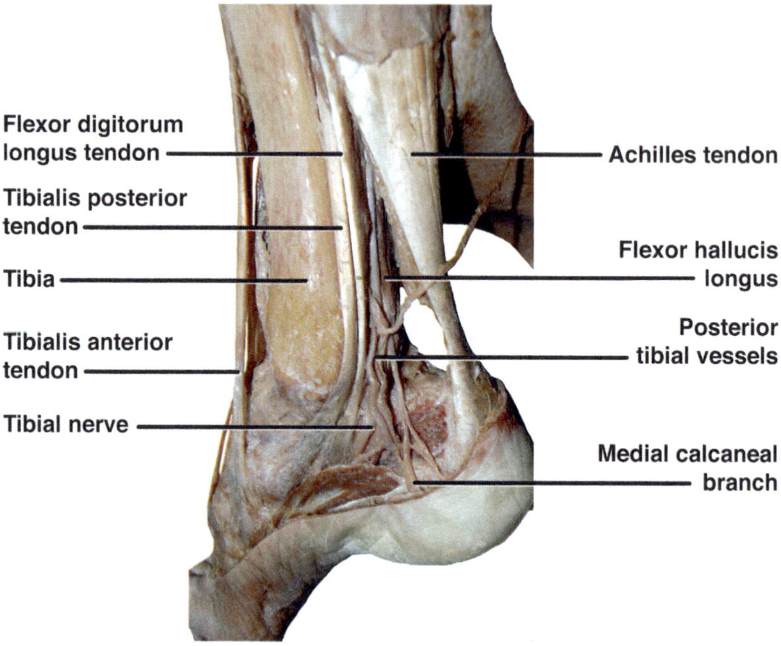

Fig. 114.1 Cadaveric dissection noting the posterior tibial artery and related anatomy. (Courtesy Prof. M. Loukas)

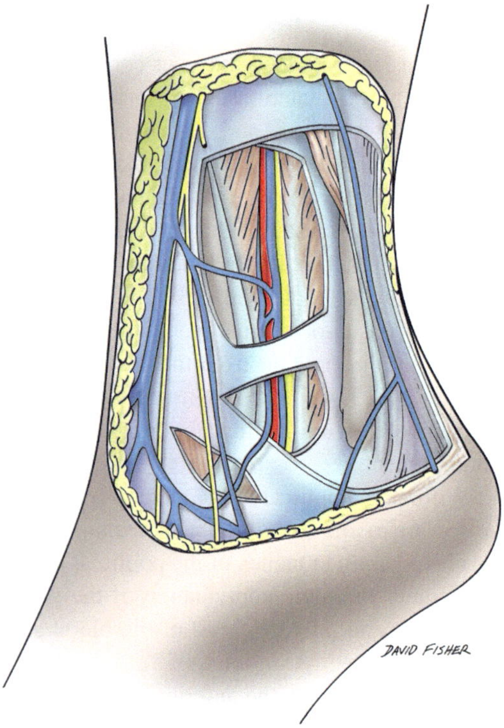

Fig. 114.2 Schematic drawing showing the superficial and deep anatomy of the posterior tibial artery (red)

tibia soleus, is the beginning of the posterior tibial artery (Chmielewski et al. 2016). As it courses distally along the deep flexor muscles of the shin, it is predominantly covered by a fascial layer and the triceps surae muscle (Chmielewski et al. 2016). In the upper segment of the leg, it lies along the tibialis posterior muscle; however, as it descends, it enters a groove between the flexor digitorum longus medially and the flexor hallucis longus laterally (Chmielewski et al. 2016). As it approaches the ankle joint, the posterior tibial artery migrates superficially and medially alongside the Achilles tendon, covered only by a small amount of fatty tissue and a superficial and deep lamina of crural fascia (Chmielewski et al. 2016). It enters the medial malleolar canal between the tendons of the tibialis posterior and flexor digitorum longus anteriorly and the flexor hallucis longus posteriorly (Chmielewski et al. 2016). At the medial aspect of the calcaneus, the posterior tibial artery gives off its terminal branches, the lateral and medial plantar arteries (Figs. 114.3, 114.4 and 114.5) (Chmielewski et al. 2016). The posterior tibial

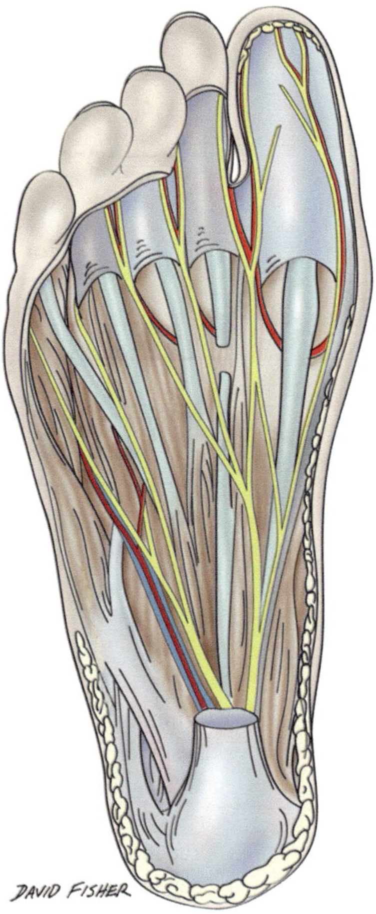

Fig. 114.3 Drawing of the superficial anatomy of the sole noting branches of the medial and lateral plantar arteries

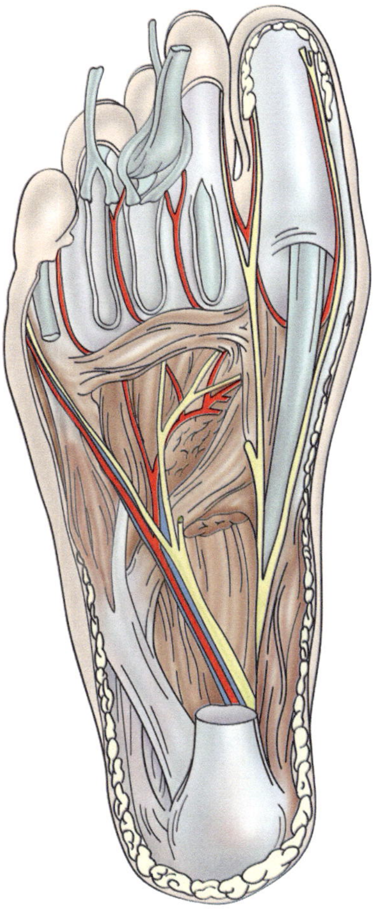

Fig. 114.4 Drawing of the deeper anatomy of the sole, highlighting primarily the lateral plantar artery

artery has several branches, mainly the fibular artery, which arises distal to the fibular neck approximately 5 cm from the popliteal bifurcation (Chmielewski et al. 2016). The other arteries that arise from the posterior tibial artery are the circumflex fibular, the nutrient, muscular, perforating, communicating, medial malleolar, calcaneal, and terminal branches of the medial and lateral plantar arteries.

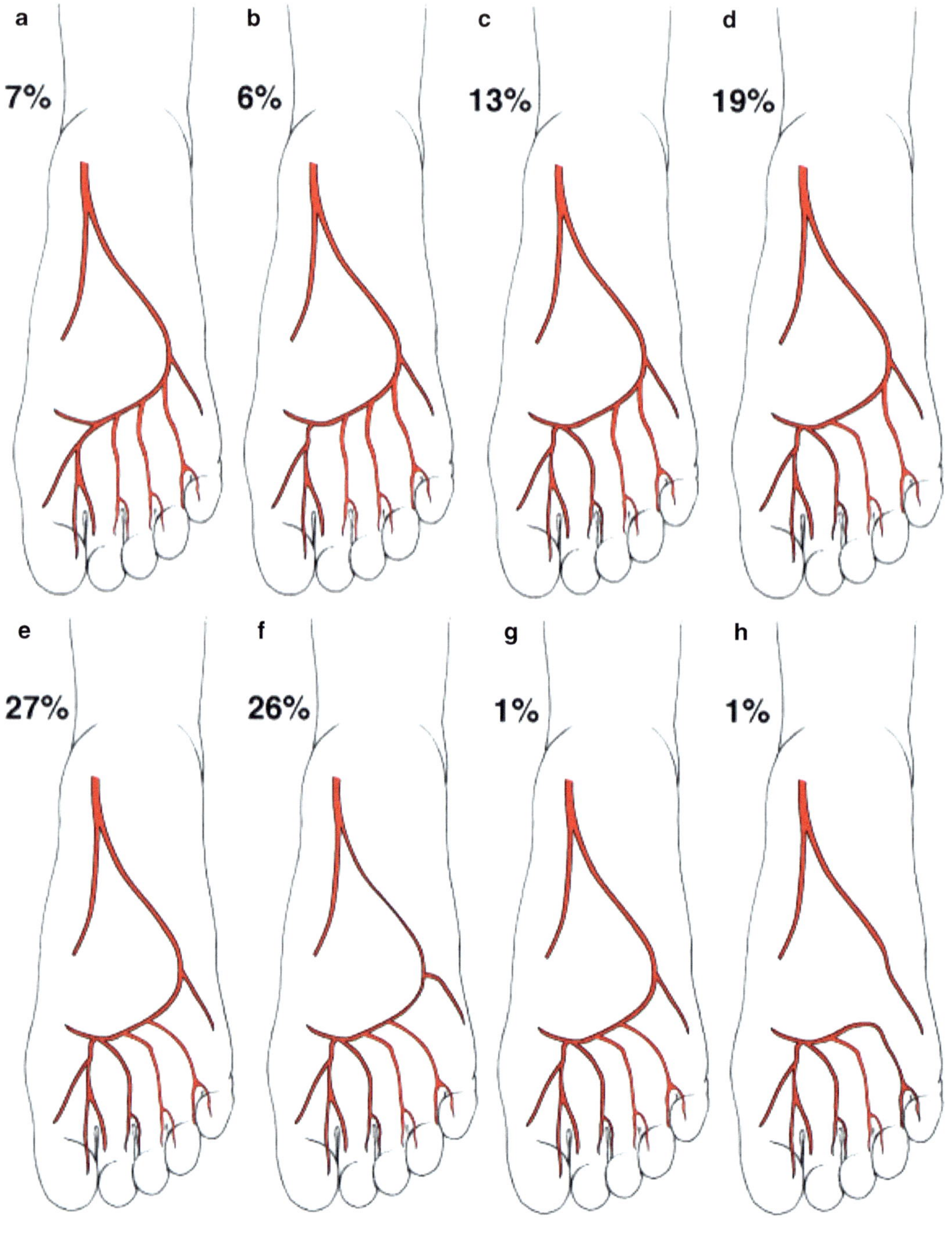

Fig. 114.5 Anatomical variations of the lateral plantar artery and their prevalence

114.3 Histology

The posterior tibial artery is a muscular (distributing) artery. Its tunica intima comprises an inner layer of endothelium surrounded by subendothelial connective tissue and a prominent internal elastic lamina. The intima is surrounded by tunica media, mostly circumferentially arranged smooth muscle, minimal elastin, and a thick external elastic lamina. The tunica adventitia is the outermost wall of organized, dense, irregular connective tissue with multiple layers of elastic tissue.

The posterior tibial vein is a medium vein with less distinct layers. It has a similar intima media that contains few layers of smooth muscle and outer adventitia.

114.4 Variations

Variations of the popliteal branching pattern arise from various embryological abnormalities. In 1985, Lippert and Pabst created the first classification system, modified by Kim et al. in 1989, to give us the system most commonly used today (Lippert and Pabst 1985; Kim et al. 1989). Kim et al. describe three types of branching variations and various subtypes:

Type I. Normal level of branching proximal to the lower border of the popliteus muscle (96.1% in 27 studies) (Tomaszeuski et al. 2017).
  A. The anterior tibial artery is the first branch, followed by the tibiofibular artery, which bifurcates into the fibular and posterior tibial arteries (92.6%).
  B. The anterior, posterior, and fibular arteries arise within 0.5 cm of each other without a tibiofibular trunk (2.4%).
  C. The posterior tibial is the first branch, followed by the tibiofibular trunk, which bifurcates into the fibular and anterior tibial arteries (1.1%).

Type II. High division of the popliteal artery (3.9% in 27 studies) (Tomaszeuski et al. 2017).
  A. The anterior tibial artery begins at or above the knee joint (2.5%).
    (a) Normal course of proximal anterior tibial (2.2%).
    (b) Medial initial curve in anterior tibial (0.3%).
  B. The posterior tibial begins at or above the knee joint. Common trunk for fibular and anterior tibial (1.2%).
  C. The fibular arises at or above the knee joint. Common trunk for anterior and posterior tibial (0.2%).

Type III. Hypoplastic or aplastic branching with altered distal supply (3.0% in 17 studies) (Tomaszeuski et al. 2017).
  A. Hypoplastic-aplastic posterior tibial. Distal posterior tibial replaced by fibular (1.7%).
  B. Hypoplastic-aplastic posterior and anterior tibial. Dorsalis pedis replaced by fibular (1.2%).
  C. Hypoplastic-aplastic posterior and anterior tibial. Posterior tibial and dorsalis pedis replaced by fibular. Termed the Peronea Arteria Magna (0.1%).

Anand et al. and Longo et al. have each suggested a modification of the Kim classification, adding another class (IV) or subclass (IIID) that includes aberrant or fibular artery insufficiency (Anand et al. 2006; Longo et al. 2014). In a unique case study by Slaba et al., the anterior tibial artery had a high origin in the femorotibial space and a tibiofibular trunk greater than 5 cm with early bifurcation of the fibular artery into the anterior and posterior branches (Slaba et al. 2007).

Variations in the origin and course of the posterior tibial artery have been well documented.

It can be hypoplastic, absent, or distally replaced by the fibular artery. If it is absent or hypoplastic, the plantar arteries arise from the fibular artery via compensatory hyperplasia (Portilho et al. 2016). The prevalence of this variation is roughly 0–5.1% (Chmielewski et al. 2016; Adachi 1928; Day and Orme 2006; Kil and Jung 2009). It can arise proximal to the anterior tibial artery branch or from a common trunk, pierce the interosseus membrane and replace or join the anterior tibial artery, arise above the tibial plateau at the knee joint, branch directly from the popliteal, or supply the common digital arteries (Kil and Jung 2009; Day and Orme 2006; Kim et al. 1989; Sanders and Alston 1986; Karaolanis et al. 2015; Bergman et al. 1988).

In the area between the calcaneus and medial malleolus of the ankle, a system for classifying the relationship of the posterior tibial artery to the tibial nerve has been described. In type I, the posterior tibial artery is medial to the tibial nerve; in type II, the posterior tibial artery is anterior to it;

in type III, the artery is lateral to it; and in type IV, the artery is between the medial and lateral plantar nerves (Kim et al. 2015).

114.5 Imaging

The posterior tibial artery and vessels of the lower extremity have been evaluated by contrast arteriography, specifically digital subtraction angiography (DSA), which has been considered the gold standard for many years (Pomposelli 2010). However, more recently, magnetic resonance angiography (MRA) and computerized tomographic angiography (CTA) have become viable alternatives, being less invasive and with image quality similar to that of DSA (Pomposelli 2010). All three imaging modalities have advantages and disadvantages regarding the patient's clinical presentation and comorbidities (Pomposelli 2010).

Vascular ultrasound in the lower extremity is a noninvasive, diagnostic modality using sound waves to reveal blood flow through vessels and is a mainstay of vascular imaging, including B-mode, continuous wave, pulsed wave Doppler, and duplex ultrasound (Varaki et al. 2018).

In addition to vascular ultrasound, the ankle-brachial index (ABI) is a simple, noninvasive method for assessing arterial perfusion. It has a high sensitivity and specificity but is not highly accurate in specific patient populations, such as the elderly, diabetics, and patients with renal disease due to largely incompressible vessels (Varaki et al. 2018). Other similar methods are segmental blood pressure measurements used to locate stenosis or occlusion and the toe brachial index (Varaki et al. 2018).

114.6 Pathology

Atherosclerosis is often responsible for occlusion or stenosis within the posterior tibial artery. Risk factors include high cholesterol, hypertension, smoking, diabetes, and obesity. The pathogenesis of atherosclerosis has been well established. The endothelial wall of the artery undergoes injury, followed by the overproduction of cellular mediators, leading to the formation of a fibrous plaque. This plaque then restricts blood flow, with consequences ranging from intermittent claudication to major tissue loss.

Aneurysms are divided into true aneurysms, dilating all three vessel wall layers, and pseudoaneurysms, which typically involve the tunica media and adventitia. Of the two, pseudoaneurysms are more common in the posterior tibial artery. They can be attributable to trauma, iatrogenic causes, connective tissue pathology, atherosclerosis, infection, fibromuscular dysplasia, Behcet disease, or hemophilia, or they can occur spontaneously (Liu et al. 2020; Sagar and Button 2014). The etiologies of true aneurysms include degenerative changes, mycotic infection, polyarteritis nodosa, possible syphilitic infection, or can be unknown (Sagar and Button 2014).

Thromboangiitis obliterans is typically a vasculitis that affects the small and medium vessels of the upper and lower extremities. It is a disease primarily found in young smokers, characterized by peripheral artery occlusion in the extremities. Histopathologically, this disease is defined by three separate phases: acute, intermediate, and end. During the acute phase, the lesions include an occlusive, highly cellular, inflammatory thrombus with the possible presence of polymorphonuclear leukocytes, microabscesses, and multinucleated giant cells (Arkkila 2016). The intermediate stage of the disease is marked by a progression to loosely organized thrombi in arteries and veins (Arkkila 2016). Finally, those thrombi become more organized, and there is fibrosis in the end-stage vessels (Arkkila 2016).

Eosinophilic vasculitis is characterized by eosinophilic infiltration with intra- and extravascular granuloma formation associated with a zone of centralized eosinophilic necrosis, histiocytes, multinucleated giant cells, and neutrophils (Khoury et al. 2014).

Vasculitis associated with antiphospholipid syndrome is usually a parallel phenomenon with no causal relationship (Lally and Sammaritano 2015). Typically, these two conditions occur in patients with systemic lupus erythematosus (SLE) or other connective tissue disease, with

80% of cases involving small vessels (Lally and Sammaritano 2015). Owing to the inflammatory nature of the disease, histopathology reveals endothelial cell activation and production of inflammatory cytokines, with the recruitment of neutrophils and monocytes to the vascular lumen (Lally and Sammaritano 2015).

114.7 Clinical and Surgical Significance

The posterior tibial artery can undergo atherosclerotic changes. These changes lead to peripheral arterial disease, in which the arterial lumen becomes occluded, decreasing blood flow and poor perfusion. Treatment is initially aimed at lifestyle modifications and medications, but surgical intervention is sometimes needed, such as angioplasty, thrombus removal, and bypass surgery.

No standardized treatment of true posterior tibial artery aneurysms has been established because cases are rare; however, treatment options should be based on size, symptoms, and the presence of a thrombus. For small and asymptomatic aneurysms with no thrombus, conservative therapy with observation can be offered. However, Sagar and Button (2014) recommend that all infrapopliteal aneurysms be surgically treated owing to the risk of embolization, thrombosis, and rupture (Sagar and Button 2014). Surgical treatment can include aneurysm excision with interposition of vein graft, ligation of the posterior tibial artery, and primary end-to-end anastomosis of this artery (Katz et al. 1992; Jenyo 1987; Tshomba et al. 2006; Sagar and Button 2014).

As with true aneurysms, treatment of pseudoaneurysms is based on clinical presentation, size, symptoms, and comorbidities. Small pseudoaneurysms can simply be observed; however, larger or symptomatic masses need intervention. Currently, endovascular intervention is preferred over open surgery for minimizing complications such as bleeding, hematoma, and arteriovenous fistula formation (Liu et al. 2020). Open surgery can be used for emergency conditions such as a ruptured or massive pseudoaneurysm if soft tissue decompression is necessary or if other treatment options have failed (Liu et al. 2020). Regardless of the mode of access, treatment is either repair or ligation with end-to-end anastomosis or venous autograft bypass (Liu et al. 2020).

Deep posterior chronic exertional compartment syndrome occurs in athletes who perform repetitive motions such as running, hockey, soccer, or cycling. Most often caused by exercise, the pain is typically found deep in the patient's calf muscle and is relieved by rest. Cramps, altered skin sensation, or restless legs at night can also accompany this pain (Winkes et al. 2020). Conservative treatment includes rest, physical therapy, orthotics, icing, chiropractic, acupuncture, and transcutaneous nerve stimulation, with varied results. Operative treatment consists of fasciotomy (Winkes et al. 2020). There have been cases in which patients with absence or deficiency of the posterior arterial system mimic chronic exertional compartment syndrome, so this must be in the differential diagnosis (Lavery et al. 2016).

Compartment syndrome of the deep and posterior compartments is rare for the four leg compartments. Fractures account for 65% of cases and are the most common cause (Taylor et al. 2012). Other causes include tight casts or dressings, extravasation of intravenous infusion, circumferential burns, postischemic swelling, crush injuries, bleeding disorders, and arterial injury (Mabvuure et al. 2012). Clinical presentation typically includes pain on passive stretch (the most sensitive finding), pallor, paresthesia, pulselessness, poikilothermia, and paralysis. Resting compartment pressure greater than 30 mmHg or "Delta P" less than 300 mmHg confirms the diagnosis. The gold standard of time to treatment is within 6 h of ischemia, after which necrosis occurs. Treatment of acute compartment syndrome is a medical emergency and requires fasciotomy.

Treatment of posterior arterial vasculitis can include both medical and surgical options. Medical therapy is focused on treating the underlying cause of the disease; surgical interventions can consist of arterial revascularization and bypass surgery to reduce the symptoms of ischemia and prevent further complications (Arkkila 2016).

References

Adachi B (1928) Das Arteriensystem Der Japaner, 2nd edn, Kyoto

Anand R, Mourouzis C, Wilbourn M, Tilley E, Brennan PA (2006) An unreported variation of the course of fibular artery during fibula free flap harvest. Br J Oral Maxillofac Surg 45(7):588–589

Arkkila P (2016) Thromboangiitis obliterans (Buerger's disease). Orphanet J Rare Dis 1:14. https://doi.org/10.1186/1750-1172-1-14

Bergman R, Thompson S, Afifi A, Saadeh F (1988) Compendium of human anatomic variation: text, atlas, and world literature. Urban & Schwarzenberg, Baltimore

Chmielewski P, Warchol L, Mroz I, Malczak M, Jaworek J, Mizia E, Walocha E et al (2016) Blood vessels of the shin – posterior tibial artery – anatomy – own studies and review of the literature. Folia Med Cracov 56(3):5–9

Day CP, Orme R (2006) Popliteal artery branching patterns – an angiographic study. Clin Radiol 61:696–699

Jenyo MS (1987) Silent posteriortibial artery aneurysm. Report ofacase and review of literature. J Cardiovasc Surg 28(4):456–459

Karaolanis G, Galyfos G, Karanikola E, Palla V, Filis K (2015) Absence of clinical and hemodynamic consequences due to posterior tibial artery congenital aplasia. Case Rep Vasc Med 82:1094. https://doi.org/10.1155/2015/821094

Katz SG, Kohl RD, Razack N (1992) Bilateral infrapoplitealartery aneurysms. Ann Vasc Surg 6(2):168–170. https://doi.org/10.1007/BF02042740

Khoury P, Grayson P, Klion A (2014) Eosinophils in vasculitis: characteristic and roles in pathogenesis. Nat Rev Rheumatol 10(8):474–483

Kil SW, Jung GS (2009) Anatomical variations of the popliteal artery and its tibial branches: analysis in 1242 extremities. Cardiovasc Intervent Radiol 2:135–139

Kim D, Orron DE, Skillman JJ (1989) Surgical significance of popliteal arterial variants. A unified angiographic classification. Ann Surg 210:776–781

Kim D, Kim Y, Han S (2015) Topography of human ankle joint: focused on posterior tibial artery and tibial nerve. Anat Cell Biol 48:130–137

Lally L, Sammaritano L (2015) Vasculitis in antiphospholipid syndrome. Rheum Dis Clin N Am 41:109–123

Lavery KP, Parcells BW, Hosea T (2016) Posterior tibial arterial system deficiency mimicking chronic exertional compartment syndrome: a case report. J Bone Joint Surg Case Connect 6(3):e72

Lippert H, Pabst R (1985) Arterial variations in man: classification and frequency. J.F. Bergmann-Verlag, Munchen

Liu K, Mu L, Yan S, Ma R, Liu B (2020) Spontaneous pseudoaneurysm of the posterior tibial artery successfully treated by open surgery: a case report and review of the literature. Medicine 99(31):e21523

Longo B, Sorotos M, Nicolotti M, Santanelli di Pompeo F (2014) Retrospective analysis of incidence of fibular artery hypoplasia in 101 free fibula transfers and new classifcation of popliteal branch. Injury 45:394–398

Mabvuure NT, Malahia M, Hindocha S, Khan W, Juma A (2012) Acute compartment syndrome of the limbs: current concepts and management. Open Orthop J 6:535–534

Pomposelli F (2010) Arterial imaging in patients with lower extremity ischemia and Diabetes mellitus. J Vasc Surg 52:81S–91S

Portilho P, de Oliveira Leite TF, Cardoso R, Pires LA, Silva JG, Chagas C (2016) Clinical features of a rare anatomical variation of the posterior tibial and fibular arteries. J Vasc Bras 15(3):234–238

Sagar J, Button M (2014) Posterior tibial artery aneurysm: a case report with review of literature. BMC Surg 14:37

Sanders R, Alston G (1986) Variations and anomalies of the popliteal and tibial arteries. Am J Surg 152(5):531–534. https://doi.org/10.1016/0002-9610(86)90222-9

Senior HD (1919) The development of teh arteries of the human lower extremity. Am J Anat 25(1):54–95. https://doi.org/10.1002/aja.1000250105

Slaba S, Khalil SA, Younan T, Nassar-Salaba J, Khoury S, Kheir C (2007) Unusual variation of popliteal arterial branches: 4 axes by early division of the fibular artery. J Mal Vasc 32(4–5):212–215

Taylor RM, Sullivan MP, Mehta S (2012) Acute compartment syndrome: obtaining diagnosis, providing treatment, and minimizing medicolegal risk. Curr Rev Musculoskelet Med 5(3):206–213

Tomaszeuski KA, Popieluszko P, Graves MJ, Pekala PA, Henry BM, Roy J, Hsieh WC, Walocha JA (2017) The evidence-based surgical anatomy of the popliteal artery and the variations in its branching patterns. J Vasc Surg 65(2):521–529

Tshomba Y, Papa M, Marone EM, Kahlberg A, Rizzo N, Chiesa R (2006) A true posterior tibial artery aneurysm – a case report. Vasc Endovasc Surg 40(3):243–249. https://doi.org/10.1177/153857440604000311

Varaki ES, Gargiulo GD, Penkala S, Breen PP (2018) Peripheral vascular disease assessment in the lower limb: a review of current and emerging non-invasive diagnostic methods. Biomed Eng Online 17:61

Winkes M, van Eerten P, Scheltinga M (2020) Deep posterior chronic exertional compartment syndrome as a cause of leg pain. Unfallchirurg 123:3–7. https://doi.org/10.1007/s00113-019-0665-1

115 Fibular Artery

Kemdi Okafor

115.1 Introduction

The fibular (peroneal) artery (Fig. 115.1) arises from the tibiofibular trunk and supplies the muscles of the lateral compartment of the leg (peroneus brevis and peroneus longus) (Susan 2015; Eovaldi and Sharma 2019). It courses in between the tibia and the fibula along the interosseous membrane (Susan 2015; Eovaldi and Sharma 2019).

115.2 Embryology

At the 14 mm embryo stage (42–45 days gestational age), the dense vascular supply of the knee starts to develop. Thereby, complex vascular remodeling takes place, which shifts the deep popliteal segment of the axial artery into the final position dorsal to the popliteus muscle (Eovaldi and Sharma 2019). Distal to the knee, the interosseous segment of the axial artery develops a perforating artery into the extensor compartment of the leg (ramus perforans cruris), the anlage of the anterior tibial artery. Simultaneously with the regression of the interosseous artery, new vessels arise by angiogenic mechanisms and form the anlagen of the posterior tibial and fibular arteries (Eovaldi and Sharma 2019).

The sciatic artery is a branch of the internal iliac artery and is responsible for supplying blood to the early bundle of the lower limb. In the eighth week of fetal development, the femoral artery, a branch of the external iliac artery, connects to the sciatic artery and becomes the main vessel supplying blood to the lower limbs. The sciatic artery regresses proximally and becomes the inferior gluteal artery in adults. The primitive middle and distal segments of the sciatic artery persist to form the definitive popliteal and fibularis arteries. The anterior tibial artery arises as a branch of the popliteal artery. In contrast, the posterior tibial artery is formed by an anastomosis between the early distal femoral artery and the popliteal artery. Complete vascularization of the lower limb is established by the end of the 12th week of pregnancy (Mauro et al. 1988).

115.3 Anatomy/Topography

The fibular artery arises from the posterior tibial artery, usually about 2.5 cm below the popliteus muscle, but it may occasionally occur up to 8 cm distal to the popliteus. From its origin point, the fibular artery descends obliquely along the medial crest of the fibula. It usually courses within the flexor hallucis longus muscle or in a fibrous canal formed between the tibialis poste-

K. Okafor (✉)
Tulane University School of Medicine,
New Orleans, LA, USA
e-mail: kokafor@tulane.edu

S. J. Bordes, Jr. et al. (eds.), *The Clinical Anatomy of the Vascular System*,
https://doi.org/10.1007/978-3-031-78326-5_115

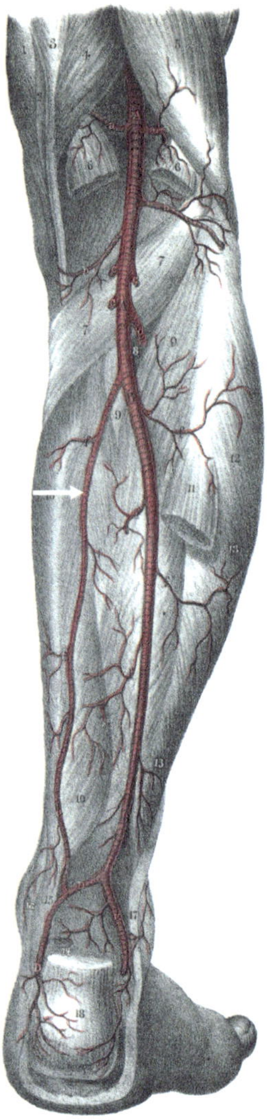

Fig. 115.1 Drawing of the fibular (arrow) and posterior tibial arteries. (After Quain and Wilson)

rior and flexor hallucis longus muscles. Proximally, the fibular artery is covered by the soleus muscle and the transverse intermuscular septum, while the flexor hallucis longus muscle covers its distal portion. As it reaches the tibiofibular syndesmosis, the fibular artery ends by dividing into its terminal branches. Along its course, the fibular artery gives off several branches: muscular, nutrient, perforating, communicating, and calcaneal branches. Muscular branches arise as several short branches that supply the muscles of the deep posterior and lateral compartments of the leg. These include the popliteus, soleus, tibialis posterior, and flexor hallucis longus muscles. The nutrient artery of fibula arises from the fibular artery roughly 7 cm after its origin and pierces the fibula at its medial crest, near the middle third of the bone shaft, providing it with endosteal supply. Perforating branches arise as several branches, usually around 5–7. One branch is usually of a larger caliber and pierces the distal interosseous membrane proximal to the lateral malleolus to enter the anterior compartment of the leg. Here, it anastomoses with the anterior lateral malleolar branch of the anterior tibial artery. The perforating branch then passes to the dorsum of the foot where it supplies the tarsus and anastomoses with the lateral tarsal artery. Other perforating branches pierce the deep fascia and supply the skin of the posterolateral leg. Posterior lateral malleolar branch surrounds the lateral malleolus and anastomoses with other malleolar arteries, contributing to the lateral malleolar vascular network. The communicating branch anastomoses with a communicating branch of the posterior tibial artery about 5 cm proximal to the ankle. Calcaneal branches arise as terminal branches of the fibular artery and pass to the lateral side of the calcaneus to anastomose with the calcaneal branches of the posterior tibial artery. They supply the skin over the lateral side of the calcaneal tendon and the calcaneus (Moore et al. 2014; Standring 2016; Tubbs et al. 2016).

115.4 Variations

The popliteal artery (PA) arises as a continuation of the femoral artery after it passes through the adductor hiatus. The PA runs in the popliteal fossa, and it normally splits at the lower border of the popliteus muscle in the anterior tibial artery (ATA) and the common junction of the posterior tibial and fibular arteries. The ATA passes into the anterior compartment (extensors) of the leg, leaving the posterior compartment (flexors) between the tibia and the fibula through the upper opening of the interosseous membrane. The common junction of the posterior tibial and fibular arteries, however, maintains its position in the posterior compartment and splits into the posterior tibial artery (PTA) and the fibular artery (FA). The PA provides blood not only for the knee, including the joint capsule and ligaments, but also for areas of the leg and foot (Bergman et al. 2015).

Anatomical variations of the branching patterns of the ATA, PTA, and FA were evaluated in 100 lower limbs (47 female and 53 male). Four branching patterns of arteries were identified in the popliteal region (Olewnik et al. 2019).

Type I. The PA divides into the ATA and the common junction of the posterior tibial and fibular arteries; the latter then splits into the FA and the PTA. This was the most common type observed in 72 cadavers (72%). This type was subdivided into subtype a (in which the ATA had a larger diameter, 42 cases) and subtype b (in which the common junction of the posterior tibial and fibular arteries had a larger diameter, 30 cases). In subtype b, the diameters of the PTA and FA were significantly larger than in subtype a: 4.00 ± 1.02 vs. 3.22 ± 0.84 (p = 0.0001) for the PTA and 3.80 ± 1.19 vs. 2.73 ± 0.82 (p = 0.0001) for the FA (Olewnik et al. 2019).

Type II. The PA divides into the ATA and PTA, and then the ATA artery gives rise to the FA. Type II was present in eight cases (8%) (Olewnik et al. 2019).

Type III. The PA trifurcates into the ATA, PTA, and FA. Type III was visualized in 12 specimens (12%). The presence of a type III PA may be particularly problematic in this sense. In the case of a hypoplastic or aplastic PTA, the FA exhibits compensatory hyperplasia and supplies the vascular bed of the PTA with blood. In this case, the FA of the foot gives rise to the lateral plantar artery, but the medial plantar artery is usually absent (Zheng et al. 2016; Olewnik et al. 2019).

Type IV. The PA divides into the ATA and FA. The PTA is aplastic. This type was seen in eight specimens (8%) (Olewnik et al. 2019).

The present study shows that the diameter of the PTA was significantly smaller in type I than in the other types (II, III), and the diameter of the FA in type I was significantly smaller than in type III (Olewnik et al. 2019).

Other Variants

The peroneal magnus artery is a vascular variant of the leg. In patients with a peroneal magnus artery, the anterior and posterior tibial arteries are smaller than normal, and the peroneal magnus artery is larger than normal fibular arteries. The peroneal magnus artery provides most blood supply to the leg and foot. A peroneal magnus artery is seen in up to 5% of the population (Gaillard et al. 2018).

115.5 Pathology

In plastic surgery, graft harvest can lead to increased complications at the donor site. Foot ischemia is possible following harvest surgery or with atherosclerotic disease or trauma that would otherwise lead to ischemia in the presence of normal 3-vessel runoff. For preharvest imaging, MR angiography is performed as distal pulses may be normal in peroneal artery dominance (Rosson and Singh 2005; Hacking and Ali 2022).

115.6 Clinical and Surgical Significances

Peripheral arterial disease (PAD), caused by atherosclerosis, is the primary clinical issue associated with the fibular artery. Mild cases of PAD correlate with mild leg pain with exertion. Moderate cases of PAD demonstrate severe pain with walking, and severe PAD presents with leg pain at rest (De Luccia et al. 2014).

115.7 Complications

The fibular artery is injured less frequently than the popliteal and tibial arteries, because of its anatomical relationships. Pseudoaneurysm of the fibular artery due to a blunt injury is a rare complication. Early diagnosis can be confirmed by angiography and when less-invasive procedures cannot be applied, a posterior approach is an attractive surgical option in the management of proximal fibular artery traumatic lesions (Tzilalis et al. 2006).

References

Bergman RA, Afifi AK, Miyauchi R (2015) Anatomy atlases: illustrated encyclopedia of human anatomic variation – anatomical variation, radiology anatomy, anatomy atlas. http://www.anatomyatlases.org/AnatomicVariants/AnatomyHP.shtml

De Luccia N, Queiroz AB, Mulatti GC, Ferreira Espírito Santo FR, Sassaki Neto PI, Schneidwind KD (2014) Lateral approach to the peroneal artery without resection of the fibula for lower limb revascularization. J Vasc Surg 59(3):857–859

Eovaldi BJ, Sharma S (2019) Anatomy, bony pelvis and lower limb, peroneal artery. In: StatPearls [Internet]. StatPearls Publishing

Gaillard J, Bourcheix LM, Masquelet AC (2018 Aug) Perforators of the fibular artery and suprafascial network. Surg Radiol Anat 40(8):927–933

Hacking C, Ali A (2022) Peroneal magnus artery. Reference article, Radiopaedia.org. Accessed 26 Jan 2022. https://doi.org/10.53347/rID-45842

Mauro M, Jaques P, Moore M (1988) The popliteal artery and its branches: embryologic basis of normal and variant anatomy. Am J Roentgenol 150:435–437. https://doi.org/10.2214/ajr.150.2.435

Moore KL, Dalley AF, Agur AMR (2014) Clinically oriented anatomy, 7th edn. Lippincott Williams & Wilkins, Philadelphia

Olewnik Ł, Łabętowicz P, Podgórski M et al (2019) Variations in terminal branches of the popliteal artery: cadaveric study. Surg Radiol Anat 41:1473–1482. https://doi.org/10.1007/s00276-019-02262-3

Rosson GD, Singh NK (2005) Devascularizing complications of free fibula harvest: peronea arteria magna. J Reconstr Microsurg 21(08):533–538. https://doi.org/10.1055/s-2005-922432

Standring S (2016) Gray's anatomy, 41st edn. Elsevier Churchill Livingstone, Edinburgh

Susan S (2015) Gray's anatomy: the anatomical basis of clinical practice. Elsevier

Tubbs RS, Shoja MM, Loukas M, Bergman RA (2016) Bergman's comprehensive encyclopedia of human anatomic variation. Wiley Blackwell, Hoboken

Tzilalis VD, Karliaftis K, Tsakiris A, Lazarides MK (2006 Sep-Oct) Peroneal artery pseudo-aneurysm following blunt injury. Acta Chir Belg 106(5):622–624. https://doi.org/10.1080/00015458.2006.11679968

Zheng M, Chen C, Qiu Q, Wu C (2016) Ultrasound in diagnosis of anatomical variation of anterior and posterior tibial arteries. Med Ultrason 18:64–69

116 Arteries of the Foot

Patrick J. McCarty

116.1 Anatomy

The anterior and posterior tibial arteries supply blood to the foot and arise from the popliteal artery and course within their respective leg compartments. The anterior tibial artery continues distally to the ankle as the dorsalis pedis artery (Fig. 116.1), which then gives rise to the medial and lateral tarsal, arcuate, and dorsal metatarsal arteries before passing into the proximal first intermetatarsal space. The posterior tibial artery terminates as two main branches, the smaller medial plantar artery and the larger lateral plantar artery (Fig. 116.2). The medial plantar artery courses along the medial aspect of the foot toward the base of the first metatarsal. It gives rise to muscular branches supplying the abductor hallucis and flexor digitorum brevis and a superficial branch before coursing toward the medial aspect of the great toe and anastomosing with a branch from the first plantar metatarsal artery. The lateral plantar artery courses laterally toward the base of the fifth metatarsal. Turning medially along with the deep branch of the lateral plantar nerve, it gives rise to a digital branch to the fifth toe. It then continues toward the space between the first and second metatarsals to join the dorsalis pedis and form the deep plantar arch. This

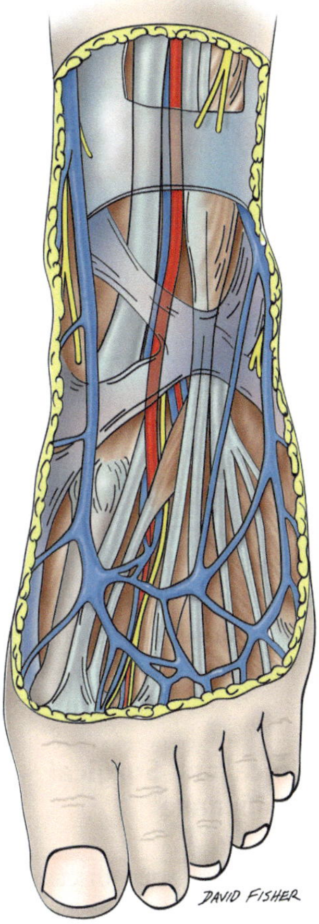

Fig. 116.1 Schematic drawing of the dorsum of the foot illustrating the dorsalis pedis, a continuation of the anterior tibial artery (red) on to the foot

P. J. McCarty (✉)
Tulane University School of Medicine,
New Orleans, LA, USA
e-mail: pmccarty2@tulane.edu

S. J. Bordes, Jr. et al. (eds.), *The Clinical Anatomy of the Vascular System*,
https://doi.org/10.1007/978-3-031-78326-5_116

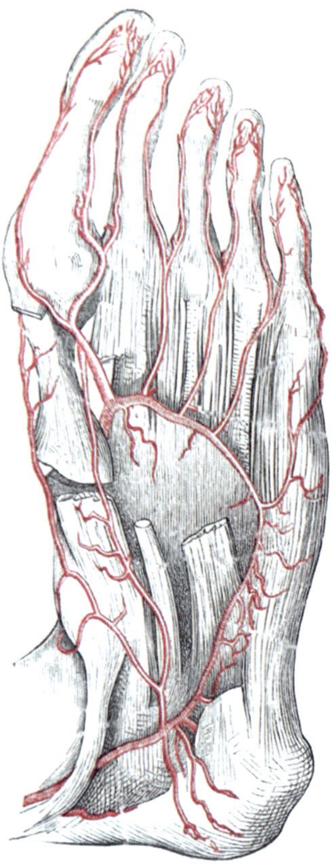

Fig. 116.2 Schematic drawing of the dorsum of the foot illustrating the dorsalis pedis, a continuation of the anterior tibial artery (red) on to the foot

arch is situated superficial to the interosseous muscles but deep to the adductor hallucis muscle and supplies much of the skin, fascia, and muscles in the sole through perforating and plantar metatarsal branches.

116.2 Clinical and Surgical Significance

Lower limb peripheral artery disease (PAD) is the third leading cause of atherosclerotic morbidity behind coronary heart disease and stroke (Criqui et al. 2021). PAD is progressive. It can lead to significant functional limitations and impaired quality of life and contribute to the development of additional complications and comorbidities. Critical limb ischemia (CLI), characterized by ulceration and ischemic pain at rest, affects up to 11% of patients with PAD and is associated with significant morbidity and mortality (Menard et al. 2016). When stratified by the level of cardiovascular disease (CVD) risk factors, PAD is independently related to future cardiovascular events and mortality (Criqui et al. 1992) and has been shown to predict future CVD events (Fowkes et al. 2008; Criqui et al. 2010).

Lower limb PAD is atherosclerotic disease and obstruction of the peripheral arterial vasculature from the aortoiliac to the pedal arteries (Criqui et al. 2021). PAD is the manifestation of an inflammatory atherogenic cascade that begins with endothelial cell injury and dysfunction and leads to leukocyte infiltration, lipid particle entrapment, oxidation, and ultimately, atheromatous plaque formation.

Individuals with diabetes mellitus (DM) are at increased risk of developing PAD, and those with both PAD and DM have worse outcomes, including increased risk for CLI, amputation, and mortality (Ding et al. 2018; Jude et al. 2001). DM's characteristic inflammatory and metabolic derangements have numerous harmful effects on the peripheral vasculature. The pathogenesis of PAD follows the same inflammatory atherogenic cascade irrespective of the presence of DM. However, advanced glycation end products (AGEs) accelerate this process in patients with DM. AGEs induce pro-inflammatory processes within the vascular compartment through direct and receptor-mediated effects, including releasing cytokines and generating reactive oxygen species (Singh et al. 2001). The extent of PAD is typically more severe in patients with DM, as indicated by more extensive, multilevel, and multivessel lesions that characteristically affect the infrapopliteal/tibial arteries (Hinchliffe et al. 2016; Graziani et al. 2007). In addition to more severe lesions, vascular compromise against the background of a chronic pro-inflammatory environment also contributes to diminished wound healing capabilities and increased susceptibility to infection through decreased immune cell function and delivery to the site of infection (Pastar et al. 2018; Hinchliffe et al. 2016). The severity of PAD in patients with diabetic foot ulcers increases the risk of first major amputation; mild, moderate, and severe PAD make the risk 20-fold, 34-fold, and 62-fold higher, respectively, than no PAD (Morbach et al. 2012). Thus, the presence and severity of PAD influence the prognosis of diabetes-related foot ulcers, but there is also an important relationship between PAD and the development of such ulcers (Hinchliffe et al. 2020; Pastar et al. 2018).

Microvascular dysfunction has also been implicated in the progression of PAD, particularly in individuals with DM, where cutaneous microvascular dysfunction can contribute to the development of necrosis in diabetic foot ulcers (Pastar et al. 2018; Roustit et al. 2018). Recent studies have also shown that diabetic retinopathy is more strongly associated with the incidence of PAD and CLI than coronary heart disease and stroke (Yang et al. 2020; Beckman et al. 2019).

Current guidelines suggest the best medical therapy and lifestyle modification as the first-line treatment for patients with PAD and claudication (Gerhard-Herman et al. 2017; Conte and Pomposelli 2015). In patients with lifestyle-limiting claudication and CLI, the standard of care is surgical revascularization to improve limb perfusion. Although the decision for open versus endovascular revascularization depends significantly on several physician and patient factors (Farber et al. 2022), endovascular techniques have supplanted open bypass procedures in the United States (Egorova et al. 2010; Menard et al. 2016; Farber et al. 2022). Endovascular techniques, which include balloon angioplasty, stent placement, and atherectomy, have been associated with less periprocedural morbidity and mortality than surgical bypass. However, robust comparison between these management strategies and their outcomes remains an active area of inquiry, particularly in individuals who are candidates for either approach (Menard et al. 2016; Criqui et al. 2021). The prospective Best Endovascular versus Best Surgical Therapy in Patients with CLTI (BEST-CLI) trial was undertaken to clarify the comparative effectiveness of these two approaches. The results highlighted the value of a good-quality bypass vein/conduit. In this group, the surgical bypass was associated with a 32% lower risk of major adverse limb events or death than endovascular revascularization (Farber et al. 2022). If there was no adequate bypass conduit, the results did not differ significantly between the two initial revascularization approaches (Farber et al. 2022). The results also revealed a higher incidence of significant reintervention in the endovascular cohort than in the surgical bypass cohort (23.5% vs. 9.2%) (Farber et al. 2022). Secondary revascularization procedures have been associated with worse outcomes (Jones et al. 2013).

References

Beckman JA, Duncan MS, Damrauer SM, Wells QS, Barnett JV, Wasserman DH, Bedimo RJ, Butt AA, Marconi VC, Sico JJ et al (2019) Microvascular disease, peripheral artery disease, and amputation. Circulation 140:449–458. https://doi.org/10.1161/CIRCULATIONAHA.119.040672

Conte MS, Pomposelli FB (2015) Society for Vascular Surgery practice guidelines for atherosclerotic occlusive disease of the lower extremities management of asymptomatic disease and claudication. Introduction. J Vasc Surg 61(3 suppl):1S. https://doi.org/10.1016/j.jvs.2014.12.006

Criqui MH, Langer RD, Fronek A, Feigelson HS, Klauber MR, McCann TJ, Browner D (1992) Mortality over a period of 10 years in patients with peripheral arterial disease. N Engl J Med 326:381–386. https://doi.org/10.1056/NEJM199202063260605

Criqui MH, McClelland RL, McDermott MM, Allison MA, Blumenthal RS, Aboyans V, Ix JH, Burke GL, Liu K, Shea S (2010) The ankle-brachial index and incident cardiovascular events in the MESA (Multi-Ethnic Study of Atherosclerosis). J Am Coll Cardiol 56:1506–1512. https://doi.org/10.1016/j.jacc.2010.04.060

Criqui MH, Matsushita K, Aboyans V, Hess CN, Hicks CW, Kwan TW, Mcdermott MM, Misra S, Ujueta F (2021) Lower extremity peripheral artery disease: contemporary epidemiology, management gaps, and future directions: a scientific statement from the American Heart Association. Circulation 144:e171. https://doi.org/10.1161/cir.0000000000001005

Ding N, Kwak L, Ballew SH, Jaar B, Hoogeveen RC, Ballantyne CM, Sharrett AR, Folsom AR, Heiss G, Salameh M et al (2018) Traditional and nontraditional glycemic markers and risk of peripheral artery disease: the Atherosclerosis Risk in Communities (ARIC) study. Atherosclerosis 274:86–93. https://doi.org/10.1016/j.atherosclerosis.2018.04.042

Egorova NN, Guillerme S, Gelijns A, Morrissey N, Dayal R, McKinsey JF, Nowygrod R (2010) An analysis of the outcomes of a decade of experience with lower extremity revascularization including limb salvage, lengths of stay, and safety. J Vasc

Surg 51(878–885):885.e1. https://doi.org/10.1016/j.jvs.2009.10.102

Farber A, Menard MT, Conte MS, Kaufman JA, Powell RJ, Choudhry NK, Hamza TH, Assmann SF, Creager MA, Cziraky MJ, Dake MD, Jaff MR, Reid D, Siami FS, Sopko G, White CJ, Van Over M, Strong MB, Villarreal MF, Mckean M, Azene E, Azarbal A, Barleben A, Chew DK, Clavijo LC, Douville Y, Findeiss L, Garg N, Gasper W, Giles KA, Goodney PP, Hawkins BM, Herman CR, Kalish JA, Koopmann MC, Laskowski IA, Mena-Hurtado C, Motaganahalli R, Rowe VL, Schanzer A, Schneider PA, Siracuse JJ, Venermo M, Rosenfield K (2022) Surgery or endovascular therapy for chronic limb-threatening ischemia. N Engl J Med 387:2305–2316. https://doi.org/10.1056/nejmoa2207899

Fowkes FG, Murray GD, Butcher I, Heald CL, Lee RJ, Chambless LE, Folsom AR, Hirsch AT, Dramaix M, deBacker G, Wautrecht JC, Kornitzer M, Newman AB, Cushman M, Sutton-Tyrrell K, Fowkes FG, Lee AJ, Price JF, d'Agostino RB, Murabito JM, Norman PE, Jamrozik K, Curb JD, Masaki KH, Rodríguez BL, Dekker JM, Bouter LM, Heine RJ, Nijpels G, Stehouwer CD, Ferrucci L, McDermott MM, Stoffers HE, Hooi JD, Knottnerus JA, Ogren M, Hedblad B, Witteman JC, Breteler MM, Hunink MG, Hofman A, Criqui MH, Langer RD, Fronek A, Hiatt WR, Hamman R, Resnick HE, Guralnik J, McDermott MM (2008) Ankle brachial index combined with Framingham Risk Score to predict cardiovascular events and mortality: a meta-analysis. JAMA 300(2):197–208. https://doi.org/10.1001/jama.300.2.197

Gerhard-Herman MD, Gornik HL, Barrett C, Barshes NR, Corriere MA, Drachman DE, Fleisher LA, Fowkes FG, Hamburg NM, Kinlay S et al (2017) 2016 AHA/ACC guideline on the management of patients with lower extremity peripheral artery disease: a report of the American College of Cardiology/American Heart Association Task Force on Clinical Practice Guidelines. Circulation 135:e726–e779. https://doi.org/10.1161/CIR.0000000000000471

Graziani L, Silvestro A, Bertone V et al (2007) Vascular involvement in diabetic subjects with ischemic foot ulcer: a new morphologic categorization of disease severity. Eur J Vasc Endovasc Surg 33(4):453–460

Hinchliffe RJ, Brownrigg JRW, Andros G, Apelqvist J, Boyko EJ, Fitridge R, Mills JL, Reekers J, Shearman CP, Zierler RE, Schaper NC (2016) Effectiveness of revascularization of the ulcerated foot in patients with diabetes and peripheral artery disease: a systematic review. Diabetes Metab Res Rev 32:136–144

Hinchliffe RJ, Forsythe RO, Apelqvist J, Boyko EJ, Fitridge R, Hong JP, Katsanos K, Mills JL, Nikol S, Reekers J, Venermo M, Zierler RE, Schaper NC (2020) Guidelines on diagnosis, prognosis, and management of peripheral artery disease in patients with foot ulcers and diabetes (IWGDF 2019 update). Diabetes Metab Res Rev 36:e3276. https://doi.org/10.1002/dmrr.3276

Jones DW, Schanzer A, Zhao Y et al (2013) Growing impact of restenosis on the surgical treatment of peripheral arterial disease. J Am Heart Assoc 2(6):e000345. https://doi.org/10.1161/jaha.113.000345

Jude EB, Oyibo SO, Chalmers N, Boulton AJ (2001) Peripheral arterial disease in diabetic and nondiabetic patients: a comparison of severity and outcome. Diabetes Care 24:1433–1437. https://doi.org/10.2337/diacare.24.8.1433

Menard MT, Farber A, Assmann SF, Choudhry NK, Conte MS, Creager MA, Dake MD, Jaff MR, Kaufman JA, Powell RJ, Reid DM, Siami FS, Sopko G, White CJ, Rosenfield K (2016) Design and rationale of the best endovascular versus best surgical therapy for patients with critical limb ischemia (BEST-CLI) trial. J Am Heart Assoc 5(7):e003219. https://doi.org/10.1161/JAHA.116.003219

Morbach S, Furchert H, Gröblinghoff U, Hoffmeier H, Kersten K, Klauke GT, Klemp U, Roden T, Icks A, Haastert B, Rümenapf G, Abbas ZG, Bharara M, Armstrong DG (2012) Long-term prognosis of diabetic foot patients and their limbs: amputation and death over the course of a decade. Diabetes Care 35(10):2021–2027. https://doi.org/10.2337/dc12-0200

Pastar I, Ojeh N, Glinos GD, Stojadinovic O, Tomic-Canic M (2018) Physiology and pathophysiology of wound healing in diabetes, in contemporary diabetes. Contemp Diabetes:109–130. https://doi.org/10.1007/978-3-319-89869-8_7

Roustit M, Loader J, Baltzis D, Zhao W, Veves A (2018) Microvascular changes in the diabetic foot, in: contemporary diabetes. Contemp Diabetes:173–188. https://doi.org/10.1007/978-3-319-89869-8_10

Singh R, Barden A, Mori T et al (2001) Advanced glycation end-products: a review. Diabetologia 44:129–146. https://doi.org/10.1007/s001250051591

Yang C, Kwak L, Ballew SH, Jaar BG, Deal JA, Folsom AR, Heiss G, Sharrett AR, Selvin E, Sabanayagam C et al (2020) Retinal microvascular findings and risk of incident peripheral artery disease: an analysis from the Atherosclerosis Risk in Communities (ARIC) Study. Atherosclerosis 294:62–71. https://doi.org/10.1016/j.atherosclerosis.2019.10.012

117 Veins of the Thigh

Wesley Ho, Aditya Vinjamuri, and Allyson Segall

117.1 Embryology

Current hypotheses date the development of primitive veins of the lower limb bud at the 8–11 mm stage of embryo development (Uhl 2015). These primitive veins include the marginal, fibular, and axial veins.

Following the development of the primitive veins, the sciatic (axial), femoral (preaxial), and postaxial nerves appear and function as angio-guiding nerves (Uhl and Gillot 2007). Schwann cells surrounding them secrete vascular endothelial growth factor (VEGF) to promote the development of venous plexuses around the nerves (Uhl 2015). The venous plexus surrounding the femoral nerve coalesces to form the femoral vein and great saphenous vein (GSV), and that surrounding the postaxial nerve forms the small saphenous vein (SSV) (Uhl and Gillot 2007). The primitive axial vein continues to grow and serves as the main pathway for venous drainage in the lower limb during early development (Uhl 2015). By the 12th week of gestation, it regresses to form the venous arcade of the sciatic nerve, and the femoral vein replaces it as the main pathway for venous drainage of the lower limb (Uhl 2015; Uhl et al. 2010).

117.2 Anatomy and Topography

Veins in the thigh can be categorized as superficial and deep. Superficial veins such as the great saphenous vein (GSV) are encased superficially by saphenous fascia and deeply by muscular fascia within this saphenous compartment. The deep veins of the thigh include the femoral and deep femoral veins and are located deep to the muscular fascia. They communicate with superficial veins via perforating veins that pierce the muscular fascia. Veins in the same category, superficial or deep, are joined by communicating veins (Caggiati et al. 2002; Bergan and Bunke-Paquette 2014).

117.3 Femoral Vein

The femoral vein (Fig. 117.1) continues as the popliteal vein as it traverses the popliteal fossa and pierces the adductor magnus muscle through the adductor hiatus. It functions as the main venous drainage of the lower limb. As it ascends within the adductor canal towards the femoral triangle, the surrounding adductor canal itself is bounded anterolaterally by the vastus medialis muscle and posteromedially by the adductor longus and adductor magnus muscles (Standring 2016). The adductor canal roof overlying the femoral vein superficially is formed by the vasto-adductor membrane, a thick band of fascia (Tubbs

W. Ho · A. Vinjamuri · A. Segall (✉)
Tulane University School of Medicine, New Orleans, LA, USA
e-mail: asegall@tulane.edu

S. J. Bordes, Jr. et al. (eds.), *The Clinical Anatomy of the Vascular System*,
https://doi.org/10.1007/978-3-031-78326-5_117

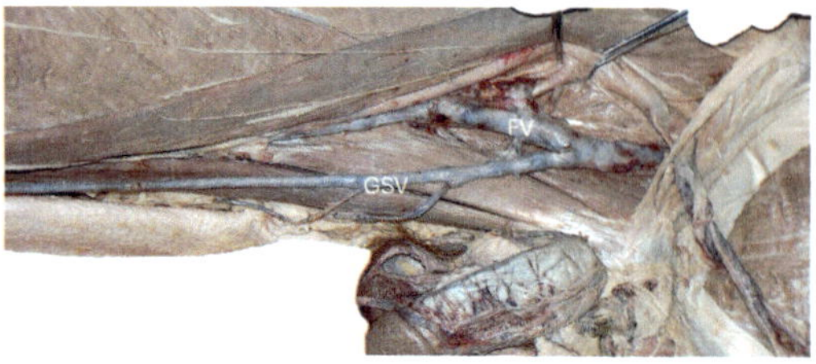

Fig. 117.1 Cadaveric dissection noting the femoral and great saphenous veins in the right thigh. (Courtesy Prof. M. Loukas)

et al. 2007). Throughout its course, the femoral vein travels alongside the femoral artery. It lies posterolateral to the femoral artery in the distal portion of the adductor canal and is positioned posterior to it in the proximal portion. At the base of the femoral triangle, it lies medial to the femoral artery (Standring 2016). It continues its ascent as the femoral vein and subsequently drains into the external iliac vein.

The femoral veins can have between one and six valves. Valves are most often seen within 5 cm of the inguinal ligament inside the femoral vein and within 3 cm of the deep femoral vein tributary within the femoral vein (Moore et al. 2011). The mean diameter of the femoral vein is greatest within 1–5 cm proximal to the level of the saphenofemoral junction and decreases to 6 cm distally (Keiler et al. 2018).

The literature frequently refers to the femoral vein as the superficial femoral vein. This is a misnomer, as the femoral vein is a deep vein (Caggiati et al. 2002).

117.4 Deep Femoral Vein

The deep femoral vein drains blood from the posterior and medial muscles of the thigh (Kachlik et al. 2012). It typically accompanies the deep femoral artery, coursing superiorly and medially to terminate in the femoral vein 8 cm below the inguinal ligament (Ba et al. 2019). It has a distal connection to the popliteal vein and a proximal connection to the inferior gluteal vein, and it receives drainage from the medial and lateral circumflex femoral veins (Standring 2016). When the femoral vein is obstructed, the deep femoral vein can provide collateral circulation, allowing for venous drainage from its distal connections in the popliteal vein (Raju and Fredericks 1991).

117.5 Obturator Vein

The obturator vein originates proximally in the adductor muscles of the thigh and enters the pelvis through the obturator canal to terminate in the internal iliac vein (Kachlik et al. 2012). It is typically accompanied by the obturator artery and nerve. In the thigh, it communicates with the medial and lateral circumflex femoral veins (Phlebology 2010).

117.6 Great Saphenous Vein (GSV) and Tributaries

The GSV (Fig. 117.1) is a superficial vein that continues in the thigh from its origin anterior to the medial ankle. It ascends on the medial aspect until it joins the femoral vein at the saphenofemoral junction, linking the venous drainage of the superficial and deep compartments of the thigh. In the saphenofemoral junction, it receives drainage from the superficial epigastric (SEV), superficial circumflex iliac (SCIV), and superficial external pudendal (SEPV) veins (Kachlik et al. 2010). The SEV typically drains into the GSV at a mean distance of 16.9 mm distal to the orifice of the GSV (Mühlberger et al. 2009). The SCIV drains into the GSV at a mean distance of 10.8 mm proximal to the orifice, while the SEPV joins it at a mean distance of 16.9 mm distal to it (Mühlberger et al. 2009). The SEV, SCIV, and SEPV drain into the GSV in 78%, 83%, and 90% of cases, respectively (Mühlberger et al. 2009). In a minority of cases, the aforementioned veins drain directly into the femoral vein (Mühlberger et al. 2009). At the proximal end of the GSV, there is a terminal valve near the saphenofemoral junction and a preterminal valve 15–25 mm below (Uhl 2015).

The GSV can also receive drainage from the anterior accessory saphenous, posterior accessory saphenous, anterior thigh circumflex, and the posterior thigh circumflex veins. There is significant variation in the presence of these GSV tributaries (Mühlberger et al. 2009).

117.7 Imaging

Ultrasound, or ultrasonography, of the lower limb is commonly used to detect venous thromboembolisms, identify injured venous valves and varicosities, assess flow in venous bypass grafts, and guide catheter placement. Prior to the advent of ultrasound imaging, contrast venography was the main modality for imaging veins; it involved using X-rays after injection of contrast material. There are two main modalities of ultrasound imaging: compression ultrasound and duplex ultrasound with color and spectral Doppler. Compression ultrasound (CUS) involves applying compression with an ultrasound probe over the veins, usually for DVT screening, to assess the compressibility of the veins (Fox and Bertoglio 2011). Two-point CUS, commonly performed, involves compression of the femoral and popliteal veins down to their branching points; whole-leg CUS examines both proximal and distal veins (Fox and Bertoglio 2011). Duplex ultrasound combines B-mode ultrasound and a directional posed Doppler, allowing varicosities to be located and the degree of venous reflux to be quantified (Thibault et al. 1990). Analysis of Doppler output allows the direction and volume of blood flow to be assessed. The addition of color Doppler enhances the visualization of blood flow by color-coding Doppler shifts, while spectral Doppler provides a graphical representation of blood flow over time. Color and spectral Doppler can also be useful when images are more difficult to obtain, as when patients with obesity or anatomical abnormalities are being screened (Fox and Bertoglio 2011). For DVT screening, duplex ultrasound from thigh to ankle with color and spectral Doppler are recommended when possible (Needleman et al. 2018). However, limitations in the availability of duplex ultrasound equipment and technicians can prevent timely screening, in which case extended CUS from the femoral vein all the way down to the popliteal vein is recommended (Needleman et al. 2018).

Different positions can confer certain advantages depending on the purpose of the evaluation. During an ultrasound examination, the patient can be situated in the reverse Trendelenburg position by placing them supine with the legs at a lower elevation than the head, increasing blood flow and vasodilation in the lower limb for optimal imaging of the deep veins (Lee et al. 2017). Deep vein visualization in the medial thigh can be further enhanced by externally rotating the patient's hip and providing flexion to the knee (Lee et al. 2017). One study also showed that femoral vein diameter increased significantly during Valsalva, making it a useful maneuver for the patient to perform during ultrasound-guided femoral vein cannulation (Beddy et al. 2006). When superficial veins are to be evaluated, an erect position is recommended over the reverse Trendelenburg position because it is difficult to assess reflux in the latter (Lee et al. 2017). Compared to other imaging modalities, ultrasound is advantageous in many situations owing to its low cost, portability, and lack of ionization radiation.

Computed tomography venography (CTV), although less often used, is another imaging modality for veins of the thigh. CTV allows venous networks to be visualized three-dimensionally and the angulated courses of veins, unusual varicose veins, and atypical inflows to the deep venous system to be seen more clearly (Sato et al. 2011). The information provided by CTV can be useful for planning a surgical course, and it can be a useful adjunct to duplex ultrasound by narrowing the area that must be assessed (Sato et al. 2011). Despite its proven utility, CTV requires contrast injections, which patients can find unpleasant, as well as ionizing radiation.

Another area where CTV can be advantageous over ultrasound is concurrent evaluation for pulmonary embolisms (PEs) and DVT (Goodman 1999; Garcia-Bolado and Del Cura n.d.). After helical CT pulmonary angiography to detect possible PE, CTV from the inferior vena cava (IVC) to the popliteal vein can be conducted within 3 min without additional contrast material (Goodman 1999). Another significant advantage of CTV is the ability to detect direct causes of

DVT and to check for multiple other pathologies when a radiological study is ordered for leg edema (Chang et al. 2014). One study demonstrated the value of CTV for detecting muscle tears, hematomas, and ganglion cysts as causes of leg edema while concurrently ruling out DVT (Chang et al. 2014). Comparing the efficacy of CTV with Doppler ultrasound for diagnosing DVT, another study observed good values for specificity (95.0%) and NPV (93.2%), but less satisfactory values for sensitivity (58.5%) and PPV (66.7%) (Garcia-Bolado and Del Cura n.d.). While Doppler ultrasound is typically advantageous over CTV, several factors, such as obesity and uncompressible air in the bowels, can impede proper visualization of veins in the abdomen and pelvis (Arnoldussen et al. 2013). For the primary purpose of diagnosing DVT in the lower extremities, and if iliac vein or IVC involvement is not suspected, the authors concluded that Doppler ultrasound remains the premier diagnostic modality over CTV (Garcia-Bolado and Del Cura n.d.).

Magnetic resonance venography (MRV) has also been explored as an alternative to Doppler ultrasound. Like CTV, MRV is potentially advantageous because it can diagnose or exclude multiple potential pathologies with symptoms similar to DVT (Spritzer n.d.). It is similarly useful for detecting obstructions in the iliac veins and the IVC and has a further advantage over CTV in visualizing detailed vein wall or intraluminal changes (Arnoldussen et al. 2013). Unlike CTV, MRV uses neither ionizing radiation nor iodine contrast; it uses a gadolinium-based contrast, which has fewer complications (Arnoldussen et al. 2013).

117.8 Variations

Femoral Vein Duplication

In one retrospective study of 445 patients assessed for varicose veins by bilateral lower limb CT venography, the femoral vein was multiple in 190 of the 890 studied limbs (21%). Of the 190 multiplications solely affecting the femoral vein, 187 were classified as duplications (Park et al. 2011). There was multiplication involving both the femoral and the popliteal vein in 6% of the studied limbs (Park et al. 2011). The medial channels of the femoral vein duplications were typically larger than the lateral channels (Park et al. 2011). In most of the duplications, the duplicated vein divided, and the constituent channels also rejoined, posterior to the accompanying femoral artery (Park et al. 2011). Less frequently, the vein divided or rejoined anterior to the femoral artery (Park et al. 2011). In another study evaluating patients who were asymptomatic or symptomatic for DVT, those with duplicated femoral veins had a significantly greater incidence of DVT than those with single femoral veins (Liu et al. 1986). Because femoral vein duplications are possible, care must be taken during imaging of the lower limb when evaluating for DVT (Quinn and Vandeman 1990; Screaton et al. 1998).

Axiofemoral Trunk and Deep Femoral Trunk

In up to 1% of cases, the primitive axial vein continues to provide the main venous drainage after early development rather than involuting to allow the femoral vein to take over (Uhl et al. 2010). When this axiofemoral trunk persists, the femoral vein is hypoplastic and reduced to the size of a collateral vein (Uhl et al. 2010).Axiofemoral trunks have been documented to drain to the deep femoral vein or directly to the femoral vein (Uhl et al. 2010).

In up to 2% of cases, hypoplasia of the femoral vein can make the deep femoral vein the main venous drainage of the lower limb (Uhl et al. 2010). This deep femoral trunk, also referred to as an axial transformation of the deep femoral vein, is connected distally to the popliteal vein through a profunda-popliteal collateral (Raju et al. 1998). This collateral can be enlarged rapidly if the femoral vein is obstructed, but only 9% of these axial transformations feature total occlusion of the femoral vein (Raju et al. 1998). Most cases feature only a profunda-popliteal junction or a slight enlargement of the deep femoral vein, smaller than the femoral vein (Raju et al. 1998).

In 36% of cases, the enlarged deep femoral vein is larger than the femoral vein, which is not totally occluded (Raju et al. 1998). This transformed axial vein can be mistaken for the femoral vein on radiological scans, though it can be distinguished by its more lateral course in the thigh overlying the femoral shaft (Raju et al. 1998). Venous reflux is common in both the obstructed femoral vein and the deep femoral trunk, and this can merit concomitant valve reconstruction to maximize venous outflow (Raju et al. 1998).

Giacomini Vein and Proximal Extensions of the Small Saphenous Vein

The SSV is typically described as draining into the popliteal vein in the popliteal fossa, terminating before the level of the thigh (Kachlik et al. 2010). However, in 1873, Carlo Giacomini described a number of variants of the SSV that extended proximally into the thigh. Some of the variants were described as terminating in the GSV or connecting to it through an anastomosis; these variants with an inter-saphenous connection are known as Giacomini veins (Kachlik et al. 2010; Georgiev et al. 2003; Caggiati et al. 2005). In most cases, a Giacomini vein terminates in the popliteal vein while providing anastomotic connections to the GSV (Georgiev et al. 2003). Less commonly, the Giacomini vein terminates directly in the GSV or in the short head of the biceps femoris muscle while providing anastomotic branches to the GSV (Georgiev et al. 2003). Superficially, the Giacomini vein courses between the biceps femoris muscle laterally and the semitendinosus and semimembranosus muscles medially. More deeply, it courses between the adductor magnus, biceps femoris, and semimembranosus muscles (Delis et al. 2004). Two to three valves are typical where the Giacomini vein meets the GSV, preventing reflux from the GSV to the SSV and allowing only upward flow (Georgiev et al. 2003).

Giacomini also described a number of other SSV variants in the thigh with no inter-saphenous anastomotic branch to the GSV. Compared to the Giacomini vein, these other proximal extensions of the SSV drain venous branches from the posterior and superior thigh down towards the saphenopopliteal junction, with valves allowing only downward flow (Georgiev et al. 2003). When these valves become incompetent, paradoxical upward reflux can ascend from the saphenopopliteal junction (Georgiev et al. 2003). Relative to the GSV and SSV, though, these proximal extensions of the SSV are less susceptible to varicosities (Delis et al. 2004).

Some literature on the Giacomini vein uses a more inclusive definition, describing it only as a proximal extension of the SSV with or without an inter-saphenous anastomosis (Delis et al. 2004). There are proximal extensions of the SSV in up to 70.4% of limbs, 49.5% of them terminating in the GSV, and 45.3% in the deep venous system (Delis et al. 2004).

Great Saphenous Vein Variations

The tributaries received by the GSV show considerable variation.

An anterior (AASV) and a posterior (PASV) accessory saphenous vein can accompany the GSV. The AASV and PASV ascend in parallel with the GSV and are located anteriorly and posteriorly to it, respectively, in a fascial compartment of the thigh (Cavezzi et al. 2006). These two accessory veins can receive drainage from the anterior and posterior thigh circumflex veins, respectively, though those circumflex veins can drain directly into the GSV (Caggiati et al. 2002). The circumflex veins ascend obliquely to drain the lateral venous system, a system of subcutaneous veins on the lateral thigh (Caggiati et al. 2002; Kachlik et al. 2010).

There is an AASV in 51% of cases, typically draining into the GSV from the lateral aspect, 20.5 mm distal to the orifice of the GSV where it meets the femoral vein (Mühlberger et al. 2009). The PASV typically drains into the GSV at an average of 73.9 cm distal to the orifice (Mühlberger et al. 2009). More frequently, though, the SCIV, SEV, SEPV, AASV, and PASV join the GSV as a common trunk formed by some combination of the aforementioned veins. Up to 69 different combinations of single-vein tributaries and common trunks at the saphenofemoral junction were observed in one study (Mühlberger et al. 2009).

Duplicated GSVs have also been described in the literature, occurring in 1.8% of duplex ultrasound examinations (Kockaert et al. 2012). Although a duplicated GSV is easily mistaken for an AASV, duplex GSVs run in the same plane and drain the same cutaneous territory, while an AASV is typically smaller and drains a different cutaneous territory (Chen and Prasad 2009). Three duplex GSV patterns have been observed: duplication with a common junction with the femoral vein, duplication with separate junctions, and a GSV that bifurcates and reconvenes before joining the femoral vein (Chen and Prasad 2009).

Persistent Sciatic Vein

A persistent sciatic vein is a rare congenital anomaly, most common in patients with Klippel-Trenaunay syndrome (Kachlik et al. 2012). Klippel-Trenaunay syndrome is typically characterized by capillary malformations resembling a port wine stain, soft tissue and bone hypertrophy, and lateral varicosities (Glovkzki and Driscoll 2007). In normal development, this vein regresses to form the sciatic arcade around the sciatic nerve by the end of the 12th week, allowing the femoral vein to become the main venous drainage of the thigh (Uhl 2015). The persistent sciatic vein can function as the main venous drainage when the femoral vein is occluded, though it can be present even when the femoral vein is patent (Cherry et al. 1996).

Three main variants of persistent sciatic veins have been described. In approximately 38% of cases, the persistent sciatic vein originates from the popliteal vein or nearby tributaries and traverses the entire length of the thigh and buttock to terminate in the internal iliac vein (Cherry et al. 1996). Less common variants traverse either the upper thigh and buttock only or the middle thigh and distal thigh only (Cherry et al. 1996). Approximately 90% of cases of persistent sciatic veins are unilateral (Cherry et al. 1996).

117.9 Pathology, Clinical Significance, and Surgical Significance

Deep Vein Thrombosis

DVT is a type of venous thromboembolism that affects the deep veins, typically in the lower extremities. Interruption of blood flow is thought to facilitate the concentration of platelet aggregates, thrombin, and fibrin monomers that would otherwise be removed owing to high shear forces at the vessel walls, leading to fibrin polymerization (Lowe 2003). DVT in the lower extremities originates near the valves, perhaps because the structure of the venous valve pockets promotes the interaction of platelets and clotting enzymes (Lowe 2003; Aird 2007). In the event of vessel injury, release of prothrombotic von Willebrand factor, and loss of membrane thrombomodulin, an anticoagulant produced by endothelial cells is believed to promote thrombosis (Blann 2003). The hypercoagulable state associated with DVT has various causes, including genetic disorders such as factor V Leiden thrombophilia and malignancies (Thomas 2001). Surgery is considered a major risk factor for DVT owing to possible iatrogenic injury to the venous intima, disruption of blood flow while positioning limbs, or immobility of the patient during postoperative recovery (Selby and Geerts 2000). Placement of central venous catheters in the femoral vein also entails considerable risk for DVT (Trottier et al. 1995).

In most nonpregnant patients diagnosed with DVT, thrombi are detected in the proximal veins of the lower limb with concurrent calf vein thrombosis (Cogo et al. 1993). In contrast, there is a greater incidence of DVT isolated to the proximal veins without calf vein involvement in pregnant women (Chan et al. 2010). Signs of DVT can include unilateral edema or pain in the affected limb, but there are no signs and symptoms in more than 50% of cases (Messmore et al. n.d.).

A clinical model developed by Wells et al. can be used to stratify pretest probabilities for DVT, accounting for signs and symptoms, risk factors, and other potential diagnoses. When used in tandem with an ultrasound examination of the lower limb, there is less need for serial or invasive testing in patients with low pretest probabilities for DVT (Wells et al. 1995). Guidelines published by the American Society of Hematology recommend a D-dimer test, if available, for excluding DVT in a patient population with low pretest probability. However, a positive D-dimer test alone is not sufficient for diagnosis of DVT, regardless of pretest probability (Lim et al. 2018).

Subcutaneous administration of low-molecular-weight heparin (LMWH), an anticoagulant, can be used for the initial management of DVT, and outpatient therapy can be appropriate if there are no additional risk factors (Douketis 2005; Koopman et al. 1996; Levine et al. 1996). Warfarin, a vitamin K antagonist, can be prescribed concurrently with LMWH for use on a longitudinal basis (Holbrook et al. 2012). Inpatient treatment is necessary if, for example, there is extensive DVT, symptomatic PE, or increased risk for bleeding, or there are major comorbidities (Douketis 2005). For patients with symptomatic DVT, there is a high risk for recurrent venous thromboembolism and post-thrombotic syndrome (Prandoni et al. 1996). In lieu of systemic anticoagulant therapy for DVT, surgical thrombectomy or pharmacological catheter-directed thrombolysis can decrease the risk of post-thrombotic syndrome (Casey et al. 2012). If the thrombus detaches from the deep vein and migrates to the right side of the heart, a PE in a pulmonary artery can result (Giordano et al. 2017).

Superficial Vein Thrombosis (SVT)

Superficial vein thrombosis (SVT) is a type of venous thromboembolism that affects superficial veins such as the GSV and SSV. Most cases of SVT occur in the lower extremities (Decousus and Leizorovicz 2005).Although SVT has historically been considered benign, documented progression of SVT into DVT and the associated risk necessitates serious attention to this condition (Hill et al. 2008). In patients with SVT, thrombi can migrate from the GSV at the level of the thigh or from the saphenofemoral junction into the femoral vein, resulting in DVT (Chengelis et al. 1996). Thrombi can also migrate from the GSV in the thigh through perforating veins to reach the femoral vein near the adductor hiatus (Chengelis et al. 1996). In one study, 19% of patients with DVT had concomitant SVT on ultrasound (Chengelis et al. 1996). A history of clinical SVT is associated with a 6.3-fold increase in risk for developing DVT and a 3.9-fold increase in risk for PE (van Langevelde et al. 2011). On average, it takes 6.3 days for SVT to progress to DVT (Chengelis et al. 1996).

Many risk factors for SVT are shared with DVT. These include, but are not limited to, prior venous thromboembolism, pregnancy, use of oral contraceptives, immobilization, and recent surgery (Leon et al. 2005). Signs and symptoms of SVT typically include pain, tenderness, swelling, warmth, erythema, and a palpable cord along the trajectory of the superficial veins (Messmore et al. n.d.). Diagnosis of SVT can be confirmed via ultrasound of the lower limb (Lutter et al. 1991).

Recommendations by the American College of Chest Physicians advise a prophylactic dose of fondaparinux, a synthetic anticoagulant, over 45 days in patients diagnosed with SVT at least 5 cm long (Kearon et al. 2012). LMWH can be considered as an alternative, but fondaparinux is recommended as the primary choice (Kearon et al. 2012). Although surgical options such as ligation of the GSV at the saphenofemoral junction or ligation combined with stripping of the affected vein have been used, anticoagulants are associated with fewer complications and less subsequent DVT or PE (Sullivan et al. 2001). Surgical treatment for SVT should be considered when anticoagulants are contraindicated (Sullivan et al. 2001).

Varicosities

Varicose veins and spider veins are two types of cosmetically undesirable leg veins that share the same etiology. Varicose veins are subcutaneous veins that are dilated and torturous. Spider veins are located superficially in the dermis and are typically narrower. In both cases, increased intravenous pressure leads to dilation and thickening of the venous walls. As the lumen widens, valves cease to function properly, and reverse blood flow ensues. Venous hypertension can also spread to more distal veins as the incompetent valves cause a reversal of blood flow in the superficial venous system (Thibault et al. 1990).

Ligation and stripping together have been considered the gold standard approach for treating varicosities in the GSV (van den Bremer and Moll 2010). The GSV is ligated just below the saphenofemoral junction, followed by a short strip of the GSV to just below the knee to avoid injuring the saphenous nerve (van den Bremer and Moll 2010). Stripping reduces the risk of recurrent varicose veins, although neovascularization remains a potential cause of recurrence at the site of ligation even after GSV stripping (Jones et al. 1996). Besides ligation and stripping, other modalities of varicose vein treatment have recently become more popular, such as ambulatory conservative hemodynamic treatment for venous insufficiency (Conservatrice Hémodynamique de l'Insuffisance Veineuse en Ambulatoire (CHIVA)), ultrasound-guided foam sclerotherapy (UGFS), radiofrequency ablation (RFA), and endovascular laser ablation (EVLA).

In CHIVA, the objective is to restore venous flow while simultaneously preserving the GSV (Faccini et al. 2019). Under local anesthesia, escape points, defined as sources of reflux such as an incompetent saphenofemoral junction, are serially ligated (Faccini et al. 2019). In some cases, CHIVA restores upward blood flow in the GSV through the saphenofemoral junction; in others, blood flows downwards nonpathogenic and enters the deep venous system via perforating veins (Faccini et al. 2019). A significant advantage of CHIVA is the absence of nerve injury, which remains a concern in stripping and thermal ablation (Faccini et al. 2019). Another advantage is the preservation of the GSV, ensuring its continued availability for a number of vital procedures (Faccini et al. 2019). Compared to stripping, CHIVA reduces the recurrence of varicose veins and the occurrence of bruising (Bellmunt-Montoya et al. 2015). Despite its advantages, it is described as having a rigorous learning curve, demanding a thorough understanding of venous hemodynamics for success (Faccini et al. 2019).

UGFS, a minimally invasive technique, involves injecting foam formed from a liquid sclerosant combined with air. The foam sclerosant is better than a liquid sclerosant because it displaces the venous blood and is not diluted when injected, allowing for controlled sclerosis of the venous segment targeted (Eckmann 2009). Ultrasound is used throughout the procedure to map the varicose veins and guide venous access for the foam injection. Typical foam sclerosants used in UGFS are polidocanol in Europe and sodium tetradecyl sulfate in the United States (Nijsten et al. 2009). UGFS is useful for primary varicose veins in both the GSV and SSV and recurrent varicosities after traditional surgery (Nijsten et al. 2009). UGFS is less effective for saphenous trunks over 6–8 mm wide than for smaller veins (Coleridge 2009). Recent studies comparing UGFS against surgery, RFA, and EVLA show that it has a higher rate of residual reflux and failure of occlusion than other approaches. However, it is associated with less postoperative pain and faster recovery times than surgery and EVLA (Rasmussen et al. 2011).

RFA, another minimally invasive technique for varicose vein treatment, involves advancing a catheter with electrodes towards the saphenofemoral junction. The electrodes make direct contact with the venous wall and emit radiofrequency energy, heating the local tissue up to 90 °C to obliterate the surrounding vein (Nijsten et al. 2009). Gradually, the catheter is withdrawn, allowing consecutive venous segments to be ablated. RFA is performed under ultrasound guidance, and manual compression is typically applied over the vein to ensure the necessary contact between the catheter and the venous wall (Nijsten et al. 2009). RFA is difficult for veins with diameters over 12 mm because direct contact between the electrodes and the venous segments is required (Nijsten et al. 2009). RFA has comparable, if not better, outcomes than surgery with less postoperative pain (Rasmussen et al.

2011; Subramonia and Lees 2010; Helmy ElKaffas et al. 2011).

EVLA is the last of the commonly used minimally invasive techniques for treating varicose veins, and it has many similarities to RFA. The veins are mapped using ultrasound, and a guidewire is introduced to position a laser fiber that affects thermal ablation while being progressively retracted. Linear primary truncal varicosities at least 5 mm in diameter are considered ideal for EVLA (Nijsten et al. 2009). EVLA is superior to RFA for treating varicosities in a GSV at least 10 mm in diameter since direct venous contact is not needed (Mese et al. 2015). Comparative studies of EVLA against UGFS and surgery show that EVLA yields better occlusion rates and has comparable outcomes to surgery, if not better (Venermo et al. 2016; Vähäaho et al. 2018; Biemans et al. 2013). Patients receiving EVLA typically experienced more severe postoperative pain than those receiving RFA, although there was no significant difference in clinical outcomes (Rasmussen et al. 2011; Shepherd et al. 2010; Nordon et al. 2011; Goode et al. 2010). However, one study using a laser source with a wavelength of 1470 nm, which increased from the usual 810–980 nm, observed less postoperative pain and comparable clinical outcomes in patients receiving EVLA (Bozoglan et al. 2017).

Central Venous Catheter Placement in the Femoral Vein

The femoral vein can be a candidate for central venous catheter placement in critically ill patients in place of the subclavian or jugular vein (Williams et al. 1991). Femoral vein catheterization entails a significantly greater risk of infectious or thrombotic complications than subclavian venous catheterization in patients under intensive care (Merrer et al. 2001). However, subclavian and jugular venous access can be impractical or limited for several reasons, such as active infection at the desired venous access site, anatomical distortion at the site, and localized vascular injury (Kolikof et al. 2022). Furthermore, the insertion of a central line can interrupt chest compression efforts during cardiopulmonary resuscitation, potentially compromising survival (Kern 2003). Femoral venous catheterization provides a safe route for rapid intravenous fluid infusion when hypovolemic shock and cardiac arrest are treated (McGee and Gould 2003; Getzen and Pollak 1979; Mangiante et al. 1988). Femoral venous catheterization has also been successful in burn patients who have burn eschar overlying the vascular access routes for subclavian and jugular venous catheterization (Purdue and Hunt 1986).

Real-time ultrasound guidance is recommended during femoral vein catheterization to decrease the risk of inadvertent arterial catheterization and the amount of time that would be involved in a landmark-oriented approach (Hilty et al. 1997). A 90% success rate has been observed in femoral venous catheterization with real-time ultrasound guidance in the CPR setting (Hilty et al. 1997). Owing to the proximity of the femoral artery to the femoral vein, care should be taken during venous catheterization to avoid inadvertent puncture of the femoral artery.

117.10 Peripherally Inserted Central Catheters

Peripherally inserted central catheters (PICCs) provide long-term intravenous access via a percutaneous route. In adults, PICCs are typically placed in peripheral veins of the upper extremities, such as the basilic, brachial, or cephalic vein. However, there are significant advantages in lower limb PICC placement in pediatric and neonatal populations. A comparative study of upper and lower limb PICC placements in neonates demonstrated an overall lower rate of complications when PICCs were placed in the GSV or SSV (Hoang et al. 2008). Furthermore, lower limb PICCs in neonates remained functional for longer than upper limb PICCs. Another study demonstrated the successful use of PICCs in the GSV at the level of the thigh and the calf in the pediatric population with a lower complication rate than other PICC placement sites (Aria et al. 2014). Advantages of PICC placement in the GSV rather than the femoral vein include the lack of risk for accidental puncture of the femoral artery, the lack of need for tunneling, and the opportunity to spare the femoral vein for future use (Aria et al. 2014). PICC placement in the GSV above the knee provides a larger vessel diameter for venous access and is recommended by the study authors over an approach from below the knee (Aria et al. 2014). The GSV has also been identified as a useful PICC placement site in adults when subclavian or internal jugular venous access is unavailable or contraindicated, as in some cases of end-stage renal failure or in patients receiving intravenous chemotherapy (Du et al. 2008; Treiman and Silberman 1993).

117.11 Saphenous Vein in Coronary Artery Bypass and Interposition Grafting

The GSV is a candidate vessel for coronary artery bypass grafting (CABG) to relieve coronary artery obstruction. Current recommendations from The Society of Thoracic Surgeons support the internal thoracic artery (ITA) as the first choice for bypassing the left anterior descending artery (LAD), followed by the right ITA or radial artery if a secondary arterial graft is needed (Aldea et al. 2016). The 10-year patency for saphenous vein grafts (SVGs) is 61% versus 85% for ITA grafts (Goldman et al. 2004). However, the ITA is not always suitable for grafting, e.g., if there is recipient vessel stenosis, injury to the ITA during harvesting, hemodynamic instability, or ventricular hypertrophy or enlargement (Geha and Baue 1979). SVGs can also be superior to ITA grafts for bypassing right coronary arteries (RCAs) with less than 70% stenosis (Sabik et al. 2005).

A larger recipient vessel predicts 10-year patency for SVGs; 88% of SVGs in >2.0 mm recipient vessels remained patent recipient vessels over 10 years (Goldman et al. 2004). When harvesting the GSV for CABG, a no-touch technique is vital for improving graft function and patency; the vein is removed with a surrounding tissue pedicle with no mechanical distension (Souza et al. 2002). More recently, though, endoscopic harvesting of veins has been popularized because there is less postoperative pain, reduced risk for wound infection, and improved patient mobility (Krishnamoorthy et al. 2017). Prescription of aspirin and statins after CABG has been recommended to improve the short-term and long-term patency of the graft, respectively (Kim et al. 2013).

117.12 Saphenous Venous Grafts in Peripheral Artery Disease

The GSV is also used as an SVG for surgical revascularization during the treatment of peripheral artery disease (PAD). PAD is characterized by the narrowing of arteries and is typically caused by atherosclerotic changes in the lower extremities. Progressive worsening of PAD can eventually lead to critical limb ischemia (CLI), entailing a high risk for limb loss, myocardial infarction, and stroke (Norgren et al. 2007). Before revascularization options such as bypass surgery are pursued, PAD is typically addressed with statins and antiplatelet therapy, along with lifestyle modifications such as smoking cessation and increased exercise (Kullo and Rooke 2016). However, thanks to continual advances in endovascular options, bypass surgery for PAD has decreased by 42% over 10 years, while endovascular intervention has become three times as common (Goodney et al. 2009). Surgical bypass is currently recommended when endovascular options have failed, if there are anatomical limitations to an endovascular option, or if PAD has progressed to CLI (Kullo and Rooke 2016).

In comparative studies of other autologous conduits, such as the basilic or cephalic vein or SSV against the GSV, GSVs outperformed the other conduits for 3-year patency and 3-year limb salvage rates in patients with critical limb ischemia (Arvela et al. 2012). When the GSV is examined as a potential graft, it should be noted that smaller veins (<3.0 mm diameter) have significantly lower early patency rates than larger veins (Towne et al. 1991; Schanzer et al. 2007). Longer grafts (>60 cm) are also associated with a lower 1-year primary patency than short grafts

(<40 cm) (Schanzer et al. 2007). Accordingly, short bypasses with >3.5 cm diameter GSV conduits are recommended for lower limb bypass procedures in patients with CLI (Schanzer et al. 2007). Compared to prosthetic materials such as polytetrafluoroethylene (PTFE), autologous saphenous vein grafts give superior results for below-knee and above-knee femoropopliteal bypasses (Klinkert et al. 2004). A systematic review of studies comparing venous grafts versus PTFE observed a 2-year patency of 81% in venous grafts versus 67% in PFTE grafts, and the use of SVGs over prosthetic grafts was recommended even in patients with limited life expectancy (Klinkert et al. 2004). After infrainguinal bypass surgery, treatment with oral anticoagulants is recommended to minimize risk for graft occlusion and ischemic events (Dutch Bypass Oral Anticoagulants or Aspirin (BOA) Study Group 2000).

More recently, cryopreserved saphenous vein (CSV) allografts have been explored as alternatives to an autogenous GSV in infrainguinal bypass grafts. The results suggest that CSVs are worth considering in patients with ischemic tissue loss who have no available autogenous conduit (O'Banion et al. 2017). However, long-term patency remains a concern for CSVs, with a 37% one-year primary patency rate (O'Banion et al. 2017). The GSV has also proved useful in SVGs in several neurosurgical cases involving intracranial aneurysms and coronary artery stenosis. However, the popularity of the technique has diminished with improved endovascular therapies (Friedman and Piepgras 2003).

117.13 Lower Limb Vascular Access in Hemodialysis

An arteriovenous fistula (AVF) can be formed surgically to link an artery to a vein and provide a reliable access site for hemodialysis. Typically, veins of the upper extremities are selected for the AVF. However, exhaustion of access sites owing to infections, prolonged placement of PICCs and central venous catheters, and placement of pacemakers warrants consideration of other venous conduits such as the femoral vein (Gradman et al. 2001). Two main approaches to using the femoral vein in AVFs were studied initially: transposition of the femoral vein (tFV) and formation of a composite loop fistula that combines PTFE and a transposed femoral vein (Gradman et al. 2001).

In tFV, the femoral vein is divided distally and freed to the level of its confluence with the deep femoral vein (Gradman et al. 2001). The femoral vein is transposed deep to the femoral artery, and the surgeon tunnels it laterally to the subcutaneous level before tunneling it back to the femoral artery between the adductor tendon and the sartorius muscle (Gradman et al. 2001). The procedure begins similarly by tunneling the femoral vein laterally to the subcutaneous level to form a composite loop fistula (Gradman et al. 2001). A PTFE fistula is implanted at the femoral artery and then tunneled deep beneath the skin towards the transposed femoral vein to create an end-to-end anastomosis (Gradman et al. 2001).

One-year primary patency rates of the tFV have ranged from 65% to 91%, while 1-year secondary patency rates have ranged from 84% to 94% (Gradman et al. 2001, 2005; Bourquelot et al. 2012; Farber et al. 2020; Hazinedaroğlu et al. 2004). tFV has demonstrated significantly better patency rates than femoral loop grafts for hemodialysis (Hazinedaroğlu et al. 2004). Furthermore, the low risk of access-related infections in tFV provides an advantage over other dialysis modalities such as prosthetic arteriovenous grafts, Hemodialysis Reliable Outflow (HeRO) grafts, and tunneled line catheters (Farber et al. 2020). Despite these advantages, surgeons should be mindful of the risks of wound infection and ischemia with tVF; wound complication rates as high as 28% have been reported (Gradman et al. 2001; Farber et al. 2020). Four potential risk factors have been identified that can predispose patients receiving tFV to ischemic events: initiation of access on the distal femoral artery, arterial occlusive disease, postoperative compartment syndrome, and a conduit size mismatch between the vein and artery (Gradman et al. 2005). In summary, when appropriately screened for risk factors, it should be considered for lower limb venous access when upper limb access is unavailable (Bourquelot et al. 2012; Farber et al. 2020; Gradman et al. 2005).

The GSV has also been explored as an option for lower limb hemodialysis through a saphenous vein loop AVF (Pierre-Paul et al. 2004). In this surgery, the GSV from the medial thigh is mobilized, transected distally, and placed in a loop configuration subcutaneously on the anterior thigh before forming an end-to-end anastomosis with the femoral artery (Pierre-Paul et al. 2004). However, poor patency rates, high complication rates, and the propensity to develop stenosis in the AVF loop argue against the use of the GSV when other options are available for lower limb bypass (Antoniou et al. 2009).

References

Aird WC (2007) Vascular bed-specific thrombosis. J Thromb Haemost 5(Suppl 1):283–291. https://doi.org/10.1111/j.1538-7836.2007.02515.x

Aldea GS, Bakaeen FG, Pal J et al (2016) The Society of Thoracic Surgeons clinical practice guidelines on arterial conduits for coronary artery bypass grafting. Ann Thorac Surg 101(2):801–809. https://doi.org/10.1016/j.athoracsur.2015.09.100

Antoniou GA, Lazarides MK, Georgiadis GS, Sfyroeras GS, Nikolopoulos ES, Giannoukas AD (2009) Lower-extremity arteriovenous access for haemodialysis: a systematic review. Eur J Vasc Endovasc Surg 38(3):365–372. https://doi.org/10.1016/j.ejvs.2009.06.003

Aria DJ, Vatsky S, Kaye R, Schaefer C, Towbin R (2014) Greater saphenous venous access as an alternative in children. Pediatr Radiol 44(2):187–192. https://doi.org/10.1007/s00247-013-2794-y

Arnoldussen CWKP, De Graaf R, Wittens CHA, De Haan MW (2013) Value of magnetic resonance venography and computed tomographic venography in lower extremity chronic venous disease. Phlebology 28(1_suppl):169–175. Accessed 16 May 2022. https://journals.sagepub.com/doi/10.1177/0268355513477785

Arvela E, Venermo M, Söderström M, Albäck A, Lepäntalo M (2012) Outcome of Infrainguinal single-segment great saphenous vein bypass for critical limb ischemia is superior to alternative autologous vein bypass, especially in patients with high operative risk. Ann Vasc Surg 26(3):396–403. https://doi.org/10.1016/j.avsg.2011.08.013

Ba B, Kanté A, Touré T et al (2019) Anatomic dissection of the femoral vein at the Bamako Anatomy Laboratory. Forensic Med Anat Res 07(04):76. https://doi.org/10.4236/fmar.2019.74012

Beddy P, Geoghegan T, Ramesh N et al (2006) Valsalva and gravitational variability of the internal jugular vein and common femoral vein: ultrasound assessment. Eur J Radiol 58(2):307–309. https://doi.org/10.1016/j.ejrad.2005.11.005

Bellmunt-Montoya S, Escribano JM, Dilme J, Martinez-Zapata MJ (2015) CHIVA method for the treatment of chronic venous insufficiency. Cochrane Database Syst Rev 2015(6):CD009648. https://doi.org/10.1002/14651858.CD009648.pub3

Bergan JJ, Bunke-Paquette N (eds) (2014) The vein book, 2nd edn. Oxford University Press

Biemans AAM, Kockaert M, Akkersdijk GP et al (2013) Comparing endovenous laser ablation, foam sclerotherapy, and conventional surgery for great saphenous varicose veins. J Vasc Surg 58(3):727–734.e1. https://doi.org/10.1016/j.jvs.2012.12.074

Blann AD (2003) How a damaged blood vessel wall contibutes to thrombosis and hypertenasion. Pathophysiol Haemost Thromb 33(5–6):445–448. https://doi.org/10.1159/000083843

Bourquelot P, Rawa M, Van Laere O, Franco G (2012) Long-term results of femoral vein transposition for autogenous arteriovenous hemodialysis access. J Vasc Surg 56(2):440–445. https://doi.org/10.1016/j.jvs.2012.01.068

Bozoglan O, Mese B, Eroglu E, Ekerbiçer HC, Yasim A (2017) Comparison of endovenous laser and radiofrequency ablation in treating varices in the same patient. J Lasers Med Sci 8(1):13–16. https://doi.org/10.15171/jlms.2017.03

Caggiati A, Bergan JJ, Gloviczki P, Jantet G, Wendell-Smith CP, Partsch H (2002) Nomenclature of the veins of the lower limbs: an international interdisciplinary consensus statement. J Vasc Surg 36(2):416–422. https://doi.org/10.1067/mva.2002.125847

Caggiati A, Bergan JJ, Gloviczki P, Eklof B, Allegra C, Partsch H (2005) Nomenclature of the veins of the lower limb: extensions, refinements, and clinical application. J Vasc Surg 41(4):719–724. https://doi.org/10.1016/j.jvs.2005.01.018

Casey ET, Murad MH, Zumaeta-Garcia M et al (2012) Treatment of acute iliofemoral deep vein thrombosis. J Vasc Surg 55(5):1463–1473. https://doi.org/10.1016/j.jvs.2011.12.082

Cavezzi A, Labropoulos N, Partsch H et al (2006) Duplex ultrasound investigation of the veins in chronic venous disease of the lower limbs—UIP consensus document. Part II. Anatomy. Eur J Vasc Endovasc Surg 31(3):288–299. https://doi.org/10.1016/j.ejvs.2005.07.020

Chan WS, Spencer FA, Ginsbergm JS (2010) Anatomic distribution of deep vein thrombosis in pregnancy. CMAJ 182(7):657–660. https://doi.org/10.1503/cmaj.091692

Chang JH, Lee HJ, Kwon JH et al (2014) Usefulness of the computed tomography venography for evaluation of leg edema including deep vein thrombosis in rehabilitation patients. Ann Rehabil Med 38(6):812–820. https://doi.org/10.5535/arm.2014.38.6.812

Chen SSH, Prasad SK (2009) Long saphenous vein and its anatomical variations. Australas J Ultrasound Med 12(1):28–31. https://doi.org/10.1002/j.2205-0140.2009.tb00004.x
Chengelis DL, Bendick PJ, Glover JL, Brown OW, Ranval TJ (1996) Progression of superficial venous thrombosis to deep vein thrombosis. J Vasc Surg 24(5):745–749. https://doi.org/10.1016/S0741-5214(96)70007-1
Cherry KJ, Gloviczki P, Stanson AW (1996) Persistent sciatic vein: diagnosis and treatment of a rare condition. J Vasc Surg 23(3):490–497. https://doi.org/10.1016/S0741-5214(96)80016-4
Cogo A, Lensing AWA, Prandoni P, Hirsh J (1993) Distribution of thrombosis in patients with symptomatic deep vein thrombosis: implications for simplifying the diagnostic process with compression ultrasound. Arch Intern Med 153(24):2777–2780. https://doi.org/10.1001/archinte.1993.00410240085010
Coleridge SP (2009) Sclerotherapy and foam sclerotherapy for varicose veins. Phlebology 24(6):260–269. https://doi.org/10.1258/phleb.2009.009050
Decousus H, Leizorovicz A (2005) Superficial thrombophlebitis of the legs: still a lot to learn. J Thromb Haemost 3(6):1149–1151. https://doi.org/10.1111/j.1538-7836.2005.01445.x
Delis KT, Knaggs AL, Khodabakhsh P (2004) Prevalence, anatomic patterns, valvular competence, and clinical significance of the Giacomini vein. J Vasc Surg 40(6):1174–1183. https://doi.org/10.1016/j.jvs.2004.09.019
Douketis JD (2005) Treatment of deep vein thrombosis. Can Fam Physician 51(2):217–223
Du L, Redmond K, Johnstone S, De Leacy M, Harper J (2008) Saphenous vein peripherally inserted central catheters: technique, indications and safety issues. J Med Imaging Radiat Oncol 52(1):68–71. https://doi.org/10.1111/j.1440-1673.2007.01914.x
Dutch Bypass Oral Anticoagulants or Aspirin (BOA) Study Group (2000) Efficacy of oral anticoagulants compared with aspirin after infrainguinal bypass surgery (The Dutch Bypass Oral anticoagulants or Aspirin study): a randomised trial. Lancet 355(9201):346–351. https://doi.org/10.1016/S0140-6736(99)07199-8
Eckmann DM (2009) Polidocanol for endovenous microfoam sclerosant therapy. Expert Opin Investig Drugs 18(12):1919–1927. https://doi.org/10.1517/13543780903376163
Faccini FP, Ermini S, Franceschi C (2019) CHIVA to treat saphenous vein insufficiency in chronic venous disease: characteristics and results. J Vasc Bras 18:e20180099. https://doi.org/10.1590/1677-5449.009918
Farber A, Cheng TW, Nimmich A et al (2020) Femoral vein transposition is a durable hemodialysis access for patients who have exhausted upper extremity options. J Vasc Surg 71(3):929–936. https://doi.org/10.1016/j.jvs.2019.07.062
Fox JC, Bertoglio KC (2011) Emergency physician performed ultrasound for DVT evaluation. Thrombosis 2011:938709. https://doi.org/10.1155/2011/938709
Friedman JA, Piepgras DG (2003) Current neurosurgical indications for saphenous vein graft bypass. Neurosurg Focus 14(3):1–3. https://doi.org/10.3171/foc.2003.14.3.2
Garcia-Bolado A, Del Cura JL (n.d) CT venography vs ultrasound in the diagnosis of thromboembolic disease in patients with clinical suspicion of pulmonary embolism | SpringerLink. Emerg Radiol 14:403–409. Accessed 15 May 2022. https://link.springer.com/article/10.1007/s10140-007-0654-5
Geha AS, Baue AE (1979) Early and late results of coronary revascularization with saphenous vein and internal mammary artery grafts. Am J Surg 137(4):456–463. https://doi.org/10.1016/0002-9610(79)90114-4
Georgiev M, Myers KA, Belcaro G (2003) The thigh extension of the lesser saphenous vein: from Giacomini's observations to ultrasound scan imaging. J Vasc Surg 37(3):558–563. https://doi.org/10.1067/mva.2003.77
Getzen LC, Pollak EW (1979) Short-term femoral vein catheterization: a safe alternative venous access? Am J Surg 138(6):875–878. https://doi.org/10.1016/0002-9610(79)90314-3
Giordano NJ, Jansson PS, Young MN, Hagan KA, Kabrhel C (2017) Epidemiology, pathophysiology, stratification, and natural history of pulmonary embolism. Tech Vasc Interv Radiol 20(3):135–140. Accessed 26 Apr 2022. https://pubmed.ncbi.nlm.nih.gov/29029707/
Glovkzki P, Driscoll DJ (2007) Klippel–Trenaunay syndrome: current management. Phlebology 22(6):291–298. Accessed 14 May 2022. https://journals.sagepub.com/doi/abs/10.1177/026835550702200611
Goldman S, Zadina K, Moritz T et al (2004) Long-term patency of saphenous vein and left internal mammary artery grafts after coronary artery bypass surgery: results from a Department of Veterans Affairs Cooperative Study. J Am Coll Cardiol 44(11):2149–2156. https://doi.org/10.1016/j.jacc.2004.08.064
Goode SD, Chowdhury A, Crockett M et al (2010) Laser and radiofrequency ablation study (LARA study): a randomised study comparing radiofrequency ablation and endovenous laser ablation (810nm). Eur J Vasc Endovasc Surg 40(2):246–253. https://doi.org/10.1016/j.ejvs.2010.02.026
Goodman LR. 1999 plenary session: Friday imaging symposium: CT diagnosis of pulmonary embolism and deep venous thrombosis – PubMed. Radiographics 20(4):1201–1205. Accessed 15 May 2022. https://pubmed.ncbi.nlm.nih.gov/10903711/
Goodney PP, Beck AW, Nagle J, Welch HG, Zwolak RM (2009) National trends in lower extremity bypass surgery, endovascular interventions, and major amputations. J Vasc Surg 50(1):54–60. https://doi.org/10.1016/j.jvs.2009.01.035
Gradman WS, Cohen W, Haji-Aghaii M (2001) Arteriovenous fistula construction in the thigh with transposed superficial femoral vein: our initial experience. J Vasc Surg 33(5):968–975. https://doi.org/10.1067/mva.2001.115000

Gradman WS, Laub J, Cohen W (2005) Femoral vein transposition for arteriovenous hemodialysis access: improved patient selection and intraoperative measures reduce postoperative ischemia. J Vasc Surg 41(2):279–284. https://doi.org/10.1016/j.jvs.2004.10.039

Hazinedaroğlu SM, Tüzüner A, Ayli D, Demirer S, Duman N, Yerdel MA (2004) Femoral vein transposition versus femoral loop grafts for hemodialysis: a prospective evaluation. Transplant Proc 36(1):65–67. https://doi.org/10.1016/j.transproceed.2003.11.031

Helmy ElKaffas K, ElKashef O, ElBaz W (2011) Great saphenous vein radiofrequency ablation versus standard stripping in the management of primary varicose veins-a randomized clinical trial. Angiology 62(1):49–54. https://doi.org/10.1177/0003319710380680

Hill SL, Hancock DH, Webb TL (2008) Thrombophlebitis of the great saphenous vein–recommendations for treatment. Phlebology 23(1):35–39. Accessed 1 May 2022. https://journals.sagepub.com/doi/10.1258/phleb.2007.007023

Hilty WM, Hudson PA, Levitt MA, Hall JB (1997) Real-time ultrasound–guided femoral vein catheterization during cardiopulmonary resuscitation. Ann Emerg Med 29(3):331–337. Accessed 3 May 2022. https://www.sciencedirect.com/science/article/abs/pii/S0196064497703445

Hoang V, Sills J, Chandler M, Busalani E, Clifton-Koeppel R, Modanlou HD (2008) Percutaneously inserted central catheter for total parenteral nutrition in neonates: complications rates related to upper versus lower extremity insertion. Pediatrics 121(5):e1152–e1159. Accessed 4 May 2022. https://pubmed.ncbi.nlm.nih.gov/18390957/

Holbrook A, Schulman S, Witt DM et al (2012) Evidence-based management of anticoagulant therapy. Chest 141(2 Suppl):e152S–e184S. https://doi.org/10.1378/chest.11-2295

Jones L, Braithwaite BD, Selwyn D, Cooke S, Earnshaw JJ (1996) Neovascularisation is the principal cause of varicose vein recurrence: results of a randomised trial of stripping the long saphenous vein. Eur J Vasc Endovasc Surg 12(4):442–445. https://doi.org/10.1016/S1078-5884(96)80011-6

Kachlik D, Pechacek V, Baca V, Musil V (2010) The superficial venous system of the lower extremity: new nomenclature. Phlebol J Venous Dis 25(3):113–123. https://doi.org/10.1258/phleb.2009.009046

Kachlik D, Pechacek V, Musil V, Baca V (2012) The deep venous system of the lower extremity: new nomenclature. Phlebol J Venous Dis 27(2):48–58. https://doi.org/10.1258/phleb.2011.010081

Kearon C, Akl EA, Comerota AJ et al (2012) Antithrombotic therapy for VTE disease. Chest 141(2 Suppl):e419S–e494S. https://doi.org/10.1378/chest.11-2301

Keiler J, Schulze M, Claassen H, Wree A (2018) Human femoral vein diameter and topography of valves and tributaries: a post mortem analysis: femoral vein dimensions, tributaries, and valves. Clin Anat 31(7):1065–1076. https://doi.org/10.1002/ca.23224

Kern KB (2003) Limiting interruptions of chest compressions during cardiopulmonary resuscitation. Resuscitation 58(3):273–274. https://doi.org/10.1016/S0300-9572(03)00269-7

Kim FY, Marhefka G, Ruggiero NJ, Adams S, Whellan DJ (2013) Saphenous vein graft disease: review of pathophysiology, prevention, and treatment. Cardiol Rev 21(2):101–109. https://doi.org/10.1097/CRD.0b013e3182736190

Klinkert P, Post PN, Breslau PJ, van Bockel JH (2004) Saphenous vein versus PTFE for above-knee femoropopliteal bypass. A review of the literature. Eur J Vasc Endovasc Surg 27(4):357–362. https://doi.org/10.1016/j.ejvs.2003.12.027

Kockaert M, de Roos KP, van Dijk L, Nijsten T, Neumann M (2012) Duplication of the great saphenous vein: a definition problem and implications for therapy. Dermatol Surg 38(1):77–82. https://doi.org/10.1111/j.1524-4725.2011.02154.x

Kolikof J, Peterson K, Baker AM (2022) Central venous catheter. In: StatPearls. StatPearls Publishing. Accessed 3 May 2022. http://www.ncbi.nlm.nih.gov/books/NBK557798/

Koopman MMW, Prandoni P, Piovella F et al (1996) Treatment of venous thrombosis with intravenous unfractionated heparin administered in the hospital as compared with subcutaneous low-molecular-weight heparin administered at home. N Engl J Med 334(11):682–687. https://doi.org/10.1056/NEJM199603143341102

Krishnamoorthy B, Critchley WR, Thompson AJ et al (2017) Study comparing vein integrity and clinical outcomes in open vein harvesting and 2 types of endoscopic vein harvesting for coronary artery bypass grafting. Circulation 136(18):1688–1702. https://doi.org/10.1161/CIRCULATIONAHA.117.028261

Kullo IJ, Rooke TW (2016) Peripheral artery disease. N Engl J Med 374(9):861–871. Accessed 7 May 2022. https://www.nejm.org/doi/10.1056/NEJMcp1507631

Lee DK, Ahn KS, Kang CH, Cho SB (2017) Ultrasonography of the lower extremity veins: anatomy and basic approach. Ultrasonography 36(2):120–130. https://doi.org/10.14366/usg.17001

Leon L, Giannoukas AD, Dodd D, Chan P, Labropoulos N (2005) Clinical significance of superficial vein thrombosis. Eur J Vasc Endovasc Surg 29(1):10–17. https://doi.org/10.1016/j.ejvs.2004.09.021

Levine M, Gent M, Hirsh J et al (1996) A comparison of low-molecular-weight heparin administered primarily at home with unfractionated heparin administered in the hospital for proximal deep-vein thrombosis. N Engl J Med 334(11):677–681. https://doi.org/10.1056/NEJM199603143341101

Lim W, Le Gal G, Bates SM et al (2018) American Society of Hematology 2018 guidelines for management of venous thromboembolism: diagnosis of venous thromboembolism. Blood Adv 2(22):3226–3256. https://doi.org/10.1182/bloodadvances.2018024828

Liu G, Ferris E, Reifsteck BM (1986) Effect of anatomic variations on deep venous thrombosis of the lower extremity. Am J Roentgenol 146(4):845–848. https://doi.org/10.2214/ajr.146.4.845

Lowe GDO (2003) Virchow's triad revisited: abnormal flow. Pathophysiol Haemost Thromb 33(5–6):455–457. https://doi.org/10.1159/000083845

Lutter KS, Kerr TM, Roedersheimer LR, Lohr JM, Sampson MG, Cranley JJ (1991) Superficial thrombophlebitis diagnosed by duplex scanning. Surgery 110(1):42–46

Mangiante EC, Hoots AV, Fabian TC (1988) The percutaneous common femoral vein catheter for volume replacement in critically injured patients. J Trauma 28(12):1644–1649. https://doi.org/10.1097/00005373-198812000-00005

McGee DC, Gould MK (2003) Preventing complications of central venous catheterization. N Engl J Med 348(12):1123–1133. https://doi.org/10.1056/NEJMra011883

Merrer J, De Jonghe B, Golliot F et al (2001) Complications of femoral and subclavian venous catheterization in critically ill patients: a randomized controlled trial. JAMA 286(6):700–707. https://doi.org/10.1001/jama.286.6.700

Mese B, Bozoglan O, Eroglu E et al (2015) A comparison of 1,470-nm endovenous laser ablation and radiofrequency ablation in the treatment of great saphenous veins 10 mm or more in size. Ann Vasc Surg 29(7):1368–1372. https://doi.org/10.1016/j.avsg.2015.03.063

Messmore HL, Bishop M, Wehrmacher WH (n.d) Acute venous thrombosis: therapeutic choices for superficial and deep veins. Postgrad Med 89(7):73–77. Accessed 1 May 2022. https://www.tandfonline.com/doi/abs/10.1080/00325481.1991.11700941

Moore HM, Gohel M, Davies AH (2011) Number and location of venous valves within the popliteal and femoral veins – a review of the literature. J Anat 219(4):439–443. https://doi.org/10.1111/j.1469-7580.2011.01409.x

Mühlberger D, Morandini L, Brenner E (2009) Venous valves and major superficial tributary veins near the saphenofemoral junction. J Vasc Surg 49(6):1562–1569. https://doi.org/10.1016/j.jvs.2009.02.241

Needleman L, Cronan JJ, Lilly MP et al (2018) Ultrasound for lower extremity deep venous thrombosis. Circulation 137(14):1505–1515. https://doi.org/10.1161/CIRCULATIONAHA.117.030687

Nijsten T, van den Bos RR, Goldman MP et al (2009) Minimally invasive techniques in the treatment of saphenous varicose veins. J Am Acad Dermatol 60(1):110–119. https://doi.org/10.1016/j.jaad.2008.07.046

Nordon IM, Hinchliffe RJ, Brar R et al (2011) A prospective double-blind randomized controlled trial of radiofrequency versus laser treatment of the great saphenous vein in patients with varicose veins. Ann Surg 254(6):876–881. https://doi.org/10.1097/SLA.0b013e318230af5a

Norgren L, Hiatt WR, Dormandy JA, Nehler MR, Harris KA, Fowkes FGR (2007) Inter-society consensus for the management of peripheral arterial disease (TASC II). J Vasc Surg 45(1):S5–S67. https://doi.org/10.1016/j.jvs.2006.12.037

O'Banion LA, Wu B, Eichler CM, Reilly LM, Conte MS, Hiramoto JS (2017) Cryopreserved saphenous vein as a last-ditch conduit for limb salvage. J Vasc Surg 66(3):844–849. Accessed 7 May 2022. https://www.jvascsurg.org/article/S0741-5214(17)30936-9/fulltext

Park EA, Chung JW, Lee W et al (2011) Three-dimensional evaluation of the anatomic variations of the femoral vein and popliteal vein in relation to the accompanying artery by using CT venography. Korean J Radiol 12(3):327–340. https://doi.org/10.3348/kjr.2011.12.3.327

Kachlik D, Pechacek V, Musil V, Baca V (2010) The venous system of the pelvis: new nomenclature. Phlebology 25(4):162–173. Accessed 19 May 2022. https://journals.sagepub.com/doi/10.1258/phleb.2010.010006

Pierre-Paul D, Williams S, Lee T, Gahtan V (2004) Saphenous vein loop to femoral artery arteriovenous fistula: a practical alternative. Ann Vasc Surg 18(2):223–227. https://doi.org/10.1007/s10016-004-0016-7

Prandoni P, Lensing AW, Cogo A et al (1996) The long-term clinical course of acute deep venous thrombosis. Ann Intern Med 125(1):1–7. https://doi.org/10.7326/0003-4819-125-1-199607010-00001

Purdue GF, Hunt JL (1986) Vascular access through the femoral vessels: indications and complications. J Burn Care Rehabil 7(6):498–500. https://doi.org/10.1097/00004630-198611000-00011

Quinn KL, Vandeman FN (1990) Thrombosis of a duplicated superficial femoral vein. Potential error in compression ultrasound diagnosis of lower extremity deep venous thrombosis. J Ultrasound Med 9(4):235–238. https://doi.org/10.7863/jum.1990.9.4.235

Raju S, Fredericks R (1991) Venous obstruction: an analysis of one hundred thirty-seven cases with hemodynamic, venographic, and clinical correlations. J Vasc Surg 14(3):305–313. https://doi.org/10.1016/0741-5214(91)90081-5

Raju S, Fountain T, Neglén P, Devidas M (1998) Axial transformation of the profunda femoris vein. J Vasc Surg 27(4):651–659. https://doi.org/10.1016/S0741-5214(98)70230-7

Rasmussen LH, Lawaetz M, Bjoern L, Vennits B, Blemings A, Eklof B (2011) Randomized clinical trial comparing endovenous laser ablation, radiofrequency ablation, foam sclerotherapy and surgical stripping for great saphenous varicose veins. Br J Surg 98(8):1079–1087. https://doi.org/10.1002/bjs.7555

Sabik JF, Lytle BW, Blackstone EH, Houghtaling PL, Cosgrove DM (2005) Comparison of saphenous vein and internal thoracic artery graft patency by coronary system. Ann Thorac Surg 79(2):544–551. https://doi.org/10.1016/j.athoracsur.2004.07.047

Sato K, Orihashi K, Takahashi S et al (2011) Three-dimensional CT venography: a diagnostic modality for the preoperative assessment of patients with varicose veins. Ann Vasc Dis 4(3):229–234. https://doi.org/10.3400/avd.oa.11.00021

Schanzer A, Hevelone N, Owens CD et al (2007) Technical factors affecting autogenous vein graft failure: observations from a large multicenter trial. J Vasc Surg 46(6):1180–1190. https://doi.org/10.1016/j.jvs.2007.08.033

Screaton NJ, Gillard JH, Berman LH, Kemp PM (1998) Duplicated superficial femoral veins: a source of error in the sonographic investigation of deep vein thrombosis. Radiology 206(2):397–401. https://doi.org/10.1148/radiology.206.2.9457192

Selby R, Geerts WH (2000) Venous thromboembolism: risk factors and prophylaxis. Semin Respir Crit Care Med 21(6):493–501. https://doi.org/10.1055/s-2000-13189

Shepherd AC, Gohel MS, Brown LC, Metcalfe MJ, Hamish M, Davies AH (2010) Randomized clinical trial of VNUS® ClosureFAST™ radiofrequency ablation versus laser for varicose veins. Br J Surg 97(6):810–818. https://doi.org/10.1002/bjs.7091

Souza DSR, Dashwood MR, Tsui JCS et al (2002) Improved patency in vein grafts harvested with surrounding tissue: results of a randomized study using three harvesting techniques. Ann Thorac Surg 73(4):1189–1195. https://doi.org/10.1016/S0003-4975(02)03425-2

Spritzer CE (n.d) Progress in MR imaging of the venous system. Perspect Vasc Surg Endovasc Ther 21(2):105–116. Accessed 16 May 2022. https://pubmed.ncbi.nlm.nih.gov/19640860/

Standring S (ed) (2016) Gray's anatomy: the anatomical basis of clinical practice, 41st edn. Elsevier Limited

Subramonia S, Lees T (2010) Randomized clinical trial of radiofrequency ablation or conventional high ligation and stripping for great saphenous varicose veins. Br J Surg 97(3):328–336. https://doi.org/10.1002/bjs.6867

Sullivan V, Denk PM, Sonnad SS, Eagleton MJ, Wakefield TW (2001) Ligation versus anticoagulation: treatment of above-knee superficial thrombophlebitis not involving the deep venous system. J Am Coll Surg 193(5):556–562. https://doi.org/10.1016/s1072-7515(01)01043-2

Thibault P, Bray A, Wlodarczyk J, Lewis W (1990) Cosmetic leg veins: evaluation using duplex venous imaging. J Dermatol Surg Oncol 16(7):612–618. https://doi.org/10.1111/j.1524-4725.1990.tb00089.x

Thomas RH (2001) Hypercoagulability syndromes. Arch Intern Med 161(20):2433–2439. https://doi.org/10.1001/archinte.161.20.2433

Towne JB, Schmitt DD, Seabrook GR, Bandyk DF (1991) The effect of vein diameter on patency of in situ grafts. J Cardiovasc Surg 32(2):192–196

Treiman GS, Silberman H (1993) Chronic venous access in patients with cancer. Selective use of the saphenous vein. Cancer 72(3):760–765. https://doi.org/10.1002/1097-0142(19930801)72:3<760::aid-cncr2820720320>3.0.co;2-j

Trottier SJ, Veremakis C, O'Brien J, Auer AI (1995) Femoral deep vein thrombosis associated with central venous catheterization: results from a prospective, randomized trial. Crit Care Med 23(1):52–59. https://doi.org/10.1097/00003246-199501000-00011

Tubbs RS, Loukas M, Shoja MM, Apaydin N, Oakes WJ, Salter EG (2007) Anatomy and potential clinical significance of the vastoadductor membrane. Surg Radiol Anat 29(7):569–573. https://doi.org/10.1007/s00276-007-0230-4

Uhl JF (2015) Focus on venous embryogenesis of the human lower limbs. Phlebolymphology 22:55–63

Uhl JF, Gillot C (2007) Embryology and three-dimensional anatomy of the superficial venous system of the lower limbs. Phlebol J Venous Dis 22(5):194–206. https://doi.org/10.1258/026835507782101717

Uhl JF, Gillot C, Chahim M (2010) Anatomical variations of the femoral vein. J Vasc Surg 52(3):714–719. https://doi.org/10.1016/j.jvs.2010.04.014

Vähäaho S, Halmesmäki K, Albäck A, Saarinen E, Venermo M (2018) Five-year follow-up of a randomized clinical trial comparing open surgery, foam sclerotherapy and endovenous laser ablation for great saphenous varicose veins. Br J Surg 105(6):686–691. https://doi.org/10.1002/bjs.10757

van den Bremer J, Moll FL (2010) Historical overview of varicose vein surgery. Ann Vasc Surg 24(3):426–432. https://doi.org/10.1016/j.avsg.2009.07.035

van Langevelde K, Lijfering WM, Rosendaal FR, Cannegieter SC (2011) Increased risk of venous thrombosis in persons with clinically diagnosed superficial vein thrombosis: results from the MEGA study. Blood 118(15):4239–4241. https://doi.org/10.1182/blood-2011-05-356071

Venermo M, Saarinen J, Eskelinen E et al (2016) Randomized clinical trial comparing surgery, endovenous laser ablation and ultrasound-guided foam sclerotherapy for the treatment of great saphenous varicose veins. Br J Surg 103(11):1438–1444. https://doi.org/10.1002/bjs.10260

Wells PS, Hirsh J, Anderson DR et al (1995) Accuracy of clinical assessment of deep-vein thrombosis. Lancet Lond Engl 345(8961):1326–1330. https://doi.org/10.1016/s0140-6736(95)92535-x

Williams JF, Seneff MG, Friedman BC et al (1991) Use of femoral venous catheters in critically ill adults: prospective study. Crit Care Med 19(4):550–553. https://doi.org/10.1097/00003246-199104000-00016

Veins of the Foot

118

Patrick J. McCarty

118.1 Anatomy

The venous drainage of the foot consists of deep and superficial systems. The plantar digital veins, originating from a plexus, merge with the dorsal digital veins to form four plantar metatarsal veins. These veins run between the metatarsals, creating the deep plantar venous arch. The lateral and medial plantar veins emerge from this arch and contribute to the short and great saphenous veins before forming the posterior tibial vein.

In the sole, superficial veins form a superficial arch at the toe roots, draining into the medial and lateral marginal veins. A particularly dense venous plexus in the heel's fat connects with the plantar venous arch and other deep veins but primarily drains into the marginal veins (Ricci et al. 2014).

The dorsal venous arch collects blood from the plantar and metatarsal digital veins. A proximal venous network receives tributaries from the deep venous system. The lateral dorsal venous arch drains into the short saphenous vein, which travels posterior to the lateral malleolus to the posterolateral leg. The medial dorsal venous arch forms a short segment, the dorsalis pedis vein, which drains into the great saphenous vein (Figs. 118.1 and 118.2).

An irregular dorsal venous network proximal to the dorsal venous arch, which receives tributaries from deep veins, connects with the venous network in the leg. This network links with the medial and lateral marginal veins on either side of the foot, formed mainly by veins from the superficial parts of the sole. The great saphenous vein drains the sole via medial marginal veins at the ankle, while the small saphenous vein connects with deep veins on the foot's dorsum.

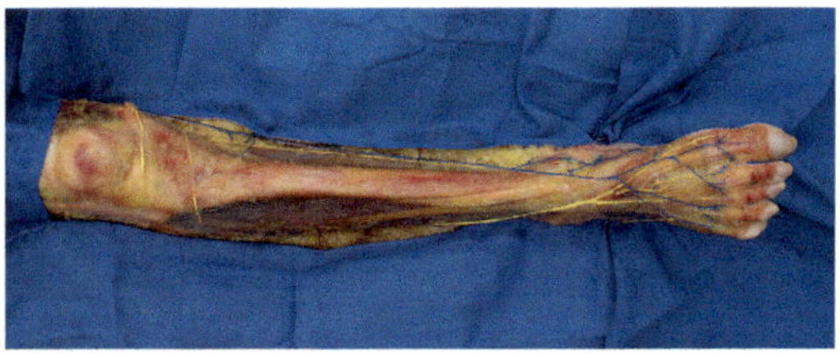

Fig. 118.1 Superficial dissection of the right-sided cadaveric lower limb. The superficial veins are colored blue, and here, the superficial venous network of the foot coalesces into the greater saphenous vein that ascends up the medial side of the lower limb to drain proximally into the femoral vein. Note the saphenous nerve traveling with the great saphenous vein in the leg

P. J. McCarty (✉)
Tulane University School of Medicine,
New Orleans, LA, USA
e-mail: pmccarty2@tulane.edu

S. J. Bordes, Jr. et al. (eds.), *The Clinical Anatomy of the Vascular System*,
https://doi.org/10.1007/978-3-031-78326-5_118

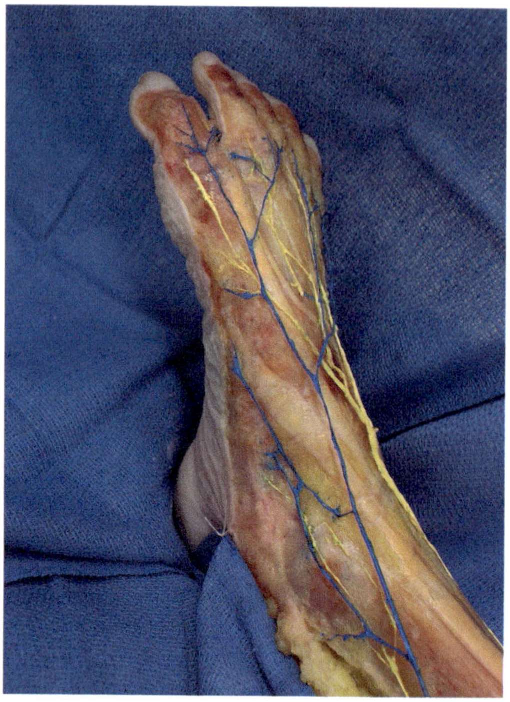

Fig. 118.2 Medial view of the image shown in Fig. 118.1. Note that branches of the superficial fibular nerve travel with the dorsal venous plexus of the foot

Reference

Ricci S, Moro L, Incalzi RA (2014) The foot venous system: anatomy, physiology and relevance to clinical practice. Dermatologic Surg 40:225–233

119 Lymphatics of the Lower Limb

Jonathon Chon Teng Chio
and Stephen J. Bordes, Jr.

Abbreviations

DVT	Deep vein thrombosis
GSV	Great saphenous vein
IL	Interleukin
iNOS	Inducible nitric oxide synthase
LECs	Lymphatic endothelial cells
MRI	Magnetic resonance imaging
SSV	Small saphenous vein
T	Thymus
VECs	Vascular endothelial cells
VEGF	Vascular endothelial growth factor
VEGF-R	Vascular endothelial growth factor receptor

119.1 Introduction

The lymphatic system has two overarching purposes: draining tissue fluid and macromolecules from the interstitial space back into the blood and mediating the immune response. This is accomplished when substances enter lymphatic vessels, which are distributed throughout the body to form a distinct vascular network that directs them toward the lymph nodes for filtration prior to entering the venous system (Ribatti and Crivellato 2010; Garza et al. 2009). The same description applies to the lower limb, which has two broad classes of lymphatic vessels (Garza et al. 2009): superficial (primary) and secondary. A major characteristic of superficial (primary) lymphatic vessels is their reliance on osmotic gradients and hydrostatic pressure differences to facilitate interstitial fluid/protein movement and absorption, as their walls lack smooth muscles. In contrast, the walls of secondary lymphatic vessels have a muscular component (albeit significantly thinner than those in arteries and veins). Another defining feature of secondary lymphatic vessels is the presence of valves that assist in afferent fluid movement. Despite the clear differences between the two systems, they cooperate with each other. The primary lymphatics collect fluid, protein, cells, and debris for transport via the secondary lymphatics, which propel fluid unidirectionally to lymph nodes and eventually the venous system (Mortimer and Rockson 2014; Petrova et al. 2004).

119.2 Embryology

Embryonic development of the lymphatic vascular system starts considerably later than that of the blood vasculature system (Wilting et al.

J. C. T. Chio
Tulane University School of Medicine,
New Orleans, LA, USA
e-mail: cchio@tulane.edu

S. J. Bordes, Jr. (✉)
Department of Surgery, Louisiana State University
Health Sciences Center, New Orleans, LA, USA

S. J. Bordes, Jr. et al. (eds.), *The Clinical Anatomy of the Vascular System*,
https://doi.org/10.1007/978-3-031-78326-5_119

2007). In the human embryo, the first lymph sacs develop in the jugular region, which is where the cranial and caudal jugular veins fuse into the common cardinal vein. There is a period of 3–4 weeks between the first appearances of vascular endothelial cells (VECs) and lymphatic endothelial cells (LECs). Studies on mammalian embryos indicate the presence of eight lymph sacs (three paired and two unpaired) (Shinaoka et al. 2020). The paired ones are the jugular, subclavian, and posterior lymph sacs, and the unpaired ones are the cisterna chyli and the retroperitoneal (mesenterial) lymph sac (Ribatti and Crivellato 2010; Shinaoka et al. 2020). The transcription factor *Prox1* is regarded as an excellent marker for LECs because it facilitates the transdifferentiation of VECs into LECs (Wilting et al. 2007; Hong et al. 2002; Petrova et al. 2002). Studies using this marker provide evidence that LECs develop from specialized parts of the venous system (Shinaoka et al. 2020). However, there is growing evidence that scattered mesenchymal cells are integrated into the growing fetal lymphatics.

Similarly, vascular endothelial growth factors (VEGF), specifically VEGF- C and D, and their receptor (VEGFR), have been used to assess lymphatic development (Ribatti and Crivellato 2010). This system relies on Prox-1 (Wigle and Oliver 1999; Wigle et al. 2002); Prox-1-deficient mice have normal vascular systems, but lymphatic development is arrested in day 10.5. Prox-1 is important as it induces the expression and permits the migration of VEGF-C^{+} LECs (Petrova et al. 2004). Furthermore, VEGF-C-deficient mice do not form lymph sacs but can be rescued by administering VEGF-C[11]. In response to VEGF-C, lymphatic cells become more sensitive to it by further increasing the expression of VEGFR-3, receptor tyrosine kinase (Syk), and substrate (SLP-76) (Ribatti and Crivellato 2010; Kriehuber et al. 2001; Abtahian et al. 2003). Regarding clinical relevance, 30% of patients with primary lymphedema (which will be described in later sections) have mutations involving the VEGF-C signaling pathway (Karkkainen et al. 2004; Connell et al. 2013).

119.3 Anatomy/Topography

A road map of the lower limb lymphatics has been determined using fluorescent lymphography (Shinaoka et al. 2020). These vessels originate beneath the dermis of each side of the toes, foot, and lateral thigh (Pan et al. 2011).

There are two major sets of superficial lymphatic vessels (Shinaoka et al. 2020). The medial vessels, comprising the posteromedial and anteromedial groups, closely follow the great saphenous vein (GSV). Originating on the dorsal surface of the foot, they travel up the anterior and posterior aspects of the medial leg along with the GSV, pass behind the medial condyle of the femur, and run along the sartorius muscle in the proximal thigh. They end in the groin, draining into the subinguinal group of inguinal lymph nodes (Fig. 119.1). Usually, 1–2 posteromedial vessels are connected to the inguinal lymph

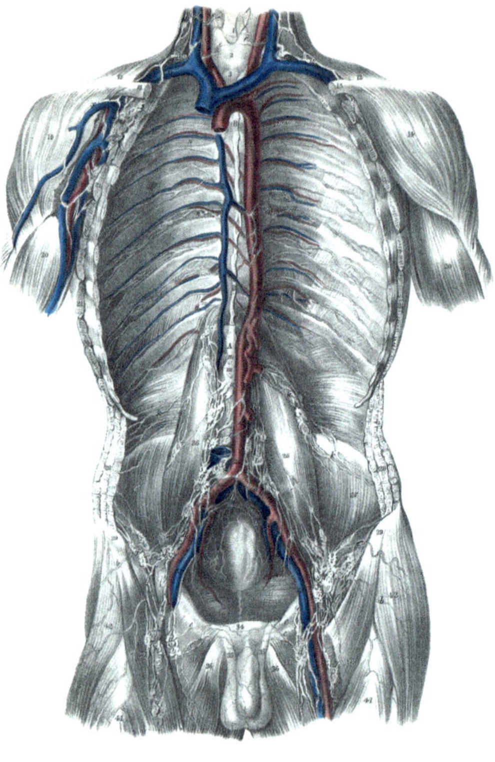

Fig. 119.1 Drawing indicating the coalescence of lower limb lymphatics in the inguinal region that then passes into the abdomen (Quain and Wilson, 1844 author's collection)

nodes. In the middle thigh, they cross the GSV to reach the lateral side. Occasionally, they branch in both the lower leg and the thigh around the knee, entering Hunter's canal, which is located at the medial aspect of the middle third of the thigh. From there, the vessels run below the deep fascia to become the deep lymphatic vessels. The other set of superficial lymphatic vessels are the anteromedial vessels, 1–5 of which are to be found around the anteromedial aspect of the ankle. These vessels run in the anteromedial aspect of the lower leg, forming a gentle sigmoidal curve parallel to the cutaneous branches of the GSV and the main trunk of the GSV in the thigh. They course in the lateral side of the GSV in the upper thigh, converging before joining their regional nodes.

Another major set of lower limb lymphatic vessels, the lateral vessels, consists of anterolateral and posterolateral groups (Shinaoka et al. 2020). In contrast to the superficial lymphatic vessels, the lateral vessels follow the small saphenous vein (SSV) to enter the popliteal nodes or ascend in front of the leg and cross just below the knee joint to join the previously described superficial lymphatic vessels. The anterolateral group of lymphatic vessels is found in the middle of the subcutaneous fat layer at the anterolateral aspect of the ankle. They travel in the anterolateral aspect of the lower leg and gradually shift toward the anteromedial aspect. The posterolateral group of vessels, as the name suggests, is found in the posterolateral aspect of the ankle level. They travel along the SSV and connect to the popliteal lymph nodes at the popliteal fossa. Occasionally, they connect to the anteromedial or anterolateral group.

A third group of lower limb lymphatic vessels, the deep lymphatic vessels, is significantly fewer in number than the superficial lymphatic vessels (Shinaoka et al. 2020). The deep vessels accompany the deep arteries of the lower leg and are found in three main groups: anterior tibial, posterior tibial, and peroneal. They are named in relation to the corresponding artery they follow to enter the popliteal lymph nodes.

After classifying the lower limb lymphatic vessels based on their anatomical layers, another way of categorizing them is based on the region of the lower limb in which they are found.

The toes contain two lymph collecting vessels that travel in the subcutaneous tissue along the mid-axial lines (Pan et al. 2011). Vessels from neighboring toes converge in the web spaces of the foot, except for those on the medial and lateral borders of the first and fifth toes. From the web spaces, 14 lymphatic vessels (on average) from the toes merge in the subcutaneous tissue of the dorsum of the foot. These vessels anastomose with each other to form a large lymphatic network. Moving superiorly, there is an average of 12 lymph collecting vessels in the subcutaneous tissue of the ankle. Distributed in the anterior aspect of the ankle between the lateral and medial malleoli, these vessels are continuous with those in the dorsum of the foot.

There is an average of 13 lymph collecting vessels distributed in the subcutaneous tissue of the leg (Pan et al. 2011). These can be divided into anteromedial, anterolateral, and posterior groups (Shinaoka et al. 2020; Pan et al. 2011). The largest vessels are mostly in the anteromedial and posterior groups (Pan et al. 2011). Anterolateral lymphatic vessels are relatively sparse, following a curved course toward the anteromedial aspect of the proximal third of the leg. In contrast, the anteromedial lymphatics are denser, follow a straighter course, and traverse adjacent to the GSV.

The subcutaneous tissue of the thigh has an average of 29 lymph-collecting vessels forming three groups, which differ in their origin and route of travel (Pan et al. 2011). The anterior group originates from the anterolateral aspect of the thigh and runs obliquely in the subcutaneous tissue before entering the lateral group of superficial inguinal lymph nodes. The medial group is continuous with vessels from the leg. They run in the subcutaneous tissue of the medial aspect of the thigh (adjacent to the GSV and its branches) and drain into the superficial inguinal lymph nodes. The posterior group originates from the posterolateral side of the thigh, running obliquely in the subcutaneous tissue of the posterior aspect of the thigh and entering the medial group of superficial inguinal lymph nodes. The largest vessels are found mostly in the anteromedial and posterior groups. The anterolateral lymphatics are again relatively sparse and travel obliquely toward the anteromedial aspect of the thigh. The anteromedial lymphatics are denser, with a straighter course, following the GSV and its branches.

There is a deep group of lymphatic vessels within the superficial femoral vascular bundle. These vessels are continuous with those from the posterior group of the leg after passing through the popliteal lymph nodes. They form 2–4 large vessels and travel adjacent to the superficial femoral artery and veins through femoral lymph nodes.

The lymphatic vessels of the lower limb drain into the inguinal nodes found in the upper aspect of the femoral triangle (de Carvalho et al. 2011). These nodes are subdivided according to their position relative to a horizontal line drawn at the level where the GSV terminates. Superficial inguinal nodes form a line directly below the inguinal ligament and receive lymph from the penis, scrotum, perineum, buttock, and abdominal wall (Tappouni et al. 2011). Approximately 10 nodes are found in the superficial fascia and parallel the course of the inguinal ligament in the upper thigh. Medially, they extend inferiorly along the terminal part of the GSV. Lymph from the superficial inguinal lymph nodes drains through vessels accompanying the femoral vessels into the external iliac nodes, associated with the external iliac artery, into the abdomen.

The subinguinal lymph nodes comprise superficial and deep groups (Bui and Bordoni 2022). Located on each side of proximal GSV, they receive lymph primarily from the superficial lateral and medial lymphatic vessels of the lower leg. The deep subinguinal nodes are most commonly found on the medial aspect of the femoral vein and receive lymph from deep lymphatic trunks associated with the femoral vessels and from those associated with the glans penis (or clitoris) and the perineum, which accompany the femoral vessels (Gvetadze and Ilkaev 2020). The deep subinguinal lymph nodes interconnect with their superficial counterparts and drain into external iliac nodes through vessels that pass along the medial side of the femoral vein as it passes under the inguinal ligament.

A third group of lower limb lymph nodes, the popliteal nodes, is embedded in the fat reserves of the popliteal fossa (Musters et al. 2011). They receive lymph from the lateral superficial vessels and drain into popliteal nodes that pass alongside the femoral vessels to empty into the deep inguinal nodes. Some popliteal lymph nodes accompany the GSV and drain into subinguinal nodes.

119.4 Imaging/Histology

First used as a diagnostic tool in 1953 by Sherman and Pergossian, lymphoscintigraphy involves interstitial injections of radioactive colloidal gold to visualize lymph nodes (Collins et al. 1989). However, the use of high dose radiation and the resulting evaluation has proved suboptimal for visualizing lymphatic channels. Despite its initial slow adoption, lymphoscintigraphy is the current reference for lymphatic imaging (Shinaoka et al. 2020; Witte et al. 2000; Maegawa et al. 2010), as it can help to exclude lymphedema as a cause of limb swelling in approximately 1/3 of patients and can differentiate between lymphedema and edema of venous origin. Assessment of radiolabeled protein (technetium 99 m-labeled colloid and albumin) allows lymphatic function, movement, drainage, and response to treatment to be measured (Mortimer 1995; Nawaz et al. 1992; Williams et al. 2000). Abnormal findings on lymphoscintigrams include interruption of lymphatic flow, collateral lymph vessels, dermal backflow, and flow and visualization of lymph nodes (Moshiri et al. 2002). These arise in the context of post-thrombotic disease, where there is reduced subfascial lymphatic flow but normal epifascial flow (Bräutigam et al. 1993). However, in lymphedema, both epifascial and subfascial lymphatics are abnormal. Therefore, evaluating flow in both compartments is key to differentiating among various diseases involving the lymphatic system.

Another method for measuring lymphatic flow is ultrasound, which reveals volumetric changes in the soft tissues (Doldi et al. 1992; Kim et al. 2013). These can include minor increases in the thickness of the dermis and subcutaneous layers, increases, decreases, or no changes in muscle mass, and structural changes. Patients with lymphedema exhibit gradual impedance of venous return, which aggravates the edema (Doldi et al. 1992). Therefore, evaluating these patients with duplex ultrasound can be particularly useful.

Single-photon emission computed tomography can be used to confirm the diagnosis and monitor treatment following disorders of the lymphatic system (Shinaoka et al. 2020; Marotel et al. 1998). Depending on the condition, physicians should look for the presence (or absence) of a honeycomb appearance, thickening of the calf skin, thickening of the subcutaneous compartment, increased fat density, and thickening of the perimuscular aponeurosis (Marotel et al. 1998; Hadjis et al. 1985; Vaughan 1990). For example, in patients with chronic venous disease, there is thickening of the skin and enlargement of the subcutaneous compartment, but no honeycomb appearance (Hadjis et al. 1985). Individuals with lipedema have an enlarged subcutaneous compartment but normal skin thickness and subfascial compartment (Hadjis et al. 1985). Patients with deep venous thrombosis exhibit an increased subcutaneous layer with signs of lymphedema, increased cross-sectional muscle area, and enlarged superficial veins (Vaughan 1990).

A fourth technique is magnetic resonance imaging (MRI), which can be used to differentiate between lymphedema, lipedema, and phlebedema (Shinaoka et al. 2020; Werner et al. 1998; Baulieu et al. 2013). Features of lymphedema on MRI include circumferential edema, increased volume of subcutaneous tissue and honeycomb pattern between muscle and subcutis, along with marked thickening of the dermis (Werner et al. 1998; Haaverstad et al. 1994; Duewell et al. 1992). Depending on the specific pathology, values can increase by different amounts: an average of 26% in patients with deep vein thrombosis and an average of 40% in chronic lymphoma patients. However, one negative aspect of MRI is the inability to differentiate between primary and secondary lymphedema (Idy-Peretti et al. 1998).

Despite the various modalities, all techniques involve intradermal or subcutaneous injection of tracer or contrast material into the foot (Shinaoka et al. 2020). The most common site of tracer injection for lymphatic imaging is the web space between the first and second toes, with the web space between the second and third toes as an alternative (Gloviczki et al. 1989). Injections into multiple toe web spaces have also been reported (Witte et al. 2000; Maegawa et al. 2010; Szuba et al. 2003).

119.5 Variations

Lymphoscintigraphy has revealed some aberrant drainage patterns of the superficial tissues in the lower extremities (Uren et al. 2000; Reynolds et al. 2007; Pan et al. 2013). Most lymphatic vessels drain into superficial inguinal lymph nodes. However, one or two lymph collecting vessels arising from the lateral side of the Achilles tendon were found in the subcutaneous tissue above the heel, traveling with the small saphenous vein (Pan et al. 2011, 2013; Kaplan 1983). These vessels drain through popliteal and femoral lymph nodes before finally entering the deep inguinal, external iliac, or inferior gluteal lymph node groups. More studies are required to determine age- and race-associated differences (Pan et al. 2013).

119.6 Pathology

Affecting 90 million people worldwide, leg lymphedema is caused by deterioration of lymphatic vessels and/or lymph nodes (Shinaoka et al. 2020; Garza et al. 2017). Edema develops when the filtration rate exceeds the capacity of the lymphatic system to manage fluid balance (Garza et al. 2017; Tiwari et al. 2003). This causes the increased volumes of interstitial and lymphatic fluids observed in 80% of lower limb lymphatic system dysfunction cases with different etiologies (valve failure, lymphatic vessel obstruction, loss of pressure gradient, or loss of smooth muscle contractility) (Szuba and Rockson

1998; Russell et al. 2019). There are two types of lymphedema. More common in females (Allen 1934), primary lymphedema is caused either by a congenital abnormality, which can be detected during the first year of life, or by dysfunction of the lymphatic system. It can further be categorized as autosomal familial (Milroy Disease) (Salem et al. 1986) or sporadic (between the ages of one and 35 years) (Haaverstad et al. 1994). Primary lymphedema is more common around menarche (Allen 1934). Patients experience swelling in the extremities, decreased coordination and mobility, and a higher risk of infections (Moshiri et al. 2002). Secondary lymphedema is due to a reduced lymph flow by an acquired cause such as trauma, recurrent infection, or malignancy (Tiwari et al. 2003). Like primary lymphedema, secondary lymphedema occurs more prominently in women and affects the distal leg first (Shinaoka et al. 2020; Cambria et al. 1993). The initial symptom is painless swelling (Tiwari et al. 2003), with progression to radial enlargement if treatment is not commenced. Treatment can be less efficacious if an underlying venous disorder impairs lymphatic function (Collins et al. 1989). Over time, the skin exhibits hyperkeratosis, papillomatosis, and skin turgor (Lewis and Wald 1984; Mortimer 2000). In extreme cases it becomes compromised, leading to lymph leakage, which increases the risk of infection. A rare complication of lymphedema is lymphangiosarcoma (Lewis and Wald 1984).

As a consequence of primary and/or secondary lymphedema, lymphatic fluid stasis leads to localized chronic inflammation (Rutkowski and Swartz 2007; Mihara et al. 2012; Zampell et al. 2012; Avraham et al. 2010). This results in extracellular matrix remodeling and fibrosis, adipose tissue differentiation, and progressive fibrosis/sclerosis, with eventual obliteration of the lymphatic vessel lumen. Inflammatory cells accumulate around the lymphatic vessels and upregulate inducible nitric oxide synthase (iNOS) expression; the increased nitric oxide levels disrupt the intrinsic nitric oxide gradient and decrease the contractility of collecting lymphatics and, therefore, lymphatic fluid transport (Avraham et al. 2010). In addition, the chronic thymus (T)-helper 2 cell–biased inflammatory response increases the expression of interleukin (IL)-4, 13, interferon-gamma, and transforming growth factor-β1 (Savetsky et al. 2015; Liao et al. 2011). This decreases the production of collateral lymphatic vessels by hindering lymphatic endothelial cell proliferation and tubule formation, migration, and function. Obesity, an independent risk factor for lymphedema, reverses lymphatic dysfunction regulated by perilymphatic NOS expression and accumulation of inflammatory T cells and macrophages (Torrisi et al. 2016).

Two other consequences of lower limb lymphatic dysfunction are lipedema and deep vein thrombosis (DVT). Lipedema, also known as lipomatosis of the leg, is described as an elastic symmetrical enlargement of both legs, the feet being spared (Rudkin and Miller 1994; Harwood et al. 1996). DVT can be due to blocked venous flow in the soleal plexus, which leads to a swollen, warm, and tender calf (Tiwari et al. 2003). A long-term consequence of DVT is postphlebitic syndrome, where there is reflux in the deep venous system leading to chronic lower limb swelling, lipodermatosclerosis, and varicose veins.

119.7 Clinical and Surgical Significance

Prior to discussing surgical intervention, it is important to list the criteria used to select patients and the potential contraindications. For patients presenting after lymphedema, if their lymphedema is at an early stage, a trial of complete decongestive therapy for 3–6 months is recommended (Akita et al. 2014; Ciudad et al. 2018; Cheng et al. 2013; Schaverien et al. 2018). Following this, surgical intervention is indicated for those with persistent lymphedema, especially with recurrent bouts of cellulitis (Akita et al. 2014; Ciudad et al. 2018; Cheng et al. 2013). For patients presenting at an advanced stage with significant pitting edema, preoperative rehabilitation using reductive complete decongestive therapy can help to optimize the conditions for surgery by transitioning toward the maintenance phase (Mikami et al. 2011; Yamamoto and Yamamoto 2007). Patients with recurrent episodes of cellulitis can benefit from prophylactic antibiotics, although surgery is contraindicated for patients with active cellulitis (Schaverien et al. 2018).

When surgical treatment was needed, the costs for patients with lymphedema were twice as much as non-lymphedema patients over the course of 2 years: $23,167 vs. $14,877, respectively (Garza et al. 2017). It can be assumed that patients will continue to incur costs related to lymphedema treatment over the course of their lifetime and the total costs associated with these treatments will be significant. In addition to the monetary impact on lymphedema patients, their condition is also associated with significant psychosocial and emotional effects.

Surgical treatments are generally regarded as effective at alleviating symptoms, reducing the risk of cellulitis and improving function and appearance in patients with lymphedema who fail to improve after nonsurgical measures (Schaverien et al. 2018). Patients should be admitted a few days prior to surgery to allow for leg elevation and compression (Zelikovski et al. 1986). There are various surgical procedures for treating lymphedema:

- Subcutaneous drainage of the lymphedema fluid through the long saphenous vein by multiperforated silicon tubes linked to a chamber by a one-way valve (Degni 1992). However, this technique is prone to blockage owing to the high protein content of the edema fluid.
- The Charles procedure, where there is radical excision of skin, subcutaneous tissue, and deep fascia with primary or staged skin grafting (Tiwari et al. 2003). A major limitation of this technique is the risk of infection and necrosis of the skin graft.
- Bypass procedures (end-to-side lymphatic-venous anastomosis) were performed in the presence of a competent venous system and intact lymphatic structures and indicated early lymphedema (Schaverien et al. 2018; Campisi et al. 1995). This technique is beneficial in reducing the leg diameter by 70% and decreasing the occurrence of cellulitis at a six-month follow-up (Gong-Kang et al. 1981; Huang et al. 1985). Notably, the efficacy results differ between primary and secondary lymphedema (Ipsen et al. 1988; Patel et al. 2015).
- Autologous interposition of vein grafting involving direct lymphatic-venous bypass (Baumeister and Siuda 1990; Baumeister et al. 1986).
- Prophylactic surgery (lymph node removal in the pelvic region in the context of malignant melanoma in order to lower the risk of lymphoceles and reduce hospital stay) (Orefice et al. 1988). However, the efficacy of this technique for treating lymphedema has not been determined (Schaverien et al. 2018; Asdourian et al. 2017).
- Vascularized lymph node transplantation, which is indicated in established lymphedema to provide new physiological function (Schaverien et al. 2018). Two main mechanisms have been demonstrated in experimental and clinical settings (Patel et al. 2015; Shesol et al. 1979; Chen et al. 1990; Suami et al. 2016; Ito et al. 2016; Cheng et al. 2014; Yan et al. 2011; Aschen et al. 2014):
  - Formation of new afferent and efferent lymphatic collateral pathways to connect the transplanted lymph nodes with lymphatic vessels in the recipient site to restore outflow. This is mediated by VEGF-C secretion from the transplanted lymph nodes.
  - Establishment of new lymphaticovenous drainage within the transplanted lymph nodes, mediated by perfusion gradients between arterial inflow and venous outflow (Chen et al. 1990; Ito et al. 2016; Cheng et al. 2014; Aschen et al. 2014).

Outcome metrics for lymphatic surgery include assessment of limb volume, incidence of cellulitis, physiological downstaging, and patient-reported outcomes (Schaverien et al. 2018). Change in limb volume is measured by limb circumferential measurements or by using a perometer to measure extracellular fluid volumes (Coroneos et al. 2019). Physiological downstaging, where the lymphedema becomes less severe, can be evaluated using the aforementioned radioisotope lymphoscintigraphy (Chang 2010; Pusic et al. 2013).

References

Abtahian F et al (2003) Regulation of blood and lymphatic vascular separation by signaling proteins SLP-76 and Syk. Science 299:247–251

Akita S et al (2014) Suitable therapy options for subclinical and early-stage lymphoedema patients. J Plast Reconstr Aesthetic Surg JPRAS 67:520–525

Allen EV (1934) Lymphedema of the extremities: classification, etiology and differential diagnosis: a study of three hundred cases. Arch Intern Med 54:606–624

Aschen SZ et al (2014) Lymph node transplantation results in spontaneous lymphatic reconnection and restoration of lymphatic flow. Plast Reconstr Surg 133:301–310

Asdourian MS et al (2017) Association between precautionary behaviors and breast cancer-related lymphedema in patients undergoing bilateral surgery. J Clin Oncol Off J Am Soc Clin Oncol 35:3934–3941

Avraham T et al (2010) Blockade of transforming growth factor-beta1 accelerates lymphatic regeneration during wound repair. Am J Pathol 177:3202–3214

Baulieu F et al (2013) Contributions of SPECT/CT imaging to the lymphoscintigraphic investigations of the lower limb lymphedema. Lymphology 46:106–119

Baumeister RG, Siuda S (1990) Treatment of lymphedemas by microsurgical lymphatic grafting: what is proved? Plast Reconstr Surg 85:64–74.; discussion 75–76

Baumeister RG, Siuda S, Bohmert H, Moser E (1986) A microsurgical method for reconstruction of interrupted lymphatic pathways: autologous lymph-vessel transplantation for treatment of lymphedemas. Scand J Plast Reconstr Surg 20:141–146

Bräutigam P, Vanscheidt W, Földi E, Krause T, Moser E (1993) The importance of the subfascial lymphatics in the diagnosis of lower limb edema: investigations with semiquantitative lymphoscintigraphy. Angiology 44:464–470

Bui T, Bordoni B (2022) Anatomy, abdomen and pelvis, inguinal lymph node. In: StatPearls. StatPearls Publishing

Cambria RA, Gloviczki P, Naessens JM, Wahner HW (1993) Noninvasive evaluation of the lymphatic system with lymphoscintigraphy: a prospective, semiquantitative analysis in 386 extremities. J Vasc Surg 18:773–782

Campisi C, Boccardo F, Tacchella M (1995) Reconstructive microsurgery of lymph vessels: the personal method of lymphatic-venous-lymphatic (LVL) interpositioned grafted shunt. Microsurgery 16:161–166

Chang DW (2010) Lymphaticovenular bypass for lymphedema management in breast cancer patients: a prospective study. Plast Reconstr Surg 126:752–758

Chen HC, O'Brien BM, Rogers IW, Pribaz JJ, Eaton CJ (1990) Lymph node transfer for the treatment of obstructive lymphoedema in the canine model. Br J Plast Surg 43:578–586

Cheng M-H et al (2013) Vascularized groin lymph node flap transfer for postmastectomy upper limb lymphedema: flap anatomy, recipient sites, and outcomes. Plast Reconstr Surg 131:1286–1298

Cheng M-H et al (2014) The mechanism of vascularized lymph node transfer for lymphedema: natural lymphaticovenous drainage. Plast Reconstr Surg 133:192e–198e

Ciudad P et al (2018) Comparison of long-term clinical outcomes among different vascularized lymph node transfers: 6-year experience of a single center's approach to the treatment of lymphedema. J Surg Oncol 117:1346–1347

Collins PS et al (1989) Abnormalities of lymphatic drainage in lower extremities: a lymphoscintigraphic study. J Vasc Surg 9:145–152

Connell FC et al (2013) The classification and diagnostic algorithm for primary lymphatic dysplasia: an update from 2010 to include molecular findings. Clin Genet 84:303–314

Coroneos CJ, Wong FC, DeSnyder SM, Shaitelman SF, Schaverien MV (2019) Correlation of L-Dex bioimpedance spectroscopy with limb volume and lymphatic function in lymphedema. Lymphat Res Biol 17:301–307

de Carvalho JPM, Patrício BF, Medeiros J, Sampaio FJB, Favorito LA (2011) Anatomic aspects of inguinal lymph nodes applied to lymphadenectomy in penile cancer. Adv Urol 2011:e952532

Degni M (1992) New technique for the subcutaneous drainage of peripheral lymphedema. Lymphology 25:182–183

Doldi SB et al (1992) Ultrasonography of extremity lymphedema. Lymphology 25:129–133

Duewell S et al (1992) Swollen lower extremity: role of MR imaging. Radiology 184:227–231

Garza AG et al (2009) Improved patient survival using a modified resuscitation protocol for out-of-hospital cardiac arrest. Circulation 119:2597–2605

Garza R, Skoracki R, Hock K, Povoski SP (2017) A comprehensive overview on the surgical management of secondary lymphedema of the upper and lower extremities related to prior oncologic therapies. BMC Cancer 17:468

Gloviczki P et al (1989) Noninvasive evaluation of the swollen extremity: experiences with 190 lymphoscintigraphic examinations. J Vasc Surg 9:683–689; discussion 690

Gong-Kang H et al (1981) Microlymphaticovenous anastomosis for treating lymphedema of the extremities and external genitalia. J Microsurg 3:32–39

Gvetadze SR, Ilkaev KD (2020) Lingual lymph nodes: anatomy, clinical considerations, and oncological significance. World J Clin Oncol 11:337–347
Haaverstad R, Nilsen G, Rinck PA, Myhre HO (1994) The use of MRI in the diagnosis of chronic lymphedema of the lower extremity. Int Angiol J Int Union Angiol 13:115–118
Hadjis NS, Carr DH, Banks L, Pflug JJ (1985) The role of CT in the diagnosis of primary lymphedema of the lower limb. AJR Am J Roentgenol 144:361–364
Harwood CA, Bull RH, Evans J, Mortimer PS (1996) Lymphatic and venous function in lipoedema. Br J Dermatol 134:1–6
Hong Y-K et al (2002) Prox1 is a master control gene in the program specifying lymphatic endothelial cell fate. Dev Dyn Off Publ Am Assoc Anat 225:351–357
Huang GK et al (1985) Microlymphaticovenous anastomosis in the treatment of lower limb obstructive lymphedema: analysis of 91 cases. Plast Reconstr Surg 76:671–685
Idy-Peretti I et al (1998) Lymphedematous skin and subcutis: in vivo high resolution magnetic resonance imaging evaluation. J Invest Dermatol 110:782–787
Ipsen T, Pless J, Frederiksen PB (1988) Experience with microlymphaticovenous anastomoses for congenital and acquired lymphoedema. Scand J Plast Reconstr Surg Hand Surg 22:233–236
Ito R, Zelken J, Yang C-Y, Lin C-Y, Cheng M-H (2016) Proposed pathway and mechanism of vascularized lymph node flaps. Gynecol Oncol 141:182–188
Kaplan WD (1983) Iliopelvic lymphoscintigraphy. Semin Nucl Med 13:42–53
Karkkainen MJ et al (2004) Vascular endothelial growth factor C is required for sprouting of the first lymphatic vessels from embryonic veins. Nat Immunol 5:74–80
Kim M et al (2013) A model to estimate the risk of breast cancer-related lymphedema: combinations of treatment-related factors of the number of dissected axillary nodes, adjuvant chemotherapy, and radiation therapy. Int J Radiat Oncol Biol Phys 86:498–503
Kriehuber E et al (2001) Isolation and characterization of dermal lymphatic and blood endothelial cells reveal stable and functionally specialized cell lineages. J Exp Med 194:797–808
Lewis JM, Wald ER (1984) Lymphedema praecox. J Pediatr 104:641–648
Liao S et al (2011) Impaired lymphatic contraction associated with immunosuppression. Proc Natl Acad Sci U S A 108:18784–18789
Maegawa J, Mikami T, Yamamoto Y, Satake T, Kobayashi S (2010) Types of lymphoscintigraphy and indications for lymphaticovenous anastomosis. Microsurgery 30:437–442
Marotel M et al (1998) Transaxial computer tomography of lower extremity lymphedema. Lymphology 31:180–185
Mihara M et al (2012) Pathological steps of cancer-related lymphedema: histological changes in the collecting lymphatic vessels after lymphadenectomy. PLoS One 7:e41126
Mikami T et al (2011) Classification of lymphoscintigraphy and relevance to surgical indication for lymphaticovenous anastomosis in upper limb lymphedema. Lymphology 44:155–167
Mortimer PS (1995) Evaluation of lymphatic function: abnormal lymph drainage in venous disease. Int Angiol J Int Union Angiol 14:32–35
Mortimer PS (2000) Swollen lower limb-2: lymphoedema. BMJ 320:1527–1529
Mortimer PS, Rockson SG (2014) New developments in clinical aspects of lymphatic disease. J Clin Invest 124:915–921
Moshiri M, Katz DS, Boris M, Yung E (2002) Using lymphoscintigraphy to evaluate suspected lymphedema of the extremities. AJR Am J Roentgenol 178:405–412
Musters GD, Kleipool RP, Bipat S, Maas M (2011) Features of the popliteal lymph nodes seen on musculoskeletal MRI in a Western population. Skeletal Radiol 40:1041–1045
Nawaz MK, Hamad MM, Abdel-Dayem HM, Sadek S, Eklof BG (1992) Lymphoscintigraphy in lymphedema of the lower limbs using 99mTc HSA. Angiology 43:147–154
Orefice S, Conti AR, Grassi M, Salvadori B (1988) The use of lympho-venous anastomoses to prevent complications from ilio-inguinal dissection. Tumori 74:347–351
Pan W-R, le Roux CM, Levy SM (2011) Alternative lymphatic drainage routes from the lateral heel to the inguinal lymph nodes: anatomic study and clinical implications. ANZ J Surg 81:431–435
Pan W-R, Wang D-G, Levy SM, Chen Y (2013) Superficial lymphatic drainage of the lower extremity: anatomical study and clinical implications. Plast Reconstr Surg 132:696–707
Patel KM, Lin C-Y, Cheng M-H (2015) From theory to evidence: long-term evaluation of the mechanism of action and flap integration of distal vascularized lymph node transfers. J Reconstr Microsurg 31:26–30
Petrova TV et al (2002) Lymphatic endothelial reprogramming of vascular endothelial cells by the Prox-1 homeobox transcription factor. EMBO J 21:4593–4599
Petrova TV et al (2004) Defective valves and abnormal mural cell recruitment underlie lymphatic vascular failure in lymphedema distichiasis. Nat Med 10:974–981
Pusic AL et al (2013) Quality of life among breast cancer patients with lymphedema: a systematic review of patient-reported outcome instruments and outcomes. J Cancer Surviv Res Pract 7:83–92
Reynolds HM et al (2007) Three-dimensional visualisation of lymphatic drainage patterns in patients with cutaneous melanoma. Lancet Oncol 8:806–812
Ribatti D, Crivellato E (2010) The embryonic origins of lymphatic vessels: an historical review. Br J Haematol 149:669–674
Rudkin GH, Miller TA (1994) Lipedema: a clinical entity distinct from lymphedema. Plast Reconstr Surg 94:841–847.; discussion 848–849

Russell PS, Hong J, Windsor JA, Itkin M, Phillips ARJ (2019) Renal lymphatics: anatomy, physiology, and clinical implications. Front Physiol 10:251

Rutkowski JM, Swartz MA (2007) A driving force for change: interstitial flow as a morphoregulator. Trends Cell Biol 17:44–50

Salem AH, Mulhim AM, Grant C, Khwaja MS (1986) Milroy's disease in a Saudi family. J R Coll Surg Edinb 31:143–146

Savetsky IL et al (2015) Th2 cytokines inhibit lymphangiogenesis. PLoS One 10:e0126908

Schaverien MV, Moeller JA, Cleveland SD (2018) Nonoperative treatment of lymphedema. Semin Plast Surg 32:17–21

Shesol BF, Nakashima R, Alavi A, Hamilton RW (1979) Successful lymph node transplantation in rats, with restoration of lymphatic function. Plast Reconstr Surg 63:817–823

Shinaoka A et al (2020) Lower-limb lymphatic drainage pathways and lymph nodes: a CT lymphangiography cadaver study. Radiology 294:223–229

Suami H, Scaglioni MF, Dixon KA, Tailor RC (2016) Interaction between vascularized lymph node transfer and recipient lymphatics after lymph node dissection-a pilot study in a canine model. J Surg Res 204:418–427

Szuba A, Rockson SG (1998) Lymphedema: classification, diagnosis and therapy. Vasc Med Lond Engl 3:145–156

Szuba A, Shin WS, Strauss HW, Rockson S (2003) The third circulation: radionuclide lymphoscintigraphy in the evaluation of lymphedema. J Nucl Med Off Publ Soc Nucl Med 44:43–57

Tappouni RF, Sarwani NI, Tice JG, Chamarthi S (2011) Imaging of unusual perineal masses. Am J Roentgenol 196:W412–W420

Tiwari A, Cheng K-S, Button M, Myint F, Hamilton G (2003) Differential diagnosis, investigation, and current treatment of lower limb lymphedema. Arch Surg Chic Ill 1960 138:152–161

Torrisi JS et al (2016) Inhibition of inflammation and iNOS improves lymphatic function in obesity. Sci Rep 6:19817

Uren RF, Thompson JF, Howland-Giles R (2000) Lymphatic drainage of the skin and breast: locating the sentinel nodes. Radiology 214:258–258

Vaughan BF (1990) CT of swollen legs. Clin Radiol 41:24–30

Werner GT, Scheck R, Kaiserling E (1998) Magnetic resonance imaging of peripheral lymphedema. Lymphology 31:34–36

Wigle JT, Oliver G (1999) Prox1 function is required for the development of the murine lymphatic system. Cell 98:769–778

Wigle JT et al (2002) An essential role for Prox1 in the induction of the lymphatic endothelial cell phenotype. EMBO J 21:1505–1513

Williams WH, Witte CL, Witte MH, McNeill GC (2000) Radionuclide lymphangioscintigraphy in the evaluation of peripheral lymphedema. Clin Nucl Med 25:451–464

Wilting J, Papoutsi M, Buttler K, Becker J (2007) Embryonic development of the lymphovascular system and tumor lymphangiogenesis. In: Leong SPL (ed) Cancer metastasis and the lymphovascular system: basis for rational therapy. Springer, pp 17–24. https://doi.org/10.1007/978-0-387-69219-7_2

Witte CL et al (2000) Advances in imaging of lymph flow disorders. Radiogr Rev Publ Radiol Soc N Am Inc 20:1697–1719

Yamamoto R, Yamamoto T (2007) Effectiveness of the treatment-phase of two-phase complex decongestive physiotherapy for the treatment of extremity lymphedema. Int J Clin Oncol 12:463–468

Yan A, Avraham T, Zampell JC, Aschen SZ, Mehrara BJ (2011) Mechanisms of lymphatic regeneration after tissue transfer. PLoS One 6:e17201

Zampell JC et al (2012) CD4(+) cells regulate fibrosis and lymphangiogenesis in response to lymphatic fluid stasis. PLoS One 7:e49940

Zelikovski A, Haddad M, Reiss R (1986) Non-operative therapy combined with limited surgery in management of peripheral lymphedema. Lymphology 19:106–108

GPSR Compliance

The European Union's (EU) General Product Safety Regulation (GPSR) is a set of rules that requires consumer products to be safe and our obligations to ensure this.

If you have any concerns about our products, you can contact us on ProductSafety@springernature.com

In case Publisher is established outside the EU, the EU authorized representative is:

Springer Nature Customer Service Center GmbH
Europaplatz 3
69115 Heidelberg, Germany

Batch number: 10732823

Printed by Printforce, the Netherlands